ARCTIC OCEAN
Svalbard (NOR.)
See inset below
THE RUSSIAN DOMAIN
RUSSIA
EUROPE
CENTRAL ASIA
KAZAKHSTAN
MONGOLIA
UZBEKISTAN
KYRGYZSTAN
TURKMENISTAN
TAJIKISTAN
GEORGIA
ARMENIA
TURKEY
AZERBAIJAN
Azores (PORT.)
Canary Is. (SP.)
TUNISIA
LEBANON
SYRIA
ISRAEL
GAZA STRIP
WEST BANK
JORDAN
IRAQ
IRAN
KUWAIT
BAHRAIN
QATAR
AFGHANISTAN
PAKISTAN
MOROCCO
ALGERIA
LIBYA
NORTH AFRICA/ SOUTHWEST ASIA
WESTERN SAHARA (MOR.)
EGYPT
SAUDI ARABIA
UNITED ARAB EMIRATES
OMAN
YEMEN
CHINA
EAST ASIA
NORTH KOREA
SOUTH KOREA
JAPAN
Kuril Is. (RUS.)
PACIFIC OCEAN
Ryukyu Islands (JAPAN)
TAIWAN
NEPAL
BHUTAN
SOUTH ASIA
INDIA
BANGLADESH
MYANMAR (BURMA)
LAOS
THAILAND
VIETNAM
CAMBODIA
PHILIPPINES
SOUTHEAST ASIA
Andaman and Nicobar Is. (INDIA)
SRI LANKA
MALDIVES
BRUNEI
MALAYSIA
SINGAPORE
INDONESIA
TIMOR-LESTE
PAPUA NEW GUINEA
Northern Mariana Is. (U.S.)
Guam (U.S.)
Wake Island (U.S.)
PALAU
FEDERATED STATES OF MICRONESIA
MARSHALL ISLANDS
NAURU
KIRIBATI
SOLOMON ISLANDS
TUVALU
VANUATU
FIJI
MAURITANIA
MALI
NIGER
CHAD
SUDAN
ERITREA
DJIBOUTI
CAPE VERDE
SENEGAL
GAMBIA
GUINEA-BISSAU
GUINEA
SIERRA LEONE
LIBERIA
BURKINA FASO
BENIN
GHANA
NIGERIA
CÔTE D'IVOIRE
TOGO
CENTRAL AFRICAN REP.
SOUTH SUDAN
ETHIOPIA
SOMALIA
CAMEROON
SÃO TOMÉ AND PRÍNCIPE
EQUATORIAL GUINEA
GABON
REP. OF THE CONGO
DEM. REP. OF THE CONGO
UGANDA
KENYA
RWANDA
BURUNDI
TANZANIA
SEYCHELLES
British Indian Ocean Territory
INDIAN OCEAN
SUB-SAHARAN AFRICA
COMOROS
Mayotte (FR.)
MALAWI
ANGOLA
ZAMBIA
MOZAMBIQUE
MADAGASCAR
MAURITIUS
Réunion (FR.)
St. Helena (U.K.)
NAMIBIA
ZIMBABWE
BOTSWANA
SWAZILAND
SOUTH AFRICA
LESOTHO
ATLANTIC OCEAN
Cocos (Keeling) Islands (AUS.)
Christmas Island (AUS.)
AUSTRALIA AND OCEANIA
AUSTRALIA
New Caledonia (FR.)
Norfolk Island (AUS.)
NEW ZEALAND
0 1,000 2,000 Miles
0 1,000 2,000 Kilometers
Kerguelen Is. (FR.)
20°W 0° 20°E 40°E 60°E 80°E 100°E 120°E 140°E 160°E
80°N 60°N 40°N 20°N 0° 20°S
0 250 500 Miles
0 250 500 Kilometers
ICELAND
Faroe Is. (DEN.)
SWEDEN
FINLAND
NORWAY
RUSSIA
ESTONIA
LATVIA
LITHUANIA
RUSSIA
North Sea
DENMARK
Baltic Sea
BELARUS
IRELAND
UNITED KINGDOM
NETHERLANDS
POLAND
BELGIUM
GERMANY
UKRAINE
Channel Islands (U.K.)
LUXEMBOURG
CZECH REP.
LIECHTENSTEIN
SLOVAKIA
ATLANTIC OCEAN
FRANCE
SWITZERLAND
AUSTRIA
HUNGARY
MOLDOVA
SLOVENIA
CROATIA
ROMANIA
MONACO
ITALY
SAN MARINO
BOSNIA AND HERZEGOVINA
SERBIA
KOSOVO
BULGARIA
Black Sea
ANDORRA
Corsica (FR.)
MONTENEGRO
MACEDONIA
PORTUGAL
SPAIN
Balearic Is. (SP.)
Sardinia (IT.)
VATICAN CITY
ALBANIA
GREECE
TURKEY
Mediterranean Sea
Sicily (IT.)
Gibraltar (U.K.)
TUNISIA
MALTA
Crete (GR.)
CYPRUS
SYRIA
LEBANON
MOROCCO
ALGERIA
20°W 0° 20°E 60°N 40°N

Europe

Diversity Amid Globalization

World Regions, Environment, Development

Sixth Edition

LES ROWNTREE
University of California, Berkeley

MARTIN LEWIS
Stanford University

MARIE PRICE
George Washington University

WILLIAM WYCKOFF
Montana State University

PEARSON

Boston Columbus Indianapolis New York San Francisco Upper Saddle River
Amsterdam Cape Town Dubai London Madrid Milan Munich Paris Montréal Toronto
Delhi Mexico City São Paulo Sydney Hong Kong Seoul Singapore Taipei Tokyo

Senior Geography Editor: Christian Botting
Senior Marketing Manager: Maureen McLaughlin
Program Manager: Anton Yakovlev
Project Manager: Sean Hale
Director of Development: Jennifer Hart
Development Editor: David Chelton
Assistant Editor: Bethany Sexton
Marketing Assistant: Nicola Houston
Media Producer: Timothy Hainley
Team Lead, Geosciences and Chemistry: Gina M. Cheselka
Production Project Manager: Connie M. Long
Full Service/Composition: Cenveo® Publisher Services
Full-Service Project Manager: Heidi Allgair/Cenveo® Publisher Services
Illustrations: International Mapping Associates
Image Lead: Maya Melenchuk
Photo Researcher: Stephen Merland, PreMediaGlobal
Text Permissions Manager: Timothy Nicholls
Text Permissions Researcher: Jenell Forschler
Design Manager: Derek Bacchus
Interior and Cover Designer: tani hasegawa
Operations Specialist: Christy Hall
Cover Photo Credit: Olaf Krügerimagebroker/SuperStock

Credits and acknowledgments borrowed from other sources and reproduced, with permission, in this textbook appear on the appropriate page within the text or on the credits page beginning on page CR-1.

Library of Congress Cataloging-in-Publication Data

Diversity amid globalization : world regions, environment, development / Les Rowntree,
Martin Lewis, Marie Price, William Wyckoff.--Sixth edition.
pages cm
Includes index.
ISBN 13: 978-0-321-91006-6
ISBN 10: 0-321-91006-0
1. Geography. I. Rowntree, Lester, 1938
G128.D58 2015
910--dc23 2013043443

About Our Sustainability Initiatives

Pearson recognizes the environmental challenges facing this planet, as well as acknowledges our responsibility in making a difference. This book has been carefully crafted to minimize environmental impact. The binding, cover, and paper come from facilities that minimize waste, energy consumption, and the use of harmful chemicals. Pearson closes the loop by recycling every out-of-date text returned to our warehouse.

Along with developing and exploring digital solutions to our market's needs, Pearson has a strong commitment to achieving carbon-neutrality. As of 2009, Pearson became the first carbon- and climate-neutral publishing company. Since then, Pearson remains strongly committed to measuring, reducing, and offsetting our carbon footprint.

The future holds great promise for reducing our impact on Earth's environment, and Pearson is proud to be leading the way. We strive to publish the best books with the most up-to-date and accurate content, and to do so in ways that minimize our impact on Earth. To learn more about our initiatives, please visit **www.pearson.com/responsibility**.

PEARSON

www.pearsonhighered.com

2 3 4 5 6 7 8 9 10—CRK—18 17 16 15 14

ISBN-10: 0-321-91006-0; ISBN-13: 978-0-321-91006-6 (Student Edition)
ISBN-10: 0-321-97241-4; ISBN-13: 978-0-321-97241-5 (Instructor's Review Copy)

Brief Contents

1 Concepts of World Geography **2**

2 Physical Geography and the Environment **50**

3 North America **82**

4 Latin America **136**

5 The Caribbean **192**

6 Sub-Saharan Africa **238**

7 Southwest Asia and North Africa **296**

8 Europe **350**

9 The Russian Domain **398**

10 Central Asia **450**

11 East Asia **494**

12 South Asia **548**

13 Southeast Asia **598**

14 Australia and Oceania **648**

Contents

About Our Sustainability Initiatives ii
Preface x
About the Authors xiii
Digital and Print Resources xiv
Book & MasteringGeography™ Walkthrough xvi

1 Concepts of World Geography 2

Converging Currents of Globalization 4

Economic Globalization 4 • Globalization and Changing Human Geographies 5

EXPLORING GLOBAL CONNECTIONS A Closer Look at Globalization 6

Geopolitics and Globalization 7 • The Environment and Globalization 8 • Controversy About Globalization 8 • Pro-globalization Arguments 8 • Critics of Globalization 10 • A Middle Position 10 • Diversity in a Globalizing World 11 • Flat and Spiky Worlds 12

Geography Matters: Environments, Regions, Landscapes 12

Areal Differentiation and Integration 13 • Regions: Formal, Functional, and Vernacular 13 • The Cultural Landscape: Space into Place 14

The Geographer's Toolbox: Location, Maps, Remote Sensing, and GIS 15

Latitude and Longitude 15 • Global Positioning Systems (GPS) 16 • Map Projections 16 • Map Scale 16 • Map Patterns and Legends 17 • Aerial Photos and Remote Sensing 17 • Geographic Information Systems (GIS) 18

Themes and Issues in World Regional Geography 19

Physical Geography and Environmental Issues: The Changing Global Environment 20

Population and Settlement: People on the Land 21

Population Growth and Change 21

WORKING TOWARD SUSTAINABILITY Managing Resources and Protecting the Environment 23

PEOPLE ON THE MOVE Migrants and Refugees 26

Global Migration and Settlement 26

CITYSCAPES The Human Habitat 28

Cultural Coherence and Diversity: The Geography of Change and Tradition 29

Culture in a Globalizing World 29

EVERYDAY GLOBALIZATION Complexities in the Common 30

Language and Culture in Global Context 31 • The Geography of World Religions 32 • Gender and Globalization 34 • Sports and Globalization 36

Geopolitical Framework: Unity and Fragmentation 37

World Political Systems 37 • The Nation-State Revisited 38 • Colonialism, Decolonialization, and Neocolonialism 39 • Global Conflict and Insurgency 40

Economic and Social Development: The Geography of Wealth and Poverty 41

More and Less Developed Countries 42 • Indicators of Economic Development 43 • Indicators of Social Development 45

Concepts of World Geography In Review 48

2 Physical Geography and the Environment 50

Geology: A Restless Earth 52

Plate Tectonics 52 • Geologic Hazards 55

Global Climates: Adapting to Change 56

Climate Controls 56

EXPLORING GLOBAL CONNECTIONS Antarctica, the Protected Continent 57

World Climate Regions 60 • Global Climate Change 61

Global Energy: The Essential Resource 66

Non-Renewable and Renewable Energy 66 • Fossil Fuel Reserves, Production, and Consumption 66 • Renewable Energy 68 • Energy Futures 69

Water: A Scarce World Resource 69

WORKING TOWARD SUSTAINABILITY Lighting Up Dark Places 70

Water Scarcity 71 • Water Sanitation 71 • Water Access 71

Bioregions and Biodiversity: The Globalization of Nature 72

Tropical Rainforests 73 • Tropical Seasonal Forests 74 • Tropical Savannas 74 • Deforestation of Tropical Forests 75 • Deserts and Grasslands 75 • Mediterranean Shrubs and Woodlands 75 • Temperate Deciduous Forests 76

EVERYDAY GLOBALIZATION The Hamburger Connection Revisited 76

Evergreen Forests 77 • Tundra 79

Physical Geography and the Environment In Review 80

3 North America 82

Physical Geography and Environmental Issues: A Vulnerable Land of Plenty 86

A Diverse Physical Setting 87 • Patterns of Climate and Vegetation 88 • The Costs of Human Modification 88 • Growing Environmental Awareness 91

WORKING TOWARD SUSTAINABILITY Bison Return to the Northern Great Plains 92

The Shifting Energy Equation 93 • Climate Change in North America 94

Population and Settlement: Reshaping a Continental Landscape 95

Modern Spatial and Demographic Patterns 95 • Occupying the Land 97 • North Americans on the Move 98 • Settlement Geographies: The Decentralized Metropolis 101 • Settlement Geographies: Rural North America 102

Cultural Coherence and Diversity: Shifting Patterns of Pluralism 103

The Roots of a Cultural Identity 104 • Peopling North America 104 • Culture and Place in North America 106

CITYSCAPES VANCOUVER 108

Patterns of North American Religion 111

EXPLORING GLOBAL CONNECTIONS West Indian Gardens in East New York 112

The Globalization of American Culture 113

EVERYDAY GLOBALIZATION The NBA Goes Global 116

Geopolitical Framework: Patterns of Dominance and Division 116

Creating Political Space 117 • Continental Neighbors 118 • The Legacy of Federalism 120 • Native Peoples and National Politics 120 • The Politics of U.S. Immigration 121 • A Global Reach 122

Economic and Social Development: Patterns of Abundance and Affluence 122

An Abundant Resource Base 122 • Creating a Continental Economy 124 • Enduring Social Issues 127 • North America and the Global Economy 130

PEOPLE ON THE MOVE Skilled Immigrants Fuel Big City Growth 132

North America In Review 134

4 Latin America 136

Physical Geography and Environmental Issues: Neotropical Diversity and Urban Degradation 141

Western Mountains and Eastern Shields 141 • River Basins and Lowlands 143 • Climate and Climate Change in Latin America 144 • Impacts of Climate Change for Latin America 147 • Environmental Issues: The Destruction of Forests 148 • Urban Environmental Challenges 151

WORKING TOWARD SUSTAINABILITY Greening Transport and Expanding Access in Bogotá 152

Population and Settlement: The Dominance of Cities 154

The Latin American City 154 • Patterns of Rural Settlement 158

CITYSCAPES Claiming the High Ground in La Paz, Bolivia 159

Population Growth and Mobility 162

Patterns of Cultural Coherence and Diversity: Repopulating a Continent 164

Decline of Native Populations 164 • Patterns of Ethnicity and Culture 166

EXPLORING GLOBAL CONNECTIONS The Catholic Church and the Argentine Pope 168

The Global Reach of Latino Culture 170

Geopolitical Framework: From Two Iberian Colonies to Many Nations 171

Iberian Conquest and Territorial Division 172

PEOPLE ON THE MOVE The Impact of U.S. Deportations on Latin America 175

Regional Organizations 176

Economic and Social Development: From Dependency to Neoliberalism 178

Development Strategies 179 • Primary Export Dependency 180

EVERYDAY GLOBALIZATION Good Morning Coffee 182

Latin America in the Global Economy 185 • Social Development 187

Latin America In Review 190

5 The Caribbean 192

Physical Geography and Environmental Issues: Paradise Undone 196

Island and Rimland Landscapes 196

EXPLORING GLOBAL CONNECTIONS Crisis Mapping in Haiti After the Earthquake 198

Caribbean Climate and Climate Change 199 • Environmental Issues 203

Population and Settlement: Densely Settled Islands and Rimland Frontiers 206

Demographic Trends 207 • The Rural–Urban Continuum 210

WORKING TOWARD SUSTAINABILITY Urban Agriculture in Havana 212

Cultural Coherence and Diversity: A Neo-Africa in the Americas 213

The Cultural Imprint of Colonialism 214 • Creolization and Caribbean Identity 217

CITYSCAPES San Juan as Cultural Crossroads of the Americas 220

Geopolitical Framework: Colonialism, Neocolonialism, and Independence 222

Life in the "American Backyard" 222 • Independence and Integration 224

PEOPLE ON THE MOVE Suriname in the Postcolonial World 226

Economic and Social Development: From Cane Fields to Cruise Ships 227

From Fields to Factories and Resorts 227 • Social Development 232

EVERYDAY GLOBALIZATION Caribbean Migrants and Health Care 235

The Caribbean In Review 236

6 Sub-Saharan Africa 238

Physical Geography and Environmental Issues: The Plateau Continent 243

Plateaus and Basins 243 • Climate and Vegetation 247 • Africa's Environmental Issues 248

WORKING TOWARD SUSTAINABILITY Can Bamboo Reduce Deforestation in Africa? 252

Climate Change and Vulnerability in Sub-Saharan Africa 254

Population and Settlement: Young and Restless 254

Population Trends and Demographic Debates 256 • Patterns of Settlement and Land Use 259 • Urban Life 262

CITYSCAPES East Africa's High-Tech Leader, Nairobi 263

Cultural Coherence and Diversity: Unity Through Adversity 265

Language Patterns 266 • Religion 268 • Globalization and African Culture 271

EXPLORING GLOBAL CONNECTIONS The Reach of Nollywood 273

Geopolitical Framework: Legacies of Colonialism and Conflict 274

European Colonization 274 • Decolonization and Independence 277 • Enduring Political Conflict 278

EVERYDAY GLOBALIZATION The African Origins of the Diamond Engagement Ring 282

Economic and Social Development: The Struggle to Develop 283

Roots of African Poverty 285 • Links to the World Economy 287

PEOPLE ON THE MOVE Chinese Merchants to Africa 289

Economic Differentiation Within Africa 290 • Measuring Social Development 291 • Women and Development 292 • Building from Within 292

Sub-Saharan Africa In Review 294

7 Southwest Asia and North Africa 296

Physical Geography and Environmental Issues: Life in a Fragile World 301

Regional Landforms 301 • Patterns of Climate 301 • Legacies of a Vulnerable Landscape 302 • Climate Change in Southwest Asia and North Africa 308

Population and Settlement: Changing Rural and Urban Worlds 309

The Geography of Population 310 • Water and Life: Rural Settlement Patterns 311

CITYSCAPES Cairo, City of Many Neighborhoods 314

Many-Layered Landscapes: The Urban Imprint 315 • A Region on the Move 317

WORKING TOWARD SUSTAINABILITY Masdar City Emerges in the Desert 319

Shifting Demographic Patterns 320

Cultural Coherence and Diversity: A Complex Cultural Mosaic 321

Patterns of Religion 321 • Geographies of Language 325 • Regional Cultures in Global Context 326

Geopolitical Framework: Never-Ending Tensions 328

The Colonial Legacy 328 • Modern Geopolitical Issues 330 • An Uncertain Political Future 335

PEOPLE ON THE MOVE The Syrian Exodus 336

Economic and Social Development: A Region of Stark Contrasts 338

Measuring Development 338 • The Geography of Fossil Fuels 339 • Regional Economic Patterns 341

EXPLORING GLOBAL CONNECTIONS Yanbu as a Global City 342

EVERYDAY GLOBALIZATION Popping Pills from Israel 344

A Woman's Changing World 345 • Global Economic Relationships 347

Southwest Asia and North Africa In Review 348

8 Europe 350

Physical Geography and Environmental Issues: Human Transformation of a Diverse Landscape 352

Landform Regions 352 • Europe's Climate 356 • Seas, Rivers, and Ports 358 • Environmental Issues: Local and Global 358 • Climate Change in Europe 360

WORKING TOWARD SUSTAINABILITY Germany's Energy Transformation 362

Population and Settlement: Slow Growth, Rapid Migration, and Widespread Urbanism 363

Low (or No) Natural Growth 363 • Migration to and Within Europe 366

PEOPLE ON THE MOVE The Roma, Europe's Largest Ethnic Minority 367

Landscapes of Urban Europe 369

CITYSCAPES Dresden and Its Symbolic Landscape 371

Cultural Coherence and Diversity: A Mosaic of Differences 372

Geographies of Language 372 • Geographies of Religion, Past and Present 374

EVERYDAY GLOBALIZATION International Students in Europe 377

European Culture in a Global Context 378

Geopolitical Framework: A Dynamic Map 380

Redrawing the Map of Europe Through War 380 • A Divided Europe, East and West 380 • The Balkans: Waking from a Geopolitical Nightmare 384 • A Europe of Small Regions? 386

Economic and Social Development: Integration and Transition 386

Europe's Industrial Revolution 386 • Rebuilding Postwar Europe 388 • Economic Disintegration and Transition in Eastern Europe 388 • Promise and Problems of the Eurozone 393

EXPLORING GLOBAL CONNECTIONS Gas, Money, and Politics in Europe's Southeast Corner 394

Europe In Review 396

9 The Russian Domain 398

Physical Geography and Environmental Issues: A Vast and Challenging Land 404

A Diverse Physical Setting 404

EXPLORING GLOBAL CONNECTIONS Russian Meteorite Fragments Go Global 408

A Devastated Environment 410

WORKING TOWARD SUSTAINABILITY Lake Baikal's Success Story 412

Climate Change and the Russian Domain 413

Population and Settlement: An Urban Domain 415

Population Distribution 415 • Regional Migration Patterns 418

CITYSCAPES Kiev, Enduring Cultural Capital 418

PEOPLE ON THE MOVE Moscow's Central Asian Communities 422

Inside the Russian City 423 • The Demographic Crisis 424

Cultural Coherence and Diversity: The Legacy of Slavic Dominance 425

The Heritage of the Russian Empire 425 • Geographies of Language 427 • Geographies of Religion 430 • Russian Culture in Global Context 431

Geopolitical Framework: Resurgent Global Superpower 433

Geopolitical Structure of the Former Soviet Union 433 • Current Geopolitical Setting 435 • Geopolitics in the South and West 436 • The Shifting Global Setting 438

Economic and Social Development: The Key Role of Energy 440

The Legacy of the Soviet Economy 440 • The Post-Soviet Economy 441 • Growing Economic Globalization 444

EVERYDAY GLOBALIZATION How the Russian Domain Shapes the Virtual World 445

Russian Domain In Review 448

10 Central Asia 450

Physical Geography and Environmental Issues: Steppes, Deserts, and Threatened Lakes 453

Central Asia's Physical Regions and Climate 453 • Central Asia's Environmental Challenges 455

EVERYDAY GLOBALIZATION Kazakhstan, Home of the Apple 461

Climate Change in Central Asia 462

Population and Settlement: Densely Settled Oases Amid Vacant Lands 463

Highland Population and Subsistence Patterns 463 • Pastoralism and Farming in the Lowlands 464 • Population Issues 465 • Urbanization in Central Asia 466

WORKING TOWARD SUSTAINABILITY Pastoralism in Mongolia 467

Cultural Coherence and Diversity: A Meeting Ground of Different Traditions 468

Historical Overview: Steppe Nomads and Silk Road Traders 468

CITYSCAPES Kabul, A Precariously Booming City 469

Contemporary Linguistic and Ethnic Geography 470 • Geography of Religion 473 • Central Asian Culture in International Context 475

Geopolitical Framework: Political Reawakening in a Power Void 476

Partitioning of the Steppes 476 • Central Asia Under Communist Rule 477 • Current Geopolitical Tensions 479 • Global Dimensions of Central Asian Geopolitics 482

PEOPLE ON THE MOVE Afghan Refugees in Iran 483

Economic and Social Development: Abundant Resources, Struggling Economies 484

Economic Development in Central Asia 484

EXPLORING GLOBAL CONNECTIONS The Heroin and Opium Trade from Afghanistan 488

Social Development in Central Asia 490

Central Asia In Review 492

11 East Asia 494

Physical Geography and Environmental Issues: Resource Pressures in a Crowded Land 498

East Asia's Physical Geography 499 • East Asia's Environmental Challenges 503

WORKING TOWARD SUSTAINABILITY Japan's Smart City Movement 506

Dams, Flooding, and Soil Erosion in China 507 • Climate Change in East Asia 510

Population and Settlement: A Realm of Crowded Lowland Basins 511

Agriculture and Settlement in Japan 511 • Agriculture and Settlement in China, Korea, and Taiwan 513 • Agriculture and Resources in Global Context 514 • Urbanization in East Asia 515

CITYSCAPES Hong Kong, the Vertical City 518

Cultural Coherence and Diversity: The Historical Influence of Confucianism and Buddhism 519

Unifying Cultural Characteristics 519 • Religious Unity and Diversity 521 • Linguistic and Ethnic Diversity 522 • East Asian Cultures in Global Context 527

Geopolitical Framework: Struggles for Regional Dominance 529

The Evolution of China 529 • The Rise of Japan 532 • Postwar Geopolitics 532 • Global Dimensions of East Asian Geopolitics 535

EXPLORING GLOBAL CONNECTIONS China Courts the Portuguese-Speaking World 535

Economic and Social Development: A Core Region of the Global Economy 537

Japan's Economy and Society 537 • The Newly Industrialized Countries 538

EVERYDAY GLOBALIZATION The East Asian Smartphone Connection 539

Chinese Development 540

PEOPLE ON THE MOVE North Korea's Hidden Migration Streams 541

Social Development in East Asia 544

East Asia In Review 546

12 South Asia 548

Physical Geography and Environmental Issues: From Tropical Islands to Mountain Rim 551

The Four Physical Subregions of South Asia 551 • Environmental Issues in South Asia 554

WORKING TOWARD SUSTAINABILITY Green Schools and Eco-Tourism in Bhutan 555

South Asia's Monsoon Climates 557 • Climate Change in South Asia 558

Population and Settlement: The Demographic Dilemma 561

The Geography of Population Expansion 561 • Migration and the Settlement Landscape 562 • Agricultural Regions and Activities 562 • Urban South Asia 566

CITYSCAPES Karachi, Pakistan's Sprawling Megacity 567

Cultural Coherence and Diversity: A Common Heritage Undermined by Religious Rivalries 568

Origins of South Asian Civilizations 569 • The Caste System 571 • Contemporary Geographies of Religion 572 • Geographies of Language 574 • South Asians in a Global Cultural Context 577

EVERYDAY GLOBALIZATION Customer Support Call Centers in India 579

Geopolitical Framework: A Deeply Divided Region 579

South Asia Before and After Independence in 1947 580 • Ethnic Conflicts and Tensions in South Asia 583

PEOPLE ON THE MOVE Migration from Bangladesh to Assam, India 584

The Maoist Challenge 586 • International Geopolitics 586

Economic and Social Development: Rapid Growth and Rampant Poverty 587

Geographies of Economic Development 588

EXPLORING GLOBAL CONNECTIONS The Troubled Tourism Industry in the Maldives 592

Globalization and South Asia's Economic Future 593 • Social Development 594

South Asia In Review 596

13 Southeast Asia 598

Physical Geography and Environmental Issues: A Once-Forested Region 602

Patterns of Physical Geography 602 • The Deforestation of Southeast Asia 605 • Fires, Smoke, and Air Pollution 607 • Climate Change in Southeast Asia 607

WORKING TOWARD SUSTAINABILITY Malaysia's National Physical Plan 608

Population and Settlement: Subsistence, Migration, and Cities 609

Settlement and Agriculture 610 • Recent Demographic Change 612 • Urban Settlement 615

Cultural Coherence and Diversity: A Meeting Ground of World Cultures 617

The Introduction and Spread of Major Cultural Traditions 617

CITYSCAPES Baguio, the "Summer Capital of the Philippines" 619

Geography of Language and Ethnicity 621 • Southeast Asian Culture in Global Context 624

EXPLORING GLOBAL CONNECTIONS Baltimore Turns to the Philippines to Recruit Teachers for Inner-City Schools 625

Geopolitical Framework: War, Ethnic Strife, and Regional Cooperation 626

Before European Colonialism 626 • The Colonial Era 626 • The Vietnam War and Its Aftermath 629 • Geopolitical Tensions in Contemporary Southeast Asia 630 • International Dimensions of Southeast Asian Geopolitics 633

PEOPLE ON THE MOVE Burmese Migrants in Thailand 634

Economic and Social Development: The Roller-Coaster Ride of Developing Economies 636

Uneven Economic Development 636 • Globalization and the Southeast Asian Economy 642 • Issues of Social Development 643

EVERYDAY GLOBALIZATION The Southeast Asian Palm Oil Connection 644

Southeast Asia In Review 646

14 Australia and Oceania 648

Physical Geography and Environmental Issues: Varied Landscapes and Habitats 652

Topography of Australia and New Zealand 652 • Australia and New Zealand's Climates 653 • Oceania's Diverse Environments 654 • Australia's Unique Plants and Animals 655 • Complex Environmental Issues 657 • Climate Change in Oceania 658

WORKING TOWARD SUSTAINABILITY Sea-Level Rise and the Future of Low Islands 660

Population and Settlement: Migration, Cities, and Empty Spaces 661

Contemporary Population Patterns 662

CITYSCAPES Perth, Australia's Boomtown 664

Historical Geography 665 • Settlement Landscapes 666 • Diverse Demographic Paths 669

Cultural Coherence and Diversity: A Global Crossroads 671

Multicultural Australia 671 • Cultural Patterns in New Zealand 673 • The Mosaic of Pacific Cultures 673

PEOPLE ON THE MOVE Australia's Asylum Problem 675

Gender Geographies 676 • Sports in Oceania 676

EVERYDAY GLOBALIZATION Hawaiian Pidgin 677

Interactions with the Larger World 678

Geopolitical Framework: A Region of Dynamic Politics 679

Roads to Independence 679

EXPLORING GLOBAL CONNECTIONS Hypercommunication Comes to Oceania 681

Land and Native Rights Issues 682 • The Strategic Pacific 684

Economic and Social Development: Increasing Ties to Asia 685

The Australian and New Zealand Economies 685 • Oceania's Economic Diversity 687 • Oceania in Global Context 690 • Continuing Social Challenges 690

Australia and Oceania In Review 692

Glossary G-1

Text and Illustration Credits CR-1

Index I-1

Preface

Diversity Amid Globalization, Sixth Edition, is an issues-oriented textbook for college and university world regional geography classes that explicitly recognizes the vast geographic changes taking place because of globalization. With this focus, we join the many scholars who consider globalization to be the most fundamental reorganization of the world's socioeconomic, cultural, and geopolitical structure since the Industrial Revolution. That premise provides the point of departure and underlying assumptions for this book. Further, as geographers, we think it essential for our readers to understand and critique two interactive themes: the consequences of converging environmental, cultural, political, and economic systems inherent to globalization and the persistence—and even expansion—of geographic diversity and differences in the face of globalization. These two opposing forces, homogenization and diversification, are reflected in our book's title, *Diversity Amid Globalization.*

New to the Sixth Edition

- ***Working Toward Sustainability*** explores sustainability projects throughout the world, emphasizing positive environmental and social initiatives.
- ***Everyday Globalization*** illustrates how globalization permeates every aspect of one's life, even the most mundane and taken-for-granted, such as one's food, clothing, cell phones, and music.
- **Quick Response (QR) code links to Google Earth Virtual Tour Videos** appear in select sidebar features, providing mobile-ready, on-the-go virtual tours of the geography and places discussed in the sidebar.
- **Chapter opening** pages introduce readers to key themes and characteristics of the regions with large panoramic photographs, a selection of visual and brief textual previews of the chapter sections, and a real-world vignette.
- **Learning Objectives** listed at the start of each chapter help students prioritize key learning goals.
- **Review questions** at the end of each major thematic section help students check comprehension as they read.
- **Visual questions** integrated with select figures give students opportunities to apply critical thinking skills and perform visual analysis.
- **In Review** end-of-chapter sections provide a highly visual summary and review of each chapter, with integrated graphics, critical thinking questions, key terms, and launching points into MasteringGeography and author blogs.
- **Quick Response (QR) code links to Author Blogs** at the end of each chapter lead readers to two blogs where authors discuss everything from current events to their travels and field research. Both blogs are graphically rich with innovative maps and photos and help extend the print book with dynamically updated information and data.

New and Updated in Chapter 1: Concepts of World Geography

- ***Geography Matters.*** New discussion of fundamental geographic concepts, including areal differentiation, regions, and the cultural landscape.
- ***Geographer's Toolbox.*** New discussion of latitude and longitude, map projections, scale, chorographic maps, aerial photos, remote sensing, and GIS.
- **Expanded, integrated treatment of globalization.** A revised presentation of globalization, including the notions of "flat" versus "spiky" worlds.
- **Demographic transition revised.** Following the lead of professional demographers, a fifth stage has been added to the traditional demographic transition model to account for the current very low natural population rates in developed countries.
- ***The Nation-State Revisited.*** A critical view of the traditional nation-state concept sets the scene for regional material on post- and neocolonial tensions, microregionalism, ethnic separatism, migrant enclaves, and multicultural nationalism.

New and Updated in Chapter 2: Physical Geography and the Environment

- **An expanded and graphically rich section on climate controls.** This expanded section explains the climate controls of solar energy, latitude, land-water interactions, global pressure and wind systems, and topography.
- **An updated and expanded section on climate change and global warming.** Drawing upon the latest data from the Intergovernmental Panel on Climate Change's *Fifth Assessment Report* (2013–2014), this section presents not just the latest data about climate change and global warming, but also the complex international negotiations on limiting CO_2 emissions.
- **A new section on global energy issues.** Linked to the previous material on climate change and global warming, this new section discusses the geography of global energy resources, both renewable and nonrenewable, including material on hydraulic fracturing ("fracking").
- **Revised and expanded material on bioregions and biodiversity.** A more detailed cartographic depiction of biomes and bioregions is complemented by a fuller discussion of the world's ecological diversity, as well as the issues faced in protecting those environments around the globe.

Thematic Organization

Diversity Amid Globalization is organized around the world geographic regions of Africa, Europe, Asia, North America, and so on. However, our 12 regional chapters depart from traditional world regional textbooks that primarily describe each individual country within the region. Instead, we have placed that material online in the MasteringGeography Study Area. This leaves us free to use five important geographic themes as the structure for each regional chapter: First is *Physical Geography and Environmental Issues,* in which we not only describe the physical geography of each region, but also environmental issues, including climate change and energy. Next is *Population and Settlement,* where we examine the region's demography, migration patterns, land use, and settlement, including cities. Our third theme, *Cultural Coherence and Diversity,* covers the traditional topics of language and religion, but also examines the ethnic and cultural tensions resulting from globalization. Gender issues and popular culture topics such as sports and music are also included in this section. The next section, covering the *Geopolitical Framework,* examines the political geography of the region, taking on such issues as postcolonial tensions, ethnic conflicts, separatism, micro-regionalism, and global terrorism. We conclude each regional chapter with a section devoted to

Economic and Social Development. Here we explore each region's economic framework at both local and global scales and examine such social issues as health, education, and gender inequalities.

These 12 regional chapters follow two substantive introductory chapters that provide the geographic fundamentals of both human and physical geography. The first chapter, "Concepts of World Geography," begins by providing readers with background on the geographic dimensions of globalization, including a section on the costs and benefits of globalization according to proponents and opponents. Next is an introduction to the discipline of geography and its major concepts, which leads into a section called "The Geographer's Toolbox," where students are informed about such matters as map-reading, cartography, aerial photos, remote sensing, and GIS. This initial chapter concludes with a discussion of the concepts and tabular data that are used throughout the regional chapters.

Chapter Two, "Physical Geography and the Environment," builds an understanding of physical geography and environmental issues with discussions of geology; environmental hazards; weather, climate, and global warming; energy; hydrology and water stress; and global bioregions and biodiversity.

Chapter Features

- **Structured learning path.** Every chapter begins with an explicit set of learning objectives to provide students with the larger context of each chapter. Review questions after each section allow students to test their learning. Each chapter ends with an innovative, graphically rich "In Review" section, where students are asked to apply what they have learned from the chapter in an active-learning framework.
- **Comparable regional maps.** Of the many maps in each regional chapter, many are constructed on the same themes and with similar data so that readers can easily draw comparisons between different regions. Most regional chapters have maps of physical geography, climate, environmental issues, population density, migration, language, religion, and geopolitical issues.
- **Other chapter maps pertinent to each region.** The regional chapters also contain many additional maps illustrating important geographic topics such as global economic issues, social development, and ethnic tensions.
- **Comparable regional data sets.** Two thematic tables in each regional chapter facilitate comparisons between regions and provide important insight into the characteristics of each region. The first table provides population data on a number of issues, including fertility rates and proportions of the population under 15 and over 65 years of age, as well as net migration rates for each country within the region. The second table presents economic and social development data for each country, including gross national income per capita, gross domestic product growth, life expectancy, percentage of the population living on less than $2 per day, child mortality rates, and the United Nations gender inequality index.
- **Sidebar essays with Google Earth Video Tours.** Each of the regional chapters has five sidebars that expand on geographic themes; to further geographic understanding, three sidebars in each chapter contain hot links to Google Earth virtual tour videos. These sidebars are:

  Cityscapes, in which text, maps, photos, and hot links to virtual tour videos are combined to convey a sense of place for a major city within each region. These sidebars also speak to the fact that our globalized world is becoming increasingly urban.

  Working Toward Sustainability sidebars feature case studies that describe sustainability projects throughout the world, emphasizing positive environmental and social initiatives and their results.

  Exploring Global Connections uses case studies to investigate the many ways in which activities in different parts of the world are linked so that students understand that in a globalized world regions are neither isolated nor discrete.

  People on the Move sidebars capture the human geography behind contemporary migration as people relocate, legally and not so legally, as they respond to the varied currents and expressions of globalization.

  Everyday Globalization sidebars illustrate the many ways that globalization permeates one's everyday life, from food, to clothing, to cell phones, to music.
- **QR links to author blogs.** These links lead readers to two blogs where authors discuss everything from current events to their travels and field research. Both blogs are graphically rich with innovative maps and photos.

Acknowledgments

We have many people to thank for the conceptualization, writing, rewriting, and production of *Diversity Amid Globalization.* First, we'd like to thank the thousands of students in our world regional geography classes who have inspired us with their energy, engagement, and curiosity; challenged us with their critical insights; and demanded a textbook that better meets their need to understand the contemporary geography of their dynamic and complex world.

Next, we are deeply indebted to many professional geographers and educators for their assistance, advice, inspiration, encouragement, and constructive criticism as we labored through the different stages of this book. Among the many who provided invaluable comments on various drafts and editions of *Diversity Amid Globalization* or who worked on supporting print or digital material are:

Gilian Acheson, *Southern Illinois University, Edwardsville*
Joy Adams, *Humboldt State University*
Dan Arreola, *Arizona State University*
Bernard BakamaNume, *Texas A&M University*
Brad Baltensperger, *Michigan Technological University*
Max Beavers, *Samford University*
Laurence Becker, *Oregon State University*
Dan Bedford, *Weber State University*
James Bell, *University of Colorado*
Katie Berchak, *University of Louisiana, Lafayette*
William H. Berentsen, *University of Connecticut*
Kevin Blake, *Kansas State University*
Mikhail Blinnikov, *St. Cloud State University*
Karl Byrand, *University of Wisconsin, Sheboygan County*
Michelle Calvarese, *California State University, Fresno*
Craig Campbell, *Youngstown State University*
G. Scott Campbell, *College of DuPage*
Elizabeth Chacko, *George Washington University*
Philip Chaney, *Auburn University*
Xuwei Chen, *Northern Illinois University*
David B. Cole, *University of Northern Colorado*
Malcolm Comeaux, *Arizona State University*
Jonathan C. Comer, *Oklahoma State University*
Catherine Cooper, *George Washington University*
Jeremy Crampton, *George Mason University*
Kevin Curtin, *University of Texas at Dallas*

James Curtis, *California State University, Long Beach*
Dydia DeLyser, *Louisiana State University*
Francis H. Dillon, *George Mason University*
Jason Dittmer, *Georgia Southern University*
Jerome Dobson, *University of Kansas*
Caroline Doherty, *Northern Arizona University*
Vernon Domingo, *Bridgewater State College*
Roy Doyon, *Ball State University*
Dawn Drake, *Missouri Western State University*
Jane Ehemann, *Shippensburg University*
Chuck Fahrer, *Georgia College and State University*
Dean Fairbanks, *California State University, Chico*
Emily Fekete, *University of Kansas*
Caitie Finlayson, *Florida State University*
Doug Fuller, *George Washington University*
Gary Gaile, *University of Colorado*
Douglas Gamble, *University of North Carolina, Wilmington*
Sherry Goddicksen, *California State University, Fullerton*
Sarah Goggin, *Cypress College*
Reuel Hanks, *Oklahoma State University*
Steven Hoelscher, *University of Texas, Austin*
Erick Howenstine, *Northeastern Illinois University*
Tyler Huffman, *Eastern Kentucky University*
Peter J. Hugil, *Texas A&M University*
Eva Humbeck, *Arizona State University*
Shireen Hyrapiet, *Oregon State University*
Drew Kapp, *University of Hawaii, Hilo*
Ryan S. Kelly, *University of Kentucky*
Richard H. Kesel, *Louisiana State University*
Rob Kremer, *Front Range Community College*
Robert C. Larson, *Indiana State University*
Alan A. Lew, *Northern Arizona University*
Elizabeth Lobb, *Mt. San Antonio College*
Catherine Lockwood, *Chadron State College*
Max Lu, *Kansas State University*
Luke Marzen, *Auburn University*
Kent Matthewson, *Louisiana State University*
James Miller, *Clemson University*
Bob Mings, *Arizona State University*
Wendy Mitteager, *SUNY, Oneonta*
Sherry D. Morea-Oakes, *University of Colorado, Denver*
Anne E. Mosher, *Syracuse University*
Julie Mura, *Florida State University*
Tim Oakes, *University of Colorado*
Nancy Obermeyer, *Indiana State University*
Karl Offen, *University of Oklahoma*
Thomas Orf, *Las Positas College*
Kefa Otiso, *Bowling Green State University*
Joseph Palis, *University of North Carolina*
Jean Palmer-Moloney, *Hartwick College*
Bimal K. Paul, *Kansas State University*
Michael P. Peterson, *University of Nebraska, Omaha*
Richard Pillsbury, *Georgia State University*
Brandon Plewe, *Brigham Young University*
Jess Porter, *University of Arkansas at Little Rock*
Patricia Price, *Florida International University*
Erik Prout, *Texas A&M University*
Claudia Radel, *Utah State University*
David Rain, *United States Census Bureau*
Rhonda Reagan, *Blinn College*
Kelly Ann Renwick, *Appalachian State University*
Craig S. Revels, *Portland State University*
Pamela Riddick, *University of Memphis*
Scott M. Robeson, *Indiana State University*
Paul A. Rollinson, *Southwest Missouri State University*
Yda Schreuder, *University of Delaware*
Kathy Schroeder, *Appalachian State University*
Kay L. Scott, *University of Central Florida*
Patrick Shabram, *South Plains College*
Duncan Shaeffer, *Arizona State University*
Dimitrii Sidorov, *California State University, Long Beach*
Susan C. Slowey, *Blinn College*
Andrew Sluyter, *Louisiana State University*
Christa Smith, *Clemson University*
Joseph Spinelli, *Bowling Green State University*
William Strong, *University of Northern Alabama*
Philip W. Suckling, *University of Northern Iowa*
Curtis Thomson, *University of Idaho*
Suzanne Traub-Metlay, *Front Range Community College*
James Tyner, *Kent State University*
Nina Veregge, *University of Colorado*
Fahui Wang, *Louisiana State University*
Gerald R. Webster, *University of Alabama*
Keith Yearman, *College of DuPage*
Emily Young, *University of Arizona*
Bin Zhon, *Southern Illinois University, Edwardsville*
Henry J. Zintambia, *Illinois State University*
Sandra Zupan, *University of Kentucky*

In addition, we wish to thank the many publishing professionals who have been involved with the project. We start with Paul F. Corey, Managing Director, General Education, Science, Technology, and Business, for his early and continued support for this book project; Senior Geography Editor and good friend Christian Botting for his professional guidance, leadership, enduring patience, and high standards; Project Manager Sean Hale and Program Manager Anton Yakovlev for their daily attention to production matters and their graceful and diplomatic interaction with four demanding and sometimes cranky authors; Development Editor David Chelton for his editorial insights, cogent suggestions, and wry humor; Assistant Editor Bethany Sexton for gracefully taking care of the many incidental tasks connected to this project; Project Manager Connie Long and *Cenveo* Production Editor Heidi Allgair for somehow turning thousands of pages of manuscript into a finished product; and *International Mapping* Senior Project Manager Kevin Lear for his outstanding work on our maps. Thanks are due as well to Nicholas Baldo for his assistance on the Asian chapters and to Marina Medina Cordero for her timely production of chapter tables.

Last, the authors want to thank that special group of friends and family who were there when we needed you most—early in the morning and late at night; in foreign countries and at home; when we were on the verge of tears and rants, but needed lightness and laughter; for your love, patience, companionship, inspiration, solace, enthusiasm, and understanding. Words cannot thank you enough: Elizabeth Chacko, Meg Conkey, Rob Crandall, Marie Dowd, Evan and Eleanor Lewis, Karen Wigen, and Linda, Tom, and Katie Wyckoff.

Les Rowntree
Martin Lewis
Marie Price
William Wyckoff

About the Authors

Les Rowntree is a Research Associate at the University of California, Berkeley, where he researches and writes about global and local environmental issues. This career change came after more than three decades teaching both Geography and Environmental Studies at San Jose State University. As an environmental geographer, Dr. Rowntree's interests focus on international environmental issues, biodiversity conservation, and human-caused global change. He sees world regional geography as a way to engage and inform students by giving them the conceptual tools needed to critically assess the contemporary world. His current research and writing projects include a natural history book on California's Coast Range and essays on Europe's environmental issues; additionally he maintains an assortment of web-based natural history, geography, and environmental blogs and websites.

Martin Lewis is a Senior Lecturer in History at Stanford University, where he teaches courses on global geography. He has conducted extensive research on environmental geography in the Philippines and on the intellectual history of world geography. His publications include *Wagering the Land: Ritual, Capital, and Environmental Degradation in the Cordillera of Northern Luzon, 1900–1986* (1992), and, with Karen Wigen, *The Myth of Continents: A Critique of Metageography* (1997). Dr. Lewis has traveled extensively in East, South, and Southeastern Asia. His current research focuses on the geography of languages. In April 2009, Dr. Lewis was recognized by *Time* magazine as one of American's most favorite lecturers.

Marie Price is a Professor of Geography and International Affairs at George Washington University. A Latin American specialist, Dr. Price has conducted research in Belize, Mexico, Venezuela, Panama, Cuba, and Bolivia. She has also traveled widely throughout Latin America and Sub-Saharan Africa. Her studies have explored human migration, natural resource use, environmental conservation, and sustainability. She is a nonresident fellow of the Migration Policy Institute, a nonpartisan think tank that focuses on migration issues, and is a Vice-President of the American Geographical Society. Dr. Price brings to *Diversity Amid Globalization* a special interest in regions as dynamic spatial constructs that are shaped over time through both global and local forces. Her publications include the co-edited book *Migrants to the Metropolis: The Rise of Immigrant Gateway Cities* (2008) and numerous academic articles and book chapters.

William Wyckoff is a Professor of Geography in the Department of Earth Sciences at Montana State University, specializing in the cultural and historical geography of North America. He has written and co-edited several books on North American settlement geography, including *The Developer's Frontier: The Making of the Western New York Landscape* (1988), *The Mountainous West: Explorations in Historical Geography* (1995) (with Lary M. Dilsaver), *Creating Colorado: The Making of a Western American Landscape 1860–1940* (1999), and *On the Road Again: Montana's Changing Landscape* (2006). His most recent book, *How to Read the American West: A Field Guide*, appeared in the Weyerhaeuser Environmental Books series and was published in 2014 by the University of Washington Press. A World Regional Geography instructor for 26 years, Dr. Wyckoff emphasizes in the classroom the connections between the everyday lives of his students and the larger global geographies that surround them and increasingly shape their future.

Digital and Print Resources

For Teachers and Students

MasteringGeography™ with Pearson eText

The Mastering platform is the most widely used and effective online homework, tutorial, and assessment system for the sciences. It delivers self-paced tutorials that provide individualized coaching, focus on course objectives, and are responsive to each student's progress. The Mastering system helps teachers maximize class time with customizable, easy-to-assign, and automatically graded assessments that motivate students to learn outside of class and arrive prepared for lecture. MasteringGeography offers:

- **Assignable activities** that include MapMaster™ interactive map activities, *Encounter* Google Earth Explorations, video activities, Geoscience Animation activities, Map Projections activities, GeoTutor coaching activities on the toughest topics in geography, Dynamic Study Modules that customize the student's learnig experience, book questions and exercises, reading quizzes, Test Bank questions, and more.
- **A student Study Area** with MapMaster™ interactive maps, videos, Geoscience Animations, web links, glossary flashcards, "In the News" RSS feeds, chapter quizzes, PDF downloads of outline maps, an optional Pearson eText including versions for iPad and Android devices, and more. The Pearson eText gives students access to the text whenever and wherever they can access the Internet. The eText pages look exactly like the printed text and include powerful interactive and customization functions, including links to the multimedia.

Television for the Environment *Earth Report* Geography Videos on DVD (0321662989)

This three-DVD set helps students visualize how human decisions and behavior have affected the environment and how individuals are taking steps toward recovery. With topics ranging from the poor land management promoting the devastation of river systems in Central America to the struggles for electricity in China and Africa, these 13 videos from Television for the Environment's global *Earth Report* series recognize the efforts of individuals around the world to unite and protect the planet.

Television for the Environment *Life* World Regional Geography Videos on DVD (013159348X)

From Television for the Environment's global *Life* series, this two-DVD set brings globalization and the developing world to the attention of any world regional geography course. These 10 full-length video programs highlight matters such as the growing number of homeless children in Russia, the lives of immigrants living in the United States and trying to aid family still living in their native countries, and the European conflict between commercial interests and environmental concerns.

Television for the Environment *Life* Human Geography Videos on DVD (0132416565)

This three-DVD set is designed to enhance any human geography course. These DVDs include 14 full-length video programs from Television for the Environment's global *Life* series, covering a wide array of issues affecting people and places in the contemporary world, including the serious health risks of pregnant women in Bangladesh, the social inequalities of the "untouchables" in the Hindu caste system, and Ghana's struggle to compete in a global market.

Geoscience Animation Library 5th Edition DVD-ROM (0321716841)

Created through a collaboration among Pearson's leading geoscience authors, this resource offers over 100 animations covering the most-difficult-to-visualize topics in physical geography, oceanography, meteorology, physical geology, and earth science. Animations are provided as Flash files and preloaded into PowerPoint® slides.

Practicing Geography by Association of American Geographers (0321811151)

This book examines career opportunities for geographers and geospatial professionals in the business, government, nonprofit, and education sectors. A diverse group of academic and industry professionals shares insights on career planning, networking, transitioning between employment sectors, and balancing work and home life. The book illustrates the value of geographic expertise and technologies through engaging profiles and case studies of geographers at work.

Teaching College Geography by Association of American Geographers (0136054471)

This two-part resource provides a starting point for becoming an effective geography teacher. Part One addresses "nuts-and-bolts" teaching issues. Part Two explores being an effective teacher in the field, supporting critical thinking with GIS and mapping technologies, engaging learners in large geography classes, and promoting awareness of international perspectives and geographic issues.

Aspiring Academics by Association of American Geographers (0136048919)

Drawing on years of research, this set of essays is designed to help graduate students and faculty start their careers in geography and social and environmental sciences. *Aspiring Academics* stresses the interdependence of teaching, research, and service—and the importance of achieving a balance of professional and personal life—while doing faculty work. Each chapter provides accessible, forward-looking advice on topics that often cause the most stress in the first years of a college or university appointment.

For Teachers

Learning Catalytics

Learning Catalytics™ is a "bring your own device" student engagement, assessment, and classroom intelligence system. With Learning Catalytics, you can:

- Assess students in real time, using open-ended tasks to probe student understanding.
- Understand immediately where students are and adjust your lecture accordingly.
- Improve your students' critical thinking skills.
- Access rich analytics to understand student performance.
- Add your own questions to make Learning Catalytics fit your course exactly.
- Manage student interactions with intelligent grouping and timing.

Learning Catalytics has grown out of 20 years of cutting-edge research, innovation, and implementation of interactive teaching and peer instruction. Available integrated with MasteringGeography.

Instructor Resource Manual (Download) (0321972422)

The *Instructor Resource Manual,* authored by Karl Byrand of the University of Wisconsin, follows the new organization of the main text. It includes a sample syllabus, chapter learning objectives, lecture outlines, a list of key terms, and answers to the textbook's review and end-of-chapter questions. Discussion questions, classroom activities, and advice on how to integrate visual supplements (including MasteringGeography and Learning Catalytics resources) are integrated throughout the chapter lecture outlines.

TestGen/Test Bank (Download) (0321972449)

TestGen is a computerized test generator that lets instructors view and edit *Test Bank* questions, transfer questions to tests, and print tests in a variety of customized formats. Authored by Elizabeth Lobb of Mount San Antonio College, this *Test Bank* includes approximately 1,500 multiple-choice, true/false, and short-answer/essay questions. Questions are correlated with the book's learning objectives, the revised U.S. National Geography Standards, chapter-specific learning outcomes, and Bloom's Taxonomy. The *Test Bank* is also available in Microsoft Word® and is importable into Blackboard.

Instructor Resource DVD (0321972600)

The *Instructor Resource DVD* provides a collection of resources to help instructors make efficient and effective use of their time. All digital resources can be found in one well-organized, easy-to-access place. The IRC DVD includes

- All textbook images as JPEGs, PDFs, and PowerPoint™ Presentations
- Pre-authored Lecture Outline PowerPoint™ Presentations that outline the concepts of each chapter with embedded art and that can be customized to fit instructors' lecture requirements
- CRS "Clicker" Questions in PowerPoint™ format, which correlate with the book's learning objectives, the U.S. National Geography Standards, chapter-specific learning outcomes, and Bloom's Taxonomy
- The TestGen software, *Test Bank* questions, and answers
- Electronic files of the *IRM* and *Test Bank*

This Instructor Resource content is also available online via the Instructor Resources section of MasteringGeography and **www.pearsonhighered.com/irc.**

For Students

Goode's World Atlas, 22nd Edition (0321652002)

Goode's World Atlas has been the world's premiere educational atlas since 1923—and for good reason. It features over 250 pages of maps, from definitive physical and political maps to important thematic maps that illustrate the spatial aspects of many important topics. The 22nd Edition includes 160 pages of new, digitally produced reference maps, as well as new thematic maps on global climate change, sea-level rise, CO_2 emissions, polar ice fluctuations, deforestation, extreme weather events, infectious diseases, water resources, and energy production.

Pearson's Encounter Series

Pearson's Encounter Series provides rich, interactive explorations of geoscience concepts through Google Earth™ activities, covering topics in regional, human, and physical geography. For those who do not use MasteringGeography, explorations are available in print workbooks and in online quizzes at **www.mygeoscienceplace.com.** Each exploration consists of a worksheet, online quizzes whose results can be emailed to instructors, along with a corresponding Google Earth™ KMZ file.

- *Encounter World Regional Geography* by Jess C. Porter (0321681754)
- *Encounter Human Geography* by Jess C. Porter (0321682203)
- *Encounter Physical Geography* by Jess C. Porter and Stephen O'Connell (0321672526)
- *Encounter Geosystems* by Charlie Thomsen (0321636996)
- *Encounter Earth* by Steve Kluge (0321581296)

Dire Predictions: Understanding Global Warming by Michael Mann and Lee R. Kump (0136044352)

Dire Predictions is appropriate for any science or social science course in need of a basic understanding of the reports from the Intergovernmental Panel on Climate Change. These periodic reports evaluate the risk of climate change brought on by humans. But the sheer volume of scientific data remains inscrutable to the general public, particularly to those who may still question the validity of climate change. In just over 200 pages, this practical text presents and expands upon the essential findings in a visually stunning and undeniably powerful way for the lay reader. Scientific findings that provide validity to the implications of climate change are presented in clear-cut graphic elements, striking images, and understandable analogies.

The world's diverse regions in context

Conveying a strong sense of place and global context, this contemporary approach to world regional geography helps students understand the unique connections among the world's diverse regions.

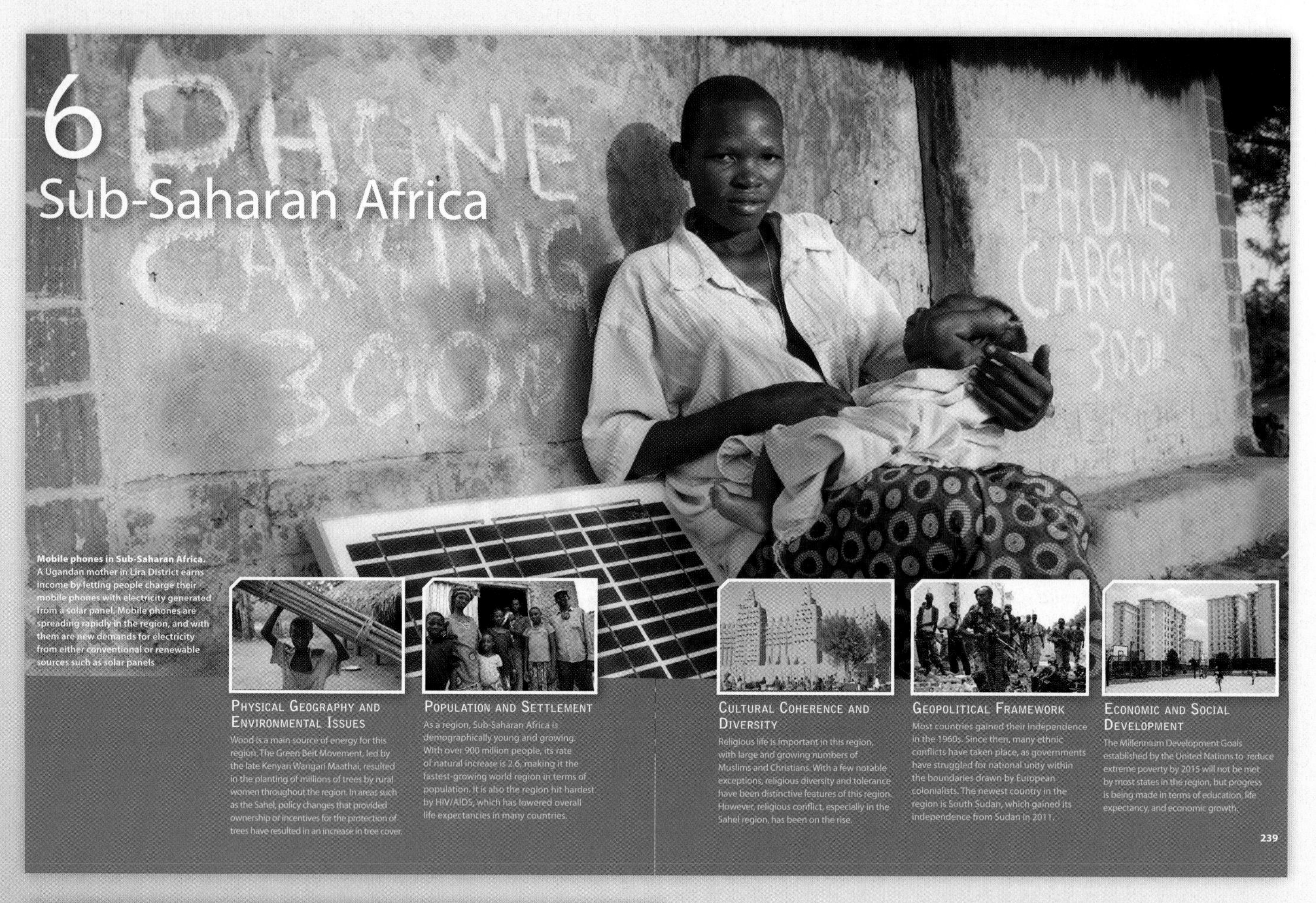

6 Sub-Saharan Africa

Mobile phones in Sub-Saharan Africa. A Ugandan mother in Lira District earns income by letting people charge their mobile phones with electricity generated from a solar panel. Mobile phones are spreading rapidly in the region, and with them are new demands for electricity from either conventional or renewable sources such as solar panels

Physical Geography and Environmental Issues
Wood is a main source of energy for this region. The Green Belt Movement, led by the late Kenyan Wangari Maathai, resulted in the planting of millions of trees by rural women throughout the region. In areas such as the Sahel, policy changes that provided ownership or incentives for the protection of trees have resulted in an increase in tree cover.

Population and Settlement
As a region, Sub-Saharan Africa is demographically young and growing. With over 900 million people, its rate of natural increase is 2.6, making it the fastest-growing world region in terms of population. It is also the region hit hardest by HIV/AIDS, which has lowered overall life expectancies in many countries.

Cultural Coherence and Diversity
Religious life is important in this region, with large and growing numbers of Muslims and Christians. With a few notable exceptions, religious diversity and tolerance have been distinctive features of this region. However, religious conflict, especially in the Sahel region, has been on the rise.

Geopolitical Framework
Most countries gained their independence in the 1960s. Since then, many ethnic conflicts have taken place, as governments have struggled for national unity within the boundaries drawn by European colonialists. The newest country in the region is South Sudan, which gained its independence from Sudan in 2011.

Economic and Social Development
The Millennium Development Goals established by the United Nations to reduce extreme poverty by 2015 will not be met by most states in the region, but progress is being made in terms of education, life expectancy, and economic growth.

239

▲ **NEW! Chapter opening** pages introduce readers to key themes and characteristics of the regions with large panoramic photographs, a selection of visual and brief textual previews of the chapter's themes and sections, followed by a real-world vignette.

Exploring **Global Connections**

Crisis Mapping in Haiti After the Earthquake

In response to the 2010 Haitian earthquake, social media, humanitarian organizations, and crisis mappers joined forces in a new and unique way that has changed how governments and civil societies will respond to complex humanitarian crises in the future. One of the leaders in the crisis-mapping movement is Patrick Meier, who was a key player in assembling the crisis-mapping team for Haiti. In 2013, Patrick Meier was a National Geographic Emerging Explorer where he blogged about the Haitian experience.

Crisis Mapping is the leveraging of mobile devices (texts and tweets), Web-based applications, participatory maps, satellite imagery, and crowd-sourced event data for rapid responses to complex humanitarian crises. Humanitarian workers need precise, real-time information that localities in crisis are often unable to provide. Working through an African-created platform called Ushahidi, crisis mappers assembled at Tufts University, just outside of Boston, Massachusetts, gathered tweets and text messages from Haitians (with translations provided by Haitians living in the U.S.). New global connections were forged, resulting in maps used by first responders that saved lives.

▼ **Figure 5.1.1 Crisis Mapping for Port-au-Prince** A portion of the map created using Open Street Map in the days after the Haitian earthquake. The circles represent the number of individual reports for that particular areas.

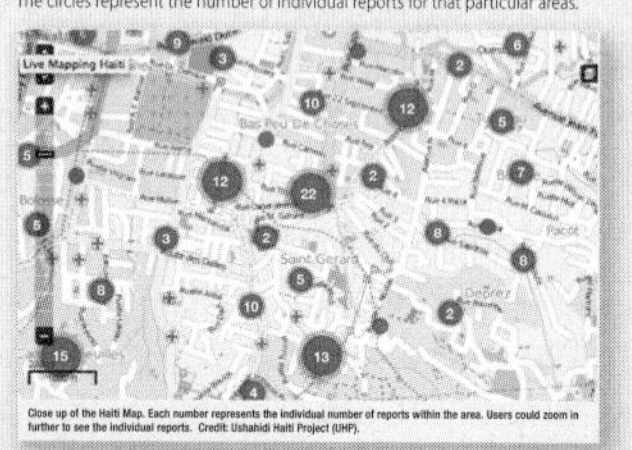

Close up of the Haiti Map. Each number represents the individual number of reports within the area. Users could zoom in further to see the individual reports. Credit: Ushahidi Haiti Project (UHP).

Two free and open-source mapping platforms were critical in moving crisis mapping forward: Ushahidi and Open Street Map. Ushahidi was developed by African bloggers who sought to report on postelection violence in Kenya in 2008 that was not covered by the media. Ushahidi (Swahili for "witness") relies on a Google Web-based map interface that plots acts of violence sent by crowd-sourced text messages. In the case of Haiti, Open Street Map was incorporated into the platform to allow for the construction of an extremely detailed and interactive map that people could use in the field and drill down to individual reports (Figure 5.1.1). Key to the success of the project was the creation of a team of crisis mappers (initially students at Tufts University) and translators who scanned for tweets. Later, through collaboration with Haiti's largest mobile phone provider, a texting number was set up so that anyone in Haiti could text urgent needs. As thousands of texts poured in, the Haitian population in the United Sates was mobilized to translate the texts from Haitian Creole to English so that the mappers could add the geo-referenced information to the map. As the real-time map grew, so did the number of contributors and users. The U.S. Coast Guard and Marines and various humanitarian groups on the ground in Port-au-Prince relied almost exclusively on its output.

Future Crisis Mapping Since the Haiti experience with crisis mapping, similar efforts have been used in response to earthquakes in Chile and Russia. An organization of crisis-mapping volunteers has formed to respond to future events. As Patrick Meier likes to say: "To map the world is to know it. But to map the world live is to change it before it's too late."

Source: Adapted from www.newswatch.nationalgeographic.com-How Crisis Mapping Saved Lives in Haiti, July 2, 2012.

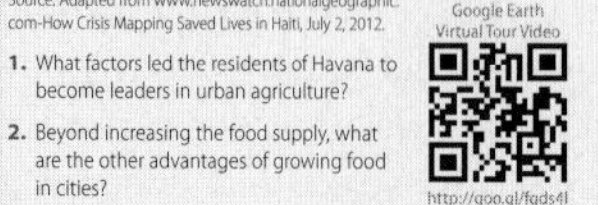

1. What factors led the residents of Havana to become leaders in urban agriculture?
2. Beyond increasing the food supply, what are the other advantages of growing food in cities?

Google Earth Virtual Tour Video
http://goo.gl/fqds4l

◄ Updated **Exploring Global Connections** case studies reinforce the theme of globalization by illustrating the interesting and sometimes unexpected interconnections between regions. Examples include the Catholic Church and the Argentine Pope; Crisis mapping in Haiti after the earthquake of 2010; the heroin and opium trade from Afghanistan; Russian meteorite fragments going global; and many others.

Explore critical and contemporary issues

A focus on critical and socially-conscious topics—sustainability, gender issues, globalization impacts, global climate change—engage and involve students on multiple levels.

► **NEW! Working Toward Sustainability** features show diverse applications of how sustainability initiatives apply to people, groups, and settlements in different places and at different scales, emphasizing positive environmental and social initiatives and their outcomes. Examples include Japan's smart city movement; green schools and eco-tourism in Bhutan; Germany's energy transformation; Lake Balkal's Success Story; and many others.

Working Toward **Sustainability**

Lake Baikal's Success Story

Lake Baikal, located in southern Siberia (see **Figure 9.2**), has become one of Russia's most important settings for protecting its vast natural environment and for developing sustainable economic activities such as ecotourism. Remarkably, the lake contains about 20 percent of Earth's unfrozen fresh surface water. Not only is the lake almost 400 miles (644 kilometers) long, but it is also 5300 feet (1600 meters) deep, occupying a structural rift in the continental crust (**Figure 9.2.1**). It remains home to a large array of unique (or *endemic*) species found nowhere else on the planet, including the world's only freshwater seal.

A Threatened Treasure Lake Baikal suffered during the later Soviet period. Large pulp and paper mills were located along the lakeshore in the 1950s and 1960s because abundant forests were nearby and the lake's amazingly pure water was useful in producing high-quality wood fibers. Unfortunately, these industries discharged pollutants into the lake and into the surrounding atmosphere. With factory discharges, the lake's purity rapidly declined. However, things have improved since the early 1990s. Stricter regulations have reduced industrial pollution. Indeed, the lake has become *the* national "poster child" of the Russian environmental movement. In 1996, the lake became a UNESCO World Heritage Site, and three years later the Russian government formally created legislation designed to protect the lake.

Recently, the lake became the center of attention as Russia planned to expand a major Siberian oil pipeline linking Russian resources to East Asian markets. High oil prices have encouraged the Russians to make large new investments in their petroleum industry, but many environmentalists feared that these growing global demands for oil might have destructive local consequences for Lake Baikal. In 2006, major protests and petition-signing drives opposed the planned pipeline's close proximity to the north shore of the lake. The initiative caught the attention of Russian President Putin, who dramatically ordered that the pipeline be directed farther away from the lake's fragile ecosystem.

▲ **Figure 9.2.1 Lake Baikal** Southern Siberia's Lake Baikal is one of the world's largest deep-water lakes. Industrialization devastated water quality after 1950 as pulp and paper factories poured wastes into the lake. Recent cleanup efforts have helped, but environmental threats remain.

1. Examine the maps of agricultural production (**Figure 9.5**), population (**Figure 9.14**) and industrial zones (**Figure 9.39**) in relation to Lake Baikal. How do these patterns help explain the relatively pristine character of the lake today?
2. Follow the Trans-Siberian Railroad in Google Earth along Lake Baikal's southern shoreline. What are the dominant features visible on your trip?

Google Earth Virtual Tour Video

http://goo.gl/JDxwFv

Expanded coverage of gender issues, food, art, music, film, and sports brings these high-interest cultural topics to the forefront.

Everyday **Globalization**

How the Russian Domain Shapes the Virtual World

It's a tough virtual world out there, especially when tanks, pirates, spacecraft, and battleships threaten us on every front.

The Russian Connection As every American college student knows, the video game and online gaming landscape has changed dramatically in the last 30 years, since Russian Alexie Pajitnov invented Tetris at the Soviet Academy of Sciences in 1984 (it has sold over 125 million copies on 30 different platforms). Less apparent, however, is the enduring connection between the Russian domain and the multi-billion-dollar video-gaming industry. Russian (1C Company), Belorussian (Wargaming.net), and Ukrainian (Persha Studia) software companies have all played pivotal roles in shaping the world's virtual landscapes. This includes outmaneuvering pirates in the Caribbean (*Age of Pirates*), organizing massive online tank-based clan wars (with more than 190,000 players online simultaneously) somewhere on the world map (*World of Tanks*), and completing house-to-house searches on the European battlefield during World War II (*Men of War*) (**Figure 9.5.1**).

Regional Advantages How did this region become so central in creating the virtual worlds shared today by hundreds of millions of gamers? Part of the answer is no doubt Soviet-era investments in pioneering computer technology and software development, much of it linked with the Cold War. Add to this a generation of sophisticated, technically trained computer geeks such as Pajitnov (who now lives in Washington State and works for an American software gaming company), who were well positioned to master the programming challenges of the budding industry. The Russian domain also offers a less expensive and less regulated environment where programmers have enjoyed considerable intellectual freedom beneath the radar of the bureaucracy.

Russia's Microsoft Although Belarus and Ukraine software developers certainly participate in the industry, Russia dominates the game. Boris Nuraliev was one of the corporate founders of the movement in the early 1990s. He created the 1C Company—often called Russia's Microsoft—which moved from the rather ordinary world of business software into the more extraordinary world of gaming (*Theater of War, Kings Bounty: The Legend, Pacific Fighters*, etc.). Today the company, based in Moscow, employs almost 1000 people (including 250 internal game developers) and is the largest game publisher and developer in the region. The company is also the most visible participant in the annual KRI (Russian Game Developers) Conference (begun in 2003), which is *the* place to be if you want to know the latest about online gaming, anti-piracy initiatives, or virtual worlds coming soon to a screen near you.

You might ponder the cultural significance of this massive Russian participation in the creation of our virtual worlds (**Figure 9.5.2**). Think about the landscapes we navigate, the strategic challenges we face, and the fascinating mix of fact (historical tanks and weapons, battle settings, and costumes are meticulously and accurately displayed) and fancy that makes up that world. Not surprisingly, a little bit of traditional Russian culture also gets passed our way. Just listen to the melody most associated with Tetris: Nikolai Nekrasov wrote the poem titled *Korobeiniki* in 1861, and the verse later became a Russian folk song that all Tetris enthusiasts have heard endless times as they skillfully maneuver their tetrominoes into place.

▲ **Figure 9.5.1 Belorussian Tanks Roll into California** The World of Tanks, a popular multiplayer online game, is being promoted in this publicity shot at the 2013 E3 Video Game Expo in Los Angeles. The game is produced by Wargaming.net, one of many Belorussian companies specializing in this global industry.

▲ **Figure 9.5.2 Russia's Blossoming Virtual World** These Russian youngsters eagerly explore the gaming cyberworld at the 2009 GameWorld interactive entertainment exhibition in Moscow.

► **NEW! *Everyday Globalization*** features illustrate how globalization permeates every aspect of one's life—even the most ordinary and taken-for-granted, such as health care, food, education, cell phones, and video games.

Structured to facilitate learning

Each regional chapter is organized into five thematic sections— Physical and Environmental Geography, Population and Settlement, Cultural Coherence and Diversity, Geopolitical Framework, and Economic and Social Development—to encourage cross-regional comparisons and highlight issues in today's globalized world. Each chapter now also includes a new active learning path to help students engage with important concepts and check their understanding.

▶ **NEW! Learning Objectives** listed at the start of each chapter help students prioritize key learning goals.

LEARNING OBJECTIVES

After reading this chapter you should be able to:

- Describe how the region's fragile, often arid setting shapes the region's contemporary environmental challenges.
- Explain how latitude and topography produce the region's distinctive patterns of climate.
- Describe four distinctive ways in which people have learned to adapt their agricultural practices to the region's arid environment.
- Summarize the major forces shaping recent migration patterns within the region.
- List the major characteristics of Islam and its key patterns of diffusion.

▶ **NEW! Review Questions** at the end of each major thematic section help students check their comprehension of the material as they read.

REVIEW

7.1 Describe the climatic changes you might experience as you travel from the eastern Mediterranean coast to the highlands of Yemen. What are some of the key climatic variables that explain these variations?

7.2 Discuss five important human modifications of the Southwest Asian and North African environment, and assess whether these changes have benefited the region.

▶ **NEW! Quick Response (QR) code links to Google Earth Virtual Tour Videos** appear in select sidebar features, providing mobile-ready, on-the-go virtual tours of the places discussed in the sidebar.

1. Find Cairo on Google Earth, and examine parts of the old city as well as new suburban developments to the east. Describe three key visual differences you can detect between these old and new settlement patterns.
2. Find a work of literature (novel, short story, poem) focused on an urban setting in your region, and identify a passage (such as the one by Mahfouz) that captures a local sense of place.

Google Earth Virtual Tour Video

http://goo.gl/D38KO1

▶ **NEW! Visual Questions** integrated into key figures in each chapter section give students opportunities to apply critical thinking skills and visual analysis.

▲ **Figure 7.43 Development Issues in Southwest Asia and North Africa: Childhood Mortality** Wealthier nations such as Israel and the United Arab Emirates have very low rates of childhood mortality, but poor countries such as Sudan, Morocco, and Iraq continue to struggle with very high rates. **Q: Why might it be argued that childhood mortality is a good measure of development?**

▶ **NEW! In Review** end-of-chapter features provide a highly visual summary and review of each chapter, with integrated graphics, critical thinking questions, key terms, Quick Response code links to the author blogs, and a launching point into MasteringGeography.

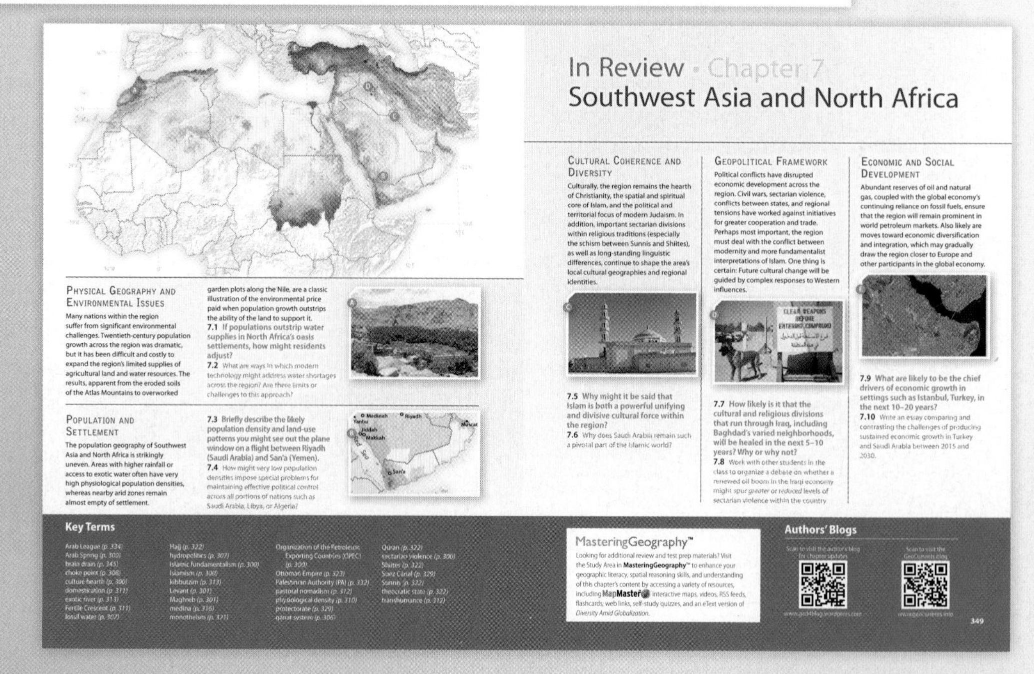

Visualize the world's places and people

A pedagogically-oriented cartography program provides many of the same thematic maps in each regional chapter. This system allows students to compare and contrast concepts and data both within and between regions. Large-format photos, satellite and remote-sensed imagery, paired population pyramids, and other visualizations of current data help students experience and understand the world's diverse regions.

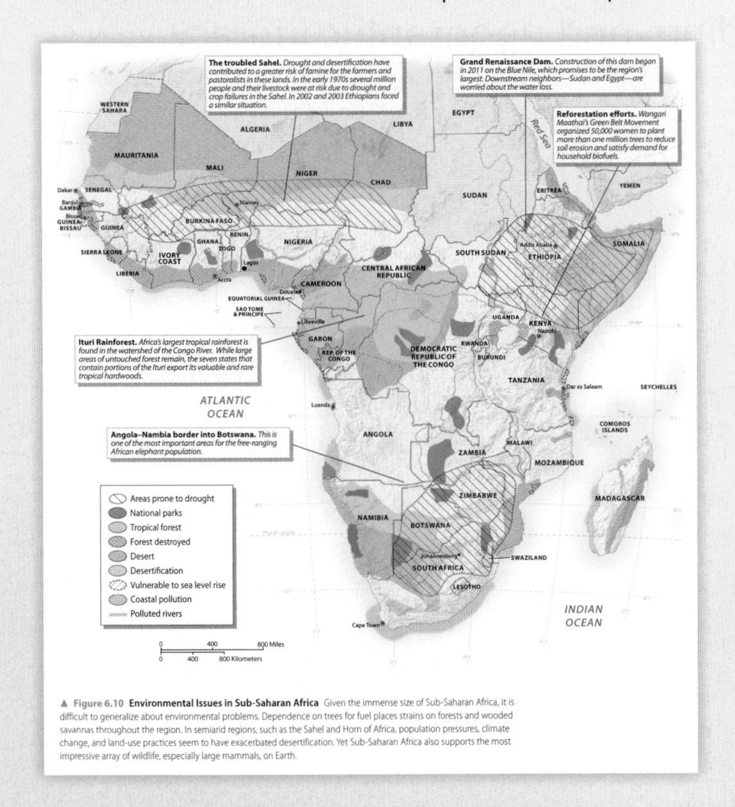

▲ **Figure 6.10 Environmental Issues in Sub-Saharan Africa** Given the immense size of Sub-Saharan Africa, it is difficult to generalize about environmental problems. Dependence on trees for fuel places strains on forests and wooded savannas throughout the region. In semiarid regions, such as the Sahel and Horn of Africa, population pressures, climate change, and land-use practices seem to have exacerbated desertification. Yet Sub-Saharan Africa also supports the most impressive array of wildlife, especially large mammals, on Earth.

▼ **Figure 10.31 Qinghai–Tibet Railway** China is engaged in a massive program to extend roads and railroads into the western half of the country. The main connection with the Tibetan Plateau is the recently completed Qinghai–Tibet Railway. This railway has resulted in a large increase in tourism in Tibet. **Q: Why have the railroad tracks been elevated so high above the valley floor?**

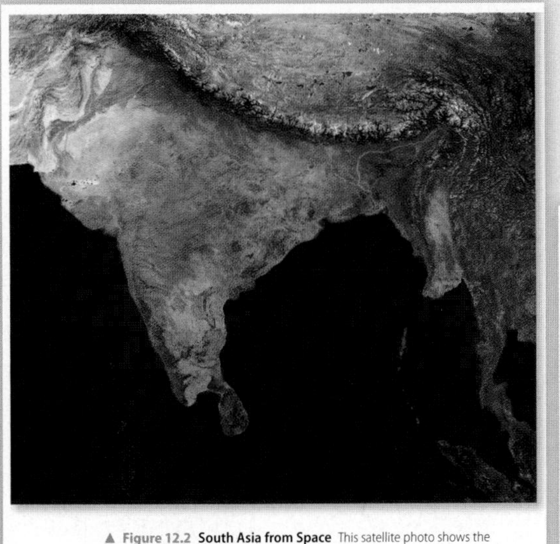

▲ **Figure 12.2 South Asia from Space** This satellite photo shows the four physical subregions of South Asia, from the snow-clad Himalayas in the north to the islands of the south. The irrigated lands of the Indus River Valley in Pakistan are clearly visible in the upper left.

▼ **Figure 13.13 Population Pyramids of Singapore and the Philippines** (a) Singapore has a very low birth rate and an aging population; as a result, its government encourages the immigration of skilled workers. (b) The Philippines, by contrast, has a high birth rate and a young population.

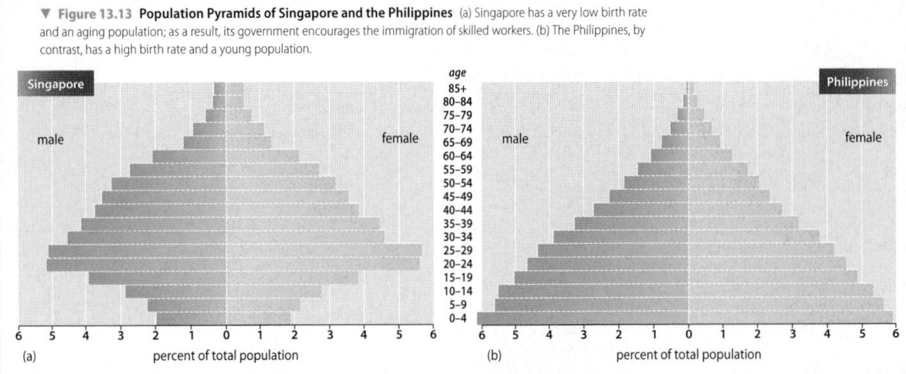

MasteringGeography™

MasteringGeography delivers engaging, dynamic learning opportunities—focusing on course objectives and responsive to each student's progress—that are proven to help students absorb world regional course material and understand difficult geographic concepts.

Improve spatial reasoning, map reading, and geographic literacy

MapMaster™ is a powerful tool that presents assignable layered thematic and place name interactive maps at world and regional scales for students to test their geographic literacy and spatial reasoning skills.

▶ **MapMaster Layered Thematic Interactive Map Activities** act as a mini-GIS tool, allowing students to layer various thematic maps to analyze spatial patterns and data at regional and global scales and answer customizable multiple-choice and short-answer questions organized by region and theme. Includes zoom and annotation functionality, and hundreds of map layers with current data from sources such as the U.S. Census, United Nations, CIA, World Bank, and Population Reference Bureau.

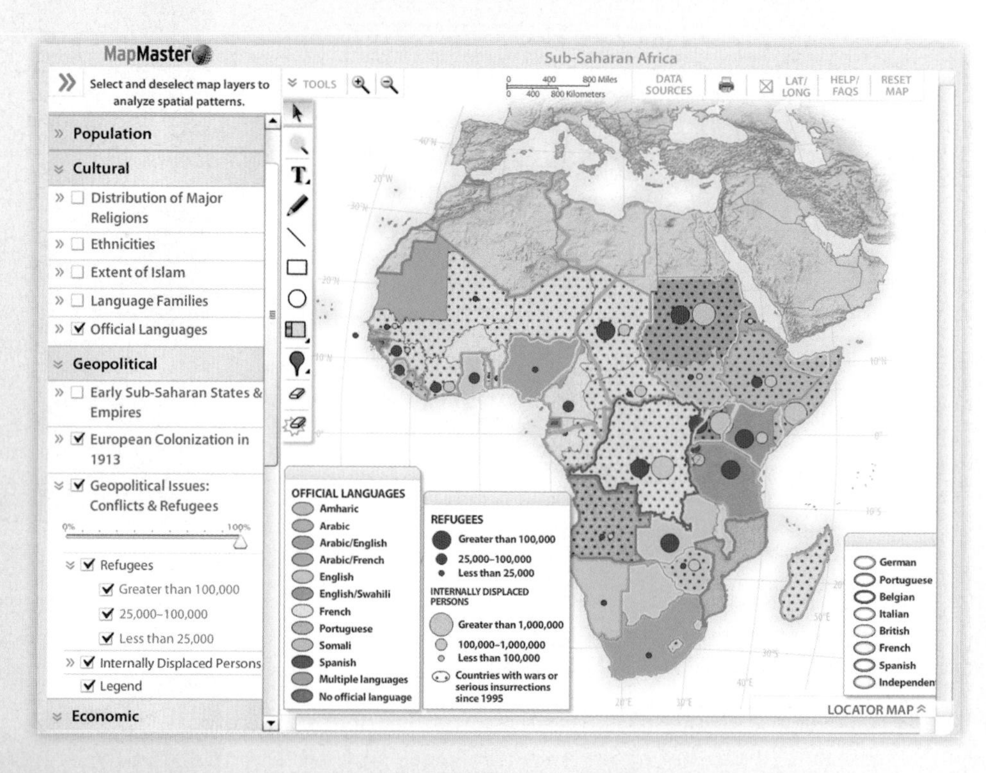

▶ **MapMaster Place Name Interactive Map Activities** have students identify place names of political and physical features at regional and global scales, explore select recent country data from the CIA World Factbook, and answer associated customizable assessment questions.

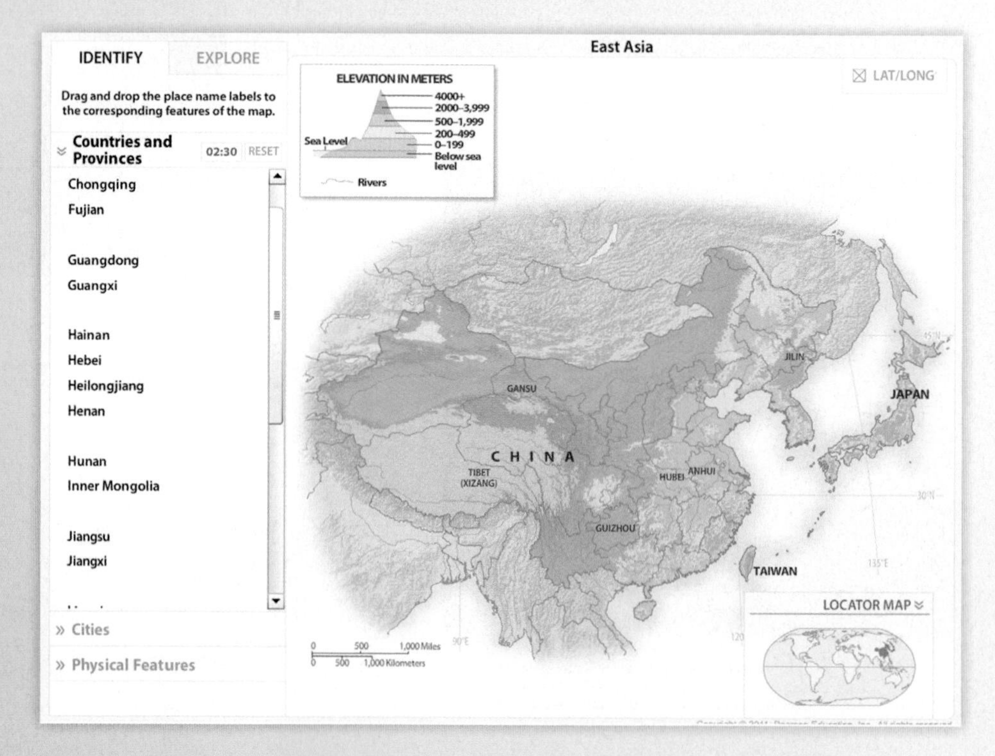

Develop a sense of place and understanding of process

◀ **Encounter Activities** provide rich, interactive explorations of geoscience concepts through Google Earth™ activities, exploring a range of topics in world regional geography. Dynamic assessment includes questions related to core world regional geography concepts. All explorations include corresponding Google Earth KMZ media files, and questions include hints and specific wrong-answer feedback to coach students towards mastery of the concepts.

▶ **Geography videos** provide students with a sense of place and allow them to explore a range of locations and topics. Covering issues of economy, development, globalization, climate and climate change, culture and more, all videos are available with customizable multiple choice questions, with hints and wrong-answer feedback, allowing teachers to test students' understanding and application of concepts.

▼ **Thinking Spatially and Data Analysis &** NEW! **GeoTutor** activities enable students to master the toughest concepts and develop spatial reasoning and critical thinking skills by identifying and labeling features from maps, illustrations, graphs, and charts. Students then examine related data sets, answering multiple-choice and increasingly higher-order conceptual questions, which include hints and specific wrong-answer feedback.

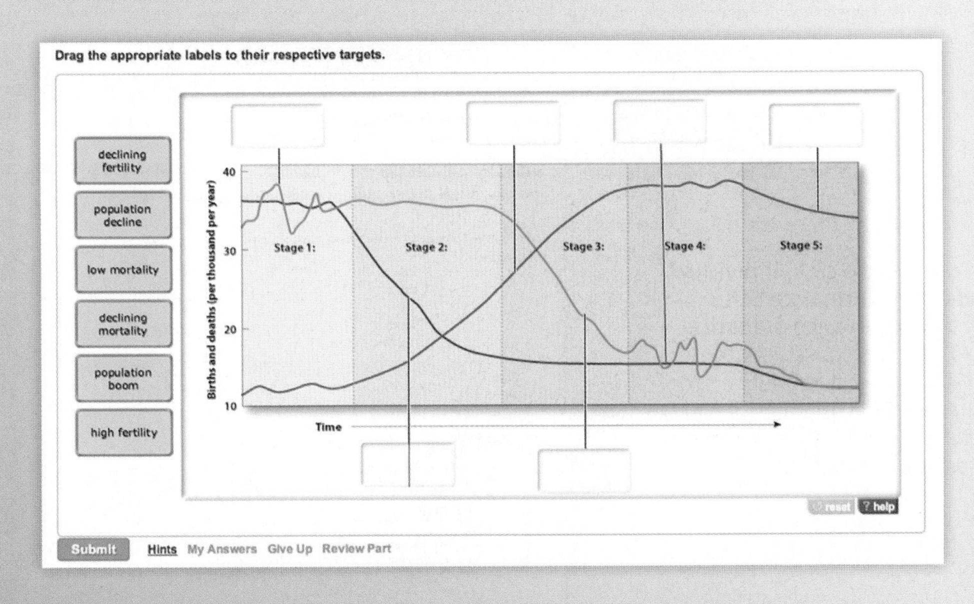

Dynamic Study Modules
Personalize each student's learning experience with Dynamic Study Modules. Created to allow students to study on their own and be better prepared to achieve higher scores on their tests. Mobile app available for iOS and Android devices for study on the go.

Student Study Resources in **MasteringGeography** include:

- MapMaster™ interactive maps
- Geography videos
- Select Geoscience Animations
- Practice quizzes
- "In the News" RSS feeds
- Glossary flashcards
- Optional Pearson eText and more

Callouts to MasteringGeography appear at the end of each chapter to direct students to extend their learning beyond the textbook.

MasteringGeography™

With the Mastering gradebook and diagnostics, you'll be better informed about your students' progress than ever before. Mastering captures the step-by-step work of every student—including wrong answers submitted, hints requested, and time taken at every step of every problem—all providing unique insight into the most common misconceptions of your class.

Quickly monitor and display student results

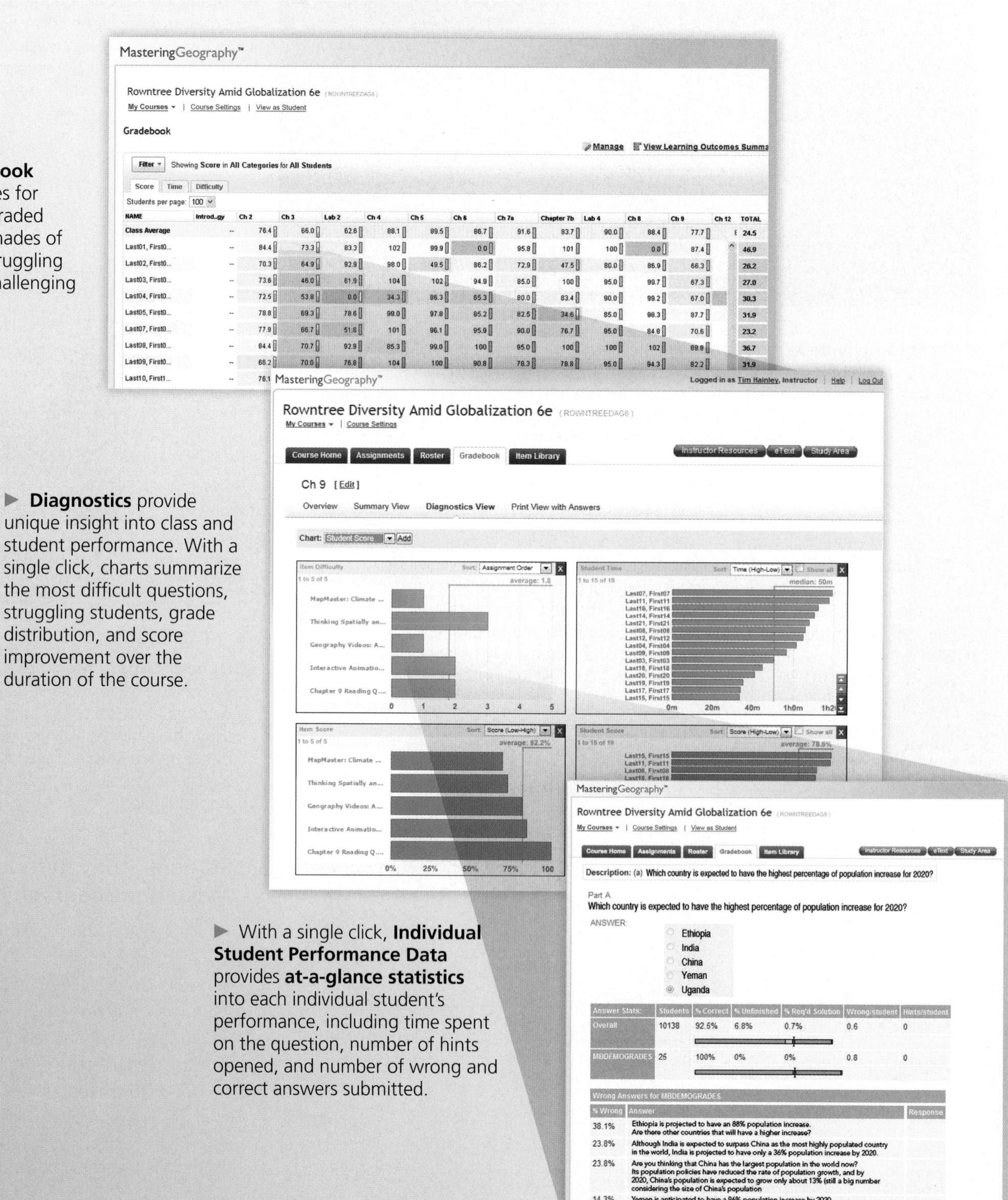

▶ The **Gradebook** records all scores for automatically graded assignments. Shades of red highlight struggling students and challenging assignments.

▶ **Diagnostics** provide unique insight into class and student performance. With a single click, charts summarize the most difficult questions, struggling students, grade distribution, and score improvement over the duration of the course.

▶ With a single click, **Individual Student Performance Data** provides **at-a-glance statistics** into each individual student's performance, including time spent on the question, number of hints opened, and number of wrong and correct answers submitted.

Easily measure student performance against your Learning Outcomes

Learning Outcomes

MasteringGeography provides quick and easy access to information on student performance against your learning outcomes and makes it easy to share those results.

- Quickly add your own learning outcomes, or use publisher-provided ones, to track student performance and report it to your administration.
- View class and individual student performance against specific learning outcomes.
- Effortlessly export results to a spreadsheet that you can further customize and/or share with your chair, dean, administrator, and/or accreditation board.

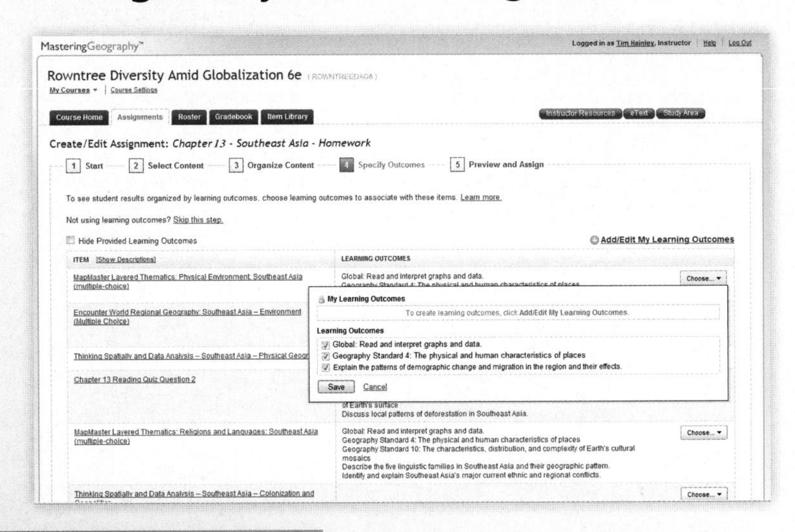

Easy to customize

Customize publisher-provided items or quickly add your own. MasteringGeography makes it easy to edit any questions or answers, import your own questions, and quickly add images, links, and files to further enhance the student experience.

Upload your own video and audio files from your hard drive to share with students, as well as record video from your computer's webcam directly into MasteringGeography—no plug-ins required. Students can download video and audio files to their local computer or launch them in Mastering to view the content.

learning | catalytics

NEW! Learning Catalytics is a "bring your own device" student engagement, assessment, and classroom intelligence system. With Learning Catalytics you can:

- Assess students in real time, using open-ended tasks to probe student understanding.
- Understand immediately where students are and adjust your lecture accordingly.
- Improve your students' critical-thinking skills.
- Access rich analytics to understand student performance.
- Add your own questions to make Learning Catalytics fit your course exactly.
- Manage student interactions with intelligent grouping and timing.

Learning Catalytics is a technology that has grown out of twenty years of cutting edge research, innovation, and implementation of interactive teaching and peer instruction. Available integrated with MasteringGeography.

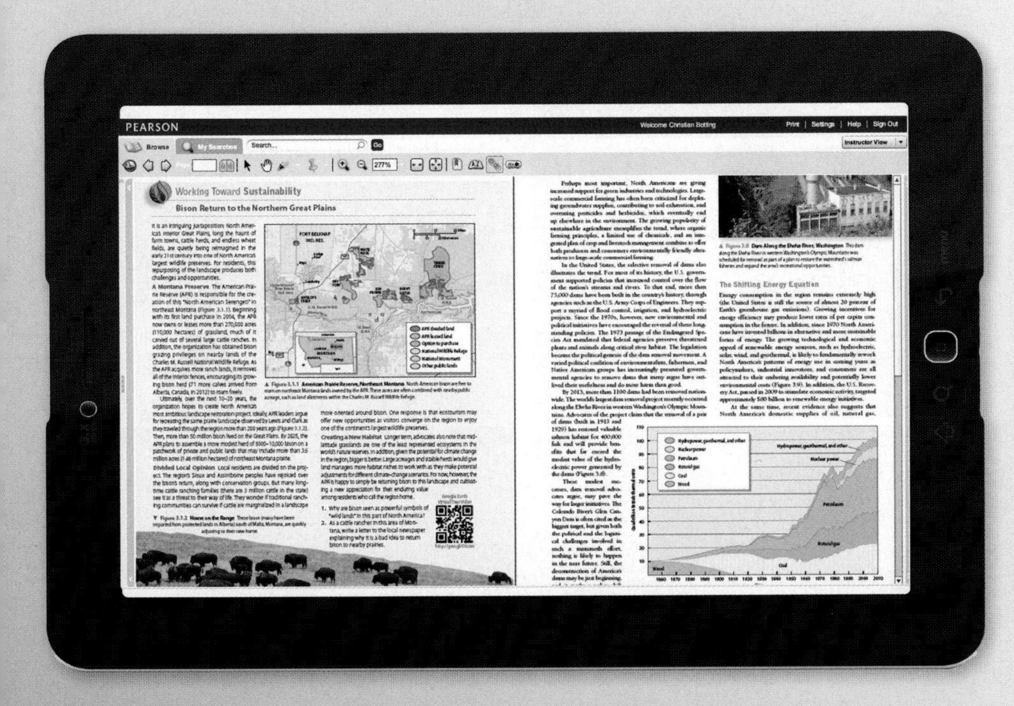

Pearson eText gives students access to ***Diversity Amid Globalization*, Sixth Edition** whenever and wherever they can access the Internet. The eText pages look exactly like the printed text, and include powerful interactive and customization functions. Users can create notes, highlight text in different colors, create bookmarks, zoom, click hyperlinked words and phrases to view definitions, and view as a single page or as two pages. Pearson eText also links students to associated media files, enabling them to view videos as they read the text, and offers a full-text search and the ability to save and export notes. The Pearson eText also includes embedded URLs in the chapter text with active links to the Internet.

The Pearson eText app is a great companion to Pearson's eText browser-based book reader. It allows existing subscribers who view their Pearson eText titles on a Mac or PC to additionally access their titles in a bookshelf on the iPad or an Android tablet either online or via download.

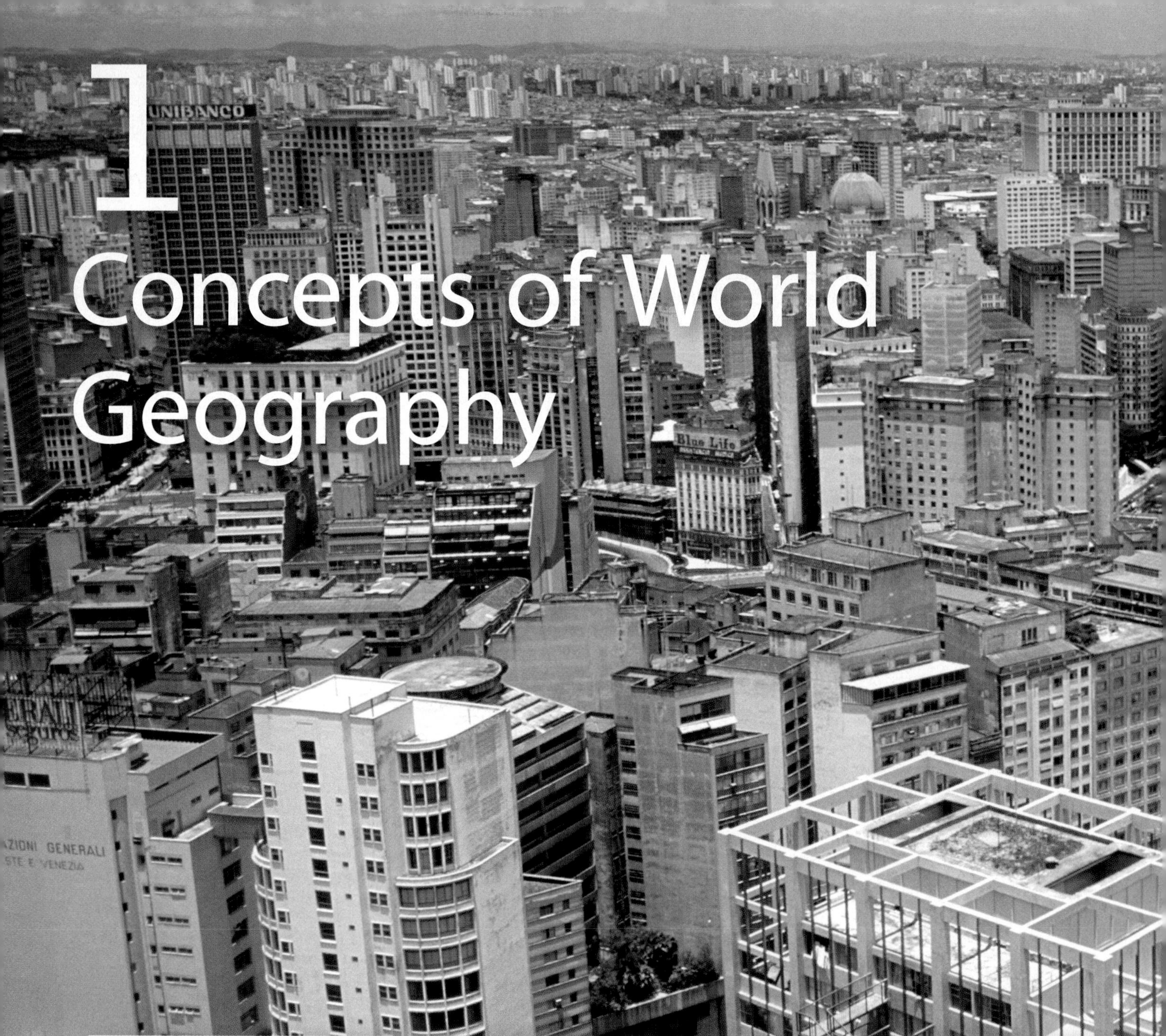

1 Concepts of World Geography

Converging Currents of Globalization

Although economic forces may drive many aspects of globalization, the effects are found in all aspects of land and life, with profound changes to world environments, cultures, settlement, demography, and geopolitics.

Geographer's Toolbox

Geography is the spatial science that describes and explains the world's physical and human environments. To do this geographers use a variety of tools such as maps, aerial photos, satellite images, global positioning systems (GPS), and geographic information systems (GIS).

Population and Settlement

While high birth rates characterize some parts of the world (Africa is an example), in many other areas (such as North America and Europe) natural growth rates are very low, thus migration becomes a major factor for demographic change.

Sao Paulo, Brazil. Home to 20 million people, Sao Paulo is the largest city in Brazil, Latin America, and the southern hemisphere. It also ranks among the world's 10 largest metropolitan areas, and illustrates how Earth recently became an urban world, with most of the its population now living in towns and cities. Sao Paulo is also a hub of Latin America's economic life, actively connected to the world economy through the globalization of commerce.

Cultural Coherence

Globalization creates a dynamic, ever-changing world cultural geography as some people take up new ways while others retreat farther into traditional cultures.

Geopolitical Framework

The last several decades have seen rapid geopolitical change linked to globalization. Not only have new countries appeared but within existing countries regionalism and ethnic separatism are causing major changes.

Economic and Social Development

Economic globalization has created new world trade patterns and centers of wealth, but not for all people in all places. Instead, critics say, economic and social disparities have actually increased the differences between rich and poor.

Converging Currents of Globalization

One of the most important challenges facing the world in the 21st century is associated with **globalization**—the increasing interconnectedness of people and places through converging economic, political, and cultural activities. Once-distant regions and cultures are now increasingly linked through commerce, communications, and travel. Although earlier forms of globalization existed, especially during Europe's colonial period, the current degree of planetary integration is stronger than ever. In fact, many observers argue that contemporary globalization is the most fundamental reorganization of the world's socioeconomic structure since the Industrial Revolution.

Pundits say globalization is like the weather: It's everywhere, all the time. It is a ubiquitous part of our lives and landscapes that is both beneficial and harmful, depending on our needs and point of view.

Economic activities may be the major driving force behind globalization, but the consequences affect all aspects of land and life: Cultural patterns, political arrangements, environmental conditions, and social development are all undergoing profound change. Because natural resources are now global commodities, the planet's physical environment is also affected by globalization. Financial decisions made thousands of miles away now affect local ecosystems and habitats, often with far-reaching consequences for Earth's health and sustainability.

Pundits say globalization is like the weather: It's everywhere, all the time. It is a ubiquitous part of our lives and landscapes that is both beneficial and harmful, depending on our needs and point of view. Some people in some places embrace the changes brought about by globalization, whereas others resist and push back, seeking refuge in traditional habits and places. As a result, the handmaiden of globalization is **diversity**: a tension between the global and the local. In Asian philosophy, *yin* and *yang* are polar opposites, yet what are seemingly contrary are actually interconnected and interdependent. Indeed, this is the case with the diversity amid globalization that makes up contemporary world regional geography.

These immense and widespread global changes make understanding our contemporary world a challenging, yet necessary task. World regional geography is central to this task because of its integration of environmental, cultural, political, and economic themes and topics (see *Exploring Global Connections: A Closer Look at Globalization*).

Economic Globalization

Most scholars agree that the major component of globalization is the economic reorganization of the world. Although different forms of a world economy have existed for centuries, a well-integrated, truly global economy is primarily the product of the past several decades. The attributes of this system, while familiar, are worth stating:

- Global communication systems that link all regions and most people on the planet instantaneously (Figure 1.1)
- Transportation systems capable of moving goods quickly by air, sea, and land
- Transnational business strategies that have created global corporations more powerful than many sovereign nations
- New and more flexible forms of capital accumulation and international financial institutions that make 24-hour trading possible
- Global agreements that promote free trade
- Market economies and private enterprises that have replaced state-controlled economies and services
- An abundance of planetary goods and services that have arisen to fulfill consumer demand (real or imagined) (Figure 1.2)

LEARNING OBJECTIVES

After reading this chapter you should be able to:

- Identify the different components of globalization, including their controversial aspects.
- List several ways in which globalization is changing world geographies.
- Describe the conceptual framework of world regional geography.
- Summarize the major tools used by geographers to study Earth's surface.
- Explain the concepts and metrics used to document changes in global population and settlement patterns.
- Describe the themes and concepts used to study the interaction between globalization and the world's cultural geographies.
- Explain how different aspects of globalization have interacted with global geopolitics from the colonial period to the present day.
- Identify the concepts and data important to documenting changes in the economic and social development of more and less developed countries (MDCs and LDCs).

▲ **Figure 1.1 Global Communications** A fundamental component of globalization is the opening up of global communications through TV, the Internet, computers, and cell phones. In many parts of the world, for example, people use cell phones for doing business and personal finance, as this farmer is doing in India.

- Economic disparities between rich and poor regions and countries that drive people to migrate, both legally and illegally, in search of a better life
- An army of international workers, managers, and executives who give this powerful economic force a human dimension

As a result of this global reorganization, economic growth in some areas of the world has been unprecedented during recent decades; China is a good example. However, not everyone has gained from economic globalization, nor have all world regions shared equally in the benefits. Globalization is often touted as universally beneficial through trickle-down economics, but evidence is mounting that this process is happening neither in all places nor for all peoples. Additionally, the global recession of 2008–2010 demonstrated that economic interconnectivity can also increase economic vulnerability, as illustrated by the precipitous decline in Hawaii's tourist trade as the economies of both Japan and the United States went flat at the same time. Currently, economic recovery in the United States is affected by the slowing economies of both China and Europe.

Globalization and Changing Human Geographies

Economic changes also trigger cultural changes. The spread of a global consumer culture, for example, often accompanies globalization and frequently creates deep and serious social tensions between traditional cultures and a new, global outlook. Global TV, movies, Facebook, Twitter, and videos implicitly promote Western culture, which is then imitated by millions throughout the world, causing friction with traditional values and lifestyles.

Fast-food franchises are changing—some would say corrupting—traditional diets, with explosive growth in most of the world's cities. Although this change may seem harmless to North Americans because of its familiarity, it is not only an expression of the deep cultural changes the world is experiencing through globalization, but also generally unhealthy and environmentally destructive. The expansion of the cattle industry, for example, as a result of the new global demand for beef is doing serious environmental damage to tropical rainforests.

Although the media give much attention to the rapid spread of Western consumer culture, nonmaterial culture is also becoming more dispersed and homogenized through globalization. Language is an obvious example—American tourists in far-flung places are often startled to hear locals speaking an English full of Hollywood clichés. However, far more than speech is involved, as social values also are dispersed globally. Changing expectations about human rights, the role of women in society, and the intervention of nongovernmental organizations are also expressions of globalization that may have far-reaching effects on cultural change.

It would be a mistake, however, to view cultural globalization as a one-way flow that spreads from the United States and Europe into the corners of the world. In actuality, when U.S. popular culture spreads abroad, it is typically melded with local cultural traditions in a process known as *hybridization*. The resulting cultural hybrids, such as hip-hop and rap music or Asian food, can themselves resonate across the planet, adding yet another layer to globalization.

In addition, ideas and forms from the rest of the world are having a great impact on U.S. culture (Figure 1.3). The growing internationalization of American food, the multiple languages spoken in the United States, and the fact that 98 percent of our clothing is imported are all expressions of globalization within the United States.

▼ **Figure 1.2 Global Shopping Malls** Once found only in suburban North America, shopping malls have now spread worldwide through economic globalization. This mall is in Bangkok, Thailand.

Exploring **Global Connections**

A Closer Look at Globalization

Globalization comes in many shapes and forms as it connects far-flung people and places. Many of these interactions are common knowledge, such as the global reach of multinational corporations. Others are more complex and sometimes rather surprising. Who would expect to find Australian firefighters dowsing California wildfires as they migrate between Southern and Northern Hemisphere fire seasons? Would you predict that South Korean investors are buying up land in Madagascar to raise more cattle to supply Asian markets with more choice beefsteaks?

Indeed, global connections are ubiquitous and often complex—so much so that an understanding of the many different shapes, forms, and scales of these interactions is a key component of the study of global geography. To complement that study, each chapter of this book contains an *Exploring Global Connections* sidebar, which presents a case study drawn from a wide variety of topics.

In Chapter 8, for example, a case study illustrates how the current banking crisis in the divided country of Cyprus (located in the eastern Mediterranean) revealed not only that rich Russians had long been stashing their money in Cypriot banks, but also that any proposed solution to the Cyprus fiscal crisis is deeply entangled with the messy politics of global gas and oil. Other examples include the global linkage between the Philippines and the Baltimore, Maryland, school district, where Filipino teachers make up 10 percent of that city's teaching force (Chapter 13); how Antarctica is protected (Chapter 2); West Indian gardens in New York City (Chapter 3); and the Afghanistan opium and heroin trade (Chapter 10) (**Figure 1.1.1**). Many of these sidebars include Google Earth virtual tour videos.

1. Come up with an example of the complicated linkages of globalization based upon your own experiences. For example, what food from another part of the world did you buy today, and how did it get to your store?
2. Now choose a foreign place in a completely different part of the world, either a cit then discuss how globalization affects the lives of people in that place.

Google Earth Virtual Tour Video

http://goo.gl/5uPpKb

▼ **Figure 1.1.1 Afghan Farmer in His Poppy Field** Globalization connects the world community in both expected and surprising ways, from international banking, to climate change, to the Afghan drug trade.

Globalization also has a clear demographic dimension. Although international migration is not new, increasing numbers of people from all parts of the world are now crossing national boundaries, legally and illegally, temporarily and permanently (Figure 1.4). Migration from Latin America and Asia has drastically changed the demographic configuration of the United States, whereas migration from Africa and Asia has transformed western Europe. Countries such as Japan and South Korea, which have long been perceived as ethnically homogeneous, now have substantial immigrant populations. Even several relatively poor countries, such as Nigeria and Ivory Coast, have large numbers of immigrants coming from even poorer countries, such as Burkina Faso and Mali. Although international migration is restricted by a huge array of laws—much more so, in fact, than the movement of goods or capital—it is rapidly increasing, propelled by the uneven economic development associated with globalization.

▼ **Figure 1.3 Global Culture in the United States** While many think that globalization is the one-way spread of North American and European socioeconomic traits into the developing world, one needs only to look around their own neighborhood to find expressions of global culture within the United States, such as this Thai restaurant in Las Vegas, Nevada.

A significant criminal element is also a component of globalization, including terrorism (discussed later in this chapter), drugs, pornography, slavery, and prostitution. Illegal narcotics, for example, are definitely a global commodity (Figure 1.5). Some of the most remote parts of the world, such as the mountains of northern Burma, are thoroughly integrated into the circuits of global exchange through the

▲ **Figure 1.4 Global Migration** Globalization—in its many different forms—is connected to the largest migration in human history as people are drawn to centers of economic activity in hopes of a better life. But along with the pull forces that lure people to new places are the forces of civil strife, environmental deterioration, and economic collapse that push migrants out of their homelands. This photo is of a truckload of African migrants crossing the Sahara to the Mediterranean shore where many will attempt to illegally enter Europe through Spain or Italy. **Q: What international groups are found in your city?**

production of opium that is central to the world heroin trade. Even many areas that do not directly produce drugs are involved in their global sale and shipment. Many Caribbean countries have seen their economies become reoriented to drug transshipments and the laundering of drug money. Prostitution, pornography, and gambling have also emerged as highly profitable global businesses. Over the past decades, for example, parts of eastern Europe have become major sources of both pornography and prostitution, finding a lucrative, but morally questionable niche in the new global economy.

Geopolitics and Globalization

Globalization also has important geopolitical components. To many, an essential dimension of globalization is that it is not restricted by territorial or national boundaries. For example, the creation of the United Nations (UN) following World War II was a step toward creating an international governmental structure in which all nations could find representation. The simultaneous emergence of the Soviet Union as a military and political superpower led to a rigid division into Cold War blocs that slowed further geopolitical integration. However, with the peaceful end of the Cold War in the late 1980s and early 1990s, the former communist countries of eastern Europe and the Soviet Union were opened almost immediately to global trade and cultural exchange, which have changed those countries immensely (Figure 1.6).

Further, there is a strong argument that globalization—almost by definition—has weakened the political power of

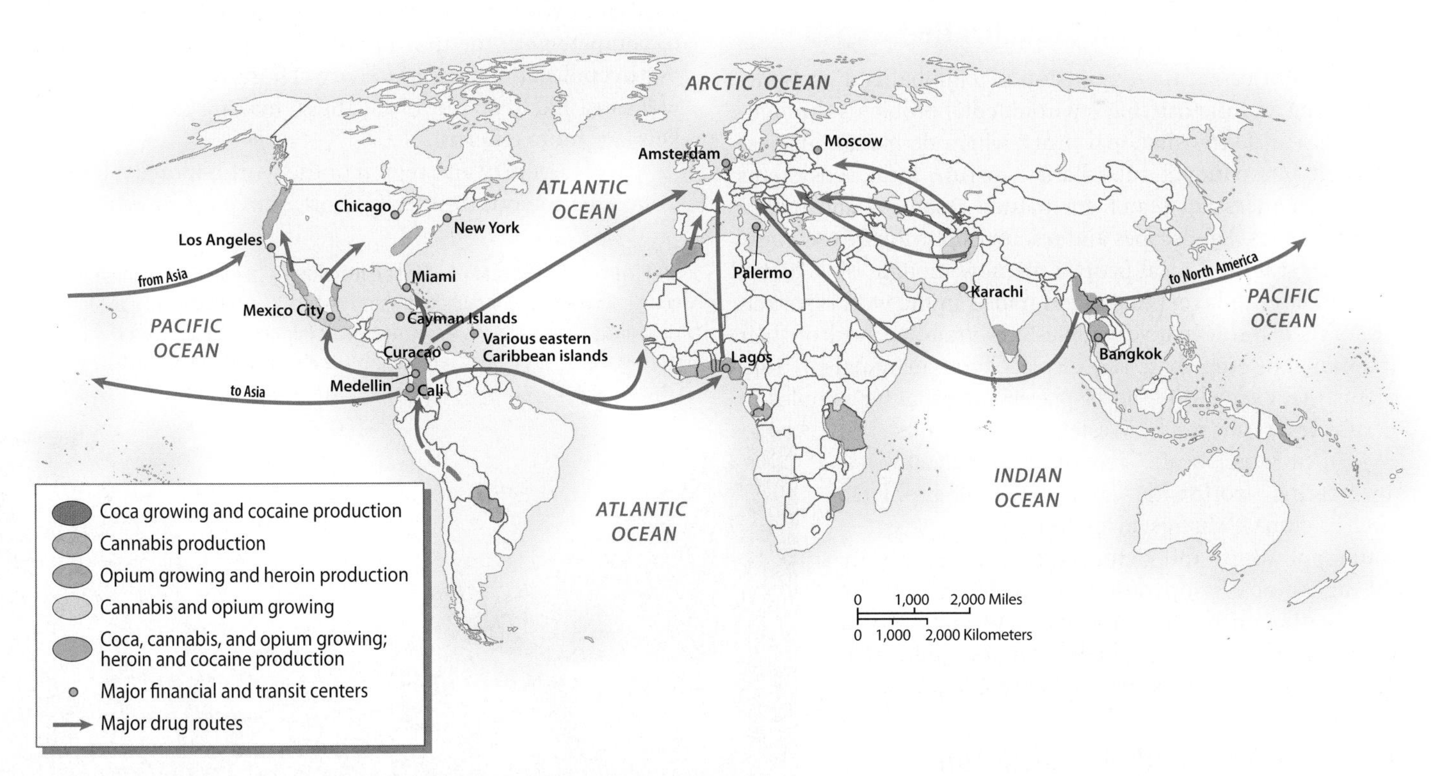

▲ **Figure 1.5 The Global Drug Trade** The cultivation, processing, and transshipment of coca (cocaine), opium (heroin), and cannabis (marijuana) are global issues. The most important cultivation centers are Colombia, Mexico, Afghanistan, and northern Southeast Asia, and the major drug financing centers are located mostly in the Caribbean, the United States, and Europe. In addition, Nigeria and Russia play significant roles in the global transshipment of illegal drugs.

▲ **Figure 1.6 End of the Cold War** The peaceful end of the Cold War in 1990 greatly facilitated global economic expansion and jump-started cultural and political globalization. Here Germans celebrate the opening of the Berlin Wall that divided East and West Berlin from August 1961 to November 1989.

individual states by strengthening the power of regional economic and political organizations, such as the European Union and the World Trade Organization (WTO). In some world regions, a weakening of traditional state power has resulted in stronger local and separatist movements, as illustrated by the turmoil on Russia's southern borders and the plethora of separatist organizations in Europe.

The Environment and Globalization

As we mentioned, the expansion of a globalized economy is creating and intensifying environmental problems throughout the world. Transnational firms, which do global business through international subsidiaries, disrupt local ecosystems with their incessant search for natural resources and manufacturing sites. Landscapes and resources previously used only by small groups of local peoples are now thought of as global commodities to be exploited and traded in the world marketplace. As a result, native peoples are often deprived of their traditional resource base and displaced into marginal environments. An example is the displacement of native peoples in Peru's upper Amazon by Western oil companies.

On a larger scale, economic globalization is aggravating worldwide environmental problems such as climate change, air pollution, energy issues, water pollution, and deforestation. Yet it is only through global cooperation, as evidenced by the UN treaties on biodiversity protection and global warming, that these problems can be addressed. These topics are discussed further in Chapter 2.

Controversy About Globalization

Globalization, and especially its economic aspect, is one of today's most contentious issues. Supporters believe that it results in a greater economic efficiency that will eventually result in rising prosperity for the entire world. In contrast, critics claim that globalization largely benefits those who are already prosperous, leaving most of the world poorer than before as the rich and powerful exploit the less fortunate.

Economic globalization is generally applauded by corporate leaders and economists, and it has substantial support among the leaders of both major political parties in the United States. Beyond North America, moderate and conservative politicians in most countries generally support free trade and other aspects of economic globalization. Opposition to economic globalization is widespread in the labor and environmental movements, as well as among many student groups worldwide. Hostility toward globalization is sometimes deeply felt, as massive protests at World Bank and WTO meetings have made obvious (Figure 1.7).

Pro-globalization Arguments

Advocates argue that globalization is a logical and inevitable expression of contemporary international capitalism and that it benefits all nations and all peoples. Economic globalization can work wonders, they contend, by enhancing competition, increasing the flow of capital to poor areas, and encouraging the spread of beneficial new technologies and ideas. As countries reduce their barriers to trade, inefficient local industries are forced to become more efficient in order to compete with the new flood of imports, thereby enhancing overall national productivity. Those that cannot adjust will most likely go out of business, making the global marketplace more efficient.

Every country and region of the world, moreover, ought to be able to concentrate on those activities for which it is

▼ **Figure 1.7 Protests Against Globalization** Meetings of international groups such as the World Trade Organization (WTO) and International Monetary Fund (IMF) commonly draw large numbers of protesters against economic globalization. This group of protesters is at a recent meeting of the WTO in Geneva, Switzerland.

best suited in the global economy. Enhancing such geographic specialization, the pro-globalizers argue, creates a more efficient world economy. Such economic restructuring is made increasingly possible by the free flow of capital to those areas that have the greatest opportunities. By making access to capital more readily available throughout the world, economists contend, globalization should eventually result in a certain global **economic convergence**, implying that the world's poorer countries will gradually catch up with the more advanced economies.

The American journalist and author Thomas Friedman, one of the first to write about globalization, argues that the world has not only shrunk, but also become economically "flat," so that financial capital, goods, services, and workers can flow freely from place to place. Friedman also describes the great power of the global "electronic herd" of bond traders, currency speculators, and fund managers who either direct money to or withhold it from developing economies, resulting in economic winners and losers (Figure 1.8).

▲ **Figure 1.8 The Electronic Herd** One component of globalization is the rapid movement of capital within the global economic system, creating financial hotspots and stampedes as money moves quickly from place to place. This electronic herd is in the Hong Kong stock exchange.

The pro-globalizers also strongly support the large multinational organizations that facilitate the flow of goods and capital across international boundaries. Three such organizations are particularly important: the World Bank, the International Monetary Fund (IMF), and the WTO. The primary function of the World Bank is to make loans to poor countries so that they can invest in infrastructure and build more modern economic foundations. The IMF is concerned with making short-term loans to countries that are in financial difficulty—those having trouble, for example, making interest payments on the loans that they had previously taken. The WTO, a much smaller organization than the other two, works to reduce trade barriers between countries to enhance economic globalization. It also tries to mediate between countries and trading blocs that are engaged in trade disputes (Figure 1.9).

To support their claims, pro-globalizers argue that countries that have been highly open to the global economy

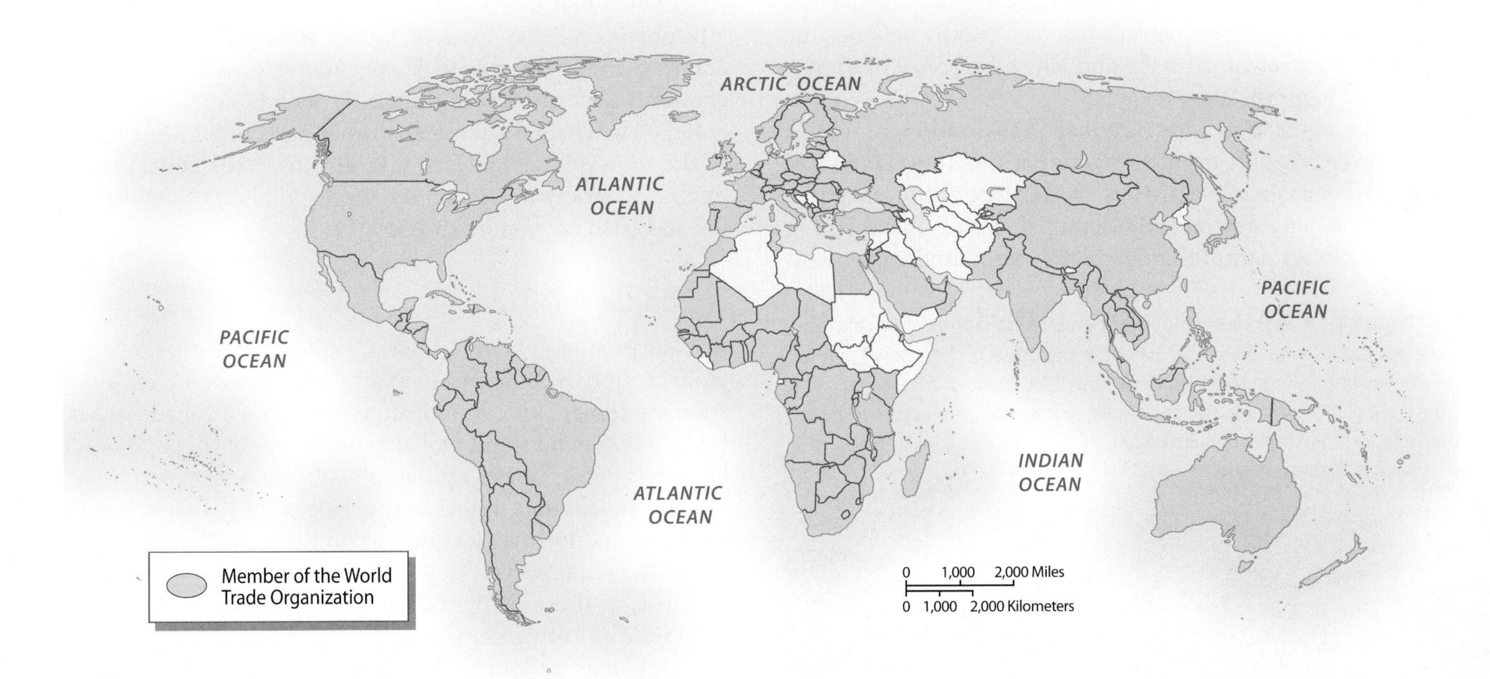

▲ **Figure 1.9 World Trade Organization** One of the most powerful institutions of economic globalization is the World Trade Organization (WTO), which was created in 1995 to oversee trade agreements, encourage open markets, enforce trade rules, and settle disputes. The WTO currently consists of 154 member countries. In addition to these member countries, more than 30 states have "observer status," including Iran and Iraq.

have generally had much more economic success than those that have isolated themselves by seeking self-sufficiency. The world's most isolated countries, Burma (Myanmar) and North Korea, have become economic disasters, with little growth and rampant poverty, whereas those that have opened themselves to global forces in the same period, such as Singapore and Thailand, have seen rapid growth and substantial reductions in poverty.

Critics of Globalization

Virtually all of the claims of the pro-globalizers are strongly contradicted by the critics of globalization. Opponents often begin by arguing that globalization is not a "natural" process. Instead, it is the product of an explicit economic policy promoted by free-trade advocates, capitalist countries (mainly the United States, but also Japan and the countries of Europe), financial interests, international investors, and multinational firms.

Further, because the globalization of the world economy is creating greater inequity between rich and poor, the trickle-down model of developmental benefits for all people in all regions has yet to be validated. On a global scale, the richest 20 percent of the world's people consume 86 percent of the world's resources, whereas the poorest 80 percent use only 14 percent. The growing inequality of this age of globalization is apparent on both global and national scales. Globally, the wealthiest countries have grown much richer over the past two decades, while many of the poorest countries have lost ground. Nationally, even in developed countries such as the United States, the wealthiest 1 percent of the population has reaped almost all of the gains that globalization has offered, while the remaining 99 percent has seen real income decline as wages have remained static and jobs have been lost to outsourcing (Figure 1.10).

Opponents also contend that globalization promotes free-market, export-oriented economies at the expense of localized, sustainable activities. World forests, for example, are increasingly cut for export timber, rather than serving local needs. As part of their economic structural adjustment package, the World Bank and the IMF often encourage developing countries to expand their resource exports so they have more hard currency to make payments on their foreign debts. This strategy, however, usually leads to overexploitation of local resources. Opponents also note that the IMF often requires developing countries to adopt programs of fiscal austerity that entail substantial reductions in public spending for education, health, and food subsidies. By adopting such policies, critics warn, poor countries end up with even more impoverished populations than before.

Furthermore, anti-globalizers contend that the "free-market" economic model commonly promoted for developing countries is not the one that Western industrial countries used for their own economic development. In Germany, France, and even to some extent the United States, governments historically have played a strong role in directing investment, managing trade, and subsidizing chosen sectors of the economy.

Those who challenge globalization also worry that the entire system—with its instantaneous transfers of vast sums of money over nearly the entire world on a daily basis—is inherently unstable. The British author and noted critic of globalization John Gray, for example, argues that the same "electronic herd" that Thomas Friedman applauds is a dangerous force because it is susceptible to "stampedes." International managers of capital tend to panic when they think their funds are at risk; when they do so, the entire intricately linked global financial system can quickly become destabilized, leading to a crisis of global proportions. The rapid downturn of the global economy in 2008 seems to support that assertion.

Even when the "herd" spots opportunity, trouble may still ensue. As vast sums of money flow into a developing country, they may create a speculatively inflated **bubble economy** that cannot be sustained. Such a bubble economy emerged in Thailand and many other parts of Southeast Asia in the mid-1990s. Analysts have also used the concept of a bubble economy to explain the tragic collapse of the Icelandic and Irish economies in 2009.

▼ **Figure 1.10 U.S. Unemployment and Globalization** One criticism of globalization is that the United States has lost jobs as commerce has moved offshore to lower-wage countries. While true to some extent, this job loss is also the result of other kinds of change in world and domestic economies. These job seekers are in Rochester Hills, Michigan.

A Middle Position

Not surprisingly, many experts argue that both the anti-globalization and the pro-globalization stances are exaggerated. Friedman, the American reporter mentioned earlier, says, "Those who think globalization is either all good or all bad don't get it," meaning that globalization is so pervasive, is so complex, and has so many aspects to it that are both negative and positive, it is unwise to limit your views with a biased generalization.

In fact, those in the middle ground tend to argue that economic globalization is indeed unavoidable. They further contend that, although globalization holds both promises and pitfalls, it can be managed, at both the national and the international levels, to reduce economic inequalities and protect the natural environment. These experts stress the need for strong, yet efficient national governments, supported by international institutions (such as the UN, World

▲ **Figure 1.11 Diversity Amid Globalization** Although much of globalization results in geographic and cultural homogeneity, geographic uniqueness and cultural diversity still persists, as shown in this photo of Masai women in a village in Kenya, Africa.

Bank, and IMF) and globalized networks of environmental, labor, and human rights groups.

Unquestionably, globalization is one of the most important issues of the day—and certainly one of the most complicated. This book does not pretend to resolve the controversy, nor does it take a position, but it does encourage readers to reflect on these critical points as they apply to different world regions.

Diversity in a Globalizing World

As globalization increases, many observers foresee a world far more uniform and homogeneous than today's. The optimists among them imagine a universal global culture uniting all humankind into a single community untroubled by war, ethnic strife, or resource shortage—a global utopia of sorts.

A more common view, however, is that the world is becoming blandly homogeneous as different places, peoples, and environments lose their distinctive character and become indistinguishable from their neighbors. This, too, is an exaggerated view, for the world is still a highly diverse place (Figure 1.11). We still find marked differences in culture (language, religion, architecture, foods, and many other attributes of daily life), economy, and politics—as well as in the physical environment. Such diversity is so vast that it cannot readily be extinguished, even by the most powerful forces of globalization. Diversity may be difficult for a society to live with, but it also may be dangerous to live without. Nationality, ethnicity, cultural distinctiveness—all are the legitimate legacy of humanity. If this diversity is blurred, denied, or repressed through global homogenization, humanity loses one of its defining traits.

In fact, globalization often provokes a strong reaction on the part of local people, making them all the more determined to maintain what is distinctive about their way of life. Thus, globalization is understandable only if we also examine the diversity that continues to characterize the world and, perhaps most important, the tension between these two forces: the homogenization of globalization, on the one hand, and the reaction against it in terms of protecting cultural and political diversity, on the other.

The politics of diversity also demand increasing attention as we try to understand worldwide tensions over terrorism, ethnic separateness, regional autonomy, and political independence. Groups of people throughout the world seek self-rule of territory they can call their own. Today most wars are fought *within* countries, not *between* them. As a result, our interest in geographic diversity takes many forms and goes far beyond simply celebrating traditional cultures and unique places. People have many ways of making a living throughout the world, and it is important to recognize this fact as the globalized economy becomes increasingly focused on mass-produced retail goods. Furthermore, a stark reality of today's economic landscape is unevenness: While some people and places prosper, others suffer from unrelenting poverty. This, unfortunately, is also a form of diversity amid globalization (Figure 1.12).

▼ **Figure 1.12 The Landscape of Economic Diversity** The geography of diversity takes many expressions. One of these is economic unevenness, as depicted in this photo from New Delhi, India, where squatter settlements of the poor contrast with the high-rise office buildings and apartment houses of the more affluent.

Flat and Spiky Worlds

Mentioned earlier was Friedman's notion that the globalized world has become increasingly flat in socioeconomic terms. This term is a metaphor for the ability of financial capital and production to flow easily from one place to another, changing locations to take advantage of technological innovation and labor costs and developing new products that can be shipped and sold anywhere in the world, both physically and digitally. Examples abound that illustrate Friedman's notion of global flatness, such as the way Silicon Valley firms have created a 24-hour workday by drawing upon skilled engineers half a world away in South Asia, who continue working on projects during California's nighttime. Although Friedman says his notion of a flat world has been overblown and generalized beyond his initial intentions, the fact is that his best-selling book *The World Is Flat* (and its sequels) captured the public's imagination, thereby providing the public with a handy metaphor for capturing the essence of globalization.

Predictably, as with any popular idea, criticism and alternatives have been proposed to Friedman's flat-world notion. Most notably, Richard Florida, in his book *The Rise of the Creative Class,* argues that the world is not flat at all, but is instead mountainous and spiky, consisting of peaks and valleys that alternatively encourage and inhibit the flow of ideas and goods around the globe. Some locations are privileged (the peaks), whereas others are not (the valleys), thus producing an uneven socioeconomic topography of winners and losers (Figure 1.13).

Clearly, both metaphors, flat and spiky, are valuable in describing the complexities of today's globalized world. We have attempted to capture that complexity by titling this book *Diversity Amid Globalization* because both are equally important in the study of world regional geography.

▼ **Figure 1.13 Spiky World** This map of eastern hemisphere urban agglomerations, which combines density and city size, conveys Richard Florida's notion of a spiky world of innovation centers (cities, usually). This contrasts with Thomas Friedman's earlier contention that globalization has made the economic world flat, with all locations theoretically able to participate in world trade. The different metaphors of "flat" versus "spiky" are helpful tools in exploring the forces and patterns of globalization.

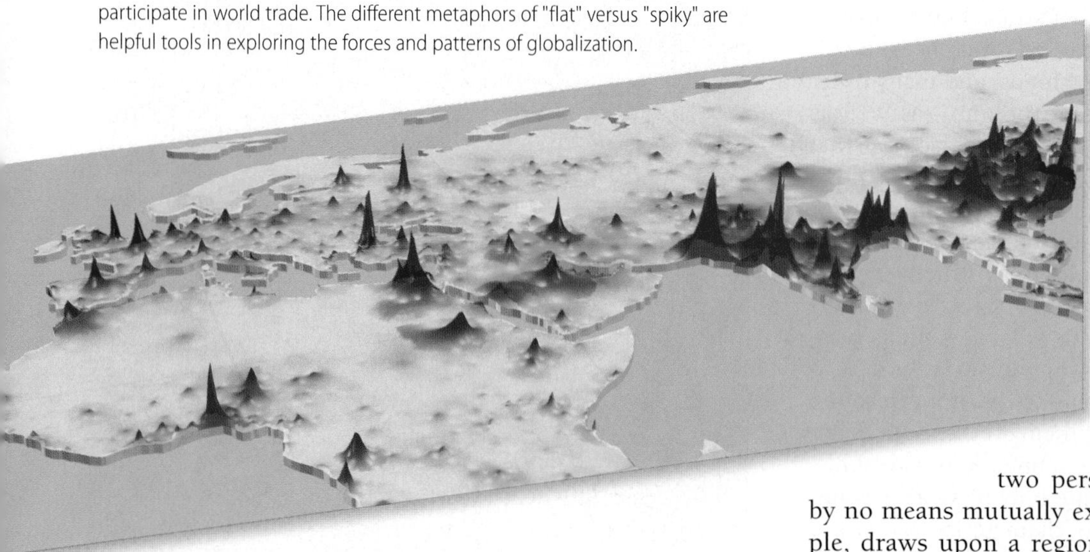

REVIEW

1.1 Describe and explain five components of economic globalization.
1.2 What is the relationship between the end of the Cold War in 1990 and economic globalization?
1.3 Summarize three elements of the controversy about globalization.
1.4 What are the characteristics of "flat" and "spiky" worlds?

Geography Matters: Environments, Regions, Landscapes

Geography is one of the most fundamental sciences, a discipline awakened and informed by a long-standing human curiosity about our surroundings and the world. The term **geography** has its roots in the Greek words for "describing the Earth," and this discipline has been central to all cultures and civilizations as they explore the world. In a simplistic way, geography can be compared to history: Historians describe and explain what has happened over time, whereas geographers primarily describe and explain the world's spatial dimensions and how Earth differs from place to place. Of course, geographers—particularly historical geographers—also document geographical changes through time (just as some historians do spatial analyses).

Given the broad scope of geography, it is no surprise that geographers have different conceptual approaches to investigating the world. At the most basic level, geography can be broken into two complementary pursuits: *physical* and *human geography*. Physical geography examines climate, landforms, soils, vegetation, and hydrology, whereas human geography concentrates on the spatial analysis of economic, social, and cultural systems.

A physical geographer, for example, studying the Amazon Basin of Brazil, might be interested primarily in the ecological diversity of the tropical rainforest or the ways in which the destruction of that environment changes the local climate and hydrology. A human geographer, in contrast, would focus on the social and economic factors explaining the migration of settlers into the rainforest or the tensions and conflicts over resources between new migrants and indigenous peoples.

Another conceptual division is that between focusing on a specific topic or theme and analyzing a place or a region. The first approach is referred to as *thematic* or *systematic geography,* whereas the second is called *regional geography*. These two perspectives are complementary and by no means mutually exclusive. This textbook, for example, draws upon a regional scheme for its overall organization, dividing the globe into 12 separate world regions.

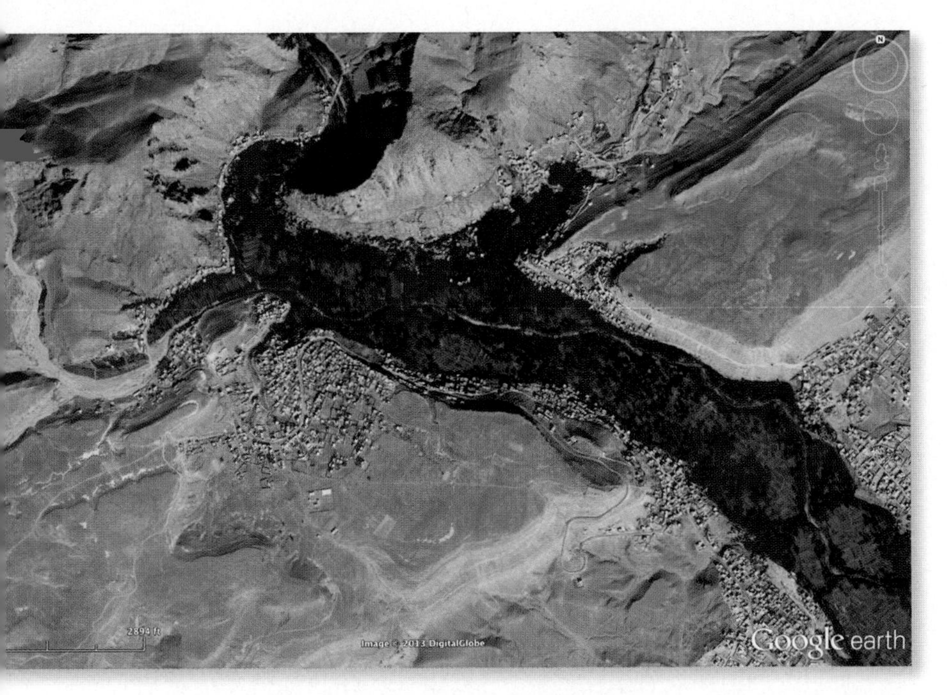

▲ **Figure 1.14 Areal Differentiation** This satellite photo of oasis villages on the southern slope of Morocco's Atlas Mountains is a classic illustration of areal differentiation, or of how landscapes can differ significantly within short distances. The dark green band are irrigated date palm and vegetable fields, watered by rivers that rise in the high mountains, then flow southward into the Sahara Desert. Since irrigated fields near the rivers are precious land, the village settlements are nearby in the dry areas.

It then presents each chapter thematically, examining the topics of physical geography and environmental issues, population and settlement, cultural coherence and diversity, geopolitical framework, and economic and social development in a systematic way. In doing so, each chapter combines four kinds of geography: physical, human, thematic, and regional.

Areal Differentiation and Integration

As a spatial science, geography is charged with the study of Earth's surface. A central component of that responsibility is describing and explaining the differences that distinguish one piece of the world from another. The geographical term for this is **areal differentiation** (*areal* means "pertaining to area"). Why is one part of Earth humid and lush, while another, just a few hundred kilometers away, is an arid desert (Figure 1.14)?

Geographers are also interested in the connections between different places and how they are linked. This theme is one of **areal integration**, or the study of how places interact with one another. An example is the analysis of how and why the economies of Singapore and the United States are closely intertwined, even though the two countries are situated in entirely different physical, cultural, and political environments. Questions of areal integration are becoming increasingly important because of the new global linkages inherent to globalization.

Global and Local All scientific inquiry has a sense of scale, whatever the discipline. In biology, some scientists study the smaller units of cells, genes, or molecules, while others take a larger view, analyzing plants, animals, or whole ecosystems. Geographers also work at different scales. One may concentrate on the analysis of a local landscape—perhaps a single village in southern China—whereas another might focus on the broader regional picture, examining all of southern China. Other geographers do research on a still larger global scale, perhaps studying emerging trade networks between southern India's center of information technology in Bangalore and North America's Silicon Valley or investigating how India's monsoon might be connected to and affected by the Pacific Ocean's El Niño.

But even though geographers may be working at different scales, they never lose sight of the interactivity and connectivity among local, regional, and global scales. They will note, for example, the ways that the village in southern India might be linked to world trade patterns or how the late arrival of the monsoon could affect agriculture and food supplies in different parts of India.

Regions: Formal, Functional, and Vernacular

The human intellect seems driven to make sense of the universe by lumping phenomena together into categories that emphasize similarities. Biology has its taxa of living organisms, history marks off eras and periods of time, and geology classifies epochs of Earth history. Geography, too, organizes information about the world, by compressing it into units of spatial similarity called **regions**.

Sometimes, the unifying threads of a region are physical, such as climate and vegetation, resulting in a regional designation like the *Sahara Desert* or the *Amazonian rainforest*. Other times, the threads are more complex, combining economic and cultural traits, as in the use of the term *Corn Belt* for parts of the central United States. People commonly compress large amounts of information into stereotypes, and in a way a geographic region is just that—a spatial stereotype for a portion of Earth that has some special signature or characteristic that sets it apart from other regions.

Geographers designate three types of regions: formal, functional, and vernacular (Figure 1.15). **Formal regions** take their name from the fact that these regions are defined by some aspect of physical form, such as a mountain range, valley, or climate. Cultural features can also be used to define formal regions. An example is the area where a certain language is spoken or a specific religion dominates. Many of the maps in this book denote formal regions. In contrast, a **functional region** is one where a certain activity (or cluster of activities) takes place. The earlier example of America's Corn Belt fits this terminology because it forms a region where a specific economic activity dominates. The

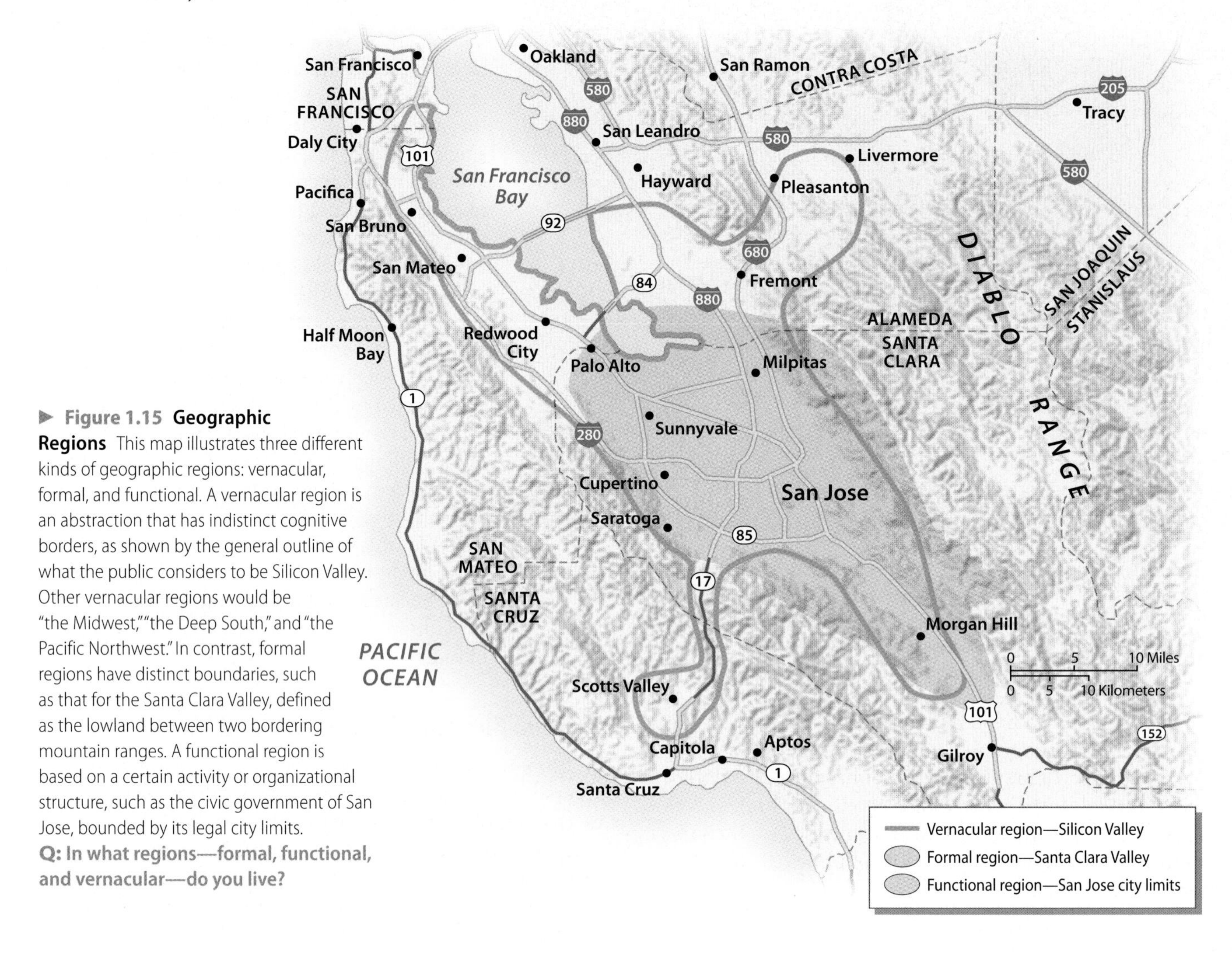

▶ **Figure 1.15 Geographic Regions** This map illustrates three different kinds of geographic regions: vernacular, formal, and functional. A vernacular region is an abstraction that has indistinct cognitive borders, as shown by the general outline of what the public considers to be Silicon Valley. Other vernacular regions would be "the Midwest," "the Deep South," and "the Pacific Northwest." In contrast, formal regions have distinct boundaries, such as that for the Santa Clara Valley, defined as the lowland between two bordering mountain ranges. A functional region is based on a certain activity or organizational structure, such as the civic government of San Jose, bounded by its legal city limits.
Q: In what regions—formal, functional, and vernacular—do you live?

Megalopolis of the eastern United States (Boston to Washington, DC) is another example, as are newspaper circulation areas and the spatial dimension of a sports team's fan base. (Think of the line somewhere in the Midwest between Chicago and St. Louis that divides baseball and football fans of each city.) Last, **vernacular regions** are defined solely in people's minds as spatial stereotypes that have no visible boundaries in the physical landscape. Examples abound: the South, the Midwest, Silicon Valley, New England, and so on.

The Cultural Landscape: Space into Place

Humans transform space into distinct places that are unique and heavily loaded with meaning and symbolism. This diverse fabric of *placefulness* is of great interest to geographers because it tells us much about the human condition throughout the world. Places can tell us how humans interact with nature and among themselves; where there are tensions and where there is peace; where people are rich and where they are poor.

A common tool for the analysis of place is the concept of the **cultural landscape**, which is, simply stated, the visible, material expression of human settlement, past and present. Thus, the cultural landscape is the tangible expression of the human habitat. It visually reflects the most basic human needs—shelter, food, and work. Additionally, the cultural landscape acts to bring people together (or keep them apart) because it is a marker of cultural values, attitudes, and symbols. Because cultures vary greatly around the world, so do cultural landscapes (Figure 1.16).

Increasingly, however, we see the uniqueness of places being eroded by the homogeneous landscapes of globalization—shopping malls, fast-food outlets, business towers, theme parks, and industrial complexes. Understanding the forces behind the spread of these landscapes is important because they tell us much about the expansion of global economies and cultures. Although a modern shopping mall in Hanoi, Vietnam, may seem familiar to someone from North America, this new landscape represents yet another component of globalized world culture that has been implanted into a once remote and distinctive city.

▲ **Figure 1.16 The Cultural Landscape** Humans, through their subsistence needs and cultural values, change the natural environmental into cultural landscapes. In this photo of an Austrian alpine village, the native forest has become essentially domesticated as pasture grassland has been increased and tree species changed to better serve timber and fuel needs. Also, the unique farmstead architecture expresses local customs and preferences.

REVIEW

1.5 Explain the difference between areal differentiation and areal integration.
1.6 How do functional regions differ from formal and vernacular regions?
1.7 How is the concept of the cultural landscape related to areal differentiation?

The Geographer's Toolbox: Location, Maps, Remote Sensing, and GIS

Geographers use many different tools to analyze the world. Today's digital tools offer geographers and other scientists an array of analytic methods not imagined just decades ago. Often, however, these new computer-based tools are combined with tried and true concepts long used by people to locate themselves in world space.

Latitude and Longitude

To navigate your way through your daily tasks, you generally use a mental map of *relative locations* that locate specific places in terms of their relationship to other landscape features. The shopping mall is near the highway, perhaps, or the college campus is along the river. In contrast, map makers use *absolute location*—often called a mathematical location—which draws upon a universally accepted coordinate system providing every place on Earth with a specific numerical address based upon latitude and longitude. The absolute location for the Geography Department at the University of Oregon, for example, has the mathematical address of 44 degrees, 02 minutes, and 42.95 seconds north and 123 degrees, 04 minutes, and 41.29 seconds west. This is written 44° 02' 42.95" N and 123° 04' 41.29" W.

Lines of latitude, called **parallels**, run east–west around the globe and are used to locate places north and south of the equator, which is 0 degrees latitude. In contrast, lines of longitude, referred to as **meridians,** run from the north pole, located at 90 degrees north latitude, to the south pole, located at 90 degrees south latitude. Longitude locates places east or west of the **prime meridian**, located at 0 degrees longitude at the Royal Naval Observatory in Greenwich, England (just east of London) (Figure 1.17). The equator itself divides the globe into northern and southern hemispheres, whereas the prime meridian divides the world into eastern and western hemispheres; these two east-west hemispheres meet at 180 degrees longitude in the western Pacific Ocean. The International Date Line, where each new solar day begins, lies along much of 180 degrees longitude, deviating where necessary to ensure that small Pacific island nations remain on the same calendar day.

Each degree of latitude measures 60 nautical miles or 69 land miles (111 km) and is made up of 60 minutes, each of which is 1 nautical mile (1.15 land miles). Each minute has 60 seconds of distance, each of which is approximately 100 feet (30.5 meters).

From the equator, parallels of latitude are used to mathematically define the tropics: the Tropic of Cancer at 23.5 degrees north and the Tropic of Capricorn at 23.5 degrees south. These lines of latitude denote where the Sun is directly overhead at noon on the solar solstices in June and December. Similarly, the Arctic and Antarctic circles, at

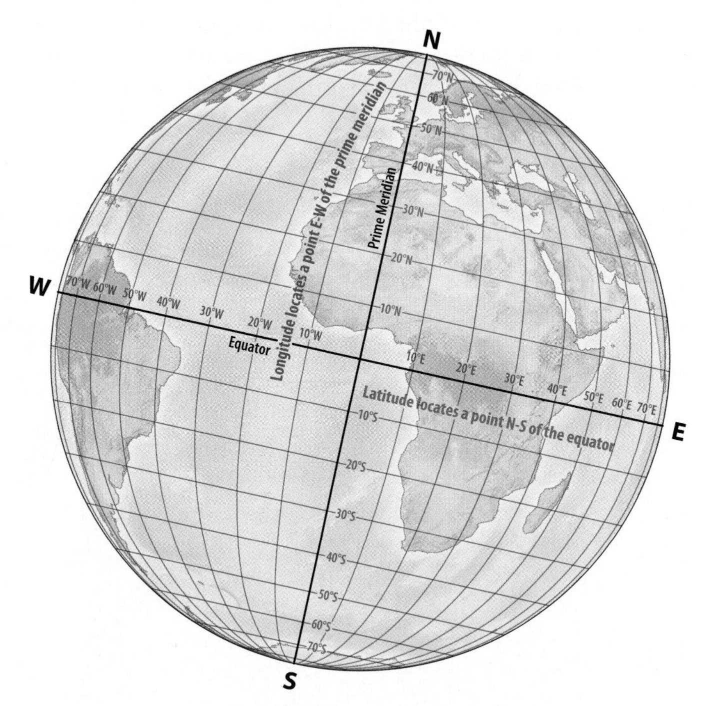

▲ **Figure 1.17 Latitude and Longitude** Latitude locates a point between the equator and the poles and is designated as so many degrees north or south. Longitude locates a point east or west of the prime meridian, which is located at the Royal Observatory in Greenwich, England, just east of London.
Q: What is the latitude and longitude of your school?

66.5 degrees north and south latitude, respectively, mathematically define where these areas experience 24 hours of sunlight on the summer solstice and 24 hours of complete darkness on the winter solstice.

Global Positioning Systems (GPS)

Historically, precise measurements of latitude and longitude were determined by a complicated method of celestial navigation, based upon the observer's location relative to the Sun, Moon, planets, and stars. Today, though, absolute location on Earth (or in airplanes above Earth's surface) is achieved through satellite-based **global positioning systems (GPS)**. These systems use time signals sent from a specific location to a satellite and back to the GPS receiver to calculate precise coordinates of latitude and longitude. These systems were first used by the U.S. military in the 1960s and were then made available to the public in the later decades of the 20th century. Today GPS guide airplanes across the skies, ships across the oceans, private autos on the roads, and hikers through wilderness areas, to name only a few of many uses. True GPS plot locations accurately with 3 feet (or a meter) on Earth's surface. Although smartphones have locational systems built into their software, most cell phones use a ground-based system of triangulation from cell phone towers, which is not quite as accurate as satellite-based GPS.

Map Projections

Because the world is spherical, mapping the globe on a flat piece of paper creates inherent distortions in the latitudinal, or north–south, depiction of Earth's land and water areas. Cartographers (those who make maps) have tried to limit these distortions by using various **map projections**, which are the different ways maps are projected onto a flat service. Historically, the Mercator projection was the projection of choice for maps used for oceanic exploration. However, just a brief look at the inflated landmasses for Greenland and Russia shows its weakness in accurate depiction of high-latitude land areas (Figure 1.18). Over time, cartographers have created literally hundreds of different map projections in their attempts to find the best and most accurate way of mapping the world.

We won't go into the details of this vexatious quest, but in the last several decades cartographers have generally used the Robinson projection for their maps and atlases. In fact, several professional cartographic societies tried unsuccessfully in 1989 to actually ban projections such as the Mercator because of their spatial distortions. Like many other professional publications, in this book we use only the Robinson projection for our maps.

Map Scale

All maps must reduce the area being mapped to a smaller piece of paper. This reduction involves the use of **map scale**, or the mathematical ratio between the map and the surface area being mapped. Many maps note their scale as a ratio or fraction between a unit on the map and the same unit in the area being mapped. An illustration is 1:63,360 or 1/63,360, which means that 1 inch on the map represents 63,360 inches on the land surface; thus, the scale is 1 inch equals 1 mile. Although 1:63,360 (1 inch equals 1 mile) is a convenient mapping scale to understand, the amount of surface area that can be mapped and fitted on a common-sized sheet of paper at this scale is limited to about 20 square miles. But at this scale mapping a larger area, say 100 square miles would produce a much larger, unwieldy map. Therefore, the ratio must be changed to a larger

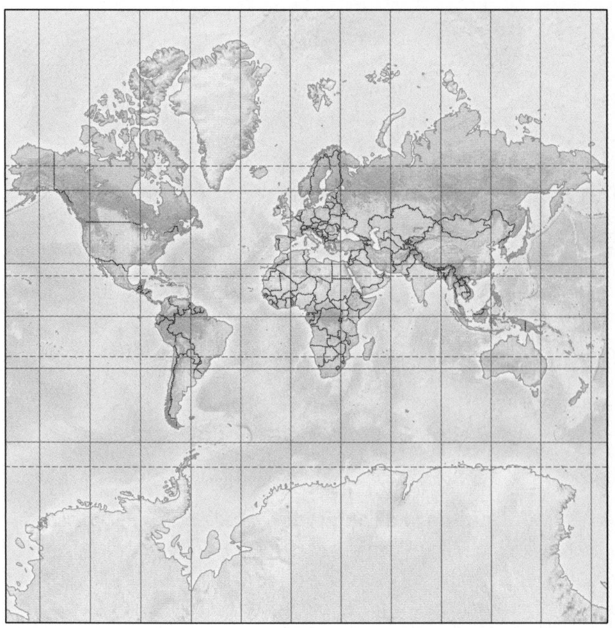

(a)

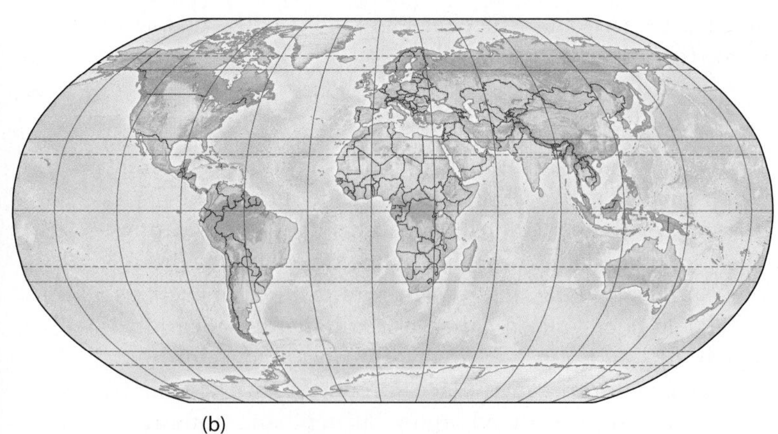

(b)

▲ **Figure 1.18 Map Projections** Cartographers have long struggled with how best to accurately map the world given the inherent distortions when transferring features on a round globe to a flat piece of paper. Early map makers commonly used the Mercator projection (a) which distorts features in the high latitudes, but worked fairly well for seagoing explorers. The map on the right (b) is the Robinson projection, which was developed in the 1960s and is now the industry standard because it minimizes cartographic distortion.

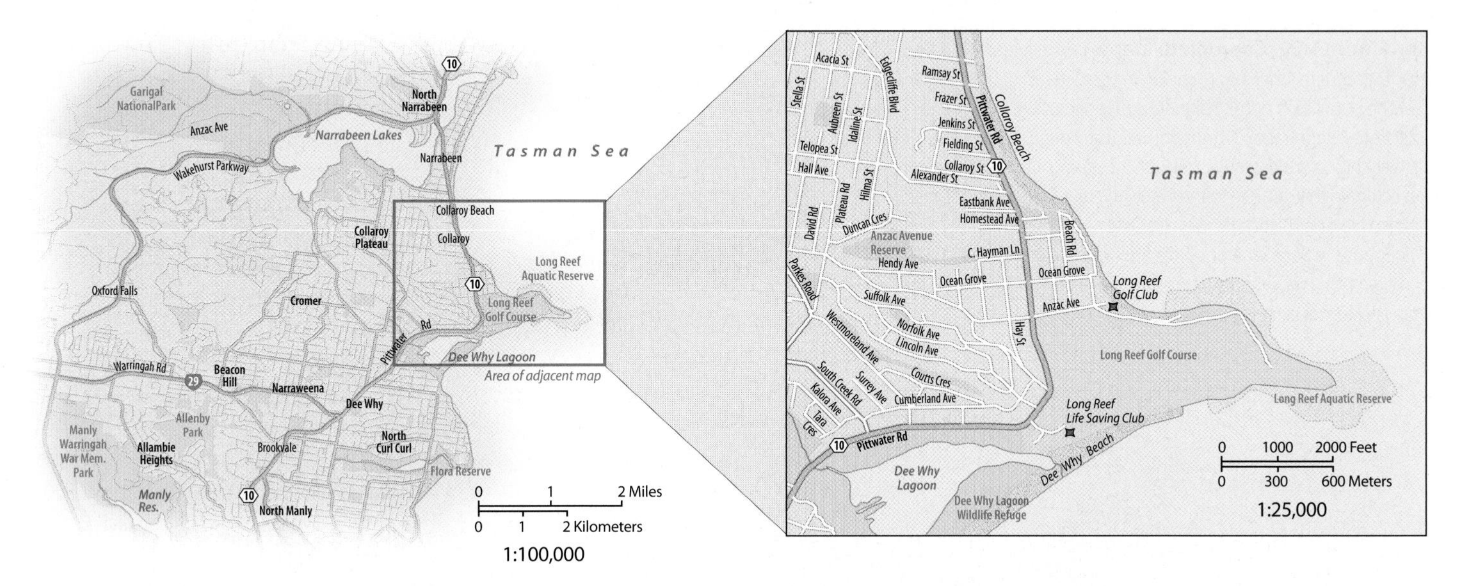

▲ **Figure 1.19 Small and Large Scale Maps** A portion of Australia's east coast north of Sydney is mapped at two scales, one (on the left) at a small scale and other (on the right) at a large scale. Note the differences in distance depicted on the linear scales of the two different maps. There is more close-up detail in the large-scale map, but it covers only a small portion of the area mapped at a small scale.

number, such as 1:316,800. This ratio signifies that 1 inch on the map now represents 5 miles (8 km) of distance on land.

Based upon the **representative fraction**, which is the cartographic term for the numerical value of map scale, maps are categorized as having either large or small scales (Figure 1.19). It may be easy to remember that large-scale maps make landscape features like rivers, roads, and cities *larger*, but because the features are larger, the maps must cover *smaller* areas. Conversely, small-scale maps cover *larger* areas, but to do so, these maps must make landscape features *smaller*. A bit harder to remember is that the larger the second number of the representative fraction—the 63,360 in the fraction 1/63,360 or the 100,000 in the fraction 1/100,000, for example—the smaller the scale of the map.

Map scale is probably easiest to interpret when it is simply portrayed in a **graphic or linear scale**, which visually depicts in a horizontal bar distance units such as feet, meters, miles, or kilometers. Most of the maps in this book are small-scale maps of large areas; thus, the graphic scale is in miles and kilometers. You can measure distances between two points on the map by making two tick marks on a piece of paper held next to the points and then measuring the distance between the two marks on the linear scale.

Map Patterns and Legends

Maps come in a wide array of colors and patterns, which depict everything from the most basic representation of topographic and landscape features to complicated patterns of population, migration, economic conditions, and so forth. Whether the map is a simple *reference map* that shows the location of certain features or a *thematic map* that displays more complicated spatial phenomena, the map legend provides the details by explaining the different map patterns.

Many maps in this book are **choropleth maps**, which map different levels of intensity of data, such as per capita income or population density, placed within discrete spatial units, such as countries, cities, counties, or cultural regions (Figure 1.20). Along with choropleth data, many of our maps contain flow arrows that depict the movement of people or the flow of trade goods.

Aerial Photos and Remote Sensing

Although maps are a primary tool of geography, much can be learned about Earth's surface by deciphering patterns on photographs taken from airplanes, balloons, or satellites. Originally, these photographs were only in black and white, but today color aerial photographs are common.

Even more information about Earth comes from electromagnetic images referred to as **remote sensing**, taken from aircraft or satellites. This technology has many scientific applications, including monitoring the loss of rainforests, tracking the biological health of crops and woodlands, and even measuring changes in ocean surface temperatures. It is also central to national defense issues, such as monitoring troop movements or the building of missile sites in hostile countries. In simple terms, aerial photographs are merely photographs taken from balloons, airplanes, or satellites, whereas remote sensing gathers electromagnetic data that then must be processed and interpreted by computer software to produce images of Earth's surface.

The Landsat satellite program launched by the United States in 1972 is a good example of both the technology

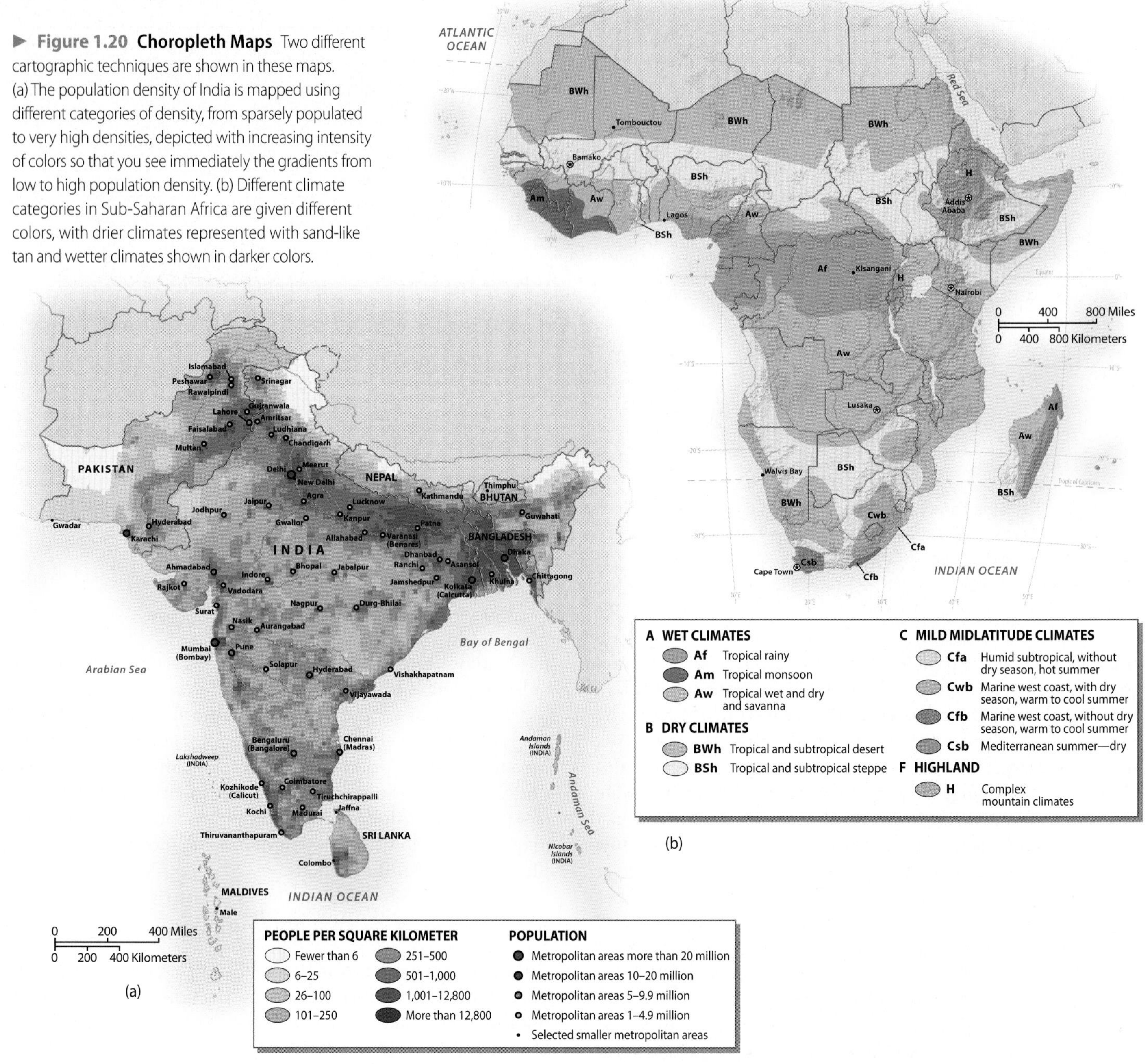

▶ **Figure 1.20 Choropleth Maps** Two different cartographic techniques are shown in these maps. (a) The population density of India is mapped using different categories of density, from sparsely populated to very high densities, depicted with increasing intensity of colors so that you see immediately the gradients from low to high population density. (b) Different climate categories in Sub-Saharan Africa are given different colors, with drier climates represented with sand-like tan and wetter climates shown in darker colors.

and the uses of remote sensing. These satellites collect data simultaneously in four broad bands of electromagnetic energy, from visible through near-infrared wavelengths, that is reflected or emitted from Earth. Once these data are processed by computers, they display a range of images, as illustrated in Figure 1.21. The resolution on Earth's surface ranges from areas 260 feet (80 meters) square down to 98 feet (30 meters) square.

Landsat 4 and 5 satellites pass from north to south over the equator at an altitude of 440 miles (700 km) each day at about 10 A.M., which allows detection of change to the environment on a continual basis. Of course, cloud cover often compromises the continuous coverage of many parts of the world.

Geographic Information Systems (GIS)

Vast amounts of computerized data from different sources, such as maps, aerial photos, remote sensing, and census tracts, are brought together in **geographic information systems (GIS)**. The resulting spatial databases are used to analyze a wide range of resource problems. Conceptually, GIS can be thought of as a computer system for producing a series of overlay maps showing spatial patterns and relationships (Figure 1.22). A GIS map, for example, might combine a conventional map with data on toxic waste sites, local geology, groundwater flow, and surface hydrology to determine the source of pollutants appearing in household water systems.

Although the earliest GIS dates back to the 1960s, it is only in the last several decades—with the advent of desktop computer systems and remote sensing technology—that GIS has become absolutely central to geographic problem solving. It has a central role in city planning, environmental science, earth science, and real estate development, to name only a few of the many activities using these systems.

▼ **Figure 1.21 Remote Sensing of the Dead Sea** This NASA satellite image of the Dead Sea, the lowest spot on Earth at 1300 feet (400 meters) below sea level, uses false-color remote sensing to capture different elements of the environment. Black is deeper water, with light blue showing shallow waters. The green areas along the shoreline are irrigated crops, whereas the white areas are salt evaporation ponds.

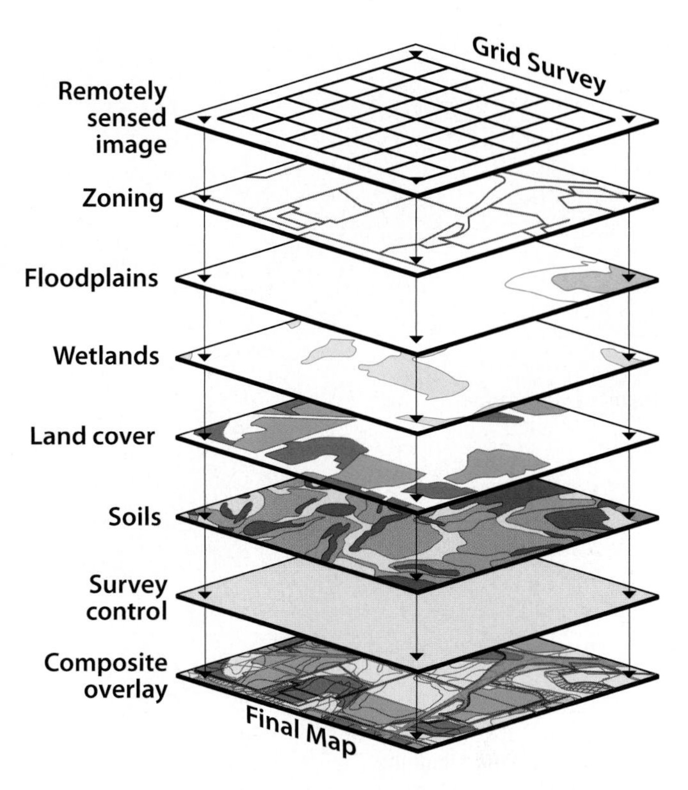

▲ **Figure 1.22 GIS Layers** Geographic information systems (GIS) maps usually consist of many different layers of information that can be viewed and analyzed either separately or as a composite overlay. This illustration is of a typical environmental planning map where different physical features (such as wetlands and soils) are combined with zoning regulations.

REVIEW

1.8 Explain the difference between latitude and longitude.
1.9 What does a map's scale tell us? Describe three different ways map scale is portrayed.
1.10 What is a choropleth map?
1.11 What is remote sensing? Give an example.

Themes and Issues in World Regional Geography

Following two introductory chapters, this book adopts a regional perspective, grouping all of Earth's countries into a framework of 12 world regions (Figure 1.23). We begin with a region familiar to most of our readers—North America—and then move on to Latin America, the Caribbean, Africa, the Middle East, Europe, Russia, and the different regions of Asia, before concluding with Australia and Oceania. Each of the 12 regional chapters employs the same five-part thematic structure—physical geography and environmental issues, population and settlement, cultural coherence and diversity, geopolitical framework, and economic and social development. The concepts and data central to each theme are discussed in the following sections.

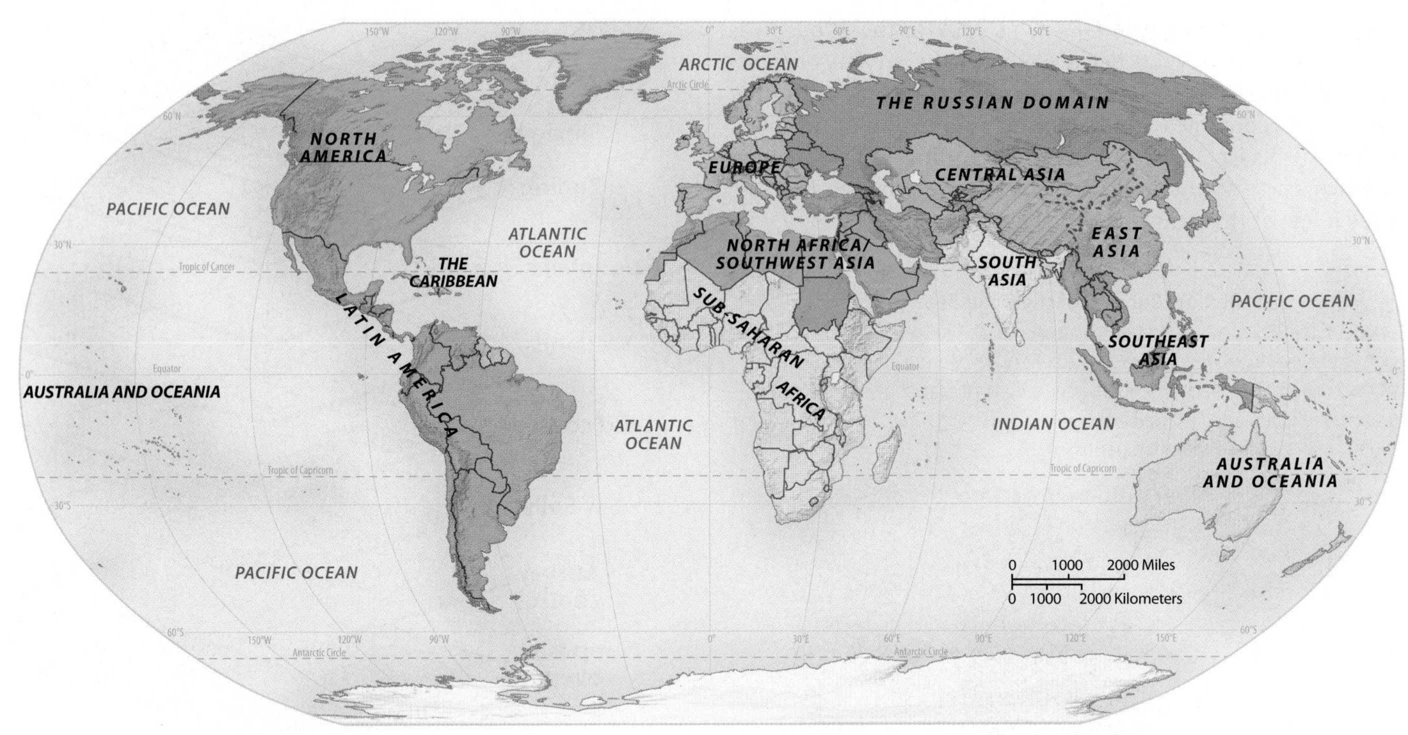

▲ **Figure 1.23 World Regions** These regions are the basis for the 12 regional chapters in this book. Countries or areas within countries that are treated in more than one chapter are designated on the map with a striped pattern. For example, western China is discussed in both Chapter 10, on Central Asia, and Chapter 11, on East Asia. Also, three countries on the South American continent are discussed as part of the Caribbean region because of their close cultural similarities to the island region.

Physical Geography and Environmental Issues: The Changing Global Environment

Chapter 2 provides background on world physical and environmental geography, outlining the global elements fundamental to human settlement—geology, climate, energy, hydrology, and vegetation. In the regional chapters, the physical geography sections explain the environmental issues relevant to each world region, covering topics such as climate change, sea-level rise, acid rain, energy issues, tropical rainforest destruction, and wildlife conservation. These environmental issues sections are not simply a list of problems, but also cover plans and policies developed to resolve those issues (see *Working Toward Sustainability: Managing Resources and Protecting the Environment*).

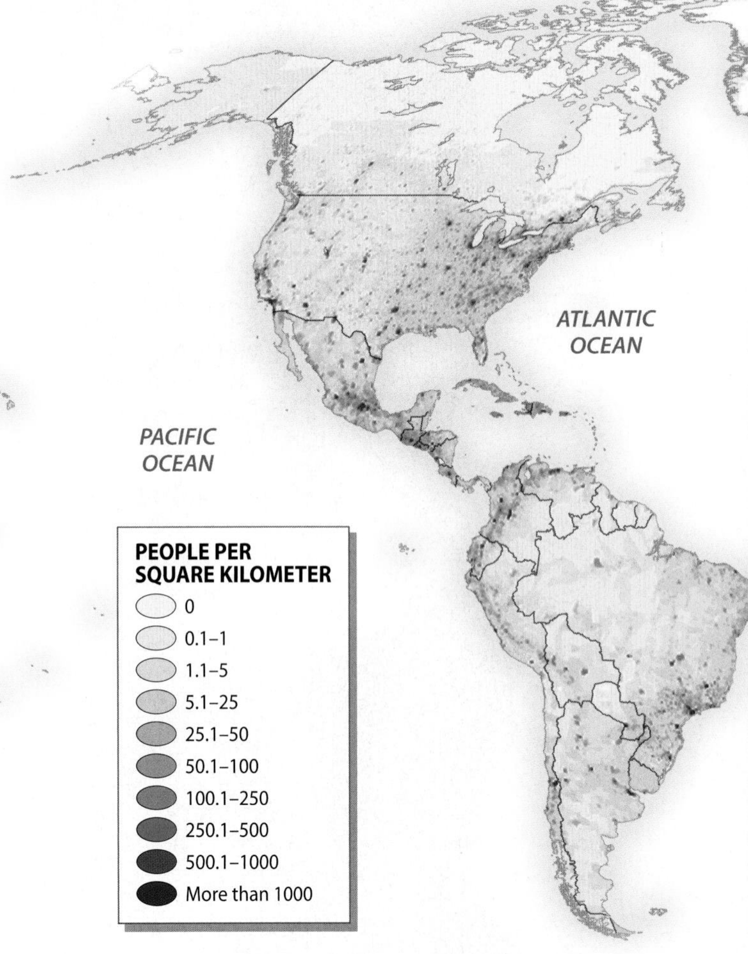

► **Figure 1.24 World Population** This map emphasizes the different population densities in the areas of the world. East Asia stands out as the most populous region, with high densities in Japan, Korea, and eastern China. The second most populous region is South Asia, dominated by India, which is second only to China in total population. In North Africa and Southwest Asia, population clusters are often linked to the availability of water for irrigated agriculture, as is apparent with the population cluster along the Nile River. Higher population densities in Europe, North America, and other countries are usually associated with large cities, their extensive suburbs, and nearby economic activities.

Population and Settlement: People on the Land

Currently, Earth has more than 7.2 billion people, with demographers forecasting an increase to 9.6 billion by 2050. Most of that increase will take place in Africa, particularly in Sub-Saharan Africa (Figure 1.24). In contrast, the European and Russian regions will probably shrink in population size, illustrating the complexity of population issues worldwide. Some countries are trying to slow population growth, but others face an uncertain future with either very slow or nonexistent population growth.

Population is obviously a very complex topic, but several points may help to focus the issues:

- As mentioned, very different rates of population growth occur throughout the world. Some countries, such as India, are growing rapidly; others, such as Italy and many other European countries, have essentially no natural growth at all. Any population growth results from in-migration.
- The current rate of population growth is now half the peak rate experienced in the 1960s. At that time, talk of "the population bomb" and "population explosions" was common, warning that the world either was very close to or had already passed a sustainable carrying capacity. Still, even with slower growth, about 20 percent of the world's population is undernourished.
- Population planning takes many forms, from the fairly rigid one- or two-child policies of late-20th-century China to the family-friendly policies of no-growth countries attempting to increase their natural birth rates (Figure 1.25).
- Not all attention should be focused on natural growth because migration is increasingly the root cause of population growth in today's world. Although most international migration is driven by a desire for a better life in the richer regions of the developed world, millions of migrants are refugees from civil strife, political persecution, and environmental disasters.
- The greatest migration in human history is going on now, as millions of people move from rural to urban environments. As of 2009, a landmark was reached when demographers estimated that more than half the world's population already lives in towns and cities.

Population Growth and Change

Because of the central importance of population growth, each regional chapter in this book includes a table of population data for the countries in that region (Table 1.1). Although at first glance these statistics may seem daunting, this information is crucial to understanding the population geography of the regions.

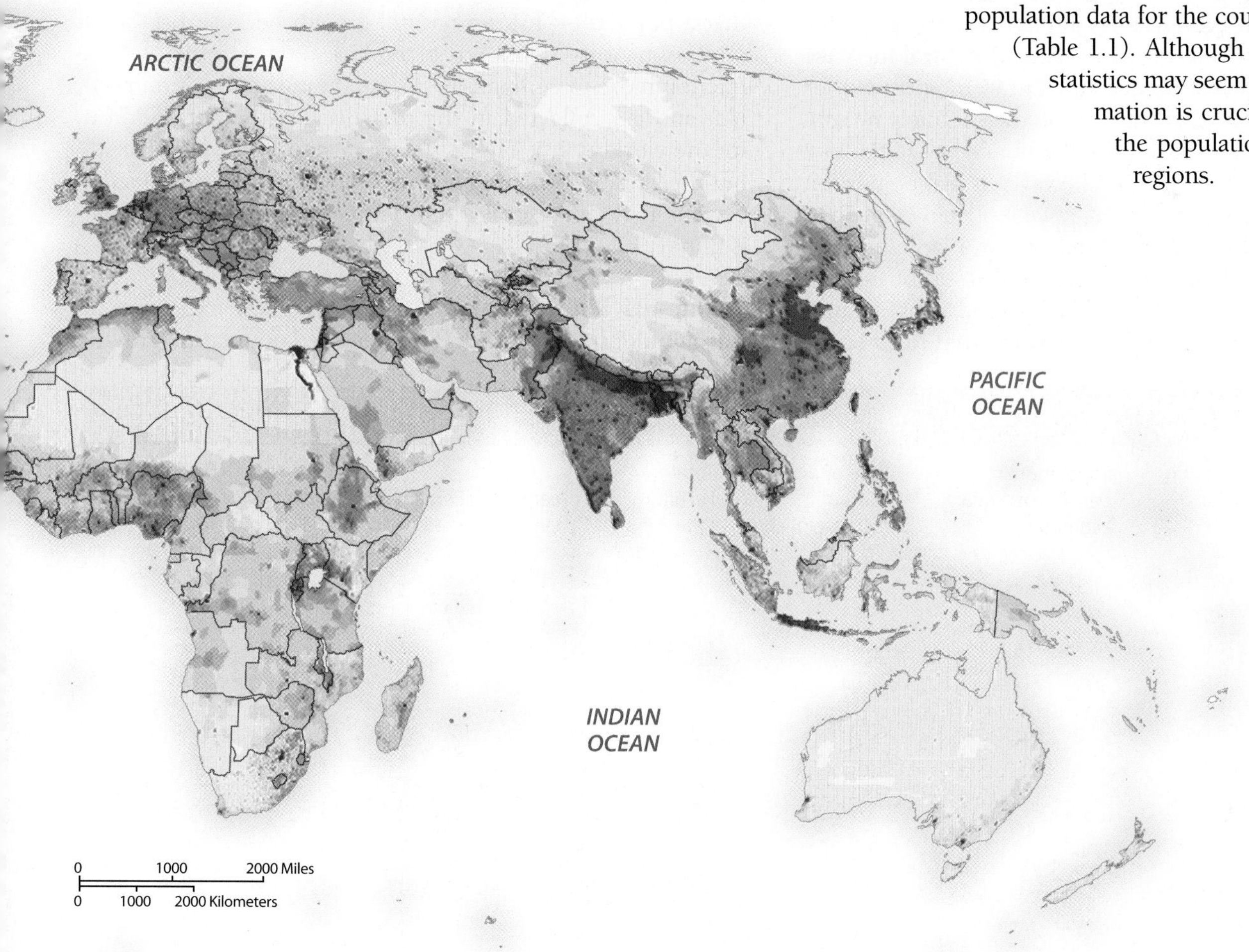

▲ **Figure 1.25 Family Planning** Many countries with fast-growing populations attempt to slow growth through government clinics like this one in Agra, India, that offer women advice on family planning matters.

Natural Population Increase A common starting point for measuring demographic change is the **rate of natural increase (RNI)**, which provides the annual growth rate for a country or region as a percentage. This statistic is produced simply by subtracting the number of deaths from the number of births in a given year. Important to remember, however, is that population gains or losses through migration are not considered in the RNI.

Also, instead of using raw numbers for a large population, demographers divide the gross numbers of births and deaths by the total population, thereby producing a number per 1000 of the population. This is referred to as either the *crude birth rate* or the *crude death rate*. For example, in 2013 the crude birth rate for the whole world was **20** per 1000, with a crude death rate of 8 per 1000. Thus, the natural growth rate was 12 per 1000. Converting that figure to a percentage produces the RNI; therefore, the global RNI in 2013 was 1.2 percent per year.

Because birth rates vary greatly among peoples and cultures (and between countries and regions of the world), rates of natural increase also vary greatly. In Africa, for example, several countries have crude birth rates of more than 40 per 1000 people. Because in these countries death rates are generally less than 11 per 1000, the rates of natural increase are greater than 3.3 percent per year, which are the highest population growth numbers found anywhere in the world.

Total Fertility Rate Although the crude birth rate gives some insight into current conditions in a country, demographers place more emphasis on the **total fertility rate (TFR)** to predict future growth. The TFR is an artificial and synthetic number that measures the fertility of a statistically fictitious, yet average group of women moving through their childbearing years. If women marry early and have many children over a long span of years, the TFR is a high number. Conversely, if data show that women marry late and have few children, the number is correspondingly low (Figure 1.26). Important to note is that any number less than 2.1 implies that a population has no natural growth because it takes a minimum of two children to replace their parents, with a fraction more to compensate for infant mortality. From population data collected in the past decade, the current TFR for the world is 2.5. This is the average for the whole world, but the variability among regions is striking. To illustrate, the current TFR for Africa is 4.8, whereas in no-growth Europe it is only 1.6.

Young and Old Populations One of the best indicators of the momentum (or its lack) for continued population growth

Table 1.1 Population Indicators

Country	Population (millions) 2013	Population Density (per square kilometer)	Rate of Natural Increase (RNI)	Total Fertility Rate	Percent Urban	Percent < 15	Percent > 65	Net Migration (Rate per 1000)
China	1,357.4	142	0.5	1.5	53	16	9	0
India	1,276.5	388	1.5	2.4	31	30	6	0
United States	316.2	33	0.5	1.9	81	19	14	2
Indonesia	248.5	130	1.5	2.6	50	29	5	−1
Brazil	195.5	23	0.9	1.8	85	25	7	0
Pakistan	190.7	230	2.3	3.8	35	37	4	−2
Nigeria	173.6	189	2.8	6.0	50	44	3	0
Bangladesh	156.6	1,087	1.5	2.3	26	31	5	−3
Russia	143.5	8	0.0	1.7	74	16	13	2
Japan	127.3	337	−0.2	1.4	91	13	25	1

Sources: Population Reference Bureau, *World Population Data Sheet, 2013.*

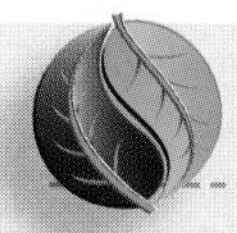

Working Toward **Sustainability**

Managing Resources and Protecting the Environment

The idea of sustainability seems to be everywhere, with much talk about sustainable cities, agriculture, forestry, businesses, corporations—even sustainable lifestyles. The list seems endless at times, and with so many different uses of the word, it's appropriate to revisit its original definitions.

The dictionary tells us that *sustainable* has two main roots. The first meaning is to endure and be able to maintain something at a certain level so that it lasts. The second definition refers to something that can be upheld or defended, such as a *sustainable idea* or *action*. Resource management has long used terms such as *sustained-yield forestry* to refer to timber practices where the amount of tree harvesting is attuned to the natural rate of forest growth so that the resource is not exhausted, but is able to renew itself over time.

Moral and ethical dimensions were added to this traditional usage in 1987 when the UN World Commission on Environment and Development addressed the complicated relationship between economic development and environmental deterioration. The commission stated that "sustainable development is development that meets the needs of the present without compromising the ability of future generations to meet their own needs." This cautionary message expands the notion of sustainability from a narrow focus on managing a specific resource, such as trees or grass, to include the whole range of human "needs," in both the present and the future. Fossil fuels, for example, are finite, so we should not consume them greedily in the present without considering their availability for future generations. Similar cautions apply to the sustainable uses of all other resources—air, water, soil, genetic biodiversity, wildlife habitats, farming, and so on.

Achieving the sustainable use of a specific resource, however, can be extremely difficult because it requires knowing the total amount of the resource in question and the current rate of consumption and then estimating the needs of future generations. These challenges have given rise to the new field of *sustainability science*, which emphasizes measuring and quantifying these factors. Because of these measurement difficulties, many researchers suggest that sustainability is better thought of as a process, rather than an achievable state.

In the following chapters, we explore the many different ways people are thinking about and working toward environmental and resource sustainability worldwide. Examples include the use of small-scale solar lighting in African villages (Chapter 2); the return of bison to North America's northern Great Plains (Chapter 3) (**Figure 1.2.1**); Japan's Smart City movement (Chapter 11); and the plight of Oceania's Low Islands as vulnerability increases because of global warming (Chapter 14). Many of these sidebars have links to Google Earth video tours.

1. Does your college or community have a sustainability plan? If so, what are the key elements?
2. How might the concept of sustainability differ for a college or university in, say, India or China? Look on the Internet to see what you can learn about sustainability programs in foreign colleges.

Google Earth Virtual Tour Video

http://goo.gl/08UbVj

▼ **Figure 1.2.1 Home on the Range** These bison (many have been imported from protected lands in Alberta) south of Malta, Montana, are quickly adjusting to their new home.

is the youthfulness of a population, since this shows the proportion of a population about to enter the prime reproductive years. The common statistic for this measure is the percentage of a population under age 15. Currently, the global average is that 26 percent of the population is younger than age 15. However, in fast-growing Africa, that figure is 41 percent, with several African countries approaching 50 percent.

◀ **Figure 1.26 Total Fertility Rate** Birth rates and death rates vary widely around the world. Fertility rates result from an array of variables, including state family-planning programs and the level of a woman's education. This family lives in the state of Rajasthan, India.

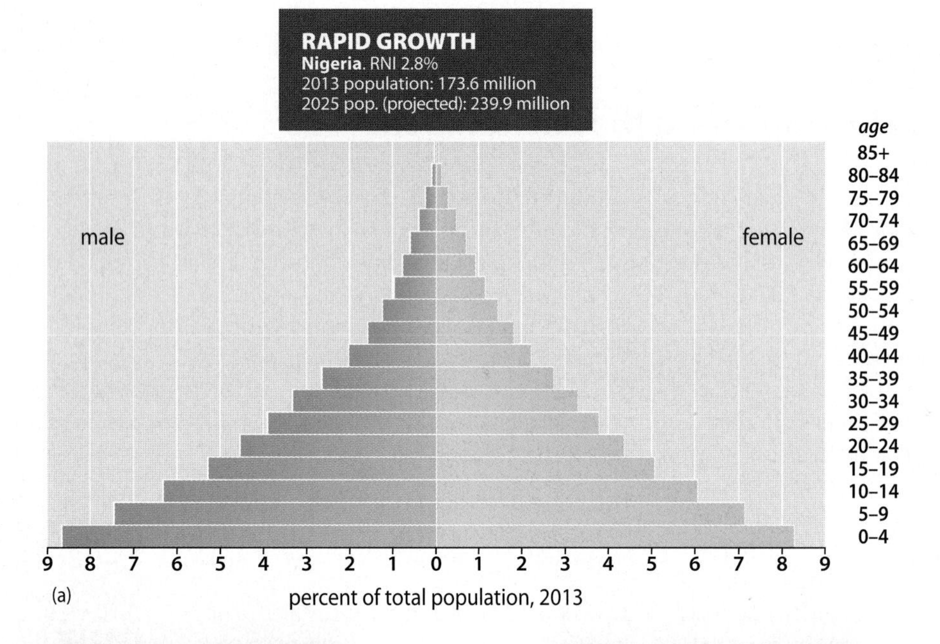

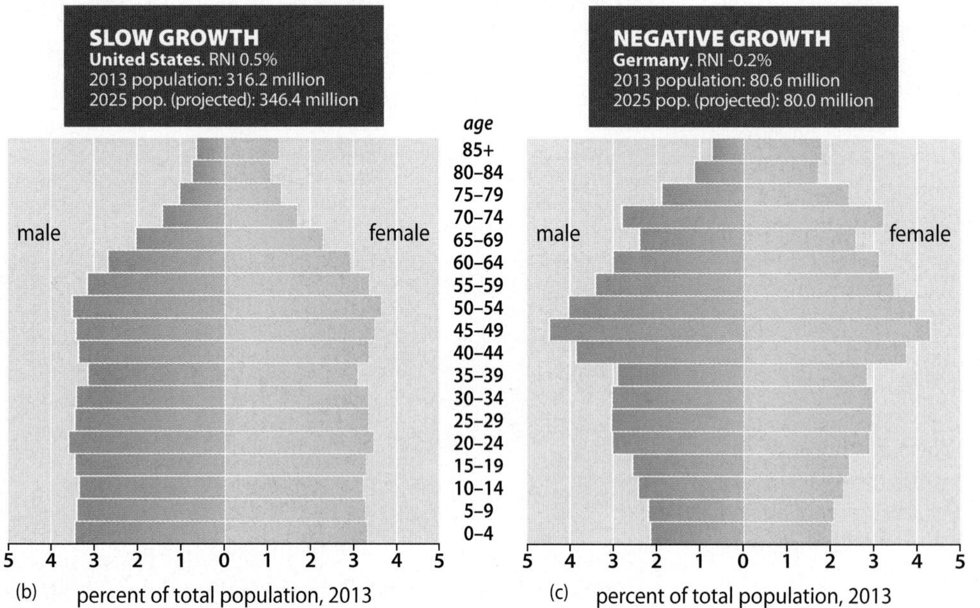

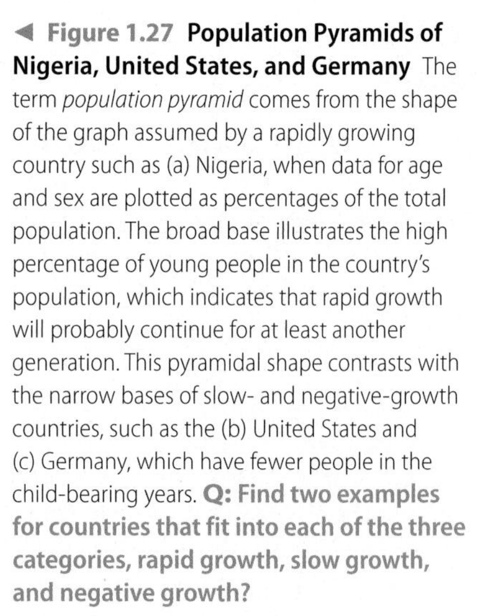
◄ **Figure 1.27 Population Pyramids of Nigeria, United States, and Germany** The term *population pyramid* comes from the shape of the graph assumed by a rapidly growing country such as (a) Nigeria, when data for age and sex are plotted as percentages of the total population. The broad base illustrates the high percentage of young people in the country's population, which indicates that rapid growth will probably continue for at least another generation. This pyramidal shape contrasts with the narrow bases of slow- and negative-growth countries, such as the (b) United States and (c) Germany, which have fewer people in the child-bearing years. **Q: Find two examples for countries that fit into each of the three categories, rapid growth, slow growth, and negative growth?**

This strongly suggests that rapid population growth in Africa will continue for at least another generation, despite the tragedy of the AIDS epidemic. In contrast, Europe has only 16 percent of its population under 15, and North America has 19 percent.

The other end of the age spectrum is also important, and it is measured by the percentage of a population over age 65. This number is useful for inferring the needs of a society in providing social services for its senior citizens and pensioners. In Japan, for example, about a quarter of the population is over the age of 65.

Population Pyramids The best graphical indicator of a population's age and gender structure is the **population pyramid**. This graph depicts the percentage of a population (or, in some cases, the raw number) that is male or female in different age classes, from young to old (Figure 1.27). If a country has higher numbers of young people than old, the graph has a broad base and a narrow tip, thus taking on a pyramidal shape that commonly forecasts rapid population growth. In contrast, slow-growth or no-growth populations are top-heavy, with a larger number of seniors than people of younger age.

Not only are population pyramids useful for comparing different population structures around the world at a given point in time, but also they can capture the structural changes of a population in time if it transitions from fast to slow growth. Population pyramids are also useful for displaying gender differences within a population, showing whether or not there is a disparity in the numbers of males and females. In the mid-20th century, for example, population pyramids for those countries that fought in World War II (such as the United States, Germany, France, and Japan) showed a distinct deficit of males, indicating those lost to warfare. Similar patterns are found today in those countries experiencing widespread conflict and civil unrest.

Cultural preferences for one sex or another, such as the preference for male infants in China and India, also show up in population pyramids. Because of their usefulness in showing different population structures, comparative population pyramids are found throughout the regional chapters of this book.

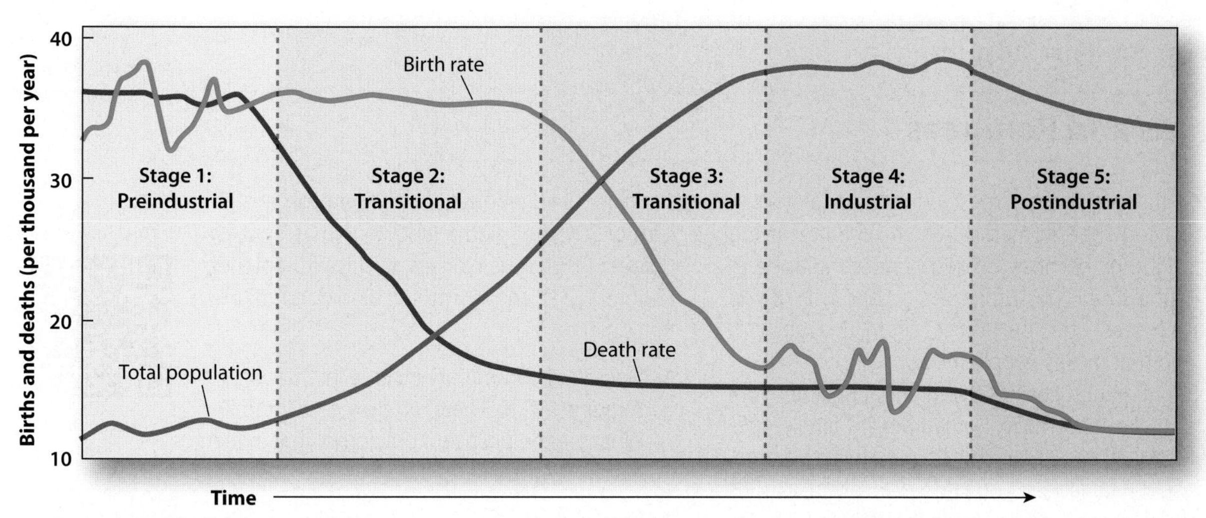

▲ **Figure 1.28 Demographic Transition** As a country goes through industrialization, its population moves through the five stages in this diagram, referred to as the *demographic transition*. In Stage 1, population growth is low because high birth rates are offset by high death rates. Rapid growth takes place in Stage 2, as death rates decline. Stage 3 is characterized by a decline in birth rates. The transition was initially thought to end with low growth once again in Stage 4, resulting from a relative balance between low birth rates and low death rates. But with a large number of developed countries now showing no natural growth, demographers have recently added a fifth stage to the traditional demographic transition model, one that shows no or even negative natural growth.

Life Expectancy Another demographic indicator that contains information about health and well-being in a society is **life expectancy**, which is the average length of a life expected at the birth of a typical male or female in a specific country. Because a large number of social factors—such as health services, nutrition, and sanitation—influence life expectancy, these data are often used as an indicator of the level of social development in a country. In this book, we use life expectancy as a social indicator; thus, these data are found in the economic and social development tables instead of in the population data tables.

Not surprisingly, because social conditions vary widely around the world, so do life expectancy figures. In general, though, life expectancy has been increasing over the decades, implying that the conditions supporting life and longevity are improving. To illustrate, in 1975 the average life expectancy figure for the world was 58 years, whereas today it **70**. In Sub-Saharan Africa, however, life expectancy has changed very little over the last 30 years because of the HIV/AIDS epidemic. As a result, the life expectancy for the region is just about the same (56) as it was in 1975 (52).

The Demographic Transition The historical record suggests that population growth rates slow down over time in a similar way from country to country. More specifically, in Europe, North America, and Japan, population growth slowed as countries became increasingly industrialized and urbanized. From these historical data, demographers generated the **demographic transition model**, a conceptualization that tracked the changes in birth rates and death rates over time. Originally, this model had four stages; however, a fifth stage is now commonly added to characterize the fact that many countries have slowed even further to a no-growth point (Figure 1.28).

In the demographic transition model, Stage 1 is characterized by both high birth rates and high death rates, resulting in a very low rate of natural increase. Historically, this stage is associated with Europe's preindustrial period, a time that predated common public health measures such as sewage treatment, an understanding of disease transmission, and the most fundamental aspects of modern medicine. Not surprisingly, death rates were high and life expectancy was short. Tragically, these conditions are still found today in some parts of the world.

In Stage 2, death rates fall dramatically, while birth rates remain high, thus producing a rapid rise in the RNI. In both historical and contemporary times, this decrease in death rates is commonly associated with the development of public health measures and modern medicine. Additionally, one of the assumptions of the demographic transition model is that these health services become increasingly available only after some degree of economic development and urbanization takes place.

However, even as death rates fall and populations increase, it takes time for people to respond with lower birth rates, which happens only in Stage 3. This, then, is the second transitional stage, in which people become aware of the advantages of smaller families in an urban and industrial setting, contrasted with the earlier need for large families in rural, agricultural settings or where children worked at industrial jobs (both legally and illegally).

People on the Move

Migrants and Refugees

Globalization has led to one of the largest migrations in human history. People are leaving their homes in search of better living conditions elsewhere, either because of desperate conditions at home—what geographers call "push forces"—or because of the lure of a better life elsewhere—the "pull forces." Often, though, it's a combination of the two that makes people move.

A third factor in the migration chain is the informational network people draw upon to make their move. Sometimes it's family connections, as people follow relatives who have successfully migrated. Other times a paid agent or labor contractor is involved. Stories abound, unfortunately, of migrant agents (called *coyotes* in Spanish) who, after taking the migrants' money, abandon them on the open seas or in the Mexican desert, or worse.

Accurate data on migration are notoriously difficult to gather because people often migrate without documentation. However, the UN says that currently at least 190 million people are legally living in a country different from where they were born. Although most of those people have moved to cities in the developed Western world, considerable migration has also occurred to magnet cities in the developing world, places such as Delhi, Dubai, Rio de Janeiro, and Mexico City. Furthermore, that UN statistic does not include those people who stay within their own country, yet leave home, like the highland peasants of Bolivia moving to La Paz or a farmer from Iowa cashing in his farm to take a job in Atlanta.

Nor does that statistic count the estimated 35 million people who are classified as refugees, those who have fled violence or natural disasters. Or the untold numbers who are not in established refugee camps, but instead are making do somehow on their own.

Because the human geography of global migration wears many different faces and tells many different stories, each of our chapters sheds light on this complicated process of migration through our *People on the Move* sidebars. Examples are the case of European Roma (also known as gypsies; Chapter 8); Chinese workers in Africa (Chapter 6) (**Figure 1.3.1**); and North Korea's hidden migration (Chapter 11).

Google Earth Virtual Tour Video

http://goo.gl/82kUAu

1. Does your community include international migrants? If so, where did they come from? What are the push and pull forces that influen tional decisions?
2. Now choose a foreign city in either Europe or Asia and, using the Internet, collect information on its international migrant population.

▼ **Figure 1.3.1 Chinese Merchant in Africa** Push or pull forces often compel people to leave their homeland in search of better opportunity elsewhere.

Then, in Stage 4, a low RNI results from a combination of low birth rates and very low death rates. Until recently, this stage was assumed be the static end point of change of a developing, urbanizing population. However, as mentioned earlier, that does not seem to be the case. In many highly urbanized developed countries, particularly those in Europe and Russia, the death rate now exceeds the birth rate. As a result, the RNI falls below a replacement level, as attested to by a negative number. This negative growth state argues for the addition of a fifth stage to the traditional demographic transition model.

Remember, though, that the RNI is just that—the rate of natural increase. Thus, it does not include a country's growth or loss from migration. In the United States, for example, although the RNI is below the replacement level, the overall population continues to grow because of immigration from other countries. The same is true of many other developed countries, particularly those in Europe. This topic is discussed further in Chapter 8.

Global Migration and Settlement

Never before in human history have so many people been on the move. Today more than 190 million people live outside the country of their birth and thus are officially designated as migrants by international agencies. Much of this international migration is directly linked to the new globalized economy because half of the migrants live either in the developed world or in developing countries with vibrant industrial, mining, or petroleum extraction economies. In the oil-rich countries of Kuwait and Saudi Arabia, for example, the labor force is composed primarily of foreign migrants. In total numbers, fully one-third of the world's migrants live in seven industrial countries: Japan, Germany, France, Canada, the United States, Italy, and the United Kingdom (see *People on the Move: Migrants and Refugees*).

Moreover, most of these migrants have moved to cities; in fact, 20 percent of migrants live in just 20 world cities. Further, because industrial countries usually have very low birth rates, immigration accounts for a large proportion of their population growth. For example, about one-third of the annual growth in the United States is due to in-migration.

However, not all migrants move for economic reasons. War, persecution, famine, and environmental destruction cause people to flee to safe havens elsewhere. Accurate data on refugees are often difficult to obtain for several reasons (such as individuals not legally crossing international boundaries and countries deliberately obscuring the number for political reasons), but UN officials estimate that some 35 million people should be considered refugees. More than half of these are in Africa and western Asia (Figure 1.29).

Net Migration Rates The amount of immigration (in-migration) and emigration (out-migration) is measured by the **net migration rate**, a statistic that indicates whether more people are entering or leaving a country. A positive figure means the population is growing because of in-migration, whereas a negative number means more people are leaving than arriving. As with other demographic indicators, the net migration rate is provided for the number of migrants per 1000 of a base population. To illustrate, the net migration rate for the United States is 49 per 1000 people, whereas Canada, which receives even more immigrants than the United States, has a net migration rate of 2. In contrast, Mexico, the source of many migrants to North America, has a net migration rate of −2.

Some of the highest net migration rates are found in countries that depend heavily on migrants for their labor force. These include Qatar, with a net migration rate of **49**, and Kuwait, at 37. Countries with the highest negative migration rates are Tonga, –17; Samoa, –17; Micronesia, –15; and several Caribbean islands with rates close to –10.

Population Density The average number of people per unit of area (square mile or square kilometer) is referred to as **population density**. This statistic conveys important information about settlement patterns and landscape in a specific area or country. In Table 1.1, you can see the striking difference between the high population density of India and the much lower figure for the United States. Flying over these two countries and looking down at the settlement patterns explains this contrast (Figure 1.30). Much of the United States is covered by farms covering hundreds of acres, with houses and barns several miles from their neighbors. In contrast, the landscape of India is made up of small villages, distant from each other by only a mile or so. This results in a population density in India three times higher than that in the United States. Japan has a settlement pattern similar to that of India, even though it is primarily an urban, industrial country. Bangladesh, one of the most densely settled countries in the world, must somehow squeeze its large and rapidly growing population into a limited amount of dry land built by the deltas of two large rivers, the Ganges and the Brahmaputra.

Because population densities differ considerably between rural and urban areas, the gross national figure can be a bit misleading. Many of the world's largest cities, for example, have densities of more than 30,000 people per square mile (10,300 per square km), with the central areas of Mumbai (Bombay) and Shanghai easily twice as dense because of the prevalence of high-rise apartment buildings. In contrast, most

▼ **Figure 1.29 Global Refugees** The United Nations estimates that some 35 million people are refugees from war and civil unrest, as illustrated by this Afghan family at a refugee camp near Peshawar, Pakistan

▲ **Figure 1.30 Contrasting Settlement Densities in the United States and India** In the United States, the population density is 33 per square kilometer, whereas it is 388 per kilometer in India. The relatively low density for the United States results from a vast area of dispersed housing, as illustrated in the bottom photo from Iowa, while India (top photo) has denser settlement in both cities and countryside.

Cityscapes

The Human Habitat

Most of the world's population lives in cities. Thus, knowing more about the world's global urban habitats is an important part of world regional geography. After all, even though cities share some traits, they also differ considerably. Few people would say, for example, that Los Angeles is like New York City or that Seattle is like New Orleans. Further, the economic and social roles cities play in the world can also differ considerably. Mexico City and Shanghai have about the same population (around 25 million), but their global economic and political roles are very different.

To further our understanding of world urbanism, each regional chapter contains a "Cityscapes" sidebar and Google Earth video tour of a prominent place in that part of the world. Vancouver, British Columbia; La Paz, Bolivia; Cairo, Egypt (**Figure 1.4.1**); Kabul, Afghanistan; Perth, Australia; and Hong Kong, China, are examples of the different cities featured. As you read about and tour these cities, keep the following facts in mind.

Although cities are old (the earliest date back to 4000 BCE), urbanism—when most of a country's population lives in cities—is new. England is thought to be the first country to have most of its population living in cities, an event that happened late in the 19th century as a result of that country's industrialization. The United States became an urban nation later, sometime between 1910 and 1920. China, the world's largest country, became predominantly urban only in the last several years, but will become increasingly urban with the government's plan to uproot and move 250 million rural people into cities by 2025.

Cities look different from one another because of what is called *urban morphology*, which is, simply put, the brick-and-mortar physical landscape of a city. Older parts of the city look different from the new because of street patterns, building architecture, and even the construction materials used. Consequently, we can learn much from studying the visual landscape of a city. Some of the earliest continually inhabited cities are found in Lebanon, Iraq, China, and Pakistan; although the most ancient parts of these cities are long gone, you can still find historical differences in the urban morphology.

Also important is deciphering where different *urban functions*, or activities, take place within the city. Where is the business district, and how does it contrast with downtowns in other cities in other parts of the world? Or consider the residential areas: Are they low or high density, or both? What about the illegal (yet often tolerated) squatter settlements where the poor often reside? The answers to these questions often tell us much about the culture and historical background of the people who live there.

Google Earth Virtual Tour Video

http://goo.gl/Cd3NKB

1. What city that you have visited has the most unique and memorable cityscape? Why?
2. Find a major city in another part of the world; and discuss its urban morphology and cityscape.

▶ **Figure 1.4.1 The Cityscape of Cairo, Egypt** The largest city in the Africa and the Arab world, Cairo has an iconic and expansive landscape.

North American cities have densities of fewer than 10,000 people per square mile (3800 per square km), due largely to the cultural preference for single-family dwellings on individual urban lots.

An Urbanizing World The focal points of the contemporary, globalizing world are the cities—the fast-paced centers of deep and widespread economic, political, and cultural change. Because of this vitality and the options cities offer to impoverished and uprooted rural peoples, they are also magnets for migration. The scale and rate of growth of some world cities are absolutely staggering. Estimates are that between natural growth and in-migration, Mumbai (Bombay) will add over 7 million people by 2020. Assuming growth is constant throughout the period (perhaps a questionable assumption), this would mean that the urban area would add over 10,000 new people each week. The same projections would have Lagos, Nigeria, which currently has the highest annual growth of any megacity, adding almost 15,000 per week (see *Cityscapes: The Human Habitat*).

As mentioned, based upon data on the **urbanized population**, which is the percentage of a country's population living in cities, at least half the world's population now lives in cities. Demographers predict that the world will be 60 percent urbanized by 2025.

Tables in this book's regional chapters include data on the urbanization rate for each country. To illustrate, more than 80 percent of the populations of Europe, Japan, Australia, and the United States live in cities. Generally speaking, most countries with such high rates of urbanization are also highly industrialized because manufacturing tends to cluster around urban centers. In contrast, the urbanized rate for developing countries is usually less than 50 percent, with figures closer to 40 percent not uncommon. Urbanization figures also show where there is high potential for urban migration. If the urbanized population is relatively small, as in Zimbabwe (Africa), where only **39** percent of the population lives in cities, the probability for high rates of urban migration in the next decades is high.

REVIEW

1.12 How is the rate of natural increase calculated? Give an example.
1.13 What is the total fertility rate?
1.14 Describe and explain the demographic transition model.
1.15 How is a population pyramid constructed, and what kind of information does it convey?

Cultural Coherence and Diversity: The Geography of Change and Tradition

Social scientists often say that culture binds together the world's diverse social fabric. If this is true, one glance at the daily news suggests this complex global tapestry could be unraveling because of widespread cultural tensions and conflict. As noted earlier, with the recent rise of global communication systems (satellite TV, films, videos, etc.), stereotypical Western culture is spreading at a rapid pace. Although some societies accept these new cultural influences willingly, others resist and push back against what they perceive as cultural imperialism with protests, censorship, and even terrorism. The geography of cultural coherence and diversity, then, entails an examination of both tradition and change, of tensions and conflict, of global patterns and unique local custom, as well as the formation of blended cultural hybrids.

Culture in a Globalizing World

Given the dynamic changes connected with globalization, traditional definitions of culture must be stretched somewhat to provide a viable conceptual framework. A very basic definition provides a starting point. **Culture** is learned, not innate, and is behavior held in common by a group of people, empowering them with what is commonly called a "way of life" (Figure 1.31).

▶ **Figure 1.31 Culture as a Way of Life** The clothing and appearance of this young couple give clue to their membership in Brooklyn Hipster culture.

In addition, culture has both abstract and material dimensions. Speech, religion, ideology, livelihood, and value systems are part of culture, but so are technology, housing, foods, and music. These varied facets of culture are relevant to the study of world regional geography because they tell us much about the way people interact with their environment, with one another, and with the larger world. Not to be overlooked is that culture is dynamic and ever changing, not static. Thus, culture is a process, not a condition—an abstract, yet useful concept that is constantly adapting to new circumstances. As a result, there are always tensions between the conservative, traditional elements of a culture and the newer forces promoting change (see *Everyday Globalization: Complexities in the Common*).

When Cultures Collide Cultural change often takes place within the context of international tensions. Sometimes, one cultural system will replace another; at other times, resistance by one group to another's culture will stave off change. More commonly, however, a newer, hybrid form of culture results from an amalgamation of two cultural traditions. Historically, colonialism was the most important perpetuator of these cultural collisions; today, though, globalization in its varied forms is a major vehicle of cultural tensions and change (Figure 1.32).

▼ **Figure 1.32 Culture Clash in Goa, India** A beach-loving western tourist gets better acquainted with a sacred symbol of Hindu culture.

Everyday **Globalization**

Complexities in the Common

Globalization is so ubiquitous that it's often taken for granted and unnoticed. Your clothing is an example. Chances are good that what you're wearing was made in a foreign country, since 98 percent of all U.S. apparel is imported. Your clothing was probably made in China, but perhaps in Bangladesh, Thailand, Haiti, Mexico, or India, all of which are major manufacturing centers for the world's clothing. Even some of the "Made in the U.S.A." clothing might be pushing the truth a bit by being produced in the U.S. commonwealth countries of Puerto Rico and the Northern Mariana Islands in the far western Pacific. However, if you paid $300 or so for your jeans, they could be made in the United States, most probably in Los Angeles, where 30 different apparel firms turn out designer jeans.

The point is not about what you're wearing, but rather that globalization is not only about multinational corporations doing business all over the world. Globalization is everywhere in your daily life, from what you eat to what you wear to the smartphone in your hand to the coffee you drink. Chances are that whatever it is involves an interesting and complex world geography.

We illustrate this idea with examples in each of the regional chapters: how your hamburger may or may not come from the tropical rainforest (Chapter 2); why the National Basketball Association is so successful globally (Chapter 3); the geography behind diamond engagement rings (Chapter 6); international students in Europe (Chapter 8) (**Figure 1.5.1**); and why you're talking to someone in India about your credit card (Chapter 12), to name just a few.

1. Identify a commonplace item or activity in your life that has an interesting backstory involving globalization.
2. How has globalization changed higher education in the United States?

Google Earth Virtual Tour Video

http://goo.gl/XMxG87

▼ **Figure 1.5.1 Students at the University of Tübingen, Germany** Globalization has accelerated the growth of students studying abroad.

The active promotion of one cultural system at the expense of another is called **cultural imperialism**. Although many expressions of cultural imperialism still exist today, the most severe examples occurred in the colonial period. In those years, European cultures spread worldwide, often overwhelming, eroding, and even replacing indigenous cultures. During this period, Spanish culture spread widely in South America, French culture diffused into parts of Africa and Southeast Asia, and British culture overwhelmed South and Southwest Asia. New languages were mandated, new educational systems were implanted, and new administrative institutions replaced the old. Foreign dress styles, diets, gestures, and organizations were added to existing cultural systems. Many vestiges of colonial culture are still evident today (Figure 1.33). In India, the makeover was so complete that pundits are fond of saying, with only slight exaggeration, that "the last true Englishman will be an Indian."

Today's cultural imperialism is seldom linked to an explicit colonizing force, but more often comes as a fellow traveler with economic globalization. Though many expressions of cultural imperialism carry a Western (even U.S.) tone—such as McDonald's, MTV, KFC, Marlboro cigarettes, and the widespread use of English as the dominant language of the Internet—these facets result more from a search for new consumer markets than from deliberate efforts to spread modern U.S. culture throughout the world.

The reaction against cultural imperialism is **cultural nationalism**. This is the process of protecting and defending a cultural system against diluting or offensive cultural expressions, while at the same time actively promoting national and local cultural values. Often cultural nationalism takes the form of explicit legislation or official censorship that simply outlaws unwanted cultural traits. Examples of cultural nationalism are common. France has long fought the Anglicization of its language by banning "Franglais" in official governmental language, thereby exorcising commonly used words such as *weekend, downtown, chat,* and *happy hour.* France has also sought to protect its national music and film industries by legislating that radio DJs play a certain percentage of French songs and artists each broadcast day (40 percent currently). Similarly, many Muslim countries limit Western cultural influences by restricting or censoring international TV, an element they consider the source of many undesirable cultural influences. Most Asian countries as well are increasingly protective of their cultural values, and many are demanding changes to tone down the sexual content of MTV and other international TV networks.

Cultural Hybrids As mentioned, a common product of cultural collision is the blending of forces to form a new, synergistic form of culture in a process called **cultural syncretism** or **hybridization**. To characterize India's culture as British, for example, is to grossly oversimplify and exaggerate England's colonial influence. Instead, Indians have adapted many British traits to their own circumstances, infusing them with their own meanings. India's use of English, for example, has produced a unique form of "Indlish" that often befuddles visitors to South Asia. Nor should we forget that India has added many words to our English vocabulary—*khaki, pajamas, veranda,* and *bungalow,* among others. Clearly, both the Anglo and the Indian cultures have been changed by the British colonial presence in South Asia. Other examples of cultural hybrids abound: Australian-rules football, hip-hop music, fusion cuisine, Tex-Mex fast food, and so on.

Language and Culture in Global Context

Language and culture are so intertwined that often language is the major characteristic that differentiates and defines one cultural group from another (Figure 1.34). Furthermore, because language is the primary means for communication, it folds together many other aspects of cultural identity, such as politics, religion, commerce, and customs. Language is fundamental to cultural cohesiveness and distinctiveness, for language not only brings people together, but also sets them apart from nonspeakers of that language. Therefore, language is an important component of national or ethnic identity, as well as a means for creating and maintaining boundaries for group and regional identity.

Because most languages have common historical (and even prehistorical) roots, linguists have grouped the thousands of languages spoken throughout the world into a handful of language families. This is simply a first-order grouping of languages into large units, based on common ancestral speech. For example, about half of the world's people speak languages of the Indo-European family, a large group that includes not only European languages such as English and Spanish, but also Hindi and Bengali, the dominant languages of South Asia.

Within language families, smaller units also give clues to the common history and geography of peoples and cultures. *Language branches and groups* (also called *subfamilies*) are closely related subsets within a language family, usually sharing similar sounds, words, and grammar. Well known are the similarities between German and English and between French and Spanish.

Additionally, individual languages often have very distinctive *dialects* associated with specific regions and places. Think of the distinctive differences, for example, among British, North American, and Australian English or the city-specific dialects that set apart New Yorkers from residents of Dallas, Berliners from inhabitants of Munich, Parisians from villagers of rural France, and so on.

When people from different cultural groups cannot communicate directly in their native languages, they often agree on a third language to serve as a common tongue, a **lingua franca**. Swahili has long served that purpose for speakers of the many tribal languages of eastern Africa, and French was historically the lingua franca of international politics and diplomacy. Today English is increasingly the common language of international communications, science, and air transportation.

The world is becoming increasingly multilingual as people find it necessary to communicate in several languages. However, in general terms, the languages with the largest numbers of native speakers are Mandarin, spoken

▼ **Figure 1.33 Cultural Imperialism in Colonial Architecture** German architecture in Namibia, Africa, provides clues to the former colonial relationship between that country and Germany.

▶ **Figure 1.34 World Language Families** Most languages of the world belong to a handful of major language families. About 50 percent of the world's population speaks a language belonging to the Indo-European language family, which includes not only languages common to Europe, but also major languages in South Asia, such as Hindi. They are in the same family because of their linguistic similarities. The next largest family is the Sino-Tibetan family, which includes languages spoken in China, the world's most populous country. **Q: What languages, other than English, are spoken in your community?**

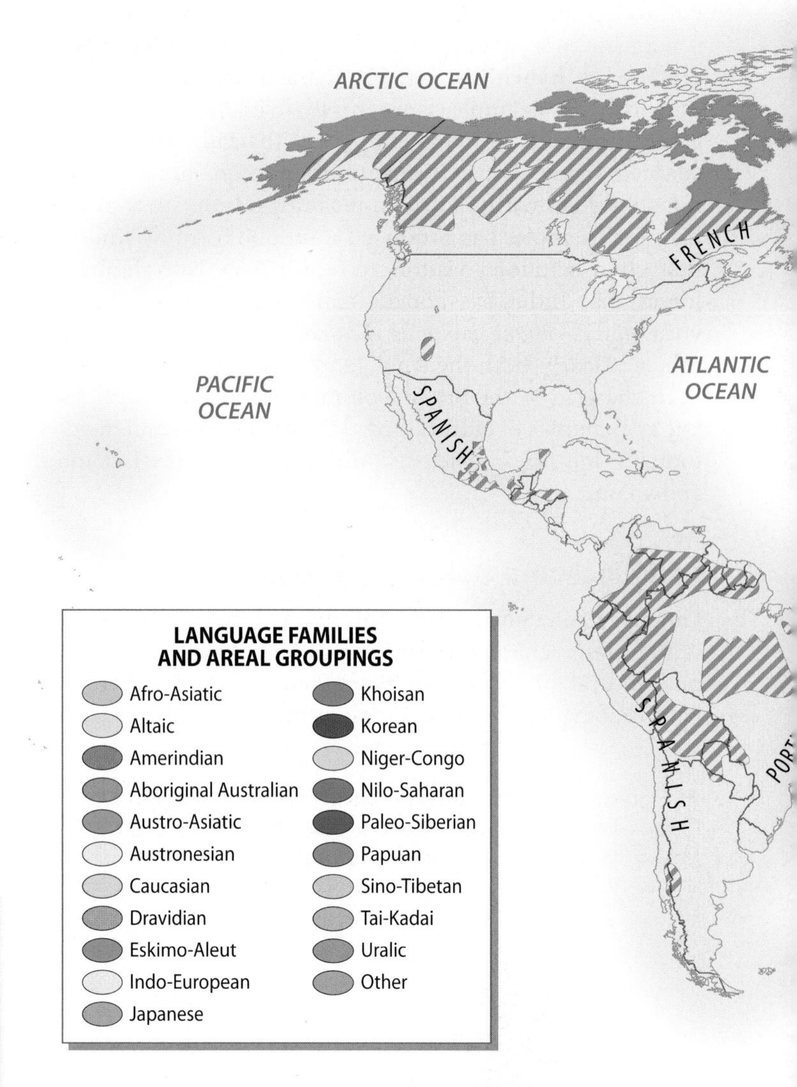

by about 13 percent of the world's population; Spanish, 6 percent; and English, also around 6 percent, followed by Hindi and Arabic, each with 5 percent (Figure 1.35).

The Geography of World Religions

Another important defining trait of cultural groups is religion (Figure 1.36). Indeed, in this era of a comprehensive global culture, religion is becoming increasingly important in defining cultural identity. Recent ethnic violence based upon religious differences in far-flung places such as the Balkans, Iraq, Syria, and Indonesia illustrates the point.

Universalizing religions, such as Christianity, Islam, and Buddhism, attempt to appeal to all peoples, regardless of location or culture. These religions usually have a proselytizing or missionary program that actively seeks new converts throughout the world. In contrast are **ethnic religions**, which are identified closely with a specific ethnic, tribal, or national group. Judaism and Hinduism, for example, are usually regarded as ethnic religions because they normally do not actively seek new converts; instead, people are born into ethnic religions.

▼ **Figure 1.35 Mandarin and English** This road sign in Shanghai, China, displays two of the world's most popular languages, Mandarin, spoken by about 12 percent of the world's population, and English, the global language of commerce, transportation, and science.

Christianity, because of its universalizing ethos, is the world's largest religion in both areal extent and number of adherents. Although fragmented into separate branches and churches, Christianity as a whole has 2.1 billion adherents, encompassing about one-third of the world's population. The largest numbers of Christians are found in Europe, Africa, Latin America, and North America.

Islam, which has spread from its origins on the Arabian Peninsula east to Indonesia and the Philippines, has about 1.3 billion members. Although not as severely fragmented

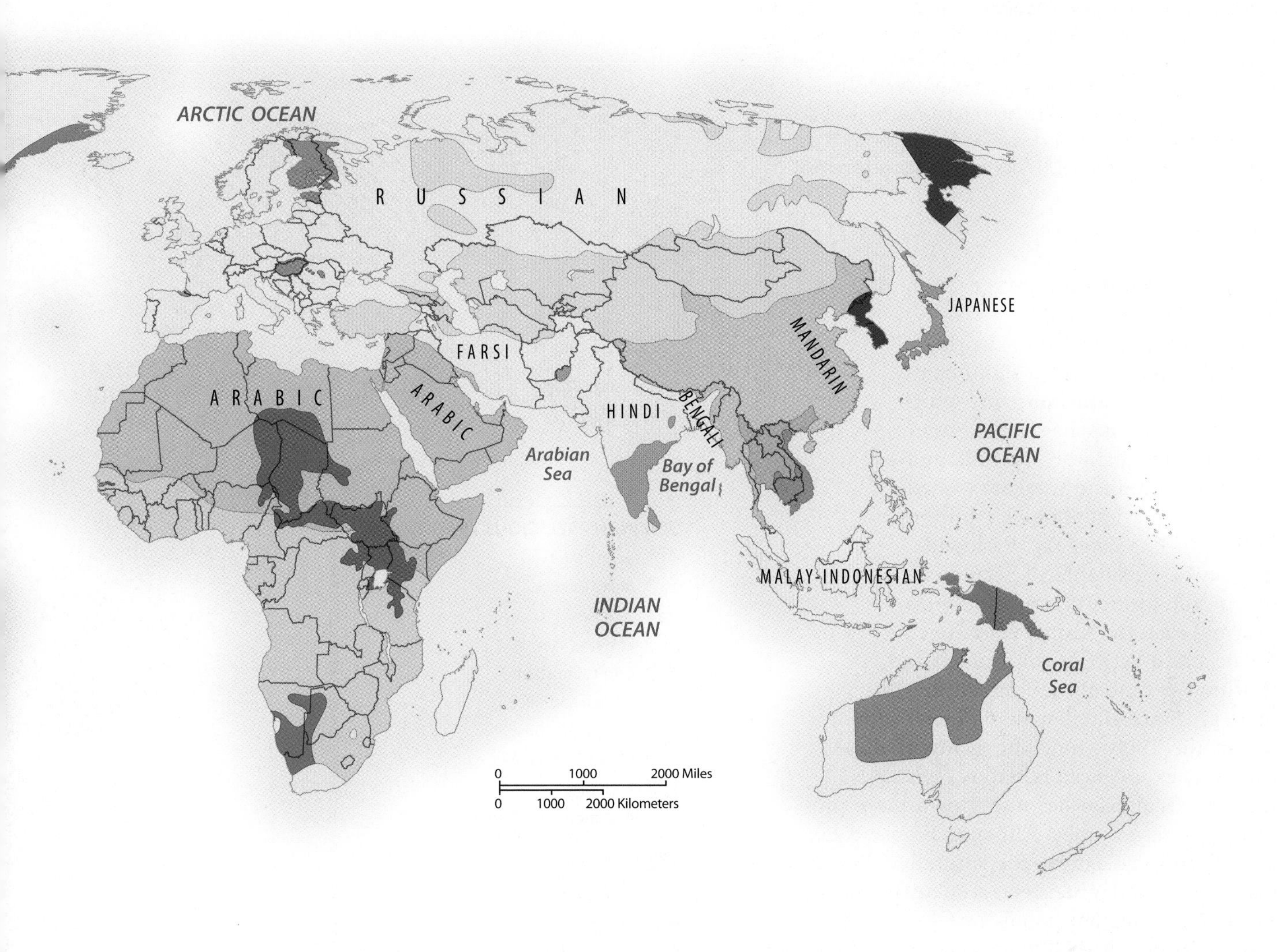

as Christianity, Islam should not be thought of as a homogeneous religion because it is also split into separate groups. One of the two major branches is **Shi'a Islam**, which constitutes about 11 percent of the total Islamic population and represents a majority in Iran and southern Iraq. The other branch is the more dominant **Sunni Islam**, which is found from the Arab-speaking lands of North Africa to Indonesia. Probably in response to Western influences connected to globalization, both of these forms of Islam are currently experiencing fundamentalist revivals in which proponents are interested in maintaining purity of faith, separate from these Western influences.

Judaism, the parent religion of Christianity, is also closely related to Islam. Although tensions are often high between Jews and Muslims, these two religions, along with Christianity, actually share historical and theological roots in the Hebrew prophets and leaders. Judaism today numbers about 14 million adherents, with the population having lost perhaps one-third of its historic population due to the systematic extermination of Jews during World War II.

Hinduism, which is closely linked to India, has about 900 million adherents. Outsiders often regard Hinduism as polytheistic because Hindus worship many deities. Most Hindus argue, however, that all of their faith's gods are merely representations of different aspects of a single divine, cosmic unity. Historically, Hinduism is linked to the caste system, with its segregation of peoples based on ancestry and occupation. Today, however, because India's democratic government is committed to reducing the social distinctions among castes, the connections between religion and caste are now much less explicit than in the past.

Buddhism, which originated as a reform movement within Hinduism about 2500 years ago, is widespread in Asia, extending from Sri Lanka to Japan and from Mongolia to Vietnam (Figure 1.37). Buddhism has two major branches: *Theravada,* found throughout Southeast Asia and Sri Lanka, and *Mahayana,* found in Tibet and East Asia. In its spread, Buddhism came to coexist with other faiths in certain areas, making it difficult to accurately estimate the number of its adherents. Estimates of the

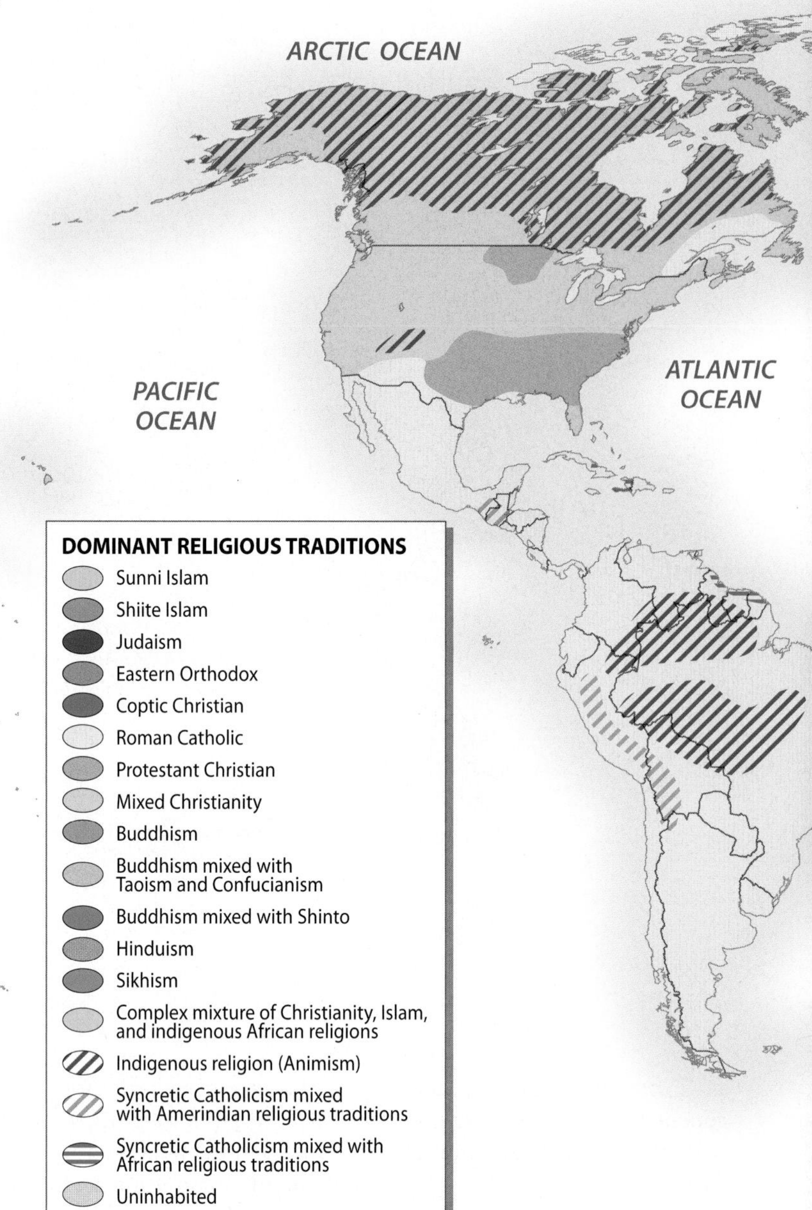

▶ **Figure 1.36 Major Religious Traditions** This map shows the regions dominated by major religions throughout the world. For most people, religious tradition is a major component of cultural and ethnic identity. Christians of different sorts account for about 34 percent of the world's population, but this religious tradition is highly fragmented. Within Christianity, there are about twice as many Roman Catholics as Protestants. Islam accounts for about 20 percent of the world's population; Hindus make up about 14 percent.

total Buddhist population range from 350 million to 900 million people.

Finally, in some parts of the world, religious practice has declined significantly, giving way to **secularism**, in which people consider themselves either nonreligious or outright atheistic. Though secularism is difficult to measure, social scientists estimate that about 1.1 billion people fit into this category worldwide. Perhaps the best example of secularism comes from the former communist lands of Russia and eastern Europe, where overt hostility occurred between government and church from the time of the Russian Revolution of 1917. Since the demise of Soviet communism in the 1990s, however, many of these countries have experienced religious revivals.

Recently, secularism has also grown more pronounced in western Europe. Although France is historically and to some extent still culturally a Roman Catholic country, today, between secularism and an increase of the immigrant population there are possibly more people attending Muslim mosques on Fridays than attending Christian churches on Sundays. Japan and the other countries of East Asia are also noted for their high degree of secularization.

Gender and Globalization

Culture includes not just that ways people speak or worship, but also the embedded systems that influence behavior and values. **Gender** is a sociocultural construct linked to the values and traditions of specific cultural groups, which then differentiate the characteristics of the two biological sexes, male and female. Central to this concept is the notion of **gender roles**, which are the cultural guidelines that define appropriate behavior for each gender within a specific context. In a traditional tribal or ethnic group, for example, gender roles might rigidly define the difference between women's work (which often consists of domestic tasks) and men's work (which is often done outside the home). Similarly, gender roles might guide all other social behaviors within a group, such as

◀ **Figure 1.37 Buddhist Landscapes** An array of buildings—temples, monasteries, and shrines—produces a distinctive landscape throughout Asia, as is illustrated by this photo from Chiang Mai in northern Thailand.

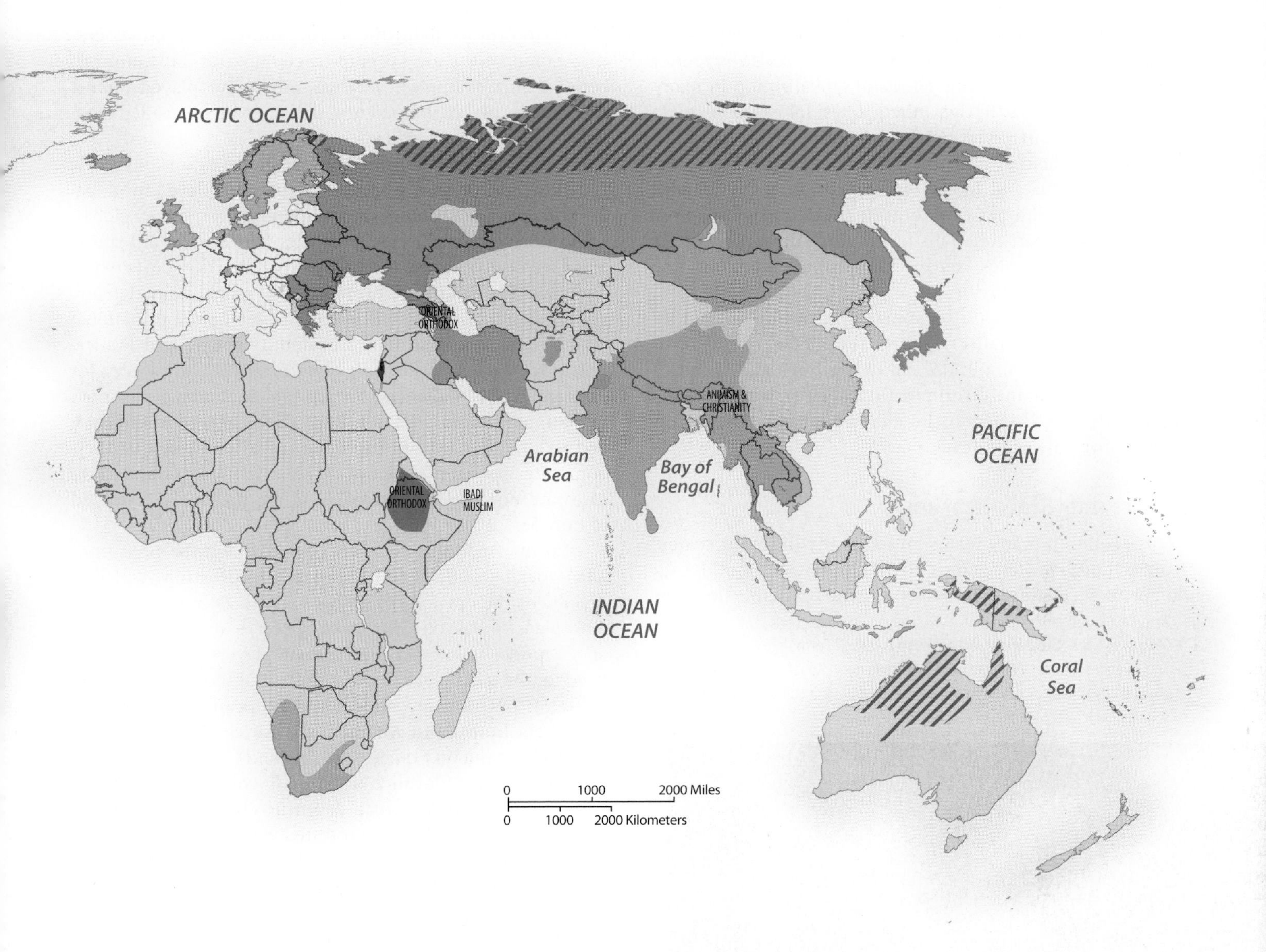

child rearing, education, marriage, and even recreational activities (Figure 1.38).

The explicit and often rigid gender roles of a traditional social unit contrast greatly with the less rigid, more implicit, and often flexible gender roles of a large, modern, urban industrial society. More to the point, globalization in its varied expressions is causing significant changes to traditional gender roles throughout the world. Global apparel firms, for example, often locate sewing factories in areas of the less developed world, such as Bangladesh. They do this to gain economic advantage by paying low wages to female workers, who are often the first generation to work outside the home. However, because of traditional gender roles, men do not always pick up the domestic tasks formerly done by these new factory workers. Instead, the women often put in what is commonly called a double day, returning to their homes from the factory to then spend hours of work doing cooking, child rearing, and housecleaning.

▼ **Figure 1.38 Traditional Dress and Gender Roles** Women's gender roles, much more so than men's, are reinforced by appropriate dress and traditional clothing, as shown here for these Muslim women in Zanzibar, Tanzania.

In contrast to providing new work opportunities for women in less developed countries, there is also a gender dimension to the economic effects of globalization in many developed countries. In the United States, for example, male workers have suffered more from unemployment than have females as industrial and technology jobs have been outsourced to China and India. To compensate, many females who previously did not work outside their homes have now taken full- and part-time jobs to provide support for their families. Many males, meanwhile, have taken on new roles in domestic activities.

Globalization has also spread the notion of gender equality (and inequality) around the globe, calling into question and exposing those cultural groups and societies that practice blatant discrimination against women. This topic is discussed later in the chapter within the section on measures of social development.

Sports and Globalization

Sports come in many forms and at many different scales, from village soccer games to Olympic and World Cup competition. However, all sports activities, whether local or global, inform us about the world's diverse cultural geography. Some sports are specific to certain cultural traditions (cricket, for example), whereas others are played almost universally, like basketball and soccer (which is called football in most of the world).

▼ **Figure 1.39 Globalization of Auto Racing** Formula 1 Grand Prix auto racing, a sport historically linked with Europe, is now global and corporate, as attested to by the logos covering the helmet and driving suit of Sebastian Vettel of the Red Bull racing team.

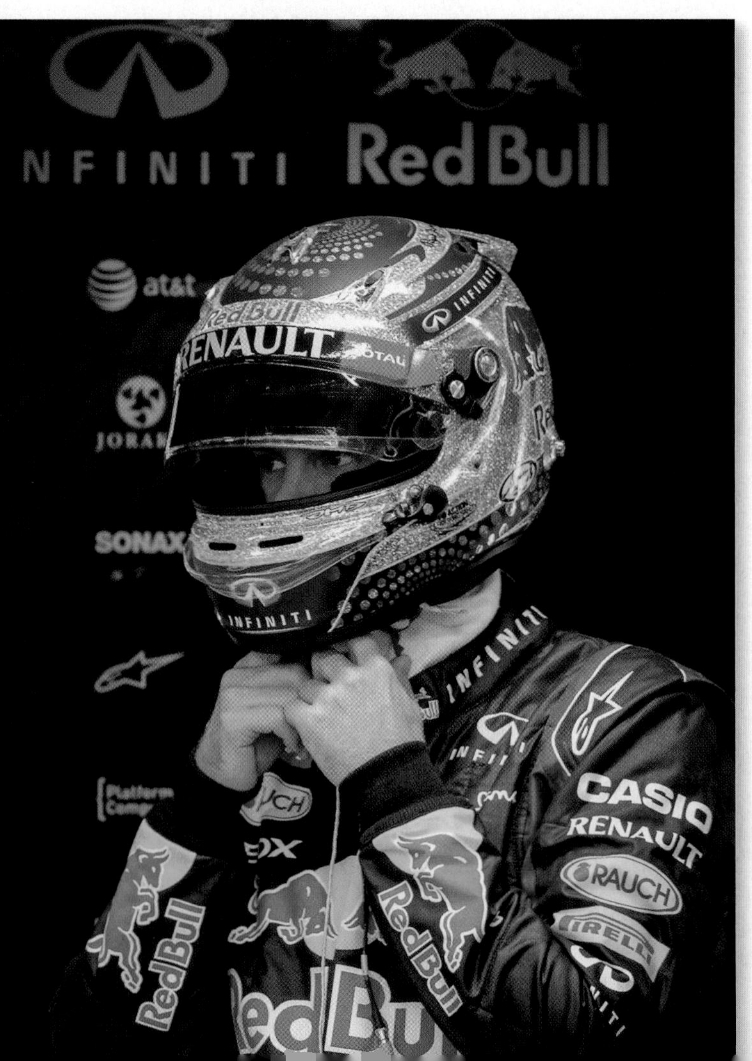

Further, the relationship between globalization and sports is a close one, primarily because of the roles played by global TV and corporate sponsorship. Important, too, is how globalization has eroded the historical ties between certain sports and specific geographic locales by expanding the sport's world audience. For example, Formula 1 auto racing was originally closely tied to Europe, with racing teams linked to national auto manufacturers in Italy, England, Germany, and France. It is now a true world sport that competes with soccer for huge global TV audiences. Racing cars are no longer painted in national colors (red for Italy, dark green for England, and so on), but instead carry the colors and logos of their corporate sponsors (Figure 1.39). Further, Grand Prix races are now held worldwide and are no longer restricted to Europe.

European soccer, too, has gone through the process of what social scientists call a **de-territorialization**, which is defined as the erosion of ties between an activity and a specific place. Early on, European soccer teams had regional and national linkages because their players came from, or lived in, the team's locality or country. But those ties lessened in the 1970s as more and more European teams brought in players from Latin America and Africa. Although some countries (namely, France and England) could make their claim to a player's locality based upon colonial ties to those African and Latin American countries, Italy, which historically had limited colonies, protested vehemently, ending the charade that soccer players must have local or national ties to their teams. Today, however, for World Cup competition, FIFA, soccer's international governing board, demands that players not born in or a citizen of the country they're playing for demonstrate a "clear connection" to that country. You could call this regulation an attempt at re-territorialization of the global sport of soccer.

American Sports Abroad Of the major North American sports, basketball, has become the most globalized, with many international players at both college and professional levels, and a large National Basketball Association (NBA) audience abroad. In 2013, NBA teams played games in China, Taiwan, the Philippines, Brazil, and England, and the 2013 NBA championship series was broadcast in 47 different languages and televised to every country in the world. Also noteworthy is not just that the NBA built a huge sports complex on the outskirts of Beijing, but also that official NBA apparel and shoes are sold in 2200 Adidas stores in China, drawing upon a market of over a billion people (Figure 1.40).

Although Major League Baseball (MLB) teams have the largest number of international players, with those being born outside of the United States making up almost 30 percent of MLB rosters, baseball has less of a global reach than

▲ **Figure 1.40 The NBA in China** Through the efforts of the NBA, American basketball has become one of the most globalized sports. This photo is of one of the over two thousand NBA-sanctioned stores in China.

NBA basketball. Its audiences are concentrated primarily in Japan, Taiwan, Mexico, and Latin America.

The National Football League (NFL) struggles on the international scene despite staging games in Japan, Mexico, England, and Germany. Some sports experts argue that, compared to soccer's nonstop 45 minutes of action (per half), American football is too slow and hard to understand because of its frequent pauses and timeouts.

Global Sports TV Despite the NBA's success internationally, at a global level the largest three TV sports audiences are those for soccer, Formula 1 auto racing, and cricket, the last being boosted into third place because of the huge numbers of viewers in the former British colonies of South and Southeast Asia. Soccer, though, is dominant in all world regions, with almost 50 percent of the world's population watching the 2010 World Cup games from South Africa. Formula 1 auto racing audiences vary from season to season, depending on where the races are held and—equally important, it seems—the nationalities of the leading drivers.

Eurosports is the global TV leader with daily sports broadcasts in 20 languages to 60 countries. Even U.S.-based ESPN broadcasts its iconic Sports Center in 5 languages, seven days a week, to Europe and Latin America.

REVIEW

1.16 Define cultural hybridization, and give an example.

1.17 Define gender roles, and illustrate your answer with examples from both modern and traditional cultures.

1.18 What are the two subcategories of Islam? Describe the geographies for each.

1.19 What is a lingua franca? Provide two examples.

Geopolitical Framework: Unity and Fragmentation

The term **geopolitics** is used to describe the close link between geography and politics. More specifically, geopolitics focuses on the interactivity between political power and territory at all scales, from the local to the global. Unquestionably, one of the global characteristics of the last several decades has been the speed, scope, and character of political change in various regions of the world; thus, discussions of geopolitics are central to world regional geography.

World Political Systems

The world's complex geopolitical fabric is composed of vastly different governmental systems, ranging from dictatorships to democracies. The type of government often determines the country's internal and external socioeconomic policies as well. Complicating matters is that many governments mix together different political and economic forms. Contemporary China is a good example, with its unique blend of communism and capitalism.

Who Has Power? A good place to begin is with the question of whether governmental power rests with a country's citizens or with one person or a small authoritarian group. Dictatorships form one end of the spectrum. A military dictatorship, as the name suggests, consists of rule by a handful of high military officials. Today the world map of dictatorships is fluid and ever changing, yet a distinct cluster of dictatorships holds power in Africa and Central Asia (Figure 1.41). Dictators are often associated with totalitarian rule, where the government controls all aspects of life in that country. An extreme form of totalitarian rule is fascism, illustrated by the historical cases of Hitler's rule of Nazi Germany (1933–1945) and Franco's regime in Spain (1936–1975).

Monarchies, led by a queen or king, take many different forms today. A monarch may be simply a ceremonial figurehead or may actually hold some—or even all—governmental power. In extreme cases, such as in Saudi Arabia, the absolute monarch (in this case, a king) holds ultimate authority and is inseparable from the government. At the other end of the spectrum are the ceremonial monarchs of several European countries, including the United Kingdom, who generally do not influence governmental activity.

In democracies, political power rests with the people and their elected representatives. This political system evolved historically as an alternative to the divine rule of

▼ **Figure 1.41 The Fall of a Dictator** U.S. Marines pull down a statue of the former Iraqi dictator, Saddam Hussein, in April 2003 shortly after entering Baghdad, Iraq, during the second Gulf War.

monarchs and/or autocratic rule by dictators. In Western society, democracy can be traced back to Athens, Greece, around 550 BCE. Common to contemporary democracies is an array of different political parties, representing various points of view and all competing for the public's support.

However, as contradictory as it seems, some countries refer to themselves as democracies, yet have only a single political party, offering a slate of preapproved candidates. This single-party democracy is commonly associated with communist governments, both historically (the Soviet Union, East Germany, and Cold War eastern Europe) and currently (Vietnam, Cuba, and China).

Governments and Socioeconomic Systems The role of government in a country's socioeconomic system, like politics itself, takes many forms. In fact, countries tend to blend together elements of three distinct socioeconomic systems: capitalism, communism, and socialism. This has led to some confusion about what each system means.

The essence of capitalism is private ownership of property, goods, and businesses, with the means of production also owned privately, by individuals or corporations. Further, capitalism supports the notion of a free market, where forces of supply and demand set prices and governmental economic policies are minimal. Capitalism traces its beginnings back to western Europe in the Middle Ages, and today it dominates the world economy.

Communism, in contrast, is based on the ownership of all property and goods by the working class, with restricted private ownership. In practice, the governments of communist countries, such as the former Soviet Union, China, and Cuba, assumed ownership of most factories, utilities, and services. This socialistic phase (see below) was considered temporary and transitional in communist societies, with workers' cooperatives being the ultimate owners. Although communal ownership has a long tradition in many societies, communism as both a political and an economic system traces its contemporary roots to the Russian Revolution, which created the Soviet Union in the early 20th century.

Socialism differs from communism in that there is no pretense of worker ownership; instead, the government owns industry and utilities. Unlike communism, private ownership of property and noncritical economic activities is allowed. Additionally, a socialist government provides the public with health, welfare, and education services. Socialism and communism both trace their heritage to Karl Marx and mid-19th-century economic and social theorists, but socialism today is less about political rule (although socialist political parties abound in Europe) and is more concerned with public policy. Unlike communism, which considers itself to be mutually exclusive with capitalism (except in China), today socialist health, welfare, and educational policies are found in many capitalist economies throughout Europe.

The Nation-State Revisited

A map of the world consists of an array of about 200 countries, ranging in size from the microstates like Vatican City and Andorra to the huge geopolitical expanses of Russia, the United States, Canada, and China. All of these countries are regulated by governmental systems, ranging from democratic to autocratic. Commonly, these different forms of government share a concern with sovereignty, which can be defined geopolitically as the ability (or the inability) of a government to control activities within its borders.

This notion of sovereignty is closely linked to the concept of the **nation-state**. In this hyphenated term, *nation* describes a large group of people with shared sociocultural traits, such as language, religion, and shared identity. The word *state* refers to a political entity that has clearly delimited boundaries, that controls its internal space, and that is recognized by other political entities. Historically, France and England are often cited as the archetypal examples of nation-states. Contemporary countries such as Albania, Egypt, Bangladesh, Japan, and the two Koreas are more modern examples of countries that show close overlap between nation and state. The related term *nationalism* is the sociopolitical expression of identity and allegiance to the shared values and goals of the nation-state.

Globalization, however, has weakened the vitality of the nation-state concept because today most of the world's countries are questionable fits with the traditional definition of nation-state. International migration, for example, has led to large populations of ethnic minorities within a country that do not necessarily share the national culture of the majority. In England, for example, large numbers of South Asians form their own communities, speak their own languages, have their own religions, and dress to their own standards. Similarly, France is home to a mosaic of peoples from its former colonial lands in Africa and Asia.

▼ **Figure 1.42 Ethnic Separatism** A major aspect of contemporary geopolitics is the way ethnic groups are demanding recognition, autonomy, and often independence from larger political units. These Basque women in southwestern France are protesting the outlawing of a Basque youth group that French and Spanish authorities suspected of aiding Basque terrorists.

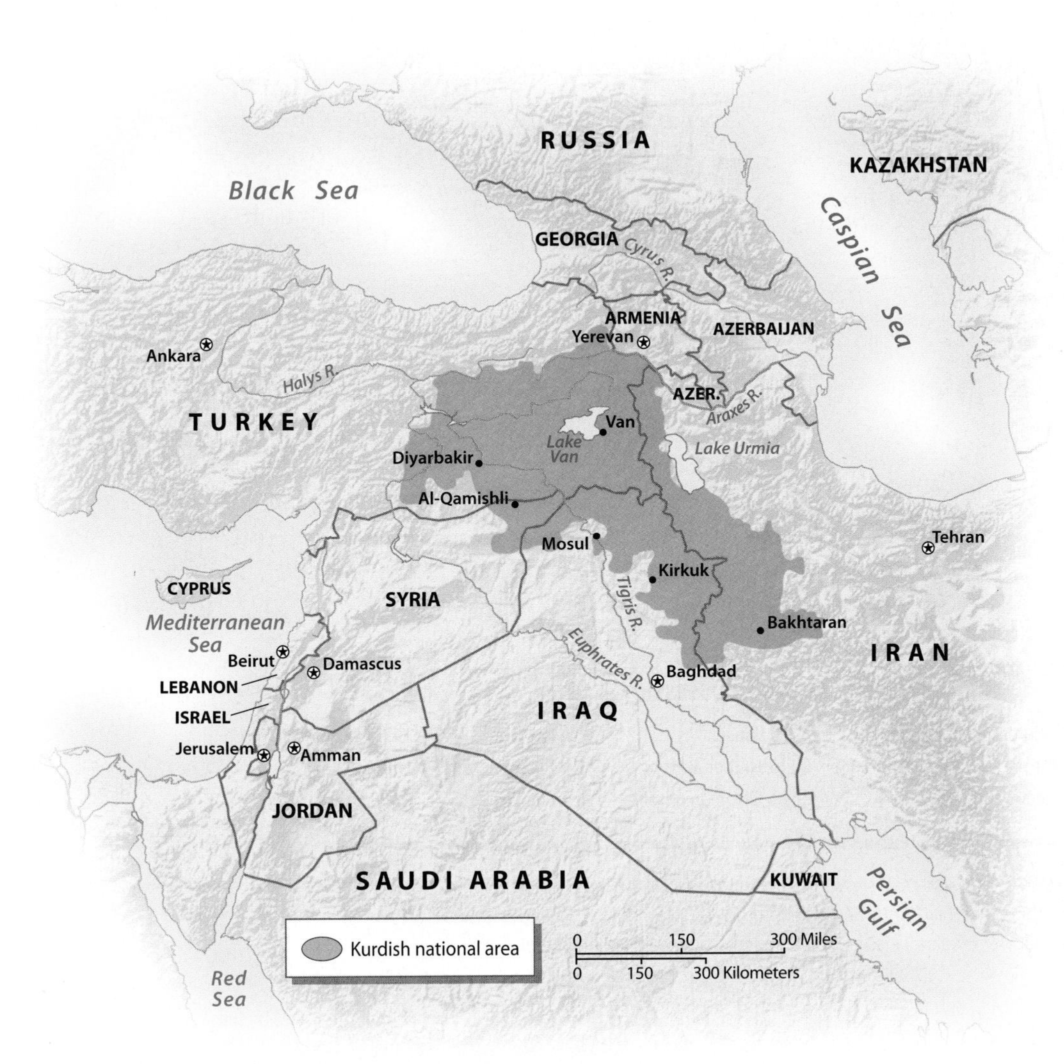

▲ **Figure 1.43 A Nation Without a State** Not all nations or large cultural groups control their own political territories. As this map shows, the Kurdish people of Southwest Asia occupy a large cultural territory that lies in four different political states—Turkey, Iraq, Syria, and Iran. As a result of this political fragmentation, the Kurds are considered a minority in each of these four countries. **Q: What kinds of issues result from the Kurds lacking a political state?**

In North America, Canada and the United States also have large immigrant populations who are legal citizens of the political state, yet who have changed the very nature of the national culture by their presence. The fact that states such as the United States, Canada, and Germany officially embrace cultural diversity, with policies declaring themselves as multicultural states, underscores these changes.

Decentralization and Devolution Residing within many nation-states are groups of people who seek autonomy from the central government and argue for the right to govern themselves. This autonomy can range from the simple decentralization of power, removed from a central government, to smaller governmental units, as is the case with states in the United States and departmental government in France. At the far end of the spectrum is outright political separation and full governmental autonomy, referred to as devolution. As an illustration, the citizens of Scotland will vote in late 2014 on the issue of full separation from England. Other separatist movements are found among French-speaking people of Quebec Province in Canada, the Catalonians and Basques of Spain, and the more radical groups of native Hawaiians who seek autonomy from the United States (Figure 1.42).

Not to be overlooked is the fact that political organizations have eclipsed the power of traditional political states. This is certainly the case for the 28 member states of the European Union, a topic discussed in Chapter 8. Finally, some cultural groups lack political voice and representation due to the way political borders have been drawn. In Southwest Asia, the Kurdish people have long been considered a nation without a state because they are divided by political borders among Turkey, Syria, Iraq, and Iran (Figure 1.43).

Colonialism, Decolonialization, and Neocolonialism

One of the overarching themes in world geopolitics is the waxing and waning of European colonial power in South America, Asia, and Africa. **Colonialism** refers to the formal establishment of rule over a foreign population. A colony has no independent standing in the world community, but instead is seen only as an appendage of the colonial power. The historical Spanish presence and rule over parts of the United States, Mexico, and South America is an example. Generally speaking, the main period of colonialization by European countries was from 1500 through the mid-1900s, with the major players being England, Belgium, the Netherlands, Spain, and France (Figure 1.44).

Decolonialization refers to the process of a colony's gaining (or, more correctly, regaining) control over its own territory and establishing a separate, independent government. As was the case with the Revolutionary War

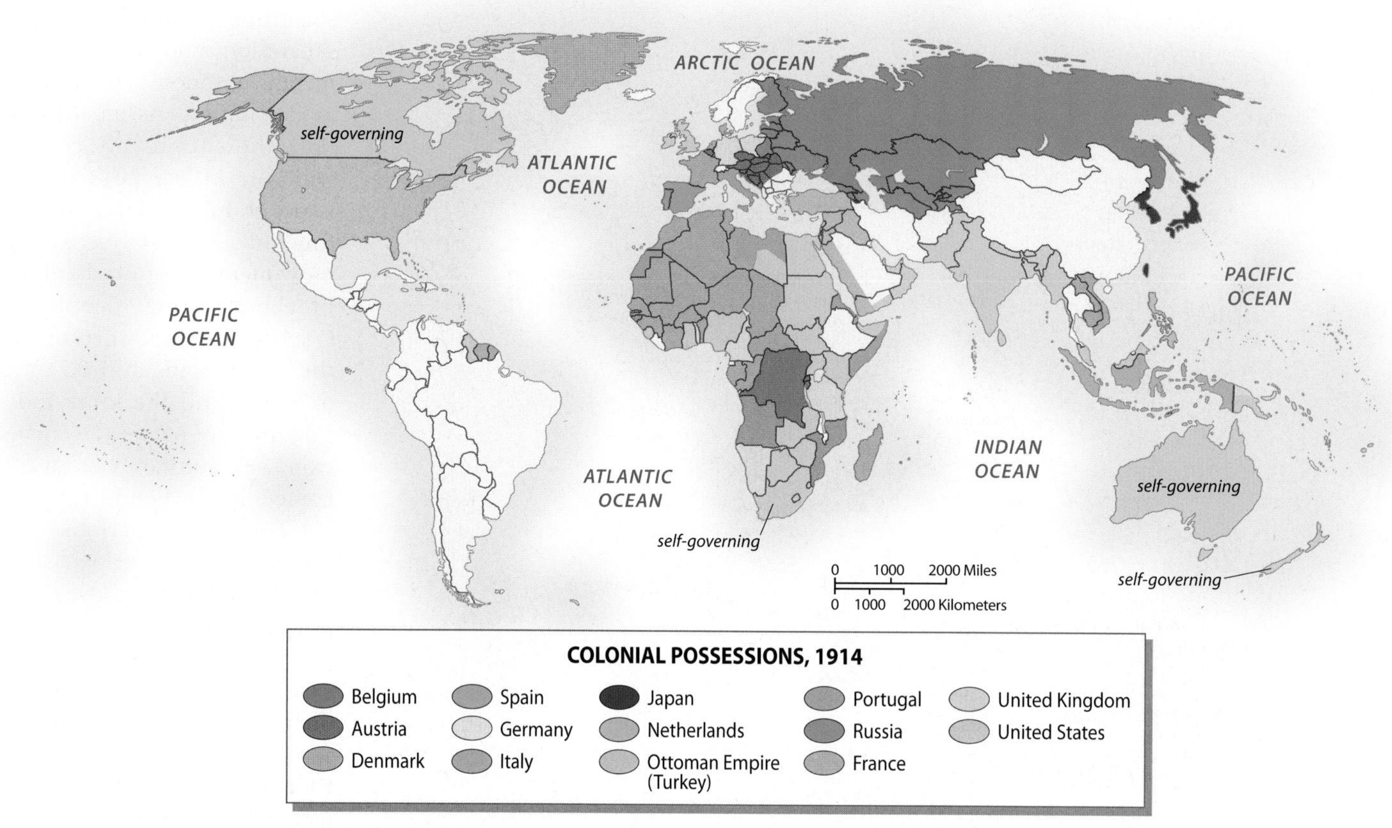

▲ **Figure 1.44 The Colonial World, 1914** This world map shows the extent of colonial power and territory just prior to World War I. At that time, most of Africa was under colonial control, as were Southwest Asia, South Asia, and Southeast Asia. Australia and Canada were very closely aligned with England. Also note that in Asia, Japan controlled colonial territory in Korea and northeastern China, which was known as Manchuria at that time.

in the United States, this process often involves violent struggle. Similar wars of independence were common during the 19th century in Latin America and, in the mid-20th century, increasingly in Southeast Asia and Africa. Consequently, most European colonial powers recognized the inevitable and began working toward peaceful disengagement from their colonies. In the late 1950s and early 1960s, for example, Britain and France granted independence to their former African colonies. This period of European decolonialization was symbolically completed in 1997 when England turned over Hong Kong to China.

However, decades and even centuries of colonial rule are not easily erased. The influences of colonialism are still commonly found in the culture, government, educational systems, and economic life of the former colonies. Examples are the many contemporary manifestations of British culture in India and the continuing Spanish influences in South America.

In the 1960s, the term **neocolonialism** came into popular usage to characterize the many ways that newly independent states, particularly those in Africa, felt the continuing control of Western powers, especially in economic and political matters. To receive financial aid from the World Bank, for example, former colonial countries were often required to revise their internal economic structures to become better integrated with the emerging global system. This economic restructuring may seem warranted from a global perspective, but the dislocations caused at national and local scales led critics of globalization to characterize these external influences as no better than the formal control by historical colonial powers—hence the currency of the term *neocolonialism*.

Global Conflict and Insurgency

As mentioned earlier, challenges to a centralized political state or authority have long been part of global geopolitics as rebellious and separatist groups seek independence, autonomy, and territorial control. These actions are referred to as **insurgency**. Armed conflict has often been part of this process; the American and Mexican revolutions were both successful wars for independence fought against European colonial powers. **Terrorism**, which can be defined as violence directed at nonmilitary targets, has also been common, albeit to a far lesser degree than today.

Until the September 2001 terrorist attacks on the United States by Al Qaeda, terrorism was usually directed at specific local targets and committed by insurgents with focused goals. The Irish Republican Army (IRA) bombings in Great Britain and Basque terrorism in Spain are illustrations. The attacks on the World Trade Center and the Pentagon (as well as the thwarted attack on the Capitol), however, went well beyond conventional geopolitics as a small group of religious extremists attacked the symbols of Western culture, finance, and power (Figure 1.45). Experts believe the goal of the Al Qaeda terrorists was not to disrupt world commerce and politics, but to make a statement about the strength of their own convictions and power. Regardless of motives, those acts of terrorism underscore the need to expand our conceptualization of the linkages between globalization and geopolitics.

More to the point, many experts argue that global terrorism is both a product of and a reaction to globalization. Unlike earlier geopolitical conflicts, the geography of global terrorism is not defined by a war between well-established political states. Instead, the Al Qaeda terrorists appear to belong to a web of small, well-organized cells located in many different countries. These cells are linked in a decentralized network that provides guidance, financing, and political autonomy and that has used the tools and means of globalization to its advantage. Members communicate instantaneously via mobile phones and the Internet. Transnational members travel between countries quickly and frequently. The network's activities are financed through a complicated array of holding companies and subsidiaries that traffic in a range of goods, including honey, diamonds, and opium. To recruit members and political support, the network feeds on the unrest and inequities (real and imagined) resulting from economic globalization. The network's terrorist acts then target symbols of those modern global values and activities it opposes. Even though the 9/11 attacks focused terror on the United States, the casualties and resultant damage were international, as citizens from more than 80 countries were killed in the World Trade Center tragedy.

Although Al Qaeda is the most visible and possibly the most threatening of contemporary global terrorist groups, the U.S. State Department names 45 groups on its list of foreign terrorist organizations. Most of these groups are clustered in North Africa and Southwest and Central Asia, but this list also includes insurgent groups in all other regions of the world.

The military responses to global terrorism and insurgency involve several components, ranging from the neutralization of terrorist activities, known as counterterrorism, to **counterinsurgency**. This is a more complicated, multifaceted strategy that combines military warfare with social and political service activities, designed to win over the local population and deprive insurgents of a political base. Counterinsurgency activities include, first, clearing and then holding territory held by insurgents and then, building schools, medical clinics, and a viable (and legal) economy. Often these nonmilitary activities are referred to as nation building, since the goal is to replace the separatist insurgency with a viable social, economic, and political fabric more complementary to the larger geopolitical state. This was the strategy employed by the United States recently in Iraq and Afghanistan.

▶ **Figure 1.45 Global Terrorism** The September 11, 2001, attacks by Al Qaeda terrorists on the World Trade Center and the Pentagon and the thwarted plan to destroy the U.S. Capitol Building resulted in more than 3000 deaths. These attacks were not targeted at local places of dispute, but were aimed at global symbols of Western power.

REVIEW

1.20 Describe the socioeconomic differences among communism, socialism, and capitalism.

1.21 Why is it common to use two different concepts—nation and state—to describe political entities?

1.22 Explain the differences between colonialism and neocolonialism.

1.23 Describe the differences between counterterrorism and counterinsurgency.

Economic and Social Development: The Geography of Wealth and Poverty

The pace of global economic change and development has accelerated dramatically in the past several decades. It increased rapidly at the start of the 21st century and then slowed precipitously in 2008 as the world fell into an economic recession (Figure 1.46). Since then, the world economy has returned partially and irregularly to a postrecession state. If nothing else, this recent global recession and its unsteady recovery have brought into focus once again the overarching question of whether the benefits of economic globalization outweigh the negative aspects. Responses vary

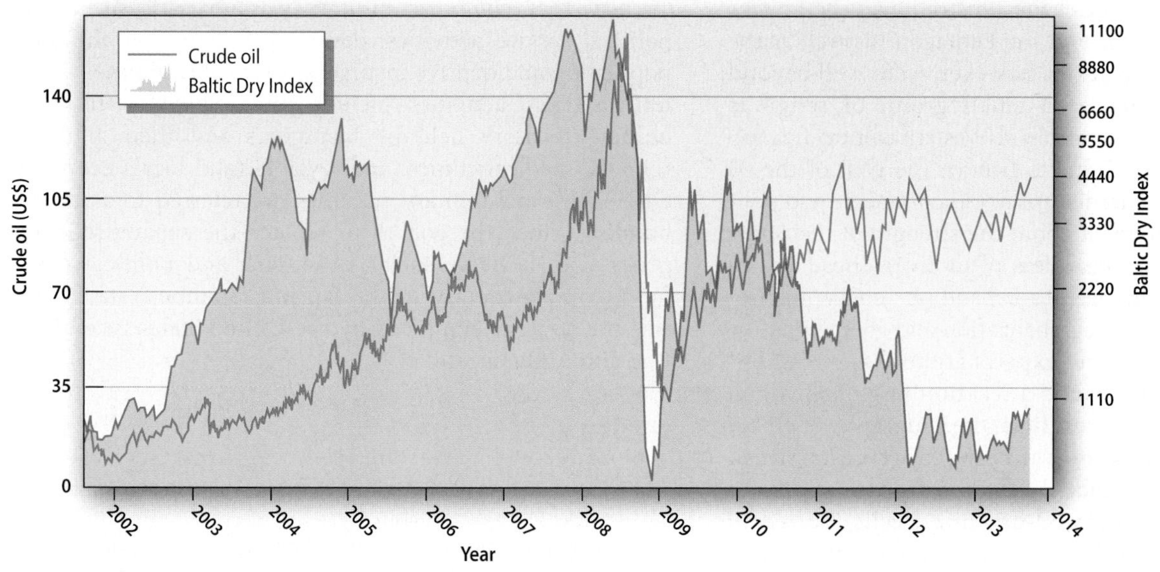

▲ **Figure 1.46 The Global Recession** Two indicators of global economic activity—the price of crude oil and the Baltic Dry Index (BDI), which is linked to world shipping activity—show the slow growth of the global economy since 2000 and then the sudden contraction in 2009. More recent data show the slow recovery from the 2009 collapse.

considerably, depending on one's point of view, occupation, career aspirations, socioeconomic status, and so on. However, one message resounds clearly: Anyone attempting to understand the contemporary world needs a basic understanding of global economic and social development. To that end, each of our regional chapters contains a substantive section on that topic, drawing upon the concepts discussed below.

Economic development is commonly accepted as desirable because it generally brings increased prosperity to people, regions, and nations. Following conventional thinking, this economic development usually translates into social improvements such as better health care, improved educational systems, and more progressive labor practices. One of the most troubling expressions of global economic growth, however, has been the geographic unevenness of prosperity and social improvement. That is, although some regions and places in the world prosper, others languish and, in fact, apparently fall further behind the more developed countries. As a result, the gap between rich and poor regions has actually increased over the past several decades in many areas. This economic and social unevenness has, unfortunately, become one of the signatures of globalization. Today, according to the World Bank, almost half the people in the world live on less than $2 per day, the commonly accepted definition of poverty (Figure 1.47). That percentage has decreased over the last decade, but perhaps even more disturbing is that the number of people living under extreme poverty, on less than $1.25 per day, has actually increased recently.

▼ **Figure 1.47 Living on Less than $2 per Day** The World Bank uses $2 a day as their definition of extreme poverty, and recent estimates place about half the world's population in that category. These people survive by picking through the rubbish dumps surrounding Manila, Philippines.

These inequities are problematic because of their inseparable interaction with political, environmental, and social issues. For example, political instability and civil strife within a nation are often driven by the economic disparity between a poor periphery and an affluent industrial core—between the haves and have-nots. Such instability throughout a country can strongly influence international economic interactions.

More and Less Developed Countries

Until the 20th century, economic development was centered in North America, Japan, and Europe, with most of the rest of the world gripped in poverty. This uneven distribution of economic power led scholars to devise a **core–periphery model** of the world. According to this scheme, the United States, Canada, western Europe, and Japan constituted the global economic core, centered in the Northern Hemisphere, whereas most of the areas in the Southern Hemisphere made up a less developed periphery. Although oversimplified, this core–periphery dichotomy does contain some truth. All the G8 countries—

the exclusive club of the world's major industrial nations, presently made up of the United States, Canada, France, England, Germany, Italy, Japan, and Russia—are located in the Northern Hemisphere. (China—which is unquestionably an industrial power, located in the Northern Hemisphere—is currently excluded from the G8.) In addition, many critics postulate that the developed countries achieved their wealth primarily by exploiting the poorer countries of the southern periphery, historically through colonial relationships and today with various forms of neocolonialism.

Following this core–periphery model, much has been made of "north–south tensions," a phrase implying that the rich and powerful countries of the Northern Hemisphere are still at odds with the poor and less powerful countries to the south. However, this model demands revision because over recent decades the global economy has grown much more complicated. A few former colonies in the Southern Hemisphere—most notably, Singapore—have become very wealthy. A few Northern Hemisphere countries—notably, Russia—have experienced very uneven economic growth since 1989, with some parts of the country actually experiencing economic decline. Also, the highly developed Southern Hemisphere countries of Australia and New Zealand never fit into the north–south division. For these reasons, many global experts conclude that the designation *north–south* is now outdated and should be avoided. We agree.

The *third world* is another term often erroneously used as a synonym for the developing world. This phrase implies a low level of economic development, unstable political organizations, and a rudimentary social infrastructure. Historically, the term was part of the Cold War vocabulary used to describe countries that were independent and not part of either the capitalist first world or the communist second world, dominated by the Soviet Union and China. In short, in its original sense the term *third world* spoke to a political and economic orientation (capitalist versus communist), not to the level of economic development. Today, because the Soviet Union no longer exists and China has changed its economic orientation considerably, the term *third world* has lost its original political meaning. Therefore, in this book we avoid that term. Instead, we prefer relational terms that capture a complex spectrum of economic and social development—**more developed country (MDC)** and **less developed country (LDC)**. This global pattern of MDCs and LDCs can be inferred from a map of gross national income (Figure 1.48), one of several indicators commonly used to assess development and economic wealth.

Indicators of Economic Development

The terms *development* and *growth* are often used interchangeably when referring to international economic activities. There is, however, value in keeping them separate. *Development* has both qualitative and quantitative dimensions. Common dictionary definitions use phrases such as "expanding or realizing potential" and "bringing gradually to a fuller or better state." When we talk about economic development, then, we usually imply structural changes, such as a shift from agricultural to manufacturing activity that also involves changes in the allocation of labor,

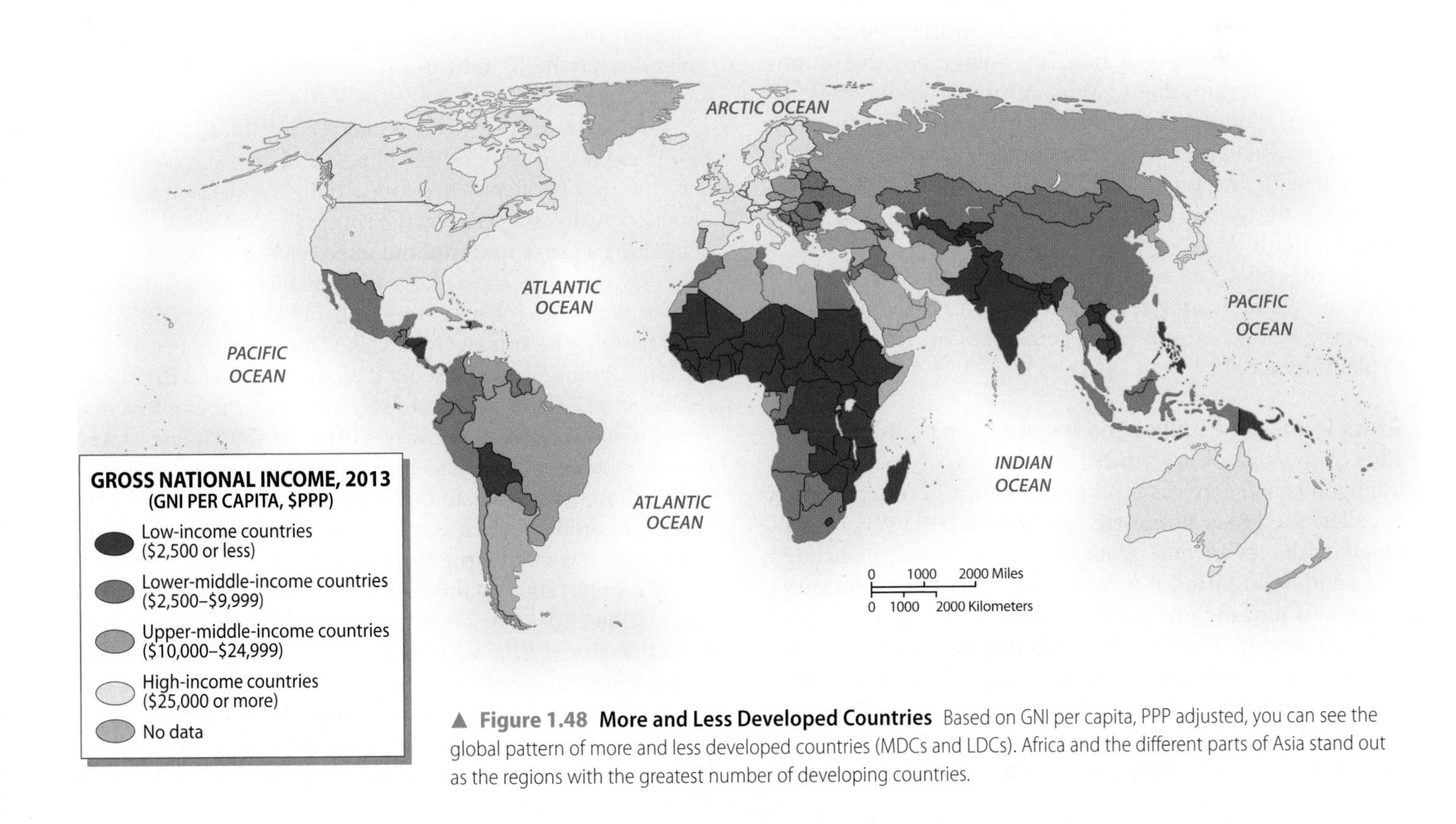

▲ **Figure 1.48 More and Less Developed Countries** Based on GNI per capita, PPP adjusted, you can see the global pattern of more and less developed countries (MDCs and LDCs). Africa and the different parts of Asia stand out as the regions with the greatest number of developing countries.

Table 1.2 Development Indicators

Country	GNI per capita, PPP 2011	GDP Average Annual %Growth 2000–11	Human Development Index (2013)[1]	Percent Population Living Below $2 a Day	Life Expectancy (2013)[2]	Under Age 5 Mortality Rate (1990)	Under Age 5 Mortality Rate (2011)	Adult Literacy (% ages 15 and older) (Male/Female)	Gender Inequality Index (2013)[3,1]
China	8,390	10.8	.699	27.2	75	49	15	97/91	0.213
India	3,640	7.8	.554	68.7	66	114	61	75/51	0.610
United States	48,820	1.6	.937	–	79	11	8	--/--	0.256
Indonesia	4,500	5.4	.629	46.1	70	82	32	96/90	0.494
Brazil	11,420	3.8	.730	10.8	74	58	16	90/90	0.447
Pakistan	2,870	4.9	.515	60.2	66	122	72	69/40	0.567
Nigeria	2,290	6.8	.471	84.5	52	214	124	72/50	–
Bangladesh	1,940	6.0	.515	76.5	70	139	46	61/52	0.518
Russia	20,410	5.1	.788	<2	70	27	12	100/99	0.312
Japan	35,330	0.7	.912	–	83	6	3	--/--	0.131

[1]1United Nations, *Human Development Report,* 2013.

[2]Population Reference Bureau, *World Population Data Sheet,* 2013.

[3]Gender Equality Index—A composite measure reflecting inequality in achievements between women and men in three dimensions: reproductive health, empowerment and the labor market that ranges between 0 and 1. The higher the number, the greater the inequality.

Sources: World Bank, *World Development Indicators, 2013.*

capital, and technology. Along with these changes are assumed improvements in standard of living, education, and political organization. The structural changes experienced by Southeast Asian countries such as Thailand and Malaysia in the past several decades capture this process.

Growth, in contrast, is simply the increase in the size of a system. The agricultural or industrial output of a country may grow, as it has for India in the past decade, and this growth may—or may not—have positive implications for development. Many growing economies, in fact, have actually experienced increased poverty with economic expansion. When something grows, it gets bigger; when it develops, it improves. Critics of the world economy often say that we need less growth and more development.

In this book, each of the regional chapters includes a table of economic and development indicators (Table 1.2). However, a few introductory comments are necessary to explain these data.

Gross Domestic Product and Income The traditional measure of the size of a country's economy is the value of all final goods and services produced within its borders, which is called the **gross domestic product (GDP)**. When combined with net income from outside its borders through trade and other forms of income, this constitutes a country's **gross national income (GNI)** (formerly referred to as gross national product [GNP]). Although the term is widely used, GNI ignores nonmarket economic activity, such as bartering or household work, and also does not take into account ecological degradation or depletion of natural resources. For example, if a country were to clear-cut its forests—an activity that would probably limit future economic growth if forest resources were in short supply—this resource usage would actually increase the GNI for that particular year. Diverting educational funds to purchase military weapons might also increase a country's GNI in the short run, but the economy would likely suffer in the future because of its less-well-educated population. In other words, GNI is a snapshot of a country's economy at a specific moment in time, not a reliable indicator of continued vitality or social well-being.

GNI data vary widely among countries and are commonly expressed in mind-boggling numbers such as billions and trillions of dollars. Therefore, a more useful comparison is obtained by dividing GNI by the country's population, thereby generating a **gross national income (GNI) per capita** figure. This way we can compare large and small economies in terms of how they may (or may not) be benefiting the population. For example, the annual GNI for the United States is over $15 trillion. Dividing that figure by the population of 316 million results in a GNI per capita of $48,820. Japan, in contrast, has a total GNI about half the size of that of the United States; however, because Japan also has a much smaller population of 127 million, its unadjusted GNI per capita is about $48,150. Thus, we could conclude that the two different economies are comparable (which is true to a degree).

An important qualification to these GNI per capita data is the concept of adjustment through **purchasing power parity (PPP)**, which takes into account the strength or weakness of local currencies (Figure 1.49). When not adjusted by PPP, GNI data are based on the market exchange rate for a country's national currency compared to the U.S. dollar. As a result, the GNI data might be inflated or

▲ **Figure 1.49 PPP and Fluctuating Local Currencies** Purchasing power parity (PPP) takes into consideration the strength or weakness of local currencies. This photo is from Hong Kong when the 2008 Wall Street failures led to runs on international banks with heavy U.S. investments.

undervalued, depending on the strength or weakness of that currency. If the Japanese yen were to fall overnight against the dollar as a result of currency speculation, Japan's GNI would correspondingly drop, despite the fact that Japan had experienced no real decline in economic output. Because of these possible distortions, the PPP adjustment provides a more accurate sense of the local cost of living. To illustrate, when Japan's GNI per capita is adjusted for PPP, which takes out the inflationary factor, the figure is $35,330, somewhat lower than the unadjusted GNI per capita figure.

Economic Growth Rate A country's rate of economic growth is measured by the average annual growth of its GDP over a five-year period, a statistic called *GDP average annual percent growth*. Looking at Table 1.2, you can see that the average growth rates for developing countries such as China, India, and Nigeria are considerably higher than those of the developed countries such as the United States and Japan. This difference in economic growth rates is expected given that developing countries are just that—developing. We would, therefore, expect a higher annual growth rate from a developing country than from a mature, developed economy like that of the United States.

Indicators of Social Development

Although economic growth is a major component of development, equally important are the conditions and quality of human life. As noted earlier, the standard assumption is that economic development will spill over into the social infrastructure, leading to improvements in public health, gender equity, and education. Unfortunately, even the briefest glance at the world reveals that poverty, disease, illiteracy, and gender inequity are still widespread, despite a growing global economy. Even in China, which has experienced unprecedented economic growth in recent decades, almost one-third of the population is still impoverished; in Pakistan, two-thirds live below the poverty

▼ **Figure 1.50 Human Development Index** This map depicts the most recent rankings assigned to four categories that make up parts of the Human Development Index (HDI). In the numerical tabulation, Norway, Australia, the United States, and New Zealand have the highest rankings, while several African countries are lowest in the scale.
Q: Why does Argentina rank higher than Bolivia on the HDI?

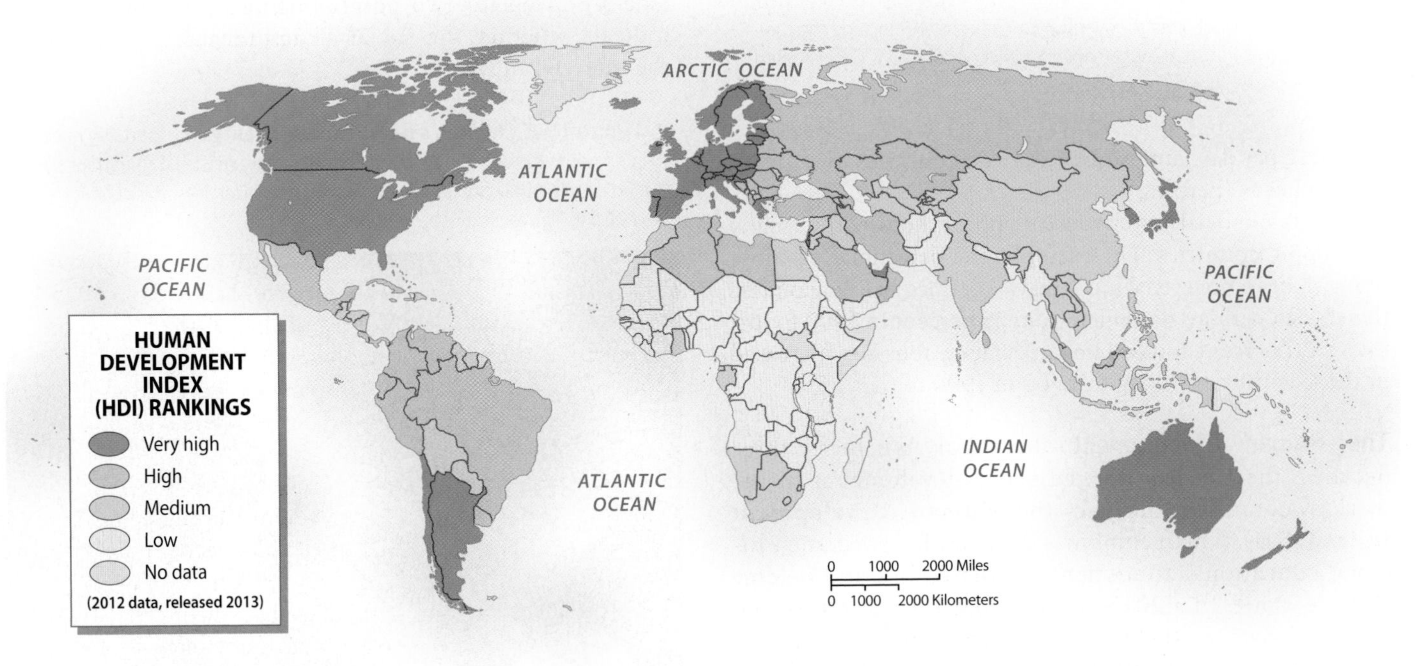

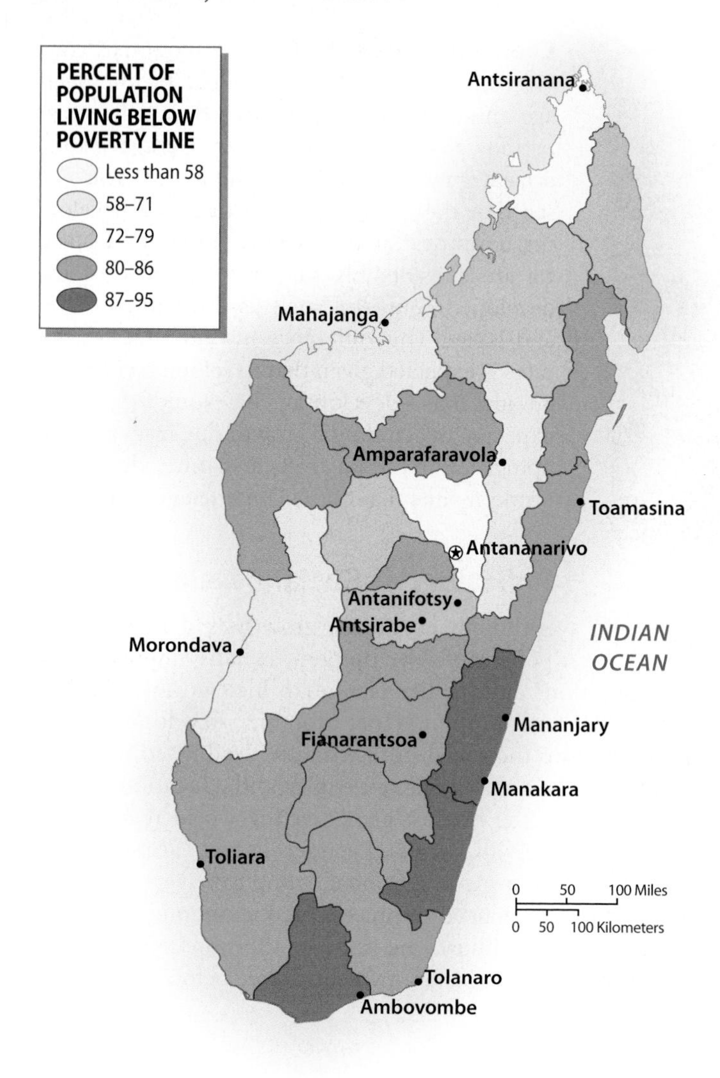

▲ **Figure 1.51 The Landscape of Poverty** As is true of most other countries, the distribution of poverty within Madagascar is uneven, with clusters of abject poverty contrasting with regions that are less poor. This map shows that the highest rates of poverty are in the central highlands and the eastern and southern coasts, where the country's population is concentrated and the density is highest

line of $2 per day; and in some African countries, that figure approaches 90 percent.

Hints of social improvement appear, however, in many developing countries. For example, the percentage of those living in deep poverty, which the UN defines as living on less than $1.25 per day, has fallen from 28 percent in 1990 to just under 20 percent today. Life expectancy, too, has increased in those countries, rising from 60 in 1990 to 65.

The Human Development Index For the past three decades, the UN has tracked social development in the world's countries through the **Human Development Index (HDI)**, which combines data on life expectancy, literacy, educational attainment, gender equity, and income (Figure 1.50). In a 2013 analysis, the 186 countries that provided data to the UN are ranked from high to low, with Norway achieving the highest score, Australia in second place, and the United States in third place. At the lowest end of the HDI are a handful of African countries, including Niger, the Democratic Republic of the Congo, Mozambique, and Chad.

Although the HDI has been criticized for using national data that overlook the diversity of development within a country, overall the HDI conveys a reasonably accurate sense of a country's human and social development. Thus, we include HDI data in our social development table for each regional chapter.

Poverty and Mortality As noted earlier, the international definition of *poverty* is living on less than $2 per day, with the category of *deep poverty* defined as existing on less than $1.25 per day. Although the cost of living varies greatly around the world, the UN has found that these definitions work well for measuring both poverty and its associated social conditions. Poverty data are usually presented at the country level, but the UN and other agencies are also attempting to compile data at a local scale in order to better understand the uneven economic and social landscape within a country. A recent mapping of poverty in Madagascar illustrates how poverty levels differ within the country (Figure 1.51).

Another widely used indicator of social development is data on *under age 5 mortality,* which is the number of children in that age bracket who die per 1000 of the general population. Aside from the tragedy of infant death, child mortality also reflects the wider conditions of a society, such as the availability of food, health services, and public sanitation. If those factors are lacking, children under age five suffer most; therefore, their death rate is taken as an indication of whether a country has the necessary social infrastructure to sustain life (Figure 1.52). In the social development tables throughout this book, child mortality data are given for two points in time, 1990 and 2011, to indicate whether the social structure has improved over the intervening years.

▼ **Figure 1.52 Children's Health and Mortality** The mortality rate of children under the age of 5 is an important indicator of social conditions such as food supply, public sanitation, and public health services. This child is being examined in Nguyen Province, Vietnam.

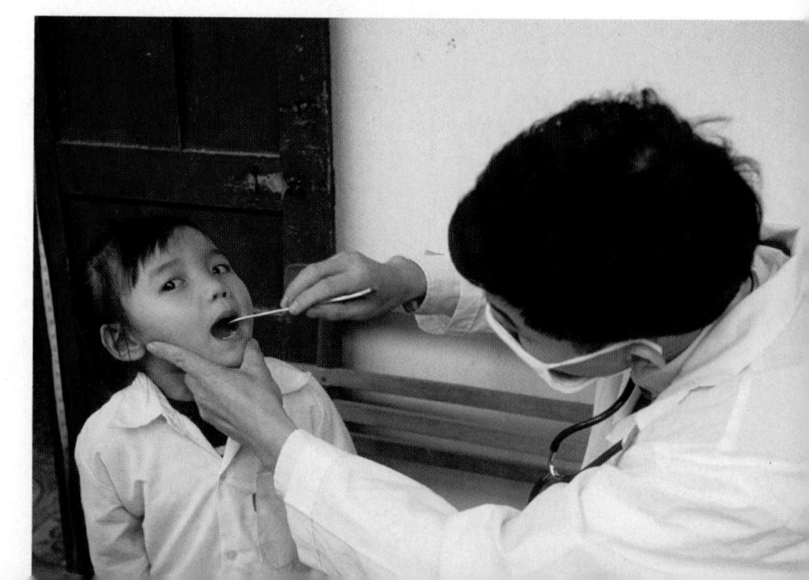

▲ **Figure 1.53 Women and Literacy** Gender inequities in education lead to higher rates of illiteracy for women. However, when there is gender equity in education, female literacy has several positive outcomes in a society. For example, educated women have a higher participation rate in family planning, which usually results in lower birth rates. These women are in a literacy class for women and girls in a refugee camp outside of Mogadishu, Somalia.

Adult Literacy Reading and writing are crucial skills in this world, yet current data show that about 20 percent of the world's adult population (those 15 years of age and over) lack those skills. Of those who cannot read or write, two-thirds are women. The greatest disparity between male and female literacy is in South Asia, an artifact of long-standing cultural favoritism toward males. Some caution, however, is needed when looking at global literacy data since governments of developed nations have been known to inflate their literacy data to improve their social development picture. Further, literacy rates do not comment on the quality of education in a country, which can be notoriously poor in some countries.

Gender Equity Discrimination against women takes many forms, from not allowing them to vote to discouraging school attendance (Figure 1.53). Given the importance of this topic, the United Nations calculates **gender equity** (and inequality) among countries in order to measure the relative position of women to men in terms of employment, empowerment, and reproductive health (in terns of maternal mortality and adolescent fertility). The UN index ranges from 0 to 1, which expresses the highest level of gender inequality. Sweden, for example, has the lowest score for inequality with 0.049, while Yemen has one of the highest at 0.769. The UN gender inequality scores are found in the development indicators tables in the regional chapters.

Of interest is that in some cases, a country may register reasonably high on the HDI, which is positive, yet also be given a relatively high gender inequality score (which is not so good) by the UN. Saudi Arabia, for example, is a rich country that uses assets from its oil resources to provide many social benefits to its citizens, which explains its high HDI ranking. At the same time, its conservative Muslim culture produces a high gender inequality rating. Readers are advised to look carefully at the development indicators data for these kinds of contradictions and inconsistencies.

REVIEW

1.24 Explain the difference between GDP and GNI.

1.25 What is PPP, and why is it useful?

1.26 How does the UN measure gender inequity? Explain why this is a useful metric for social development.

1.27 What are some reasons for the differences in adult literacy rates around the world?

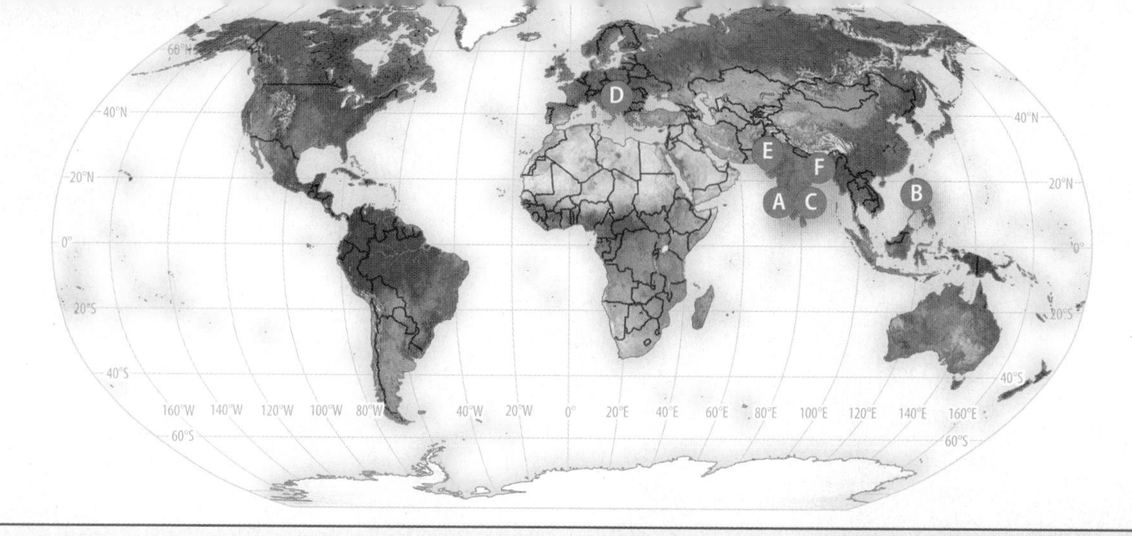

Converging Currents of Globalization

Globalization appears everywhere, all the time, affecting all aspects of world geography with its economic, cultural, and political connectivity (or lack of it). However, despite fears that globalization will produce a homogeneous and flat world, a great deal of diversity (or spikyness) is still apparent—often as economic unevenness, other times as pushback against globalization.

1.1 Why are so many U.S. call centers located in India?

1.2 How has globalization affected—for better or worse—the local economy in your area?

Geographer's Toolbox

Geography is responsible for describing and explaining Earth's varied landscapes and environments. This can be done conceptually in many different ways, by physical or human geography and either topically or regionally—by using a combination of all these approaches. Geographers use a variety of tools, from paper maps to computer models, drawing upon information gathered on the ground and by satellites high above Earth's surface, at all different scales, large and small.

1.3 Using the landscape as your guide, describe the economy and social structure of this village on the island of Luzon, Philippines.

1.4 How are geographic tools used by city and county planners in your community?

Population and Settlement

Human populations around the world are growing either quickly or slowly, as a function of varying birth and death rates (natural growth), along with widely different patterns of in- and out-migration. Urbanization is also a major factor in settlement patterns as people continue to move from rural to urban locales.

1.5 What are the reasons for the differences in population density in this portion of southern India?

1.6 Do a brief study of the population situation in your area of the country in terms of its natural growth, out-migration, in-migration, and resulting growth or decline. Then explain why your area's population is either growing or declining.

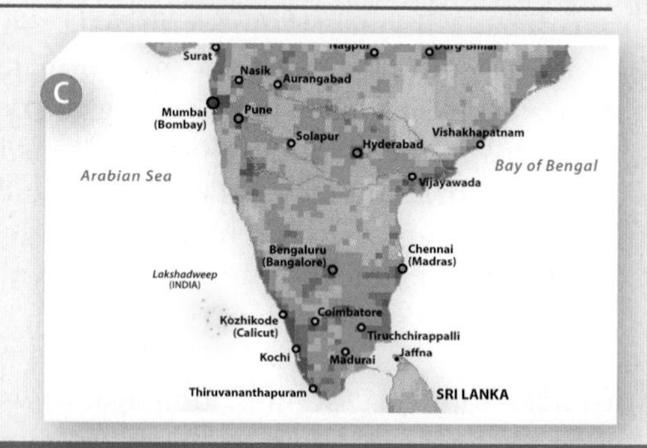

Key Terms

areal differentiation *(p. 13)*
areal integration *(p. 13)*
bubble economy *(p. 10)*
choropleth map *(p. 17)*
colonialism *(p. 39)*
core–periphery model *(p. 42)*
counterinsurgency *(p. 41)*
cultural imperialism *(p. 30)*
cultural landscape *(p. 14)*
cultural nationalism *(p. 30)*
cultural syncretism or hybridization *(p. 31)*
culture *(p. 29)*
decolonialization *(p. 39)*
demographic transition model *(p. 25)*
de-territorialization *(p. 36)*
diversity *(p. 4)*
economic convergence *(p. 9)*
ethnic religion *(p. 32)*
formal region *(p. 13)*
functional region *(p. 13)*
gender *(p. 34)*
gender equity *(p. 47)*
gender roles *(p. 34)*
geographic information systems (GIS) *(p. 18)*
geography *(p. 12)*
geopolitics *(p. 37)*
globalization *(p. 4)*
global positioning systems (GPS) *(p. 16)*
graphic or linear scale *(p. 17)*
gross domestic product (GDP) *(p. 44)*
gross national income (GNI) *(p. 44))*
gross national income (GNI) per capita *(p. 44)*
Human Development Index (HDI) *(p. 46)*
insurgency *(p. 40)*
less developed country (LDC) *(p. 43)*
life expectancy *(p. 25)*
lingua franca *(p. 31)*
map projection *(p. 16)*
map scale *(p. 16)*
meridians (lines of longitude) *(p. 15)*
more developed country (MDC) *(p. 43)*
nation-state *(p. 38)*
neocolonialism *(p. 40)*
net migration rate *(p. 27)*
parallels (lines of latitude) *(p. 15)*
population density *(p. 27)*
population pyramid *(p. 24)*
prime meridian *(p. 15)*
purchasing power parity (PPP) *(p. 44)*
rate of natural increase (RNI) *(p. 22)*
region *(p. 13)*
remote sensing *(p. 17)*
representative fraction *(p. 17)*
secularism *(p. 34)*
Shi'a Islam *(p. 33)*
Sunni Islam *(p. 33)*
terrorism *(p. 40)*
total fertility rate (TFR *(p. 22)*)
universalizing religion *(p. 32)*
urbanized population *(p. 28)*
vernacular region *(p. 14)*

In Review • Chapter 1
Concepts of World Geography

Cultural Coherence and Diversity

Culture is learned behavior and includes a wide range of both tangible and intangible behaviors and objects, from language to house architecture, from gender roles to sports. Because of globalization, the world's cultural geography is rapidly changing, producing new cultural hybrids in many places. In other places, people resist change with different kinds of cultural nationalism that protect (or even resurrect) traditional ways of life.

D

1.7 This photo was taken in downtown Prague, in the Czech Republic. What are the signs of global versus traditional culture?

1.8 Give five examples of cultural hybridization in your local area.

Geopolitical Framework

Varying political systems, ranging from dictatorships to democracies, provide the world with a dynamic geopolitical framework that is stable in some places and filled with tension and violence in others. As a result, the traditional concept of the nation-state is challenged by separatism, insurgency, and even terrorism.

E

1.9 What are the reasons behind terrorist attacks in Pakistan? What groups are responsible, and what are their goals?

1.10 Working in a small group, choose an African country that was a European colony in the early 20th century, and trace its geopolitical geography and history over the last century.

Economic and Social Development

Although proponents of economic globalization argue that all people in all places gain from expanded world commerce, that does not seem to be the case. Instead, there appear to be winners and losers—places that profit, while others lose out—resulting in a world geography of economic disparity. Social development of health care and education also remains highly uneven.

F

1.11 Many clothing factories have sprung up in Bangladesh in the last decade. Why? Make a list of the positive and negative aspects of this development from a local perspective.

1.12 In the latest Human Development Index study, the United States ranks number 3 in the world, behind Norway and Australia. Why doesn't the United States rank higher?

MasteringGeography™

Looking for additional review and test prep materials? Visit the Study Area in **MasteringGeography™** to enhance your geographic literacy, spatial reasoning skills, and understanding of this chapter's content by accessing a variety of resources, including **MapMaster™** interactive maps, videos, RSS feeds, flashcards, web links, self-study quizzes, and an eText version of Diversity Amid Globalization.

Authors' Blogs

Scan to visit the author's blog for chapter updates

www.gad4blog.wordpress.com

Scan to visit the GeoCurrents blog

www.geocurrents.info

2 Physical Geography and the Environment

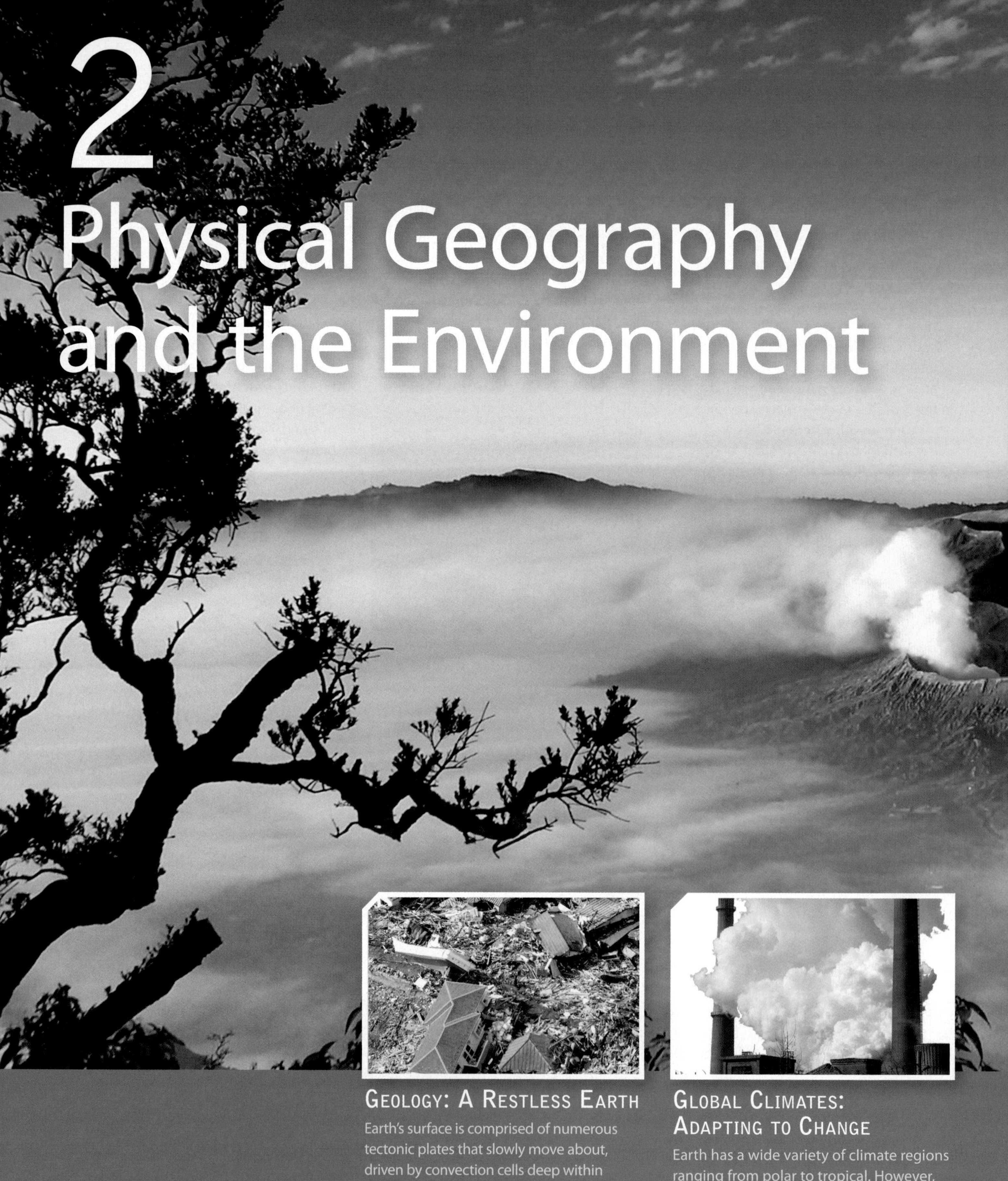

Geology: A Restless Earth

Earth's surface is comprised of numerous tectonic plates that slowly move about, driven by convection cells deep within the mantle. These movements cause earthquakes and volcanoes as they shape Earth's topography.

Global Climates: Adapting to Change

Earth has a wide variety of climate regions ranging from polar to tropical. However, these climates may be changing because of human-caused global warming, with problematic consequences..

Mt. Bromo in the morning. The volcanic landscape of Bromo Tengger Semeru National Park in East Java, Indonesia illustrates the dramatic diversity of Earth's physical geography. The area became a national park in 1982.

Global Energy: The Essential Resource

Fossil fuels—oil, coal, and natural gas—currently dominate the world's energy usage, while renewable energy sources—water, wind, solar, and biomass—remain primarily an alternative for the future.

Water: A Scarce World Resource

Clean freshwater is a necessity of life, but world supplies are scarce and dwindling, resulting in water stress and shortages in many areas of the world.

Bioregions and Biodiversity: The Globalization of Nature

A cloak of vegetation that varies greatly from place to place covers Earth, creating diverse ecologies. Humans, however, are changing these ecologies rapidly through various activities.

◄ **Figure 2.1 The World's Physical Geography** A fundamental step for studying world geography is to learn about Earth's physical geography—its geology, climate, energy resources, hydrology, and biodiversity. This scene is of a glacial-carved fjord on Baffin Island, Canada. In the background is Davis Strait, which separates North America from Greenland.

Earth's Physical Geography

http://earthobservatory.nasa.gov/IOTD/

The immense physical diversity of the global environment, with its varied climates, deep oceans, towering mountain ranges, dry deserts, and wet tropics, makes Earth unique in our solar system. Other planets are too warm for life (Venus) or too cold (Mars), but Earth is the Goldilocks planet, having just the right temperature range for life. In turn, life forms of all different sorts—plant, animal, and human—have all interacted with the environment to produce the diversity of landscapes and habitats that make Earth the human home. Thus, a necessary starting point for the study of world regional geography is knowing more about Earth's physical environment—its geology, climate, energy resources, hydrology, and diverse life forms (Figure 2.1).

A necessary starting point for the study of world regional geography is knowing more about Earth's physical environment.

Geology: A Restless Earth

Geologic processes shape Earth's surface and provide character to its diverse landscapes, creating a physical fabric of mountains, hills, valleys, and plains. This geologic foundation is also central to a wide array of human activities, such as agriculture, mining, transportation, and settlement patterns. Not to be overlooked, however, is that the geologic environment also presents serious challenges in the form of earthquakes, landslides, and volcanic eruptions. These geologic complexities make understanding the physical processes shaping Earth's landscapes crucial to appreciating its geography.

Plate Tectonics

The starting point for understanding most geologic processes is **plate tectonics**, a geophysical theory that Earth's outer layer consists of large geologic platforms, or plates, that move slowly across its surface. Driving these tectonic plates is a heat exchange deep within Earth; Figure 2.2 illustrates this complicated process.

On top of these tectonic plates sit continents and ocean basins. Note that continents are not identical to the underlying tectonic plates. Instead, most continents and ocean basins commonly straddle several different tectonic plates. This is important because most earthquakes and volcanoes are found along these plate boundaries. Figure 2.3 shows those relationships.

In western North America, for example, coastal California lies atop two different tectonic plates—the Pacific

LEARNING OBJECTIVES

After reading this chapter you should be able to:

- Describe those aspects of tectonic plate theory responsible for shaping Earth's surface.
- Identify those parts of the world where earthquakes and volcanoes are hazardous to human settlement and explain why casualty rates from those hazards differ from place to place.
- List the factors that control the world's weather and climate.
- Describe the major characteristics and locations of the world's major climate regions.
- Explain the greenhouse effect and how it relates to anthropogenic global warming.
- Summarize the major issues underlying the international controversy over reducing emissions of greenhouse gases.
- Define the concept of fossil fuel proven reserves and explain the socioeconomic factors that influence the amount of fossil fuel reserves.
- Describe the world geography of fossil fuel production and consumption.
- List the advantages and disadvantages of different kinds of renewable energy.
- Identify the causes of global water stress.
- Describe the characteristics and distribution of the world's major bioregions.
- Explain the reasons behind deforestation in both tropical and higher-latitude forests.

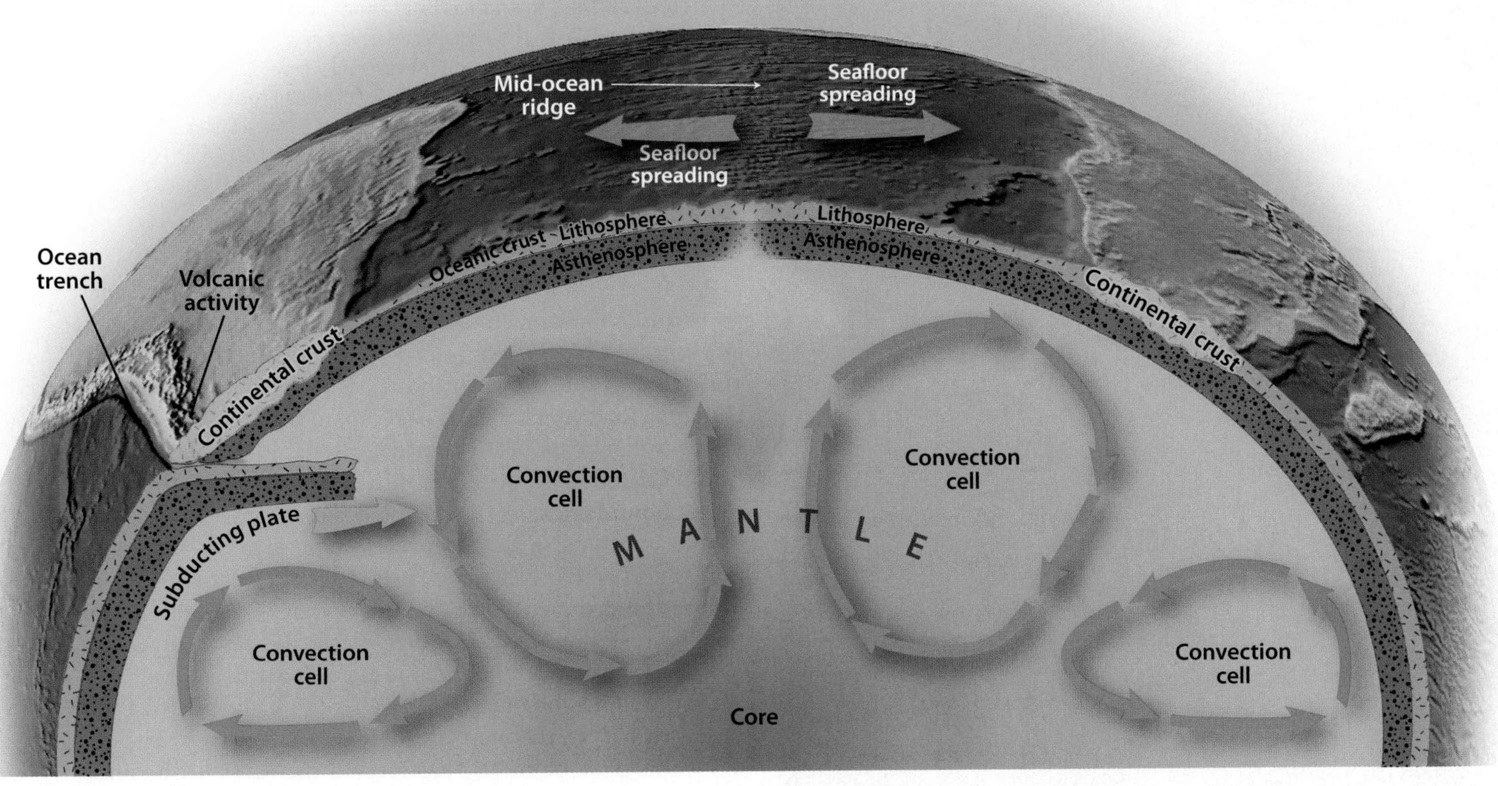

▲ **Figure 2.2 Plate Tectonics** According to plate tectonics theory, large convection cells circulate molten rock in different directions within Earth's mantle. In the crust, the slow movement of the cells drags tectonic plates away from the mid-oceanic ridges, resulting in the collision of plates in convergent plate boundaries. Although the rate of movement differs for each convection cell, in general it is only a few inches (or centimeters) per year.

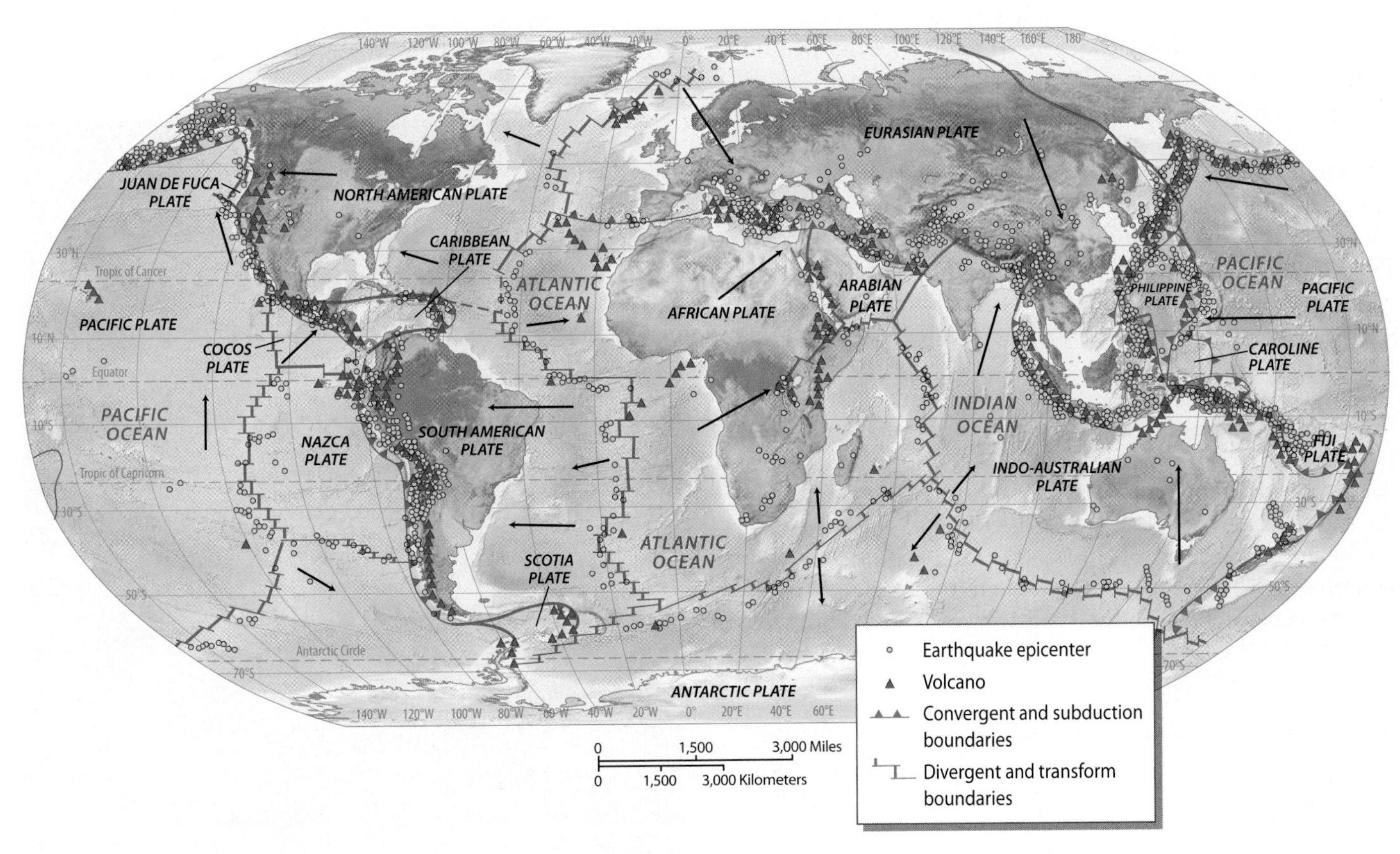

▲ **Figure 2.3 Tectonic Plate Boundaries, Earthquakes, and Volcanoes** This map of the world shows the close relationship between tectonic plate boundaries and the world distribution of earthquakes and volcanoes. Note, particularly, the so-called Pacific Rim of Fire that includes the eastern coasts of Asia and the western coasts of both North and South America.

◀ **Figure 2.4 The San Andreas Fault Zone** This fault zone, which is actually 40 miles (64 km) wide, may be the transform boundary where the North American and Pacific plates slide past each other. More recently, however, some geologists believe that the small Sierran Microplate lies east of the San Andreas Fault, relocating the actual border of the North American Plate just east of the Sierra-Nevada Mountains. This photo is of the main fault trace of the San Andreas in the dry Carrizo Plain area halfway between San Francisco and Los Angeles in the California Coast Ranges.

and the North American plates—on a former **colliding plate boundary**. These two tectonic plates converged about 35 million years ago, forced together by convection cells circulating in different directions deep within Earth's mantle. The infamous San Andreas Fault traversing coastal California forms the plate boundary. Originally formed by colliding plates, in more recent geologic times the San Andreas Fault became a transform fault, where the eastern edge of the Pacific Plate moves laterally northward at a rate of several inches each year, pushing sideways against the North American Plate. The nearness of San Francisco and Los Angeles to the San Andreas Fault makes these two urban areas—like many others around the world—vulnerable to destructive earthquakes (Figure 2.4).

In tectonic collision zones, one tectonic plate often sinks below another, creating a **subduction zone**. Such a region is characterized by deep trenches where the ocean floor has been pulled downward by regional tectonic warping. Subduction zones exist off the western coast of South America, off the northwest coast of North America, offshore of eastern Japan, and near the Philippines, where the Mariana Trench is the deepest point of the world's oceans at 35,000 feet (10,700 meters) below the surface. These subduction zones are also the locations of Earth's most powerful earthquakes, as witnessed by the magnitude 8.8 quake in Chile in 2010 and the magnitude 9.0 earthquake (with an accompanying tsunami) that devastated coastal Japan in 2011 (Figure 2.5).

In other parts of the world, tectonic plates move away from each other in opposite directions, forming **divergent plate boundaries**. As these plates diverge, magma from Earth's interior can flow between them, creating mountain ranges and active volcanoes. In the North Atlantic, Iceland lies on the divergent plate boundary that bisects the Atlantic Ocean (Figure 2.6). In other places, however, divergent boundaries form deep depressions, called rift valleys, such as that occupied by the Red Sea between northern Africa and Saudi Arabia. To the west of the Red Sea, in Africa, a splinter of this plate boundary has created the African rift valley, part of which forms the Olduvai Gorge, where tectonic activity has preserved in volcanic soils the earliest traces of our human ancestors.

Geologic evidence suggests that some 250 million years ago, all the world's plates were tightly consolidated into a supercontinent centered on present-day Africa. Over time, this

▶ **Figure 2.5 Japanese Earthquake and Tsunami** On March 11, 2011, at 2:46 P.M. local time a magnitude 9.0 undersea subduction zone earthquake struck the northeastern coast of Japan. Its epicenter was located 43 miles (70 km) east of the Oshika Peninsula, and it generated a massive tsunami wave that reached heights of 133 feet (40.5 meters) and traveled 6 miles (10 km) inland, causing about 20,000 deaths. This photo shows the devastation in Miyagi Prefecture.

large area, called Pangaea, was broken up as convection cells moved the tectonic plates apart. A hint of this former continent can be seen in the jigsaw-puzzle fit of South America with Africa and of North America with Europe.

Tectonic plate theory explains many of the world's large mountain ranges, but it cannot account for all highlands. Although both the Himalayan and the Alpine mountain ranges were created by the forces of colliding tectonic plates, many mountain ranges today are far removed from current tectonic boundaries. In North America, the Rocky Mountains serve as an example, as do the Ural Mountains in the center of Russia. Two possibilities explain these interior mountain ranges. One is that they were formed on the edges of tectonic plates when the global arrangement of plates was very different from what we see today. Alternatively, geologic stresses and strains along tectonic boundaries rippled across large distances from those boundaries, creating mountain ranges in plate interiors.

Geologic Hazards

Although floods and tropical storms typically take a higher toll of human life each year, earthquakes and volcanoes can have major effects on human settlement and activities. An estimated 20,000 people died in March 2011 from the combination of an earthquake and a tsunami in coastal Japan, and the year before (January 2010) over 230,000 people were killed in a magnitude 7.0 earthquake in Haiti. The vastly different effects of these two quakes underscore the fact that vulnerability to geologic hazards differs considerably around the world, depending on local building standards, population density, housing traditions, and the effectiveness of search, rescue, and relief organizations.

In addition to earthquakes, volcanic eruptions are found along most tectonic plate boundaries and can also cause major destruction. In 1985, for example, a volcanic eruption resulted in 23,000 deaths in Colombia, South America. In some cases, eruptions can be predicted days in advance, which usually provides enough time for evacuation. During the 1991 eruption of Mt. Pinatubo in the Philippines, 60,000 people were evacuated, although 800 did die in the disaster. Because volcanoes usually generate warning activity before they erupt, the loss of life from volcanoes is generally a fraction of that from earthquakes. More to the point, in the 20th century an estimated 75,000 people were killed by volcanic eruptions, whereas approximately 1.5 million died in earthquakes.

Unlike earthquakes, volcanoes provide some benefits to people. In Iceland, New Zealand, and Italy, geothermal activity produces energy to heat houses and power factories. In other parts of the world, such as the islands of Indonesia, volcanic ash has enriched soil fertility for agriculture. Additionally, local economies benefit from tourists attracted by scenic volcanic landscapes in such places as Hawaii, Japan (Figure 2.7), and the Pacific Northwest.

REVIEW

2.1 What are the three kinds of tectonic plate boundaries?

2.2 What drives tectonic plate movement?

2.3 Where are most of the world's earthquakes and volcanoes? Why are they located where they are?

◀ **Figure 2.6 Iceland's Divergent Plate Boundary** Volcanic activity is common in Iceland because of the divergent tectonic border that bisects the island (and the North Atlantic Ocean). Here people watch the 2010 Eyjafjallajökull eruption, which was most notable for the major disruption its volcanic ash caused to both trans-Atlantic and intra-Europe air traffic.

◀ **Figure 2.7 Mt. Fuji** The highest mountain in Japan at 12,389 feet (3776 meters), Mt. Fuji is a typical Pacific Rim of Fire stratovolcano, much like Mt. Ranier near Seattle, Washington. Although Mt. Fuji last erupted in 1708, it is considered an active volcano, lying just 60 miles (100 km) southwest of the Tokyo metropolitan area of more than 36 millon people.

Global Climates: Adapting to Change

Many, if not most, human activities are closely linked to weather and climate, from farming, which depends on balanced conditions of moisture and warmth to produce food, to city workers, who suffer from heat waves and disruptive snow storms. Moreover, much of the world's landscape diversity results from the many different ways in which people adapt to local weather and climate. For example, some desert areas in California are covered with high-value irrigated agriculture, producing vegetables for the global marketplace year-round. In contrast, most of the world's arid regions are barren and support very little farming. In addition, severe weather events such as storms and droughts often ripple through interconnected world trade and food systems with major repercussions. When drought strikes Russia's grain belt or Africa's Sahel, the socioeconomic consequences are felt throughout the world.

Aggravating these interconnections is the fact that the world's climates are presently changing because of global warming. Just what the future holds is not entirely clear, but even if the long-term forecast has some uncertainty, there is little question that all forms of life—including humans—will have to adjust to vastly different climatic conditions by the middle of the 21st century (see *Exploring Global Connections: Antarctica, the Protected Continent*).

Climate Controls

The world's climates differ considerably from place to place and seasonally, with highly varying temperature and precipitation (rain and snow) patterns. Most of these differences can be explained by a set of physical processes that influence weather and climate throughout the world. Let's examine each of these factors.

Solar Energy Both Earth's surface and the atmosphere immediately above it are heated by energy from the Sun, making solar energy the most important factor affecting world climates. Not only does this energy explain the differences in temperature between warm equatorial zones and the cold polar regions, but also it drives other important processes, including global pressure systems, winds, and ocean currents.

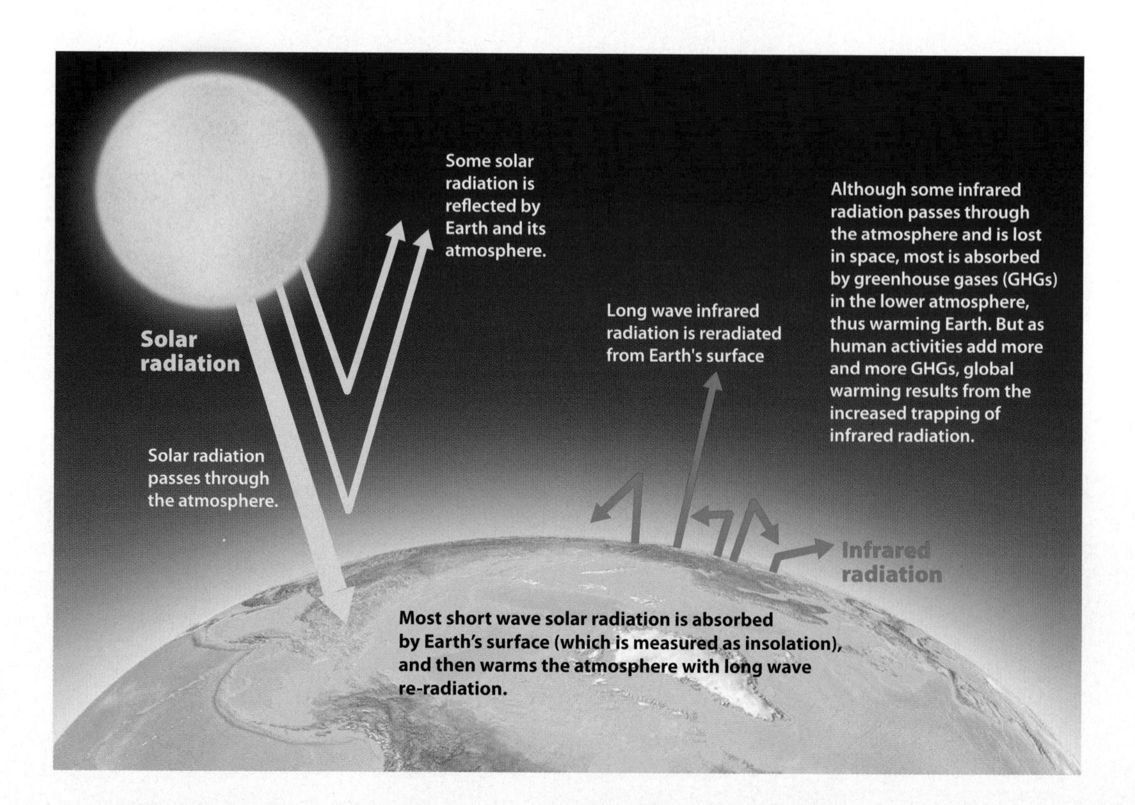

▶ **Figure 2.8 Solar Energy and the Greenhouse Effect** The greenhouse effect is the trapping of solar radiation in the lower atmosphere, resulting in a warm envelope surrounding Earth.

Exploring **Global Connections**

Antarctica, the Protected Continent

Globalization may bring both good and bad: positives for some, problems for others. However, one of globalization's unquestionable bright spots is Antarctica. The international community came together in a peaceful and cooperative spirit to make sure that this unique continent was protected for the common good. Even more amazing was that this international agreement occurred in 1959, during the darkest days of the Cold War between the world's superpowers: the Soviet Union (now Russia) and the United States. At that time, 12 countries (including both superpowers) signed the Antarctica Treaty System (ATS), which protects Antarctica from territorial claims, mining, and military bases and also establishes the land as a center for international scientific research. Today the ATS has 50 full member countries and 22 associate members.

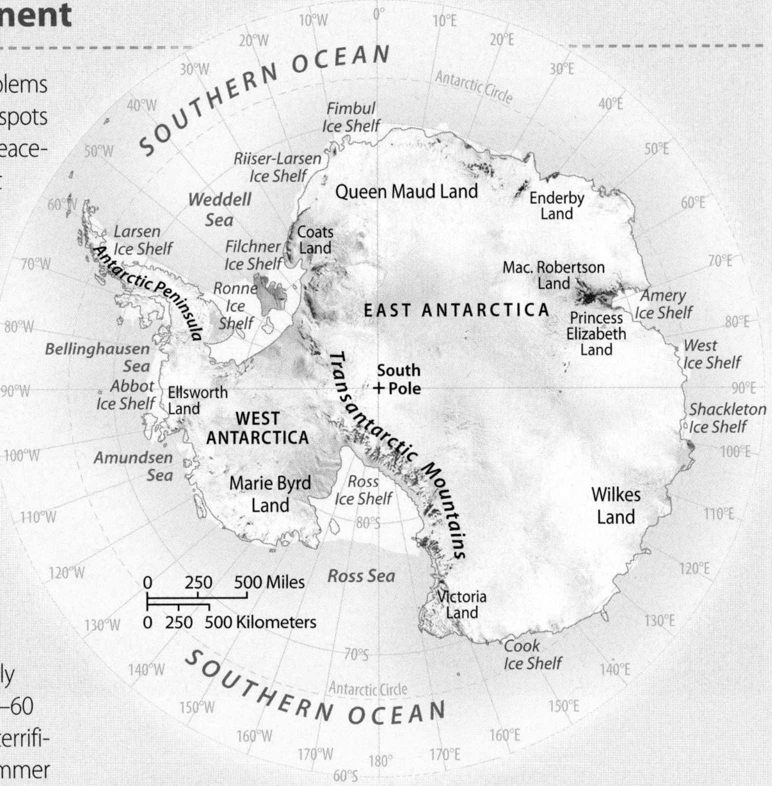

▲ **Figure 2.1.1 Antarctica** This map of Antarctica shows the East and West Antarctica Ice Sheets that make up much of the continent, as well as the Transantarctic Mountains separating the two ice sheets. McMurdo Station, the American Antarctic research center is southeast of the Ross Ice Shelf, on the small point north of Victoria Land.

Physical Geography Antarctica, which includes the South Pole, is the world's fifth largest continent. By comparison, Antarctica is twice as large as Australia. Not only is this southern continent huge, but also many other superlatives apply: it's the coldest, driest, windiest, and highest continent. The average yearly temperature at the South Pole is –70°F (–57°C), and on average only 6.5 inches (16.5 cm) of precipitation fall each year. Winds of 50–60 miles per hour (80–95 km per hour) blow constantly, resulting in terrifically low wind-chill factors. Only on the Antarctic Peninsula do summer temperatures get above freezing.

The geography of Antarctica is fairly simple: Two major ice sheets dominate, one in the east and the other in the west, separated by a large mountain range. In addition, several giant coastal ice shelves extend into nearby ocean waters (Figure 2.1.1). Ice caps cover most of Antarctica with an average depth of 7900 feet (2400 meters), but reaching 16,500 feet (5000 meters) in the thickest part. Mt. Vinson is the highest point above sea level at 16,144 feet (4892 meters).

The Laboratory on Ice Another unique feature of Antarctica is that it has no indigenous people—no natives have ever lived there. In fact, the only familiar life forms on the entire continent are penguins. This makes Antarctica cleaner and purer than any other place on Earth—perfect for investigating atmospheric science without the effects of air pollution.

Shortly after the ATS Treaty was signed, the British Antarctica Survey began to monitor atmospheric ozone, resulting in cooperative science on the infamous ozone hole. This is a depletion of the ozone layer in the upper atmosphere, which normally protects life on Earth from too much harmful ultraviolet radiation from the Sun. Scientists found a hole in the ozone layer over Antarctica, growing at an alarming rate due to reactions with certain chemicals that drifted into the upper atmosphere. This research led to another landmark in international cooperation, the Montreal Protocol of 1989, which banned ozone-depleting sprays and refrigerants. It is still upheld today as a model of international scientific and legal cooperation.

Today much of the international science being done in Antarctica is focused on global warming and climate change. Ice cores from the continent's gigantic ice sheets, for example, provide invaluable information about past climates going back some 800,000 years, almost seven times longer than the ice cores extracted from Greenland.

Not only does Antarctica provide clues to Earth's past climates, but also the continent is a critical player in contemporary global-warming and climate change scenarios because of the potential for melting and collapse of the West Antarctica Ice Sheet (Figure 2.1.2). Should that happen, it could raise Earth's sea levels by 10 feet (33 meters) or more, devastating coastal settlements throughout the world. However, such a disaster would not happen overnight, for scientists think that, at the current rate of warming, it may take at least several decades for this to occur.

Google Earth Virtual Tour Video

http://goo.gl/nU1gfO

1. What's the highest elevation in Antarctica? Is it part of an ice sheet?
2. How do scientists measure melting of the Antarctic ice sheets?

▲ **Figure 2.1.2 West Antarctica Ice Sheet** The edge of the ice shelf in the Ross Sea, with the Transantarctic mountains in the background.

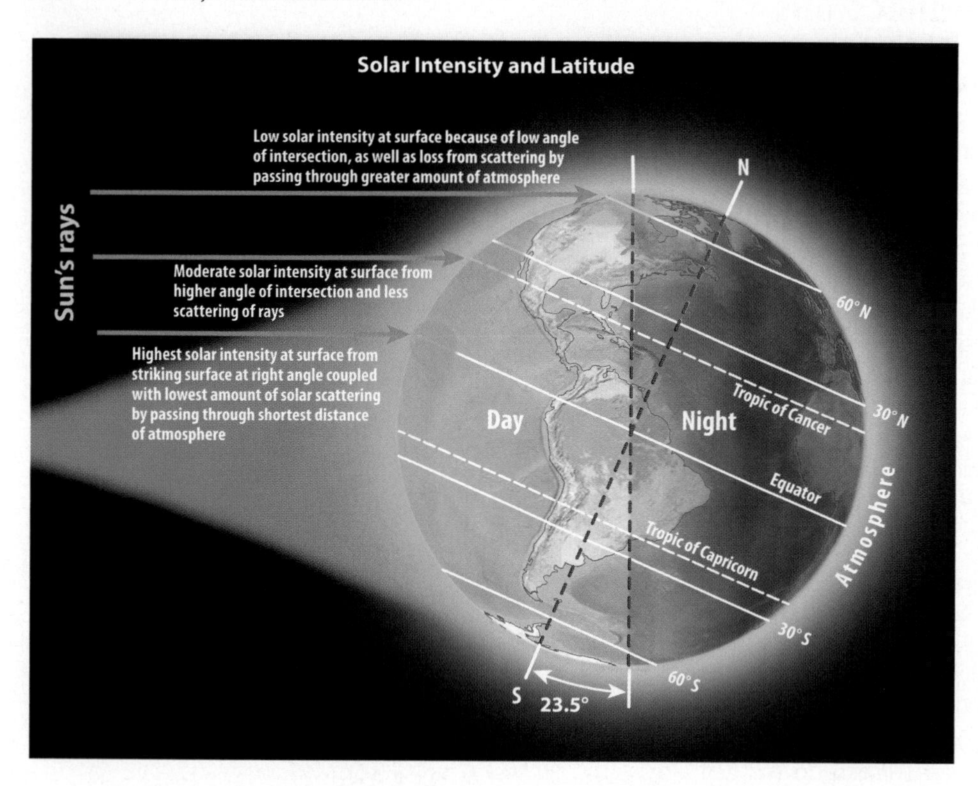

◀ **Figure 2.9 Solar Intensity and Latitude** Because of Earth's curvature, solar radiation is more effective at warming the surface in the tropics than in higher latitudes.

Incoming short-wave solar energy, called **insolation**, passes through the atmosphere and is absorbed by Earth's land and water surfaces. As these surfaces warm, they reradiate heat back into the lower atmosphere as infrared, long-wave energy. This energy, in turn, is absorbed by water vapor and atmospheric gases such as carbon dioxide (CO_2) in the lower atmosphere. This absorbed energy produces the envelope of warmth that makes life possible on our planet. There is some similarity between this heating process and the way a garden greenhouse traps sunlight to warm the structure's interior; therefore, this natural process of atmospheric heating is called the **greenhouse effect** (Figure 2.8). Were it not for this process, Earth's climate would average about 60°F (33°C) colder, resulting in conditions much like Mars.

Latitude Because of the curvature of the globe, insolation strikes Earth at a true right angle only in the tropics. This means that each ray of solar energy is more intense in the tropics than at higher latitudes north and south of the equator. As a result, the Sun is more effective at heating Earth's surface at the equator than at higher latitudes (Figure 2.9). You can feel the difference in solar intensity when standing in direct sunlight near the equator in Singapore, contrasted to the welcome warmth provided by the summer sun in an Arctic Circle city like Tromso, Norway.

Not only does this difference in solar intensity result in warm tropical climates that contrast with cooler climates in the middle or high latitudes, but also this difference in solar intensity builds up heat in the equatorial regions. This heat is then redistributed away from the tropics through physical processes such as global pressure and wind systems, ocean currents, tropical typhoons, hurricanes, and even midlatitude storms (Figure 2.10).

Interactions Between Land and Water Because land and water differ in their abilities to absorb and reradiate insolation, the global arrangement of oceans and land areas is a major influence on world climates (Figure 2.11). Basically, land areas heat and cool faster than do bodies of water. This explains why the temperature extremes of hot summers and cold winters, such as experienced in the interior United States and Canada, are always found away from coastal areas. These differences are also found at smaller scales, within hundreds of miles of each other. For example, the average high July temperature in San Francisco, on the California coast, is 60°F (15.6°C), whereas in Sacramento, the inland state capital—only 80 miles away—the average high temperature in July is 92.4°F (33.5°C).

The term **continentality** describes inland climates with hot summers and cold, snowy winters, such as those found in interior North America and Europe. In contrast, **maritime climates** are those close to the ocean, with moderate temperatures in

▼ **Figure 2.10 "Superstorm Sandy," October 2012** Hurricane Sandy merged with a mid-latitude frontal system moving across North American to become what the media called a "superstorm" as it spread its destruction over an 1100 mile (1770 km) wind field, making it the largest hurricane recorded. Traveling first across the Caribbean then up the East Coast, Sandy caused at least 286 deaths and over $68 billion in damage.

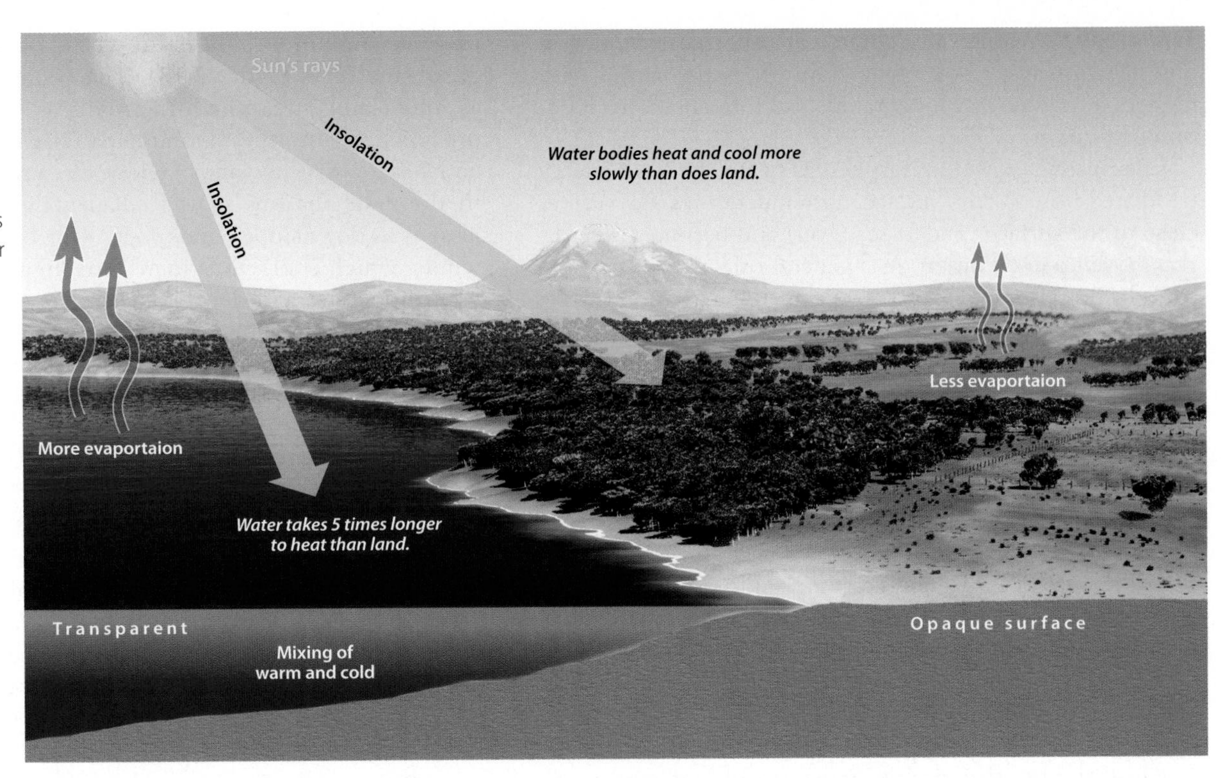

▶ **Figure 2.11 Differential Heating of Land and Water** Because land heats and cools faster than do bodies of water, temperatures are usually both warmer in the summer and colder in the winter inland than on the coasts.

both summer and winter. Western Oregon and the British Isles are good examples of regions with maritime climates.

Global Pressure Systems The uneven heating of Earth due to latitudinal differences and the arrangement of oceans and continents produces a regular pattern of high- and low-pressure cells. These cells drive the movement of the world's wind and storm systems. The interaction between high- and low-pressure systems in the North Pacific, for example, produces storms that are carried by the jet stream onto the North American continent from west to east in both winter and summer. Similar processes in the North Atlantic also produce winter and summer weather for Europe.

Farther south, in the subtropical zones, large oceanic cells of high pressure cause very different conditions. These high-pressure cells expand during the warm summer months because of the subsidence of warm air from the equatorial regions. As they enlarge, the cells produce the warm, rainless summers of Mediterranean climate areas in Europe and California. In the equatorial zone itself, summer weather spawns the strong tropical storms known as typhoons in Asia and hurricanes in North America and the Caribbean.

Global Wind Patterns High- and low-pressure systems also produce global wind patterns at local, regional, and global scales. As a rule, winds flow away from high-pressure cells and toward low-pressure cells (just as water flows from high to low elevations). This explains the monsoon in India, for example, which arrives in June as moisture-laden air masses flow northward from high pressure over the Indian Ocean. They travel across the South Asian subcontinent toward the low-pressure area of northern India and Tibet, where summer heat has created a massive low-pressure system. In the winter, wind flow is opposite: As high pressure builds over these inland areas with winter cooling, winds flow outward from cold Tibet and the snowy Himalayas toward low pressure over the warm Indian Ocean (see Chapter 12 for more detail). Similar seasonal pressure and wind regimes are found in many other parts of the world as well. One example is the U.S. Southwest, where the summer monsoon brings moisture to the arid lands of Nevada, Arizona, and New Mexico (Figure 2.12).

▼ **Figure 2.12 Summer Monsoon in the U.S. Southwest** As the U.S. Southwest warms in June and July, this heating creates thermal lows that draw in moist air from the Gulfs of Mexico and California, resulting in cloudiness (as shown here), thunderstorms, and much-needed rainfall.

Topography Weather and climate are affected by topography in two ways: Cooler temperatures are associated with higher elevations, and topography also influences precipitation patterns.

Because the lower atmosphere is heated by solar energy reradiated from Earth's surface, air temperatures are warmer close to the surface (and at sea level) and become cooler as you move up in elevation. As a general rule, the atmosphere cools by 3.5°F for every 1000 feet gained in elevation (in the metric system, the cooling rate is 1°C per 100 meters). This is called the **adiabatic lapse rate**, which is the rate of cooling with increasing altitude within the lower atmosphere. Thus, on a hot summer day in Phoenix, AZ, elevation 1100 feet (335 meters), the temperature often reaches 100°F (37.7°C). Just 140 miles away, in the mountains of northern Arizona at 7100 feet (2160 meters) near Flagstaff, the temperature is usually 21°F lower at a pleasant 79°F (26°C). The difference of 21°F equals 6 (thousands of feet difference in elevation) times 3.5, the adiabatic lapse rate. Cooler temperatures, then, are always found at higher elevations. This explains why in mountainous areas, the precipitation that falls as rain in the lowlands will probably fall as snow in the uplands.

Topography also wrings moisture out of storms by forcing moving air masses to cool as they are forced up and over mountain ranges, in what is known as the **orographic effect** (Figure 2.13). The cooler air cannot hold as much moisture, which then falls as precipitation. This process explains the common pattern of wet mountains and nearby dry lowlands. These dry areas are said to be in the **rain shadow** of the adjacent mountains. Rainfall lessens as downslope winds warm (the opposite of upslope winds, which cool), thus increasing an air mass's ability to retain moisture and deprive nearby lowlands of precipitation. Rain shadow areas are common in the mountainous areas of western North America, Andean South America, and many parts of South and Central Asia.

World Climate Regions

Even though the world's weather and climate vary greatly from place to place, we can use general similarities in temperature, moisture, and seasonality to make maps of global climate regions (Figure 2.15). Before going further, though, it is important to note the difference between these two terms. **Weather** is the short-term, day-to-day expression of atmospheric processes: Weather can be rainy, cloudy, sunny, hot, windy, calm, or stormy, all within a short time period. As a result, weather is measured at regular intervals each day (usually hourly). These data are then compiled over a 30-

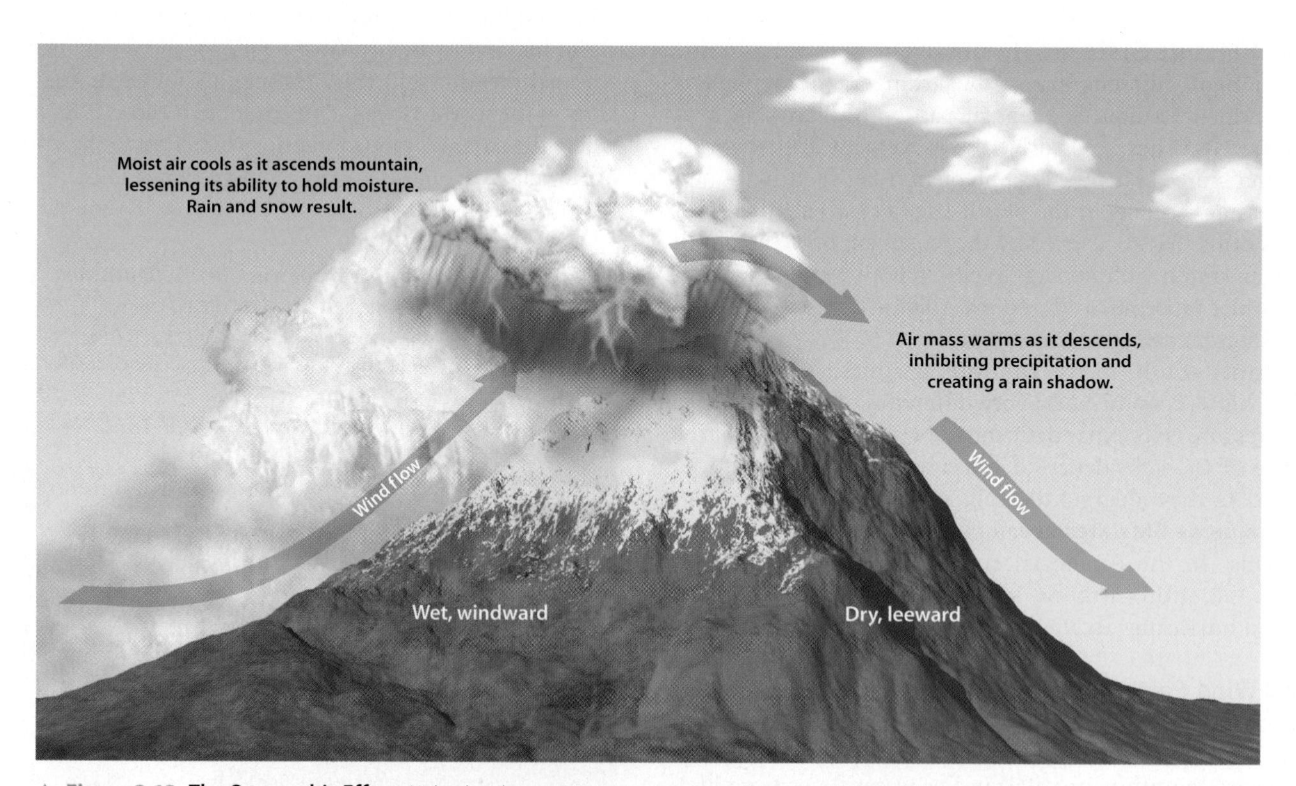

▲ **Figure 2.13 The Orographic Effect** Upland and mountainous areas are usually wetter than the adjacent lowland areas because of the orographic effect, whereby rising air is cooled and loses its ability to hold moisture as it flows up and over mountains, resulting in rain and snowfall. In contrast, the leeward or downwind side of the mountains is usually drier because warming air increases its ability to hold moisture. **Q: After reviewing the concept of the orographic effect, look at a map of the world and list at least five different areas where a rain shadow would be found.**

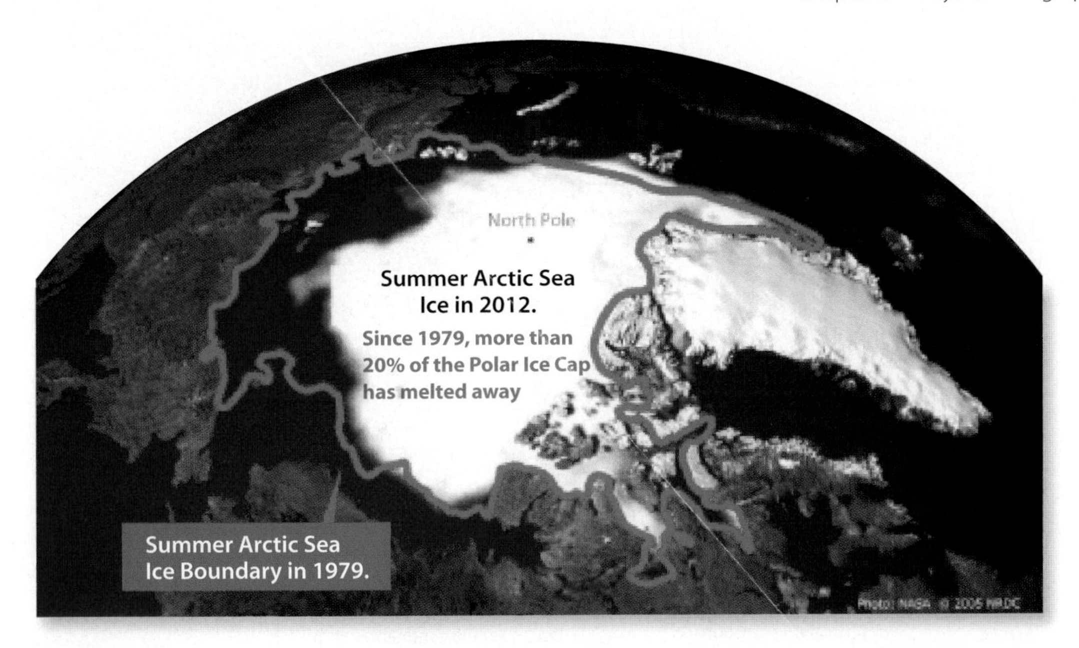

◄ **Figure 2.14 Loss of Arctic Sea Ice Through Global Warming** This schematic illustrates the effect of global warming and climate change on the extent of Arctic Sea Ice. The red line marks the extent of sea ice at the end of summer in 1979, while the white area maps the reduced distribution of ice at the end of summer 2012. What is not mapped is the reduction in snow and ice cover for Greenland, which has also been significant.

year period to generate statistical averages that describe the common meteorological conditions of a specific place. From these averages, we can place the weather of a specific locale in the worldwide scheme of climate regions. Simply stated, then, *weather* is the short-term expression of atmospheric processes, and **climate** is the long-term average from daily weather measurements. As pundits like to say, climate is what you expect and weather is what you get.

Knowing the climate type for a specific part of the world not only conveys a clear sense of average rainfall and temperatures, but also allows inferences about a wider array of human activities. If an area is categorized as desert, then we infer that rainfall is so limited that any agricultural activities require irrigation. In contrast, when an area is characterized as a tropical monsoon climate, we know it has warm temperatures and adequate amounts of rainfall for certain kinds of farming.

We use a standard scheme of climate types throughout this book, and each regional chapter contains a map showing the different climates of the area. In addition, these maps contain **climographs**, which are graphic representations of monthly average high and low temperatures, along with monthly precipitation amounts for a specific location. Figure 2.15 shows climographs for Tokyo, Japan, and Cape Town, South Africa. Two lines for temperature data are presented on each climograph: The upper line plots average high temperatures for each month, while the lower line shows average low temperatures. Besides these temperature lines, climographs contain bar graphs depicting average monthly precipitation. Not only is the total amount of rainfall and snowfall important, but also the seasonality of precipitation provides valuable information about the link between moisture and agricultural growing seasons.

Global Climate Change

Human activities, primarily those connected with economic development and industrialization, are changing the world's climate in ways that have significant consequences for all living organisms, whether plants, animals, or humans. More specifically, **anthropogenic** (human-caused) pollution of the lower atmosphere is increasing the natural greenhouse effect so that worldwide **global warming** is taking place. This warming, in turn, could lead to climate changes with major consequences. Rainfall patterns, for example, could change, so that existing agricultural production in traditional breadbasket areas such as the U.S. Midwest and Canadian prairies is threatened. Low-lying coastal settlements in places like Florida and Bangladesh could be flooded as sea levels rise from warming oceans and melting polar ice caps. An increase in searing heat waves could cause higher human death tolls in the world's cities, with water becoming an increasingly scarce resource in many areas of the world. Although the world's nations recognize the seriousness of climate change, only minor progress has been made in the last decades in crafting international agreements that could mitigate the problem.

Causes of Global Warming and Climate Change Before going further, it would be useful to clarify what we mean by global warming, as contrasted to the closely related term **climate change**. Global warming refers not just to an increase of Earth's average temperature, but also to the many implications of that warming, such as rising sea levels, warming oceans, changes in plant and animal behaviors, and so forth. Related, but more specific, climate change is just that: a documented or predicted change in the temperature, winds, and precipitation patterns that differs from the current (or very recent) climate.

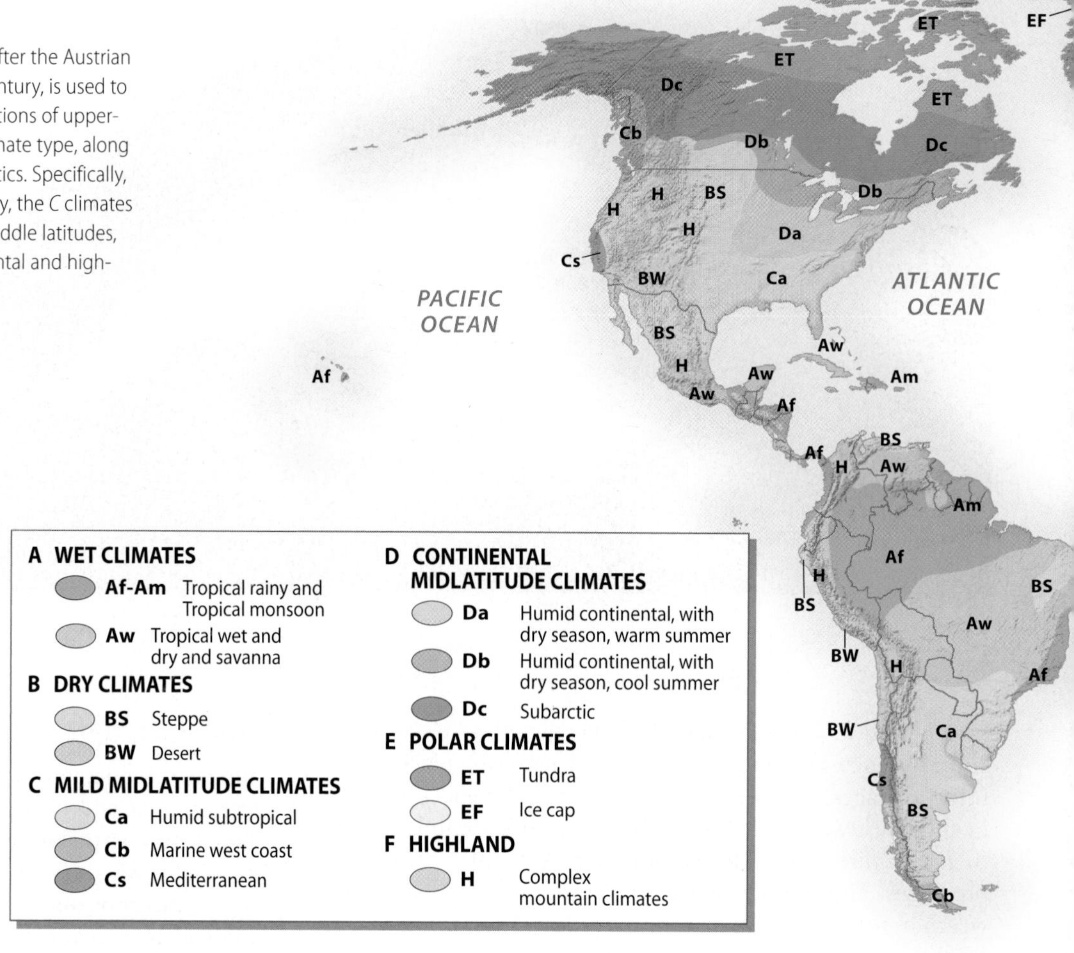

▶ **Figure 2.15 Climates of the World**
A standard scheme, called the *Köppen system,* after the Austrian geographer who devised it in the early 20th century, is used to describe the world's diverse climates. Combinations of upper- and lower-case letters describe the general climate type, along with precipitation and temperature characteristics. Specifically, the *A* climates are tropical, the *B* climates are dry, the *C* climates are generally moderate and are found in the middle latitudes, and the *D* climates are associated with continental and high-latitude locations.

In the last 20,000 years, for example, Earth's climate has warmed naturally from the last glacial or ice age to the climate that our great-grandparents knew around 1870 or so. Within that general warming trend were also some notable fluctuations, such as the Little Ice Ages from about 1550 to around 1850 CE. However, since 1870, human-caused global warming has caused climates to change to warmer average temperatures. The year 2012, to illustrate, was among the 10 warmest years on record (based upon the world's average temperature) and the 36th year in a row exceeding long-term averages.

As mentioned earlier, the natural greenhouse effect provides Earth with a warm atmospheric envelope that supports life, and this warmth comes from the trapping of incoming and outgoing solar radiation by an array of natural greenhouse gases (GHGs)—water vapor, carbon dioxide (CO_2), methane (CH_4), and ozone (O_3). Although the composition of these natural GHGs has varied somewhat over long periods of geologic time, they have been relatively stable since the last ice age ended 20,000 years ago (more detail on GHGs is found online in *MasteringGeography*).

However, with the widespread consumption of coal and petroleum associated with global industrialization, a huge increase has occurred in atmospheric carbon dioxide and methane (byproducts of burning oil and coal). This increase has greatly magnified the natural greenhouse effect and produced global warming. More specifically, in 1860 atmospheric CO_2 was 280 parts per million (ppm); today it is greater than 390 ppm (Figure 2.16). More troubling is that, unless global emissions are reduced considerably in the very near future, the CO_2 level will reach 450 ppm by 2020, a level at which climate scientists project that Earth's climate will change irrevocably. Although the complexity of the global climate system leaves some uncertainty about exactly how the world's climates may change, climate scientists using high-powered computer models are reaching consensus on what can be expected from continued global warming. Basically, computer models predict that average global temperatures will increase almost 4°F (2°C) by 2020, a temperature change that is the same magnitude as the cooling that caused ice-age glaciers to cover much of Europe and North America 30,000 years ago. If emissions continue at

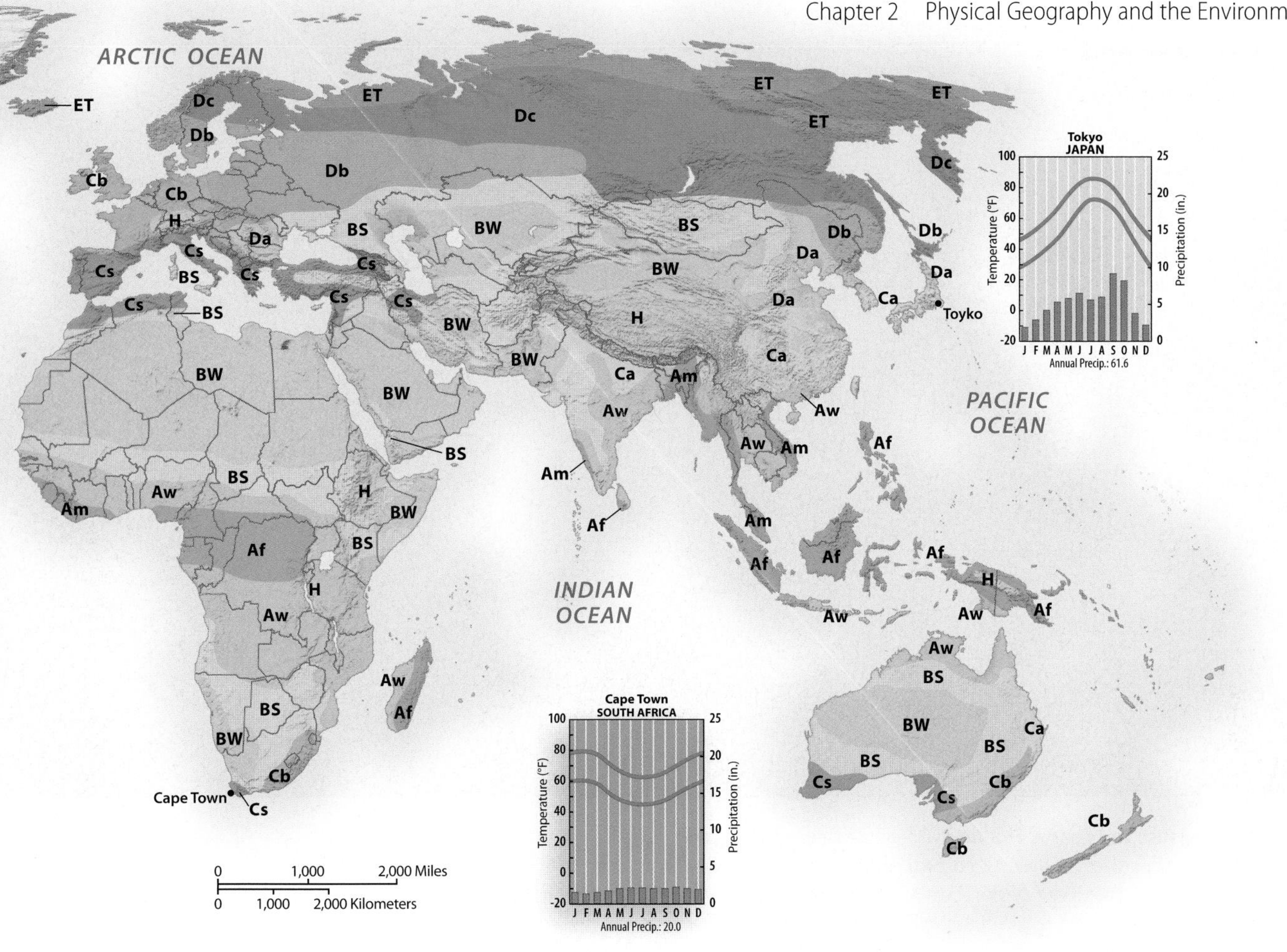

the current rate, this temperature increase is projected to double by 2100. As noted earlier, the consequences of future global warming are indeed sobering, with widespread coastal flooding from sea-level rise, worldwide droughts that impact food systems, widespread water shortages, devastating heat waves, more frequent wildfires, and catastrophic animal and plant extinctions (Figure 2.17).

The International Debate on Limiting Emissions The international debate on limiting GHGs emissions has passed through two phases in the last 30 years and is now in a crucial third stage, one that could determine the extent of global warming and climate change experienced for the rest of the century.

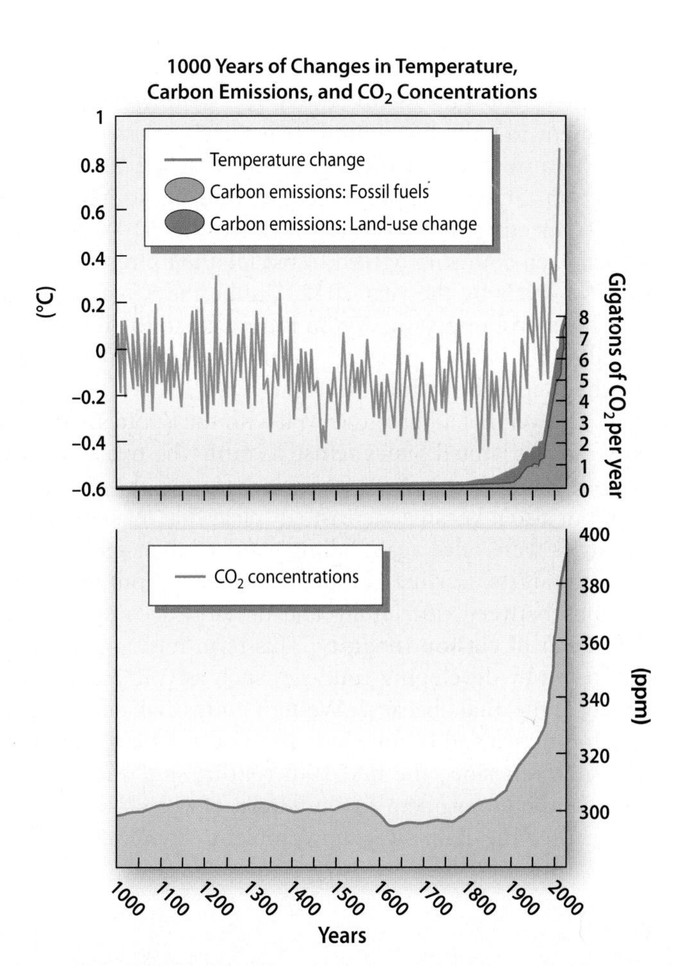

▶ **Figure 2.16 Increase in CO_2 and Temperature** These two graphs show the relationship between the rapid increase of CO_2 in the atmosphere and the associated rise in average annual temperature for the world. The graphs go back 1000 years and show both CO_2 and temperature to have been relatively stable until the recent industrial period, when the burning of fossil fuels (coal and oil) began on a large scale.

▲ **Figure 2.17 Sea-Level Rise and Coastal Flooding** One consequence of global warming will be a rise in the world's sea level due to a combination of polar ice cap melting and the thermal expansion of warmer ocean water. Forecasts are that, at the current rate of climate warming, sea levels will rise about 4 feet (1.4 meters) by the year 2100, causing considerable flooding of low-lying coastal areas such as this region of the Netherlands near Amsterdam. Other densely settled coastal areas throughout the world will also be vulnerable.

The first phase began at the UN-sponsored 1992 Earth Summit in Rio de Janeiro, when 167 countries signed an agreement to voluntarily limit their GHGs emissions. However, because none of the signatories reached its emission reduction targets, a more formal second phase began with a 1997 meeting in Kyoto, Japan. Here the 30 Western industrialized countries agreed to reduce their emissions back to 1990 levels by the year 2012. Unlike the Rio agreement, which was voluntary, the Kyoto Protocol had the force of international law, with penalties for those countries not reaching their emission reduction targets.

However, not all has gone well with the Kyoto Treaty. Not only did the United States refuse to ratify the treaty because of political concerns about possible injury to the U.S. economy, but also it became increasingly clear that most industrial countries were once again falling short of their emission reduction targets. Further complications arose from increasing tensions between developing and developed countries over the matter of **carbon inequity**. This term refers to the position taken by developing countries such as China and India, which argue that, because Western industrial countries in North America and Europe have been burning large amounts of fossil fuels since the mid-19th century and because CO_2 stays in the atmosphere for hundreds of years, these countries caused the global-warming problem. Therefore, they say, North America and Europe should fix global warming before requiring emerging economies to reduce their emissions. Because of this carbon inequity concern, the Kyoto Treaty of 1997 did not include emission reduction targets for China, India, Brazil, or any other developing country.

By 2008, China's annual GHGs emissions surpassed those of the United States, which until that year had historically been the world's largest emitter (Figure 2.18). Adding to the controversy was fast-growing India's stated ambition to follow China's example of rapid industrialization. Brazil, too, has become a problem because of rapidly increasing emissions from deforestation of the Amazon rainforest and from its expansive livestock economy. (Trees in the rainforest absorb carbon dioxide when growing and emit it when cut and burned; livestock emit methane gas.)

With the Kyoto Protocol scheduled to expire in 2012, the third phase of the international global-warming emission debate began in December 2011, with a meeting of 194 countries in Durban, South Africa. At that conference, a conceptual road map was agreed upon that guides current discussions on limiting emissions. The key points are as follows:

- The Kyoto Protocol was extended beyond its 2012 expiration date to 2015, which is now the target date for a completely new international agreement.

▼ **Figure 2.18 Recent CO_2 Emission Trends for the Five Largest Emitters** China's yearly CO_2 emissions continue to grow as hundreds of new coal-fired power plants come online. In contrast, emissions in the United States have stabilized recently because more power plants have switched from coal to natural gas as the price of that cleaner source of energy has become increasingly competitive. The flat line of Russia's emissions after the collapse of the Soviet Union is somewhat of a mystery and may be a reporting problem. In contrast, Germany's flat line is a result of an increasing reliance on renewable energy—namely, wind and solar. India's every-increasing emissions continue to be of concern. **Q: Note the difference in the trajectories over the last 20 years of China and India contrasted with those of the United States, Germany, and Russia. Explain the differences.**

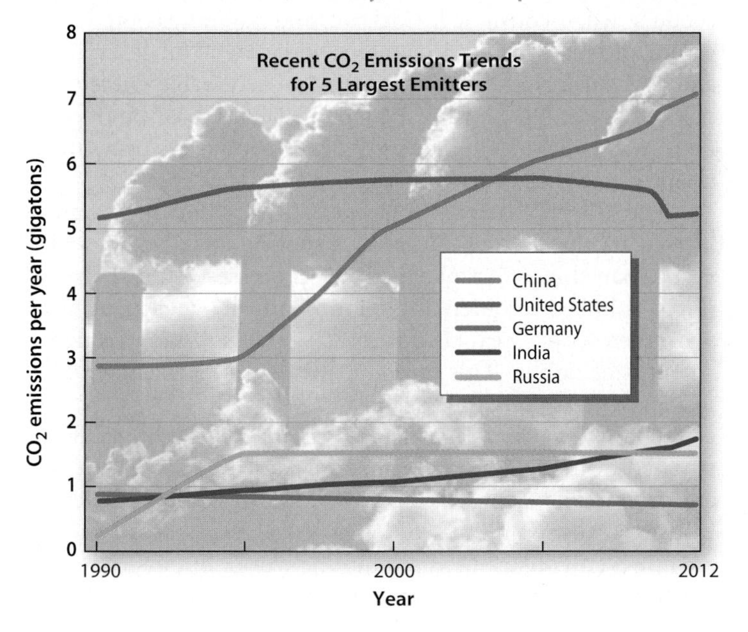

- Unlike Kyoto, this new agreement must include all countries, developed and developing; this clearly includes the large CO_2 emitters China and India, as well as those countries like Brazil whose emission levels are projected to grow significantly in the next decades. Not surprisingly, this component of the Durban Agreement is extraordinarily contentious because of China and India's strong opposition to emission reduction targets. Whether these two giant countries will lessen their resistance over the next several years is crucial to future global-warming agreements (Figure 2.19).
- By 2015, new emission reduction goals will be set for all the world's countries, and these will take effect no later than 2020.
- A \$100 billion Green Fund was created to help less developed countries (LDCs) meet the challenges of global warming. This fund will finance a wide range of activities, including renewable energy development and flood control structures for areas threatened by sea-level rise.
- A Reducing Emissions from Deforestation and Degradation (REDD) program was established to finance preservation of the world's forests by paying developing countries to reduce the cutting of their forests and instead treat them as carbon storehouses.

Will the Durban Agreement and the 2015 agreement significantly reduce the global-warming problem? Probably not, since, as noted earlier, the current rate of global CO_2 emissions will most likely reach 450 ppm by the time the new Durban reduction targets take effect. As a result, the amount of CO_2 in the atmosphere will probably reach the point where—even if all emissions were halted overnight—the world's climate processes will be irrevocably changed.

REVIEW

2.4 What is the difference between insolation and reradiation?

2.5 How and why has the natural greenhouse effect been changed by human activities?

2.6 What are the differences between continental and maritime climates? What causes these differences?

2.7 What issues are hindering international efforts to reduce carbon emissions?

▼ **Figure 2.19 China's Global-Warming Emissions** China surpassed the United States as the largest emitter of GHGs in 2008. A major component of this global-warming pollution is CO_2 emissions from coal combustion. China is the world's largest consumer of coal, which is used primarily to generate electricity from coal-fired power plants. Because of this growing need, experts predict that China's emissions could double in the next 20 years, a worrisome prospect that will have profound effects on the world's climate.

Global Energy: The Essential Resource

The world runs on energy. In the biological world, the Sun's energy provides the driving force behind life. In our human world, the situation is more complicated because most of the modern world's energy comes from a finite supply of fossil fuels that, as discussed in the previous section, pollute the atmosphere when consumed. As a result, understanding of energy resources is fundamental to comprehending the world's geography.

To start with basics, let's first distinguish between energy and power. Energy is the capacity to do work of any sort, such as heating a house, moving a car, or generating electricity. Power is the rate at which energy is used over a specific period of time. As an illustration, imagine a dam across a river, trapping and storing vast amounts of water behind the barrier. The water lying still in the lake behind the dam is considered to be a form of stored energy, but as it flows downhill through the dam's turbines, it generates electricity over some time period and becomes a source of power (Figure 2.20).

Energy is measured in units such as joules, calories, or British thermal units (BTUs), whereas power is measured in units such as the kilowatt-hours seen on household electricity bills. Larger amounts of energy, like those found in oil fields or coal mines, are measured in equivalents to millions of barrels of oil or tons of coal; large amounts of power usage are measured in units such as thousands of barrels of oil per day or millions of tons of coal per year.

Non-Renewable and Renewable Energy

Energy resources are categorized as either **non-renewable** or **renewable**. Non-renewable energy is consumed at a higher rate than it is replenished, which is indeed the case with the world's oil, coal, uranium, and natural gas resources. In contrast, renewable energy depends on natural processes that are being constantly renewed; water, wind, and solar energy are examples. Currently, most of the world is powered by non-renewable energy, with renewable energy powering just 3 percent of the world's needs.

More specifically, fossil fuels provide 87 percent of the world's power. Of the three major fossil fuel groups, more oil is used than coal, generating 33 percent of the world's power, whereas coal provides 30 percent (Figure 2.21). Global usage of coal, however, is increasing at oil's expense because currently coal is cheaper than oil and, further, its price is more stable than that of oil. Until recently coal usage was increasing at oil's expense because it was cheaper; however at this writing (late 2013), natural gas is replacing both coal and oil in North America and parts of Europe because gas is both cleaner and cheaper.

Another distinction is that between "dirty" and "clean" energy, based upon the amount of atmospheric emissions they produce when consumed. Coal, oil, and natural gas produce carbon dioxide and methane (along with several other GHGs) and are considered "dirty energy." Most renewable energy is pollution-free. Of the three major fossil fuels, coal produces more CO_2 emissions than does oil, with natural gas being the cleanest, emitting 60 percent less CO_2 than does coal.

The reason fossil fuels give off CO_2 is that they were formed by the decomposition of organisms over millions of years, shaped by an array of geologic processes. Because their origin is linked to different life forms (mainly ancient plants and microscopic sea creatures), fossil fuels contain high amounts of carbon. This carbon is then emitted as carbon dioxide (CO_2) when burned as fuel.

The global distribution of fossil fuels is highly differentiated from place to place because these energy resources are products of complex tectonic forces reshaping ancient landscapes over millions of years. As a result, fossil fuels are not evenly distributed around the world, but instead are clustered into specific geologic formations, scattered differentially around the globe.

Fossil Fuel Reserves, Production, and Consumption

Table 2.1 shows the varied world geography of energy reserves, production, and consumption—who's got the buried resources, who's mining and producing them, and who the major consumers are of coal, oil, and gas. More details on this complex global energy geography appear in each of the regional chapters of this book.

A word of explanation is necessary for the category of proven fossil fuel reserves. Because a high degree of uncer-

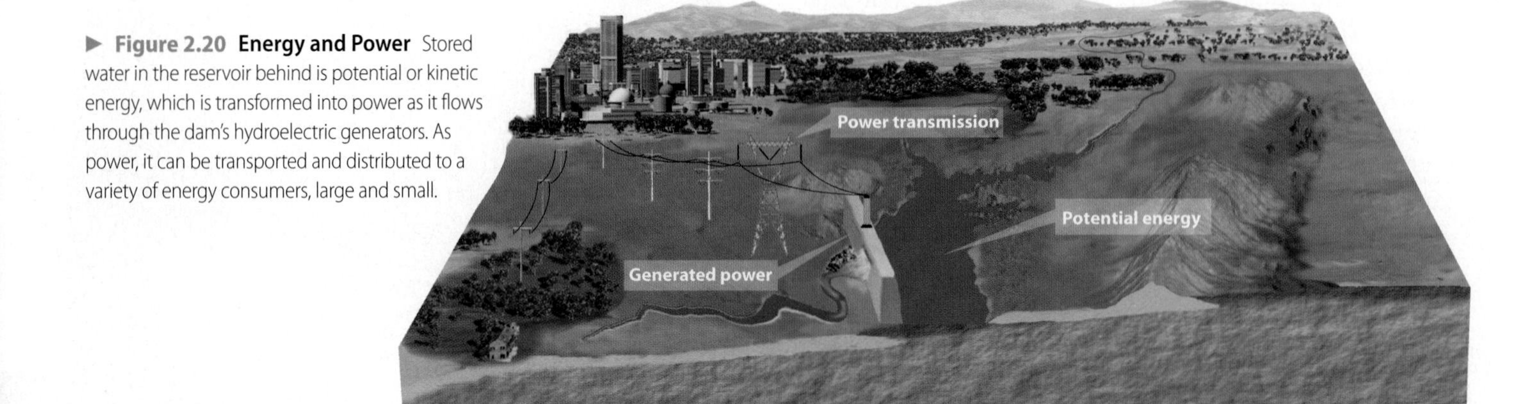

▶ **Figure 2.20 Energy and Power** Stored water in the reservoir behind is potential or kinetic energy, which is transformed into power as it flows through the dam's hydroelectric generators. As power, it can be transported and distributed to a variety of energy consumers, large and small.

▲ **Figure 2.21 Wyoming Coal Train** Wyoming is the largest coal-producing state in the U.S., mining 40 percent of the country's coal, which is then shipped by huge coal trains to the east and west coasts where it's exported to Europe and Asia. Because it produces more pollution than natural gas or petroleum, coal's future as a fuel is uncertain in the U.S. **Q: List several reasons why coal is moved long distances in the United States? What are the costs and benefits of this at a national scale?**

tainty and technological difficulty is involved with fossil fuel mining, the energy industry uses the concept of **proven reserves** of oil, coal, and gas to refer to deposits that are possible to mine and distribute under current economic and technological conditions. An important component of this definition is "current," since economic, regulatory, and technological conditions can change rapidly, expanding or even reducing the amount of a resource deemed feasible to mine and produce at a given moment in time. If, for example, the price of oil on the global market is high, then oil reserves with relatively high costs of drilling can be produced at a profit. If the market price of oil is low, then only easily accessible reserves qualify as "proven."

To illustrate, within the last 10 years the price of a barrel of crude oil (the standard unit) has ranged from a low of $37.71 in December 2008 to its current price around $100. When the global economy stalled in 2008, the price of oil fell drastically because of low worldwide demand. This led to the closing of many production sites, along with a revision and reduction in what were considered at the time to be oil's proven reserves. Now, with higher oil prices, it's economically feasible to produce more expensive oil, such as that stored in North Dakota's shale formations, which reportedly costs $60 a barrel to mine, so consequently, those proven reserves have once again expanded.

Not only do the market prices for fossil fuels change over time, but also drilling and mining technologies change and generally improve. These, too, force revision of the

Table 2.1 Fossil Fuel Facts

Proven Reserves	World Share	Production	World Share	Consumption	World Share
Oil		**Oil**		**Oil**	
Venezuela	17.9%	Saudi Arabia	13.2%	United States	20.5%
Saudi Arabia	16.1%	Russia	12.8%	China	11.4%
Canada	10.6%	United States	8.8%	Japan	5.0%
Iran	9.1%	Iran	5.2%	India	4.0%
Iraq	8.7%	China	5.1%	Brazil	3.0%
Coal		**Coal**		**Coal**	
United States	27.6%	China	49.5%	China	50%
Russia	18.2%	United States	14.1%	United States	13.5%
China	13.3%	Australia	5.8%	India	7.9%
Australia	8.9%	India	5.6%	Japan	3.2%
India	7.0%	Indonesia	5.1%	South Africa	2.5%
Natural Gas		**Natural Gas**		**Natural Gas**	
Russia	21.4%	United States	20.0%	United States	21.5%
Iran	16.9%	Russia	18.5%	Russia	13.2%
Qatar	15.9%	Canada	4.9%	Iran	4.7%
Turkmenistan	11.7%	Iran	4.6%	China	4.0%
United States	4.1%	Qatar	4.5%	Japan	3.3%
				Nuclear Energy	
				United States	31.4%
				France	16.7%
				Japan	6.2%
				South Korea	5.7%
				Canada	3.6%

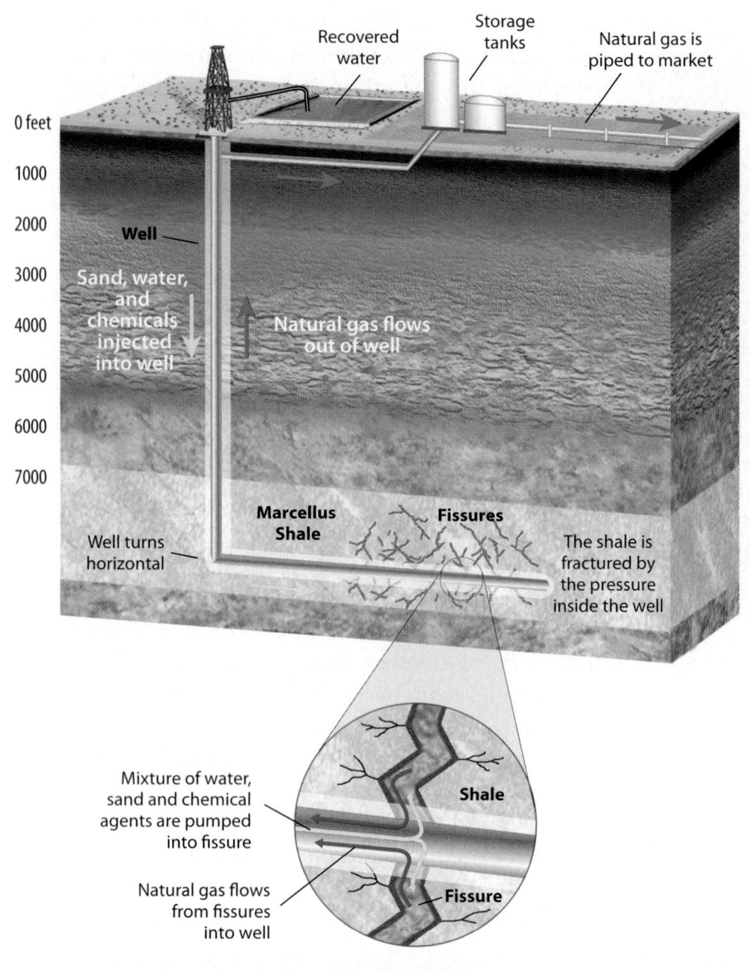

▲ **Figure 2.22 Hydraulic Fracturing** This graphic illustrates the different components of hydraulic fracturing (fracking) for oil and gas. What it does not show are the environmental issues associated with fracking, issues that include the high amount of freshwater used, the possibility of polluting local ground water, disposal issues of waste water, local noise and nuisance issues, and the possibility fracking could cause more earthquakes in certain geologic structures.

proven energy reserves. A good example is the recent development of **hydraulic fracturing**, or **"fracking,"** a relatively new oil and gas mining technique that forces oil and gas out of shale rock (Figure 2.22). If fracking proves successful (and there are some doubts about this), it might dramatically alter world energy supplies, trade, and politics. We will return to this topic later in this section.

Renewable Energy

As mentioned earlier, renewable energy is naturally replenished at a faster rate than it is consumed. Wind and solar power are good examples. The usual list of renewable energy sources includes not just wind and solar power, but also hydroelectric, geothermal (from Earth's interior heating), tidal currents, and biofuels, which use the carbon in plants as their power source. We should also add experimental efforts, such as growing a diesel-like fuel from algae, to the list of renewable energy sources.

Nevertheless, despite the widespread availability of sun, wind, and possible biofuels, renewable energy provides only a fraction of the world's power, at 3 percent. Some countries do make considerable use of renewables in their power grid. Iceland, for example, is blessed with bountiful supplies of both water and thermal resources and generates fully 95 percent of its power needs from renewable sources. In the United States, renewable energy accounts for about 13 percent of power needs, with most (7.7 percent) resulting from hydroelectricity. Many individual states within the United States have mandated goals for achieving a certain percentage of power from renewables, usually ranging from 10 to 30 percent (Figure 2.23).

Of the large industrial economies, Germany leads the way with over 20 percent of its power coming from renewables, primarily because of extensive wind and solar power stations. China, the world's largest producer of energy (mainly from coal, thus its high CO_2 emissions), reportedly drives a quarter of its massive economy with renewable energy. Unlike Germany, most of China's renewable energy comes from hydropower, although the country is rapidly expanding its wind and solar facilities.

Issues with Renewable Energy Despite the attractiveness of wind and solar power, significant issues must be resolved before their usage can expand. Both wind and solar, for example, are intermittent sources of power, since the sun does not always shine or the wind blow. To compensate for these lulls, wind and solar facilities are usually mandated to have gas-fired generators as backups to produce power at a moment's notice. At a smaller scale, for households or even small villages, wind- and solar-charged batteries usually take up the slack when the wind or sun falters.

The intermittent nature of large-scale wind and solar generation also requires that national power grids perform a tricky balancing act between energy supply and demand. Power surges generated by sunny and windy periods must somehow be assimilated into a power grid built long ago on the assumption of steady and continual inputs of power from gas- or coal-fired power plants. Germany, for example, often finds its power grid overloaded with power during windy days and has to take extreme measures to siphon off the power before blowing out a national circuit. As a result, and because the country is committed to expanding its array of renewable energy, Germany is now completely rebuilding its national power grid so it can balance more effectively the surges and lulls in input from its vast array of solar and wind generators.

Finally, developing and implementing renewable power can be more costly than fossil fuels because renewables lack the same degree of economic subsidies and tax incentives as those enjoyed by oil, coal, gas, and nuclear. At a global level, it has been estimated that traditional fossil fuels receive six times the financial support through governmental incentives that renewables do. When new technologies serve the public interest, they usually receive considerable economic support from national governments, and while this has happened in several notable cases (China and Germany), these subsidies need to become more widespread before renewable energy can compete on a level playing field with fossil fuels.

◄ **Figure 2.23 Wind Turbines in California** Wind power generates about 4 percent of U.S. power, and is the second largest source of renewable energy behind hydropower. Currently, Texas produces the most wind power, with California in second. This wind farm is in the Tehachapi Pass area of southern California.

Energy Futures

Global energy demand is forecast to increase 40 percent by 2030 as developing economies in China, India, and Brazil industrialize further and a billion new people are added to the world. Not to be overlooked is the goal of providing power to the 25 percent of the world's population that currently lacks access to an energy source (see *Working Toward Sustainability: Lighting Up Dark Places*). Energy demand in the developed world (the United States, Europe, Japan) will increase at a much slower rate than in the developing world, and perhaps even level off, because of advancements in energy efficiency. In coming years, cars, trucks, and airplanes will go farther on less fuel, reducing energy demand from the transportation sector. Buildings will also become more energy efficient, lessening energy demands for heating, cooling, and lighting.

Forecasts assume that fossil fuel reserves are adequate to meet future energy demands, but the expansion of renewable technologies may actually decrease the demand for coal, oil, and natural gas. Also important is whether unconventional drilling techniques, such as hydraulic fracturing, will continue to expand. In the United States, which has extensive deposits of the shale oil and gas ripe for fracking, this means first finding solutions to the many environmental issues associated with this new technology. If this is done (and opinions vary on whether these environmental issues can be resolved satisfactorily), the United States might find itself in the enviable position of satisfying most of its energy needs with domestic oil and gas instead of relying on its traditional import partners—Mexico, Canada, Saudi Arabia, and Venezuela.

China and Russia also have extensive oil and gas shales that could be exploited by fracking, and if they, too, are successful, this could also bring widespread changes to the current global energy geography. China, as the second largest importer of crude oil after the United States, most certainly has a vested interest in expanding its domestic supply of oil and gas. Russia's position as the world's largest exporter of natural gas, with considerable markets in Europe and China, is less clear beyond its obvious need to retain its current export advantage. Also to be considered is the future of oil-exporting countries like Saudi Arabia and Iraq if natural gas from fracking becomes widely available at competitive prices. Could that mean an end to oil's regime as the world's preferred energy source? And if so, what are the implications for Mideast oil kingdoms? We discuss these topics in more detail in the regional chapters.

REVIEW

2.8 What are fossil fuel proven reserves?

2.9 List the top three countries using the most oil, coal, and natural gas.

2.10 What are some of the problems with renewable energy?

2.11 How might fracking change the world?

Water: A Scarce World Resource

Water is central to human life and our supporting activities—agriculture, industry, transportation, even recreation. Yet water is unevenly distributed around the world, being plentiful in some areas while distressingly scarce in others. Water problems are not simply due to varied global climates that produce wet or dry conditions; they are also caused by complex factors such as political control over river basins and storage facilities. In addition, different cultures and different economies have highly varied water usage patterns.

At first glance, Earth is indeed the water planet, with more than 70 percent of its surface area covered by oceans. As a result, 97 percent of the total global water budget is saltwater, with only 3 percent freshwater. Of that small amount of freshwater, almost 70 percent is locked up in polar ice caps and mountain glaciers. Additionally, groundwater, which is often difficult to access, accounts for almost 30 percent of the world's freshwater. This leaves less than 1 percent of the world's water in more accessible surface rivers and lakes.

Another way to conceptualize this limited amount of freshwater is to think of the total global water supply as 100 liters, or 26 gallons. Of that amount, only 3 liters (0.8 gallons) is freshwater; and of that small supply, only a mere 0.003 liters, or about half a teaspoon, is readily available to humans.

Water planners use the concept of **water stress** and scarcity to map where water problems exist and also to predict where future problems will occur (Figure 2.24). Water stress data are generated by calculating the amount of freshwater available in relation to current and future population. Northern Africa stands out as a region of high water stress;

Working Toward **Sustainability**

Lighting Up Dark Places

About a quarter of the world's population lives in the dark, mainly in Sub-Saharan Africa and South Asia. These rural people are way off the power grid, making do with candles, kerosene lamps, or perhaps an extension cord hooked up to a village diesel generator that runs for an hour or two each day to meet their most basic energy needs. The social and economic costs of living in the dark are high: Education suffers because students can't study after dark; community health is poor without refrigerated vaccines and modern diagnostic instruments; household water supplies and crop irrigation must be provided by hand or animal labor pumps; and so on.

Here Comes the Sun Recently, village and household solar power systems have been making a big difference in lighting up these dark places (**Figure 2.2.1**). For example, in the village of Ruhilra, Uganda, electric lights were switched on for the first time after a solar micro-grid was installed at eight points around the village, each serving 20 households. Even though sunshine is plentiful, the system stores a three-day charge in batteries to guard against cloudy periods. Households on the grid now spend about $5 a month for their energy needs, about the same amount they . . . as they traditionally paid for candles and kerosene. Also, subscribers can pay with their cell phones, a billing and payment method now common in rural Africa. This new project was accomplished by Shared Solar, a nonprofit research group at Columbia University in New York City.

The Barefoot College Another group bringing solar power to poor rural peoples is the Social Work and Research Centre—commonly known as the Barefoot College—in Tilonia, Rajasthan, India. Since 2005, more than 140 women from Africa, many of them illiterate grandmothers, have trained at the Barefoot College. In six months, these women learned how to fabricate, install, and maintain solar-powered household lighting systems, thus becoming solar engineers and technicians who have transformed the lives of over 2000 families in the first self-sufficient and self-reliant, solar-electrified villages in Africa. This program has reached remote, poor, rural villages in 25 countries in Africa, Asia, and Latin America, flying poor women to India so that they can return home and install solar power systems to electrify their own villages.

Each household agrees to pay $5–$10 per month for the solar lighting, roughly what they used to spend on kerosene, candles, and flashlight batteries. Over time, these modest fees pay for the solar system, replacement parts, and also a modest monthly salary for the Barefoot Engineer who maintains the system (Figure 2.2.2).

Google Earth
Virtual Tour Video

http://goo.gl/8e2Odm

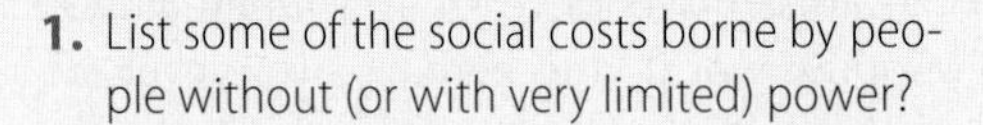

1. List some of the social costs borne by people without (or with very limited) power?
2. What other alternatives than solar power do rural Africans have for lighting their villages?

▼ **Figure 2.2.1 Small Scale Solar in Africa** This family shows off the solar panel that brings light and electricity to their home in the village of Kwanza Sul in Angola.

▼ **Figure 2.2.2 Barefoot College Solar Technicians** African women watch their teacher demonstrate how to maintain a village solar energy system during a class at the Barefoot College in Tilonia, India.

hydrologists predict that three-quarters of Africa's population will experience water shortages by 2025. Other problem areas are China, India, much of Southwest Asia, and even several countries in humid Europe. Although global warming may actually increase rainfall in some parts of the world, scientists forecast that climate change will probably aggravate global water problems. Three areas of concern occupy water planners: scarcity, sanitation, and access.

Water Scarcity

Currently, about half the world's population lives in areas where water shortages are common. As the population in these areas increases, these water problems will become even more acute. Also, since 70 percent of the world's freshwater usage is for agriculture, food production will probably decline as water becomes increasingly scarce.

Water Sanitation

Where clean water is not available, people use polluted water for their daily needs, resulting in a high rate of sickness and even death. More specifically, the UN reports that over half of the world's hospital beds are occupied by people suffering from illnesses linked to contaminated water, and more people die each year from polluted water than are killed in all forms of violence, including wars. This toll from polluted water is particularly high for infants and children, who have not yet developed any sort of resistance to or tolerance for contaminated water. The UN Children's Fund (UNICEF) reports that nearly 4000 children die each day from unsafe water and lack of basic sanitation facilities.

Water Access

By definition, when a resource is scarce, access is problematic, and these hardships take many forms. Women and children, for example, often bear the burden of providing water for family use, and this can mean walking long distances to pumps and wells and then waiting in long lines to draw water. The result is that their daily time allowed for other activities, such as school or work, is severely curtailed (Figure 2.25). Given the amount of human labor involved in providing water for crops, it is not surprising that some studies have shown that in certain areas, people expend as many calories of energy irrigating their crops as they gain from the food itself.

Ironically, some recent international efforts to increase people's access to clean water have gone astray and have actually aggravated access problems instead. Historically, domestic water supplies have been public resources, organized and regulated—either informally by common consent or more formally as public utilities—resulting in free or low-cost water. In recent decades, however, the World Bank and the International Monetary Fund have promoted the privatization of water systems as a condition for providing loans and economic aid to developing countries. The agency's goals have been laudable, trying to ensure that water is clean and healthful. However, the means have been controversial because, typically, the international engineering firms that have upgraded rudimentary water systems by installing modern water treatment and delivery technology have increased the costs of water delivery to recoup their investment. Although the people may now have access to cleaner and more reliable water, in many cases the price is

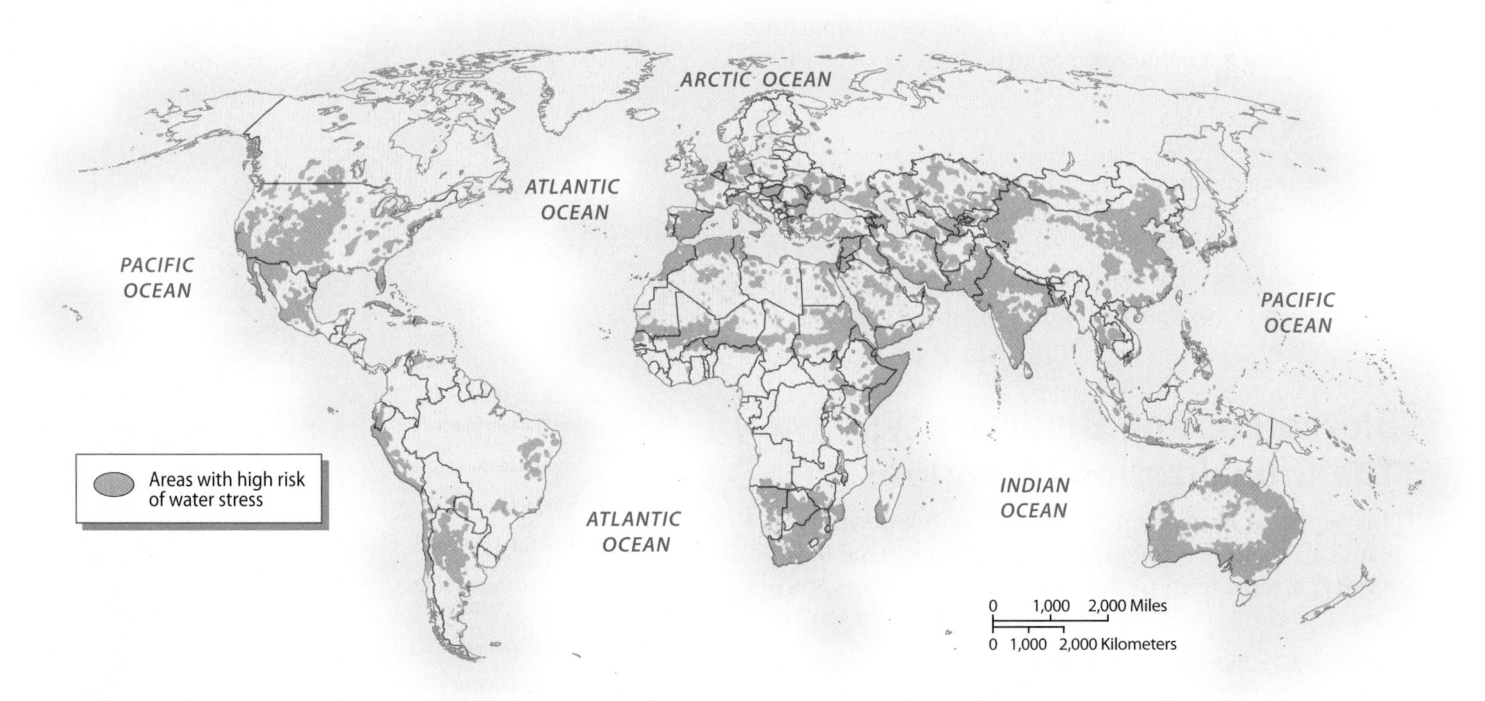

▲ **Figure 2.24 Global Water Stress** Water planners use the concept of water stress for those regions where water is—or will be—in short supply because of the combination of high water usage, low water supply, and the forecast population growth of an area. **Q: Why do several European countries in humid climate areas face water stress?**

▲ **Figure 2.25 Women and Water** In many parts of the world, women and young girls spend much of their day providing water for their homes and villages. For young girls, this task often interferes with their schooling. These women and girls are fetching water in the Punjab region of India.

higher than they can afford, forcing them to either do without or go to other unreliable and polluted sources.

In Cochabamba, Bolivia, for example, the privatization of the water system in 2000 resulted in a 35 percent increase in water costs. In response, the people rebelled and rioted, with demonstrations that became tragically violent. Eventually, the water system was returned to public control. Reportedly, however, today half the city's population is still without a reliable water source.

REVIEW

2.12 How much water is there on Earth, and how available is it for human usage? Use in your answer the concept that Earth's water budget is just 100 liters.

2.13 What are the three major issues that cause water stress?

2.14 Where in the world are the most severe areas of water stress?

Bioregions and Biodiversity: The Globalization of Nature

One aspect of Earth's uniqueness is the rich diversity of plants and animals covering its continents. This **biodiversity**, a term used to describe species richness, can be thought of as the "green glue" that binds together life, land, and atmosphere. Humans are very much a part of this interaction. Not only are we evolutionary products of a specific **bioregion**—a local or regional assemblage of plants and animals—but also our human prehistory includes the domestication of specific plants and animals that led to modern agriculture.

Our human presence, however, is taking a toll on nature by destroying habitat and hunting animals. As a result, biologists estimate that we're losing several dozen species every day, causing an extinction crisis that could see 50 percent of Earth's species gone by mid-century unless we change our ways and take measures to preserve threatened plant and animal species.

Here we provide a brief overview of the most important bioregions. We begin with the warm equatorial tropics and then move poleward, describing the dry desert and grassland regions, followed by two different kinds of temperate forests (Figure 2.26). Note that these bioregions are closely correlated to the climate regions discussed earlier, since the major traits of a climate region—temperature, precipitation, and seasonality—are the same factors influencing the distribution of flora and fauna making up a particular bioregion.

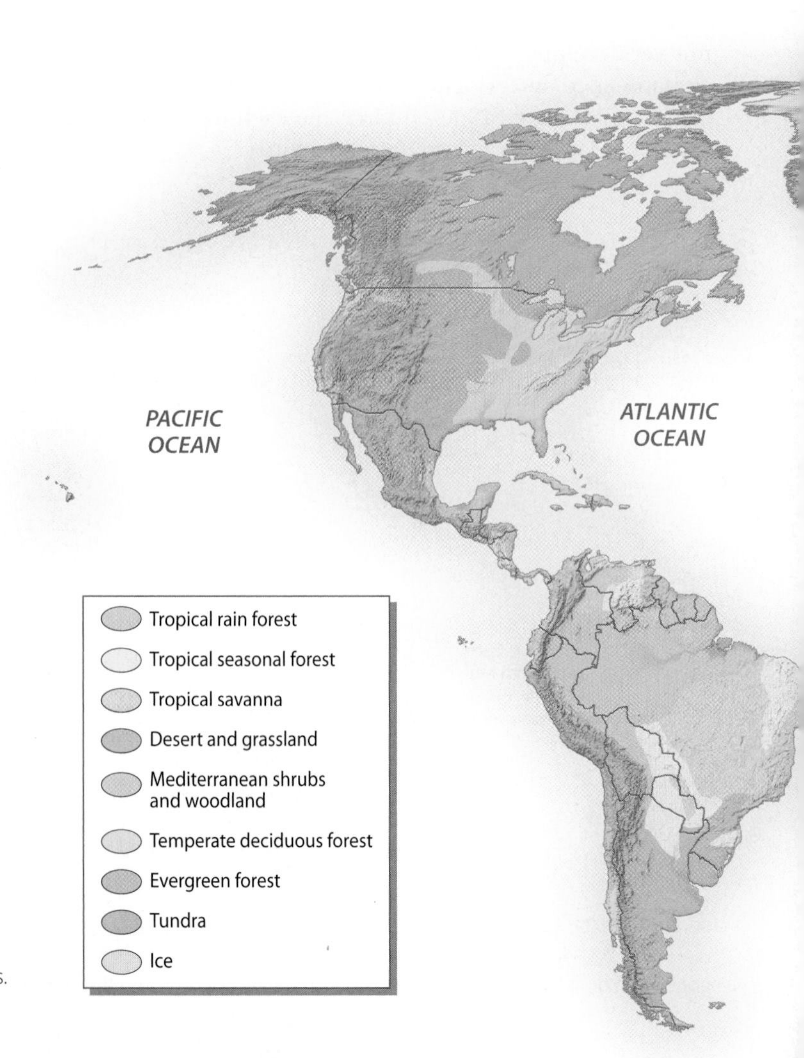

► **Figure 2.26 Bioregions of the World** Although global vegetation has been greatly modified by clearing land for agriculture and settlements and by cutting forests for lumber and paper pulp, there is still a recognizable pattern to the world's bioregions, ranging from tropical forests to arctic tundra. Each bioregion has its own array of ecosystems, containing plants, animals, and insects.

◀ **Figure 2.27 Tropical Rainforest** As fragile as they are diverse, tropical rainforest environments feature a complex, multilayered canopy of vegetation. Plants on the forest floor are well adapted to receiving very little direct sunlight.

Tropical Rainforests

Tropical rainforests are found along the equator in the Af climate region, with its high average annual temperatures, abundant sunlight, and copious rainfall occurring throughout the year. Annual rainfall amounts of 70–150 inches (180–380 cm) are common.

This bioregion covers about 7 percent of the world's land area (roughly the size of the contiguous United States) in Central and South America, Sub-Saharan Africa, Southeast Asia, and Australia and on many tropical Pacific islands.

More than half of the world's known plant and animal species live in the tropical rainforest bioregion, making it the bioregion with the highest biodiversity.

The dense tropical forest vegetation is usually arrayed in three distinct levels that are adapted to decreasing amounts of sunlight, from the treetops to the darker forest floor (Figure 2.27). The tallest trees, around 200 feet (60 meters) high, receive open sunlight; the middle level, around 100 feet (30 meters) high, gets filtered sunlight; and the lowest level, the forest floor, is where plants survive with very little direct sunlight. Even though much organic material accumulates on the forest floor in the form of falling leaves, rainforest soils are not particularly well suited to intensive agriculture because most nutrients are absorbed by extensive tree root systems.

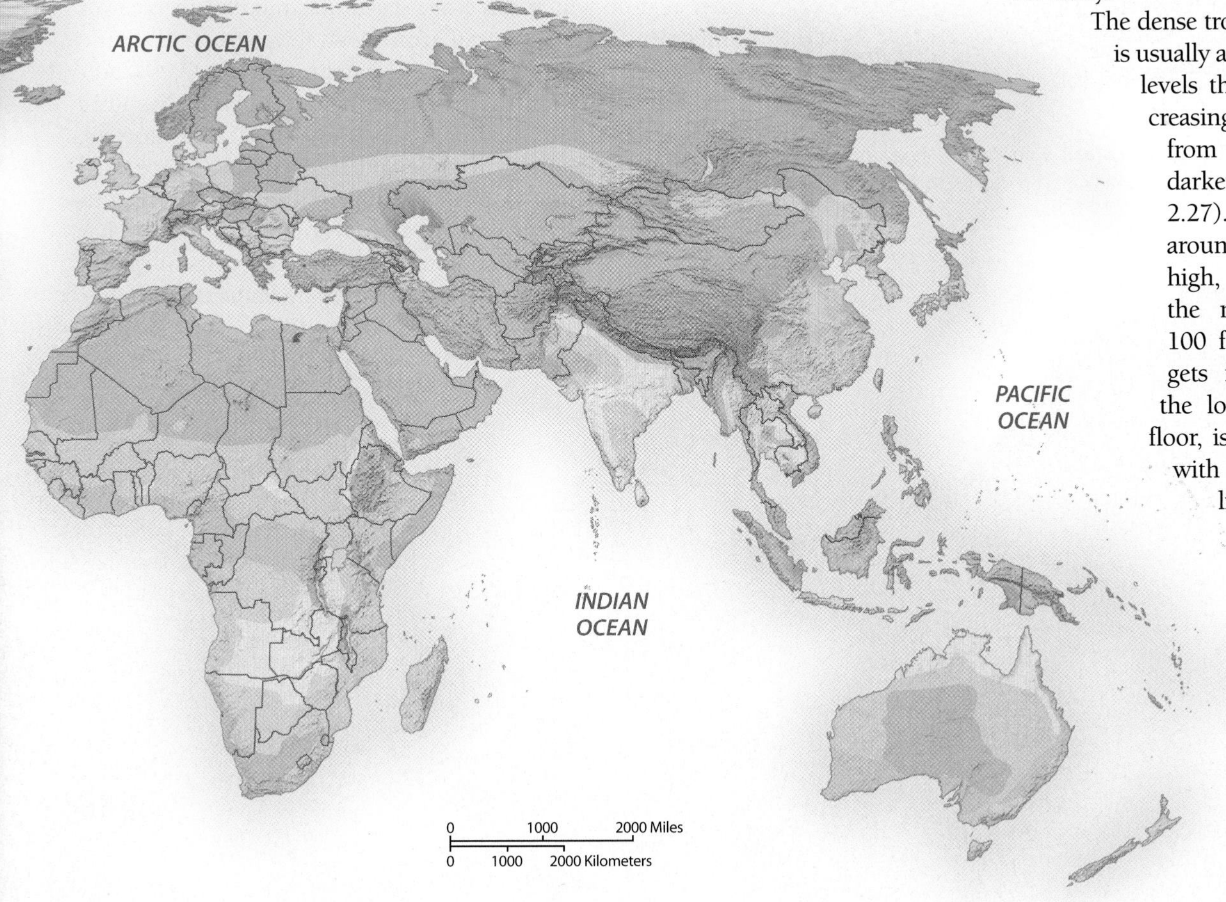

◀ **Figure 2.28 Tropical Seasonal Forest** Unlike the true tropical rainforest that receives rainfall throughout the year, the tropical seasonal forest has a distinct dry season, during which many of the trees and bushes lose their leaves and hibernate until the rains return. This photo is from the Panna National Park in India.

Tropical Seasonal Forests

Bracketing the true tropical rainforest, north and south of the equator, are the tropical seasonal forests, found in those climates that have a distinct dry season, typically four months or so. This is the Am or monsoonal climate region. Because of seasonal droughts, rainfall is only about half of that found in the true rainforest, resulting in a more open landscape with trees spaced farther apart than where rain falls throughout the year (Figure 2.28). Additionally, seasonal forest trees are often **deciduous**, meaning they shed their leaves during the harsh dry season in order to slow or completely halt growth. With a pronounced dry season, this bioregion is also subject to regular fires, both natural and human-set, a process that thins the forest even further by killing young trees and expanding the grass cover. Humans often set these fires to make this bioregion more suitable for grazing animals and farming.

Tropical Savannas

The **tropical savanna** bioregions are found in the Aw climate region, which has a dry season lasting half the year or longer. In these equatorial areas, even fewer trees are found in a landscape dominated by grasslands (Figure 2.29). Annual rainfall amounts are commonly around 50 inches (125 cm), which is a third of what falls in the tropical rainforests. As is true for the tropical seasonal forests, humans have used fire to expand the tropical savanna and make it more suitable for grazing and farming. This change has been so extensive that ecologists are hard-pressed to differentiate between a *climatic*

▼ **Figure 2.29 Tropical Savanna in Brazil** With a much longer dry season, tropical savanna has fewer trees and more open grasslands compared to seasonal forests.

savanna, created by natural climatic conditions, and a *derived savanna,* produced by human manipulation.

Deforestation of Tropical Forests

Tropical forests are being devastated at an unprecedented rate, creating a crisis that tests our political, economic, and ethical will. Although deforestation rates differ from region to region, each year an area about the size of Wisconsin or Pennsylvania is denuded of trees. Geographically, almost half of this activity is in the Amazon Basin of South America; however, deforestation appears to be occurring faster in Southeast Asia, where reportedly logging is taking place at rates three times faster than in the Amazon (Figure 2.30).

Besides destroying the habitat of crucial and highly diverse plants and animals, the other major concern is that tropical deforestation releases huge amounts of CO_2 into the atmosphere. Current estimates suggest that fully 20 percent of all human-caused GHGs emissions result from cutting and burning tropical forests.

Driving tropical deforestation is the recent globalization of commerce in international wood products. Currently, about one-half of all tropical forest timber is destined for China, where much of the wood is used for throwaway items such as chopsticks and newspapers. Another factor contributing to the rapid destruction of tropical rainforests is the world's seemingly insatiable appetite for beef. Cattle species originally bred to survive in the hot weather of India are now raised on grassland pastures created worldwide by converting tropical forests and savannas into rangeland (see *Everyday Globalization: The Hamburger Connection Revisited*).

More recently, tropical forests have been cleared away to make room for palm oil plantations in response to a growing demand for this popular of cooking oil. Also, tropical forest areas are often thought of as a settlement solution for rapidly growing populations of developing countries. In these places, national policies allow settlers to clear and homestead tropical forestlands as a way (real or imagined) of relieving pressure on overcrowded cities.

▲ **Figure 2.30 Tropical Forest Destruction in a new cacao plantation on Borneo Island, East Malaysia** There are many ill effects resulting from tropical rainforest destruction, including the loss of plants and animals, the loss of homeland for native peoples, and the release of large amounts of global-warming GHGs.

Deserts and Grasslands

Poleward of the tropics lie large areas of arid and semiarid climates (BW and BS climate regions) where the world's extensive deserts and steppe grasslands are found. In fact, fully one-third of Earth's land area qualifies as either true desert, where annual rainfall is less than 5 inches (13 cm), or steppe, where the yearly rainfall is between 5 and 10 inches (13 and 25 cm).

Grassy plants appear in these semiarid areas, forming a lush cover during the wet season. In North America, the midsections of both Canada and the United States are covered by grasslands known as **prairie**, characterized by thick, long grasses. In other parts of the world, such as Central Asia, Russia, and Southwest Asia, shorter, less dense grasslands form **steppes**.

The boundary between desert and grassland has always varied naturally because of fluctuations in rainfall. During wet periods, grasslands might expand, only to contract once again during drier years. The transition zone between the two is a precarious environment for people, as the United States learned during the 1930s, when the semiarid grasslands of the western prairie turned into the notorious "Dust Bowl." At that time, thousands of farmers watched their fields become devastated by wind erosion and drought—a disaster that led to an exodus of people from these once-productive lands. Farming these semiarid lands may lead to **desertification**, which is the creation of truly arid desert lands in what were formerly grasslands (Figure 2.31). This has happened on a large scale throughout the world—in Africa, Australia, Asia, and North America.

Mediterranean Shrubs and Woodlands

In the Mediterranean climate regions (Cs), where annual precipitation exceeds 10 inches (25 cm), a prolonged summer-season drought of three or four months produces a unique array of grasses, shrubs, and trees. This is the Mediterranean shrub and woodland bioregion, found in only five areas of the world—around the Mediterranean Sea; in parts of South Africa, Australia, Chile, and along the U.S. West Coast (mainly in California, but also extending into southern Oregon). In these bioregions, the typical landscape is a combination of grasslands; low, drought-resistant shrubs (called *chaparral* in Spain and California and *maquis* in southern France); and

Everyday Globalization

The Hamburger Connection Revisited

Did your fast-food hamburger come from the rainforest? Or, if you don't eat meat, might your tofu burger have started life on cutover lands in the Brazilian tropics? The reason for these questions is that the major cause of destruction in Brazil's rainforest is clearing land to create cattle pastures or to grow soybeans. (Figure 2.3.1).

The Hamburger Connection Back in 1981, as public awareness was growing about tropical rainforest issues, Norman Myers, a resource ecologist at Oxford University, wrote a compelling article documenting how North America's hamburger hunger was responsible for rainforest destruction. In what became widely known as the "the hamburger connection," Myers reported that much of Central America's tropical forest was cleared to create pasture for livestock that became hamburger meat in North American fast-food outlets. Public response was immediate and intense, with widespread boycotts and picketing of fast-food eateries.

▼ **Figure 2.3.1 Cattle in the Brazilian Rainforest** Zebu cattle, a South Asian breed suited to the heat of tropical rainforests, graze on a new pasture created by cutting away and burning rainforest vegetation. Burning the vegetation enriches the soil with nutrients stored in the tree trucks, but also adds considerable amounts of greenhouse gases to the atmosphere.

Fast Food Today But that was decades ago. What about today? Unfortunately, a recent report from Brazil's Center for International Forestry Research suggests strongly that the hamburger connection is still alive and well, since Brazil is now the world's largest exporter of beef, most of which originates on pastures carved out of the Amazon rainforest. Brazil also exports large amounts of soybeans, which are also grown on freshly cut rainforest tracts. Some of those exports come here. Despite millions of cattle on U.S. pastures and in U.S. feedlots and despite the fact that the United States exports beef to Japan, South Korea, and Europe (generally, the United States exports the choicer and more expensive cuts of beef), the United States does import some beef. Most of these meat imports come from Australia and Canada, but also from New Zealand, Mexico, and Brazil.

Does the beef that we import today come from cut-down rainforest? Probably, in the case of imports from Brazil (about 5 percent of total U.S. beef imports) and quite possibly for meat from Mexico (about 12 percent); however, it is far less likely for beef coming from Australia (about 31 percent) and very unlikely for beef from Canada and New Zealand, since those countries don't have any tropical rainforest. Overall, then, the chance of your fast-food burger coming from rainforest pastures is pretty small. Be careful, though, if you stop for a hamburger in Mexico City, Rio de Janeiro, Beijing, or Tokyo, since fast-food outlets in countries other than the United States are reportedly far less picky about sourcing their hamburger.

What about tofu and other soybean products—do they come from the rainforest? Probably not, since the United States grows huge amounts of soybeans (mainly to feed livestock). Brazil is indeed a major soybean exporter, but most of its export soybeans go to China. Thus, it's highly unlikely that tofu sold in the United States came from rainforest soybeans.

1. What are the causes behind deforestation in Brazil?
2. Compare the rainforest cutting patterns between Novo Progresso and Itaituba, Brazil. Is there a difference between cutting for cattle pasture contrasted to logging for timber? Where is the most cattle pasture?

Google Earth Virtual Tour Video

http://goo.gl/G1XT9Q

woodlands consisting of oak and pine (Figure 2.32). In the late 19th century, Australia's fast-growing eucalyptus trees were introduced into these Mediterranean bioregions and are now so widespread that they completely dominate many summer-dry landscapes throughout the world.

Temperate Deciduous Forests

This bioregion is associated with the widespread temperate C climate, where precipitation falls year-round in amounts of 30–60 inches (75–150 cm), summers are warm, and winters are cold, yet no month averages below freezing. In North America, temperate deciduous forests are the major habitat from the Gulf Coast to New England, as well as in parts of the Midwest; in Europe, they were the natural vegetation from the British isles eastward to Germany, as well as in the more temperate parts of Russia and Asia.

As mentioned earlier, a *deciduous* tree loses its leaves during the harsh dry season to slow metabolic activity. In the temperate deciduous forest, this occurs during the win-

▲ **Figure 2.31 Desertification in China** A worker builds biological barriers using hay to stabilize sand dunes and prevent desertification in the Tengger Desert, Gansu province, China. Located in the arid northwest, the area is surrounded by the encroaching Tengger and Badain Jaran deserts, and this area is one of the major sources of China's sandstorms. The Chinese government has spent billions of dollars trying to prevent further expansion of the desert by introducing vegetation and planting trees.

ter, when sunlight is scarce and temperatures are too cold for the tree's nutrient circulation system. To make up for a period of winter hibernation, deciduous trees usually have large leaves to take full advantage of photosynthesis in the warm season. These leaves drop as winter approaches, often providing a colorful display as the leaves wither and die (Figure 2.33).

Common tree species in temperate deciduous forests are maple, oak, elm, ash, and beech. The interior cellular structures of these broadleaf species make them difficult to mill for lumber; thus, they are grouped into the category of *hardwoods*. This category contrasts with the evergreen, needle-leaf trees (discussed below), called the *softwoods*, which are the preferred lumber species in most areas of the world.

▲ **Figure 2.32 California Coastal Grassland and Shrubs** California's state flower, the poppy, is seen in the foreground of this photo from the Big Sur coastal area, with grassland, chaparral shrubs, and oak woodlands filling in the landscape. This photo was taken two years after a large wildfire had burned the area, illustrating how rapidly Mediterranean vegetation recovers from fire.

Because the fallen tree leaves of the deciduous forest produce a rich soil, this bioregion has been cleared extensively for agriculture in North America and Europe, leaving only remnants in areas that were previously richly forested.

Evergreen Forests

In the colder-winter D climate regions, where temperatures average below freezing for at least one month and annual precipitation is around 40 inches (100 cm), evergreen, needle-leaf trees replace the deciduous broadleaf trees because they are better adapted to cold temperatures and low sunlight. In this bioregion, the dominant tree species are fir, pine, and spruce. However, several deciduous species, such as the cold-loving aspen and alder, are also found among the evergreens, thus providing splashes of autumnal color in the otherwise monotonous tracts of evergreen.

In North America, evergreen, needle-leaf forests are found north of the temperate deciduous forests, from New England to the upper Midwest and then west to the Pacific Coast. They are also found in all mountainous environments of western North America, from the Mexican border northward. In Canada and Alaska, these forests are called **boreal** forests, referring to their near-arctic location. Evergreen forests dominate in continental and northern Europe, as well as in the mountains, and then form vast forest tracts eastward through cold-win-

▲ **Figure 2.33 New England Deciduous Forests** As they prepare to hibernate during the long, cold winter by dropping their leaves, deciduous trees such as these in New Hampshire often put on a colorful display. As winter weather approaches, trees shut down their nutrient systems and starve their foliage, resulting in a variety of colors.

ter Asia, where the Russian word **taiga** is commonly applied (Figure 2.34).

In the needle-leaf forests of western North America, the struggle between timber harvesting and environmental groups remains contentious. Timber interests argue that increased cutting is the only way to meet the high domestic demand for lumber and other wood products, whereas environmental groups are concerned about the protection of habitat for endangered species like the northern spotted owl. This conflict has led the government to close large tracts of evergreen forest to commercial logging (Figure 2.35). Further complicating the future of western forests are global market forces. Many Japanese and Chinese timber firms pay premium prices for logs cut from U.S. and Canadian forests, outbidding domestic firms for these scarce resources. Adding yet another element of controversy, many of these trees are cut from public lands in both Canada and the United States. This aspect of globalization raises troublesome questions about whether North American public forests should be cut to meet the lumber and wood-product needs of foreign countries.

The Siberian *taiga* forest is a resource that may become an attractive source of income for Russia's hard-pressed economy. Some observers argue that, if lumber from the Siberian forests is put on the market for global trade, it will reduce logging pressure on North America's western forests, which could make it easier to enact and enforce comprehensive environmental protection in the United States and Canada. On the other hand, cutting this extensive Siberian forest would release massive amounts of CO_2 and methane into the atmosphere, increasing Russia's global-warming emissions.

Tundra

The **tundra** bioregion appears in two versions: the expansive *arctic tundra* of the far northern hemisphere and, sharing many similar traits, the *alpine tundra* found at high elevations in mountainous areas worldwide. In both cases, the tundra landscape is primarily treeless because

▼ **Figure 2.34 Evergreen Trees** Evergreen needle-leaf trees are adapted to cold temperatures. They have small needles (in contrast to the larger leaves of broadleaf deciduous trees) and are able to retain them year-round (again in contrast to broadleaf deciduous trees that lose their leaves and shut down during the harsh winter season). In terms of evolution, needle-leaf trees are thought to be an older—even archaic—plant form contrasted to deciduous trees. This evergreen needle-leaf forest is in the Austrian Alps.

◄ **Figure 2.35 Clear-Cut Forest** Commercial logging in Washington's Olympic Peninsula has dramatically reshaped this landscape. Throughout the Pacific Northwest, environmental lobbies have successfully restricted logging to protect habitat for endangered species and recreation. **Q: After examining this photo carefully, discuss the cost and benefits—both economically and ecologically—of clear-cutting.**

of a very short growing season lasting only a month or two. Moisture is also limited, with arctic tundra areas annually receiving the equivalent of only 6–10 inches (15–25 cm) of precipitation, which falls mostly as snow. As a result of these severe conditions, the tundra bioregions consist primarily of low shrubs, reindeer moss, sedge, and grasses (Figure 2.36). For a brief time in summer, a colorful flower display breaks the monotony of this otherwise bleak landscape.

Despite the low biotic productivity of the tundra bioregion, it stores vast amounts of methane (as do all boggy environments), which is being increasingly released into the atmosphere because of global warming. Because methane is 20 times more effective than CO_2 at trapping atmospheric heat, the thawing of the tundra is a very worrisome factor of the global-warming picture.

REVIEW

2.15 How and why do the three tropical bioregions differ?

2.16 What are the causes of tropical forest deforestation?

2.17 What is desertification?

▼ **Figure 2.36 Tundra North of the Arctic Circle in Norway** High in the mountains and close to the poles where harsh winter seasons inhibit tree growth, the tundra is dominated by grasses and shrubs adapted to very short summers and very long winters.

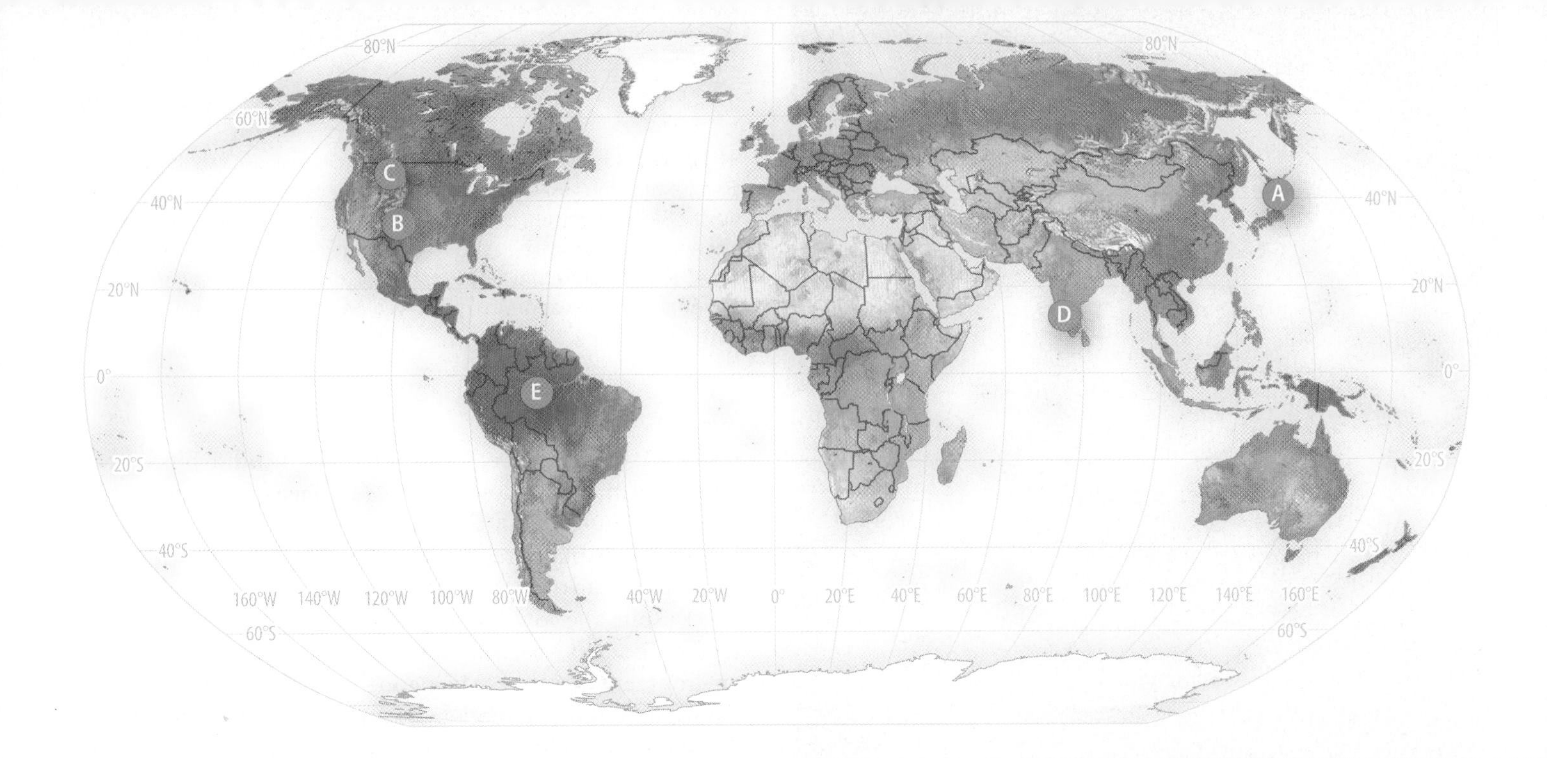

Geology: A Restless Earth

The arrangement of tectonic plates on Earth is responsible for diverse global landscapes. The motion of these plates also causes earthquake and volcanic hazards that threaten the safety of millions of people, particularly in the large cities of North America, Asia, and South America around the Pacific Rim.

2.1 Make a list of cities with populations larger than half a million that are located near tectonic plate boundaries, thus are vulnerable to damage from a major earthquake. Also note those cities where a nearby volcanic explosion could be a hazard.

2.2 Choose one of those cities then go on the Internet to gather information about how that city is reducing its earthquake or volcanic hazard vulnerability in terms of planning and disaster preparation.

Global Climates: Adapting To Change

Current global warming and climate change results from the pollution of the atmosphere by greenhouse gases (GHGs) and is a byproduct of human activities, both past and present. Historically, the developed countries of Europe and North America were the major GHGs producers; today, however, the developing economies of China and India have also become major polluters.

2.3 What causes tornados and where in North America are they most common? How do people and communities prepare for the tornado hazard?

2.4 Make a list of other climate hazards in other parts of the world and then investigate how people in those areas adapt to those threats.

Key Terms

adiabatic lapse rate *(p. 60)*
anthropogenic *(p. 61)*
biodiversity *(p. 72)*
bioregion *(p. 72)*
boreal forest *(p. 79)*
carbon inequity *(p. 64)*
climate *(p. 61)*
climate change *(p. 61)*
climograph *(p. 61)*
colliding plate boundary *(p. 54)*
continentality *(p. 58)*
deciduous forest *(p. 74)*
desertification *(p. 75)*
divergent plate boundary *(p. 54)*
global warming *(p. 61)*
greenhouse effect *(p. 58)*
hydraulic fracturing (fracking) *(p. 68)*
insolation *(p. 58)*
maritime climate *(p. 58)*
non-renewable energy *(p. 66)*
orographic effect *(p. 60)*
plate tectonics *(p. 52)*
prairie *(p. 75)*
proven reserves *(p. 67)*
rain shadow *(p. 60)*
renewable energy *(p. 66)*
steppe *(p. 75)*
subduction zone *(p. 54)*
taiga *(p. 79)*
transform fault *(p. 54)*
tropical savanna *(p. 74)*
tundra *(p. 79)*
water stress *(p. 71)*
weather *(p. 60)*

In Review • Chapter 2
Physical Geography and the Environment

Global Energy: The Essential Resource

Fossil fuels (oil, coal, and natural gas) dominate the world's energy picture, with renewable energy (wind, solar, hydro, and biomass) currently providing only a fraction of the world's energy needs. Energy demands from China, India, and other developing economies will continue to rise, thereby worsening global warming pollution.

2.5 What are the possible environmental effects from fracking in this area? Compare those to the possible benefits for landowners and municipalities. Which is greater?

C

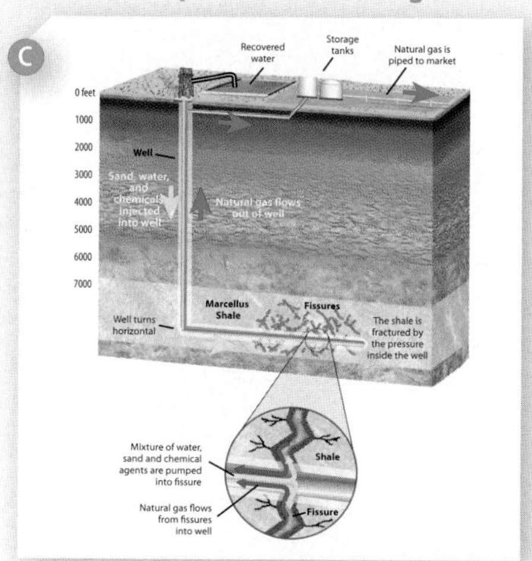

2.6 Currently the United States relies on vast amounts of imported oil to meets its energy needs. While some say that fracking could end this dependence of foreign oil, others say not. Acquaint yourself with this debate, and then make a case for one side or the others.

Water: A Scarce Resource

Water, a necessity for all life, is becoming an increasingly scarce resource in the world and will cause serious water stress problems in Sub-Saharan Africa, Southwest Asia, and western North America.

D

2.7 In what climate region is this dam and reservoir? Given that climate, why are there water problems in this area?

2.8 Consult the Internet to become acquainted with the controversy over large dams in South Asia by listing their benefits and liabilities for different groups of people (city dwellers, large farmers, subsistence farmers, etc).

Bioregions and Biodiversity: The Globalization of Nature

Plants and animals throughout the world face an extinction crisis because of habitat destruction from a variety of human activities. Tropical forests are a focus of these problems because they contain the most plant and animal species of any bioregion, yet they are threatened by the world's demands for wood products, cattle ranching, and resettlement of expanding human populations.

E

2.9 Describe the different bioregions shown on this aerial photo of the northeastern Amazon in Brazil.

2.10 Discuss the specific human activities that might be changing these bioregions and what effects these might have on specific plants and animals.

MasteringGeography™

Looking for additional review and test prep materials? Visit the Study Area in **MasteringGeography**™ to enhance your geographic literacy, spatial reasoning skills, and understanding of this chapter's content by accessing a variety of resources, including **MapMaster**™ interactive maps, videos, RSS feeds, flashcards, web links, self-study quizzes, and an eText version of *Diversity Amid Globalization*.

Authors' Blogs

Scan to visit the author's blog for chapter updates

www.gad4blog.wordpress.com

Scan to visit the GeoCurrents blog

www.geocurrents.info

3 North America

Harnessing the Winds of Texas. The high plains of North Texas are one of the continent's best locations for harnessing low-cost wind energy. The Brazos Wind Farm is co-owned by Mitsui Tokyo (Japan) and Shell Wind Energy (the Netherlands).

Physical Geography and Environmental Issues

Stretching from Texas to the Yukon, the North American region is home to an enormously varied natural setting and to an environment that has been extensively modified by human settlement and economic development.

Population and Settlement

Settlement patterns in North American cities reflect the diverse needs of an affluent, highly mobile population. The region's sprawling suburbs are designed around automobile travel and mass consumption, whereas many traditional city centers struggle to redefine their role within the decentralized metropolis.

Cultural Coherence and Diversity

Cultural pluralism remains strong in North America. Currently, more than 46 million immigrants live in the region, more than double the total in 1990. The tremendous growth in the numbers of Hispanic and Asian immigrants since 1970 has fundamentally reshaped the region's cultural geography.

Geopolitical Framework

Cultural pluralism continues to shape political geographies in the region. Immigration policy remains hotly contested in the United States, and Canadians confront persistent regional and native peoples' rights issues.

Economic and Social Development

North America's economy continues to recover from the harsh economic downturn of 2007–2010. Enduring examples of poverty and many social issues related to gender equity, aging, and health care challenge both Canada and the United States in the second decade of the 21st century.

It is a long and lonely drive from Lubbock, Texas, to rural Borden County (population 641), 50 miles south. This is a land of endless horizons and isolated farms and ranches. The local high school is so small that it has trouble fielding a six-man football team. These high and windswept prairies of north-central Texas seem a world away from North America's globalized economy. But appearances can be deceiving. Even in Borden County, global capital is reshaping the landscape and creating new economic opportunities. Change is (literally) in the air: Since the 2003 completion of the Brazos Wind Farm, Borden County (and neighboring areas) now produces enough electricity (through a network of 160 one-megawatt turbines) to power 48,000 Texas homes. This setting marks the first place in the United States where Mitsui Tokyo, a Japanese multinational company, has invested in wind power. It manufactured the turbines and owns half of the project. The other half of the wind farm reflects investments from another part of the world: Dutch-owned Shell Wind Energy partners with Mitsui to make sure the turbines keep on turning.

By any measure of multinational corporate investment and global trade, the region plays a dominant role that far outweighs its population of 350 million residents.

The Brazos Wind Farm reminds us that globalization, often in subtle ways, has fundamentally refashioned many portions of the North American landscape. A walk down any busy street in Toronto, Tucson, or Toledo reveals how international products, foods, culture, and economic connections shape the everyday scene. Large foreign-born populations also provide direct links to every part of the world. Tourism brings in millions of additional foreign visitors and billions of dollars, which are spent everywhere from Las Vegas to Disney World. In more subtle ways, North Americans see globalization in their everyday lives. They eat ethnic foods, enjoy the sounds of salsa and Senegalese music, and surf the Internet from one continent to the next.

Globalization is a two-way street, and North American capital, culture, and power are ubiquitous. By any measure of multinational corporate investment and global trade, the region plays a dominant role that far outweighs its population of 350 million residents. North American consumer goods, information technology, and investment capital circle the globe. In addition, North American foods and popular culture are diffusing globally at a rapid pace. North American music, cinema, and fashion have also spread rapidly around the world. Soaring downtown skylines from Mumbai to Beijing increasingly resemble their North American counterparts.

The North American region encompasses the United States and Canada, a culturally diverse and resource-rich region that has seen unparalleled human modification and economic development over the past two centuries (Figure 3.1). The result is one of the world's most affluent regions, where two highly urbanized, mobile, and rapidly globalizing societies have the highest rates of resource consumption on Earth. Indeed, the region superbly exemplifies a **postindustrial economy** that is shaped by modern technology, by innovative financial and information services, and by a popular culture that dominates both North America and the world beyond.

Politically, North America is home to the United States, the last remaining global superpower. Such status brings the country onto center stage in times of global tensions, whether they are in the Middle East, South Asia, or West Africa. In addition, North America's largest metropolitan area of New York City (22 million people) is home to the United Nations (UN) and other global political and financial institutions.

North of the United States, Canada is the other political unit within the region. Although slightly larger in area than the United States (3.83 million square miles [9.97 million square km] versus 3.68 million square miles [9.36 million square km]), Canada's population is only about 10 percent that of the United States.

The United States and Canada are commonly referred to as "North America," but that regional terminology can be confusing. Some geography textbooks call the region "Anglo America" because of its close and abiding connections with

LEARNING OBJECTIVES

After reading this chapter you should be able to:

- Describe North America's major landforms and climate regions.
- Identify key environmental issues facing North Americans in the 21st century and describe how these relate to the region's resource base and economic development.
- Explain the major ways in which people have modified the North American environment.
- Summarize the three most important periods of European settlement in North America.
- Identify major migration flows in North American history.
- Explain the processes that shape contemporary urban and rural settlement patterns.
- List the five phases of immigration shaping North America and describe the recent importance of Hispanic and Asian immigration.
- Provide examples of how cultural globalization has shaped the region.
- Describe how the United States and Canada developed distinctive federal political systems and identify each nation's current political challenges.
- Describe the role of key factors in explaining why economic activities are located where they are in North America.

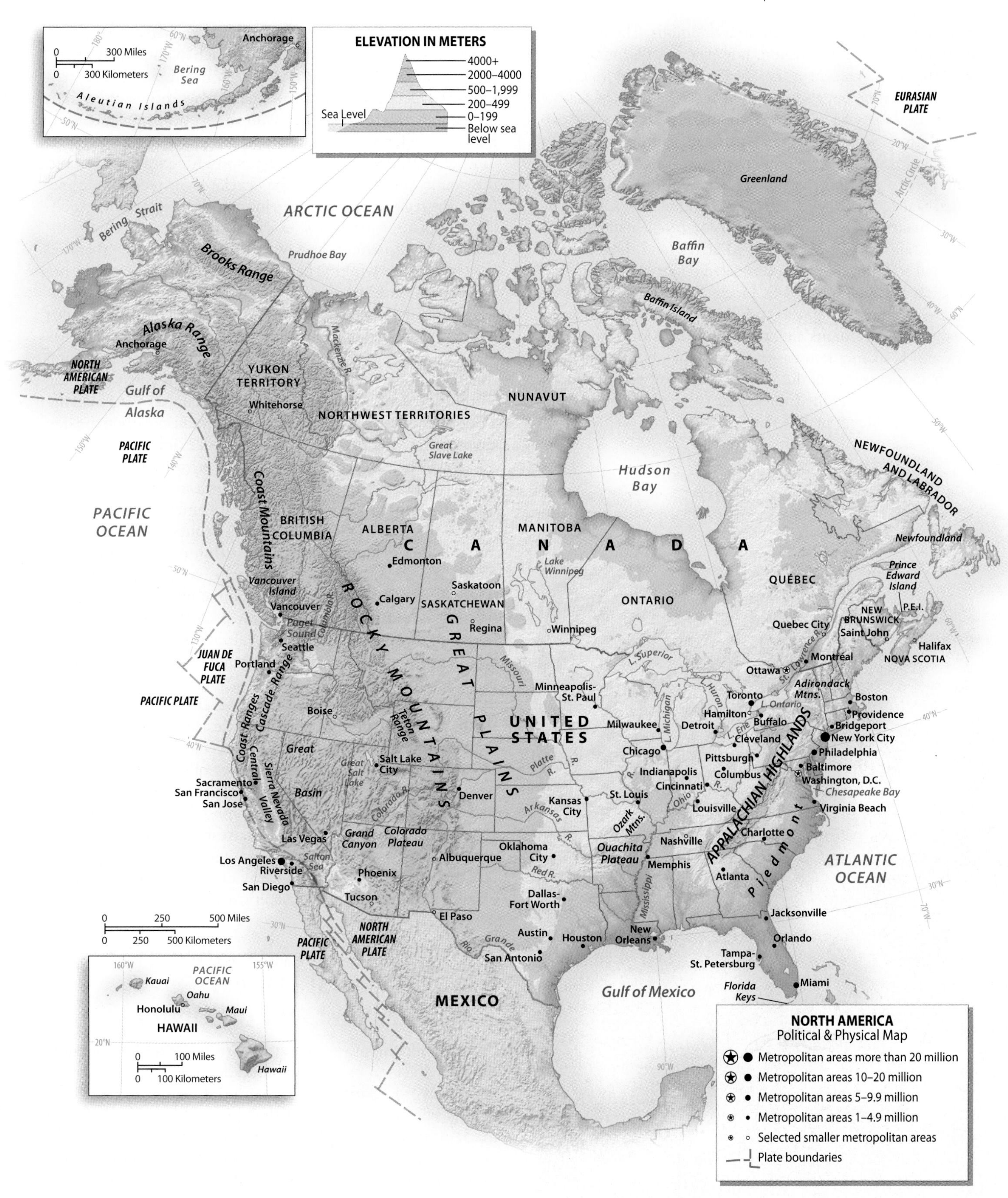

▲ **Figure 3.1 North America** North America plays a pivotal role in globalization. The region also contains one of the world's most highly urbanized and culturally diverse populations. With 350 million people and extensive economic development, North America is also one of the largest consumers of natural resources on the planet.

Britain and its Anglo-Saxon cultural traditions. The increasing cultural diversity of the region, however, has discouraged the use of this term in recent years. The term *North America* is more culturally neutral, but also has problems. As a physical feature, the North American continent commonly includes Mexico, Central America, and often the Caribbean. Culturally, however, the United States–Mexico border seems a better dividing line, although the growing Hispanic presence in the southwestern United States, as well as ever-closer economic links across the border, makes even that regional division problematic. In addition, Hawaii is a part of the United States (and thus included in Chapter 3), but it is also considered a part of Oceania (and thus discussed in Chapter 14). Finally, Greenland (population 56,000), which often appears on the North American map, is actually an autonomous country within the Kingdom of Denmark and is mainly known for its valuable, but diminishing, ice cap. Our coverage of the "North American" region concentrates on Canada and the United States, two of the world's largest and most affluent nation-states.

▼ **Figure 3.2 Toronto's Caribana Parade** Toronto's varied cultural mix is celebrated during the annual Caribana Parade through the city. Powerful forces of globalization have reshaped the cultural and economic geographies of dozens of North American cities.

Widespread abundance and affluence characterize North America. The region is extraordinarily rich in natural resources, such as navigable waterways, good farmland, fossil fuels, and industrial metals. The modern scene displays the results of combining that rich resource base with the acquisitive nature of its European colonizers and an accelerating pace of technological innovation. Indeed, contemporary North America displays both the bounty and the price of the development process. On one hand, the region shares the benefits of modern agriculture, globally competitive industries, excellent transport and communications infrastructures, and two of the most highly urbanized societies in the world. The cost of abundance, however, has been high: European settlers all but eliminated native populations, and later populations logged forests, converted grasslands into farms, eroded precious soils, threatened numerous species with extinction, diverted great rivers, and often wasted natural resources. Today, although home to only about 5 percent of the world's population, the region consumes about 25 percent of the world's commercial energy budget and produces carbon dioxide emissions at a per person rate more than 12 times that of India.

Nevertheless, economic growth has vastly improved the standard of living for many North Americans, who enjoy high rates of consumption and varied urban amenities that are the envy of the developing world. Internet connections, sushi restaurants, and shopping malls are within easy reach of most North American residents. Amid this material abundance, however, are persistent disparities in income and in the quality of life. Poor rural and inner-city populations still struggle to match the affluence of their wealthier neighbors, and poverty continues, despite the unprecedented economic growth of the second half of the 20th century.

North America's unique cultural character also defines the region. The cultural characteristics that hold this region together include a common process of colonization, a heritage of Anglo dominance, and a shared set of civic beliefs in representative democracy and individual freedom. However, the history of the region has also juxtaposed Native Americans, Europeans, Africans, and Asians in fresh ways, and the results are two societies unlike any other (Figure 3.2). Adding to the mix is a popular culture that today exerts a powerful homogenizing influence on North American society.

Physical Geography and Environmental Issues: A Vulnerable Land of Plenty

North America's physical and human geographies are enormously diverse. In the past decade, the region has also witnessed a dizzying array of natural disasters and environmental hazards that suggest the close connections between the region's complex physical setting and its human population. In 2012, for example, a widespread drought gripped much of the North American interior in the spring and summer; the record-setting heat had major implications for farmers and food prices. In the fall, much of the East Coast (with major impacts from Virginia to New England)

▲ **Figure 3.3 Coastal Damage from Hurricane Sandy** When Hurricane Sandy made landfall in New Jersey in October, 2012, the storm inflicted heavy damage on this portion of the Atlantic City boardwalk.

was slammed by Hurricane Sandy (Figure 3.3). The so-called superstorm (formed by the merging of tropical and midlatitude storm systems) rearranged the coastline of New Jersey, flooded lower Manhattan, and produced $50–$60 billion in damage. For millions of coastal residents, Sandy was yet another reminder (along with Katrina in the Gulf of Mexico in 2005) that such settings are extraordinarily vulnerable to natural disasters, especially in a time of increasingly dense coastal development and the looming possibilities of global climate change.

A Diverse Physical Setting

North America's complex landscape is dominated by vast interior lowlands bordered by more mountainous topography in the western portion of the region (see Figure 3.1). In the eastern United States, extensive coastal plains stretch from southern New York to Texas and include a sizable portion of the lower Mississippi Valley. The Atlantic coastline is complex and is made up of drowned river valleys, bays, swamps, and low barrier islands (Figure 3.4). The nearby Piedmont region, which is a transition zone between nearly flat lowlands and steep mountain slopes, consists of rolling hills and low mountains that are much older and less easily eroded than the lowlands. West and north of the Piedmont are the Appalachian Highlands, an internally complex zone of higher and rougher country, reaching altitudes from 3000 to 6000 feet (915 to 1830 meters). Farther to the southwest, Missouri's Ozark Mountains and the Ouachita Plateau of northern Arkansas resemble portions of the southern Appalachians. Much of the North American interior is a vast lowland extending east–west from the Ohio River Valley across the Great Plains and north–south from west central Canada to the coastal lowlands near

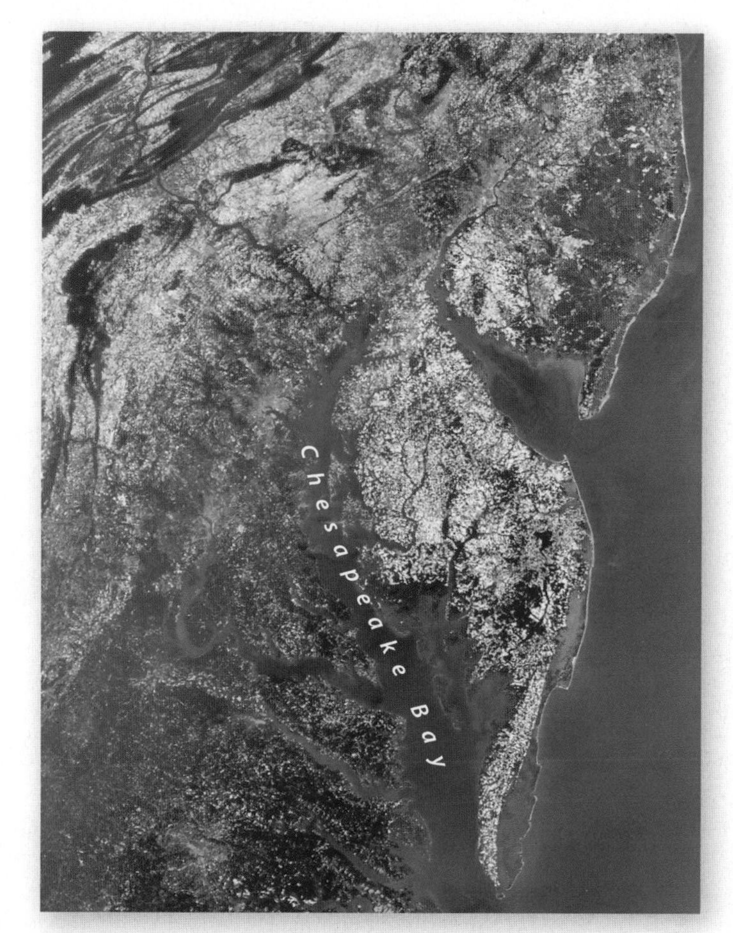

▲ **Figure 3.4 Satellite Image of Chesapeake Bay** This view of the Mid-Atlantic coast reveals the intricate and vulnerable low-lying shoreline of the Chesapeake Bay (lower center). The coast sector is characterized by drowned river valleys, barrier islands, and sandy beaches. The Piedmont zone and Appalachian Highlands appear to the northwest.

the Gulf of Mexico. Glacial forces, particularly north of the Ohio and Missouri rivers, have actively carved and reshaped the landscapes of this lowland zone.

In the West, mountain-building (including large earthquakes and volcanic eruptions), alpine glaciation, and erosion produce a visually spectacular regional topography quite unlike that of eastern North America. The Rocky Mountains reach more than 10,000 feet (3000 meters) in height and stretch from Alaska's Brooks Range to northern New Mexico's Sangre de Cristo Mountains (Figure 3.5). West of the Rockies, the Colorado Plateau is characterized by highly colorful sedimentary rock eroded into spectacular buttes and mesas. Nevada's sparsely settled basin and range country features north–south-trending mountain ranges alternating with structural basins with no outlet to the sea. North America's western border is marked by the mountainous and rain-drenched coasts of southeast Alaska and British Columbia; the Coast Ranges of Washington, Oregon, and California; the lowlands of the Puget Sound (Washington), Willamette Valley (Oregon), and Central Valley (California); and the complex uplifts of the Cascade Range and Sierra Nevada.

Patterns of Climate and Vegetation

North America's climates and vegetation are highly diverse, mainly due to the region's size, latitudinal range, and varied terrain (Figure 3.6). Much of eastern North America south of the Great Lakes is characterized by a long growing season, 30–60 inches (75–150 cm) of precipitation annually, and a deciduous broadleaf forest (later cut down and replaced by crops). From the Great Lakes north, the coniferous evergreen or **boreal forest** dominates the continental interior. Near Hudson Bay and across harsher northern tracts, trees give way to **tundra**, a mixture of low shrubs, grasses, and flowering herbs that grow briefly in the short growing seasons of the high latitudes.

Drier continental climates found from west Texas to Alberta feature large seasonal ranges in temperature and unpredictable precipitation that averages 10–30 inches (25–75 cm) annually (see climographs for Dallas and Cheyenne in Figure 3.6). The soils of much of this region are fertile and originally supported **prairie** vegetation, dominated by tall grasslands in the East and by short grasses and scrub vegetation in the West. Western North American climates and vegetation are greatly complicated by the region's many mountain ranges. The Rocky Mountains and the intermontane interior experience the typical seasonal variations of the middle latitudes, but patterns of climate and vegetation are greatly modified by the effects of topography. Farther west, marine west coast climates dominate north of San Francisco, whereas a dry-summer Mediterranean climate occurs across central and southern California. (Compare the climographs for Los Angeles and Vancouver in Figure 3.6.)

The Costs of Human Modification

North Americans have modified their physical setting in many ways. Processes of globalization and accelerated urban and economic growth have transformed North America's landforms, soils, vegetation, and climate. Globalization has brought many benefits to North America, but with the accompanying urbanization, industrialization, and heightened consumption, the region is also paying an environmental price for its affluence. Indeed, problems such as acid rain, nuclear waste storage, groundwater depletion, and toxic

▼ **Figure 3.5 Rocky Mountains** This spectacular lake in Alberta's Banff National Park reveals the characteristic signatures of alpine glaciation that are found in many portions of the Rocky Mountain region, both in the United States and in Canada.

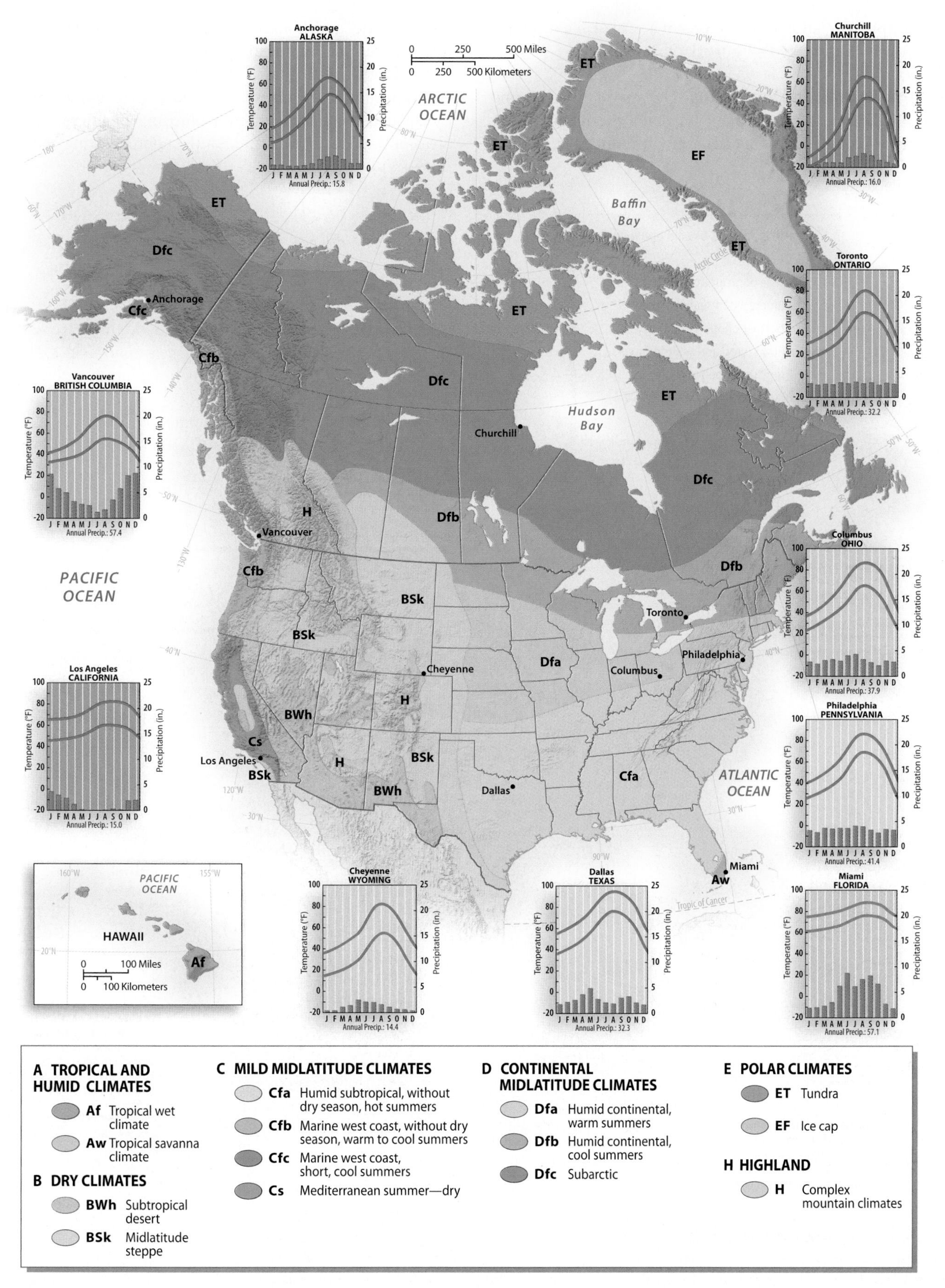

▲ **Figure 3.6 Climate of North America** North American climates include everything from tropical savanna (Aw) to tundra (ET) environments. Most of the region's best farmland and densest settlements lie in the mild (C) or continental (D) midlatitude climate zones.

chemical spills are all manifestations of a way of life unimaginable only a century ago (Figure 3.7).

Transforming Soils and Vegetation The arrival of Europeans on the North American continent impacted the region's flora and fauna as countless new species were introduced, including wheat, cattle, and horses. As the number of settlers increased, forest cover was removed from millions of acres. Grasslands were plowed under and replaced with grain and forage crops not native to the region. Widespread soil erosion was increased by unsustainable farming and ranching practices, and many areas of the Great Plains and South suffered lasting damage.

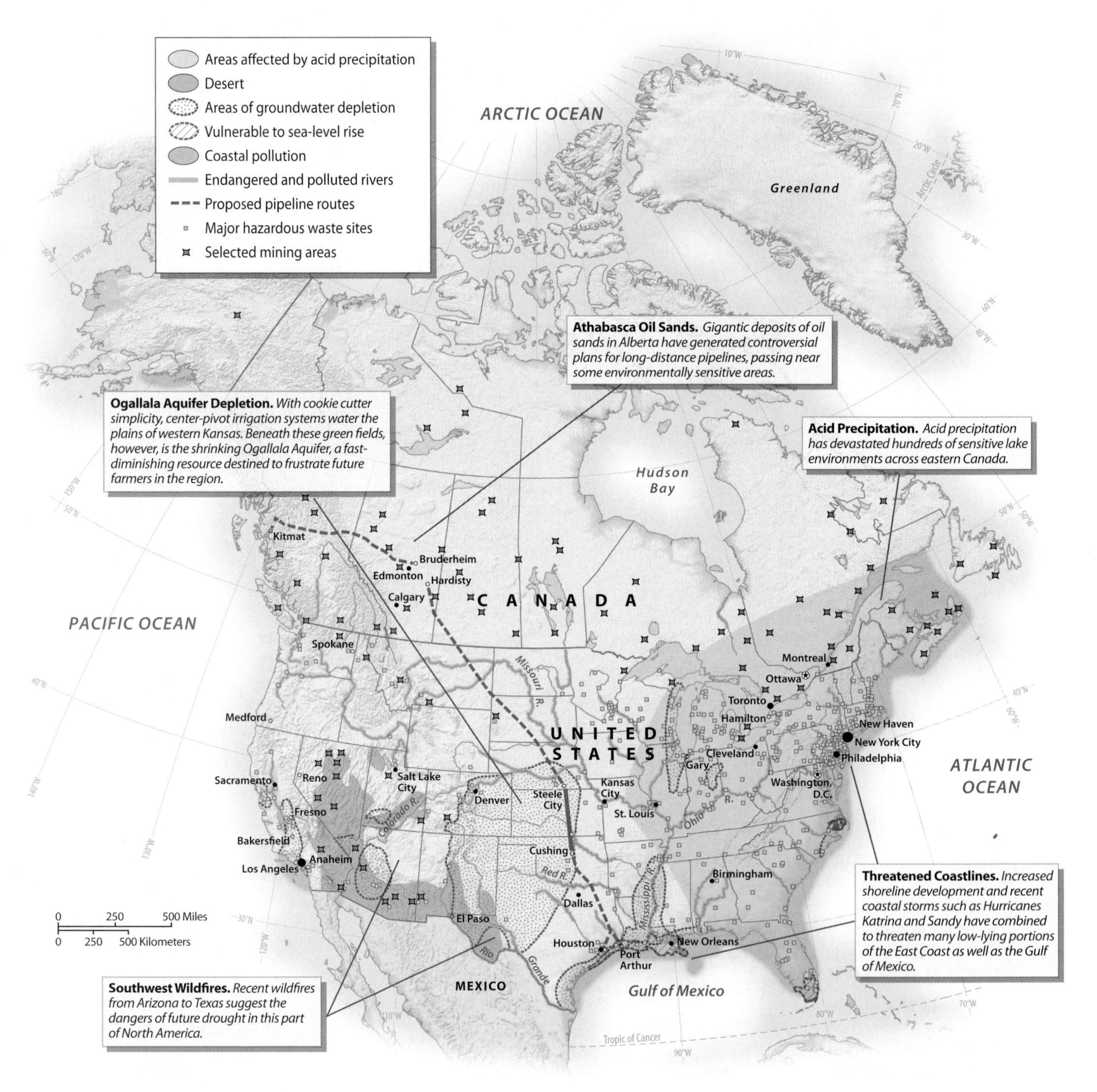

▲ **Figure 3.7 Environmental Issues in North America** Many environmental issues threaten North America. Acid rain damage is widespread in regions downwind from industrial source areas. Elsewhere, widespread water pollution, cities with high levels of air pollution, and zones of accelerating groundwater depletion pose health dangers and economic costs to residents of the region. Since 1970, however, both Americans and Canadians have become increasingly responsive to the dangers posed by these environmental challenges.

Managing Water North Americans also consume huge amounts of water. In the United States, although conservation efforts and technology have slightly reduced per capita rates of water use over the past 25 years, city dwellers still use an average of more than 175 gallons daily. Indirectly, through other forms of food and industrial consumption, Americans may consume more than 1400 gallons of water per day. Many places in North America are threatened by water shortages. Metropolitan areas such as New York City struggle with outdated municipal water-supply systems.

Beneath the Great Plains, the waters of the Ogallala Aquifer are being depleted. Central-pivot irrigation systems have steadily lowered water tables across much of the region by as much as 100 feet (30 meters) in the past 50 years. The costs of pumping are rising steadily, and 50 percent of the area's irrigated land may see wells run dry by 2020. Farther west, California's complex system of water management is a reminder of that state's large demands within a setting that remains prone to periodic drought.

Water quality is also a major issue. North Americans are exposed to water pollution every day, and even environmental laws and guidelines, such as the U.S. Clean Water Act or Canada's Green Plan, have not eliminated the problem. In 2010, the *Deepwater Horizon* rig explosion and leaking oil well in the Gulf of Mexico quickly became one of North America's greatest environmental disasters. The inability of British Petroleum (BP) and the U.S. government to quickly control the flow of oil allowed tens of millions of gallons to escape into the Gulf, damaging sea and bird life, sensitive coastal ecosystems, and the region's resource and tourist economies. The full environmental and human consequences of the spill will take years to assess. Meanwhile, the federal government has slowed the pace of granting new offshore drilling permits, while onshore leasing activities have expanded.

Elsewhere, North America's varied fisheries illustrate the complexities of continental water-resource management. The case of the Asian carp suggests how unpredictable the process can be. Imported from East Asia by southern catfish farmers in the 1970s, the carp removed algae and suspended matter from ponds in the lower Mississippi Valley. But many carp escaped their pens during large floods in the early 1990s. Twenty years later, carp have migrated northward and are now knocking on the door of the Great Lakes as they discover artificial ship canals that link the lakes with the Mississippi system. The Army Corps of Engineers has spent millions to prevent an invasion that could decimate the Great Lakes fishing industry.

Eastern Canada's cod fisheries, British Columbia's coastal salmon industry, and many of Alaska's freshwater and saltwater fishing grounds have all seen their annual catches decline dramatically in recent years. Overfishing, deadly infections from commercial fish-farming operations, and global climate shifts have all been blamed for the declines. New England fisheries have had steep quotas imposed on their catch so that fish stocks can grow in the Georges Bank and Gulf of Maine areas. Data released in 2009 by the United States Geological Survey (U.S.G.S.) are also troubling. Fish sampled in 291 streams in the United States revealed widespread mercury contamination in the nation's freshwater fish population, much of it related to coal-burning power plants.

Altering the Atmosphere North Americans modify the very air they breathe; in doing so, they change local and regional climates, as well as the chemical composition of the atmosphere. For example, built-up metropolitan areas create **urban heat islands**, in which development associated with cities often produces nighttime temperatures some 9 to 14°F (5 to 8°C) warmer than those of nearby rural areas. At the local level, industries, utilities, and automobiles contribute carbon monoxide, sulfur, nitrogen oxides, hydrocarbons, and particulates to the urban atmosphere. Some of the region's worst offenders are U.S. cities such as Houston and Los Angeles, but Canadian cities such as Toronto, Hamilton, and Edmonton also experience significant problems of air quality. In 2009, the U.S. Environmental Protection Agency (EPA) issued a report identifying almost 600 urban American neighborhoods in which elevated cancer rates were associated with air pollution. High-impact localities such as Los Angeles or the Illinois suburbs of St. Louis might have pollution-related cancer rates that are 500 times higher than those for rural, clean-air localities in settings such as central Montana.

On a broader scale, North America is plagued by **acid rain**, caused by industrially produced sulfur dioxide and nitrogen oxides in the atmosphere. The excess acidity can damage forests, poison lakes, and kill fish. Many acid rain producers are located in the Midwest and southern Ontario, where industrial plants, power-generating facilities, and motor vehicles contribute atmospheric pollution. Prevailing winds transport the pollutants and deposit damaging acid rain and snow across the Ohio Valley, Appalachia, the northeastern United States, and eastern Canada (see Figure 3.7).

Air pollution is also going global, freely drifting into Canada and the United States on prevailing winds. One recent estimate suggests at least 30 percent of the region's ozone comes from beyond its borders. Both China and Mexico are major contributors of airborne pollutants.

Growing Environmental Awareness

Many environmental initiatives in the United States and Canada have addressed local and regional problems. For example, both public and private initiatives have preserved and restored wildlife habitats across the United States and Canada (see *Working Toward Sustainability: Bison Return to the Northern Great Plains*). Conservation practices also have greatly reduced water- and wind-related soil erosion in North America, although experts estimate that one-third of U.S. and one-fifth of Canadian cropland are still at high to severe risk of future erosion. Tougher air-quality standards have also selectively reduced emissions in many North American cities. Similarly, the U.S. Superfund program (begun in 1980) and Canada's Environmental Protection Act (CEPA, begun in 1988) have significantly cleaned up hundreds of North America's toxic waste sites.

Working Toward **Sustainability**

Bison Return to the Northern Great Plains

It is an intriguing juxtaposition: North America's interior Great Plains, long the haunt of farm towns, cattle herds, and endless wheat fields, are quietly being reimagined in the early 21st century into one of North America's largest wildlife preserves. For residents, this repurposing of the landscape produces both challenges and opportunities.

A Montana Preserve The American Prairie Reserve (APR) is responsible for the creation of this "North American Serengeti" in northeast Montana (Figure 3.1.1). Beginning with its first land purchase in 2004, the APR now owns or leases more than 270,000 acres (110,000 hectares) of grassland, much of it carved out of several large cattle ranches. In addition, the organization has obtained bison grazing privileges on nearby lands of the Charles M. Russell National Wildlife Refuge. As the APR acquires more ranch lands, it removes all of the interior fences, encouraging its growing bison herd (71 more calves arrived from Alberta, Canada, in 2012) to roam freely.

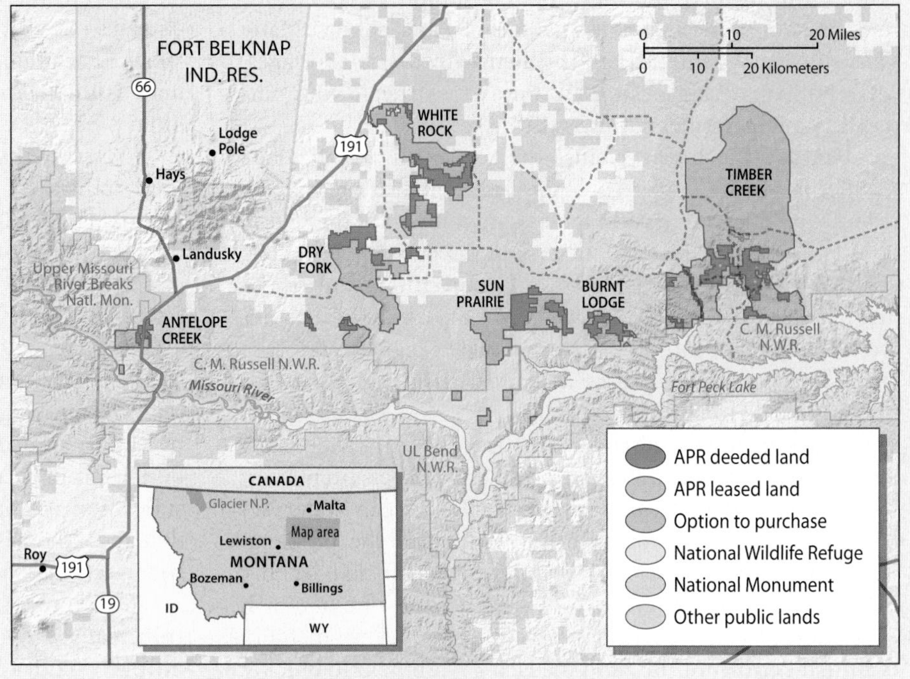

▲ **Figure 3.1.1 American Prairie Reserve, Northeast Montana** North American bison are free to roam on northeast Montana lands owned by the APR. These acres are often combined with nearby public acreage, such as land allotments within the Charles M. Russell Wildlife Refuge.

Ultimately, over the next 10–20 years, the organization hopes to create North America's most ambitious landscape restoration project. Ideally, APR leaders argue for recreating the same prairie landscape observed by Lewis and Clark as they traveled through the region more than 200 years ago (Figure 3.1.2). Then, more than 50 million bison lived on the Great Plains. By 2025, the APR plans to assemble a more modest herd of 5000–10,000 bison on a patchwork of private and public lands that may include more than 3.6 million acres (1.46 million hectares) of northeast Montana prairie.

Divided Local Opinion Local residents are divided on the project. The region's Sioux and Assiniboine peoples have rejoiced over the bison's return, along with conservation groups. But many long-time cattle ranching families (there are 3 million cattle in the state) see it as a threat to their way of life. They wonder if traditional ranching communities can survive if cattle are marginalized in a landscape more oriented around bison. One response is that ecotourism may offer new opportunities as visitors converge on the region to enjoy one of the continent's largest wildlife preserves.

▼ **Figure 3.1.2 Home on the Range** These bison (many have been imported from protected lands in Alberta) south of Malta, Montana, are quickly adjusting to their new home.

Creating a New Habitat Longer term, advocates also note that mid-latitude grasslands are one of the least represented ecosystems in the world's nature reserves. In addition, given the potential for climate change in the region, bigger is better: Large acreages and sizable herds would give land managers more habitat niches to work with as they make potential adjustments for different climate-change scenarios. For now, however, the APR is happy to simply be returning bison to this landscape and cultivating a new appreciation for their enduring value among residents who call the region home.

1. Why are bison seen as powerful symbols of "wild lands" in this part of North America?
2. As a cattle rancher in this area of Montana, write a letter to the local newspaper explaining why it is a bad idea to return bison to nearby prairies.

Google Earth Virtual Tour Video

http://goo.gl/Olczso

Perhaps most important, North Americans are giving increased support for green industries and technologies. Large-scale commercial farming has often been criticized for depleting groundwater supplies, contributing to soil exhaustion, and overusing pesticides and herbicides, which eventually end up elsewhere in the environment. The growing popularity of **sustainable agriculture** exemplifies the trend, where organic farming principles, a limited use of chemicals, and an integrated plan of crop and livestock management combine to offer both producers and consumers environmentally friendly alternatives to large-scale commercial farming.

In the United States, the selective removal of dams also illustrates the trend. For most of its history, the U.S. government supported policies that increased control over the flow of the nation's streams and rivers. To that end, more than 75,000 dams have been built in the country's history, through agencies such as the U.S. Army Corps of Engineers. They support a myriad of flood control, irrigation, and hydroelectric projects. Since the 1970s, however, new environmental and political initiatives have encouraged the reversal of these long-standing policies. The 1973 passage of the Endangered Species Act mandated that federal agencies preserve threatened plants and animals along critical river habitat. The legislation became the political genesis of the dam removal movement. A varied political coalition of environmentalists, fishermen, and Native American groups has increasingly pressured governmental agencies to remove dams that many argue have outlived their usefulness and do more harm than good.

By 2013, more than 1100 dams had been removed nationwide. The world's largest dam removal project recently occurred along the Elwha River in western Washington's Olympic Mountains. Advocates of the project claim that the removal of a pair of dams (built in 1913 and 1929) has restored valuable salmon habitat for 400,000 fish and will provide benefits that far exceed the modest value of the hydroelectric power generated by the dams (Figure 3.8).

These modest successes, dam removal advocates argue, may pave the way for larger initiatives. The Colorado River's Glen Canyon Dam is often cited as the biggest target, but given both the political and the logistical challenges involved in such a mammoth effort, nothing is likely to happen in the near future. Still, the deconstruction of America's dams may be just beginning, and it marks a policy shift intended to restore streams to some of their former environmental vitality.

▲ **Figure 3.8 Dam Along the Elwha River, Washington** This dam along the Elwha River in western Washington's Olympic Mountains was scheduled for removal as part of a plan to restore the watershed's salmon fisheries and expand the area's recreational opportunities.

The Shifting Energy Equation

Energy consumption in the region remains extremely high (the United States is still the source of almost 20 percent of Earth's greenhouse gas emissions). Growing incentives for energy efficiency may produce lower rates of per capita consumption in the future. In addition, since 1970 North Americans have invested billions in alternative and more sustainable forms of energy. The growing technological and economic appeal of **renewable energy sources**, such as hydroelectric, solar, wind, and geothermal, is likely to fundamentally rework North America's patterns of energy use in coming years as policymakers, industrial innovators, and consumers are all attracted to their enduring availability and potentially lower environmental costs (Figure 3.9). In addition, the U.S. Recovery Act, passed in 2009 to stimulate economic activity, targeted approximately $80 billion to renewable energy initiatives.

At the same time, recent evidence also suggests that North America's domestic supplies of oil, natural gas,

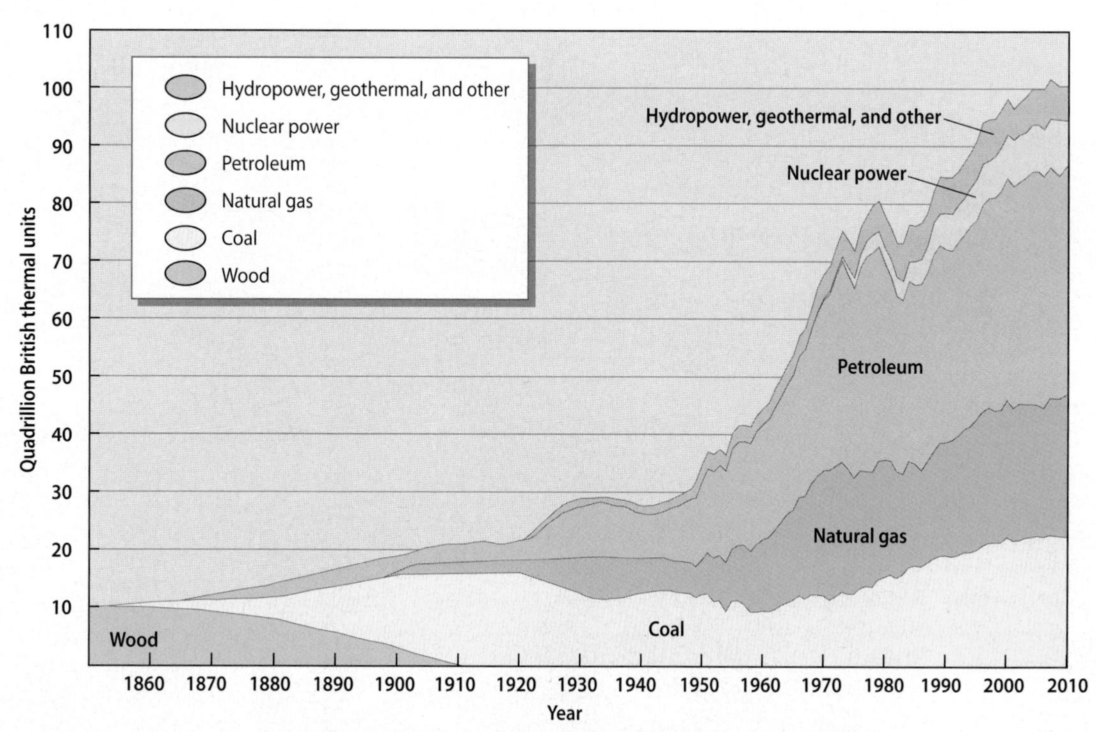

▲ **Figure 3.9 U.S. Energy Consumption** The growing popularity of fossil fuels is evident in U.S. energy consumption during the late 19th century as coal, oil, and then natural gas supplanted wood consumption. **Q: Approximately what percentage of U.S. energy consumption is currently accounted for by nuclear power and renewable energy sources?**

coal-based resources, and tar sands may offer several more decades of relatively abundant fossil fuel–based energy resources. New discoveries and drilling technologies have shifted the continent's energy equation. For example, recent estimates suggest that the Bakken formation in North Dakota and Montana, thanks to new oil extraction methods, may one day produce more than 20 billion barrels of oil, making it one of the planet's truly great energy reserves (on par with Alaska's North Slope field). The 2012 *World Energy Outlook* issued by the International Energy Agency predicted that the United States will become a major exporter of natural gas by 2020. Canada alone has the world's third largest proven oil reserves: About 170 billion barrels of oil can be recovered just from its rich oil sands. Overall, even with its high rates of consumption, North America may be an oil export region by 2035.

Still, many environmental issues complicate the clean development of North America's untapped fossil fuels. Many resources exist in pristine natural areas such as Alaska's Arctic National Wildlife Refuge. Moving fossil fuels also involves huge investments and risks. Plans for transporting large volumes of coal from the Rocky Mountains to the Pacific Coast for export to Asia have stirred protests across that region. Controversial energy pipelines, such as the Keystone XL project (from Alberta to the Gulf of Mexico) and the Northern Gateway Pipeline (from Alberta to the Pacific Coast), designed to tap into Canada's rich Athabasca oil sands, also illustrate the tensions between increasing energy production (and job creation) and the potentially dangerous environmental consequences of moving fossil fuels long distances (see Figures 3.7 and 3.10). In addition, the growing use of hydraulic fracturing, or **fracking**, drilling technologies (which inject a mix of water, sand, and chemicals underground in order to release natural gas) in settings from North Dakota to Pennsylvania may lead to polluted groundwater and potentially hazardous environmental conditions for nearby residents.

Climate Change in North America

Global climate change has already profoundly reshaped North America. High-latitude and alpine environments are particularly vulnerable to warming. Changes in arctic temperatures, sea ice, and sea levels have increased coastal erosion, affected migrating whale and polar bear populations, and stressed traditional ways of life for Eskimo and Inuit populations. At the same time, a more ice-free Arctic Ocean also opens up potential for more commercial

▼ **Figure 3.10 Canadian Tar Sands** This aerial view from near Fort McMurray suggests the massive scale of development associated with Alberta's tar sands projects.

(a)

(b)

▲ **Figure 3.11 Glaciers in Retreat** With prospects for further global warming, the outlook is bleak for many of North America's alpine glaciers. This pair of images records change at Grinnell Glacier, in Montana's Glacier National Park, between (a) 1940 and (b) 2006.

shipping and resource development in that high-latitude zone.

The cumulative long-term consequences of climate change for North Americans are enormous. After the unparalleled warmth in 2012 (it was the hottest year on record in the United States), many experts predict that more records are likely in the near future. The accompanying droughts and wildfires cost billions. Many coastal localities, especially low-lying zones in the arctic, along the East Coast, and in the Gulf of Mexico, are also vulnerable to rising sea levels. Although some experts disagree about precise connections, the large number of destructive hurricanes (such as Katrina in 2005 and Sandy in 2012) may also be connected to shifting climate patterns and warmer ocean temperatures.

Basic redistributions of plants, animals, and crops are already under way. For example, high-latitude distributions of tundra and permafrost environments may shift more than 300 miles (480 km) poleward by 2050, dramatically impacting wildlife populations and human settlements in these regions. In western North American mountains, expanding mountain pine beetle populations are rapidly infesting lodgepole and whitebark pine forests. Many of North America's spectacular alpine glaciers are rapidly disappearing, often with important implications for downstream fisheries and populations (Figure 3.11). For northern U.S. and Canadian farmers, longer growing seasons may open new possibilities for agriculture, but weeds and plant pathogens will migrate north with the crops. The Great Lakes region may also become wetter by the end of the century.

REVIEW

3.1 Describe North America's major landform regions and climates, and suggest ways in which the region's physical setting has shaped patterns of human settlement.

3.2 Identify the key ways in which humans have transformed the North American environment since 1600.

3.3 Identify four key environmental problems that North Americans face in the early 21st century.

Population and Settlement: Reshaping a Continental Landscape

The North American landscape is the product of human settlement that extends back in time for at least 12,000–25,000 years. The pace of change for much of that period was modest and localized, but the last 400 years have witnessed an extraordinary transformation as Europeans, Africans, and Asians arrived in the region, disrupted Native American peoples, and created dramatically new patterns of human settlement. Today 350 million people live in the region, and they are some of the world's most affluent and highly mobile populations (Table 3.1).

Modern Spatial and Demographic Patterns

Metropolitan clusters dominate North America's population geography, producing strikingly uneven patterns of settlement across the region (Figure 3.12). In Canada, about 90 percent of the population is found within 100 miles (160 km)

Table 3.1 Population Indicators

Country	Population (millions), 2013	Population Density (per square kilometer)	Rate of Natural Increase (RNI)	Total Fertility Rate	Percent Urban	Percent <15	Percent >65	Net Migration (rate per 1000)
Canada	35.3	4	0.4	1.6	80	16	15	7
United States	316.2	33	0.5	1.9	81	19	14	2

Source: Population Reference Bureau, *World Population Data Sheet, 2013.*

► **Figure 3.12 Population of North America** North America's geography of population reveals a strikingly clustered pattern of large cities interspersed with more sparsely settled zones. Notable concentrations are found on the eastern seaboard between Boston and Washington, DC; along the shores of the Great Lakes; and across the Sun Belt from Florida to California.

of the U.S. border. Within this broad region, Canada's "Main Street" corridor contains most of that nation's urban population, led by the cities of Toronto (5.8 million) and Montreal (3.9 million). The federal capital of Ottawa (1.3 million) and the industrial center of Hamilton (700,000) are also within the Canadian urban corridor. **Megalopolis**, the largest settlement agglomeration in the United States, includes the Washington, DC/Baltimore area (8.6 million), Philadelphia (6.5 million), New York City (22 million), and the greater Boston metropolitan area (7.6 million) (Figure 3.13). Beyond these two national core areas, other sprawling urban centers cluster around the southern Great Lakes (Chicago, 9.7 million), in various parts of the South (Dallas, 7.4 million), and along the Pacific Coast (Los Angeles, 17.9 million; Vancouver, 2.4 million).

Both Canada and the United States keep track of their populations by taking a census every 10 years. Most recently (in 2010 for the United States and 2011 for Canada), mailed questionnaires and census takers fanned out across the North American continent, gathering basic information on household size, age, and a variety of other social and economic characteristics. For the United States, millions of bilingual forms were also utilized for the first time, in an effort to increase the response rate in Latino communities across the country.

In both countries, census data are used to make key geographical decisions. For example, census tallies are used to adjust election districts and to apportion tax dollars more efficiently. In the United States, more than $400 billion are allocated annually according to census data to help fund services such as schools, hospitals, job-training centers, senior centers, and infrastructure improvement. Businesses also use the census as North America's largest market-research survey. It helps restaurants and retailers plan locations for new stores and can suggest what products and services might do well in particular neighborhoods. In addition, the censuses are one of the most widely used tools of spatial analysis by North American human geographers interested in solving a variety of demographic, social, and economic problems. Information gathered in 2010 and 2011 is used as benchmark data for thousands of experts examining a myriad of urban, economic, and political issues in the region.

Population Change over Time Historically, North America's population has increased greatly since the beginning of European colonization. Before 1900, high rates of natural increase produced large families. In addition, waves of foreign immigration swelled settlement, a pattern that continues today. In Canada, a population of fewer than 300,000 Indians and Europeans in the 1760s grew to an impressive 3.2 million a century later. For the United States, a late colonial (1770) total of around 2.5 million increased more than 10-fold to more than 30 million by 1860. Both countries saw even higher rates of immigration in the late 19th and 20th centuries, although birth rates gradually fell after 1900. After World War II, birth rates rose once again in both countries, resulting in the "baby boom" generation born between 1946 and 1965.

Today, however, as in much of the developed world, rates of natural increase in North America are below 1 percent annually, and the overall population is growing older, particularly in states such as Iowa (Figure 3.14). Still, the region attracts many immigrants: More than 46 million foreign-born migrants now live in North America. These growing numbers, along with higher birth rates among immigrant populations (exemplified by Texas), recently have led demographic experts to increase long-term population projections for the 21st century (Figure 3.14). Indeed, new predictions by the UN that by 2050 the region's population will reach 464 million (423 million in the United States and 41 million in Canada) may prove conservative.

Occupying the Land

When Europeans began occupying North America more than 400 years ago, they were not settling an empty land. North America was populated for at least 12,000–25,000 years by peoples as culturally diverse as the Europeans who conquered them. Native Americans arrived in North America from northeast Asia in multiple migratory waves

▼ **Figure 3.13 New York City** Midtown Manhattan's cluster of high-rise buildings remains one of North America's most dramatic urban skylines.

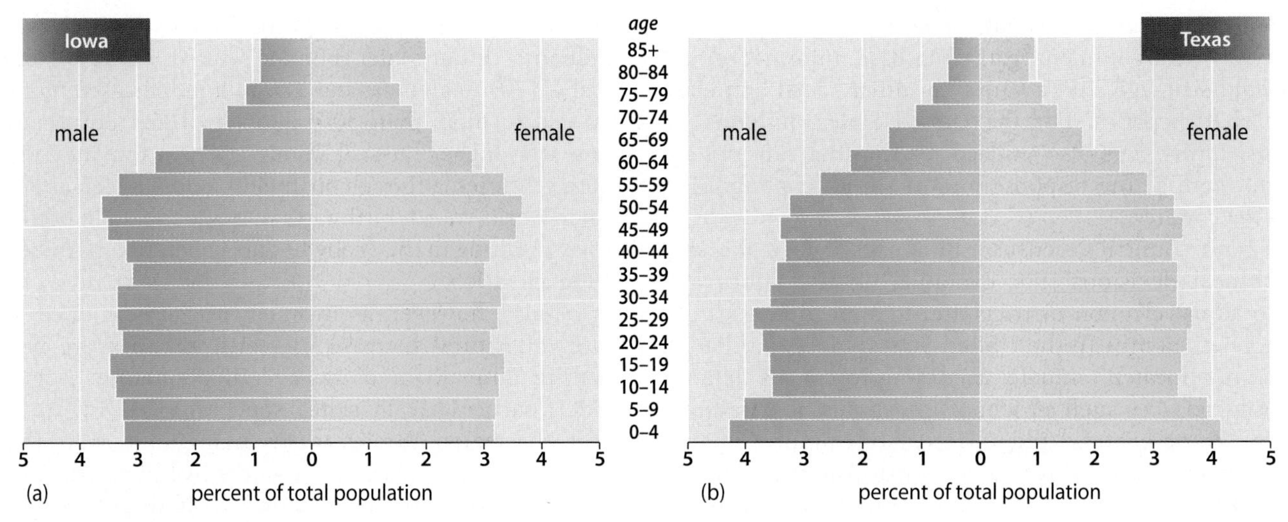

▲ **Figure 3.14 Population Pyramids of Iowa and Texas** (a) Iowa's aging population stands in contrast to the larger proportion of young people in (b) Texas, reflecting that state's higher birth rate and sizable influx of young immigrants.
Q: How is the baby boom generation reflected in the Iowa pyramid?

and became broadly distributed across the region, adapting in diverse ways to its many natural environments. Cultural geographers estimate Native American populations in 1500 CE at 3.2 million for the continental United States and another 1.2 million for Canada, Alaska, Hawaii, and Greenland. In many areas, European diseases, wars, and economic disruptions reduced these Native American populations by more than 90 percent as contacts increased.

North America's native peoples met many different fates. Some populations were exterminated by disease and war. Many others were expelled from their homelands and relocated on reservations, both in Canada and in the United States. For example, as a consequence of the Indian Removal Act of 1830, thousands of Native Americans were driven out of the Southeast in the notorious "Trail of Tears" trek to Indian Territory (later Oklahoma). Some Native Americans also intermingled with European populations, losing parts of their cultural identity in the process. Today the majority of the region's native peoples actually live in cities, often far removed from the land of their ancestors.

What replaced Native North Americans? The first stage of a dramatic new settlement geography began with a series of European colonies, mostly within the coastal regions of eastern North America (Figure 3.15). Established between 1600 and 1750, these regionally distinct societies were anchored in the north by the French settlement of the St. Lawrence Valley and extended south along the Atlantic Coast, including separate English colonies. Scattered developments along the Gulf Coast and in the Southwest also appeared before 1750.

The second stage in the Europeanization of the North American landscape took place between 1750 and 1850. It was highlighted by settlement of much of the better agricultural land within the eastern half of the continent. After the American Revolution (1776) and a series of Indian conflicts, pioneers surged across the Appalachians. They found much of the Interior Lowlands region almost ideal for agricultural settlement. Much of southern Ontario, or Upper Canada, was also opened to widespread development after 1791.

The third stage in North America's settlement expansion picked up speed after 1850 and continued until just after 1910. During this period, most of the region's remaining agricultural lands were settled by a mix of native-born and immigrant farmers. Farmers were challenged and sometimes defeated by drought, mountainous terrain, and short northern growing seasons. In the American West, settlers were attracted by opportunities in California, the Oregon country, Mormon Utah, and the Great Plains. In Canada, thousands occupied southern portions of Manitoba, Saskatchewan, and Alberta. Gold and silver discoveries led to initial development in areas such as Colorado, Montana, and British Columbia's Fraser Valley.

Incredibly, in a mere 160 years, much of the North American landscape was occupied as expanding populations sought new land to settle and as the global economy demanded resources to fuel its growth. It was one of the largest and most rapid transformations of the landscape in human history. This European-led advance forever reshaped North America in its own image, and in the process it also changed the larger globe in lasting ways by creating a "New World" destined to reshape the Old.

North Americans on the Move

From the mythic days of Davy Crockett and Calamity Jane to the 20th-century sojourns of John Steinbeck and Jack Kerouac, North Americans have been on the move. Indeed, almost one in every five Americans moves annually, suggesting that residents of the region are quite willing to change residence in order to improve their income or their quality of life. Although interregional population flows are complex in both the United States and Canada, several trends dominate the picture.

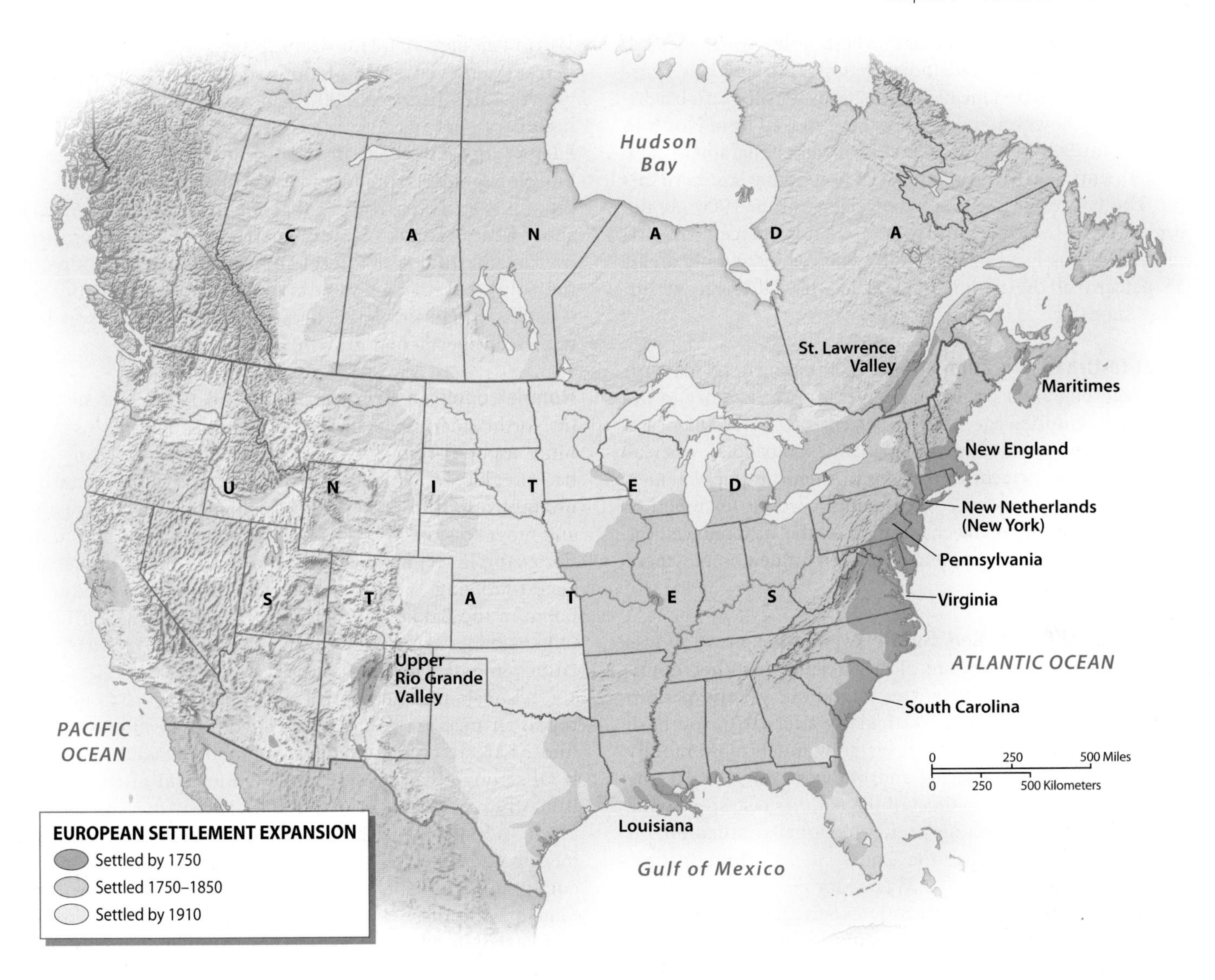

▲ **Figure 3.15 European Settlement Expansion** Sizable portions of North America's East Coast and the St. Lawrence Valley were occupied by Europeans before 1750. The most remarkable surge of settlement occurred during the next century, as Europeans opened vast areas of land and dramatically disrupted Native American populations.

Westward-Moving Populations The most persistent regional migration trend in North America has been the tendency for people to move west. By 1990, more than half the population of the United States lived west of the Mississippi River, a dramatic shift from colonial times. Since 1990, some of the fastest-growing areas have been in the states of the American West (including Arizona and Nevada), as well as in the western Canadian provinces of Alberta and British Columbia. This move was fueled by new job creation in high-technology, energy, and service industries, as well as by the region's scenic, recreational, and retirement attractions. With new population, demands for water in this arid region continue to grow. Also, the area's increasing political and economic power is redefining its place in national affairs. In the United States, the 2010 census data paved the way for the West to gain power in the House of Representatives and play a larger role in national elections.

The economic slowdown of 2007–2010 hit the region hard. The contraction of construction and leisure-time industries in settings such as Las Vegas, Phoenix, and Tucson slowed growth in these metropolitan areas to their lowest rates in decades, and home values in some neighborhoods declined by more than 50 percent. Recently, however, a slowly recovering economy in many desirable western settings suggests that the enduring appeal of "going west" is not likely to end anytime soon.

Black Exodus from the South African Americans have also generated distinctive patterns of interregional migration. Most blacks remained economically tied to the rural South after the Civil War. Conditions changed, however, in the early 20th century. Many African Americans migrated because of declining demands for labor in the agricultural South and growing industrial opportunities in the North and West.

Migrants ended up in cities where jobs were located. Boston, New York, Philadelphia, Detroit, Chicago, Los Angeles, and Oakland became key destinations for southern blacks. Since 1970, however, more blacks have moved from North to South. Sun Belt jobs and federal civil rights guarantees now attract many northern urban blacks to growing southern cities. The net result is still a major change from 1900: At the beginning of the last century, more than 90 percent of African Americans lived in the South, whereas today only about 55 percent of the nation's 45 million blacks reside within those states.

Rural-to-Urban Migration Another continuing trend in North American migration has taken people from the country to the city. Two centuries ago, only 5 percent of North Americans lived in urban areas (cities of more than 2500 people), whereas today about 80 percent of the North American population is urban. Shifting economic opportunities account for much of the transformation: As mechanization on the farm reduced the demand for labor, many young people left for new employment opportunities in the city.

Growth of the Sun Belt South Twentieth-century moves to the American South are clearly related to other dominant trends in North American migration, yet the pattern deserves closer inspection. Particularly after 1970, southern states from the Carolinas to Texas grew much more rapidly than states in the Northeast and Midwest. Since 2010, the South has been home to most of the nation's fastest-growing counties and has remained a key regional destination for domestic migrants within the United States, adding people even more rapidly than the West. Florida, Texas, Georgia, and North Carolina have recently experienced sizable population gains, with migrants heading for job-rich suburbs, as well as high-amenity retirement locations. Dallas–Fort Worth (23 percent), Houston (26 percent), and Atlanta (24 percent) enjoyed some of the nation's fastest metropolitan growth rates between 2000 and 2010, replacing previous highfliers such as Las Vegas and Phoenix (Figure 3.16). Factors that have contributed to the South's growth are its buoyant economy, modest living costs, adoption of air conditioning, attractive recreational opportunities, and appeal to snow-weary retirees. Movements have been selective, however; many rural agricultural and mountain counties within the South have seen few new residents, while amenity-rich coastal settings and job-generating metropolitan areas have witnessed spectacular growth.

Nonmetropolitan Growth During the 1970s, certain areas in North America beyond its large cities began to see significant population gains, including many rural settings that had previously lost population. Selectively, this pattern of **nonmetropolitan growth**, in which people leave large cities and move to smaller towns and rural areas, continues today. For example, Williston, North Dakota, is one of the region's fastest-growing smaller cities, thanks to the production boom in the Bakken oil fields in the northern Great Plains. Other smaller centers have proven attractive to retiree populations in both Canada and the United States.

A substantial number of migrants attracted to nonmetropolitan areas are younger, so-called *lifestyle migrants*. They find or create employment in affordable smaller cities and rural settings that are rich in amenities and often removed from the perceived problems of urban America. In fact, recent nonmetropolitan population growth has exceeded metropolitan growth in many western states. Other smaller communities outside the West, such as Mason City, Iowa; Mankato, Minnesota; and Traverse City, Michigan, also are seen as desirable destinations for migrants interested in downsizing from their metropolitan roots.

▼ Figure 3.16 Downtown Houston Houston's energy-based economy has helped make it one of North America's fastest-growing cities since 1990.

Settlement Geographies: The Decentralized Metropolis

North America's settlement landscape reflects the population movements, shifting regional economic fortunes, and technological innovations of the last century. The ways in which settlements are organized on the land—the actual appearance of cities, suburbs, and farms—as well as the very ways in which North Americans socially construct their communities, have changed greatly in the past century. Today's cloverleaf interchanges, sprawling suburbs, outlet malls, and theme parks would have struck most 1900-era residents as utterly extraordinary.

Settlement landscapes of North American cities boldly display the consequences of **urban decentralization**, in which metropolitan areas sprawl in all directions and suburbs take on many of the characteristics of traditional downtowns. Although both Canadian and U.S. cities have experienced decentralization, the impact is particularly profound in the United States, where inner-city problems, poor public transportation, widespread automobile ownership, and fewer regional-scale planning initiatives have encouraged many middle-class urban residents to move beyond the central city. Even beyond North America, observers note a globalization of urban sprawl: Many Asian, European, and Latin American cities are taking on attributes of their North American counterparts as they experience similar technological and economic shifts. Indeed, suburban Walmarts, semiconductor industrial parks, and shopping malls, so familiar in Seattle and Albuquerque, may become increasingly common sights on the peripheries of Kuala Lumpur or Mexico City.

Historical Evolution of the City in the United States

Changing transportation technologies decisively shaped the evolution of the city in the United States (Figure 3.17). The pedestrian/horsecar city (pre-1888) was compact, essentially limiting urban growth to a 3- or 4-mile-diameter ring around downtown. The invention of the electric trolley in 1888 expanded the urbanized landscape farther into new "streetcar suburbs," often 5 or 10 miles from the city center. A star-shaped urban pattern resulted, with growth extending outward along and near the streetcar lines. The biggest technological revolution came after 1920 with the widespread adoption of the automobile. The automobile city (1920–1945) promoted the growth of middle-class suburbs beyond the reach of the streetcar and added even more distant settlement in the surrounding countryside. Following World War II, growth in the outer city (1945 to the present) promoted more decentralized settlement along freeways and commuter routes as built-up areas appeared 40 to 60 miles from downtown.

Urban decentralization also reconfigured land-use patterns, producing metropolitan areas today that are strikingly different from their counterparts of the early 20th century. In the city of 1920, urban land uses were generally organized in rings around a highly focused central business district (CBD) that contained much of the city's retailing and office functions. Residential districts beyond the CBD were added as the city expanded, with higher-income groups seeking more desirable locations on the outside edge of the urbanized area.

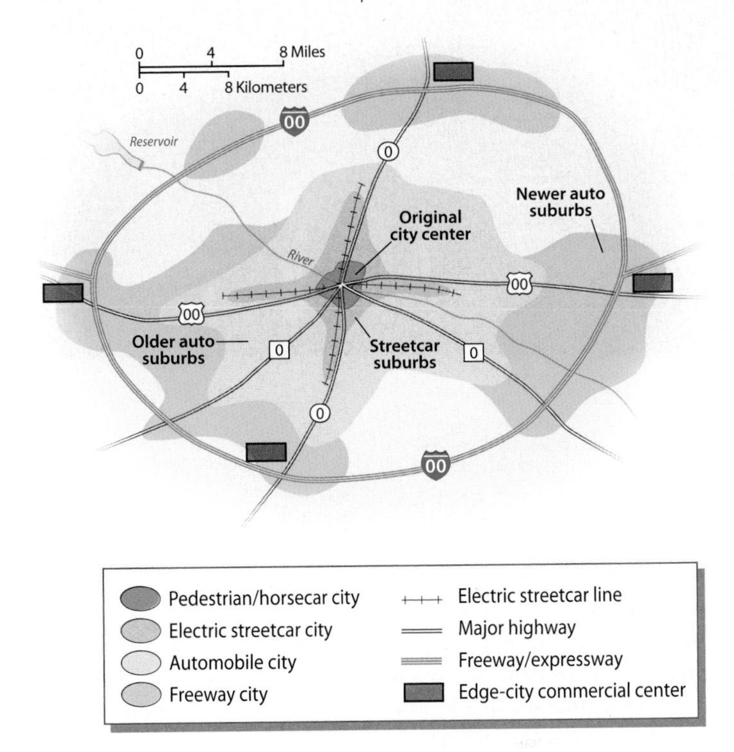

▲ **Figure 3.17 Growth of the American City** Many U.S. cities became increasingly decentralized as they moved through eras dominated by the pedestrian/horsecar, electric streetcar, automobile, and freeway. Each era left a distinctive mark on metropolitan America, including the recent growth of edge cities on the urban periphery.

The modern pattern has shifted. Today's suburbs feature a mix of peripheral retailing (such as commercial strips, shopping malls, and big-box stores), industrial parks, office complexes, and entertainment facilities. This larger, peripheral node of activity, called an **edge city**, has fewer functional connections with the central city than it has with other suburban centers. Southern California's Costa Mesa office and retailing district, located south of Los Angeles, is an excellent example of an edge-city landscape on the expanding periphery of a North American metropolis (Figure 3.18).

The Consequences of Sprawl The rapid evolution of the North American city continues to transform the urban landscape and those who live in it. As suburbanization accelerated in the 1960s and 1970s, many inner cities, especially in the Northeast and Midwest, suffered losses in population, increased levels of crime and social disruption, and a shrinking tax base, which often brought them to the brink of bankruptcy. Today inner-city poverty rates average almost three times those of nearby suburbs. Unemployment rates remain above the national average. Central cities in the

◀ Figure 3.18 **Costa Mesa, California** North America's edge-city landscape is nicely illustrated by Costa Mesa, California. Far from downtown Los Angeles, this sprawling complex of suburban offices and commercial activities reveals how and where many North Americans will live their lives in the 21st century.

United States are also places of racial tension, the product of decades of discrimination, segregation, and poverty. Although the number of middle-class African Americans and Hispanics is growing, many exit the central city for the suburbs, further isolating the urban underclass that remains behind. In Detroit, for example, about 85 percent of inner-city residents are black, and the city's poverty rate is about triple the national norm.

Amid these challenges, inner-city landscapes are also experiencing a selective renaissance. Referred to as **gentrification**, the process involves the displacement of lower-income residents of central-city neighborhoods by higher-income residents, the rehabilitation of deteriorated inner-city landscapes, and the construction of new shopping complexes, entertainment attractions, or convention centers in selected downtown locations. The older and more architecturally diverse housing of the central city is also a draw, combining specialty shops and restaurants for a cosmopolitan urban clientele with residential opportunities for upscale singles who wish to live near downtown. Seattle's Pioneer Square, Toronto's Yorkville district, and Baltimore's Harborplace exemplify how such public and private investments shape the central city. Recent census data show that selected central-city settings (such as in Chicago, San Francisco, and New York City) continue to post healthy population growth as new residents relocate to these vibrant, convenient downtown settings.

Many city planners and developers involved in such efforts are advocates of **new urbanism**, an urban design movement stressing higher-density, mixed-use, pedestrian-scaled neighborhoods where residents might be able to walk to work, school, and local entertainment. Pittsburgh's recent urban renaissance offers an affordable housing market, an older highly skilled workforce, and mixed-use neighborhoods such as the SouthSide Works development, a 34-acre assortment of residences, offices, and stores created on the site of an old steel plant on the Monongahela River.

The suburbs are also changing. The edge-city construction of new corporate office centers, fashion malls, and industrial facilities has created true "suburban downtowns" in suburbs that are no longer simply bedroom communities for central-city workers. Indeed, such localities have been growing players in the continent's globalization process. Many of North America's key internationally connected corporate offices (IBM, Microsoft), industrial facilities (Boeing, Cisco, Oracle), and entertainment complexes (Walt Disney World, Universal Studios, and the Las Vegas Strip) are now in such settings, and they are intimately tied to global information, technology, capital, and migration flows.

The edge-city lifestyle has also transformed both Canada and the United States into a continent of suburban commuters in which people live in one suburb and work in another. The average daily commute now exceeds 30 miles (48 km) per person in cities such as Atlanta, Georgia, and Birmingham, Alabama. Longer term, North Americans are likely to continue their outward drift, creating a vast suburban periphery where the boundary between country and city blurs in a mosaic of clustered housing developments, shopping complexes, and remaining open space (Figure 3.19).

Settlement Geographies: Rural North America

Rural North American landscapes trace their origins to early European settlement. Over time, these immigrants from Europe showed a clear preference for a dispersed rural settlement pattern as they created new farms on the North

▶ Figure 3.19 **Life on the Urban Periphery** New homes mingle with rolling grasslands in suburban Douglas County, Colorado, miles south of Denver's central business district. Today Douglas County is a mix of upscale residential developments, commercial nodes, and open space.

▲ **Figure 3.20 Iowa Settlement Patterns** The regular rectangular look of this Iowa town and the nearby rural setting reveals the North American penchant for simplicity and efficiency. In the United States, the township-and-range survey system stamped such predictable patterns across vast portions of the North American interior.

American landscape. In portions of the United States settled after 1785, the federal government surveyed and sold much of the rural landscape. Surveys were organized around the simple, rectangular pattern of the federal government's township-and-range survey system, which offered a convenient method of dividing and selling the public domain in 6-mile-square townships (Figure 3.20). Canada developed a similar system of regular surveys that stamped much of southern Ontario and the western provinces with a strikingly rectilinear character.

Commercial farming and technological changes further transformed the settlement landscape. Railroads opened corridors of development, provided access to markets for commercial crops, and helped to establish towns. By 1900, several transcontinental lines spanned North America, radically transforming the farm economy and the pace of rural life. After 1920, however, even greater change accompanied the arrival of the automobile, farm mechanization, and better rural road networks. The need for farm labor declined with mechanization, and many smaller market centers became unnecessary, as farmers equipped with automobiles and trucks could travel farther and faster to larger, more diverse towns.

Today many areas of rural North America face population declines as they adjust to the changing conditions of modern agriculture. Both U.S. and Canadian farm populations fell by more than two-thirds during the last half of the 20th century. Typically, a fewer number of farms (but larger in acreage) dot the modern rural scene, and many young people leave the land to obtain employment elsewhere. The visual record of abandonment offers a painful reminder of the economic and social adjustments that come from population losses. Weed-choked driveways, empty farmhouses, roofless barns, and the empty marquees of small-town movie houses tell the story more powerfully than any census or government report.

Elsewhere, rural settings show signs of growth. Some places begin to experience the effects of expanding edge cities. Other growing rural settings lie beyond direct metropolitan influence, but are seeing new populations who seek amenity-rich environments removed from city pressures. These trends are shaping the settlement landscape from British Columbia's Vancouver Island to Michigan's Upper Peninsula. Newly subdivided land, numerous real estate offices, and the presence of new espresso bars are all signs of growth in such surroundings.

REVIEW

3.4 Describe the dominant North American migration flows during the 20th century.

3.5 Describe the principal patterns of land use within the modern U.S. metropolis. Include a discussion of (a) the central city and (b) the suburbs/edge city. How have forces of globalization shaped North American cities?

Cultural Coherence and Diversity: Shifting Patterns of Pluralism

North America's cultural geography is both globally dominant and internally pluralistic. History and technology have produced a contemporary North American cultural force that is second to none in the world. Many people outside the United States speak of cultural imperialism when they describe the global dominance of American popular culture, which they often see as threatening the vitality of other cultural values. Yet North America is also a home for many different peoples who retain part of their traditional cultural identities and celebrate their pluralistic roots.

The Roots of a Cultural Identity

Powerful forces formed a common dominant culture within North America. Both the United States (1776) and Canada (1867) became independent from Great Britain, but the two countries remained closely tied to their Anglo roots. Key Anglo legal and social institutions solidified the common set of core values that many North Americans shared with Britain and, eventually, with one another. Traditional Anglo beliefs emphasized representative government, separation of church and state, liberal individualism, privacy, pragmatism, and social mobility. From those shared foundations, particularly within the United States, consumer culture blossomed after 1920, producing a common set of experiences oriented around convenience, consumption, and the mass media.

However, North America's cultural unity coexists with pluralism: the persistence and assertion of distinctive cultural identities. Closely related is the concept of **ethnicity**, the shared cultural identity held by a group of people with a common background and history. For Canada, the early and enduring French colonization of Quebec and the lasting power of the country's native peoples complicate its modern cultural geography. Canadians face the challenge of creating a truly multicultural society where issues of language and political representation are central concerns. Within the United States, given its unique immigration history, a greater diversity of ethnic groups exists, and differences in cultural geography are often found at both local and regional scales.

Peopling North America

North America is a region of immigrants. Quite literally, global-scale migrations made possible the North America we know today. Decisively displacing Native Americans in most portions of the region, immigrant populations created a new cultural geography of ethnic groups, languages, and religions. Early migrants often had considerable cultural influence, despite minuscule numbers. Over time, varied immigrant groups and their changing destinations produced a culturally diverse landscape. Also varying among groups was the pace and degree of **cultural assimilation**, the process in which immigrants were absorbed by the larger host society.

Migration to the United States In the United States, variations in the number and source regions of migrants produced five distinctive chapters in the country's history (Figure 3.21). In Phase 1 (prior to 1820), English influences dominated. Slaves, mostly from West Africa, contributed additional cultural influences in the South. In Phase 2 (1820–1870), Northwest Europe served as the main source region of immigrants. The emphasis, however, shifted away from English migrants. Instead, Irish and Germans dominated the flow and provided more cultural variety.

As Figure 3.21 shows, immigration reached a much higher peak around 1900, when almost 1 million foreigners entered the United States *annually*. During Phase 3 (1870–1920), the majority of immigrants were southern and eastern Europeans. Political

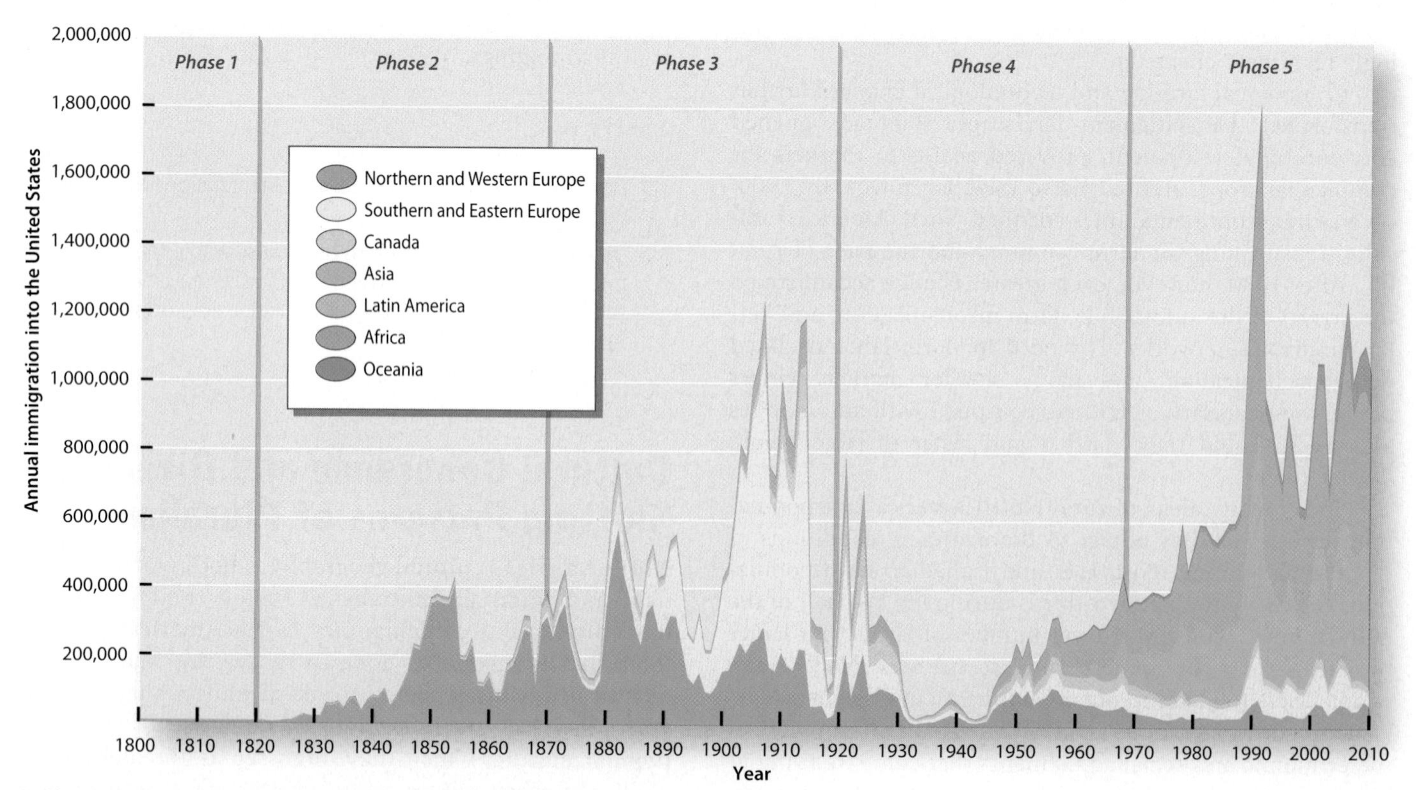

▲ **Figure 3.21 U.S. Immigration, by Year and Group** Annual immigration rates peaked around 1900, declined in the early 20th century, and then surged again, particularly since 1970. The source areas of these migrants have also shifted. Note the decreased role of Europeans currently versus the growing importance of Asians and Latin Americans.

Q: What were dominant source regions of the U.S. population in 1850, 1910, and 1980? Why did they change?

► **Figure 3.22 Alley Murals, Mission District, San Francisco, California** One of the Bay Area's largest Hispanic communities features dozens of street-side murals that help define the area's cultural identity. A portrait of César Chávez celebrates the importance of the farm labor-union movement within the state.

strife and poor economies in Europe existed during this period. News of available land and expanding industrialization in the United States offered an escape from such difficult conditions. By 1910, almost 14 percent of the nation was foreign-born. Very few of these immigrants, however, targeted the job-poor U.S. South, creating a cultural divergence that still exists.

Between 1920 and 1970 (Phase 4), more immigrants came from neighboring Canada and Latin America, but overall totals fell sharply, a function of more restrictive federal immigration policies (the Quota Act of 1921 and the National Origins Act of 1924), the Great Depression, and the disruption caused by World War II.

Since 1970 (Phase 5), the region has witnessed a sharp reversal in numbers, and now annual arrivals surpass those of the early 20th century (see Figure 3.21). Most legal migrants since 1970 originated in Latin America or Asia. In 2000, about 60 percent of immigrants were Hispanics and only 20 percent were Asians, but by 2010 the balance shifted: In that year, 36 percent of immigrants were from diverse Asian settings, while only about 30 percent were Hispanic. The post-1970 surge was made possible by economic and political instability abroad, a growing postwar American economy, and a loosening of immigration laws (the Immigration Acts of 1965 and 1990 and the Immigration Reform and Control Act of 1986). Undocumented immigration, particularly from Mexico, rose after 1970, but since 2008 the pace has slowed appreciably, mostly because of fewer job opportunities in the United States, as well as an increased number of U.S. border patrol agents. Today the United States is home to about 11–12 million undocumented immigrants.

The nation's Hispanic population continues to grow (Figure 3.22). An estimated 12 million Mexican-born residents (about 10 percent of Mexico's population) now live in the United States. In the next 25 years, most of the projected increase in the U.S. Hispanic population will be fueled by births within the country, rather than by new immigrants. Almost half of U.S. Hispanics live in California (27 percent of California's population is foreign-born) or Texas, but they are increasingly moving to other areas (Figure 3.23). States such as South Carolina, Alabama, Wisconsin, Georgia, Kansas, and Arkansas have witnessed dramatic increases in Hispanic populations. Many settlements across the Great Plains are also home to Hispanic immigrants, who bring new churches, taquerías, and school-aged children to once-dying communities.

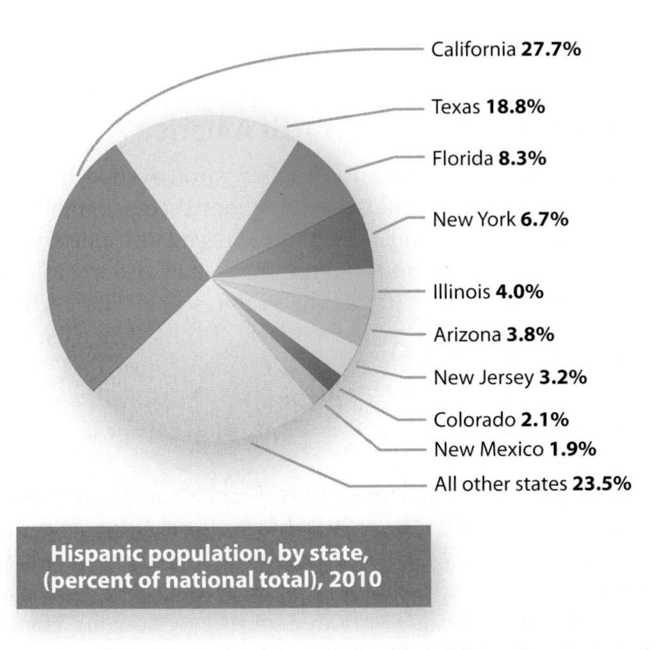

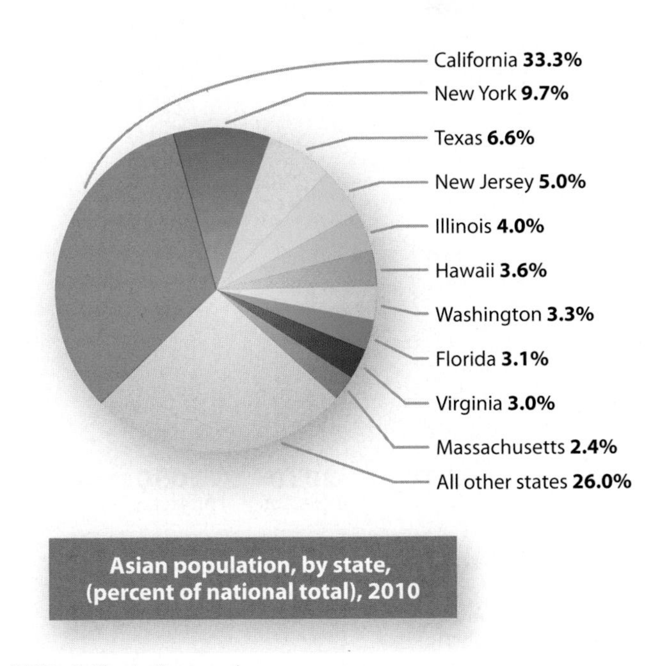

▲ **Figure 3.23 Distribution of U.S. Hispanic and Asian Populations, by State, 2010** California, Texas, and Florida claim more than half of the nation's Hispanic population, and California alone is still home to one-third of the country's Asian population. **Q: Outside of the West and South, why do you think New York, New Jersey, and Illinois are also important destinations for these immigrants?**

The cultural implications of these Hispanic migrations are profound: Today the United States is the fifth largest Spanish-speaking nation on Earth.

In percentage terms, migrants from Asia constitute the fastest-growing immigrant group, and various Asian ethnicities, both native and foreign-born, account for 6 percent of the U.S. population. Chinese is the third most common spoken language in the United States (behind English and Spanish). California remains a key entry point for migrants and is home to one-third of the nation's Asian population, whereas Hawaii has the highest percentage of Asian immigrants (Figure 3.23). Asian migrants often move to large cities such as Los Angeles, San Francisco, and New York City. Beyond these key gateway cities, a diverse array of Asian immigrants is also moving to growing communities in Washington, DC, Chicago, Seattle, and Houston. The largest Asian groups in the United States include Chinese (3.8 million), Filipino (3.2 million), Asian Indian (2.8 million), Vietnamese (1.7 million), and Korean (1.6 million). Although some Asian immigrants live in poverty, on average, Asian Americans tend to have higher household incomes and are better educated than the overall U.S. population.

The future cultural geography of the United States will be dramatically redefined by these recent immigration patterns (Figure 3.24). By 2050, Asians may total almost 10 percent of the U.S. population, and almost one American in three will be Hispanic. Indeed, it is likely that the U.S. non-Hispanic white population will achieve minority status by that date.

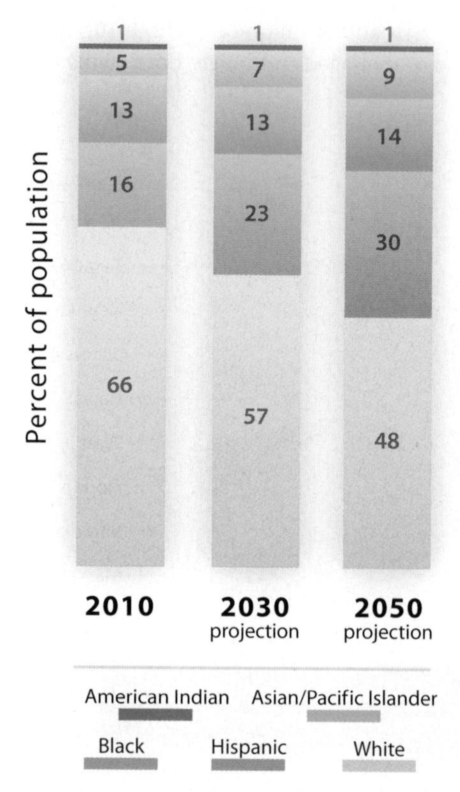

▲ **Figure 3.24 Projected U.S. Ethnic Composition, 2010 to 2050** By the middle of the 21st century, almost one in three Americans will be Hispanic, and non-Hispanic whites will achieve minority status amid an increasingly diverse U.S. population.

Finney County, Kansas, suggests how U.S. communities are changing. Set deep in the heartland, far from any international border or port city, this southeastern Kansas county entered "majority-minority" status in 2008, meaning its proportion of non-Hispanic whites is now under 50 percent. Garden City, the largest community in the county, has long been home to Hispanic populations who have worked in the agricultural sector, especially in the area's growing meatpacking industry. But recently, thousands of Somali and Southeast Asian immigrants have been added to the mix, drawn to the area's steady work and modest cost of living. The result is a complex cultural mosaic that suggests how immigration is now reworking the cultural geography of every corner of the country.

The Canadian Pattern The peopling of Canada included early French arrivals who concentrated in the St. Lawrence Valley. After 1765, many migrants came from Britain, Ireland, and the United States. Canada then experienced the same surge and reorientation in migration flows seen in the United States around 1900. Between 1900 and 1920, more than 3 million foreigners moved to Canada, an immigration rate far higher than for the United States given Canada's much smaller population. Eastern Europeans, Italians, Ukrainians, and Russians were the most important nationalities in these later movements. Today about 60 percent of Canada's recent immigrants are Asians, and its 20 percent foreign-born population is among the highest in the developed world. In Toronto, the city's 44 percent foreign-born population reveals a slight bias toward European backgrounds. On Canada's west coast, Vancouver (38 percent foreign-born) has been a key destination for Asian immigrants, particularly Chinese (see *Cityscapes: Vancouver*).

Culture and Place in North America

Cultural and ethnic identity is often strongly tied to place. North America's cultural diversity is expressed geographically in two ways. First, similar people congregate near one another and derive meaning from the territories they occupy in common. Second, culture marks the visible scene: The everyday landscape is filled with the artifacts, habits, language, and values of different groups. Boston's Italian North End simply looks and smells different from nearby Chinatown, and rural French Quebec is a world away from a Hopi village in Arizona.

Persistent Cultural Homelands French-Canadian Quebec superbly exemplifies the **cultural homeland**: It is a culturally distinctive nucleus of settlement in a well-defined geographical area, and its ethnicity has survived over time, stamping the cultural landscape with an enduring personality (Figure 3.25). Overall, about 23 percent of Canadians are French, but about 80 percent of the population of Quebec speaks French, and language remains the cultural glue that unites the homeland. Indeed, policies adopted after 1976 strengthened the

French language within the province by requiring French instruction in the schools and by mandating national bilingual programming by the Canadian Broadcasting Corporation (CBC). Ironically, many Quebecois feel that the greatest cultural threat may come not from Anglo-Canadians, but rather from recent immigrants to the province. Southern Europeans or Asians in Montreal, for example, show little desire to learn French, preferring instead to put their children in English-speaking private schools.

Another well-defined cultural homeland is the Hispanic Borderlands (see Figure 3.25). It is similar in geographical magnitude to French-Canadian Quebec and significantly larger in total population, but not specifically linked to a single political entity such as a state or province. Historical roots of the homeland are deep, extending back to the 16th century, when Spaniards opened the region to the European world. The homeland's historical core is in northern New Mexico, including Santa Fe and much of the surrounding rural hinterlands. A rich legacy of Spanish place-names, earth-toned Catholic churches, and traditional Hispanic settlements dots the rolling highlands of northern New Mexico and southern Colorado. From California to Texas, other historical sites, place-names, missions, and presidios also reflect the Hispanic heritage.

Unlike Quebec, however, massive 20th-century migrations from Latin America brought an entirely new wave of Hispanic settlement to the Southwest. Over 54 million Hispanics now live in the United States, with more than half in California, Texas, and Florida combined. Indeed, in 2015, Hispanics will likely outnumber non-Hispanic whites in California.

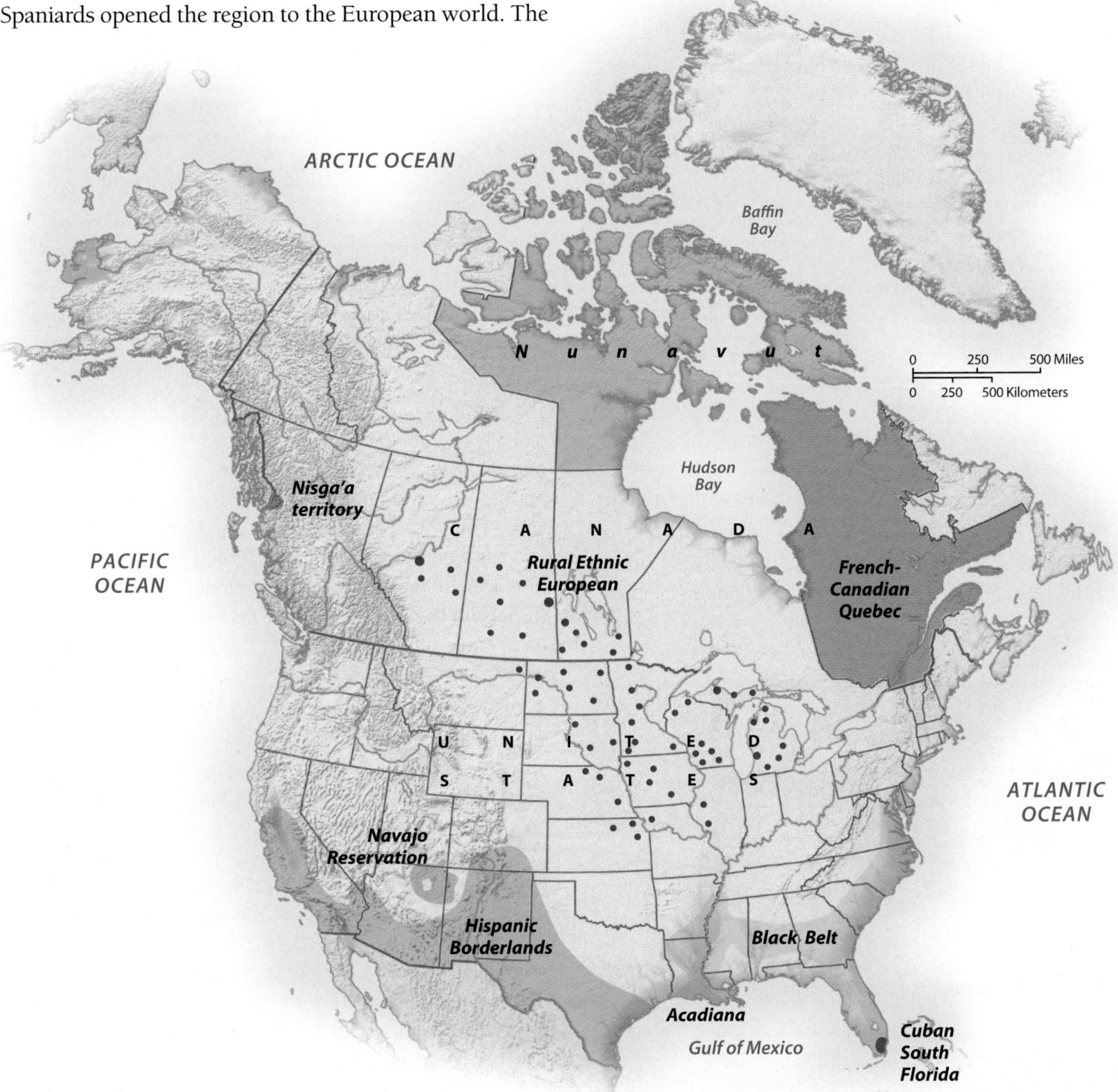

▲ **Figure 3.25 Selected Cultural Regions of North America** From northern Canada's Nunavut to the Southwest's Hispanic Borderlands, different North American cultural groups strongly identify with traditional local and regional homelands. Shaded portions of the map display a sampling of these regions across North America.

Cityscapes

VANCOUVER

Southern British Columbia's lower Fraser River Valley has attracted human settlement for thousands of years. Coastal Salish-speaking peoples were drawn to the Burrard Peninsula's strategic location and its accessibility to nearby natural resources. Today little remains of this indigenous habitation, but the modern city of Vancouver (metropolitan population, 2.4 million) is still defined by the area's spectacular natural setting (Figure 3.2.1). The steep, forested Coast Ranges dramatically define the city's northern edge, and much of the rest of the metropolitan area is never far from water. Its mild maritime climate (palm trees grow here) and amenity-rich surroundings have repeatedly made Vancouver one of the world's "most livable cities" according to many surveys. The city also proved to be an attractive host for the 2010 Winter Olympics. However, fame comes with a price: Given its setting and its enduring popularity as a place to live, Vancouver is Canada's most densely settled major city, its daily traffic jams can rival those of Seattle and San Francisco, and its costly housing makes it one of North America's least affordable cities.

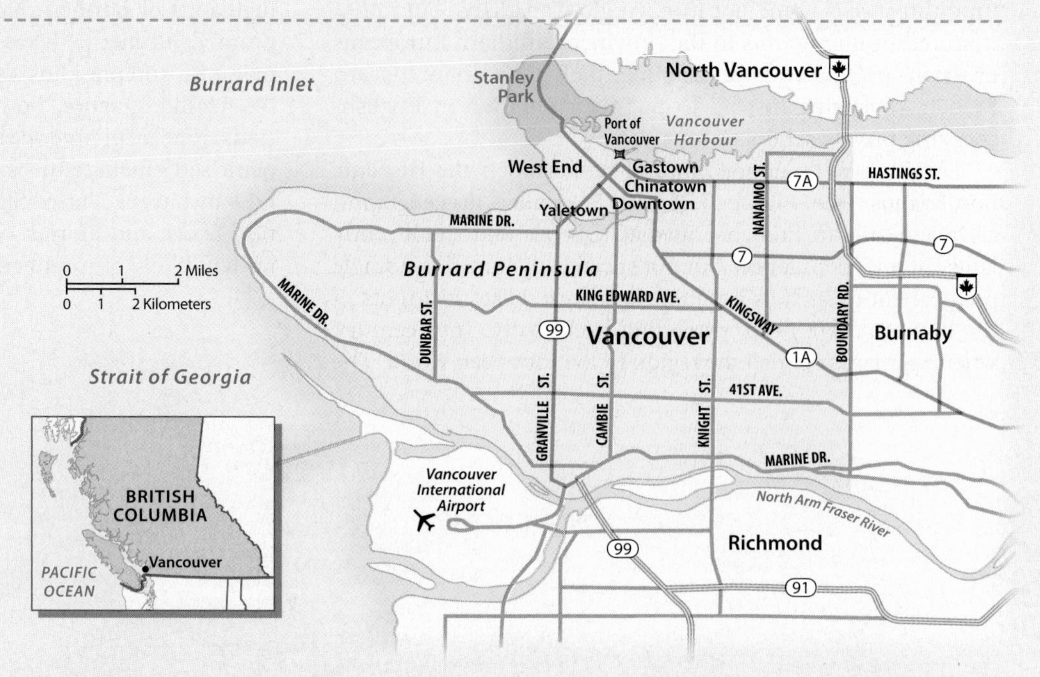

▲ **Figure 3.2.1 Vancouver** With a metropolitan population of 2.4 million people, Vancouver is Canada's most densely settled major city. It is home to a spectacular natural setting and an ethnically diverse population.

Growing Global Connections Much of Vancouver's urban landscape and place identity are oriented to its enduring economic and cultural connections to the nearby Pacific Ocean. Its role as a globally oriented gateway city was cemented in 1886 when builders of the Canadian Pacific Railway chose the spot as the Pacific terminus of their transcontinental line. More than 125 years later, Vancouver's economic fortunes remain anchored to long-distance trade: The city's port is Canada's largest (and the fourth largest in North America). China is now the port's biggest trading partner, recently surpassing the United States. Over two dozen major marine terminals handle some of the world's largest ships and busiest container-cargo traffic.

A City of Neighborhoods A closer look at Vancouver reveals a city of many neighborhoods and diverse cultures. Vancouver's Pacific linkages are richly revealed in its ethnic landscapes: Almost 30 percent of the city's residents are of Chinese heritage, with important communities located east of downtown (*Chinatown*), as well as in the nearby southern suburb of *Richmond* (about 60 percent Chinese) (Figure 3.2.2). Many recent immigrants came from Hong Kong (thus the name *Hongcouver*) in the 1980s and 1990s, in anticipation of that city's formal return to Mainland China in 1997. In addition, Vancouver is home to sizable South Asian (6 percent) and Filipino (5 percent) communities. Elsewhere, the urban landscape offers great examples of an older, once-gritty warehouse district that has been stylishly gentrified (*Yaletown*) and one of North America's largest gay communities (the *West End neighborhood*), as well as a carefully

◄ **Figure 3.2.2 Chinatown** Vancouver's traditional Chinatown is located just east of downtown.

preserved historic district complete with cobblestone streets (*Gastown*) located along the waterfront just northeast of downtown (**Figure 3.2.3**). The city's largest open space is also distinctive: With great forethought, Stanley Park was created in 1888, and its 1000 acres (400 hectares) of forests, waterfront vistas, and recreational facilities make it one of North America's largest urban parks (**Figure 3.2.4**).

1. Vancouver's Chinatown and suburban Richmond are both home to large numbers of Chinese residents. How might the cultural landscapes of these two settings reflect this ethnic signature, and how might they differ?
2. Explore the streets of Vancouver's Gastown neighborhood. Identify examples of specific landscape features that likely reflect efforts to create a distinctive sense of place and time in this historic neighborhood.

Google Earth
Virtual Tour Video

http://goo.gl/QaUj8O

▲ **Figure 3.2.3 Gastown** The city's historic Gastown neighborhood has been carefully preserved and is a major draw for visitors.

▼ **Figure 3.2.4 Stanley Park** One of the largest urban parks in North America, Stanley Park offers residents welcome relief from the stress of city life.

Within the homeland, Hispanics have created a distinctive Borderlands culture that mixes many elements of Latin and North America. These newer migrants augment the rural Hispanic presence in agricultural settings such as the lower Rio Grande Valley in Texas and the Imperial and Central valleys in California. Cities such as San Antonio, Phoenix, and Los Angeles also play leading roles in expressing the Hispanic presence within the Southwest. Regionally distinctive Latin foods and music add internal cultural variety to the region. New York City, Chicago, and Miami serve as key points of Hispanic cultural influence beyond the homeland.

African Americans also retain a cultural homeland, but it has diminished in intensity because of out-migration (see Figure 3.25). Reaching from coastal Virginia and North Carolina to East Texas, the Black Belt is a zone of African-American population remaining from the cotton South, when a vast majority of American blacks resided within the region. Today, although many blacks have left for cities, dozens of rural counties in the region still have large black majorities. Blacks account for more than one-quarter of the populations of Mississippi (38 percent), Louisiana (33 percent), South Carolina (29 percent), Georgia (32 percent), and Alabama (27 percent). More broadly, the South is home to many black folk traditions, including music such as black spirituals and the blues, which have now become popular far beyond their rural origins. Beyond the South, African Americans have created large, enduring communities in mostly urban settings in the Northeast, Midwest, and West. Regrettably, even though the rural neighborhoods of the black homeland differ greatly in density and appearance from African-American urban neighborhoods outside the South, poverty plagues both types of communities.

A second rural homeland in the South is Acadiana, a zone of enduring Cajun culture in southwestern Louisiana (see Figure 3.25). This homeland was created in the 18th century when French settlers were expelled from eastern Canada (an area around the Bay of Fundy known as Acadia) and relocated to Louisiana. Nationally popularized today through their food and music, the Cajuns have a lasting attachment to the bayous and swamps of southern Louisiana.

Native American Signatures Native American populations are also strongly tied to their homelands. Indeed, many native peoples maintain intimate relationships with their surroundings, weaving elements of the natural environment together with their material and spiritual lives. Over 5 million Indians, Inuits, and Aleuts live in North America, and they claim allegiance to more than 1100 tribal bands. Although many Native Americans now live in cities, they still retain close contact with their homelands. Place-names, landscape features, and family ties cement this connection between people and place.

Particularly in the American West and the Canadian and Alaskan North, native peoples also control sizable reservations, although less than 25 percent of the region's overall native populations reside on reservations. The largest block of native-controlled land in the lower 48 states is the Navajo Reservation in the Southwest. About 300,000 people claim allegiance to the Navajo Nation. To the north, Canada's self-governing Nunavut Territory (population about 35,000) is another reminder of the enduring presence of native cultural influence and emergent political power within the region (see Figure 3.25).

Although these homelands preserve traditional ties to the land, they are also settings for pervasive poverty, health problems, and increasing cultural tensions (Figure 3.26a). Within the United States, many Native American groups have taken advantage of the special legal status of their reservations and have built gambling casinos and tourist facilities that bring in much-needed capital, but also challenge traditional lifestyles (Figure 3.26b).

(a)

(b)

◄ **Figure 3.26 Native American Landscapes**
(a) A Navajo girl and her grandfather stand in front of the family home in the Navajo Nation. (b) Near Albuquerque, New Mexico, the Isleta Pueblo Indian Reservation operates the Isleta Resort and Casino, providing many new jobs for local native residents.

A Mosaic of Ethnic Neighborhoods North America's cultural mosaic is also enlivened by smaller-scale ethnic signatures that shape both rural and urban landscapes. For example, distinctive rural communities that range from Amish settlements in Pennsylvania to Ukrainian neighborhoods in southern Saskatchewan add cultural variety. When much of the agricultural interior was settled, immigrants often established close-knit communities. Among others, German, Scandinavian, Slavic, Dutch, and Finnish neighborhoods took shape, held together by common origins, languages, and religions. Although many of these ties weakened over time, rural landscapes of Wisconsin, Minnesota, the Dakotas, and the Canadian prairies still display some of these cultural imprints. Folk architecture, distinctive settlement patterns, ethnic place-names, and the simple elegance of rural churches selectively survive as signatures of cultural diversity upon the visible scene of rural North America.

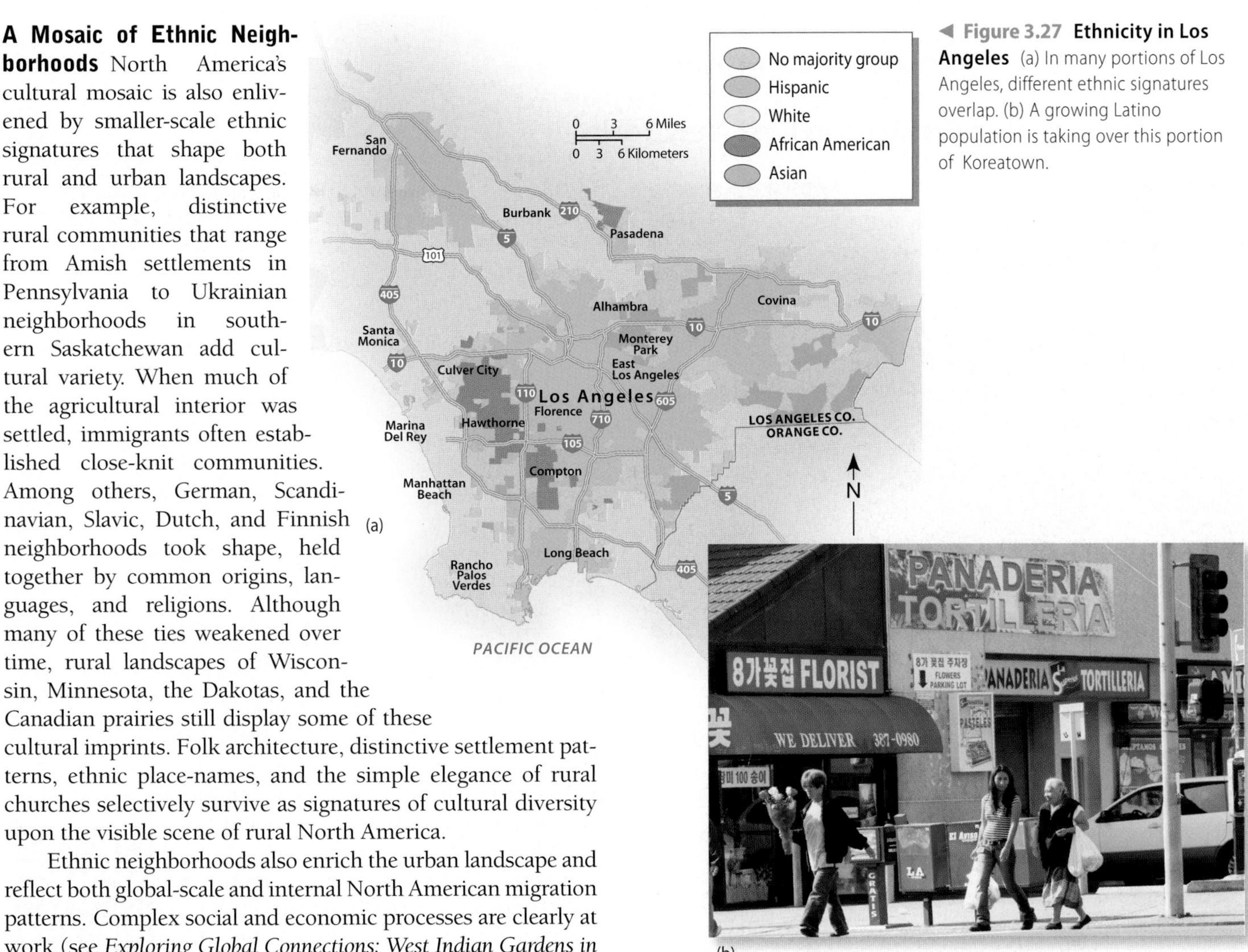

◀ **Figure 3.27 Ethnicity in Los Angeles** (a) In many portions of Los Angeles, different ethnic signatures overlap. (b) A growing Latino population is taking over this portion of Koreatown.

Ethnic neighborhoods also enrich the urban landscape and reflect both global-scale and internal North American migration patterns. Complex social and economic processes are clearly at work (see *Exploring Global Connections: West Indian Gardens in East New York*). Historically, employment opportunities fueled population growth in North American cities, but the cultural makeup of the incoming labor force varied, depending on the timing of the economic expansion and the relative accessibility of an urban area to different cultural groups. The ethnic geography of Los Angeles exemplifies the interplay of the economic and cultural forces at work (Figure 3.27a). Since most of its economic expansion took place during the 20th century, its ethnic geography reflects the movements of more recent migrants. African-American communities on the city's south side (Compton and Inglewood) represent the legacy of black population movements out of the South. Hispanic (East Los Angeles) and Asian (Alhambra and Monterey Park) neighborhoods are a reminder that about 40 percent of the city's population is foreign-born (Figure 3.27b).

Particularly in the United States, ethnic concentrations of nonwhite populations increased in many cities during the 20th century as whites exited for the perceived safety of the suburbs. In terms of central-city population, African Americans make up more than 60 percent of Atlanta, whereas Los Angeles is now more than 40 percent Hispanic (almost 5 million Hispanics reside in Los Angeles County, greater than the population of Costa Rica). Ethnic concentrations are also growing in the suburbs of some U.S. cities: Southern California's Monterey Park has been called the "first suburban Chinatown," and growing numbers of middle-class African Americans and Hispanics shape suburban neighborhoods in metropolitan settings such as Atlanta and San Antonio.

Patterns of North American Religion

Many religious traditions also shape North America's human geography. Reflecting its colonial roots, Protestantism is dominant within the United States, accounting for about 60 percent of the population (Figure 3.28). In some settings, hybrid American religions sprang from broadly Protestant roots. By far the most successful are the Latter-Day Saints (LDS or Mormons), regionally concentrated in Utah and Idaho and claiming more than 6 million North American members. Although many traditional Catholic neighborhoods have lost population in the urban Northeast, Catholic numbers are growing in the West and South, reflecting both domestic migration patterns and higher rates of Hispanic immigration and births. Strong Protestant religious traditions enliven many African-American communities, both in the rural South and in dozens of urban settings where churches serve as critical centers of community identity and solidarity.

Exploring Global Connections

West Indian Gardens in East New York

◀ **Figure 3.3.1 Community Garden, East New York** A volunteer examines the crops at the United Community Center garden, a member of East New York Farms that produces about 4,000–5,000 lbs. of fruits and vegetables in the local area.

Take a summer stroll down East New York's Schenck Avenue, or peer into the backyards along Bradford Street. Pause to enjoy the small, densely planted garden plots wedged in between and behind the aging two-story homes and brownstones (Figure 3.3.1). These small, carefully tended urban green spaces are a reminder of New York City's vibrant community garden program (the Green Thumb project has 600 gardens throughout the city), now the nation's largest.

Growing Your Own East New York is a part of the city located in the borough of Brooklyn. Its gardens offer a glimpse into the much larger world outside New York, where the residents of this gritty, working-class neighborhood came from. Look closely at the vine, root, bush, and tree crops. Even better, explore the local Farmers Market. Since 1998, the Farmers Market has offered an amazingly diverse collection of West Indian and African crops, all grown in nearby gardens (Figure 3.3.2). Recent immigrants from Caribbean islands (Puerto Rico, Jamaica, Trinidad, etc.) have brought their seeds and plants with them, along with the know-how to make them grow in East New York's vacant lots. Today about 16,000 people with a West Indian heritage call the neighborhood home.

What grows in the 60 community garden plots? Many of the plants, such as the yard-long bean (a variety of string bean), first came to the Caribbean from West Africa, brought to the islands as slave populations were forcibly removed from their native land. You can also find the green leafy dasheen bush (the "coco" of Jamaican cuisine), which is used to make a variety of dishes, including a flavorful gumbo. Don't neglect the peppers: One local resident has counted more than 42 varieties, including a new breed that originated in East New York. Other specialties include sour gherkins, bitter melons, and the lablab (a bean from Trinidad). The neighborhood's cultural mix is as rich as the menu. The Nehemiah 10 garden is worked by immigrants from Grenada, Guyana, Trinidad, Panama, Puerto Rico, St. Vincent, Belize, Ghana, Nigeria, and Egypt.

Garden Supporters The gardens and the Farmers Market are supported by several local community institutions. In 2006, the East New York Farms Project (an initiative of United Community Centers) received support from Green Thumb to expand local green space and make micro loans available to private entrepreneurs. It also dedicated the "Hands and Heart Garden," designed to cultivate interest in farming. A nearby Youth Farm sponsors educational programs and also encourages young people in the neighborhood to continue the grow-your-own tradition. In addition, Brooklyn's Weeksville Heritage Center has received a seed grant to develop a breeding program for West Indian and African plants, all to be shared with East New York gardeners.

1. Given your cultural background and locality, what crops might you grow in your own backyard garden?
2. In your home community (or near your local college/university), cite examples of community garden and farmers market initiatives that encourage the local production and consumption of food. What specialty crops are featured?

Google Earth Virtual Tour Video

http://goo.gl/9rKwwt

▼ **Figure 3.3.2 Farmers Market, East New York** The diverse global offerings found at the East New York Farmers Market attract both nearby residents and visitors from Manhattan and beyond.

Within Canada, almost 40 percent of the population is Protestant, with the United Church of Canada claiming large numbers of followers (see Figure 3.28). Roman Catholicism is important in regions that received large numbers of Catholic immigrants. French-Canadian Quebec is a bastion of Catholic tradition and makes Canada's population (43 percent) distinctly more Catholic than that of the United States (24 percent).

Millions of other North Americans practice religions outside of the Protestant and Catholic traditions or are unaffiliated with traditional religions. Orthodox Christians congregate in the urban Northeast, where many Greek, Russian, and Serbian Orthodox communities were established between 1890 and 1920. The telltale domes of Ukrainian Orthodox churches still dot the Canadian prairies of Alberta, Saskatchewan, and Manitoba. More than 5 million Jews live in North America, concentrated in East and West Coast cities. In the United States, the rapidly growing organization known as the Nation of Islam also has a strong urban orientation, reflecting its appeal to many economically dispossessed African Americans. Many other Muslims (6 million), Buddhists (1 million), and Hindus (1 million) also live in the United States. Only about 8 percent of people in the United States classify themselves as nonbelievers, but a recent survey showed that 30 percent of the population claimed to have a largely secular lifestyle in which religion was rarely practiced.

The Globalization of American Culture

North America's culture is becoming more global at the same time that global cultures are becoming more North American (influenced particularly by the United States). The process of cultural globalization itself is becoming more complex. No longer can we think of simple flows of foreign influences into North America or the juggernaut of U.S. cultural dominance invading every traditional corner of the globe. In the 21st century, the story of cultural globalization is increasingly a mix of influences that flow in many directions at once and that feature new hybrid cultural creations.

North Americans: Living Globally More than ever before, North Americans in their everyday lives are exposed to people from beyond the region. With more than 46 million foreign-born migrants living across the region, diverse global influences are free to mingle in new ways. In addition,

▼ **Figure 3.28 Religions of North America** Although many portions of North America feature great religious diversity, selected regions are dominated by Roman Catholicism or various Protestant denominations. Portions of rural Utah and Idaho dominated by the Mormon faith display some of the West's highest concentrations of any single religion.

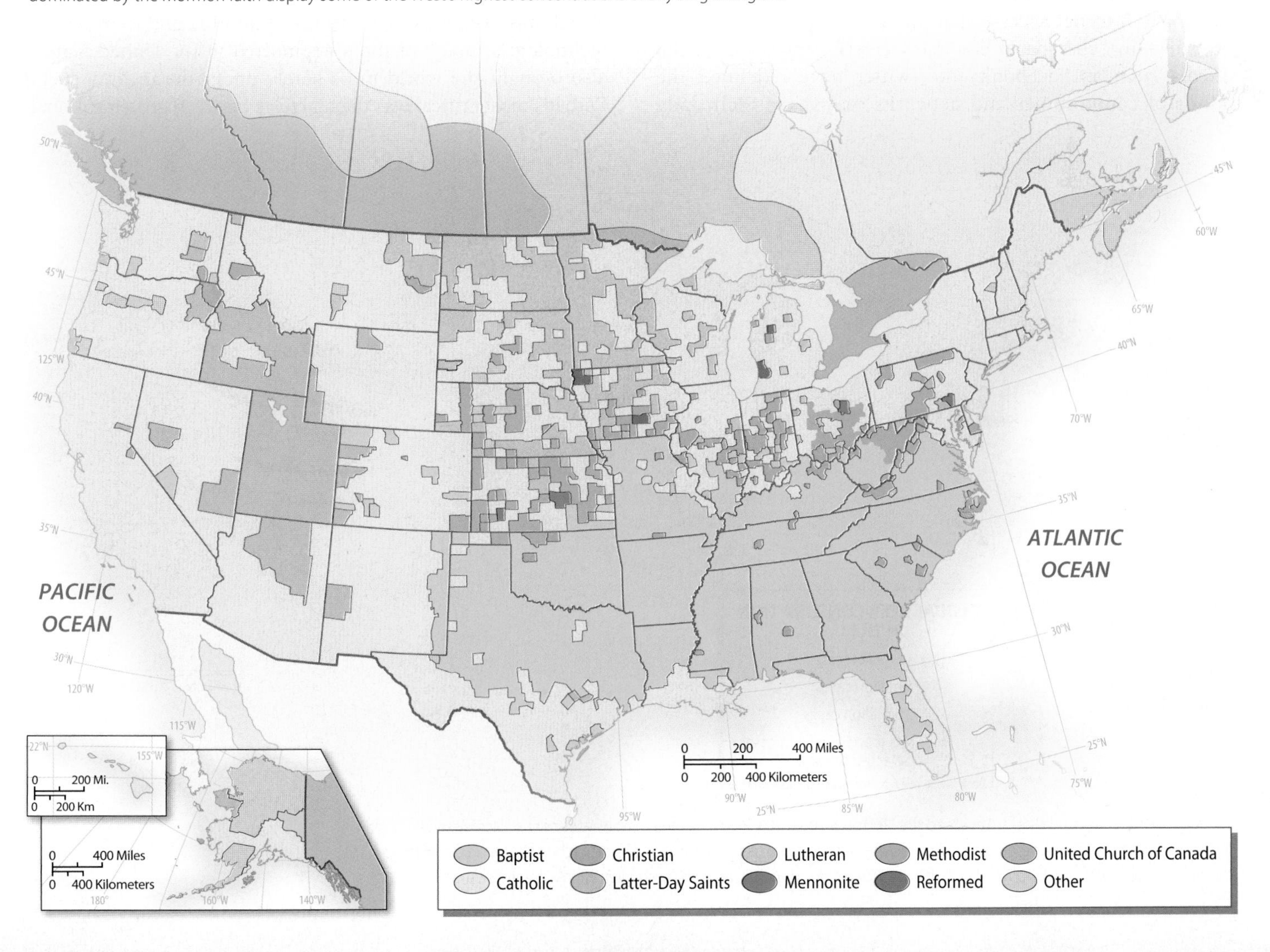

tens of thousands of foreigners arrive daily, mostly as tourists. In 2012, the United States recorded more than 60 million international visitors in the country, many arriving from neighboring Canada and Mexico. At American colleges and universities, more than 760,000 international students (more than half are from Asia) add a global flavor to the ordinary curriculum. Canada experiences a similar saturation of such global influences, with the U.S. presence particularly dominant.

Globalization presents cultural challenges for North Americans. In the United States, one key issue revolves around the English language, which some have described as the "social glue" holding the nation together. Since 1980, the continuing flow of non-English-speaking immigrants into the country has sharpened the debate over the role English should play in U.S. culture. Evidence indicates that North America's immigrants are learning English more rapidly than ever before, seeing it as a powerful way to gain entrance into the economic mainstream, both in the United States and in Canada. The growing popularity of **Spanglish**, a hybrid combination of English and Spanish spoken by Hispanic Americans, also illustrates the complexities of North American globalization. Spanglish includes interesting hybrids such as *chatear,* which means "to have an online conversation."

North Americans are going global in other ways as well. By 2014, the vast majority of Americans and Canadians will have Internet access, opening the door for far-reaching journeys in cyberspace. For many North Americans, social media such as Facebook and Twitter have redefined the kinds of communities and networks that shape their daily lives. North Americans also travel more widely. For example, about 15 percent of U.S. college students now participate in "study abroad" opportunities.

Within North America, the popularity of ethnic restaurants has peppered the region with a bewildering variety of Cuban, Ethiopian, Basque, and Pakistani eateries. Americans consume more than 125 million cases of imported beer annually (Figure 3.29). In another example of globalization, Heineken, famed for its fine Dutch brew, now owns the rights to resell in the United States both the Tecate and the Dos Equis brands, two popular Mexican beers. In fashion, *Gucci, Brioni,* and *Prada* are household words for millions who keep their eyes on European styles. The beat of German techno bands, Gaelic instrumentals, and Latin rhythms is an increasingly seamless part of daily life. Indeed, from acupuncture and massage therapy to soccer and New Age religions, North Americans are tirelessly borrowing, adapting, and absorbing the larger world around them.

The Global Diffusion of U.S. Culture In a parallel fashion, U.S. culture has forever changed the lives of billions of people beyond the region. Although the economic and military power of the United States was notable by 1900, after World War II the country's popular culture reshaped global human geographies in fundamental ways. The Marshall Plan and Peace Corps initiatives exemplified the growing presence of the United States on the world stage, even as European colonialism waned. Rapid improvements in global transportation and information technologies, many of them engineered in the United States, also brought the world more surely under the region's spell. Perhaps most critical was the marriage between growing global

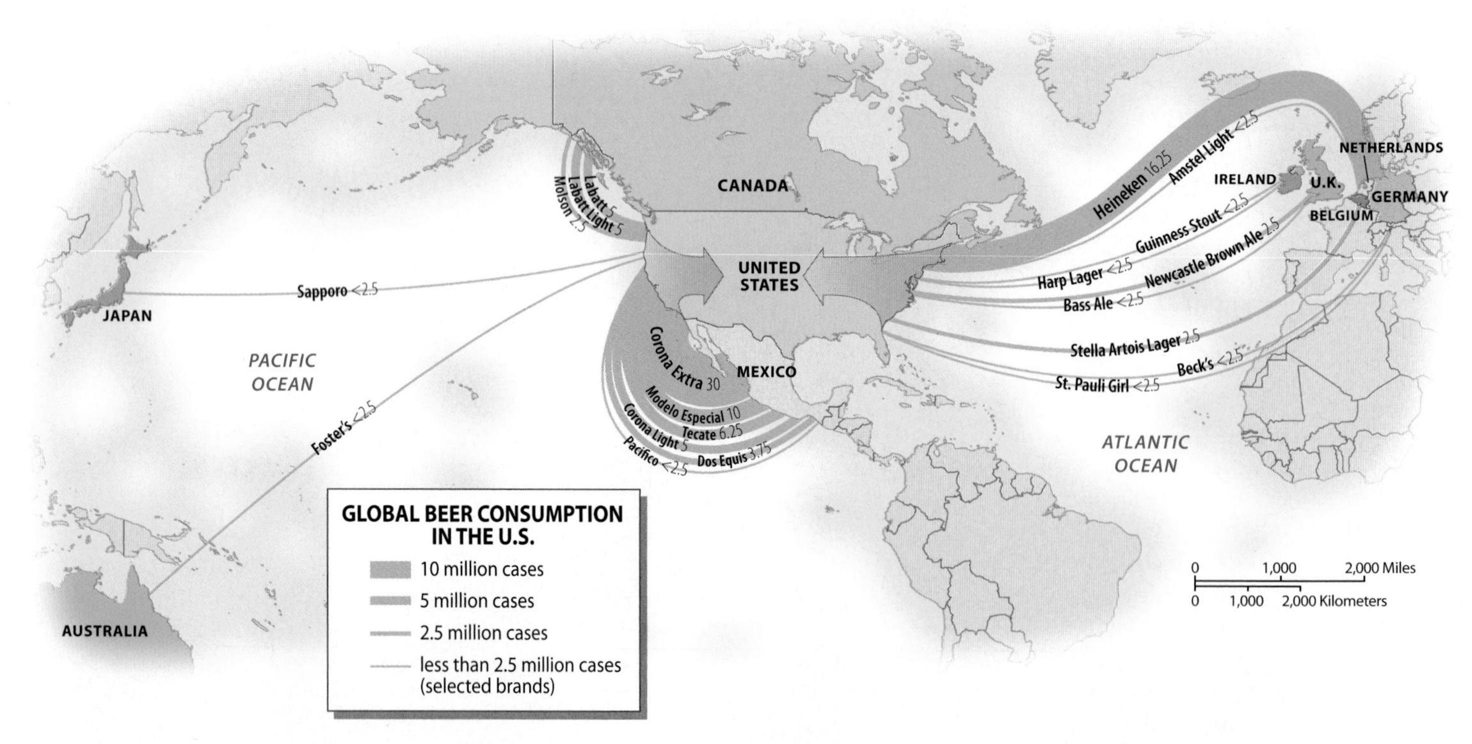

▲ **Figure 3.29 Annual Beer Imports to the United States, 2010** Whether they are aware of it or not, North Americans are increasingly eating and drinking globally. Rising beer imports, including many more expensive foreign brands, exemplify the pattern. The nation's beer drinkers know no bounds to their thirsts. Mexico dominates, along with varied European, Asian, and Australian producers.

demands for consumer goods and the rise of the multinational corporation, which was superbly structured to meet and cultivate those needs.

The results of these connections are not a simple Americanization of traditional cultures or a single, synthesized global culture shaped in the U.S. image. Still, millions of people, particularly the young, are strongly attracted by the North American emphasis on individualism, consumption, youth, and mobility. The wide popularity of English-language teaching programs in places from China to Cuba is testimony to the cultural power of the United States. Indeed, English is already an official language in more than 60 countries around the world. Global information flows illustrate the country's cultural influence on the world. Companies such as Time Warner and Walt Disney dominate multiple entertainment media. Book and magazine publishing is thriving in the United States, and an expanding international appetite has risen for everything from technical journals to romance novels and science fiction. The United States also dominates television; as television sets, cable networks, satellite dishes, and Internet download options spread, so do U.S. sitcoms, CNN, and MTV.

▲ **Figure 3.30 Japan's Love Affair with Baseball** North America's passion for baseball has readily leaped across the Pacific. Today many Japanese follow their favorite teams on both sides of the ocean. In this view, fans of the Yakult Swallows Japan League baseball team sing and hold umbrellas overhead as one of their team's players hits a home run at Tokyo's Jingu Stadium.

The United States shapes contemporary culture around the world. In the built landscape, central-city skylines become indistinguishable from one another, suburban apartment blocks take on a global sameness, and one airport hotel looks the same as another eight time zones away. Global corporate advertising, distribution networks, and mass consumption bring Cokes and Big Macs to Moscow and Beijing, golf courses to Thai jungles, and Mickey and Minnie Mouse to Tokyo and Paris. Western-style business suits have become the professional uniform of choice, while T-shirts and jeans offer standardized global comfort on days away from work.

However, U.S. cultural control has not gone unchallenged, illustrating the varied consequences of globalization. Hollywood's dominance within the global film industry has declined dramatically as filmmakers have built their own movie businesses in India, Latin America, China, and elsewhere. As worldwide use of the Internet has grown, the online global dominance of English-speaking users has dramatically declined, from more than 71 percent in 1998 to only 27 percent in 2010. Not surprisingly, given the rapid diffusion of the Internet and China's growing influence (more than 1 billion Internet users are in Asia alone), Mandarin may soon surpass English as the leading global language of Internet users.

Active resistance to U.S. cultural influence is also notable. For example, Canadian government agencies routinely chastise their radio, television, and film industries for letting in too many U.S. cultural influences. The French also criticize U.S. dominance in such media as the Internet. Elsewhere, Iran has banned satellite dishes and many U.S. films, although illegal copies of top box-office hits often find their way through national borders.

The Globalization of North American Sports Increasingly, major North American sports (hockey, baseball, basketball, football) are transcending national boundaries. At the same time, sports popular in other parts of the world (particularly soccer) are becoming integral elements of North American culture. The so-called *global media sports complex* has faciltated this cultural and economic transformation: Today media companies broadcast sporting events by satellite and cable TV around the globe. Huge investments in league franchises, individual players, stadiums, and related sport-equipment companies, along with growing mobility among professional athletes, have all made it easier for North American sports to go global.

Undoubtedly, North American sports are reshaping the larger world. American-style golf courses are being built in exotic settings such as the United Arab Emirates (now home to the Dubai Desert Classic). American baseball is a passion from the Caribbean to Japan, where American teams have played exhibition games since 2000 (Figure 3.30). Interest in American-style football has also expanded: The National Football League regularly plays exhibition or regular-season games in settings such as London, Toronto, and Mexico City. Although the NFL Europe League (1995–2007) failed financially, it signaled the growing visibility of the sport in countries such as Germany, Great Britain, and the Netherlands.

In the 1992 Olympics, the electrifying performance of the "Dream Team" (led by Michael Jordan) also brought American-style basketball onto the world stage in a bigger way than ever before. Today the sport has a huge following in settings as diverse as Serbia, China, and Argentina (where viewers watch National Basketball Association [NBA] games with a passion and home-grown leagues have blossomed). In fact, since 2008, Chinese planners have mandated the building of basketball courts in every Chinese village, an initiative no doubt spurred by the popularity of Yao Ming in the NBA.

Similarly, North American sports have been transformed by the rest of the world. Within the region, Canadian ice hockey has expanded to become a truly continental sport. Soccer has also emerged on North American shores. Since 1994, the American Youth Soccer Organization has multiplied its presence from Iowa to Alabama. Millions of North American children now participate in league competition.

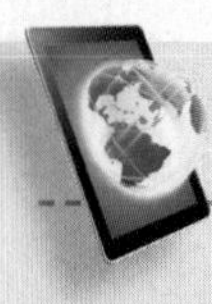

Everyday **Globalization**

The NBA Goes Global

The Shanghai Sharks knew they had a special player. Yao Ming, their talented 7 ft 6 in. star, took them to the Chinese Basketball Association (CBA) finals three times between 1999 and 2002 (in the third year, they won the national championship). Later in 2002, however, things changed forever, for both Chinese and American basketball. Yao entered the American NBA draft and was selected by the Houston Rockets. He became one of the biggest sports celebrities in basketball, both in China and in the United States.

▲ **Figure 3.4.1 International Players in North American Professional Basketball, 2013** This map shows the national origins of active NBA players born outside the United States. Tall, sharpshooting immigrants from Europe and Latin America dominate the pattern, but émigrés from Tanzania to Turkey have been fitted with NBA jerseys.

The popularity of professional soccer teams in many American cities, as well as the growing North American interest in World Cup Soccer events, also illustrates the trend. More broadly, the number of foreign players on professional American teams continues to grow rapidly. In the interconnected world of the 21st century, great talent from Nigeria to Nicaragua can more easily be discovered and imported to North America. Baseball teams feature truly globalized rosters, and many of the region's leading basketball players grew up practicing layups in another part of the world (see *Everyday Globalization: The NBA Goes Global*).

REVIEW

3.6 What are the distinctive eras of immigration in U.S. history, and how do they compare with those of Canada?

3.7 Identify four enduring North American cultural regions, and describe their key characteristics.

Geopolitical Framework: Patterns of Dominance and Division

In disarmingly simple fashion, North America's political geography brings together two of the world's largest countries. The creation of these political entities was the outcome of complex historical processes that might have created quite a different North American map. Once established, the two countries have coexisted in a close relationship of mutual economic and political interdependence. President John F. Kennedy summarized the links in a speech to the Canadian parliament in 1962: "Geography has made us neighbors, history has made us friends, economics has made us partners, and necessity has made us allies." That cozy continental relationship has not been without its tensions, and some persist today. In addition, both countries have had to deal with fundamental internal political complexities that have not only tested the limits of their federal structures, but also challenged their very existence as states.

Going Global Over the next 10 years, Yao exemplified a trend that has increasingly shaped the look of American basketball as talented players from around the globe have been attracted to the fame and fortune of North American professional sports (**Figure 3.4.1**). Just switch on the latest college or professional game or examine the roster of some of the NBA's best teams (recently, the San Antonio Spurs had no fewer than eight foreign players).

Much of the league's inspiration to go global came from David Stern, the commissioner of the NBA from 1984 to 2014. Stern oversaw the league's continental expansion into Canada, promoted televised NBA games overseas (the NBA now broadcasts in 43 languages), and facilitated the North American importation of talented international players, at both the college and the professional levels. Stern also recognized the larger forces at work: In an increasingly interconnected world, it was only logical that a popular North American sport such as basketball could be successfully marketed in other countries and that talent from those countries might ultimately add to the American game.

An International Game Between 2007 (60 players from 28 countries) and 2013 (87 players from 37 countries), the NBA's growing global harvest has been impressive. Recently, some of the league's top players included brothers Marc and Pau Gasol (Spain), Dirk Nowitski (Germany), Hedo Türkoğlu (Turkey), and Manu Ginóbili (Argentina) (**Figure 3.4.2**). The global flow of players destined for the North American court is dominated by Europe (France [11] and Spain [6] are key source countries), but the reach for talent extends deep into Latin America, Sub-Saharan Africa, and China (Figure 3.4.1). The trend seems likely to continue as the global popularity of the game grows and as lucrative NBA salaries grow.

Finally, one expression of the game's interconnected and global presence is revealing: A recent roster of the Shanghai Sharks includes Gilbert Arenas (who grew up in California), D. J. White (Alabama), and Elijah Milsap (Alabama), former North American players who now put up 15-foot jump shots for a very different audience.

► **Figure 3.4.2 Global Giants Clash in the NBA** Players from Spain (Pau Gasol-#16) and Germany (Dirk Nowitski-#41) meet on the court in this NBA contest between the Los Angeles Lakers and the Dallas Mavericks.

Creating Political Space

The United States and Canada sprang from very different political roots. The United States broke cleanly and violently from England. The American Revolution fostered a powerful sense of nationalism that sped the process of spatial expansion and contributed toward the creation of a continental, indeed global, power. By contrast, Canada was a country of convenience, born from a peaceful, incremental separation from Britain and then assembled as a collection of distinctive regional societies that only gradually acknowledged their common political destiny.

Uniting the States Turning back the clock to the early 18th century reveals a political geography very much in the making. Beyond scattered frontiers of European settlement lay vast domains of Native American–controlled political space. Although boundaries were not formally surveyed or mapped, native peoples carved up the continent in an elaborate geography of homelands, allied territories, and enemy terrain. The creation of the United States replaced this continental division of political space with another.

With amazing rapidity, European nations and then the United States imposed their own political boundaries across the region. The 13 English colonies, sensing their common destiny after 1750, finally united two decades later and clashed violently with their colonial parent. By the 1790s, the young nation's political claims had reached the Mississippi River; the new republic was busily coercing land cessions from native peoples; and the Ordinance of 1787 provided a template for western territory and state formation that served as the model of expansion for the next century. Soon the Louisiana Purchase (1803) nearly doubled the country's size, creating a new political domain that was as vast as it was unexplored. By mid-century, Texas had been annexed; treaties with Britain had secured the Pacific Northwest;

◄ **Figure 3.31 Satellite Image of the Great Lakes** North America's Great Lakes region features one of the most environmentally complex political boundaries in the world. Canada and the United States share responsibility (at a variety of local, state/provincial, and federal levels) for managing the ecological health of the five Great Lakes (from west to east: Superior, Michigan, Huron, Erie, and Ontario). **Q: Which of the Great Lakes define international borders, and what states and provinces do they separate?**

and an aggressive war with Mexico had captured much of the Southwest. The territorial acquisition of Alaska (1867) and Hawaii (1898) eventually rounded out the present political domain of 50 states.

Assembling the Provinces Canada was created under quite different circumstances. The modern pattern of provinces was assembled in a slow and uncertain fashion. After the American Revolution, England's remaining territorial claims in the region came under the control of colonial administrators in British North America. The Quebec Act of 1774 allowed for continued French settlement in the St. Lawrence Valley and provided the initial template for governing the region. Soon, however, Anglo settlers near Lakes Ontario and Erie pressed for more local colonial representation. The result was the Constitutional Act of 1791, which divided the colony into Upper Canada (Ontario) and Lower Canada (Quebec). Frustration with that system led to the Act of Union in 1840, thereby reuniting the two Canadas. In 1867, the British North America Act united the provinces of Ontario, Quebec, Nova Scotia, and New Brunswick in an independent Canadian Confederation. This peaceful separation from the mother country also guaranteed to Quebec special legal and cultural privileges that set the stage for some of the modern power struggles within the country.

Once created, the Canadian Confederation grew in piecemeal fashion, more out of geographical convenience than from any compelling nationalism at work to unite the northern portion of the continent. Within a decade, the Northwest Territories (1870), Manitoba (1870), British Columbia (1871), and Prince Edward Island (1873) joined Canada, and the continental dimensions of the country took shape. Soon, the Yukon Territory (1898) separated from the Northwest Territories; Alberta and Saskatchewan gained provincial status (1905); and Manitoba, Ontario, and Quebec were enlarged (1912) north to Hudson Bay. Newfoundland finally joined in 1949. The creation of Nunavut Territory (1999), carved from the Northwest Territories, represents the latest change in Canada's political geography.

Continental Neighbors

Geopolitical relationships between Canada and the United States have always been close: Their common 5525-mile (8900-km) boundary requires both nations to pay close attention to one another. During the 20th century, the two countries lived largely in political harmony with one another.

Sharing the Great Lakes The Great Lakes, in particular, have been a setting for remarkable continental cooperation (Figure 3.31). In 1909, the Boundary Waters Treaty created the International Joint Commission, an early step in the common regulation of cross-boundary issues involving Great Lakes water resources, transportation, and environmental quality. The St. Lawrence Seaway (1959) opened the Great Lakes region to better global trade connections. With the signing of the Great Lakes Water Quality Agreement (1972) and the U.S.–Canada Air Quality Agreement (1991), the two nations have joined more formally in cleaning up Great Lakes pollution and in reducing acid rain in eastern North America. In 2012, these environmental agreements were updated, opening the way to new cooperative cleanup efforts between the two countries.

Close Trading Connections The two nations are also key trading partners. The United States receives about 73 percent of Canada's exports and supplies 63 percent of its imports. Conversely, Canada is the United States' most important trading partner, accounting for roughly 20 percent of its exports and 15 percent of its imports. A landmark agreement reached in 1989 was the signing of the bilateral (two-way) Free Trade Agreement (FTA). Five years later, the larger **North American Free Trade Agreement (NAFTA)** extended the alliance to Mexico. Paralleling the success of the European Union (EU), NAFTA has forged the world's largest trading bloc, including more than 450 million consumers and a huge free-trade zone that stretches from beyond the Arctic Circle to Latin America. However, NAFTA has also posed challenges for the region: U.S. jobs have been lost to Mexico, and Mexico's economy has become increasingly dependent on demand from the United States, creating problems in times of declining consumption in North America.

Continuing Conflicts Political conflicts occasionally still divide North Americans (Figure 3.32). Environmental issues produce cross-border tensions, especially when environmental degradation in one nation affects the other. In addition to the intricate trans-boundary mingling of the Great Lakes, other regional water issues are common, since so many drainage systems cross the border. For example, Canada has protested North Dakota's plans to control the north-flowing Red River (which leads into Manitoba), while some Montana residents are nervous that Canadian logging and mining interests in British Columbia will increase pollution on the south-flowing North Flathead River.

More generally, tighter U.S. regulations since 2009 have made it more difficult to cross the border in either direction. Reflecting security concerns in the United States, the world's longest "open border" now sees more surveillance drones and border agents than ever before. New regulations demand that persons crossing the border present a passport or other approved form of identification, just as they would on the border with Mexico. Many people in Canada cite the new rules as potentially harmful to tourism between the two countries.

Agricultural and natural resource competition has also caused periodic controversy between the two neighbors. The appearance of mad cow disease in Canadian livestock curtailed exports to the United States and elsewhere. Furthermore, when the disease appeared in U.S. cattle in 2003,

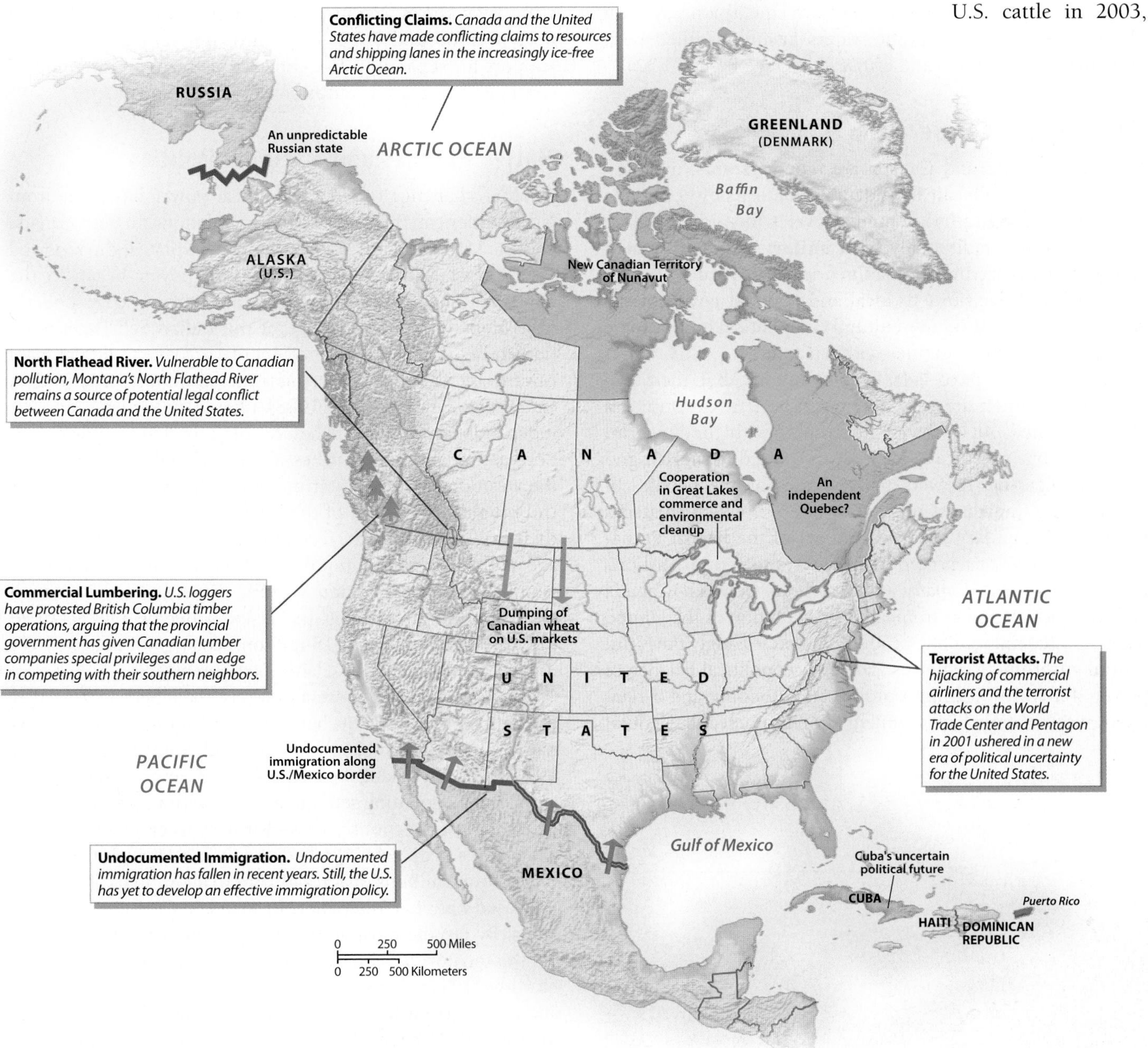

▲ **Figure 3.32 Geopolitical Issues in North America** Although Canada and the United States share a long and peaceful border, many political issues still divide the two countries. In addition, internal political conflicts cause tensions, particularly in multicultural Canada.

Canadian sources were suspected, raising tensions between the two nations. Problems have periodically developed when Canadian wheat and potato growers were accused of dumping their products into U.S. markets, thus depressing prices and profits for U.S. farmers. Similar issues have arisen in the logging industry, although a 2006 agreement signed between the two countries has lessened tensions over that issue.

In the far north, the two countries also disagree on the maritime boundary between the Yukon and the state of Alaska. In addition, the United States does not agree with the assertion that a potential Northwest Passage opening across a more ice-free Arctic Ocean would essentially be within Canada's territorial waters. From their perspective, in a warmer world, Canadian officials are equally uneasy with the prospect of "their" arctic waters becoming filled with unregulated oil tankers and cruise ships (see Figure 3.32).

The Legacy of Federalism

The United States and Canada are **federal states**, meaning that both nations allocate considerable political power to units of government beneath the national level. Other nations, such as France, have traditionally been **unitary states**, in which power is centralized at the national level. Federalism leaves many political decisions to local and regional governments and often allows distinctive cultural and political groups to be recognized as distinct entities within a country.

Both nations have federal constitutions, but their origins and evolution are very different. The U.S. Constitution (1787), created out of a violent struggle with the powerful British nation, specifically limited centralized authority, giving all unspecified powers to the states or the people. In contrast, the Canadian Constitution (1867), which created a federal parliamentary state, was an act of the British Parliament. Originally, it reserved most powers to central authorities and maintained many political links between Canada and the British Crown. Ironically, the evolution of the United States as a federal republic produced an increasingly powerful central government, whereas Canada's geopolitical balance of power shifted toward more provincial autonomy and a relatively weak national government. For example, the federal government largely controls U.S. public lands, but in Canada provincial authorities retain power over public Crown lands.

Quebec's Challenge The political status of Quebec remains a major issue within Canada (see Figure 3.32). Economic disparities between the Anglo and French populations have reinforced cultural differences between the two groups, with the French Canadians often suffering when compared with their wealthier neighbors in Ontario. Beginning in the 1960s, a separatist political party in Quebec (the Parti Quebecois) increasingly voiced French Canadian concerns. When the party won provincial elections in 1976, it declared French the official language of Quebec. Formal provincial votes over the question of remaining within Canada were held in 1980 and 1995: Both measures failed. Since then, support for separation has ebbed in favor of a more modest strategy of increased "autonomy" within Canada.

Native Peoples and National Politics

Another challenge to federal political power has come from North American Indian and Inuit populations, in both Canada and the United States. Within the United States, the renewed assertion of Native American political power began in the 1960s and marked a decisive turn away from earlier policies of assimilation. Since passage of the Indian Self-Determination and Education Assistance Act of 1975, the trend has been toward increased Native American autonomy. The Indian Gaming Regulatory Act (1988) offered potential economic independence for many tribes. By 2012, Indian gaming operations nationally netted tribes about $27 billion annually. In the western American interior, where Native Americans control roughly 20 percent of the land, tribes are also solidifying their hold on resources, reacquiring former reservation acreage, and participating in political interest groups, such as the Native American Fish and Wildlife Society and the Council of Energy Resource Tribes. In Alaska, native peoples acquired title to 44 million acres (18 million hectares) of land in 1971 under the Alaska Native Claims Settlement Act.

In Canada, even more ambitious challenges to a weaker centralized government have yielded dramatic results. As natives pressed their claims for land and political power in the 1970s, Canada established the Native Claims Office (1975) and began negotiating settlements with various groups, particularly within the country's vast northern interior. Agreements with native peoples in Quebec, Yukon, and British Columbia turned over millions of acres of land to aboriginal control and increased native participation in managing remaining public lands. By far, the most ambitious agreement has been to create the territory of Nunavut out of the eastern portion of the Northwest Territories in 1999 (see Figure 3.32). Nunavut is about 85 percent Inuit and is the largest territorial/provincial unit within Canada (Figure 3.33). Its creation represents a new level of native self-government in North America, par-

◄ **Figure 3.33 Life in Nunavut** This woman navigates the unpaved streets of Pond Inlet on an ATV. The small hamlet is a rugged outpost on the northern end of Baffin Island, part of Canada's Nunavut Territory.

ticularly significant in a part of the world witnessing rapid climate change. Recently, agreements between the federal Parliament and British Columbia tribes have initiated a similar move toward more native self-government in that western province. The 6400 members of the Nisga'a tribe, for example, now control a 770-square-mile (1992-square-km) portion of the Nass River Valley near the Alaska border (see Figure 3.25). Elsewhere, recent disputes over land claims in Ontario (near Toronto) and in oil-rich Alberta are reminders that Canadians, much like their neighbors to the south, are still struggling with defining the political status of their native populations. In early 2013, more coordinated First Nations protests by Canadian tribes charged the federal government with failing to resolve treaties, fund native programs, and guarantee environmental oversight of native lands.

The Politics of U.S. Immigration

Immigration policies have long been hotly contested within the United States. Four key issues remain at the center of the debate. First, there are ongoing disagreements concerning the overall numbers of legal immigrants that should be allowed into the country. Some groups, such as the Federation for American Immigration Reform (FAIR), suggest that sharply reduced numbers of immigrants would protect American jobs and allow for a more gradual assimilation of existing foreigners. Other groups, such as the Cato Institute, take the opposite position, proposing to loosen existing restrictions on immigrants in a move to spur economic growth and business expansion.

A second major issue, particularly along the border with Mexico, is how to tighten up on the daily flow of undocumented immigrants. Since the terrorist attacks in 2001, many have argued that the country's wide-open southern border is a national security issue. Recent federal legislation has mandated an increased number of border patrol agents, and more than 20,000 officers now monitor the boundary. More than 700 miles (1125 km) of fencing also have been built or improved (Figure 3.34). Others argue such measures are pointless and merely sour relations with Mexico. Meanwhile, hundreds of thousands of foreigners are apprehended along the border annually, although the economic recession of the late 2000s has reduced the numbers since 2008.

Third, an epidemic growth of drug-related violence near the border has soured relations between the two countries. Mexico remains the leading source of methamphetamine, heroin, and marijuana for the United States and is a key transit nation for northward-bound cocaine originating in South America. In addition, more than 30,000 deaths, mostly in northern Mexico, were tied to the drug business between 2007 and 2012. Some of that violence has spilled north into places such as El Paso and Phoenix. Many American officials worry that the Mexican government has fundamentally lost effective political control of its northern border.

Finally, a great deal of political debate has taken place about how to deal fairly with millions of existing undocumented workers within the United States. Some policymakers have advocated stricter felony-level penalties for these immigrants, but most politicians have proposed loosening requirements for citizenship or some form of amnesty to enable undocumented immigrants to more easily enter the mainstream of American society. Initial hopes for a nationwide "Dream Act" (first proposed in 2001), which would grant

▼ **Figure 3.34 International Border** North America's southwestern landscape is boldly divided by an increasingly hardened international border that separates the United States and Mexico.

conditional legal residency for immigrants, were met with hostile responses from some officials, especially in Arizona, where new laws passed in 2010 (Arizona Senate Bill 1070) made it easier to stop and question suspected undocumented residents. Finally, after President Obama was reelected in 2012 with a huge plurality of Hispanic voters, national consensus emerged among Democrats and Republicans to address the issue in 2013. New legislative proposals focused on tightening the border, but also on providing a "pathway to citizenship" for undocumented workers and their children.

A Global Reach

The geopolitical reach of the United States, in particular, has taken its influence far beyond the bounds of North America. The Monroe Doctrine (1824) asserted that U.S. interests were hemispheric and transcended national boundaries, but not until after 1895 did the United States accelerate its global expansion. Two principal settings served as early laboratories for political imperatives in the United States. In the Pacific, the United States claimed the Philippines as a prize of the Spanish-American War (1898), and further annexations that year of Guam and the Hawaiian Islands began the country's 20th-century dominance of the region. In Central America and the Caribbean, the growing role of the American military between 1898 and 1916 shaped politics in Cuba, Puerto Rico, Panama, Nicaragua, Haiti, Mexico, and elsewhere. Further, the country's role in World War I raised its stakes in European affairs.

The 1920s and 1930s briefly returned the United States to isolationist policies, but World War II and its aftermath forever redefined the country's role in world affairs. Victorious in both the Atlantic and the Pacific theaters, postwar America emerged from the conflict as the world's dominant political power. Quickly, however, a resurgent Soviet Union challenged the United States, and the Cold War began in the late 1940s. In response, the Truman Doctrine promised aid to struggling postwar economies and actively challenged communist expansion in Europe and elsewhere. The United States also fashioned multinational political and military agreements, such as those establishing the North Atlantic Treaty Organization (NATO) and the Organization of American States (OAS), which were designed to cast a broad umbrella of U.S. protection across much of the noncommunist world. Violent conflicts in Korea (1950–1953) and Vietnam (1959–1975) pitted U.S. political interests against communist attempts to extend their Asian dominance beyond the Soviet Union and China. Tensions also ran high in Europe as the Berlin Wall crisis (1961) and nuclear weapons deployments by NATO- and Soviet-backed forces brought the world closer to another global war. The Cuban missile crisis (1962) reminded Americans that traditional political boundaries provide little defense in a world uneasily brought closer together by technologies of potential mass destruction.

Even as the Cold War gradually receded during the late 1980s, the global political reach of the United States continued to expand. A few examples suggest the pattern. Interventionist policies in Central America favored regimes friendly to the United States. President Carter's successful Middle East Peace Treaty between Israel and Egypt (1979) guaranteed a continuing diplomatic and military presence in the eastern Mediterranean. When Iraq's Saddam Hussein threatened Persian Gulf oil supplies in 1990, the United States led a UN coalition to contain the aggression. In the late 1990s, Serbian aggression within Kosovo prompted an American- and NATO-led intervention, which included major air attacks on the Serbian capital of Belgrade (1999) and a peacekeeping presence (with the UN) in the disputed area of Kosovo. Recent controversial wars in Iraq (2003–2011) and Afghanistan (from 2001) and an annual defense budget ($680 billion in 2011) nearly as much as the defense budgets of the rest of the world combined suggest that the United States will continue playing a key, highly visible role in the world's political affairs.

The geographic distribution of the U.S. military has also changed dramatically since 2000. Military planners argue that in the future, less emphasis should be given to housing large numbers of troops in relatively friendly foreign "hub" settings, such as Germany, South Korea, and Japan. A growing number of the U.S. military's 1.4 million active-duty personnel appear headed for more spartan assignments, located near source regions for terrorists and in arenas of potential conflict. From the drug- and rebel-filled jungles of Colombia to the terrorist training camps of the Philippines, the region of growing global deployment also includes much of central and northern Africa, the Middle East, Central Asia, and Southeast Asia. In addition, the growing use of drones in warfare may have long-term impacts on where U.S. troops are stationed in the future. With drones, for example, recent bombing runs in Afghanistan were "flown" by pilots stationed half a world away at air force installations in the Nevada desert.

REVIEW

3.8 How do the political origins of the United States and Canada differ, and what issues divide these nations today?

3.9 What are the four key elements surrounding U.S. immigration policy?

Economic and Social Development: Patterns of Abundance and Affluence

Along with its global political clout, North America possesses the world's most powerful economy and its most affluent population. Its 350 million people consume huge quantities of global resources, but also produce some of the world's most sought-after manufactured goods and services. North America's size, geographic diversity, and resource abundance have all contributed to the region's global dominance in economic affairs. More than that, however, the region's human capital—the skills and diversity of its population—has enabled North Americans to achieve high levels of economic development (Table 3.2).

An Abundant Resource Base

North America is blessed with a varied storehouse of natural resources. The region's climatic and biological diversity, its soils and terrain, and its abundant energy, metals, and forest resources have provided a variety of raw materials for development. Indeed, the direct extraction of natural resources still

Table 3.2 Development Indicators

Country	GNI per Capita PPP, 2011	GDP Average Annual % Growth (2000–2011)	Human Development Index (2011)[1]	Percent Population Living Below $2 a Day	Life Expectancy (2013)[2]	Under Age 5 Mortality Rate (1990)	Under Age 5 Mortality Rate (2010)	Adult Literacy (% ages 15 and older) Male/Female	Gender Inequality Index (2011)[3,1]
Canada	39,660	1.9	.911	–	81	8	6	–/–	0.119
United States	48,820	1.6	.937	–	79	11	8	–/–	0.256

[1]United Nations, *Human Development Report, 2013.*

[2]Population Reference Bureau, *World Population Data Sheet, 2013.*

[3]Gender Inequality Index—A composite measure reflecting inequality in achievements between women and men in three dimensions: reproductive health, empowerment and the labor market that ranges between 0 and 1. The higher the number, the greater the inequality.

Sources: World Bank, *World Development Indicators, 2013.*

makes up 3 percent of the U.S. economy and more than 6 percent of the Canadian economy. Some of these North American resources are then exported to global markets, while other raw materials are imported to the region.

Opportunities for Agriculture North Americans have created one of the most efficient food-producing systems in the world, and agriculture remains a dominant land use across much of the region (Figure 3.35). Farmers practice highly

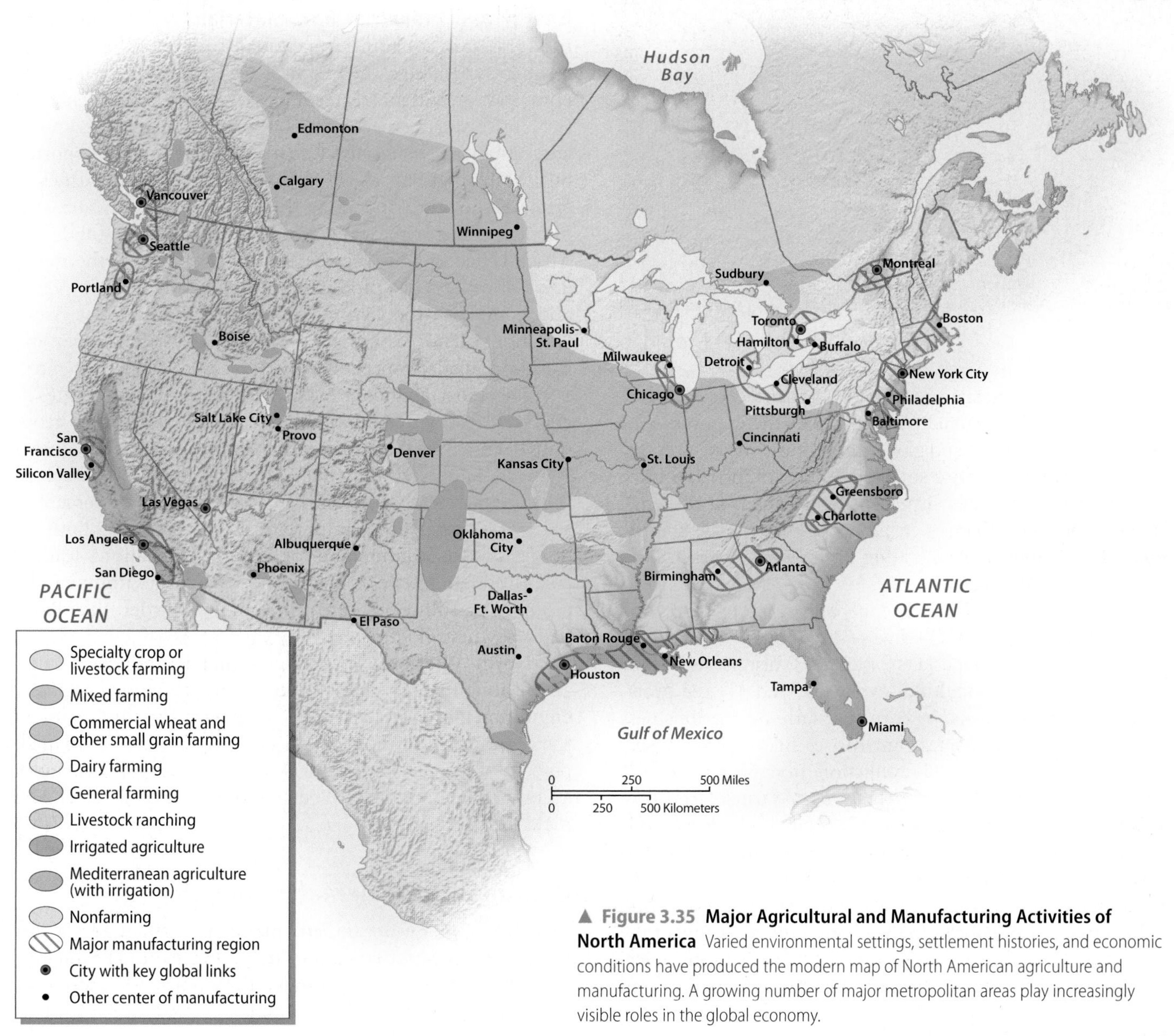

▲ **Figure 3.35 Major Agricultural and Manufacturing Activities of North America** Varied environmental settings, settlement histories, and economic conditions have produced the modern map of North American agriculture and manufacturing. A growing number of major metropolitan areas play increasingly visible roles in the global economy.

◀ **Figure 3.36 Ginseng Farm, Marathon County, Wisconsin** Wealthy Chinese consumers are demanding high-quality American ginseng, and they are reshaping global patterns of trade in this age-old commodity. This ginseng farm in Marathon County, Wisconsin, exports its crop to China and also illustrates the labor-intensive nature of this agricultural operation.

commercialized, mechanized, and specialized agriculture (Figure 3.36). The system also emphasizes the importance of efficient transportation and global agricultural markets. Today agriculture employs only a small percentage of the labor force in both the United States (1 percent) and Canada (2 percent). At the same time, changes in farm ownership have sharply reduced the number of operating units, while average farm sizes have steadily risen.

The geography of North American farming represents the combined impacts of (1) diverse environments; (2) varied continental and global markets for food; (3) historical patterns of settlement and agricultural evolution; and (4) the growing role of **agribusiness**, or corporate farming. Agribusiness involves large-scale business enterprises that control closely integrated segments of food production, from farm to grocery store. In the Northeast, dairy operations and truck farms take advantage of their nearness to major cities in Megalopolis and southern Canada. Corn and soybeans dominate the Midwest and western Ontario, where a tradition of mixed farming combines the growing of feed grains with the production and fattening of livestock. To the south, only remnants of the old Cotton Belt remain, largely replaced by varied subtropical specialty crops; poultry, catfish, and livestock production; and commercial logging. Farther west, extensive, highly mechanized commercial grain-growing operations stretch from Kansas to Saskatchewan and Alberta. Depending on surface water and groundwater resources, irrigated agriculture across western North America also offers opportunities for farming. Indeed, California's agricultural output, nourished particularly by large agribusiness operations in the irrigated Central Valley, accounts for more than 10 percent of the nation's farm economy.

Energy and Industrial Raw Materials North Americans produce and consume huge quantities of other natural resources. Although the region is well endowed with a variety of energy and metals resources, the scale and diversity of the North American economy have led to the need to import additional raw materials. Canada is a net energy exporter, but the United States still imports substantial quantities of crude oil, much of it coming from the Americas (Venezuela, Mexico, and Canada), the Middle East (Saudi Arabia), and Africa (Nigeria and Angola). Within the region, the major areas of oil and gas production are the Gulf Coast of Texas and Louisiana; the Central Interior, including West Texas, Oklahoma, and eastern Kansas; Alaska's North Slope; and Central Canada (especially Alberta).

The most abundant fossil fuel in the United States is coal, but its relative importance in the overall energy economy declined in the 20th century as industrial technologies changed and environmental concerns grew. Nonetheless, the country's 400-year supply of coal reserves (27 percent of the world's total) is an important energy resource, both for domestic consumption and for export. The nation's leading coal-producing state is Wyoming, where environmentally cleaner low-sulfur coals are mined in the Powder River Basin and elsewhere (Figure 3.37).

North America also remains a major producer of metals, although global competition, rising extraction costs, and environmental concerns pose challenges for this sector of the economy. Still, the region is endowed with more than 20 percent of the world's copper, lead, and zinc reserves, and it accounts for more than 20 percent of global gold, silver, and nickel production.

Creating a Continental Economy

The timing of European settlement in North America was critical in its rapid economic transformation. The region's abundant resources came under the control of Europeans possessing new technologies that reshaped the landscape and reorganized its economy. By the 19th century, North Americans actively contributed to those technological changes. In addition, new natural resources were developed in the interior, and new immigrant populations arrived in large numbers. In the 20th century, although natural resources remained important, industrial innovations and more jobs in the service sector added to the economic base and extended the country's global reach.

Connectivity and Economic Growth Dramatic improvements in North America's transportation and communication systems laid the foundation for urbanization, industrialization, and the commercialization of agriculture. Indeed, the region's economic success was a function of its **connectivity**, or how well its different locations became linked with one another through improved transportation and communications networks. Those links greatly facilitated the potential for interaction between places and dramatically reduced the cost of moving people, products, and information over long distances. Before 1830, North American connectivity gradually improved with the help of better roads, but it still took a week to travel from New York City to Ohio. For commercial traffic, canal construction also facilitated the movement of bulky goods. More than 1000 miles (1600 km) of canals crisscrossed the eastern United States, including the Erie Canal (completed in 1825), which linked New York City and its shipping port with the North American interior.

Tremendous technological breakthroughs revolutionized North America's economic geography between 1830 and 1920. By 1860, more than 30,000 miles (48,000 km) of railroad track had been laid in the United States, and the network grew to more than 250,000 miles (400,000 km) by 1910. Farmers in the Midwest and Plains found ready markets for their products in cities hundreds of miles away. Industrialists collected raw materials from faraway places, processed them, and shipped manufactured goods to their final destinations. The telegraph brought similar changes to information: Long-distance messages flew across eastern North America by the late 1840s, and 20 years later undersea cables linked the region to Europe, another milestone in the process of globalization.

North America's transportation and communications systems were modernized further after 1920. Automobiles, mechanized farm equipment, paved highways, commercial air links, national radio broadcasts, and dependable transcontinental telephone service reduced the cost of travel and communication across the region. After World War II, continental connectivity also benefited from the St. Lawrence Seaway between the Atlantic Ocean and Great Lakes (1959), vast improvements in jet airline connections, and the increasing prevalence of television. Perhaps most important, the region has taken the lead in the global information age, integrating computer, satellite, telecommunications, and Internet technologies in a web of connections that assists the flow of knowledge both within the region and beyond. Since 1990, North America's transformative role in information technologies has been one of its most profound, innovative contributions to globalization and global economic change.

The Sectoral Transformation Changes in employment structure signaled North America's economic modernization just as surely as its increasingly interconnected society did. The **sectoral transformation** refers to the evolution of a nation's labor force from one dependent on the *primary* sector (natural resource extraction) to one with more employment in the *secondary* (manufacturing or industrial), *tertiary* (services), and *quaternary* (information processing) sectors. For example, with agricultural mechanization, demands

▼ **Figure 3.37 Black Thunder Coal Mine** Wyoming's Black Thunder Mine, located in the Powder River Basin, is one of the largest coal producers in North America.

▲ Figure 3.38 **Gulf Coast Petroleum Refining** Petroleum-related manufacturing has transformed many Gulf Coast settings. Much of Houston's 20th-century growth was fueled by the dramatic expansion of oil-related industries. The port of Houston remains a major center of North America's refining and petrochemical operations.

for primary-sector workers drop and are replaced by new opportunities in the growing industrial sector. In the 20th century, new services (trade, retailing) and information-based activities (education, data processing, research) created other employment opportunities. Today the tertiary and quaternary sectors employ more than 70 percent of the labor force in both Canada and the United States. These recent trends also reveal the tangible imprint of globalization on the North American labor force. Much of North America's sustained growth in the tertiary and quaternary sectors is directly tied to the ability of those industries to export innovations in services, financial management, and information processing to a worldwide clientele.

Regional Economic Patterns North America's industries show important regional patterns. **Location factors** are the varied influences that explain *why* an economic activity is located where it is. Many influences, both within and beyond the region, shape patterns of economic activity. Patterns of industrial location illustrate the concept (see Figure 3.35). The historical manufacturing core includes Megalopolis (Boston, New York, Philadelphia, and Baltimore), southern Ontario (Toronto and Hamilton), and the industrial Midwest. The region's proximity to *natural resources* (farmland, coal, and iron ore); increasing *connectivity* (canals and railroad networks, highways, air traffic hubs, and telecommunications centers); a ready supply of *productive labor;* and a growing national, then global, *market demand* for its industrial goods encouraged continued *capital investment* within the core.

Traditionally, the core has dominated in the production of steel, automobiles, machine tools, and agricultural equipment and played a critical role in producer services such as banking and insurance. In the last half of the 20th century, industrial- and service-sector growth shifted to the South and West. Cities of the South's Piedmont manufacturing belt (Greensboro to Birmingham) grew after 1960, partly because lower labor costs and Sun Belt amenities attracted new investment. For example, the North Carolina "research triangle" area encompassing Raleigh, Durham, and Chapel Hill has emerged to become the nation's third largest biotech cluster behind California and Massachusetts. The Gulf Coast industrial region is strongly tied to nearby fossil fuels that provide raw materials for its many energy-refining and petrochemical industries (Figure 3.38).

The varied West Coast industrial region stretches from Vancouver, British Columbia, to San Diego, California (and beyond into northern Mexico), and it demonstrates the increasing importance of Pacific Basin trade. Large aerospace operations in the West also suggest the role of *government spending* as a location factor. Silicon Valley is one of North America's leading regions of manufacturing exports. Its proximity to Stanford, Berkeley, and other universities demonstrates the importance of *access to innovation and research* for many fast-changing high-technology industries.

Silicon Valley's location also shows the advantages of *agglomeration economies,* in which many companies with similar and often integrated manufacturing operations locate near one another (Figure 3.39). Smaller places such as Provo, Utah, and Austin, Texas, also specialize in high-technology industries and demonstrate the role of *lifestyle amenities* in shaping industrial location decisions, both for entrepreneurs and for the skilled workers who need to be attracted to such opportunities.

▲ **Figure 3.40 Gated America** Emerald Bay is a high-end gated community that offers luxury living in this upscale portion of Laguna Beach, California.

Enduring Social Issues

Profound economic and social problems shape the human geography of North America. Even with its continental wealth, great differences persist and have increased between rich and poor. High per capita incomes in the United States and Canada fail to reveal the differences in wealth within the two countries (see Table 3.2). Broader measures of social well-being suggest significant disparities in health care and education. Race, particularly within the United States, continues to be an issue of overwhelming importance. In addition, both nations face problems associated with gender inequity and aging populations. One consequence of globalization is that many of these economic and social challenges are increasingly defined beyond the region. Poverty in the rural American South may be related to low Asian wage rates, for example, and a viral outbreak in Hong Kong might be only a plane flight away from suburban Vancouver.

Wealth and Poverty The global economic downturn of the late 2000s rippled through the economy of both the United States and Canada. Unemployment levels soared between 2008 and 2011 and hovered between 8.5 and 10 percent. In the United States, young and poor African Americans and Hispanics typically fared the worst, with much higher rates of unemployment and lower overall incomes. Real estate values and home ownership rates fell as many people lost their homes to foreclosure. Poverty rates rose for the first time in years. Tax revenues in both nations fell. States such as California, Nevada, and Arizona, once home to real estate and construction booms, witnessed some of the most dramatic economic declines. By 2013, unemployment fell slightly in both the United States and Canada, but many families still struggled with stagnant wages and rising health-care and education costs.

The distribution of wealth and poverty varies widely across the United States and Canada. Elite northeastern suburbs, gated California neighborhoods, upscale shopping malls, and posh alpine ski resorts are all expressions of private and exclusive landscape settings that characterize wealthier North American communities (Figure 3.40). Many of America's wealthiest communities are suburbs on the edge of large metropolitan areas. Just outside Washington, DC, the area of Fairfax County, Virginia, is one of the nation's wealthiest counties; similar settings are found in suburban Connecticut, New Jersey, and Maryland, as well as on the peripheries of dozens of southern and western cities. Resort and retirement communities are havens for the rich as well: Palm Beach, Florida, and Aspen, Colorado, have some of the nation's most desirable real estate and costliest housing.

▼ **Figure 3.39 Silicon Valley** The high-technology industrial landscape of California's Silicon Valley contrasts sharply with the look of traditional manufacturing centers. Here similar industries form complex links, benefiting from their proximity to one another and to nearby universities such as Stanford and Berkeley.

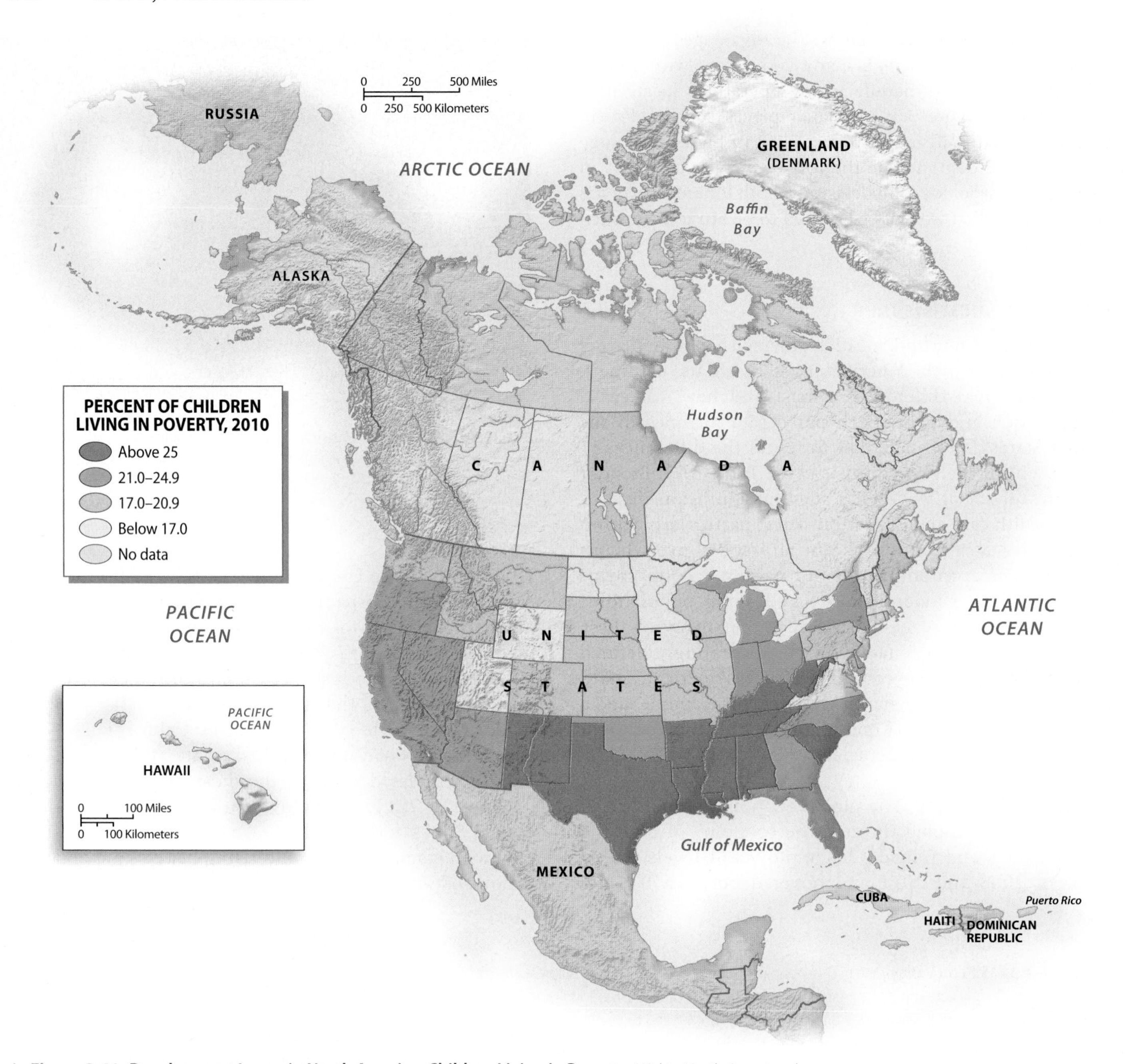

▲ **Figure 3.41 Development Issues in North America: Children Living in Poverty** Within North America, the South, Appalachia, and portions of the Southwest feature higher regional rates of childhood poverty. Overall, 22 percent of U.S. children and 14 percent of Canadian children are in families living at or below the poverty level. **Q: What might explain high childhood poverty in the Deep South? What about California and the Southwest?**

In terms of average income, the Northeast and West remain the richest regions in the United States. In Canada, Ontario, Alberta, and British Columbia are the country's wealthiest provinces, with Vancouver's high housing prices vying with those of San Francisco.

In contrast, substandard housing, abandoned and foreclosed property, aging infrastructure, and unemployed workers are reminders of the gap between rich and poor. Recently, poverty levels in the region rose in response to the economic recession of 2007–2010. Approximately 13–16 percent of the U.S. and Canadian populations live in poverty. Poverty rates for children (22 percent of children in the United States and about 14 percent of children in Canada live in poverty) vary considerably across North America and act as one regional measure of overall economic development (Figure 3.41). Children born into poor families are markedly less prepared for school at age five, are less likely to graduate high school or college, and are more likely to be convicted of a crime while still of school age.

Poor populations are still clustered in a variety of geographical settings. The problems of the rural poor remain major regional social issues in the Canadian Maritimes, Appalachia, the Deep South, the Southwest, and agricultural California. Most poor people in the United States, however, live in central-city locations, and links between ethnicity and poverty are strong in these communities. Nationally, about 27 percent of the country's African-American and Hispanic populations live below the poverty line.

Access to Education Education is also a major public policy issue in Canada and the United States. Although political parties differ in their approach, most public officials agree that more investment in education can only improve North America's chances for competing successfully in the global marketplace. Canada has witnessed steady improvement in its school dropout rate since the early 1990s. By 2012, U.S. high schools reported their highest graduation rates (over 75 percent) in decades, although dramatically lower numbers in many poor urban neighborhoods and rural areas suggest ongoing challenges in these settings. In addition, race plays a key role: American whites are two or three times more likely than African Americans or Hispanics to hold a college degree. Another challenge, particularly in the United States, is debating an effective national education policy in the face of a very strong tradition of local control of education.

Gender Equity Since World War II, both the United States and Canada have seen great improvements in the role that women play in society. However, the **gender gap** is yet to be closed when it comes to differences in salary, working conditions, and political power. Women comprise more than half of the North American workforce and are often more educated than men, but they still earn only about 80 cents for every dollar that men earn. Measured by corporate power, women still play modest roles. A 2011 survey reveals that women comprise only 16 percent of corporate board membership of large publicly held companies in the United States (versus 35 percent in Norway, for example). Women also head the vast majority of poorer single-parent families in the United States, and almost 40 percent of all births in the country are to unwed mothers. Canadian women, particularly single mothers who work full-time, are also greatly disadvantaged, averaging only about 65 to 70 percent of the salaries of Canadian men.

Although women have played critical roles in deciding recent national elections (they made up an essential component of President Obama's 2008 and 2012 presidential victories), political power remains largely in male hands. Canadian women have voted since 1918 and U.S. women since 1920, but females in the early 21st century remain significant minorities in both the Canadian Parliament (25 percent in 2012) and the U.S. Congress (18 percent in 2013).

Aging and Health-Care Issues Aging and health care are also key concerns within a region of graying baby boomers. A recent report on aging in the United States predicted that 20 percent of the nation's population will be older than 65 by 2050. Today the most elderly senior citizenry (age 85+) constitutes more than 12 percent of all seniors, and they are the fastest-growing part of the population, particularly among women. Poverty rates are also higher for seniors. With fewer young people to support their parents and grandparents, officials debate the merits of reforming social security programs. Whatever the outcome of such debates, the geographical consequences of aging are already abundantly clear. Whole sections of the United States—from Florida to southern Arizona—have become increasingly oriented around retirement (Figure 3.42). Communities cater to seniors with special assisted-living arrangements, health-care facilities, and recreational opportunities.

Health care remains a key issue in both countries. Both systems are costly by global standards: Canadians spend about 12 percent of their gross domestic product (GDP) on health care, and costs are even higher in the United States (over 15 percent of GDP). For several decades, Canada has offered an enviable system of government-subsidized universal health care to its residents (who pay higher taxes to fund it). In the United States, the Patient Protection and Affordable Care Act was signed into law in 2010 and has gradually moved that nation toward more universal coverage, still largely within a system of private insurers. Even with ongoing issues related to costs and access, the long life spans and low childhood mortality rates (see Table 3.2) of the region suggest that both countries reap many rewards from these modern health-care systems, offering the latest in high technology.

The rising incidence of chronic diseases associated with aging (heart disease, cancer, and stroke are the three leading causes of death) will continue to pressure both health-care systems. In addition, hectic lives, often oriented around fast food and more meals eaten out (the average American consumed 603 more calories daily in 2000 than 20 years earlier), have contributed to rapidly growing rates of obesity. On a typical day, more than 30 percent of American children and adolescents report eating fast food. The long-term results are sobering: Almost two-thirds of adult Americans are overweight, and the trend is contributing to higher rates of heart disease and diabetes.

Chronic alcoholism is also widespread and costly. For example, in the United States, about half of college students who drink engage in harmful binge drinking, more than 150,000 students develop alcohol-related health problems annually, and large numbers of students report being assaulted (about 700,000 per year) or sexually abused (about 100,000 per year) by other students who have been drinking.

Another critical health-care issue has been the care and treatment of the region's 1.2 million HIV/AIDS victims. The price of the disease will be broadly borne in the 21st century,

▲ **Figure 3.42 Tomorrow's Baby Boom Landscape?** Hundreds of golf resorts and retirement communities have been built across North America's Sun Belt since 1980 and now cater increasingly to the baby boom generation. This is an aerial view of Sun City, Arizona.

but particularly among poorer black (44 percent of AIDS cases in the United States) and Hispanic (20 percent) populations.

North America and the Global Economy

Together with Europe and East Asia, North America plays a pivotal role in the global economy. In prosperous times, the region benefits from global economic growth, but in periods of international instability, globalization means that the region is more vulnerable to economic downturns. These links mean more to the region than abstract trade flow and foreign investment statistics. Increasingly, North American workers and localities find their futures directly tied to export markets in Latin America, the rise and fall of Asian imports, or the pattern of global investments in U.S. stock and bond markets.

The region is home to a growing number of truly "global cities" that serve as key connecting points and decision-making centers in the world economy (see Figure 3.35). New York City is the largest, although smaller metropolitan areas such as Toronto, Chicago, Los Angeles, and Seattle are emerging as pivotal urban players on the global stage. Centers such as Miami (links to Latin America) and Las Vegas (a global tourist destination) have carved out their own niches, as well. The global status of these urban centers has had profound local consequences. Life changes as thousands of new jobs are created; suburbs grow with a rush of new and ethnically diverse migrants; and a new, more cosmopolitan culture replaces older regional traditions (see *People on the Move: Skilled Immigrants Fuel Big City Growth*).

The United States, with Canada's firm support, played a formative role in creating much of this new global economy and in shaping many of its key institutions. In 1944, allied nations met at Bretton Woods, New Hampshire, to discuss economic affairs. Under U.S. leadership, the group set up the International Monetary Fund (IMF) and the World Bank, giving these global organizations the responsibility for defending the world's monetary system and making key postwar investments in infrastructure. The United States also spurred the creation (1948) of the General Agreement on Tariffs and Trade (GATT). Renamed the **World Trade Organization (WTO)** in 1995, its 159 member states are dedicated to reducing global barriers to trade. In addition, the United States and Canada participate in the **Group of Eight (G8)**—a collection of economically powerful countries (including Japan, Germany, Great Britain, France, Italy, and Russia)—which confers regularly on key global economic and political issues.

Patterns of Trade North America is prominent in both the sale and the purchase of goods and services within the international economy. Both countries import diverse products from many global sources, and the post-1990 growth of the Asian trade (particularly with China, South Korea, and southeast Asia) has fundamentally changed the North American economy.

Dominated overwhelmingly by the United States, Canada imports large quantities of manufactured parts, vehicles, computers, and foodstuffs. For the United States, imports continue to grow, creating a persistent global trade deficit for the country. Canada, Mexico, China, Japan, and world oil exporters supply the United States with a diversity of raw materials, low-cost consumer goods, and high-quality vehicles and electronics products (Figure 3.43).

Outgoing trade flows suggest what North Americans produce most cheaply and efficiently. Canada's exports include large quantities of raw materials (energy, metals, grain, and wood products), but manufactured goods are becoming increasingly important, particularly in its pivotal trade with the United States. Since 1994, trade initiatives with the Pacific Rim have offered Canadians new opportunities for export growth. The United States also enjoys many lucrative global economic ties, and its geography of exports reveals particularly strong links to other portions of the more developed world. Sales of automobiles, aircraft, computer and telecommunications equipment, entertainment, financial and tourism services, and food products contribute to the nation's flow of exports.

Patterns of Global Investment Patterns of capital investment and corporate power place North America at the center of global money flows and economic influence. Given its relative stability, the region attracts huge inflows of foreign capital, both as investments in North American stocks and bonds and as foreign direct investment (FDI) by international companies. For Canada, U.S. wealth and proximity have meant that 80 percent of foreign-owned corporations in the country are based in the United States. In the United States, sustained economic growth and supportive government policies have encouraged large foreign investments, particularly since the late 1970s. Today the United States is the largest destination of foreign investment in the world, and the average annual pay of manufacturing-related jobs originating from these investments is more than $70,000, more than 30 percent higher than the average pay across the entire workforce.

The impact of U.S. investments in foreign stock markets also suggests how flows of outbound capital are transforming the way business is done throughout the world. Aging U.S. baby boomers have poured billions of pension fund and investment dollars into Japanese, European, and "emerging" stock markets such as Brazil, Russia, and China. In addition, U.S. investments in foreign countries flow through direct investments made by multinational corporations based in the United States.

However, the geography of 21st-century multinational corporations is changing, illustrating three recent shifts in broader patterns of globalization. These shifts have important consequences for North Americans. First, traditional American-based multinational corporations are adopting a new, more globally integrated model. For example, IBM now has more than 50,000 employees in India.

▼ **Figure 3.43 Container Shipping, Port of Seattle** Major North American ports such as Seattle are key links in facilitating global trade between the region and the rest of the world. Standard-sized container-shipping modules can easily be stored, stacked, and moved in such settings.

People on the Move

Skilled Immigrants Fuel Big City Growth

Broad questions surrounding immigration continue to foster debate in both Canada and the United States, but one economic trend is beyond dispute: The region's growing influx of skilled immigrants is providing a powerful economic stimulus for many of North America's largest and most dynamic cities (**Figure 3.5.1**). Part of the story is simply the numerical impact of immigrant populations: The New York City, Los Angeles, San Francisco, and Boston metropolitan areas, for example, would all have actually lost population between 2000 and 2010 without substantial numbers of new immigrants. The economic consequences of these movements are also huge. William

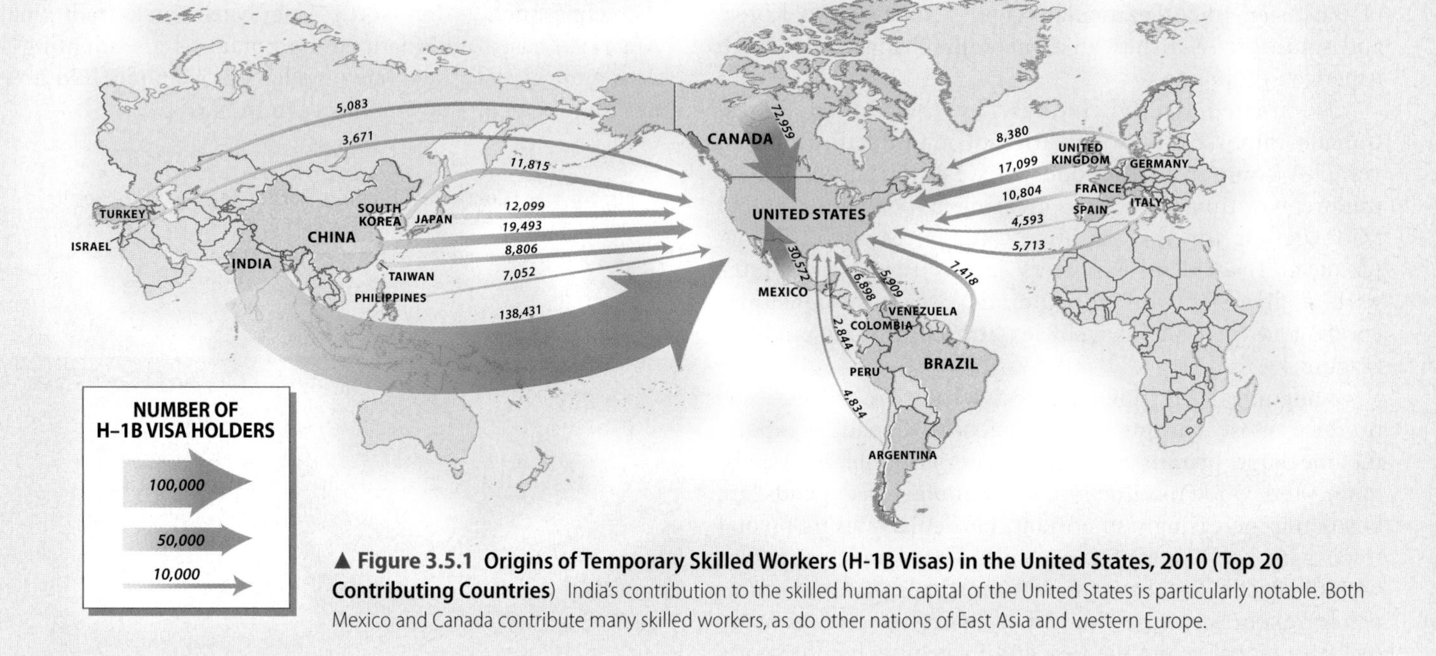

▲ **Figure 3.5.1 Origins of Temporary Skilled Workers (H-1B Visas) in the United States, 2010 (Top 20 Contributing Countries)** India's contribution to the skilled human capital of the United States is particularly notable. Both Mexico and Canada contribute many skilled workers, as do other nations of East Asia and western Europe.

Second, a growing array of multinational corporations based elsewhere in the world—especially in places such as China, India, Russia, and Latin America—is buying up companies and assets once controlled by North American or European capital. Brazilian investors, for example, spend billions annually in buying overseas assets (especially in North America). In Asia, China's Lenovo purchased IBM's huge personal computer business within that country. Similarly, Indian multinational companies are buying hundreds of foreign assets based in the more developed world. One sign of the times can be measured in the home country of the world's largest multinational companies. Measured by their market value (2012), only 4 of the top 10 global corporations are based in North America (Exxon Mobil, Microsoft, Walmart Stores, and Apple). Only one is European (Royal Dutch Shell), whereas five others are now based in East Asia or Australia (PetroChina, ICBC, BHP Billiton, China Mobile, and Samsung).

Third, many of these same multinational companies are making huge investments of their own in other portions of the less developed world, from Africa to Southeast Asia, bypassing North American control altogether. Today more than one-third of FDI in emerging market nations comes from other emerging market nations. Simply put, the late-20th-century top-down model of multinational corporate control and investment, traditionally based in North America, Europe, and Japan, is being replaced by a more globally distributed model of corporate control. This new model has many origins, many destinations, and new patterns of labor, capital, production, and consumption.

Frey, a demographer with the Brookings Institution, notes that "a lot of cities rely on immigration to prop up their housing market and prop up their economies."

Foreign Spark Plugs Many immigrants become urban entrepreneurs, starting new businesses that cater to their own communities, as well as to larger metropolitan populations (**Figure 3.5.2**). Whether it is the Chinese in Vancouver or the Cubans in Miami, immigrants in many of North America's largest, most global cities have made huge capital and human investments in their adopted communities. Almost 30 percent of the Korean-born and 20 percent of the Iranian-born populations in the United States are self-employed, a strong indicator of business ownership. A recent report issued by the Center for an Urban Future argues that "immigrants have been the entrepreneurial spark plugs of cities from New York to Los Angeles." Is it any surprise that first-generation immigrants created 22 of the 100 fastest-growing companies in Los Angeles or that 25 percent of all U.S. engineering and technology firms founded between 1995 and 2005 were headed by immigrant entrepreneurs?

Granting Visas Additional statistics gathered by the U.S. Department of Homeland Security also point to the unique contributions of highly skilled immigrants. So-called H-1B visas are granted to special "temporary skilled workers" to encourage computer programmers, doctors, and other professionals to work in the United States. More than 400,000 such visas were issued in 2010, enabling these individuals to work within the United States, adding immeasurably to that country's creative human capital. New federal legislation proposed in 2013 aimed at substantially increasing the number of available H-1B visas, as well as making it easier for college-educated foreign students to remain in the country to work and eventually become citizens.

The geography of the top 20 contributing countries to the H-1B visa program offers yet another snapshot of the economic impact of globalization and is a powerful reminder of another way in which the United States benefits in the process (see **Figure 3.5.1**). Predictably, many ties are to European nations, but the linkages with India (particularly through the computer, software, and electronics industries) are impressive. Continental connections to nearby Mexico and Canada are also notable, along with growing links to East Asian nations such as China, South Korea, Japan, the Philippines, and Taiwan. The patterns offer a powerful reminder that the economic evolution of both the United States and Canada remains intimately connected to skilled immigrant populations and that the region's large urban centers will continue to be a destination for these creative and talented individuals.

▲ **Figure 3.5.2 Immigrant Entrepreneurs** These Vietnamese merchants own the Tien Hung Complete Oriental Foods and Gifts store in Orlando, Florida.

North Americans have certainly experienced direct consequences from these shifts in global capitalism. For example, reaction has been growing in the United States against corporate **outsourcing**, a business practice that transfers portions of a company's production and service activities to lower-cost settings, often located overseas. In addition, millions of jobs in manufacturing, textiles, semiconductors, and electronics have effectively migrated to settings such as China, India, and Mexico, since those countries offer low-cost, less regulated settings for production, both for local and for foreign firms. The results are complex: North American consumers benefit from buying cheap imports, but they may find their own jobs are threatened in the corporate restructurings that make such bargains possible.

REVIEW

3.10 What is the sectoral transformation, and how does it help explain economic change in North America?

3.11 Cite five types of location factors, and illustrate each with examples from your local economy.

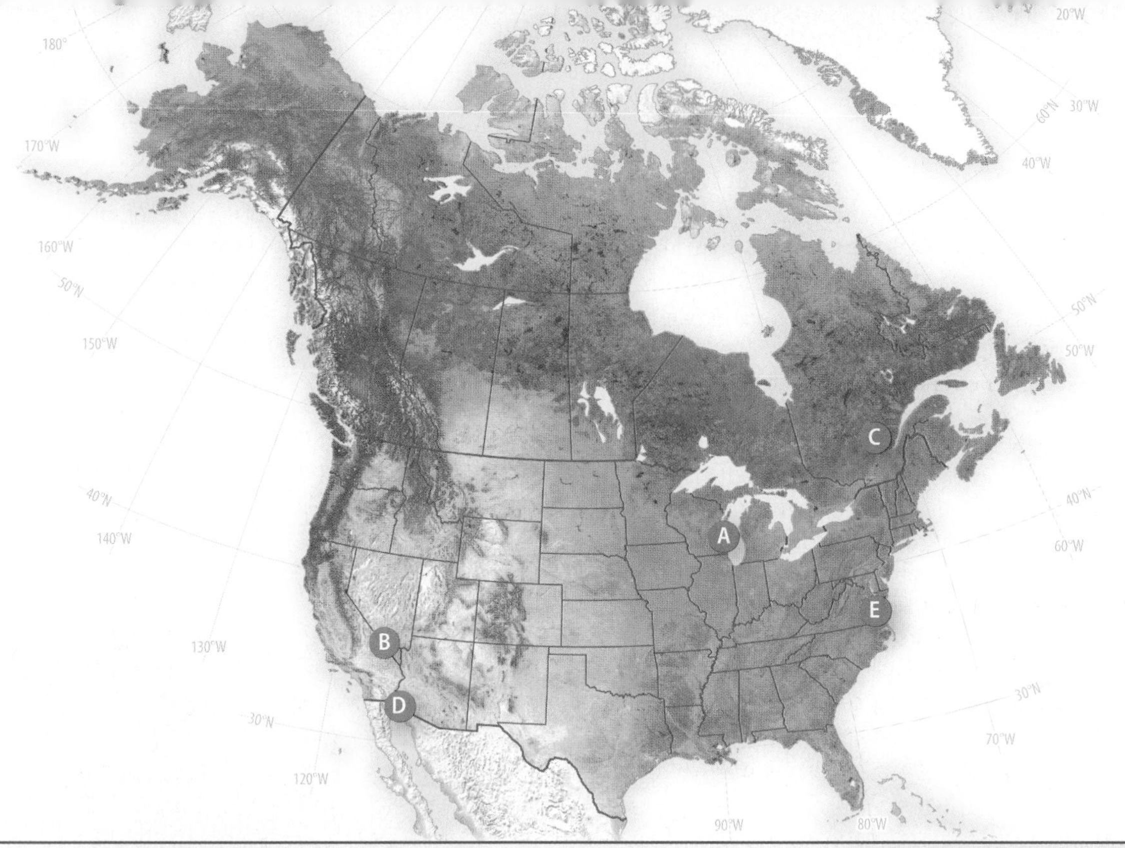

Physical Geography and Environmental Issues

North Americans have reaped the natural abundance of their region, and in the process they have transformed the environment, created a highly affluent society, and extended their global economic, cultural, and political reach. North America's affluence has come with a considerable price tag, and today the region faces significant environmental challenges, including soil erosion, acid rain, and air and water pollution.

3.1 The yellow squares on the map indicate major hazardous waste sites. Why are so many sites concentrated along major rivers and near the Great Lakes?

3.2 Can you identify key hazardous waste sites in your area? What are the sources of the waste? Have these sites been cleaned up?

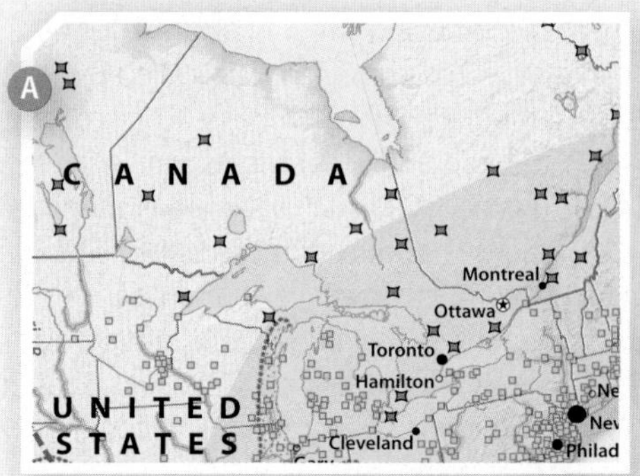

Population and Settlement

In a remarkably short time, a unique mix of varied cultural groups from around the world has contributed to the settlement of a huge and resource-rich continent that is now one of the world's most urbanized regions.

3.3 What are some of the reasons for the rapid growth of Las Vegas since 1980?

3.4 In this harsh desert (especially vulnerable to future drought), describe a path to sustainability for maintaining southern Nevada's population in the next 50–100 years.

Key Terms

acid rain *(p. 91)*
agribusiness *(p. 124)*
boreal forest *(p. 88)*
connectivity *(p. 125)*
cultural assimilation *(p. 104)*
cultural homeland *(p. 106)*
edge city *(p. 101)*
ethnicity *(p. 104)*
federal state *(p. 120)*
fracking *(p. 94)*
gender gap *(p. 129)*
gentrification *(p. 102)*
Group of Eight (G8) *(p. 131)*
location factors *(p. 126)*
Megalopolis *(p. 96)*
new urbanism *(p. 101)*
nonmetropolitan growth *(p. 100)*
North American Free Trade Agreement (NAFTA) *(p. 118)*
outsourcing *(p. 133)*
postindustrial economy *(p. 84)*
prairie *(p. 88)*
renewable energy sources *(p. 93)*
sectoral transformation *(p. 125)*
Spanglish *(p. 114)*
sustainable agriculture *(p. 93)*
tundra *(p. 88)*
unitary state *(p. 120)*
urban decentralization *(p. 101)*
urban heat island *(p. 91)*
World Trade Organization (WTO) *(p. 131)*

In Review • Chapter 3
North America

Cultural Coherence and Diversity

North Americans produced two societies that are closely intertwined, but that face distinctive national political and cultural issues. In Canada, the nation's identity remains problematic as it works through the persistent challenges of its multicultural character and the costs and benefits of its proximity to its dominating continental neighbor.

3.5 Why does one still find large numbers of French speakers in the Canadian province of Quebec?

3.6 Is the French language likely to retain its cultural vitality in Quebec over the next century? What key challenges does it face?

Geopolitical Framework

Canada and the United States enjoy a close political relationship, but several issues—often related to environmental quality, resource claims, and trade—still produce tensions between the two countries.

3.7 What are the key characteristics of a political "borderlands" zone such as this one along the Mexico/California border?

3.8 Immigration remains a key issue for North America. Organize a class debate on the pros and cons of sharply curtailing immigration in the future. What about opening up the region to even larger flows of immigrants?

Economic and Social Development

North America displays great regional affluence, but enduring issues of poverty, health care, and gender equity continue to challenge both countries in the 21st century.

3.9 What national and global economic trends are illustrated by edge-city settings such as Tysons Corner, Virginia?

3.10 Identify an edge city or peripheral suburban shopping area near you. What economic activities are emphasized in these settings?

Authors' Blogs

Scan to visit the author's blog for chapter updates

www.gad4blog.wordpress.com

Scan to visit the GeoCurrents blog

www.geocurrents.info

4 Latin America

Puerto Maldonado Bridge. The newly constructed Puerto Maldonado Bridge in the Peruvian Amazon spans the Madre de Dios River. The bridge is a major infrastructural feature of the Interoceanic Highway which connects Atlantic ports in Brazil with Pacific ports in Peru by traversing the Andes mountains and the Amazon forest. A symbol of the region's surge in natural resource exports to Asia and Europe, many worry that the new highway will further tropical deforestation and undermine indigenous livelihoods.

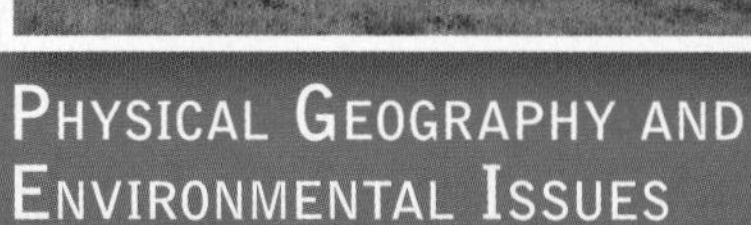

Physical Geography and Environmental Issues

Tropical forests in Latin America, especially in the Amazon Basin, are one of the planet's greatest reserves of biological diversity. How this diversity will be managed is a critical question, especially with increasing pressure to extract mineral wealth, build roads and dams, and convert forests into farms or pasture.

Population and Settlement

Latin America is the most urbanized region of the developing world, with 78 percent of the population living in cities. Four megacities (10 million people or more) are found here. Yet it is also a region with high rates of emigration, especially to North America.

Cultural Coherence and Diversity

Amerindian activism is on the rise in Latin America. Indigenous people from Central America to the Andes are finding their political voice by demanding territorial and cultural recognition. These demands are occurring while Latin America is on a global stage due to Brazil hosting the World Cup in 2014 and the Olympics in 2016.

Geopolitical Framework

As Latin American governments mark 200 years of independence from Spain, most are fully democratic. Recent elections in the region have seen liberal democrats and populists gain power, promising to reduce income inequality through government programs. Women are also becoming political actors in Latin America, holding nearly one-quarter of all seats in national parliaments.

Economic and Social Development

Economic growth, increased trade, and fewer people in extreme poverty are all positive trends for the region, but serious income inequality persists. Government programs such as Brazil's Bolsa Familia have sought to address both social and economic development for poor families. Meanwhile, heightened violence and insecurity, especially in Central America, have more people living in fear.

One of the iconic symbols of trade in the Americas is the Panama Canal, which marks its 100th anniversary in 2014. In 2015, Panama inaugurates a new set of locks to service larger vessels. Yet as big as the Panama Canal is, the biggest infrastructure project in the region is the recently completed Interoceanic Highway, which extends some 3400 miles.

Road crews worked day and night, blasting rock and laying pavement, so that this east–west transcontinental link could traverse Amazonian rainforest and Andean mountain passes. Part manifest destiny and part access to resources, plans have existed for decades to construct this road—but only recently did Brazil have the capital and Peru the will to complete it. A key element of this road is the new bridge in Puerto Maldonado that crosses the Madre de Dios River. The river is half a mile wide at this point, but is only a tributary of the mighty Amazon. After leaving this lowland area, the road divides into two main branches that cross the Andes and serve three ports, two of which—Ilo and Matarani—already exist along Peru's arid coast.

The economics of moving soy from the Brazilian Amazon over the high Andean passes for shipping across the Pacific are hard to justify. But the road also provides access to remote areas filled with gold, lumber, and future farms and pasture. Completion of the Interoceanic Highway also signals the region's growing trade with China to the west and Europe to the east. However, critics see this new highway as opening up pristine and biologically diverse rainforest for exploitation. Such large-scale infrastructure projects that facilitate trade and resource extraction are a reflection of neoliberal forces at work in Latin America. (As we will see later in detail, neoliberalism refers to policies of free trade, foreign investment, and expansion of the private sector.)

The modern states of Latin America are multiethnic, with distinct indigenous and immigrant profiles and very different rates of social and economic development. Beginning with Mexico and extending to the tip of South America, Latin America's regional coherence stems largely from its shared colonial history. More than 500 years ago, the Iberian countries of Spain and Portugal began their conquest of the Americas. Iberia's mark is still visible throughout Latin America: Officially, two-thirds of the inhabitants speak Spanish, and the rest speak Portuguese. Iberian architecture and town design add homogeneity to the colonial landscape. The largest concentration of Catholics worldwide is in Latin America, and in 2013 an Argentine priest became Pope Francis, the first Pope from the Americas.

These European traits blended with those of different Amerindian peoples. The Indian presence remains especially strong in Bolivia, Peru, Ecuador, Guatemala, and southern Mexico, where large and diverse indigenous populations maintain their native languages, dress, and traditions. After the initial Iberian conquest, other cultural groups were added to this mix of indigenous and Iberian peoples. The legacy of slavery imparted a strong African influence, primarily on the coasts of Colombia and Venezuela and throughout Brazil. In the 19th and 20th centuries, new waves of settlers came from Spain, Italy, Germany, Japan, and Lebanon. The result is one of the world's most racially mixed regions. Given the significant outflows of emigrants from Latin America today, people from this region are found throughout the world, but especially in North America, Europe, and Japan.

The modern states of Latin America are multiethnic, with distinct indigenous and immigrant profiles and very different rates of social and economic development.

The concept of Latin America as a distinct region has been popularly accepted for nearly a century. The boundaries of this region are straightforward, beginning at the Rio Grande (called the *Rio Bravo* in Mexico) and ending at Tierra del Fuego (Figure 4.1). French geographers are credited with coining the term *Latin America* in the 19th century to distinguish the Spanish- and Portuguese-speaking republics of the Americas plus Haiti from the English-speaking territories. There is nothing particularly "Latin" about the area, other than the predominance of romance languages. The term stuck because it was vague enough to be inclusive of different colonial histories, while also offering a clear cultural boundary from Anglo-America, the region referred to as North America in this book.

This chapter describes Latin America, consisting of the Spanish- and Portuguese-speaking countries of Central and South America, including Mexico. This division emphasizes the important Indian and Iberian influences affecting mainland Latin America and separates it from the unique colonial and demographic history of the Caribbean, discussed in Chapter 5.

LEARNING OBJECTIVES

After reading this chapter you should be able to:

- Explain the relationships among elevation, climate, and agricultural production in Latin America, especially in tropical highland areas.
- Identify the major environmental issues of Latin America and how countries are addressing them.
- Summarize the demographic issues impacting this region, such as rural-to-urban migration, urbanization, smaller families, and emigration.
- Describe the cultural mixing of European and Amerindian groups in this region and indicate where Amerindian cultures thrive today.
- Explain the global reach of Latino culture through immigration, sport, music, and television.
- Describe the Iberian colonization of the region and how it affected the formation of today's modern states.
- Identify major Amerindian groups today and their efforts toward territorial and political recognition.
- Identify the major trade blocs in Latin America and how they are influencing development.
- Summarize the significance of primary exports from Latin America, especially agricultural commodities, minerals, wood products, and fossil fuels.
- Identify the important energy sources for Latin America and how they have shifted since 1970.
- Describe the neoliberal economic reforms that have been applied to Latin America and how they have influenced the region's development.

▲ **Figure 4.1 Latin America** Roughly equal in size to North America, Latin America supports a larger population and far greater ecological diversity. The 17 countries included in this region share a history of Iberian colonization. Three-quarters of the region's 560 million people live in cities, making it the most urbanized region of the developing world. It is noted for its production of primary exports and manufactured goods, although rates of economic development vary greatly among states.

Roughly equal in area to North America, Latin America has a much larger and faster-growing population of 560 million people. Its most populous state, Brazil, has 200 million people, making it the fifth largest country (by population) in the world. The next largest state, Mexico, has a population of 118 million, and one of its citizens, telecom mogul Carlos Slim, is widely considered one of the richest men in the world. Collectively, many Latin American states fall into the middle-income category and support a significant middle class. But national debt, currency devaluation, and at times triple-digit inflation triggered grave economic hardships throughout the region, especially during the 1980s and early 1990s. It is estimated that 12 percent of the people in the region live on less than $2 per day. In the 1990s, the percentage living on less than $2 per day was nearly twice as large, which shows that efforts to reduce extreme poverty are working.

Through colonialism, immigration, and trade, the forces of globalization have been embedded in the Latin American landscape. The early Spanish Empire concentrated on extracting precious metals, sending galleons laden with silver and gold across the Atlantic. The Portuguese became prominent producers of dyewoods, sugar products, gold, and, later, coffee. In the late 19th and early 20th centuries, exports to North America and Europe fueled the region's economy. Most countries specialized in one or two products: bananas and coffee, meats and wool, wheat and corn, petroleum and copper. Such a primary export tradition, according to Latin American economists, led to an unhealthy economic dependence. They argued in the 1960s that Latin American economies were too specialized and faced unequal terms of trade that inhibited overall development.

Since then, the countries of the region have industrialized and diversified their production, but they continue to be major producers of primary goods for North America, Europe, and East Asia. Today neoliberal policies that encourage foreign investment, export production, and privatization have been adopted by many states. These policies exemplify the current impact of economic globalization on Latin America. The results are mixed, with some states experiencing impressive economic growth, but increased disparity between rich and poor. Intraregional trade within Latin America has been stimulated by **Mercosur** (the Southern Common Market found in South America), as well as by the impact of the North American Free Trade Agreement (NAFTA) (see Chapter 3) and the **Central American Free Trade Association (CAFTA).** These associations are indicators of heightened economic integration in the hemisphere.

Despite the region's growing industrial capacity, extractive industries continue to prevail, in part because of the area's impressive natural resources. Latin America is home to Earth's largest rainforest, the greatest river by volume, and massive reserves of natural gas, oil, gold, and copper. With its vast territory, its tropical location, and its relatively low population density (Latin America has half the population of India in nearly seven times the area), the region is also recognized as one of the world's great reserves of biological diversity. How this diversity will be managed in the face of global demand for natural resources is an increasingly important question for the countries of this region.

Unlike most areas of the developing world today, Latin America is decidedly urban. Prior to World War II, most people lived in rural settings and worked as farmers. Today three-quarters of Latin Americans are city dwellers. Even more startling is the number of **megacities.** São Paulo, Mexico City, Buenos Aires, and Rio de Janeiro all have more than 10 million inhabitants (Figure 4.2). In addition, more than 40 cities have at least 1 million residents. Residents of these cities aspire for global recognition as they compete for political summits, foreign tourists, and major cultural and sporting events. A dozen Brazilian cities hosted the World Cup in 2014, and Rio de Janeiro hosts the Olympics in 2016.

▼ **Figure 4.2 Megacity Buenos Aires** This capital city of over 13 million inhabitants is the economic and cultural hub of Argentina. The bustling ceremonial boulevard, 9 de Julio Avenue, cuts through the downtown and commemorates Argentine Independence Day (July 9). The obelisk built in the early 20th century is an iconic structure for the city.

Physical Geography and Environmental Issues: Neotropical Diversity and Urban Degradation

Much of the region is characterized by its tropicality. Travel posters of Latin America showcase verdant forests and brightly colored parrots. The diversity and uniqueness of the **neotropics** (tropical ecosystems in the Western Hemisphere) have long been attractive to naturalists eager to understand their distinct flora and fauna. It is no accident that Charles Darwin's theory of evolution was inspired by his two-year journey in tropical America. Even today scientists throughout the region work to understand complex ecosystems, discover and protect new species, conserve genetic resources, and interpret the impact of human settlement, especially in neotropical forests.

Not all of the region is tropical. Important population centers extend below the Tropic of Capricorn, most notably Buenos Aires and Santiago. Much of northern Mexico, including the city of Monterrey, is north of the Tropic of Cancer. Yet Latin America's tropical climate and vegetation prevail in popular images of the region. Given the territory's large size and relatively low population density, Latin America has not experienced the same levels of environmental degradation witnessed in Europe and East Asia.

Huge areas of Latin America still remain relatively untouched, supporting an incredible diversity of plant and animal life. Throughout the region, national parks offer some protection to unique communities of plants and animals. A growing environmental movement in countries such as Costa Rica and Brazil has yielded both popular and political support for "green" initiatives. In short, Latin Americans enter the 21st century with a real opportunity to avoid many of the environmental mistakes seen in other regions of the world. At the same time, global market forces are driving governments to exploit minerals, fossil fuels, forests, and soils. The region's biggest natural resource challenge is to balance the economic benefits of extraction with the ecological soundness of conservation. Another major challenge is to improve the environmental quality of Latin American cities.

Western Mountains and Eastern Shields

Latin America is a region of diverse landforms, including high mountains and extensive upland plateaus. The movement of tectonic plates explains much of the region's basic topography, including the formation of its geologically young and tectonically active western mountain ranges (see Figure 4.1). In contrast, the Atlantic side of South America is characterized by humid lowlands interspersed with large upland plateaus called **shields.** Across these lowlands meander some of the great rivers of the world, including the Amazon.

Historically, the most important areas of settlement in tropical Latin America were not along the region's major rivers, but across its shields, plateaus, and fertile intermontane basins. In these localities, the combination of arable land, mild climate, and sufficient rainfall produced the region's most productive agricultural areas and its densest settlements. The Mexican Plateau, for example, is a massive upland area ringed by the Sierra Madre Mountains, with the Valley of Mexico located at the southern end of the plateau. Similarly, the elevated and well-watered basins of Brazil's southern mountains provide an ideal setting for agriculture. These especially fertile areas are able to support high population densities, so it is not surprising that the two largest cities in the region, Mexico City and São Paulo, emerged in these settings.

The Andes Beginning in northwestern Venezuela and ending at Tierra del Fuego, the Andes are relatively young mountains that extend nearly 5000 miles (8000 km). Created by the collision of oceanic and continental plates, the mountains are a series of folded and faulted sedimentary rocks with intrusions of crystalline and volcanic rock. The subduction of the Nazca Plate under the South American Plate has produced an impressive chain of some 30 peaks higher than 20,000 feet (6000 meters). Due to the violent and complex origins of the Andes, many rich veins of precious metals and minerals are found here. The initial economic wealth of many Andean countries came from mining silver, gold, tin, copper, and iron.

The Andes are still forming, so active volcanism and regular earthquakes are common in this zone. This was made clear in February 2010 when an 8.8-magnitude earthquake struck near the Chilean city of Concepción, killing some 400 people and unleashing tsunami warnings across the Pacific. The earthquake was so powerful that geologists estimate that the entire city of Concepción moved 10 feet (3 meters) to the west. Because the epicenter was not near a major population center and Chilean buildings are engineered to withstand earthquakes, the death toll was remarkably light given the intensity of the event. Even so, it is estimated that nearly 1.5 million Chileans were temporarily displaced by this natural disaster.

Given the length of the Andes, the mountain chain is typically divided into northern, central, and southern components. In Colombia, the northern Andes actually split into three distinct mountain ranges before merging near the border with Ecuador. High-altitude plateaus and snow-covered peaks distinguish the central Andes of Ecuador, Peru, and Bolivia. The Andes reach their greatest width here. Of special interest is the treeless high plain of Peru and Bolivia, called the **Altiplano.** The floor of this elevated plateau ranges from 11,800 feet (3600 meters) to 13,000 feet (4000 meters) in altitude, and it has limited usefulness for grazing. Two high-altitude lakes, Titicaca on the Peruvian and Bolivian border and the smaller Poopó in Bolivia, are located in the Altiplano, as are many mining sites (Figure 4.3). The southern Andes are shared by Chile and Argentina.

▼ **Figure 4.3 Bolivian Altiplano** The Altiplano is an elevated plateau straddling the Bolivian and Peruvian Andes. Below is picturesque Laguna Canapa in Bolivia with the Andean peaks towering in the background. This high and windswept land is home to many Amerindian peoples.

◀ **Figure 4.4 Mexico's Mesa Central** Mexico's elevated central plateau has long been the demographic and agricultural core of the county. This image shows a variety of agave grown in Jalisco that is used for the country's tequila production. Tequila, a traditional drink in Mexico, has a growing export market.

The highest peaks of the Andes are found in the southern Andes, including the highest peak in the Western Hemisphere, Aconcagua, at almost 23,000 feet (7000 meters). South of Santiago, Chile, the mountains are lower and the chain less compact. The impact of glaciation is most evident in the southernmost extension of the Andes.

The Uplands of Mexico and Central America The Mexican Plateau and the Volcanic Axis of Central America are the most important Latin American uplands in terms of long-term settlement. Most major cities of Mexico and Central America are located here. The Mexican Plateau is a large, tilted block that has its highest elevations, about 8000 feet (2500 meters), in the south around Mexico City, and its lowest, just 4000 feet (1200 meters), at Ciudad Juárez. The southern end of the plateau, the Mesa Central, contains several flat-bottomed basins interspersed with volcanic peaks that have long been significant areas for agricultural production (Figure 4.4). It also contains Mexico's megalopolis—a concentration of the largest population centers, such as Mexico City and Puebla. Across the Mexican Plateau are rich seams of silver, copper, and zinc. The quest for silver drove much of the economic activity of colonial Mexico.

Along the Pacific coast of Central America lies a chain of volcanoes that stretches from Guatemala to Costa Rica. The Volcanic Axis of Central America is a handsome landscape of rolling green hills, elevated basins with sparkling lakes, and conical volcanic peaks. More than 40 volcanoes are found here, many of them still active. Their legacy is a rich volcanic soil that yields a wide variety of domestic and export crops. Most of Central America's population is also concentrated in this zone, in the capital cities or the surrounding rural villages. The bulk of the agricultural land is tied up in large holdings that yield export products including beef, cotton, and coffee. Yet most of the farms are small subsistence properties that produce corn, beans, squash, and assorted fruits. A major rift (a large crustal fracture) lies east of the Volcanic Axis in Nicaragua. In this low-lying valley are Lakes Managua and Nicaragua, the largest in Central America.

The Shields South America has three major shields—large upland areas of exposed crystalline rock that are similar to upland plateaus found in Africa and Australia. (Two are discussed here, while the third, the Guiana shield, will be discussed in Chapter 5.) The Brazilian shield is the larger and more important in terms of natural resources and settlement. Far from a uniform land surface, the Brazilian shield covers much of Brazil from the Amazon Basin in the north to the Plata Basin in the south. It is studded with isolated low ranges and flat-topped plateaus in the north. In southeastern Brazil, a series of mountains (Serra da Mantiqueira and Serra do Mar) reach elevations of 9000 feet (2700 meters). In between these ranges are elevated basins that offer a mild climate and fertile soils. In one of these basins is the city of São Paulo, the largest urban conglomeration in South America.

The Paraná basalt plateau, located on the southern end of the Brazilian shield, is celebrated for its fertile red soils (*terra roxa*), which yield coffee, oranges, and soybeans. The basalt plateau, much like the Deccan plateau in India, is an ancient lava flow that resulted from the breakup of Gondwanaland. So fertile is this area that the economic rise of São Paulo is attributed to the expansion into this area of commercial agriculture.

The vast low-lying Patagonian shield lies in the southern tip of South America. Beginning south of Bahia Blanca and extending to Tierra del Fuego, the region to this day remains sparsely settled and hauntingly beautiful. It is treeless, covered by scrubby steppe vegetation, and home to wildlife such as the condor and guanaco (Figure 4.5). Sheep were introduced to Patagonia in the late 19th century, spurring a wool boom. More recently, offshore oil production has renewed the economic importance of this area.

▼ **Figure 4.5 Patagonian Wildlife** Native to South America, guanacos thrive on the thin steppe vegetation found throughout Patagonia. Their numbers have fallen dramatically due to hunting and competition with introduced livestock.

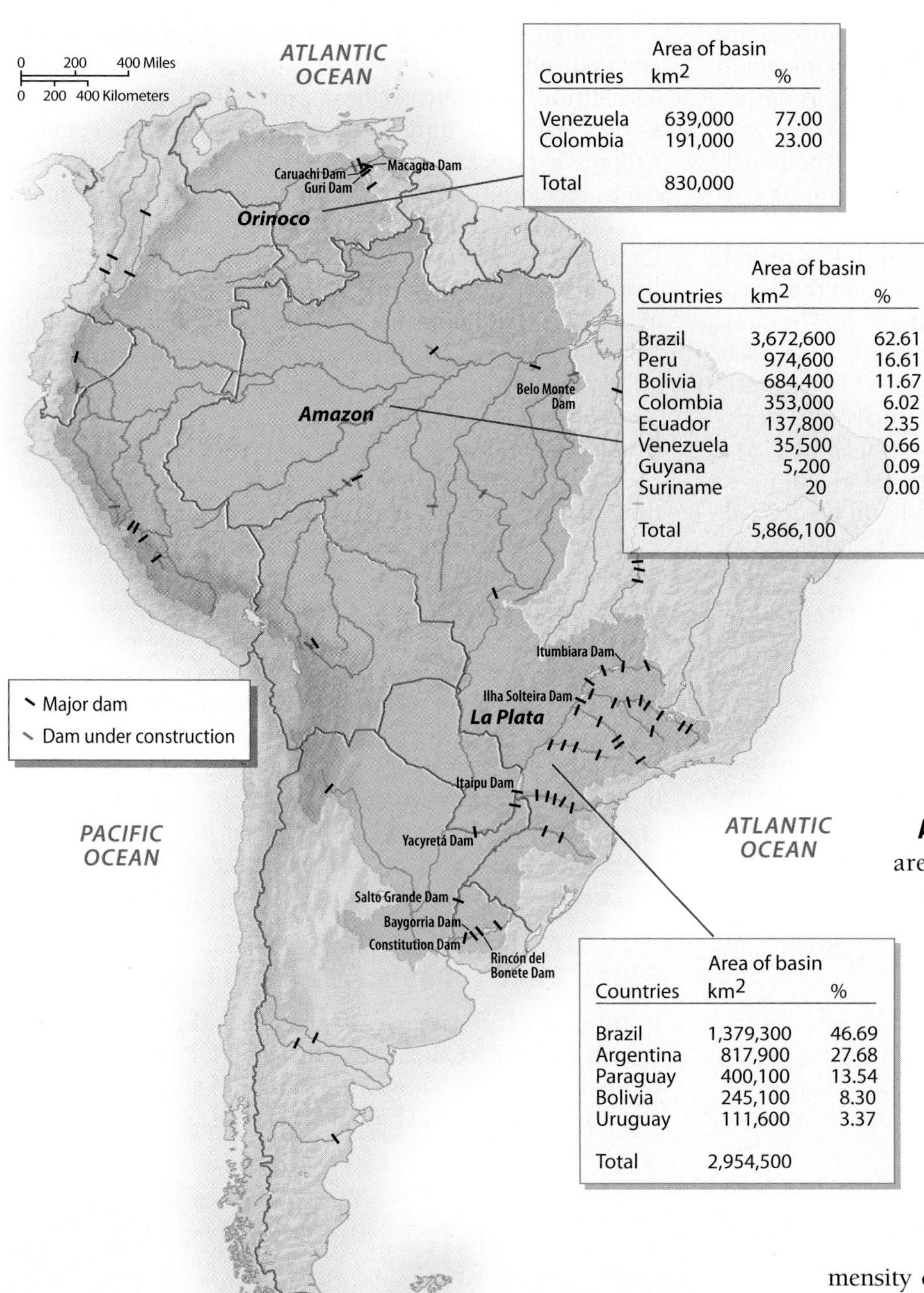

Countries	Area of basin km^2	%
Venezuela	639,000	77.00
Colombia	191,000	23.00
Total	830,000	

Countries	Area of basin km^2	%
Brazil	3,672,600	62.61
Peru	974,600	16.61
Bolivia	684,400	11.67
Colombia	353,000	6.02
Ecuador	137,800	2.35
Venezuela	35,500	0.66
Guyana	5,200	0.09
Suriname	20	0.00
Total	5,866,100	

Countries	Area of basin km^2	%
Brazil	1,379,300	46.69
Argentina	817,900	27.68
Paraguay	400,100	13.54
Bolivia	245,100	8.30
Uruguay	111,600	3.37
Total	2,954,500	

◀ **Figure 4.6 South American River Basins and Dams** The three great river basins of the region are the Amazon, Plata, and Orinoco. The Amazon Basin covers 6 million square kilometers, including portions of eight countries, but the majority of the basin is within Brazil. The Amazon is the largest river system in the world in terms of volume of water and area. It is currently experiencing a boom in dam construction to provide for Brazil's energy needs. The Plata Basin drains nearly 3 million square kilometers across five countries and is intensely farmed. It also contains Latin America's largest hydroelectric dam, Itaipú. The Orinoco Basin is shared by Venezuela and Colombia, covering nearly 1 million square kilometers. **Q: What are the benefits and costs of reliance upon hydroelectricity?**

River Basins and Lowlands

Three great river basins drain the Atlantic lowlands of South America: the Amazon, Plata, and Orinoco (Figure 4.6). Within these basins are vast interior lowlands, less than 600 feet (200 meters) in elevation, which lie over young sedimentary rock. From north to south, they are the Llanos, the Amazon lowlands, the Pantanal, the Chaco, and the Pampas. With the exception of the Pampas, most of these lowlands are sparsely settled and offer limited agricultural potential except for grazing livestock. Yet the pressure to open new areas for settlement and to exploit natural resources has created pockets of intense economic activity in the lowlands. Areas such as the Amazon and the Chaco have witnessed marked increases in resource extraction, soy cultivation, dam construction, and settlement since the 1970s.

Amazon Basin The Amazon drains an area of roughly 2.4 million square miles (6.6 million square km), making it the largest river system in the world by volume and area and the second longest by length. Everywhere in the basin, annual rainfall is more than 60 inches (150 cm), and in many places it is more than 80 inches (200 cm). The basin's largest city, Belém, averages close to 100 inches (250 cm) of rainfall a year. Although there is no real dry season, there are definitely drier and wetter times of year, with August and September the driest months. The immensity of this watershed and its hydrologic cycle is underscored by the fact that 20 percent of all freshwater discharged into the oceans comes from the Amazon.

Since the Amazon Basin draws from eight countries, this watershed is an ideal network to integrate the northern half of South America. In fact, the river is navigable to Iquitos, Peru, some 2100 miles (3600 km) upstream. Ironically, compared with the other great rivers of the world, settlement in the basin continues to be sparse, in large part due to the poor quality of the forest soils. The best soils are found on the floodplain, where natural levees reach heights of 20 feet (6 meters). Alluvium deposited on these levees for thousands of years has made them extremely fertile, as well as safe from normal flooding. Consequently, most of the older settlements, such as Manaus, are found on the levees. Active colonization of the Brazilian portion of the Amazon since the 1960s has boosted the population. Over 20 million people live in the Brazilian Amazon, which is about 10 percent of the country's total population, and the Amazonian states are increasing at nearly 4 percent a year. The development of the basin, most notably through towns, roads, dams, farms, and mines, is constantly changing what was viewed as a vast tropical wilderness just a half century ago. The Brazilian government has plans to build 30 new dams in its portion of the Amazon to meet the country's growing energy demands and to facilitate resource extraction in this region. Perhaps the most contested dam is Belo Monte on the Xingu River (a tributary of the Amazon) (Figure 4.7). The project, when completed, will be the third largest hydroelectric dam in the world, generating more than 11,000 megawatts of electricity. It includes two dams, two canals, two reservoirs, and a system of dikes. Yet the dam has its critics, as a relatively pristine river will be radically transformed, up to 20,000 people will be displaced, and forest will be flooded. The Belo Monte project has galvanized both domestic and international movements to stop construction. The project has experienced setbacks and court-ordered stoppages, but as of 2013 it continues to be built.

Plata Basin The region's second largest watershed begins in the tropics and discharges into the Atlantic in the midlatitudes. Three major rivers make up this system: the Paraná, the Paraguay, and the Uruguay. The Paraguay River and its tributaries drain the eastern Andes of Bolivia, the Brazilian shield, and the Chaco. The Paraná primarily drains the Brazilian uplands before the Paraguay River joins it in northern Argentina. The Paraná and the considerably smaller Uruguay empty into the Rio de la Plata estuary, which begins north of Buenos Aires.

▼ **Figure 4.7 Amazonian Dam** An early phase in the construction of the Belo Monte Hydroelectric Project on the Xingu River in Brazil is shown here. Located in Para state near the town of Altamira, when completed Belo Monte will be the third largest dam in the world.

Unlike the Amazon, much of the Plata Basin is now economically productive through large-scale mechanized agriculture, especially soybean production. Arid areas such as the Chaco and inundated lowlands such as the Pantanal support livestock. The Plata Basin contains several major dams, including Latin America's largest hydroelectric plant, the Itaipú Dam on the Paraná, which generates electricity for all of Paraguay and much of southern Brazil. Only China's Three Gorges Dam and hydroelectric plant is larger. As agricultural output in this watershed grows, sections of the Paraná River are being canalized and dredged to enhance the river's capacity for barge and boat traffic.

Orinoco Basin The third largest river basin by area is the Orinoco in northern South America. The Orinoco River meanders through much of southern Venezuela and part of eastern Colombia, giving character to a tropical grassland called the Llanos. Although it is only one-sixth the size of the Amazon watershed, its discharge roughly equals that of the Mississippi River. Like the Amazon, this basin is home to very few individuals; 90 percent of Venezuela's population lives north of the basin. With the exception of the industrial developments between Ciudad Guayana and Ciudad Bolívar, cities are few. Much of the Orinoco drains the Llanos, which are inundated by several feet of water during the rainy season. Since the colonial era, these grasslands have supported large cattle ranches. Although cattle are still important, the Llanos have also become a dynamic area of petroleum production for both Colombia and Venezuela.

Climate and Climate Change in Latin America

In tropical Latin America, average monthly temperatures in settings such as Managua (Nicaragua), Quito (Ecuador), or Manaus (Brazil) show little variation (Figure 4.8). Precipitation patterns do vary, however, and create distinct wet

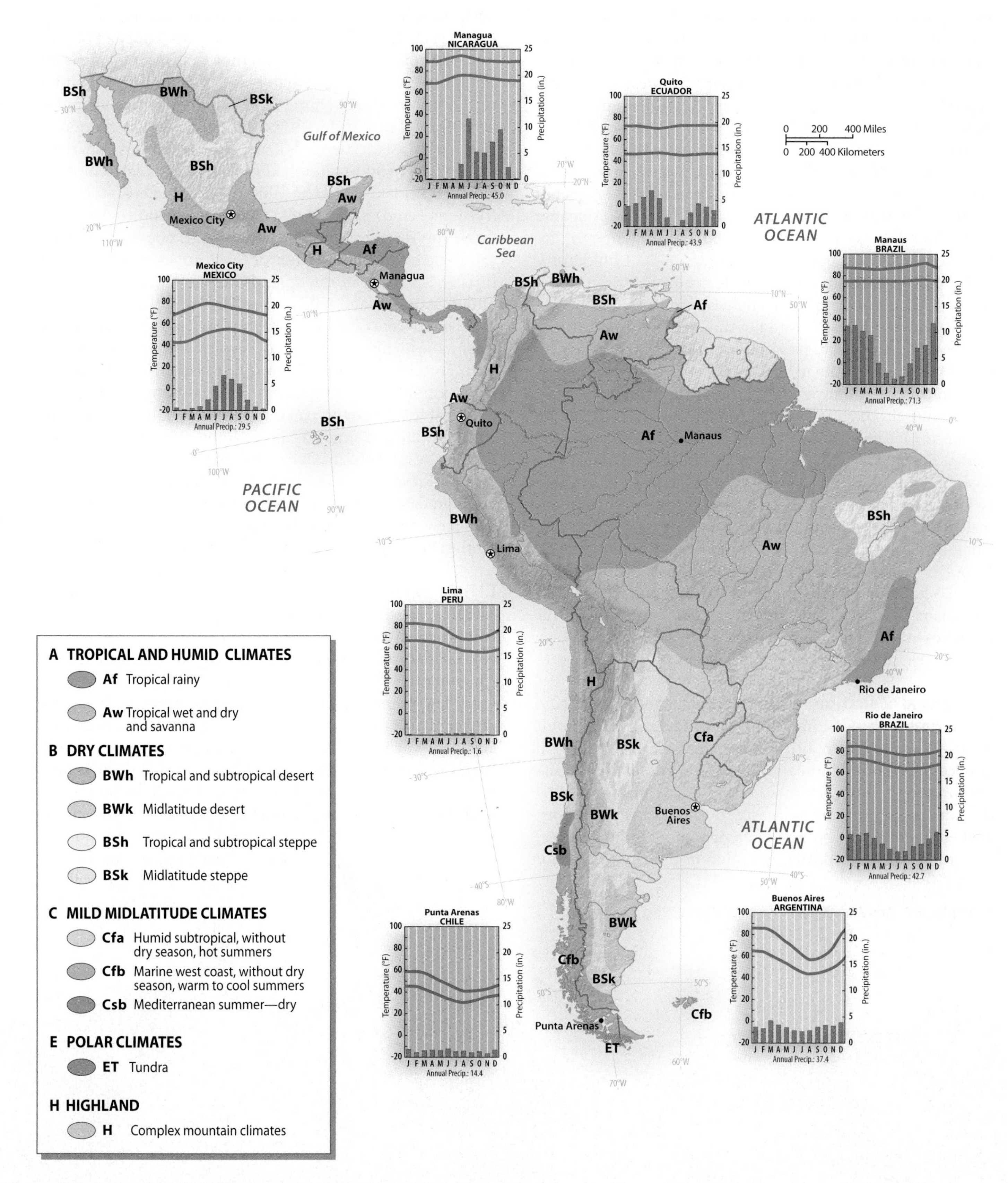

▲ **Figure 4.8 Climate of Latin America** Latin America includes the world's largest rainforest (Af) and driest desert (BWh), as well as nearly every other climate classification. Latitude, elevation, and rainfall play important roles in determining the region's climates. Note the contrast in rainfall patterns between humid Quito and arid Lima.

and dry seasons. In Managua, for example, January is typically a dry month, and June is a wet one. The tropical lowlands of Latin America, especially east of the Andes, are usually classified as tropical humid climates that support forest or savanna, depending on the amount of rainfall. The region's desert climates are found along the Pacific coasts of Peru and Chile, Patagonia, northern Mexico, and the Bahia of Brazil. Because of the extreme aridity of the Peruvian coast, a city such as Lima, Peru, which is clearly in the tropics, averages only 1.5 inches (4 cm) of annual rainfall. Some sections of the Atacama Desert of Chile get no measurable rainfall (Figure 4.9). Yet the discovery of resources such as nitrates in the 19th century and copper in the 20th century made this hyper-arid region a source of conflict among Chile, Bolivia, and Peru.

Midlatitude climates, with hot summers and cold winters, prevail in Argentina, Uruguay, and parts of Paraguay and Chile. Of course, the midlatitude temperature shifts in the Southern Hemisphere are the inverse of those in the Northern Hemisphere, meaning cold Julys and warm Januarys. Chile's climate is a mirror image of that of the west coast of Mexico and the United States; the Atacama Desert in the north is much like Baja California, a Mediterranean dry summer around Santiago is similar to that in Los Angeles, and a marine west coast climate with no dry season south of Concepción is like the climate along the coasts of Oregon and Washington. In the mountain ranges, complex climate patterns result from changes in elevation. To appreciate how humans adapt to tropical mountain ecosystems, it is important to understand the concept of **altitudinal zonation**—the relationship between cooler temperatures at higher elevations and changes in vegetation.

Altitudinal Zonation First described in the scientific literature by Prussian naturalist Alexander von Humboldt in the early 1800s, altitudinal zonation has practical applications that are intimately understood by all the region's native inhabitants. Humboldt systematically recorded declines in temperature as he ascended to higher elevations, a phenomenon known as the **environmental lapse rate.** According to Humboldt's description of the environmental lapse rate, temperature declines approximately 3.5°F for every 1000 feet in elevation, or 6.5°C for every 1000 meters. Humboldt also noted changes in vegetation by elevation, demonstrating that plant communities common to the midlatitudes could thrive in the tropics at higher elevations. These different altitudinal zones are commonly referred to as the *tierra caliente* (hot land) from sea level to 3000 feet (900 meters); the *tierra templada* (temperate land) at 3001 to 6000 feet (900 to 1800 meters); the *tierra fría* (cold land) at 6001 to 12,000 feet (1800 to 3600 meters); and the *tierra helada* (frozen land) above 12,000 feet (3600 meters). Exploitation of these zones allows agriculturists, especially in the uplands, access to a great diversity of domesticated and wild plants (Figure 4.10).

The concept of altitudinal zonation is most relevant for the Andes, the highlands of Central America, and the Mexican Plateau. For example, traditional Andean farmers might use the high pastures of the Altiplano for grazing llamas and alpacas, the tierra fría for potato and quinoa production, and the lower temperate zone to produce corn. All the great indigenous civilizations, especially the Incas and the Aztecs, systematically extracted resources from these zones, thus ensuring a diverse and abundant resource base. Yet these complex ecosystems are extremely fragile and have become important areas of research for the effects of climate change in the tropics.

El Niño One of the most studied weather phenomena in Latin America, called **El Niño** (referring to the Christ child), occurs when a warm Pacific current arrives along the normally cold coastal waters of Ecuador and Peru in December, around Christmastime. This change in ocean temperature happens every few years and produces torrential rains, signaling the arrival of an El Niño year. The 2009–2010 El Niño was especially bad for Latin America; scores of people were killed by floods or storms attributed to El Niño–related disturbances. Devastating floods occurred in Peru and Brazil. In Peru, heavy rains and flooding damaged the railroad leading to the ancient Incan site of Machu Picchu, temporarily limiting access to this popular tourist destination until the railroad could be rebuilt.

▼ **Figure 4.9 Atacama Desert** One of the driest places on earth with almost no vegetation, many visitors liken it to a moonscape. Yet, the soils of the Atacama contain a wealth of copper and nitrates. Below is an image of the Valley of the Moon in northern Chile.

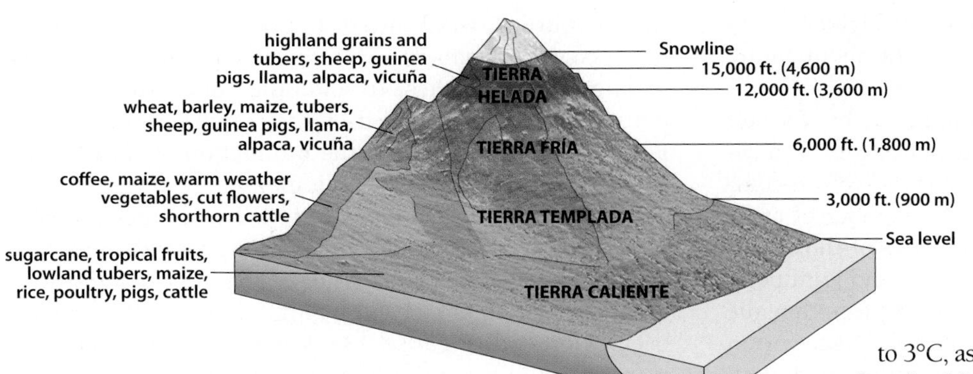

◄ **Figure 4.10 Altitudinal Zonation** Tropical highland areas support a complex array of ecosystems. In the *tierra fría* zone (6000 to 12,000 feet, or 1800 to 3700 meters), for example, midlatitude crops such as wheat and barley can be grown. This diagram depicts the range of crops and animals found at different elevations in the Andes.

El Niño's impacts vary, however, across Latin America. While the Pacific coast of South and North America experienced record rainfall in the 1997–1998 El Niño, Colombia, Venezuela, northern Brazil, Central America, and Mexico battled drought. In addition to crop and livestock losses, estimated to be in the billions of dollars, hundreds of brush and forest fires left their mark on the region's landscape. In addition, extreme weather events unrelated to El Niño, such as hurricanes and their associated heavy rain, flooding, and landslides, affect Central America and Mexico. In April of 2013, dozens died in Argentina when over 12 inches of rain fell in one night and flash flooding occurred as the Plata River spilled over its banks.

Impacts of Climate Change for Latin America

Latin America represents about 8 percent of the world's population and produces about 6 percent of global greenhouse gas (GHG) emissions. The rate of growth of GHG emissions in Latin America is dramatically lower than in all other regions of the developing world, except for Sub-Saharan Africa. The region's relatively low emissions can be explained by lower average energy consumption, higher reliance on renewable energy (especially hydropower and biofuels), and greater dependence on public transportation. The burning of forest and brush, a common practice in the region, does produce spikes in carbon dioxide (CO_2) emissions, but the regrowth of vegetation also absorbs vast amounts of CO_2.

Global warming has both immediate and long-term implications for Latin America. Of greatest immediate concern is how climate change is influencing agricultural productivity, water availability, changes in the composition and productivity of ecosystems, and incidence of vector-born diseases such as malaria and dengue fever. Changes attributable to global warming are already apparent in higher elevations, making these concerns more pressing. For example, coffee growers in the Colombian Andes have seen a decline in productivity over the past five years, which they attribute to higher temperatures and longer dry spells. The long-term effects of global climate change on lowland tropical forest systems are less clear: For example, some areas may experience more rainfall, others less. Other long-term impacts, such as rising sea level, will not cause the same levels of displacement in Latin America as predicted for the Caribbean or Oceania.

Climate change research indicates that highland areas are particularly vulnerable to global warming. Tropical mountain systems are projected to experience increased temperatures of 1 to 3°C, as well as lower rainfall. This will raise the altitudinal limits of various ecosystems, affecting the range of crops and arable land available to farmers and ranchers. Research over the past 50 years has documented the dramatic retreat of Andean glaciers—some no longer exist, and others will cease to exist in the next 10–15 years. Although this is a visible indicator of global warming, it also has pressing human repercussions. Many Andean villages, as well as metropolitan areas such as La Paz, Bolivia, get much of their water from glacial runoff. A major Bolivian glacier, Chacaltaya, has lost 80 percent of its area in the last 20 years and is virtually gone (Figure 4.11).

(a) Chacaltaya Glacier in 1996, was small but still existed.

(b) Chacaltaya Glacier in 2006, has only a small remaining area on the far right of the image.

▲ **Figure 4.11 Andean Glacial Retreat** The rapid disappearance of Chacaltaya Glacier in the Bolivian Andes is a visible reminder of warming conditions in high altitude tropical zones. Seasonal melt from this glacier, which has long been important to water supply in La Paz, Bolivia, has declined dramatically.

Thus, as average temperatures increase in the highlands and glaciers recede, there is widespread concern about future supplies of drinking water.

Another immediate concern brought on by warmer temperatures is the sudden rise in dengue fever, a mosquito-borne virus. It was once considered relatively uncommon in highland Latin America, but the number of cases has risen sharply in the past few years. Tens of thousands now suffer from its fever, headache, nausea, and joint pain. In rare cases, external and internal bleeding from dengue fever can result in death. The World Health Organization has estimated that dengue is now widespread in more than 100 tropical and subtropical countries around the world, but the sudden rise in cases in Latin America suggests that warmer highland temperatures have placed millions more at risk.

Scientists are not yet sure whether or how global warming is impacting the frequency and strength of El Niño cycles. If El Niño cycles intensify and occur more often as a result of global warming, increased flooding in western South America and a declining fishery (which is supported by nutrient upwelling in the normally cold currents) off the coast of Peru and Chile will result. Moreover, some evidence indicates that hurricane intensity will increase as ocean temperatures warm with climate change, heightening the impacts of these natural disasters. The effects of global warming on El Niño and extreme storm events are the subject of ongoing research, but are of particular concern to the Latin American region.

Environmental Issues: The Destruction of Forests

Perhaps the environmental issue most commonly associated with Latin America is deforestation (Figure 4.12). The Amazon Basin and portions of the eastern lowlands of Central America and Mexico still maintain unique and impressive stands of tropical forest. But other woodland areas, such as the Atlantic coastal forests of Brazil and the Pacific forests of Central America, have nearly disappeared as a result of agriculture, settlement, and ranching. The coniferous forests of northern Mexico are also falling, in part because of a bonanza for commercial logging stimulated by the NAFTA. In Chile, the ecologically unique evergreen rainforest (the Valdivian forest) in the midlatitudes is being cleared for wood chip exports to Asia.

The loss of tropical rainforests is most critical in terms of biological diversity. Tropical rainforests account for only 6 percent of Earth's landmass, but at least 50 percent of the world's species are found in this biome. Moreover, the Amazon Basin contains the largest undisturbed stretches of rainforest in the world. Unlike Southeast Asian forests, where hardwood extraction drives forest clearance, Latin American forests are usually seen as an agricultural frontier that state governments divide in an attempt to give land to the landless or reward political cronies. Thus, forests are cut and burned, with settlers and politicians carving them up to create permanent settlements, slash-and-burn plots, or large cattle ranches. In addition, some tropical forest cutting has been motivated by the search for gold (Brazil, Peru, and Costa Rica) and the production of coca leaf for cocaine (Peru, Bolivia, and Colombia).

Brazil has incurred more criticism than other countries for its Amazon forest policies. During the past 40 years, close to 20 percent of the Brazilian Amazon has been deforested. In states such as Rondônia, close to 60 percent of the state has been deforested (Figure 4.13). What most alarms environmentalists and forest dwellers (Indians and rubber tappers) is the dramatic increase in the rate of rainforest clearing since 2000, estimated at nearly 8000 square miles (20,000 square km) per year. The increased rates of deforestation in the Brazilian Amazon are due to the expansion of industrial mining and logging, the growth in corporate farms, the development of new road networks and dams, the incidence of human-ignited wildfires, and continued population growth. Under the Advance Brazil program started in 2000, some $40 billion will go to new highways, railroads, gas lines, hydroelectric projects, power lines, and river canalization projects that will reach into remote areas of the basin. The Brazilian government, under President Lula da Silva, has created 150,000 square kilometers of new conservation areas, many of them alongside the "arc of deforestation"—a swath of agricultural development along the southern edge of the Amazon Basin where the worst deforestation has occurred (see Figure 4.12). However, Brazil's Forest Code, as revised in 2012, has reduced the amount of "forest reserve" that private landholders must maintain. Many conservationists fear this will lead to more forest clearing and fragmentation.

Grassification The conversion of tropical forest into pasture, called **grassification,** is another practice that has contributed to deforestation. Particularly in southern Mexico, Central America, and the Brazilian Amazon, a hodgepodge of development policies from the 1960s through the 1980s encouraged deforestation to make room for cattle. The preference for ranching as a status-conferring occupation seems to be a carryover from Iberia. The image of the *vaquero* (cowboy) looms large in the region's history. Even poor farmers appreciate the value of having livestock. They know that cattle can be quickly sold for cash, something like having a savings account on hand.

Many natural grasslands are suitable for grazing in Latin America, such as the Llanos in Colombia and Venezuela and the Chaco and Pampas in Argentina. However, the rush to convert forest into pasture made ranching a scourge on the land. Even in cases in which domestic demand for beef has increased, ranching in remote tropical frontiers is seldom economically self-sustaining. The conversion of tropical forest to pasture is especially dramatic in the Central American countries of Guatemala, Costa Rica, and Panama, where huge tracts of forest have been

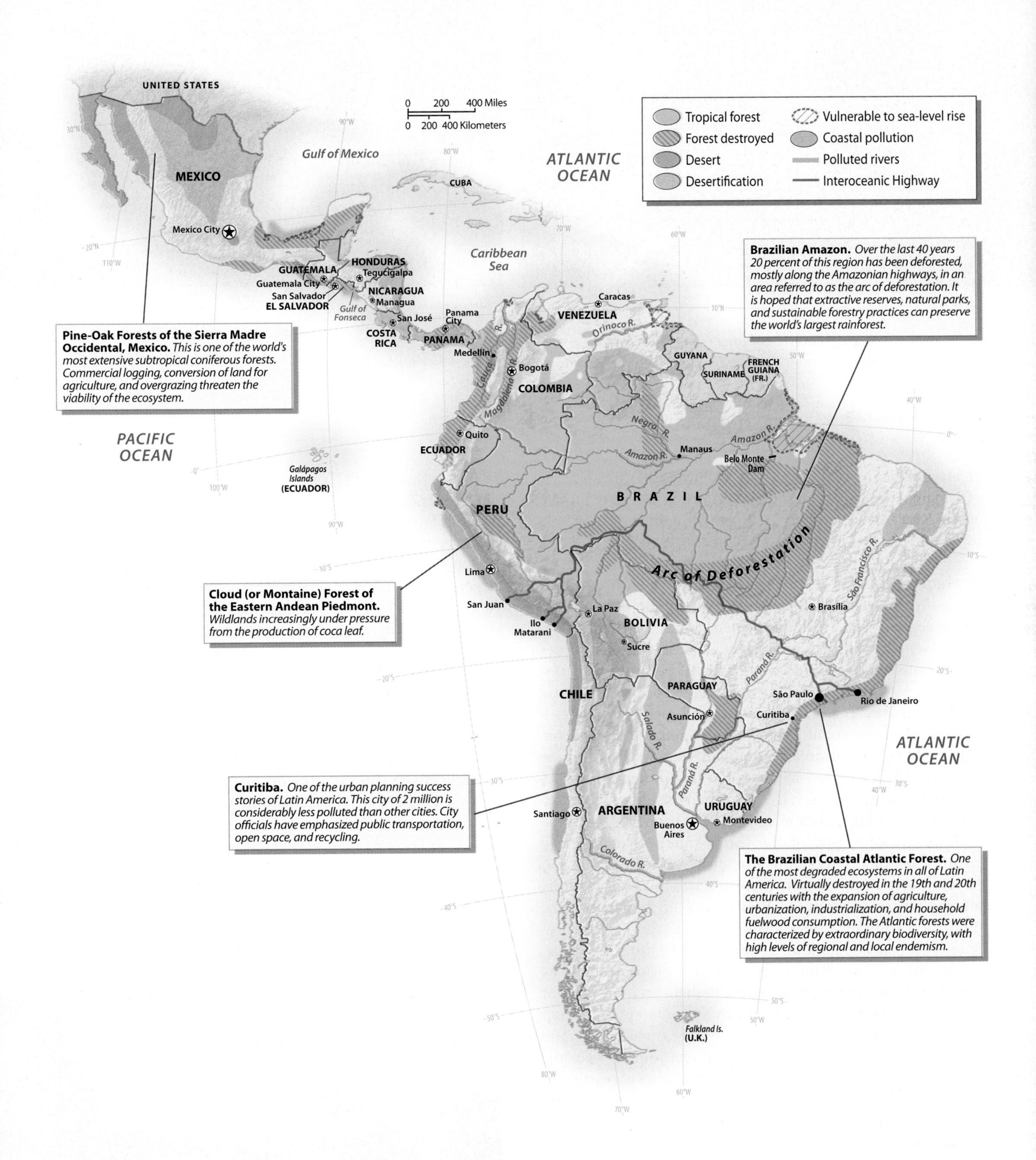

▲ **Figure 4.12 Environmental Issues in Latin America** Tropical forest destruction, desertification, water pollution, and poor urban air quality are some of the pressing environmental problems facing Latin America. Still present, however, are vast areas of tropical forest, supporting a wealth of genetic and biological diversity.

(a) July 30, 2000

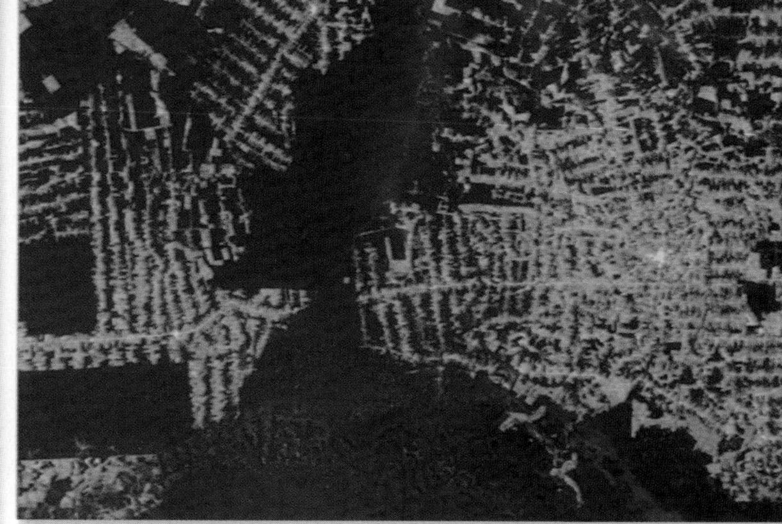

(b) August 2, 2010

▲ **Figure 4.13 Tropical Forest Settlement in the Amazon** These satellite images of Rondônia, Brazil, illustrate the dramatic change in forest cover in just 10 years near the settlement of Buritis and road BR-364. Intact forest is dark green, whereas cleared areas are light green (crops) or tan (bare ground). Typically, the first clearings appear off of roads, forming a fishbone pattern. Over time, as more forest is cleared and settlements grow, the fishbone pattern collapses into a mosaic of pasture, farmland, and forest fragments.

cleared and converted to pasture (Figure 4.14). In the case of Panama, the Guna (an indigenous group formerly known as the Kuna) have banned cattle ranching in the Guna Yala territory in an effort to protect their forest lands.

Protecting Lands for Future Generations Latin America has more nationally protected lands than any other developing region. The areas designated as national parks, nature reserves, wildlife sanctuaries, and scientific reserves with limited public access went from 10 percent of the territory in 1990 to 20 percent in 2010, according to World Bank estimates. Brazil's protected land went from just 9 percent of the national territory to 26 percent in 20 years. Although conservationists complain that many of these areas are "paper parks" with limited real protection, many countries in the region have used the conservation of forests and other lands as a means to attract tourists.

Costa Rica is one of the Latin American pioneers in creating national parks and promoting ecotourism. In the 1970s, Costa Rican conservationist Mario Boza successfully lobbied for the creation of national parks in response to the rampant forest destruction occurring to expand coffee production, banana plantations, and cattle pasture. By 1990, almost 20 percent of the territory had been protected—by then nearly all the unprotected lands had been cleared for agriculture or settlement. Given Costa Rica's impressive natural beauty, Pacific and Caribbean beaches, volcanoes, and biodiversity, about 2 million international tourists visit the country each year (Figure 4.15). The parks are accessible to Costa Ricans as well, at a reduced fee, whereas tourists pay higher park entrance fees to support conservation and park maintenance.

Problems on Agricultural Lands The pressure to modernize agriculture has produced a series of environmental problems. As peasants were encouraged to adopt new hybrid varieties of corn, beans, and potatoes, an erosion of genetic diversity occurred. Efforts to preserve dozens of native domesticates are under way at agricultural research centers in the central Andes and Mexico. Nonetheless, many useful native plants may have been lost. Modern agriculture also depends on chemical fertilizers and pesticides that eventually run off into surface streams and groundwater. Consequently, many rural areas suffer from contamination of local water supplies. Even more troublesome is the direct exposure of farmworkers to toxic agricultural chemicals. Mishandling of pesticides and fertilizers can lead to exposure, resulting in rashes and burns. In some areas, such as Sinaloa, Mexico, a rise in serious birth defects parallels the widespread application of chemicals.

Soil erosion and fertility decline occur in all agricultural areas. Certain soil types in Latin America are particularly

▼ **Figure 4.14 Converting Forest into Pasture** Cattle graze in northern Guatemala's Petén region. Clearing of this tropical forest lowland began in the 1960s and continues today. Ranching is a status-conferring occupation in Latin America with serious ecological costs. The beef produced from this region is for domestic and export markets.

▲ **Figure 4.15 Costa Rican National Park** Hugging the Pacific Coast, the tropical forest and beaches of Manuel Antonio National Park make it a popular destination for Costa Ricans as well as international tourists. The pressures to develop tropical coasts are real which makes creating protected areas an urgent need.

vulnerable to erosion, most notably the volcanic soils and the reddish *oxisols* found in the humid tropical lowlands. The productivity of the Paraná basalt plateau in Brazil, for example, has declined over decades due to the ease with which these volcanic soils erode and the failure to apply soil conservation methods. By contrast, the oxisols of the tropical lowlands can quickly degrade into a baked claypan surface when the natural cover is removed, making permanent agriculture nearly impossible. Ironically, the consolidation of the large-scale modern farms in the basins and valleys of the highlands tends to push peasant subsistence farmers into marginal areas. On these hillside farms, gullies and landslides reduce productivity. Lastly, the sprawl of Latin American cities consumes both arable land and water, eliminating some of the region's best farmland.

Urban Environmental Challenges

For most Latin Americans, air pollution, water availability and quality, and garbage removal are the pressing environmental problems of everyday life. Consequently, many environmental activists from the region focus their efforts on making urban environments cleaner by introducing "green" legislation and calling people to action. In this most urbanized region of the developing world, city dwellers do have better access to water, sewers, and electricity than their counterparts in Asia and Africa. Moreover, the density of urban settlement seems to encourage the widespread use of mass transportation; both public and private bus and van routes make getting around cities fairly easy. However, the usual environmental problems that come from dense urban settings ultimately require expensive remedies, such as new power plants and modernized sewer and water lines. The money for such projects is never enough, due to currency devaluation, inflation, and foreign debt. Because many urban dwellers tend to reside in unplanned squatter settlements, servicing these communities with utilities after they are built is difficult and costly.

Air Pollution Air pollution is a concern for most major cities (see *Working Toward Sustainability: Greening Transport and Expanding Access in Bogotá*). However, this is especially true

Working Toward **Sustainability**

Greening Transport and Expanding Access in Bogotá

Most major Latin American cities are over four centuries old and were designed for pedestrians and carriage traffic, not automobiles. As these cities exploded in size in the 20th century, many observers lamented how automobile dependence had destroyed urban life in Latin America. An infusion of cars took over public space, contaminated air, and saddled people with slow commutes. Yet innovative leaders in Latin American cities dreamed of something different.

Mass Transit Buses For the 9 million people in Bogotá, Colombia, a new urban phase began in 2000 with the opening of a rapid-transit bus system called TransMilenio. TransMilenio replicates some aspects of the highly regarded transit system in Curitiba, Brazil, which has long been considered Latin America's "Green City." Using large-capacity articulated buses with dedicated bus lanes and rapid-loading platforms, the bright red buses now dominate the city's main arteries (Figure 4.1.1). Today about 1400 buses are in use, plus several hundred smaller feeder buses, moving 1.5 million people per day. Older, more polluting buses were pulled from the streets. Extensive coverage, the use of smart cards that integrate transfers from one bus to another, and a fixed fare of less than a dollar make this an accessible system. Rapid-transit bus systems with dedicated lanes are also much less expensive than subways or light rail, which is an important consideration for developing countries.

TransMilenio was just one component of a broader vision for the city that focused on increased social integration, improved mobility, enhanced public space, and a human scale of design. Guided by Mayor Enrique Peñalosa in the late 1990s, the plan emphasized creating new pedestrian zones, revitalized parks and sidewalks, bike paths, and a more integrative and efficient public transport system. Better public transport made driving in private cars less attractive. Moreover, the city imposed restrictions on private cars. Based on license plate numbers, two days a week drivers could not use their cars during rush hour.

▼ **Figure 4.1.1 TransMilenio in Bogotá** Gliding along the busy arteries of Bogotá are large red atriculated buses that make up the TransMileno system. Operating along designated lanes and with large fast loading platforms, the system moves 1.5 million people a day.

▲ **Figure 4.1.2 Bogotá's Bikers** A cyclist on a bike path pedals past a nun. Abundant bike lanes make Bogotá Latin America's most bike-friendly city. In 2003 the mayor introduced the "day without cars" campaign to demonstrate how this congested city could function without automobiles.

Mile-High Biking Bogotá's planners intentionally included bicycles in the transportation system. In the past decade, bike lanes have been built throughout the city, and bike stations are found at suburban TransMilenio bus stops (Figure 4.1.2). Bogotá is at an elevation of 8600 feet (2600 meters), but it is relatively flat with spring-like temperature year round, so the bicycle is a practical and clean form of urban transport. Today Bogotá is considered Latin America's most bike-friendly city—and one of the most bike-friendly big cities in the world. Biking is a popular sport in Colombia, and the practice of closing some streets for weekend biking has existed for decades. Yet to reinforce the value of biking and walking, Bogotá held its first Car Free Day in 2000 in which no private cars and trucks could be used in the city. This weekday event was so popular that it is now held annually.

So what is Bogotá like today? Congestion, especially in the downtown, has been greatly reduced. Also, the city's air quality has visibly improved. Bicycles are definitely more prevalent, and the city is filled with lively and popular public spaces.

Google Earth Virtual Tour Video

http://goo.gl/NqzYx7

1. Consider the density of Bogotá. Why would a bus system work well here?
2. What are the advantages of designated bus lanes?

for the capitals of Santiago and Mexico City. A combination of geographical factors (basin settings) and meteorological factors (winter inversion layers), along with dense human settlement and automobile dependence, has led to these two cities having some of the highest recorded concentrations of particulate matter and ozone, major contributors to air pollution. Air pollution is not just an aesthetic issue—the health costs of breathing such contaminated air are real, as shown by elevated death rates due to heart disease, asthma, influenza, and pneumonia. The burden of air pollution is not evenly distributed among city residents, as the elderly, the very young, and the poor are more likely to suffer the negative health effects of contaminated air. Fortunately, both cities have taken steps to address this vexing problem.

Santiago, a prosperous city of nearly 7 million in Chile's Central Valley, has an elevation of 1700 feet (520 meters). Although not as high as Mexico City, this basin setting regularly produces thermal inversions, when warm air traps a layer of cold air near the surface. This trapped surface layer becomes filled with engine exhaust, industrial pollution, garbage, and even fecal matter (Figure 4.16). The inversion layers happen year round, but can be especially bad in the winter months from May through August, often forcing schools to suspend all sports when smog emergencies are called. Santiago officials began addressing this problem in the late 1980s by restricting vehicular traffic. On a given weekday, 20 percent of all buses, taxis, and cars are restricted from driving, based on license plate numbers. During smog emergencies, up to 40 percent of vehicles can be restricted from driving. In addition, older buses were replaced by a new fleet of cleaner-running buses with greater carry capacity. These buses, combined with the city's subway system, can move over 2 million riders a day. By 2010, air quality had noticeably improved, and public support for restrictive measures and public transport has solidified.

The smog of Mexico City has been so bad that most modern-day visitors have no idea that mountains surround them. Air quality has been a major issue for Mexico City since the 1960s, driven in part by the city's unusually high rate of growth. (Between 1950 and 1980, the city's annual rate of growth was 4.8 percent.) It is difficult to imagine a better setting for creating air pollution. The city sits in a bowl 7400 feet (2200 meters) above sea level, and thermal inversions form regularly.

▼ **Figure 4.16 Air Pollution in Santiago** Smog blankets Santiago, with the Andes in the background. During the winter months (May through August), thermal inversion layers form that trap pollutants near ground level, causing a spike in pollution-related health problems. By reducing vehicular traffic and greatly expanding public transportation, the city has experienced improved air quality.

Steps were finally taken in the late 1980s to reduce emissions from factories and cars. Unleaded gas is now widely available for the 4 million cars in the metropolitan area, and cars manufactured for the Mexican market must have catalytic converters. Also, some of the worst polluting factories in the Valley of Mexico have closed. In the last few years, the mayor of Mexico City has expanded a low-emissions bus system, eliminating thousands of tons of carbon monoxide. In 2007, the decision was made to close the elegant Paseo de Reforma to traffic on Sunday mornings and open it to bike riders. This change was so popular that now bike lanes have been introduced to some downtown areas in an effort to encourage bike ridership. For longer commutes, a suburban train system is being built that will complement the existing subway system. The payoff is real: Mexico City no longer ranks among the top polluted cities in the world, and it appears to have cut most of its pollutants by at least half.

Water Mexico City's other significant environmental problem is water. When Vicente Fox was president of Mexico, he declared water (both scarcity and quality) not just a problem for the capital, but also a national security issue for the entire country. Ironically, it was the abundance of water that made Mexico's central valley attractive for settlement initially. Large shallow lakes once filled the valley, but over the centuries most were drained to expand agricultural land. As surface water became scarce, wells were dug to tap the basin's massive freshwater aquifer. Today approximately 70 percent of the water used in the metropolitan area is drawn from this aquifer. Troubling evidence has appeared that the aquifer is being overdrawn and at risk of contamination, especially in areas where unlined drainage canals can leak pollutants into the surrounding soil, which then leach into the aquifer. To reduce reliance on the aquifer, the city now pumps water nearly a mile uphill from more than 100 miles (160 km) away.

Andean cities such as Bogotá, Quito, and La Paz are increasingly experiencing water scarcity and rationing. Some of this is due to increased demands put on aging water systems as a result of population growth. However, changes in precipitation patterns due to El Niño years or global climate change make these large urban centers especially vulnerable. La Paz, Bolivia, for example, gets much of its water from glacial runoff. However, as noted earlier, a major Bolivian glacier, Chacaltaya, has lost 80 percent of its area in the past 20 years. Thus, as average temperatures increase in the highlands and glaciers recede, there is widespread concern about future drinking-water supplies in this metropolitan area of nearly 2 million people.

REVIEW

4.1 Describe the major ecosystems in Latin America and how humans have adapted to and modified these different ecosystems.

4.2 Summarize some of the major environmental issues impacting this region and how different countries have tried to address them.

Population and Settlement: The Dominance of Cities

Latin America did not have great river-basin civilizations like those in Asia. In fact, the great rivers of the region are surprisingly underutilized as areas of settlement or corridors for transportation. The major population clusters of Central America and Mexico are in the interior plateaus and valleys, whereas the interior lowlands of South America are relatively empty. Historically, the highlands supported most of the region's population during the pre-Hispanic and colonial eras, although archeological and geographical research has shown greater pre-Hispanic lowland settlement in the Amazon and Central American lowlands than previously thought. In the 20th century, population growth and migration to the Atlantic lowlands of Argentina and Brazil, along with continued growth of Andean coastal cities such as Guayaquil, Barranquilla, and Maracaibo, have reduced the demographic significance of the highlands. Major highland cities such as Mexico City, Guatemala City, Bogotá, and La Paz still dominate their national economies, but the majority of large cities are on or near the coasts (Figure 4.17).

Like the rest of the developing world, Latin America experienced dramatic population growth in the 1960s and 1970s. In 1950, its population totaled 150 million people, which equaled the population of the United States at that time. By 1995, the population had tripled to 450 million; in comparison, the United States reached 300 million in 2006. Latin America outpaced the United States because its birth rate remained consistently higher as infant mortality rates dropped and life expectancy soared. In 1950, Brazilian life expectancy was only 43 years; by the 1980s, it was 63, and by 2013, it was 74. In fact, between 1950 and 1980, most countries in the region experienced a 15- to 20-year improvement in life expectancy, which pushed up growth rates. Four countries account for over 70 percent of the region's population: Brazil with 195 million, Mexico with 118 million, Colombia with 48 million, and Argentina with 41 million (see Table 4.1).

During the 1980s, population growth rates in Latin America suddenly began to slow, and by the 1990s most countries reported rates of less than 2 percent. By 2012, the regional rate of natural increase was 1.2 percent. This relatively sudden shift surprised demographers, who had predicted in 1985 that the region's population would reach 750 million in 2025. Today's projection is for only 620 million by that date. One of the reasons for this fertility decline is the shift to urban living, which tends to reduce family size.

The Latin American City

A quick glance at the population map of Latin America shows a concentration of people in cities. The movement from rural areas to cities has been one of the most significant demographic shifts in the region, beginning in earnest in the 1950s. In 1950, only one-quarter of the region's

► Figure 4.17 **Population of Latin America** The concentration of population in urban and coastal settlements is evident in this map. Population density in central and southern Mexico, as well as Central America, is quite high. In South America, the majority of people live on or near the coasts, leaving the interior of the continent lightly populated.

population was urban; the rest lived in small villages and the countryside. Today the pattern is reversed, with three-quarters of the population living in cities. In the most urbanized countries, such as Argentina, Chile, Uruguay, and Venezuela, more than 85 percent of the population lives in cities (see Table 4.1). This preference for urban life is attributed to cultural as well as economic factors. Under Iberian rule, people residing in cities had higher social status and greater economic opportunity. Initially, only Europeans were allowed to live in the colonial cities, but this exclusivity was not strictly enforced. Over the centuries, colonial cities became the hubs for transportation and communication, making them the primary centers for economic and political activities.

Latin American cities are noted for high levels of **urban primacy,** a condition in which a country has a primate city three to four times larger than any other city in the country. Examples of primate cities are Lima, Caracas, Guatemala City, Santiago, Buenos Aires, and Mexico City. Primacy is often viewed as a liability, as too many national resources are concentrated into one urban center. In an effort to decentralize, some governments have intentionally built new cities far from existing primate cities (for example, Ciudad Guayana in Venezuela and Brasília in Brazil). Despite these efforts, the tendency toward primacy remains. Moreover, the growth of urbanized regions that include several major cities has inspired the label of *megalopolis* for three areas in Latin America. Emerging megalopolises include Mexico City–Puebla–Toluca–Cuernavaca on the Mesa Central, the Niterói–Rio de Janeiro–Santos–São Paulo–Campinas

Table 4.1 Population Indicators

Country	Population (millions), 2013	Population Density (per square kilometer)	Rate of Natural Increase (RNI)	Total Fertility Rate	Percent Urban	Percent < 15	Percent > 65	Net Migration (Rate per 1000)
Argentina	41.3	15	1.1	2.4	93	25	11	−1
Bolivia	11.0	10	1.9	3.2	67	35	5	0
Brazil	195.5	23	0.9	1.8	85	25	7	−0
Chile	17.6	23	0.9	1.9	87	22	10	0
Colombia	48.0	42	1.4	2.3	76	28	6	−1
Costa Rica	4.7	92	1.2	1.9	73	25	7	3
Ecuador	15.8	56	1.7	2.7	67	32	6	−0
El Salvador	6.3	300	1.2	2.2	65	31	7	−8
Guatemala	15.4	142	2.6	3.9	50	41	4	−2
Honduras	8.6	76	2.2	2.9	52	38	4	−2
Mexico	117.6	60	1.5	2.2	78	30	6	−5
Nicaragua	6.0	46	1.9	2.6	58	34	5	−5
Panama	3.9	51	1.5	2.6	75	29	7	1
Paraguay	6.8	17	1.8	2.9	62	33	5	−1
Peru	30.5	24	1.5	2.6	75	30	6	−3
Uruguay	3.4	19	0.4	2.0	94	22	14	−2
Venezuela	29.7	33	1.7	2.4	89	29	6	0

Source: Population Reference Bureau, *World Population Data Sheet, 2013.*

axis in southern Brazil, and the Rosario–Buenos Aires–Montevideo–San Nicolás corridor in Argentina and Uruguay's lower Rio Plata Basin (see Figure 4.17).

Urban Form Latin American cities have a distinct urban morphology that reflects both their colonial origins and their contemporary growth (Figure 4.18). Usually, a clear central business district (CBD) exists in the old colonial core. Radiating out from the CBD is older middle- and lower-class housing found in the zones of maturity and *in situ* accretion. In this model, residential quality declines as you move from the center to the periphery. The exception is the elite spine, a newer commercial and business strip that extends from the colonial core to newer parts of the city. Along the spine are superior services, roads, and transportation. The city's best residential zones, as well as shopping malls, are usually on either side of the spine. Close to the elite residential sector, a limited area of middle-class housing is typically found. Most major urban centers also have a *periférico* (a ring road or beltway highway) that circumscribes the city. Industry is located in isolated areas of the inner city and in larger industrial parks outside the ring road.

In outer rings of the city (sometimes straddling the periférico) is a zone of peripheral squatter settlements where many of the urban poor live in the worst housing. Services and infrastructure are extremely limited: Roads are unpaved, water is often trucked in, and sewer systems are nonexistent. The dense ring of squatter settlements (variously called *ranchos, favelas, barrios jovenes,* or *pueblos nuevos*) that encircle Latin American cities reflect the speed and intensity with which these zones were created. The squatter settlements are also found in disamenity zones near the core of the city. These are settings such as steep hillsides or narrow gorges prone to flooding that are considered too risky for formal housing. They can also be polluted industrial areas or even garbage dumps where the very poor reside because they might find employment there and/or a place to live. In some cities, more than one-third of the population lives in these self-built homes of marginal or poor quality. These kinds of dwellings are recognizable throughout the developing world, yet the practice of building homes on the "urban frontier" has a longer history in Latin America than in most Asian and African cities. The combination of a rapid inflow of migrants, the inability of governments to meet pressing housing needs, and the eventual official recognition of many of these neighborhoods with land titles and utilities meant that this housing strategy was rarely discouraged. In cities as diverse as Medellin, Colombia, and Caracas, Venezuela, planners have become creative in addressing the needs of urban settlers on steep hillsides by introducing gondolas to link shanty settlements with the rest of the city. Rio de Janeiro installed a six-station gondola line running over a group of favelas known as the *Complexo do Alemão* in 2011. Residents now have a 15-minute ride in the gondola

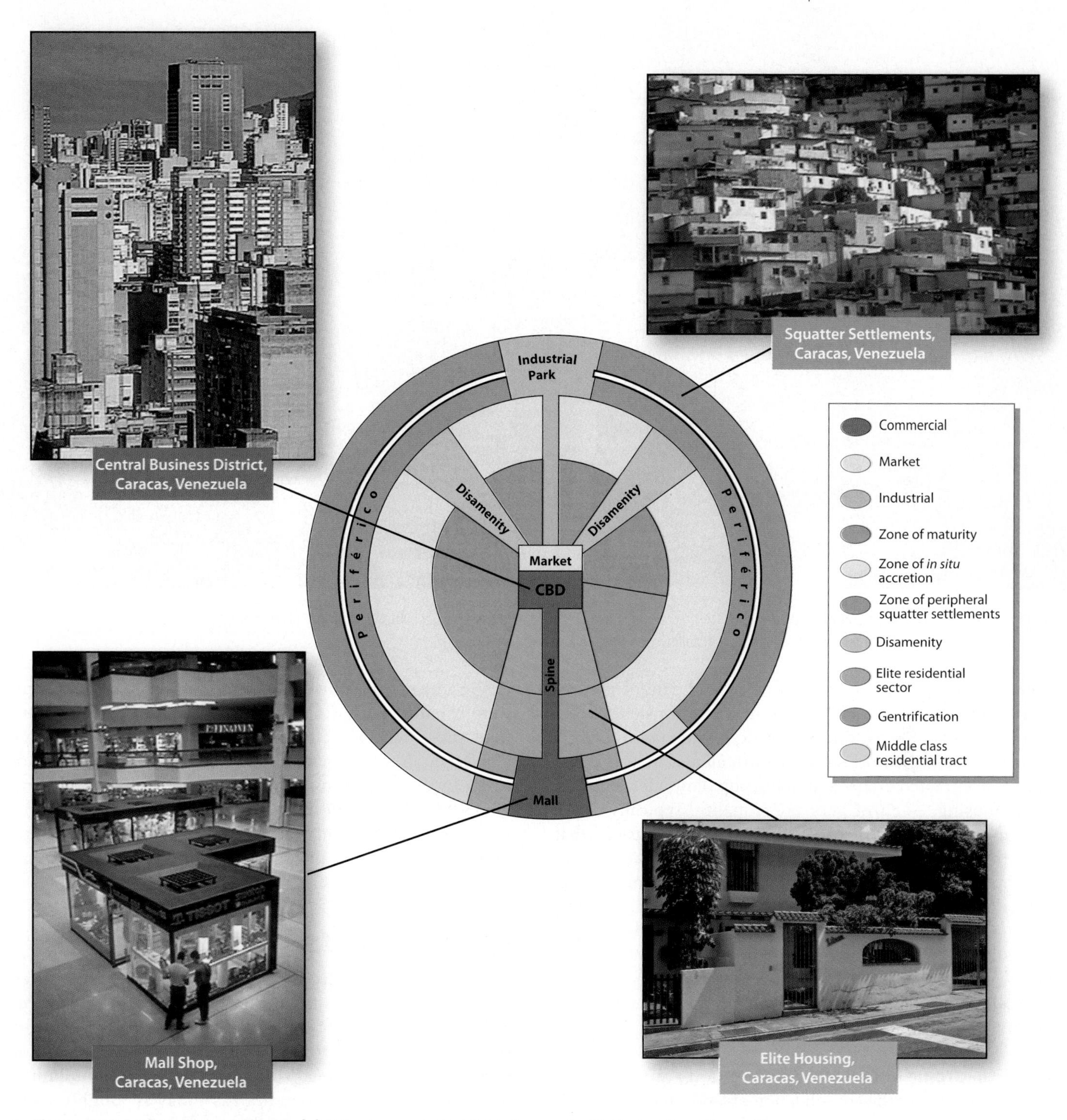

▲ **Figure 4.18 Latin American City Model** This urban model highlights the growth of Latin American cities and the class divisions within them. The central business district (CBD), elite spine, and residential sectors may have excellent services and utilities, but life in the zone of peripheral squatter settlements is much more difficult. In many Latin American cities, one-third of the population resides in squatter settlements. **Q: How does the Latin American city model compare with that of North America in terms of where the rich and the poor live? What are the factors that drive urban growth?**

versus an over-an-hour-long hike to the rail station (Figure 4.19).

Among the inhabitants of these neighborhoods, the **informal sector** is a fundamental force that houses, services, and employs them. Definitions of the informal sector are much debated. The term usually refers to the economic sector that relies on self-employed, low-wage jobs (such as street vending, shoe shining, and artisan manufacturing), which are virtually unregulated and untaxed. Some scholars include as part of the informal sector illegal activities

▲ **Figure 4.19 Gondolas Over Rio de Janeiro** Urban planners have creatively addressed issues of access in some of the city's densely settled favelas (informal settlements) by installing gondolas to improve transport.

such as drug smuggling, prostitution, and sale of contraband items such as illegally copied movie or music CDs and tapes. One of the most interesting expressions of informality is the housing in the squatter settlements. In arid Lima, Peru, an estimated 40 percent of the population lives in self-built housing, often of very poor quality. Typically, these settlements begin as illegal invasions of open spaces that are carefully planned and timed to avoid the risk of eviction by city authorities. If the hastily built communities go unchallenged, squatters steadily improve their houses. The creation of these landscapes reflects a conscious and organized effort on the part of the urban poor, many of whom have rural origins, to make a place for themselves in Latin America's cities (see *Cityscapes: Claiming the High Ground in La Paz, Bolivia*).

Rural-to-Urban Migration As conditions in rural areas deteriorated due to the consolidation of lands, mechanization of agriculture, and increased population pressure, peasants began to pour into the cities of Latin America in a process referred to as **rural-to-urban migration.** The strategy of rural households sending family members to the cities for employment as domestics, construction workers, artisans, and vendors has been well established since the 1960s. Once in the cities, rural migrants generally found conditions better, especially in terms of access to education, health care, electricity, and clean water.

It was not poverty alone that drove people out of rural areas, but individual choice and an urban preference. Migrants believed in, and often realized, greater opportunities in cities, especially the capital cities. Those who came were usually young (in their twenties) and better educated than those who stayed behind. Women slightly outnumbered men in this migrant stream. The move itself was made easier by extended kin networks formed by earlier migrants who settled in discrete areas of the city and aided new arrivals. The migrants maintained their links to their rural communities by periodically sending remittances and making return visits.

Patterns of Rural Settlement

Although the majority of Latin Americans live in cities, some 125 million people do not. Throughout the region, a distinct rural lifestyle exists, especially among peasant subsistence farmers. In Brazil alone, more than 35 million people live in rural areas. Interestingly, the absolute number of people living in rural areas today is roughly equal to the number in the 1960s. Yet rural life has changed dramatically. In addition to subsistence agriculture, highly mechanized, capital-intensive farming occurs in most rural areas. The links between rural and urban areas are much improved, with the result that rural folks are less isolated. Also, as international migration increases, many rural communities are directly connected to cities in North America and Europe, with immigrants sending back remittances and supporting hometown associations. This is especially evident in rural Mexico and Central America. Much like the region's cities, the rural landscape is divided by extremes of poverty and wealth. The root of social and

Cityscapes

Claiming the High Ground in La Paz, Bolivia

La Paz, quite literally, takes your breath away. Most visitors first experience the city by landing at the International Airport in El Alto (the Heights) at 13,500 feet (4100 meters) above sea level. Unless you live in Tibet, this elevation is usually a shock to the system. The local custom of sipping coca tea is suggested to help with nausea, headaches, and difficulty breathing, but most people acclimate after resting for a day or two.

In truth, there are two cities in metropolitan La Paz: One is the colonial Nuestra Senora de La Paz with about 1 million people, and the other is the majority Aymara (an Amerindian group) city of El Alto, incorporated 30 years ago and also with about 1 million people. These two parts straddle the rim of the Altiplano—La Paz rests in a steep-sided bowl below the rim, while El Alto spreads out in neat rectilinear lines across the high, flat, windswept plain of the Altiplano. Above it all looms snow-covered Illimani, a 21,000-foot (6500-meter) Andean peak (Figure 4.2.1).

La Paz: Shelter in the Valley La Paz was settled in the 16th century as a critical administrative point for the shipment of silver from Potosi, Bolivia, to Lima, Peru. The Spanish chose this setting as a shelter from the extreme climate of the Altiplano. Today La Paz is the administrative capital of Bolivia (Sucre is the constitutional capital), so the president resides here, as do many government agencies and embassies. In the CBD of La Paz are glistening high-rises, both upscale apartments and office buildings, but the entire city is rimmed by brick shanty towns built into the steep hillsides of the city's bowl-shaped valley. Descending the valley to the south some 3000 feet (900 meters), you encounter the posh neighborhoods of the southern zone—La Paz's elite spine. Here the well-off have settled in suburban neighborhoods, attracted to the lower elevation and warmer daytime temperatures. The laborers who work in these neighborhoods often travel by collective buses from El Alto, which is at least an hour-long trip—and a major elevation change.

▼ **Figure 4.2.1 Vertical La Paz** This mountain city with Illimani in the background was settled by the Spanish in the 16th century. Today the metropolitan area has 2 million people.

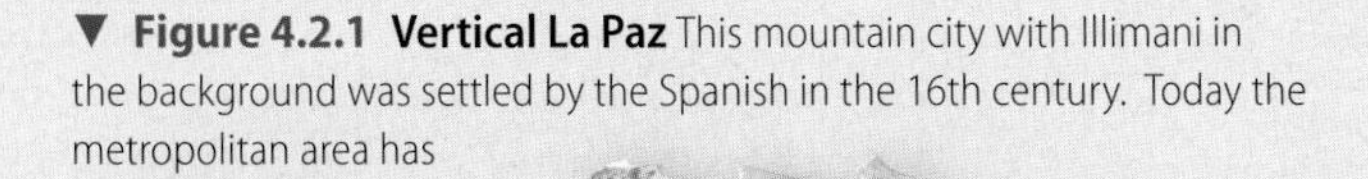

▲ **Figure 4.2.2 El Alto** This mostly Aymara city that developed on the altiplano overlooking La Paz has emerged as a vital center for Amerindian activism. In this image Aymara union members rally in El Alto's streets.

El Alto: On the Highland Plain The city of El Alto is remarkable in its homogeneity and newness—it is Bolivia's fastest-growing city. The city surrounded the airport and then spread out in a grid pattern of one- and two-story brick homes that blend with the muted vegetation of the high plain. This mostly Aymara city began to form after Bolivia's agrarian reform in the 1950s. Aymara peasants made their way to the outskirts of La Paz, and by the 1970s the city was growing daily as more peasants settled among their kin. El Alto incorporated in 1987 in an effort to distance itself from La Paz and better control its resources. It is a dynamic commercial center for the Aymara. The most affluent merchants have built wildly mirrored and painted multistoried buildings that are places of business and residence, as well as a testament to their success (Figure 4.2.2). It is also a definitely Amerindian city, whose inhabitants have used their strategic high ground to shut down La Paz with roadblocks in protest against neoliberal reforms. The political muscle of El Alto became most evident with the election of President Evo Morales in 2005—Bolivia's first Aymara leader.

1. How does the landscape of El Alto contrast with that of southern zone of La Paz?
2. What environmental, cultural, or economic factors might explain the way that metropolitan La Paz grew?

Google Earth Virtual Tour Video

http://goo.gl/GG04z1

economic tension in the countryside is the uneven distribution of arable land.

Rural Landholdings Historically, the control of land in Latin America was the basis for political and economic power. Colonial authorities granted large tracts of land to the colonists, who were also promised the service of Indian laborers as part of the *encomienda* system regulating the land. These large estates typically took up the best lands along the valley bottoms and coastal plains. The owners were often absentee landlords, spending most of their time in the city and relying on a mixture of hired, tributary, and slave labor to run their rural operations. Passed down from one generation to the next, many estates can trace their ownership back a century or more. The allocation of large blocks of land to one owner also denied peasants access to land, so they were forced to labor on the estates. This entrenched practice of maintaining large estates is referred to as **latifundia.**

Although the pattern of estate ownership is well documented, peasants have always farmed small plots for their subsistence. This practice, called **minifundia,** can lead to permanent or shifting cultivation. Small farmers typically plant a mixture of crops for subsistence, as well as for trade. Peasant farmers in Colombia or Mexico, for example, grow corn, fruits, and various vegetables alongside coffee bushes that produce beans for export. Strains on the minifundia system occur when rural populations grow and land becomes scarce, forcing farmers to divide their properties into smaller and less-productive parcels or seek out new parcels on steep slopes.

Much of the turmoil in 20th-century Latin America surrounded the issue of land ownership, with peasants demanding its redistribution through the process of **agrarian reform.** Governments have addressed these concerns in different ways. The Mexican Revolution in 1910 yielded a system of communally held lands called *ejidos*. In the 1950s, Bolivia crafted agrarian reform policies that led to the expropriation of estate lands and their reallocation to small farmers. As part of the Sandinista revolution in Nicaragua in 1979, lands were expropriated from the political elite and converted into collective farms. In 2000, President Hugo Chavez ushered in a new era of agrarian reform in Venezuela. In 2006, Bolivian President Evo Morales introduced an agrarian reform program aimed at giving land title to indigenous communities in the eastern lowlands, often far from their ancestral territories. These programs have met with resistance and, at times, have proven to be costly politically. Eventually, the path chosen by most governments was to make frontier lands available to land-hungry peasants. The opening of tropical frontiers, especially in South America, was a widely practiced strategy that changed national settlement patterns and began waves of rural-to-rural and even urban-to-rural migration.

Agricultural Frontiers The expansion into agricultural frontiers serves several purposes: providing peasants with land, tapping unused resources, and shoring up political boundaries. Several frontier colonization efforts in South America are noteworthy. In addition to settlement along Brazil's Trans-Amazon Highway, Peru developed its Carretera Marginal (Marginal Highway) in an effort to lure colonists into the cloud and rain forests of eastern Peru. Most recently, the completion of the Interoceanic Highway linking Peru and Brazil has opened up new areas of settlement in the Amazonian territories of these countries (Figure 4.20). In Bolivia,

▼ **Figure 4.20 Amazon Mining Town** The small gold mining town of Mazuco, Peru is part of resource boom that is bringing in settlers and bringing down forest. This community is also served by the new Interoceanic Highway.

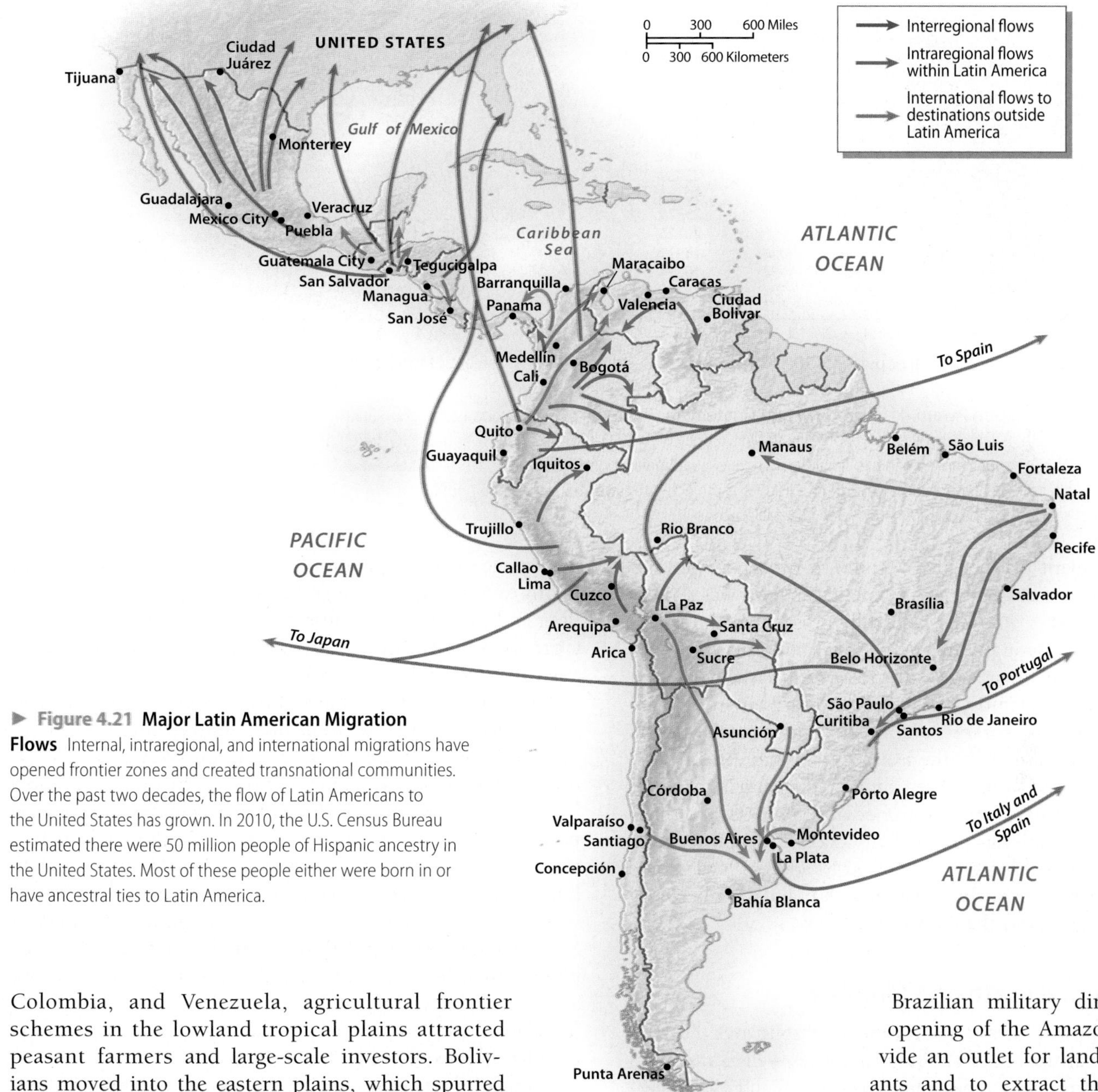

▶ Figure 4.21 **Major Latin American Migration Flows** Internal, intraregional, and international migrations have opened frontier zones and created transnational communities. Over the past two decades, the flow of Latin Americans to the United States has grown. In 2010, the U.S. Census Bureau estimated there were 50 million people of Hispanic ancestry in the United States. Most of these people either were born in or have ancestral ties to Latin America.

Colombia, and Venezuela, agricultural frontier schemes in the lowland tropical plains attracted peasant farmers and large-scale investors. Bolivians moved into the eastern plains, which spurred the growth and importance of the city of Santa Cruz. Mexico sent colonists, some displaced by dam construction, into the forests of Tehuantepec. Guatemala developed its northern Petén region. El Salvador had no frontier left, but many desperately poor Salvadorans poured into the neighboring states of Honduras and Belize in search of land. In short, although the dominant demographic trend has been a rural-to-urban movement, an important rural-to-rural flow has turned previously virgin areas into agricultural zones (see the purple arrows in Figure 4.21).

The opening of the Brazilian Amazon for settlement was the most ambitious frontier colonization scheme in the region. In the 1960s, Brazil began its frontier expansion by constructing several new Amazonian highways, a new capital (Brasília), and state-sponsored mining operations. The Brazilian military directed the opening of the Amazon to provide an outlet for landless peasants and to extract the region's many resources. However, the generals' plans did not deliver as intended. Throughout the basin, thin forest soils were incapable of supporting permanent agricultural colonies and, in the worst cases, degraded into baked claylike surfaces devoid of vegetation. Government-promised land titles, agricultural subsidies, and credit were slow to reach small farmers, even if they were fortunate enough to be given land on *terra roxa*, a nutrient-rich purple clay soil. Instead, a disproportionate amount of money went to subsidizing large cattle ranches through tax breaks and improvement deals, where "improved" land meant cleared land.

Because of the many competing factions (miners, loggers, ranchers, peasants, and corporate farmers) and the uncertainty of land title and political authority, the Amazon

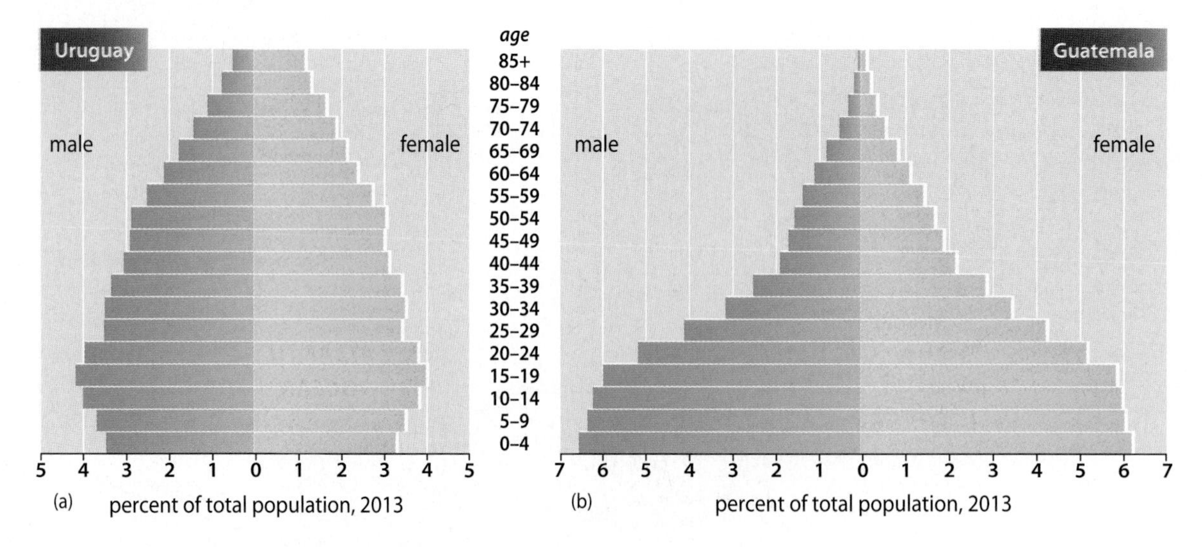

▲ **Figure 4.22 Population Pyramids of Uruguay and Guatemala** These two pyramids contrast the population structure of (a) the more developed and demographically stable Uruguay with that of (b) the youthful and rapidly growing Guatemala. The average Uruguayan woman has 2 children, whereas Guatemalan women have a total fertility rate (TFR) of 3.9 children. Due to differences in natural increase and TFR, Uruguay is projected to have about the same population size in 2050, but Guatemala is projected to be twice as large, with 27.4 million people.

can be a violent place. Chico Mendes, a rubber tapper and internationally recognized environmental activist, was slain in 1988 by a rancher who was eventually prosecuted and jailed. Mendes's death brought an international outpouring of support, and eventually a large extractive reserve was created in Acre state in his memory. Yet violence toward environmental activists, religious workers, and organizers of rural workers is extremely common. In the last 20 years, more than 1000 Amazonian activists have been murdered in Brazil. Most of the murders occur due to resource or land conflicts, and those who order the killings are seldom successfully prosecuted. The Brazilian government has vowed to address lawlessness in the Amazon; meanwhile, the region's population is nearly 10 times larger today than it was in 1960.

Population Growth and Mobility

The high growth rates in Latin America throughout the 20th century are attributed to natural increase, as well as immigration in the early part of the century. The 1960s and 1970s were decades of tremendous growth, resulting from high fertility rates and increasing life expectancy. In the 1960s, for example, a Latin American woman typically had six or seven children. By the 1980s, family sizes were half as big. Today the total fertility rate (TFR) for the region is 2.2, which is only slightly higher than replacement value (see Table 4.1). Several factors explain this trend: more urban families, which tend to be smaller than rural ones; increased participation of women in the workforce; higher education levels of women; state support of family planning; and better access to birth control. Even the more rural countries with a high percentage of Amerindians are experiencing smaller families—Bolivia's TFR is 3.3, and Ecuador's is 2.5.

Even with family sizes shrinking—and in the cases of Chile and Costa Rica, falling below replacement rates—there is built-in potential for continued growth because of the relative demographic youth of these countries. The average percentage of the population below age 15 is 28 percent. In North America, that same group is 19 percent of the population, and in Europe it is just 16 percent. This means that a proportionally larger segment of the population has yet to enter into its childbearing years.

The population pyramids of two countries, Uruguay and Guatemala, contrast the profile of a country with a stable population size with that of a demographically growing state (Figure 4.22). Uruguay is a small, but prosperous country, with a high Human Development Index ranking and relatively little poverty. Uruguayan women average two children, which is slightly below replacement level. Life expectancy is also high, but most population projections have the country growing very slowly between now and 2050. In contrast, Guatemala has a wide-based population pyramid and is considerably poorer. Total fertility rates have declined in Guatemala, but are still considered high at 3.6. Due to the youthfulness of the population and the increase in life expectancy, Guatemala's population is expected to double between 2010 and 2050.

In addition to natural increase, waves of immigrants into Latin America and migrant streams within Latin America have influenced population size and patterns of settlement. Beginning in the late 19th century, new immigrants from Europe and Asia added to the region's size and ethnic complexity. Important population shifts within countries have also occurred in recent decades, as witnessed by the growth of Mexican border towns and the demographic expansion of the Bolivian plains. In an increasingly globalized economy, even more Latin Americans live and work outside the region, especially in the United States and Europe.

European Migration After Latin American countries gained independence from Iberia in the 19th century, their new leaders sought to develop economically through immigration. Firmly believing that "to govern is to populate," many countries set up immigration offices in Europe to attract hardworking peasants to till the soils and "whiten" the **mestizo** (people of mixed European and Amerindian ancestry) population. The Southern Cone countries of Argentina, Chile, Uruguay, and southern Brazil were the most successful in attracting European immigrants from the 1870s until the depression of the 1930s. During this period, some 8 million Europeans arrived (more than came during the entire colonial period), with Italians, Portuguese, Spaniards, and Germans the most numerous. Some of this immigration was state-sponsored, such as the nearly 1 million laborers (including entire families) brought to the coffee estates surrounding São Paulo at the start of the 20th century. Other migrants came seasonally, especially the Italian peasants who left Europe in the winter for agricultural work during the Argentine summer and were thus nicknamed "the swallows." Still others paid their own passage, intending to settle permanently and prosper in the growing commercial centers of Buenos Aires, São Paulo, Montevideo, and Santiago.

Asian Migration Less well known than the European immigrants to Latin America are the Asian immigrants, who arrived during the late 19th and 20th centuries. Although considerably fewer, over time they established an important presence in the large cities of Brazil, Peru, Argentina, and Paraguay. Beginning in the mid-19th century, the Chinese and Japanese who settled in Latin America were contracted to work on the coffee estates in southern Brazil and the sugar estates and guano mines of Peru. Over time, Asian immigrants became prominent members of society. A son of Japanese immigrants, Alberto Fujimori was president of Peru from 1990 to 2000.

Between 1908 and 1978, a quarter-million Japanese immigrated to Brazil; today the country is home to more than 1.3 million people of Japanese descent (Figure 4.23). Initially, most Japanese were landless laborers, yet by the 1940s they had accumulated enough capital so that three-quarters of the migrants had their own land in the rural areas of São Paulo and Paraná states. As a group, the Japanese have been closely associated with the expansion of soybean and orange production. Today Brazil leads the world in exports of orange juice concentrate, with most of the oranges grown on Japanese-Brazilian farms. Increasingly, second- and third-generation Japanese have taken professional and commercial jobs in Brazilian cities; many of them have married outside their ethnic group and have lost their fluency in Japanese. South America's economic turmoil in the 1990s encouraged many ethnic Japanese to emigrate to Japan in search of better opportunities. Nearly one-quarter of a million ethnic Japanese left South America in the 1990s (mostly from Brazil and Peru) and now work in Japan.

▶ **Figure 4.23 Japanese Brazilians** Brazilian youth of Japanese ancestry perform in Curitiba, Brazil, to mark the 100th anniversary of Japanese immigration to Brazil in 2008. In 1908, the first Japanese immigrants arrived as agricultural workers, choosing Brazil as a destination after countries such as the United States and Canada had banned Japanese immigration. Today there are over 1.3 million ethnic Japanese in Brazil, especially in the states of São Paulo and Parana, and they have distinguished themselves as major farmers and urban professionals.

The latest Asian immigrants are from South Korea. Unlike their predecessors, most of the Korean immigrants came with enough capital to invest in small businesses, and they settled in cities, rather than in the countryside. According to official South Korean statistics, 120,000 Koreans emigrated to Paraguay between 1975 and 1990. Although many have stayed in Paraguay, a pattern has appeared of secondary immigration to Brazil and Argentina. Recent Korean immigrants in São Paulo have created more than 2500 small businesses. Unofficial estimates of the number of Koreans living in Brazil range from 40,000 to 120,000. As a group, they are decidedly commercial in orientation and urban in residence; their cities of choice are Asunción and Ciudad del Este in Paraguay, São Paulo in Brazil, and Buenos Aires in Argentina.

Latino Migration and Hemispheric Change Migration within Latin America and between Latin America and North America has had a significant impact on sending and receiving communities alike. Within Latin America, international migration is shaped by shifting economic and political realities. Thus, Venezuela's oil wealth during the 1960s and 1970s attracted 1–2 million Colombian immigrants who tended to work as domestics or agricultural laborers. Argentina has long been a destination for Bolivian and Paraguayan laborers. And, of course, farmers in the United States have depended on Mexican laborers for over a century (see Figure 4.21).

Political turmoil also sparked waves of international migrants. Chilean intellectuals fled to neighboring countries in the 1970s when General Pinochet wrested power from the socialist government led by Salvador Allende. Nicaraguans likewise fled when the socialist Sandanistas came to power in 1979. The bloody civil wars in El Salvador and Guatemala sent waves of refugees into neighboring countries, such as Mexico and the United States. Violence and low-intensity conflict have internally displaced over 2.5 million Colombians in the past two decades, and official statistics suggest another 3 million Colombians live abroad. With democratization on the rise in the region, many of today's immigrants are classified as economic migrants, not political asylum seekers.

Presently, Mexico is by far the largest country of origin of legal immigrants to the United States. The 2010 U.S. Census recorded 50 million Hispanics in the United States, two-thirds of whom claimed Mexican ancestry or Mexico as their birthplace. Mexican labor migration to the United States dates back to the late 1800s, when relatively unskilled labor was recruited to work in agriculture, mining, and railroads. This practice was formalized in the 1940s through the 1960s with the *Bracero* program, which granted temporary employment residence to 5 million Mexican laborers. Yet when that program ended with a major reform of immigration laws in 1965, most of the *braceros* returned to Mexico. Later flows of Mexicans were made up of legal permanent residents, temporary agricultural workers, and undocumented laborers. Mexican immigrants are most concentrated in California and Texas, but increasingly they are found throughout the country. Although Mexicans continue to have the greatest presence among Latinos in the United States, the number of immigrants from El Salvador, Guatemala, Nicaragua, Colombia, Ecuador, and Brazil has steadily grown. The Census Bureau estimates that by 2050, nearly 30 percent of the U.S. population will be Hispanic (see Figure 3.23). Most of this Hispanic population will have ancestral ties with peoples from Latin America and the Caribbean (see Chapter 5 on Caribbean migration).

Today Latin America is a region of emigration rather than immigration. The majority of states have negative rates of annual net migration, which means they are losing more people each year than they are gaining through immigration. Mexico has an annual rate of –2.3 per 1000, and El Salvador's rate is –7.3. By comparison, the average rate of net migration for the developing world as a whole is a much smaller figure of –0.5 (see Table 4.1).

Both skilled and unskilled workers from Latin America are an important source of labor in North America, Europe, and Japan. Many of these immigrants send monthly **remittances** (monies sent back home) to sustain family members. In 2008, it was estimated that immigrants sent nearly $70 billion to Latin America. By 2012, that figure had declined to $64 billion, which shows the lingering impact of the economic recession on both remittances and migrant flows. Most of this money came from workers in the United States, but Latino immigrants in Spain, Portugal, Japan, Canada, and Italy also sent money back to the region.

REVIEW

4.3 What are the historical and economic explanations for urban dominance and urban primacy in Latin America?

4.4 How have policies such as agrarian reform and frontier colonization impacted the patterns of settlement and primary resource extraction in the region?

4.5 Demographically, Latin America has grown much faster than North America. What factors contribute to faster growth, and is this growth likely to continue?

Cultural Coherence and Diversity: Repopulating a Continent

The Iberian colonial experience (1492 to the 1800s) imposed a political and cultural coherence on Latin America that makes it recognizable today as a world region. Yet this was not a simple transplanting of Iberia across the Atlantic. Instead, a complex process unfolded in which European and Indian traditions blended as indigenous groups were incorporated into either the Spanish or the Portuguese Empire. In some areas, such as southern Mexico, Guatemala, Bolivia, Ecuador, and Peru, Amerindian cultures have shown remarkable resilience, as evidenced by the survival of indigenous languages. Yet the prevailing pattern is one of forced assimilation in which European religion, languages, and political organization were imposed on the surviving fragments of native society. Later, other cultures—especially 10 million African slaves—added to the cultural mix of Latin America, the Caribbean, and North America. The legacy of the African slave trade will be examined in greater detail in Chapters 5 and 6. For Latin America, perhaps the single most important factor in the dominance of European culture was the demographic collapse of native populations.

Decline of Native Populations

It is difficult to grasp the enormity of cultural change and human loss due to this encounter between the Americas and Europe. Throughout the region, archaeological sites are reminders of the complexity of Amerindian civilizations prior to European contact. Dozens of stone temples found throughout Mexico and Central America, where the Mayan and Aztec civilizations flourished, attest to the ability of these societies to thrive in the area's tropical forests and upland plateaus. The Mayan city of Tikal flourished in the lowland forests of Guatemala, supporting tens of thousands, before its mysterious collapse centuries before the arrival of Europeans (Figure 4.24). In the Andes, stone terraces built by the Incas are still being used by Andean farmers. Cuzco—the core of the great Incan empire—and the Incan site of Machu Picchu are testaments to Incan ingenuity. The Spanish, too, were impressed by the sophistication and wealth they saw around them, especially in Tenochtitlán, where Mexico City is today. Tenochtitlán was the political and ceremonial center of the Aztecs, supporting a complex metropolitan area with some 300,000 residents. The largest city in Spain at the time was considerably smaller.

The most telling figures of the impact of European expansion are demographic. Experts believe that the precontact Americas had 54 million inhabitants; by comparison, western Europe in 1500 had approximately 42 million. Of the 54 million, about 47 million were in what is now Latin America, and the rest were in North America and the Caribbean. The region had two major population centers: one in central Mexico with 14 million people and the other in the central Andes (highland Peru and Bolivia) with nearly

◀ Figure 4.24 **Tikal, Guatemala** This ancient Mayan city, located in the lowland forests of the Petén, was part of a complex network of cities located in the Yucatan and northern Guatemala. At its height, Tikal supported over 100,000 residents before its collapse in the late 10th century. Today Tikal is a major tourist destination.

12 million. By 1650, after a century and a half of colonization, the indigenous population was one-tenth its precontact size. The human tragedy of this population loss is difficult to comprehend. The relentless elimination of 90 percent of the indigenous population was largely caused by epidemics of influenza and smallpox, but warfare, forced labor, and starvation due to a collapse of food production systems also contributed to the rapid population decline.

Interestingly, one legacy of the Amerindian collapse was an extensive recovery of forest, wildlife, and soils due to reduced human pressure. Geographer William Denevan has argued that a **pristine myth** has been perpetuated about Latin America as sparsely settled wilderness when the Spanish arrived in 1492. In fact, there is tremendous evidence that the native population cleared forest, created grasslands, built urban centers, and experienced serious problems with soil erosion. By 1750, as the colonial period was ending, the extensive human presence throughout Latin America was much less evident than it was when the Spanish first arrived. The population low point for Amerindians was in 1650, but the tragedy continued throughout the colonial period and to a much lesser extent continues today. After the indigenous population began its slow recovery in the central Andes and central Mexico, there were still tribal bands in southern Chile (the Mapuche) and Patagonia (the Araucania) that experienced the ravages of disease three centuries after Columbus landed. Even now, the isolation of some Amazonian tribes has made them vulnerable to disease.

The Columbian Exchange Historian Alfred Crosby likens the contact period between the Old World (Europe, Africa, and Asia) and the New World (the Americas) as an immense biological swap, which he terms the **Columbian exchange.** According to Crosby, Europeans benefited greatly from this exchange, and Amerindian peoples suffered the most from it. On both sides of the Atlantic, however, the introduction of new diseases, peoples, plants, and animals forever changed the human ecology.

Consider, for example, the introduction of Old World crops. The Spanish brought their staples of wheat, olives, and grapes to plant in the Americas. Wheat did surprisingly well in the highland tropics and became a widely consumed grain over time. Grapes and olive trees did not fare as well, but eventually grapes were produced commercially in the temperate zones of South America. The Spanish grew to appreciate the domestication skills of Indian agriculturalists, who had developed valuable starch crops such as corn, potatoes, and bitter manioc, as well as condiments such as hot peppers, tomatoes, pineapple, cacao, and avocados. Corn never became a popular food for Europeans, but many African peoples adopted it as a vital staple. After initial reluctance, Europeans and Russians widely consumed the potato as a basic food. Domesticated in the highlands of Peru and Bolivia, the humble potato has an impressive ability to produce a tremendous volume of food in a very small area, especially in cool climates. This root crop is credited with driving Europe's rapid population increase in the 18th century, when peasant farmers from Ireland to Russia became increasingly dependent on it as a basic food. This potato dependence also made them vulnerable to potato blight, a fungal disease that emerged in the 19th century and came close to unraveling Irish society.

Tropical crops transferred from Asia and Africa reconfigured the economic potential of the region. Sugarcane, an Asian transfer, became the dominant cash crop of the Caribbean and the Atlantic tropical lowlands of South America. With sugar production came the importation of millions of African slaves. Coffee, a later transfer from East Africa, emerged as one of the leading export crops throughout Central America, Colombia, Venezuela, and Brazil in the 19th century. Pasture grasses introduced from Africa enhanced the forage available to livestock. Rice, from Africa and Asia, also became a critical addition to Latin American diets.

The movement of Old World animals across the Atlantic had a profound impact on the Americas. Initially, these animals hastened Indian decline by introducing animal-borne

◀ **Figure 4.25** A Guna woman with her four children in front of her home on one of the San Blas islands in Panama. The Guna (formerly known as the Kuna) have maintained their territory of Guna Yala in Eastern Panama since the 1920s.

diseases and by producing feral offspring that consumed everything in their paths. However, the utility of domesticated swine, sheep, cattle, and horses was eventually appreciated by native survivors. Draft animals were adopted, as was the plow, which facilitated the preparation of soil for planting. Wool became a very important fiber for indigenous communities in the uplands. Slowly, pork, chicken, and eggs added protein and diversity to the staple diets of corn, potatoes, and cassava. With the major exception of disease, many transfers of plants and animals ultimately benefited both worlds. Still, it is clear that the ecological and material basis for life in Latin America was completely reworked through the exchange process initiated by Columbus.

Amerindian Survival and Political Recognition Presently, Mexico, Guatemala, Ecuador, Peru, and Bolivia have the largest indigenous populations. Not surprisingly, these areas had the densest native populations at contact. Indigenous survival also occurs in isolated settings where the workings of national and global economies are slow to penetrate. The isolated Miskito Coast of Honduras is home to Miskito, Pech, and Garífuna. In eastern Panama, the Guna and Emberá are present. In the Brazilian state of Roraima in the northern Amazon, some 15,000 indigenous Pemong speakers organized in 2004 to create the Raposa/Serra do Sol Indian reservation. In these relatively isolated areas, small groups of people have managed to maintain a distinct way of life despite the pressures to assimilate.

In many cases, Indian survival comes down to one key resource—land. Indigenous peoples who are able to maintain a territorial home, formally through land title or informally through long-term occupancy, are more likely to preserve a distinct ethnic identity. Because of this close association between identity and territory, native peoples are increasingly insisting on a recognized space within their countries. Some of Panama's indigenous groups have organized indigenous territories called *comarcas* where they assert local authority and have limited autonomy. The comarca of Guna Yala, on the Caribbean coast of eastern Panama, is the recognized territory of some 40,000 Guna (Figure 4.25). These efforts to define indigenous territory are seldom welcomed by the state, but they are occurring throughout the region.

From Amazonia to the highlands of Chiapas, many native groups are demanding formal political and territorial recognition as a means to redress centuries of injustice. The Zapatista rebellion in southern Mexico began on January 1, 1994, in Chiapas, the day NAFTA took effect. Although Zapatista supporters—largely Amerindian peasants—are mostly interested in access to land and basic services, their movement reflects a general concern about how increased foreign trade and investment hurts rural peasants. In January 2000, thousands of Indians, allied with dissident army officers, forced the resignation of Ecuadorian President Jamil Mahuad. Two years later, the Indian movement in Ecuador helped to elect Lucio Gutiérrez, one of the dissident army colonels. Bolivia witnessed the organized protests of Amerindians who inhabit the city of El Alto, overlooking La Paz. Angry with their pro-mining and pro-trade president, the Indians set up roadblocks in 2004 that cut off La Paz from supplies. In the end, indigenous-led protests forced two presidents to resign in two years and led to the election of Bolivia's first Amerindian president, Evo Morales, in 2005. Such experiences show the greater involvement of organized Indian groups in national politics and, some might argue, a deepening of democracy. Still others express concern that ethnically driven politics could lead to national fragmentation.

Patterns of Ethnicity and Culture

The Indian demographic collapse enabled Spain and Portugal to refashion Latin America into a European likeness. Yet instead of a neo-Europe rising in the tropics, a complex ethnic blend evolved. Beginning with the first years of contact, unions between European sailors and Indian women began the process of racial mixing that over time became a defining feature of the region. The courts of Spain and Portugal officially discouraged racial mixing, but were unable to prevent it. Spain, which had a far larger native population to oversee than did the Portuguese in Brazil, became obsessed with the matter of race and maintaining racial purity among its colonists. An elaborate classification system was constructed to distinguish emerging racial castes. Thus, in Mexico in the 18th century a Spaniard and an Indian union resulted in a *mestizo* child. A child of a mestizo and a Spanish woman was a *castizo*. However, the children from a castizo woman and a Spanish man were considered Spanish in Mexico, but a quarter mestizo in Peru. Likewise, *mulattoes* were the progeny of European and African unions, and *zambos* were the offspring of Africans and Indians.

After generations of intermarriage, such a classification system collapsed under the weight of its complexity, and four broad categories resulted: *blanco* (European ancestry), *mestizo* (mixed ancestry), *indio* (Indian ancestry), and *negro* (African ancestry). The blancos (or Europeans) continue to be well represented among the elites, yet the vast majority of people are of mixed racial ancestry. Dia de la Raza, the region's observance of Columbus Day, recognizes the emergence of a new mestizo race as the legacy of European conquest. Throughout Latin America, more than other regions of the world, miscegenation—or racial mixing—is the norm, which makes the process of mapping racial or ethnic groups especially difficult.

Languages Roughly two-thirds of Latin Americans are Spanish speakers, and one-third speak Portuguese. These colonial languages were so prevalent by the 19th century that they were the unquestioned languages of government and instruction for the newly independent Latin American republics. In fact, until recently many countries actively discouraged, and even repressed, Indian tongues. It took a constitutional amendment in Bolivia in the 1990s to legalize native-language instruction in primary schools and to recognize the country's multiethnic heritage (more than half the population is Amerindian, and Quechua, Aymara, and Guaraní are widely spoken) (Figure 4.26).

▲ **Figure 4.26 Languages of Latin America** The dominant languages of Latin America are Spanish and Portuguese. Nevertheless, there are significant areas in which indigenous languages persist and, in some cases, are recognized as official languages. Smaller language groups exist in Central America, the Amazon Basin, and southern Chile. **Q: What does this language map tell us about the patterns of Amerindian survival and endurance in Latin America?**

Because Spanish and Portuguese dominate, there is a tendency to neglect the influence of indigenous languages in the region. Mapping the use of indigenous languages, however, reveals important pockets of Indian resistance and survival. In the central Andes of Peru, Bolivia, and southern Ecuador, more than 10 million people still speak Quechua and Aymara, along with Spanish. In Paraguay and lowland Bolivia, there are 4 million Guaraní speakers, and in southern Mexico and Guatemala at least 6–8 million speak Mayan languages. Small groups of native-language speakers are found scattered throughout the sparsely settled interior of South America and the more isolated forests of Central America, but many of these languages have fewer than 10,000 speakers.

Blended Religions Like language, the Roman Catholic faith appears to have been imposed upon the region without challenge. Most countries report 90 percent or more of their population as Catholic. Every major city has dozens of churches, and even the smallest hamlet maintains a graceful church on its central square. In some countries, such as El Salvador and Uruguay, a sizable portion of the population attends Protestant evangelical churches, but the Catholic core of this region is still intact (see *Exploring Global Connections: The Catholic Church and the Argentine Pope*).

Exactly how native peoples absorbed the Christian faith is unclear. Throughout Latin America, **syncretic religions,** or blends of different belief systems, enabled pre-Hispanic religious practices to be folded into Christian worship. These blends took hold and endured, in part because Christian saints were easy surrogates for pre-Christian gods and because the Catholic Church tolerated local variations in worship as long as the process of conversion was under way. The Mayan practice of paying tribute to spirits of the underworld seems to be replicated today in Mexico and Guatemala via the practice of building small cave shrines to favorite Catholic saints and leaving offerings of fresh flow-

Exploring **Global Connections**

The Catholic Church and the Argentine Pope

In 2013, a new Pope was selected, and for the first time the spiritual leader of nearly 1.2 billion Roman Catholics was born in the Americas. Pope Francis, formerly Bishop Jorge Mario Bergoglio, is the son of

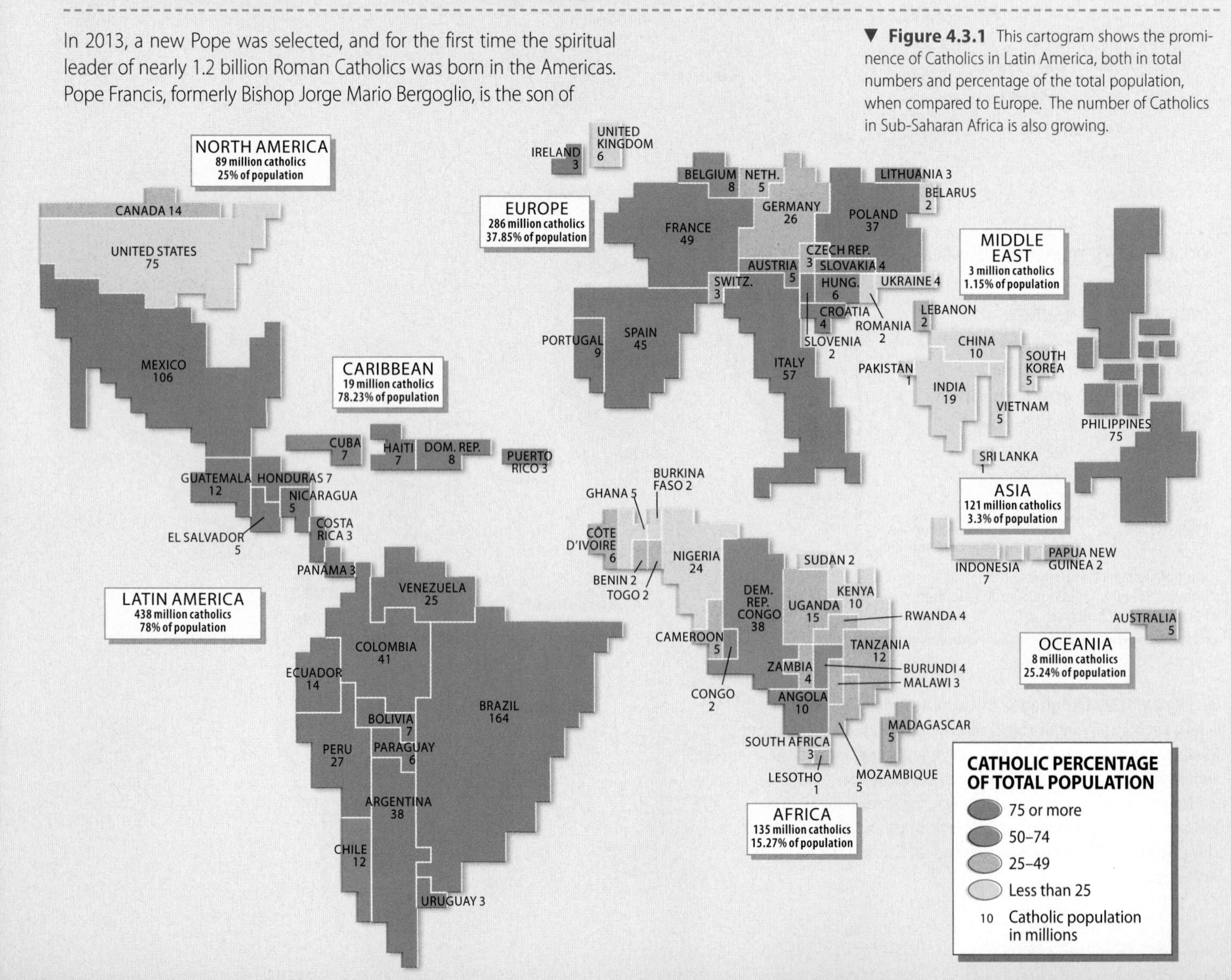

▼ **Figure 4.3.1** This cartogram shows the prominence of Catholics in Latin America, both in total numbers and percentage of the total population, when compared to Europe. The number of Catholics in Sub-Saharan Africa is also growing.

ers and fruits. One of the most celebrated religious icons in Mexico is the Virgin of Guadalupe—a dark-skinned virgin seen by an Indian shepherd boy in the 16th century—who became the patron saint of Mexico.

Syncretic religious practices also evolved and endured among African slaves. By far the greatest concentration of slaves was in the Caribbean, where slaves were used to replace the indigenous population, which was wiped out by disease (see Chapter 5). Within Latin America, the Portuguese colony of Brazil received the most Africans—at least 4 million. In Brazil, where the volume and the duration of the slave trade were the greatest, the transfer of African-based religious systems is most evident. West African–based religious systems such as Batuque, Umbanda, Candomblé, and Shango are often mixed with, or ancillary to, Catholicism and are widely practiced in Brazil. So accurate were some of these religious transfers that it is common to have Nigerian priests journey to Brazil to learn forgotten traditions. In many parts of southern Brazil, Umbanda is as popular with people of European ancestry as with Afro-Brazilians. Typically, a person becomes familiar with Umbanda after falling victim to a magician's spell by having some object of black magic buried outside his or her home. To regain control of his or her life, the victim needs the help of a priest or priestess.

The syncretic blend of Catholicism with African traditions is most obvious in the celebration of carnival, Brazil's most popular festival and one of the major components of Brazilian national identity. The three days of carnival, known as the Reign of Momo, combine Christian Lenten beliefs with pagan influences and feature African musical traditions epitomized by the rhythmic samba bands. Although the street festival was banned for part of the 19th century, Afro-Brazilians in Rio de Janeiro resurrected it in the 1880s with nightly parades, music, and dancing. Within 50 years, the street festival had given rise to formalized samba schools and helped break down racial barriers. By the 1960s, carnival became an important symbol for

Italian immigrants and was born in Buenos Aires, Argentina. As a religious leader and a member of the Jesuit Order, he earned a reputation for his humility and devotion to the poor. Now as leader of the Catholic Church, he overseas a vast global network of churches, schools, missions, and clergy that look to his guidance from Rome.

Shifting Catholic Demographics Pope Francis's selection brings to the fore the gradual, but dramatic demographic shift in the world's Catholic population. In 1900, the clear majority of the world's Catholics resided in Europe, but today less than one-quarter of them live there. For some time, the demographic core of the Catholic Church has been in Latin America. Worldwide, Brazil has the largest Catholic population (150 million), followed by Mexico (106 million), the Philippines (75 million), and then the United States (75 million)—helped in part by the large influx of immigrants from Latin America (**Figure 4.3.1**). Half the world's Catholics are in the Americas. Sub-Saharan Africa—particularly the Democratic Republic of the Congo and Nigeria—is another region where the numbers of Catholics are growing quickly, due in part to missionary work and colonial legacies.

Influence of Liberation Theology The global influence of Latin American Catholics became evident in the 1960s when the message of **liberation theology** began to take hold. Inspired by the region's widespread poverty and the social conditions of racial prejudice and unequal distribution of wealth that perpetuated it, a Peruvian priest, Gustavo Gutiérrez, coined the term. By the 1970s, liberation theology was strongly embraced by clergy in Brazil, Uruguay, and El Salvador as well. Critics of liberation theology complained that it was inspired more by Marxism than by the teachings of the Bible. In 1980, the Bishop of El Salvador, Oscar Romero, was assassinated while celebrating mass because of his outspoken support of the poor and criticism of the government. In his death, he became a martyr for the plight of the poor in Latin America. Not all Catholics in Latin America adopted liberation theology, but most acknowledge that its origins are rooted in the profound inequalities that have existed in the region.

Although the Pope is foremost a spiritual leader, he is clearly one with political clout. Many Latin Americans are thrilled at the selection of Pope Francis and the recognition it implies. The president of Argentina was quick to visit him at the Vatican, asking for his intervention on behalf of Argentina with regard to the disputed Falkland/Malvinas Islands (**Figure 4.3.2**)—a dispute he is unlikely to engage. The church also faces serious challenges, such as declining numbers of priests and troubling sexual abuse scandals being litigated in Europe and North America. Ironically, the future of the Catholic Church may lie more in the Southern Hemisphere than the northern one, especially with a messenger such as Pope Francis and a focus on the needs of the poor.

▼ **Figure 4.3.2 Pope Francis with Argentine President Kirchner** Pope Francis met with Argentine President Cristina Fernandez de Kirchner in the Vatican soon after he became Pope in 2013. Born in Argentina, Pope Francis is the first leader of the Catholic Church from the Americas.

▲ **Figure 4.27 Carnival in Rio de Janeiro** Samba schools, such as this one, compete each year during Carnival for the best costumes and music. Rio de Janeiro's Carnival is a spectacle that draws thousands of revelers.

Brazil's multiracial national identity. Today the festival—which is most associated with Rio de Janeiro—draws thousands of participants from all over the world (Figure 4.27).

The Global Reach of Latino Culture

Latin American culture, vivid and diverse as it is, is widely recognized throughout the world. Whether it is the sultry pulse of the tango or the fanaticism with which Latinos embrace soccer as an art form, aspects of Latin American culture have been absorbed into the global world culture. A dramatic example of the reach of Latino culture appears every Saturday on the Univision television network, based in Miami. From there, a charismatic Chilean, Don Francisco, hosts *Sabado Gigante*. This 3-hour-long Spanish variety show is viewed in 40 countries and draws a weekly audience of 100 million viewers. Don Francisco, whose real name is Mario Kreutzberger, began his show in Chile in 1962 and moved it to Miami in 1986; as of 2012, it was the longest-running television variety show in history.

In the arts, Latin American writers such as Gabriel García Marquez, Jorge Luis Borges, Octavio Paz, and Isabel Allende have obtained worldwide recognition. In terms of popular culture, musical artists such as Colombia's Shakira and Brazil's hip-hop samba singer Max de Castro reach international audiences. Through music, literature, and even *telenovelas* (soap operas), Latino culture is being transmitted to an eager worldwide audience.

Telenovelas Popular nightly soap operas are a mainstay of Latin American television. These tightly plotted series are filled with intrigue and double dealing. Unlike their counterparts in the United States, they end, usually after 100 episodes. Once standard fare for the working class, many telenovelas take hold and absorb an entire nation. During particularly popular episodes, the streets are noticeably calm as millions of people tune in to catch up on the lives of their favorite heroines. Brazil, Venezuela, and Mexico each produce scores of telenovelas, but the Mexican ones are international mega-hits.

Televisa, a Mexican production agency, has aggressively marketed its inventory of soap operas to an eager global public. Mexican telenovelas are avidly watched in countries as diverse as Croatia, Russia, China, South Korea, Iran, the United States, and France, as well as throughout Latin America. Predictably scripted as Mexican Cinderella stories, these sagas of poor underclass women (often domestics) falling in love with members of the elite, battling jealous rivals, and ultimately emerging triumphant seem to resonate with fans around the world. In addition to their broad appeal, telenovelas are big business, perhaps Mexico's largest cultural export. While Hollywood and Mumbai grind out movies for theaters, much of Mexico's entertainment industry is geared toward producing this popular home art form.

Soccer Perhaps the quintessential global sport, soccer has a fanatical following throughout much of the world. Yet it is in Latin America, and especially in South America, where *fútbol* is considered a cultural necessity. Still largely a male game, young boys and men are constantly seen on fields, beaches, and blacktops playing soccer, especially in late afternoons and on weekends. This is beginning to change in some countries as women take up the sport, especially in Brazil. The great soccer stadiums of Buenos Aires (Bombonera) and Rio de Janeiro (Maracaña) are regarded as shrines to the game. Many individuals use the victories and losses of their national soccer teams as the important chronological markers of their lives.

Pelé, the Brazilian soccer phenomenon of the 1960s and 1970s, introduced the free-flowing acrobatic style that became known as "the beautiful game." Today Latino soccer stars such as Argentine Lionel Messi and Brazilian Kaká

▼ **Figure 4.28 Maracaña Stadium** Rio de Janeiro's soccer stadium, a sports icon for fans around the world, will be the site of the 2016 Olympic Opening Ceremonies. The final match of the 2014 World Cup is scheduled to be held here as well.

play for corporate clubs in Europe and earn millions. Latin Americans also fill up the few slots allotted to foreign players on the U.S. Major League Soccer teams. Yet the dream of many Latin American soccer players is to be on the national team and bring home the World Cup. Visit any Latin American country when its team is playing a World Cup qualifying match, and the streets are eerily quiet. Of the 19 World Cups awarded between 1930 and 2010, South American teams have won 9. In 2014, Brazil hosts the World Cup, with the final match to be held in the recently renovated Maracaña stadium (Figure 4.28). Somehow, as if by magic, it is believed that a World Cup victory will make things better, improve the economy, and even reduce crime. In short, soccer is regarded with near religious significance. And as Latin Americans emigrate (both as players and as laborers), they bring their enthusiasm for the sport with them.

National Identities Viewed from the outside, the region displays considerable homogeneity; yet distinct national identities and cultures flourish in Latin America. Since the early days of the republics, countries celebrated particular elements from their pasts when creating their national histories. In the case of Brazil, the country's interracial characteristics were highlighted to proclaim a new society in which the color lines between Europeans and Africans ceased to matter—although racism against Afro-Brazilians does persist. Mexico celebrated the architectural and cultural achievements of its Aztec predecessors, while at the same time forging an assimilationist strategy that discouraged surviving indigenous culture and language.

Musical and dance traditions evolved and became emblematic of these new societies: The tango in Argentina, caporales in Bolivia, the vallenato and cumbia in Colombia, the mariachi in Mexico, and the samba in Brazil are easily distinguished styles that are representative of distinct national cultures (Figure 4.29). Literature also reflects the distinct identities found in Latin America.

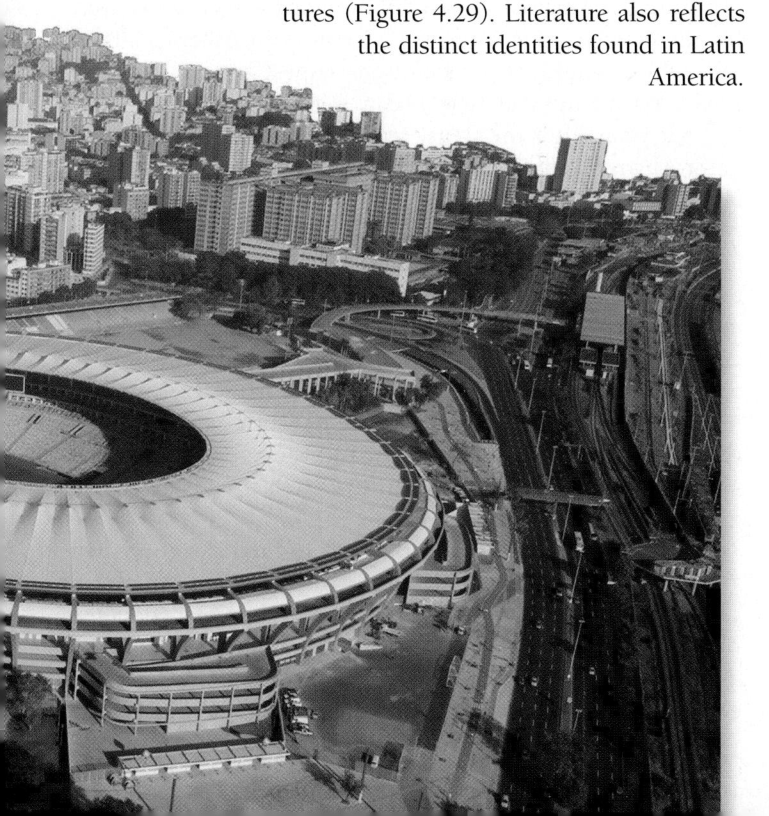

▲ **Figure 4.29 Argentine Tango** Tango is the signature dance and music of Buenos Aires, Argentina. Today tango clubs are popular with tourists who want to see the dance and even learn how to do it.

Writers such as Isabel Allende, Gabriel García Marquez, Mario Vargas Llosa, Carlos Fuentes, and Jorge Amado situate their stories in their native countries and, in so doing, celebrate the unique characteristics of Chileans, Colombians, Peruvians, Mexicans, and Brazilians. Distinct political cultures also evolved, which at times led to expansionist policies that brought neighbors into conflict.

REVIEW

4.6 What factors contributed to racial mixing in Latin America, and where are the areas of strongest Amerindian survival?

4.7 What are the cultural legacies of Iberia in Latin America, and how are they expressed?

Geopolitical Framework: From Two Iberian Colonies to Many Nations

Latin America's colonial history, more than its present condition, unifies this region geopolitically. For the first 300 years after the arrival of Columbus, Latin America was a territorial prize sought by various European countries, but effectively settled by Spain and Portugal. By the early 19th century, the independent states of Latin America had formed, but they continued to experience foreign influence and, at times, overt political pressure, especially from the United States. At other times, a more neutral hemispheric vision of American relations and cooperation has held sway, represented by the formation of the **Organization of American States (OAS).** The present organization was chartered in 1948, but its origins date back to 1889. Yet there is no doubt that U.S. policies toward trade, economic assistance, political development, and at times military intervention are often seen as undermining the sovereignty of these states.

Today the geopolitical influence of the United States in the region is declining, especially in South America.

▶ **Figure 4.30 Shifting Political Boundaries of Latin America** The evolution of political boundaries in Latin America began with the Treaty of Tordesillas, which gave much of the Americas to Spain and a slice of South America to Portugal. The larger Spanish territory was gradually divided into viceroyalties and audiencias, which formed the basis for many modern national boundaries. The 1830 borders of these newly independent states were far from fixed. Bolivia would lose its access to the coast; Peru would gain much of Ecuador's Amazon; and Mexico would be stripped of its northern territory by the United States.

In many South American countries, trade with the European Union, China, and Japan is as important as, if not more important than, trade with the United States. For example, Brazil's largest trading partner is now China. Brazilian influence in the region and the world is rising. In 2013, it was the sixth largest economy, and it is considered one of the rapidly advancing BRIC countries in the world (BRIC stands for Brazil, Russia, India, and China). As a sign of its growing economic clout, Brazilian Roberto Azevêdo took over the leadership of the World Trade Organization in 2013.

Within Latin America, cycles of intraregional cooperation and antagonism have occurred. Neighboring countries have fought over territory, closed borders, imposed high tariffs, and cut off diplomatic relations. Even today a dozen long-standing border disputes in Latin America have the potential to erupt into conflict. The last 20 years have witnessed a revival in the trade bloc concept, with the formation of Mercosur (the Southern Common Market) in South America, NAFTA in North America and Mexico, and CAFTA, the Central American Free Trade Association. In 2008, Brazil proposed the formation of the **Union of South American Nations (UNASUR)**, which includes all the states of South America except French Guiana. Modeled more like the European Union than a free trade association, UNASUR marks a significant change in the region's political geography, but its long-term impact remains to be seen.

Iberian Conquest and Territorial Division

When Christopher Columbus claimed the Americas for Spain, the Spanish became the first active colonial agents in the Western Hemisphere. In contrast, the Portuguese presence in the Americas was the result of the **Treaty of Tordesillas** in 1493–1494. At that time, Portuguese navigators had charted much of the coast of Africa in an attempt to find an ocean route to the Spice Islands (Moluccas) in Southeast Asia. With the help of Columbus, Spain sought a western route to the Far East. When Columbus landed in the Americas, Spain and Portugal asked the Pope to settle how these new territories should be divided. Without consulting other European powers, the Pope divided the Atlantic world in half—the eastern half, containing the African continent, was awarded to Portugal, and the western half, with most of the Americas, was given to Spain. The line of division established by the treaty actually cut through the eastern edge of South America, placing it under Portuguese control. The treaty was never recognized by the French, English, or Dutch, who also asserted territorial claims in the Americas, but it did provide the legal apparatus for the creation of a Portuguese territory in America—Brazil—which would later become the largest and most populous state in Latin America (Figure 4.30).

Six years after the treaty was signed, Portuguese navigator Alvares Cabral inadvertently reached the coast of Brazil on a voyage to southern Africa. The Portuguese soon realized that this territory was on their side of the Tordesillas line. Initially, they were unimpressed by what Brazil had to offer; there were no spices or major indigenous settlements. Over time, they came to appreciate the utility of the coast as a provisioning site, as well as a source for brazilwood, used to produce a valuable dye. Portuguese interest in the territory intensified in the late 16th century, with the development of sugar estates and the expansion of the slave trade, and in the 17th century, with the discovery of gold in the Brazilian interior.

Spain, in contrast, aggressively pursued the conquest and settlement of its new American territories from the very start. After discovering little gold in the Caribbean, by the mid-16th century Spain's energy was directed toward developing the silver resources of central Mexico and the central Andes (most notably Potosí in Bolivia). Gradually,

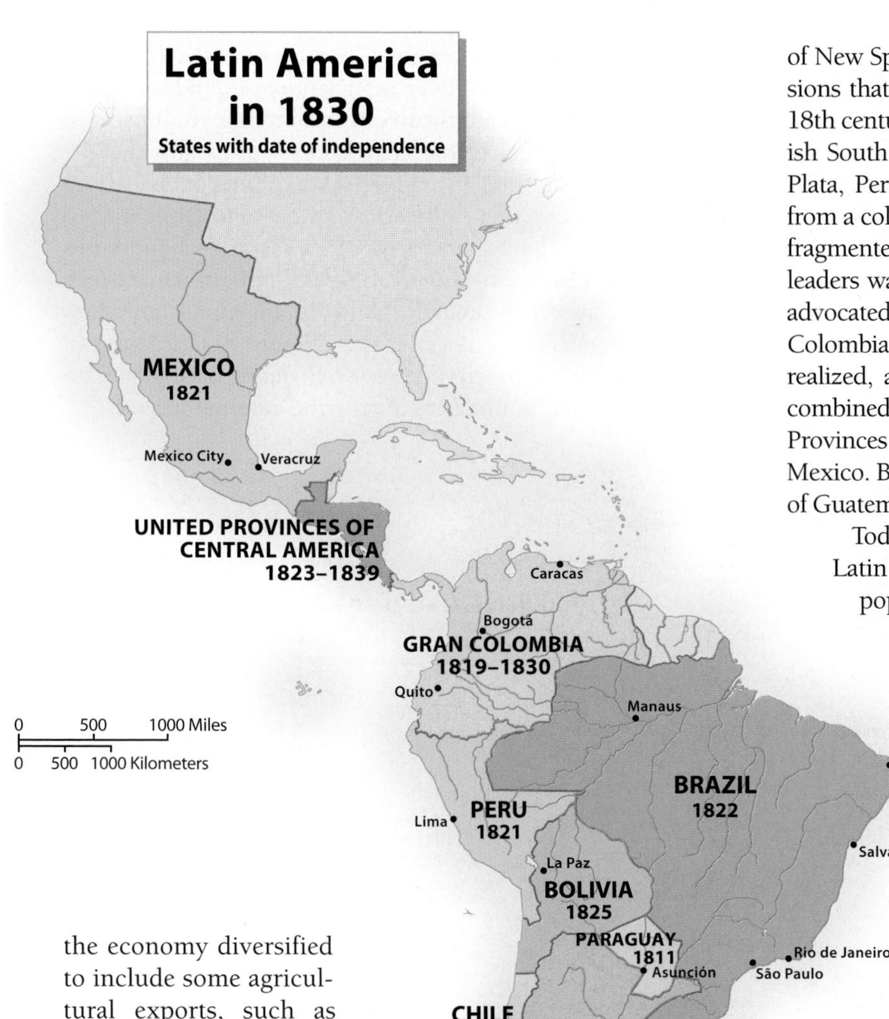

the economy diversified to include some agricultural exports, such as cacao (for chocolate) and sugar, as well as a variety of livestock. In terms of foodstuffs, the colonies were virtually self-sufficient. Some basic manufactured items, such as crude woolen cloth and agricultural tools, were also produced, but, in general, manufacturing was forbidden in the Spanish American colonies to keep them dependent on Spain.

Revolution and Independence Not until the rise of revolutionary movements between 1810 and 1826 was Spanish authority on the mainland challenged. Ultimately, elites born in the Americas gained control, displacing the representatives of the crown. In Brazil, the evolution from Portuguese colony to independent republic was a slower and less violent process that spanned eight decades (1808–1889). In the 19th century, Brazil was declared a separate kingdom from Portugal with its own monarch, and later it became a republic.

The territorial division of Spanish and Portuguese America into administrative units provided the legal basis for the modern states of Latin America (see Figure 4.30). The Spanish colonies were first divided into two viceroyalties (the administrative units of New Spain and Peru), and within these were various subdivisions that later became the basis for the modern states. (In the 18th century, the viceroyalty of Peru, which included all of Spanish South America, was divided to form three viceroyalties: La Plata, Peru, and New Granada.) Unlike Brazil, which evolved from a colony into a single republic, the former Spanish colonies fragmented in the 19th century. Prominent among revolutionary leaders was Venezuelan-born Simon Bolívar (Figure 4.31), who advocated his vision for a new and independent state of Gran Colombia. For a short time (1822–1830), Bolívar's vision was realized, as Colombia, Venezuela, Ecuador, and Panama were combined into one political unit. Similarly, in 1823 the United Provinces of Central America was formed to avoid annexation by Mexico. By the 1830s, this union also broke apart into the states of Guatemala, Honduras, El Salvador, Nicaragua, and Costa Rica.

Today the former Spanish mainland colonies include 16 Latin American states plus 3 Caribbean ones, with a total population of over 400 million. If the Spanish colonial territory had remained a unified political unit, it would now have the third largest population in the world, following China and India.

Persistent Border Conflicts As the colonial administrative units turned into states, it became clear that the territories were not clearly delimited, especially the borders that stretched into the sparsely populated interior of South America. This would later become a source of conflict as new states struggled to demarcate their territorial boundaries. Numerous border wars erupted in the 19th and 20th centuries, and the map of Latin America has been redrawn many times. Some of the more noted conflicts were the War of the Pacific (1879–1882), in which Chile expanded to the north and Bolivia lost its access to the Pacific; warfare between Mexico and the United States in the 1840s, which resulted in the present

▼ **Figure 4.31 Simon Bolívar** Heroic likenesses of Simon Bolívar (the Liberator) are found throughout South America, especially in his native country of Venezuela. This statue stands in the central plaza of the Andean city of Mérida, Venezuela.

border under the Treaty of Hidalgo (1848); and the War of the Triple Alliance (1864–1870), the bloodiest war of the postcolonial period, which occurred when Argentina, Brazil, and Uruguay allied themselves to defeat Paraguay in its claim to control the upper Paraná River Basin. It is estimated that this conflict resulted in the reduction of Paraguay's adult male population by nine-tenths. Sixty years later, the Chaco War (1932–1935) resulted in a territorial loss for Bolivia in its eastern lowlands and a gain for Paraguay. In the 1980s, Argentina lost a war with Great Britain over control of the Falkland, or Malvinas, Islands in the South Atlantic. As recently as 1998, Peru and Ecuador skirmished over a disputed boundary in the Amazon Basin.

Outright war in the region is less common than inactive, but unresolved disputes over international boundaries. Any of a dozen dormant claims can erupt from time to time given the political climate between neighbors. For example, every March in Bolivia, the Dia del Mar (Day of the Sea) is held to recognize the day that Bolivia lost its coast to Chile and to inspire support for regaining it. Other disputed boundaries include the Venezuela–Guyana border, the Peruvian eastern lowlands claimed by Ecuador, and the maritime boundary between Venezuela and Colombia (Figure 4.32). Many of these disputes are based on territorial claims dating back to poorly delimited boundaries during the colonial period.

▲ **Figure 4.32 Geopolitical Issues in Latin America** Of the five economic trade blocs depicted, Mercosur and NAFTA are the most dynamic. As UNASUR develops, it could fold Mercosur and the Andean Community into one common market. Members of the Central American Common Market signed an agreement in 2004 to form CAFTA (Central American Free Trade Association), which also includes the Dominican Republic. **Q: How could the growth and strength of trade blocs impact how Latin America functions as a region?**

The Mexico–U.S. border is the most fortified border of the Americas. Since 1996, the United States has poured resources into more fencing, surveillance, and biometric forms of identification for all recipients of visas and green cards. Part of the rationale is to prevent illegal border crossings—which have declined dramatically since the late 2000s—in an effort cut down on the number of undocumented people in the United States. Ironically, scholars who track the flow of undocumented entrants say that intense border security may have had the unintended consequence of reducing circular labor flows because recrossing the border has become so difficult and expensive. As a result, undocumented laborers who used to leave the United States are now staying. The other strategy has been to remove or deport undocumented migrants. Since 2000, over 3 million people have been removed from the United States, which is more removals than the combined total from 1893 to 2000. (See *People on the Move: The Impact of U.S. Deportations on Latin America.*)

People on the Move

The Impact of U.S. Deportations on Latin America

Much of the immigration debate in the United States focuses on the 11 million undocumented immigrants. Less attention is paid to the nearly 3 million Latin Americans who have been forcibly removed from the United States since 2000. However, more removals occurred from 2000 to 2010 than in the entire 107-year period preceding that decade, according to Department of Homeland Security data. The surge in removals does not count people turned away or detained when trying to cross the border; the deportees are people who have lived and worked in the United States, often for many years, and were caught up in the deportation machine that has been created through integrated databases, biometric data (such as digital fingerprints), and tougher local enforcement. The vast majority of the deported were sent to Latin America, mostly Mexico (72 percent), followed by Guatemala, Honduras, and El Salvador. Hence, the removal of so many migrants is a politically charged and deeply personal issue for people in Latin America. A recent Pew survey estimated that one in three Latinos in the United States knows someone who has been deported or detained in the last 12 months (**Figure 4.4.1**).

▲ **Figure 4.4.1 Immigration Reform Rally** Latinos gather in front of the U.S. Capitol to rally for immigration reform in the spring of 2013. Record numbers of deportations have driven Latinos to mobilize for change.

Personal Calamity This unintended return happens suddenly and without warning, like a lightning strike. A migrant can be stopped for a traffic violation or caught up in a workplace raid and suddenly be channeled back to his or her country of origin (often after months of detention). Rather than making a triumphant return with new possessions, financial resources for investment, and even one's entire family, the unintended returnee has little to show for years of toil abroad. Moreover, the process of being forcibly removed often results in financial calamity: Remittances are no longer sent, mortgages cannot be paid, and savings are rapidly depleted. The research on involuntary returns to Mexico shows loss of income, family strain, social stigma, and the difficulties of reintegration, especially for young children who have grown up in the United States and must now be integrated into Mexican schools.

Forced removals have always taken place, especially of criminal deportees. But the recent growth in this involuntary movement is due to noncriminal deportations (people removed because they did not have the right to stay in the United States). Mexico eclipses all other countries by its sheer number of deportations, 2.2 million. Living within the United States are 12 million Mexican-born individuals and some 35 million people of Mexican ancestry. When comparing the size of the Mexican-born population in the United States with the number of removals, nearly one in five foreign-born Mexicans in the United States has been forcibly removed in the 2000s.

National Calamity The economic and social impact of unintended return may be greater on the Central American states of Honduras and Guatemala. In the 2000s, deportations to Guatemala increased fivefold, and those to Honduras increased fourfold. The reason for the greater impact is that the Hondurans and Guatemalans removed from the United States made up a greater proportion of the remittance-sending population from those states. And remittances are a critical source of income for households in these countries. At a time when these countries are also suffering from a surge in crime, the rise in deportations is an added strain. Unless comprehensive immigration reform is passed, forced removal of the undocumented is likely to remain a contentious issue between Latin America and the United States.

▲ **Figure 4.33 Student Protests in Chile** Chilean student leaders, including Camila Vallejo (center), lead a march of some 50,000 students in Santiago over the growing costs of higher education in Chile.

The Trend Toward Democracy Most of the 17 countries in Latin America have celebrated, or will soon celebrate, their bicentennials. Compared with most of the rest of the developing world, Latin Americans have been independent for a long time. Yet political stability is not a hallmark of the region. Among the countries in the region, some 250 constitutions have been written since independence, and military takeovers have been alarmingly frequent. In 2012, after the disputed removal of Paraguayan President Lugo from office, the other members of Mercosur suspended that country's membership in the group. Since the 1980s, however, the trend has been toward democratically elected governments, the opening of markets, and broader popular participation in the political process. Where dictators once outnumbered elected leaders, by the 1990s each country in the region had a democratically elected president. (Cuba, the one exception, will be discussed in Chapter 5.)

Democracy may not be enough for the millions frustrated by the slow pace of political and economic reform. In survey after survey, Latin Americans register their dissatisfaction with politicians and governments. Many of the democratic leaders have also been free-market reformers who are quick to eliminate state-backed social safety nets, such as food subsidies, government jobs, and pensions. Many of the poor and middle class have grown doubtful about whether this brand of neoliberal policy can make their lives better. Even in prosperous Chile, widespread student protests in 2011–2013 demanding a more equitable society and subsidized higher education have left politicians scrambling. One of the Chilean student leaders is a charismatic geography student, Camila Vallejo, who has gained international recognition for her ability to mobilize university students and shut down a city (Figure 4.33). Under such conditions, it is not surprising the left-leaning politicians have won presidential elections in Brazil, Bolivia, Nicaragua, Ecuador, Peru, and Venezuela. These leaders have not rolled back neoliberal trade reforms, but many have tried to improve social services and attempted to reduce income inequalities.

Regional Organizations

At the same time that democratically elected leaders struggle to address the pressing needs of their countries, political developments at the supranational and subnational levels pose new challenges to their authority. The most discussed **supranational organizations** (governing bodies that include several states) are the trade blocs, the newest one being UNASUR. **Subnational organizations** (groups that represent areas or people within a state) form along ethnic or ideological lines and can support organized crime such as drug trafficking. Subnational organizations can have positive or destabilizing impacts. Examples include native groups that seek territorial recognition (such as the Guna in Panama), insurgent groups (such as the FARC [Revolutionary Armed Forces of Colombia] or the Zapatistas in Mexico) that have challenged the authority of the state, and more recently drug cartels such as the *Zetas* and *Sinaloa,* who have terrorized Mexican society with extreme violence since 2006.

Trade Blocs Beginning in the 1960s, regional trade alliances were attempted in an effort to foster internal markets and reduce trade barriers. The Latin American Free Trade Association (LAFTA), the Central American Common Market, and the Andean Community have existed for decades, but their ability to influence economic trade and growth is limited at best. In the 1990s, Mercosur and NAFTA emerged as supranational structures that could influence development (see Figure 4.32). For Latin America, the lessons from Mercosur in particular led Brazil to propose UNASUR in 2008, uniting virtually all of South America.

NAFTA took effect in 1994 as a free-trade area that would gradually eliminate tariffs and ease the movement of goods among the member countries (Mexico, the United States, and Canada). NAFTA has increased intraregional trade, but has provoked considerable controversy about costs to the environment and to employment (see Chapter 3). NAFTA did prove, however, that a free-trade area combining industrialized and developing states was possible. In 2004, the United States, five Central American countries—Guatemala, El Salvador, Nicaragua, Honduras, and Costa Rica—and the Dominican Republic signed CAFTA. CAFTA, like NAFTA, aims to increase trade and reduce tariffs among member countries. The treaty became fully ratified in 2009, but much debate surrounds the question of whether such a treaty will lead to more economic development in Central America.

Mercosur was formed in 1991 with Brazil and Argentina, the two largest economies in South America, and the smaller states of Uruguay and Paraguay as members. Since its formation, trade among these countries has grown tremendously, so much so that Chile, Bolivia, Peru, Ecuador, and Colombia have joined the group as associate members and Venezuela received ratification as a full member in 2012. This success is significant in two ways: It reflects the growth of these economies and the willingness to put aside old rivalries (especially long-standing antagonisms between Argentina and Brazil) for the economic benefits of cooperation.

In 2008, Brazil initiated the formation of the 12-member UNASUR. Some see this as an assertion of Brazil's greater political and economic clout in the region, since it is Latin America's largest country and the sixth largest economy in the world. UNASUR includes all countries in South America, minus French Guiana, which is a territory of France. It has formally organized with a permanent secretariat, and it has responded to political crises in Bolivia (2008), Ecuador (2010), and Paraguay (2012). Significantly, unlike NAFTA or CAFTA, UNASUR was a Brazilian-led effort, not one led by the United States. Brazil's lead in this initiative underscores its larger geopolitical ambitions to influence South American development and to secure a permanent seat on the United Nations Security Council. Brazil's ambitions have not always been well received by its neighbors, but the strengthening of UNASUR suggests a shift in the region's geopolitical alignment and the potential to form a common market more like the European Union.

Insurgencies and Drug Cartels Guerilla groups such as the FARC in Colombia have controlled large territories of their countries through the support of those loyal to the cause, along with theft, kidnapping, and violence. The FARC, along with the ELN (National Liberation Army), gained wealth and weapons through the drug trade. The level of violence in Colombia escalated further with the rise of paramilitary groups—armed private groups that terrorize those sympathetic to insurgency. The paramilitary groups have been blamed for hundreds of politically motivated murders each year. As many as 2.5 million Colombians have been internally displaced by violence since the late 1980s, most fleeing rural areas for towns and cities. Fortunately, the situation has improved considerably in the last decade. Under President Uribe, the police presence has increased throughout the state, and negotiations with insurgencies have stopped, ultimately reducing their power. Still, Colombia remains the world's largest producer of cocaine, followed by Peru and Bolivia.

Drug cartels and gangs in states as diverse as Mexico, Guatemala, El Salvador, Honduras, and Brazil have been blamed for increases in violence and lawlessness. The spike in violence and corruption in Mexico has been especially destabilizing. Profiting from the illegal production and/or shipment of cocaine, marijuana, methamphetamine, and heroin, the cartels generate billions of dollars. Beginning in 2006, the Mexican government brought in the army to quell the violence, kidnapping, and intimidation brought on by cartel groups, especially in the border region, but now extending throughout Mexico and into Central America (Figure 4.34). The Mexican government reported some 70,000 cartel-related murders from 2006 to 2012 and many thousands more "disappered" persons. Finding ways to stem the violence (rather than the flow of drugs) was one of the biggest issues in the 2012 Mexican election. Mexico's newly elected President

▲ **Figure 4.34 The Influence of Drug Cartels in Mexico** This map suggests the major areas of influence of Mexican Cartels in 2011. Cartel influence is fluid and groups often compete for territories. Los Zetas and the Sinaloa cartels control the drug trade in large swaths of Mexico, and their influence crosses into Central America as well.

Enrique Peña is shifting his focus more toward police work and judicial reform to reduce violence rather than maintaining the military focus.

Tragically, the most violent states in Latin America are in Central America. Honduras, El Salvador, and Guatemala have some of the highest homicide rates in the world outside of war zones. These three impoverished countries have long been in the transshipment zone of cocaine produced in Colombia, Peru, and Bolivia. More recently, however, Mexico's Sinaloa, Gulf, and Zetas cartels have been active in the isthmus, paying locals in drugs, creating drug-processing centers, and, in the process, driving up the murder rate.

REVIEW

4.8 How did the colonization of Latin America by Iberia lead to the formation of the modern states of Latin America?

4.9 How are trade blocs reshaping the region's geopolitics and development?

Economic and Social Development: From Dependency to Neoliberalism

Most Latin American economies fit into the broad middle-income category set by the World Bank. Clearly part of the developing world, Latin American people are much better off than those in Sub-Saharan Africa, South Asia, and most of China. Still, the economic contrasts are sharp both between states and within them. Generally, the Southern Cone countries (including southern Brazil and excluding Paraguay) and Mexico are the richest states. The poorest countries in terms of per capita purchasing power parity (PPP) are Nicaragua, Bolivia, Honduras, and Guatemala. Although per capita incomes in Latin America are well below levels of developed countries, the economies are growing, and the region has witnessed steady improvements in various social indicators, such as life expectancy, child mortality, and literacy. This is reflected in the relatively strong rankings of Latin American countries in the Human Development Index, most notably Chile (Table 4.2).

The economic engines of Latin America are the two largest countries, Brazil and Mexico. According to the International Monetary Fund, Brazil's economy in 2011 was the 6th largest in the world, and Mexico's was the 14th largest, based on gross domestic product. The region has also seen a reduction in extreme poverty, as the percentage of people living on less than $2 per day dropped from 22 percent in 1999 to 12 percent in 2008, in part because of conditional cash transfer programs such as Brazil's Bolsa Familia.

Table 4.2 Development Indicators

Country	GNI per capita, PPP 2011	GDP Average Annual % Growth 2000–11	Human Development Index (2011)[1]	Percent Population Living Below $2 a Day	Life Expectancy (2013)[2]	Under Age 5 Mortality Rate (1990)	Under Age 5 Mortality Rate (2011)	Adult Literacy (% ages 15 and older) Male/Female	Gender Inequality Index (2011)[3,1]
Argentina	15,570	5.6	.811	<2	76	28	14	98/98	0.380
Bolivia	4,890	4.2	.675	24.9	67	120	51	96/87	0.474
Brazil	11,420	3.8	.730	10.8	74	11	8	--/--	0.256
Chile	16,330	4.1	.819	2.7	79	19	9	99/98	0.360
Colombia	9,560	4.5	.719	15.8	74	34	18	93/93	0.459
Costa Rica	11,860	4.8	.773	6.0	79	17	10	96/96	0.346
Ecuador	8,510	4.8	.724	10.6	75	52	23	93/90	0.442
El Salvador	6,640	2.0	.680	16.9	72	60	15	82/87	0.441
Guatemala	4,760	3.6	.581	26.3	71	78	30	81/70	0.539
Honduras	3,820	4.4	.632	29.8	73	55	21	85/85	0.483
Mexico	15,390	2.1	.775	4.5	77	49	16	94/92	0.382
Nicaragua	3,730	3.2	.599	31.7	74	66	26	78/78	0.461
Panama	14,510	7.2	.780	13.8	77	33	20	95/93	0.503
Paraguay	5,390	4.1	.669	13.2	72	53	22	95/93	0.472
Peru	9,440	6.2	.741	12.7	74	75	18	95/85	0.387
Uruguay	14,640	4.0	.792	<2	76	23	10	98/98	0.367
Venezuela	12,430	4.4	.748	12.9	75	31	15	96/95	0.466

[1]United Nations, *Human Development Report, 2013.*

[2]Population Reference Bureau, *World Population Data Sheet, 2013.*

[3]Gender Inequality Index—A composite measure reflecting inequality in achievements between women and men in three dimensions: reproductive health, empowerment and the labor market that ranges between 0 and 1. The higher the number, the greater the inequality.

Source: World Bank, *World Development Indicators, 2013.*

Development Strategies

The path toward economic development in Latin America has been a volatile one. In the 1960s, Brazil, Mexico, and Argentina all seemed poised to enter the ranks of the developed world. Multilateral agencies such as the World Bank and the Inter-American Development Bank loaned money for big development projects: continental highways, dams, mechanized agriculture, and power plants. All sectors of the economy were radically transformed. Agricultural production increased with the application of "green revolution" technology and mechanization. State-run industries reduced the need for imported goods, and the service sector ballooned as a result of new government and private-sector jobs. In the end, most countries in Latin America made the transition from predominantly rural and agrarian economies, dependent on one or two commodities, to more economically diversified and urbanized countries with mixed levels of industrialization.

The modernization dreams of Latin American countries were trampled in the 1980s, when debt, currency devaluation, hyperinflation, and falling commodity prices undermined the aspirations of the region. By the 1990s, most Latin American governments had radically changed their economic development strategies. State-run national industries and tariffs were jettisoned for policy reforms that emphasized privatization, direct foreign investment, and free trade, collectively labeled **neoliberalism.** Through tough fiscal policy, increased trade, privatization, and reduced government spending, most countries saw their economies grow and poverty decline. In aggregate, the economies of the region averaged an annual growth rate of 4 percent from 2000 to 2010. However, sporadic economic downturns have made these neoliberal policies highly unpopular with the masses, at times causing major political and economic turmoil. In particular, the value of increased trade and direct foreign investment has been criticized as benefiting a minority of the people in the region, while not adequately addressing the region's long-standing problem of gross income disparities.

Maquiladoras and Foreign Investment The growth in foreign investment and the presence of foreign-owned factories are examples of neoliberalism. The Mexican assembly plants called **maquiladoras,** which line the border with the United States, are characteristic of manufacturing systems in an increasingly globalized economy. These plants began to be constructed in the 1960s as part of a border industrialization program. The Mexican government allowed the duty-free import of machinery, components, and supplies from the United States to be used for manufacturing goods for export back to the United States. Initially, all products had to be exported, but changes in the law in 1994 now allow up to half of the goods to be sold in Mexico. Today more than 3000 maquiladoras exist, employing over 1 million people, who assemble automobiles, consumer electronics, and apparel. Since 2003, China has become a favorite destination for the labor-intensive assembly work that Mexico has specialized in for the last three decades. As Mexican wages have gone up, some companies have relocated factories to East Asia. Northern Mexico is still an attractive location, but competition from China and even Central America may erode Mexico's various locational and structural advantages.

Considerable controversy on both sides of the border surrounds this form of industrialization. Organized labor in the United States complains that well-paying manufacturing jobs are being lost to low-cost competitors, whereas environmentalists decry serious industrial pollution resulting from lax government regulation. Mexicans worry that these plants are poorly integrated with the rest of the economy and that many of the factories choose to hire young, unmarried women because they are viewed as docile laborers (Figure 4.35). In the border city of Ciudad Juarez, for many years women outnumbered men in factory jobs, and this led to many women outearning men. And, shockingly, beginning in the 1990s, a surge in female homicides occurred there, which went unprosecuted and generated popular protests by women's groups. With NAFTA, foreign-owned manufacturing plants are no longer restricted to the border zone and are increasingly being built near the population centers of Monterrey, Puebla, and Veracruz. Aguascalientes, in central Mexico, has emerged as the country's auto city, although most of the cars produced there are by foreign companies and are destined for export.

Other Latin American states are attracting foreign companies through tax incentives and low labor costs. Assembly plants in Honduras, Guatemala, and El Salvador are drawing foreign investors, especially in the apparel industry.

► **Figure 4.35 Mexican Maquiladora Workers** These women manufacture car radios at Delphi Delco Electronics in Matamoros, Mexico. Delphi, which makes parts for General Motors cars, has about 11,000 Mexican workers in seven factories near Matamoros. Many maquiladoras rely upon women for tedious and demanding assembly work. Matamoros is across the border from Brownsville, Texas, near the Gulf of Mexico.

A recent report from El Salvador claims that not one of its apparel factories has a union. Making goods for American labels such as the Gap, Liz Claiborne, and Nike, many Salvadoran garment workers complain they do not make a living wage, work 80-hour weeks, and face mandatory pregnancy tests. With the signing of the CAFTA agreement, Central American states are hopeful that more foreign investment will flow into their countries and that wages and conditions will improve.

The situation in Costa Rica, which has been a major chip manufacturer for Intel since 1998, is quite different. With a well-educated population, low crime rate, and stable political scene, Costa Rica is now attracting other high-tech firms. Hopeful officials claim that Costa Rica is transitioning from a banana republic (bananas and coffee were the country's long-standing exports) to a high-tech manufacturing center. As a result, the Costa Rican economy averaged 5 percent annual growth from 2000 to 2010. Uruguay is another example of a small country with a well-educated population that has recently emerged as the leader in Latin American **outsourcing** operations. Outsourcing, most commonly associated with India, is the practice of moving service jobs such as tech support, data entry, and programming to cheaper locations. Partnered with the Indian multinational company Tata, in the last few years TCS Iberoamerica in Uruguay has created the largest outsourcing operation in the region. Uruguay takes advantage of being in a similar time zone as the eastern United States. While India's top engineers sleep, Uruguayan engineers and programmers can serve their customers from Montevideo.

▼ **Figure 4.36 Peruvian Street Vendors** A women sells vegetables on a street in Huancayo, Peru. Street vending is a common economic activity of the informal sector in Latin American cities.

The Entrenched Informal Sector Even in prosperous Montevideo, Uruguay, a short drive to the urban periphery shows large neighborhoods of self-built housing filled with street traders and family-run workshops. Such activities make up the informal sector, the provision of goods and services without the benefit of government regulation, registration, or taxation. Most people in the informal economy are self-employed and receive no wages or benefits except the profits they clear. The most common informal activities are housing construction (in many cities, as many as half of all residents live in self-built housing), manufacturing in small workshops, street vending, transportation services (messenger services, bicycle delivery, and collective taxis), garbage picking, street performing, and even line-waiting (Figure 4.36). These activities are legal. Illegal informal activities also exist: drug trafficking, prostitution, and money laundering, for example. The vast majority of people who rely on informal livelihoods produce legal goods and services.

No one is sure how big this economy is, in part because separating formal activities from informal ones is difficult. Visitors to Lima, Belém, Guatemala City, or Guayaquil could easily get the impression that the informal economy *is* the economy. From self-help housing that dominates the landscape to hundreds of street vendors that crowd the sidewalks, it is impossible to avoid. There are advantages in the informal sector—hours are flexible, children can work with their parents, and there are no bosses. Peruvian economist Hernando de Soto even argues that this most dynamic sector of the economy should be encouraged and offered formal lines of credit. As important as this sector may be, however, widespread dependence on it signals Latin America's poverty, not its wealth. It reflects the inability of the formal economies of the region, especially in industry, to absorb labor. For millions of urban dwellers, formal employment that offers benefits, safety, and a living wage is still a dream. With nowhere else to go, the numbers of the informally employed are substantial.

Primary Export Dependency

Historically, Latin America's abundant natural resources were its wealth. In the colonial period, silver, gold, and sugar generated tremendous riches for the colonists. With independence in the 19th century, the region began a series of export booms to an expanding world market, including commodities such as bananas, coffee, cacao, grains, tin, rubber, copper, wool, and petroleum. One of the legacies of this export-led development was a tendency to specialize in one or two major commodities, a pattern that continued into the 1950s. During that decade, Costa Rica earned 90 percent of its export earnings from bananas and coffee; Nicaragua earned

▲ **Figure 4.37 Soy Production in Brazil** Fartura Farm in the state of Mato Grosso, Brazil, embodies the large-scale industrial agriculture that has transformed much of South America into one of the world's largest producers and exporters of soy products.

70 percent from coffee and cotton; 85 percent of Chilean export income came from copper; half of Uruguay's export income came from wood. Even Brazil generated 60 percent of its export earnings from coffee in 1955; by 2000, coffee accounted for less than 5 percent of the country's exports, yet Brazil remained the world leader in coffee production (see *Everyday Globalization: Good Morning Coffee*).

The economies of the region have industrialized and diversified. Since the 1990s, however, an increased demand from Asia for primary exports (from fossil fuels to metals and soybeans) has driven up commodity prices, creating another boom in primary export commodities. Although many Latin America states are riding this wave, from oil-rich Venezuela to copper-laden Chile, there is concern that Latin America may once again become too reliant on its bountiful natural resources.

Agricultural Production Since the 1960s, the trend in Latin America has been to mechanize agriculture and expand the range of crops produced. Nowhere is this more evident than in the Plata Basin, which includes southern Brazil, Uruguay, northern Argentina, Paraguay, and eastern Bolivia. Soybeans, used for oil and animal feed, transformed these lowlands in the 1980s and 1990s. Brazil is now the second largest producer of soy in the world (following the United States) and is already the world's largest exporter of soy. Argentina is the third largest producer, and production is still increasing. Between the late 1990s and 2010, soy production tripled in Argentina. The speed with which the Plata and Amazon basins are being converted into soy fields alarms many; it is eliminating forest and savanna and negatively impacting biodiversity and GHGs. But with soy prices high, the rush to plant continues (Figure 4.37). In addition to soy, acres of rice, cotton, and orange trees, as well as the more traditional wheat and sugar, continue to be planted in the Plata Basin.

Similar large-scale agricultural frontiers exist along the piedmont zone of the Venezuelan Llanos (mostly grains), the Pacific slope of Central America (cotton and some tropical fruits), and the Central Valley of Chile and the foothills of Argentina (wine and fruit production). In northern Mexico, water supplied from dams along the Sierra Madre Occidental has turned the valleys in Sinaloa into intensive producers of fruits and vegetables for consumers in the United States. The relatively mild winters in northern Mexico allow growers to produce strawberries and tomatoes during the winter months.

In each of these cases, the agricultural sector is capital-intensive and dynamic. By using machinery, high-yielding hybrids, chemical fertilizers, and pesticides, many corporate farms are extremely productive and profitable. What these operations fail to do is employ many rural people, which is especially problematic in countries where a quarter or more of the population depends on agriculture for its livelihood. Thus, the modernization of agriculture has left behind many subsistence producers who make up the ranks of "Latin America's" most impoverished people. Interestingly, a few traditional Amerindian foods, such as quinoa, are gaining consumers thanks to a growing appetite for organic and healthy foods. Recently, Peru and Bolivia have experienced a boom in quinoa production and exports, with much of the crop being grown in small and medium-sized highland farms. Bolivian President Evo Morales even declared 2012 the "year of quinoa" in an effort to promote traditional "Indian" foods.

Mining and Forestry The mining of silver, zinc, copper, iron, bauxite, and gold is an economic mainstay for many countries in the region. Moreover, many commodity prices reached record levels in the last decade, boosting foreign exchange earnings. Chile is the world leader in copper production, far outproducing the next two largest producers, Peru and the United States.

Everyday Globalization

Good Morning Coffee

For many Americans, the day begins with coffee. More coffee is consumed in the United States than in any other country, for an annual per capita consumption rate of 10 pounds. This is not the world's highest per capita consumption—in many European countries, consumption is twice as high—but the U.S. market is huge. Americans have been swigging coffee in large quantities since the 19th century, yet the United States is not a coffee producer. Consequently, we have looked to the tropics, particularly Latin America, to get our morning fix.

Coffeenomics Coffee looms large in the history of Latin America. For over a century, Brazil has been the world leader in coffee production. With nearly 200 million people, it is also a major consumer as well. Up until 1990, Colombia was the second largest global producer, but in the past two decades Colombia's production has declined and the output of two Southeast Asian countries—Vietnam and Indonesia—has surpassed it (**Figure 4.5.1**). Today Latin America accounts for 58 percent of the world's coffee production, down from 65 percent in 1990. Over 145 million 60-kilo bags were produced in the 2012–2013 harvest. Much of the exported coffee is bound to Europe, North America, and Japan.

In the early 20th century, many Latin American countries earned half of their export earnings from coffee in a classic pattern of primary export dependence. Yet today in Brazil, coffee accounts for just 2 percent of that country's exports. In smaller economies, coffee accounts for a larger proportion of export value—20 percent in Honduras and 12 percent in Guatemala. There are dozens of

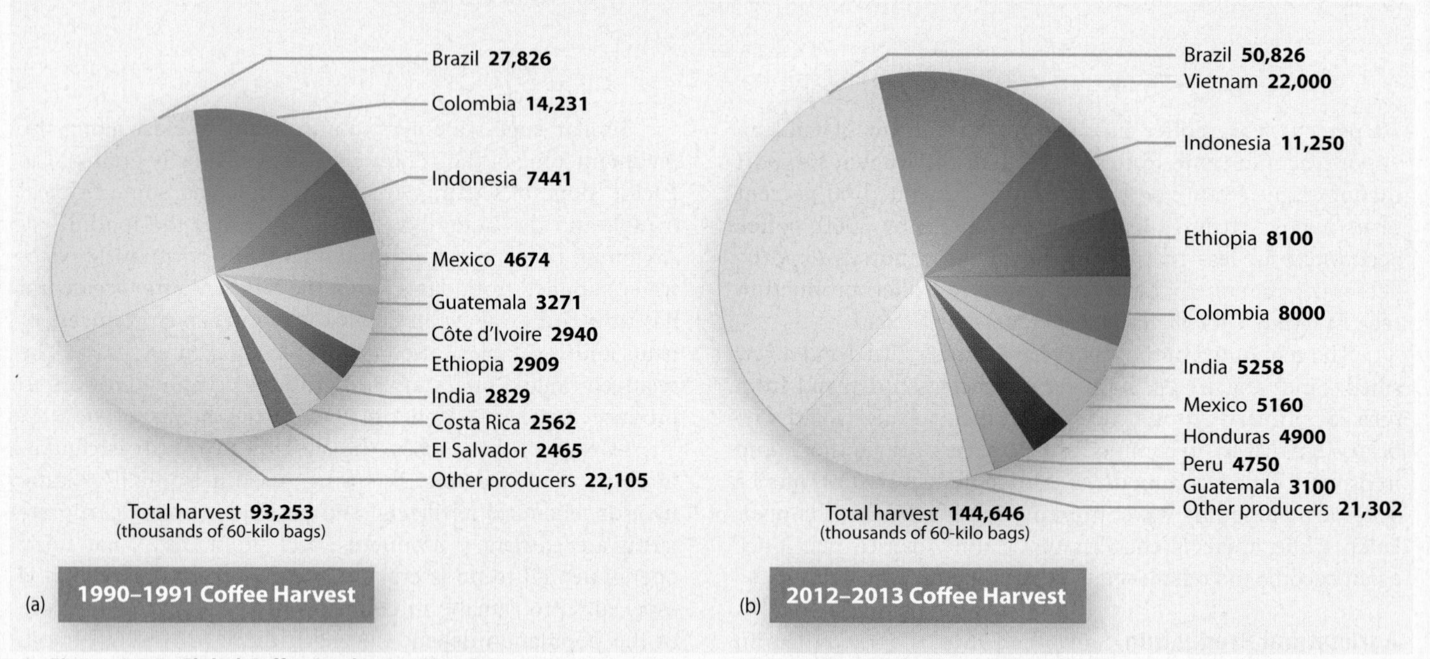

▲ **Figure 4.5.1 Global Coffee Production Trends** Latin America is still the region that produces the most coffee, with Brazil leading the world in total harvest. Yet in the last 20 years, Vietnam, Indonesia and Ethiopia have all seen dramatic increases in production, while Colombia's harvests have declined.

Mexico and Peru were top silver producers in 2012, but Chile, Bolivia, and Argentina were all top-10 producers. Peru was also Latin America's top gold producer in 2012.

Lithium metal is gaining worldwide interest. This soft, silver-white metal is used for making lightweight batteries, like those in cell phones and laptops. It is also a key metal for electric car batteries. Today the largest producer of lithium is Chile, but the world's largest reserves are in Bolivia, under the Salar de Uyuni in the Altiplano. These reserves are so immense Bolivia has been dubbed the Saudi Arabia of lithium. But it remains to be seen how and under what terms this critical resource will be extracted from this remote region.

Like agriculture, mining has become more mechanized and less labor-intensive. Even Bolivia, a country long dependent on tin production, cut 70 percent of its miners from the payrolls in the 1990s. The measure was part of a broad-based austerity program, yet it suggests that the majority of the miners were not needed. Similarly, the vast copper mines of northern Chile are producing record amounts of copper with few miners. In contrast, gold mining continues to be labor-intensive, offering employment for thousands of prospectors.

Logging is another important, and controversial, extractive activity in the region. Ironically, many of the forest areas cleared for cattle were not systematically harvested.

▲ **Figure 4.5.2 Harvesting Colombian Coffee** Two coffee pickers in Risaralda Department, harvest coffee by hand. Noted for its quality coffee, harvests in Colombia have declined in the last decade due to climate change and rural violence.

coffee-producing countries, but the top 10 countries accounted for 85 percent of production in 2012–2013. The International Coffee Organization estimates that coffee growing employs 26 million people on small farms and large estates. World exports are valued at $15 billion.

Coffee Farming Coffee is an ideal commodity for global trade because it does not spoil. Coffee is a notoriously volatile crop, with price fluctuations that bring prosperity and ruin from one season to the next. Presently, prices are high, so more coffee bushes are planted, which triggers oversupply and downward price pressure. Unlike other major agricultural exports, coffee is labor-intensive. The berries are handpicked; depulping, washing, and drying the beans for shipment is also labor-intensive (Figure 4.5.2). Small family farms do not produce great wealth, but in the highland tropics coffee is often the most profitable legal crop farmers can grow. It is a critical commodity for rural livelihoods.

Increasingly, **fair trade** coffee is promoted as a means to ensure that small growers receive a higher price for their crop by cutting out some of the middlemen. Coffee farmers are also finding that the use of organic methods lowers their costs. An added benefit of shade-grown and organic coffee is that it helps maintain economically productive and biologically diverse habitats. Even major outlets now market fair trade and organic coffee. When consumers are made aware of the connection between their daily cup of coffee and the people who produce it, rural livelihoods in Latin America can be more economically and environmentally sustainable.

Google Earth
Virtual Tour Video

http://goo.gl/qkSqo2

1. After exploring images of coffee producing areas, what are the the differences between the large coffee estates in Brazil with the smaller ones in Honduras?
2. What are the environmental and economic advantages of growing organic coffee?

More often than not, all but the most valuable trees were burned. Logging concessions are commonly awarded to domestic and foreign timber companies, which export boards and wood pulp. These one-time arrangements are seen as a quick means for foreign exchange, particularly if prized hardwoods such as mahogany are found. Logging can mean a short-term infusion of cash into a local economy. Yet rarely do long-term conservation strategies exist, making this system of extraction unsustainable. Interest is growing in certification programs that designate wood products that have been produced sustainably. This is due to consumer demand for certified wood, mostly in Europe. Unfortunately, such programs are small, and the lure of profit usually overwhelms the impulse to conserve for future generations.

Several countries rely on plantation forests of introduced species of pines, teak, and eucalyptus to supply domestic fuelwood, pulp, and board lumber. These plantation forests grow single species and fall far short of the complex ecosystems occurring in natural forests. Nonetheless, growing trees for paper or fuel reduces the pressure on other forested areas. Leaders in plantation forestry are Brazil, Venezuela, Chile, and Argentina. Considered Latin America's economic star in the 1990s, Chile relied on timber and wood chips to boost its export earnings. Thousands of hectares of nonnative trees (eucalyptus and pine) have been planted,

▲ **Figure 4.38 Chilean Wood Chips** A mountain of wood chips awaits shipment to Japanese paper mills from the southern Chilean port of Punta Arenas. The exploitation of wood products from both native and plantation forests supports Chile's booming export economy. Increasingly, these wood exports are bound for markets in East Asia.

systematically harvested, and cut into boards or chipped for wood pulp (Figure 4.38). Japanese capital is heavily involved in this sector of the Chilean economy. The recent expansion of the wood chip business, however, has led to a dramatic increase in the logging of native forests.

The Energy Sector The oil-rich nations of Venezuela, Mexico, and now Brazil are able to meet their own fuel needs and also earn vital state revenues from oil exports. In 2011, Mexico was the 8th largest producer of oil in the world, Brazil was 10th, and Venezuela was 13th. Of the three, Venezuela is the most dependent on revenues from oil, earning up to 90 percent of its foreign exchange from petroleum and natural gas products. The largest new oil discovery in recent years has been off the coast of Brazil. In the past decade, oil production in Brazil has doubled, moving Brazil into the ranks of oil exporters and drawing increased foreign investment. Although Latin American oil producers do not receive as much attention as those in the Middle East, they are significant. Venezuela was one of the five founding members of the Organization of the Petroleum Exporting Countries (OPEC), and Ecuador joined the group in 1973. Latin America has the second largest reserves of oil outside the Middle East, which has, by far, the largest known reserves.

Natural gas production is also on the rise in this region. Venezuela and Bolivia have the largest proven reserves of natural gas in Latin America, but Mexico and Argentina are by far the largest producers. In recent years, Argentina's natural gas production has been boosted by new finds in Patagonia. In 2012, Argentina made headlines when President Christina Fernandez de Kirchner seized majority control of YPF, the major producer of oil and natural gas in Argentina, from a Spanish-owned company, claiming frustration over unnecessary declines in output. Argentina is especially dependent upon natural gas for its urban markets and exports; the production declines produced shortages and price increases in the domestic liquefied natural gas market that required action.

In the area of biofuels, Brazil offers a story of sweet success. In the 1970s, when oil prices skyrocketed, then oil-poor Brazil decided to convert its abundant sugarcane harvest into ethanol. Over the years, even when oil prices plummeted, Brazil continued to invest in ethanol, building mills and a distribution system that delivered ethanol to gas stations. One of Brazil's major technological successes was inventing flex-fuel cars that run on any combination of ethanol and gasoline. At the time, Brazil was motivated by its limited oil reserves, but today, with interest in biofuels growing as a way to reduce CO_2 emissions, Brazil's support of its ethanol program looks visionary.

Due to these innovations, Latin America's energy mix has changed considerably since the 1970s (Figure 4.39). Forty years ago, 60 percent of the region's energy was supplied by oil and 20 percent by wood fuel. By 2010, the region's energy supply was more diverse and cleaner, most notably in the growth in natural gas, hydroelectricity, and biofuels (or bagasse). That said, Latin America's energy consumption has also increased fivefold from 1970 to 2010 due to population increase, urban growth, improved transportation, and greater economic activity.

▼ **Figure 4.39 Latin America's Energy Mix, 1970–2010** As the region's energy sources have increased and diversified, there is much less reliance on wood, and more reliance on natural gas. **Q: What would explain the decline in wood fuel for this region?**

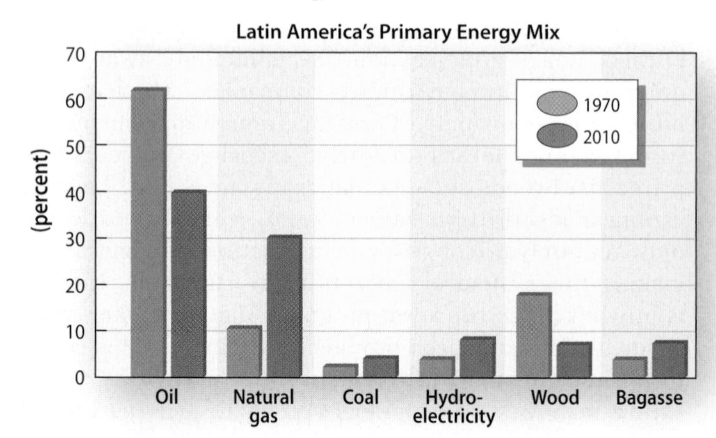

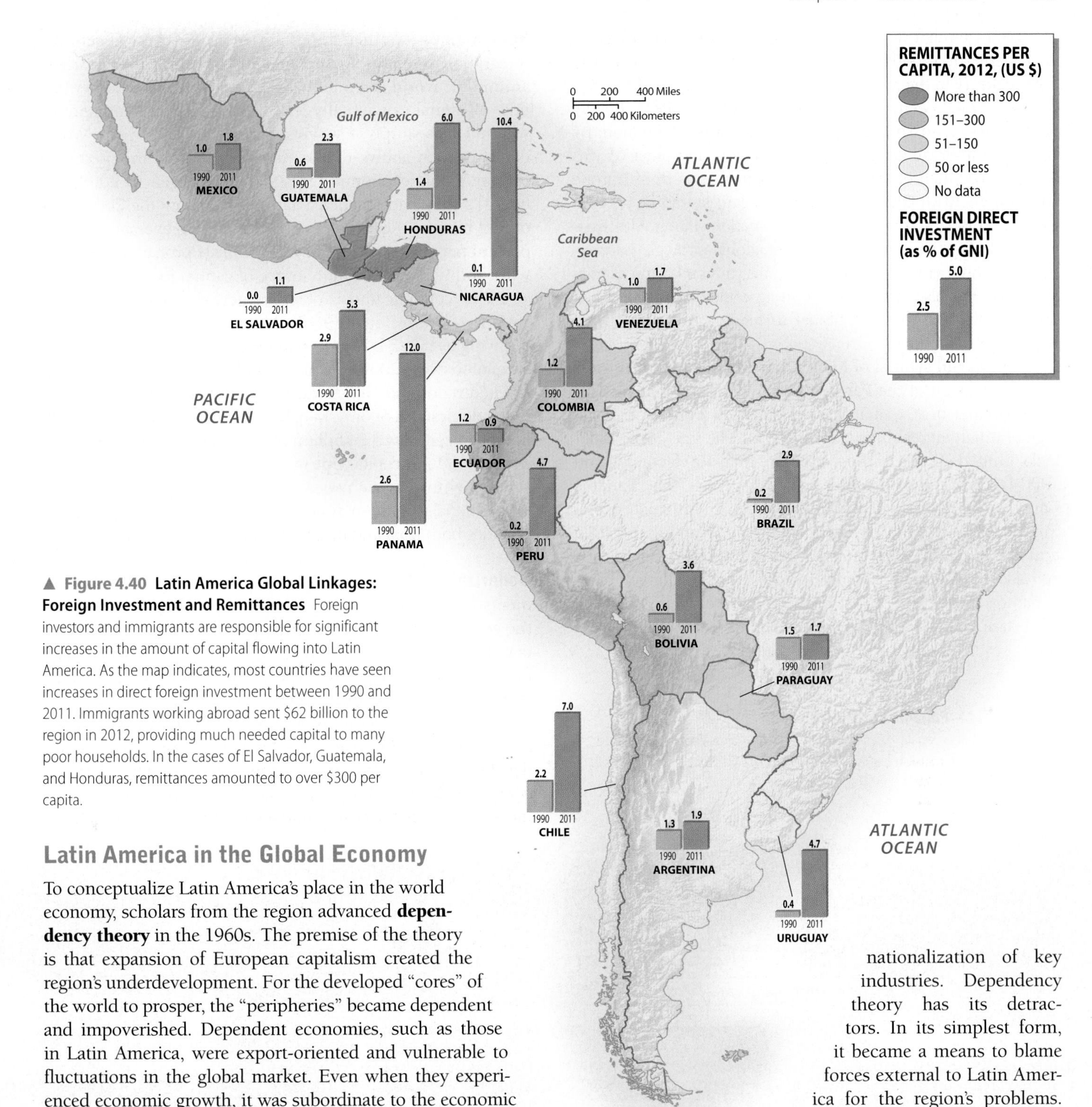

▲ **Figure 4.40 Latin America Global Linkages: Foreign Investment and Remittances** Foreign investors and immigrants are responsible for significant increases in the amount of capital flowing into Latin America. As the map indicates, most countries have seen increases in direct foreign investment between 1990 and 2011. Immigrants working abroad sent $62 billion to the region in 2012, providing much needed capital to many poor households. In the cases of El Salvador, Guatemala, and Honduras, remittances amounted to over $300 per capita.

Latin America in the Global Economy

To conceptualize Latin America's place in the world economy, scholars from the region advanced **dependency theory** in the 1960s. The premise of the theory is that expansion of European capitalism created the region's underdevelopment. For the developed "cores" of the world to prosper, the "peripheries" became dependent and impoverished. Dependent economies, such as those in Latin America, were export-oriented and vulnerable to fluctuations in the global market. Even when they experienced economic growth, it was subordinate to the economic demands of the core (North America and Europe).

Economists who accepted this interpretation of Latin America's history were convinced that economic development could occur only through self-sufficiency, growth of internal markets, agrarian reform, and greater income equality. In short, they argued for strong state intervention and an uncoupling from the core economies of western Europe and North America. This view partially influenced policies such as import substitution industrialization (developing a country's industrial sector by making imported products extremely expensive and domestic manufactured goods cheaper) and nationalization of key industries. Dependency theory has its detractors. In its simplest form, it became a means to blame forces external to Latin America for the region's problems. Implicit in dependency theory is also the notion that the path to development taken by Europe and North America cannot be easily replicated. This was a radical idea for its time.

Latin America's century-long dependence on North America, especially the United States, as its major trading partner is still evident, especially with Mexico and Central America. However, growing trade with—as well as investment from—Europe and East Asia suggests that a more complex and less U.S.-dependent pattern of trade is emerging. Latin America is linked to the world economy in ways other than trade. Figure 4.40 shows the changes in foreign direct investment (FDI) as a percentage of gross domestic product (GDP) from 1990 to 2011. For nearly every country in the region, the value of FDI in terms of the percentage of GDP went up, except for Ecuador. In 1995, Brazil's FDI was less than $5 billion, and Mexico's was $9.5 billion. By 2011, FDI in Brazil had soared to $71 billion, and Mexico's was $21 billion. Much of this foreign investment was from Europe and Asia. In 2012, China became Brazil's largest trading partner. China and Brazil—the so-called BRIC nations along with Russia and India—have recognized their global strategic partnership. Not only is Brazil exporting grains, minerals, and energy resources, but also it has signed an agreement that will allow Brazilian airplane maker Embraer to manufacture and sell its regional jets in China.

A powerful symbol of Latin America's role in global trade is the expansion of the Panama Canal. Built and managed by Panamanians, in 2015 a new set of locks will allow for the passage of much larger ships (Figure 4.41). The size of the older locks limited vessels to those no more than 100 feet wide and 1000 feet long, the so-called Panamex vessels. The Panama Canal Authority, which manages the canal, anticipates total cargo levels increasing by 3 percent a year, doubling the 2005 tonnage by 2025. Ports around North America and Europe are also retrofitting to adjust for the arrival of larger post-Panamex ships. The timing of this expansion was especially important, considering the potential for competition from Arctic sea routes opening up due to global warming.

▼ **Figure 4.41 Panama Canal Expansion** A cruise ship on the Pacific side of the Panama Canal passes the construction area for the new and wider locks. When the expansion project is complete in 2015, the canal will be able to accommodate more and larger ships.

Remittances Another important indicator that reflects the integration of Latin American workers into labor markets around the world is remittances. Scholars debate whether this flow of capital can actually lead to sustained development or whether it is simply a survival strategy of last resort. World Bank research shows that remittances sharply dropped during the global economic recesssion that began in 2008, but by 2012 they had reached $62 billion (below the 2008 peak of $69 billion). Remittances are growing on average by 5–10 percent per year, so that many economists project remittances will continue to be a major source of captial for the region.

The economic impact of remittances shown on a per capita basis is real (see Figure 4.40). Mexico is the regional leader, receiving over $23 billion in remittance income in 2012 (which is equal to over $200 per capita). But for smaller countries such as El Salvador and Honduras, remittances contribute far more to the domestic economy. El Salvador, a country of about 6 million people, received $4 billion in remittances in 2012, which is nearly $630 per capita. For many Latinos, remittances are the surest way to alleviate poverty, although they depend upon an international migration system that is constantly changing and includes both legal and illegal channels of movement.

Dollarization During the 1990s, as Latin American governments faced various financial crises, many began to consider the economic benefits of **dollarization,** a process by which a country adopts—in whole or in part—the U.S. dollar as its official currency. In a totally dollarized economy, the U.S. dollar becomes the only medium of exchange, and the country's national currency ceases to exist. This was the radical step taken by Ecuador in 2000 to address the dual problems of currency devaluation and hyperinflation rates of more than 1000 percent annually. El Salvador adopted dollarization in 2001 as a means to reduce the cost of borrowing money. Dollarization is not a new idea; back in 1904, Panama dollarized its economy the year after it gained independence from Colombia. Until 2000, however, Panama was the only fully dollarized state in Latin America.

A more common strategy in Latin America is limited dollarization, in which U.S. dollars circulate and are used alongside the country's national currency. Limited dollarization exists in many countries around the world, but most notably in Latin America. Since the economies of Latin America are prone to currency devaluation and hyperinflation, limited dollarization is a type of insurance. Many banks in Latin America, for example, allow customers to maintain accounts in dollars to avoid the problem of capital flight should a local currency be devalued. Other countries keep their national currency, but peg its value one-for-one to the dollar; this was the innovative strategy adopted by Argentina in 1991, although it led to a serious financial crisis in 2001 and was eventually stopped. Dollarization, partial or full, tends to reduce inflation, eliminate fears of currency devaluation, and reduce the cost of trade by eliminating currency conversion costs.

Dollarization has its drawbacks. The obvious one is that a country no longer has control of its monetary policy, making it reliant on the decisions of the U.S. Federal Reserve. Foreign governments do not have to ask permission to dollarize their economies. At the same time, the United States insists that all its monetary policies be based exclusively on domestic considerations, regardless of the impact such decisions may have on foreign countries. The political impact of eliminating a national currency is serious. The case of Ecuador is instructive. In 1999, when President Jamil Mahuad announced his plan to dollarize the economy to head off hyperinflation, he was quickly forced out of office by a coalition of military and Indian activists. When Vice President Gustavo Naboa became president and the economic situation worsened, the country's political leadership went ahead with dollarization. In short, dollarization may help in a time of economic duress, but it is not a popular policy.

Social Development

Over the past three decades, Latin America has experienced marked improvements in life expectancy, child survival, and educational equity. One telling indicator is the steady decline between 1990 and 2010 in mortality rates for children below age five (Table 4.2). This indicator is important because an increase in the number of children younger than five years surviving suggests that basic nutritional and health-care needs are being met. We can also conclude that resources are being used to sustain women and their children. Despite economic downturns, the region's social networks have been able to lessen the negative effects on children.

A combination of government policies and grassroots and nongovernmental organizations (NGOs) plays a fundamental role in contributing to social well-being. In the past few years, conditional cash transfer programs, such as **Bolsa Familia,** have reduced extreme poverty. Poor Brazilian families who qualify for Bolsa Familia receive a monthly check from the state, but are required to keep their children in school and take them to clinics for health checkups. Such programs have both the immediate impact of giving poor families cash and the long-term impact of improving the educational attainment and health care of their children. Mexico has adopted a similar program. For states with far fewer resources than Brazil or Mexico, international humanitarian organizations, church organizations, and community activists provide many services that state and local governments cannot. Catholic Relief Services and Caritas, for example, work with rural poor throughout the region to improve their water supplies, health care, and education. Other groups lobby local governments to build schools and recognize squatters' claims. Grassroots organizations also develop cooperatives that market everything from sweaters to cheeses.

Other important gauges for social development are life expectancy, gender educational equity, and access to improved water sources. In aggregate, 84 percent of the people in the region have access to an adequate amount of water from an improved source; slightly more girls receive education than boys (Figure 4.42); and life expectancy (men and women) is 74 years (see Table 4.2). Masked by this aggregate data are extreme variations between rural and urban areas, between regions, and along racial and gender lines.

Within Mexico and Brazil, tremendous internal differences exist in socioeconomic indicators. The northeastern part of Brazil lags behind the rest of the country in every social indicator. The country has a literacy rate of over 85 percent, but in the northeast it is only 60 percent. Moreover, within the northeast, literacy for city residents is 70 percent, but for rural residents it is only 40 percent. In Mexico, the levels of poverty are highest in the more Indian south. In contrast, Mexico City and the states of Nuevo Leon (Monterrey is the capital), Quintana Roo (home to Mexico's largest resort, Cancun), and Campeche have the highest GDP per capita. Noting this north–south economic divide, former President Calderón explained, "there is one Mexico more like North America and another Mexico more like Central America. It is a very clear challenge for me to make them more alike." All countries have spatial inequities regarding income and availability of services, but the contrasts tend to be sharper in the developing world. In the cases of Mexico and Brazil, it is hard to ignore ethnicity and race when explaining these patterns.

Race and Inequality There is much to admire about race relations in Latin America. The complex racial and ethnic mix that was created in Latin America fostered tolerance for diversity. That said, Indians and blacks are more likely to be counted among the poor of the region. More than ever, racial discrimination is a major political issue in Brazil. Reports of organized killings of street children, most of them Afro-Brazilian, make headlines. For decades, Brazil put forward its vision of a color-blind racial democracy. True, residential segregation by race is rare in Brazil, and interracial marriage is common, but certain patterns of social and economic inequity seem best explained by race.

Assessing racial inequities in Brazil is problematic. The Brazilian census asks few racial questions, and all are based on self-classification. In the 2000 census, less than 11 percent of

▲ **Figure 4.42 School Children in Panama** Uniformed public school children walk to school in Panama City. Latin American states have seen steady improvements in youth literacy, with 97 percent of youth (between the ages of 15 and 24) being literate. Per capita expenditures on education and access to postsecondary education still lag behind levels in Europe and North America.

the population called itself black. Some Brazilian sociologists, however, claim that more than half the population is of African ancestry, making Brazil the second largest "African state" after Nigeria. Racial classification is always highly subjective and relative, but certain patterns support the existence of racism. Evidence from northeastern Brazil, where Afro-Brazilians are the majority, shows death rates approaching those of some of the world's poorest countries. Throughout Brazil, blacks suffer higher rates of homelessness, landlessness, illiteracy, and unemployment. To address this problem, various affirmative action measures have been implemented (along with the Bolsa Familia program). From federal ministries to public universities, various quota systems are being tried to improve the condition of Afro-Brazilians.

In areas of Latin America where Indian cultures are strong, indicators of low socioeconomic position are also present. In most countries, areas where native languages are widely spoken invariably correspond with areas of persistent poverty. In Mexico, the Indian south lags behind the booming northern states and Mexico City. Prejudice is embedded in the language. To call someone an *indio* (Indian) is an insult in Mexico. In Bolivia, women who dress in the Indian style of full, pleated skirts and bowler hats are called *cholas,* a descriptive term referring to the rural mestizo population that suggests backwardness and even cowardice. No one of high social standing, regardless of skin color, would ever be called a *chola* or *cholo*.

It is difficult to separate status divisions based on class from those based on race. From the days of conquest, being European meant an immediate elevation in status over the Indian, African, and mestizo populations. Class awareness is very strong. Race does not necessarily determine a person's economic standing, but it certainly influences it. Most people recognize the power of the elite and envy their lifestyle.

A growing middle class also exists that is formally employed, aspires to own a home and car, and strives to give its children a university education. The vast majority of people, however, are the working poor who struggle to meet basic food, shelter, clothing, and transport needs. These class differences are evident in the landscape. Go to any large Latin American city, and you will find handsome suburbs, country clubs, and trendy shopping centers. High-rise luxury apartment buildings with beautiful terraces offer all the modern amenities, including maids' quarters. The elite and the middle class even show a preference for decentralized suburban living and dependence on automobiles, as do North Americans. Yet near these same residences are shantytowns where urban squatters build their own homes, create their own economy, and eke out a living.

The Status of Women Many contradictions exist with regard to the status of women in Latin America. Many Latina women work outside the home. In most countries, the formal figures are between 30 and 40 percent of the workforce, not far off from many European countries, but lower than in the United States. Legally speaking, women can vote, own property, and sign for loans, although they are less likely to do so than men, reflecting the patriarchal (male-dominated) tendencies in the society. Even though Latin America is predominantly Catholic, divorce is legal, and family planning is promoted. In most countries, however, abortion remains illegal.

Overall, access to education in Latin America is good, compared to other developing regions, and thus illiteracy rates tend to be low. The rates of adult illiteracy are slightly higher for women than for men, but usually by only a few percentage points. Throughout higher education in Latin America, male and female students are equally represented today. Consequently, women are regularly employed in the fields of education, medicine, and law.

The biggest changes for women are the trends toward smaller families, urban living, and educational parity with men. These factors have greatly improved the participation of women in the labor force. In the countryside, however, serious inequalities remain. Rural women are less likely to be educated and tend to have larger families. In addition, they are often left to care for their families alone, as hus-

▶ **Figure 4.43 Development Issues in Latin America: Female Participation in Politics** Women are increasingly playing vital roles in Latin American politics. On average, 25 percent of seats in national parliaments are held by women. (In contrast, the figure for the United States is 18 percent.) In Brazil, only 9 percent of the seats were held by women, but in Mexico and Argentina 37 percent were held by women in 2012. Moreover, six countries have or have had women presidents. The most recently elected female president is Dilma Rousseff of Brazil. (World Bank Development Indicators, 2013)

bands leave in search of seasonal employment. In most cases, the conditions facing rural women have been slow to improve.

Women are increasingly playing an active role in politics. In 1990, Nicaragua elected the first woman president in Latin America, Violeta Chamorro, the owner of an opposition newspaper. Nine years later, Panamanians voted Mireya Moscoso into power. In 2005, South America had its first woman president when Dr. Michelle Bachelet, a pediatrician and single mother, took the oath of office in Chile. Brazilian President Dimla Rousseff took office in 2011, so that the sixth largest economy in the world is run by a woman. As shown in Figure 4.43, many Latin American states have larger percentages of women in their national parliaments than does the United States, which has only 18 percent. In contrast, Argentina, Costa Rica, Ecuador, Mexico and Nicaragua had over 30 percent of their national parliament seats held by women in 2012.

Across the region, women and indigenous groups are active organizers and participants in cooperatives, small businesses, and unions and are elected to national office. In a relatively short period, they have won a formal place in the economy and a political voice. Moreover, evidence suggests that this trend will continue.

REVIEW

4.10 How has the export of primary products (food, fiber, and energy) shaped the economies of Latin America?

4.11 What explains some of the positive indicators of social development in Latin America?

Physical Geography and Environmental Issues

Compared to Europe and Asia, this region is still rich in natural resources and relatively lightly populated. Yet as populations continue to grow and trade in natural resources increases, concern for the state of the environment rises. The relentless cutting of tropical forest and the construction of new dams are of particular concern. In terms of urban environments, Latin American cities have made strides in improving air quality.

4.1 How have changes in urban transportation planning made Latin American cities greener?

4.2 Provide some examples of resource conservation in Latin America. Are these practices working?

Population and Settlement

Unlike in other developing areas, three-quarters of Latin Americans live in cities. This shift started early and reflects a cultural bias toward urban living with roots in the colonial past. The cities are large and combine aspects of the formal industrial economy along with the informal one. Grinding poverty in rural areas drives people to live in cities or to emigrate in search of employment abroad.

4.3 Explain where this area might be with regard to the Latin American urban model (Figure 4.18).

4.4 Why has Latin America become a region of emigration? Is this pattern likely to continue?

Key Terms

agrarian reform *(p. 160)*
Altiplano *(p.* 141*)*
altitudinal zonation *(p. 146)*
Bolsa Familia (marked as a key term on *p. 187*)
Central American Free Trade Agreement (CAFTA) *(p. 140)*
Columbian Exchange *(p. 165)*
dependency theory *(p. 185)*
dollarization *(p. 186)*
El Niño *(p. 146)*
environmental lapse rate *(p.* 146*)*
fair trade *(p. 183)*
grassification *(p. 148)*
informal sector *(p. 157)*
latifundia *(p. 160)*
liberation theology *(p. 169)*
maquiladora *(p. 179)*
megacity *(p. 140)*
Mercosur *(p. 140)*
mestizo *(p. 163)*
minifundia *(p. 160)*
neoliberalism *(p. 179)*
neotropics *(p. 141)*
Organization of American States (OAS) *(p. 171)*
outsourcing *(p. 180)*
pristine myth *(p. 165)*
remittance *(p. 164)*
rural-to-urban migration *(p. 158)*
shield *(p. 141)*
subnational organization *(p. 176)*
supranational organization *(p. 176)*
syncretic religion *(p. 168)*
Treaty of Tordesillas *(p. 172)*
Union of South American Nations (UNASUR) *(p. 172)*
urban primacy *(p. 155)*

In Review • Chapter 4
Latin America

Cultural Coherence and Diversity

Latin America and the Caribbean were the first world regions to be fully colonized by Europe. In the process, perhaps 90 percent of the native population died from disease, cruelty, and forced resettlement. The slow demographic recovery of native peoples and the continual arrival of Europeans and Africans resulted in an unprecedented level of racial and cultural mixing in Latin America. Today Amerindian activism is on the rise as indigenous groups seek territorial and political recognition.

4.5 What factors explain the language patterns in this area of Latin America?

4.6 How do religious practices in this region reflect both globalization and diversity?

Geopolitical Framework

Most Latin American states have been independent for 200 years, yet during that time they experienced dependent political and trade relations with Europe and North America that limited the region's overall development. Today Latin America, especially Brazil, is exerting more geopolitical influence. Within Latin America, new political actors are emerging—from indigenous groups to women—who are challenging old ways of doing things.

4.7 What international agreements have shaped land-use patterns in this area?

4.8 How might the evolution of UNASUR impact development within South America?

Economic and Social Development

Latin American governments were early adopters of neoliberal economic policies. Some states prospered, whereas others faltered, sparking popular protests against the negative effects of neoliberalism and globalization. Since 2000, most states have experienced steady economic growth, driven in part by higher commodity prices. In addition, extreme poverty has declined, and social indicators of development are improving.

4.9 What commodity is this, and how is its rise in production changing agricultural practices and patterns of trade?

4.10 Do you think neoliberal policies are increasing or decreasing income inequality in the region?

MasteringGeography™

Looking for additional review and test prep materials? Visit the Study Area in **MasteringGeography**™ to enhance your geographic literacy, spatial reasoning skills, and understanding of this chapter's content by accessing a variety of resources, including **MapMaster** interactive maps, videos, RSS feeds, flashcards, web links, self-study quizzes, and an eText version of *Diversity Amid Globalization.*

Authors' Blogs

Scan to visit the author's blog for chapter updates

www.gad4blog.wordpress.com

Scan to visit the GeoCurrents blog

www.geocurrents.info

5 The Caribbean

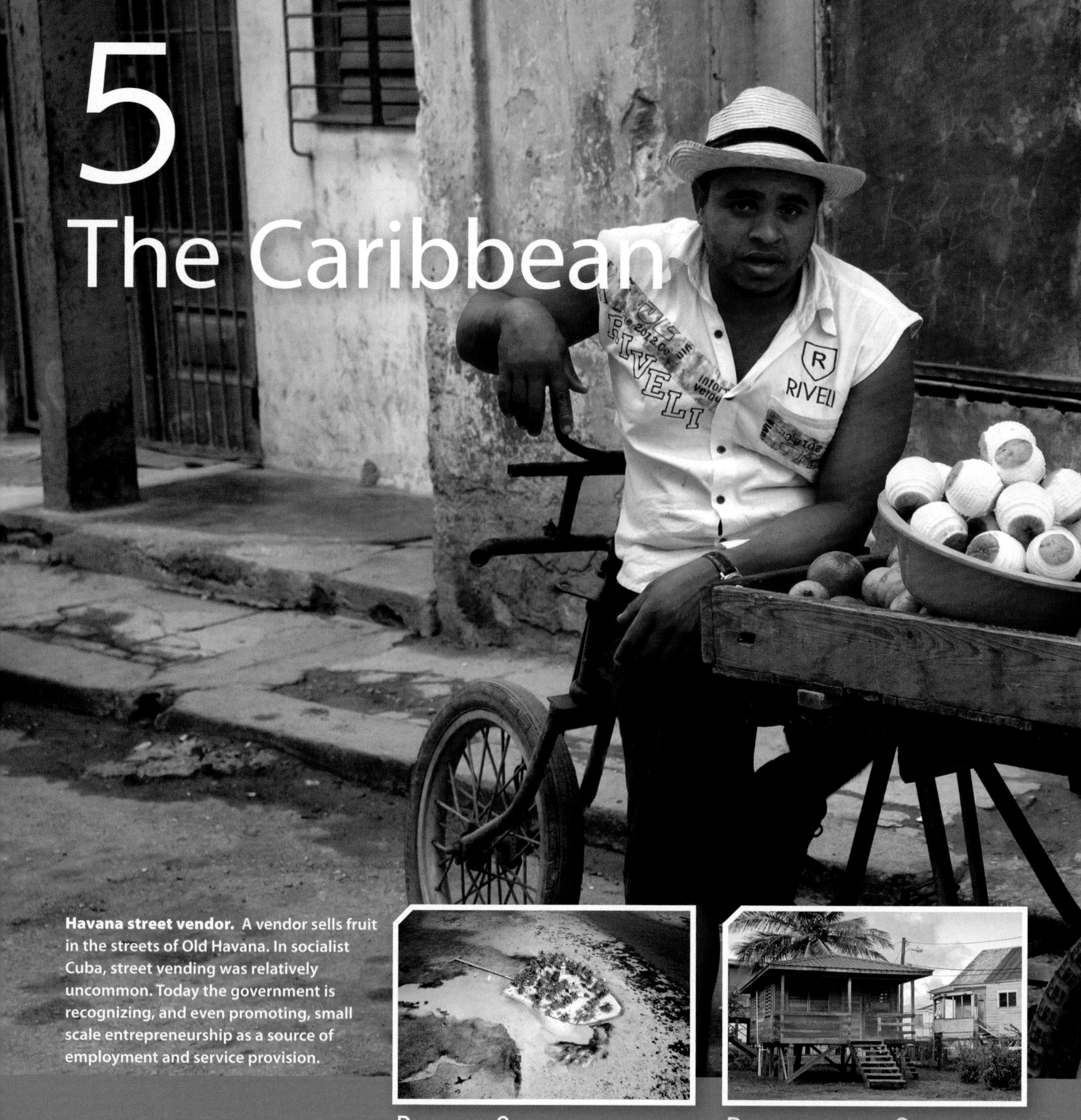

Havana street vendor. A vendor sells fruit in the streets of Old Havana. In socialist Cuba, street vending was relatively uncommon. Today the government is recognizing, and even promoting, small scale entrepreneurship as a source of employment and service provision.

Physical Geography and Environmental Issues

Climate change threatens the Caribbean, with the potential for stronger and more frequent hurricanes, loss of territory due to rising sea level, and destruction of coral reefs. In 2010, a devastating earthquake flattened the capital of Haiti, initiating the region's worst natural disaster in decades.

Population and Settlement

Having experienced its demographic transition, the region now has slow population growth. In addition, large numbers of Caribbean people have emigrated from the region, leaving in search of economic opportunity and sending back billions in dollars.

Cultural Coherence and Diversity

Creolization—the blending of African, European, and Amerindian elements—has resulted in many unique Caribbean expressions of culture, such as carnival, reggae, and steel drum bands.

Geopolitical Framework

The first area of the Americas to be extensively explored and colonized by Europeans, the region has seen many rival European claims and, since the early 20th century, has experienced strong U.S. influence. Many Caribbean territories have been independent states for less than 50 years.

Economic and Social Development

Environmental, locational, and economic factors make tourism a vital component of this region's economy, particularly in Puerto Rico, Cuba, the Dominican Republic, Jamaica, and the Bahamas. Offshore manufacturing and banking are also significant in the region's modern economic development.

The largest country in the Caribbean is Cuba. Yet as a socialist country where the state controls virtually all aspects of the economy, it is also an anomaly. Short on cash and needing to cut state payrolls, Cuba is presently experimenting with private entrepreneurship for the first time in decades. ***Cuentapropistas*** (private entrepreneurs) are filling Cuba's cities and towns with small restaurants, barbershops, family-run hotels, hardware stores, and even religious article shops (for both Christian and Santería worshippers). Small private businesses were banned early on in Fidel Castro's revolution, but a major reform by his brother, President Raul Castro, in 2010 has seen a flourishing of small restaurants, hotels, street vendors, and services that yield higher incomes for the owners, as well as tax revenue for government. In a dramatic reversal in policy, Cubans can obtain licenses for 181 lines of work and, for the first time, hire employees. Private restaurants are now found throughout the island, where once state-run cafeterias dominated. These simple establishments, serving take-out food or offering dining for less than a dozen people, would be considered ordinary elsewhere in the Caribbean, but for Cubans they are transformative. So, too, is the ability to rent rooms to visitors, which enables close interaction between Cubans and a growing numbers of tourists. New businesses, such as bicycle and shoe repair, furniture production, music shops, and nail salons, are an increasingly popular form of employment, and *cuentapropistas* usually earn higher salaries than government workers. The declared goal is to have 1.8 million workers in the private sector by 2015, roughly one-third of the workforce. Although this goal is not likely to be met, it shows that Cuba is remaking the island's economy so that it more closely resembles its Caribbean neighbors.

The Caribbean was the first region of the Americas to be extensively explored and colonized by Europeans. Yet its modern regional identity is unclear—often merged with Latin America, but also viewed as apart from it.

The Caribbean was the first region of the Americas to be extensively explored and colonized by Europeans. Yet its modern regional identity is unclear—often merged with Latin America, but also viewed as apart from it. Today the region is home to 43 million inhabitants scattered across 26 countries and dependent territories. They range from the small British dependency of the Turks and Caicos, with 12,000 people, to the island of Hispaniola, with nearly 20 million. In addition to the Caribbean Islands, Belize of Central America and the three Guianas—Guyana, Suriname, and French Guiana—of South America are often included as part of the Caribbean. For historical and cultural reasons, the peoples of these mainland states identify with the island nations and are thus included in this chapter (Figure 5.1).

Historically, the Caribbean was a battleground between rival European powers competing for territorial control of these remarkably profitable tropical lands. In the early 1900s, the United States took over as the dominant geopolitical force in the region, regularly sending troops and maintaining what some have called a neocolonial presence. As in many developing areas, external control of the Caribbean produced highly dependent and inequitable economies. Extreme inequalities were accentuated by the area's reliance upon slave labor, plantation agriculture, and the prosperity that sugar promised in the 18th century. Over time, other products began to have economic importance, as did the international tourist industry. Increasingly, governments have sought to diversify their economies by expanding nontraditional exports (such as flowers, nuts, and processed fruits), banking services, information processing services, gambling, and manufacturing to reduce the region's dependence on agriculture and tourism.

The basis for treating the Caribbean as a distinct area lies within its particular cultural and economic history. Culturally, this region can be distinguished from the largely Iberian-influenced mainland of Latin America because of its more diverse European colonial history and its strong African imprint, due to the region's historical reliance upon slavery. In terms of economic production, the dominance of export-oriented plantation agriculture explains many of the social, economic, and environmental patterns in the region.

LEARNING OBJECTIVES

After reading this chapter you should be able to:

- Differentiate between island and rimland environments and the environmental issues that affect these areas.
- Summarize the demographic shifts in the Caribbean as population growth slows, settlement in cities intensifies, emigration abroad continues, and a return migration begins.
- Explain why European colonists so aggressively sought control of the Caribbean and why independence in the region came about more gradually than in neighboring Latin America.
- Identify the demographic and cultural implications of the massive transfer of African peoples to the Caribbean and the creation of a neo-African society in the Americas.
- Describe how the Caribbean is linked to the global economy through offshore banking, emigration, and tourism.
- Understand the limited energy resources of the region and why renewable energy and exploration of offshore sites are important for Caribbean economic development.
- Explain the difference between social and economic development and suggest reasons why the Caribbean does better in social indicators of development, compared to economic indicators.

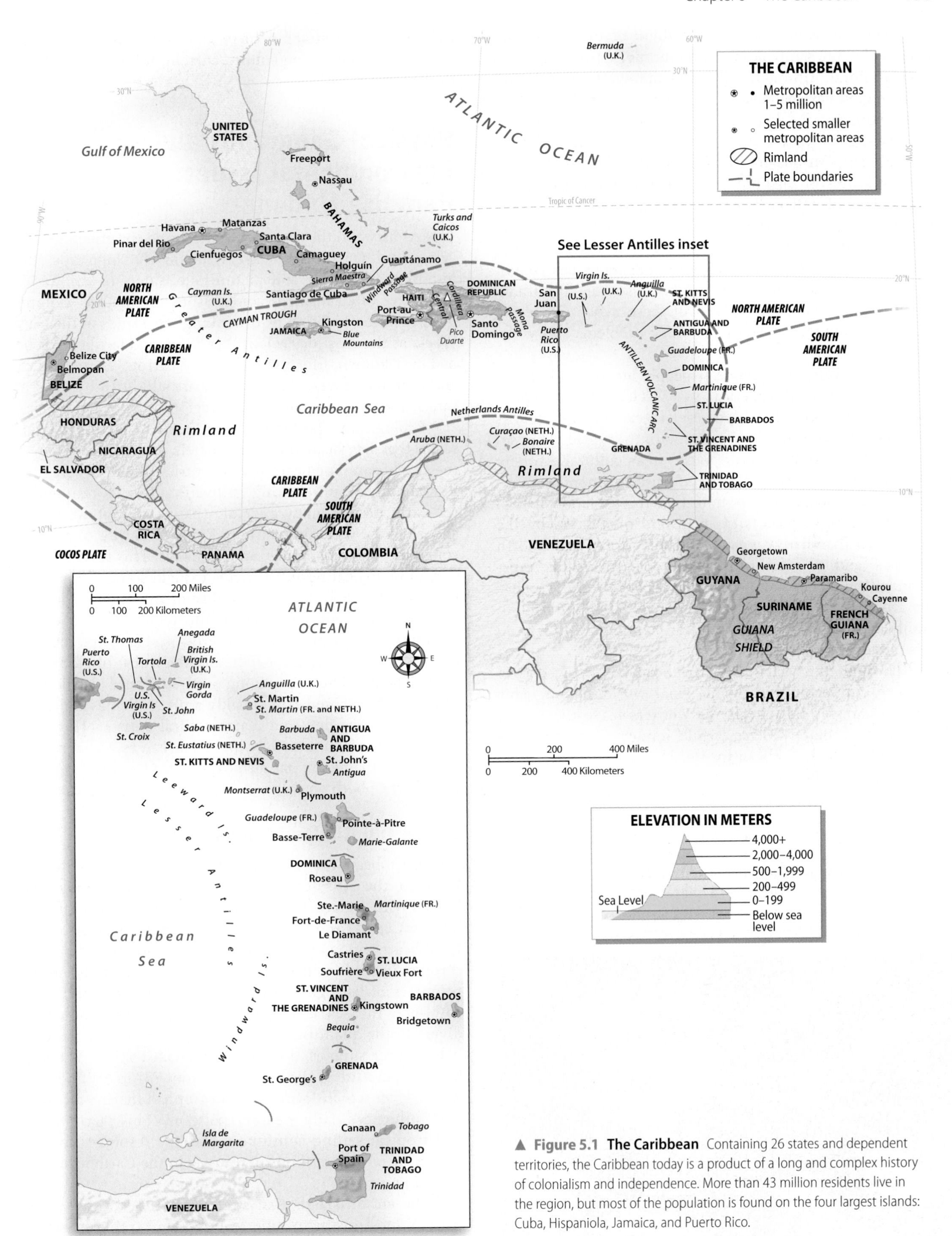

▲ **Figure 5.1 The Caribbean** Containing 26 states and dependent territories, the Caribbean today is a product of a long and complex history of colonialism and independence. More than 43 million residents live in the region, but most of the population is found on the four largest islands: Cuba, Hispaniola, Jamaica, and Puerto Rico.

Generally, when most people think of the Caribbean, images of white sandy beaches and turquoise tropical waters come to mind. Although still a developing area, most states in the region have achieved life expectancies in the 70s, low child mortality, and high rates of literacy. Millions of tourists visiting the Caribbean view it as an international playground. The Caribbean also plays a role in global finance, hosting scores of offshore financial institutions that manage billions of dollars. However, another Caribbean exists, which is far poorer and economically more dependent than the one portrayed on travel posters. Haiti, by many measures the poorest country in the Western Hemisphere, has 10 million people. The Dominican Republic also has 10 million, and Cuba has 11 million. Each of these major Caribbean nations suffers from serious economic problems and widespread poverty.

The majority of Caribbean people are poor, living in the shadow of North America's vast wealth. The concept of **isolated proximity** has been used to explain the region's unusual and contradictory position in the world. The *isolation* of the Caribbean sustains the area's cultural diversity (Figure 5.2), but also explains its limited economic opportunities. Caribbean writers note that this isolation fosters a strong sense of place and an inward orientation by the people. Yet the relative *proximity* of the Caribbean to North America (and, to a lesser extent, Europe) ensures its transnational connections and economic dependence. For example, Dominican workers abroad sent over $3 billion back to family and friends in the Dominican Republic, who rely on this money for sustenance.

Through the years, the Caribbean has evolved as a distinct, but economically marginal world region. This status expresses itself today as workers leave the region in search of better wages, while foreign companies are attracted to the Caribbean for its cheaper labor costs. The economic well-being of most Caribbean countries is precarious. Despite such uncertainty, an enduring cultural richness and an attachment to place are present here that may explain a growing countercurrent of immigrants back to the region.

▼ **Figure 5.2 Carnival Drummer** A steel pan drummer performs while his drum cart is pushed through the streets during carnival in Port of Spain, Trinidad. Steel drums were created in Trinidad in the 1940s from discarded oil drums from a U.S. military base. They have become an iconic sound for the region.

Physical Geography and Environmental Issues: Paradise Undone

Tucked between the Tropic of Cancer and the equator, with year-round temperatures averaging in the high 70s, the hundreds of islands and picturesque waters of the Caribbean have often inspired comparisons to paradise. Columbus began the tradition by describing the islands of the New World as the most marvelous, beautiful, and fertile lands he had ever known, filled with flocks of parrots, exotic plants, and friendly natives. Writers today are still lured by the sea, sands, and swaying palms of the region.

Ecologically speaking, it is difficult to picture a landscape that has been more completely altered than that of the Caribbean. For nearly five centuries, the destruction of forests and the unrelenting cultivation of soils resulted in the extinction of many endemic (native) Caribbean plants and animals, including various kinds of shrubs and trees, songbirds, large mammals, and monkeys. This severe depletion of biological resources helps explain some of the present economic and social instability of the region. Most of the environmental problems in the region are associated with agricultural practices, soil erosion, excessive reliance on wood and charcoal for fuel, and the threat of global climate change. The devestating Port-au-Prince earthquake of 2010 underscored how quickly a place such as Haiti can become undone due to a major natural disaster. However, because many countries rely upon tourism as a vital source of income, the region has also experienced a growth in protected areas, both on the land and in maritime locales.

Island and Rimland Landscapes

The Caribbean Sea itself—that body of water enclosed between the Antillean islands (the arc of islands that begins with Cuba and ends with Trinidad) and the mainland of Central and South America—links the states of the Caribbean region. Historically, the sea connected people through its trade routes and sustained them with its marine resources of fish, green turtle, manatee, lobster, and crab. Although the sea is noted for its clarity and biological diversity, the quantities of any one species are not great, so it has never supported large commercial fishing. The surface temperature of the sea ranges from 73° to 84°F (23° to 29°C), over which forms a warm tropical marine air mass that influences daily weather patterns. This warm water and tropical setting continue to be key resources for the region, attracting millions of tourists to the Caribbean each year (Figure 5.3).

The arc of islands that stretches across the sea is the region's most distinguishing feature. The Antillean islands

◄ **Figure 5.3 Caribbean Sea** Noted for its calm, turquoise waters, steady breezes, and treacherous shallows, the Caribbean Sea has both sheltered and challenged sailors for centuries. This aerial photograph shows English Caye off the coast of Belize.

are divided into two groups: the Greater and Lesser Antilles. The **rimland** (the Caribbean coastal zone of the mainland) includes Belize and the Guianas, as well as the Caribbean shoreline of Central and South America stretching between them (refer to Figure 5.1). In contrast to the islands, the rimland has low population densities.

Most of the islands, with the exception of Cuba, are on the Caribbean tectonic plate, wedged between the South American and North American plates. Generally, this is not one of the most tectonically active zones, although earthquakes and volcanic eruptions do happen. On January 12, 2010, a 7.0 earthquake leveled Haiti's capital of Port-au-Prince in one of the most tragic natural disasters to strike the Caribbean.

The Enriquillo Fault, near the densely settled and extremely poor capital city, had been inactive for more than a century. When it violently shifted, the epicenter of the resulting earthquake was just a few miles away from the city (Figure 5.4). The disaster affected nearly 3 million people, as homes were rendered unsafe and water and electricity supplies were disrupted. Shockingly, over 200,000 people died as a result of the earthquake, and 1 million people were made homeless. Many of the government agencies that would have assisted in the relief response were also destroyed. The tragedy of the Haitian earthquake was compounded by the state's poverty and corruption, as most buildings were not built to standards that could withstand an earthquake of this magnitude. The international community, as well as the large diaspora of Haitians living abroad, immediately offered financial aid and assistance. Even several years after the earthquake, rebuilding efforts have been slow, and tens of thousands remain in tent cities (see *Exploring Global Connections: Crisis Mapping in Haiti After the Earthquake*).

Greater Antilles The four large islands of Cuba, Jamaica, Hispaniola (shared by Haiti and the Dominican Republic), and Puerto Rico make up the **Greater Antilles.** These islands have most of the region's population, arable lands, and large mountain ranges. Given the popular interest in the Caribbean coasts, it still surprises many people that Pico Duarte on the Dominican Republic is more than 10,000 feet (3000 meters) tall, Jamaica's Blue Mountains top 7000 feet (2100 meters), and Cuba's Sierra Maestra is more than 6000 feet (1800 meters) tall. The mountains of the Greater Antilles were of little economic interest to plantation owners, who preferred the coastal plains and valleys. However, the mountains were an important refuge for runaway slaves and subsistence farmers and thus figure prominently in the cultural history of the region.

The best farmlands are found in the central and western valleys of Cuba, where a limestone base contributes to the formation of a fertile red clay soil (locally called *matanzas*) and a gray or black soil called *rendzinas* (also found in Antigua, Barbados, and lowland Jamaica). The rendzinas soils, which consist of a gravelly loam with a high organic content—ideal for sugar production—are actively exploited for agriculture wherever found. Surprisingly, given the area's agricultural

▼ **Figure 5.4 Earthquake in Haiti** The earthquake that struck Port-au-Prince on January 12, 2010, was one of the worst natural disasters in the region's history, killing over 200,000 people. In this satellite image, Port-au-Prince (in grey and white) rests just north of the Enriquillo Fault line and the nearby epicenter of the 7.0-magnitude quake.

Exploring **Global Connections**

Crisis Mapping in Haiti After the Earthquake

In response to the 2010 Haitian earthquake, social media, humanitarian organizations, and crisis mappers joined forces in a new and unique way that has changed how governments and civil societies will respond to complex humanitarian crises in the future. One of the leaders in the crisis-mapping movement is Patrick Meier, who was a key player in assembling the crisis-mapping team for Haiti. In 2013, Patrick Meier was a National Geographic Emerging Explorer where he blogged about the Haitian experience.

Crisis Mapping is the leveraging of mobile devices (texts and tweets), Web-based applications, participatory maps, satellite imagery, and crowd-sourced event data for rapid responses to complex humanitarian crises. Humanitarian workers need precise, real-time information that localities in crisis are often unable to provide. Working through an African-created platform called Ushahidi, crisis mappers assembled at Tufts University, just outside of Boston, Massachusetts, gathered tweets and text messages from Haitians (with translations provided by Haitians living in the U.S.). New global connections were forged, resulting in maps used by first responders that saved lives.

▼ **Figure 5.1.1 Crisis Mapping for Port-au-Prince** A portion of the map created using Open Street Map in the days after the Haitian earthquake. The circles represent the number of individual reports for that particular areas.

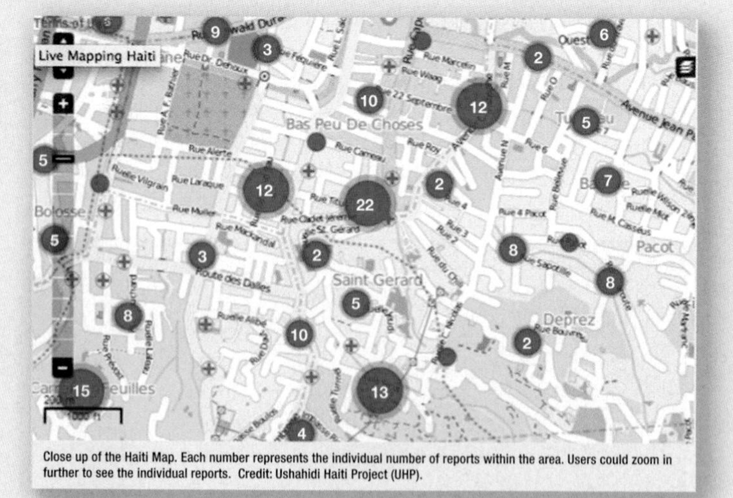

Close up of the Haiti Map. Each number represents the individual number of reports within the area. Users could zoom in further to see the individual reports. Credit: Ushahidi Haiti Project (UHP).

Two free and open-source mapping platforms were critical in moving crisis mapping forward: Ushahidi and Open Street Map. Ushahidi was developed by African bloggers who sought to report on postelection violence in Kenya in 2008 that was not covered by the media. Ushahidi (Swahili for "witness") relies on a Google Web-based map interface that plots acts of violence sent by crowd-sourced text messages. In the case of Haiti, Open Street Map was incorporated into the platform to allow for the construction of an extremely detailed and interactive map that people could use in the field and drill down to individual reports (**Figure 5.1.1**). Key to the success of the project was the creation of a team of crisis mappers (initially students at Tufts University) and translators who scanned for tweets. Later, through collaboration with Haiti's largest mobile phone provider, a texting number was set up so that anyone in Haiti could text urgent needs. As thousands of texts poured in, the Haitian population in the United Sates was mobilized to translate the texts from Haitian Creole to English so that the mappers could add the geo-referenced information to the map. As the real-time map grew, so did the number of contributors and users. The U.S. Coast Guard and Marines and various humanitarian groups on the ground in Port-au-Prince relied almost exclusively on its output.

Future Crisis Mapping Since the Haiti experience with crisis mapping, similar efforts have been used in response to earthquakes in Chile and Russia. An organization of crisis-mapping volunteers has formed to respond to future events. As Patrick Meier likes to say: "To map the world is to know it. But to map the world live is to change it before it's too late."

Source: Adapted from www.newswatch.nationalgeographic.com-How Crisis Mapping Saved Lives in Haiti, July 2, 2012.

1. What factors led the residents of Havana to become leaders in urban agriculture?
2. Beyond increasing the food supply, what are the other advantages of growing food in cities?

Google Earth Virtual Tour Video

http://goo.gl/fqds4l

orientation, many of the soils are nutrient-poor, heavily leached, and acidic. These poor, *ferralitic* soils are found in the wetter areas where crystalline base rock exists (as in parts of Hispaniola, the Guianas, and Belize). They are characterized by heavy accumulations of red and yellow clays and offer little potential for permanent intensive agriculture.

Lesser Antilles The **Lesser Antilles** form a double arc of small islands stretching from the Virgin Islands to Trinidad. Smaller in size and population than the Greater Antilles, they were important early footholds for rival European colonial powers. The islands from St. Kitts to Grenada form the inner arc of the Lesser Antilles. These mountainous islands, with peaks ranging from 4000 to 5000 feet (1200 to 1500 meters), have volcanic origins. In this subduction zone, the heavier North and South American plates sink beneath the Caribbean Plate, producing volcanic activity and earthquakes. Erosion of the island peaks and the accumulation of ash from eruptions have created small pockets of arable soils, although the steepness of the terrain limits agricultural development. The latest round of volcanic activity began in July 1995 on Montserrat. A series of volcanic eruptions of ash and rock took several lives and forced most of the island's 10,000 inhabitants to nearby islands and even to London. Plymouth, the capital, was abandoned in 1997, although interim government buildings now exist on the northwest corner of the island and some residents have returned.

Just east of this volcanic arc are the low-lying islands of Barbados, Antigua, Barbuda, and the eastern half of Guadeloupe. Covered in limestone that overlays volcanic rock, these lands were inviting for agriculture, especially sugarcane. Barbados was the experimental hub for colonial British sugar production in the Caribbean; innovations from that intensely cultivated island diffused throughout the region. Trinidad and Tobago are on the South American Plate and consist of sedimentary rather than volcanic rock. These islands include alluvial soils and, more important, sedimentary basins that contain oil and natural gas reserves.

The Rimland Unlike the rest of the Caribbean, the rimland states of Belize and the Guianas still contain significant amounts of forest cover. As on the islands, agriculture in these states is closely tied to local geology and soils. Much of low-lying Belize is limestone. Sugarcane dominates in the drier north, whereas citrus is produced in the wetter central portion of the state. The Guianas, however, are characterized by the rolling hills of the Guiana Shield. The shield's crystalline rock explains the area's overall poor soil quality. Most agriculture in the Guianas occurs on the narrow coastal plain, where sugar and rice are produced. Timber continues to be an important export for these rimland states. Metal extraction (bauxite and gold) also is vital to the economies of Guyana and Suriname. French Guiana, which is an overseas territory of France, relies mostly on French subsidies, but exports shrimp and timber. It also is home to the European Space Center at Kourou (Figure 5.5).

▲ **Figure 5.5 Kourou, French Guiana** The European Space Agency launches rockets from it center in Kourou, French Guiana. This French territory, near the equator and on the coast, makes an ideal launching site. In this photo an unmanned Ariane rocket is prepared for launch.

Caribbean Climate and Climate Change

Much of the Antillean islands and rimland receives more than 80 inches (200 cm) of rainfall annually, which is enough to support tropical forests. Average temperatures are typically highs of 80°F and lows of 70°F (27°C and 21°C) (Figure 5.6). Seasonality in the Caribbean is defined by changes in rainfall more than temperature. Although some rain falls throughout the year, the rainy season is from July to October. This is when the Atlantic high-pressure cell is farthest north and easterly winds generate moisture-laden and unstable atmospheric conditions that sometimes yield hurricanes. In Belize City and Havana, October is the wettest month. In Bridgetown and San Juan, the wettest month is November (see Figure 5.6). During the slightly cooler months of December through March, rainfall declines. This time of year corresponds with the peak tourist season.

The Guianas have a different rainfall cycle. These territories, on average, receive more rain than the Antillean islands. In Cayenne, French Guiana, an average of 126 inches (320 cm) falls each year (see Figure 5.6). Unlike the Antilles, the Guianas experience a brief dry period in late summer (September to October). Also, January tends to be a wet period for the mainland, while it is a dry time for the islands. Climatically, the Guianas also are distinguishable from the rest of the region because they are not affected by hurricanes.

Hurricanes Each year several **hurricanes** pound the Caribbean, as well as Central and North America, with

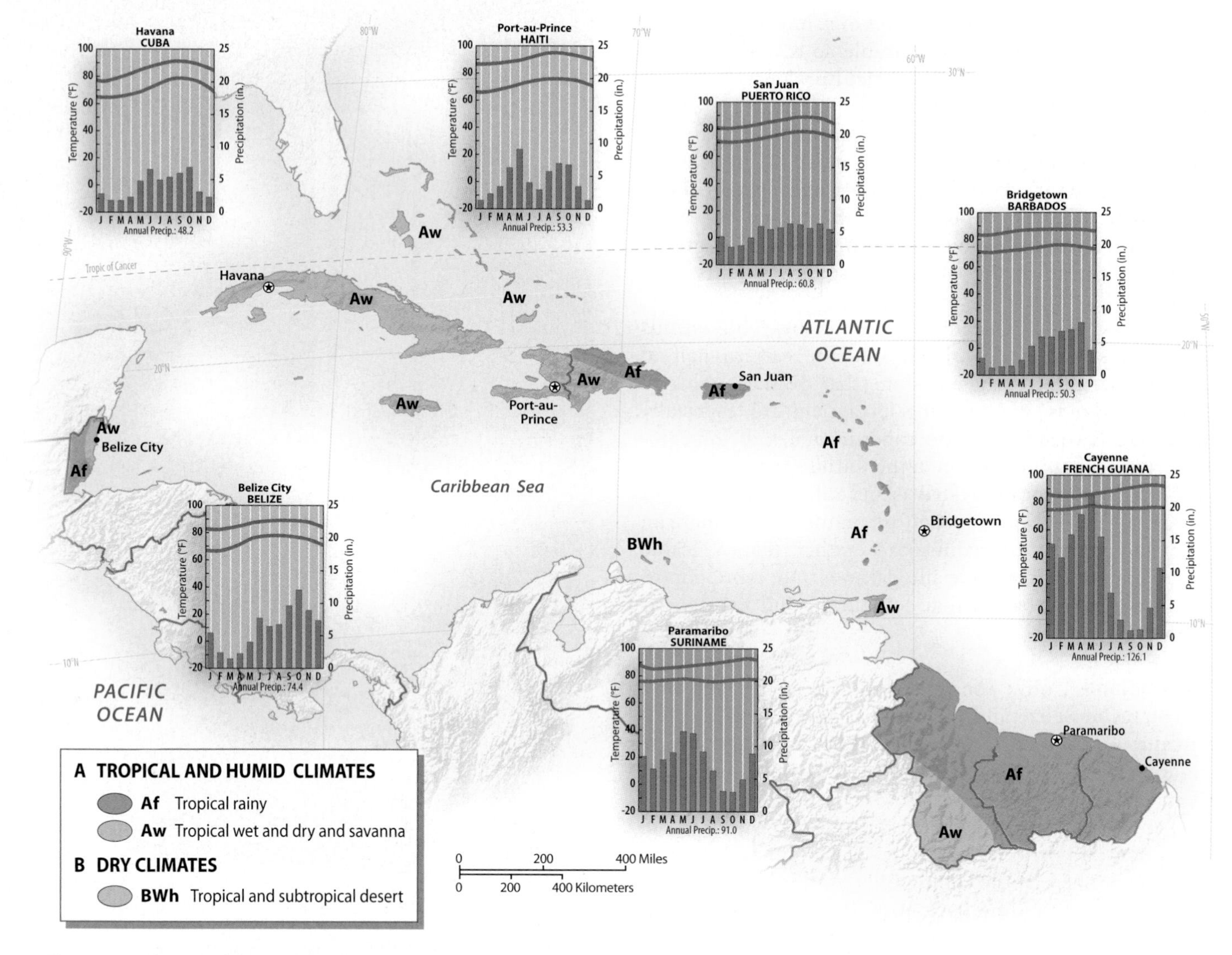

▲ **Figure 5.6 Climate of the Caribbean** Most of the region is classified as having either a tropical wet (Af) or a tropical savanna (Aw) climate. Temperature varies little over the year, as shown by the relatively straight temperature lines. Important differences in total rainfall and the timing of the dry season distinguish different localities.

heavy rains and fierce winds. Beginning in July, westward-moving low-pressure disturbances form off the coast of West Africa and pick up moisture and speed as they move across the Atlantic Ocean. Usually no more than 100 miles across, these disturbances achieve hurricane status when wind speeds reach 74 miles per hour. Hurricanes may take several paths through the region, but they typically enter through the Lesser Antilles. They then arc north or north-west and collide with the Greater Antilles, Central America, Mexico, or southern North America before moving to the northeast and dissipating in the Atlantic. The hurricane zone lies just north of the equator on both the Pacific and Atlantic sides of the Americas. Typically, a half dozen to a dozen hurricanes form each season and move through the region, causing limited damage.

Exceptions do occur, but most longtime residents of the Caribbean have felt the full force of at least one major hurricane in their lifetimes. The destruction caused by these storms is not just from the high winds, but also from the heavy downpours, which can cause severe flooding and deadly coastal storm surges. In 2005, Hurricane Dennis struck Cuba twice, flattening houses, downing power lines, and dropping more than 40 inches of rain in some areas before moving north (Figure 5.7). Cuba's primary citrus-growing areas were hit hard, drastically reducing harvests that year. In 2010, Antigua in the Lesser Antilles was hit hard by Hurricane Earl, which destroyed homes and caused serious flooding.

The National Hurricane Center in Miami tracks and predicts hurricanes in the Atlantic, the Caribbean, the Gulf of Mexico, and the eastern Pacific. During hurricane season, surface reports and satellite data are constantly monitored to detect and track tropical depressions that may achieve hurricane status. The center uses a combination of sophisticated models, geostationary satellites, and actual measurements of the approaching hurricanes made through specially equipped aircraft that fly into hurricanes before they approach land. Figure 5.8 shows the actual paths of three hurricanes over the course of two weeks in 12-hour intervals. A particular storm can take an irregular jog or have periods of slower or faster movement through an area. One reason Hurricane Mitch (in 1998) was so deadly was that it slowed down and lingered over a small portion of Central America for nearly a week, dropping huge amounts of rain.

Many atmospheric scientists and geographers believe that we have entered a more active hurricane period. Since 1995, a change in the multiyear cycle of sea surface temperatures in the Atlantic Ocean may be contributing to more-intense tropical depressions. With settlement and population growth intensifying in coastal zones of North America and the Caribbean, major storms are likely to do more damage. Apart from deaths resulting from Hurricanes Mitch and Katrina, fatalities from hurricanes over the last 30 years have declined because of better forecasting of hurricane movement, resulting in initiation of evacuation procedures at least 24 hours ahead of a storm's landing. Improved forecasting saves lives, but it cannot reduce the damage to crops, forests, or infrastructure. An unfortunately timed storm can destroy a banana harvest or shut down a resort for a season or more. The sheer force of a storm can radically change the landscape as well, turning a palm-lined sandy beach into a barren rocky shore covered with debris.

Climate Change Of all the issues facing the Caribbean, one of the most difficult to address is climate change. The Caribbean has not been a major contributor of greenhouse gases (GHGs), but this maritime region is extremely vulnerable to the negative impacts of climate change, including sea-level rise, increased intensity of storms, variable rainfall leading to both floods and droughts, and loss of biodiversity (both in forests and in coral reefs). The scientific consensus is that climate change could cause a sea-level rise of 3–10 feet (1–3 meters) in this century. In terms of land loss due to inundation, the low-lying Bahamas would be the most affected country in the region—it would lose nearly 30 percent of

▼ **Figure 5.7 Hurricane Dennis Strikes Cuba** A Cuban man looks at what remains of his house in the village of Casilda after Hurricane Dennis pounded the island with drenching rains and devastating winds. Each season hurricanes roll through the Caribbean. In many climate change scenarios, the intensities of hurricanes in this region are expected to increase.

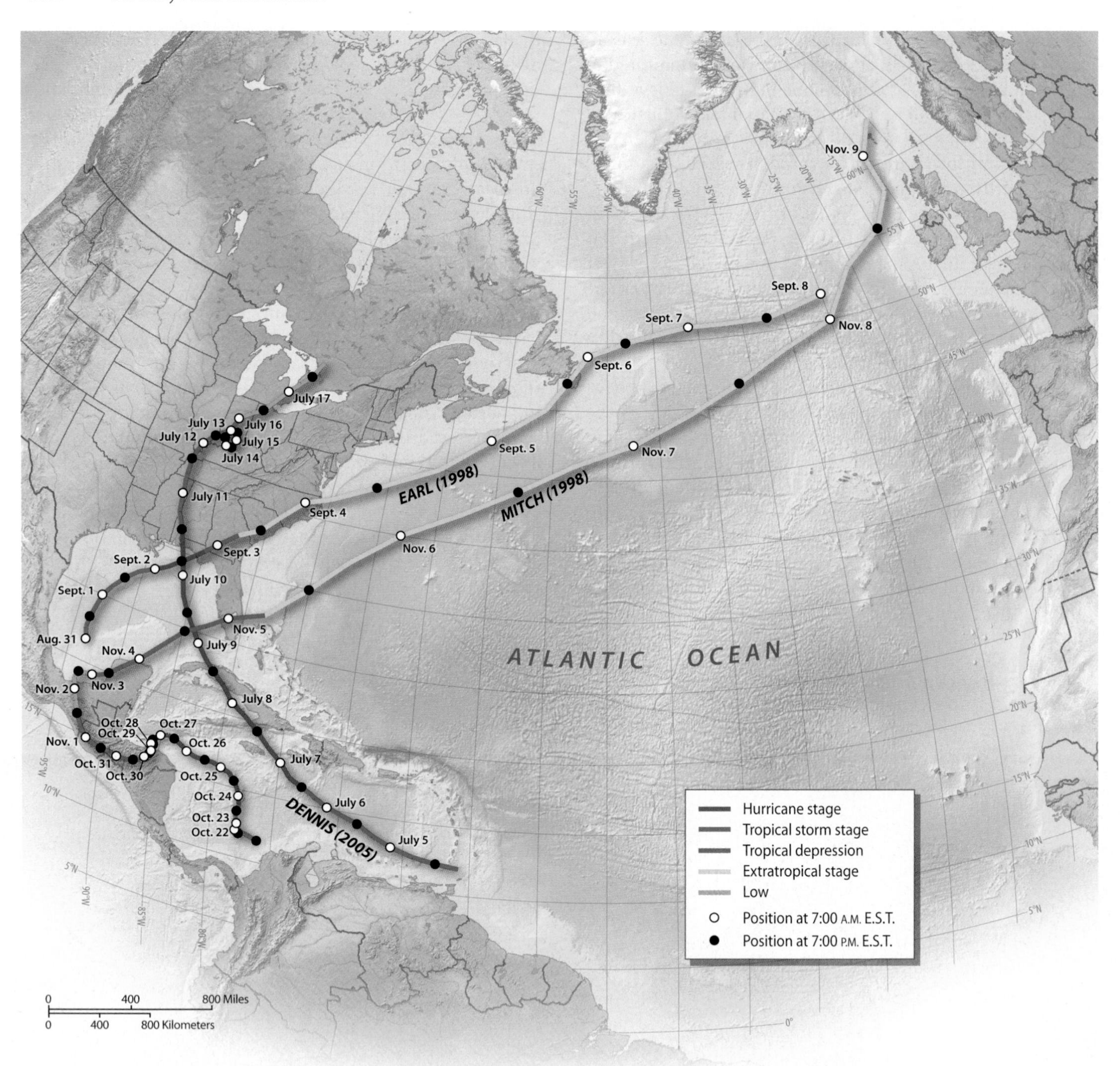

▲ **Figure 5.8 Tracking Paths of Hurricanes** The path, duration, and intensity of Hurricanes Dennis, Earl, and Mitch are shown as they arc across the Caribbean to the North Atlantic.

its land with a 10-foot (3-meter) sea-level rise. In terms of people affected by inundation, Suriname, French Guiana, Guyana, Belize, and The Bahamas would be the most severely impacted: A 3-foot (1-meter) sea-level rise would be devastating because most of the population lives near the coast. With a sea-level rise of 10 feet (3 meters), 30 percent of Suriname's population and 25 percent of Guyana's would be displaced (see the areas most affected by sea-level rise in Figure 5.9).

In addition to land loss and population displacement due to sea-level rise, other concerns include changes in rainfall patterns, leading to declines in agricultural yields and freshwater supplies, and increases in storm intensity—especially hurricanes—that cause destruction of infrastructure and other problems. All these changes would negatively affect tourism and thus the gross domestic income of countries in the region. Some of the worst-case scenarios are catastrophic.

In terms of biodiversity, continued warming of ocean temperatures will further negatively affect the Caribbean's coral reefs, which are the most biologically diverse ecosystems of the marine world. These reefs, particularly those of the rimland, are already threatened by water pollution and subsistence fishing practices. Recently, evidence is mounting of coral bleaching and die-off due to higher sea

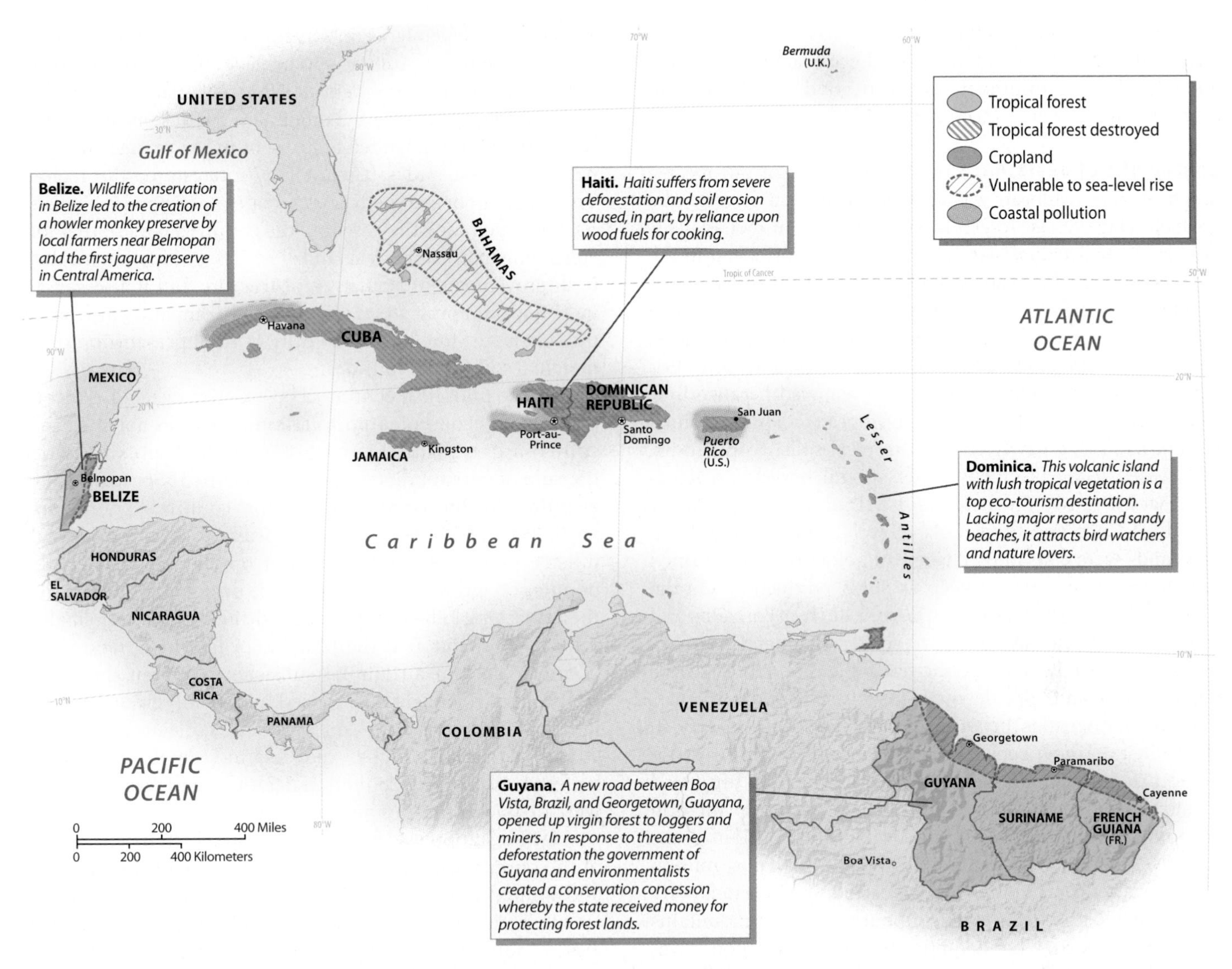

▲ **Figure 5.9 Environmental Issues in the Caribbean** It is hard to imagine a region in which the environment has been so completely transformed. Most of the island forests were removed long ago for agriculture or fuel, and soil erosion is a chronic problem. Coastal pollution is serious around the largest cities and industrial zones. The forest cover of the rimland states, however, is largely intact and is attracting the interest of environmentalists.

temperatures. Coral reefs are diverse and productive ecosystems that function as nurseries for many marine species. Healthy reefs, along with mangrove swamps and wetlands, also serve as barriers to protect populated coastal zones. As the reefs become more ecologically vulnerable, so, too, do the human populations that depend on the many benefits that the reefs provide.

Throughout the Caribbean, protecting the environment and preparing for the effects of climate change are considered fundamental to the region's economic livelihood. In fact, the Caribbean Community and Common Market (CARICOM), a regional organization that we will discuss later, has monitored the threat of climate change for over a decade. To address the issue of GHGs regionally, Guyana entered into an innovative agreement with Norway in 2009. Norway provided Guyana with an initial payment of $30 million into its Reducing Emissions from Deforestation and Forest Degradation (REDD) development fund. In 2013, the German government contributed $15 million toward assisting Guyana in a low-carbon development strategy that focuses resources on enhanced management of protected lands.

Environmental Issues

Climate change is both a medium- and a long-term concern for the Caribbean region. However, other environmental issues, such as soil erosion and deforestation, have preoccupied the region due to its long-standing dependence upon agriculture. Also, as the Caribbean has become more urbanized and more reliant upon tourism, governments have realized that protection of local ecosystems is not just good for the environment, but also good for the overall economy of the region (see Figure 5.9).

Legacy of Deforestation Prior to the arrival of Europeans, much of the Caribbean was covered in tropical forests. The great clearing of these forests began on European-owned plantations on the smaller islands of the eastern Caribbean in the 17th century and spread westward. The island forests were removed not only to make room for sugarcane, but also to provide the fuel necessary to turn the cane juice into sugar, as well as to provide lumber for buildings, fences, and ships. Primarily, however, tropical forests were removed because they were seen as unproductive; the European colonists valued cleared land. The newly exposed tropical soils easily eroded and ceased to be productive after several harvests, a situation that led to two distinct land-use strategies. On the larger islands of Cuba and Hispaniola, as well as on the mainland, new lands were constantly cleared and older ones abandoned or fallowed in an effort to keep up sugar production. On the smaller islands, where land was limited, such as Barbados and Antigua, labor-intensive efforts to conserve soil and maintain fertility were employed. In either case, the island forests were replaced by a landscape devoted to crops for world markets.

The connections between deforestation and poverty in Haiti illustrate how social and economic inequities contribute to unsound environmental practices, which, in turn, lead to more poverty. In the late 18th century, Haiti's plantation economy yielded vast wealth for several thousand Europeans (mostly French), who exploited the labor of half a million African slaves. During the colonial period, the lowlands were cleared to make way for cane fields, but most of the mountains remained forested. The inequities of Haiti's plantation economy led to the world's first successful slave uprising, followed by Haitian independence in 1804. Even though Haiti was the second state in the Americas to achieve independence, social and economic inequities persisted. Most Haitians were subsistence farmers, planting food crops along with some coffee and cacao. In 1915, the United States sent in the marines to quell political unrest and safeguard U.S. interests in the region. These U.S. troops and advisors remained for nearly 20 years, during which time they rewrote the country's laws so that foreign companies could own land. The U.S. occupation resulted in improvements in Haitian infrastructure, but it also caused increased land pressure as foreign sugar companies purchased the best lands, forcing the majority of rural peasants onto the marginal soils of the hillsides.

▼ **Figure 5.10 Deforestation Challenges** Hillsides that were once forested become denuded as trees are cleared for agriculture and fuel production, especially in Haiti. **Q: Which side of the yellow border is Haiti? How would you compare land use in Haiti versus that in the Dominican Republic?**

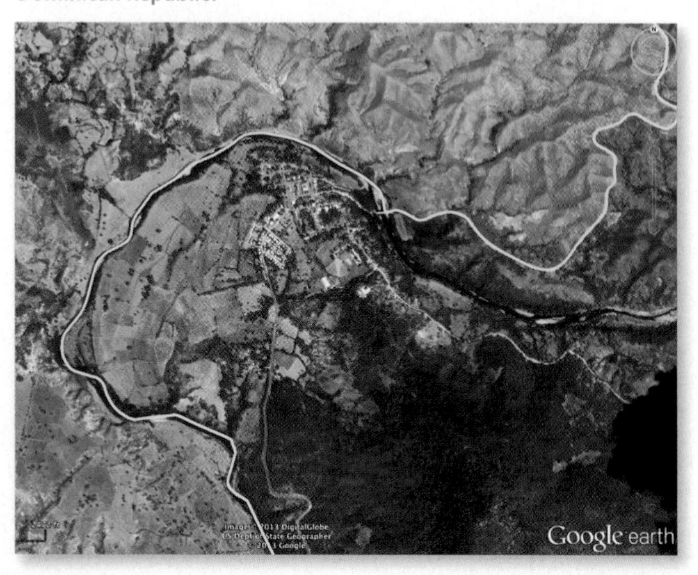

By the mid-20th century, a destructive cycle of environmental and economic impoverishment was established that still confronts Haiti (Figure 5.10). While Haiti was under the rule of corrupt dictators from 1957 to 1986, Haiti's elite benefited as the conditions for the country's poor worsened. Half of Haiti's people are peasants who work small hillside plots and seasonally labor on large estates. As the population grew, people sought more land. They cleared the remaining hillsides, subdivided their plots into smaller units, and abandoned the practice of fallowing land in an effort to eke out an annual subsistence. When the heavy tropical rains came, the exposed and easily eroded mountain soils washed away. As sediment collected in downstream irrigation ditches and behind dams, agriculture suffered, electricity production declined, and water supplies were degraded throughout the country.

Deforestation was further aggravated by the dependence of the population on wood for fuel. Because of their poverty and limited electricity supplies, most Haitians use charcoal (made from trees) to cook meals and heat water. By the late 1990s, only an estimated 3 percent of Haiti remained forested. Ongoing political turmoil hampers reforestation programs, and the effects of forest removal have only worsened. In less than a lifetime, hills that were once covered in forest now support shrubs and grasses.

Although Haiti has lost most of its forest cover, on Jamaica and Cuba nearly one-third of the land is still forested. About 40 percent of the Dominican Republic is forested, as is more than half of Puerto Rico. For these more-forested islands, the decline in agricultural production overall has allowed forests to recover. In the case of Puerto Rico, which is a territory of the United States, the creation of national forests—such as El Yunque on the eastern side of the island—led to greater protection of forests and better conservation of biological diversity (Figure 5.11).

Conservation Efforts In general, biological diversity and stability are less threatened in the rimland states than in the rest of the Caribbean. Thus, current conservation efforts could produce important results. Even though much of Belize was selectively logged for mahogany in the 19th and 20th centuries, healthy forest cover still supports a diversity of mammals, birds, reptiles, and plants. Public awareness of the negative consequences of deforestation is also greater now. Many protected areas have been established

◀ **Figure 5.11 El Yunque National Forest** Puerto Rico's largest remaining rainforest is in El Yunque National Forest. Visitors enjoy the park's hiking trails and its rich biological diversity.

in Belize. In the mid-1980s, villagers in Bermudian Landing, Belize, established a community-run sanctuary for black howler monkeys (locally referred to as baboons). The villagers banded together to maintain habitat for the monkeys and commit to land management practices that accommodate this gregarious species. The success of the project has resulted in tourists visiting the villages to see these indigenous primates up close (Figure 5.12). In 1986, a Jaguar Preserve was established in the Cockscomb Basin in southern Belize, the first of its kind in Central America.

In Guyana, 77 percent of the land is forested, and in Suriname 94 percent is in forest cover, according to 2010 World Bank figures. Yet these relatively pristine interior forests are becoming a battleground among conservationists, indigenous peoples, and developers. For over a decade, the Guyanese and Surinamese governments have made the wood-processing industry a priority by encouraging private investment and granting concessions to companies from Malaysia, Europe, and China for logging and sawmill operations. A new dry-season highway traverses the length of Guyana, connecting Boa Vista, Brazil, with Georgetown, Guyana. Not only does this road improve trade between Guyana and Brazil, but also it provides access to the forest and mineral resources (chiefly gold) in southern Guyana. Although the governments in Guyana and Brazil are encouraged by the economic possibilities of this road, an alliance of conservationists (local and foreign) and indigenous peoples established the region's first conservation concession in 2002, whereby the government of Guyana receives money for not clearing the concession lands. Meanwhile, as the pace of logging and mining accelerates in Suriname, the government is meeting resistance from its maroon population, made up of escaped slaves who have lived in the forests for more than 200 years. They claim legal rights to the land based on old treaties with the Dutch, but many of these land claims are in dispute.

Slowly, the territorial waters surrounding the Caribbean nations have gained protection, although more could be done. Here again, Belize has been a leader in creating over a dozen marine reserves and national parks along its barrier reef and outer atolls. The country has also created a substantial coastal wildlife sanctuary to protect mangrove swamps. The Caribbean island of Bonaire, which attracts large numbers of scuba divers, maintains the Bonaire Marine Park, recognized as one of the most effectively managed marine reserves in the region.

Urban Environmental Issues The growth of Caribbean cities has led to environmental issues concerning water quality and proper waste disposal. The urban poor are the most vulnerable to health problems associated with overtaxed or nonexistent sewage services. Almost everyone in Caribbean cities has access to improved water sources, typically through piped water to homes or shared neighborhood faucets. Several dams exist on the larger islands to supply water, while some of the smaller islands rely on expensive desalination plants.

Access to improved sanitation facilities (from protected pit latrines to flush toilets with a sewerage connection) is more uneven. According to the World Bank, 80 percent of

▼ **Figure 5.12 Protecting Habitat and Wildlife** Tourists visit the community "Baboon" Sanctuary in Bermudian Landing, Belize. The sanctuary is a community-run project to preserve the habitat and increase the number of black howler monkeys (locally referred to as baboons). The sanctuary, established in 1985, attracts domestic and foreign visitors.

Dominicans and Jamaicans have access to improved sanitation, compared with just 17 percent of Haitians. The expense of improving basic urban infrastructure, especially sewerage, is considerable, but slowly this infrastructure is being developed and improved. As tourism, offshore manufacturing, and a growing urban population demand more water and produce more waste, Caribbean countries are being forced to make hard decisions about their infrastructure. In addition to being a public health concern, water contamination from improper waste disposal poses serious economic problems for states dependent on tourism. Governments are caught between a desire to fix the problem before tourists notice it and a tendency not to discuss it at all. Doing nothing, however, may not be an option for countries that rely heavily on tourism. In the more industrialized Puerto Rico, it has been estimated that half the country's coastline is unfit for swimming, mostly due to contamination from sewage.

REVIEW

5.1 What environmental issues currently impact the Caribbean? Describe the risks and possible solutions.

5.2 Describe the locational, environmental, and climatic factors that together help make the Caribbean a major international tourist destination.

Population and Settlement: Densely Settled Islands and Rimland Frontiers

In the Caribbean, the population density is generally quite high and, as in neighboring Latin America, increasingly urban. Eighty-five percent of the region's population is concentrated on the four islands of the Greater Antilles (Figure 5.13). Add to this Trinidad and Tobago's 1.3 million and

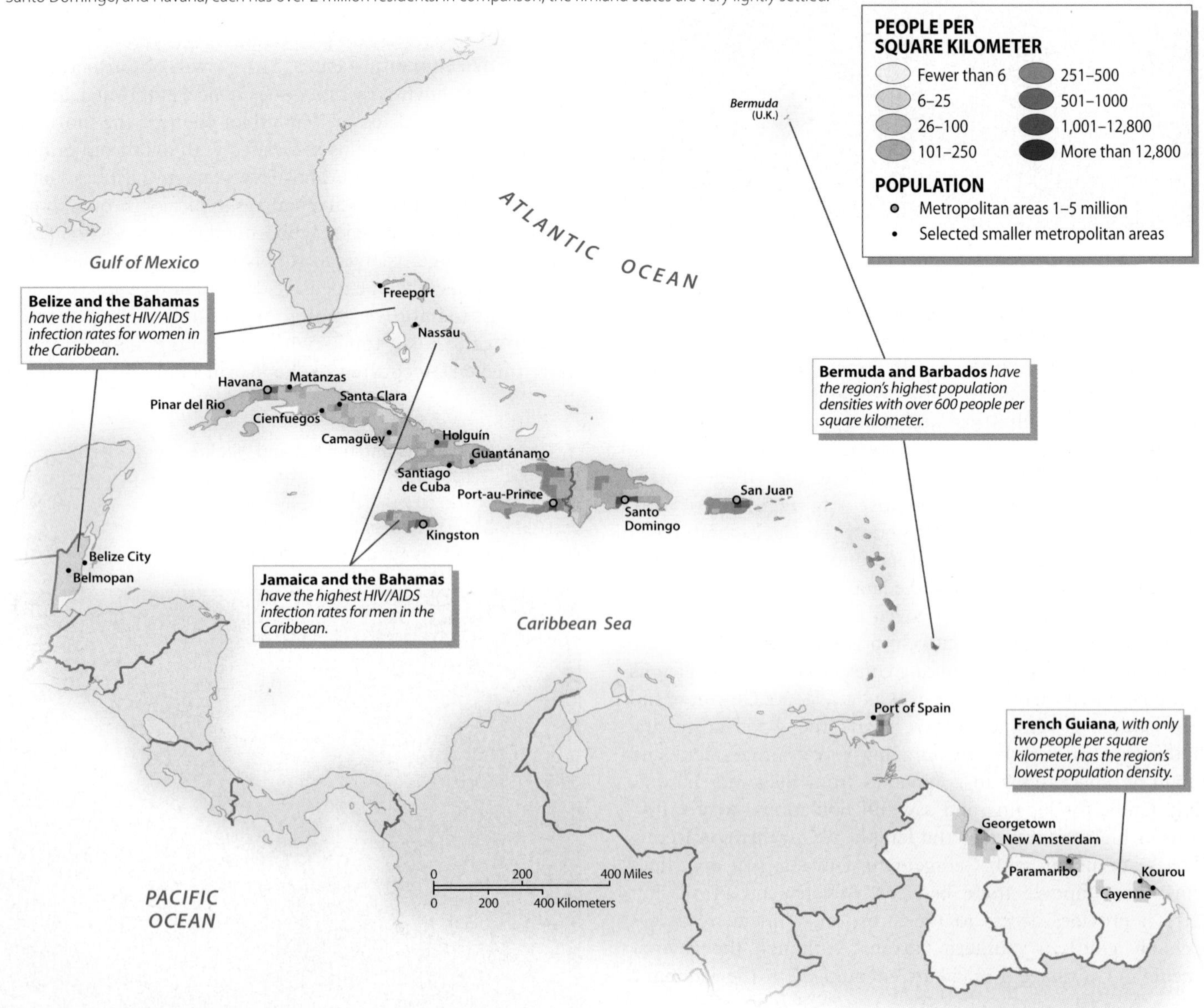

▼ **Figure 5.13 Population of the Caribbean** The major population centers are on the islands of the Greater Antilles. The tendency here, as in the rest of Latin America, is toward greater urbanism. The largest metropolitan areas in the region are San Juan, Santo Domingo, and Havana; each has over 2 million residents. In comparison, the rimland states are very lightly settled.

Guyana's 800,000, and most of the population of the Caribbean is accounted for by six countries and one U.S. territory (Puerto Rico). Of these Puerto Rico has the greatest population density with 410 people per square kilometer (1066 per square mile), followed by Haiti with 376 people per square kilometer (978 per square mile).

In absolute numbers, few people inhabit the Lesser Antilles; nevertheless, some of these microstates are densely settled. The small island of Barbados is an extreme example. With only 166 square miles (430 square kilometers) of territory, it has 1,530 people per square mile (589 people per square kilometer). Bermuda, which is one-third the size of the District of Columbia, has nearly 1300 people per square kilometer. Population densities on St. Vincent, Martinique, and Grenada, while not as high, are still more than 700 people per square mile (270 people per square kilometer). If you take into consideration the scarcity of arable land on some of these islands, it is clear that access to land is a basic resource problem for many inhabitants of the Caribbean. The growth in the region's population, coupled with its scarcity of land, has forced many people into the cities or abroad. It also has forced many Caribbean states to be net importers of food.

In contrast to the islands, the mainland territories of Belize and the Guianas are lightly populated; Guyana averages 10 people per square mile (4 people per square kilometer), Suriname only 8 (3 per square kilometer), and Belize 36 (15 per square kilometer). These areas are sparsely settled in part because the relatively poor quality and accessibility of arable land made them less attractive to colonial enterprises.

Demographic Trends

Prior to European contact with the New World, diseases such as smallpox, influenza, and malaria did not exist in the Americas. As discussed in Chapter 4, these diseases contributed to the demographic collapse of Amerindian populations. In the Caribbean, epidemics spread quickly, and within 50 years of Columbus's arrival, the indigenous

Table 5.1 Population Indicators

Country	Population (millions) 2013	Population Density (per square kilometer)	Rate of Natural Increase (RNI)	Total Fertility Rate	Percent Urban	Percent <15	Percent >65	Net Migration (Rate per 1000)
Anguilla*	0.02	173	–	1.8	100	24	–8	13
Antigua and Barbuda	0.1	199	0.8	2.1	30	26	7	0
Bahamas	0.3	25	0.7	1.7	84	26	6	6
Barbados	0.3	589	0.4	1.8	44	22	12	1
Belize	0.3	15	1.8	2.6	45	35	4	5
Bermuda*	0.07	1286	–	2.0	100	18	16	2
Cayman*	0.05	203	–	1.9	100	19	11	15
Cuba	11.3	102	0.4	1.8	75	17	13	–4
Curacao	0.2	348	0.5	2.1	–	20	14	19
Dominica	0.1	94	0.5	2.0	67	22	11	–6
Dominican Republic	10.3	211	1.6	2.6	67	31	6	–3
French Guiana	0.2	3	2.3	3.4	76	35	4	–1
Grenada	0.1	324	0.8	2.0	39	27	7	–8
Guadeloupe	0.4	238	0.6	2.2	98	21	15	–6
Guyana	0.8	4	1.4	2.6	28	37	3	–8
Haiti	10.4	376	1.7	3.5	53	36	4	–4
Jamaica	2.7	247	0.8	2.1	52	29	8	–6
Martinique	0.4	349	0.4	1.9	89	19	16	–5
Montserrat*	0.005	51	–	1.3	14	26	6	0
Puerto Rico	3.6	410	0.3	1.6	99	19	15	–8
St. Kitts and Nevis	0.1	210	0.6	1.8	32	23	8	1
St. Lucia	0.2	316	0.9	2.0	18	25	9	1
St. Vincent and the Grenadines	0.1	279	1.1	2.2	49	26	7	–9
Suriname	0.6	3	1.1	2.3	70	28	7	–2
Trinidad and Tobago	1.3	261	0.6	1.8	14	21	9	–2
Turks and Caicos*	0.05	50	–	1.7	93	22	4	15

Source: Population Reference Bureau, *World Population Data Sheet, 2013.*

*Additional data from the CIA *World Factbook*, 2013

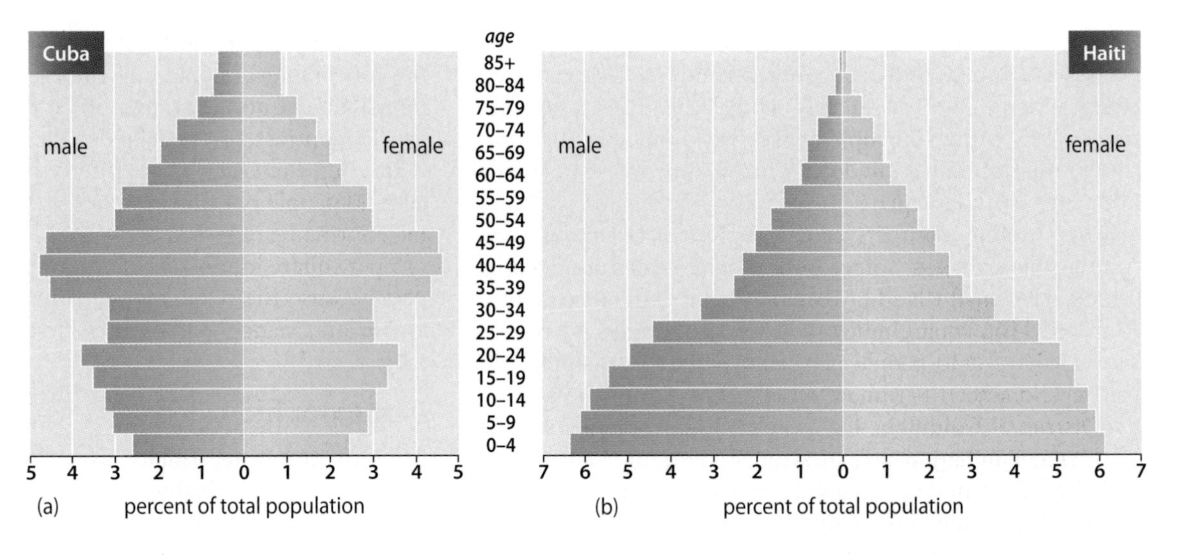

▶ Figure 5.14 Population Pyramids of Cuba and Haiti Although neighbors, Cuba and Haiti have extremely different population profiles. (a) Cuba's population is stable and older, with a notable decline in family size. (b) Haiti's population is much younger and growing, which is reflected in its broad-based pyramid.

population was virtually gone. Only the name *Caribbean* suggests that a Carib people once inhabited the region. Initially, European planters experimented with white indentured labor to work on sugar plantations. However, newcomers from Europe were especially vulnerable to malaria in the lowland Caribbean; typically, half died during the first year of settlement. Those that survived were considered "seasoned." In contrast, Africans had prior exposure to malaria and thus some immunity. They, too, died from malaria, but at much lower rates. This is not to argue that malaria caused slavery in the region, but it did strengthen the economic case for it.

During the years of slave-based sugar production, mortality rates were extremely high because of disease, inhumane treatment, and malnutrition. Consequently, the only way population levels could be maintained was through the continual importation of African slaves. With the end of slavery in the mid- to late 19th century and the gradual improvement of health and sanitary conditions on the islands, natural population increase began to occur. In the 1950s and 1960s, many states achieved peak growth rates of 3.0 or higher, causing population totals and densities to soar. Over the past 30 years, however, growth rates have steadily come down and stabilized. As noted earlier, the current population of the Caribbean is 44 million. However, the population is now growing at an annual rate of 1.1 percent, and projected population in 2025 is 47 million (see Table 5.1).

Fertility Decline and Longer Lives The most significant demographic trends in the Caribbean are the decline in fertility and the increase in life expectancy. Cuba and Puerto Rico have the region's lowest rates of natural increase (0.4). In socialist Cuba, due to the education of women, combined with the availability of birth control and abortion, the average woman has 1.8 children (compared to 2.1 in the United States). Yet in capitalist Puerto Rico, low rates of natural increase have also been achieved, along with a total fertility rate of 1.6. In general, educational improvements, urbanization, and a preference for smaller families have contributed to slower growth rates. Even states with relatively high total fertility rates, such as Haiti, have seen a decline in family size. Haiti's total fertility rate fell from 6.0 in 1980 to 3.5 in 2013.

Figure 5.14 provides a stark contrast in the population profiles of Cuba and Haiti. Although both are poor Caribbean countries, Haiti has the more classic, broad-based pyramid of a developing country, where more than one-third of the population is under the age of 15. Also, there are very few old people, due to the relatively low life expectancy (62 years) in Haiti. In contrast, Cuba's population pyramid is more diamond-shaped, bulging in the 35- to 49-year-old age cohort and tapering down after that. Here the impact of the Cuban revolution and socialism is evident. Family size came down sharply after education improved and modern contraception became readily available. With better health care, Cuba's population also lives longer, having nearly the same life expectancy as those in the United States (78 years). Cuba has 13 percent of its population over 65 and just 17 percent under 15; thus, it has an extremely low rate of natural increase, similar to many developed countries in the world.

The Rise of HIV/AIDS Although nowhere near the infection rates in Sub-Saharan Africa (see Chapter 6), about 1 percent of the Caribbean population between the ages of 15 and 49 had HIV/AIDS in 2012, with women having slightly higher infection rates than men. After Sub-Saharan Africa, the prevalence of HIV in the Caribbean is second in the world. Reflecting global patterns, the main transmission route is through heterosexual sex. In 2009, an estimated 17,000 people became infected with HIV and 12,000 people died of AIDS in the Caribbean.

In Haiti, one of the earliest locations where AIDS was detected, 1.5 percent of men and 2.3 percent of women between the ages of 15 and 49 are infected with the virus. The Bahamas has the highest infection rate in the region. Here, too, a higher percentage of women (3.7 percent) are infected compared to men (2.4 percent). Both Belize and Jamaica have infection rates higher than the regional average; in Belize, nearly 3 percent of women and 2 percent of men are infected, whereas in Jamaica men have twice the infection rate (2.3 percent) of women. The rate of HIV/AIDS infection in the Caribbean has come down in the past few years, but it is still high, making the disease an important regional issue.

Various factors led to the spread of the disease. In the 1980s and 1990s, limited information was available about safe sexual practices, and a stigma existed against discussing infection and prevention. In addition, many Caribbean islands have tourist-based economies that often contribute to the growth of prostitution. Many Caribbean countries have taken steps to educate their populations about the spread of HIV/AIDS. In 2001, the Pan-Caribbean Partnership Against HIV/AIDS (PANCAP) formed to help prevent the spread of the disease and alleviate suffering by taking a regionwide approach. PANCAP was effective in negotiating lower costs for antiretroviral drugs. Through state and regional efforts, mother-to-child transmission prevention is common, condoms are now widely available, and testing is easily done. Nearly every country has launched educational campaigns to bring infection rates down. Cuba, which witnessed a surge in both tourism and prostitution in the 1990s, has maintained a very low infection rate of 0.1 percent among its 15- to 49-year-old population. Education programs and an effective screening and reporting system have kept Cuba's infection rates down.

Emigration Driven by the region's limited economic opportunities, a pattern of emigration to other Caribbean islands, North America, and Europe began in the 1950s. For more than 50 years, a **Caribbean diaspora**—the economic flight of Caribbean peoples across the globe—has defined existence and identity for much of the region (Figure 5.15).

▼ **Figure 5.15 Caribbean Diaspora** Emigration has long been a way of life for Caribbean peoples. With relatively high education levels, but limited professional opportunities, migrants from the region head to North America, Great Britain, France, and the Netherlands. Intraregional migrations between Haiti and the Dominican Republic and between the Dominican Republic and Puerto Rico also occur.

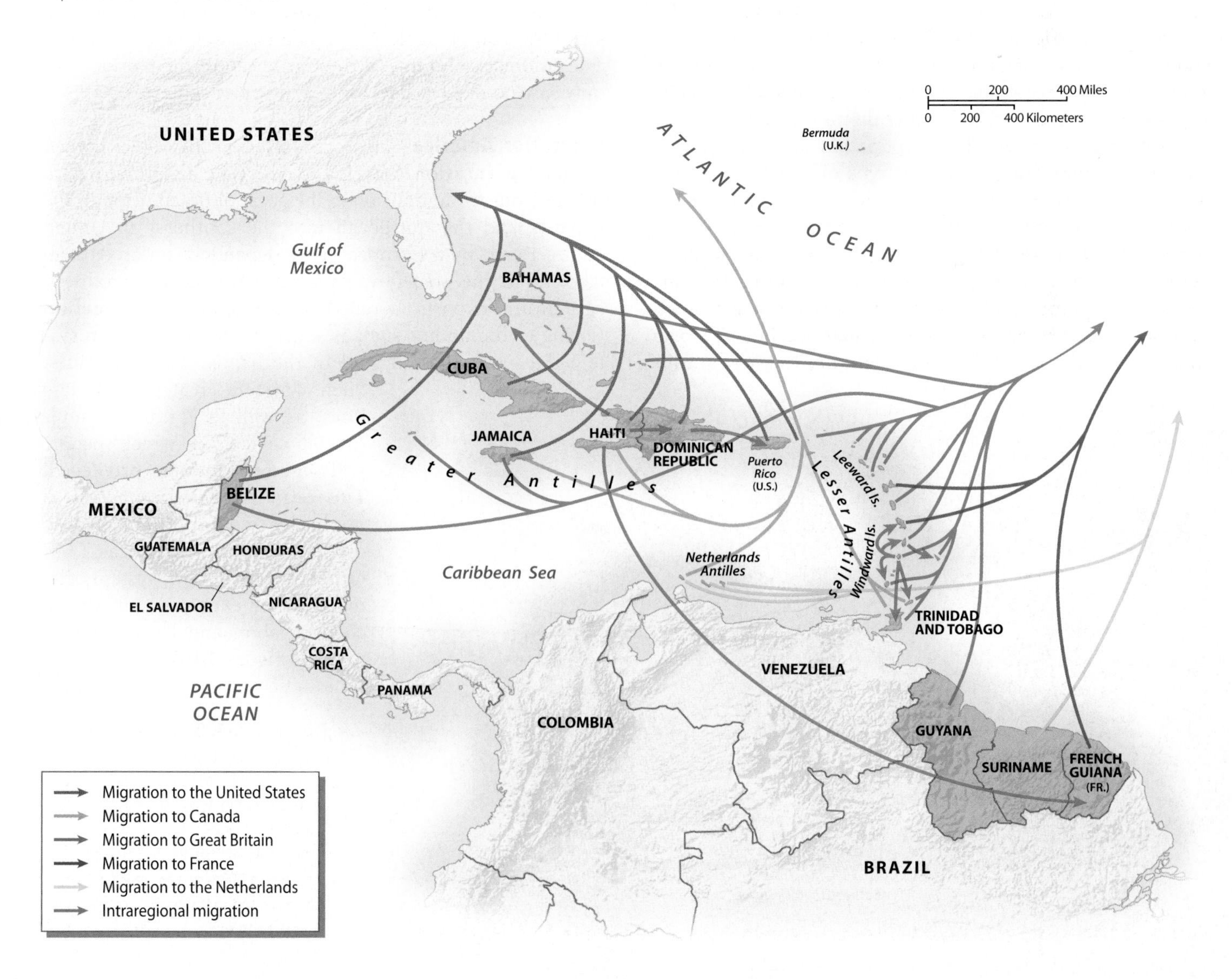

Barbadians generally choose England, most settling in the London suburb of Brixton with other Caribbean immigrants. In contrast, one out of every three Surinamese has moved to the Netherlands, with most residing in Amsterdam. As for Puerto Ricans, only slightly more live on the island than reside on the U.S. mainland. In the 1980s, roughly 10 percent of Jamaica's population legally emigrated to North America (some 200,000 to the United States and 35,000 to Canada). Cubans have made the city of Miami their destination of choice since the 1960s. Today they are a large percentage of that city's population, and since the mid-1980s nearly all of Miami's mayors have been Cuban-born.

Intraregional movements also are important. Perhaps one-fifth of all Haitians do not live in their country of birth. Their most common destination is the neighboring Dominican Republic, followed by the United States, Canada, and French Guiana. Dominicans are also on the move; the vast majority come to the United States, settling in New York City, where they are the single largest immigrant group. Others, however, simply cross the Mona Passage and settle in Puerto Rico. As a region, the Caribbean has one of the highest annual rates of net migration in the world at –3.0 per thousand. That means for every 1000 people in the region, 3 leave annually. Individual countries have much higher rates, such as Guyana and Grenada at –8 per 1000 and Jamaica at –6 per 1000 (see Table 5.1). The economic implications of this labor-related migration are significant and will be discussed later.

Most migrants, with the exception of Cubans, are part of a **circular migration** flow. In this type of migration, a man or woman typically leaves children behind with relatives in order to work hard, save money, and return home. Other times a **chain migration** begins, in which one family member at a time is brought over to the new country. In some cases, large numbers of residents from a Caribbean town or district send migrants to a particular locality in North America or Europe. Thus, chain migration can account for the formation of immigrant enclaves. Caribbean immigrants have increasingly practiced **transnational migration**—the straddling of livelihoods and households between two countries. Dominicans are probably the most transnational of all the Caribbean groups. They regularly move back and forth between two islands: Hispaniola and Manhattan. Dominican President Leonel Fernandez was first elected in 1996 for a four-year term and was reelected in 2004 and in 2008. He grew up in New York City, still holds a green card, and has said he intends to return when his presidential term is over.

The Rural–Urban Continuum

Initially, plantation agriculture and subsistence farming shaped Caribbean settlement patterns. Low-lying arable lands were dedicated to export agriculture and controlled by wealthy colonial landowners. Only small amounts of land were set aside for subsistence production. Over time, villages of freed or runaway slaves were established, especially in remote areas of the interior. But the vast majority of people continued to live on estates as owners, managers, or slaves. Cities were formed to serve the administrative and social needs of the colonizers, but most were small, containing a small fraction of a colony's population, and often defensive. The colonists who linked the Caribbean to the world economy saw no need to develop major urban centers.

Plantation America Anthropologist Charles Wagley coined the term **plantation America** to designate a cultural region that extends from midway up the coast of Brazil through the Guianas and the Caribbean into the southeastern United States. Ruled by a European elite dependent on an African labor force, this society was primarily coastal and produced agricultural exports. It relied upon **monocrop production** (a single commodity, such as sugar) under a plantation system that concentrated land in the hands of elite families. Such a system created rigid class lines, as well as forming a multiracial society in which people with lighter skin were privileged. The term *plantation America* is not meant to describe a race-based division of the Americas, but rather a production system that relied upon export commodities, coerced labor, and limited access to land (Figure 5.16).

◀ **Figure 5.16 Tobacco Plantation** This woodcut from the 1840s depicts slaves harvesting tobacco in Cuba while a white supervisor looks on smoking a cigar. Commodities such as tobacco and sugar were profitable but the work was arduous. Several million Africans were enslaved and forcibly relocated to the region to produce these commodities.

▲ **Figure 5.17 Santo Domingo Metro** Passengers load onto metro cars in downtown Santo Domingo, Dominican Republic. The Metro, which opened in 2009, received technical support from the Metro in Madrid, Spain.

Even today, the structure of Caribbean communities reflects the plantation legacy. Many of the region's subsistence farmers are descendants of former slaves who continue to work their small plots and seek seasonal wage-labor on estates. The social and economic patterns generated by slavery still mark the landscape. Rural communities tend to be loosely organized; labor is transient; and small farms are scattered on available pockets of land. Because men have tended to leave home for seasonal labor, matriarchal family structures and female-headed households are common.

Caribbean Cities The mechanization of agriculture, offshore industrialization, and rapid population growth caused a surge in rural-to-urban migration in the 1960s. Cities have grown accordingly, and today 66 percent of the region is classified as urban. Of the large islands, Puerto Rico is the most urban, and Haiti is the least (see Table 5.1). Caribbean metropolitan areas are not large by world standards, as only five have more than 1 million residents: Santo Domingo, Havana, Port-au-Prince, San Juan, and Kingston. Three were laid out by the Spanish, one by the French, and one by the English.

Like their counterparts in Latin America, the Spanish Caribbean cities were laid out on a grid with a central plaza. Vulnerable to raids by rival European powers and pirates, these cities were usually walled and extensively fortified. The oldest continually occupied European city in the Americas is Santo Domingo in the Dominican Republic, settled in 1496. Today it is a metropolitan area of 2.9 million. Merengue—a fast-paced, highly danceable music that originated in the Dominican Republic—is the soundtrack that pulses through the metropolis day and night. As rural migrants poured into the city over the last four decades in search of employment and opportunity, the city steadily grew. In 2009, a high-speed Metro opened in Santo Domingo with one line and 16 stations; more lines are planned to link the downtown with the suburbs and reduce the crushing traffic (Figure 5.17). The country has experienced solid growth in the 2000s, but there is still inadequate housing, electricity, employment, and schooling for a large portion of Santo Domingo's residents. Some critics argue that an expensive underground Metro was ill-advised given the city's other pressing needs. Yet it is also a sign of both big-city status and modernity that Dominicans have embraced.

The second largest city in the region is metropolitan San Juan, estimated at 2.6 million. It, too, has a renovated colonial core that is dwarfed by the modern sprawling city, which supports the island's largest port. San Juan is the financial, political, manufacturing, and tourism hub of Puerto Rico. With its highways, high rises, shopping malls, and ever-present shoreline, it is an interesting blend of Latin American, North American, and Caribbean urbanism.

Havana emerged as the most important colonial city in the region, serving as a port for all incoming and outgoing Spanish galleons. Strategically situated on Cuba's north coast at a narrow opening to a natural deep-water harbor, Havana became an essential city for the Spanish empire. Consequently, Old Havana possesses a handsome collection of colonial architecture, especially from the 18th and 19th centuries, and is a UNESCO World Heritage Site. The modern city is more sprawling, with a mix of Spanish colonial and Soviet-inspired concrete apartment blocks. It is also a city that had to reinvent itself when subsidies from the former Soviet Union stopped flowing (see *Working Toward Sustainability: Urban Agriculture in Havana*). Americans are still banned from tourist travel to Cuba, but Havana's streets are filled with tourists from other parts of the world.

Working Toward Sustainability

Urban Agriculture in Havana

Many cities around the world have seen a renewed interest in urban gardening as a way to build community unity, reduce food insecurity, improve nutrition, create income opportunities, and enhance urban environments by converting brown spaces into green ones. Cuba is a global leader in urban agriculture, and these farming efforts are especially evident in metropolitan Havana. Scattered throughout this city of 2 million are thousands of small and large plots where urban residents are producing vegetables on raised beds; harvesting fruit trees; and raising rabbits, chickens, and goats for meat, eggs, and milk (Figure 5.2.1). Although the Cuban context is unique—a socialist planned economy with fixed prices and limited exposure to market forces—some of the successes of Havana farmers are transferable to other cities in the world.

▼ **Figure 5.2.1 Urban Farmer** A worker harvests lettuce from an urban farm in Havana.

Gardening by Necessity In 1989, the Cuban government officially recognized the potential for urban gardens as a means to address the pressing food shortages provoked by drastic cuts in food and energy supports from the Soviet Union. Within a couple of years after the Soviet collapse, average per capita daily caloric intake in Cuba plummeted by about 1000 calories (from 2800 to 1800). Cubans needed a creative, fast, and durable solution to their food problem. The Ministry of Agriculture responded by creating the first coordinated urban agriculture program, providing access to land, especially small urban lots; extension services for training and research; supply stores; and sales outlets. In addition to government actions, nongovernmental organizations from Germany, Canada, and the United States were consulted for best urban agricultural practices and innovative organic techniques. From the start, intensive organic farming techniques and the use of biological agents for pest control were emphasized—in part, due to the expense of imported fertilizers and pesticides.

By the early 1990s, it became clear that the entire agricultural system in Cuba required radical reform. Large state-run farms that typically grew sugarcane or citrus were partitioned into Basic Units of Cooperative Production. The producers on these smaller farms have use rights to the land for an indefinite period and the freedom to choose the crops they grow and to sell their products at market prices. In cities, residents interested in growing food on empty lots or other open spaces (say, public parks) have free use of the land as long as they keep it in production. An Urban Agriculture Department was formed to change city laws so that gardeners would have

Other colonial powers left their mark on the region's cities. For example, Paramaribo, the capital of Suriname, has been described as a tropical, tulipless extension of Holland. In the British colonies, a preference for wooden whitewashed cottages with shutters is evident. Yet the British and French colonial cities tended to be unplanned afterthoughts; these port cities were built to serve the needs of the rural estates, rather than the needs of all residents. Most of them have grown dramatically over the last half century. No longer small ports for agricultural exports, cities such as Bridgetown, Barbados, and Point-a-Pitre, Guadeloupe, increasingly are oriented to welcoming cruise ships and sun-seeking tourists.

Caribbean cities and towns do have their charms and reflect a variety of cultural influences. Throughout the region, houses are often simple structures (made of wood, brick, or stucco), raised off the ground a few feet to avoid flooding and painted in pastels (Figure 5.18). Most people still get around by foot, bicycle, or public transportation; neighborhoods are filled with small shops and services that are within easy walking distance. Streets are narrow, and the pace of life is markedly slower than in North America and Europe. Even when space is tight in town, most settlements are close to the sea and its cooling breezes. An afternoon or evening stroll along the waterfront is a common activity.

REVIEW

5.3 What are the major demographic trends for this region, and what factors explain these patterns?

5.4 How did the long-term reliance on a plantation economy influence patterns of settlement in the Caribbean?

▶ **Figure 5.2.2 Urban Gardens in Havana** In central Havana, near the memorial to Cuban poet José Martí, there are numerous urban gardens, as well as public parks. Can you identify three different garden areas?

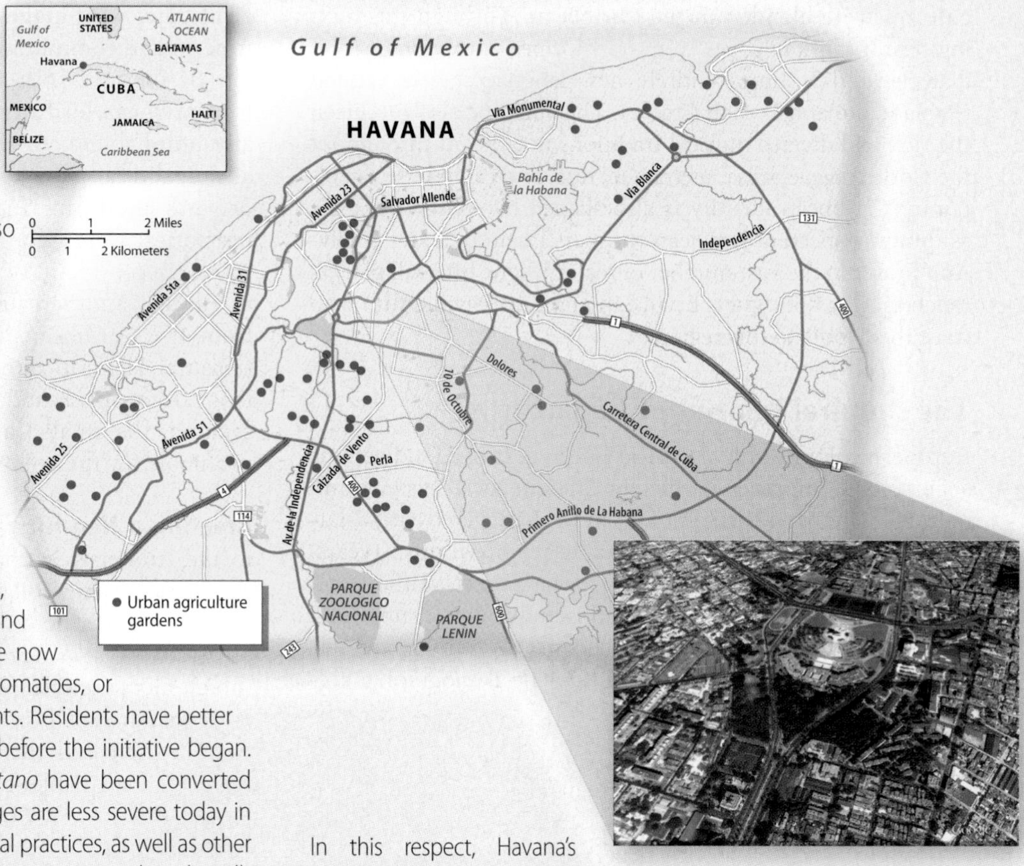

legal priority for all unused space and also to set up consulting centers in each of the administrative districts of the city (**Figure 5.2.2**). These consulting centers are a key innovation of the program, as they provide tools, seeds, compost, and advice for a population that before these reforms was unaccustomed to farming.

Havana as a Garden City These efforts have transformed Havana, improving access to a quantity and variety of fresh foods, creating new forms of self-employment, and forging new green spaces. Empty plots are now filled with raised beds growing eggplants, tomatoes, or strawberries that are tended by local residents. Residents have better access to fresh eggs, milk, and meats than before the initiative began. Even portions of Havana's *Parque Metropolitano* have been converted into food-producing plots. The food shortages are less severe today in Havana because of urban and rural agricultural practices, as well as other market-based reforms that have stimulated economic growth and small-scale entrepreneurship. Interestingly, a new form of tourism has emerged as international visitors book special tours to view the agricultural innovations found in Havana.

It is possible that, if Cuba becomes more of a market-driven economy, then new pressures will arise to convert urban farms into other uses. Yet today many city governments beyond Cuba are looking at urban farming as a way to make cities greener and the citizens within them healthier. In this respect, Havana's urban gardeners have much to share with a rapidly urbanizing world.

1. What factors led the residents of Havana to become leaders in urban agriculture?
2. Beyond increasing the food supply, what are the other advantages of growing food in cities?

Google Earth Virtual Tour Video
http://goo.gl/NiDBtY

▼ **Figure 5.18 Belize City Cottage** Residents of Belize City build their wooden cottages on stilts as protection against flooding. Shuttered cottages with metal roofs are typical throughout the English Caribbean.

Cultural Coherence and Diversity: A Neo-Africa in the Americas

Linguistic, religious, and ethnic differences abound in the Caribbean. A score of former European colonies, millions of descendants of ethnically distinct Africans and indentured workers from India and China, and isolated Amerindian communities on the mainland challenge any notion of cultural coherence.

Common historical and social processes hold the region together. European colonies, with their plantation-based economies, produced similar social structures throughout the region. The imprint of more than 7 million African slaves, creating a **neo-Africa** in the Americas in which African peoples and customs prevailed, reflects the important linkages between the Caribbean and the wider Atlantic world.

Last, in a process called **creolization,** African and European cultures were blended in the Caribbean. Through this mixing, European languages were transformed into vibrant local dialects, and at times entirely new languages were created (Papiamento and French Creole). This melding also produced the rich and diverse musical traditions now heard throughout the world: reggae, salsa, merengue, rara, and calypso. Contemporary Caribbean identity is also shaped by sports, from the astounding speed of Jamaican sprinters Usain Bolt and Shelly-Ann Fraser to the vast number of major league baseball players (such as Alex Rodriguez, Ervin Santana, and David Ortiz) who trace their roots to this region.

The Cultural Imprint of Colonialism

European colonization of the Caribbean destroyed indigenous societies and imposed completely different social systems and cultures. The arrival of Columbus in 1492 triggered a devastating chain of events that depopulated the region within 50 years. A combination of Spanish brutality, enslavement, warfare, and disease reduced the densely settled islands, which supported up to 3 million Caribs and Arawaks, into an uninhabited territory ready for the colonizer's hand. The demographic collapse of Amerindian populations occurred throughout the Americas (see Chapter 4), but the death rates were highest in the Caribbean. Only fragments of Amerindian communities survive, mostly on the rimland.

By the mid-16th century, as rival European states vied for Caribbean territory, the lands they fought for were virtually uninhabited. In many ways, this simplified their task, as they did not have to acknowledge indigenous land claims or work amid Amerindian societies. Instead, the Caribbean territories were reorganized to serve a plantation-based production system, overseen from fortified port cities such as Havana and San Juan. The Spanish galleon trade sailed from the Caribbean back across the Atlantic, laden with treasure. The critical missing element was labor. Once slave labor from Africa, and later indentured labor from Asia, was secured, the small Caribbean colonies became remarkably profitable for the colonists.

Creating a Neo-Africa African slaves were first introduced to the Americas in the 16th century, partly in response to the demographic collapse of the Amerindians. The flow of slaves continued into the 19th century. This forced migration of mostly West Africans to the Americas was only part of a much more complex **African diaspora**—the forced removal of Africans from their native area. The slave trade also crossed the Sahara to include North Africa and linked East Africa with a slave trade in the Middle East (see Chapter 6). The best-documented slave route is the transatlantic one; at least 10 million Africans landed in the Americas, and it is estimated that another 2 million died en route. More than half of these slaves were sent to the Caribbean (Figure 5.19).

This influx of slaves, combined with the extermination of nearly all the native inhabitants, recast the Caribbean as the area with the greatest concentration of African transfers in the Americas. The African source areas extended from Senegal to Angola, and slave

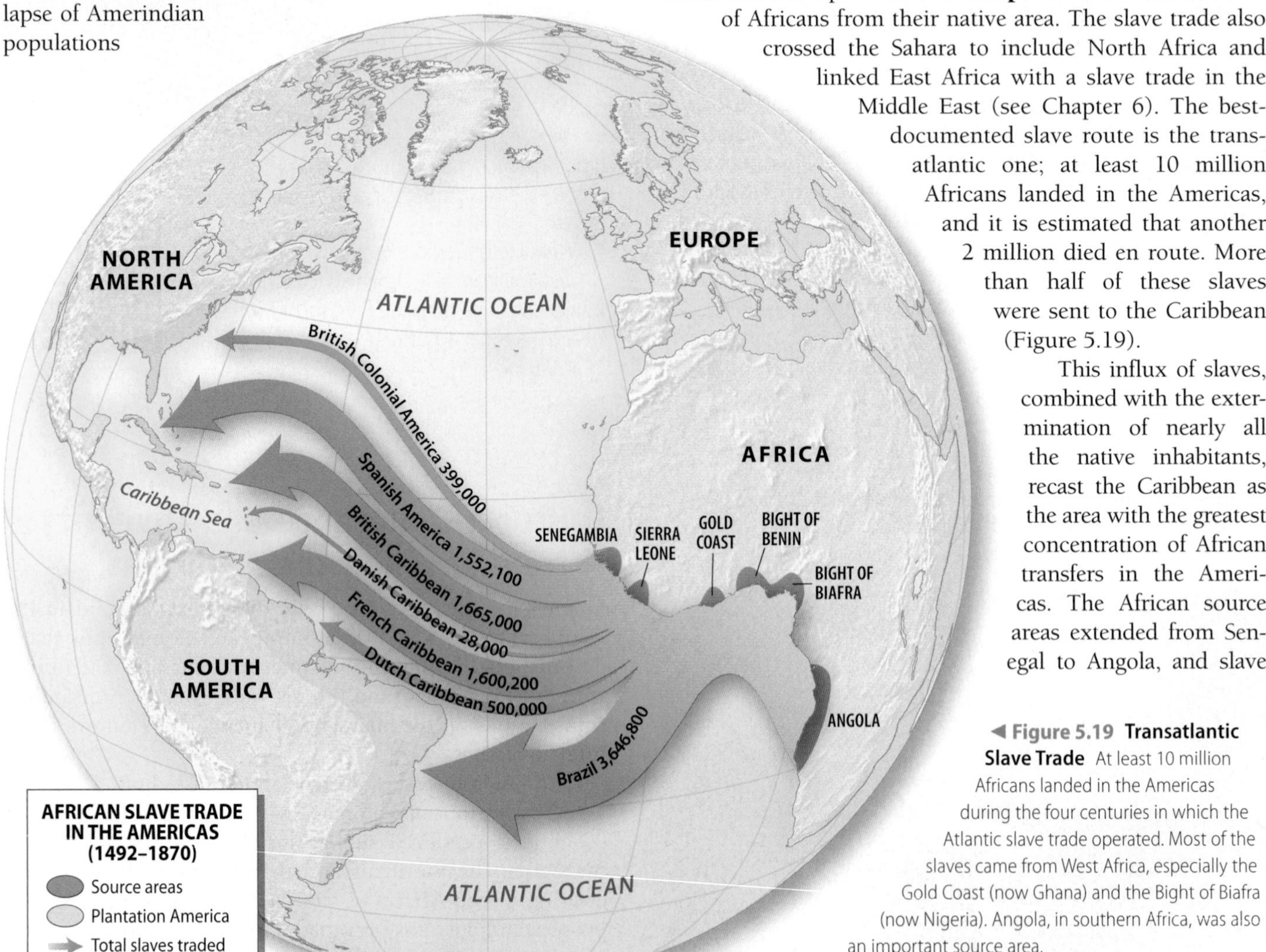

Figure 5.19 Transatlantic Slave Trade At least 10 million Africans landed in the Americas during the four centuries in which the Atlantic slave trade operated. Most of the slaves came from West Africa, especially the Gold Coast (now Ghana) and the Bight of Biafra (now Nigeria). Angola, in southern Africa, was also an important source area.

▲ **Figure 5.20 Maroons in Suriname** A woman with white clay rubbed on her face dances at the Black People's Day celebration in Paramairbo, Suriname. Surinamese Maroons (whose heritage is from runaway slave communities) recognize the first Sunday in January as Black People's Day to express their shared cultural heritage.

purchasers intentionally mixed tribal groups in order to dilute ethnic identities. Consequently, intact transfer of African religions and languages into the Caribbean did not occur; instead, languages, customs, and beliefs were blended.

Maroon Societies Communities of runaway slaves—termed **maroons** in English, *palenques* in Spanish, and *quilombos* in Portuguese—offer the most compelling examples of African cultural diffusion across the Atlantic. Hidden settlements of escaped slaves existed wherever slavery was practiced. While many of these settlements were short-lived, others have endured and allowed for the survival of African traditions, especially farming practices, house designs, community organization, and language.

The maroon communities of Suriname and French Guiana still manifest clear links to West Africa. They form the largest maroon population in the Western Hemisphere. Whereas other maroon societies were gradually assimilated into local populations, to this day the maroons of Suriname maintain a distinct identity. These runaways fled the Dutch coastal plantations in the 17th and 18th centuries, forming settlements along rivers amid the interior rainforest. Six distinct maroon tribes formed, ranging in size from a few hundred to 20,000. Clear connections to West African cultural traditions persist, including religious practices, crafts, patterns of social organization, agricultural systems, and even dress (Figure 5.20). Living relatively undisturbed for 200 years, these rainforest inhabitants fashioned a rich ritual life for themselves, involving oracles, spirit possession, and witch doctors.

More recently, pressures to modernize and extract resources have placed the maroons in direct conflict with the state and private investors. The maroons in Suriname have been directly affected by the construction of dams, gold mining operations, and logging concessions. From 1986 until 1992, a civil war raged between the maroons and the Creole-run military, in which hundreds of maroons were killed and villages destroyed. Although peace was brokered in 1992, the maroons continue to fight for legal recognition of ancestral claims to land and resources.

African Religions Linked to maroon societies, but more widely diffused, is the transfer of African religious and magical systems to the Caribbean. These patterns, another reflection of neo-Africa in the Americas, are most closely associated with northeastern Brazil and the Caribbean. In Chapter 4, we discussed how millions of Brazilians practice the African-based religions of Umbanda, Macuba, and Candomblé, along with Catholicism. Likewise, Afro-religious traditions in the Caribbean have evolved into unique forms that have clear ties to West Africa.

The most widely practiced religions are Voodoo (also Vodoun) in Haiti, Santería in Cuba, and Obeah in Jamaica. These religions have their own priesthood and unique patterns of worship. Their impact is considerable; the father

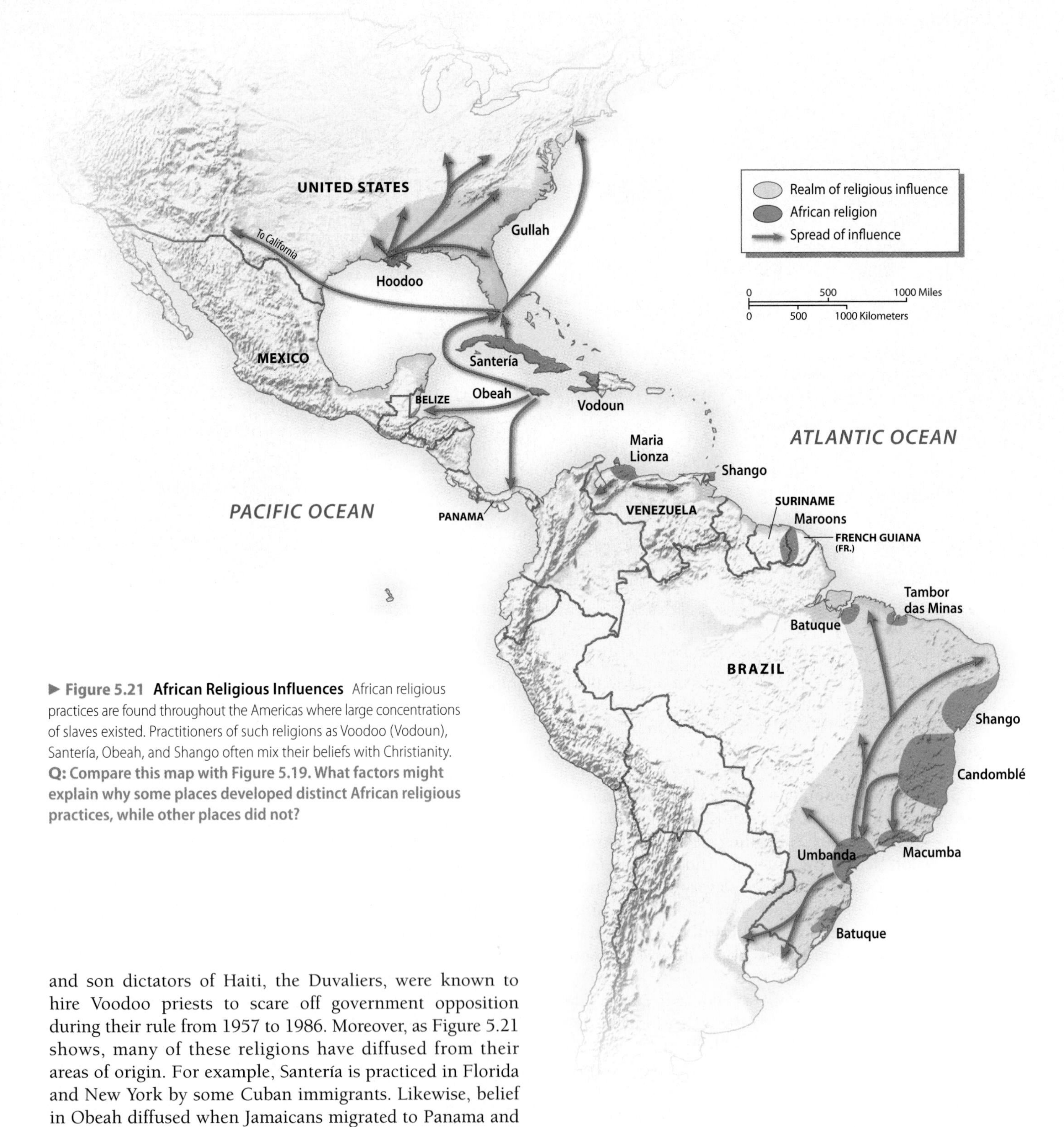

▶ **Figure 5.21 African Religious Influences** African religious practices are found throughout the Americas where large concentrations of slaves existed. Practitioners of such religions as Voodoo (Vodoun), Santería, Obeah, and Shango often mix their beliefs with Christianity. **Q: Compare this map with Figure 5.19. What factors might explain why some places developed distinct African religious practices, while other places did not?**

and son dictators of Haiti, the Duvaliers, were known to hire Voodoo priests to scare off government opposition during their rule from 1957 to 1986. Moreover, as Figure 5.21 shows, many of these religions have diffused from their areas of origin. For example, Santería is practiced in Florida and New York by some Cuban immigrants. Likewise, belief in Obeah diffused when Jamaicans migrated to Panama and Los Angeles.

Indentured Labor from Asia By the mid-19th century, most colonial governments in the Caribbean had begun to free their slaves. Fearful of labor shortages, they sought **indentured labor** (workers contracted to labor on estates for a set period of time, often several years) from South, Southeast, and East Asia.

The legacy of these indentured arrangements is clearest in Suriname, Guyana, and Trinidad and Tobago. In Suriname, a former Dutch colony, more than one-third of the population is of South Asian descent, and 16 percent is Javanese (from Indonesia, another former Dutch colony). Guyana and Trinidad were British colonies, and most of their contract labor came from India. Today half of Guyana's population and 40 percent of Trinidad and Tobago's claim South Asian ancestry. Hindu temples are found in the cities and villages, and many families speak Hindi at home. The current prime minister of Trinidad and Tobago, Kamla Persad-Bissessar, is of Indian ancestry (Figure 5.22).

Most of the former English colonies have Chinese populations of not more than 2 percent. Once these East Asian immigrants fulfilled their agricultural contracts, they often became merchants and small-business owners, positions they still hold in Caribbean society.

Creolization and Caribbean Identity

Creolization refers to the blending of African, European, and even some Amerindian cultural elements into the unique sociocultural systems found in the Caribbean. The Creole identities that have formed over time are complex; they illustrate the cultural and national identities of the region. Today Caribbean writers (V. S. Naipaul, Derek Walcott, and Jamaica Kincaid), musicians (Bob Marley, Ricky Martin, and Juan Luis Guerra), and artists (Trinidadian costume designer Peter Minshall) are internationally regarded. Collectively, these artists are representative of their individual islands and of Caribbean culture as a whole.

The story of the Garifuna people illustrates creolization at work. Now settled along the Caribbean rimland from the southern coast of Belize to the northern coast of Honduras, the Garifuna (formerly called the *Black Carib*) are descendants of African slaves who speak an Amerindian language. Unions between Africans and Carib Indians on the island of St. Vincent produced an ethnic group that was predominantly African, but spoke an Indian language. In the late 18th century, Britain forcibly resettled some 5000 Garifuna from St. Vincent to the Bay Islands in the Gulf of Honduras. Over time, the Garifuna settled along the Caribbean coast of Central America, living in isolated fishing communities. In addition to maintaining an Indian language, the Garifuna are the only group in Central America who regularly eat bitter manioc—a root crop common in lowland tropical South America. It is assumed that they acquired their taste for manioc from their exposure to Carib culture. The Afro-Indian blend that the Garifuna manifest is unique, but the process of creolization is recognizable throughout the Caribbean, especially in language and music.

Figure 5.22 South Asian Influences Trinidad and Tobago's prime minister, Kamla Persad-Bissessar, attends a celebration to mark the 165th anniversary of Indian Arrival Day in Trinidad. The prime minister is of Indian descent.

Language The dominant languages in the region are European: Spanish (25 million speakers), French (11 million), English (7 million), and Dutch (0.5 million) (Figure 5.23). However, these figures tell only part of the story. In Cuba, the Dominican Republic, and Puerto Rico, Spanish is the official language, and it is universally spoken. As for the other countries, colloquial variants of the official language exist, especially in spoken form, which can be difficult for a nonnative speaker to understand. In some cases, completely new languages have emerged. In the islands of Aruba, Bonaire, and Curaçao, Papiamento (a trading language that blends Dutch, Spanish, Portuguese, English, and African languages) is the *lingua franca,* with usage of Dutch declining. In Suriname, the vernacular language is Sranan Tongo (an amalgam of Dutch and English with many African words). Similarly, French Creole or *patois* in Haiti has constitutional status as a distinct language. In practice, French is used in higher education, government, and the courts, but patois (with clear African influences) is the language of the street, the home, and oral tradition. Most Haitians speak patois, but only the formally educated know French.

With the independence of Caribbean states from European colonial powers in the 1960s, Creole languages became politically and culturally charged with national meaning. Most formal education is taught using standard language forms, but the richness of vernacular expression and its ability to instill a sense of identity are appreciated. As linguists began to study these languages, they found that, although the vocabulary came from Europe, the syntax or semantic structure had other origins, notably from African language families. Locals rely on their ability to switch from standard to vernacular forms of speech. Thus, a Jamaican can converse with a tourist in standard English and then switch to a Creole variant when a friend walks by, effectively excluding the outsider from the conversation. This ability to switch is evident in many cultures, but is widely used in the Caribbean.

Music The rhythmic beats of the Caribbean might be the region's best-known product. This small area is the

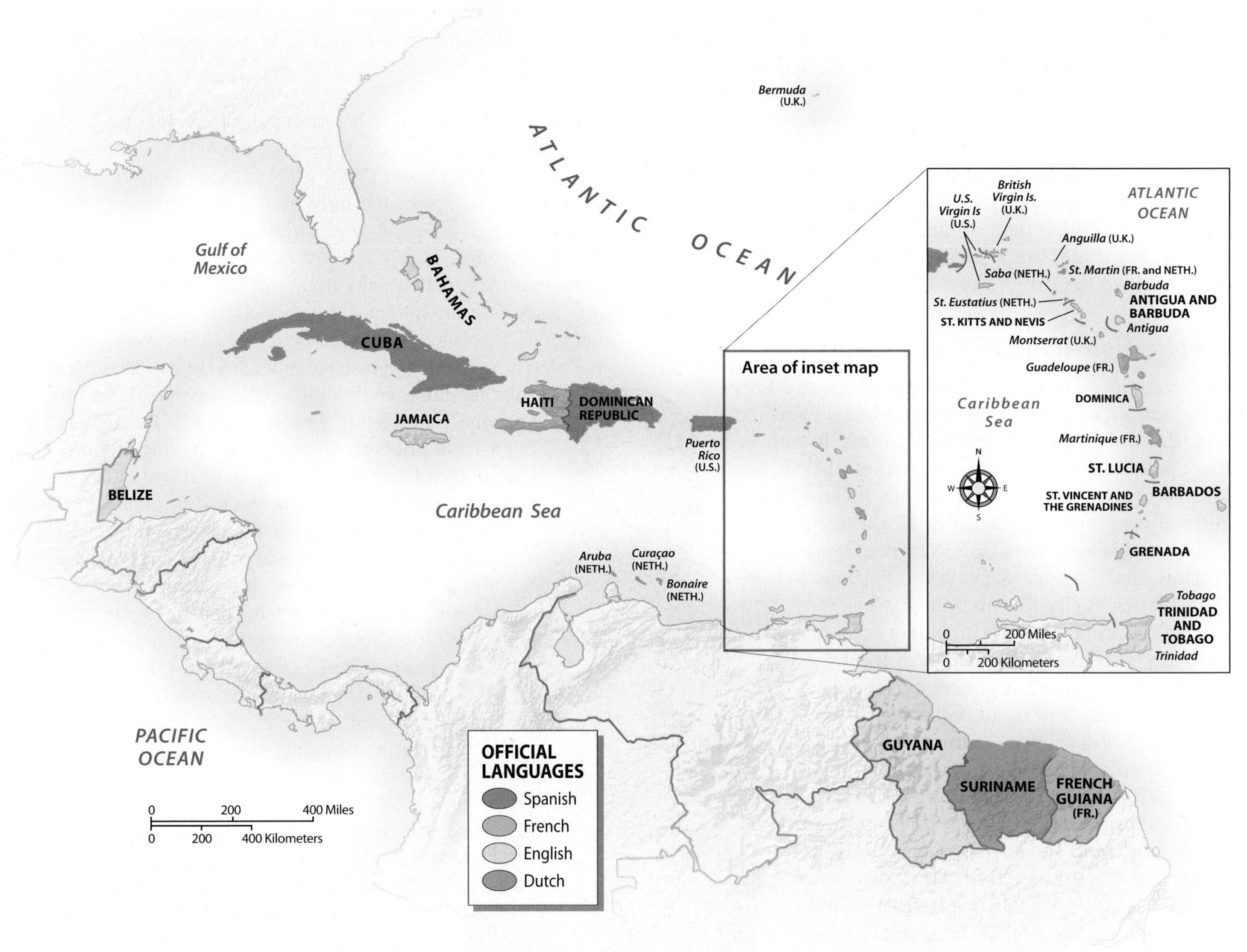

▲ **Figure 5.23 Languages of the Caribbean** Since this region has no significant Amerindian population (except on the mainland), the dominant languages are European: Spanish (25 million), French (11 million), English (7 million), and Dutch (0.5 million). However, many of these languages have been creolized, making it difficult for outsiders to understand them.

birthplace of reggae, calypso, merengue, rumba, zouk, and scores of other musical forms. The roots of modern Caribbean music reflect a combination of African rhythms with European forms of melody and verse. These diverse influences, coupled with a long period of relative isolation, sparked distinct local sounds. As circulation among Caribbean inhabitants increased, especially during the 20th century, musical traditions were grafted onto each other, but characteristic sounds remained.

The famed steel pan drums of Trinidad were created from oil drums discarded from a U.S. military base there in the 1940s. The bottoms of the cans are pounded with a sledge hammer to create a concave surface that produces different tones. During carnival (a pre-Lenten celebration), racks of steel pans are pushed through the streets by dancers while the drummers play (see Figure 5.2, page 196). So skilled are these musicians that they even perform classical music, and government agencies encourage troubled teens to learn steel pan.

The eclectic sound and the ingenious rhythms make Caribbean music very popular. It is much more than good dancing music; it is closely tied to Afro-Caribbean religions and is used in political protest. In Haiti, rara music mixes percussion instruments, saxophones, and bamboo trumpets, while weaving in funk and reggae bass lines (Figure 5.24). The songs are always performed in French Creole and typically celebrate Haiti's African ancestry and the use of Voodoo. The lyrics often address difficult issues, such as political oppression or poverty. Consequently, rara groups and other musicians have been banned at times from performing and

▲ **Figure 5.24 Haiti's Rara Music** Performed in procession, rara music is sung in patois. Considered the music of the poor, it is used to express risky social commentary. This rara band performs at a folk festival in Washington, D.C.

even forced into exile—most notably, folk singer Manno Charlemagne, who later returned to Haiti and was elected mayor of Port-au-Prince in the 1990s.

When Jamaican Bob Marley and the Wailers crashed the international music scene in the 1970s with their soulful reggae sound, it was the lyrics about poverty, injustice, and freedom that people identified with, making Marley one of the developing world's first pop superstars. Marley sang of his life in the Kingston ghetto of Trenchtown. He was a devout Rastafarian who believed that Jah was the living force, that New World Africans should look to Africa for a prince to emerge (determined to be Haile Selassie of Ethiopia), and that *ganja* (marijuana) should be consumed regularly. It was Marley's political voice as peacemaker, however, that touched so many lives. His first hit, "Simmer Down," was written to quell street violence that had erupted in Kingston in 1964. Other songs, such as "Get Up, Stand Up" and "No Woman No Cry," had a message of social unity and freedom from oppression that resonated in the 1970s. Commercial success never dulled Marley's political edge. He was wildly popular in Africa, and before his death in 1981, one of his last concerts was in Zimbabwe to mark its independence.

Sports: From Baseball to Béisbol Latin Americans are known for their love of soccer, but baseball is the dominant sport for much of the Caribbean. A byproduct of early U.S. influence in the region, baseball is the sport of choice in Cuba, Puerto Rico, and the Dominican Republic (see *Cityscapes: San Juan as Cultural Crossroads of the Americas*). Even in socialist Cuba, baseball is embraced with a fervor that would humble many U.S. fans. However, the Dominican Republic sends more players to the major leagues than any other country outside of the United States. In 2012, 28 percent of all major league baseball players were foreign-born, with Dominican players accounting for 11 percent of all major leaguers, followed by Venezuela with 8 percent. Remarkably, international players account for half of the minor league rosters, where Caribbean players also dominate.

The Dominican Republic became a talent pipeline for major league baseball due to a complex mix of talent, economic inequality, and greed. This small country has produced many baseball legends, and over the decades franchises have invested millions of dollars in training camps there. In the past two decades, however, Dominican pride in its baseball prowess has been tinged by the realities of a merciless feeder system that depends on impoverished kids, performance-enhancing drugs, fake documents, and scouts who skim a percentage of the signing bonuses. Still, the reality is that more and more young boys, who can sign contracts at age 16, see their future in baseball, rather than schooling. Even a modest signing bonus of $10,000 to $20,000 can build a nice home for a teen's family (Figure 5.25).

San Pedro de Macoris, not far from Santo Domingo, epitomizes this field of dreams. This humble sugarcane town has produced many baseball legends. It is a place of cane fields, kids on bicycles with bats and gloves, sugarcane factories, dusty baseball diamonds, and large homes of former players, such as George Bell, Pedro Guerrero, and Sammy Sosa. These houses are silent testaments to what is possible with baseball.

▼ **Figure 5.25 Caribbean Baseball** A young Cuban batter takes aim during a pick-up game in rural Cuba. Several Caribbean states have adopted baseball as their national sport—most notably, the Dominican Republic and Cuba.

Cityscapes

San Juan as Cultural Crossroads of the Americas

The 500-year-old city of San Juan is emblematic of Caribbean, Latin American, and U.S. cultures. Its stunning Caribbean setting, natural harbor, attractive beaches, and tropical climate make it an inviting place to live and to vacation; San Juan has several resorts and casinos, especially along Isla Verde near its international airport. Spanish is the dominant language and Catholicism the main religion. Puerto Rico is also a commonwealth of the United States, so it has familiar U.S. institutions, it uses the dollar, and everyone born in Puerto Rico can freely travel to the mainland. Most Puerto Ricans are bilingual in Spanish and English. Yet because San Juan is a fusion of all three regions, it often does not fit easily into any of them.

City with a History Founded by Spain in the early 1500s, its natural harbor on the island's Atlantic coast made the city a critical staging area for the transport of silver to Spain. The imposing fort of El Morro ("The Wall") controlled the harbor and protected the crown's treasure and its representatives (**Figure 5.3.1**). The fort, along with Old San Juan, was declared a World Heritage Site in 1983. Consequently, many of the historic buildings have been renovated and converted into expensive homes, hotels, shops, museums, and restaurants (**Figure 5.3.2**). Most of the former residents of Old San Juan have relocated to other parts of the city. In the historic core, you are more likely to meet someone from a cruise ship than a native of the old town.

Contemporary San Juan Since Puerto Rico was ceded to the United States in 1898 after the Spanish-American War, America's cultural imprint is everywhere. From its auto-dependent society to its highway system signage and fast-food restaurants, many elements of San Juan are like other U.S. cities, such as Miami or Tampa. San Juan's residents are usually bilingual, and many have spent time on the mainland. They love their baseball and are politically engaged. Even though residents of the island cannot vote in U.S. federal elections, the 4 million Puerto Ricans on the mainland can, which might explain why President Obama made San Juan a campaign stop during the 2012 election.

San Juan is also a Caribbean place. Culturally a mix of European and African peoples, the city's residents love Salsa and La Bomba, musical styles with clear Afro-Antillean roots. The sea is always present, and people head to the beaches on weekends to hang out under the palms, enjoy a cool drink, and eat hardy creole food. Home to one of the largest cruise ship ports in the Caribbean, San Juan's people are welcoming, in part because so much of their economy depends upon tourism. Yet they are also divided over their political future; residents range in opinion from those who want independence from the United States to those who want to become the 51st state. For the near future, however, Puerto Rico's position as commonwealth and San Juan's place as cultural crossroads are certain.

▲ **Figure 5.3.2 Old San Juan** Tourists roam the cobbled streets of Old San Juan. Because this port city was named a UNESCO World Heritage Site, many of its 18th- and 19th-century structures have been handsomely restored.

1. What aspects of San Juan's relative location and physical setting made it an important city for Spanish colonists?
2. How is San Juan a cultural blend of Caribbean, Latin American, and U.S. cultures?

Google Earth Virtual Tour Video

http://goo.gl/wm1TOH

► **Figure 5.3.1 San Juan Fort** Tourists visit Castillo San Felipe del Morro, a historic fortress that protected the harbor of Old San Juan from invasion.

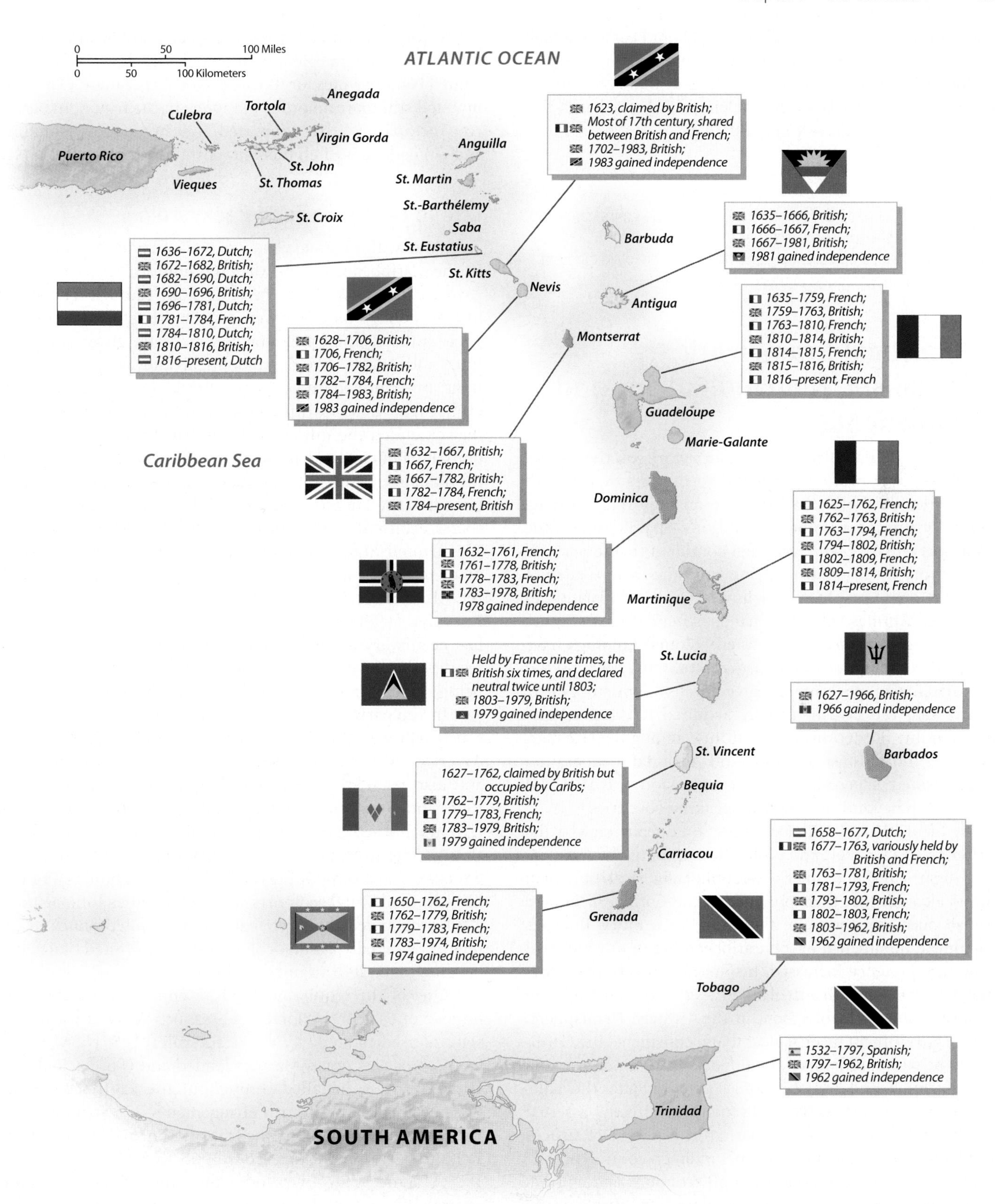

▲ **Figure 5.26 Changing Colonial Powers** The Lesser Antilles had several changes in colonial affiliation. The French and British traded islands such as Tobago, Grenada, Dominica, and Guadeloupe several times. Many of these territories gained their independence in the 1960s through the 1980s.

In an effort to clean up baseball's image, Major League Baseball has officials in the Dominican Republic investigating drug use and fraudulent papers. However, as long as there are families pushing their teenage boys and a talent pool that delivers, this transnational system is self-perpetuating.

REVIEW

5.5 What kinds of neo-African influences exist in the Caribbean, and how do they express themselves?

5.6 What is meant by creolization, and how does it explain different cultural patterns found in the Caribbean?

Geopolitical Framework: Colonialism, Neocolonialism, and Independence

Caribbean colonial history is a patchwork of rival powers dueling over profitable tropical territories. By the 17th century, the Caribbean had become an important proving ground for European colonial ambitions. Spain's grip on the region was tentative, and rivals felt confident that they could win territory by gradually moving from the eastern edge of the sea to the west. Many territories, especially islands in the Lesser Antilles, changed hands several times (Figure 5.26). In a few instances, contested colonial holdings have produced contemporary border disputes. Only recently did the Guatemalan government give up its claim to Belize, arguing that the British illegally acquired it. Also, several long-standing border disputes continue among the Guianas. However, indications of newfound regional cooperation have appeared, especially with the expansion of CARICOM membership beyond the English-speaking countries.

In the 17th and 18th centuries, Europeans viewed the Caribbean as a strategic and profitable region in which to produce sugar, rum, and spices. Geopolitically, rival European powers also felt that their presence in the Caribbean limited Spanish authority there. By the mid-19th century, Europe's geopolitical dominance in the Caribbean began to wane, just as the U.S. presence increased. Inspired by the **Monroe Doctrine**, which claimed that the United States would not tolerate European military involvement in the Western Hemisphere, the U.S. government made it clear that it considered the Caribbean to be within its sphere of influence. Even though several English, Dutch, and French colonies persisted after this date, the United States indirectly (and sometimes directly) asserted its control over the region, ushering in a period of **neocolonialism**. (Neocolonialism is the indirect control of one country or region by another through economic and cultural domination, rather than by direct military or political control as occurs under colonialism.)

Today, in an increasingly global age, even neocolonial interest can be short-lived or sporadic. The Caribbean has not attracted the level of private foreign investment seen in other regions. Moreover, as the Caribbean's strategic importance in a post–Cold War era fades, new geopolitics is shaping the region. Taiwan began wooing small Caribbean islands in the 1990s with strategic investments, in the hopes of winning United Nations votes for its cause. Not surprisingly, China has invested still more amounts of money in the region, in part to convince nations that supported Taiwan, such as Dominica and Grenada, to switch their support to China.

Life in the "American Backyard"

To this day, the United States exerts considerable influence in the Caribbean, which was commonly referred to as the "American backyard" in the early 20th century. The stated foreign policy objectives were to free the region from European authority and encourage democratic governance. Yet time and again, American political and economic ambitions undermined those goals. President Theodore Roosevelt made his priorities clear with imperialistic policies that extended the influence of the United States beyond its borders. Policies and projects such as the construction of the Panama Canal and the maintenance of open sea-lanes benefited the United, States but did not necessarily support social, economic, or political gains for the Caribbean people. The United States later offered benign-sounding development packages such as the Good Neighbor Policy (1930s), the Alliance for Progress (1960s), and the Caribbean Basin Initiative (1980s). The Caribbean view of such initiatives has been wary at best. Rather than feeling liberated, many residents believe that one kind of political dependence was being traded for another—colonialism for neocolonialism.

In the early 1900s, the role of the United States in the Caribbean was overtly military and political. The Spanish-American War (1898) secured Cuba's freedom from Spain and also resulted in Spain's ceding the Philippines, Puerto Rico, and Guam to the United States; the latter two are still U.S. territories. The U.S. government also purchased the Danish Virgin Islands in 1917, renaming them the U.S. Virgin Islands and developing the harbor of St. Thomas. French, English, and Dutch colonies were tolerated as long as these allies recognized the supremacy of the United States in the region. Avowedly against colonialism, the United States had become much like an imperial force.

One of the requirements of an empire is the ability to impose one's will, by force if necessary. When a Caribbean state refused to abide by U.S. trade rules, U.S. Navy vessels would block its ports. Marines landed and U.S.-backed governments were installed throughout the Caribbean Basin. These were not short-term engagements: U.S. troops occupied the Dominican Republic from 1916 to 1924, Haiti from 1913 to 1934, and Cuba from 1906 to 1909 and 1917 to 1922 (Figure 5.27). Even today the United States maintains several important military bases in the region, including Guantánamo in eastern Cuba. There is greater reluctance to commit troops in the area now, but as recently as 1994 and 2004, U.S. troops were sent to Haiti to suppress political violence and prevent a mass exodus of Florida-bound refugees. Also, after the Haitian earthquake in 2010, U.S. naval vessels and troops were deployed to assist in the relief effort.

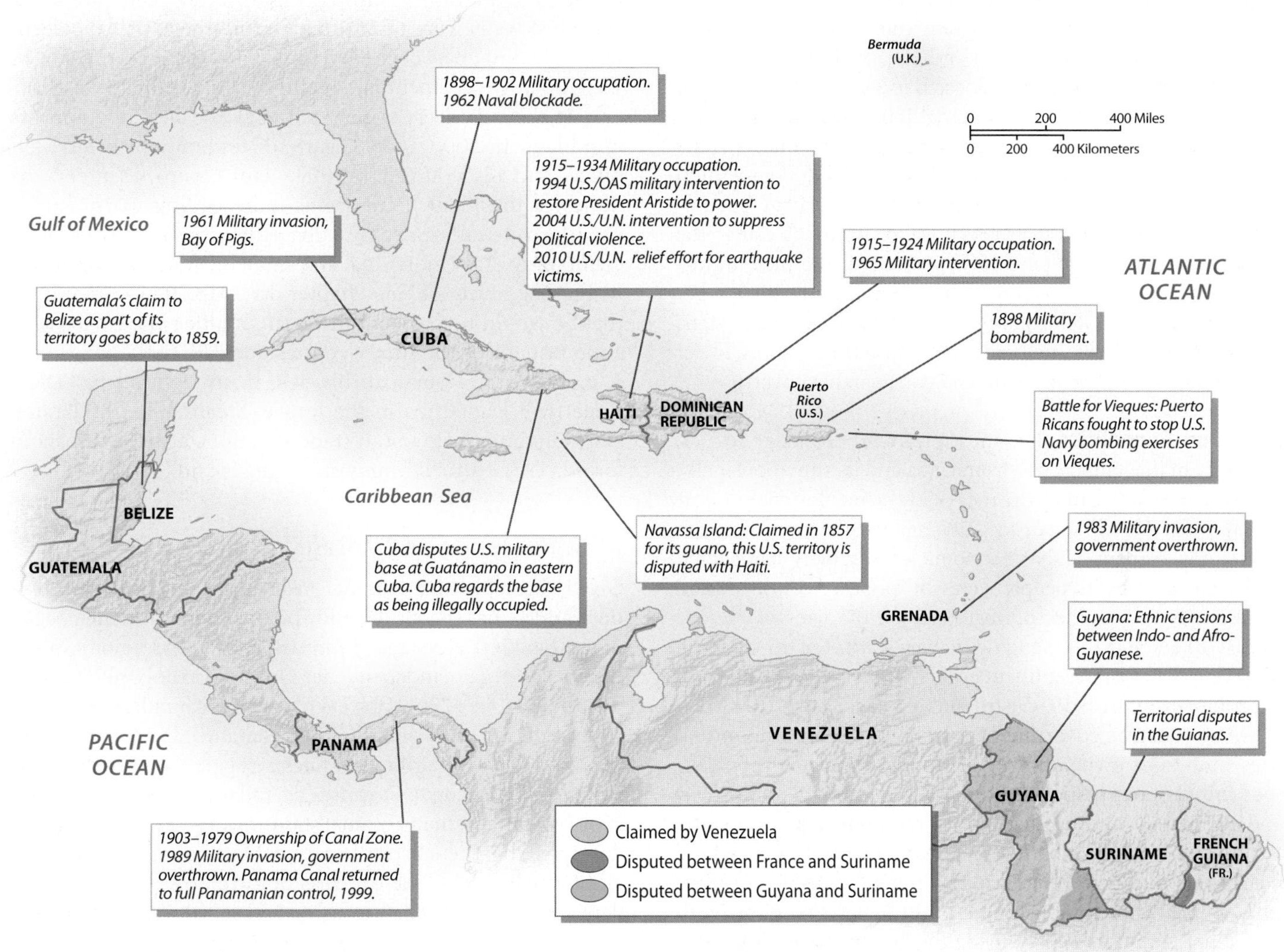

▲ **Figure 5.27 Geopolitical Issues in the Caribbean** The Caribbean was labeled the geopolitical backyard of the United States, and U.S. military occupation was a common occurrence in the first half of the 20th century. Border and ethnic conflicts also exist, most notably in the Guianas.

Many critics of U.S. policy in the Caribbean complain that business interests overwhelm democratic principles when foreign policy is determined. For example, U.S. banana companies settled the coastal plain of the Caribbean rimland and operated as if they were independent states. Sugar and rum manufacturers from the United States bought the best lands in Cuba, Haiti, and Puerto Rico. Meanwhile, truly democratic institutions remained weak, and there was little improvement in social development. True, exports increased, railroads were built, and port facilities were improved; but levels of income, education, and health remained abysmally low throughout the first half of the 20th century.

The Commonwealth of Puerto Rico Puerto Rico is both within the Caribbean and apart from it because of its status as a commonwealth of the United States. Throughout the 20th century, various Puerto Rican independence movements sought to uncouple the island from the United States. Even today residents of the island are divided about their political future. At the same time, Puerto Rico depends on U.S. investment and welfare programs; U.S. food stamps are a major source of income for many Puerto Rican families. Commonwealth status also means that Puerto Ricans can freely move between the island and the U.S. mainland, a right they actively assert. In other ways, Puerto Ricans symbolically manifest their independence; for example, they support their own "national" sports teams and send a Miss Puerto Rico to international beauty pageants. The dispute over the use of Vieques Island (a small island off the east coast) as a naval testing ground for bombing exercises became a flashpoint in U.S.-Puerto Rican relations. Ultimately, the political pressure from Puerto Ricans led President George W. Bush to close the facility in 2003. The island now supports an upscale resort.

Beginning in the 1950s, Puerto Rico led the Caribbean in the transition from an agrarian economy to an industrial one. For some U.S. officials, Puerto Rico became the model for the rest of the region. Puerto Rican President Muñoz Marín championed an industrialization program called "Operation Bootstrap." Drawn by tax incentives and cheap labor, hundreds of U.S. textile and apparel firms relocated to Puerto Rico. Over the next two decades, 140,000 industrial jobs were added, resulting in a marked increase in per capita gross national income (GNI). In the 1970s, when Puerto Rico faced stiff competition from Asian apparel manufacturers, the government encouraged petrochemical and pharmaceutical plants to relocate to the island. By the 1990s, Puerto Rico was one of the most industrialized places in the region, with a significantly higher per capita income than its neighbors. Yet it still shows many signs of underdevelopment, including rampant out-migration, lower rates of educational attainment, and considerable poverty.

Cuba and Geopolitics The most profound challenge to U.S. authority in the region came from Cuba and its superpower ally, the former Soviet Union. In the 1950s, a revolutionary effort led by Fidel Castro began in Cuba against the pro-American Batista government. Cuba's economic productivity had soared under Batista, but its people were still poor, uneducated, and increasingly angry. The contrast between the lives of average cane workers and the foreign elite was stark. Castro tapped a deep vein of Cuban resentment against six decades of American neocolonialism. In 1959, Castro took power.

After Castro's government nationalized American industries and took ownership of all foreign-owned properties, the United States responded by refusing to buy Cuban sugar and ultimately ending diplomatic relations with the state. Various U.S. trade embargoes against Cuba have existed for nearly five decades. What sealed Cuba's fate as a geopolitical enemy was its establishment of diplomatic relations with the Soviet Union in 1960, during the height of the Cold War. With the Soviet Union financially and militarily backing Castro, a direct U.S. invasion of Cuba was too risky. Instead, CIA-trained paramilitaries attempted the Bay of Pigs invasion in the spring of 1961, but failed. The fall of 1962 produced one of the most dangerous episodes of the Cold War when Soviet missiles were discovered on Cuban soil. Ultimately, the Soviet Union removed its weapons; in return, the United States promised not to invade Cuba.

Even with the end of the Cold War, when Cuba lost its financial support from the Soviet Union, it managed to reinvent itself by growing its tourism sector and courting foreign investment, especially from Spain. Castro and the late Venezuelan President Hugo Chavez also became close political allies. In 2004, they signed an important exchange agreement in which Cuba provided Venezuela with doctors, while Venezuela shipped much-needed oil to Cuba. In the long term, the Cuban government hopes that offshore oil will provide vital economic and natural resources.

In the territorial waters off the north coast of Cuba, the U.S. Geological Survey estimates there may be 4–5 billion barrels of undiscovered, but technically recoverable crude oil. This would be the largest new oil find in the Caribbean. The Cuban government has delimited its territorial waters into blocks that oil companies from Angola to Vietnam have leased for exploratory drilling (Figure 5.28). As of late 2012, three test wells were unproductive, and the expensive deepwater drilling has been temporarily suspended. However, the Cuban government is hopeful that they will strike oil.

At the very least, a new political era for Cuba appears imminent. In 2008, Fidel Castro, 82 and in poor health, left office, and his younger brother, Raúl Castro, assumed the duties of president. Raúl seems more willing to encourage small private enterprise in Cuba as he expanded licenses to cuentapropistas (individual and small businesses), described in the beginning of this chapter. In 2013, Raúl announced that he would not seek a third term in office and praised the appointment of the first vice president, Miguel Díaz-Canel. A much younger man in his 50s, many believe that Díaz-Canel will lead Cuba in the future. Meanwhile, the United States maintains its tough trade sanctions against Cuba and forbids U.S. tourists from visiting the island.

Independence and Integration

Given the repressive colonial history of the Caribbean, it is no wonder that the struggle for political independence began more than 200 years ago. Haiti was the second colony in the Americas to gain independence, in 1804 (the United States was the first, in 1776). However, the political independence of many states in the region has not guaranteed economic independence. Many Caribbean states struggle to meet the basic needs of their people. Surprisingly, today some Caribbean territories maintain their colonial status as an economic asset. For example, the French territories of Martinique, Guadeloupe, and French Guiana are overseas departments of France; residents have full French citizenship and social welfare benefits.

Independence Movements Haiti's revolutionary war began in 1791 and ended in 1804. Spanish, French, and British forces were involved, as well as factions within Haiti that had formed along racial lines. During this conflict, the island's population was cut in half by casualties and emigration; ultimately, the former slaves became the rulers. Independence, however, did not allow this jewel of the French Caribbean to prosper. Plantation America watched in horror as Haitian slaves used guerrilla tactics to gain their freedom. Fearing that other colonies might follow Haiti's lead, plantation owners in other countries were on guard for the slightest hint of revolt. For its part, Haiti did not become a leader in liberation. Slowed by economic and political problems, it was shunned by the European powers and never embraced by the states of the Spanish mainland when they became independent in the 1820s.

Several revolutionary periods followed in the 19th century. In the Greater Antilles, the Dominican Republic finally gained independence in 1844 after wresting control of the territory from Spain and Haiti. Cuba and Puerto Rico were freed from Spanish colonialism in 1898, but their independence was compromised by greater U.S. involvement. The British colonies also faced revolts, especially in the 1930s, yet it was not until the 1960s that independent states emerged from the English Caribbean. First, the larger colonies of Jamaica,

Trinidad and Tobago, Guyana, and Barbados gained their independence. Other British colonies followed throughout the 1970s and early 1980s. Suriname, the only Dutch colony on the rimland, became an autonomous territory in 1954, but remained part of the Kingdom of the Netherlands until 1975, when it declared itself an independent republic (see *People on the Move: Suriname in the Postcolonial World*).

Present-Day Colonies Britain still maintains several crown colonies in the region: the Cayman Islands, the Turks and Caicos, Anguilla, Montserrat, and Bermuda. The combined population of these islands is about 120,000 people, yet their standard of living is high, due in part to their specialization in the recently developed industry of offshore financial services. French Guiana, Martinique, and Guadeloupe are each departments of France and thus, technically speaking, not colonies. Together, they total 1 million people. The Dutch islands in the Caribbean are considered autonomous countries that are part of the Kingdom of the Netherlands. Curaçao, Bonaire, St. Maartin, Saba, and St. Eustatius make up the federation of the Netherlands Antilles. Aruba left the federation in 1986 and governs without its influence. In 2010 the federation dissolved completely with Curaçao and St. Maarten becoming independent countries and the other islands remaining with the Netherlands. Together, the population of the Dutch islands is a quarter of a million people.

Limited Regional Integration Perhaps the most difficult task facing the Caribbean is to increase economic integration. Scattered islands, a divided rimland, different

▼ **Figure 5.28 Energy Geopolitics and Offshore Drilling in Cuba** The territorial waters off Cuba's northwest shore are believed to contain oil and have been divided into blocks for test drilling. The map shows a variety of foreign companies that have leased blocks, none of which are American firms. **Q: Which countries/companies are leasing blocks, and which are not? What environmental impact might Cuban deep-water drilling have on the United States?**

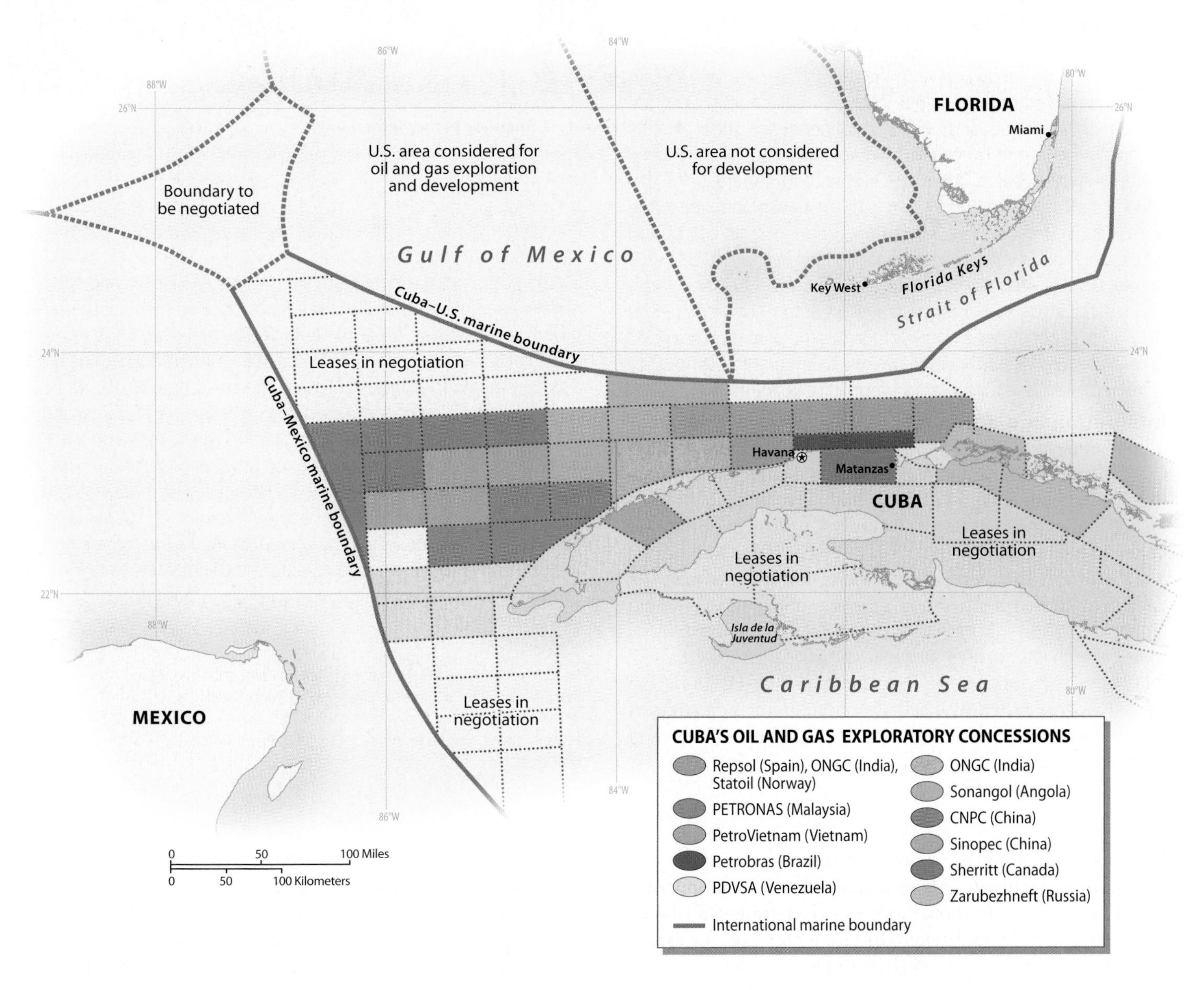

People on the Move

Suriname in the Postcolonial World

The former Dutch colony of Suriname is in northern South America, with close ties to the Caribbean region. The country is about the same size as Florida, but has only half a million residents. Its transition from Dutch colony to self-governing territory in the 1950s and to independent state in 1975 is similar to the experience of other Caribbean states. However, unlike other Caribbean states, as Suriname began its political transition to independence, nearly one-third of Suriname's population emigrated to the Netherlands to take advantage of Dutch citizenship. The majority of those migrants stayed in the Netherlands, where 330,000 first- and second-generation Surinamese now reside. Though migration to the Netherlands has slowed, the story of Surinamese settlement and integration into Dutch society is instructive.

▲ **Figure 5.4.1 Surinamese in the Netherlands** A participant wearing the clothes of the former Dutch colonists attends the Keti Koti festival in Amsterdam which commemorates the abolition of slavery in the Dutch colonies. Ethnic Surinamese established this event.

Two-Wave Migration The largest migration happened in the 1970s; a first wave came prior to independence and a second wave in the late 1970s, before the Dutch government required visitors to have a visa to travel to the Netherlands. Prior to independence, the schools mirrored the Dutch educational system, so students knew the Dutch language, history, and culture, which would appear to help them with the transition to life in the Netherlands. Unfortunately, they arrived during a difficult period when a global oil crisis and rampant unemployment strained Dutch society. Legally recognized as Dutch citizens, the Afro- and Indo-Surinamese arrivals in cities such as Amsterdam, Rotterdam, and the Hague experienced extremely high unemployment, depended upon state welfare, and seemed unlikely to integrate. In short, they were perceived as a postcolonial problem population.

Integration into Society By the 1990s, however, the image of the Surinamese in the Netherlands had grown more positive. Dutch people perceive the Surinamese as more like them (due to their language skills and background), even though they are racially and ethnically distinct (**Figure 5.4.1**). Moreover, they are better integrated into Dutch society when compared to the Turks and Moroccans, the country's other major immigrant groups. The rates of unemployment among Surinamese migrants, although higher than those of the native Dutch, have come down. Also, there are high rates of intermarriage, so that nearly half of the second-generation Surinamese had married a native Dutch person by 2001. Lastly, in terms of self-identification, three-quarters of Surinamese saw themselves as either Dutch or equally Dutch and Surinamese. Consequently, return migration to Suriname is very low, although many send remittances there and maintain contact with family members.

The postcolonial exodus from Suriname was dramatic and had a negative demographic and social impact, since so many of the territory's most educated people left. Today, however, Suriname is experiencing a minor migration boom as investment from China has also contributed to a surge in Chinese migrants (some 50,000, which is 10 percent of the country's population). In parts of Paramaraibo, the capital, Chinese food stores or restaurants can be found on nearly every block. China may even be the top provider of aid to Suriname, including military assistance, construction of low-income housing, and renewable energy. Chinese laborers brought to Suriname for new infrastructure projects have stayed (some legally, others not). They are drawn to the potential resources, the lightly settled territory, and the vision that this territory is a gateway to both the Caribbean and South America.

Source: Adapted from Mies van Niekerk, "Second Generation Caribbeans in the Netherlands: Different Migration Histories, Diverging Trajectories," *Journal of Ethnic and Migration Studies* 33, no. 7 (2007): 1063–1081; and Simon Romero, "With Aid and Migrants, China Expands Its Presence in a South American Nation," *New York Times*, April 10, 2011.

languages, and limited economic resources hinder the formation of a meaningful regional trade bloc. It is more common to see economic cooperation between groups of islands with a shared colonial background than between, for example, former French and English colonies.

During the 1960s, the Caribbean began to experiment with regional trade associations as a means to improve its economic competitiveness. The goal of regional cooperation was to improve employment rates, increase intraregional trade, and ultimately reduce economic dependence.

The countries of the English Caribbean took the lead in this development strategy. In 1963, Guyana proposed an economic integration plan with Barbados and Antigua. In 1972, the integration process intensified with the formation of the **Caribbean Community and Common Market (CARICOM).** Representing the former English colonies, CARICOM proposed an ambitious regional industrialization plan and the creation of the Caribbean Development Bank to assist the poorer states. CARICOM also oversees the University of the West Indies, with campuses in Trinidad, Jamaica, and Barbados. As important as this trade group is as an institutional symbol of collective identity, it has produced limited improvements in intraregional trade.

Today CARICOM has 13 full-member states—all of the English Caribbean and French-speaking Haiti. Other dependencies, such as Anguilla, Bermuda, the Turks and Caicos, the Cayman Islands, and the British Virgin Islands, are associate members. As CARICOM's membership grows, so does its services. In 2005, CARICOM passports began to be issued to facilitate travel between member nations. Also, during the 2007 Cricket World Cup, a CARICOM visa was issued to facilitate travel between the nine host countries for this global sporting event.

The dream of regional integration as a way to produce a more stable and self-sufficient Caribbean has yet to be realized. One scholar of the region argues that a factor limiting this regional integration is a "small-islandist ideology." For example, islanders tend to keep their backs to the sea, oblivious to the needs of neighbors. At times, such isolationism results in suspicion, distrust, and even hostility toward nearby states. Yet economic necessity dictates engagement with partners outside the region. Thus, this peculiar status of isolated proximity unfolds in the Caribbean, expressing itself in uneven social and economic development trends.

REVIEW

5.7 Which countries have had colonial or neocolonial influences in the Caribbean, and why have they engaged with the region?

5.8 What are the obstacles to Caribbean political or economic integration?

Economic and Social Development: From Cane Fields to Cruise Ships

Collectively, the population of the Caribbean, although poor by U.S. standards, is economically better off than most of Sub-Saharan Africa, South Asia, and even China. Despite periods of economic stagnation in the Caribbean, social gains in education, health, and life expectancy are significant (Table 5.2). Historically, the Caribbean's links to the world economy were tropical agricultural exports, yet several specialized industries—such as tourism, offshore financial services, and assembly plants—have challenged the dominance of agriculture. These industries grew because of the region's proximity to North America and Europe, the availability of cheap labor, and the implementation of policies that created a nearly tax-free environment for foreign-owned companies. Unfortunately, growth in these sectors does not employ all the region's displaced rural workers, so the lure of jobs in North America and Europe is still strong.

From Fields to Factories and Resorts

Agriculture used to dominate the economic life of the Caribbean. However, decades of turbulent commodity prices and decline in preferential trade agreements with former colonial states have produced more hardship than prosperity. Ecologically, the soils are overworked, and there are no frontiers into which to expand production, except for areas of the rimland. Moreover, agricultural prices have not kept pace with rising production costs, so wages and profits remain low. With the exception of a few mineral-rich territories, such as Trinidad, Guyana, Suriname, and Jamaica, most countries have systematically tried to diversify their economies, relying less on their soils and more on manufacturing and services.

Comparing export figures over time demonstrates the shift away from monocrop dependence. In 1955, Haiti earned more than 70 percent of its foreign exchange through the export of coffee; by 1990, coffee accounted for only 11 percent of its export earnings. Similarly, in 1955 the Dominican Republic earned close to 60 percent of its foreign exchange through sugar, but 35 years later sugar earned less than 20 percent of the country's foreign exchange, and pig iron exports surpassed those of sugar.

Sugar and Coffee The economic history of the Caribbean cannot be separated from the production of sugarcane. Even relatively small territories such as Antigua and Barbados yielded fabulous profits because there was no limit to the demand for sugar in the 18th century. Once considered a luxury crop, it became a popular necessity for European and North American laborers by the 1750s. It sweetened tea and coffee and made jams a popular spread for stale bread. In short, it made the meager and bland diets of ordinary people tolerable, and it also boosted caloric intake. Distilled into rum, sugar produced a popular intoxicant. Though it is hard to imagine today, individual consumption of a pint of rum a day was not uncommon in the 1800s.

Sugarcane is still grown throughout the region for domestic consumption and export. Its economic importance has declined, however, mostly because of increased competition from corn and sugar beets grown in the midlatitudes. The Caribbean and Brazil are the world's major sugar exporters. Until 1990, Cuba alone accounted for more than 60 percent of the value of world sugar exports, and the country earned 80 percent of its foreign exchange through sugar production. However, Cuba's dominance in sugar exports had more to do with its subsidized and guaranteed markets in eastern Europe and the Soviet Union than with exceptional productivity. Since 1990, the value of the Cuban sugar harvest has plummeted.

Coffee is planted in the mountains of the Greater Antilles. Haiti has been the most dependent on coffee, relying on peasant sharecroppers to tend the plants and harvest the beans. For other countries, coffee is a valued specialty commodity. Beans harvested in the Blue Mountains of Jamaica, for example, fetch two to three times the going price for similar highland coffee grown in Colombia. Puerto Rico and Cuba also are trying

Table 5.2 Development Indicators

Country	GNI per capita, PPP 2011	GDP Average Annual % Growth 2000–11	Human Development Index (2011)[1]	Percent Population Living Below $2 a Day	Life Expectancy (2013)[2]	Under Age 5 Mortality Rate (1990)	Under Age 5 Mortality Rate (2011)	Adult Literacy (% ages 15 and older) Male/Female	Gender Inequality Index (2011)[3,1]
Anguilla	12,200*	–		–	81*	–	–	--/--	–
Antigua and Barbuda	17,900	2.7	.760	–	76	27	8	98/99	–
Bahamas	29,790	0.5	.794	–	75	22	16	--/--	0.3
Barbados	19,000	0.9	.825	–	75	18	20	--/--	0.3
Belize	6,090	3.7	.702	22.0	73	44	17	--/--	0.4
Bermuda	69,900*	1.9	–	–	81*	–	–	--/--	–
Cayman	43,800*	–	–	–	81*	–	–	99/99	–
Cuba	10,200*	6.1	.780	–	78	13	6	100/100	0.3
Dominica	13,000	3.5	.745	–	74	17	12	--/--	–
Dominican Republic	9,420	5.7	.702	9.9	73	58	25	89/90	0.5
French Guiana	–	–	–	–	79	–	–	--/--	–
Grenada	10,350	2.2	.770	–	73	21	13	--/--	–
Guadeloupe	–	–	–	–	80	–	–	--/--	–
Guyana	3,460	2.5	.636	18.0	66	63	36	--/--	0.4
Haiti	1,180	0.7	.456	77.5	62	143	70	53/45	0.5
Jamaica	9,300*	–	.730	5.4	73	35	18	82/91	0.4
Martinique	–	–	–	–	81	–	–	--/--	–
Montserrat	8,500*	–	–	–	73*	–	–	--/--	–
Netherlands Antilles	–	–	–	–	77*	–	–	--/--	–
Puerto Rico	16,300*	0.0	–	–	79	–	–	90/91	–
St. Kitts and Nevis	16,470	2.5	.745	–	75	28	7	--/--	–
St. Lucia	11,220	2.9	.725	40.6	75	23	16	--/--	–
St. Vincent and the Grenadines	10,440	3.1	.733	–	72	27	21	--/--	–
Suriname	7,730	4.9	.684	27.2	71	52	30	95/94	0.4
Trinidad and Tobago	24,350	5.6	.760	13.5	71	37	28	99/98	0.3
Turks and Caicos	29,100*	–	–	–	79*	–	–	99/98*	–

[1]United Nations, *Human Development Report, 2013.*

[2]Population Reference Bureau, *World Population Data Sheet, 2013.*

[3]Gender Inequality Index—A composite measure reflecting inequality in achievements between women and men in three dimensions: reproductive health, empowerment and the labor market that ranges between 0 and 1. The higher the number, the greater the inequality. * Additional data from the *CIA World Fackbook,* 2013.

Source: World Bank, *World Development Indicators, 2013.*

to develop a niche in the gourmet coffee market. An important production distinction with coffee, in contrast to sugar, is that it is mostly grown on small farms and sold to buyers or delivered to cooperatives. Typically, farmers plant other crops between the coffee bushes so that they can meet their subsistence needs, as well as produce a cash crop. Even with this self-provisioning system, peasants often seasonally abandon their farms for work elsewhere as laborers.

The Caribbean also has other agricultural commodities. Several small states of the Lesser Antilles reply upon banana exports, although reduced preferential trade with the European Union has hurt this industry. There is also specialty production in nontraditional export commodities such as flowers or spices. A not inconsiderable cash crop is marijuana; illegal, but tolerated, it is grown for local consumption, along with some for export.

Energy Needs and Innovations The Caribbean economy depends largely upon imported oil for its energy needs. With the exception of Trinidad and Tobago, which exports oil and liquefied natural gas, the Caribbean states are net importers

▶ **Figure 5.29 Wind Farm in Puerto Rico** In 2012, Puerto Rico inaugurated the St. Isabel Wind Farm on the island's south coast near the city of Ponce. Puerto Rico plans to grow its renewable energy sector with wind and reduce its dependence upon oil. **Q: Considering the dominant wind patterns of the Caribbean, where else might wind power play a role in the region's energy production?**

of oil and highly dependent upon foreign sources. Venezuela has been one of the major providers of crude oil for Caribbean states—sometimes at below market prices, as in the case of Cuba. The Caribbean does have some oil refineries that process crude oil shipments into petroleum for domestic consumption and even some for export. But as oil prices climb, surges in food and energy costs make the small economies of this region vulnerable.

Not surprisingly, Caribbean nations have a growing interest in renewable energy. In many ways, wind energy has long been important for the Caribbean economy. After all, the entire colonial enterprise depended on the trade winds to move commodities and people across the Atlantic. But commercial wind energy is still relatively new. Puerto Rico just opened a major new wind farm on its southern coast near the city of Ponce (Figure 5.29). The Puerto Rican government intends to have 12 percent of its energy come from renewable sources by 2015. Similarly, the Los Cocos wind farm in the Dominican Republic is a major investment in wind power to serve the country's growing electricity needs. The potential for solar power is also excellent for this region.

Assembly-Plant Industrialization Another important Caribbean development strategy has been to invite foreign investors to set up assembly plants and thus create jobs. This was first tried successfully in Puerto Rico in the 1950s and was copied throughout the region. During Puerto Rico's "Operation Bootstrap," island leaders encouraged U.S. investment by offering cheap labor, local tax breaks, and, most importantly, federal tax exemptions (something only Puerto Rico can do because of its special status as a commonwealth of the United States). Initially, the program was a tremendous success, and by 1970 nearly 40 percent of the island's gross domestic product (GDP) came from manufacturing. Today 25 percent of the male labor force and 11 percent of the female labor force in Puerto Rico are employed in industry, and this sector accounts for nearly half of the island's GDP. However, competition from other states with even lower wages and the U.S. Congress's decision in 1996 to phase out many of the tax exemptions may threaten Puerto Rico's ability to maintain its specialized industrial base.

Through the creation of **free trade zones (FTZs)**—duty-free and tax-exempt industrial parks for foreign corporations—the Caribbean is an increasingly attractive location for assembling goods for North American consumers. The Dominican Republic took advantage of tax incentives and guaranteed access to the U.S. market offered through the Caribbean Basin Initiative. The Dominican Republic now has 50 FTZs, with the majority clustered around the outskirts of Santo Domingo and Santiago, the country's two largest cities (Figure 5.30).

◀ **Figure 5.30 Free Trade Zones in the Dominican Republic** A sign of globalization is the increase in duty-free and tax-exempt industrial parks in the Caribbean. The Dominican Republic, which is also a member of the Central American Free Trade Association, has 50 FTZs with foreign investors from the United States, Canada, South Korea, and Taiwan.

Firms from the United States and Canada are the most frequent investors in these zones, followed by Dominican, South Korean, and Taiwanese companies. Traditional manufacturing on the island was tied to sugar refining and rum production, whereas production in the FTZs focuses on garments and textiles. These manufacturing centers now account for three-quarters of the country's exports.

The growth in manufacturing depends on national and international policies that support export-led development through foreign investment. Certainly, new jobs are being created, and national economies are diversifying in the process, but critics believe that foreign investors gain more than the host countries. Because most goods are assembled from imported materials, there is little development of national suppliers. Although wages are often higher than local averages, they are still low compared to those in the developed world—sometimes just a few dollars a day.

Offshore Banking and Online Gambling The rise of **offshore banking** in the Caribbean is most closely associated with The Bahamas, which began this industry back in the 1920s. Offshore banking centers appeal to foreign banks and corporations by offering specialized services that are confidential and tax exempt. Places that provide offshore banking make money through registration fees, not taxes. The Bahamas was so successful in developing this sector that by 1976, the country was the third largest banking center in the world. Its dominance began to decline because competitors from the Caribbean, Hong Kong, and Singapore appeared and because there was no longer a tax advantage in arranging large international loans offshore. Concerns about corruption and laundering of drug money also hurt the islands' financial status in the 1980s, and major reforms were introduced to reduce the presence of funds gained from illegal activities. By 1998, The Bahamas' ranking among global financial centers had dropped to 15th. Still, offshore banking remains an important part of the Bahamian economy. In the 1990s, the Cayman Islands emerged as the region's leader in financial services. With a population of 50,000, this crown colony of Britain has more registered companies than inhabitants and a per capita income purchasing power parity of $54,000. In 2010, the Cayman Islands was the fifth largest banking center in the world after New York, London, Hong Kong, and Tokyo.

It is estimated that $20–$30 trillion are hidden in offshore tax havens all over the world; the Caribbean is just one location where corporations and rich individuals park their money. Each of the offshore banking centers in the Caribbean tries to develop special financial services to attract clients, such as banking, functional operations, insurance, or trusts. Bermuda, for example, is the global leader in the reinsurance business, which makes money from underwriting part of the risk of other insurance companies (Figure 5.31). The Caribbean is an attractive location for such services because of its closeness to the United States (home of many of their registered firms), client demand for these services in different countries, and the steady improvement in telecommunications that make this industry possible. The resource-poor islands of the region see providing financial services as a way to bring foreign capital to state treasuries. Envious of the economic success of The Bahamas, Bermuda, and the Cayman Islands, countries such as Antigua, Aruba, Barbados, and Belize have also developed offshore sectors. For example, in 2012 Barbados was the preferred tax haven for Canadians.

As offshore financial services were expanding in the 1990s in the Caribbean, Southeast Asia, the Pacific, and Europe (Liechtenstein and Jersey), international efforts sought to curb money laundering by threatening the suspension of privacy provisions whenever criminal activities were suspected. After the terrorist attacks on the United States in September 2001, interest in promoting know-your-customer laws as a means of tracking offshore assets of terrorist organizations was renewed. The result has been a more stringent regulatory environment that is making offshore banking less attractive to newcomers. Grenada recently announced that it is no longer a site for offshore banking.

Online gambling is the newest industry for the microstates of the Caribbean. Antigua and St. Kitts were the leaders of the region, beginning legal online gambling services in 1999. Other states soon followed; as of 2003, Dominica, Grenada, Belize, and the Cayman Islands had gambling domain sites. In 2007, the World Trade Organization deemed

◀ **Figure 5.31 Financial Services in Bermuda** Front Street in Hamilton, Bermuda, is a reflection of the territory's ties to the United Kingdom and its prosperity. Tourism and financial services in the reinsurance business explain Bermuda's wealth.

restrictions imposed on overseas Internet gambling sites by the United States to be illegal. The tiny nation of Antigua is currently seeking $3 billion from the United States as compensation for lost revenue due to illegal restrictions placed on Antigua's business.

Meanwhile, sensing a lucrative business opportunity, efforts to legalize Internet gambling in the United States moved into full gear. By 2013, the governors of Delaware, New Jersey, and Nevada all signed laws to allow online gambling in their states. Other cash-strapped states may find the potential revenues from online gambling impossible to resist. For the Caribbean, however, this would have negative repercussions for their Internet gambling business.

Tourism Environmental, locational, and economic factors converge to support tourism in the Caribbean. The earliest visitors to this tropical sea admired its clear and sparkling turquoise waters. By the 19th century, wealthy North Americans were fleeing winter to enjoy the healing warmth of the Caribbean during its dry season. Developers later realized that the simultaneous occurrence of the Caribbean dry season and the Northern Hemisphere winter was ideal for beach resorts. By the 20th century, tourism was well established, with both destination resorts and cruise lines. By the 1950s, the leader in tourism was Cuba, and The Bahamas was a distant second. Castro's rise to power, however, eliminated this sector of the island's economy for nearly three decades and opened the door for other islands to develop tourism economies.

Six countries or territories hosted two-thirds of the 21 million international tourists who came to the Caribbean in 2011: the Dominican Republic, Puerto Rico, Cuba, The Bahamas, Jamaica, and Aruba (Figure 5.32). Puerto Rico saw its tourist sector begin to grow with commonwealth status in 1952. San Juan is now the largest home port for cruise lines and the second largest cruise-ship port in the world in terms of total visitors. The Bahamas attributes most of its

▼ **Figure 5.32 The Caribbean's Global Linkages: International Tourism** The Caribbean is directly linked to the global economy through tourism. Each year more than 21 million tourists come to the islands, mostly from North America, Latin America, and Europe. The most popular destinations are the Dominican Republic, Puerto Rico, Cuba, Jamaica, The Bahamas, and Aruba.

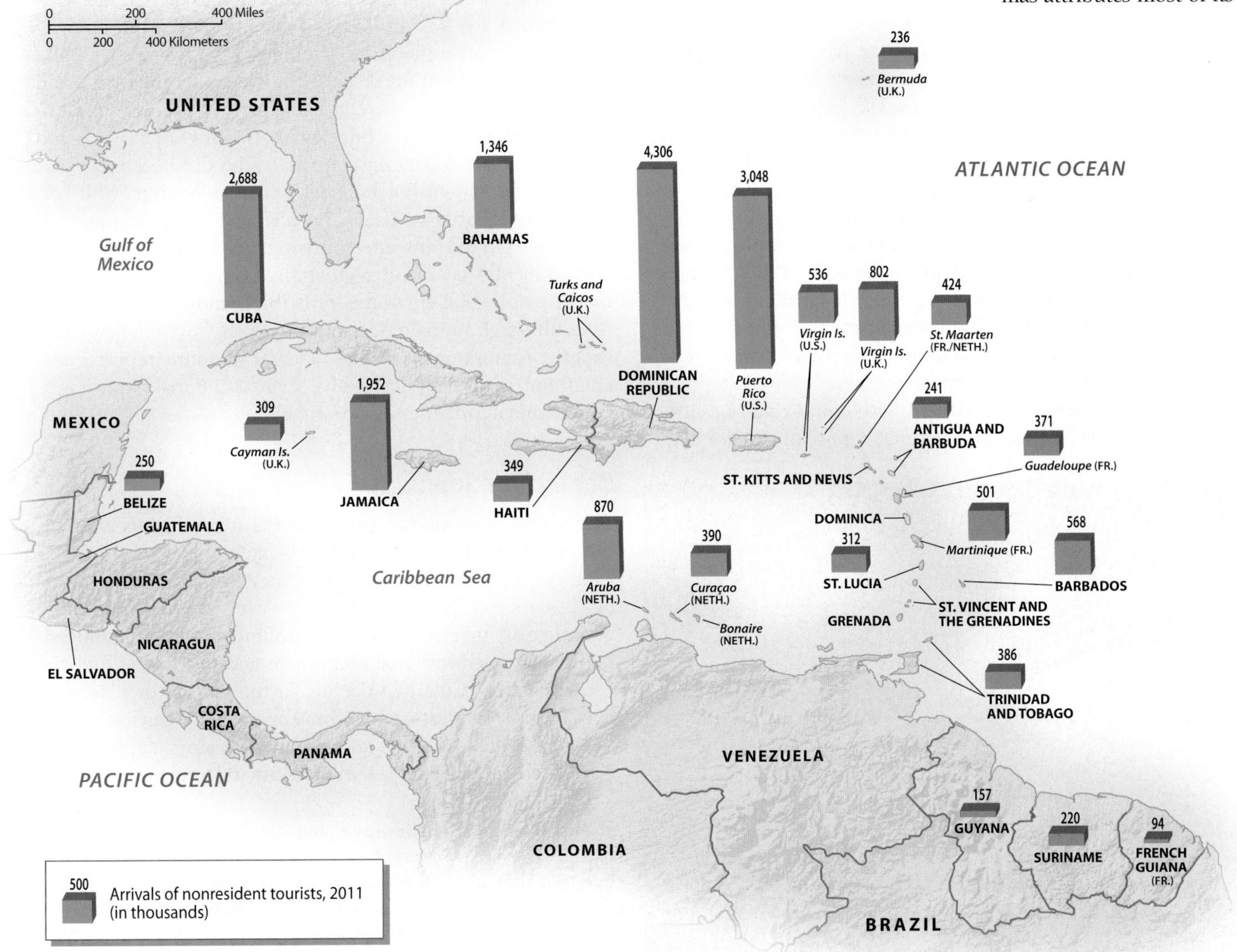

economic development and high per capita income to tourism. With 1.4 million stay-over visitors in 2011 and almost as many cruise-ship passengers, The Bahamas is another major hub for tourism in the region. Some 30 percent of the Bahamian population is employed in tourism, and tourism represents nearly half the country's GDP (Figure 5.33).

The Dominican Republic is the region's largest tourist destination, receiving over 4 million visitors in 2011, many of them Dominican nationals who live in the United States. Since 1980, tourist receipts have increased 20-fold, making tourism the leading foreign-exchange earner at more than $4 billion. Jamaica has become similarly dependent on tourism for hard currency. The vast majority of tourists to Jamaica are from the United States and the United Kingdom. While drawing nearly 2 million stay-overs in 2011, Jamaica was a port of call for another half-million cruise-ship passengers. Tourism receipts for Jamaica totaled $2 billion in 2011.

After years of neglect, Cuba has revived tourism in an attempt to earn badly needed hard currency. Tourism represented less than 1 percent of the national economy in the early 1980s. By 2011, 2.7 million tourists (mostly Canadians and Europeans) poured onto the island bring in $2.5 billion in tourism receipts. Conspicuous in their absence are travelers from the United States, forbidden to travel to Cuba because of the U.S.-imposed sanctions. With U.S. investors out of the picture, Spanish and other European investors are busy building up Cuba's tourist capacity, anticipating the day when the U.S. ban will be lifted.

As important as tourism is for the larger islands, it is often the principal source of income for smaller ones. The Virgin Islands, Barbados, the Turks and Caicos, and, recently, Belize all greatly depend on international tourists. To show how quickly this sector can grow, consider this example: When Belize began promoting tourism in the early 1980s, it had just 30,000 arrivals per year. An English-speaking country close to North America, Belize specialized in ecotourism that showcased its interior tropical forests and coastal barrier reef. By the mid-1990s, the number of land-based tourists topped 300,000, and tourism was credited with employing one-fifth of the workforce. Belize City became a port of call for day visitors from cruise ships in 2000, making it the fastest-growing tourist port in the Caribbean. Yet the influx of day visitors to this impoverished coastal town of 60,000 has done little to improve the city's infrastructure or high unemployment.

For more than four decades, tourism has been the foundation of the Caribbean economy. However, this regional industry has grown more slowly in recent years, compared to other tourism destinations in the Middle East, southern Europe, and even Central America. It seems that Americans are favoring domestic destinations, such as Hawaii, Florida, and Las Vegas, or are going to more "exotic" localities, such as Costa Rica. European tourists also seem to be staying closer to home or venturing to new locations, such as Dubai on the Persian Gulf or Goa in India. A destination's reputation can suddenly deteriorate as a result of social or natural forces. Increasingly, foreign tourists are opting to experience the Caribbean from the decks of cruise ships, rather than land-based resorts. This trend undermines the local benefits of tourism, directing capital to large cruise lines, rather than island economies.

Tourism-led growth has detractors for other reasons. It is subject to the overall health of the world economy and current political affairs. Thus, if North America experiences a recession or international tourism declines due to heightened fears of terrorism, the flow of tourist dollars to the Caribbean dries up. Where tourism is on the rise, local resentment may build as residents confront the disparity between their own lives and those of the tourists. There is also a serious problem of **capital leakage,** which is the huge gap between gross receipts and the total tourist dollars that remain in the Caribbean. Because many guests stay in hotel chains or on cruise ships with corporate headquarters outside the region, leakage of profits is inevitable. On the plus side, tourism tends to promote stronger environmental laws and regulation. Countries quickly learn that their physical environment is the foundation for success. Also, although tourism does have its costs (higher energy and water consumption, as well as demand for more imports), it is environmentally less destructive than traditional export agriculture and at present more profitable.

Social Development

The record of economic growth in the region is inconsistent, but measures of social development are generally strong. For example, Caribbean peoples have an average life expectancy of 72 years (see Table 5.2). Literacy levels are high, and there is near parity in terms of school enrollment by gender. Indeed, high levels of educational attainment and out-migration have contributed to a marked decline in the natural increase rate over the past 30 years, which hovers around 1 percent.

These demographic and social indicators explain why Caribbean nations fare well in the Human Development Index

◄ **Figure 5.33 Caribbean Cruise Ship** A ship from Carnival Cruise Lines, a British- and American-owned company, anchors at Grand Turk Island in the Turks and Caicos. Tourism is vital to many Caribbean states, but most cruise ships are owned by companies outside the region and offer relatively little direct employment for Caribbean workers.

(Figure 5.34). All ranked states (territories are not ranked) are in the high and medium human development categories. The island nation of Barbados, ranked 38th in the world in 2013, is in the very high human development category. Among the Caribbean states, Haiti has the lowest ranking, at 161st in the world, placing it in the category of low human development. Figure 5.34 also shows that many of these well-ranked states, especially Jamaica, Guyana, and St. Kitts and Nevis, have significant annual per capita flows of **remittances** (monies sent back home by migrants working overseas) entering the economy. It has been argued that remittances have become extremely important in boosting the overall level of social and economic development in the region. Despite real social gains, many inhabitants are chronically underemployed, poorly housed, and perhaps overly dependent on foreign remittances. For rich and poor alike, the temptation to leave the region in search of better opportunities remains.

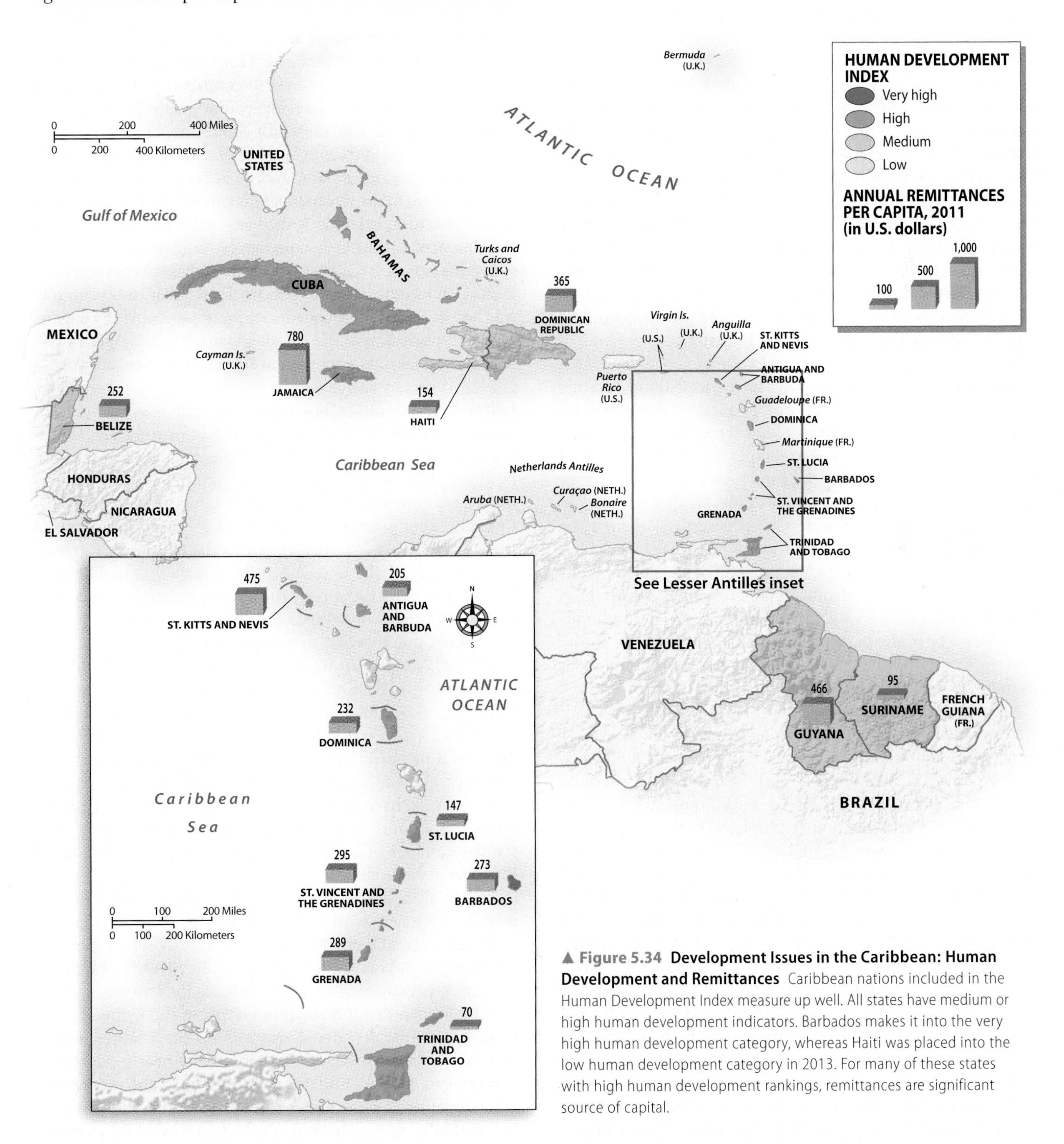

▲ **Figure 5.34 Development Issues in the Caribbean: Human Development and Remittances** Caribbean nations included in the Human Development Index measure up well. All states have medium or high human development indicators. Barbados makes it into the very high human development category, whereas Haiti was placed into the low human development category in 2013. For many of these states with high human development rankings, remittances are significant source of capital.

Status of Women The matriarchal (female-dominated) basis of Caribbean households is often singled out as a distinguishing characteristic of the region. The rural custom of men leaving home for seasonal employment tends to nurture strong and self-sufficient female networks. Women typically run the local street markets. With men absent for long periods of time, women tend to make household and community decisions. Although giving women local power, this position does not always confer status. In rural areas, female status is often undermined by the relative exclusion of women from the cash economy—men earn wages, while women provide subsistence.

As Caribbean society urbanizes, more women are being employed in assembly plants (the garment industry, in particular, prefers to hire women), in data-entry firms, and in tourism. With new employment opportunities, female labor-force participation has surged; in countries such as Barbados, Haiti, Jamaica, Puerto Rico, and Trinidad and Tobago, more than 40 percent of the workforce is female. Increasingly, women are the principal earners of cash, and they are more likely than men to complete secondary education. There are also signs of greater political involvement by women. In recent years, Jamaica, Dominica, Trinidad and Tobago, and Guyana have all had women prime ministers.

Education Many Caribbean states have excelled in educating their citizens. Literacy is the norm, and the expectation is for most people to receive at least a high school degree. In many respects, Cuba's educational accomplishments are the most impressive given the size of the country and its high illiteracy rates in the 1960s. Today nearly all adults are literate. Hispaniola is the obvious contrast to Cuba's success. Although the Dominican Republic has made strides in improving adult literacy (88 percent of adults are literate), half of all Haitian adults are illiterate. Political stability and economic growth have helped the Dominican Republic better its social conditions over the past decade. In fact, many Haitians crossed the border into the Dominican Republic because conditions, although far from ideal, are much better than in their homeland.

Education is expensive for these nations, but it is considered essential for development. Ironically, many states express frustration about training professionals for the benefit of developed countries, a phenomenon called **brain drain.** Brain drain occurs throughout the developing world, especially between former colonies and the mother countries. In the early 1980s, Jamaica's prime minister complained that 60 percent of his country's newly university-trained workers left for the United States, Canada, and Britain, representing a subsidy to these economies far greater than the foreign aid Jamaica received from them. A study by the World Bank of skilled migrants revealed that 40 percent of Caribbean immigrants living abroad were college educated. In countries such as Guyana, Grenada, Jamaica, St Vincent and the Grenadines, and Haiti, over 80 percent of the college-educated population will emigrate. No other region in the world has this many educated people leaving. To be fair, some of these migrants moved when they were young and received their education in North America and Europe. Still, other immigrants, especially health professionals, received their educations in the Caribbean and were recruited abroad because of higher wages and better opportunities. Given the small population of many Caribbean territories, each professional person lost to emigration can negatively impact local health care, education, and enterprise (see *Everyday Globalization: Caribbean Migrants and Health Care*).

The Brain Gain However, although the outflow of professionals continues to be high, many countries are experiencing a return migration of Caribbean peoples who left for work in North America and Europe in the 1960s and 1970s and are now returning after decades overseas. This **brain gain,** some argue, offers the potential for returnees to contribute to the social and economic development of a home country with the experiences they have gained abroad. Throughout the English-speaking Caribbean, crates filled with household possessions are arriving, new houses are being built, and associations of returnees are being formed. Many returnees are finally living a long-deferred dream of return, bringing with them cash, new skills, distinct life experiences, and new expectations. Sometimes, they bring adult children with them who have never lived in the Caribbean but are "returning" nevertheless to a place that they were told was home. Economic inequalities may drive people to emigrate in search of work, but the emotional attachment to places persists and explains, in part, the Caribbean emigrants' return.

Labor-Related Migration Given the region's high educational rates and limited employment opportunities, Caribbean countries have seen their people emigrate for decades. Besides influencing regional economies, this strategy impacts community and household structures as well. Historically, labor circulated within the Caribbean in pursuit of the sugar harvest—Haitians went to the Dominican Republic; residents of the Lesser Antilles journeyed to Trinidad. During the construction of the Panama Canal in the early 1900s, thousands of Jamaicans and Barbadians journeyed to the Canal Zone to work, and many remained.

After World War II, better transportation and political developments in the Caribbean produced a surge of migrants to North America. This trend began with Puerto Ricans going to New York in the early 1950s and intensified in the 1960s with the arrival of nearly half a million Cubans. Since then, large numbers of Dominicans, Haitians, Jamaicans, Trinidadians, and Guyanese have also migrated to North America, typically settling in Miami, New York, Los Angeles, and Toronto. Substantial numbers of Caribbean migrants are also in the United Kingdom, France, and the Netherlands.

Crucial in this exchange of labor from south to north is the counterflow of cash remittances. Immigrants are expected to send something back, especially when immediate family members are left behind. Collectively, remittances add up; it is estimated that $3.5 billion are sent annually to the Dominican Republic by immigrants in the United States, making remittances the country's second leading source of income. Jamaicans and Haitians remit nearly $2 billion annually to their countries. Governments and individuals alike depend on these transnational family networks. Families carefully select the household member most likely to

Everyday **Globalization**

Caribbean Migrants and Health Care

If you live on the East Coast of the United States and have ever been to a hospital or clinic or have ever required home health care, it is likely that someone from the Caribbean assisted you. In the United States, 16 percent of all health-care professionals were foreign-born in 2010. Women account for three-quarters of these immigrant health-care providers. They are doctors, nurses, technicians, and aides who overwhelmingly come from the developing world.

The Brain Drain Even though the Caribbean is demographically small, many of its health-care professionals work abroad (**Figure 5.5.1**). The Caribbean provides as many health-care workers to the United States as does all of Latin America, which is far larger. The overall quality of Caribbean education, its relatively low wages, and the fact that many people from the region speak English make the move to the United States attractive. In addition, various temporary and permanent visa categories created by the U.S. government have facilitated this move and allowed skilled laborers to be actively recruited. For the individuals and households involved, working overseas makes financial sense. But for the Caribbean region, it often results in chronic shortages of health-care personnel. Such an exodus epitomizes the brain drain.

One World Bank study found that the Caribbean lost 30 percent of its physicians from 1991 to 2004 due to emigration. About 300,000 to 500,000 Caribbean-born health-care professionals work in the United States. More than half of these people work as nursing, psychiatric, or home health aides—positions that are lower paid, have lower status, and are often the most physically demanding. Another third of Caribbean health-care workers are technicians or registered nurses (**Figure 5.5.2**). They may work anywhere in the United States, but are concentrated in the Northeast (especially metropolitan New York) and Florida.

▼ **Figure 5.5.1 Caribbean Health-Care Workers** A Jamaican nurse with her colleagues in a Massachusetts hospital.

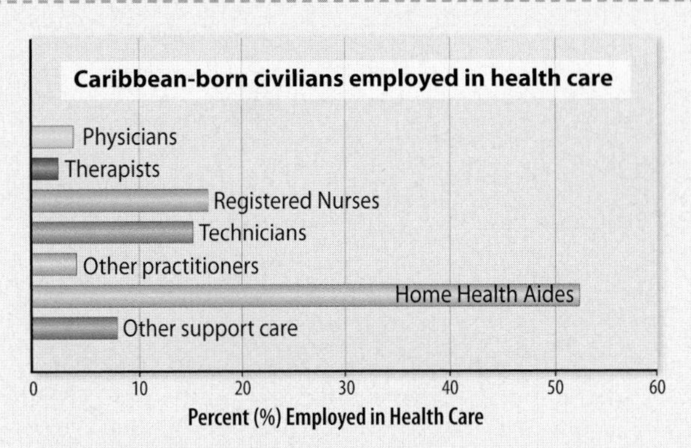

▲ **Figure 5.5.2 Caribbean Health-Care Employment** Caribbean health-care workers are over represented in the areas of home health aides, registered nurses and technicians.

Global Care Industry The rise of Caribbean health-care workers—and foreign-born health-care providers in general—is indicative of both the feminization of migration and the globalization of care industries. Working as a nurse or health-care aide is difficult work with long and irregular hours. Increasingly, the native-born U.S. population is unwilling to seek this employment, allowing women from places such as the Caribbean, the Philippines, and West Africa to fill these positions. Feminist scholars have argued that the care industry (from health-care workers to child-care and elder-care workers) has increasingly become the domain of immigrant women of color throughout North America and Europe. This segmentation of the labor market is driven by a complex mix of income inequalities, racial and gender preferences, and the demands and relatively low status of care work in general.

For the Caribbean, this means that now slightly more women migrate to the United States than men. These women often are reliable senders of remittances, which are used for everyday needs and perhaps savings or investments. Yet it is also true that these female migrants may leave their own families and children behind, due to the demands of the work and the stipulations of the visas. Many Caribbean governments worry about this trend, but it is not one that can be easily addressed.

succeed abroad in the hope that money will flow back and a base for future immigrants will be established. A Caribbean nation in which no one left would soon face crisis.

Labor-related migration is a standard practice for tens of thousands of households in the region, as well as a clear expression of the globalization of labor flows. Still, it is unlikely that migrants and their remittances can challenge the larger forces that make migration attractive in the first place. The impact of emigration and remittances is experienced primarily at the household level and is still too fragmented to represent a national force for development.

REVIEW

5.9 As the Caribbean has shifted out of agricultural dependence, what other economic sectors have emerged in the region?

5.10 What explains the relatively high levels of social development in the Caribbean given the region's relative poverty?

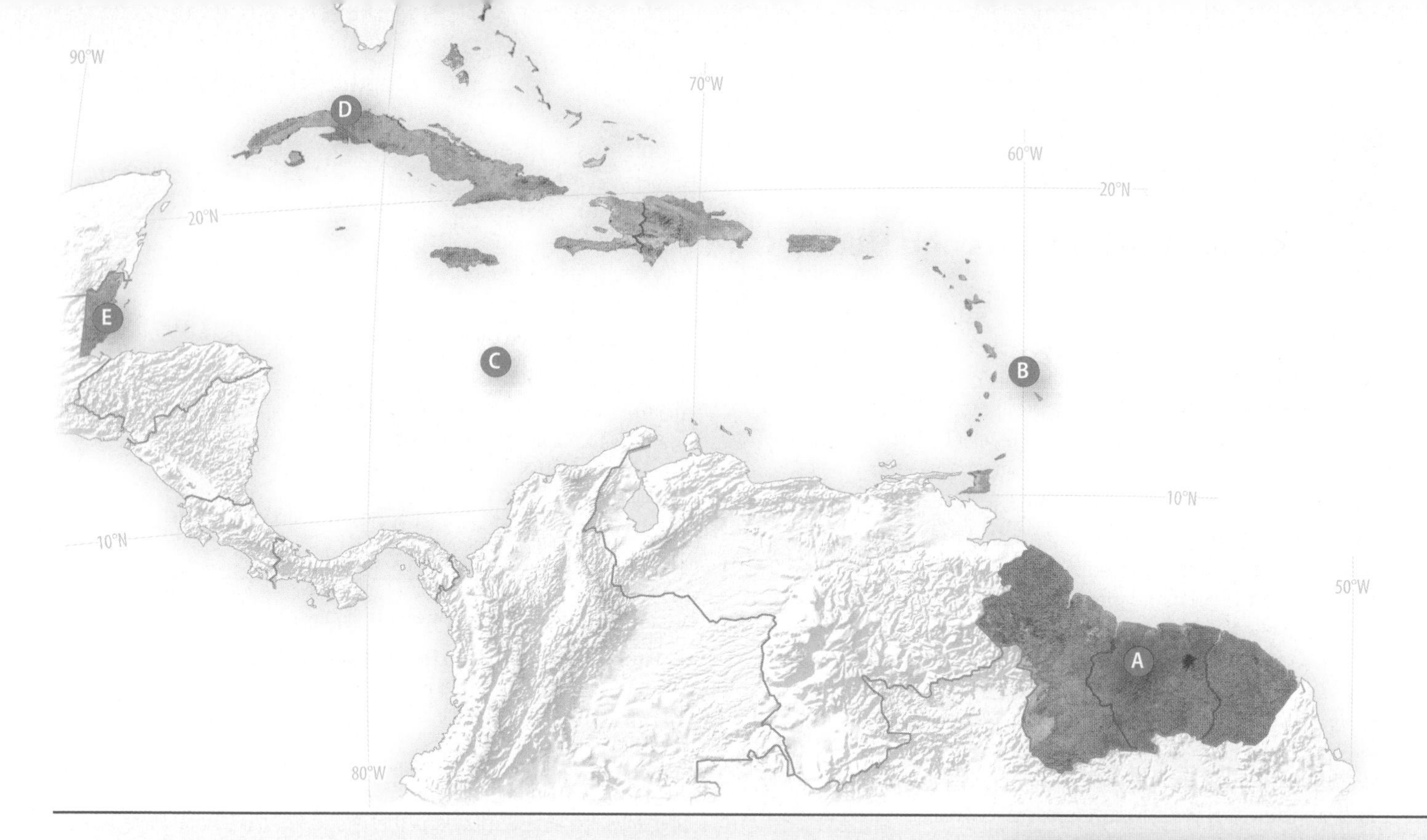

Physical Geography and Environmental Issues

This tropical region has been exploited to produce export commodities such as sugar, coffee, and bananas. The region's warm waters and mild climate attract millions of tourists. Yet serious problems with deforestation, soil erosion, and water contamination have degraded urban and rural environments. Global warming poses a serious threat to the region, with the likelihood of more-intense hurricanes and sea-level rise.

5.1 What factors contributed to the maintenance for forest in this Caribbean state?

5.2 How will global warming impact the Caribbean over the next century?

Population and Settlement

Population growth in the Caribbean has slowed over the past two decades; the average woman now has two to three children. Life expectancy is quite high, as are literacy rates. Most Caribbean people live in cities, but Caribbean cities are not large by world standards. The Caribbean region is noted for high rates of emigration, especially among the highly skilled who settle in North America and Europe.

5.3 What factors might explain the slow population growth in the Caribbean country of Barbados?

5.4. Figure 5.15 shows migration flows within and beyond the Caribbean. What are the reasons behind particular destination preferences?

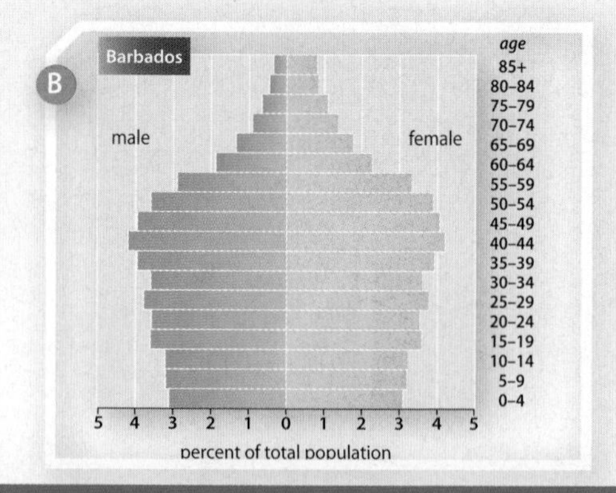

Key Terms

African Diaspora *(p. 214)*
brain drain *(p. 234)*
brain gain *(p. 234)*
capital leakage *(p. 232)*
Caribbean Community and Common Market (CARICOM) *(p. 227)*
Caribbean Diaspora *(p. 209)*
chain migration *(p. 210)*
circular migration *(p. 210)*
creolization *(p. 214)*
crisis mapping *(p. 198)*
cuentapropista (p. 194)
free trade zone (FTZ) *(p. 229)*
Greater Antilles *(p. 197)*
hurricane *(p. 199)*
indentured labor *(p. 216)*
isolated proximity *(p. 196)*
Lesser Antilles *(p. 198)*
maroon *(p. 215)*
monocrop production *(p. 210)*
Monroe Doctrine *(p. 222)*
neo-Africa *(p. 213)*
neocolonialism *(p. 222)*
offshore banking *(p. 230)*
plantation America *(p. 210)*
remittance *(p. 233)*
rimland *(p. 197)*
transnational migration *(p. 210)*

In Review • Chapter 5
The Caribbean

Cultural Coherence and Diversity

The Caribbean was forged through European colonialism and the labor of millions of Africans. The blending of African and European elements, referred to as creolization, has resulted in many unique cultural expressions in music, language, and religion. Others view the Caribbean as a neo-Africa, in which African peoples, cultures, and even some agricultural practices dominate, especially in isolated maroon communities. The Amerindian population, which was considerable at contact, was virtually eliminated in this region.

5.5 What religious practices transferred from Africa to the Caribbean, and how did they diffuse?

5.6 What explains the language diversity in the Caribbean?

Geopolitical Framework

Geopolitical Framework

Today the region contains 26 independent countries and territories. Even with the end of the Cold War, Cuba is still a geopolitical hot spot in the region. The United States still maintains its trade embargo with Cuba, but there are signs of political and economic change coming to the Caribbean's largest country. While U.S. influence is strong in this region, other countries such as Venezuela, China, and the United Kingdom also exert influence in the Caribbean.

5.7 This protected harbor was vital for Spain's control of the Caribbean. Where is this location and why was this port city important for Spain?

5.8 Why have U.S. actions in the region been considered neocolonial? What other countries are influencing the region today?

Economic and Social Development

In terms of development, the Caribbean has gradually shifted from being an exporter of primary agricultural resources (especially sugar) to a service and manufacturing economy. Employment opportunities in assembly plants, tourism, and offshore banking have replaced jobs in agriculture. The region's strides in social development, especially in education, health, and the status of women, distinguish it from other developing areas.

5.9 What environmental, economic, and locational factors contribute to the strength of the tourism economy in the Caribbean?

5.10 Is reliance upon remittances a sign of the Caribbean's isolation from or integration into the global economy?

MasteringGeography™

Looking for additional review and test prep materials? Visit the Study Area in **MasteringGeography**™ to enhance your geographic literacy, spatial reasoning skills, and understanding of this chapter's content by accessing a variety of resources, including **MapMaster**™ interactive maps, videos, RSS feeds, flashcards, web links, self-study quizzes, and an eText version of *Diversity Amid Globalization*.

Authors' Blogs

Scan to visit the author's blog for chapter updates

www.gad4blog.wordpress.com

Scan to visit the GeoCurrents blog

www.geocurrents.info

6 Sub-Saharan Africa

Mobile phones in Sub-Saharan Africa. A Ugandan mother in Lira District earns income by letting people charge their mobile phones with electricity generated from a solar panel. Mobile phones are spreading rapidly in the region, and with them are new demands for electricity from either conventional or renewable sources such as solar panels

Physical Geography and Environmental Issues

Wood is a main source of energy for this region. The Green Belt Movement, led by the late Kenyan Wangari Maathai, resulted in the planting of millions of trees by rural women throughout the region. In areas such as the Sahel, policy changes that provided ownership or incentives for the protection of trees have resulted in an increase in tree cover.

Population and Settlement

As a region, Sub-Saharan Africa is demographically young and growing. With over 900 million people, its rate of natural increase is 2.6, making it the fastest-growing world region in terms of population. It is also the region hit hardest by HIV/AIDS, which has lowered overall life expectancies in many countries.

Cultural Coherence and Diversity

Religious life is important in this region, with large and growing numbers of Muslims and Christians. With a few notable exceptions, religious diversity and tolerance have been distinctive features of this region. However, religious conflict, especially in the Sahel region, has been on the rise.

Geopolitical Framework

Most countries gained their independence in the 1960s. Since then, many ethnic conflicts have taken place, as governments have struggled for national unity within the boundaries drawn by European colonialists. The newest country in the region is South Sudan, which gained its independence from Sudan in 2011.

Economic and Social Development

The Millennium Development Goals established by the United Nations to reduce extreme poverty by 2015 will not be met by most states in the region, but progress is being made in terms of education, life expectancy, and economic growth.

The highland East African country of Uganda is filled with promise. Its rolling hills, rich soils, and abundant rainfall provide the basis for the region's agricultural productivity. Eighty-five percent of the country's 3X million people live in rural settings such as the Lira District, where subsistence farming is a way of life and families are large—the average woman has six children. Promising signs include an economy that has averaged over 7 percent growth for over a decade, the decline in violence associated with the Lord's Resistance Army, the widespread availability of mobile phones, and the discovery of substantial oil reserves that have yet to be exploited.

Many countries have reduced infant mortality, expanded basic education, and increased food production in the last two decades. One of the most transformational changes has been the rapid diffusion of cell phones in the region.

Yet the challenges are real as well. Half of the country's population is under 15 and needs to be educated. Infrastructure is lacking; just 15 percent of Ugandans are connected to the power grid, and improved roads are rare. One way that Uganda's rural energy needs are being met is through off-the-grid solar energy. Rural households typically use kerosene or candles for lighting and wood fuel for cooking. Yet with the rise of cell phone use, rural people are turning to inexpensive solar panel kits to produce the relatively small amounts of electricity needed for evening light and cell phone recharging. The solar panel kits promoted by Uganda's Energy for Rural Transformation program can pay for themselves in five months. The panels are cleaner and safer than kerosene and last for 10 years, and households can earn money by offering phone-recharging services, as shown in the opening photo. A solar market is growing in rural settings like Uganda because for now it is the most cost-effective source of off-the-grid energy. Just as the people of Sub-Saharan Africa are finding innovative uses for cellular technology, they are also becoming innovators in adapting solar technology to the small-scale needs of rural households.

In general, Africa south of the Sahara is poorer and more rural, and its population is much younger when compared to Latin American and the Caribbean. Over 900 million people reside in this region, which includes 48 states and 1 territory (Reunion off the coast of Madagascar). Demographically, this is the world's fastest-growing region (with a 2.6 percent rate of natural increase); in most countries, nearly half the population (43 percent) is younger than 15 years old. Income levels are extremely low: 70 percent of the population lives on less than $2 per day. Life expectancy is only 55 years. Such statistics and the all-too-frequent negative headlines about violence, disease, and poverty might lead to despair. Yet this is also a region of resilience, where many Africans are optimistic about the future. Local and international nongovernmental organizations, diaspora-led groups, and various government agencies are improving the quality of life in many parts of the region. In the process, many countries have reduced infant mortality, expanded basic education, and increased food production in the past two decades. One of the most transformational changes has been the rapid diffusion of cell phones, along with innovative applications that improve communication, digital payment for goods, and information sharing. In addition, since 2000 most of the national economies of the region are growing much faster than their population growth, which is a positive indicator.

Sub-Saharan Africa—that portion of the African continent lying south of the Sahara Desert—is a commonly accepted world region (Figure 6.1). The unity of this region has to do with similar livelihood systems and a shared colonial experience. No common religion, language, philosophy, or political system has ever united the area. Instead, loose cultural bonds developed from a variety of lifestyles and idea systems that evolved here. The impact of outsiders also helped to determine the region's identity. Slave traders from Europe, North Africa, and Southwest Asia treated Africans as chattel; up until the mid-1800s, millions of Africans were taken from the region and sold into slavery. In the late 1800s, the entire African continent was divided by European colonial powers, imposing political boundaries that, for the most part, remain to this day. In the postcolonial period, which began in the 1960s, Sub-Saharan African countries faced many of the same economic and political challenges.

LEARNING OBJECTIVES

After reading this chapter you should be able to:

- List the characteristics that make Sub-Saharan Africa a distinct world region.
- Summarize the major ecosystems in the region and how humans have adapted to living in them.
- Describe the factors that have made wildlife conservation and tourism important aspects of the region's economy.
- Explain the region's rapid demographic growth and describe the differential impact of HIV/AIDS upon the region.
- Describe the relationship between ethnicity and conflict in this region and the strategies for maintaining peace.
- Assess the roots of African poverty and explain why many of the fastest-growing economies in the world today are in Sub-Saharan Africa.
- List the major resources of the region, especially metals and fossil fuels, and describe how they are impacting the region's development.
- Summarize various cultural influences of African peoples within the region and globally.

▲ **Figure 6.1 Sub-Saharan Africa** Africa south of the Sahara includes 48 states and 1 territory. This vast region of rainforest, tropical savanna, and desert is home to 900 million people. Much of the region consists of broad plateaus ranging from 1600 to 6500 feet (500 to 2000 meters) in elevation. Although the population is growing rapidly, the overall population density of Sub-Saharan Africa is low. Considered one of the least developed regions of the world, it remains an area rich in natural resources.

When setting this particular regional division, the major question is how to treat North Africa. Some scholars argue for treating the African continent as one world region because the Sahara has never formed a complete barrier between the Mediterranean north and the rest of the African landmass. Regional organizations such as the African Union are modern expressions of this continental unity. However, North Africa is generally considered more closely linked, both culturally and physically, to Southwest Asia. Arabic is the dominant language and Islam the dominant religion of North Africa. Consequently, North Africans feel more closely connected to the Arab hearth in Southwest Asia than to the Sub-Saharan world.

In this chapter, we focus on the states south of the Sahara. We preserve political boundaries when delimiting the region so that the Mediterranean states of North Africa, along with Western Sahara and Sudan, are discussed with Southwest Asia. Before the independence of South Sudan from Sudan in 2011, Sudan was Africa's largest state by area. In the more populous and powerful north, Muslim leaders crafted an Islamic state that is culturally and politically oriented toward North Africa and Southwest Asia. The new country of South Sudan, however, has more in common with the Christian and animist groups in Sub-Saharan Africa. South Sudan, along with the Sahelian states of Mauritania, Mali, Niger, and Chad, form the northern boundary of the countries discussed in this chapter.

▼ **Figure 6.2 Subsistence Farmers in Tanzania** A family from Iringa, Tanzania, pose in their garden. Many African households grow food and tend animals for their own consumption, selling any surplus at local markets.

The region is culturally complex, with dozens of languages spoken in some large states. Consequently, most Africans understand and speak several languages. Ethnic identities do not conform to the political divisions of Africa, sometimes resulting in deadly ethnic conflict, such as in Rwanda in the mid-1990s and Nigeria today. Nevertheless, throughout the region peaceful coexistence among distinct ethnic groups is the norm. The cultural significance of European colonizers cannot be ignored: European languages, religions, educational systems, and political ideas were adopted and modified. Yet the daily rhythms of African life are often far removed from the industrial or postindustrial world. Most Africans engage in subsistence and cash-crop agriculture (Figure 6.2). Women in particular are charged with tending crops and procuring household necessities. Male roles revolve around tending livestock and participating in the public life of the village and market.

The influence of African peoples outside the region is great, especially considering that human origins are traceable to this part of the world. In historic times, the legacy of the slave trade resulted in the transfer of African peoples, religious systems, and even musical traditions throughout the Western Hemisphere. Even today, African-based religious systems are widely practiced in the Caribbean and Latin America, especially in Brazil. American jazz, Brazilian samba, and Cuban rumba would not exist but for the influence of Africans. The popularity of "world music" has created a large international audience for performers such as Senegal's Youssou N'Dour and South Africa's Ladysmith Black Mambazo. Nigeria is the movie capital of the region and is second only to India in terms of annual films produced. Through music and the arts, Sub-Saharan Africa has exerted a significant influence on global culture.

African economies are growing: In 2012, over one-third of the countries saw growth rates of 6 percent or higher. This growth is attributable to high commodity prices, robust domestic demand, rising exports, and steady remittance flows. Yet Sub-Saharan Africa's overall economy is marginal when compared with the rest of the world. According to the World Bank, the region's economic output in 2011 amounted to less than 2 percent of global output, even though the region contains 13 percent of the world's population. Moreover, the gross national product of just one country, South Africa, accounts for one-third of the region's total economic output. The region's persistent poverty and the unnecessary vulnerability of its people have drawn the concern of the global community.

Many scholars feel that Sub-Saharan Africa has benefited little from its integration (both forced and voluntary) into the global economy. Slavery, colonialism, and export-oriented mining and agriculture served the needs of consumers outside the region, while failing to improve domestic food supplies, infrastructure, and standards of living.

Foreign assistance in the postcolonial years initially improved agricultural and industrial output, but also led to mounting foreign debt and corruption, which over time have undercut the region's economic gains. Ironically, many of the scholars and politicians who worry about the lack of benefit

▲ **Figure 6.3 Victoria Falls** The Zambezi River descends over Victoria Falls. A fault zone in the African plateau explains the existence of a 360-foot (110-meter) drop. The Zambezi has never been important for navigation, but it is a vital supply of hydroelectricity for Zimbabwe, Zambia, and Mozambique.

from economic integration also worry that negative global attitudes about the region have produced a pattern of neglect. The idea of debt forgiveness for Africa's poorest states is steadily being applied as a strategy to improve access to basic services and education. Private capital investment in Sub-Saharan Africa is growing, although it lags behind that in other developing regions. Much of the new foreign investment has gone into extracting oil and mineral wealth in states such as South Africa, Nigeria, Angola, Chad, and Equatorial Guinea. This direct investment comes from a range of countries, from traditional investor states such as the United States, France, and the United Kingdom to important newcomers such as China and Malaysia.

The past few years have witnessed a surge in philanthropic outreach to the region. Rock star Bono of U2 and the Bill and Melinda Gates Foundation have led the One Campaign, directing millions of dollars to support health care, disease prevention, education, and poverty reduction in Sub-Saharan Africa and other developing areas. Many observers believe that through a combination of internal reforms, better governance, foreign assistance, and foreign investment in infrastructure and technology, social and economic gains are possible for Sub-Saharan Africa. Such investments in social development are paying off, as a dramatic decline in child mortality was achieved between 1990 and 2011.

Physical Geography and Environmental Issues: The Plateau Continent

Sub-Saharan Africa is the largest landmass straddling the equator. It is vast in scale, and its physical environment is remarkably beautiful. Called the *plateau continent,* the African interior is dominated by extensive uplifted areas that resulted from the breakup of **Gondwana**, an ancient mega-continent that included Africa, South America, Antarctica, Australia, Madagascar, and the Arabian Peninsula. Some 250 million years ago, it began to split apart through the forces of continental drift. As this process unfolded, the African landmass experienced a series of continental uplifts that left much of the area with vast elevated plateaus. The highest areas are found on the eastern edge of the continent, where the Great Rift Valley forms a complex upland area of lakes, volcanoes, and deep valleys. In contrast, lowlands prevail in West Africa (see Figure 6.1).

The landscape of Sub-Saharan Africa offers a palette of intense colors: deep red soils studded with bright green food crops; the blue tropical sky; golden savannas that ripple with the movement of animal herds; dark rivers meandering through towering rainforests; and sun-drenched deserts. Amid this beauty, however, are relatively poor soils, persistent tropical diseases, and frequent droughts. Large areas exist with great potential for agricultural development, especially in southern Africa, and throughout the continent significant water resources, biodiversity, and mineral wealth abound.

Plateaus and Basins

A series of plateaus and elevated basins dominates the African interior and explains much of the region's unique physical geography. Generally, elevations increase toward the south and east of the continent. Most of southern and eastern Africa lies well above 2000 feet (600 meters), and sizable areas sit above 5000 feet (1500 meters). These areas are typically referred to as High Africa; Low Africa includes West Africa and much of Central Africa. The high plateaus in countries such as Kenya, Zimbabwe, and Angola are noted for their cooler climates and relatively abundant moisture. Steep escarpments form where the plateaus abruptly end, as illustrated by the majestic Victoria Falls on the Zambezi River (Figure 6.3). Much of southern Africa is rimmed by a landform called the **Great Escarpment** (a steep cliff separating the coastal lowlands from the plateau uplands), which begins in southwestern Angola and ends in northeastern South Africa. South Africa's Drakensberg Range (with elevations reaching 10,000 feet, or 3100 meters) rise up from the Great Escarpment. Because of this landform, coastal plains tend to be narrow, with few natural harbors, and river navigation is impeded by a series of falls. Such landforms proved to be a barrier for European colonial settlement in the interior of the continent and, in part, explain the prolonged colonizing period.

Though Sub-Saharan Africa is an elevated landmass, it has few significant mountain ranges. The one extensive area of mountainous topography is in Ethiopia, which lies in the northern portion of the Rift Valley zone. Yet even there the dominant features are high plateaus intercut with deep valleys, rather than actual mountain ranges. Receiving heavy rains in the wet season, the Ethiopian Plateau is densely settled and forms the headwaters of several important rivers—most notably, the Blue Nile, which joins the White Nile at Khartoum, Sudan.

A discontinuous series of volcanic mountains, some of them quite tall, is associated with the southern half of the Rift Valley. Kilimanjaro at 19,000 feet (5900 meters) is the continent's tallest mountain (Figure 6.4), and nearby Mount Kenya (17,000 feet, or 5200 meters) is the second tallest. The Rift Valley itself reveals the slow, but inexorable progress of geological forces. Eastern Africa is slowly being torn away from the rest of the continent, and within some tens of millions of years it will form a separate landmass. Such motion has already produced a great gash across the uplands of eastern Africa, much of which is occupied by elongated and extremely deep lakes (most notably, Nyasa, Malawi, and Tanganyika). In central East Africa, this rift zone splits into two separate valleys, each of which is flanked by volcanic uplands. Between the eastern and western rifts lies a bowl-shaped depression, the center of which is filled by Lake Victoria—Africa's largest body of water. Not surprisingly, some of the densest areas of settlement are found amid the fertile and well-watered soils that border the Rift Valley.

Watersheds Africa south of the Sahara conspicuously lacks the broad, alluvial lowlands that influence patterns of settlement throughout other regions. The four major river systems are the Congo, Nile, Niger, and Zambezi. Smaller rivers—such as the Orange in South Africa; the Senegal, which divides Mauritania and Senegal; and the Limpopo in Mozambique—are locally important, but drain much smaller areas. Ironically, most people think of this region as suffering from water scarcity and tend to discount the size and importance of the watersheds (or catchment areas) that these river systems drain.

The Congo River (or Zaire) is the largest watershed in the region in terms of drainage and volume of flow. It is second only to South America's Amazon River in terms of annual flow. The Congo flows across a relatively flat basin that lies more than 1000 feet (300 meters) above sea level, meandering through Africa's largest tropical forest, the Ituri (Figure 6.5). Entry from the Atlantic into the Congo Basin is prevented by a series of rapids and falls, making the Congo River only partially navigable. Despite these limitations, the Congo River has been the major corridor for travel within the Republic of the Congo and the Democratic Republic of the Congo (formerly Zaire); the capitals of both countries, Brazzaville and Kinshasa, rest on opposite sides of the river.

The Nile River, the world's longest, is the lifeblood of Egypt and Sudan. Yet this river originates in the highlands of the Rift Valley zone and is an important link between North and Sub-Saharan Africa. The Nile begins in the lakes of the Rift Valley zone (Victoria and Edward) before descending into a vast wetland in South Sudan known as the Sudd. The Sudd is

▶ **Figure 6.4 Mount Kilimanjaro** Rising above the tropical plateau and capped in snow, Africa's highest peak is a popular destination for tourists who aspire to reach its lofty heights.

▲ **Figure 6.5 Congo River** Africa's largest river by volume, the mighty Congo River flows through the Ituri rainforest in the Democratic Republic of the Congo.

one of the great wetlands of the world. It averages over 30,000 square kilometers, but can expand to four times that size during the rainy season. Navigation in this region is tricky, but settlers have long been drawn to this area's abundant water supply and aquatic life (Figure 6.6). Agricultural development projects in the 1970s increased the agricultural potential of the Sudd, especially its peanut crop. Unfortunately, three decades of civil war in South Sudan ravaged this area, turning farmers and herders into refugees and undermining the productive capacity of this important ecosystem. One of the development objectives in the new nation of South Sudan is to harness the potential of the Sudd. A more extensive discussion of the Nile River appears in Chapter 7.

For the energy-poor nation of Ethiopia, harnessing the hydroelectric potential of the Blue Nile, a major tributary of the Nile, is key to the country's energy needs. The country opened the Tekeze Dam in 2009 on a tributary of the Nile in northern Tigray Province to provide water for irrigation and electricity. In 2011, the ambitious Grand Renaissance Dam project began on the Blue Nile. When completed, it will be the largest hydroelectric project in the region and supply much of Ethiopia's energy needs. Because the dam is located near the border of Sudan, Ethiopia's downstream neighbors, Sudan and Egypt, are concerned about how this project will affect their overall supply of water. Ethiopia's response is that hydropower along the Blue Nile is an essential and sustainable approach to begin supplying the energy needs of the country's 90 million people.

Like the Nile, the Niger River is the critical source of water for two otherwise arid countries: Mali and Niger. Originating in the humid Guinea highlands, the Niger flows first to the northeast and then spreads out to form a huge inland delta in Mali before making a great bend southward at the margins of the Sahara near Gao. On the banks of the Niger River are the capitals of Mali (Bamako) and Niger (Niamey), as well as the historic city of Tombouctou (Timbuktu). After flowing through the desert, the Niger River returns to the humid lowlands of Nigeria, where the Kainji Reservoir temporarily blocks its flow to produce electricity for Africa's most populous state.

The considerably smaller Zambezi River originates in Angola and flows east, spilling over an escarpment at Victoria Falls and finally reaching Mozambique and the Indian Ocean. More than other rivers in the region, the Zambezi has been a major supplier of commercial energy for decades. Two of Sub-Saharan Africa's early hydroelectric installations are here: The Kariba Dam is on the border of Zambia and Zimbabwe, and the Cabora Bassa Dam is in Mozambique.

Soils With a few major exceptions, Sub-Saharan Africa's soils are relatively infertile. Generally speaking, fertile soils are young soils, deposited in recent geological time by rivers, volcanoes, glaciers, or windstorms. In older soils—especially those located in moist tropical environments—natural processes tend to wash out most plant nutrients over time. Over most of Sub-Saharan Africa, the agents of soil renewal have largely been absent; the region has few alluvial lowlands where rivers periodically deposit fertile silt, and it did not experience significant glaciation in the last ice age, as did North America.

Portions of Sub-Saharan Africa are, however, noted for their natural soil fertility, and not surprisingly these areas support denser settlement. Some of the most fertile soils are in the Rift Valley, enhanced by the volcanic activity associated with the area. The population densities of rural Rwanda and Burundi, for example, are partially explained by the highly productive volcanic soils. The same can be said for highland Ethiopia, which supports the region's second largest population, with nearly 90 million people. The Lake Victoria

▼ **Figure 6.6 Settlement in the Sudd** The Sudd in South Sudan is one of the region's largest wetlands, and during the rainy season the Nile River can expand the wetlands to four times it's size.

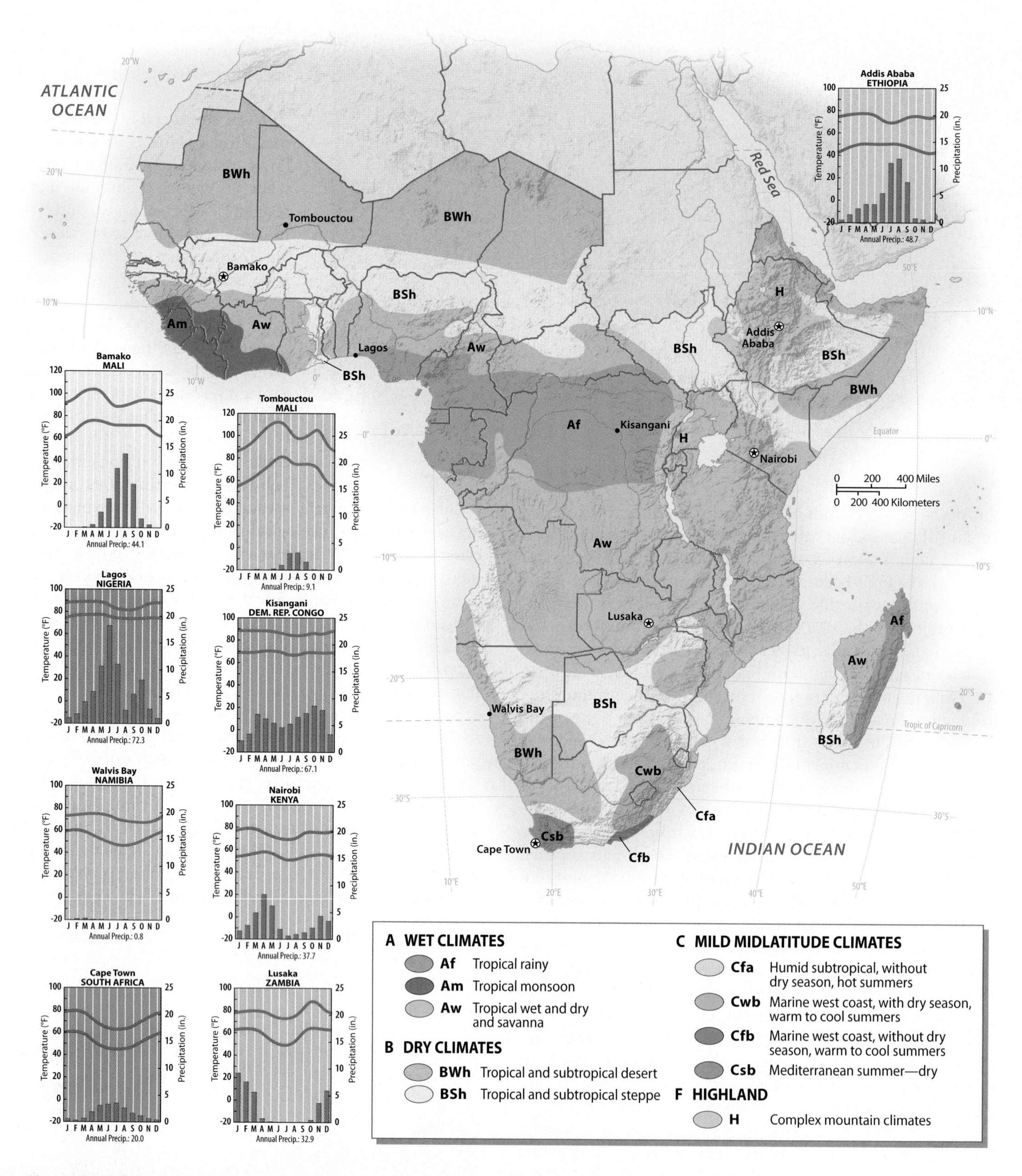

▲ **Figure 6.7 Climate of Sub-Saharan Africa** Much of the region lies within the tropical humid and tropical dry climatic zones; thus, the seasonal temperature changes are not great. Precipitation, however, varies significantly. Compare the distinct rainy seasons in Lusaka and Lagos: Lagos is wettest in June, and Lusaka receives most of its rain in January. Although West and Central Africa have important tropical forests, much of the territory is tropical savanna.

lowlands and central highlands of Kenya also are noted for their sizable populations and productive agricultural bases.

The drier grasslands and semidesert areas have a soil type called *alfisols*. High in aluminum and iron, these red soils have greater fertility than comparable soils found in wetter zones. This helps to explain the tendency of farmers to plant in drier areas, such as the Sahel, even though they risk exposure to drought. With irrigation, many agronomists suggest that the southern African countries of Zambia and Zimbabwe could greatly increase commercial grain production on these soils.

Climate and Vegetation

Sub-Saharan Africa lies in the tropical latitudes. Beginning just north of the Tropic of Cancer, crossing the equator, and extending past the Tropic of Capricorn in the south, it is the largest tropical landmass on the planet. Only the far south of the continent extends into the subtropical and temperate belts. Much of the region averages high temperatures from 70° to 80°F (22° to 28°C) year-round. The seasonality and amount of rainfall, more than temperature, determine the different vegetation belts that characterize the region. As Figure 6.7 shows, Addis Ababa, Ethiopia, and Walvis Bay, Namibia, have similar average temperatures, but Addis Ababa is in the moist highlands and receives nearly 50 inches (127 cm) of rainfall annually, whereas Walvis Bay lies on the Namibian Desert and receives less than 1 inch (2.5 cm).

To understand the relationship between climate and vegetation in Sub-Saharan Africa (see Figure 6.7), imagine a series of concentric vegetation belts that begin in the western equatorial zone as forest (Am and Af), followed by woodlands and grasslands (Aw), semidesert (BSh), and finally desert (BWh). The montane zones of East Africa, Cameroon, Guinea, and South Africa (especially along the Drakensberg Range) exhibit the effects of altitudinal zonation discussed in Chapter 4. Capturing more rainfall than the surrounding lowlands, these mountains often support unique flora, with large numbers of native species. The only midlatitude climates of the region are found in South Africa. The southwestern corner of South Africa contains a small zone of Mediterranean climate that is comparable to that of California or Spain and is noted for its production of fine wine (Csb). The eastern coast of South Africa has a moist subtropical climate, not unlike that of Florida (Cfa).

Tropical Forests The core of Sub-Saharan Africa is remarkably moist. The world's second largest expanse of humid equatorial rainforest, the Ituri, lies in the Congo Basin, extending from the Atlantic coast of Gabon two-thirds of the way across the continent, including the Republic of the Congo and northern portions of the Democratic Republic of the Congo (Zaire). The conditions here are constantly warm to hot, and precipitation falls year-round (see the climograph for Kisangani in Figure 6.7).

Commercial logging and agricultural clearing have degraded the western and southern fringes of this vast forest, but much of the northeastern section is still intact. Considering the high rates of tropical deforestation in Southeast Asia and Latin America, the Central African case is a pleasant exception. Major national parks such as Okapi and Virunga have been created in the Democratic Republic of the Congo. Virunga National Park—one of the oldest in Africa—is on the eastern fringe of the Ituri forest and is home to the endangered mountain gorillas. Poor infrastructure and political chaos in the Democratic Republic of the Congo over the past 20 years have made large-scale logging impossible, but they have also made conservation difficult. Due to regional conflict, parks such as Virunga have been repeatedly taken over by rebel groups, and park rangers have been killed; poaching has become a means of survival for people in the region. In the future, it seems likely that Central Africa's rainforest, and the wildlife within it, could suffer the same kind of degradation experienced in other equatorial areas. As deforestation proceeds elsewhere in the world, the trees of equatorial Africa become increasingly valuable and hence more vulnerable.

Savannas Wrapped around the Central African rainforest belt in a great arc lie Africa's vast tropical wet and dry savannas. Savannas are dominated by a mixture of trees and tall grasses in the wetter zones immediately adjacent to the forest belt and shorter grasses with fewer trees in the drier zones. North of the equatorial belt, rain generally falls only from May to October. The farther north one travels, the less the total rainfall and the longer the dry season. Climatic conditions south of the equator are similar, only reversed, with the wet season occurring between October and May and precipitation generally decreasing toward the south (see the climograph for Lusaka in Figure 6.7). A larger area of wet savanna exists south of the equator, with substantial woodlands in southern portions of the Democratic Republic of the Congo, Zambia, northern Zimbabwe, and eastern Angola. These savannas also are a critical habitat for the region's large fauna (Figure 6.8).

▶ **Figure 6.8 African Savannas** Buffalo gather at a watering hole in Zambezi National Park in Zimbabwe. The savannas of southern Africa are a noted habitat for the region's larger mammals, including buffalo, elephant, zebra, and lions.

◄ **Figure 6.9 Namib Desert** A hiker explores the coastal dunes of the Namib Desert, which can reach 300 feet in height. Namibia has the region's lowest population density; much of the country is desert.

Africa's Environmental Issues

The prevailing perception of Africa south of the Sahara is one of environmental scarcity and degradation, no doubt fostered by televised images of drought-ravaged regions and starving children. Single explanations such as rapid population growth or colonial exploitation cannot fully capture the complexity of Africa's environmental issues or the ways that people have adapted to living in marginal ecosystems with debilitating tropical diseases. Because much of Sub-Saharan Africa's population is rural, earning its livelihood directly from the land, sudden environmental changes can have devastating effects on household income and food consumption. As Figure 6.10 illustrates, **desertification**—the expansion of desert-like conditions as a result of human-induced degradation—and deforestation are commonplace. Sub-Saharan Africa is also vulnerable to drought, most notably in the Horn of Africa, parts of southern Africa, and the Sahel. Many scientists fear that droughts will come more often and be prolonged with global warming.

As Sub-Saharan cities grow in size and importance, urban environments increasingly face problems of air and water pollution, as well as sewage and waste disposal. At the same time, wildlife tourism is an increasingly important source of foreign exchange for many African states. Throughout the region, national parks have been created in an effort to strike a balance between humans' and animals' competing demands for land.

Deserts The vast extent of tropical Africa is bracketed by several deserts. The Sahara, the world's largest desert and one of its driest, spans the landmass from the Atlantic coast of Mauritania all the way to the Red Sea coast of Sudan. A narrow belt of desert extends to the south and east of the Sahara, wrapping around the **Horn of Africa** (the northeastern corner that includes Somalia, Ethiopia, Djibouti, and Eritrea) and pushing as far as eastern and northern Kenya. An even drier zone is found in southwestern Africa. In the striking red dunes of the Namib Desert of coastal Namibia, rainfall is a rare event, although temperatures are usually mild (Figure 6.9). Inland from the Namib lies the Kalahari Desert. Most of the Kalahari is not dry enough to be classified as a true desert because it receives slightly more than 10 inches (25 cm) of rain a year. Its rainy season, however, is brief. Most of the precipitation is immediately absorbed by the underlying sands. Surface water is thus scarce, giving the Kalahari a desertlike aspect for most of the year.

The Sahel and Desertification The **Sahel** is a zone of ecological transition between the Sahara in the north and the wetter savannas and forest of the south (Figure 6.10). In the 1970s, the Sahel became a symbol for the dangers of unchecked population growth and human-induced environmental degradation when a relatively wet period came to an abrupt end. Six years of drought (1968–1974) were followed by a second prolonged drought during the mid-1980s, ravaging the land. During these droughts, rivers in the area diminished, and desertlike conditions began to move south. Unfortunately, tens of millions of people lived in this area, and farmers and pastoralists whose livelihoods had come to

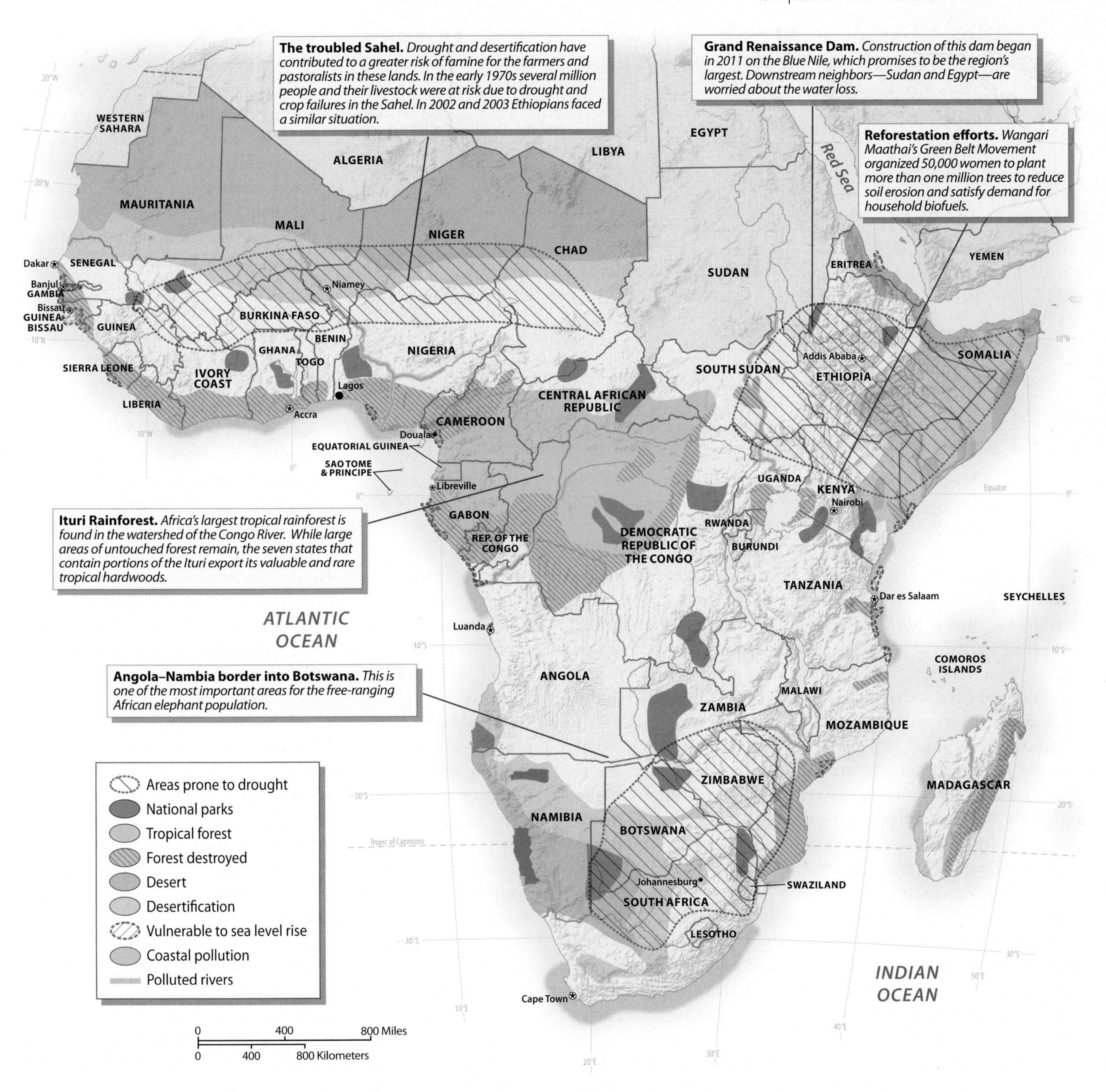

▲ **Figure 6.10 Environmental Issues in Sub-Saharan Africa** Given the immense size of Sub-Saharan Africa, it is difficult to generalize about environmental problems. Dependence on trees for fuel places strains on forests and wooded savannas throughout the region. In semiarid regions, such as the Sahel and Horn of Africa, population pressures, climate change, and land-use practices seem to have exacerbated desertification. Yet Sub-Saharan Africa also supports the most impressive array of wildlife, especially large mammals, on Earth.

depend on the more abundant precipitation of the relatively wet period were temporarily forced out.

Life in the Sahel depends on a delicate balance of limited rain, drought-resistant plants, and a pattern of animal **transhumance** (the movement of animals between wet-season and dry-season pasture). What appears to be desert wasteland in April or May is transformed into a lush garden of millet, sorghum, and peanuts after the drenching rains of June. Relatively free of the tropical diseases found in the wetter zones to the south, the Sahel also has soils that are quite fertile, which helps to explain why people continue to live there despite the unreliable rainfall patterns (Figure 6.11).

▲ **Figure 6.11 Sahel in Bloom** A woman prepares millet grains grown near the city of Maradi, Niger. The soils of the Sahel are fertile, and peasant farmers can produce a surplus when adequate rain falls. Yet in times of drought, crop failures can lead to famine in this region.

Considerable disagreement continues over the basic causes of desertification and drought in the Sahel. Has it been due to too many humans degrading their environment, or unsound settlement schemes encouraged by European colonizers, or a failure to understand global atmospheric cycles? Certainly, human activity in the region greatly increased in the mid-20th century, making the case for human-induced desertification more likely. But parts of the Sahel were important areas of settlement long before European colonization.

The main practices cited in the desertification of the Sahel are the expansion of agriculture and overgrazing, leading to the loss of natural vegetation and declines in soil fertility. For example, French colonial authorities forced villagers to grow peanuts as an export crop, a policy continued by the newly independent states of the region. However, peanuts tend to deplete several key soil nutrients, which means that peanut fields are often abandoned after a few years as cultivators move on to fresh sites. Harvesting this crop also turns up the soil at the onset of the dry season, leading to accelerated wind erosion of valuable topsoil.

Overgrazing by livestock, another traditional product of the region, has also been implicated in Sahelian desertification. Development agencies, hoping to increase livestock production, introduced deep wells into areas that previously had been unused by herders through most of the year. The new supplies of water, in turn, allowed year-round grazing in places that, over time, could not withstand it. Large barren circles around each new well began to appear even on satellite images.

Importantly, some Sahelian areas are experiencing some vegetative recovery thanks to simple actions taken by farmers, a change in government policy, and better rainfall. In the Sahelian portion of Niger, local agronomists have recently documented an unanticipated increase in tree cover over the past 35 years. More interesting still, increases in tree cover have occurred in some of the most densely populated rural areas. After a drought in 1984, farmers began to actively protect saplings instead of clearing them from their fields, including the nitrogen-fixing *goa* tree, which had disappeared from many villages. During the rainy season, the goa tree loses its leaves, so it does not compete with crops for water or sun. The leaves themselves fertilize the soil. Sahelian farmers also use branches, pods, and leaves from trees for fuel and for animal fodder.

Until the 1990s, all trees were considered property of the state of Niger, thus giving farmers little incentive to protect them. Since then, the government has recognized the value of allowing individuals to own trees. Not only can farmers sell branches, pods, or fruit, but also they can conserve them to ensure sustainable rural livelihoods. The villages that protect their trees are much greener than those that do not. In the village of Moussa Bara, where regeneration has been successful, not one child died of malnutrition in the famine of 2005. The Sahel is still poor and prone to drought, but as the case of Niger shows, relatively simple conservation practices can have a positive impact.

Deforestation Although Sub-Saharan Africa still contains extensive forests, much of the region is either grasslands or agricultural lands that were once forest. Lush forests that existed in places such as highland Ethiopia were long ago reduced to a few remnant patches. Throughout history, local populations have relied on such woodlands for their daily needs. Tropical savannas, which cover large portions of the region to the north and south of the tropical rainforest zone, are dotted with woodlands. For many people of the region, deforestation of the savanna woodlands is of greater local concern than the commercial logging of the rainforest. This is because of the importance of **biofuels**—wood or charcoal used for household energy needs, especially cooking—as the leading source of energy for many rural settlements. Loss of woody vegetation has resulted in extensive hardship, especially for women and children who must spend many hours a day looking for wood.

In some countries, village women have organized into community-based nongovernmental organizations (NGOs) to plant trees and create green belts to meet ongoing fuel needs. One of the most successful efforts is that in Kenya, led by Wangari Maathai. Maathai's Green Belt Movement has more than 50,000 members, mostly women, organized into 2000 local community groups. Since the movement's beginning in 1977, millions of trees have been successfully planted. In those areas, village women now spend less time collecting fuel, and local environments have improved. Kenya's success has drawn interest from other African countries, spurring a Pan-African Green Belt Movement largely organized through NGOs interested in biofuel generation, protection of the environment, and the empowerment of women. In 2004, Professor Maathai was awarded a Nobel Peace Prize for her contribution to sustainable development, democracy, and peace. She died in 2011, but the Green Belt Movement remains a powerful force in the region.

Destruction of tropical rainforests for logging is most evident in the fringes of Central Africa's Ituri (refer to Figure 6.10). Given the vastness of this forest and the relatively small number of people living there, however, it is less threatened than other forest areas. Two smaller rainforests—one along the Atlantic coast from Sierra Leone to western Ghana and the other along the eastern coast of the island of Madagascar—have nearly disappeared. These rainforests have been severely degraded by commercial logging and agricultural clearing. Madagascar's eastern rainforests, as well as its western dry forests, have suffered serious degradation in the past three decades. Deforestation in Madagascar is especially worrisome because the island forms a unique environment with a large number of native species—most notably, the charismatic lemurs. In order to address the biofuel needs in areas of deforestation, NGOs are experimenting with the introduction of bamboo. (See *Working Toward Sustainability: Can Bamboo Reduce Deforestation in Africa?*)

Energy Issues The people in this region suffer from serious energy shortages. At the same time, foreign investors are actively developing the region's supplies of oil and natural gas, mostly for export. Many Sub-Saharan states have oil and natural gas; major producers such as Nigeria and Angola are even members of OPEC (Figure 6.12). More recently, countries such as Ivory Coast, Tanzania, and Mozambique have developed their natural gas reserves for domestic consumption and export. Yet for most Africans, wood and charcoal (labeled as combustible renewables) account for the majority of total energy production. Figure 6.12 shows the 21 states in which the World Bank estimates the percentage that these biofuels contribute to national energy production. Even though Angola is a major oil producer, more than half of the country's national energy supply comes from biofuels. In Nigeria,

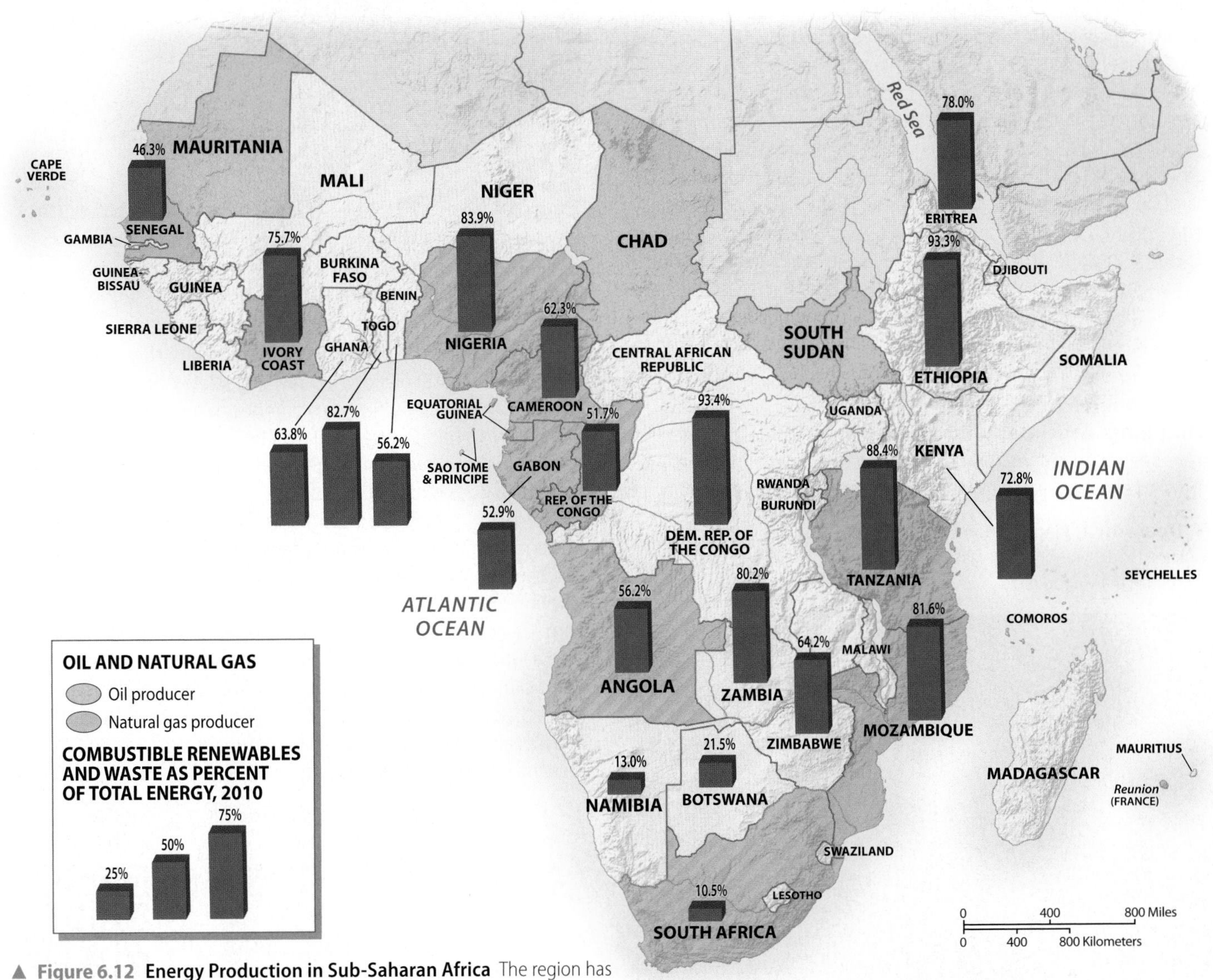

▲ **Figure 6.12 Energy Production in Sub-Saharan Africa** The region has many states that produce oil and natural gas. Two of the region's largest producers, Nigeria and Angola, are members of OPEC. Yet many states still receive the majority of their total energy from burning wood and agricultural waste. **Q: A few states in this figure are much less dependent upon biofuels. Why might that be?**

Working Toward **Sustainability**

Can Bamboo Reduce Deforestation in Africa?

For the near future, wood fuels will be a major contributor to Africa's energy needs. The question remains, what can be done to reduce the pressure on the region's forests beyond planting more trees, as has been done by the Green Belt Movement? Some argue that growing bamboo is a solution. Bamboo grows on every continent and in a tremendous variety of climates. The International Network for Bamboo and Rattan (INBAR) has several pilot programs in Ethiopia and Ghana to promote bamboo as an alternative to wood and charcoal. Several species of bamboo are native to Sub-Saharan Africa, but the plant has never been commercially important in the region (**Figure 6.1.1**). In contrast, bamboo is widely used in China and India as a cooking fuel, as a construction material, and for furniture and flooring. There are many areas of Africa where bamboo could grow well for household consumption and commercial use.

Advantages of Bamboo Bamboo is a fast-growing grass, and once it matures, after five to six years, it can be harvested annually for the decades-long life of the plant. In contrast, a hardwood tree may take 20–30 years to mature, and once harvested, a new one must be planted. Bamboo burns more cleanly than wood, and it can be converted into charcoal, a preferred cooking fuel. Beyond fuel needs, bamboo is a perennial plant that helps prevent soil erosion, especially when planted on hillsides or along riverbanks. More importantly, it does not consume that much water, which is also a critical resource in drought-prone Ethiopia. Throughout Asia, bamboo is commercially harvested as a lightweight, but remarkably strong building and flooring material used mostly for domestic markets. In addition, bamboo floors and furniture are marketed as a sustainable or "green" alternative to hardwoods in many developed countries.

Changing Attitudes and Landscapes Biofuel alternatives have been tried before in Africa. In the early 20th century, the fast-growing eucalyptus tree from Australia was considered the "miracle" tree in highland Ethiopia. Yet over the decades, this tree has been maligned because of how rapidly it depletes groundwater supplies. This does not happen with bamboo, so what is preventing the bamboo revolution in this region? The challenge is both to introduce people to new technologies and to persuade them to change traditional ways of doing things.

The INBAR project provides bamboo seedlings, trains people to manage bamboo forests, and teaches villagers to build kilns to convert bamboo into charcoal. Some of the technology, such as energy-efficient cooking stoves that are fueled by bamboo charcoal, comes from China, and cultural resistance persists, especially in Ghana, where wood fuel is more available. Also, bamboo won't grow in the driest climates of the region, which is often where biofuels are most needed. But bamboo is being adopted iin countries such as Ethiopia and Mozambique as an inexpensive and sustainable solution to meet pressing demands for fuel and construction materials (**Figure 6.1.2**).

▲ **Figure 6.1.2 Building with Bamboo** This young boy in Mozambique prepares bamboo poles for construction material.

▼ **Figure 6.1.1 Planting Bamboo for Biofuels** To address the region's fuel needs, many believe that fast growing bamboo would be an better alternative than wood fuels.

1. Which countries in the region would benefit from increased bamboo production?
2. Besides fuel, what are the other uses of bamboo?

Source: Based on Tina Rosenberg's "In Africa's Vanishing Forests, the Benefits of Bamboo," *New York Times*, March 13, 2012.

Google Earth Virtual Tour Video

http://goo.gl/1PaV3K

▲ **Figure 6.13 Oil Pollution in the Niger Delta** Thousands of Nigerians engage in the practice of hacking into oil pipelines to steal crude oil, refine it, and then sell it locally or abroad. This informal practice leaves the delta horribly polluted and cuts deeply into Nigeria's national oil production.

biofuels account for 84 percent of the national energy supply. In large countries such as Ethiopia or the Democratic Republic of the Congo, the figure is over 90 percent. This is why energy production places tremendous strain on forests and vegetation in the region; it is also why so many countries are developing alternatives such as hydroelectricity and solar power. Another environmental issue for the majority who rely upon biofuels is the smoke that fills homes and causes respiratory problems.

Developing oil and natural gas reserves is not a sure path to economic development, and some have even called it a curse for Sub-Saharan Africa. In the case of Nigeria, politicians and oil executives have prospered from oil profits. But in the Niger Delta, where oil was first extracted over 50 years ago, many places are without roads, electricity, and schools. Moreover, careless and unregulated oil extraction has grossly degraded the delta ecosystem (Figure 6.13). As geographer Michael Watts has observed about the delta, oil has been "a dark tale of neglect and unremitting misery." Not all oil production leads to such misery. Yet Nigeria is a cautionary tale about the limits of the ability of oil to foster development.

Wildlife Conservation Sub-Saharan Africa is famous for its wildlife. No other region of the world has such abundance and diversity of large mammals. The survival of wildlife here reflects, to some extent, the historically low human population density and the fact that sleeping sickness and other diseases have kept people and their livestock out of many areas. In addition, many African peoples have developed various ways of successfully coexisting with wildlife, and about 12 percent of the region is included in nationally protected lands.

However, as is true elsewhere in the world, wildlife is declining in much of Sub-Saharan Africa. The most noted wildlife reserves are in East Africa (Kenya and Tanzania) and southern Africa (South Africa, Zimbabwe, Namibia, and Botswana). These reserves are vital for wildlife protection and are major tourist attractions. Wildlife reserves in southern Africa now seem to be the most secure. In fact, elephant populations are considered to be too large for the land to sustain in countries such as Zimbabwe. Yet throughout the region, population pressure, political instability, and poverty make the maintenance of large wildlife reserves difficult. Poaching is a major problem, particularly for rhinoceroses and elephants; the price of a single horn or tusk in distant markets represents several years' wages for most Africans (Figure 6.14). There is a market for ivory in East Asia, especially in Japan. In China, powdered rhino horn is used as a traditional medicine, whereas in Yemen rhino horn is prized for dagger handles. Some wildlife experts contend that herds could be reduced to prevent overgrazing and that the ivory and rhino horn should be legally sold in the international market in order to generate revenue for further conservation. Not surprisingly, many environmentalists disagree strongly.

In 1989, a worldwide ban on ivory trade was imposed as part of the Convention on International Trade in Endangered Species (CITES). Although several African states, such as Kenya, lobbied hard for the ban, others, such as Zimbabwe, Namibia, and Botswana, complained that their herds were growing and the sale of ivory helped to pay for conservation efforts. Conservationists

▼ **Figure 6.14 South African Wildlife** A white rhinoceros grazes the savannas of Kruger National Park in South Africa, one of the region's oldest wildlife parks. A protected and endangered species, white rhinos are found in South Africa, Zimbabwe, Botswana, Angola, and Kenya.

feared that lifting the ban would bring on a new wave of poaching and illegal trade. However, in the late 1990s, the ban was lifted so that some southern African states could sell down their inventories of elephant ivory confiscated from poachers, and limited sales have continued. The last legal auction of elephant ivory was in 2008; officials have been reluctant to hold more auctions, as there has been a corresponding spike in poaching. The ivory controversy shows how differences in animal distribution in the region, global markets, and international conservation policies are impacting the long-term survival of some 400,000 elephants in Sub-Saharan Africa's protected areas.

Climate Change and Vulnerability in Sub-Saharan Africa

Global climate change poses extreme risks for Sub-Saharan Africa due to the region's poverty, recurrent droughts, and overdependence on rain-fed agriculture. Sub-Saharan Africa is the lowest emitter of greenhouse gases in the world, but it is likely to experience greater-than-average human vulnerability to global warming because of the region's limited resources to both respond and adapt to environmental change. The areas most vulnerable are arid and semiarid regions such as the Sahel and the Horn of Africa, some grassland areas, and the coastal lowlands of West Africa and Angola.

Climate change models suggest that parts of highland East Africa and equatorial Central Africa may receive more rainfall in the future. Thus, some lands that are currently marginal for farming might become more productive. These effects are likely to be offset, however, by the decline in agricultural productivity in the Sahel, as well as in the grasslands of southern Africa, especially in Zambia and Zimbabwe. Drier grassland areas could deplete wildlife populations, which are a major factor behind the growing tourist economy. As in Latin America, higher temperatures in the tropics may result in the expansion of vector-borne diseases such as malaria and dengue fever into the highlands, where they have until now been relatively rare. Given the relatively high elevations in the region, the negative consequences of a rising sea level would be mostly felt on the West African coast (Senegal, Gambia, Sierra Leone, Nigeria, Cameroon, and Gabon).

Even without the threat of climate change, famine stalks many areas of Africa. The Famine Early Warning Systems (FEWS) network is in place to monitor food insecurity throughout the developing world, but especially in Sub-Saharan Africa. **Food insecurity** measures both daily food intake and irreversible coping strategies—for example, selling assets such as livestock or machinery—that will lead to food consumption gaps. By tracking rainfall, vegetation cover, food production, food prices, and conflict, the network maps food insecurity along a continuum from food secure to famine. In 2005, crops in Niger failed due to drought, and an estimated 2.5 million people were vulnerable to famine, but early warning systems were taken seriously, and relief efforts resulted in relatively low mortality rates. Figure 6.15 shows the status of food insecurity in West Africa during an average year (2013). Food insecurity is minimal throughout much of the Sahel, but northern Mali and northeastern Nigeria reached crisis levels, in part due to conflict in these areas. Even with early warning, not all famines can be averted. Between October 2010 and April 2012, nearly a quarter of a million Somalis died, more than half of whom were under five years of age. A combination of severe drought, massive loss of livestock, and conflict-related insecurity interfered with humanitarian assistance efforts.

▼ **Figure 6.15 Map of Food Insecurity in West Africa** Anticipating areas of food insecurity, based upon the timing and amount of rainfall and changes in vegetation/crop cover, is the mission of FEWS network. FEWS has existed since the late 1980s to map areas of potential famine, especially in the Sahelian region of Sub-Saharan Africa.
(Famine Early Warning System (FEWS) Network, April to June 2013 Report, U.S. Agency for International Development)

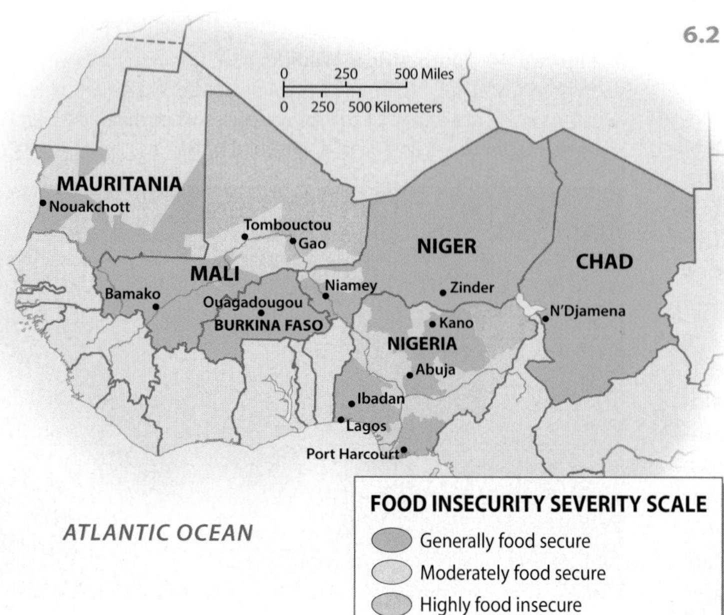

REVIEW

6.1 What environmental factors help explain the patterns of human settlement in the region?

6.2 Summarize the factors that make Sub-Saharan Africans especially vulnerable to climate change.

Population and Settlement: Young and Restless

Sub-Saharan Africa's population is growing quickly. By 2050, the projected population for Africa south of the Sahara is 2 billion people, more than double the current population. It is also a very young population, with 43 percent of the people younger than age 15, compared to just 17 percent for more developed countries. Only 3 percent of the region's population is over 65 years of age, whereas in Europe 16 percent is over 65. Families tend to be large, with a woman having an average of five children (Table 6.1). However, high child and maternal mortality rates also exist, although child mortality rates have declined

Table 6.1 Population Indicators

Country	Population (millions) 2013	Population Density (per square kilometer)	Rate of Natural Increase (RNI)	Total Fertility Rate	Percent Urban	Percent < 15	Percent >65	Net Migration (Rate per 1000)
Angola	21.6	17	3.2	6.3	59	48	2	1
Benin	9.6	86	2.9	5.2	45	43	3	–0
Botswana	1.9	3	0.7	2.7	24	34	4	2
Burkina Faso	18.0	66	3.1	6.0	27	46	2	–2
Burundi	10.9	391	3.2	6.2	11	44	2	1
Cameroon	21.5	45	2.7	5.1	42	43	3	–1
Cape Verde	0.5	128	1.5	2.4	63	31	6	–9
Central African Republic	4.7	8	3.2	6.2	39	40	4	0
Chad	12.2	10	3.6	7.0	22	49	2	–2
Comoros	0.8	354	2.3	4.3	28	42	3	–3
Congo	4.4	13	2.8	5.0	64	42	3	–1
Dem. Rep. of Congo	71.1	30	2.9	6.3	34	45	3	–0
Djibouti	0.9	40	2.0	3.7	77	34	4	–4
Equatorial Guinea	0.8	27	2.3	5.1	39	39	3	6
Eritrea	5.8	49	3.1	4.9	21	43	2	2
Ethiopia	89.2	81	2.6	4.8	17	44	3	–0
Gabon	1.6	6	2.2	4.1	86	39	5	1
Gambia	1.9	167	3.3	5.8	57	46	2	–2
Ghana	26.1	109	2.4	4.2	52	39	4	–0
Guinea	11.8	48	2.6	5.1	35	43	3	–0
Guinea-Bissau	1.7	46	2.5	5.0	44	42	3	–2
Ivory Coast	21.1	66	2.2	5.0	51	42	3	–1
Kenya	44.2	76	2.7	4.5	24	42	3	–0
Lesotho	2.2	74	1.2	3.1	28	37	4	–2
Liberia	4.4	39	3.3	5.7	48	43	3	2
Madagascar	22.5	38	2.8	4.6	33	43	3	–0
Malawi	16.3	138	2.9	5.6	16	46	3	0
Mali	15.5	12	3.1	6.1	35	48	3	–3
Mauritania	3.7	4	2.6	4.8	41	40	3	–1
Mauritius	1.3	636	0.4	1.4	42	21	7	–0
Mozambique	24.3	30	3.0	5.9	31	45	3	–0
Namibia	2.4	3	1.9	3.2	38	37	3	–2
Niger	16.9	13	3.8	7.6	18	50	3	0
Nigeria	173.6	189	2.8	6.0	50	44	3	–0
Reunion	0.8	337	1.2	2.4	94	25	9	0
Rwanda	11.1	422	2.9	4.7	19	45	2	–1
São Tomé and Principe	0.2	195	3.1	4.6	63	42	4	–2
Senegal	13.5	69	3.1	5.0	47	44	3	–2
Seychelles	0.1	205	1.1	2.4	54	20	7	–3
Sierra Leone	6.2	87	2.0	4.9	41	42	3	0
Somalia	10.4	16	3.2	6.8	38	48	3	–4
South Africa	53.0	43	1.0	2.4	62	30	5	4
South Sudan	9.8	15	2.4	5.1	18	43	3	16
Swaziland	1.2	71	1.6	3.5	21	38	3	–1
Tanzania	49.1	52	3.1	5.4	27	45	3	–1
Togo	6.2	109	2.6	4.7	38	42	3	–0
Uganda	36.9	153	3.5	6.2	16	49	2	–1
Zambia	14.2	19	3.3	5.9	39	47	3	–1
Zimbabwe	13.0	33	2.2	3.8	39	41	4	0

Source: Population Reference Bureau, *World Population Data Sheet, 2013.*

substantially in the past two decades. The most troubling indicator for the region is its low life expectancy, which dropped to 50 years in 2008 (in part due to the AIDS epidemic), but is currently estimated at 55 years. Life expectancy in other developing nations is much better: India's is 65 years and China's is 75 years. The growth of cities is also a major trend. In 1980, an estimated 23 percent of the population lived in cities; now the figure is 37 percent.

Behind these demographic facts lie complex differences in settlement patterns, livelihoods, belief systems, and access to health care. Although the region is experiencing rapid population growth, Sub-Saharan Africa is not densely populated. The entire region holds 900 million persons—roughly half the population that is crowded into the much smaller land area of South Asia. In fact, the overall population density of the region (38 people per square kilometer or 98 people per square mile) is comparable to that of the United States (33 people per square kilometer or 85 people per square mile). Just six states account for over half of the region's population: Nigeria, Ethiopia, the Democratic Republic of the Congo, South Africa, Tanzania, and Kenya (see Table 6.1). Some states have very high population densities (such as Rwanda or Mauritius), whereas others are sparsely settled. Namibia and Botswana have just 3 people per square kilometer.

Crude population density is an imperfect indicator of whether a country is overpopulated. Geographers are often more interested in **physiological density**, or the number of people per unit of arable land. The physiological density in Chad, where only 3 percent of the land is arable, is much higher than its crude population density of 9 people per square kilometer. Perhaps a more telling indicator of population pressure and potential food shortages is **agricultural density**, the number of farmers per unit of arable land. Because the majority of people in Sub-Saharan Africa earn their livings from agriculture, agricultural density indicates the number of people who directly depend on each arable square kilometer. The agricultural density of many Sub-Saharan countries is 10 times greater than their crude population density.

Population Trends and Demographic Debates

African demography is a dynamic and controversial field of study. Some believe that Sub-Saharan Africa could support many more people than it presently does, especially if large investments were made in agricultural development. Pessimists, however, argue that the region is a demographic time bomb and that, unless fertility rapidly declines, Sub-Saharan Africa will face massive famines in the near future. The majority of African states officially support lowering rates of natural increase and are promoting modern contraception practices, both to reduce family size and to protect people from sexually transmitted diseases.

The demographic profile of the region is changing. One positive change is the decline in child mortality due to greater access to primary health care. Gone are the days when 1 in 5 children did not live past his or her fifth birthday. Today the child mortality figure is closer to 1 in 10—still high by world standards, but a considerable improvement. Also, life expectancy figures bottomed out in the 2000s due to the devastating impact of HIV/AIDS and are now on the rise. Finally, like populations in other world regions, Africans are moving to cities, which in most cases tends to result in declines in fertility. The family size in South Africa, one of the more urbanized large countries, is half the regional average.

Figure 6.16 compares the population pyramids of Ethiopia and South Africa. Ethiopia has the classic broad-based pyramid of a demographically growing and youthful country. In Ethiopia, most women have five children, and the rate of natural increase is 2.4 percent. There are nearly even numbers of men and women, and only 3 percent of the people are over the age of 65 (life expectancy is 59). In contrast, the South African population pyramid tapers

▼ **Figure 6.16 Population Pyramids of Ethiopia and South Africa** These two pyramids show the contrasting demographic profiles of (a) the more rural and rapidly growing Ethiopia and (b) the more urbanized and slower-growing South Africa. A lost generation of South African women in their 30s is also evident, due to the disproportionate impact of HIV/AIDS on women in South Africa.

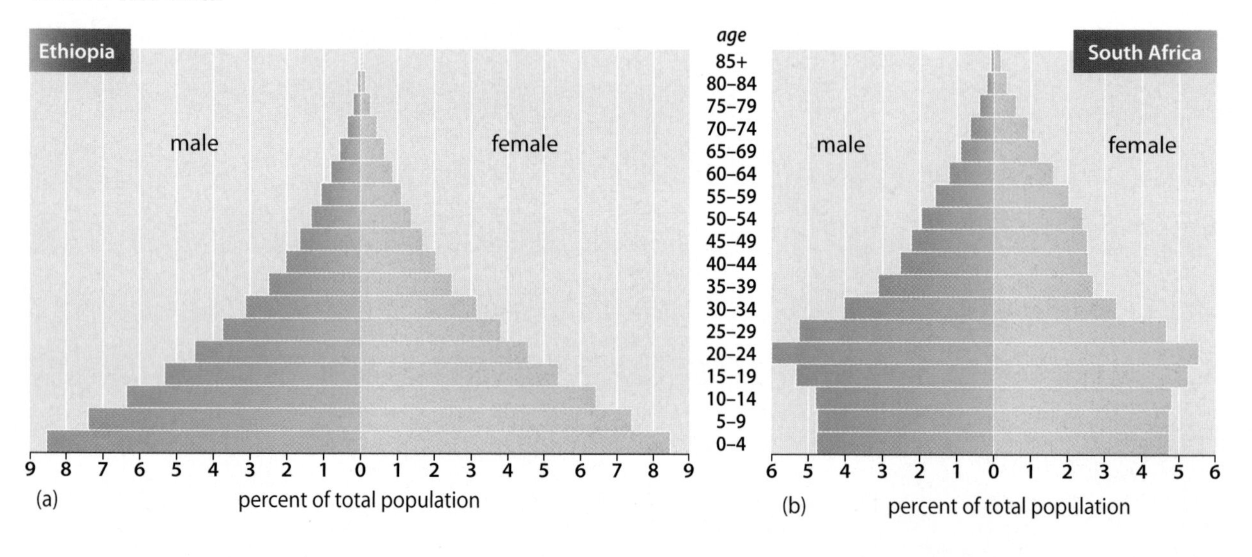

▲ **Figure 6.17 Large Families** Large families are still common in Sub-Saharan Africa. The average total fertility rate for the region is five children per woman.

down, reflecting the country's smaller family size (the average woman has two or three children). One unusual aspect in this graphic is the smaller number of women in their 30s and early 40s when compared to men. This is due to the disproportionate impact of AIDS on women in Africa (which will be discussed later). South Africa has more people over the age of 65 (5 percent), but its life expectancy is only 54 years, less than that of Ethiopia. This, too, is the legacy of the AIDS epidemic, which hit southern Africa with deadly force in the 1990s, but, thankfully, infection rates are now on the decline. Even with the region's low life expectancy, the population is still growing, due to large family size and improvements in child survival rates.

Family Size A continued preference for large families is the basis for the region's demographic growth. In the 1960s, many areas in the developing world had a total fertility rate (TFR) of 5.0 or higher. Today Sub-Saharan Africa, at 5.1, is the only region with such a high TFR. A combination of cultural practices, rural lifestyles, child mortality, and economic realities encourages large families (Figure 6.17). Yet the average family sizes are coming down; as recently as 1996, the regional TFR was 6.0.

Throughout Sub-Saharan Africa, large families guarantee a family's lineage and status. The everyday realities of rural life make large families an asset. Children are an important source of labor; from tending crops and livestock to gathering fuel, they add more to the household economy than they take. Even now, most women marry young, typically when they are teenagers, which increases their opportunities to have children. Demographers often point to the limited formal education available to women as another factor contributing to high fertility. Religious affiliation seems to have little bearing on the region's fertility rates; Muslim, Christian, and animist communities all have similarly high birth rates.

Government policies toward family size have shifted dramatically in the past four decades. During the 1970s, population growth was not perceived as a problem by many African governments; in fact, many equated limiting population size with a neocolonial attempt to slow regional development. Beginning in the 1980s, a shift in national policies occurred. For the first time, government officials argued that smaller families and slower population growth were needed for social and economic development. Following the United Nations International Conference on Population and Development in Cairo, Egypt, African governments announced their intent to bring natural increase rates down to 2.0 and to increase the rate of contraceptive use to 40 percent by 2020. They may reach their goal: As of 2012, the regional rate of natural increase was 2.6 percent, and nearly one-quarter of married women used some method of contraception. Other factors are bringing down the growth rate. As African states slowly become more urban, there is a corresponding decline in family size—a pattern seen throughout the world. Tragically, declines in natural increase and life expectancy were also occurring as a result of AIDS.

The Impact of AIDS and Malaria on Africa Now in its fourth decade, HIV/AIDS has been one of the deadliest epidemics in modern human history, yet it is beginning to decline. At the 2012 International AIDS Conference, the consensus was that HIV infection is still a dangerous, but now a treatable disease. This is especially welcome news for Sub-Saharan Africa, where two-thirds of the 34 million people living with HIV/AIDS are found.

The virus is thought to have originated in the forests of the Congo, possibly crossing over from chimpanzees

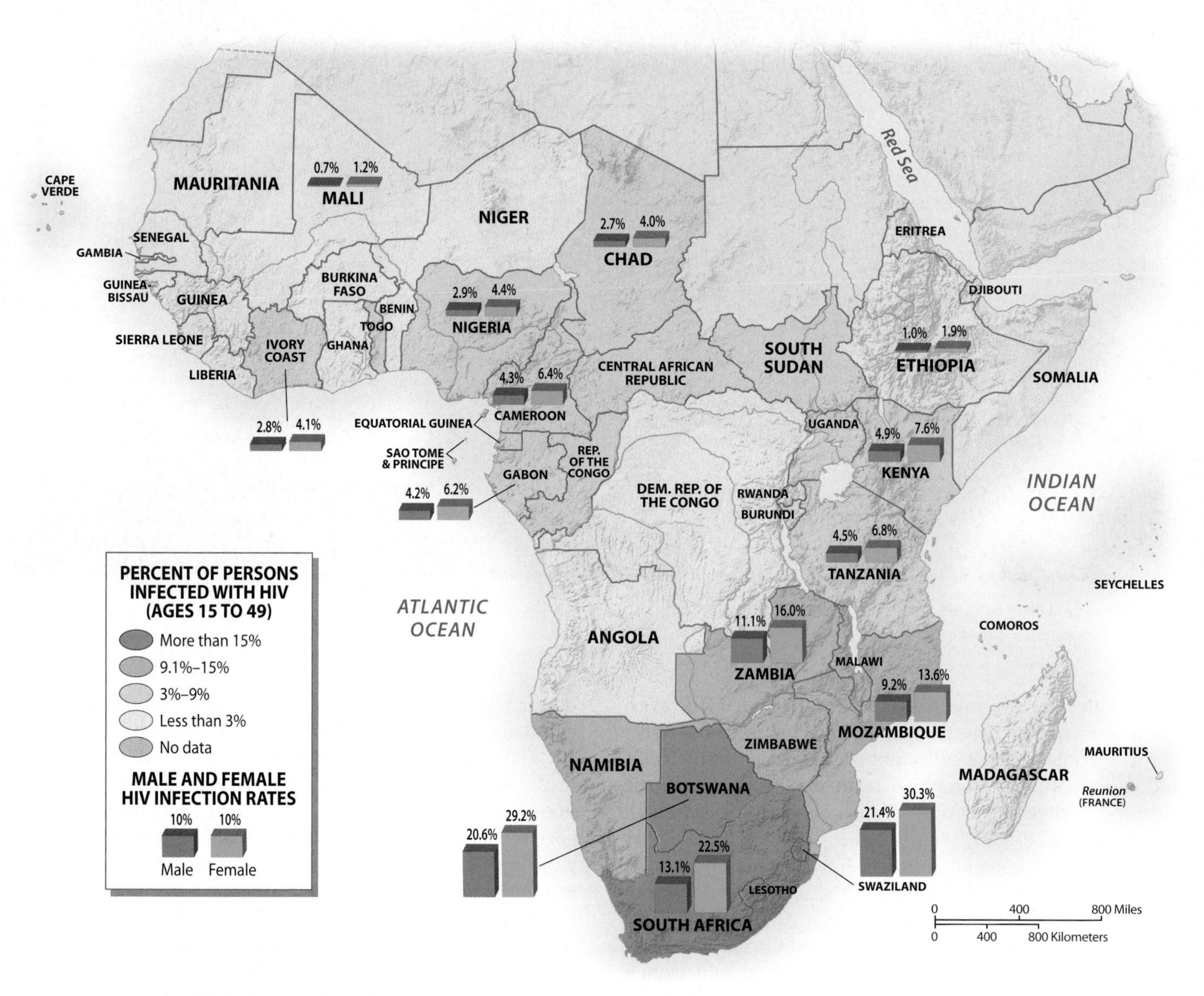

▲ **Figure 6.18 HIV Prevalence in Sub-Sahara Africa** Two-thirds of the world's people infected with HIV/AIDS are in Sub-Saharan Africa. Infection rates in 2012 were highest in southern Africa, especially in Botswana, South Africa, Swaziland, and Lesotho. Infection rates are generally higher among women than men. In Swaziland, for example, 30 percent of women are infected with the virus, whereas only 21 percent of men are infected.
(Data from the Population Reference Bureau, World Population Data Sheet, 2012)

to humans sometime in the 1950s. Yet it was not until the 1980s that the impact of the disease was widely felt. In Sub-Saharan Africa, as in much of the developing world, the disease is transmitted by unprotected heterosexual behaviors. Mother-to-child transmission through the birthing process or breast-feeding also contributes to the number of new cases, although strides have been made in reducing these numbers through prenatal testing and treatment. A long-term pattern of seasonal male labor migration helped to spread the disease. So, too, did lack of education, inadequate testing early on in the epidemic, and the disempowerment of women. Consequently, women bear a disproportionate burden of the HIV/AIDS epidemic. They account for approximately 60 percent of the HIV infections, and they are usually the caregivers for those who are infected. Until the late 1990s, many African governments were unwilling to acknowledge publicly the severity of the situation or to discuss frankly the measures necessary for prevention.

Southern Africa is ground zero for the AIDS epidemic that is ravaging the region (Figure 6.18). The eight countries with the highest HIV prevalence are all located there. In South Africa, the most populous state in southern Africa, 5.7 million people (nearly one in five people age 15–49) are infected with HIV/AIDS. The rate of infection in neighboring Botswana is 25 percent (down from a staggering 36 percent in 2001) for the same age group. Infection rates in other African states are lower, but still high by world standards. In Kenya, it is estimated that 6 percent of the 15–49 age group has HIV or AIDS. By comparison, only 0.6 percent of the same age group in North America is infected.

The social and economic implications of this epidemic have been profound. Life expectancy rates tumbled—in a few

▲ **Figure 6.19 HIV/AIDS Activism in South Africa** School children march through Cape Town, South Africa, to celebrate the anniversary of the establishment of a program to distribute anti-retroviral drugs to AIDS patients. The program run by Doctors without Borders began distribution of these life-prolonging drugs in 2001.

places even dropping to the early 40s. AIDS typically hits the portion of the population that is most economically productive. Time lost to care for sick family members and the outlay of workers' compensation benefits have reduced economic productivity and overwhelmed public services in hard-hit areas. The disease makes no class distinctions: Countries are losing both peasant farmers and educated professionals (doctors, engineers, and teachers). Many countries struggle to care for millions of children orphaned by AIDS.

After three devastating decades, there are finally hopeful signs. Prevention measures are widely taught, and treatment with a mix of available drugs means that HIV infection is now manageable and not a death sentence. Due to international financial support and national outreach efforts, Sub-Saharan Africa now has many more health facilities that offer HIV testing and counseling. It is estimated that over 1 million South Africans now receive antiretroviral drugs, which are prolonging life. Prevention services in prenatal settings in 2011 allowed nearly half of all pregnant HIV-positive women to get antiretroviral drugs to prevent transmission of the virus to their babies. Changes in sexual practices, driven by educational campaigns, more condom use, and higher rates of male circumcision, have prevented hundreds of thousands of new cases (Figure 6.19).

Malaria Malaria has been a scourge for this region for centuries. Transmitted from infected individuals to others via the anopheles mosquito, malaria causes high fever, severe headache, and in the worst cases, death. The World Health Organization estimates that 200 million people experience malaria each year, resulting in 600,000 deaths. The majority of infections and deaths occur in Sub-Saharan Africa. Since 2000, African governments, NGOs, and foreign aid sources have spent increasing amounts of money to reduce the threat of infection. Presently, a malaria vaccine does not exist, but research in this area is promising. Medication to prevent infection helps in many cases, but is not reliable over the long term. Insecticide spraying is also used, but mosquito resistance to the chemicals occurs. The most effective tool to reduce infection has been the distribution of insecticide-treated mosquito nets to millions of African homes. That, along with rapid diagnostic tests and access to medication once infected, has cut infections and related deaths by one-third since 2000.

Malaria and poverty are closely related in Sub-Saharan Africa, with many of the poorest tropical countries experiencing the higher rates of infection. The areas hardest hit by malaria are West and Central Africa. The Democratic Republic of the Congo and Nigeria account for 40 percent of malaria deaths globally. Forty percent of all malaria cases occur in these two countries plus India.

Patterns of Settlement and Land Use

Because of the dominance of rural settlements in Sub-Saharan Africa, people are widely scattered throughout the region (Figure 6.20). Population concentrations are highest in West Africa, highland East Africa, and the eastern half of South Africa. The first two areas have some of the region's best soils, and indigenous systems of permanent agriculture developed there. In South Africa, the more densely settled east results from an urbanized economy based on mining, as well as the forced concentration of black South Africans into eastern homelands.

West Africa is more heavily populated than most of Sub-Saharan Africa, although the actual distribution pattern is patchy. Density in the far west, from Senegal to Liberia, is moderate. This area is characterized by broad lowlands with decent soils, and in many areas the cultivation of rice has enhanced agricultural productivity. Greater concentrations of people are found along the Gulf of Guinea, from southern Ghana through southern Nigeria, and again in northern Nigeria along the southern fringe of the Sahel. Nigeria is moderately to densely settled through most of its extensive territory; with over 170 million inhabitants, it stands as the demographic core of Sub-Saharan Africa. The next largest country, Ethiopia, has a population of nearly 90 million.

As more Africans move to cities, patterns of settlement are evolving into clusters of higher concentration. Towns that were once small administrative centers for colonial elites grew into major cities. The region has one megacity, Lagos, although the actual size of metropolitan Lagos is unclear. The United Nations (UN) estimates this urban agglomeration at over 11 million people, but other sources suggest it is nearing 20 million. The country of Nigeria does not have an official figure for the metropolitan area, but recognizes that the city is growing quickly. Throughout the continent, African cities are growing faster than rural areas. But before examining the Sub-Saharan urban scene, a more detailed discussion of rural subsistence is needed.

Agricultural Subsistence The staple crops over most of Sub-Saharan Africa are millet, sorghum, rice, and corn (maize), as well as a variety of tubers and root crops such as yams. Irrigated rice is widely grown in West Africa and Madagascar. Geographer Judith Carney in her book *Black Rice* documents how African slaves introduced rice cultivation to the Americas. Corn, in contrast, was introduced to Africa from the Americas by the slave trade and quickly grew to become a basic food. In higher elevations of Ethiopia and

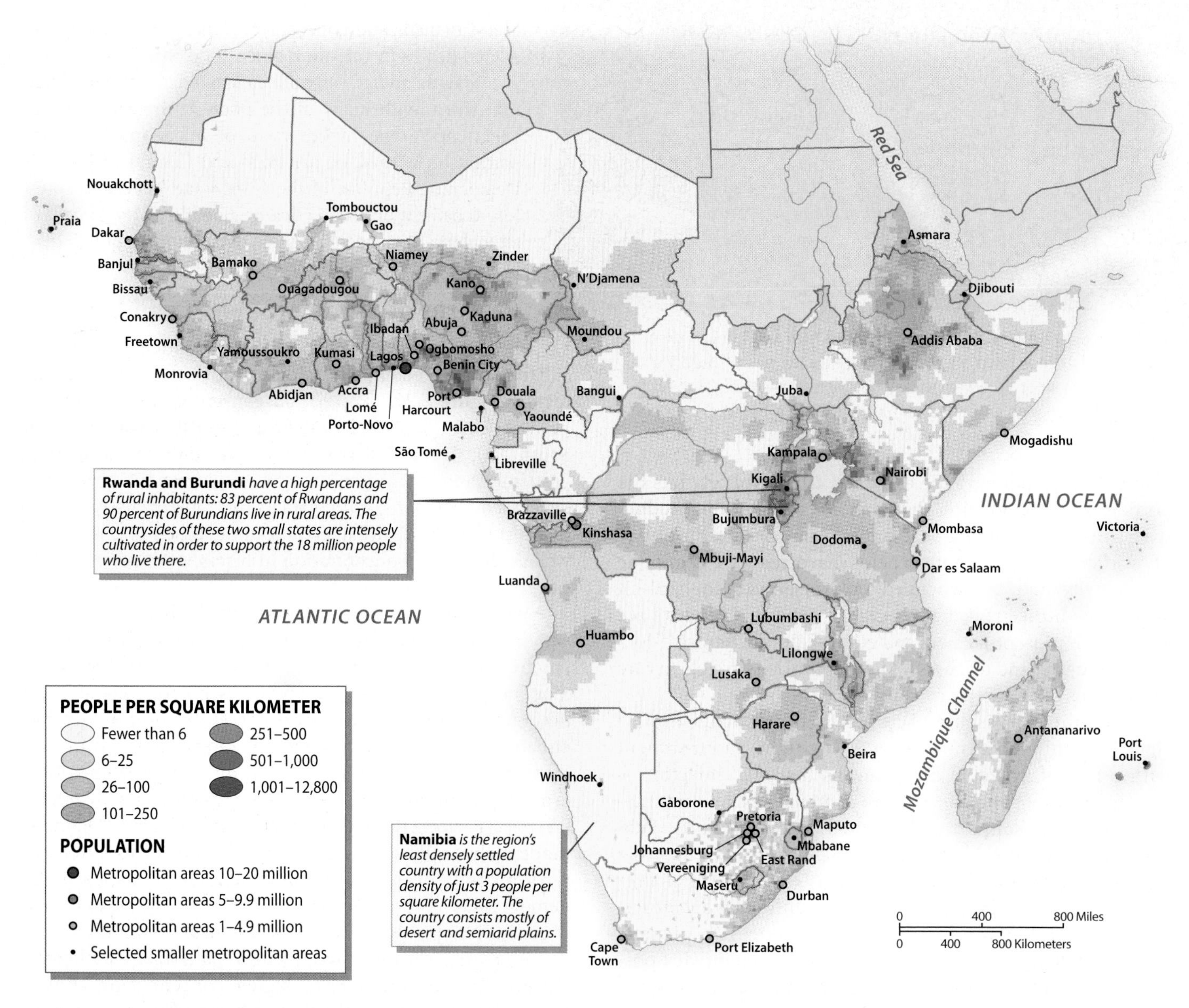

▲ **Figure 6.20 Population of Sub-Saharan Africa** The majority of people in Sub-Saharan Africa live in rural areas. However, some of these rural zones—such as West Africa and the East African highlands—are densely settled. Major urban centers, especially in South Africa and Nigeria, support millions. Lagos, Nigeria, is the one megacity in the region (with over 10 million people), but some three dozen cities have more than 1 million residents. **Q: What factors contribute to the extremely low population density in the southwest corner of the continent?**

South Africa, wheat and barley are grown. Intermixed with subsistence foods are a variety of export crops—coffee, tea, rubber, bananas, cocoa, cotton, and peanuts—that are grown in distinct ecological zones and often in some of the best soils.

In areas that support annual crop yields, population densities are greater. In parts of humid West Africa, for example, the yam became the king of subsistence crops. The Ibos' mastery of yam production allowed them to procure more food and live in denser permanent settlements in the southeastern corner of present-day Nigeria. Much of traditional Ibo culture is tied to the arduous tasks of clearing the fields, tending the delicate plants, and celebrating the harvest.

Over much of the continent, African agriculture remains relatively unproductive, and rural population densities tend to be low. Amid the poorer tropical soils, growing crops usually entails shifting cultivation, or **swidden**. This process involves burning the natural vegetation to release fertilizing ash and planting crops such as maize, beans, sweet potatoes, bananas, papaya, manioc, yams, melon, and squash. Each plot is temporarily abandoned once its source of nutrients has been exhausted. Swidden cultivation often is a very finely tuned adaptation to local environmental conditions, but it is unable to support high population densities. Women are the subsistence farmers of the region, producing for their household needs, as well as for local and foreign markets.

Export Agriculture Agricultural exports, whether from large estates or small producers, are critical to the economies of many states. If African countries are to import the modern goods and energy resources they require, they must sell their own products on the world market. Because the region has

▲ **Figure 6.21 Kenyan Floriculture** Women cut roses for export at a greenhouse in Naivasha, Kenya. Floriculture is a vital contributor to Kenya's agricultural exports and has grown dramatically in the past two decades. One-third of the cut flowers sold in Europe, especially roses, come from Kenya.

few competitive industries, the bulk of its exports are primary products derived from farming, mining, and forestry.

In the densely settled country of Rwanda, most farms are small, but the highland volcanic soil is ideal for growing coffee. However, for decades the country's coffee production languished, and farmers earned very little for their low-quality beans. Yet across Rwanda's hillsides were older varieties of coffee plants with high value in today's premium coffee market. Farmers just needed a better way to prepare the beans and market them. In an effort to rebuild the country after the ethnic genocide in the 1990s, the government targeted improving the quality of the country's coffee through the formation of cooperatives that would bring ethnic groups together and improve farmers' incomes. Through community-run cooperatives, the quality of their washed and sorted beans improved. So, too, did their ability to bargain with major buyers such as Starbucks and Green Mountain, leaving out the middleman. For many small farmers in Rwanda, their premium coffee is now a source of pride. The cooperatives have taken off, and farmers, many of them war widows, have seen their incomes improve.

Several African countries rely heavily on one or two export crops. Coffee, for example, is vital for Ethiopia, Kenya, Rwanda, Burundi, and Tanzania. Peanuts have historically been the primary source of income in the Sahel, whereas cotton is tremendously important for the Central African Republic and South Sudan. Ghana and Ivory Coast have long been the world's main suppliers of cacao (the source of chocolate), Liberia produces plantation rubber, and many farmers in Nigeria specialize in palm oil. The export of such products can bring good money when commodity prices are high, but when prices collapse, as they periodically do, economic hardship follows.

Nontraditional agricultural exports that depend upon significant capital inputs and refrigerated air transport have emerged in the last two to three decades. One industry is floriculture for the plant and cut flower industry. Here the highland tropical climate of Kenya, Ethiopia, and South Africa is advantageous. After Colombia in South America, Kenya is the largest exporter of cut flowers in the world, and most of its exports go to Europe (Figure 6.21). Similarly, the European market for fresh vegetables and fruits in the winter is being met by some African producers in West and East Africa.

Pastoralism Animal husbandry (the care of livestock) is extremely important in Sub-Saharan Africa, particularly in the semiarid zones. Camels and goats are the principal animals in the Sahel and the Horn of Africa, but farther south cattle are king (Figure 6.22). Many African peoples have traditionally specialized in cattle raising and are often tied into

▼ **Figure 6.22 Pastoralists in the Horn of Africa** Two Ethiopian girls walk along a herd of camels. Camels are well suited for the arid conditions of eastern Ethiopia and Somalia. Pastoralists rely on their camels for milk, transport, and trade.

mutually beneficial relationships with neighboring farmers. Such **pastoralists** typically graze their stock on the stubble of harvested fields during the dry season and then move them to drier uncultivated areas during the wet season, when the pastures turn green. Farmers thus have their fields fertilized by the manure of the pastoralists' stock, while the pastoralists find good dry-season grazing. At the same time, the nomads can trade their animal products for grain and other goods of the sedentary world. Several pastoral peoples of East Africa, such as the Masai of the Tanzanian-Kenyan borderlands, are noted for their extreme reliance on cattle and general (but never complete) independence from agriculture. The Masai derive a large percentage of their nutrition from drinking a mixture of milk and blood. The blood is obtained by periodically tapping the animal's jugular vein, a procedure that evidently causes little harm.

Large expanses of Sub-Saharan Africa have been off-limits to cattle because of infestations of **tsetse flies**, which spread sleeping sickness to cattle, humans, and some wildlife. In environments containing brush or woodland (which is necessary for tsetse fly survival), wild animals that harbor the disease, but are immune to it, could be present in large numbers, but cattle simply could not be raised. Some evidence suggests that tsetse fly infestations dramatically increased in the late 1800s, greatly harming African societies dependent on livestock, but benefiting wildlife populations. At present, tsetse fly eradication programs are reducing the threat, and cattle ranching is spreading into areas where it was previously unsuitable. This process is beneficial for African livestock, but it may endanger the continued survival of many wild animals. When people move their cattle into new areas in large numbers, wildlife almost inevitably declines.

Sub-Saharan Africa thus presents a difficult environment for raising livestock because of the virulence of its animal diseases. In the tropical rainforests of Central Africa, cattle have never survived well, and the only domestic animal that thrives is the goat. Raising horses, moreover, has historically been feasible only in the Sahel and in South Africa. As we shall see later, the disease environment of tropical Africa has presented a variety of problems for humans as well.

Urban Life

Sub-Saharan Africa and South Asia are the least urbanized regions in the world, although in both regions cities are growing at twice the national growth rates. More than one-third of Sub-Saharan people live in urbanized areas. One of the consequences of this surge in city living is urban sprawl. Rural-to-urban migration, industrialization, and refugee flows are forcing the cities of the region to absorb more people and use more resources. As in Latin America, the tendency is toward urban primacy, the condition in which one major city is dominant and at least three times larger than the next largest city. For example, Luanda, the capital of Angola, was built for half a million people, but now has over 5 million residents and is projected to have 9 million by 2025. Most of the city's inhabitants live in the vast slums that surround the modernizing urban core. In such rapidly growing places, city officials struggle to build enough roads and provide electricity, water, trash collection, and employment for all these people.

European colonialism greatly influenced urban form and development in the region. Although a very small percentage of the population lived in cities, Africans did have an urban tradition prior to the colonial era, Ancient cities, such as Axum in Ethiopia, thrived 2000 years ago. Similarly, in the Sahel, prominent trans-Saharan trade centers, such as Tombouctou (Timbuktu) and Gao, have existed for more than a millennium. In East Africa, an urban mercantile culture emerged that was rooted in Islam and the Swahili language. The prominent cities of Zanzibar, Tanzania, and Mombasa, Kenya, flourished by supporting a trade network that linked the East African highlands with the Persian Gulf (see *Cityscapes: East Africa's High-Tech Leader, Nairobi*). The stone ruins of Great Zimbabwe in southern Africa are testimony to the achievements of stone working, metallurgy, and religion achieved by Bantu groups in the 14th century. West Africa, however, had the most developed precolonial urban network, reflecting both indigenous and Islamic traditions. It also supports some of the region's largest cities today.

West African Urban Traditions The West African coastline is dotted with cities, from Dakar, Senegal, in the far north (the westernmost city in Africa) to Lagos, Nigeria, in the east. Half of Nigerians live in cities, and in 2011 the country had eight metropolitan areas with populations of more than 1 million. Historically, the Yoruba cities in southwestern Nigeria have been the best documented. Developed in the 12th century, cities such as Ibadan were walled and gated, with a palace encircled by large rectangular courtyards at the city center. An important center of trade for an extensive hinterland, Ibadan also was a religious and political center. Lagos was another Yoruba settlement. Founded on a coastal island on the Bight of Benin, most of the modern city has spread onto the nearby mainland. Its coastal setting and natural harbor made this relatively small, indigenous city attractive to colonial powers. When the British took control in the mid-19th century, the city's size and importance grew.

In 1960, Lagos was a city of only 1 million, but today it has at least 11 million residents, and some estimate it to be nearly twice as large. Unable to keep up with the surge of rural migrants, the streets of Lagos are clogged with traffic; for those living on the city's periphery, 3- and 4-hour commutes (one way) are common. For many, the informal sector (unregulated services and trade) provides employment (Figure 6.23). Crime is another major problem. The chances of being attacked on the streets or robbed in one's home are quite high, even though most windows are barred and houses are fenced. To deal with this problem, Lagos has the highest density of police in Nigeria, but the weak social bonds between urban migrants, widespread poverty, and the gap between rich and poor seem to encourage lawlessness.

Most West African cities are hybrids, combining Islamic, European, and national elements such as mosques, Victorian architecture, and streets named after independence leaders. Accra is the capital city of Ghana and home to 2.6 million people. Originally settled by the Ga people in the 16th century,

Cityscapes

East Africa's High-Tech Leader, Nairobi

Nairobi, the capital of Kenya, is considered the hub for transportation, finance, and communication for East Africa. This metropolis of 3.5 million people just south of the equator spreads across a high plateau at nearly 6000 feet (1800 meters). The elevation supports a pleasant daytime temperature in the 70s and cool evenings. The city's origins are colonial; the British built it to resupply the Mombasa-to-Kampala railroad. But the city quickly emerged as the administrative hub of the British colony. Elegant homes were built on the city's western hills, along with golf courses and a horse-racing track.

When the city became the capital of independent Kenya in 1963, it was reworked. The city's main boulevard was renamed Kenyatta Avenue for the country's first president. It would later intersect with Moi Avenue, named for the country's second president. The modest skyline is anchored by the cylindrical Kenyatta International Conference Center. From its upper floors, you can watch giraffes stride through Nairobi National Park, the only game reserve in the world to border a major city and a prime tourist attraction (Figure 6.2.1). However, if Kenyans have their way, Nairobi may become the high-tech hub for the continent with the development of Konza City.

The Silicon Savannah Nairobi's rise as a high-tech superstar came suddenly. High-speed Internet became widely available only in the last six years, but it has transformed the way the city works. It is estimated that half the city uses the Internet, and many start-up companies are being created here, especially along Ngong Road in the western hills, where business incubator centers such as I-Hub receive financial backing from Google. As cell phones have become ubiquitous, so have many innovative applications. The majority of the city's adult population now uses M-Pesa, a phone-based money transfer service, to pay for everything from street food to rides on the city's privately owned minibuses, called *matatus*. Kenya's development of innovative mobile applications far outpaces that of the United States. Ushahidi, a crowd-mapping phone application developed in Kenya in 2008 to monitor post-election violence, has been used to map earthquake victims in Haiti and the 2012 U.S. elections.

Nairobi has many college-educated Kenyans and relatively high unemployment, so 20-something Nairobi residents are turning to digital entrepreneurship. Hence, the government established a 5000-acre technology park called Konza City, some 60 kilometers south of Nairobi. It is hoped that when completed in 2025, this city and technology hub—dubbed the Silicon Savannah—will lead the continent in mobile and Internet innovations and entrepreneurship.

Nairobi's Underside Despite such robust embrace of technology, Nairobi still has many unemployed and impoverished residents. Close to the downtown and bordering a golf course is the slum of Kibera, with some 200,000 residents (Figure 6.2.2). One of the oldest and most studied slums in East Africa, it is experiencing the dual process of upgrading and resident relocation. Still, this is a place where garbage lines the streets, crime is rampant, and housing is crude and crowded. Indeed, a true testament to Nairobi's ability to embrace the power of new technology is to provide the residents of Kibera with the same access to mobile and Internet technologies that other city residents share.

▲ **Figure 6.2.2 Kibera Slum** One of the city's largest slums, improving basic infrastructure in this densely settled area in challenging.

▼ **Figure 6.2.1 Downtown Nairobi** A flock of sacred ibis flies above downtown Nairobi in the late afternoon. Kenya's bustling capital city is a regional hub for all of East Africa and aspires to be the high-tech center of the continent. The tall, cylindrical building at the left is the Kenyatta International Conference Center.

1. Where does the population density appear to be greatest in the city?
2. How is Kenyan national identity expressed in the Nairobi landscape?

Google Earth Virtual Tour Video

http://goo.gl/OLvVen

◄ **Figure 6.23 Downtown Lagos, Nigeria** A sea of shoppers, vendors, and buses fills the streets of the Oshodi Market, the commercial hub of Lagos. Lagos is the largest city in Sub-Saharan Africa, with at least 11 million inhabitants.

it became a British colonial administrative center by the late 1800s. Here again, the modern city is being transformed through neoliberal policies introduced in the 1980s that attracted international corporations. A growth in foreign investment in financial and producer services led to the creation of a "Global Central Business District (CBD)" on the east side of the city, away from the "National CBD." Here foreign companies clustered in areas with secure land title, new modern roads, parking, and access to the airport. Upper-income gated communities have also formed near the Global CBD, with names such as Trasacco Valley, Airport Hills, and Buena Vista (Figure 6.24). Accra, like other cities in the region, is rapidly changing due to an influx of foreign capital. The result is highly segregated urban spaces that reflect a global phase in urban development, in which world market forces, rather than colonial or national ones, are driving the change.

Urban Industrial South Africa The major cities of southern Africa, unlike those of West Africa, are colonial in origin. Cities such as Lusaka, Zambia, and Harare, Zimbabwe, grew as administrative or mining centers. South Africa is one of the most urbanized states in the entire region, and it is certainly the most industrialized. The foundations of South Africa's urban economy rest largely on its incredibly rich mineral resources (diamonds, gold, chromium, platinum, tin, uranium, coal, iron ore, and manganese). Seven of its metropolitan areas have more than 1 million people; the largest of them are Johannesburg, Durban, and Cape Town.

The form of South African cities continues to be imprinted by the legacy of **apartheid**, an official policy of racial segregation that shaped social relations in South Africa for nearly 50 years. Even though apartheid was abolished in 1994, it is still evident in the landscape. Under apartheid rules, all cities were divided into residential areas according to racial categories: white, **coloured** (a South African term describing people of mixed African and European ancestry), Indian (South Asian), and African (black). Whites resided in the most appealing and spacious portions of the cities, whereas blacks were crowded into the least desired areas, forming squatter settlements called **townships.** Today blacks, coloureds, and Indians are legally allowed to live anywhere they want. Yet the economic differences among racial groups, as well as deep-rooted animosity, hinder residential integration.

Even in relatively affluent South Africa, the challenge of accommodating new urban residents is constant. The largely black community of Diepsloot, north of Johannesburg, illustrates how even planned communities can become densely settled in short order. Diepsloot is a post-apartheid project where the government divided land into lots and eventually built several thousand houses. Figure 6.25 illustrates the increased settlement density of a Diepsloot neighborhood over a decade. The relatively large lots were gradually filled in with small shacks built with scrap metal, wood, and plastic. The initial homes had access to running water and sewage, but the informal dwellings surrounding them do not. Later, small shops were added, and trees were planted, but much of the population lacks formal employment. This area, which was planned for a few thousand people, is now a community of over 150,000.

Still, Johannesburg is the African metropolitan area that most consciously aspires to global city status, a dream underscored by its hosting of soccer's World Cup in 2010. However, as apartheid ended, Johannesburg became infamous for its high crime rate. Many white-owned businesses fled the CBD for the northern suburb of Sandton (Figure 6.26). By the late 1990s, Sandton was the new financial and business hub for the city of Johannesburg. It epitomizes the modern urban face of South Africa: While it is racially mixed, affluent whites are overrepresented. When the province of Gauteng invested $3 billion in a new high-speed commuter rail called Gautrain, it was no accident that the first functioning line linked Sandton with Tambo International Airport.

▼ **Figure 6.24 Elite Accra Neighborhood** In the Legon area east of downtown Accra one finds the city's nicest homes, private schools and the University of Ghana.

(a)

(b)

▲ **Figure 6.25 Satellite Images of Diepsloot, South Africa** In less than a decade, black settlers poured into the planned community of Diepsloot, filling in single-family-home lots (a) with multiple structures. They also added businesses, planted trees and vastly increased the population density of the area (b).

REVIEW

6.3 Explain the factors that contribute to high population growth rates in this region.
6.4 Where has the impact of HIV/AIDS and malaria been most pronounced, and what are governments and aid organizations doing to fight these diseases?
6.5 What are the major rural livelihoods in this region?

Cultural Coherence and Diversity: Unity Through Adversity

No world region is culturally homogeneous, but most have been partially unified in the past by widespread systems of belief and communication. Traditional African religions, however, were largely limited to local areas, and the religions that did become widespread—namely, Islam and Christianity—are primarily associated with other world regions. A handful of African trade languages have long been understood over vast territories (Swahili in East Africa, Mandingo and Hausa in West Africa), but none spans the entire Sub-Saharan region. Sub-Saharan Africa also lacks a history of widespread political union or even an indigenous system of political relations. The powerful African kingdoms and empires of past centuries all were limited to distinct subregions of the landmass.

The lack of traditional cultural and political coherence across Sub-Saharan Africa is not surprising if you consider the region's huge size. Sub-Saharan Africa is more than four times larger than Europe or South Asia. Had foreign imperialism not impinged on the region, it is quite possible that West Africa and southern Africa would have developed into their own distinct world regions.

An African identity south of the Sahara was created through a common history of slavery and colonialism, as

▼ **Figure 6.26 Sandton, Johannesburg** Just north of Johannesburg's central business district, Sandton has emerged as the financial and business center of the new South Africa. Many businesses relocated here, fleeing high crime rates in the old central business district. With world-class shopping, hotels, a convention center, and even Nelson Mandela Square, Sandton has become the part of Johannesburg that most tourists and businesspeople experience.

well as struggles for independence and development. More telling, the people of the region often define themselves as African, especially to the outside world. No one will argue the fact that Sub-Saharan Africa is poor. And yet the cultural expressions of its people—its music, dance, and art—are joyous. Africans share a resilience and optimism that visitors to the region often comment on. The cultural diversity of the region is obvious, yet there is unity among the people, drawn from surviving many adversities.

Language Patterns

In most Sub-Saharan countries, as in other former colonies, multiple languages are used that reflect tribal, ethnic, colonial, and national affiliations. Indigenous languages, many from the Bantu subfamily, often are localized to relatively small rural areas. More widely spoken African trade languages, such as Swahili or Hausa, serve as a lingua franca over broader areas. Overlaying native languages are Indo-European (French and English) and Afro-Asiatic (Arabic) languages. Figure 6.27 illustrates the complex pattern of language families and major languages found in Africa today. Contrast the larger map with the inset that shows current "official" languages. A comparison of the two shows that most African countries are multilingual, which can be a source of tension within states. In Nigeria, for example, the official language is English, yet there are millions of Hausa, Yoruba, Igbo (or Ibo), Ful (or Fulani),

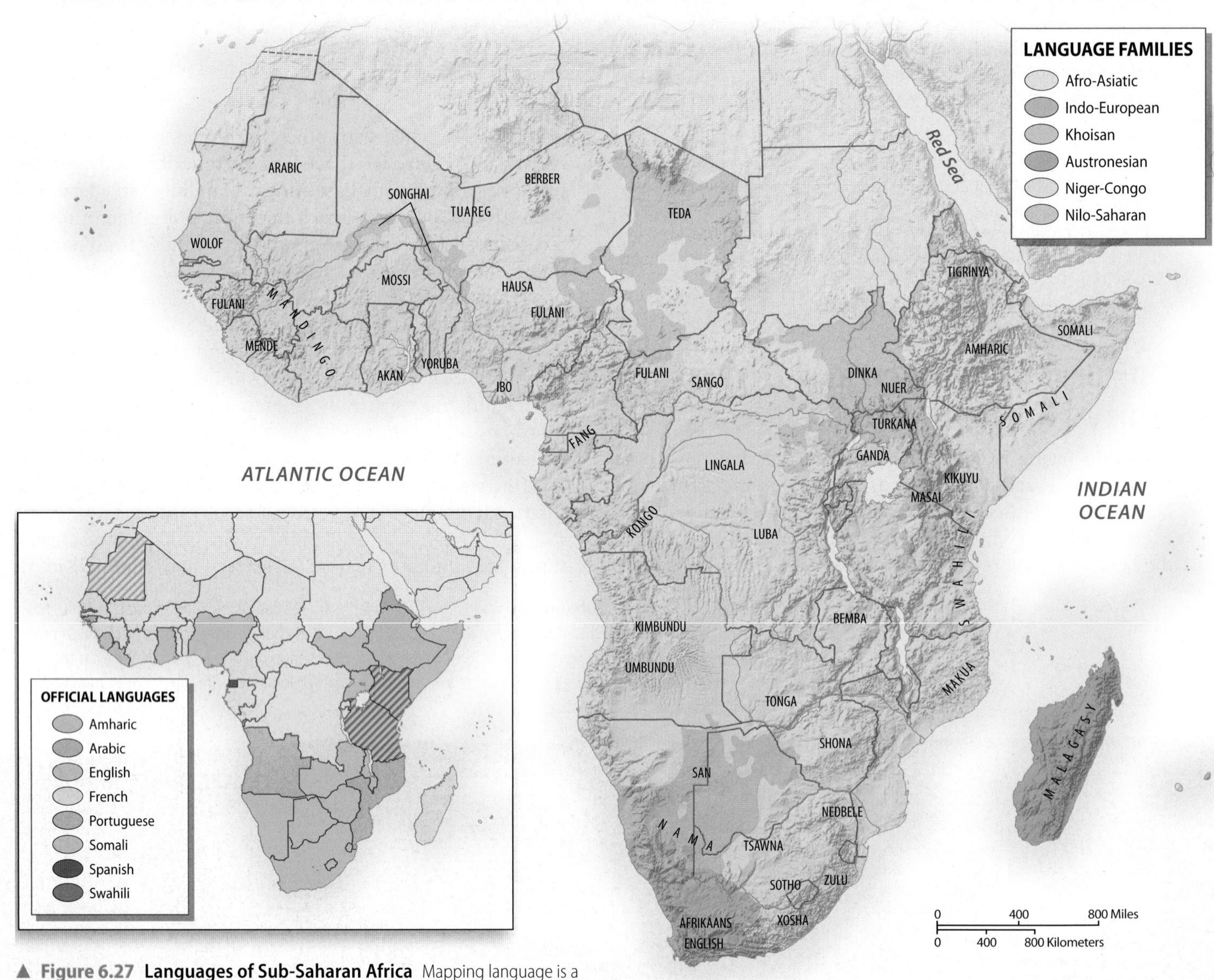

▲ **Figure 6.27 Languages of Sub-Saharan Africa** Mapping language is a complex task for Sub-Saharan Africa. There are languages with millions of speakers, such as Swahili, and there are languages spoken by a few hundred people living in isolated areas. Six language families are represented in the region. Among these families are scores of individual languages (see the labels on the map). Because most modern states have many indigenous languages, the colonial language often became the "official" language because it was less controversial than picking from among several indigenous languages. English and French are the most common official languages in the region (see inset). **Q: Considering the distribution of language families in the region, what does this pattern tell us about this region's interaction with peoples from other regions?**

and Efik speakers, as well as speakers of dozens of other languages.

African Language Groups Three of the six language groups mapped in Figure 6.27 are unique to the region (Niger-Congo, Nilo-Saharan, and Khoisan), while the other three (Afro-Asiatic, Austronesian, and Indo-European) are more closely associated with other parts of the world. Afro-Asiatic languages, especially Arabic, dominate North Africa and are understood in Islamic areas of Sub-Saharan Africa as well. Amharic in Ethiopia and Somali of Somalia are also Afro-Asiatic languages. The Malayo-Polynesian language family is limited to the island of Madagascar, which many believe was first settled by seafarers from Indonesia some 1500 years ago. Indo-European languages, especially French, English, Portuguese, and Afrikaans, are a legacy of colonialism and are widely used today.

Of the three language groups found exclusively in the region, the Niger-Congo language group is by far the most influential. This linguistic group originated in West Africa and includes Mandingo, Yoruba, Ful (ani), and Igbo, among others. Around 3000 years ago a people of the Niger-Congo stock began to expand out of western Africa into the equatorial zone. This group, called the Bantu, commenced one of the most far-ranging migrations in human history, which introduced agriculture into large areas of central and southern Africa. One Bantu group migrated east across the fringes of the rainforest to settle in the Lake Victoria basin in East Africa, where they formed an eastern Bantu core that later pushed south all the way to South Africa (Figure 6.28). Another group moved south, into the rainforest proper. The equatorial rainforest belt immediately adjacent to the original Bantu homeland had been very sparsely settled by the ancestors of the modern pygmies (a distinct people noted for their short stature and hunting skills). Pygmy groups, having entered into close trading relations with the Bantu newcomers, eventually came to speak Bantu languages as well. Although several pygmy populations have persisted to the present, their original languages disappeared long ago.

Once the Bantu migrants had advanced beyond the rainforest into the savannas and woodlands, their agricultural techniques proved highly successful and their influence expanded (see Figure 6.28). Around 650 CE, Bantu-speaking peoples reached South Africa. Over the centuries, the various languages and dialects of the many Bantu-speaking groups, which often were separated from each other by considerable distances, gradually diverged from each other. Today there are several hundred distinct languages in the Bantu subfamily of the great Niger-Congo group. All Bantu languages, however, remain closely related to each other, and a speaker of one can generally learn any other without undue difficulty.

Most individual Sub-Saharan languages are limited to relatively small areas and are significant only at the local scale. One language in the Bantu subfamily, Swahili, eventually became the most widely spoken Sub-Saharan language. Swahili originated as a trade language on the East African coast, where several merchant colonies from Arabia were established around 1100 CE. A hybrid society grew up in a narrow coastal band of modern Kenya and Tanzania, speaking a language of Bantu structure enriched with many Arabic words. Swahili became the primary language only in the narrow coastal belt, but it spread far into the interior as the language of trade. After independence, both Kenya and Tanzania

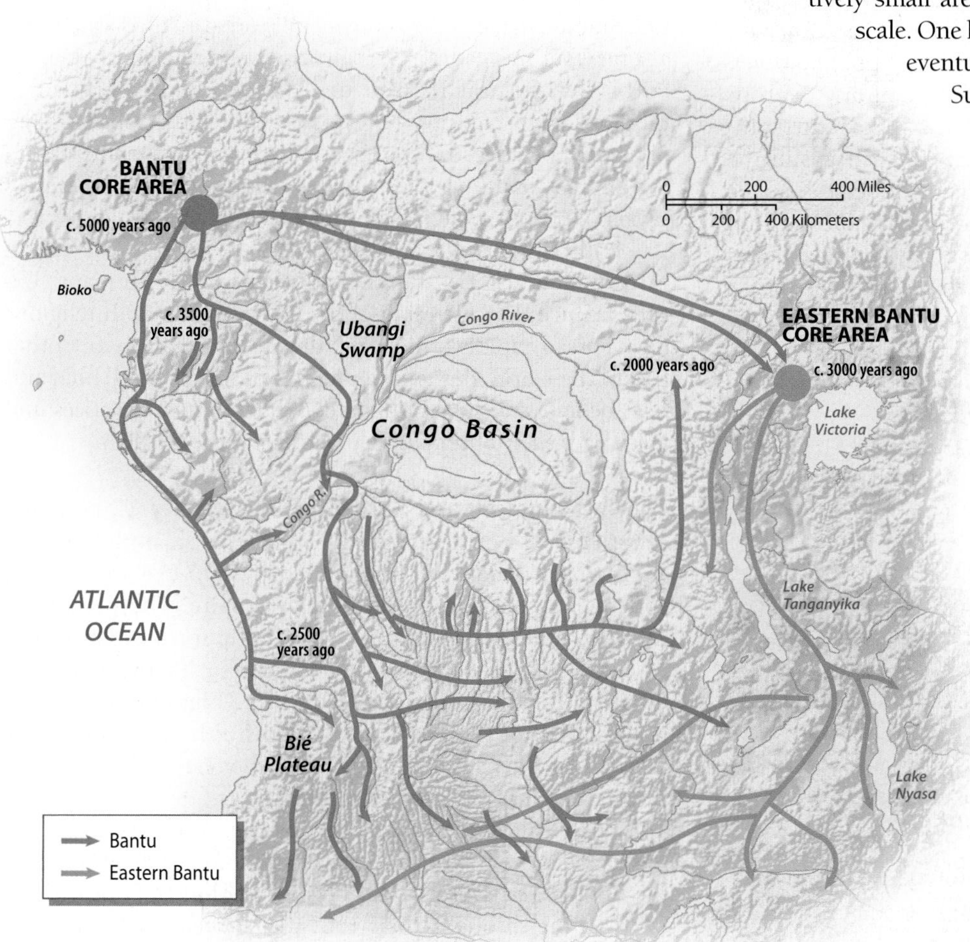

◄ **Figure 6.28 Bantu Migrations** Bantu languages, a subfamily of the Niger-Congo language family, are widely spoken throughout Sub-Saharan Africa. The out-migration of Bantu tribes from an original core in West Africa and a secondary core in East Africa helps to explain the diffusion of Bantu languages, which include Zulu, Swahili, Bemba, Shona, Lingala, and Kikuyu, among others.
(From James L. Newman, The Peopling of Africa, New Haven, CT: Yale University Press, 1995, 141)

adopted Swahili as an official language. Swahili, with an estimated 70 million speakers, is the lingua franca of East Africa. It has generated a fairly extensive literature and is often studied in other regions of the world.

Language and Identity Ethnic identity, as well as linguistic affiliation, has historically been highly unstable over much of Sub-Saharan Africa. The tendency was for new groups to form when people threatened by war fled to less settled areas, where they often mixed with peoples from other places. In such circumstances, new languages arise quickly, and divisions between groups become blurred. Nevertheless, distinct **tribes** formed that initially consisted of a group of families or clans with a common kinship, language, and definable territory. The impetus to formalize tribal boundaries came from European colonial administrators, who were eager to establish a fixed indigenous social order to control native peoples. In this process, a flawed cultural map of Sub-Saharan Africa evolved. Some tribes were artificially divided, meaningless names were applied, and territorial boundaries were often misinterpreted.

Social boundaries between different ethnic and linguistic groups have become more stable in recent years, and some individual languages have become particularly important for communication on a national scale. Wolof in Senegal; Mandingo in Mali; Mossi in Burkina Faso; Yoruba, Hausa, and Igbo in Nigeria; Kikuyu in central Kenya; and Zulu, Xhosa, and Sesotho in South Africa are all nationally significant languages spoken by millions of people (Figure 6.29). None, however, has the status of being the official language of any country. With the end of apartheid in South Africa, the country officially recognized 11 languages, although English is still the lingua franca of business and government. Indeed, a single language has a clear majority status in only a handful of Sub-Saharan countries. The more linguistically homogeneous states include Somalia (where virtually everyone speaks Somali) and the very small states of Rwanda, Burundi, Swaziland, and Lesotho.

▼ **Figure 6.29 Multilingual South Africa** A South African sign warns pedestrians not to cross a highway in three languages: English, Afrikaans, and Zulu. Throughout Sub-Saharan Africa, many people are multilingual because of the diversity of languages that exist in each state.

European Languages In the colonial period, European countries used their own languages for administrative purposes in their African empires. Education in the colonial period also stressed literacy in the language of the imperial power. In the postindependence period, most Sub-Saharan African countries have continued to use the languages of their former colonizers for government and higher education. Few of these new states had a clear majority language that they could employ, and picking any minority tongue would have aroused the opposition of other peoples. The one exception is Ethiopia, which maintained its independence during the colonial era. The official language is Amharic, although other indigenous languages are also spoken.

Two vast blocks of European languages exist in Africa today: Francophone Africa, encompassing the former colonies of France and Belgium, where French serves as the main language of administration; and Anglophone Africa, where the use of English prevails (see inset, Figure 6.27). Early Dutch settlement in South Africa resulted in the use of Afrikaans (a Dutch-based language) by several million South Africans. Portuguese is spoken in Angola and Mozambique, former colonies of Portugal. Interestingly, when South Sudan gained its independence from Sudan in 2011, it changed its official language from Arabic to English.

Religion

Indigenous African religions generally are classified as *animist*, a somewhat misleading catchall term used to classify all local faiths that do not fit into one of the handful of "world religions." Most animist religions are centered on the worship of nature and ancestral spirits, but the internal diversity within the animist tradition is vast. Classifying a religion as animist says more about what it is not than what it actually is.

Both Christianity and Islam actually entered the region early in their histories, but they advanced slowly for many centuries. Since the beginning of the 20th century, both religions have spread rapidly—more rapidly, in fact, than in any other part of the world. But tens of millions of Africans still hold animist beliefs, and many others combine animist practices and ideas with their observances of Christianity and Islam.

The Introduction and Spread of Christianity Christianity came first to northeast Africa. Kingdoms in both Ethiopia and central Sudan were converted by 300 CE—the earliest conversions outside of the Roman Empire. The peoples of Ethiopia and Eritrea adopted the Coptic form of Christianity and have thus historically looked to Egypt's Christian minority for their religious leadership (Figure 6.30). Today roughly half of the population of both Ethiopia and Eritrea follows Coptic Christianity; most of the rest are Muslim, but there are still some animist communities.

European settlers and missionaries introduced Christianity to other parts of Sub-Saharan Africa beginning in the 1600s. The Dutch, who began to colonize South Africa at this time, brought their Calvinist Protestant faith. Later European immigrants to South Africa brought Anglicanism

▲ **Figure 6.30 Eritrean Christians** Coptic Christians gather for an Easter celebration in Asmara, Eritrea.

and other Protestant creeds, as well as Catholicism. A substantial Jewish community also emerged, concentrated in the Johannesburg area. Most black South Africans eventually converted to one or another form of Christianity as well. In fact, churches in South Africa were instrumental in the long fight against white racial supremacy. Religious leaders such as Bishop Desmond Tutu were outspoken critics of the injustices of apartheid and worked to bring down the system.

Elsewhere in Africa, Christianity came with European missionaries, most of whom arrived after the mid-1800s. As was true in the rest of the world, missionaries had little success where Islam had preceded them, but they eventually made numerous conversions in animist areas. As a general rule, Protestant Christianity prevails in areas of former British colonization, while Catholicism is more important in former French, Belgian, and Portuguese territories. There are nearly 180 million Catholics in the region. In the postcolonial era, African Christianity has diversified, at times taking on a life of its own, independent from foreign missionary efforts. Increasingly active in the region are various Pentecostal, Evangelical, and Mormon missionary groups. It is difficult to map the distribution of Christianity in Africa, however, because it has spread irregularly across the entire non-Islamic portion of the region.

The Introduction and Spread of Islam Islam began to advance into Sub-Saharan Africa 1000 years ago (Figure 6.31). Berber traders from North Africa and the Sahara introduced

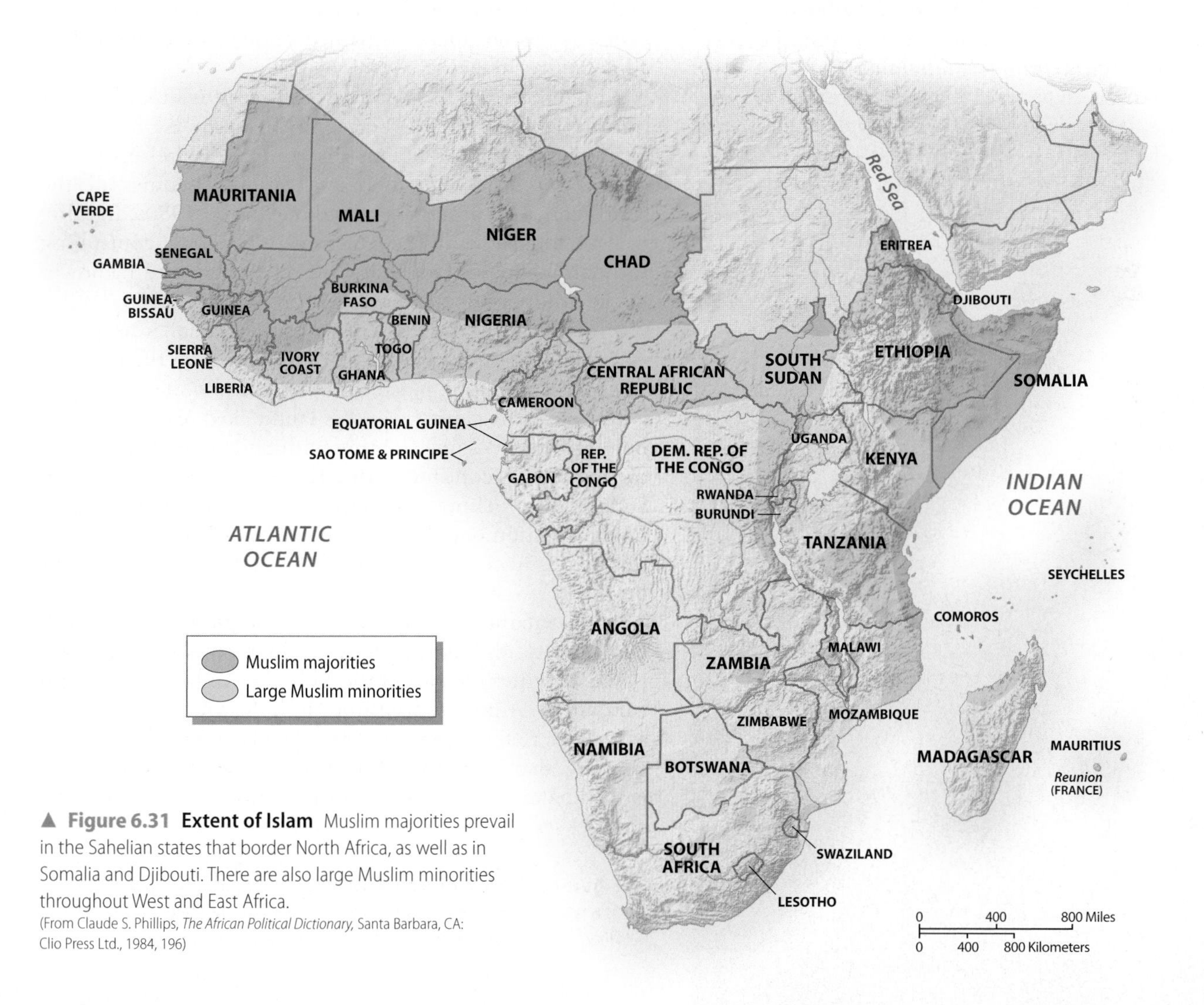

▲ **Figure 6.31 Extent of Islam** Muslim majorities prevail in the Sahelian states that border North Africa, as well as in Somalia and Djibouti. There are also large Muslim minorities throughout West and East Africa.
(From Claude S. Phillips, *The African Political Dictionary*, Santa Barbara, CA: Clio Press Ltd., 1984, 196)

the religion to the Sahel, and by 1050 the Kingdom of Tokolor in modern Senegal emerged as the first Sub-Saharan Muslim state. Somewhat later, the ruling class of the powerful Mande-speaking mercantile empires of Ghana and Mali converted as well. In the 14th century, the emperor of Mali astounded the Muslim world when he and his huge entourage made the pilgrimage to Mecca, bringing with them so much gold that they set off an inflationary spiral throughout Southwest Asia (Figure 6.32).

Mande-speaking traders, whose networks spanned the area from the Sahel to the Gulf of Guinea, gradually introduced the religion to other areas of West Africa. There, however, many peoples remained committed to animism, and Islam made slow and fitful progress. Even in the Sahel, syncretic (blended) forms of Islam prevailed through the 1700s. In the early 1800s, however, the pastoral Fulani people launched a series of successful holy wars designed to shear away animist practices and to establish pure Islam. Today orthodox Islam prevails through most of the Sahel. Farther south, Muslims are mixed with Christians and animists, but their numbers continue to grow, and their practices tend to be orthodox as well.

Interaction Between Religious Traditions The southward spread of Islam from the Sahel, coupled with the northward dissemination of Christianity from the port cities, has generated a complex religious frontier across much of West Africa. In Nigeria, the Hausa are firmly Muslim, while the southeastern Igbo are largely Christians. The Yoruba of the southwest are divided between Christians and Muslims. In the more remote parts of Nigeria, moreover, animist traditions remain vital. Despite this religious diversity, religious conflict in Nigeria has been relatively rare until recently. In 2000, seven Nigerian states imposed Muslim sharia (religious) laws, which triggered intermittent violence ever since, especially in the northern cities of Kano and Kaduna. In the last few years, an armed jihadist group called Boko Haram formed in northeastern Nigeria, escalating the levels of religious violence in the north and in the south. In response, the Nigeria military launched an offensive in 2013 to dislodge this group from the Borno region and has killed hundreds of sympathetic supporters. Still, most of West Africa's regional conflicts have been framed more along ethnic terms, rather than religious ones.

Religious conflict historically has been far more acute in northeastern Africa, where Muslims and Christians have struggled against each other for centuries. Such a clash eventually led to the creation of the region's newest state when South Sudan separated from the country of Sudan in 2011. Islam was introduced to Sudan in the 1300s by an invasion of Arabic-speaking pastoralists who destroyed the indigenous Coptic Christian kingdoms of the area. Within a few hundred years, northern and central Sudan had become completely Islamic. The southern equatorial province of Sudan, where tropical diseases and extensive wetlands prevented Arab advances, remained animist or converted to Christianity under British colonial rule.

In the 1970s, the Arabic-speaking Muslims of northern and central Sudan began to build an Islamic state. Experiencing both religious discrimination and economic exploitation, the Christian and animist peoples of the south launched a massive rebellion. In the 1980s, fighting became intense, with the government generally controlling the main towns and roads and the rebels maintaining power in the countryside. A peace was brokered in 2003, and as part of the peace agreement, southern Sudan was promised an opportunity to vote on secession from the north in 2011. The vote took place, and the new nation was formed with Juba as its capital. Yet this landlocked territory is still not at peace, as some fighting with Sudanese troops along the border has continued. In 2012, South Sudan's access to Sudan's oil pipeline was temporarily cut off, which meant the new nation could not export its oil. Clearly, conflict in this area is likely to continue as the nation-building efforts proceed.

Sub-Saharan Africa is a land of religious vitality. Both Christianity and Islam are spreading rapidly, and devotional activities are part of the daily flow of life in cities and rural areas. Animism continues to hold widespread appeal as well, so that new and syncretic forms of religious expression are also emerging. With such a diversity of faiths, it is fortunate that religion is not typically the cause of overt conflict.

◀ **Figure 6.32 West African Muslims** Residents of Djenne, Niger, gather in an open market in front of the city's great mosque. Much of the Sahelian region converted to Islam more than six centuries ago.

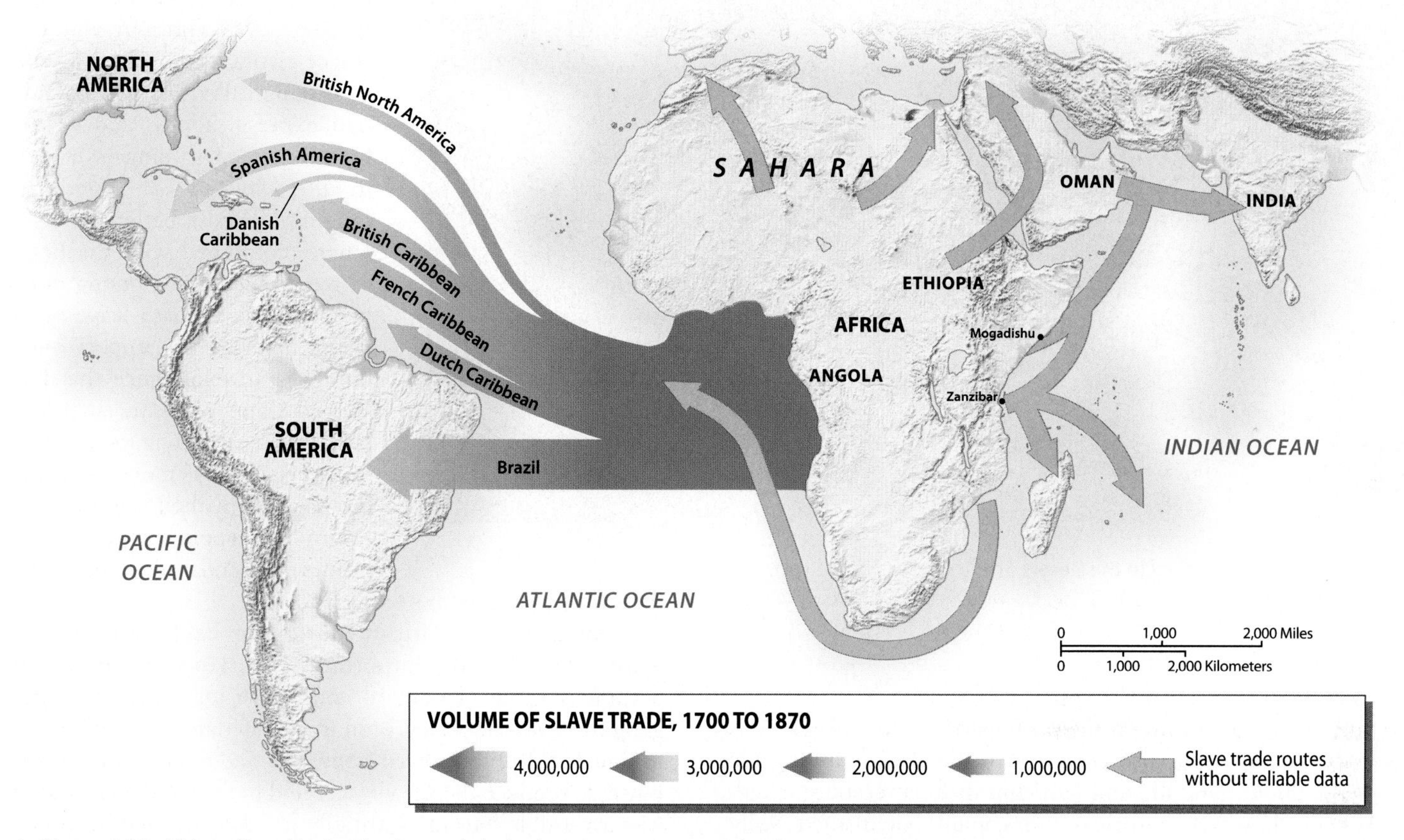

▲ **Figure 6.33 African Slave Trade** The slave trade had a devastating impact on Sub-Saharan societies. From ship logs, it is estimated that 12 million Africans were shipped to the Americas to work as slaves on sugar, cotton, and rice plantations; the majority went to Brazil and the Caribbean. Yet other slave routes existed, although the data are less reliable. Africans from south of the Sahara were used as slaves in North Africa. Others were traded across the Indian Ocean into Southwest Asia and South Asia.

Globalization and African Culture

The slave triangle that linked Africa to the Americas and Europe set in process patterns of cultural diffusion that transferred African peoples and practices across the Atlantic. Tragically, slavery damaged the demographic and political strength of African societies, especially in West Africa, from where most slaves were taken. An estimated 12 million Africans were shipped to the Americas as slaves from the 1500s until 1870 (Figure 6.33). Slavery impacted the entire region, sending Africans not just to the Americas, but also to Europe, North Africa, and Southwest Asia. The vast majority, however, worked on plantations across the Americas.

Out of this tragic displacement of people came a blending of African cultures with Amerindian and European ones. African rhythms are at the core of various musical styles, from rumba to jazz, the blues, and rock and roll. Brazil, the largest country in Latin America, is claimed to be the second largest "African state" (after Nigeria) because of the huge Afro-Brazilian population. Thus, the forced migration of Africans as slaves had a huge cultural influence on many areas of the world.

Cultural exchanges are never one-way. Foreign languages, religions, and dress were absorbed by Africans who remained in the region. With independence in the 1960s and 1970s, several states sought to rediscover their ancestral roots by openly rejecting European cultural elements. Ironically, the search for traditional religions found African scholars traveling to the Caribbean and Brazil to consult with Afro-religious practitioners about ceremonial elements lost to West Africa.

So, too, have contemporary movements of Africans influenced the cultures of many world regions. Perhaps one of the most celebrated persons of African ancestry today is President Barack Obama, the son of a Kenyan man and a woman from Kansas. Obama's heritage and upbringing embody the forces of globalization. In Kenya, he is hailed as part of the modern African diaspora—young professionals (and their offspring) who leave the continent and make their mark somewhere else. As an indicator of our global age, Kenyan President Mwai Kibaki declared a Kenyan national holiday after Obama won the U.S. presidential election in November 2008.

Popular culture in Africa, like everywhere else in the world, is a dynamic mixture of global and local influences. Kwaito, a popular musical form in South Africa, sounds a lot like rap from the United States. A closer listen, however, reveals an incorporation of local rhythms, lyrics in Zulu and Xhosa,

▲ **Figure 6.34** **Festival in the Desert** The Tuareg band Igbayen plays at the Mali Festival in the Desert. This annual winter festival in the oasis town of Essakane, Mali, draws thousands of Malian musicians, Tuareg nomads, and Western tourists.

and themes about life in the post-apartheid townships. Zola, a kwaito superstar, also costarred in the Oscar-winning film *Tsotsi,* about gang life and hardship in Johannesburg. Lagos, Nigeria, however, is Africa's film capital. Nicknamed Nollywood, Nigeria is the second most prolific producer of movies, after India. Nollywood films are popular all over Africa, but some directors have global ambitions (see *Exploring Global Connections: The Reach of Nollywood*).

Music in West Africa Nigeria is the musical center of West Africa, with a well-developed and cosmopolitan recording industry. Modern Nigerian styles such as juju, highlife, and Afro-beat are influenced by jazz, rock, reggae, and gospel, but they are driven by an easily recognizable African sound.

Farther up the Niger River lies the country of Mali. Bamako, the capital, is also a music center that has produced scores of recording artists. Many Malian musicians descend from a traditional caste of musical storytellers performing on either the traditional kora (a cross between a harp and a lute) or the guitar. The musical style is strikingly similar to that of blues from the Mississippi Delta—so much so that Ali Farka Touré, from northern Mali, was referred to as the Bluesman of Africa because of his distinctive, yet familiar, guitar work. Each January, not far from Timbuktu, music fans from West Africa and Europe gather for the Festival in the Desert. In this remote Saharan locale, a celebration of Malian music and Tuareg nomadic culture draws together Western tourists, African musicians, and nomads (Figure 6.34). Even with the recent conflict in Mali (which we will discuss later), the festival was held in January 2012. As fighting intensified, the gathering was cancelled in 2013, but is scheduled for 2014, which is a hopeful sign.

For many African emigrants, music can also be deeply nostalgic. No singer expresses this longing for home better than Cesária Evora from Cape Verde. A Grammy-winning singer, Evora is known for a style of singing called *mornas,* ballads that express sadness and yearning. Her need to sing is rooted in the emotion of *sodade*—a longing to find a better life elsewhere, combined with the hope of returning to live among one's people. This need to leave one's home in order to support one's family is an experience that has shaped Cape Verdeans since the late 1800s, when mass emigration began. As more Africans emigrate from their countries of origin in response to dire economic or political conditions, their music reflects this experience.

Contemporary African music can be both commercially and politically important. Nigerian singer Fela Kuti became a voice of political conscience for Nigerians struggling for true democracy. Born in an elite family and educated in England, Kuti borrowed from jazz, traditional, and popular music to produce the Afro-beat sound in the 1970s. The music was irresistible, but his searing lyrics also attracted attention. Acutely critical of the military government, he sang of police harassment, the inequities of the international economic order, and even Lagos's infamous traffic. Singing in English and Yoruba, his message was transmitted to a larger audience, and he became a target of state harassment. Kuti died in Lagos in 1997 from complications related to AIDS; yet his music and politics later became the subject of the awarding-winning Broadway musical *Fela!*

Pride in East African Runners Ethiopia and Kenya have produced many of the world's greatest distance runners. Abebe Bikila won Ethiopia's and Africa's first Olympic gold medal, running barefoot at the Rome games in 1960. Since then, nearly every Olympic Games has yielded medals for Ethiopia and Kenya. At the London Olympics in 2012, Kenyan runners won 11 medals and Ethiopians 7. These states, along with South Africa, were the top medal winners for Sub-Saharan Africa in London.

Running is a national pastime in Kenya and Ethiopia, where elevation—Addis Ababa sits at 7300 feet (2200 meters) and Nairobi at 5300 feet (1600 meters)—increases oxygen-carrying capacity. Past medalists Haile Gebrselassie and Derartu Tulu are national celebrities in Ethiopia, where they are idealized by the country's youth. Tulu, the first black African woman to win a gold medal in distance running, is a forceful voice for women's rights in a country where women are discouraged from putting on running shorts. In the 2012 Olympics, Ethiopian women Tiki Gelana and Tirunesh Dibaba won gold in the women's marathon and 10,000-meter race, respectively (Figure 6.35).

Exploring Global Connections

The Reach of Nollywood

Africa's undisputed film capital is Lagos, Nigeria. Referred to as Nollywood, the Nigerian movie industry currently grinds out 2500 films a year (50 films a week) and employs about 1 million people. Nollywood makes more films than Hollywood. Relying on relatively inexpensive digital video technology, most of these movies are shot in a few days and with budgets of $10,000 to $20,000. The typical themes of religion, ethnicity, corruption, witchcraft, the spirit world, violence, and injustice resonate with African audiences. They are almost always shot on location, in city streets, office buildings, and homes or in the countryside. Nollywood films can be bloody and exploitative. They can also be overtly evangelical, promoting Christianity over indigenous faiths. Many are acted in English, but there are also movies for Yoruba, Igbo, and Hausa speakers.

The DVD Market Rather than being released in theaters, most Nollywood movies go directly to DVD and are rarely viewed beyond Africa (Figure 6.3.1). The shelf life of these movies is rather short; production companies need to make back their money in a couple of weeks before pirated copies undermine profits. However, a $20,000 film can earn $500,000 in DVD sales in a couple of weeks. Consequently, distribution of films remains tightly controlled by Igbo businessmen, sometimes called the Alaba cartel for their distribution center on the outskirts of Lagos. But this is beginning to change as an African middle-class population wants to go to theaters and directors want to create higher-quality films for Nigeria and beyond.

In 2010, the President of Nigeria, Goodluck Jonathan, pledged to improve the quality of Nigerian filmmaking by making grants available to established directors. With budgets of 10 times the going rate, directors can improve the postproduction quality of their work. Also the Nigerian diaspora in Europe and North America is growing the market beyond Africa and encouraging more transnational themes. Nigerian director Tony Abulu released his film *Doctor Bello* in a dozen U.S. cities in 2013. A recipient of a Nigerian government grant, he worked in New York City and Nigeria to produce his movie about an African-American cancer specialist who turns to a Nigerian (Dr. Bello) for a nontraditional cancer treatment for a gravely ill patient. Things go badly, and the American doctor journeys to Nigeria in search of a secret elixir in a mountain hideaway. The film was not a box office hit, but it was an important step for Nigeria's maturing filmmakers who are seeking recognition beyond Africa.

▼ **Figure 6.3.1 Location Filming in Nigeria** A Nigerian film crew prepares to shoot a scene in Lagos. Nicknamed Nollywood, Nigeria produces more films each year than Hollywood.

▲ **Figure 6.3.2 Queen of Nollywood** Nigerian actress Omotola Jalade Ekeinde was recognized by Time Magazine as one of the 100 most influential people in 2013. One of Nigeria's top movie stars, she is also active in charity work around the world.

The Queen of Nollywood The reigning queen of Nollywood is Omotola Jalade-Ekeinde, popularly referred to as *OmoSexy*. This talented actress has made over 300 feature films. She is also a singer and a philanthropist known for her Youth Empowerment Program in Nigeria and her work throughout Africa for the UN (Figure 6.3.2). In the plot of *Ijé*, a recent film partially shot in Los Angeles, Omotola is charged with the murder of three men, including her husband. The story is told through her sister, who comes from Nigeria to visit her in a U.S. jail. Such transnational intrigue is part of Nollywood's trajectory as it reaches out to the world.

◄ **Figure 6.35 East African Distance Runners** Runners from Ethiopia and Kenya lead the pack in the 10,000-meter race during the 2012 London Olympics. Ethiopian Tirunesh Dibaba (third from the front) took gold, and Kenyans Sally Jepkosgei Kipyego and Vivian Jepkemoi Cheruiyot took silver and bronze.

REVIEW

6.6 What are the dominant religions of Sub-Saharan Africa, and how have they diffused throughout the region?

6.7 What are the ways in which African peoples have influenced world regions beyond Africa?

Geopolitical Framework: Legacies of Colonialism and Conflict

The duration of human settlement in Sub-Saharan Africa is unmatched by that of any other region. Evidence shows that humankind originated here, evidently evolving from a rather apelike *Australopithicus* all the way to modern *Homo sapiens*. Over the millennia, many diverse ethnic groups formed in the region. Although conflicts among these groups have occurred, cooperation and coexistence among different peoples have also continued over centuries.

Some 2000 years ago, the Kingdom of Axum arose in northern Ethiopia and Eritrea, strongly influenced by political models derived from Egypt and Arabia. The first wholly indigenous African states were founded in the Sahel around 700 CE. Kingdoms such as Ghana, Mali, Songhai, and Kanem-Bornu grew rich by exporting gold to the Mediterranean and importing salt from the Sahara, and they maintained power over lands to the south by monopolizing horse breeding and mastering cavalry warfare (Figure 6.36).

Over the next several centuries, a variety of other states emerged in West Africa. Some were large, but diffuse empires, organized through elaborate hierarchies of local kings and chiefs; others were centralized states focused on small centers of power. The Yoruba of southwestern Nigeria, for example, developed a city-state form of government, and their homeland is still one of the most urbanized and densely populated parts of Africa. The most powerful Sub-Saharan states continued to be located in the Sahel until the 1600s, when European coastal trade undercut the lucrative trans-Saharan networks. Subsequently, the focus of power moved to the Gulf of Guinea, where well-organized African states (such as Dahomey and Ashanti), as well as Europeans, took advantage of the lucrative opportunities presented by the slave trade and in the process increased their military and economic power.

Thus, prior to European colonization, Sub-Saharan Africa presented a complex mosaic of kingdoms, states, and tribal societies. With the arrival of Europeans, patterns of social organizations and ethnic relations were changed forever. As Europeans rushed to carve up the continent to serve their imperial ambitions, they set up various administrations that heightened ethnic tensions and promoted hostility. Many of the region's modern conflicts can trace their roots back to the colonial era, especially the drawing of political boundaries.

European Colonization

Unlike the relatively rapid colonization of the Americas, Europeans needed centuries to gain effective control of Sub-Saharan Africa. Portuguese traders arrived along the coast of West Africa in the 1400s, and by the 1500s they were well established in East Africa as well. Initially, the Portuguese made large profits, converted a few local rulers to Christianity, established several fortified trading posts, and acquired dominion over the Swahili trading cities of the east. They stretched themselves too thin, however, and many of their settlements failed or were lost to other colonizers. Only where a sizable population of mixed African and Portuguese descent emerged, along the coasts of modern Angola and Mozambique and on the islands of Cape Verde, could Portugal maintain power. Along the Swahili, or eastern, coast, they were eventually expelled by Arabs from Oman, who subsequently established their own mercantile empire in the area.

The Disease Factor One of the main reasons for the Portuguese failure was the disease environment of Africa.

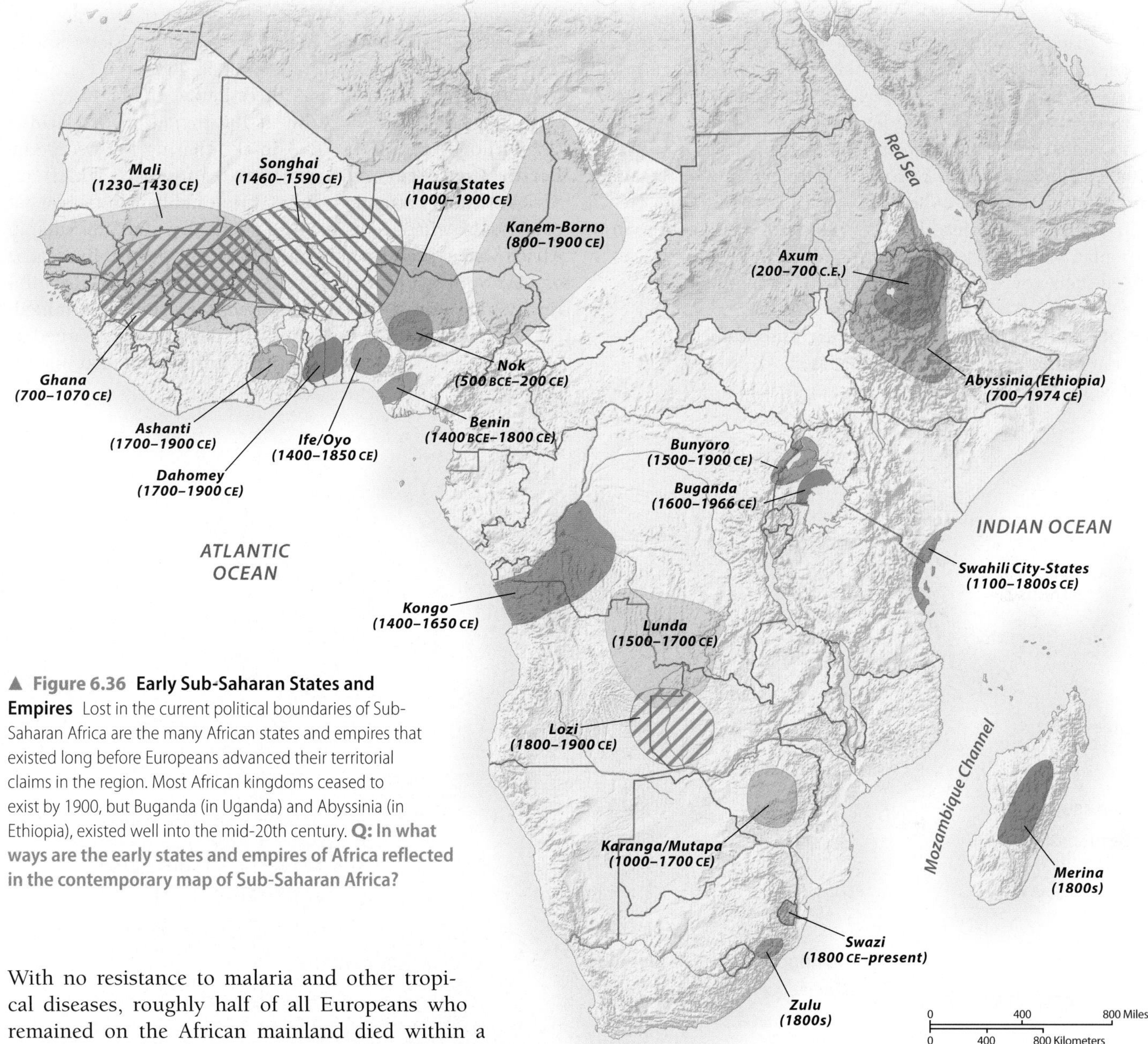

▲ **Figure 6.36 Early Sub-Saharan States and Empires** Lost in the current political boundaries of Sub-Saharan Africa are the many African states and empires that existed long before Europeans advanced their territorial claims in the region. Most African kingdoms ceased to exist by 1900, but Buganda (in Uganda) and Abyssinia (in Ethiopia), existed well into the mid-20th century. **Q: In what ways are the early states and empires of Africa reflected in the contemporary map of Sub-Saharan Africa?**

With no resistance to malaria and other tropical diseases, roughly half of all Europeans who remained on the African mainland died within a year. Protected both by their formidable armies and by the diseases of their native lands, African states were able to maintain an upper hand over European traders and adventurers well into the 1800s. Unlike the Americas, where European conquest was facilitated by the introduction of Old World diseases that devastated native populations (see Chapters 4 and 5), Sub-Saharan Africa's native diseases limited European settlement until the mid-19th century.

The hazards of malaria and other tropical diseases, such as sleeping sickness, were compensated for by the lure of profit, and other European traders soon followed the Portuguese. By the 1600s, Dutch, British, and French firms dominated the lucrative export of slaves, gold, and ivory from the Gulf of Guinea. The Dutch also established a settler colony in South Africa, safely outside of the tropical disease zone, to supply their ships bound for Indonesia. For the next 200 years, European traders came and went, occasionally building fortified coastal posts, but they almost never ventured inland and seldom had any real influence on African rulers. By exporting millions of slaves, however, they had a profoundly negative impact on African society.

In the 1850s, European doctors discovered that a daily dose of quinine would offer protection against malaria, radically changing the balance of power in Africa. Explorers immediately began to penetrate the interior of the continent, while merchants and expeditionary forces began to move inland from the coast. The first imperial claims soon followed. The French quickly grabbed power along the easily navigated Senegal River, while the British established protectorates over the indigenous states of the Gold Coast (modern-day Ghana), in which the British guaranteed protection for these territories in exchange for various trade preferences.

Also in the early 1800s, two small territories were established in West Africa so that freed and runaway slaves would have a place to return to in Africa. The territory that was to become Liberia was set up by the American Colonization Society in 1822 to settle former African-American slaves. By 1847, it was the independent and free state of Liberia. Sierra Leone served a similar function for ex-slaves from the British Caribbean, but it remained a protectorate of Britain until the 1960s. Despite the good intentions behind the creation of these territories, they, too, were colonies. Liberia, in particular, was imposed on existing indigenous groups who viewed their new "African" leaders with contempt.

The Scramble for Africa In the 1880s, European colonization of the region quickly accelerated, leading to the so-called scramble for Africa. By this time, due to the invention of the machine gun, no African state could long resist European force.

As the colonization of Africa intensified, tensions among the colonizing forces of Britain, France, Belgium, Germany, Italy, Portugal, and Spain mounted. Rather than risk war, 13 countries convened in Berlin at the invitation of German Chancellor Bismarck in 1884 in a gathering known as the **Berlin Conference**. During the conference, which no African leaders attended, rules were established as to what constituted "effective control" of a territory, and Sub-Saharan Africa was carved up and traded like properties in a game of Monopoly (Figure 6.37). Exact boundaries in the interior, which was still poorly known, were not determined, and a decade of "orderly" competition remained as imperial armies marched inland.

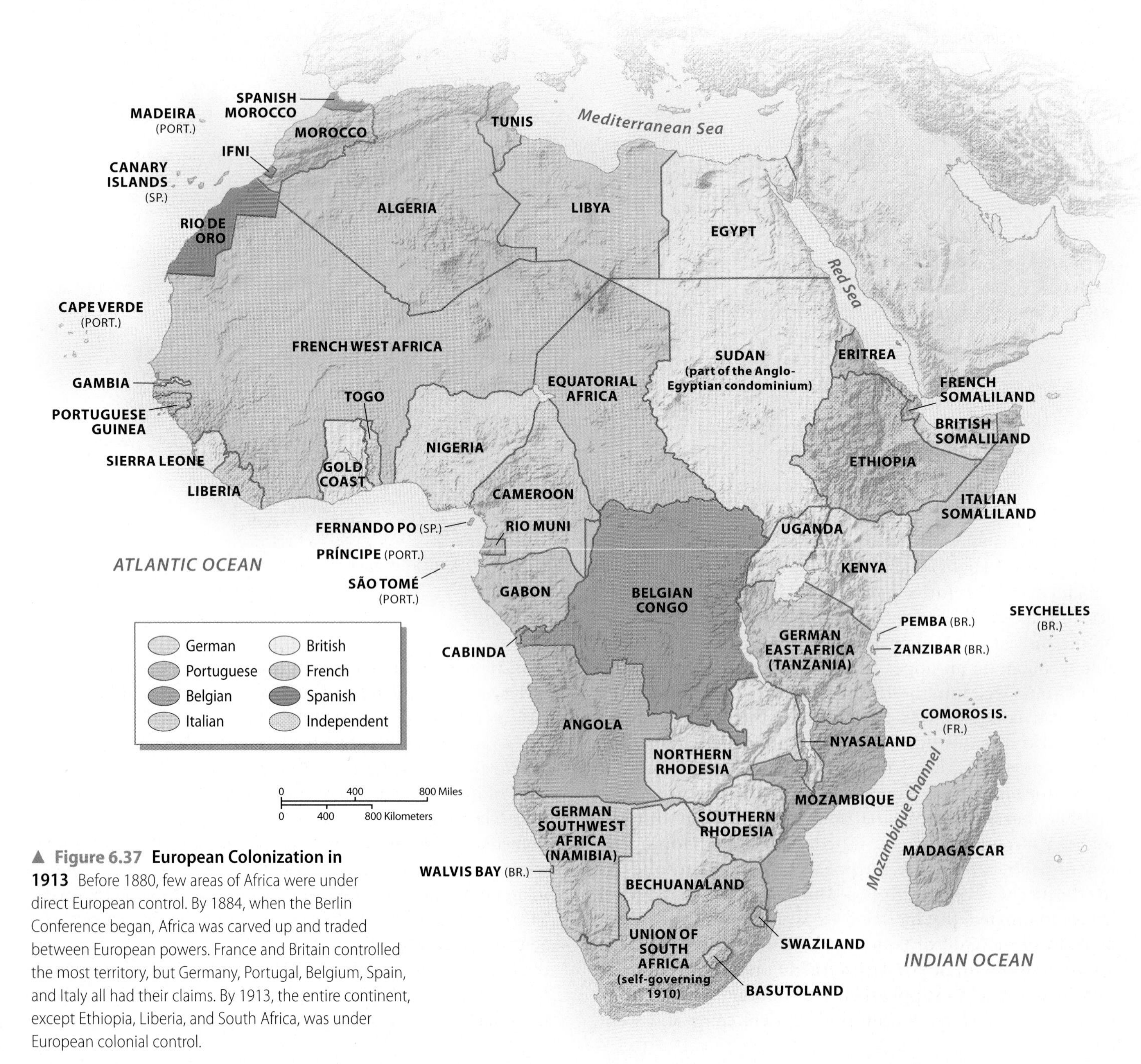

▲ **Figure 6.37 European Colonization in 1913** Before 1880, few areas of Africa were under direct European control. By 1884, when the Berlin Conference began, Africa was carved up and traded between European powers. France and Britain controlled the most territory, but Germany, Portugal, Belgium, Spain, and Italy all had their claims. By 1913, the entire continent, except Ethiopia, Liberia, and South Africa, was under European colonial control.

Although European weapons in the 1880s were far superior to anything found in Africa, several indigenous states did mount effective resistance campaigns. For example, in South Africa, Zulu warriors resisted British invasion into their lands in what has been termed the Anglo-Zulu Wars (1879–1896). Eventually, European forces prevailed everywhere except Ethiopia. The Italians had conquered the Red Sea coast and the far northern highlands (modern Eritrea) by 1890, and they quickly set their sights on the large Ethiopian kingdom, called Abyssinia, which had been vigorously expanding for several decades. In 1896, however, Abyssinia defeated the invading Italian army, earning the respect of the European powers. In the 1930s, fascist Italy launched a major invasion of the country, by this time renamed Ethiopia, to redeem its earlier defeat, and with the help of poison gas and aerial bombardment, it quickly prevailed. However, by 1942 Ethiopia had regained its freedom.

Although Germany was a principal instigator of the scramble for Africa, it lost its own colonies after suffering defeat in World War I. Britain and France then partitioned most of Germany's African empire between themselves. Figure 6.37 shows the colonial status of the region in 1913, prior to Germany's territorial loss.

While the Europeans were cementing their rule over Africa, South Africa was inching toward political independence, at least for its white population. South Africa was one of the oldest colonies in Sub-Saharan Africa, and in 1910 it became the first to obtain its political independence from Europe. However, because of its formalized system of discrimination and racism, it was hardly a symbol of liberty. Ironically, as the Afrikaners tightened their political and social control over the nonwhite population through their policy of apartheid (or "separateness"), introduced in 1948, the rest of the continent was preparing for political independence from Europe.

Decolonization and Independence

Decolonization of Sub-Saharan Africa happened rather quickly and peacefully, beginning in 1957. Independence movements, however, had sprung up throughout the continent, some dating back to the early 1900s. Workers' unions and independent newspapers became voices for African discontent and the hope for freedom. Black intellectuals, who had typically studied abroad, were influenced by the ideas of the **Pan-African Movement** led by W. E. B. Du Bois and Marcus Garvey in the United States. Founded in 1900, the movement's slogan of "Africa for Africans" encouraged a trans-Atlantic liberation effort. Nevertheless, Europe's hold on Africa remained secure through the 1940s and early 1950s, even as other colonies in South and Southeast Asia gained their independence.

By the late 1950s, Britain, France, and Belgium decided that they could no longer maintain their African empires and began to withdraw. (Italy had already lost its colonies during World War II, and Britain gained Somalia and Eritrea.) Once started, the decolonization process moved rapidly. By the mid-1960s, virtually the entire region had achieved independence. In most cases, the transition was relatively peaceful and smooth, with the exception of southern Africa.

Dynamic African leaders put their mark on the region during the early independence period. Men such as Kenya's Jomo Kenyatta, Ivory Coast's Felix Houphuët-Boigney, Tanzania's Julius Nyerere, and Ghana's Kwame Nkrumah became powerful father figures who molded their new nations (Figure 6.38). President Nkrumah's vision for Africa was the most expansive. After helping to secure independence for Ghana in 1957, his ultimate aspiration was the political unity of Africa. Although his dream was never realized, it set the stage for the founding of the Organization of African Unity (OAU) in 1963, which was renamed the **African Union (AU)** in 2002. The AU is a continent-wide organization headquartered in Addis Ababa, Ethiopia, whose main role has been to mediate disputes between neighbors. Certainly, in the 1970s and 1980s it was a constant voice of opposition to South Africa's minority rule, and the AU intervened in some of the more violent independence movements in southern Africa.

Southern Africa's Independence Battles Independence did not come easily to southern Africa. In Southern Rhodesia (modern-day Zimbabwe), the problem was the presence of some 250,000 white residents, most of whom owned large farms. Unwilling to see power pass to the country's black majority, then some 6 million strong, these settlers declared themselves the rulers of an independent, white-supremacist state in 1965. The black population continued to resist, however, and in 1978 the Rhodesian government was forced to give up power. The renamed country of Zimbabwe was, henceforth, ruled by the black majority, although the remaining whites still formed an economically privileged community. Since the mid-1990s, disputes over government land reform (splitting up the large commercial farms, mostly owned by whites, and giving the land to black farmers) and President Robert Mugabe's strongman politics have resulted in serious racial and political tensions, as well as the collapse of the country's economy.

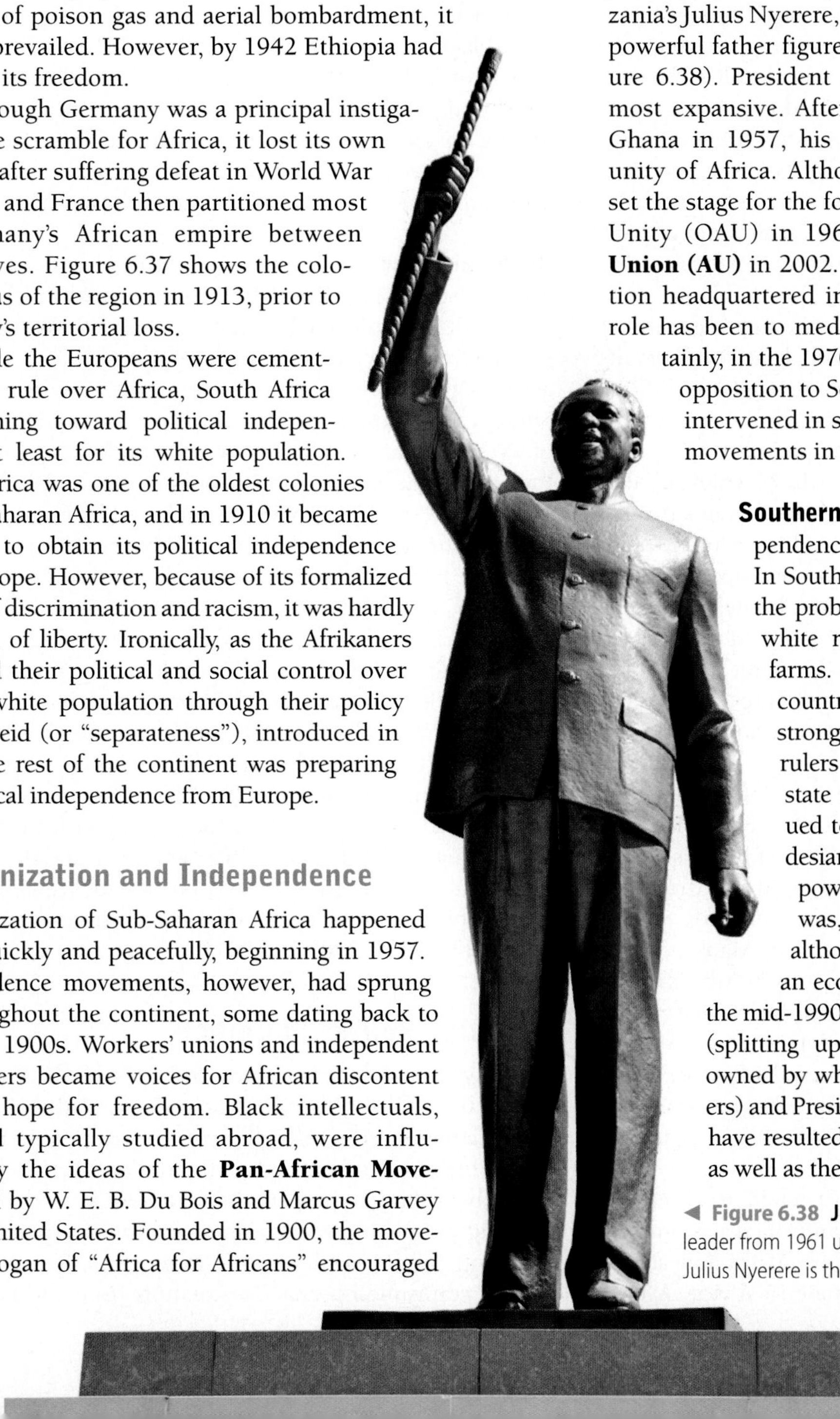

◀ **Figure 6.38 Julius Nyerere Monument** An independence leader from 1961 until he retired from the Presidency in 1985, Julius Nyerere is the founding father of Tanzania.

In the former Portuguese colonies, independence came violently. Unlike the other imperial powers, Portugal refused to relinquish its colonies in the 1960s. As a result, the people of Angola and Mozambique turned to armed resistance. The most powerful rebel movements adopted a socialist orientation and received support from the Soviet Union and Cuba. A new Portuguese government came to power in 1974, however, and it withdrew abruptly from its African colonies. At this point, Marxist regimes quickly came to power in both Angola and Mozambique. The United States, and especially South Africa, responded to this perceived threat by supplying arms to rebel groups that opposed the new governments. Fighting dragged on for nearly three decades in Angola and Mozambique. The countryside in both states became heavily laced with land mines, a threat that rural residents still face if they unknowingly walk or farm in a mined area. Efforts to clear the mines and make the land usable are ongoing, but uneven. With the end of the Cold War, however, outsiders lost interest in continuing these conflicts, and sustained efforts began to negotiate a peace settlement. Mozambique has been at peace since the mid-1990s. After several failed attempts at peace in Angola, the Angolan army signed a peace treaty with rebels in 2002 that ended a 27-year conflict in which more than 300,000 people died and 3 million Angolans were displaced.

Apartheid's Demise in South Africa While fighting continued in the former Portuguese zone, South Africa underwent a remarkable transformation. From 1948 through the 1980s, the ruling Afrikaners' National Party was firmly committed to apartheid (or separateness) that was prejudicial to nonwhite groups. Under apartheid, only whites enjoyed real political freedom, while blacks were denied even citizenship in their own country—technically, they were citizens of segregated **homelands**. Likened to reservations created for Native Americans in the United States, homelands were rural, overcrowded, and on marginal land. Moreover, to ensure the notion that every black had a homeland, some 3 million blacks were forcibly relocated into homelands during apartheid, and residence outside of a homeland was strictly regulated.

Opposition to apartheid began in the 1960s, intensifying and becoming more violent by the 1980s. Blacks led the opposition, but coloureds and Asians (who suffered severe, but less extreme, discrimination) also opposed the Afrikaner government. As international pressure mounted, white South Africans found themselves ostracized. Many corporations refused to do business there, and South African athletes (regardless of color) were banned from most international competitions, such as the Olympics and World Cup Soccer. Increasing numbers of whites also opposed the apartheid system, and many businesspeople began to believe that apartheid threatened to undermine their economic endeavors.

The first major change came in 1990, when South Africa withdrew from Namibia, which it had controlled as a protectorate since the end of World War I. South Africa now stood alone as the single white-dominated state in Africa. A few years later, the leaders of the Afrikaner-dominated National Party decided they could no longer resist the pressure for change. In 1994, free elections were held in which Nelson Mandela (1918-2013), a black leader who had been imprisoned for 27 years by the old regime, emerged as the new president. Black and white leaders pledged to put the past behind them and work together to build a new, multiracial South Africa. The homelands themselves were the first to be eliminated from the political map of the new South Africa. Since Mandela's presidency, orderly elections have been held, and South Africans have elected Thabo Mbeki for two terms (1999–2009) and Jacob Zuma in 2009.

Unfortunately, the legacy of apartheid is not so easily erased. Residential segregation is officially illegal, but neighborhoods are still sharply divided along racial lines. Under the multiracial political system, a black middle class emerged, but most blacks remain extremely poor (and most whites remain prosperous). Violent crime has increased, and rural migrants and immigrants have poured into South African cities, producing a xenophobic anti-immigrant backlash. Because the political change was not matched by a significant economic transformation, the hopes of many people are frustrated.

Regional Blocks Given the scale of the African continent, it is not surprising that as the region gained independence, groups of states formed regional organizations to facilitate intraregional exchange and development. The two most active regional organizations are the Southern African Development Community (SADC) and the Economic Community of West African States (ECOWAS). Both were founded in the 1970s, but became more prominent in the 1990s (Figure 6.39). SADC and ECOWAS are anchored by the region's two largest economies: South Africa and Nigeria, respectively.

The Economic Community of Central African States (ECCAS) was founded in the mid-1980s and is headquartered in Libreville, Gabon. Its effectiveness has been hampered by the political instability of the area. Several ECCAS members also are members of SADC, which has been a more dynamic group. The East African Community has five member states. It was founded in 1967, collapsed 10 years later, and was reinstated in 2000. In 2010, it began its own common market for goods, labor, and capital with the intention of creating a common currency within a decade. Interestingly, these trade groups have facilitated growing foreign direct investment between African countries, which nearly doubled between 2007 and 2011.

Enduring Political Conflict

Although most Sub-Saharan countries made a relatively peaceful transition to independence, virtually all of them immediately faced a difficult set of institutional and political problems. In several cases, the old authorities had done virtually nothing to prepare their colonies for independence. Lacking an institutional framework for independent government, countries such as the Democratic Republic of the Congo confronted a chaotic situation from the beginning.

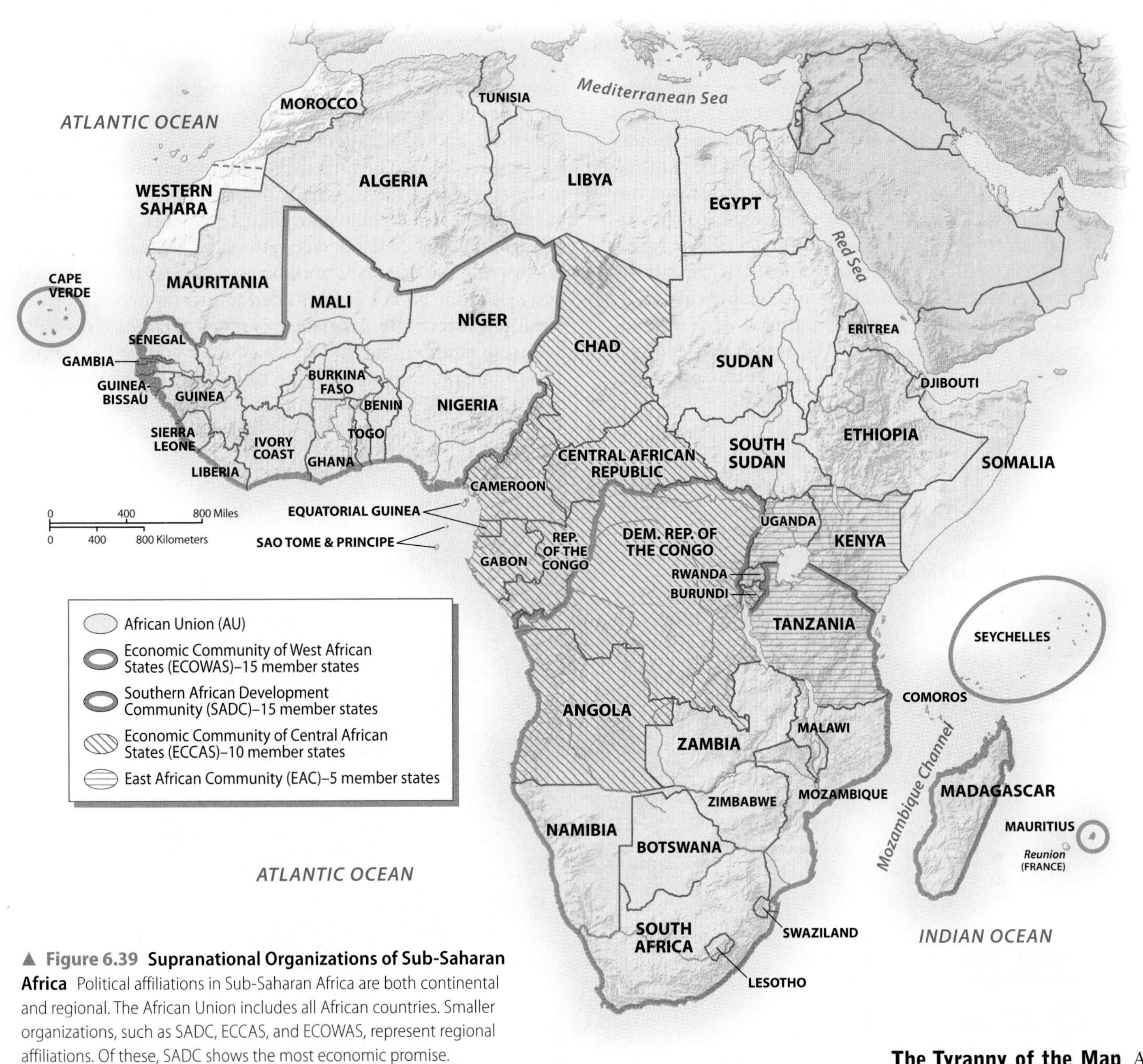

▲ **Figure 6.39 Supranational Organizations of Sub-Saharan Africa** Political affiliations in Sub-Saharan Africa are both continental and regional. The African Union includes all African countries. Smaller organizations, such as SADC, ECCAS, and ECOWAS, represent regional affiliations. Of these, SADC shows the most economic promise.

Only a handful of Congolese had received higher education, let alone been trained for administrative posts. The indigenous African political framework had been essentially destroyed by colonization, and in most cases very little had been built in its place.

Even more problematic in the long run was the political geography of the newly independent states. Civil servants could always be trained and administrative systems built, but little could be done to rework the region's basic political map. The problem was the fact that the European colonial powers had essentially ignored indigenous cultural and political patterns, both in dividing Africa among themselves and in creating administrative subdivisions within their own imperial territories.

The Tyranny of the Map All over Africa, different ethnic groups found themselves forced into the same state with peoples of different linguistic and religious backgrounds, many of whom had recently been their enemies. At the same time, some of the larger ethnic groups of the region found their territories split between two or more countries. The Hausa people of West Africa, for example, were divided between Niger (formerly French) and Nigeria (formerly British), each of which they had to share with several former rivals.

Given the imposed political boundaries, it is no wonder that many African countries struggled to generate a common sense of national identity or establish stable political institutions. **Tribalism**, or loyalty to the ethnic group rather than to the state, has emerged as the bane of African political life. Especially in rural areas, tribal identities usually supersede national ones. Because virtually all of Africa's countries inherited an inappropriate set of colonial borders, you might assume that they would have been better off redrawing a new political map based on indigenous identities. However, such a strategy was impossible, as all the leaders of the newly independent states realized. Any new territorial divisions would have created winners and losers, and thus would have resulted in more conflict. Moreover, because ethnicity in Sub-Saharan Africa was traditionally fluid, and because many groups were territorially intermixed, it would have been difficult to generate a clear-cut system of division. Finally, most African ethnic groups were considered too small to form viable countries. With these complications in mind, the new African leaders, meeting in 1963 to form the OAU, agreed that colonial boundaries should remain. The violation of this principle, they argued, would lead to pointless wars between states and endless civil struggles within them.

Despite the determination of Africa's leaders to build their new nations within existing boundaries, challenges to the states began soon after independence. Figure 6.40 shows the ethnic and political conflicts that have disabled parts of Africa since 2005. The human cost of this turmoil is several million refugees and internally displaced persons. **Refugees** are people who flee their state because of a well-founded fear of persecution based on race, ethnicity, religion, or political orientation. Nearly 3 million Africans were considered refugees at the end of 2012. Added to this figure are another 5 million **internally displaced persons (IDPs)**. IDPs have fled from conflict, but still reside in their country of origin. The Democratic Republic of the Congo has the largest number of IDPs (2.7 million), followed by Somalia (1.1 million).

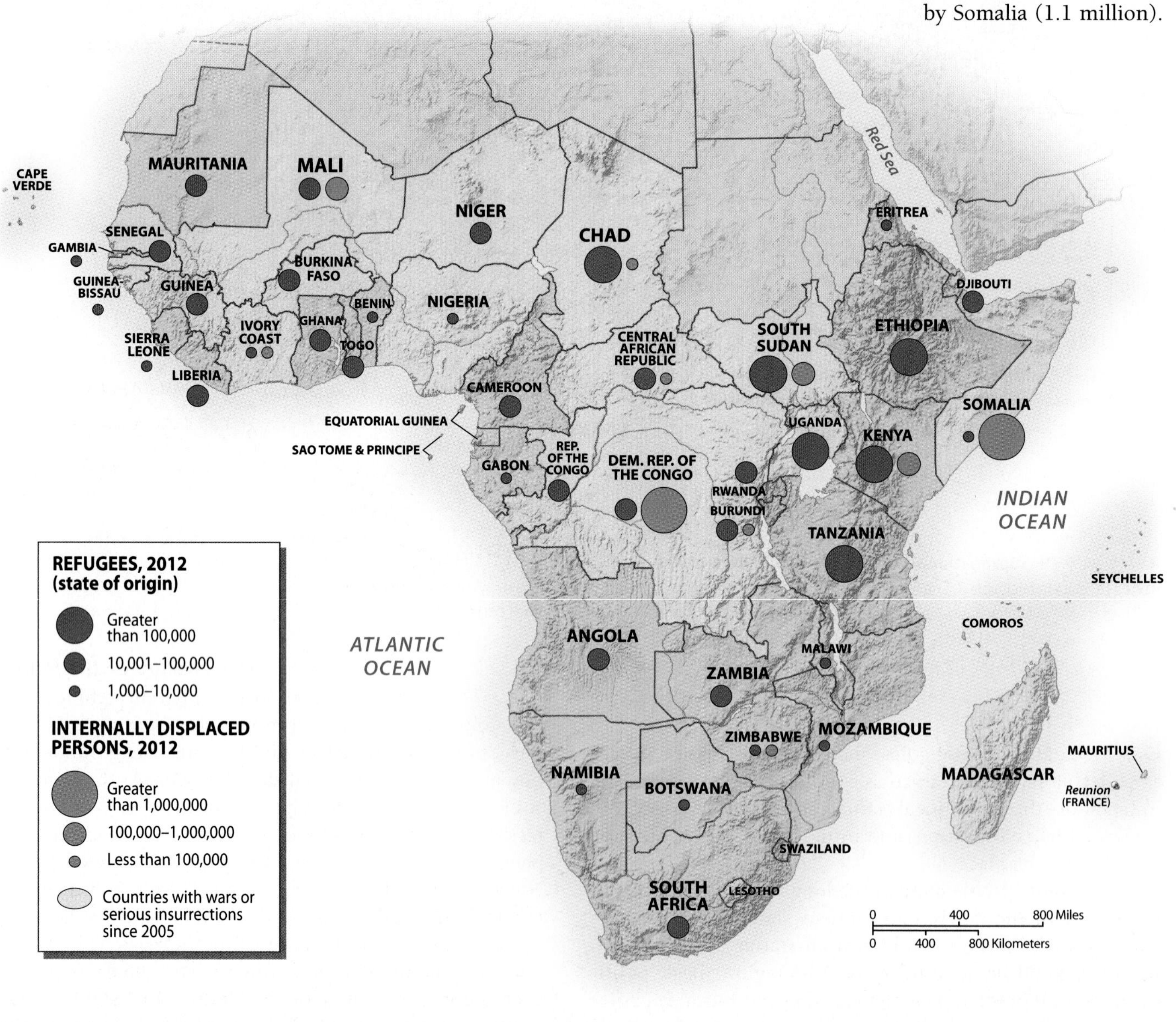

▲ **Figure 6.40 Geopolitical Issues in Sub-Saharan Africa** Several Sahelian and Central African countries have experienced wars or serious insurrections since 2005. These same states are also likely to produce refugees (and internally displaced persons). As of 2012, nearly 3 million Africans were refugees and 5 million were internally displaced.
(Data from UNHCR 2012)

These populations are not technically considered refugees, making it difficult for humanitarian NGOs and the UN to assist them. The good news is that even the number of IDPs is declining. In 2009, the IDP population was estimated at 13 million people, nearly three times the present figure. Significantly fewer IDPs and refugees today is another indicator that the impact of many of the region's prolonged conflicts from the 1990s and early 2000s is diminishing.

Ethnic Conflicts In the 1990s, nearly two-thirds of the states in the region were experiencing serious ethnic conflict and, in the case of the Rwanda, even genocide. As Figure 6.40 shows, most of the conflict since 2005 occurred in the Sahelian states and Central Africa. Fortunately, in the past few years peace has returned to Sierra Leone, Liberia, Ivory Coast, and Angola, states that produced large numbers of refugees in the 1990s.

Many attributed the cycles of violence in Sierra Leone and Liberia to the availability of diamonds as a means of financing the conflict. Although the relationship between resources and conflict is complex, the term **conflict diamonds** was employed when discussing the diamond trade in West Africa in the 1990s (see *Everyday Globalization: The African Origins of the Diamond Engagement Ring*). One result of the public concern about conflict diamonds was a certification scheme adopted in 2002 called the Kimberly Process. Its aim is to keep conflict diamonds out of the global market and thus avoid tainting the image of the diamond business. In Ivory Coast, where conflict began as violence spilled over from Liberia in 2002, a peace deal was brokered in 2007 between the New Forces rebel group in the north and the government-controlled south.

The deadliest ethnic and political conflict in the region has been in the Democratic Republic of the Congo. It is estimated that between 1998 and 2010, 5.4 million people died there, although many of the deaths were from war-induced starvation and disease, rather than bullets or machetes. In addition, half a million refugees are living outside the country and five times as many IDPs within it (see Figure 6.40). In 1996 and 1997, a loose alliance of armed groups from Rwanda (led by Tutsis) and Uganda joined forces with other militias in the Congo and marched their way across the country, installing Laurent Kabila as president. Under Kabila's rocky and ruthless leadership, which ended in assassination in 2001, rebel groups again invaded the Congo from Uganda and Rwanda and soon controlled the northern and eastern portions of the country, while the Kinshasa-based government loosely controlled the western and southern portions.

With Kabila's death, his son Joseph took power and signed a peace accord with the rebels in 2002. In 2003, rebel leaders were made part of a transitional government, and an unsteady peace was in place, with help from the UN, the AU, and Western donors. Remarkably, when elections were held in 2006, Joseph Kabila was elected president, and he was reelected in 2011. Yet Sub-Saharan Africa's largest state in terms of territory has only limited experience with democracy. Its civil service barely functions, corruption is rampant, and there are few roads and little working infrastructure for the nearly 70 million people who live there. Moreover, armed groups scattered throughout the country continue to commit serious crimes, including mass rapes, torture, and murders. Due to years of conflict, the formal economy is small, and the informal economy dominates. Consequently, foreign aid accounted for 38 percent of the country's gross national income (GNI) in 2011. The level of violence is certainly lower now, but this is still one of the region's most troubled spots.

A low-intensity ethnic conflict in northern Mali heated up in 2012 as a Tuareg-based National Movement for the Liberation of Azawad (MNLA) proclaimed independence from the Bamako-based government in the south, largely controlled by Mande-speaking groups. (Azawad is a territory in northern Mali.) The idea of Tuareg independence, or autonomy, has existed for decades, but the MNLA was formed recently, in 2011, partly by armed Tuareg fighters returning from Libya after the fall of Gaddafi's regime as part of the Arab Spring (see Chapter 7). As fighting intensified, a military coup occurred in the capital of Bamako, and President Toure was removed from office in March 2012. Leaders of the coup were dissatisfied with the state's inability to fight Tuareg rebels, emboldened by the arrival of weapons from Libya. The AU has denounced the actions and suspended Mali from the union, but the UN estimates that over 200,000 people have already been uprooted by the violence. As of 2013, a peace was negotiated between the Tuareg and the Malian government with support of UN, ECOWAS, and French troops.

Although this latest conflict seems to be driven by political events in North Africa and ethnic rivalries, other conflicts point to ecological pressures brought on by drought and/or climate change that undermine traditional patterns of resource sharing between farmers and pastoralists. There is no definitive connection between resource scarcity and ethnic clashes, but several of the region's current conflicts—from Nigeria to Mali, South Sudan to Somalia—are in semiarid zones becoming more vulnerable to drought and famine.

Secessionist Movements Problematic African political boundaries have occasionally led to attempts by territories to secede and form new states. The Shaba (or Katanga) Province in what was then the state of Zaire (now the Democratic Republic of the Congo) tried to leave its national territory soon after independence. The rebellion was crushed a couple of years after it started with the help of France and Belgium. Zaire was unwilling to give up its copper-rich territory, and the former colonialists had economic interests in the territory worth defending. Similarly, the Igbo in oil-rich southeastern Nigeria declared an independent state of Biafra in 1967. After a short, but brutal war, during which Biafra was essentially starved into submission, Nigeria was reunited.

In 1991, the government of Somalia disintegrated, and the territory has been in civil war for over 20 years. The lack of political control facilitated the rise of piracy; Somali pirates raid vessels and extort ransom from ships from the

Everyday Globalization

The African Origins of the Diamond Engagement Ring

The customary practice of sealing a marriage engagement with a diamond ring has much to do with the discovery of enormous diamond mines in southern Africa in the 1870s and the remarkable advertising efforts of DeBeers, a South African firm that held near monopolistic control of the diamond market throughout the 20th century. Yes, *a diamond is forever*, as the DeBeers ad campaign states, but the practice of demonstrating your devotion with a diamond ring is a modern convention.

Rise of the Monopoly DeBeers was founded by Cecil Rhodes, a British businessmen and imperialist, who saw firsthand the potential of the Kimberly, South Africa, mines (**Figure 6.4.1**). After the first major diamond discovery there in the 1870s, and following several mergers of smaller companies, DeBeers Consolidated Mines came into existence in 1888. The DeBeers mining empire steadily expanded to include diamond mines in Botswana, Namibia, and Canada. Throughout most of the 20th century, DeBeers controlled 90 percent of the global diamond trade.

DeBeers mines, operating with government partners, are open-pit, subsurface, and even offshore (**Figure 6.4.2**). They are all capital-intensive formal businesses that extract the diamond-carrying ore. Later, the rough diamonds need to be cut, which usually means they are brought to Antwerp, Belgium, but also to Tel Aviv, Israel, and New York City.

DeBeers' genius was threefold: It expanded the supply of quality diamonds, controlled the global market, and convinced a growing middle class that diamonds were an ideal demonstration of love. After diamond sales began to flatten in Europe in the 1920s, DeBeers began marketing in the United States, hiring Madison Avenue advertising agencies in the 1930s to promote the concept that a suitor should spend at least one month's wage on a diamond ring. The idea seems to have stuck; most marriages in the United States include a diamond ring.

Commitment Written in Bling Social historians claim that the popularity of diamond rings in the 1930s coincided with the striking down of "breach of promise to marry" laws, which allowed women to sue men for breaking off an engagement. In the convention of the day, virginity of brides was of utmost importance. Giving a woman a ring showed a suitor's commitment. And if the groom walked, the jilted bride could keep the ring as a penalty for default on the engagement. This anachronistic view of the marriage contract seems foreign today, now that women have their own careers and often outearn their partners. Clearly, diamond rings have a different cultural meaning today, but the "bling" is still a recognizable symbol that a woman is committed to another.

▲ **Figure 6.4.2 Diamond Cutter** A young diamond cutter checks a 2 carat diamond she is polishing in Gaborone, Botswana. In 2008 the government of Botswana launched its own diamond trading company in partnership with DeBeers in order to retain more jobs and income in this diamond producing country.

▼ **Figure 6.4.1 Kimberly Mine** Kimberly, South Africa, was the site of a diamond rush in the 1870s. What began as a flat-topped mountain ended as the "big hole," shown here, on the outskirts of town.

DeBeers' lock on the diamond trade began to unravel in the 1990s. One reason was that Russia became a major diamond producer, marketing outside of the DeBeers commodity chain. Also, the media grabbed hold of the idea of "conflict diamonds" in the 1990s, which began to taint the image of the diamond, especially the African diamond. Through the certification process, blood diamonds are much less of a concern today, but DeBeers' control of the business has also slipped. The company today controls 40 percent of diamond sales, although southern Africa is still a major diamond producer.

1. What are the apparent environmental differences between industrial versus artisan diamond mining?
2. Why might diamonds be an ideal resource to support armed conflict?

Google Earth Virtual Tour Video

http://goo.gl/8cG9Cq

Gulf of Aden to the Indian Ocean. The territory has been ruled by clan-based warlords and their militias, who have informally divided the country into clan-based units. **Clans** are social units that are branches of a tribe or an ethnic group larger than a family. Early in the conflict, the northern portion of the country declared its independence as a new country—Somaliland. Somaliland has a constitution, a functioning parliament, government ministries, a police force, a judiciary, and a president. The territory produces its own currency and passports. Yet no country has recognized this territory, in part because no government exists in Somalia to negotiate the secession. In 1998, neighboring Puntland also declared itself an autonomous state. Although it does not seek outright independence like Somaliland, Puntland is creating its own administration. Meanwhile, Islamic insurgents with their well-armed militias control the south around Mogadishu. In the past three years, UN peacekeeping forces, with the help of the Kenyan and Ethiopian military and U.S. drones, have been fighting the insurgents, trying to reclaim the cities. The need for stability has been exacerbated by three years of drought, creating a humanitarian emergency. In 2012, the Somalian government was able to reform, but conflict still persists. In June 2013, the Shabab, a violent Islamist group formed just to control Mogadishu, bombed a UN compound in the city (Figure 6.41).

Only two territories in the region have successfully seceded. In 1993, Eritrea gained independence from Ethiopia after two decades of civil conflict. This territorial secession is striking because Ethiopia gave up its access to the Red Sea, making it landlocked. Yet the creation of Eritrea still did not bring about peace. After years of fighting, the transition to Eritrean independence began remarkably well. Unfortunately, border disputes between the two countries erupted in 1998, resulting in the deaths of some 100,000 troops. In 2000, a peace accord was reached, and the fighting stopped. The second example is South Sudan, which gained its independence from Sudan in 2011 after some three decades of violent conflict between the largely Arab and Muslim north and the Christian and animist south. It is too early to measure South Sudan's success as an independent state, but its formation, along with that of Eritrea, may provide a model for territorial solutions in war-torn areas in the region. Still, major transformations of Africa's political map should not be expected.

▼ **Figure 6.41 Conflict in Somalia** Somali National Government soldiers patrol the streets of Mogadishu after Shebab insurgents attacked the United Nations compound in that city in June of 2013. The country has suffered through more than 20 years of conflict.

REVIEW

6.8 What are the processes behind Sub-Saharan Africa's political map, and why have there been relatively few boundary changes since the 1960s?

6.9 What are the present major conflicts in this region, and where are they occurring?

Economic and Social Development: The Struggle to Develop

By almost any measure, Sub-Saharan Africa is the poorest world region. According to World Bank estimates, two-thirds of the population live on less than $2 per day, although in 1993 the figure was 78 percent. Due to poverty and low life expectancy, nearly all the states in the region are ranked at the bottom of the Human Development Index. Although some demographically small or resource-rich states, such as Botswana, Equatorial Guinea, Mauritius, the Seychelles, and South Africa, have much higher per capita gross national income, adjusted for purchasing power parity (GNI-PPP), the regional average was about $2200 in 2011 (Table 6.2). By way of comparison, the figure for South Asia, the next poorest region, was $3300.

Table 6.2 Development Indicators

Country	GNI per capita, PPP 2011	GDP Average Annual % Growth 2000-11	Human Development Index (2011)[1]	Percent Population Living Below $2 a Day	Life Expectancy (2013)[2]	Under Age 5 Mortality Rate (1990)	Under Age 5 Mortality Rate (2011)	Adult Literacy (% ages 15 and older) Male / Female	Gender Inequality Index (2013)[3,1]
Angola	5,230	12.2	.508	70.2	51	243	158	83/58	-
Benin	1,610	3.8	.436	75.3	59	177	106	55/30	0.618
Botswana	14,550	4.0	.634	49.4	47	53	26	84/85	0.485
Burkina Faso	1,330	5.8	.343	72.6	56	208	146	37/22	0.609
Burundi	610	3.5	.355	93.5	56	183	139	73/62	0.476
Cameroon	2,330	3.2	.495	30.4	54	145	127	79/63	0.628
Cape Verde	3,980	6.2	.586	40.9	74	58	21	89/79	–
Central African Republic	810	1.3	.352	80.1	49	169	164	69/43	0.654
Chad	1,540	8.7	.340	83.3	50	208	169	45/24	–
Comoros	1,110	1.9	.429	65.0	60	122	79	80/70	–
Congo	3,240	4.5	.534	74.4	58	119	99	--/--	0.610
Dem. Rep. of Congo	340	5.6	.304	95.2	49	181	168	77/57	0.681
Djibouti	2,450	4.0	.445	41.2	61	122	90	--/--	–
Equatorial Guinea	25,620	14.4	.554	–	52	190	118	97/91	–
Eritrea	580	0.9	.351	–	62	138	68	79/58	–
Ethiopia	1,110	8.9	.396	66.0	62	198	77	49/29	–
Gabon	13,740	2.4	.683	19.6	63	94	66	92/85	0.492
Gambia	1,750	3.4	.439	55.9	58	165	101	60/40	0.594
Ghana	1,810	6.3	.558	51.8	61	121	78	73/61	0.565
Guinea	1,020	2.6	.355	69.6	56	228	126	52/30	–
Guinea-Bissau	1,230	2.5	.364	78.0	54	210	161	68/41	–
Ivory Coast	1,710	1.0	.432	46.3	50	151	115	65/47	.632
Kenya	1,710	4.4	.519	67.2	60	98	73	91/84	0.608
Lesotho	2,050	3.8	.461	62.3	48	88	86	83/96	0.534
Liberia	540	6.0	.388	94.9	60	241	78	65/57	0.658
Madagascar	950	3.2	.483	92.6	64	161	62	67/62	–
Malawi	870	5.1	.418	90.5	54	227	83	81/68	0.573
Mali	1,060	5.1	.344	78.7	54	257	176	43/20	0.649
Mauritania	2,530	5.6	.467	47.7	61	125	112	65/51	0.643
Mauritius	14,330	3.9	.737	–	73	24	15	95/86	0.377
Mozambique	960	7.5	.327	81.8	50	226	103	71/43	0.582
Namibia	6,610	4.9	.608	51.1	63	73	42	89/88	0.455
Niger	720	4.2	.304	75.2	57	314	125	43/15	0.707
Nigeria	2,290	6.8	.471	84.5	52	214	124	72/50	–
Reunion	–	–	–	–	79	–	–	–	–
Rwanda	1,270	7.9	.434	82.4	63	156	54	75/68	0.414
São Tomé and Principe	2,080	–	.525	54.2	66	96	89	94/85	–
Senegal	1,940	4.1	.470	55.2	63	136	65	62/39	0.540
Seychelles	26,280	2.9	.806	<2	73	17	14	91/92	–
Sierra Leone	1,140	6.6	.359	76.1	45	267	185	54/31	0.643
Somalia	–	–	–	–	54	180	180	–	–
South Africa	10,710	3.7	.629	31.3	58	62	47	91/87	0.462
South Sudan	–	–	–	–	54	217	121	–	–
Swaziland	6,610	2.4	.536	60.4	49	83	104	88/87	0.525
Tanzania	1,500	7.0	.476	87.9	60	158	68	79/67	0.556
Togo	1,040	2.6	.459	52.7	56	147	110	71/44	0.566
Uganda	1,310	7.7	.456	64.7	58	178	90	83/65	0.517
Zambia	1,490	5.7	.448	82.6	56	193	83	81/62	0.623
Zimbabwe	–	–5.1	.397	–	56	79	67	95/90	0.544

[1]United Nations, *Human Development Report, 2013.*

[2]Population Reference Bureau, *World Population Data Sheet, 2013.*

[3]Gender Inequality Index—A composite measure reflecting inequality in achievements between women and men in three dimensions: reproductive health, empowerment and the labor market that ranges between 0 and 1. The higher the number, the greater the inequality.

Source: World Bank, *World Development Indicators, 2013.*

The decade of the 1990s was especially difficult for Sub-Saharan Africa, with most states registering low or negative growth rates. Not only was the AIDS crisis raging, but also an economic and debt crisis of the 1980s and 1990s prompted the introduction of **structural adjustment programs**. Promoted by the International Monetary Fund (IMF) and the World Bank, structural adjustment programs typically reduce government spending, cut food subsidies, and encourage private-sector initiatives. Yet these same policies have caused immediate hardships for the poor, especially women and children, and have led to social protest, most notably in cities. The idea of debt forgiveness for Africa's poorest states has gradually gained acceptance as a strategy for reducing human suffering by redirecting monies that would have gone to debt repayment to the provision of services and the building of infrastructure. American economist Jeffrey Sachs argues that in order for the region to get out of the poverty trap, it will need substantial sums of new foreign aid and investment.

On the positive side, since 2000 there have been signs of economic growth. For most states, the average annual growth rates from 2000 to 2011 were positive. Several countries have seen average annual growth rates of 5 percent or more; Angola's annual growth rate averaged a stunning 12 percent for the decade (see Table 6.2). Although this is good news overall, such figures can be deceptive. Some are due to soaring oil prices (notably so for Angola, Chad, and Nigeria), whereas others are due to countries beginning from a very low base after years of conflict (Mozambique, Sierra Leone, and Rwanda). Still, for the first time in many years, Sub-Saharan African economies are growing at a faster rate than their populations. This is very good news.

Roots of African Poverty

In the past, outside observers often attributed Africa's poverty to its colonial history, poorly conceived development policies, corrupt governance, and/or the physical environment. For those who favored environmental explanations for poverty, the region's infertile soils, erratic patterns of rainfall, lack of navigable rivers, and virulent tropical diseases were all pointed to as reasons for underdevelopment. Most contemporary scholars, however, argue that such environmental handicaps are not prevalent everywhere throughout the region and that even where they do exist, they can be—and often have been—overcome by human labor and ingenuity. The favored explanations for African poverty today look much more to historical and institutional factors than to environmental circumstances.

Numerous scholars have singled out the slave trade for its debilitating effect on Sub-Saharan African economic life. Large areas of the region were depopulated, and many people were forced to flee into poor, inaccessible refuges. Colonization was another blow to Africa's economy. European powers invested little in infrastructure, education, and public health and were instead interested mainly in developing mineral and agricultural resources for their own benefit. Several plantation and mining zones did achieve some prosperity under colonial regimes, but strong national economies failed to develop. In almost all cases, the basic transport and communications systems were designed to link administration centers and zones of extraction directly to the colonial powers, rather than to their own surrounding areas. As a result, after achieving independence, Sub-Saharan African countries faced economic and infrastructural challenges that were as daunting as their political problems.

Since 2000, investment in infrastructure projects has grown dramatically, resulting in improvements in transport and trade throughout the region. For decades, South Africa was the only African state with a fully developed modern road network, but major road projects are occurring throughout the continent. Recently, Kenya inaugurated its first superhighway, an eight-lane, 42-kilometer road from Nairobi to Thika (Figure 6.42). Not only has the highway transformed many of the towns along its route, but also it is an expression of growing Chinese investment in this region, since the Chinese firm Wu Yi Co. did much of the engineering and construction work.

Failed Development Policies The first decade or so of independence was a time of relative prosperity and optimism for many African countries. Most of them relied heavily on the export of mineral and agricultural products, and through the 1970s commodity prices generally remained high. Some foreign capital was attracted to the region, and in many cases the European economic presence actually increased after decolonization.

▼ **Figure 6.42 Kenyan Highway** The country's first superhighway was inaugurated in 2012. Built from Nairobi to the northern town of Thika, the road will ease congestion on this heavily traveled route.

In the 1980s, as most commodity prices began to decline, foreign debt began to weigh down many Sub-Saharan countries. By the end of the 1980s, most of the region was in serious economic decline. By the early 1990s, the region's foreign debt was around $200 billion. Although low compared to that of other developing regions (such as Latin America), as a percentage of its economic output, Sub-Saharan Africa's debt was the highest in the world.

Many economists argue that Sub-Saharan African governments enacted counterproductive economic policies and thus brought some of their misery on themselves. Eager to build their own economies and reduce their dependency on the former colonial powers, most African countries followed a course of economic nationalism. Specifically, they set about building steel mills and other forms of heavy industry that were simply not competitive (Figure 6.43). Local currencies were often maintained at artificially elevated levels, which benefited the elite who consumed imported products, but undercut exports.

The largest blunders made by Sub-Saharan leaders were in agricultural and food policies in the postindependence period. The main objective was to maintain a cheap supply of staple foods in the urban areas. Yet the majority of Africans are farmers, who could not make money from their crops because prices were artificially low. Thus, they opted to grow food mainly for subsistence, rather than selling at a loss to national marketing boards. The end result was the failure to meet staple food needs at a time when the population was growing explosively. Many of these price subsidies have since been removed as part of structural adjustment policies.

Reflecting a neoliberal turn toward agricultural production, several Sub-Saharan states—namely, Uganda, Tanzania, Somalia, and Mozambique—have experienced a 21st-century land grab driven by food-insecure governments. Dutch geographer Annelies Zoomers describes food-insecure governments, such as China and the Gulf States, as those that seek to outsource their domestic food production by buying or leasing vast areas of farmland abroad for their own offshore food production. Ironically, the places that these countries invest in tend to be poor countries with their own food insecurity issues.

Corruption Although prevalent throughout the world, corruption also seems to have been particularly rampant in several African countries. Civil servants typically are not paid

▼ **Figure 6.43 Industrialization** Heavy industry, such as this iron foundry in Zambia, failed to deliver Sub-Saharan Africa from poverty. In the worst cases, these industrial enterprises were unable to produce competitive products for world and domestic markets.

▶ **Figure 6.44 Mobile Phones for Africa** A woman on her cell phone in downtown Monrovia, Liberia. Cell phone subscriptions in Sub-Saharan Africa doubled between 2007 and 2010, greatly improving communication.

a living wage and are thus virtually forced to solicit bribes. Teachers solicit informal fees from students in order to get paid. According to a recent poll of international businesspeople, Nigeria ranks as the world's most corrupt country. (Skeptical observers, however, point out that several Asian nations with highly successful economies, such as China, also are noted for high levels of corruption, so that corruption alone may not be an explanation for Africa's economic problems.)

With millions of dollars in loans and aid pouring into the region, officials at various levels were tempted to skim from the top. Some African states, such as the Democratic Republic of the Congo (DRC), were dubbed kleptocracies. A **kleptocracy** is a state in which corruption is so institutionalized that politicians and government bureaucrats siphon off a huge percentage of the country's wealth. President Mobutu of the DRC was a legendary kleptocrat. While his country was saddled with an enormous foreign debt, he reportedly skimmed several billion dollars and deposited them in Belgian banks during his presidency from 1965 to 1997.

Decade of Growth Since 2000, strong commodity prices, new infrastructure, and improved technology (mobile phone subscriptions cover most of the population) have brightened the economic prospects of the region. Over the past 10 years, real income per person grew (20 to 30 percent), whereas in the previous 20 years it actually decreased. The most optimistic views of the region see democracy strengthening, greater civic engagement, less violence, and growing investment (from within and outside the region).

It can be argued that domestic and international aid targeted toward reducing extreme poverty (as outlined in the **Millennium Development Goals**, a global UN effort to reduce extreme poverty by focusing on basic education, health care, and access to clean water) contributed to the region's economic and social development. Not all the specific goals for 2015 will be met, but the number of people in extreme poverty (living on less than $1.25 a day) has declined, more children are in school, and tremendous strides have been made to combat both HIV/AIDS and malaria. This is still the world's poorest region, with serious problems, but its connections with the world are deepening and in many cases proving beneficial.

Links to the World Economy

Sub-Saharan Africa's trade connection with the world is limited, accounting for 2 percent of global trade. The level of overall trade is low both within the region and outside it. Traditionally, most exports went to the European Union (EU), especially the former colonial powers of England and France. The United States is the second most common destination. That pattern is changing fast; China is now the single largest trading partner with the region, although collective trade between the EU nations and Sub-Saharan Africa is greater. Throughout the decade of the 2000s, China's trade with Sub-Saharan Africa grew on average 30 percent per year. During the same decade, India's and Brazil's levels of trade with Sub-Saharan Africa annually grew more than 20 percent. As of 2010, India's level of trade with Sub-Saharan Africa equaled that of the United States.

By most measures of connectivity, Sub-Saharan Africa has lagged behind other developing regions, but this is also changing quickly due to cellular and digital technology. Admittedly, fixed telephone lines are scarce; the regional average is 1 line per 100 people. Cell phone usage, however, has soared. In 2007, the World Bank estimated 23 cell phone subscriptions per 100 people in Sub-Saharan Africa; by 2011, the figure had more than doubled to 51 per 100 people. Multinational providers are now competing for mobile phone customers. Development specialists and entrepreneurs are exploring many new uses for cell phones and smartphones, with applications to secure not only micro-finance, but also educational tools and updates about health issues or weather patterns (Figure 6.44). The Internet is also facilitating the sharing of alternative and positive narratives about Sub-Saharan Africa's future. Two new websites—See Africa Differently (based in the United Kingdom) and Africa Good News (based in South Africa)—focus on the new ways in which Sub-Saharan peoples are developing their communities and engaging with the world.

Aid Versus Investment In many ways, Sub-Saharan Africa is linked to the global economy more through the flow of financial aid and loans than through the flow of goods. As Figure 6.45 shows, for several states this aid accounted for more than 20 percent of the GNI, and in the case of Liberia it was 54 percent. Most of the aid came from a handful of developed countries (see the insert in Figure 6.45). The United States provided one-quarter of the foreign aid in 2011, followed by EU institutions and then individual states, such as France, the United Kingdom, and Germany. In total, some $34 billion in foreign aid went to Sub-Saharan Africa in 2011.

Although aid is extremely important for many African states, foreign direct investment in the region substantially increased from only $4.5 billion in 1995 to $36 billion in 2012. The overall level of foreign investment remains low when compared to that of other developing regions. In recent years, the largest recipients of foreign investment were the region's major mineral and oil producers, such as South Africa, Ghana, the Democratic Republic of the Congo, Zambia, Angola, and Nigeria. China has been the leading investor in the region at a time when the United States and EU are more focused on offering aid or fighting terrorism. China wants to secure the oil and ore it needs for its massive industrial economy (see *People on the Move: Chinese Merchants in Africa*). In exchange, it offers Sub-Saharan nations money for roads, railways, housing, and schools, with relatively few strings attached. Some African leaders see China as a new kind of global partner, one that wants straight commercial relations without an ideological or political agenda. Angola, a country where China has invested heavily, is now one of China's top suppliers of oil.

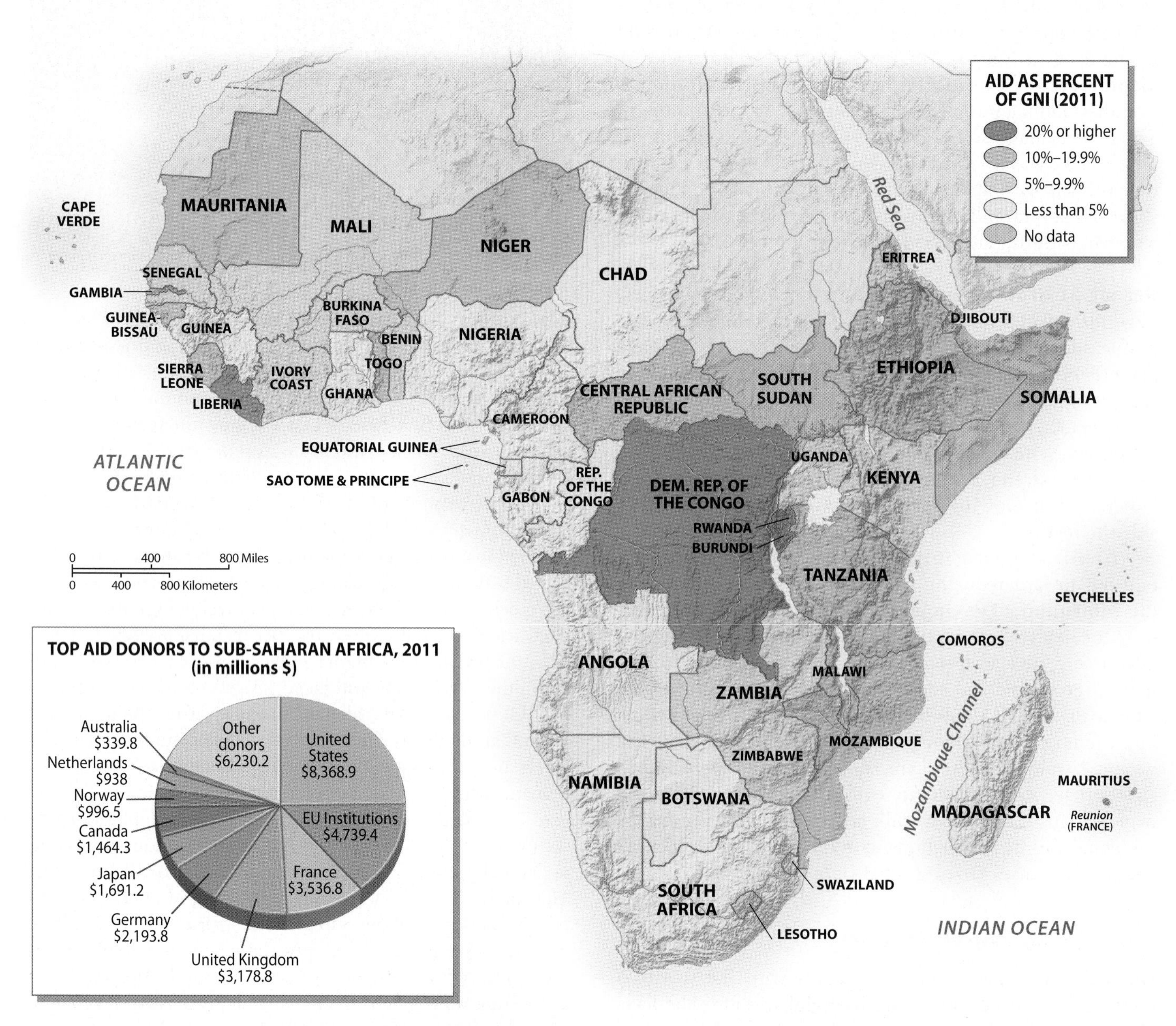

▲ **Figure 6.45 Sub-Saharan Africa Global Linkages: Aid Dependency** Many states in Sub-Saharan Africa are dependent on foreign aid as their primary link to the global economy. This figure maps aid as a percentage of GNI, which ranges from less than 1 percent to 54 percent in Liberia. **Q: Compare the most aid-dependent states in the region, with the least aid-dependent ones. What are the differences between these two groups of states?**

People on the Move

Chinese Merchants in Africa

The rise of China as the largest trading partner of and investor in Sub-Saharan Africa has generated much geopolitial discussion about China's influence in this resource-rich, but developing world region. In 2013, China's President Xi Jinping visited Tanzania, South Africa, and the Democratic Republic of the Congo, promoting the mutual benefits of Sino-African relations (Figure 6.5.1). The estimates of the number of Chinese currently working in Sub-Saharan Africa vary. At the high end, Chinese sources claim over 1 million Chinese migrants have settled in the region; the low end is a quarter that number. Most agree that the top destination is South Africa, where Chinese migrants number between 200,000 and 400,000.

Life as a Chinese Migrant in Africa Chinese government and corporate investment in the region draws attention, but less is known about the small traders who see greater economic opportunity in Africa south of the Sahara than in China. On the ground in African cities and villages, the growing number of new Chinese merchants and small businesses evokes both admiration and resentment, especially in southern Africa (Figure 6.5.2). Chinese merchants are an important component of the new migration to Africa. A recent survey of this population by a South African foundation revealed important insights into who these people are and local perceptions about them.

The majority of the Chinese emigrating to Africa in the past decade are from the southern province of Fujian, long an important source for Chinese traders and emigrants. Informal networks from Fujian facilitate this migration, as well as wholesale connections for the goods that are sold. This is not a state-directed flow, but the result of individuals and households seeing economic potential in emigration and small trade.

Generally, these are not the most educated Chinese, and southern Africa was usually not their first choice, but it was selected because of its minimal entry requirements. Also, the move makes economic sense as shopkeepers can earn perhaps three times more in southern Africa than in their home country. In interviews, Chinese migrants see themselves as working hard, forging opportunities in places neglected by others, creating new supply chains, and outcompeting local merchants. Most do not see themselves staying in their new homes—which is typical for most economic migrants. Yet they also express fears of local resentment, language difficulties, and cultural isolation as they struggle to build new businesses and lives in this region. Only Chinese traders in South Africa expressed any sense of attachment to their country of settlement.

▲ **Figure 6.5.2 Chinese Merchants in Sub-Saharan Africa** A man browses through pairs of Chinese made shoes in a shop owned by a Chinese merchant in Kampala, Uganda.

Local Resentments On the other hand, these close-knit and ethnically derived Chinese business networks are also seen as exclusionary by the Africans who purchase products from the Chinese. Also, isolated xenophobic incidents against some merchants have occurred. Although many African consumers enjoy the wider range of goods, they complain about low quality and price fixing. Rumors often circulate about underhanded deals that cheat, rob, or undermine African-owned businesses. Many Sub-Saharan governments often welcome large-scale Chinese investment, yet it seems to be local farmers, retailers, and petty traders who complain about unfair competition. In the countries of Malawi, Tanzania, Uganda, and Zambia, new rules have restricted the industries or sectors in which Chinese can operate.

▼ **Figure 6.5.1 Chinese-African Relations** China's President Xi Jinping and first lady Peng Liyuan are greeted by Tanzanian president Jakaya Kikwete and first lady Salma Kikwete at Julius Nyerere International Airport in Dar es Salaam, Tanzania. In 2013 the new Chinese president made a two-day visit to the country.

Source: Based on from *Africa in Their Words: A Study of Chinese Traders in South Africa, Lesotho, Botswana, Zambia and Angola* (2012) by Terence McNamee. Discussion Paper 2012/2013 for the Brenthurst Foundation, South Africa.

◄ **Figure 6.46 Chinese Investment in Angola** Children play on the basketball courts at the Kilambi Kiaxi housing development, a massive Chinese-built project in the suburbs of Luanda, Angola. Most of these apartments remain vacant because they are far too expensive for Angolan workers to purchase.

Yet not all investments have paid off. A massive real estate development outside of Luanda, Angola, called Nova Cidade de Kilamba, was largely unoccupied in 2013 because the cost of purchasing an apartment was well beyond the means of the average Angolan (Figure 6.46).

Debt Relief Another development strategy making a difference in the region is debt relief. The World Bank and the IMF proposed in 1996 to reduce debt levels for heavily indebted poor countries, many of which are in Sub-Saharan Africa. Most Sub-Saharan states are indebted to official creditors such as the World Bank, not to commercial banks (as is the case in Latin America and Southeast Asia). Under this World Bank/IMF program, substantial debt reduction is given to Sub-Saharan countries that the banks determine have "unsustainable" debt burdens. Mauritania, for example, spends six times more money on repaying its debts than it does on health care.

States qualify for different levels of debt relief provided they present a poverty reduction strategy. Uganda was the first state to qualify for the program, using the money it saved on debt repayment to expand primary schooling. Ghana qualified for debt relief in 2004 and received a $3.5 billion relief package. Other countries that have benefited from debt reduction are Tanzania, Mozambique, Ethiopia, Mauritania, Mali, Niger, Nigeria, Senegal, Burkina Faso, and Benin. More African countries may qualify in the future to redirect debt payments toward building infrastructure and improving basic health and education.

Economic Differentiation Within Africa

As in most other regions, considerable differences in levels of economic and social development persist in Sub-Saharan Africa. In many respects, the small island nations of Mauritius and the Seychelles have little in common with the mainland. With high per capita GNI, life expectancies averaging in the low 70s, and economies built on tourism, they could more easily fit into the Caribbean were it not for their Indian Ocean location. In contrast, two-thirds of the population in mainland Sub-Saharan Africa subsist on less than $2 per day. Also, only a few states, such as Angola, Botswana, South Africa, Swaziland, Gabon, Namibia, and Equatorial Guinea, have a per capita GNI-PPP over $5000 (see Table 6.2).

Millions of Africans are still living in extreme poverty in every state in the region. In rural areas, many of the very poor still practice subsistence farming and are barely a part of the formal economy. In the rapidly growing urban areas, massive slums are home to millions with inadequate shelter and no regular access to clean water. Most make their living through the informal economy. These are the people that the Millennium Development Goals hope to reach, although securing the resources and finding the right programs to ease this poverty are both difficult and long-term propositions.

South Africa The unchallenged economic powerhouse of Sub-Saharan Africa is South Africa. The per capita GNI-PPP of Nigeria, the next largest economy, is just one-fifth that of South Africa. Only South Africa has a well-developed and well-balanced industrial economy. It also boasts a healthy agricultural sector, and, more importantly, it is one of the world's mining superpowers. South Africa remains unchallenged in gold production and is a leader in many other minerals and precious gems, including diamonds. In the summer of 2010, South Africa hosted the World Cup, the first African country to do so, symbolizing its arrival as a developed and modern nation.

South Africa is undeniably a wealthy country by African standards, but while its white minority is prosperous by any standard, it is also a country beset by severe and widespread poverty. In the townships lying on the outskirts of the major cities and in the rural districts of the former homelands, unemployment is high and living standards marginal. Despite the end of apartheid, South Africa continues to suffer from one of the most unequal distributions of income in the world.

Oil and Mineral Producers Another group of relatively prosperous Sub-Saharan countries benefits from substantial oil and mineral reserves and small populations. The prime example is Gabon, a country of noted oil wealth that is inhabited by nearly 2 million people. Its neighbors, the Republic of the Congo and Equatorial Guinea, have also experienced steady growth in income and foreign investment due to oil finds in the 1980s and 1990s, respectively. Equatorial Guinea, a former Spanish colony with less than 1 million people, began producing significant amounts of oil in 1998. Today it has the distinction of having the region's highest per capita GNI at $25,620. Yet over a decade of oil production, these newfound revenues have not been invested in the country's citizens, but, as is often the case, seem to have fallen into the hands of a few elites. Farther south, Namibia and Botswana also have the advantage of small populations and abundant mineral resources, especially diamonds.

Over the past few years, both countries have enjoyed sound government and experienced significant economic and social development, reflected in their relatively high Human Development Index figures. Angola now rivals Nigeria as one of Sub-Saharan Africa's major oil exporters. Its per capita GNI has risen to over $5000, but 70 percent of the population still lives on less than $2 per day.

The Leaders of ECOWAS The most populous country in Africa, Nigeria is the core member of the Economic Community of West African States (ECOWAS). Nigeria has the largest oil reserves in the region and is a member of the Organization of Petroleum Exporting Countries (OPEC). Yet despite its natural resources, its per capita GNI-PPP is a relatively low $2290. It has been argued that oil money has helped to make Nigeria notoriously inefficient and corrupt. A small minority of its population has grown fantastically wealthy, more by manipulating the system than by engaging in productive activities. Eighty-three percent of Nigerians, however, remain trapped in poverty, earning less than $2 per day in 2011. Oil money led to the explosive growth of the former capital city of Lagos, which by the 1980s had become one of the most expensive—and least livable—cities in the region. As a result, the Nigerian government built a new capital city in Abuja, located near the country's center, a move that has proved tremendously expensive. In 1991, Abuja became the national capital, and it now has over 2 million residents. The southwestern corner of the country, in the Niger Delta where most of the oil is produced, has seen few of the benefits of oil production, but bears the burden of the environmental costs and related social unrest.

The second and third most populous states in ECOWAS, Ivory Coast and Ghana, are also important West African commercial centers. These states rely upon a mix of agricultural and mineral exports. In the mid-1990s, the Ivorian economy began to take off. Boosters within the country called it an emerging "African elephant" (comparing it to the successful "economic tigers" of eastern Asia). Yet a destructive civil war began in 2002, in which rebel forces controlled the northern half of the county and over half a million Ivorians were displaced. A peace agreement was signed in 2007, but the economic growth of the 1990s has yet to return. Ghana, a former British colony, began its economic recovery in the 1990s. In 2001, it negotiated with the IMF and World Bank for debt relief to reduce its nearly $6 billion foreign debt. Ghana maintained an average annual growth rate of nearly 6 percent. In 2011, Ghana also became an emerging oil producer for Africa, with offshore wells being pumped near the city of Takoradi.

East Africa Long the commercial and communications center of East Africa, Kenya experienced economic decline and political tension throughout the 1990s. From 2000 to 2011, the economy has averaged 4 percent annual growth, and per capita GNI-PPP is at $1710. Kenya boasts good infrastructure by African standards, and over 1 million foreign tourists come each year to marvel at its wildlife and natural beauty. Traditional agricultural exports of coffee and tea, as well as nontraditional exports such as cut flowers, dominate the economy. As for social indicators, Kenyan women are having fewer children now than in the late 1970s (down from an average of 8 to 4.4 children per woman), and those children are better educated and less likely to be in extreme poverty compared to other countries in the region—only 42 percent of the population lives on less than $2 per day. If Kenya can avoid political unrest due to ethnic rivalries, it could lead East Africa into better economic integration with the southern and western parts of the continent.

The political and economic indicators for Kenya's neighbors, Uganda and Tanzania, are also improving, with average annual growth rates in the 2000s of 7.7 percent and 7.0 percent, respectively (see Table 6.2). When compared with Kenya, both countries have a much higher proportion of their populations living on less than $2 a day. Both countries rely heavily upon agricultural exports and mining (especially gold). Uganda and Tanzania both benefited from debt reduction agreements that redirected funds from debt repayment to education and health care.

Measuring Social Development

By global standards, measures of social development in Africa are extremely low. Yet there are some positive trends, especially with regard to child survival and literacy, that are cause for hope (see Table 6.2). Many governments in the region have reached out to the modern African diaspora (economic migrants and refugees who now live in Europe and North America). In other parts of the world, African immigrant organizations have worked to improve schools and health care, and former emigrants have returned to invest in businesses and real estate. The economic impact of remittances in this region is small, but growing.

Child Mortality and Life Expectancy Reductions in child mortality are often an indication of improved social development because, if most children make it to their fifth birthday, it usually indicates adequate primary care and nutrition. In a country where the child mortality rate is 200 per thousand, it means that 1 out of 5 children dies before his or her fifth birthday. As Table 6.2 shows, most of the states in the region saw modest to significant improvements in child survival between 1990 and 2011. Eritrea, Liberia, Madagascar, Malawi, and Rwanda actually experienced dramatic gains in child survival rates. However, countries with prolonged conflict (e.g., Somalia and the Democratic Republic of the Congo) have seen little improvement. In 1990, the regional child mortality rate was 178 per thousand; in 2011, it was down to 109. Thus, high child mortality is still a major concern for the region, but steady reductions in child mortality are significant for African families.

Life expectancy for Sub-Saharan Africa is only 55 years. Countries hit hard by HIV/AIDS or conflict have seen life expectancies tumble into the 40s. Despite these statistics, there are indications that access to basic health care is improving, and, eventually, so will life expectancies. Keep in mind that high infant- and child-mortality figures depress overall life expectancy figures; average life expectancies for people who make it to adulthood are much better.

The causes of short life expectancy generally are related to extreme poverty, environmental hazards (such as drought), and various environmental and infectious diseases (cholera, measles, malaria, schistosomiasis, and AIDS). Often these factors work in combination. Malaria, for example, kills half a million African children each year. The death rate is also affected by poverty, undernourished children being the most vulnerable to the effects of high fevers. Tragically, diseases that are preventable, such as measles, occur when people have no access to or cannot afford vaccines. National and international health agencies, along with NGOs such as the Gates Foundation, are working to improve access to vaccines, bed nets (to prevent malaria), and primary health care. These efforts are slowly making a difference.

Meeting Educational Needs Basic education is another obstacle confronting the region. The goal of universal access to primary education is a daunting one for a region in which 43 percent of the population is less than 15 years old. The UN estimates that 75 percent of African children are enrolled in primary school, but only 23 percent of the relevant population is in secondary school (high school or its equivalent). The region is home to one-sixth of the world's children under 15, but half of the world's uneducated children. Girls are still less likely than boys to attend school. In West African countries such as Chad, Niger, and Ivory Coast, girls are decidedly underrepresented at all levels.

A renewed focus on education since 2000 has been attributed to the Millennium Development Goals. Although the region will not meet the goals set for 2015, more resources have been directed to schools from governments and from nonprofit organizations through debt relief and aid. More schools are being built across the region, and more children are attending them (Figure 6.47).

Women and Development

Development gains cannot be made in Africa unless the economic contributions of African women are recognized. Officially, many of the labors of women are not accounted for in local or national statistics. In agriculture, women account for 75 percent of the labor that produces more than half the food consumed in the region. Tending subsistence plots, taking in extra laundry, and selling surplus produce in local markets all contribute to household income. Yet because many of these activities are considered informal economic activities, they are not counted. For many of Africa's poorest people, however, the informal sector is the economy, and within this sector women dominate.

Status of Women The social position of women is difficult to measure for Sub-Saharan Africa. Female traders in West Africa, for example, have considerable political and economic power. By such measures as female labor force participation, many Sub-Saharan African countries show relative gender equality. Also, women in most Sub-Saharan societies do not suffer the kinds of traditional social restrictions encountered in much of South Asia, Southwest Asia, and North Africa; in Sub-Saharan Africa, women work outside the home, conduct business, and own property. In 2006, Ellen Johnson-Sirleaf was sworn in as Liberia's president, making her Africa's first elected female leader. In 2012, she was joined by Joyce Banda, the recently elected president of Malawi. In fact, throughout the region, women occupied 22 percent of all seats in national parliaments in 2012. In the country of Rwanda, over half the parliamentary seats are filled by women (Figure 6.48).

By other measures, however, such as the prevalence of polygamy, the practice of the "bride-price," and the tendency for males to inherit property over females, African women do suffer discrimination. Perhaps the most controversial issue regarding women's status is the practice of female circumcision, or genital mutilation. In Ethiopia, Somalia, and Eritrea, as well as parts of West Africa, the majority of girls are subjected to this practice, which is extremely painful and can have serious health consequences. Yet because the practice is considered traditional, most African states are unwilling to ban it.

Regardless of their social position, most African women still live in remote villages where educational and wage-earning opportunities remain limited and caring for large families is time-consuming and demanding labor. As educational levels increase and urban society expands—and as reduced infant mortality provides greater security—we can expect fertility in the region to gradually decrease. Governments can speed up the process by providing birth control information and cheap contraceptives—and by investing more money in health and educational efforts aimed at women. As the economic importance of women receives greater attention from national and international organizations, more programs are being directed exclusively toward them.

Building from Within

Surveys reveal that the majority of people in Sub-Saharan Africa are optimistic about their future. Considering many of the real development hurdles the region faces, this surprises outside observers. Yet considering the levels of conflict, food insecurity, and neglect that Sub-Saharan Africa experienced during the 1990s, perhaps the developments of the past decade are a cause for hope. Most African states have been independent for only 50 years. During that time,

◀ **Figure 6.47 Educating African Youth** Botswana students attend classes at a Lutheran theological seminary in Francistown, Botswana. Religious organizations, as well as public institutions, are critical in providing educational services.

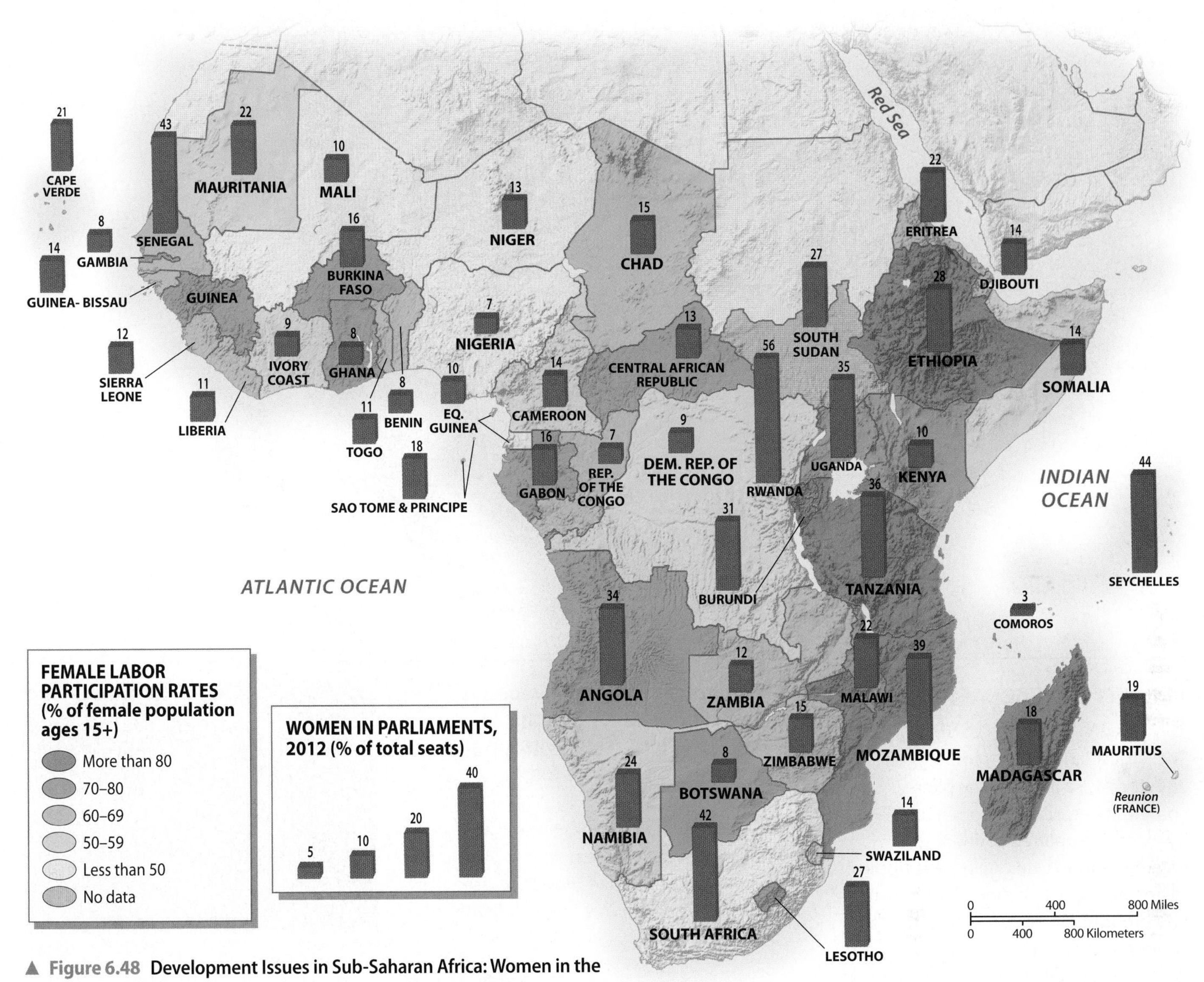

▲ **Figure 6.48 Development Issues in Sub-Saharan Africa: Women in the Workforce and Politics** Female participation in the workforce is comparable to that of developed countries, as 70 percent of women over the age of 15 are in the labor force. Another significant change for the region is the increase of women holding seats in national parliaments. The regional average in 2012 is 22 percent, but in South Africa 42 percent of parliamentary seats are held by women, and in Rwanda 56 percent of the seats are held by women. In contrast, in Africa's largest country, Nigeria, only 7 percent of the parliamentary seats are held by women.

the governance of these countries has shifted from one-party authoritarian states to multiparty democracies. Targeted aid projects, aimed at addressing particular critical indicators such as mother and child mortality or combating malaria, have proven their effectiveness. Civil society is also vigorous, from raising the status of women to supporting small businesses with micro-credit loans. Even members of the African diaspora are beginning to return and invest in their countries. The rapid adoption and adaptation of mobile phone technology in the region demonstrates the region's desire to be better connected with each other and the world.

Whether inspired by the free market, African socialism, or feminism, various internal and international organizations are finding ways to meet basic needs and innovate in the future. No doubt some of these projects will fall short of their objectives. And for areas such as the Sahel, the long-term implications of climate change may overwhelm efforts to develop and adjust to severe water shortages. Yet for many people, the message of forming local networks to solve community problems is an empowering one. More importantly, in this very young continent where nearly half the population is under 15, their orientation tends to be more toward a hopeful future.

REVIEW

6.10 What are the environmental, historical, structural, and institutional reasons offered to explain poverty in the region?

6.11 What technological and investment changes are impacting the region's social development?

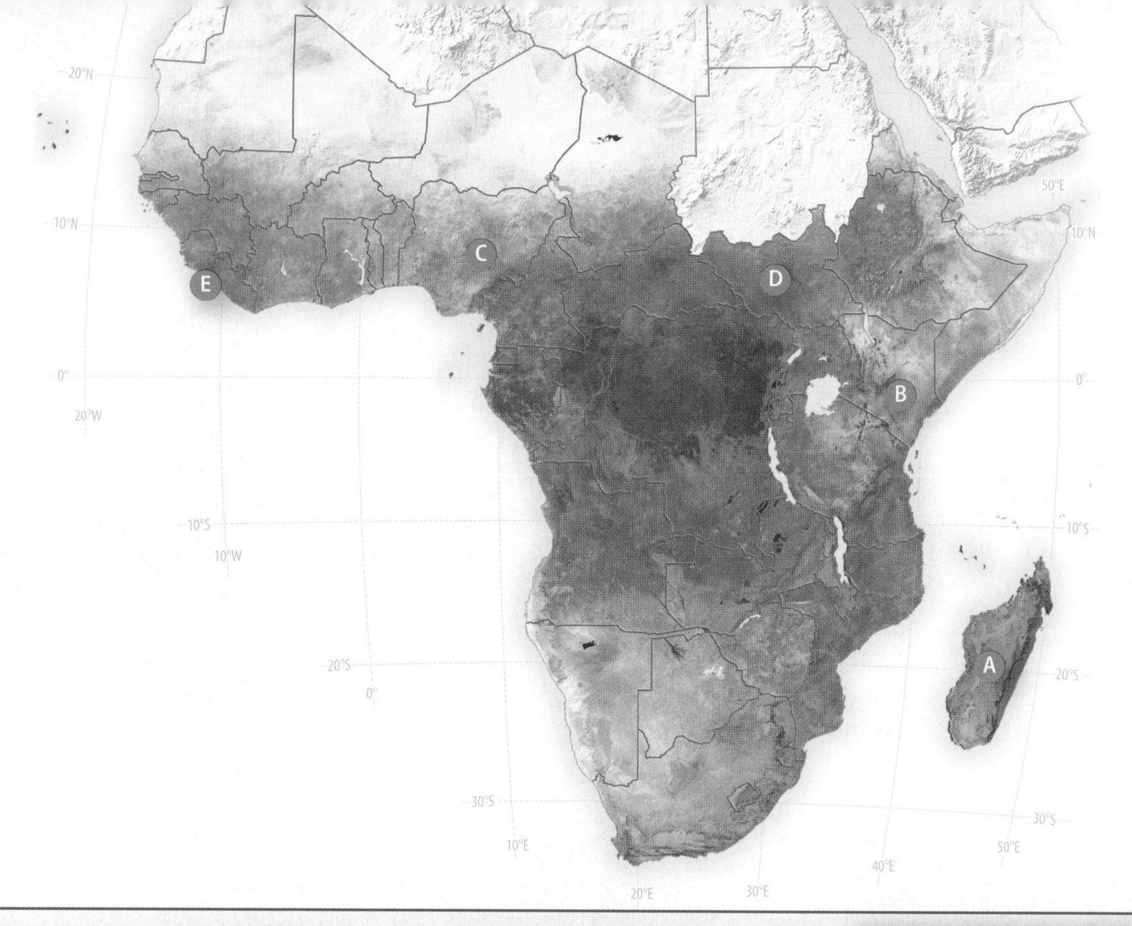

Physical Geography and Environmental Issues

The largest landmass straddling the equator, Africa is called the plateau continent because it is dominated by extensive uplifted plains. Key environmental issues facing this tropical region are desertification, deforestation, and drought. At the same time, the region supports a tremendous diversity of wildlife, especially large mammals.

6.1 This area was heavily forested in the 1950s. Looking at this image, what activities are contributing to deforestation in this area of Madagascar?

6.2 Sub-Saharan Africa is noted for its wildlife, especially large mammals. What environmental and historical processes explain the existence of so much fauna?

Population and Settlement

With nearly 900 million people, Sub-Saharan Africa is the fastest-growing region in terms of population, with the average woman having five children. Yet it is also the poorest region, with two-thirds of its people living on less than $2 a day. In addition, it has the lowest average life expectancy, at 55 years. HIV/AIDS has hit this region especially hard.

6.3 What factors might explain the density of settlement in this region of Africa?

6.4 Consider how increasing urbanization may impact the overall structure of the population in Sub-Saharan Africa.

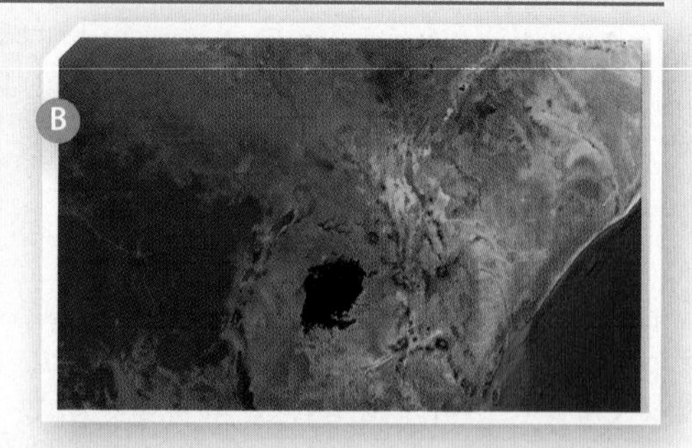

Key Terms

African Union (AU) *(p. 277)*
agricultural density *(p. 256)*
apartheid *(p. 264)*
Berlin Conference *(p. 276)*
biofuels *(p. 250)*
clan *(p. 283)*
coloured *(p. 264)*
conflict diamonds *(p. 281)*
desertification *(p. 248)*
food insecurity *(p. 254)*
Gondwana *(p. 243)*
Great Escarpment *(p. 244)*
homelands *(p. 278)*
Horn of Africa *(p. 248)*
internally displaced person *(p. 280)*
kleptocracy *(p. 287)*
Millennium Development Goals *(p. 287)*
Pan-African Movement *(p. 277)*
pastoralist *(p. 262)*
physiological density *(p. 256)*
refugee *(p. 280)*
Sahel *(p. 248)*
structural adjustment program *(p. 285)*
swidden *(p. 260)*
township *(p. 264)*
transhumance *(p. 249)*
tribalism *(p. 279)*
tribe *(p. 268)*
tsetse fly *(p. 262)*

In Review • Chapter 6
Sub-Saharan Africa

Cultural Coherence and Diversity

Culturally, Sub-Saharan Africa is an extremely diverse region, where multiethnic and multireligious societies are the norm. With a few exceptions, religious diversity and tolerance have been a distinctive feature of the region. Most states have been independent for 50 years, and in that time pluralistic, but distinct, national identities have been forged. Many African cultural expressions, such as music, dance, and religion, have been influential beyond the region.

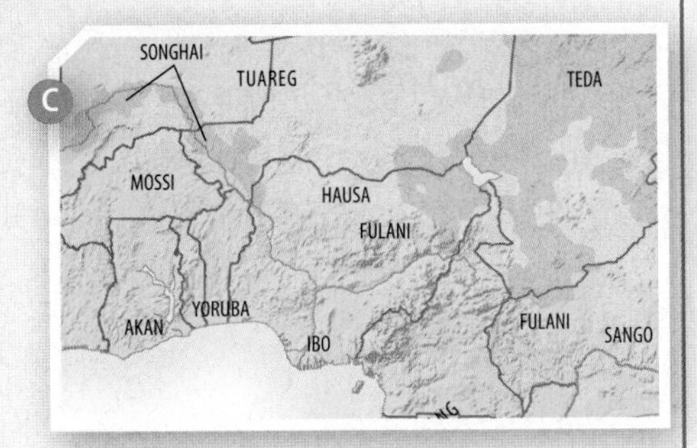

6.5 Nigeria is linguistically diverse. What do the distinct language families in this state tell us about the area's settlement?

6.6 Compare and contrast the role of tribalism in Sub-Saharan Africa with that of nationalism in Europe.

Geopolitical Framework

In the 1990s, many bloody ethnic and political conflicts occurred in the region. Fortunately, peace now exists in many conflict-ridden areas, such as Angola, Sierra Leone, and Liberia. However, ongoing ethnic and territorial disputes in Somalia, the Democratic Republic of the Congo, and South Sudan, and more recently in Mali, have produced millions of internally displaced persons and refugees.

6.7 Consider South Sudan. Which regional trade bloc do you think this new nation is likely to join?

6.8 Historically, how was Sub-Saharan Africa integrated into the global economy? Was its role similar to that of other developing regions?

Economic and Social Development

Widespread poverty is the region's most pressing concern. Since 2000, Sub-Saharan economies have grown, led in part by higher commodity prices, greater investment, debt forgiveness, and the end of some of the longest-running conflicts in the region. Social indicators of development are also improving, due to greater attention from the international community and better access to health care and drugs to fight HIV/AIDS.

6.9 What economic activity is happening in this area of Sierra Leone, and what are its economic and environmental consequences?

6.10 Compare and contrast the development model put forward by the United States and Europe with that of China. Will Chinese influence in the region alter the course of development for Sub-Saharan Africa?

Authors' Blogs

Scan to visit the author's blog for chapter updates

www.gad4blog.wordpress.com

Scan to visit the GeoCurrents blog

www.geocurrents.info

7 Southwest Asia and North Africa

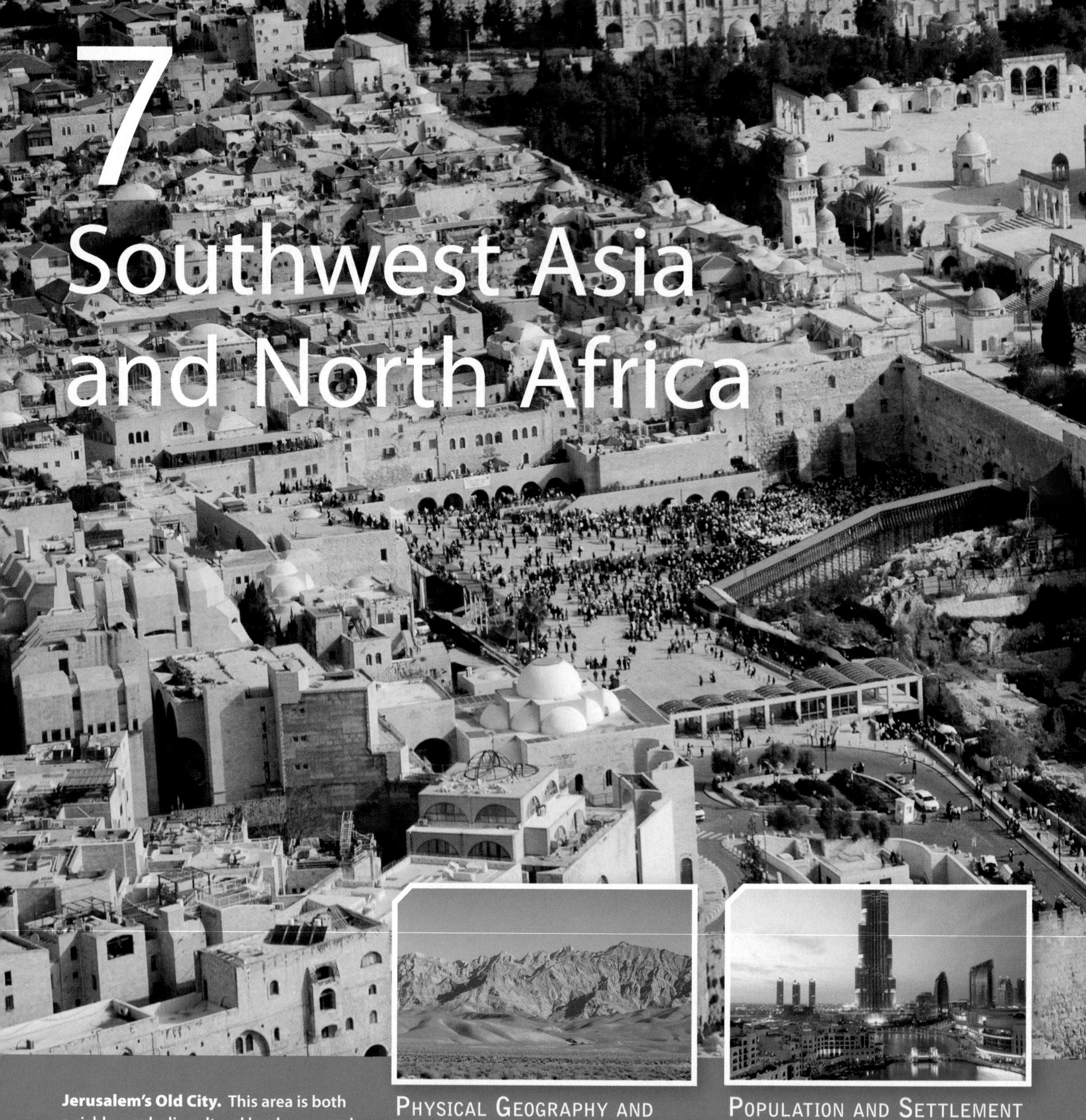

Jerusalem's Old City. This area is both a richly symbolic cultural landscape and some of the world's most hotly contested real estate.

Physical Geography and Environmental Issues

The region's vulnerability to water shortages is likely to increase in the early 21st century as growing populations, rapid urbanization, and increasing demands for agricultural land consume limited supplies.

Population and Settlement

Many settings within the region continue to see rapid population growth. These demographic pressures are particularly visible in fragile, densely settled rural zones, as well as in fast-growing large cities.

Cultural Coherence and Diversity

Islam continues to be a vital cultural and political force within the region, but increasing fragmentation within that world is leading to more culturally defined political instability.

Geopolitical Framework

Ongoing political instability has become a fact of life. The Arab Spring uprisings jolted geopolitical realities in Tunisia, Egypt, Libya, Yemen, and Bahrain. Internal instability has also produced extensive bloodshed in Syria. Prospects for peace between Israel and the Palestinians remain murky, and Iran's growing political role is seen by many as a threat both within and beyond the region.

Economic and Social Development

Unstable world oil prices and unpredictable geopolitical conditions have discouraged investment and tourism in many countries. The pace of social change, especially for women, has quickened, producing diverse regional responses.

How does a *place* become a *symbol?* Within the Middle East, the city of Jerusalem (now the Israeli capital) holds special cultural and religious significance for several groups and also stands at the core of the region's political problems. Indeed, the sacred space of this ancient Middle Eastern city remains deeply scarred and divided. Just considering the 220 acres of land within the Old City, Jews pray at the old Western Wall (the site of a Roman-era Jewish temple); Christians honor the Church of the Holy Sepulchre (the burial site of Jesus); and Muslims hold sacred religious rites in the city's eastern quarter (including the place from which the prophet Muhammad reputedly ascended to heaven). Nearby suburban communities are also contested real estate as Arab and Israeli neighborhoods (including newly built Jewish settlements) uneasily sit next to one another. Ultimately, longer-term stability within the region will at least partly depend on a political solution that recognizes the complex role of this symbolic place. Imaginative compromises in defining that political space will need to recognize its special value to both its Israeli and its Palestinian residents.

Climate, culture, and oil resources all help define this world region.

Jerusalem itself sits almost at the geographical center of a much larger and even more complex world region known as Southwest Asia and North Africa. Climate, culture, and oil resources all help define this world region. Straddling the historic meeting ground of Europe, Asia, and Africa, the region sprawls across thousands of miles of parched deserts, rugged plateaus, and oasis-like river valleys. It extends 4000 miles (6400 km) between Morocco's Atlantic coastline and Iran's eastern boundary with

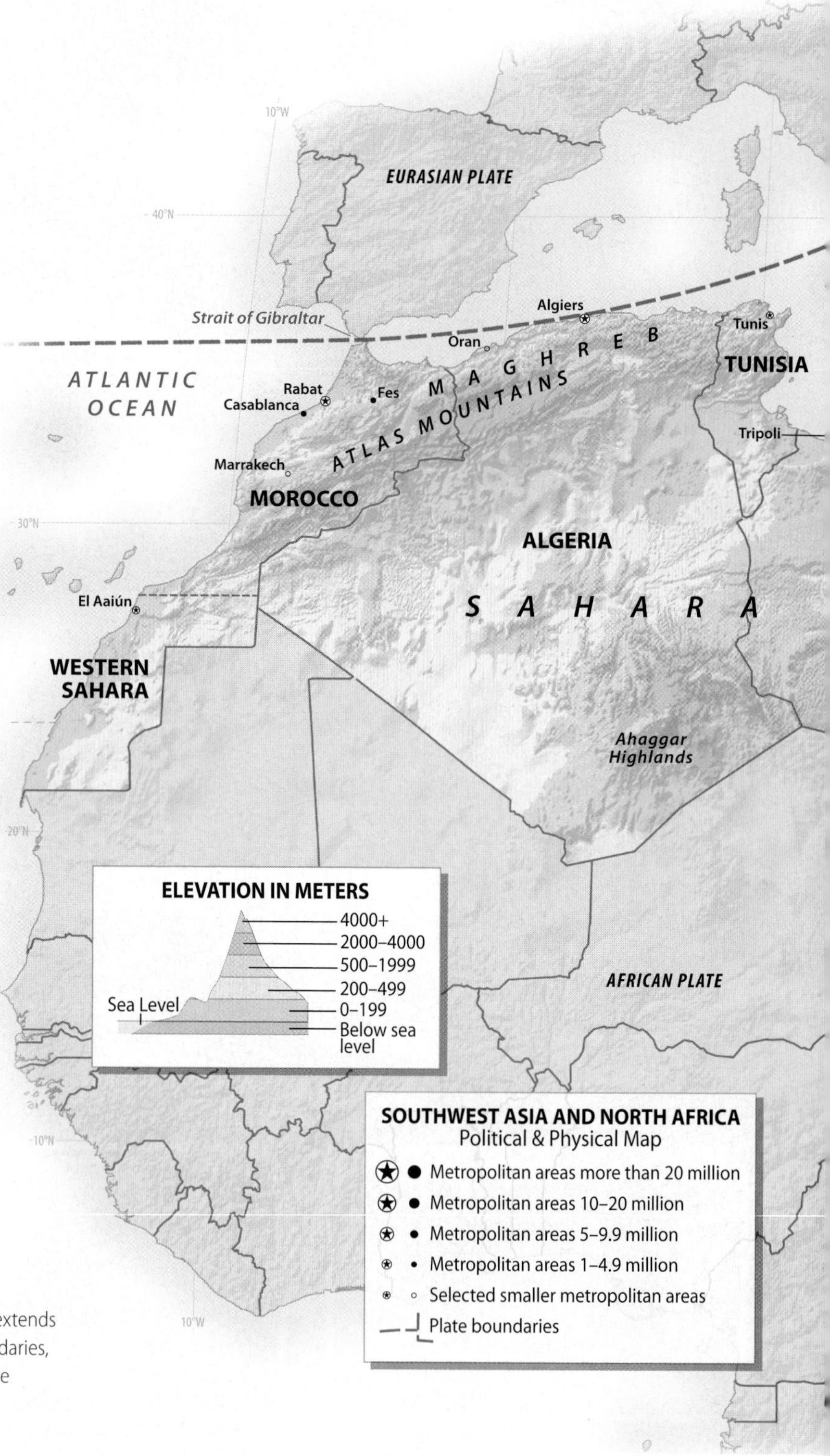

▶ **Figure 7.1 Southwest Asia and North Africa** This vast region extends from the shores of the Atlantic Ocean to the Caspian Sea. Within its boundaries, major cultural differences and globally important petroleum reserves have contributed to recent political tensions.

LEARNING OBJECTIVES

After reading this chapter you should be able to:

- Describe how the region's fragile, often arid setting shapes the region's contemporary environmental challenges.
- Explain how latitude and topography produce the region's distinctive patterns of climate.
- Describe four distinctive ways in which people have learned to adapt their agricultural practices to the region's arid environment.
- Summarize the major forces shaping recent migration patterns within the region.
- List the major characteristics of Islam and its key patterns of diffusion.
- Identify the region's dominant religions and language families.
- Describe the local impacts of the Arab Spring rebellions in different regional settings.
- Identify the role of cultural variables and sectarian differences in understanding key regional conflicts in Israel, Syria, and Iraq.
- Summarize the geography of oil and gas reserves in the region.
- Describe traditional roles for Islamic women and provide examples of recent changes.

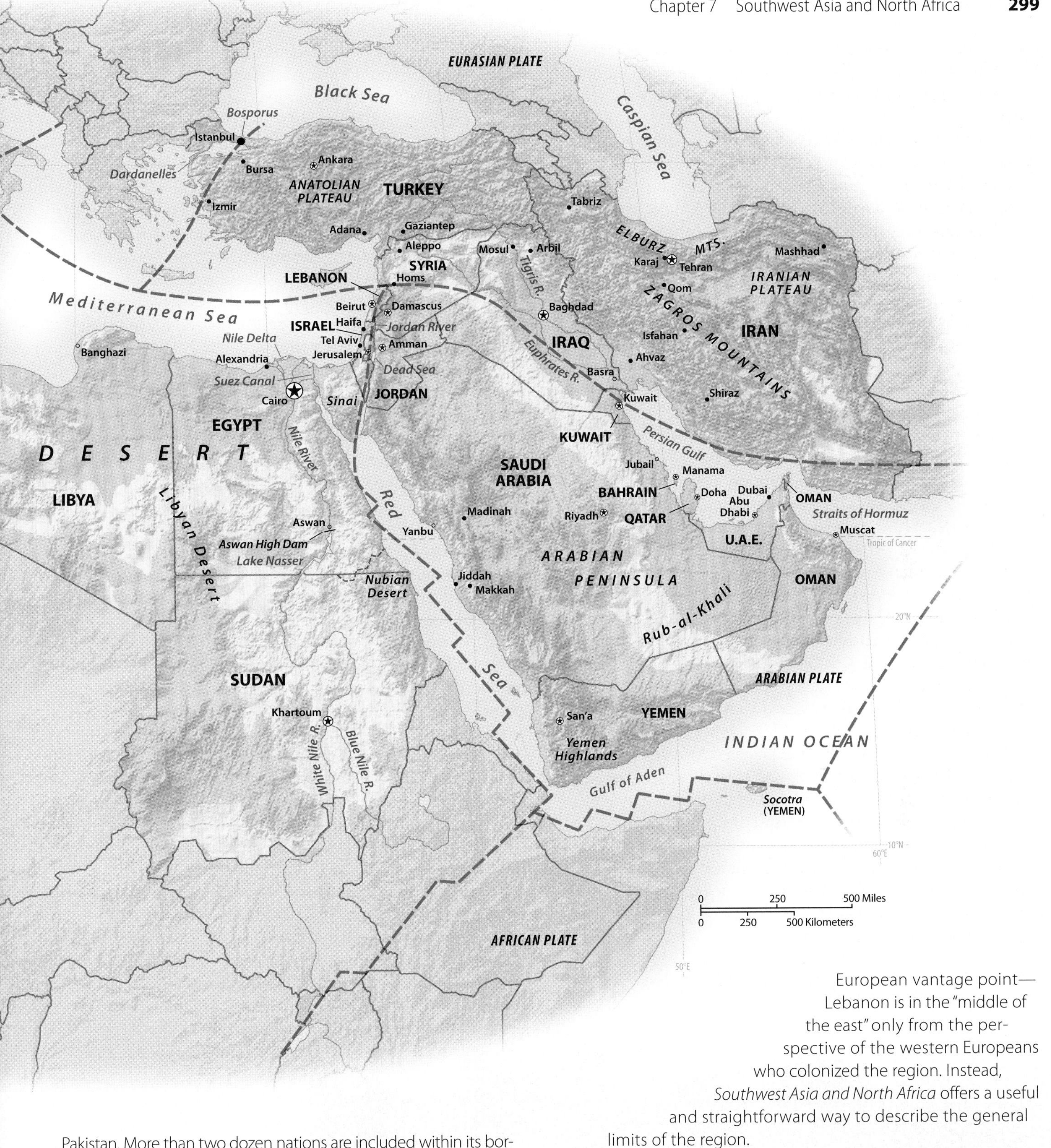

Pakistan. More than two dozen nations are included within its borders, with the largest populations found in Egypt, Turkey, and Iran (Figure 7.1). Generally, its climates are arid, although the region's diverse physical geography causes precipitation to vary considerably.

Regional boundaries and terminology defy easy definition. Often the same broad area is simply called the *Middle East,* but most experts would exclude the western parts of North Africa, as well as Turkey and Iran, from a region with this designation. In addition, the term *Middle East* carries with it a peculiarly European vantage point—Lebanon is in the "middle of the east" only from the perspective of the western Europeans who colonized the region. Instead, *Southwest Asia and North Africa* offers a useful and straightforward way to describe the general limits of the region.

Problems also arise with simply defining the geographical limits of the region. For example, a small piece of Turkey actually sits west of the Bosporus Strait, generally considered to be the dividing line between Europe and Asia (see Figure 7.1). The addition of Cyprus (off the coast of Turkey) to the European Union in 2004 effectively removed it from the region, and that country is treated in the chapter on Europe (see Chapter 8). To the northeast, the largely Islamic peoples of Central Asia share many religious and linguistic ties with Turkey and Iran, but for historical, geopolitical, and economic reasons, we have chosen to treat these groups in a separate chapter on Central Asia (see Chapter 10).

African borders also remain problematic. The conventional division of "North Africa" from "Sub-Saharan Africa" cuts through the middle of modern Mauritania, Mali, Niger, and Chad. These transitional countries are discussed in Chapter 6. Sudan's recent split also emphasizes this complexity. The new nation of South Sudan (created in 2011) is also treated in Chapter 6, whereas Sudan (an "Islamic republic") retains many ties to the Muslim world and remains in this chapter.

Culturally, diverse languages, religions, and ethnic identities have molded land and life within the region for centuries. One traditional zone of conflict is in the eastern Mediterranean, where Jewish, Christian, and Islamic peoples have yet to resolve long-standing cultural tensions and political differences, particularly as they relate to the state of Israel and various Palestinian groups within the region.

Since 2010, the so-called **Arab Spring** movements (a series of public protests, strikes, and rebellions, often facilitated by social media, that have called for fundamental governmental and economic reforms) have overthrown some governments and pressured others to accelerate political and economic reforms (Figure 7.2). However, it remains unclear what the long-term consequences of these dramatic changes will be. At the same time, cycles of **sectarian violence**—conflicts that divide people along ethnic, religious, and sectarian lines—have repeatedly plagued the region. For example, enduring differences have led to clashes between Jews and Muslims and between varying factions of Islam.

Recently, Syria, in particular, has been a setting for escalating sectarian violence as various ethnic and political factions have divided the country. Nearby Iran and its ongoing development of a nuclear capability have also increased tensions. In addition, Southwest Asia and North Africa's extraordinary petroleum resources keep the area in the global economic spotlight. The strategic value of oil has combined with ongoing ethnic and religious conflicts to produce one of the world's least stable political settings, one prone to geopolitical conflict both within and between the countries of the region.

No world region has better exemplified the theme of globalization throughout history than Southwest Asia and North Africa. It is a key global **culture hearth**—that is, a region that witnessed many cultural innovations that subsequently diffused widely to other portions of the world. As an early center for agriculture, several great civilizations, and three major world religions, the region has been a key human crossroads for thousands of years. Important trade routes have connected North Africa with the Mediterranean and Sub-Saharan Africa. Southwest Asia also has had historical ties to Europe, the Indian subcontinent, and Central Asia. As a result, innovations within the region have often spread far beyond its bounds. For example, the domestication of wheat and cattle in Southwest Asia had far-reaching global impacts. In addition, religions born within the region (Judaism, Christianity, and Islam) have shaped many other parts of the world.

Particularly within the past century, globalization also has operated in the opposite direction: The region's strategic importance has made it increasingly vulnerable to outside influences. Traditional lifestyles have been transformed. The 20th-century development of the petroleum industry, largely initiated by U.S. and European investment, has had enormous, but selective, consequences for economic development. Global demand for oil and natural gas has powered rapid industrial change within the region, defining its pivotal role in world trade. Many key members of the **Organization of the Petroleum Exporting Countries (OPEC)** are found within the region, and these countries strongly influence global prices and production levels for petroleum. Oil-rich nations such as the United Arab Emirates (UAE), Saudi Arabia, and Kuwait have been fundamentally transformed by the global fossil fuel economy, while petroleum-poor neighbors such as Jordan, Lebanon, and Tunisia have seen less dramatic impacts. Politically, petroleum resources also have elevated the region's strategic importance, adding to many traditional cultural tensions.

Islamic fundamentalism in the region advocates a return to more traditional practices within the Islamic religion. Fundamentalists in any religion advocate a conservative adherence to enduring beliefs within their creed, and they strongly resist change. A related political movement within Islam, known as **Islamism**, challenges the encroachment of global popular culture and blames colonial, imperial, and Western elements for many of the region's political, economic, and social problems. Islamists resent the role they claim the West has played in creating poverty in their world, and many Islamists advocate merging civil and religious authority and rejecting many characteristics of modern, Western-style consumer culture.

▼ **Figure 7.2 Tahrir Square** Cairo's Tahrir Square fills with demonstrators at a mass rally in November 2011.

The region's environment provides additional challenges for the populations of Southwest Asia and North Africa. The availability of water in this largely dry portion of the world has shaped the region's physical and human geographies. Biologically, the region's plants and animals must adapt to the aridity of long dry seasons and short, often unpredictable rainy periods. Human settlement is linked to water in similar ways. Whether it comes from precipitation, underground aquifers, or rivers, water has shaped settlement patterns and has placed severe limits on agricultural development across huge portions of the region. In the future, the region's growing population of 500 million people will stress these available resources even further, and water issues will no doubt increase economic and political instability.

Physical Geography and Environmental Issues: Life in a Fragile World

In the popular imagination, much of Southwest Asia and North Africa is a land of shifting sand dunes, searing heat, and scattered oases. Although examples of those stereotypes certainly can be found across the region, the actual physical setting, in both landforms and climate, is considerably more complex (see Figure 7.1). In reality, the regional terrain varies greatly, with rocky plateaus and mountain ranges more common than sandy deserts. Even the climate, although dominated by aridity, varies remarkably from the dry heart of North Africa's Sahara Desert to the well-watered highlands of northern Morocco, coastal Turkey, and western Iran. One theme is pervasive, however: A lengthy legacy of human settlement has left its mark on a fragile environment, and the entire region is faced with increasingly daunting ecological problems in the decades ahead.

Regional Landforms

A quick tour of the region reveals a surprising diversity of environmental settings and landforms (see Figure 7.1). In North Africa, the **Maghreb** region (meaning "western island") extends across the northern parts of Morocco, Algeria, and Tunisia and is dominated near the Mediterranean coastline by the Atlas Mountains. The rugged flanks of the Atlas rise like a series of islands above the narrow coastal plains to the north and the vast stretches of the lower Saharan deserts to the south (Figure 7.3). South and east of the Atlas Mountains, interior North Africa varies between rocky plateaus and extensive lowlands. In northeast Africa, the Nile River dominates the scene as it flows north through Sudan and Egypt.

Southwest Asia is more mountainous than North Africa. In the **Levant**, or eastern Mediterranean region, mountains rise within 20 miles (32 km) of the sea, and the highlands of Lebanon reach heights of more than 10,000 feet (3000 meters). Farther south, the Arabian Peninsula forms a massive tilted plateau, with western highlands higher than 5000 feet (1500 meters) gradually sloping eastward to extensive lowlands in the Persian Gulf area. North and east of the Arabian Peninsula lie the two great upland areas of Southwest Asia: the Iranian and Anatolian plateaus (*Anatolia* refers to the large peninsula of Turkey, sometimes called Asia Minor) (see Figures 7.1 and 7.4). Both of these plateaus, averaging between 3000 and 5000 feet (1000 to 1500 meters) in elevation, are geologically active and prone to earthquakes. One dramatic quake in western Turkey (1999) measured 7.8 on the Richter scale, killed more than 17,000 people, and left 350,000 residents homeless. Another quake near the Iranian city of Bam (2003) claimed more than 30,000 lives.

Smaller lowlands characterize other portions of Southwest Asia. Narrow coastal strips are common in the Levant, along both the southern (Mediterranean Sea) and northern (Black Sea) Turkish coastlines and north of the Iranian Elburz Mountains near the Caspian Sea. Iraq contains the most extensive alluvial lowlands in Southwest Asia, dominated by the Tigris and Euphrates rivers, which flow southeast to empty into the Persian Gulf. Although much smaller, the distinctive Jordan River Valley is also a notable lowland that straddles the strategic borderlands of Israel, Jordan, and Syria and drains southward to the Dead Sea (Figure 7.5).

Patterns of Climate

Although Southwest Asia and North Africa is often termed the *dry world*, a closer look reveals a more complex climatic pattern (Figure 7.6). Both latitude and altitude come into play.

▼ **Figure 7.3 Atlas Mountains** The rugged Atlas Mountains dominate a broad area of interior Morocco.

▲ **Figure 7.4 Satellite View of Turkey** This satellite image of Turkey suggests the varied, quake-prone terrain encountered across the Anatolian Plateau. The Black Sea coastline is visible near the top of the image, and the island-studded Aegean Sea borders Turkey on the west.

Aridity dominates large portions of the region. A nearly continuous belt of desert lands stretches eastward from the Atlantic coast of southern Morocco across the continent of Africa, through the Arabian Peninsula, and into central and eastern Iran. Throughout this vast dry zone, plant and animal life has adapted to extreme conditions. Deep or extensive root systems and rapid life cycles allow desert plants to benefit from the limited moisture they receive. Similarly, animals adjust by efficiently storing water, hunting nocturnally, or migrating seasonally to avoid the worst of the dry cycle.

Some of the driest conditions are found across North Africa. Away from the Atlas Mountains, the Sahara Desert dominates much of the region. Much of the central Sahara receives less than 1 inch (2.5 cm) of rain a year and can thus support only the most meager forms of vegetation. Saharan summers feature extremely hot days and warm evenings, but winters are generally pleasant, even cool at night. Only in the tropical latitudes of southern Sudan do precipitation levels rise significantly. Here summer rains produce 20–50 inches (50–125 cm) of precipitation, and tropical savannas and woodlands replace the desert vegetation to the north.

Across the Red Sea, deserts also dominate Southwest Asia. Most of the Arabian Desert is not quite as dry as the Sahara proper, although the Rub-al-Khali along Saudi Arabia's southern border is one of the world's most desolate areas. On the fringe of the summer-monsoon belt, the Yemen Highlands along the southwestern edge of the Arabian Peninsula receive much more rain than the rest of the region and thus form a more favorable site for human habitation. To the northwest, precipitation also increases slightly in the central Tigris and Euphrates river valleys, with Baghdad, Iraq, averaging just over 5 inches (13 cm) of rain annually. Another major arid zone lies across Iran, a country divided into a series of minor mountain ranges and desert basins (Figure 7.7).

Elsewhere, altitude and latitude dramatically alter the desert environment and produce a surprising amount of climatic variety. For example, the Atlas Mountains and the nearby lowlands of northern Morocco, Algeria, and Tunisia experience a distinctly Mediterranean climate in which hot, dry summers alternate with cooler, relatively wet winters. In these areas, the landscape resembles that found in nearby southern Spain or Italy (Figure 7.8). A second zone of Mediterranean climate extends along the Levant coastline into the nearby mountains and northward across large portions of northern Syria, Turkey, and northwestern Iran.

Legacies of a Vulnerable Landscape

The island of Socotra illustrates the region's fragile and vulnerable environment and suggests how processes of globalization may threaten the area's long-term ecological health. Socotra's stony slopes rise out of the shimmering waters of the Indian Ocean about 230 miles (370 km) southeast of Yemen. The island's unique natural and cultural history has recently caught the world's attention. Separated for millions of years from the mainland of the Arabian Peninsula, Socotra's environment evolved in isolation. More than 30 percent of the island's 850 plants are found nowhere else on Earth.

▼ **Figure 7.5 Jordan Valley** This panoramic view of the Jordan Valley shows a fertile mix of irrigated tree, vine, and grain crops.

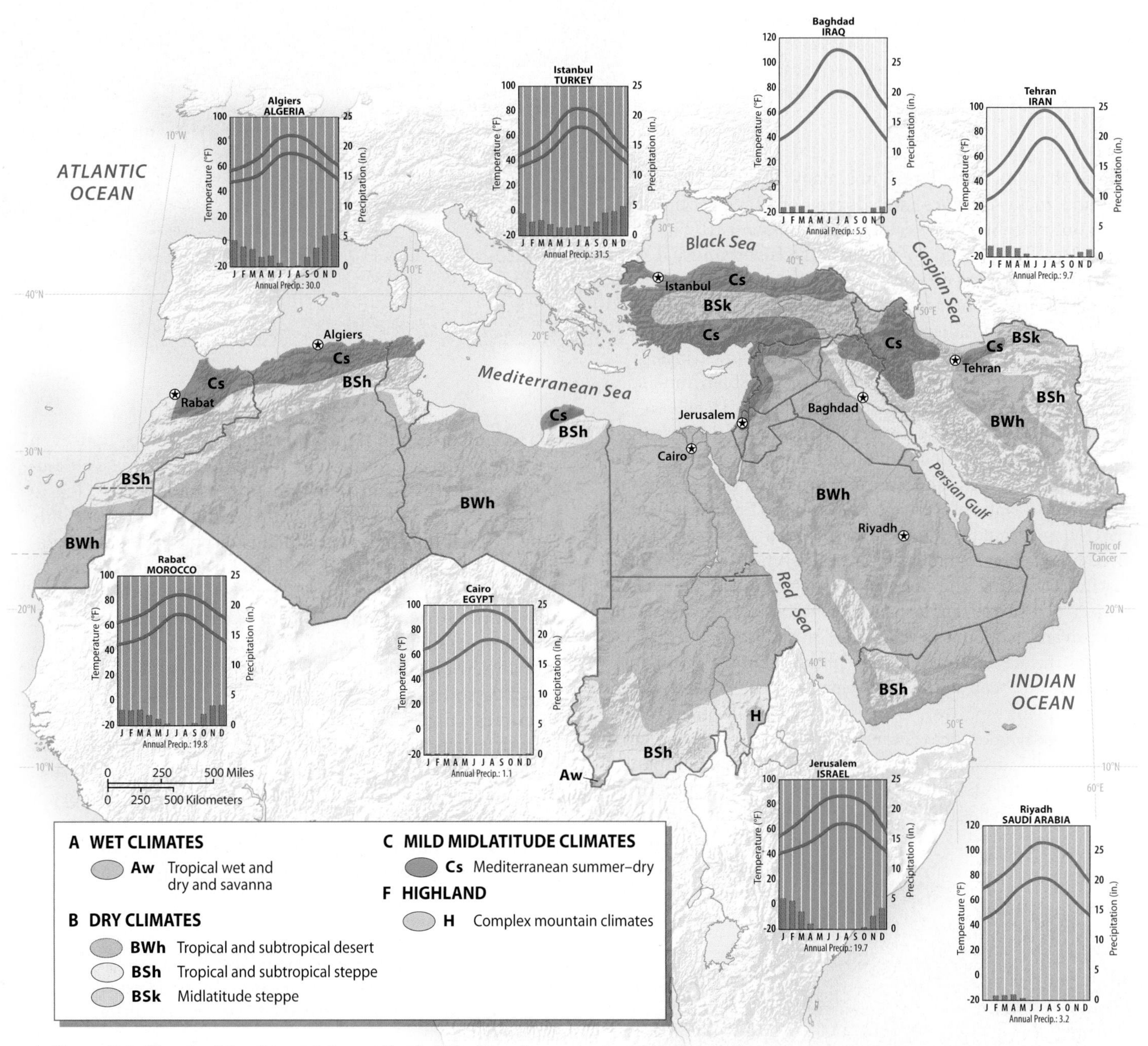

▲ **Figure 7.6 Climate of Southwest Asia and North Africa** Dry climates dominate from western Morocco to eastern Iran. Within these zones, persistent subtropical high-pressure systems offer only limited opportunities for precipitation. Elsewhere, mild midlatitude climates with wet winters are found near the Mediterranean Basin and Black Sea.

► **Figure 7.7 Arid Iran** Only sparse vegetation dots this arid scene from central Iran, a landscape characterized by isolated mountain ranges and dry interior plateaus.

▶ **Figure 7.8 Mediterranean Landscape, Northern Algeria** The Mediterranean moisture in northern Algeria produces an agricultural landscape similar to that of southern Spain or Italy. Winter rains create a scene that contrasts sharply with deserts found elsewhere in the region.

Exotic dragon's blood trees dot many of the island's dry and rocky hillsides, and dozens of other species have only recently been catalogued by botanists (Figure 7.9). Offshore, equally rare coral reefs and unusual fish populations have evolved as a part of the island's unique environmental setting in the wet–dry tropics of the northwestern Indian Ocean.

Socotra's unique environmental heritage now hangs in the balance. Global participants in the 1992 environmental summit at Rio de Janeiro established a $5 million international biodiversity project on Socotra, and Edinburgh's Royal Botanic Garden has conducted extensive studies on the island. In 2008, the island was recognized as a United Nations (UN) World Natural Heritage Site, and the European Union (EU) has officially supported the idea of preserving the area's unique biogeography. In the past several years, with funding from the UN, the Socotra Governance and Biodiversity Project has played an increasingly visible role in regulating tourism and economic development on the island.

At the same time, global pharmaceutical companies have already been allowed to harvest some of the island's unique plant species. Other plants and animals have been illegally taken by visitors. In addition, the Yemeni government has invited international petroleum companies to explore the island's offshore oil and gas potential, and it has considered a grand scheme for developing a series of luxury tourist hotels that would forever change the island's character. Environmental experts note that even well-meaning ecotourists on the island are already damaging coral reefs, and their growing numbers have led to increased firewood gathering and road construction. Only time will tell how Socotra, which is often cited as the Galápagos of the Indian Ocean, will fare in the 21st century.

The larger environmental history of Southwest Asia and North Africa contains many examples of the clever, but short-sighted legacies of its human occupants. Lengthy human settlement in a marginal land has resulted in deforestation, soil salinization and erosion, and depleted water resources (Figure 7.10). Indeed, the pace of population growth and technological change during the past century suggests that the region's vulnerable environment is destined to face even greater challenges in the 21st century.

Deforestation and Overgrazing Deforestation is an ancient problem in Southwest Asia and North Africa. Although

▼ **Figure 7.9 Socotra's Dragon's Blood Tree** Unique to Socotra, the rare dragon's blood tree reflects the island's environmental isolation. Evolving as a part of the island's ecosystem, the tree survives in the region's dry tropical climate.

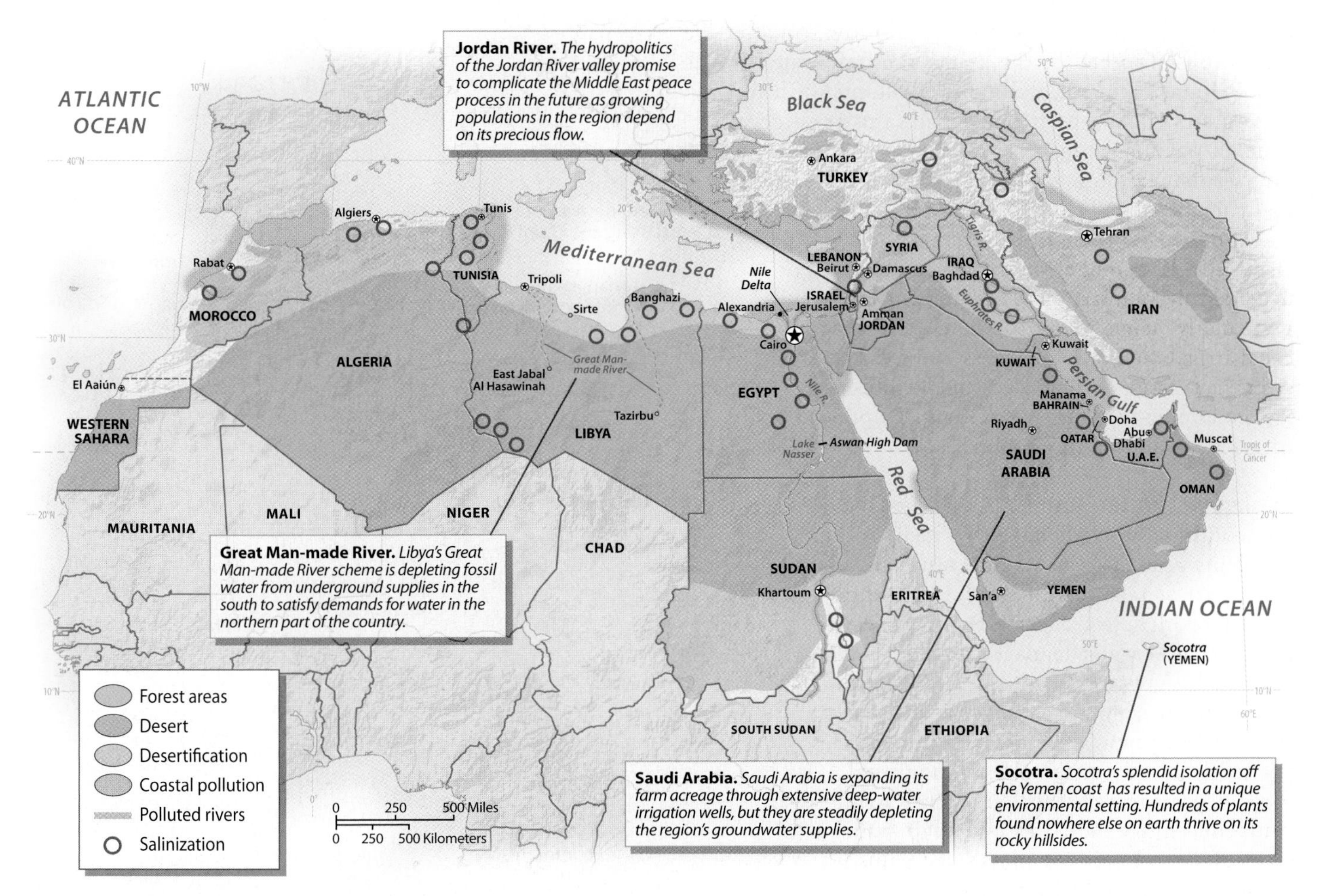

▲ **Figure 7.10 Environmental Issues in Southwest Asia and North Africa** Growing populations, pressures for economic development, and pervasive aridity combine to create environmental hazards across the region. Long human occupancy has contributed to deforestation, irrigation-induced salinization, and expanding desertification. **Q: Compare this map with Figure 7.6. What climate types are most strongly associated with desertification?**

much of the region is too dry for trees, the more humid and elevated lands that ring the Mediterranean once supported heavy forests. Included in these woodlands were the cedars of Lebanon, cut during ancient times and now reduced to a few scattered groves in a largely denuded landscape. In many settings, growing demands for agricultural land caused upland forests to be removed and replaced with grain fields, orchards, and pastures.

Human activities have conspired with natural conditions to reduce most of the region's forests to grass and scrub. Mediterranean forests grow slowly, are highly vulnerable to fire, and usually fare poorly if subjected to heavy grazing. Browsing by goats in particular often has been blamed for much of the region's forest loss, but other livestock have also impacted the vegetative cover, especially in steeply sloping semiarid settings that are slow to recover from grazing. Deforestation also has resulted in a millennia-long deterioration of the region's water supplies and in accelerated soil erosion.

Some forests survive in the mountains of northern and southern Turkey, and northern Iran also retains considerable tree cover. Scattered forests can be found in western Iran, the eastern Mediterranean, and the Atlas Mountains. Moreover, several governments have launched reforestation drives and forest preservation efforts in recent years. For example, both Israel and Syria have expanded their coverage of wooded lands since the 1980s. In nearby Lebanon, more than 5 percent of the country's total area was included in the Al-Shouf Cedar Reserve in 1996 to protect old-growth cedars, junipers, and oak forests. An even larger area (including the Al-Shouf Cedar Reserve) was declared the Shouf Biosphere Reserve by the UN in 2005 and is designed to protect the rare trees as well as endangered mammals, such as the wolf and Lebanese jungle cat.

Salinization Salinization, or the buildup of toxic salts in the soil, is another ancient environmental issue in a region where irrigation has been practiced for centuries (see Figure 7.10). The accumulation of salt in the topsoil is a common problem wherever desert lands are subjected to extensive irrigation. All freshwater contains a small amount of dissolved salt, and when water is diverted from streams into fields, salt remains in the soil after the moisture is absorbed by the plants and evaporated by the sun. In humid climates, accumulated salts are washed away by saturating rains, but in arid climates this rarely occurs. Where irrigation is practiced, salt concentrations build up over time, leading to lower crop yields and eventually to abandoned lands.

Hundreds of thousands of acres of once-fertile farmland within the region have been destroyed or degraded by salinization. The problem has been particularly acute in Iraq, where centuries of canal irrigation along the Tigris and Euphrates rivers have seriously degraded land quality. Similar conditions plague central Iran, Egypt, and irrigated portions of the Maghreb.

Managing Water Residents of the region are continually challenged by many other problems related to managing water in one of the driest portions of Earth. As technological change has accelerated, the scale and impact of water management schemes have had a growing effect on the region's environment. Sometimes the costs are justified if large new areas are brought into productive and sustainable agricultural use. In other cases, the long-term environmental price will far outweigh any immediate and perhaps short-lived gains in agricultural production.

Regional populations have been modifying drainage systems and water flows for thousands of years. The Iranian **qanat system** of tapping into groundwater through a series of gently sloping tunnels was widely replicated on the Arabian Peninsula and in North Africa. With simple technology, farmers directed the downslope flow of underground water to fields and villages where it could be efficiently utilized.

In the past half century, however, the scope of environmental change has been greatly magnified. Water management schemes have become huge engineering projects, major budget expenditures, and even political issues, both within and between countries of the region. One remarkable example is Egypt's Aswan High Dam, completed in 1970 on the Nile River south of Cairo (see Figure 7.10). Many benefits came with the dam's completion, including greatly increased storage

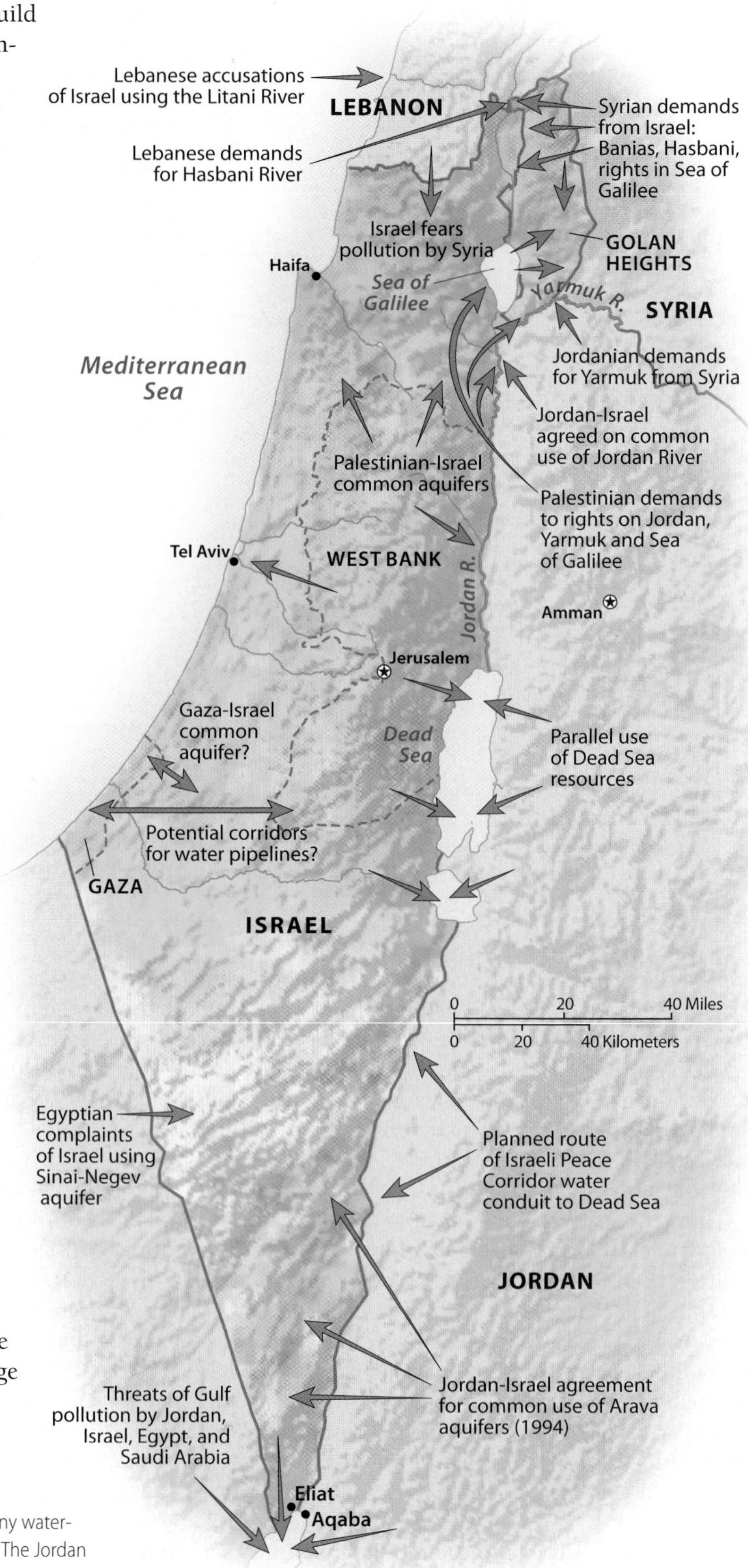

▶ **Figure 7.11 Hydropolitics in the Jordan River Basin** Many water-related issues complicate the geopolitical setting in the Middle East. The Jordan River system has been a particular focus of conflict.

capacity in the upstream reservoir. (This reservoir, consisting of Lake Nasser in Egypt and Lake Nubia in Sudan, is one of the largest human-made lakes in the world.) This promoted more year-round farming and an expansion of cultivated lands along the Nile—critical changes in a country that continues to experience rapid population growth. The dam also generates large amounts of clean electricity for the region. But the environmental costs have been high. The dam has changed methods of irrigation along the Nile, greatly increasing salinization. Fresh sediments and soil nutrients once washed across the valley floor with the annual high river flows, but more controlled irrigation has meant greatly increased inputs of costly fertilizers. Sediments that used to move downstream now accumulate behind the dam, infilling Lake Nasser. Other problems with the dam include an increased incidence of schistosomiasis (a debilitating parasitic disease spread by waterborne snails in irrigation canals) and the collapse of the Mediterranean fishing industry near the Nile Delta, an area previously nourished by the river's silt.

Israel's "Peace Corridor" project is another potentially massive engineering venture within the region. Designed to stimulate the regional economy and increase cooperation among Jordanians, Palestinians, and Israelis, the project would bring Red Sea water north into the Dead Sea. The ambitious plan calls for a 120-mile (200-km) conduit to be built from near Aqaba to the southern end of the Dead Sea (Figure 7.11). The water could be desalinated, generate hydroelectricity, and help save the Dead Sea and its major tourist resorts from literally drying up.

Elsewhere, **fossil water**, or water supplies stored underground during earlier and wetter climatic periods, has also been put to use by modern technology. Libya's "Great Manmade River" scheme taps underground water in the southern part of the country, transports it 600 miles (960 km) to northern coastal zones, and uses the precious resource to expand agricultural production in one of North Africa's driest countries (see Figure 7.10). Similarly, Saudi Arabia has invested huge sums to develop deep-water wells, allowing it to expand its food output greatly. Unfortunately, these underground supplies are being depleted much more rapidly than they are being recharged, thus limiting the long-term sustainability of such ventures.

A growing number of countries, including Saudi Arabia, Kuwait, and Israel, are also making large investments in seawater desalination plants (the region accounts for half the world's total). However, this technology consumes large amounts of energy and increases seawater salinity (Figure 7.12).

Most dramatically, **hydropolitics**, or the interplay of water resource issues and politics, has raised tensions between countries that share drainage basins. For example, in 2007, with the help of Chinese capital and engineering expertise, Ethiopia began construction of its huge Tekeze Dam project on the Nile. Already referred to by its promoters as the equivalent of "China's Three Gorges Dam in Africa," the controversial project threatens to disrupt downstream fisheries and irrigation in North Africa. Similarly, Sudan's Merowe Dam project (on another stretch of the Nile) has raised major concerns in nearby Egypt.

In Southwest Asia, Turkey's growing development of the upper Tigris and Euphrates rivers (the Southeast Anatolia Project, or GAP), complete with 22 dams and 19 power plants, has raised issues with Iraq and Syria, who argue that capturing "their" water might be considered a provocative political act. In addition, a 2013 study—based on new NASA satellite data—measured the overall water flows within the Tigris–Euphrates basin and found accelerating losses since 2003. In Iraq, as surface water flows from neighboring nations have declined, farmers have dug over 1000 new wells, severely impacting groundwater supplies and lowering regional water tables.

Hydropolitics also has played into negotiations among Israel, the Palestinians, and other neighboring states, particularly in the Golan Heights area (an important headwaters zone for multiple nations) and within the Jordan River drainage (see Figure 7.11). Israelis worry about Palestinian and Syrian pollution; nearby Jordanians argue for more water from Syria; and all regional residents must deal with the uncomfortable reality that, regardless of their political differences, they must drink from the same limited supplies of freshwater.

▼ **Figure 7.12 Desalination Plant, Israel** This facility, located south of Tel Aviv, produces about 13 percent of Israel's domestic freshwater and is one of the largest desalination plants in the world.

▲ **Figure 7.13 Straits of Hormuz** This satellite view shows the entrance to the Persian Gulf at the Straits of Hormuz, one of the region's most important choke points.

Aside from the use of water as a resource, it is also a vital transportation link in the area. The region's physical geography has produced enduring **choke points**, where narrow waterways are vulnerable to military blockade or disruption. For example, the Strait of Gibraltar (entrance to the Mediterranean), Turkey's Bosporus and Dardanelles straits, and the Suez Canal have all been key historical choke points within the region (see Figure 7.1). Iran's periodic threat to close the Straits of Hormuz (eastern end of the Persian Gulf) to world oil shipments suggests the strategic role water transportation continues to play in the region (Figure 7.13).

▼ **Figure 7.14 Alexandria, Egypt** This beachside view along northern Egypt's low-lying coastline at Alexandria could change significantly if global sea levels were to rise.

Climate Change in Southwest Asia and North Africa

Projected climate changes in Southwest Asia and North Africa will aggravate already existing environmental issues within the region. Temperature changes are predicted to have a greater impact on the region than changes in precipitation. The already arid and semiarid region will likely remain relatively dry, but warmer average temperatures are likely to have several major consequences:

- First, higher overall evaporation rates and lower overall soil moisture across the region will stress crops, grasslands, and other vegetation. Semiarid lands are particularly vulnerable, especially dryland cropping systems that cannot depend on irrigation. Even in irrigated zones, higher temperatures are likely to reduce yields for crops such as wheat and maize.
- Second, warmer temperatures will likely reduce net runoff into the region's already stressed streams and rivers, potentially reducing hydroelectric potential and water that is available for the region's increasingly urban population.
- Third, there will be a higher likelihood of extreme weather events, such as record-setting summertime temperatures. These extreme events will undoubtedly lead to more heat-related deaths as residents struggle to adapt, particularly in urban settings where many people cannot afford air conditioning.

Sea-level changes pose special threats to the Nile Delta. This portion of northern Egypt is a vast, low-lying landscape of settlements, farms, and marshland. Studies that simulate rising sea levels reveal that much of this region could be lost to inundation, erosion, or salinization. Farmland losses of more than 250,000 acres (100,000 hectares) are quite possible with even modest sea-level changes. For example, estimates are that a sea-level rise of 3.3 feet (1 meter) could affect 15 percent of Egypt's habitable land and displace 8 million Egyptians in coastal and delta settings. In the Egyptian city of Alexandria, $30 billion in losses have been projected because sea-level changes will devastate the city's huge resort industry, as well as nearby residential and commercial areas (Figure 7.14).

Some experts have also attempted to estimate the broader political and economic costs associated with potential climate changes within the region. For example, given the political instability of the Middle East, even relatively small changes in water supplies, particularly where they might involve several nations, could significantly add to the potential for conflict within the region. Projected economic implications of climate change may depend on the relative importance of agriculture within different nations. In addition, wealthier nations such as Israel and Saudi Arabia may have more available resources to plan, adjust, and adapt to climate shifts and extreme events versus poorer, less-developed countries such as Yemen, Syria, and Sudan.

REVIEW

7.1 Describe the climatic changes you might experience as you travel from the eastern Mediterranean coast to the highlands of Yemen. What are some of the key climatic variables that explain these variations?

7.2 Discuss five important human modifications of the Southwest Asian and North African environment, and assess whether these changes have benefited the region.

Population and Settlement: Changing Rural and Urban Worlds

The geography of human population across Southwest Asia and North Africa demonstrates the intimate tie between water and life in this part of the world. The pattern is complex: Large areas of the population map remain almost devoid of permanent settlement, while more moisture-favored lands suffer increasingly from problems of crowding and overpopulation (Figure 7.15).

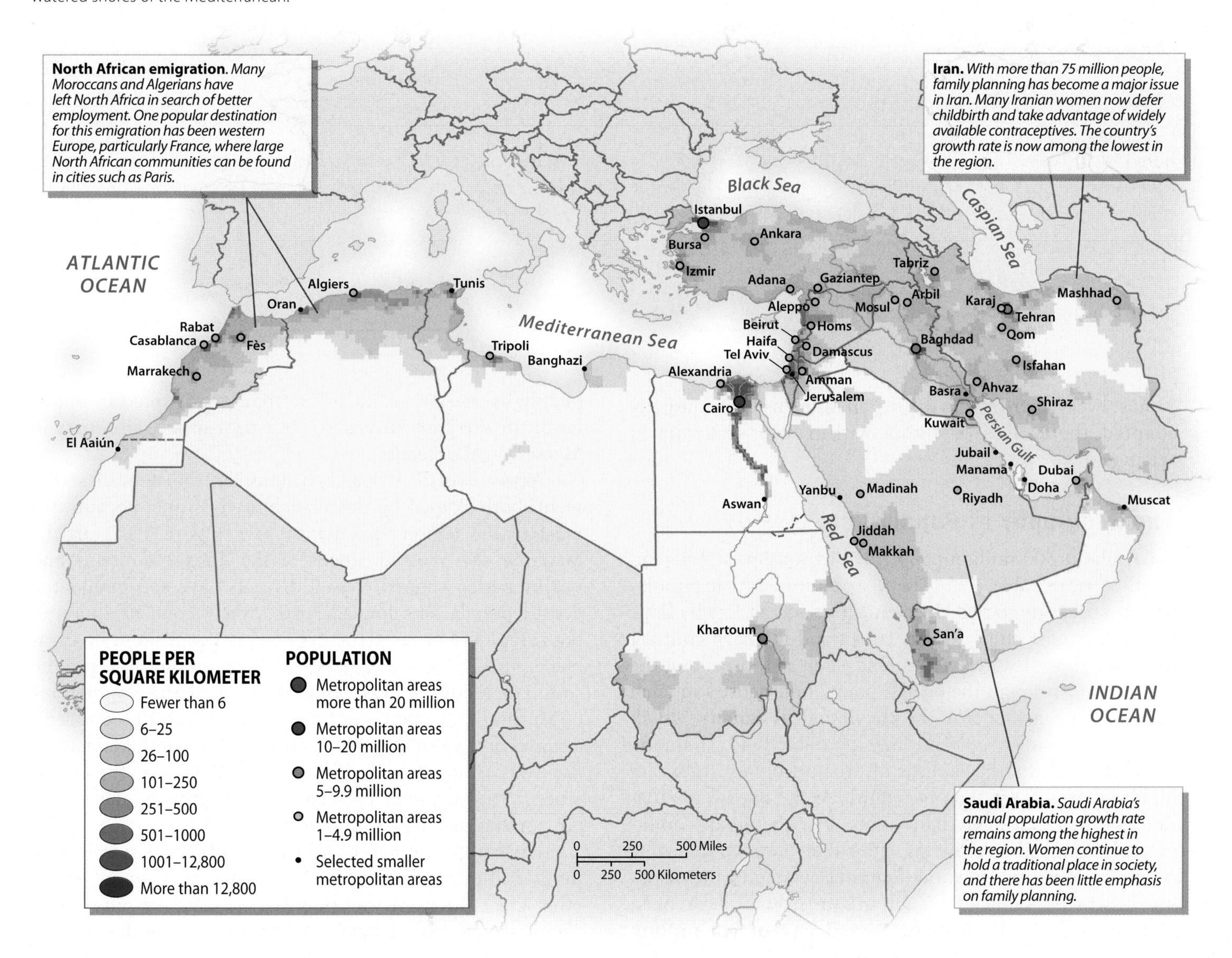

▼ **Figure 7.15 Population of Southwest Asia and North Africa** The striking contrasts are clearly evident between large, sparsely occupied desert zones and much more densely settled regions where water is available. The Nile Valley and the Maghreb region contain most of North Africa's people, whereas Southwest Asian populations cluster in the highlands and along the better-watered shores of the Mediterranean.

Table 7.1 Population Indicators

Country	Population (millions) 2013	Population Density (per square kilometer)	Rate of Natural Increase (RNI)	Total Fertility Rate	Percent Urban	Percent <15	Percent >65	Net Migration (Rate per 1000)
Algeria	38.3	16	2.2	3.0	73	28	6	–0
Bahrain	1.1	1,630	1.2	1.9	100	21	2	–40
Egypt	84.7	85	1.9	3.0	43	31	6	–1
Gaza and West Bank	4.4	734	2.9	4.1	83	41	3	–2
Iran	76.5	46	1.4	1.9	71	25	5	–1
Iraq	35.1	80	2.6	4.3	71	39	3	–1
Israel	8.1	364	1.6	3.0	91	28	10	2
Jordan	7.3	82	2.4	3.5	83	37	3	11
Kuwait	3.5	194	1.7	2.4	98	23	2	37
Lebanon	4.8	464	0.9	1.5	87	23	9	18
Libya	6.5	4	1.7	2.5	78	29	5	–6
Morocco	33.0	74	1.6	2.7	59	30	6	–3
Oman	4.0	13	1.8	2.9	75	26	3	42
Qatar	2.2	197	1.1	2.2	100	12	1	49
Saudi Arabia	30.1	14	1.8	2.9	81	30	3	2
Sudan	34.2	18	2.6	4.6	33	42	3	–4
Syria	21.9	118	2.2	3.1	54	35	4	–14
Tunisia	10.9	67	1.3	2.2	66	23	7	–1
Turkey	76.1	97	1.2	2.1	77	25	8	0
United Arab Emirates	9.3	112	1.5	1.9	83	13	0	11
Western Sahara	0.6	2	1.6	2.5	82	27	2	18
Yemen	25.2	48	2.7	4.9	29	42	3	–1

Source: Population Reference Bureau, *World Population Data Sheet, 2013.*

Almost everywhere across the region, humans have uniquely adapted themselves to living within arid or semiarid settings.

The Geography of Population

Today about 500 million people live in Southwest Asia and North Africa (Table 7.1). The distribution of that population is strikingly varied. In countries such as Egypt, large zones of almost empty desert land stand in sharp contrast to crowded, well-watered locations, such as those along the Nile River. Although the overall population density in such countries appears modest, the **physiological density**, which is the number of people per unit of arable land, is among the highest on Earth. Patterns of urban geography also are highly uneven: Less than two-thirds of the overall population is urban, but many nations are overwhelmingly dominated by huge and sprawling cities that produce the same problems of urban crowding found elsewhere in the developing world. Rates of recent urban growth have been phenomenal: Cairo, a modest-sized city of 3.5 million people in the 1960s, has more than quadrupled in population in the past 50 years (see *Cityscapes: Cairo, City of Many Neighborhoods*).

Across North Africa, two dominant clusters of settlement, both shaped by the availability of water, account for most of the region's population (see Figure 7.15). In the Maghreb, the moister slopes of the Atlas Mountains and nearby better-watered coastal districts have accommodated denser populations for centuries. Today concentrations of both rural and urban settlement extend from south of Casablanca in Morocco to Algiers and Tunis on the shores of the southern Mediterranean. Indeed, most of the populations of Morocco, Algeria, and Tunisia crowd into this crescent, a stark contrast to the almost empty lands south and east of the Atlas Mountains. Casablanca and Algiers are the largest cities in the Maghreb. They have rapidly growing metropolitan populations of 3–4 million residents each.

Farther east, much of Libya and western Egypt is very thinly settled. Egypt's Nile Valley, however, is home to the other great North African population cluster (Figure 7.16). The vast majority of Egypt's 85 million people live within

▲ **Figure 7.16 Nile Valley** This satellite image of the Nile Valley dramatically reveals the impact of water on the North African desert. Cairo lies at the southern end of the Nile Delta, where it begins to widen toward the Mediterranean Sea. The coastal city of Alexandria sits on the northwest edge of the low-lying delta. Lake Nasser is visible toward the bottom (center) of the image.

10 miles of the river. Optimistic politicians and planners are creating a second corridor of denser settlement in Egypt's "New Valley" west of the Nile by diverting water from Lake Nasser into the Western Desert.

Most Southwest Asian residents are clustered in favored coastal zones, moister highland settings, and desert localities where water is available from nearby rivers or subsurface aquifers. High population densities are found in better-watered portions of the eastern Mediterranean (Israel, Lebanon, and Syria) and Turkey. Nearby Iran is home to almost 80 million residents, but population densities vary considerably, from thinly occupied deserts in the east to more concentrated settlements near the Caspian Sea and across the more humid highlands of the northwest. Turkey's Istanbul, formerly Constantinople (13.5 million), and Iran's Tehran (12.2 million) are Southwest Asia's largest urban areas. Both have grown rapidly in recent years as rural populations gravitate toward the economic opportunities of these cities (Figure 7.17). Elsewhere, sizable populations are scattered through the Tigris and Euphrates river valleys, in the Yemen Highlands, and near oases where groundwater can be tapped to support agricultural or industrial activities.

Water and Life: Rural Settlement Patterns

Water and life are closely linked across the rural settlement landscapes of Southwest Asia and North Africa. Indeed, the diverse environments of Southwest Asia, in particular, are home to one of the world's earliest hearths of **domestication**, where plants and animals were purposefully selected and bred for their desirable characteristics. Beginning around 10,000 years ago, increased experimentation with wild varieties of wheat and barley led to agricultural settlements that later included domesticated animals, such as cattle, sheep, and goats.

Much of the early agricultural activity focused on the **Fertile Crescent**, an ecologically diverse zone that stretches from the Levant inland through the fertile hill country of northern Syria into Iraq. Between 5000 and 6000 years ago,

▼ **Figure 7.17 Istanbul, Turkey** Turkey's largest city is now home to more than 13 million people.

better knowledge of irrigation techniques and increasingly centralized political states promoted the spread of agriculture into the nearby valleys of the Tigris and Euphrates (Mesopotamia) and North Africa's Nile Valley. Since then, different peoples of the region have adapted to its environmental diversity and limitations in distinctive ways. In the process, they have practiced forms of agriculture appropriate to their settings and have left their own unique imprints upon the landscape (Figure 7.18).

Pastoral Nomadism Most common in the drier portions of the region, **pastoral nomadism** is a traditional form of subsistence agriculture in which practitioners depend on the seasonal movement of livestock for a large part of their livelihood. Arabian Bedouins, North African Berbers, and Iranian Bakhtiaris provide surviving examples of nomadism within the region. Today, however, with fewer than 10 million nomads remaining, the lifestyle is in decline, the victim of constricting political borders, reduced demand for traditional beasts of burden such as camels, competing land uses, and selective overgrazing. In addition, government resettlement programs in Saudi Arabia, Syria, Egypt, and elsewhere are actively promoting a more settled lifestyle for many nomadic groups.

The settlement landscape of pastoral nomads reflects their need for mobility and flexibility as they seasonally move camels, sheep, and goats from place to place. Near highland zones such as the Atlas Mountains and the Anatolian Plateau, nomads practice **transhumance** (the seasonal movement of animals between wet-season and dry-season pastures) by herding their livestock to cooler, greener high-country pastures in the summer and then returning them to valley and lowland settings for fall and winter grazing. Elsewhere, seasonal movements often involve huge territories of desert to support small groups of a few dozen families. In addition, nomads trade with sedentary agricultural populations, a mutually beneficial relationship in which they exchange meat, milk, hides, and wool for cereal and orchard crops available at desert oases.

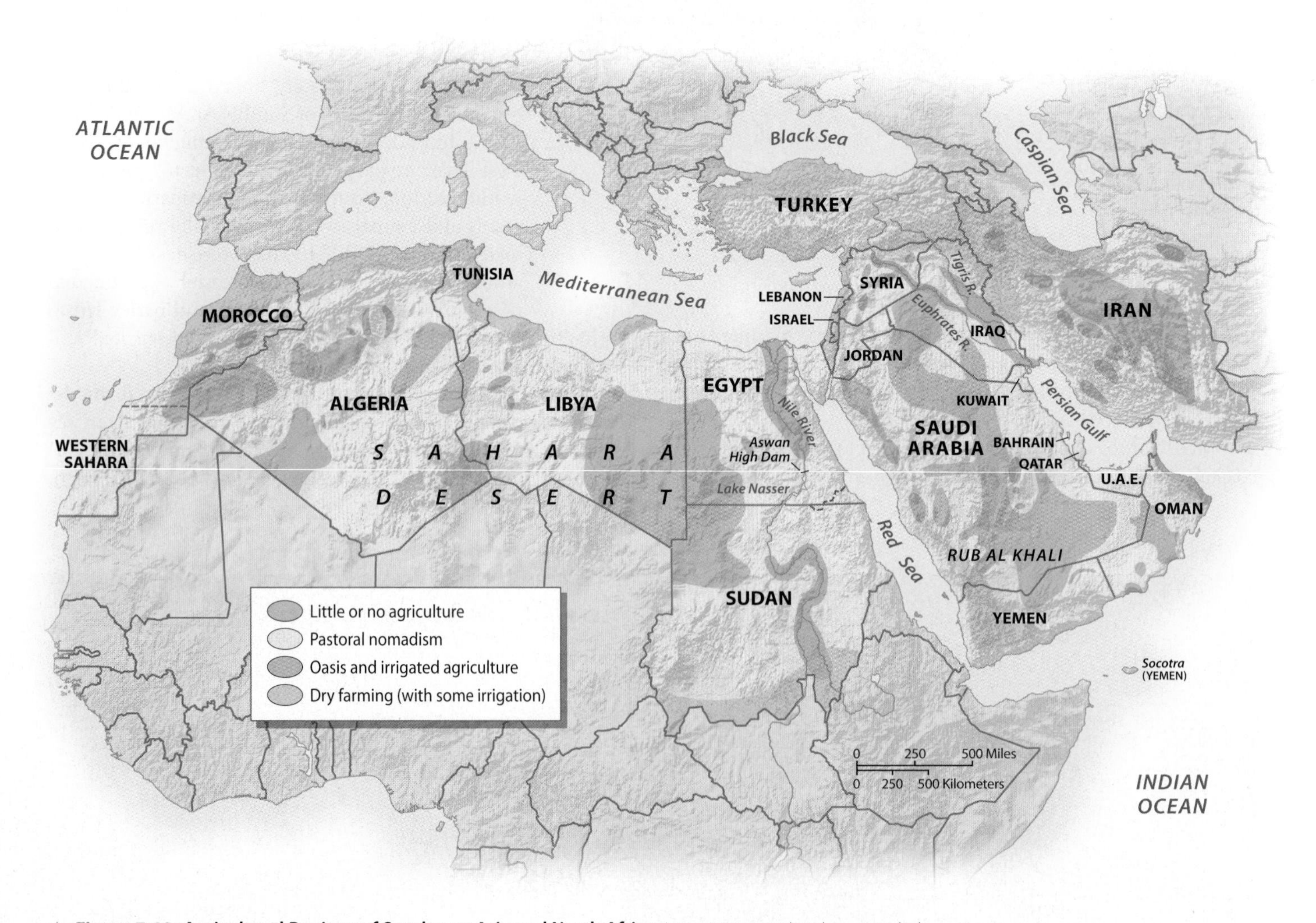

▲ **Figure 7.18 Agricultural Regions of Southwest Asia and North Africa** Important agricultural zones include oases and irrigated farming where water is available. Elsewhere, dry farming supplemented with irrigation is practiced in midlatitude settings.

▲ **Figure 7.19 Oasis Settlement** Date palms and irrigated fields shape the landscape around Tinehir, a fertile oasis settlement located in central Morocco.

activity, and both systems have large, densely settled deltas. Other linear irrigated settlements can be found near the Jordan River in Israel and Jordan, along short streams originating in North Africa's Atlas Mountains, and on the more arid peripheries of the Anatolian and Iranian plateaus. These settings, although capable of supporting sizable rural populations, also are among the most vulnerable to overuse, particularly if irrigation results in salinization.

Farming in such localities supports much higher population densities than is the case with pastoral nomadism or traditional desert oases. Fields are small, intensely utilized, and connected with closely managed irrigation systems designed to store and move water efficiently through the settlement. In Egypt, farmers of the Nile Valley live in densely settled, clustered villages near their fields; work much of their land with the same tools and technologies their ancestors used; and grow a mix of cotton, rice, wheat, and forage crops (Figure 7.20). However, rural life is also changing in such settings. New dam- and canal-building schemes in Egypt, Israel, Syria, Turkey, and elsewhere are increasing the storage capacity of river systems, which allows for more year-round agricultural activity. Higher-yielding rice and wheat varieties and more mechanized agricultural methods also have raised food output, particularly in places such as Egypt and Israel. Some of the most efficient farms in the region are associated with Israeli **kibbutzim**, which are collectively worked settlements producing grain, vegetable, and orchard crops irrigated by waters from the Jordan River and from the country's elaborate feeder canals. For example, Israel's National Water Carrier system takes water from the Sea of Galilee and moves it south and west, where intensively worked agricultural operations produce key food crops for nearby Tel Aviv and Jerusalem.

Oasis Settlements Permanent oasis settlements dot the arid landscape where high groundwater levels or modern deep-water wells provide reliable moisture in otherwise arid locales. Tightly clustered, often walled villages, their sun-baked mud houses blending into the surrounding scene, sit adjacent to small, but intensely utilized fields where underground water is carefully applied to tree and cereal crops. In more recently created oases, concrete blocks and prefabricated housing add a contemporary look. Surrounded by large zones of desert, these green islands of rural activity stand out in sharp contrast to the sand- and rock-strewn landscape.

Traditional oasis settlements (Figure 7.19) are composed of close-knit families who work their own irrigated plots or, more commonly, work for absentee landowners. Although oases are usually small, eastern Saudi Arabia's Hofuf oasis covers more than 30,000 acres (12,000 hectares). Some crops are raised for local consumption, but commercial trade has always played an important role in such settings. In the past century, the expanding world demand for products such as figs and dates has included even these remote oases in the global economy, and many products end up on the tables of hungry Europeans or North Americans. New drilling and pumping technologies, particularly in Saudi Arabia, have added to the size and number of oasis settlements. But oasis life across the region faces major challenges. Population growth, groundwater depletion, and the pressures of global cultural change threaten the economic and social integrity of these settlements.

Settlement Along Exotic Rivers For centuries, the region's densest rural settlement has been tied to its great river valleys and their seasonal floods of water and enriching nutrients. In such settings, **exotic rivers** transport precious water and nutrients from distant, more humid lands into drier regions, where the resources are utilized for irrigated farming (see Figure 7.16). The Nile and the Tigris and Euphrates rivers are the largest regional examples of such

▼ **Figure 7.20 Nile River Valley** These farmers are pulling clover for animal feed from these irrigated fields along the Nile River north of Cairo, Egypt.

Cityscapes

Cairo, City of Many Neighborhoods

▲ **Figure 7.1.1 Downtown Cairo** This aerial view of downtown Cairo shows high-rise office buildings and hotels lining the shoreline of the Nile River.

Crowded Cairo is now home to more than 15 million people, making it easily the region's largest city (Figure 7.1.1). Founded more than 1000 years ago along the shores of the Nile River, Cairo emerged as a key Islamic city in the Muslim world. Its universities cultivated Islamic scholarship, and its commercial bazaars served as a key regional crossroads between northern Africa and western Asia.

Modern Downtown Emerges Modern Cairo began taking shape in the later 19th century when Egyptian rulers reimagined major sections of downtown, inspired by the European Renaissance. This section of the city—sometimes called the Paris of the Nile—was redesigned with distinctive open spaces, grand boulevards, and key government buildings. On its southwestern edge, the area's most famous landmark is undoubtedly *Tahrir* (or Liberation) *Square,* which served as a key public rallying point in 2011 during Egypt's Arab Spring rebellions that overthrew the regime of Hosni Mubarak (see Figure 7.2).

▼ **Figure 7.1.2 Old Cairo** Cairo's narrow, twisting streets are often crowded with shoppers and pedestrians.

The Challenge of Dryland Agriculture Mediterranean climates in portions of the region permit varied forms of dryland agriculture that depend largely on seasonal moisture to support farming. These zones include the better-watered valleys and coastal lowlands of the northern Maghreb, lands along the shore of the eastern Mediterranean, and favored uplands across the Anatolian and Iranian plateaus (see Figure 7.18). A regional variant of dry farming is practiced across Yemen's terraced highlands in the moister corners of the southern Arabian Peninsula.

Often, a variety of crops and livestock surrounds the Mediterranean villages of the region. Drought-resistant tree crops are common, with olive groves, almond trees, and citrus orchards producing output for both local consumption and commercial sale. Elsewhere, favored locations support grape vineyards, whereas more marginal settings are used to grow wheat and barley or to raise forage crops to feed cattle, sheep, and goats. Vulnerability to drought and the availability of more sophisticated water management strategies are leading some Mediterranean farmers to utilize more irrigated farming, thus improving production of cotton, wheat, citrus fruits, and tobacco across portions of the region. More mechanization, crop specialization, and fertilizer use are also transforming such agricultural settings, following a pattern set earlier in nearby areas of southern Europe. One commercial adaptation of growing regional and global importance is Morocco's flourishing hashish crop. More than 200,000 acres (80,000 hectares) of cannabis are cultivated in the hill

Strolling City Streets Some of Cairo's most interesting neighborhoods are found east and south of downtown (**Figure 7.1.2**). Stroll the twisting lanes and narrow streets—crowded with shops and people—in *Old Cairo*. These southern districts grew slowly and haphazardly, and some streets are remnants of cities that even predate Cairo's official founding. The city's Coptic population and Christian churches are also concentrated here.

The *Islamic Cairo* neighborhood retains a similar feel. It is east of downtown and home to many of the city's important mosques and Islamic monuments. Egypt's greatest novelist, Naguib Mahfouz (1911–2006), captured some of the character of this area in his Cairo Trilogy of novels (*Palace Walk, Palace of Desire, Sugar Street*). In *Palace of Desire,* one of his characters, Yasin, returns to his boyhood neighborhood:

> When his feet brought him to al-Gamaliya Street, he was so choked up he felt he would die. He had not been there for eleven years...yet it remained exactly the way it had been when he was growing up. Nothing had changed. The street was still so narrow a handcart would almost block it when passing by. The protruding balconies of the houses almost touched each other overhead. The small shops resembled the cells of a beehive, they were so close together and crowded with patrons, so noisy and humming. . . . There was the same never-ending stream of pedestrian traffic. Uncle Hasan's snack shop and Uncle Sulayman's restaurant too remained just as he had known them. . . .

Dynamic Suburban Growth Many of Cairo's peripheral districts have changed dramatically in the past 50 years. Huge suburban developments west of the river (in *Giza*) and east of downtown (*Nasr City* and beyond) have spilled into former farmland and desert waste. Nearby, one of the city's prime burial grounds—the famous *City of the Dead*—is now home to almost 1 million residents (mostly poor), who live in cheap homes and apartment houses erected amid a sprawling cemetery. Another satellite community called *New Cairo City,* about 20 miles east of downtown, features planned boulevards, affluent housing, and new university facilities (**Figure 7.1.3**). However, the development's long-term prospects remain uncertain amid Egypt's current economic and political challenges.

▲ **Figure 7.1.3 New Cairo City** Recent construction east of downtown is creating a very different landscape in suburban Cairo.

Google Earth Virtual Tour Video

http://goo.gl/D38KO1

1. Find Cairo on Google Earth, and examine parts of the old city as well as new suburban developments to the east. Describe three key visual differences you can detect between these old and new settlement patterns.
2. Find a work of literature (novel, short story, poem) focused on an urban setting in your region, and identify a passage (such as the one by Mahfouz) that captures a local sense of place.

country near Ketama in northern Morocco, generating more than $2 billion annually in illegal exports (mostly to markets in Europe) (Figure 7.21).

Many-Layered Landscapes: The Urban Imprint

Cities have played a pivotal role in the region's human geography. Indeed, some of the world's oldest urban places are located in the region. Today enduring political, religious, and economic ties link the city and countryside.

A Long Urban Legacy Cities in the region traditionally have played important functional roles as centers of political and religious authority, as well as key focal points of local and long-distance trade. Urbanization in Mesopotamia (modern Iraq) began by 3500 BCE, and cities such as Eridu and Ur reached populations of 25,000 to 35,000 residents. Similar centers appeared in Egypt by 3000 BCE, with Memphis and Thebes assuming major importance amid the dense populations of the middle Nile Valley. These ancient cities were key centers of political and religious control. Temples, palaces, tombs, and public buildings dominated the urban landscapes of such settlements, and surrounding walls (particularly in Mesopotamia) offered protection from outside invasion. By 2000 BCE, however, a different kind of city was emerging, particularly along the shores of the eastern Mediterranean and at the junction points of important caravan routes. Centers such as Beirut, Tyre, and

▲ **Figure 7.21 Cannabis Fields, Morocco** Bundles of processed hash dry in the sun near Ketama, Morocco, and cannabis fields clothe the nearby hillside. Much of the region's hashish crop is bound for Europe.

Sidon, all in modern Lebanon, as well as Damascus in nearby Syria, exemplified the growing role of trade in creating urban landscapes. Expanding port facilities, warehouse districts, and commercial thoroughfares suggested how trade and commerce shaped these urban settlements, and many of these early Middle Eastern trading towns have survived to the present.

Islam also left a lasting mark because cities traditionally served as centers of Islamic religious power and education. By the 8th century, Baghdad had emerged as a religious center, followed soon thereafter by the appearance of Cairo as a seat of religious authority and expansion. Urban settlements from North Africa to Turkey felt the influences of Islam. Indeed, the Islamic Moors carried the Muslim culture to Spain, where it shaped urban centers such as Córdoba and Málaga. Islam's impact upon the settlement landscape merged with older urban traditions across the region and established a characteristic Islamic cityscape that exists to this day. Its traditional attributes include a walled urban core, or **medina**, dominated by the central mosque and its associated religious, educational, and administrative functions (Figure 7.22). A nearby bazaar, or *suq,* functions as a marketplace where products from city and countryside are traded. Housing districts feature an intricate maze of narrow, twisting streets that maximize shade and accentuate the privacy of residents, particularly women (Figure 7.23). Houses have small windows, frequently are situated on dead-end streets, and typically open inward to private courtyards, which are often shared by extended families with similar ethnic or occupational backgrounds.

More recently, European colonialism added another layer of urban landscape features in selected cities. Particularly in North Africa, coastal and administrative centers during the late 19th century added dozens of architectural features from Britain and France. Victorian building blocks, French mansard roofs, suburban housing districts, and wide European-style commercial boulevards complicated the settlement landscapes of dozens of cities, both old and new. Centers such as Algiers (French), Fes (French), and Cairo (British) vividly displayed the effects of colonial control, and many of these urban signatures remain today (see Figure 7.23).

Signatures of Globalization Since 1950, dramatic new forces have transformed the urban landscape. Cities have become key gateways to the global economy. As the region has been opened to new investment, industrialization, and tourism, the urban landscape reflects the fundamental changes taking place. Expanded airports, commercial and financial districts, industrial parks, and luxury tourist facilities all mark the imprint of the global economy.

Further, as urban centers become focal points of economic growth, surrounding rural populations are drawn to the new employment opportunities, thus fueling rapid population increases. The results are both impressive and problematic. Many traditional urban centers, such as Algiers and Istanbul, have more than doubled in size in recent years. Booming demand for homes has produced ugly, cramped high-rise

▼ **Figure 7.22 Mosque in Qom, Iran** Qom's Hazrati Masumeh Shrine is annually visited by thousands of faithful Shiites in this sacred Iranian city south of Tehran.

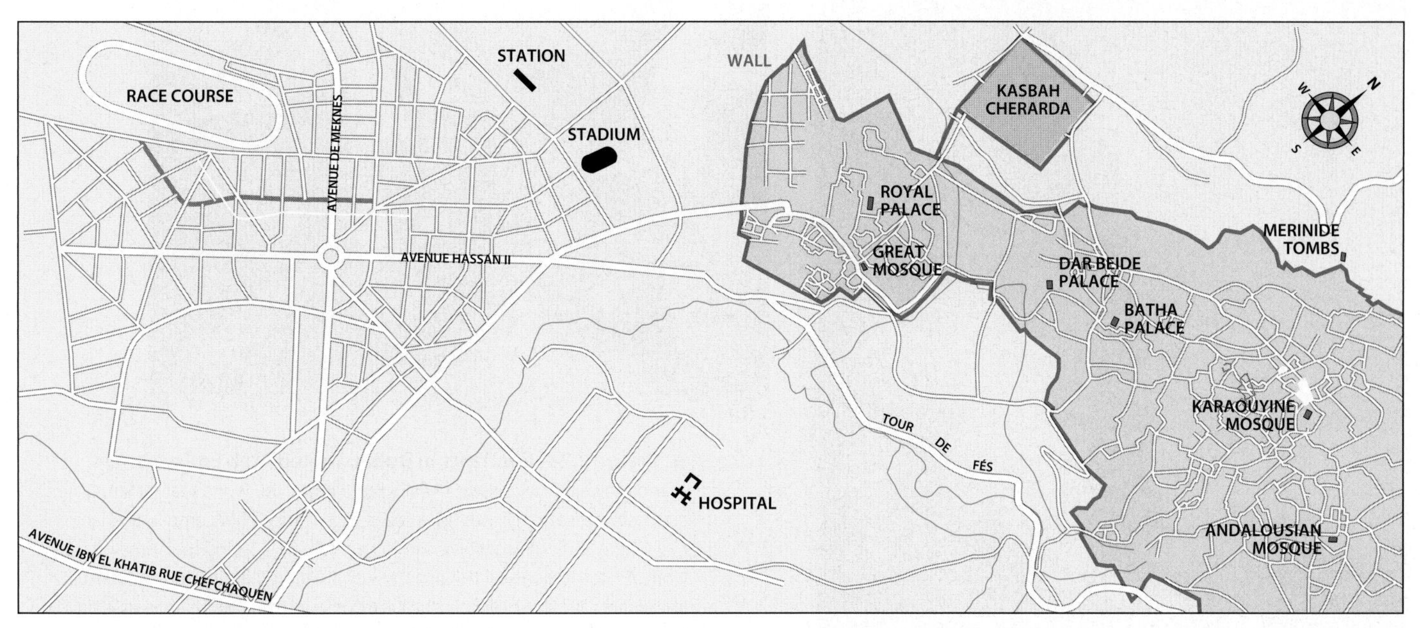

▲ **Figure 7.23 Map of Fes, Morocco** The tiny neighborhoods and twisting lanes of the old walled city reveal features of the traditional Islamic urban center. To the west, however, the rectangular street patterns, open spaces, and broad avenues suggest colonial European influences.

apartment houses in some government-planned neighborhoods. Elsewhere, sprawling squatter settlements provide little in the way of quality housing or municipal services.

Certainly, the oil-rich states of the Persian Gulf display the most extraordinary changes in the urban landscape (Figure 7.24). Before the 20th century, urban traditions were relatively weak in the area, and even as late as 1950 only 18 percent of Saudi Arabia's population lived in cities. All that changed, however, as the global economy's demand for petroleum mushroomed. Today the Saudi Arabian population is more urban than those of many industrialized nations, including the United States, and the capital city of Riyadh has grown to about 5 million people. Particularly after 1970, other cities, such as Dubai (UAE), Doha (Qatar), Kuwait City (Kuwait), and Manama (Bahrain), grew in size and took on modern Western characteristics, including futuristic architecture and new transportation infrastructure. Dubai's skyline is now home to the soaring Burj Khalifa (completed in 2010), a needle-like, 160-story building that ascends a half mile into the desert sky (Figure 7.25a). In addition, investments in petrochemical industries have fueled the creation of new urban centers, such as Jubail along Saudi Arabia's Persian Gulf coastline. Masdar City, one of the world's most carefully planned and energy-efficient cities, is also taking shape in this part of the world (see *Working Toward Sustainability: Masdar City Emerges in the Desert*).

A Region on the Move

Nomads have crisscrossed this region for ages, but entirely new patterns of migration reflect the global economy and recent political events. The rural-to-urban shift seen

▼ **Figure 7.24 Persian Gulf City of Manama, Bahrain** In this aerial view of the city, Manama's changing skyline is marked by high-rise construction projects.

(a)

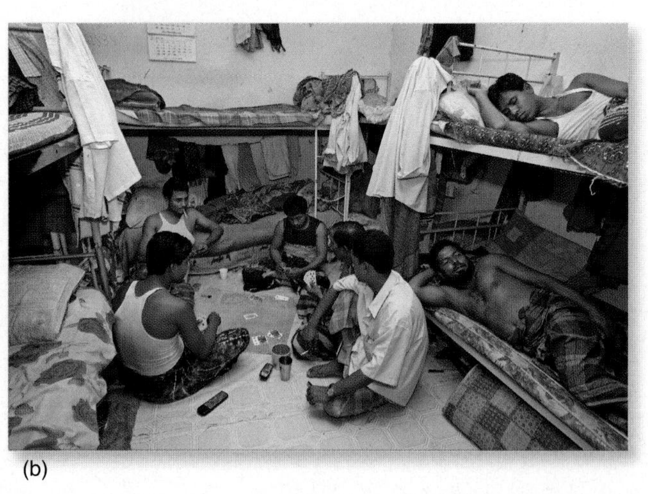

(b)

▲ **Figure 7.25 Contrasts in Dubai, United Arab Emirates** (a) The soaring tower of downtown Dubai's Burj Khalifa, the world's tallest structure, contrasts dramatically with (b), a scene in a nearby labor camp in Shariah City where South Asian workers enjoy a card game during their time off from work. A large majority of the country's population is foreign-born, and many immigrants like these work in the construction sector of the economy, building structures such as the Burj Khalifa.

widely in the less developed countries is reworking population patterns across Southwest Asia and North Africa. The Saudi Arabian example is echoed in many other countries. Cities from Casablanca to Tehran are experiencing phenomenal growth rates, spurred by in-migration from rural areas.

Foreign workers have also migrated to areas within the region that have large labor demands (Figure 7.26). In particular, Persian Gulf

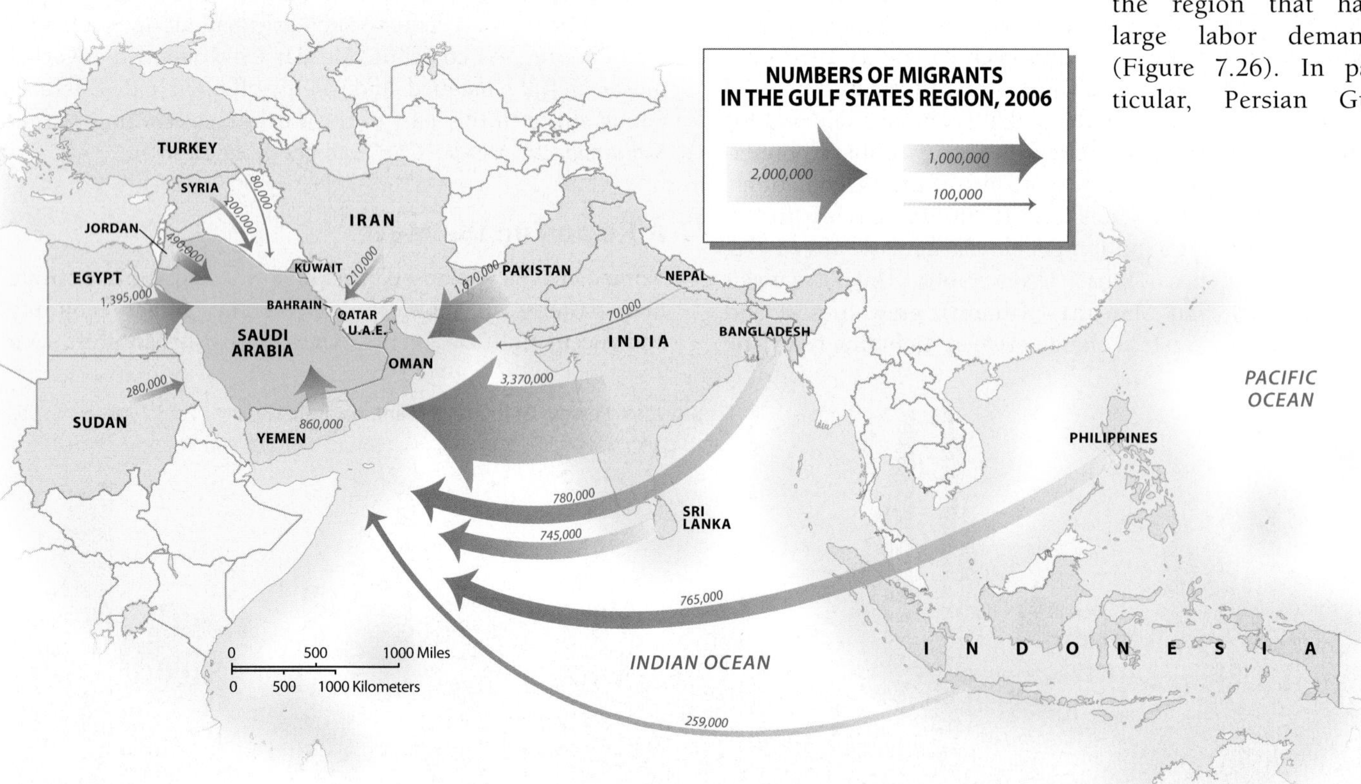

▲ **Figure 7.26 Estimated Migrant Populations in the Gulf States, 2006** The map shows the sources of immigrant populations working in the Persian Gulf states region in the early 21st century. Note the importance of nearby Arab countries, other Muslim nations, and South Asia.

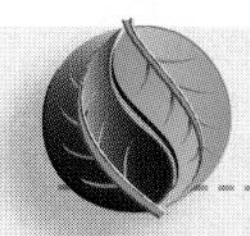

Working Toward Sustainability

Masdar City Emerges in the Desert

▲ **Figure 7.2.1 Masdar City** Masdar City is being laid out near the Abu Dhabi Airport in the United Arab Emirates.

A new walled city—reminiscent of traditional regional urban forms—is slowly taking shape in Abu Dhabi (within the UAE). Located about 11 miles (17 km) outside of the city of Abu Dhabi, Masdar City is no ordinary suburb (Figure 7.2.1). Its planners have ambitiously designed it as the planet's first completely sustainable, zero-carbon, zero-waste city. Moreover, promoters want it to be the one-stop shopping place for new ideas and technologies for designing other green cities around the 21st-century world.

Long-Term Trajectory Initial construction of Masdar City began in 2008. The project received financial assistance from the Abu Dhabi government, much of it through the Abu Dhabi Future Energy Corporation. The global financial crisis slowed development in 2009 and 2010, but the first phase of the $15–$25 billion project is now scheduled to be finished by 2015. The long-term trajectory for Masdar City sees construction continuing until 2020 or 2025. By then, the city will support 45,000 to 50,000 new residents and enough jobs to support an additional workforce of 60,000 daily commuters. The city will also be home to a large university community (led by the Massachusetts Institute of Technology–supported Masdar Institute of Science and Technology), oriented around studying and promoting green technologies (Figure 7.2.2). Most importantly, Masdar City boosters envision a huge collection of green-oriented industries (1500 businesses) and research incubators where the next generation of energy-efficient technologies will be developed and marketed to the rest of the world.

▼ **Figure 7.2.2 Masdar Institute of Science and Technology** One of the focal points of green-technology research and innovation, the Masdar Institute of Science and Technology has already attracted a growing number of students and research initiatives and has also received support from the U.S.-based Massachusetts Institute of Technology.

A Sustainable Urban Design The city's overall urban design is the brainchild of Fosters and Partners, a British firm led by innovative architect Norman Foster. Foster's dream is to make Masdar City the world's most sustainable urban place. A varied array of large-scale solar-energy projects will supply much of the city's power, in addition to wind farms and geothermal energy sources. All buildings will feature green-construction technologies. A large solar-powered desalination plant will provide much of the city's water (from the nearby Gulf), and the municipal system will recycle about 80 percent of its wastewater. Urban transportation will largely avoid gas-powered vehicles. Instead, energy-efficient mass transportation (light rail and an underground metro line) and electric vehicles are designed to offer more sustainable patterns of urban movement.

Project Critics The World Wildlife Fund, Greenpeace, and even the U.S. Department of Energy have enthusiastically endorsed the project, but Masdar City also has its critics. Some observers claim that Masdar City is simply destined to become an elite walled enclave, largely a showplace for Abu Dhabi and wealthy corporations interested in profiting from green technology. Will ordinary people in the region really benefit? Other experts doubt that Masdar City can affordably implement all of its innovative goals. For example, cost-cutting measures have largely eliminated an innovative PRT (personal rapid transit) system of pod cars originally designed to move people around the city. Still, Masdar City promises to be a well-funded and highly visible initiative that (somewhat ironically) will search for pathways to a postpetroleum future in a setting blessed with some of the world's richest reserves of fossil fuels.

1. In considering the overall urban layout of Masdar City, what elements of traditional urban design parallel the historical regional pattern of the Islamic city? What traditional features are given less emphasis?
2. Cite local examples of sustainable green building technologies on your own campus or within your community.

Google Earth Virtual Tour Video

http://goo.gl/HBWTx9

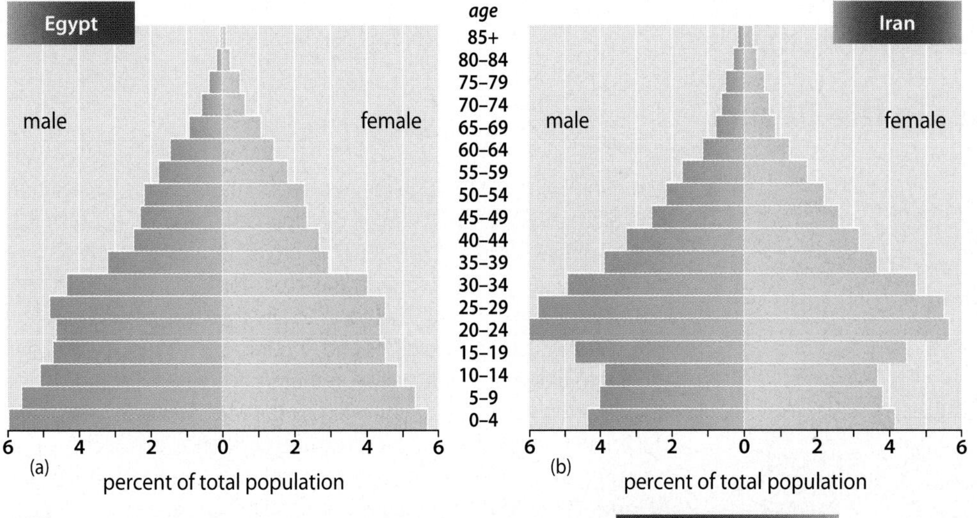

◄ **Figure 7.27 Population Pyramids of Egypt, Iran, and United Arab Emirates, 2012** Three distinctive demographic snapshots highlight regional diversity: (a) Egypt's above-average growth rates differ sharply from those of (b) Iran, where a focused campaign on family planning has reduced recent family sizes. Male immigrant laborers play a special role in skewing the pattern within (c) the United Arab Emirates. **Q: For each example, cite a related demographic or cultural issue that you might potentially find in these countries.**

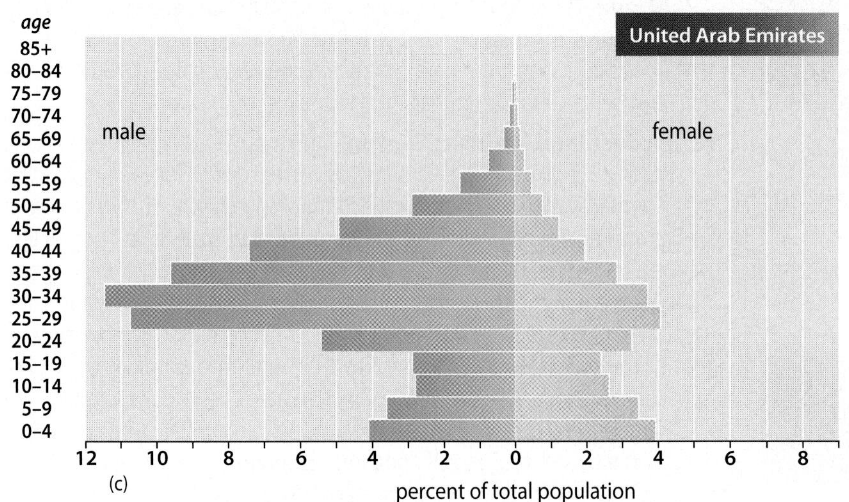

nations support immigrant workforces that often comprise large proportions of the overall population. The influx has major economic, social, and demographic implications in nations such as Saudi Arabia (27 percent of the total population is foreign workers), Kuwait (68 percent), Qatar (87 percent), and the UAE (81 percent) (Figure 7.25b). Source regions for this Gulf-nation labor force are varied, but the largest communities are from South Asia and other Muslim countries within and beyond the region (see Figure 7.26, p. 318). In Dubai (UAE), Pakistani cab drivers, Filipino nannies, and Indian shop clerks typify a foreign workforce that currently makes up a large majority of the city's 1.4 million inhabitants. Most migrants to the UAE, however, are male, creating an extraordinary gender imbalance in the country's population (Figure 7.27).

Other residents migrate to jobs elsewhere in the world. Because of its strong economy and close location, Europe is a powerful draw. More than 2 million Turkish guest workers live in Germany. Both Algeria and Morocco also have seen large out-migrations to western Europe, particularly France. Many Lebanese, often skilled workers and business owners, have also emigrated.

Political forces also encourage migration. Thousands of wealthier residents, for example, left Lebanon and Iran during the political turmoil of the 1980s and are today living in cities such as Toronto, Los Angeles, and Paris. More recent political instability has provoked other refugee movements. More than 35,000 Iraqis have resettled in the United States since 2003. Elsewhere, thousands of displaced Afghans remain in eastern Iran. In the Middle East, more than 500,000 Russian-Jewish immigrants have migrated to Israel since 1991 and have reshaped the cultural geography of that country. Syria's recent civil war also has produced growing refugee populations in neighboring countries, especially Jordan, Lebanon, and Turkey. To the south, huge numbers of people in western Sudan's unsettled Darfur region have been forced to move to dozens of refugee camps in nearby Chad.

Shifting Demographic Patterns

High population growth remains a critical issue throughout Southwest Asia and North Africa, but the demographic picture is shifting. Uniformly high growth rates in the 1960s have been replaced by more varied regional patterns. For example, women in Tunisia and Turkey now average fewer than three births, representing a large decline in total fertility rates (see Table 7.1). Various factors explain these changes. More urban, consumer-oriented populations opt for fewer children. Many Arab women now delay marriage into their middle 20s and early 30s. Family-planning initiatives are expanding in many countries. For example, programs in Tunisia, Egypt, and Iran have greatly increased access to contraceptive pills, IUDs, and condoms. Intriguingly, fundamentalist Iran has witnessed the fastest decline in fertility in the past two decades (see Figure 7.27). Fertility has fallen by more than two-thirds since the mid-1970s (from an average of 6.6 births to 1.9 births per woman). Although Iran's family-planning program was initially dismantled after the fundamentalist revolution in 1979 (it was seen as a Western idea), recent leaders have recognized the wisdom in containing the country's large population. Iran has one of the world's most successful family-planning programs, according to the Population Reference Bureau.

Still, areas such as the West Bank, Gaza, and Yemen are growing much faster than the world average. Poverty and traditional ways of rural life contribute to large rates of population increase, and even in more urban Saudi Arabia, growth rates remain near 2 percent. The increases result from high birth rates combined with low death rates. In Egypt, even though birth rates may decline, the labor market will need to absorb more than 500,000 new workers annually over the next 10 to 15 years just to keep up with the country's large youthful population (see Figure 7.27). As that population ages in the mid-21st century, it will be increasingly urban and demand jobs, housing, and social services. Growing populations will also increase demands on the region's already limited water resources.

REVIEW

7.3 Discuss how pastoral nomadism, oasis agriculture, and dryland wheat farming represent distinctive adaptations to the regional environments of Southwest Asia and North Africa. How do these rural lifestyles create distinctive patterns of settlement?

7.4 Describe the distinctive contributions of (a) Islam, (b) European colonialism, and (c) recent globalization to the region's urban landscape.

7.5 Summarize the key patterns and drivers of migration in and out of the region.

Cultural Coherence and Diversity: A Complex Cultural Mosaic

Southwest Asia and North Africa clearly define the heart of the "Islamic" and "Arab" worlds, but a surprising degree of cultural diversity characterizes the region. Muslims practice their religion in varied ways, often disagreeing profoundly on basic religious views, as well as on how much of the modern world and its mass consumer culture should be incorporated into their daily lives. In addition, diverse religious minorities complicate the region's contemporary cultural geography. Linguistically, Arabic languages form an important cultural core historically centered on the region. However, different Arab dialects can mean that a Syrian may struggle to comprehend an Algerian. In addition, many non-Arab peoples, including Persians, Kurds, and Turks, populate important homelands in the region and historically have dominated large portions of it. Understanding these varied patterns of cultural geography is essential to comprehending many of the region's political tensions, as well as appreciating why many of its residents resist processes of globalization and celebrate the lasting cultural identity of their home neighborhoods and communities.

Patterns of Religion

Religion permeates the lives of most people within the region. Its centrality stands in sharp contrast to largely secular cultures in many other parts of the world. Whether it is the quiet ritual of morning prayers or profound discussions about contemporary political and social issues, religion is part of the daily routine of most regional residents from Casablanca to Tehran. The geographies of religion—their points of origin, paths of diffusion, and patterns of modern regional distribution—are essential elements in understanding cultural and political conflicts within the region.

Hearth of the Judeo-Christian Tradition Both Jews and Christians trace their religious roots to the eastern Mediterranean, and while neither group is numerically dominant across the area, each plays a key cultural role. The roots of Judaism lie deep in the past: Abraham, an early patriarch in the Jewish tradition, lived some 4000 years ago and led his people from Mesopotamia to Canaan (modern-day Israel), near the shores of the Mediterranean. From Jewish history, recounted in the Old Testament of the Holy Bible, springs a rich religious heritage focused on a belief in one God (or **monotheism**), a strong code of ethical conduct, and a powerful ethnic identity that continues to the present. Around 70 CE, during the time of the Roman Empire, most Jews were forced to leave the eastern Mediterranean after they challenged Roman authority. The resulting forced migration, or diaspora, of the Jews took them to the far corners of Europe and North Africa. Only in the past century have many of the world's far-flung Jewish populations returned to the religion's hearth area, a process that accelerated greatly with the formation of the Jewish state of Israel in 1948.

Christianity also emerged in the vicinity of modern-day Israel and has left a lasting legacy across the region. An outgrowth of Judaism, Christianity was based on the teachings of Jesus and his disciples, who lived and traveled in the eastern Mediterranean about 2000 years ago. Although many Christian traditions became associated with European history, some forms of early Christianity remained strong near the religion's original hearth. To the south, one stream of Christian influences linked with the Coptic Church diffused into northern Africa, shaping the culture of places such as Egypt and Ethiopia. In the Levant, another group of early Christians, known as the Maronites, retained a separate cultural identity that survives today. The region remains the "Holy Land" in the Christian tradition and attracts millions of the faithful annually, as they visit the most sacred sites in the Christian world.

The Emergence of Islam Islam originated in Southwest Asia in 622 CE, forming yet another cultural hearth of global significance. Muslims can be found today from North America to the southern Philippines; however, the Islamic

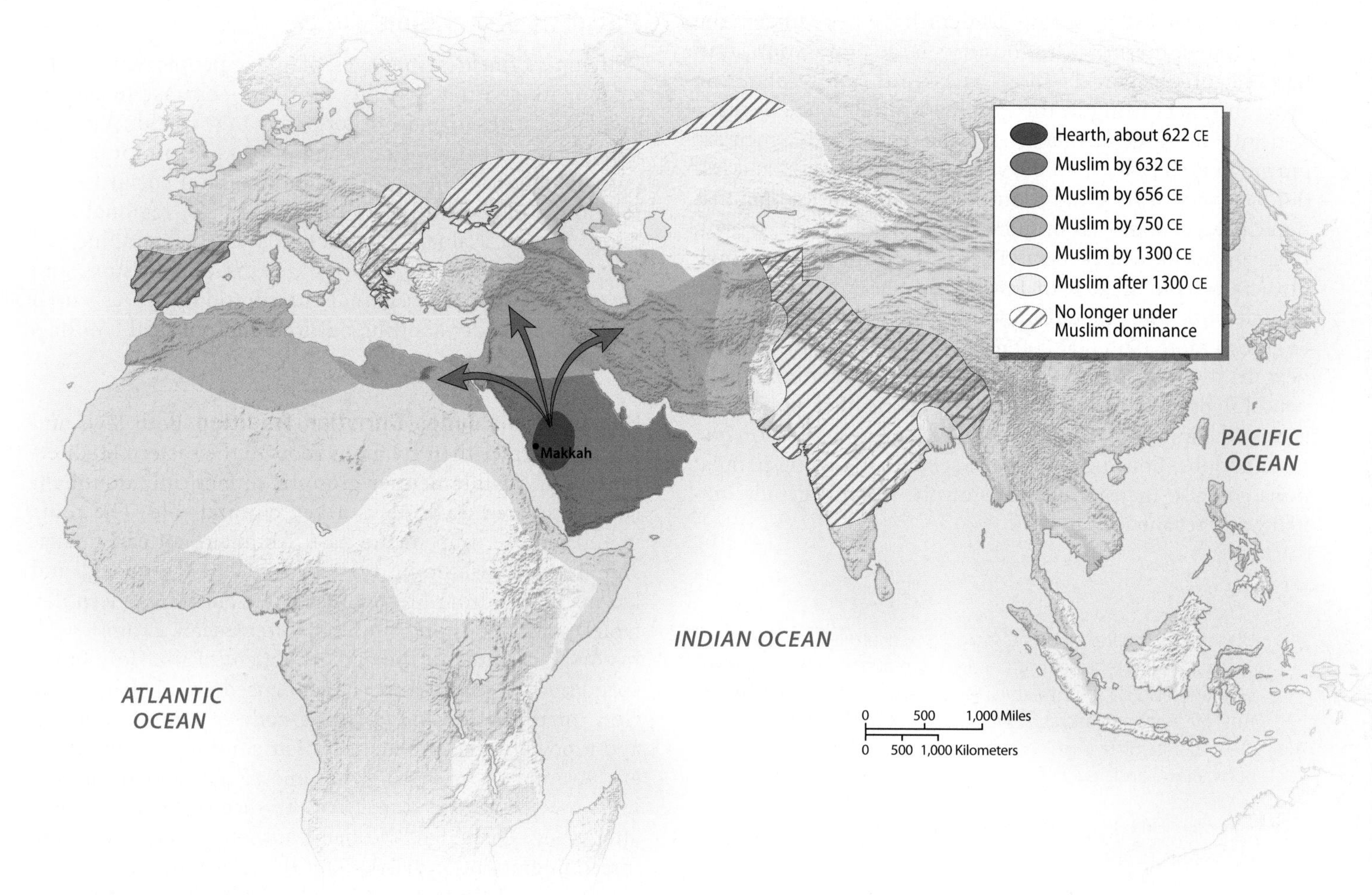

▲ **Figure 7.28 Diffusion of Islam** The rapid expansion of Islam that followed its birth is shown here. From Spain to Southeast Asia, Islam's legacy remains strongest nearest its Southwest Asian hearth. In some settings, its influence has ebbed or has come into conflict with other religions, such as Christianity, Judaism, and Hinduism.

world is still centered on its Southwest Asian origins. Most Southwest Asian and North African peoples still follow its religious teachings and moral doctrines. Muhammad, the founder of Islam, was born in Makkah (Mecca) in 570 CE and taught in nearby Madinah (Medina) (Figure 7.28). In many respects, his creed parallels the Judeo-Christian tradition. Muslims believe that both Moses and Jesus were true prophets and that both the Hebrew Bible (or Old Testament) and the Christian New Testament, while incomplete, are basically accurate. Ultimately, however, Muslims hold that the **Quran** (or Koran), a book of revelations received by Muhammad from Allah (God), represents God's highest religious and moral revelations to humankind.

The basic teachings of Islam offer an elaborate blueprint for leading an ethical and a religious life. Islam literally means "submission to the will of God," and the creed rests on five essential pillars: (1) repeating the basic creed ("There is no god but God, and Muhammad is his prophet"); (2) praying facing Makkah five times daily; (3) giving charitable contributions; (4) fasting between sunup and sundown during the month of Ramadan; and (5) making at least one religious pilgrimage, or **Hajj**, to Muhammad's birthplace of Makkah (Figure 7.29). Islam is a more austere religion than most forms of Christianity, and its modes of worship and forms of organization are generally less ornate. Muslims avoid the use of religious images, and the strictest interpretations even forbid the depiction of the human form. Followers of Islam are prohibited from drinking alcohol and are instructed to lead moderate lives, avoiding excess. Many Islamic fundamentalists still argue for a **theocratic state**, such as modern-day Iran, in which religious leaders (ayatollahs) guide public policy.

A major religious schism divided Islam early on, and the differences endure to the present. The breakup occurred almost immediately after the death of Muhammad in 632 CE. Key questions surrounded the succession of religious power. One group, now called the **Shiites**, favored passing on power within Muhammad's own family—specifically to Ali, his son-in-law. Most Muslims, later known as **Sunnis**, advocated passing power down through the established clergy. This group emerged victorious. Ali was killed, and his Shiite supporters went underground. Ever since, Sunni Islam has formed the mainstream branch of the religion, to which Shiite Islam has presented a recurring, and sometimes powerful, challenge (Figure 7.30). The Shiites argue that a successor to Ali will someday return to reestablish the pure, original form of Islam. The Shiites also are more hierarchically organized than the Sunnis. Iran's ayatollahs (Shia), for example, have overriding religious and political power.

▲ **Figure 7.29 Makkah** Thousands of faithful Muslims gather at the Grand Mosque in central Makkah, part of the pilgrimage to this sacred place that draws several million visitors annually.

In a very short period of time, Islam diffused widely from its Arabian hearth, often following camel caravan routes and Arab military campaigns as it expanded its geographical range and converted thousands to its beliefs. By the time of Muhammad's death in 632 CE, the peoples of the Arabian Peninsula were united under its banner. Shortly thereafter, the Persian Empire fell to Muslim forces, and the Eastern Roman (or Byzantine) Empire lost most of its territory to Islamic influences. By 750 CE, Arab armies had swept across North Africa, conquered most of the Iberian Peninsula (modern Spain and Portugal), and established footholds in Central and South Asia. At first, only the Arab conquerors followed Islam, but the diverse inhabitants of Southwest Asia and North Africa gradually were absorbed into the religion, although with many distinct local variants. By the 13th century, most people in the region were Muslims, while older religions such as Christianity and Judaism became minority faiths or disappeared altogether.

Between 1200 and 1500, Islamic influences expanded in some areas and contracted in others. The Iberian Peninsula returned to Christianity by 1492, although many Moorish (Islamic) cultural and architectural features remained behind and still shape the region today. At the same time, Muslims expanded their influence southward and eastward into Africa. In addition, Muslim Turks largely replaced Christian Greek influences in Southwest Asia after 1100. One group of Turks moved into the Anatolian Plateau and finally conquered the last vestiges of the Byzantine Empire in 1453. These Turks soon created the vast **Ottoman Empire** (named after one of its leaders, Osman), which included southeastern Europe (including modern-day Albania, Bosnia, and Serbia) and most of Southwest Asia and North Africa. The legacy of the Ottoman Empire was considerable. It offered a new, distinctly Turkish interpretation of Islam, and it provided a focus of Muslim

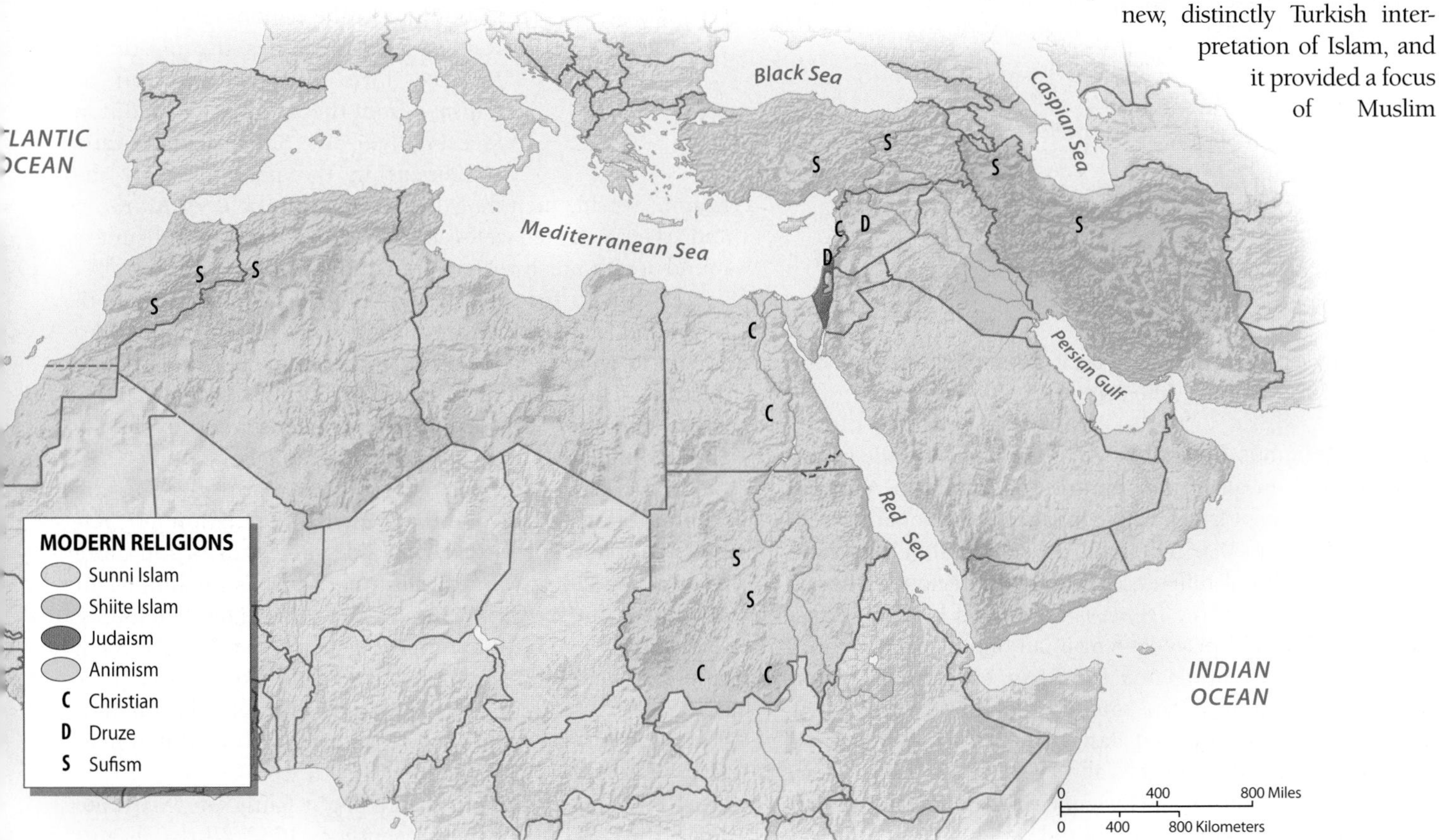

▲ **Figure 7.30 Religions of Southwest Asia and North Africa** Islam remains the dominant religion across the region. Most Muslims are tied to the Sunni branch, whereas Shiites reside in places such as Iran and southern Iraq. In some locales, however, Christianity and Judaism remain important.

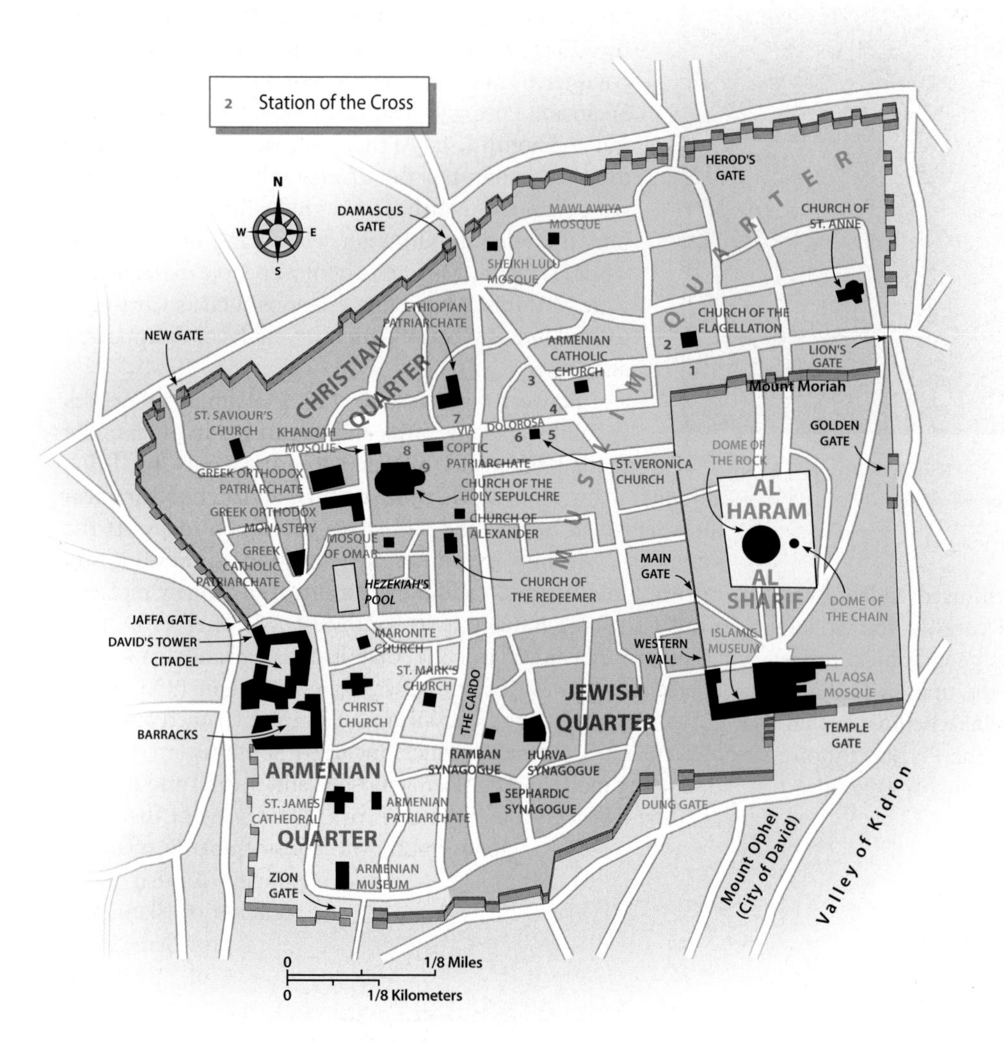

▲ **Figure 7.31 Old Jerusalem** The historic center of Jerusalem reflects its varied religious legacy. Sacred sites for Jews, Christians, and Muslims all cluster within the Old City. The Western Wall, a remnant of the ancient Jewish temple, stands at the base of the Dome of the Rock and Islam's Al Aqsa Mosque.

political power within the region until the empire's disintegration in the late 19th and early 20th centuries.

Modern Religious Diversity Today Muslims form the majority population in all of the countries of Southwest Asia and North Africa except Israel, where Judaism is the dominant religion (see Figure 7.30, p. 323). Still, divisions within Islam have created key cultural differences within the region. Although most (73 percent) of the region is dominated by Sunni Muslims, the Shiites (23 percent) remain an important element in the contemporary cultural mix. In Iraq, for example, the majority Shiite population in the southern portion of the country (around Najaf, Karbala, and Basra) set its own cultural and political course following the fall of Saddam Hussein. Many Shiites saw the departure of Hussein (who gathered most of his support from Sunnis in the central and western parts of Iraq) as an opportunity to increase their cultural and political influence within the country. Today, however, they still often find themselves in sharp conflict with the country's Sunni population.

Shiites claim a majority in nearby Iran as well, where their religious fervor has fundamentally shaped government policies since the late 1970s. In Yemen, important Shiite minorities claim the government has neglected their needs, which has led to increased political instability in that impoverished nation. In addition, they form a major religious force in Lebanon, Bahrain, Algeria, and Egypt.

Sunni elements, however, benefited from many of the Arab Spring revolts around the region and have sharpened their attacks against the Shiites. Resurgent sectarian conflicts in Iraq, growing Saudi Arabian (mainly Sunni) suspicion of Iranian intentions in the region, and Sunni–Shiite violence among European Muslims have all been recent signals of continuing animosity—both religious and political—between these major Islamic groups.

Whereas the Sunni–Shiite split is the great divide within the Muslim world, other variations of Islam are also found in the region. One division, for example, separates the mystically inclined form of Islam—known as *Sufism*—from the mainstream tradition (see Figure 7.30). Sufism is especially prominent in the peripheries of the Islamic world, including the Atlas Mountains of Morocco and Algeria. Other well-established examples of Sufism are found across northwestern Iran and portions of Turkey.

Elsewhere, the Salafists and Wahhabis, numerous in both Egypt and Saudi Arabia, are austere, conservative Sunnis who adhere to what they see as an earlier, purer form of Islamic doctrine. The Druze of Lebanon practice another variant of Islam, and they form a cohesive religious minority in the Shouf Mountains east of the Lebanese capital of Beirut.

Southwest Asia also is home to many non-Islamic communities. Israel's dominant Jewish population (77 percent) is divided between Jewish fundamentalists and more reform-minded Jews. The country also has an important Muslim minority (16 percent), and Christians (including Armenians) make up another 2 percent of the total. Such cultural diversity is certainly displayed in settings such as the Old City of Jerusalem, where well-established neighborhoods reflect these historical and cultural differences (Figure 7.31).

Neighboring Lebanon had a slight Christian (Maronite and Orthodox) majority as recently as 1950. Christian outmigration and differential birth rates, however, have created a nation that today is about 60 percent Muslim. Christians also form approximately 10 percent of Syria's population;

Iraqi Christians, concentrated mostly in the rugged northern uplands, make up about 3 percent of its population.

Important Jewish and Christian communities also have left a long legacy across North Africa. Roman Catholicism was once dominant in much of the Maghreb, but it disappeared several hundred years after the Muslim conquest. The Maghreb's Jewish population, on the other hand, remained prominent until the post–World War II period, when most of the region's Jews migrated to the new state of Israel. In Egypt, Coptic Christianity has maintained a stable presence through the centuries and today includes approximately 9 percent of the country's population. In earlier years, the Coptic community had a secure place in Egyptian society, and numerous Copts held high-level posts in government and business. Today, however, Egypt's Christians are being increasingly marginalized, and some of their communities have been put under pressure, even subjected to physical attack, by extremist Islamic elements. A smaller area of diverse, locally based animist religions is found across southern Sudan along the border with South Sudan.

Geographies of Language

Although the region is often referred to as the "Arab World," linguistic complexity creates many important cultural divisions across Southwest Asia and North Africa (Figure 7.32). The geography of language offers insights into regional patterns of ethnic identity and potential cultural conflicts that exist at linguistic borders. The language map is useful in identifying the major families found across the region, but it is important to remember that many local variations in language, representing distinctive dialects and well-defined islands of cultural and ethnic identity, are not seen on the map. For example, more than 70 separate languages are recognized in Iran, 36 in Turkey, and 23 in Iraq.

Semites and Berbers Afro-Asiatic languages dominate much of the region. Within that family, Arabic-speaking Semitic peoples can be found from Morocco to Saudi Arabia. Before the expansion of Islam, Arabic was limited to the Arabian Peninsula. Today, however, Arabic is spoken from the Persian Gulf to the Atlantic and reaches southward into Sudan, where it borders the Nilo-Saharan-speaking peoples of Sub-Saharan Africa. As the language has diffused, it has slowly diverged into local dialects. As a result, the everyday Arabic spoken on the streets of Fes, Morocco, is quite distinct from the Arabic spoken in the UAE. The Arabic language also has a special religious significance for all Muslims because it was the sacred language in which God delivered his message to Muhammad. Although most of the world's Muslims do not speak Arabic, the faithful often

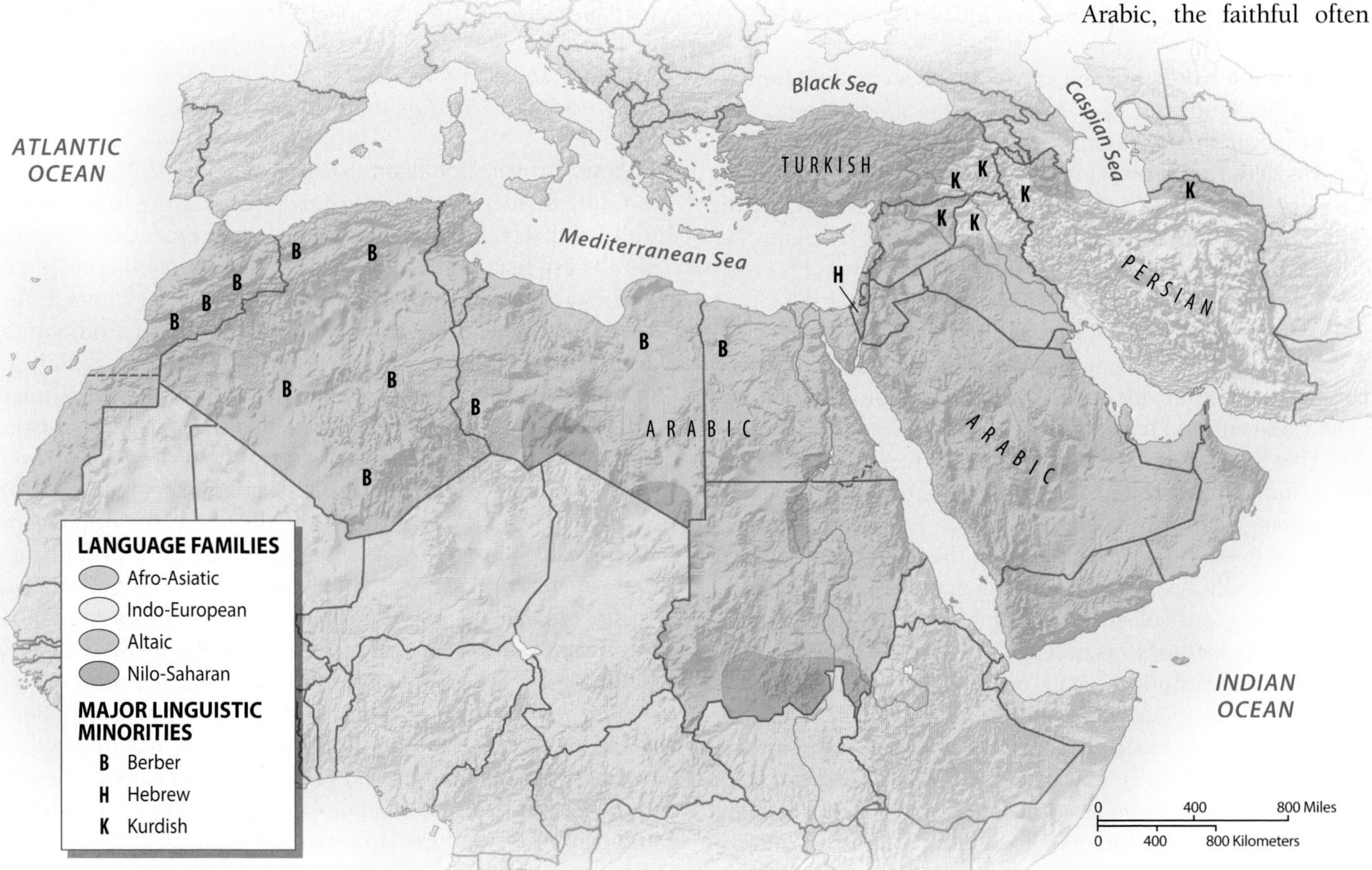

▲ **Figure 7.32 Languages of Southwest Asia and North Africa** Arabic is a Semitic Afro-Asiatic language, and it dominates the region's cultural geography. Turkish, Persian, and Kurdish, however, remain important exceptions, and such differences within the region often have had lasting political consequences. Israel's more recent reintroduction of Hebrew further complicates the region's linguistic geography. **Q: Cite examples where Islam (see Figure 7.30) dominates in non-Arabic-speaking regions.**

memorize certain prayers in the language, and many Arabic words have entered other important languages of the Islamic world. Advanced Islamic learning, moreover, demands competence in Arabic.

Hebrew, another traditional Semitic language of Southwest Asia, was recently reintroduced into the region with the creation of Israel. Hebrew originated in the Levant and was used by the ancient Israelites 3000 years ago. Today its modern version survives as the sacred tongue of the Jewish people and is the official language of Israel, although the country's non-Jewish population largely speaks Arabic. English is also widely used as a second or third language throughout the country.

While Arabic eventually spread across North Africa, several older languages survive in more remote areas. Older Afro-Asiatic tongues endure in the Atlas Mountains and in certain parts of the Sahara. Collectively known as Berber, these languages are related to each other, but are not mutually intelligible. Most Berber languages have never been written, and none has generated a significant literature. Indeed, a Berber-language version of the Quran was not completed until 1999. The decline in pastoral nomadism and pressures of modernization threaten the integrity of these Berber languages. Scattered Berber-speaking communities are found as far to the east as Egypt, but Morocco is the center of this language group, where it plays an important role in shaping that nation's cultural identity.

Persians and Kurds Although Arabic spread readily through portions of Southwest Asia, much of the Iranian Plateau and nearby mountains are dominated by older Indo-European languages. Here the principal tongue remains Persian, although, since the 10th century, the language has been enriched with Arabic words and written in the Arabic script. Persian, like other languages, has developed distinct local dialects. Today Iran's official language is called *Farsi*, which denotes the form of Persian spoken in Fars, the area around the city of Shiraz. Thus, although both Iran and neighboring Iraq are Islamic nations, their ethnic identities spring from quite different linguistic and cultural traditions.

The Kurdish speakers of northern Iraq, northern Iran, and eastern Turkey add further complexity to the regional pattern of languages. Kurdish, also an Indo-European language, is spoken by 10–15 million people in the region. Kurdish has not historically been a written language, but the Kurds do have a strong sense of shared cultural identity. Indeed, "Kurdistan" has sometimes been called the world's largest nation without its own political state because the group remains a minority in several countries of the region. In the 2003 war in Iraq, the Kurds emerged as a cohesive group in the northern part of the country and opposed Saddam Hussein. Iraqi Kurds have gained more political autonomy in a post-Saddam Iraq state and are also intent on maintaining control of important oil resources found in their portion of the country. Their leading city of Kirkuk has been called the "Kurdish Jerusalem" because its history and settlement so richly capture the group's cultural identity. Nearby Kurds in eastern Turkey, however, still complain that their ethnic identity is frequently challenged by the majority Turks, leaving some to wonder if they will attempt to join forces with their Iraqi neighbors.

The Turkish Imprint Turkish languages provide more variety across much of modern Turkey and in portions of far northern Iran. The Turkish languages are a part of the larger Altaic language family that originated in Central Asia. Turkey remains the largest nation in Southwest Asia dominated by that family. Tens of millions of people in other countries of Southwest and Central Asia speak related Altaic languages, such as Azeri, Uzbek, and Uighur. During the era of the Ottoman Empire, Turkish speakers ruled much of Southwest Asia and North Africa, but Iran is the only other large country in the region today where Turkic languages persist, particularly in the northwest part of the country.

Regional Cultures in Global Context

Many cultural connections tie the region with the world beyond. Islam links the region with a Muslim population that now lives in many different settings around the world. In addition, European and American cultural influences have multiplied greatly across the region since the mid-19th century. Colonialism, the boom in petroleum investment, and the growing presence of Western-style popular culture have had enduring impacts, from the Atlas Mountains to the Indian Ocean. These global cultural connections are nuanced and complex, linking this region with the rest of the world in fascinating, often unanticipated ways.

Islamic Internationalism Islam is geographically and theologically divided, but all Muslims recognize the fundamental unity of their religion. This religious unity extends far beyond Southwest Asia and North Africa. Islamic communities are well established in such distant places as central China, European Russia, central Africa, and the southern Philippines. Today Muslim congregations also are expanding rapidly in the major urban areas of western Europe and North America, largely through migration but also through local conversions. Islam is thus emerging as a truly global religion. Even with its global reach, however, Islam remains centered on Southwest Asia and North Africa, the site of its origins and its holiest places. As Islam expands in number of followers and geographical scope, the religion's tradition of pilgrimage ensures that Makkah will become a city of increasing global significance in the 21st century. The global growth of Islamist fundamentalism and Islamism also focuses attention on the region, where these contemporary movements recently burst upon the scene. In addition, the oil wealth accumulated by many Islamic nations is used to sustain and promote the religion. Countries such as Saudi Arabia invest in Islamic banks and economic ventures and make donations to Islamic cultural causes, colleges, and hospitals worldwide.

Globalization, Technology, and Popular Culture The region also is struggling with how its growing role in the global economy is changing traditional cultural values. European

▲ **Figure 7.33 Communicating from the Front Lines, Cairo, Egypt, 2011** A young Egyptian woman talks on a mobile phone in Cairo's Tahrir Square during demonstrations in February 2011.

colonialism left its own cultural legacy, not only in the architectural landscapes still found in the old colonial centers, but also in the widespread use of English and French among the Western-educated elite across the region. In oil-rich countries, huge capital investments have had important cultural implications, as the number of foreign workers has grown and as more affluent young people have embraced elements of Western-style music, literature, and clothing. The expansion of Islamic fundamentalism and Islamism is in many ways a reaction to the threat posed by external cultural influences, particularly the supposed evils of European colonialism, American and Israeli power, and local governments and cultural institutions that are seen as selling out to the West.

Technology also shapes cultural and political change. Particularly among the young, millions within and beyond the region found themselves linked by the Internet, cell phones, and various forms of social media during the Arab Spring uprisings of 2011 and 2012 (Figure 7.33). Cell phones, blogs, email, and tweets facilitated the flow of information that helped protesters plan events and coordinate strategies with their allies. Local videos from smartphones and pinhole cameras documented government abuse and often provided (in settings such as Syria) the only proof of widespread state-supported violence. Also, the global diffusion of this information promoted the internationalization of political discourse and made it easier to spread the word about local conflicts and to identify common threads among different protest movements. A related point is that about 60 percent of the region's population is under 30—precisely the group most inclined to use these technologies and often the people most frustrated by unresponsive governments that refuse to change their ways.

Hybrid forms of popular culture also reflect globalization. Recently, for example, both within the region and beyond, Arab hip-hop music has offered a form of artistic and political expression that represents the fusion of cultural traditions. Tune in MTV Arabia and you can hear Desert Heat (UAE), Malikah (Lebanon), or MWR (Palestine). These artists represent a generation of rappers who spit out lyrics challenging cultural and political stereotypes within the region. The global vibe is hard to miss: The African-American roots of rap remain a fixture in the strong beats that set the tone amid the Arabic ouds (lutes), clangs of Asian pop, and soulful lyrics that might emerge in Arabic, English, or French.

Widespread television viewing has also transformed the region, with viewers offered everything from Islamist religious programming to American-style reality TV. Even in conservative Iran, where satellite TV dishes are officially banned, millions of people have access to them, beaming in multicultural programming from around the world. In other settings, millions of regional viewers tune in weekly to (and share tweets with) leading Islamic televangelists such as Amr Khaled (Egyptian) or Ahmad al-Shugairi (Saudi). Building on a Christian practice pioneered a half century earlier in the United States by Billy Graham, these broadcast preachers have become regional celebrities and iconic cultural figures.

Some cultural changes among younger residents of the region are not going unchallenged. In the culturally conservative Persian Gulf states of Qatar and the UAE, a new penchant for cross-dressing among young women has produced a considerable backlash among more orthodox Muslims. Many young women in those countries (particularly on university campuses) have donned Western-style baggy trousers and wear short-cropped hair. Termed *boyats*, or "tomboys," the women have been accused of adopting a "foreign trend" in dress brought about by globalization. Their "deviant behavior" has been seen as a menace to society. The same clothing style and look that would go unnoticed on North American college campuses have provoked calls among UAE conservatives for the death penalty, "medical treatment," or reeducation workshops in femininity.

Elsewhere, conservative cultural influences are gaining strength and visibility. For example, in Tunisia, long known for its more Western-leaning, modernized brand of Islam, the growing use of headscarves and veils among conservative women has caused a stir. A ban on such traditional practices was enacted in 1981 by a Tunisian government that described it as a "sectarian form of dress." More recently, such clothing has been seen by some as promoting the political agenda of extremist Islamist elements within the country. Interestingly, European human rights advocates have come to the aid of the women as they are increasingly harassed by police to remove the scarves and sign pledges that they will not go back to wearing them. In Turkey, a growing traditional Islamist presence in the government and in the civil service bureaucracy has allowed for more religious study and prayers in that country's public school system. Science textbooks are also being changed, deemphasizing Western interpretations in favor of more traditional, religiously based explanations.

The Role of Sports Sports have also assumed a hugely important cultural role in everyday life within the region. Soccer rules the day in most countries, both as a spectator sport and as an activity that many young people enjoy. Most countries have national football (soccer) associations

▲ **Figure 7.34 Iraq Soccer Star, Younis Mahmoud** Emotions run high as Younis Mahmoud scores a goal against Kuwait in the 2013 Gulf Cup tournament.

that also participate in regional and global (FIFA) league competitions. Begun in 1974, the Union of Arab Football Associations (UAFA), headquartered in Riyadh, Saudi Arabia, offers many opportunities for regional competition, often carried on the Al Jazeera Sports network. A smaller Gulf Cup of Nations competition also sparks spirited rivalries. Large soccer stadiums are a common part of modern urban landscapes, including Kuwait City's Sabah Al-Salem Stadium, which comfortably seats more than 28,000 fans. Recently, Algeria, Tunisia, Saudi Arabia, and Iraq have supported strong teams (in 2013, Algeria and Tunisia enjoyed high global FIFA rankings). As in other parts of the world, regional players can earn superstar status, such as Iraq's goal-shooting veteran Younis Mahmoud (Figure 7.34).

REVIEW

7.6 Describe the key characteristics of Islam, and explain why distinctive Sunni and Shiite branches exist today.

7.7 Identify three examples that illustrate how modern communication and transportation technologies have promoted cultural change and globalization within the region.

Geopolitical Framework: Never-Ending Tensions

Geopolitical tensions remain very high in Southwest Asia and North Africa (Figure 7.35). In the Arab Spring rebellions, governments fell in Tunisia (the regional movement began here in late 2010), Egypt, Libya, and Yemen; widespread protests shook once-stable states such as Bahrain; and a more protracted civil war decimated Syria. Many other countries witnessed shorter, more intermittent demonstrations against state authority. To varying degrees, these uprisings focused broadly on (1) charges of widespread government corruption; (2) limited opportunities for democracy and free elections; (3) rapidly rising food prices; and (4) the enduring reality of widespread poverty and high unemployment, especially for people under 30. When you add to this mix the complex local realities of traditional tribal rivalries, ethnic identities, and religious factionalism, it is no wonder that the future paths of particular countries appear as diverse as they are uncertain.

In addition to these recent conflicts, ongoing issues include the future of Israeli-Palestinian relations, Iran's saber rattling and nuclear ambitions, and Iraq's unsteady emergence as an independent state. Some of these tensions relate to age-old patterns of cultural geography. European colonialism also contributes to the situation because modern boundaries were formed by colonial powers. Geographies of wealth and poverty also enter the geopolitical mix: Some residents profit from petroleum resources and industrial expansion, while others struggle to feed their families. Islamist elements in Iran, Egypt, Algeria, Tunisia, Turkey, Yemen, and Saudi Arabia have also changed the political atmosphere. The result is a political climate charged with tension, a region in which the sounds of bomb blasts and gunfire remain all-too-common characteristics of everyday life.

The Colonial Legacy

European colonialism arrived relatively late in Southwest Asia and North Africa, but the era had an important impact upon the modern political landscape. The Turks were one reason for Europe's late participation in imposing colonial rule. Between 1550 and 1850, much of the region was dominated by the Turkish Ottoman Empire, which expanded from its Anatolian hearth to engulf much of North Africa, as well as nearby areas of the Levant, the western Arabian Peninsula, and modern-day Iraq. Ottoman rule imposed an outside political order and economic framework that had lasting consequences. The tide began to turn, however, in the early 19th century as the European presence increased and Ottoman power ebbed. Still, it took a century for Ottoman influences to be replaced with largely European colonial dominance after World War I (1918). Although much of that direct European control ended by the 1950s, old colonial ties persist in a variety of economic, political, and cultural contexts. It is still common to encounter British English on the streets of Cairo, and French still can be heard in Algiers and Beirut. Indeed, the persistence of terms such as "Middle East" and "Levant" (a French term meaning "east" or "the Orient") within our 21st-century geographical vocabulary is a reminder of the long-standing linkages that tied this "eastern" land to western European influences.

Imposing European Power French colonial ties have long been a part of the region's history. Beginning around 1800, France became committed to a colonial presence in North

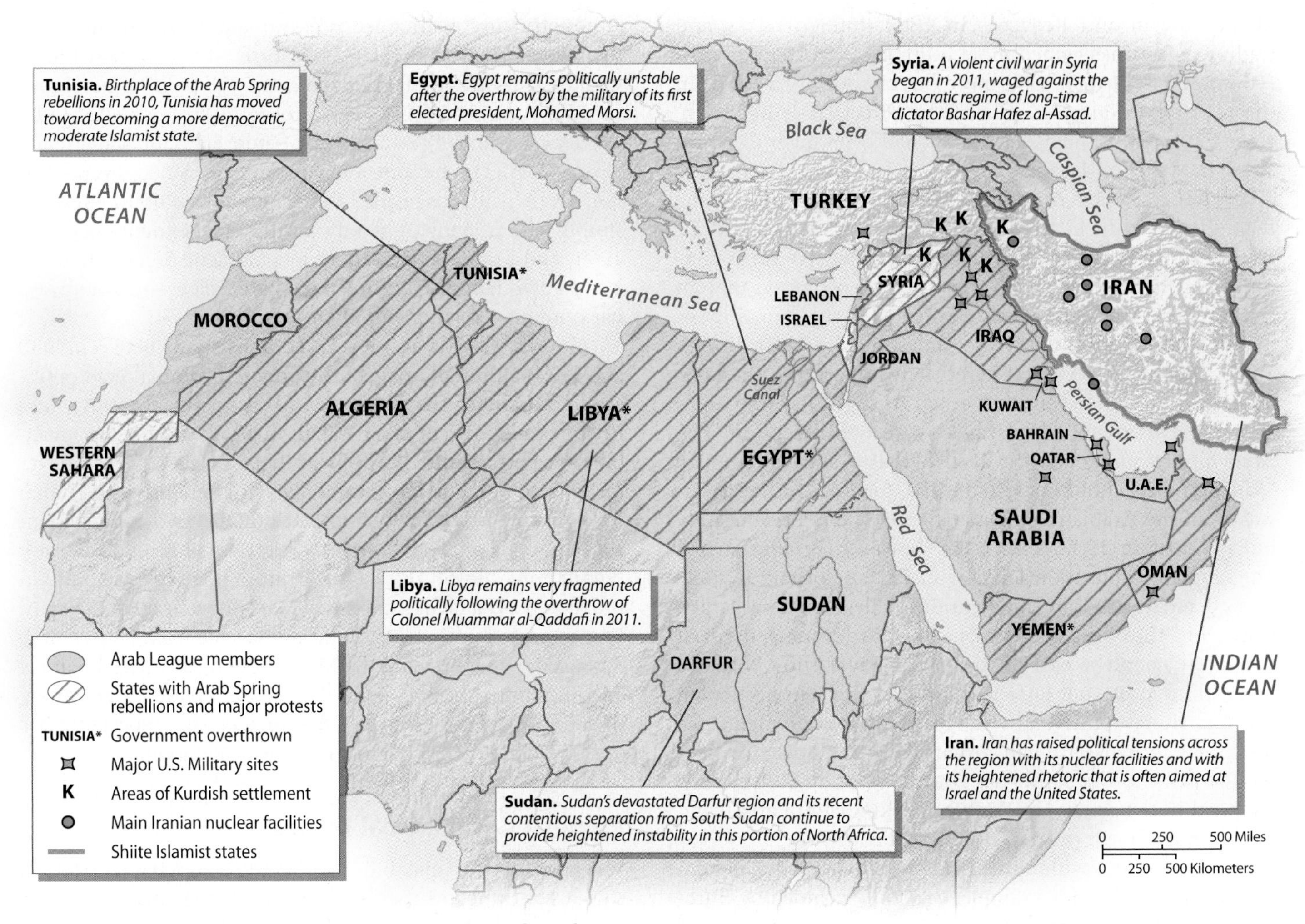

▲ **Figure 7.35 Geopolitical Issues in Southwest Asia and North Africa** Political tensions continue across much of the region. Many countries are members of the Arab League, a regional organization focused on Arab political and economic unity. The Arab Spring rebellions and Syria's violent civil war have shaped recent political changes. The Israeli-Palestinian conflict also remains pivotal.

Africa, including an expedition to Egypt by Napoleon. By the 1830s, France moved more directly into Algeria, forcing that region into its colonial sphere of influence. In the next 120 years, several million French, Italian, and other European immigrants poured into the country, taking the best lands from the Algerian people. The French government expected this territory to become an integral part of France, dominated by the growing French-speaking immigrant population. Indeed, the landscape of modern Algiers still reflects these colonial connections (Figure 7.36). France also established its influence in Tunisia (1881) and Morocco (1912), ensuring a lasting French political and cultural presence in the Maghreb. These **protectorates** in Tunisia and Morocco retained some political autonomy, but remained under a broader sphere of French influence and protection from other competing colonial powers. Finally, France's victory over the German–Ottoman Turk alliance in World War I produced additional territorial gains in the Levant, as France acquired control of the northern zone, encompassing the modern nations of Syria and Lebanon.

Great Britain's colonial fortunes also grew within the region before 1900. To control its sea-lanes to India, Britain established a series of 19th-century protectorates over the small coastal states of the southern Arabian Peninsula and the Persian Gulf. In this manner, such places as Kuwait, Bahrain, Qatar, the UAE, and Aden (in southern Yemen) were loosely incorporated into the British Empire. Nearby Egypt also caught Britain's attention. Once the European-engineered **Suez Canal** linked the

▼ **Figure 7.36 Algiers** European colonial influences still abound in the old French capital of Algiers in northern Algeria.

Mediterranean and Red seas in 1869, foreign banks and trading companies gained more influence over the Egyptian economy. The British took more direct control in 1883. In the process, Britain also inherited a direct stake in Sudan, since Egyptian soldiers and traders had been pushing south along the Nile for decades.

Another series of British colonial gains within the region came at the close of World War I. In Southwest Asia, British and Arab forces had joined to expel the Turks during the war. To obtain Arab trust, Britain promised that an independent Arab state would be created in the former Ottoman territories. At roughly the same time, however, Britain and France signed a secret agreement to partition the area. When the war ended, Britain opted to slight its Arab allies and honor its treaty with France, with one exception: The Saud family convinced the British that a smaller country (Saudi Arabia) should be established, focused on the desert wastes of the Arabian Peninsula. Saudi Arabia became fully independent in 1932. Elsewhere, however, Britain carried out its plan to partition lands with France. Britain divided its new territories into three entities: Palestine (now Israel, Gaza, and the West Bank) along the Mediterranean coast; Trans-jordan to the east of the Jordan River (now Jordan); and a third zone that later became Iraq. Iraq, in particular, was a territory contrived by the British that combined three dissimilar former Ottoman provinces. It included the centers of Basra in the south (an Arabic-speaking Shiite area), Baghdad in the center (an Arabic-speaking Sunni area), and Mosul in the north (a Kurd-dominated zone).

Other settings within the region felt more marginal colonial impacts. Libya, for example, was long regarded by Europeans as a desert wasteland. Italy, never a dominant colonial power, expelled Turkish forces from the coastal districts by 1911, and its colonial influence continued until 1947. Spain also carved out its own territorial stake within the region, gaining control over southern Morocco (now Western Sahara) in 1912. To the east, Persia and Turkey never were directly occupied by European powers. In Persia, the British and Russians agreed to establish mutual spheres of economic influence (the British in the south, the Russians in the north), while respecting Persian independence. In 1935, Persia's modernizing ruler, Reza Shah, changed the country's name to Iran.

In nearby Turkey, the old core of the Ottoman Empire was almost partitioned by European powers following World War I. After several years of fighting, however, the Turks expelled the French from southern Turkey and the Greeks from western Turkey. The key to the successful Turkish resistance was the spread of a new modern, nationalist ideology under the leadership of Kemal Ataturk. Ataturk decided to emulate the European countries and establish a culturally unified and resolutely secular state. He was successful, and Turkey was quickly able to stand up to European power.

Decolonization and Independence European colonial powers began their withdrawal from several Southwest Asian and North African colonies before World War II. By the 1950s, most of the countries in the region were independent, although many maintained political and economic ties with their former colonial rulers. In North Africa, Britain finally withdrew its troops from Sudan and Egypt in 1956. Libya (1951), Tunisia (1956), and Morocco (1956) achieved independence peacefully during the same era, but the French colony of Algeria became a major problem. Since several million French citizens resided there, France had no intention of simply withdrawing. A bloody war for independence began in 1954, and France agreed to an independent Algeria in 1962, but the two nations continued to share a close—if not always harmonious—relationship thereafter.

Southwest Asia also lost its colonial status between 1930 and 1971, although many of the imposed colonial-era boundaries continue to shape the regional geopolitical setting. Iraq became independent from Britain in 1932, but its later instability in part resulted from its artificial borders, which never recognized much of its cultural diversity. Similarly, the French division of its Levant territories into the two independent states of Syria and Lebanon (1946) greatly angered local Arab populations and set the stage for future political instability in the region. As a favor to its small Maronite Christian majority, France carved out a separate Lebanese state from largely Arab Syria, even guaranteeing the Maronites constitutional control of the national government. The action created a culturally divided Lebanon as well as a Syrian state that repeatedly has asserted its influence over its Lebanese neighbors.

Modern Geopolitical Issues

The geopolitical instability in Southwest Asia and North Africa continues in the 21st century. It remains difficult to predict political boundaries that seem certain to change as a result of negotiated settlements or political conflict. A quick regional transect from the shores of the Atlantic to the borders of Central Asia suggests how these forces are playing out in different settings early in the century.

Across North Africa Varied North African settings have recently witnessed dramatic political changes (see Figure 7.35). In Tunisia, birthplace of the Arab Spring, a moderate Islamist government was elected to replace deposed dictator Zine el-Abidine Ben Ali. Growing tensions with conservative Salafists, however, threatened the regime's stability in 2013. In addition, many Tunisian women reacted negatively to Islamist proposals to reduce their rights and to see them as "complementary" to men.

In nearby Libya, while many cheered the end of Colonel Muammar al-Qaddafi's rule in 2011, transitional government leaders and political parties were slow to establish a clear agenda and national consensus. Given the varied mix of local militias, continued outbreaks of violence, and lack of centralized authority in the country, some observers have called Libya essentially a country of fragmented city-states.

Next door, following the 2011 overthrow of Hosni Mubarak, Egypt remained politically unstable. It held parliamentary and presidential elections in 2012, ushering in a brief period of rule by the Muslim Brotherhood, led by President

Mohamed Morsi. The government adopted a new constitution, but many ordinary Egyptians felt Morsi moved forward too quickly and ignored many democratically guaranteed civil rights. Also unsettling (particularly to the nation's Coptic Christian population) was the growing visibility of fundamentalist Islamist extremists (especially among the Salafists). In the summer of 2013, the Egyptian military staged a coup, ousted Morsi, and moved the country toward another round of elections. Meanwhile, these new political uncertainties in Egypt have discouraged tourism, dampened foreign investment, and led to questions about Egypt's role in the Arab World.

Elsewhere in North Africa, Islamist political movements have also reshaped the geopolitical landscape in several states. Most notably, recent violence has rocked both Algeria and Morocco, where protesters have pressed for some of the same democratic and economic reforms they see unfolding in nearby countries. Sudan also faces daunting political issues. A Sunni Islamist state since a military coup in 1989, Sudan imposed Islamic law across the country, antagonizing both moderate Sunni Muslims and the large non-Muslim (mostly Christian and animist) population in the south. Civil war between the north and south produced more than 2 million casualties (mostly in the south) between 1988 and 2004. A tentative peace agreement was signed in 2005, which opened the way for a successful vote on independence in South Sudan. Even though the two nations officially split in 2011, tensions remain, especially focused on an oil-rich and contested southern border zone between the two countries.

In addition, Sudan's western Darfur region remains in shambles (see Figure 7.35). Ethnicity, race, and control of territory seem to be at the center of the struggle in the largely Muslim region, as a well-armed Arab-led militia group (with many ties to the central government in Khartoum) has attacked hundreds of black-populated villages, killing more than 300,000 people (through violence, starvation, and disease) and driving 2.5 million more from their homes.

The Arab-Israeli Conflict The 1948 creation of the Jewish state of Israel produced another enduring zone of cultural and political tensions within the eastern Mediterranean (Figure 7.37). Jewish migration to Palestine increased after the defeat of the Ottoman Empire in World War I. In 1917, Britain issued the Balfour Declaration, a pledge to encourage the creation of a Jewish homeland in the region.

▼ **Figure 7.37 Evolution of Israel** Modern Israel's complex evolution began with an earlier British colonial presence and a UN partition plan in the late 1940s. Thereafter, multiple wars with nearby Arab states produced Israeli territorial victories in settings such as Gaza, the West Bank, and the Golan Heights. Each of these regions continues to figure importantly in the country's recent relations with nearby states and with resident Palestinian populations.

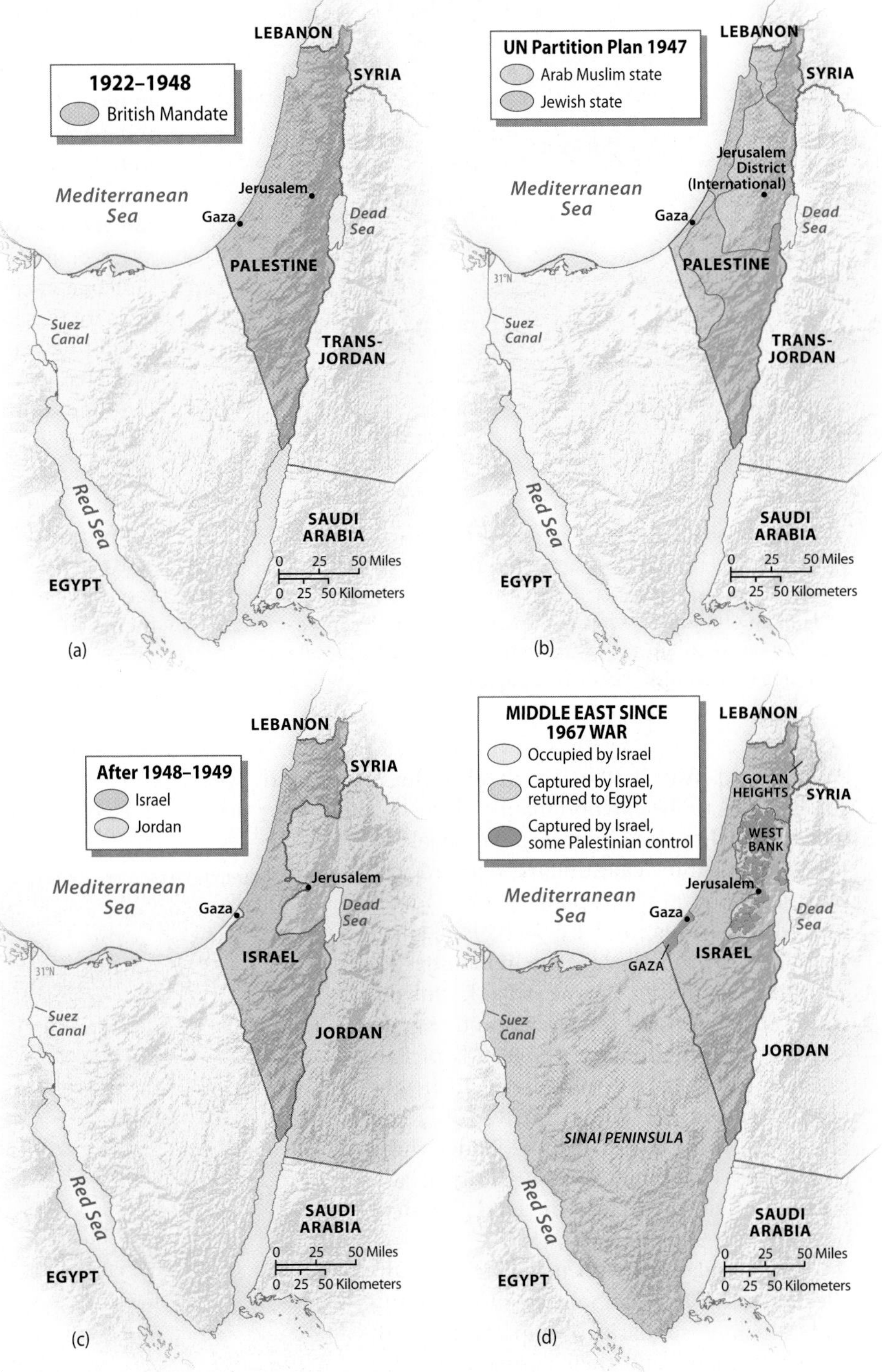

▲ **Figure 7.38 West Bank** Portions of the West Bank were returned to Palestinian control in the 1990s, but Israel has partially reasserted its authority in some of these areas since 2000, citing the increased violence in the region. **Q: Look carefully at the scale of the map. Measure the approximate distance between Jerusalem and Hebron, and find two local towns in your area that are a similar distance apart.**

After World War II, the UN divided the region into two states, one to be predominantly Jewish, the other primarily Muslim (see Figure 7.37, p. 331). Indigenous Arab Palestinians rejected the partition, and war erupted. Jewish forces proved victorious, and by 1949 Israel had actually grown in size. The remainder of Palestine, including the West Bank and the Gaza Strip, passed to Jordan and Egypt, respectively. Hundreds of thousands of Palestinian refugees fled from Israel to neighboring countries, where many of them remained in makeshift camps. Under these conditions, Palestinians nurtured the idea of creating their own state on land that had become part of Israel.

Israel's relations with neighboring countries remained poor. Supporters of Arab unity and Muslim solidarity sympathized with the Palestinians, and their antipathy toward Israel grew. Israel fought additional wars in 1956, 1967, and 1973. In territorial terms, the Six-Day War of 1967 was the most important conflict (see Figure 7.37). In this struggle against Egypt, Syria, and Jordan, Israel occupied substantial new territories in the Sinai Peninsula, the Gaza Strip, the West Bank, and the Golan Heights. Israel also annexed the eastern (Muslim) part of the formerly divided city of Jerusalem, arousing particular bitterness among the Palestinians (see Figure 7.31). A peace treaty with Egypt resulted in the return of the Sinai Peninsula in 1982, but tensions focused on other occupied territories that remained under Israeli control. To strengthen its geopolitical claims, Israel also built additional Jewish settlements in the West Bank and in the Golan Heights, further angering Palestinian residents.

Palestinians and Israelis began to negotiate a settlement in the 1990s. Preliminary agreements called for a quasi-independent Palestinian state in the Gaza Strip and across much of the West Bank. A tentative agreement late in 1998 strengthened the potential control of the ruling **Palestinian Authority (PA)** in the Gaza Strip and portions of the West Bank (Figure 7.38). But a new cycle of heightened violence erupted late in 2000 as Palestinian attacks against Jews increased and the Israelis continued with the construction of new settlements in occupied lands (especially in the West Bank) (Figures 7.39 and 7.40). In 2013, with several hundred thousand Israeli settlers already living in these West Bank settlements, the Israeli government pledged to continue its support for new construction, including developments in and near Jerusalem. Palestinian authorities continue to strongly protest these new West Bank Jewish settlements near the Israeli capital. In addition, Palestinians see their land under a growing threat of new Jewish settlements in the eastern, more rural portions of the West Bank as well.

Adding to the friction has been ongoing construction of the Israeli security barrier, a partially completed series of concrete walls, electronic fences, trenches, and

▼ **Figure 7.39 Jewish Settlement, West Bank** This new housing is part of the Jewish settlement of Givat Ze'ev located just a few miles northwest of Jerusalem.

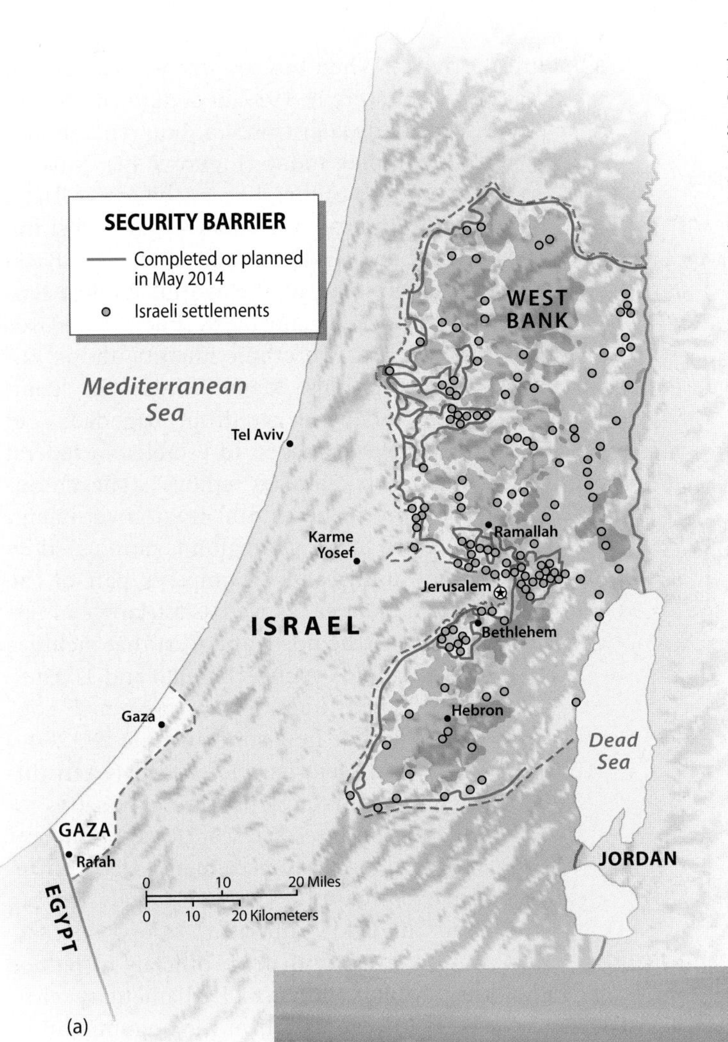

(b)

▲ **Figure 7.40 Israeli Security Barrier** (a) The map shows the completed and planned portions of the Israeli security barrier, as well as many of the Israeli settlements located in the West Bank region. (b) The photo shows a segment of the Israeli security barrier.

watchtowers designed to effectively separate the Israelis from Palestinians across much of the West Bank region (see Figure 7.40). Israeli supporters of the barrier (to be more than 400 miles long when completed) see it as the only way they can protect their citizens from suicide bombings and more terrorist attacks. Palestinians see it as a land grab, an "apartheid wall" designed to socially and economically isolate many of their settlements along the Israeli border.

Political fragmentation of the Palestinians has added further uncertainty. In 2006, control of the Palestinian government was split between the Fatah and Hamas political parties. Hamas has long been seen by many Israelis as an extremist and violent political party, whereas Fatah has shown more willingness to work peacefully with Israel. The split between these Palestinian factions became violent, with Hamas gaining effective control of the PA within Gaza and Fatah maintaining its greatest influence across the West Bank. Further complicating this political geography is the emergent Islamist movement among some Palestinians. Often, Islamists challenge the secular orientation of both Hamas and Fatah. These growing complexities within the Palestinian movement make an eventual peace treaty with Israel even more difficult to achieve.

One thing is certain: Geographical issues will remain at the center of the conflict. Israelis continue their search for secure borders to guarantee their political integrity. Most Palestinians still call for a "two-state solution," in which their autonomy is guaranteed, but a growing minority of Palestinians, frustrated with the stalemate, suggest considering a "one-state solution," in which Israel would be compelled to recognize the Palestinians as equals. Many Israelis, however, are wary of this option, seeing the fast-growing Palestinian population as a threat to a Jewish state they worked hard to create.

Instability in Syria and Iraq Elsewhere in the region, political instability in Syria erupted into civil war between 2011 and 2013. Rebel protests (mostly by Sunni Muslims) against the autocratic regime of President Bashar Hafez al-Assad (a member of the minority Alawite sect) reached a fever pitch, and government soldiers have killed thousands of civilians and used chemical weapons in a series of violent confrontations (Figure 7.41). By 2013, more than 90,000 deaths were directly related to the violence.

▼ **Figure 7.41 Violence in Homs, Syria** Smoke rises from the Baba Amr neighborhood in Homs. The Syrian city became an early focal point of rebel protests against the Assad government.

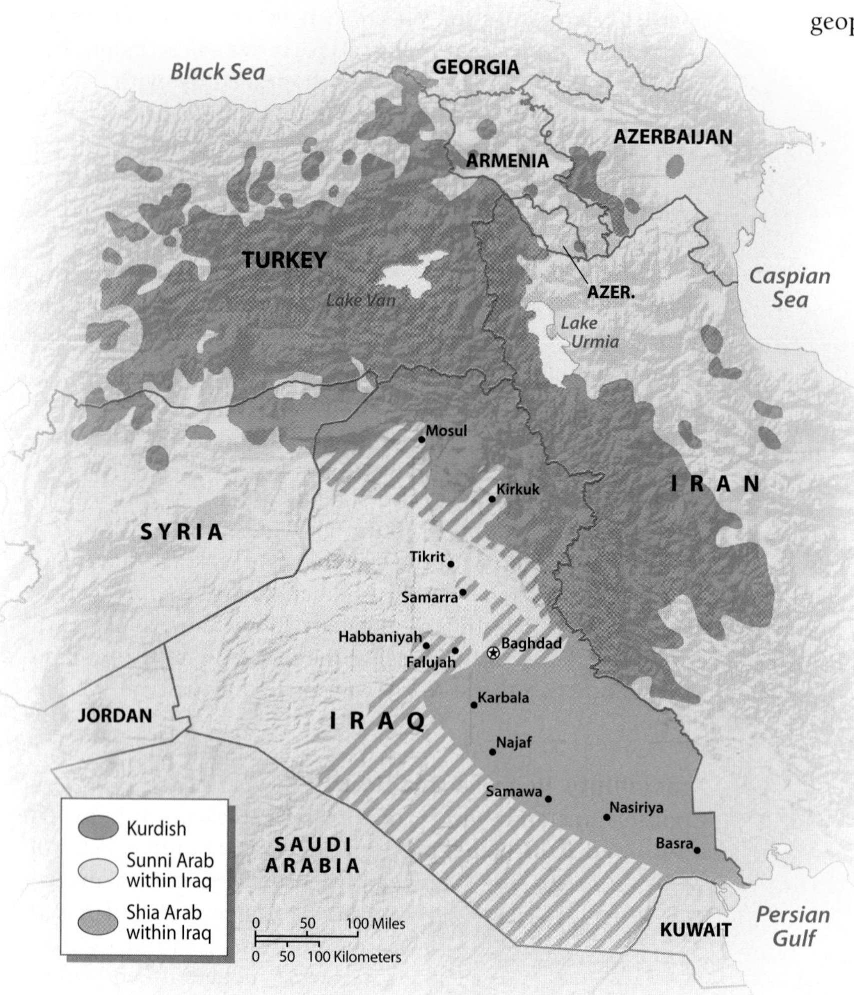

▲ **Figure 7.42 Multicultural Iraq** Iraq's complex colonial origins produced a state with varying ethnic characteristics. Shiites dominate south of Baghdad, Sunnis hold sway in the western triangle zone, and Kurds are most numerous in the north, near oil-rich Kirkuk and Mosul.

The rebels themselves are fragmented and divided: Some simply demand less corruption and a more democratic, religiously tolerant state, while others are Sunni religious extremists allied with conservative Salafists who want to impose an Islamist theocracy within the region.

The larger regional Arab community has also reacted against Assad, suspending Syria from the **Arab League** (a regional political and economic organization focused on Arab unity and development; see Figure 7.35) and urging an international solution to the crisis. For most Syrians, any semblance of normal life has ended as the civil conflict has expanded to many portions of the country. Massive numbers of Syrian refugees have fled, burdening several nearby nations (see *People on the Move: The Syrian Exodus*). Complicating matters is the fact that Assad has gained support from Iranian leaders, as well as from Iranian-supported Hezbollah (a Shiite militia group) in neighboring Lebanon.

Iraq is another multinational state born during the colonial era that has yet to escape the consequences of its geopolitical origins. When the country was carved out of the British Empire in 1932, it contained the cultural seeds of its later troubles. Iraq remains culturally complex today (Figure 7.42). Most of the country's Shiites live in the lower Tigris and Euphrates river valleys near and south of Baghdad. Indeed, the region near Basra contains some of the world's holiest Shiite shrines. In northern Iraq, the Kurds have their own ethnic identity and political aspirations. Many Kurds want complete independence from Baghdad, and they have managed to establish a federal region that already enjoys some autonomy from the central Iraqi government. A third major subregion is dominated by the Sunnis and encompases part of the Baghdad area, as well as a triangle of territory to the north and west that includes strongholds such as Falujah and Tikrit.

When Iraqi leaders assumed control of their new state on June 28, 2004, growing sectarian violence between different Iraqi factions threw portions of the nation into civil war. Rival Sunni and Shiite groups forced many Iraqis from their communities. Before largely leaving Iraq in 2011, American troops successfully worked with Iraqi officials to reduce the level of violence. Parliamentary elections, however, confirm the ongoing political uncertainties within Iraq. Both Shiite and Sunni candidates prove about equally popular, largely splitting the vote along regional lines (Shiites winning in the south and narrowly in Baghdad; Sunnis winning in central and western provinces). In 2013, growing violence between Shiite and Sunni factions appeared to be heightened by Sufi extremists and by Al Qaeda terrorists within the country. Complicating matters further, the politically and culturally distinct Kurdish Alliance remains dominant in the north.

An enduring geopolitical solution for a stable Iraqi state has yet to take shape, but it will need to recognize these cultural differences, as well as allow for an ongoing revival of the country's oil industry. By far, the country's dominant oil fields are located in Shiite- and Kurdish-controlled portions of Iraq, an uneasy fact of resource geography that has long troubled the nation's sizable Sunni population.

Politics in the Arabian Peninsula In the Persian Gulf region, tiny Bahrain, although ruled by a Sunni monarchy, is a mostly Shiite country. Violent protests also erupted there in 2011, aimed at encouraging democratic reforms, rather than overthrowing the government. After violence escalated, troops from neighboring Saudi Arabia were called in by the Bahraini government to end the conflict.

Saudi Arabia itself remains ruled by an aging conservative monarchy (the Al Saud family) unwilling to promote democratic reforms. On the surface, it supports U.S. efforts to provide stable flows of petroleum and fight terrorism, but beneath the surface, certain elements of the regime may have financed radically anti-American groups such as Al Qaeda. The Saudi people themselves, largely Sunni Arabs, are torn among an allegiance to their royal family (and the economic stability it brings); the lure of a more democratic, open Saudi society; and an enduring distrust of foreigners, particularly Westerners. Furthermore, the Sunni majority includes Wahhabi sect members, whose radical Islamist philosophy has fostered anti-American sentiment. In addition, the large number of foreign laborers and the persistent U.S. military and economic presence within the country (one of the chief complaints of former Al Qaeda leader Osama bin Laden) create a setting ripe for political instability.

In nearby Yemen, President Ali Abdullah Saleh was forced from office and elections held in 2012. Even so, calls for democratic reforms are complicated by ongoing factionalism within the country, including pockets of Shiite militants that maintain close connections with Iran and southern separatists anxious to reassert a separate political identity they enjoyed before 1990 (when Yemen emerged as a single state). Al Qaeda's growing influence (including terrorist training camps) within the country has also been noted as a threat to its political stability.

Iran Ascendant? Iran increasingly garners international attention. Islamic fundamentalism dramatically appeared on the political scene in 1978 as Shiite Muslim clerics overthrew Shah Mohammad Reza Pahlavi, an authoritarian, pro-Western ruler friendly to U.S. interests. The new leaders proclaimed an Islamic republic in which religious officials ruled both clerical and political affairs. Today Iran supports Shiite Islamist elements throughout the region (such as Hezbollah in Lebanon and Syria) and has repeatedly threatened the state of Israel. The country is also seen as a threat by moderate Arab states. Adding uncertainty has been Iran's ongoing nuclear development program, an initiative its government claims is related to the peaceful construction of power plants (see Figure 7.35). Israel, in particular, has suggested it may destroy Iranian nuclear sites before Iranian leaders have an opportunity to attack Israel with a nuclear-tipped missile. Others in the West also remain unconvinced of the government's motives and demand its program be stopped or opened to more inspections before allowing for greater international cooperation with the regime. As a result, strong economic sanctions have been imposed on Iran by the West, limiting its ability to both export oil and import desirable consumer products.

Within Iran, varied political and cultural impulses are evident. Elections in 2013 ushered in a new, more moderate president (Hassan Rouhani), who may be more inclined to push for domestic economic and political reforms. Many younger, wealthier, more cosmopolitan Iranians are hopeful that the country will become less isolated on the world stage. At the same time, most people in the country actually support the nuclear program, arguing that they have the same right as Pakistan or India to develop this resource. Hard-line religious extremists also maintain control of key positions in the country, and they continue to be harshly critical of Western sanctions and suspected interference in their country.

Tensions in Turkey Turkey has also emerged as a key geopolitical question mark. Turkey continues to find itself strategically positioned between diverse, often contradictory geopolitical forces. Many pro-Westerners within Turkey, for example, are committed to joining the EU. To do so, the country has embarked on an active agenda of reforms designed to demonstrate its commitment to democracy. Press freedoms and multiparty elections have been permitted, the role of the military has been downplayed, and minority groups (particularly the Kurds) have gained greater recognition from the central government.

At the same time, anti-Western Islamist political elements have been on the rise, linked to a growing number of terrorist bombings in the country since 2003. Deeply suspicious of the floundering EU, these factions advocate a new, more religious-based course for the country that charts an independent economic trajectory in the region and resists cozying up to Western political interests.

Recently, large public protests concerning government planning policies also erupted in violence, further threatening the country's political and economic stability.

An Uncertain Political Future

Few areas of the world pose more geopolitical questions than Southwest Asia and North Africa. Twenty years from now, the region's political map could look quite different from the one today. The region's strategic global importance increased greatly after World War II, propelled into the international spotlight by the creation of Israel, the tremendous growth in the world's petroleum economy, Cold War tensions between the United States and the Soviet Union, and, more recently, the rise of Islamic fundamentalism and Islamist political movements. Now that the Cold War between global superpowers has ended, the region is experiencing a geopolitical reorientation.

One key question that remains is, what geopolitical role should the United States play in Southwest Asia? That simple question continues to confound policymakers in the United States, as well as residents of that ever-troubled part of the world. The 2003 Iraqi invasion put large numbers of American troops in the region. Even as many of those troops have returned home, the American role remains central to the region's geopolitical future. How does the United States identify and deal with hostile elements? Should the United States go it alone in these efforts ("unilateralism") or depend on sharing the responsibilities with other countries or international organizations ("multilateralism")? Regional resistance movements directed against the United States also have grown more powerful in many countries within the

People on the Move

The Syrian Exodus

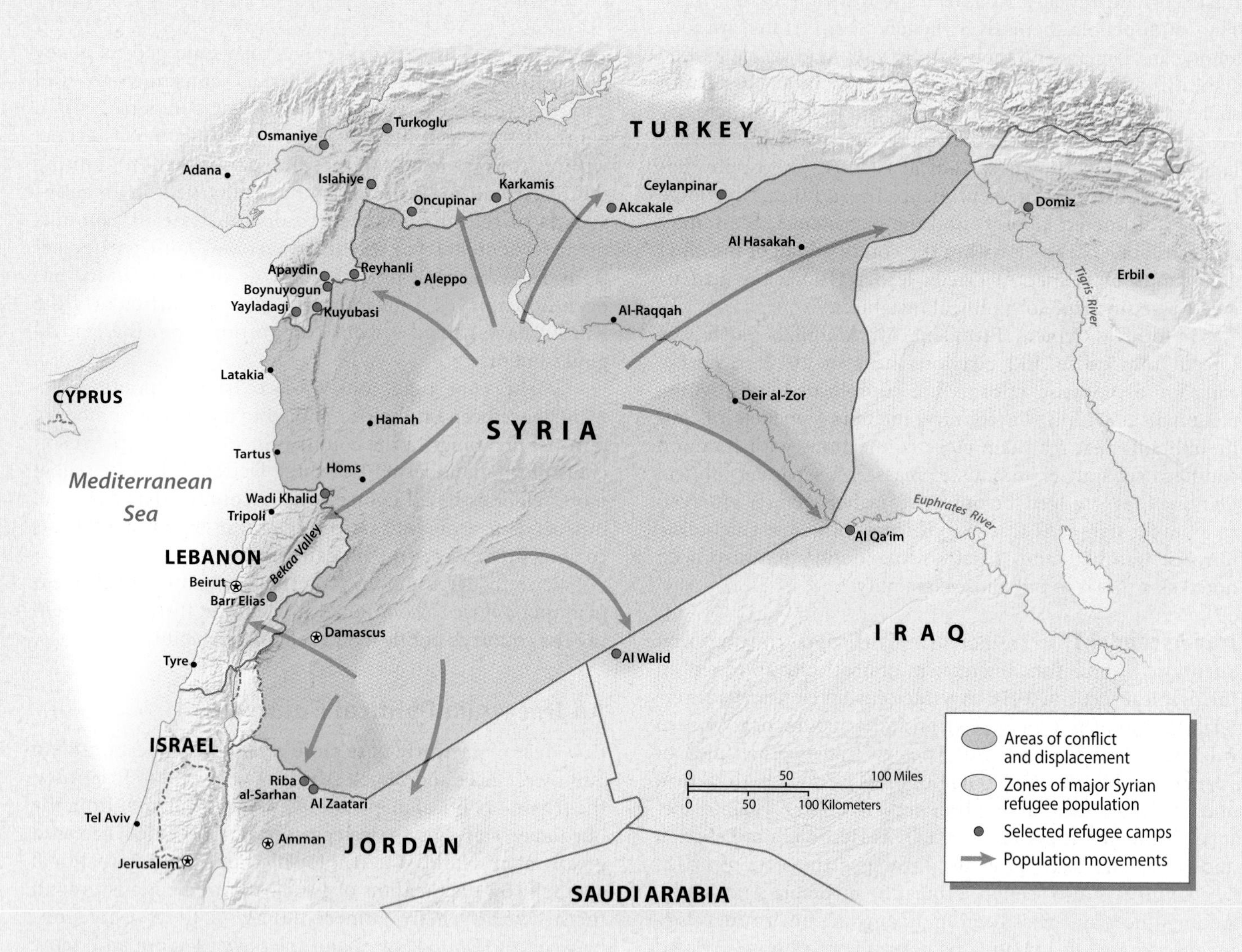

▲ **Figure 7.3.1 Syrian Refugee Zones and Selected Camps** Neighboring areas of Turkey, Jordan, Iraq, and Lebanon were inundated with refugees in 2012 and 2013 as violence escalated in the Syrian civil war.

region, drawing on growing popular suspicion of American motives and on the heated religious rhetoric of Islamist elements.

Several interrelated settings appear destined to hold continuing U.S. attention within the region:

- Whether the United States will become more involved in the Syrian conflict remains uncertain, since the rebel movements against the Assad regime have failed to arrive at their own consensus for the country's future. Still, the bloodshed within the region may demand some coordinated international response.
- Nearby Iraq also remains important, even as the country struggles to find its own, more independent political identity. Iraq's fragile political structure and its globally significant oil economy, all cast amid the country's complex ethnic and religious geography, seem likely to remain key concerns for American policymakers.
- Another key question mark is oil-rich Saudi Arabia. Of particular concern are domestic challenges to the

Sadly, political crises in the Middle East have produced millions of refugees since World War II. Recently, Syria's civil war has created a mass exodus from many war-torn portions of the country. The scale of these desperate and forced migrations is unprecedented, even for this unstable part of the world. The situation has produced a crisis that ripples throughout the entire region and also threatens to increase political instability in areas dealing with the flood of refugees.

Magnitude of the Crisis The Syrian conflict involves multiple, often divided factions that are battling government forces. The Assad regime has not hesitated to launch widespread air and ground attacks on its own people, including the use of chemical weapons. The response has been massive: In some cases, entire villages have virtually emptied out as families seek to escape the violence. By mid-2013, more than 4.3 million internally displaced persons (IDPs) had left their homes and were living elsewhere in the country. In addition, by the end of 2013, UN experts estimated that as many as 3.5 million more people were living in nearby countries, in existing towns and cities, as well as in refugee camps. Southern Turkey and northern Jordan witnessed large numbers of migrants early in the crisis, but increased violence in southwestern Syria has also propelled huge numbers of refugees into many areas of northern and eastern Lebanon (**Figure 7.3.1**). Many people have also fled to Iraq and Egypt.

Conditions in the Camps Refugee conditions in these neighboring nations have been perilous. In Jordan's Zaatari camp, for example, more than 100,000 people struggle just to survive. A sea of tents marks the site, and it has become one of the world's largest refugee camps (**Figure 7.3.2**). Crowded living conditions, power outages, food shortages, and disease have become a way of life for many Syrians who once owned homes and had comfortable middle-class lives. Local residents are also resentful of the intrusions: The flood of migrants into northern Jordan, for example, has precipitated a looming water shortage as aquifers are rapidly depleted. In settings such as northern Lebanon, where authorities (many of them sympathetic to the Assad regime) have resisted opening up new camps, existing towns and villages have also been overwhelmed with new arrivals. The refugees themselves represent diverse political interests, often resulting in conflicts among themselves and with their local hosts.

A Legacy of Refugees Whatever the immediate outcome of the Syrian crisis, the longer-term consequences of these people on the move are enduring and deeply problematic for this part of the world. They reflect a heritage of political instability and the basic human response to escape violence. Palestinian refugee camps are now decades old in various parts of the region (including Syria). Many Iraqi refugees also have yet to return to their still-troubled homeland. The human and economic costs of these crises are immense, and it appears that yet another generation of people in the region will bear them for a long time to come.

▼ **Figure 7.3.2 Zaatari Refugee Camp, Jordan** More than 100,000 refugees poured into Jordan's Zaatari refugee camp between 2012 and 2013.

all-powerful Al Saud royal family, and U.S. observers fret that a sudden change in Saudi Arabian politics would have global ramifications, particularly if oil exports are disrupted. The country remains the most important trading partner for the United States in the entire region.

- The growing geopolitical presence of Iran challenges American, Israeli, and many moderate Arab (particularly Saudi Arabian) interests within the region. Iran's regional aspirations and its always murky nuclear installations remain key question marks for this portion of the world.
- Finally, recent events in Israel, Gaza, and the West Bank continue to make global headlines. Palestinian infighting, continuing attacks on Israel, and Israeli reprisals (including the construction of its security barrier) continue to fracture and enflame the region. What is an appropriate role for the United States in tactically improving short-term relationships between Israel and its Palestinian populations and in crafting a longer-term strategic policy aimed at achieving a permanent peace in the region? How the United States responds to Southwest Asia's shifting geopolitics seems destined to shape the region for decades to come.

REVIEW

7.8 Describe the role played by the French and British in shaping the modern political map of Southwest Asia and North Africa. Provide specific examples of their lasting legacy.

7.9 Discuss the root causes behind the Arab Spring rebellions, and offer three examples of how the movement has played out politically in different ways within the region.

7.10 Explain how ethnic and political differences have shaped recent political conflicts in Syria and Iraq.

Economic and Social Development: A Region of Stark Contrasts

Southwest Asia and North Africa comprise a region of incredible wealth and discouraging poverty (Table 7.2). Some countries enjoy prosperity, due mainly to rich reserves of petroleum and natural gas, but other nations are among the world's least developed. Continuing political instability contributes to the region's struggling economy. The Arab Spring rebellions, for example, while raising prospects for longer-term political reforms, also produced shorter-term challenges as economic output and tourism declined in many places. In Iran and Syria, economic sanctions have also crippled the economy. Petroleum will no doubt figure significantly in the region's future economy, but some countries in the area have also focused on increasing agricultural output, investing in new industries, and promoting tourism to broaden their economic base.

Measuring Development

Investments in education, health care, and new employment opportunities have slowed considerably in many countries from the heady gains of the late 1970s and 1980s. Although overall social and economic conditions have improved modestly for many people, striking regional variability continues.

Table 7.2 Development Indicators

Country	GNI per capita, PPP 2011	GDP Average Annual % Growth 2000–11	Human Development Index (2011)[1]	Percent Population Living Below $2 a Day	Life Expectancy (2013)[2]	Under Age 5 Mortality Rate (1990)	Under Age 5 Mortality Rate (2011)	Adult Literacy (% ages 15 and older) Male/ Female	Gender Inequality Index (2011)[3,1]
Algeria	8,310	3.7	.713	23.6	76	66	30	81/64	0.391
Bahrain	21,200	6.3	.796	–	76	21	10	93/90	0.258
Egypt	6,120	5.0	.662	15.4	70	86	21	80/64	0.590
Gaza and West Bank	–	–	.670	< 2	73	43	22	98/92	–
Iran	11,420	5.4	.742	8.0	73	61	25	89/81	0.496
Iraq	3,750	1.1	.590	21.4	69	46	38	86/71	0.557
Israel	27,110	3.6	.900	–	82	12	4	--/--	0.114
Jordan	5,930	6.5	.700	< 2	73	37	21	96/89	0.482
Kuwait	58,720	6.0	.790	–	75	17	11	95/92	0.274
Lebanon	14,470	5.0	.745	–	79	33	9	93/86	0.433
Libya	16,800	5.4	.769	–	75	44	16	96/83	0.216
Morocco	4,880	4.8	.591	14.0	70	81	33	69/44	0.444
Oman	25,720	4.9	.731	–	76	48	9	90/81	0.340
Qatar	86,440	13.6	.834	–	78	20	8	97/95	0.546
Saudi Arabia	24,700	3.7	.782	–	74	43	9	90/81	0.682
Sudan	2,120	7.1	.414	44.1	62	123	86	80/82	0.604
Syria	5,080	5.0	.648	16.9	75	36	15	90/77	0.551
Tunisia	8,990	4.4	.712	4.3	75	51	16	86/71	0.261
Turkey	16,940	4.7	.722	4.7	74	72	15	96/85	0.366
United Arab Emirates	47,890	4.8	.818	–	76	22	7	89/91	0.241
Western Sahara	–	–	–	–	67	–	–	–	–
Yemen	2,170	3.6	.458	46.6	62	126	77	81/47	0.747

[1]United Nations, *Human Development Report, 2013.*

[2]Population Reference Bureau, *World Population Data Sheet, 2013.*

[3]Gender Inequality Index—A composite measure reflecting inequality in achievements between women and men in three dimensions: reproductive health, empowerment, and the labor market that ranges between 0 and 1. The higher the number, the greater the inequality.

Sources: World Bank, World Development Indicators, 2013.

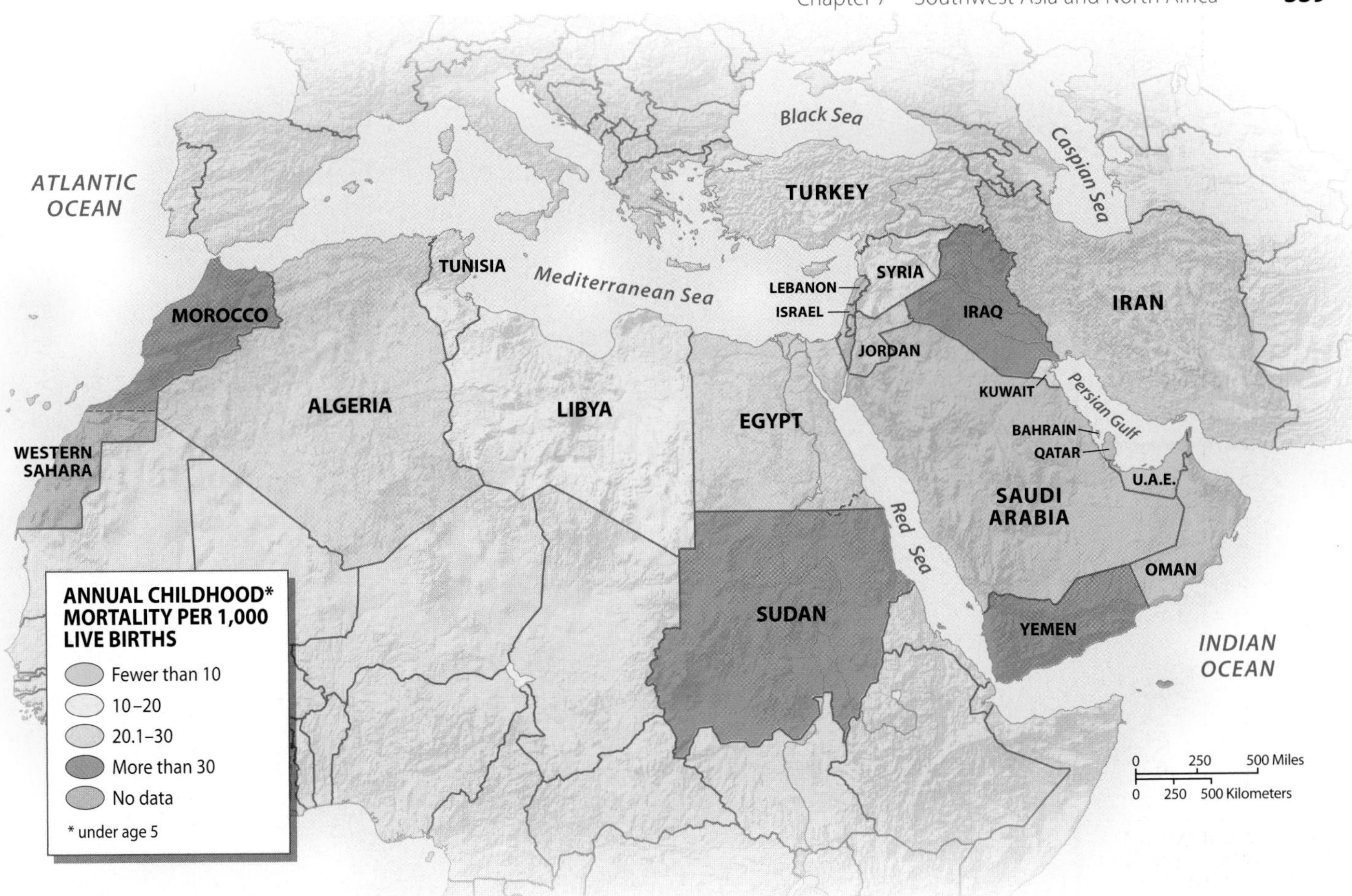

▲ **Figure 7.43 Development Issues in Southwest Asia and North Africa: Childhood Mortality** Wealthier nations such as Israel and the United Arab Emirates have very low rates of childhood mortality, but poor countries such as Sudan, Morocco, and Iraq continue to struggle with very high rates. **Q: Why might it be argued that childhood mortality is a good measure of development?**

In looking at childhood (under age 5) mortality, for example, access to modern medicine and excellent health-care systems is obvious in some nations, such as Israel (though Palestinians still receive fewer benefits) and the UAE (Figure 7.43). Elsewhere, however, nations such as Sudan, Iraq, and Morocco still have stubbornly high rates of childhood mortality, even though they have declined slightly since the early 1990s.

Development within the region also can be assessed by its access to information and how well it is connected to the rest of the world. Cell phone and Internet use, for example, vary widely across North Africa and Southwest Asia. Lower penetration of cell phone technology in some settings (Egypt, Sudan, and Iran) suggests basic investments in infrastructure remain lacking, whereas areas of very high cell phone usage (Israel, many Persian Gulf states) are correlated with greater affluence, as well as highly mobile, foreign-born populations. Internet use is also highly variable, ranging from 1 percent in Yemen to most of the population of the UAE. In some settings such as Iran, official government resistance to the Internet plays a role as well: More than 5 million websites are blocked in the country, including the use of Google Earth (the government in 2013 threatened to develop its own Iranian version).

The Geography of Fossil Fuels

The striking global geographies of oil and natural gas reveal the region's persistent importance in the world oil economy, as well as the extremely uneven distribution of these resources within the region (Figure 7.44). Higher oil prices early in the new century have once again highlighted the economic clout of this region. Saudi Arabia remains one of the major producers of petroleum in the world, and Iran, the UAE, Libya, and Algeria also contribute significantly. The region plays an important, though less dominant, role in natural gas production. The distribution of fossil fuel reserves suggests that regional supplies will not be exhausted anytime soon. Overall, with only 7 percent of the world's population, the region holds over half of the world's proven oil reserves. Saudi Arabia's pivotal position, both regionally and globally, is clear: Its 30 million residents live atop almost 20 percent of the planet's known oil supplies.

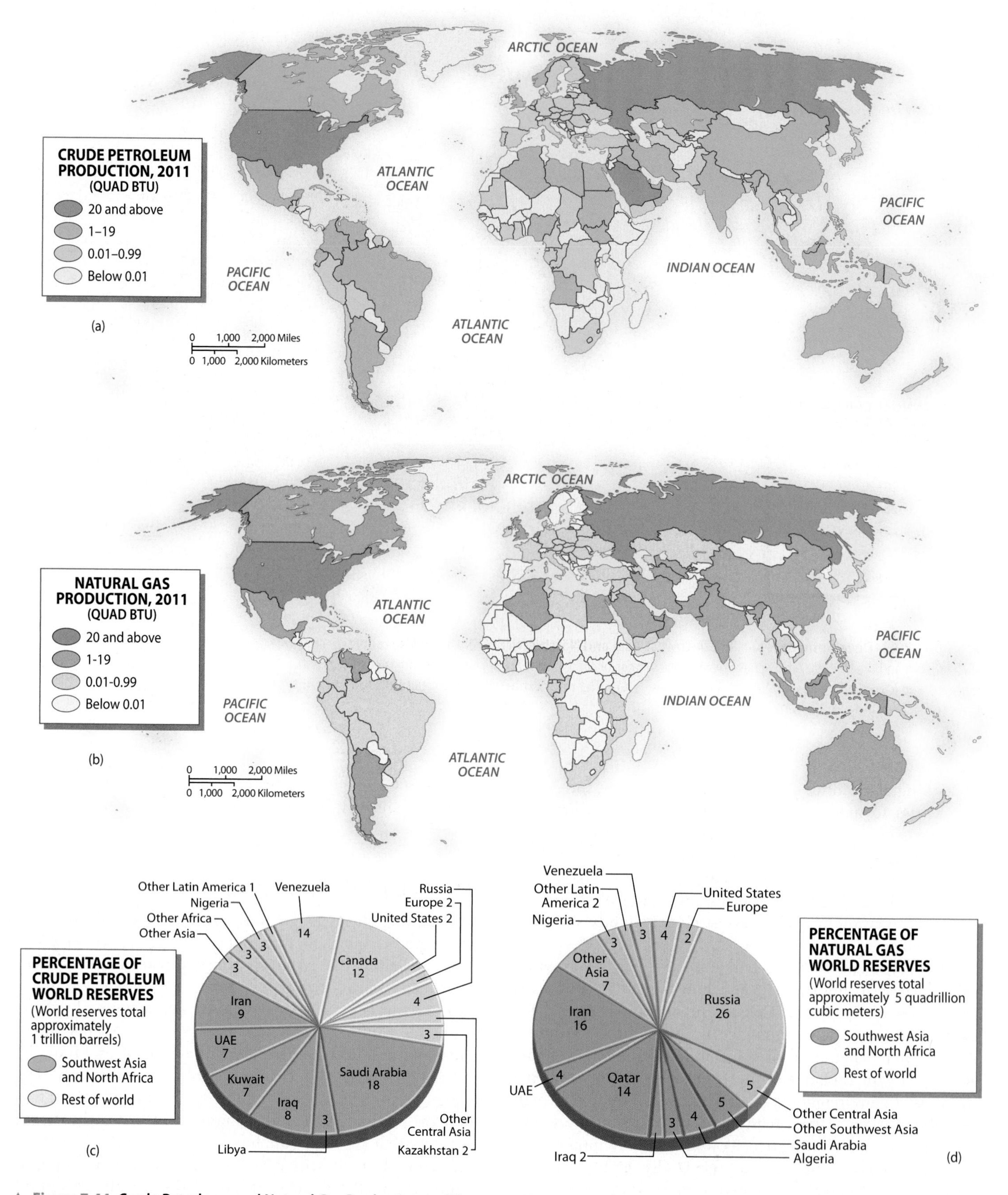

▲ **Figure 7.44 Crude Petroleum and Natural Gas Production and Reserves** The region plays a central role in (a) the global geography of both crude petroleum production as well as (b) natural gas production. Abundant regional reserves of both (c) crude petroleum and (d) natural gas suggest that the pattern will continue.

Several major geological zones supply much of the region's output of fossil fuels. The world's largest concentration of petroleum lies within the Arabian-Iranian sedimentary basin, a geological formation that extends from northern Iraq and western Iran to Oman and the lower Persian Gulf. All of the states bordering the Persian Gulf reap the benefits of oil and gas deposits within this geological basin, and it is not surprising that the world's densest concentration of OPEC members is found in the area. A second important zone of oil and gas deposits includes eastern Algeria, northern and central Libya, and scattered developments in northern Egypt. As in the Persian Gulf region, these North African fields are tied to regional processing points and to global petroleum markets by a complex series of oil and gas pipelines and by networks of technologically sophisticated oil-shipping facilities. Third, a zone of growing importance is being developed in the African nation of Sudan, including areas southwest of Khartoum. Sudan may also figure prominently in the development and transport of oil out of producing zones in nearby South Sudan.

Even with all these riches, the geography of fossil fuels is strikingly uneven across the region. Some states—even those with tiny populations (Bahrain, Qatar, and Kuwait, for example)—contain incredible fossil fuel reserves, especially when considered on a per capita basis. Many other countries, however, and millions of regional inhabitants reap relatively few benefits from the oil and gas economy. In North Africa, for example, Morocco possesses few developed petroleum reserves, and even the fruits of Sudan's blossoming wealth will likely remain very concentrated within a small segment of a large, poor population. In oil-rich Southwest Asia, the distribution of fossil fuels is also uneven: Israel, Jordan, and Lebanon all lie outside favored geological zones for either petroleum or natural gas. Although Turkey has some developed fields in the far southeast, it must import substantial supplies to meet the needs of its large and industrializing population.

Regional Economic Patterns

Remarkable economic differences characterize the region (see Table 7.2). Some oil-rich countries have prospered greatly since the early 1970s, but in many cases fluctuating oil prices, political disruptions, and rapidly growing populations have reduced prospects for economic growth. Other nations, although poor in oil and gas reserves, have seen brighter prospects through moves toward greater economic diversification. Finally, some countries in the region are subject to persistent poverty, where rapid population growth and the basic challenges of economic development combine with political instability to produce very low standards of living.

Higher-Income Oil Exporters The richest countries of Southwest Asia and North Africa owe their wealth to massive oil reserves. Nations such as Saudi Arabia, Kuwait, Qatar, Bahrain, and the UAE benefit from fossil fuel production, as well as from their relatively small populations (see Figure 7.44). Large investments in transportation networks, urban centers, and petroleum-related industries have reshaped the cultural landscape. Billions of dollars have also poured into new schools, medical facilities, low-cost housing, and modernized agriculture, significantly raising the standard of living in the past 40 years.

The Saudi petroleum-processing and -shipping centers of Jubail (on the Persian Gulf) and Yanbu (on the Red Sea) are examples of this commitment to expanding the region's economic base beyond the simple extraction of crude oil (see *Exploring Global Connections: Yanbu as a Global City*). Other oil-rich cities have cultivated a more cosmopolitan image, with globalized Dubai in the UAE appealing to both international tourists and business interests, as well as a global labor force (see Figure 7.25).

Still, problems remain. Dependence on oil and gas revenues produces economic pain in times of falling prices, such as those seen in the economic downturn between 2007 and 2010. Such fluctuations in world oil markets will inevitably continue in the future, disrupting construction projects, producing large layoffs of immigrant populations, and slowing investment in the region's economic and social infrastructure. In addition, countries such as Bahrain and Oman are faced with the problem of rapidly depleting their reserves over the next 20 to 30 years.

Lower-Income Oil Exporters Other states in the region are important secondary players in the oil trade, but different political and economic variables have hampered sustained economic growth. In North Africa, for example, Algerian oil and natural gas overwhelmingly dominate its exports, but the past 20 years have also brought political instability and increasing shortages of consumer goods. Although the country contains some excellent agricultural lands in the north, the overall amount of arable land has increased little over the past 25 years, even as the country's population has grown by more than 50 percent.

In Southwest Asia, Iraq still faces economic headwinds. Recent conflicts crippled much of Iraq's already deteriorated infrastructure, and continuing political instability is making the task of rebuilding its economy more difficult. Iraq still suffers from high unemployment. However, the country's economic leaders have implemented an ambitious plan for increasing oil output, and in 2012 Iraq produced more than 3 million barrels of oil per day for the first time in more than 20 years.

The situation in Iran remains challenging. The country's oil and gas reserves are huge, but Iran is relatively poor, burdened with a stagnating standard of living. Since 1980, fundamentalist leaders have downplayed the role of international trade in consumer goods and services, fearing the import of unwanted cultural influences. Recent international sanctions on purchases of Iranian oil (related to its nuclear development program) have severely depressed the economy. Some bright spots have

Exploring Global Connections

Yanbu as a Global City

Just what is a *global city?* This term has been used to describe many different kinds of places around the world, from North America's New York City to China's Macau. In Southwest Asia and North Africa, some of the best examples of global cities are linked with world petroleum markets and the regional developments they have spawned.

Consider Yanbu along the shore of the Red Sea in the western portion of Saudi Arabia (**Figure 7.4.1**). This city is an old trading town that linked Yemen and Egypt for centuries, but since 1975 (when it was designated as a new industrial center by the Saudis) it has blossomed into a city of almost 200,000 people. Yanbu's urban geography certainly fulfills three criteria that might be useful in defining any truly global city: (1) Yanbu's growth has been globally financed and engineered, (2) its workforce draws on a diverse immigrant population, and (3) its products end up in every corner of the world.

Global Investments and Expertise Yanbu's post-1975 rise to global prominence was initially sparked by Saudi government policy and key government investments in infrastructure (such as the country's East–West pipelines). These investments have helped make it the world's third largest hub of petroleum refining and the largest shipping port on the Red Sea (**Figure 7.4.2**). However, much of the $50 billion investment that has propelled the city's growth has come from beyond the country and often represents joint ventures between Saudi Arabia and a varied array of European, North American, and Asian petroleum, engineering, and banking interests. For example, U.S. firms such as Conoco Phillips and Jacobs Engineering have played key roles in expanding Yanbu's refinery capacity. Nearby, a large new power plant, to be completed in 2017, is a joint venture among Saudi Arabia, China's Shanghai Electric Company, and Japan's Samsung Engineering.

▲ **Figure 7.4.1 Yanbu, Saudi Arabia** This Google Earth image shows regional roads converging on Yanbu, a city of almost 200,000 people in western Saudi Arabia. The Red Sea is on the lower left.

A Global Workforce Yanbu also supports a truly global workforce. Although Saudi residents dominate the old portion of the city near downtown, much of Yanbu Al-Sina'iya (the Industrial Yanbu) supports a

appeared, however: The country benefits from energy developments in Central Asia, and literacy rates (particularly for women) have risen, reflecting a new emphasis on rural education.

Prospering Without Oil Some countries, while lacking petroleum resources, have nevertheless found paths to increasing economic prosperity. Israel, for example, supports one of the highest standards of living in the region, even with its political challenges (see Table 7.2). The Israelis and many foreigners have invested large amounts of capital to create a highly productive industrial base, which produces many products for the global marketplace (see *Everyday Globalization: Popping Pills from Israel*). The country also is emerging as a global center for high-tech computer and telecommunications products. Israel is known for its fast-paced and highly entrepreneurial business culture, which resembles California's Silicon Valley. Israel also has daunting economic problems. Its persistent struggles with the Palestinians and with neighboring states have sapped much of its potential vitality. Defense spending absorbs a large share of total gross national income (GNI), necessitating high tax rates. Despite a recent economic recovery, especially in Gaza, poverty and diverse assortment of upper-income European, Asian, and North American residents, as well as nearby working-class populations (mostly Muslims) from Africa, Turkey, the Persian Gulf region, and South and Southeast Asia. Elite residential communities and hotels often cater to wealthier foreign residents and visitors (**Figure 7.4.3**). Most lower-paid immigrant workers, however, live in marginal housing and have few legal protections.

Global Oil Markets Yanbu's outbound petroleum products circle the globe. Almost 90 percent of the country's exports are directly related to fossil fuels, and Yanbu plays a critical role in shipping both crude and refined petroleum products to diverse consumers. More than half of Saudi Arabia's crude oil exports end up in Asia—much of it in Japan, China, South Korea, and India. Another 30 percent is bound for European markets, and the remainder ends up largely in the United States (16 percent).

1. What are three characteristics that help define a *global city?*
2. Examine the corporate website of one of the world's large oil companies. Can you find examples of global investments and joint ventures that link it with key oil-producing areas in the Arabian Peninsula?

Google Earth Virtual Tour Video

http://goo.gl/Bb9cY6

▲ **Figure 7.4.2 Yanbu Petroleum Installation** Pipelines facilitate the movement of oil to and from Yanbu's large coastal refineries in western Saudi Arabia.

◀ **Figure 7.4.3 Exclusive Residential Area, Yanbu** Many foreign professionals and elite residents live in this upper-class suburban enclave.

unemployment among Palestinians also remain unacceptably high, and the gap between rich and poor within the country has widened.

Turkey also has a diversified economy, and its per capita income has advanced in the early 21st century. Lacking petroleum, Turkey produces varied agricultural and industrial goods for export. About 30 percent of the population remains employed in agriculture, and the country's principal commercial products include cotton, tobacco, wheat, and fruit. The industrial economy has grown considerably since 1980, including exports of textiles, food, and chemicals. Turkey has also gone high tech: About 44 percent of Turks use the Internet, and the country has been a fertile ground for dozens of global Internet start-up companies that connect well (many are online or virtual gaming enterprises) with younger Turks, as well as with the global economy (Figure 7.45). Turkey also remains a major tourist destination in the region, attracting more than 6 million visitors annually in recent years.

Regional Patterns of Poverty Poorer countries of the region share the problems of the less developed world. For example, Sudan, Egypt, and Yemen each face unique economic challenges. For Sudan, continuing political

Everyday **Globalization**

Popping Pills from Israel

Every year more than 2.5 *billion* prescriptions for generic pharmaceuticals are written in the United States. Few people realize how many of these drugs are actually manufactured in Southwest Asia—specifically, Israel. When you reach for that generic antibiotic (amoxicillin), pain killer (oxycodone), or anti-inflammatory (naproxen), you may well be taking medications manufactured halfway around the world (Figure 7.5.1).

Rise of the Generic Industry This global connection was made possible by the 1984 passage of the Hatch–Waxman Act (and the amendments that followed). Essentially, the legislation created the modern system of regulating and monitoring the production of generic drugs that could be sold in the United States. It also established a regular timeline for the expiration of patented pharmaceuticals, thus encouraging generic drug producers to create low-cost alternatives for many medications. The consequences have been far-reaching: American consumers have benefited from cheaper generic alternatives, and generic drug manufacturers worldwide have seen American demand for their products skyrocket.

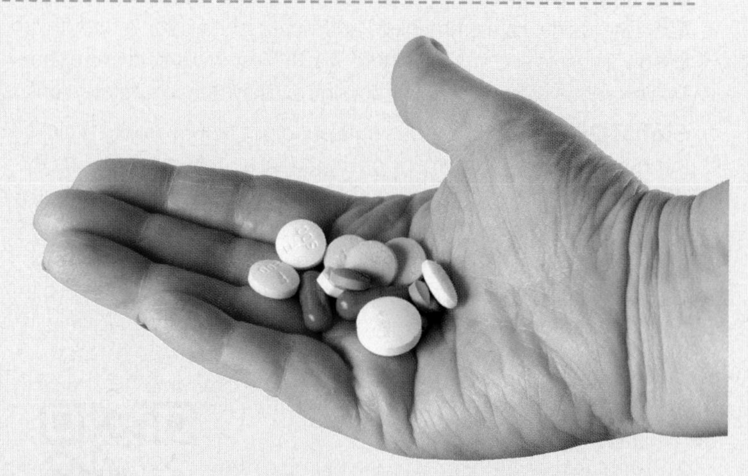

▲ **Figure 7.5.1 Generic Drugs** The multibillion- dollar market for generic drugs in the United States has attracted huge Israeli companies such as Teva Pharmaceutical Industries.

The Israeli Connection Israel has emerged as a key global player in the generic drug industry. Even before the creation of the modern state of Israel, many German Jews with pharmaceutical expertise migrated to Palestine in the early and mid-1930s. Some of these talented immigrants ended up at the Hebrew University of Jerusalem (opened in 1925), which established important research-oriented programs in microbiology, biochemistry, and bacteriology. After World War II (and the creation of the state of Israel), the country's research infrastructure grew. During the 1980s, the Israeli government also encouraged private-sector investment in biotech and the drug industry. Today Israel is home to seven research universities and a host of companies that focus on the biological sciences and innovations in the pharmaceutical industry.

▲ **Figure 7.5.2 Teva Headquarters, Petah Tikva, Israel** Employing thousands of skilled workers, Israel's Teva Pharmaceuticals Industries produces both the world's largest volume of generic drugs and a growing array of its own patented pharmaceuticals.

Teva's Success Story By far, the largest player in Israel's generic drug industry is Teva Pharmaceutical Industries. The company estimates that one in six generic prescriptions in the United States is filled with a Teva (Hebrew for "nature") product. Based in Petah Tikva, Israel, Teva went public in 1951 on the newly created Tel Aviv Stock Exchange. Today the company is the largest global manufacturer of generic pharmaceuticals, as well as an innovative producer of its own proprietary drugs (Figure 7.5.2). In the past decade, the company also has aggressively acquired other drug companies in the United States, Germany, and Japan, increasing its worldwide reach. Many other generic drug manufacturers have built factories in Israel as well, adding to the region's global dominance. The result is that Israel has emerged as one of the planet's key focal points in an industry that seems destined to grow along with the world's insatiable demand for affordable pharmaceuticals.

problems have stood in the way of progress. Political disruptions have resulted in major food shortages. The country's transportation and communications systems have seen little new investment, and secondary school enrollments remain very low. On the other hand, Sudan's fertile soils could support more farming, and its share of regional oil prospects suggests petroleum's expanding role in the economy.

Egypt's economic prospects are also unclear. Since the overthrow of the Mubarak regime in 2011, Egyptian currencies have weakened, unemployment has risen, tourism has slumped, and foreign investors have remained wary of the nation's uncertain political environment. Many Egyptians live in poverty, and the gap between rich and poor continues to widen. Illiteracy is widespread, and the country suffers from the **brain drain** phenomenon, as some of its brightest young people leave for better jobs in western Europe or the United States. Egypt's 85 million people already make it the region's most heavily populated state, and recent efforts to expand the nation's farmland have met with numerous environmental, economic, and political problems.

▲ **Figure 7.45 Turkey's High-Tech Economy** In an international joint venture, the Karsan automotive factory, based in Turkey, is producing minivans for an Italian bus company. This modern assembly plant is located in Bursa, a city in northwest Turkey

Yemen remains the poorest country on the Arabian Peninsula. Positioned far from the region's principal oil fields, Yemen's low per capita GNI puts the country on par with nations in impoverished Sub-Saharan Africa. The largely rural country relies mostly on marginally productive subsistence agriculture, and much of its mountain and desert interior lacks effective links to the outside world. Coffee, cotton, and fruits are its commercial agricultural products, and modest oil exports bring in needed foreign currency. Overall, however, high unemployment remains widespread. Recent political instability—including the growing role of Al Qaeda—also discourages foreign investment and clouds economic prospects.

A Woman's Changing World

The role of women in largely Islamic Southwest Asia and North Africa remains a major social issue. Female participation rates in the workforce are among the lowest in the world, and large gaps typically exist between levels of education for males and females. In conservative parts of the region, few women work outside the home. Even in parts of Turkey, where Western influences are widespread, it is rare to see rural women selling goods in the marketplace or driving cars in the street.

More orthodox Islamic states impose legal restrictions on the activities of women. In Saudi Arabia, for example, women are not allowed to drive, although growing protests among many younger, educated Saudi females may overturn that ban. In 2012, a well-publicized cellphone video (more than 2 million hits) by an outraged Saudi woman showed her being kicked out of a shopping mall in Riyadh for wearing nail polish. In Iran, full veiling remains mandatory in more conservative parts of the country, and police often hassle Iranian women in more liberal dress, even in larger urban areas. Generally, Islamic women lead more private lives than men: Much of their domestic space is shielded from the world by walls and shuttered windows, and their public appearances are filtered through the use of the *niqab* (face veil) or *chador* (full-body veil).

However, in some places, women's lives are changing, even within norms of more conservative Islamist societies. From Tunisia to Yemen, women widely participated in the Arab Spring rebellions, asserting their new political visibility in very public ways. The local consequences of these political and social changes are complex: In liberated Libya, young Islamic women rejoice in their freedom to wear the *niqab* in public, a practice banned under Colonel Muammar al-Qaddafi's rule. On the other hand, greater postrebellion freedoms in Egypt encouraged one young woman to pose almost naked on Facebook and Twitter, sparking controversy (and a supportive show of solidarity from a large group of Israeli women who also posed naked) and more than a million page views. Other women have been encouraged to play more active roles in political affairs, including running for office.

Algerian women demonstrate the pattern (Figure 7.46). Most studies suggest those in the younger generation are more religious than their parents and are more likely to cover their heads and drape their bodies with traditional religious clothing. At the same time, they are more likely

▲ **Figure 7.46 Algerian Women** Women now make up more than 30 percent of Algeria's National Assembly, a higher proportion of female representation than in many Western nations.

▼ **Figure 7.47 Naama Bay, Egypt** At Naama Bay, scenic El Fanar beach attracts sunbathers and snorkeling enthusiasts.

to be educated and employed. Today 70 percent of Algeria's lawyers and 60 percent of its judges are women. A majority of university students are women, and women dominate the health-care field. These new social and economic roles help explain why birth rates are declining. The Algerian case also demonstrates how complex social processes actually unfold, rarely following simple models for how "traditional" societies evolve. Women's roles are shifting throughout the region. In Sudan and Saudi Arabia, a growing number of women pursue high-level careers. Education may be segregated, but it is available. In Libya, more women than men graduate from the country's university system. In Western Sahara, Saharawi women play a leading role in the country's political fight for independence from Morocco, and their broader educational backgrounds and social freedoms (including the right to divorce their husbands) further separate them from Moroccan women. Women also have a more visible social position in Israel, except in fundamentalist Jewish communities, where conservative social customs limit women to more traditional domestic roles.

Global Economic Relationships

Southwest Asia and North Africa share close economic ties with the rest of the world. Although oil and gas remain critical commodities that dominate international economic linkages, the growth of manufacturing and tourism is also redefining the region's role in the world.

OPEC's Changing Fortunes OPEC does not control global oil and gas prices, but it still influences the cost and availability of these pivotal products. Western Europe, the United States, Japan, China, and many less industrialized countries depend on the region's fossil fuels. In the case of Saudi Arabia, for example, petroleum and its related products comprise more than 90 percent of its exports. Many oil-producing countries (such as Saudi Arabia) form partnerships with foreign corporations, accelerating the economic integration of the region with the rest of the world. Even so, the region's major energy producers are always vulnerable to global-scale recessions, such as the downturn that hit the region hard between 2007 and 2010.

Beyond key OPEC producers, other trade flows also contribute to global economic integration. Turkey, for example, ships textiles, food products, and manufactured goods to its principal trading partners: Germany, the United States, Italy, France, and Russia. Tunisia sends more than 60 percent of its exports (mostly clothing, food products, and petroleum) to nearby France and Italy. Israeli exports emphasize the country's highly skilled workforce: Products such as cut diamonds, electronics, and machinery parts are exported to the United States, western Europe, and Japan.

Regional and International Linkages Future interconnections between the global economy and Southwest Asia and North Africa may depend increasingly on cooperative economic initiatives far beyond OPEC. Relations with the EU are critical. Since 1996, Turkey has enjoyed close ties with the EU, but recent attempts at full membership in the organization have failed. Other so-called Euro-Med agreements also have been signed between the EU and countries across North Africa and Southwest Asia that border the Mediterranean Sea.

Most Arab countries, however, are wary of too much European dominance. In 2005, 17 Arab League members established the Greater Arab Free Trade Area (GAFTA), designed to eliminate all intraregional trade barriers and spur economic cooperation. In addition, Saudi Arabia plays a pivotal role in regional economic development through organizations such as the Islamic Development Bank and the Arab Fund for Economic and Social Development.

The Geography of Tourism Tourists are another link to the global economy. Traditional magnets such as ancient historical sites and globally significant religious localities draw millions of visitors annually. In addition, as the developed world becomes wealthier, a growing demand arises for recreational spots that offer beaches, sunshine, and novel entertainment. Indeed, many miles of the Mediterranean Sea, Black Sea, and Red Sea coastlines are now lined with upscale, but often ticky-tacky, landscapes of resort hotels and condominiums dedicated to serving travelers' needs. More adventurous travelers seek ecotourist activities such as snorkeling in Naama Bay on Egypt's Sinai coast (Figure 7.47) or four-wheeling among the Berbers in the Moroccan backcountry. Endangered wildlife beckons photographers and poachers hoping to catch a glimpse of a South Arabian grey wolf, a Nubian ibex, or a darting Persian squirrel.

Political instability, however, often dampens demand for travel to areas perceived to be unstable and dangerous. In the aftermath of the disruptive Arab Spring rebellions, for example, countries such as Tunisia have redoubled their efforts to boost their tourist economies, trying to assure potential visitors that all is well. Egypt, in particular, has seen tourist travel decline, and many hotels near historic sites and resort areas sit half empty. Widespread instability in Syria and Iraq has also made travel in those areas hazardous. Even Turkey's multibillion-dollar tourist economy may be hit, given growing political instability in that country.

REVIEW

7.11 Describe the basic geography of oil reserves across the region, and compare the pattern with the geography of natural gas reserves.

7.12 What different strategies for economic development have recently been employed by nations such as Saudi Arabia, Turkey, Israel, and Egypt? How successful have they been, and how do they relate to the theme of globalization?

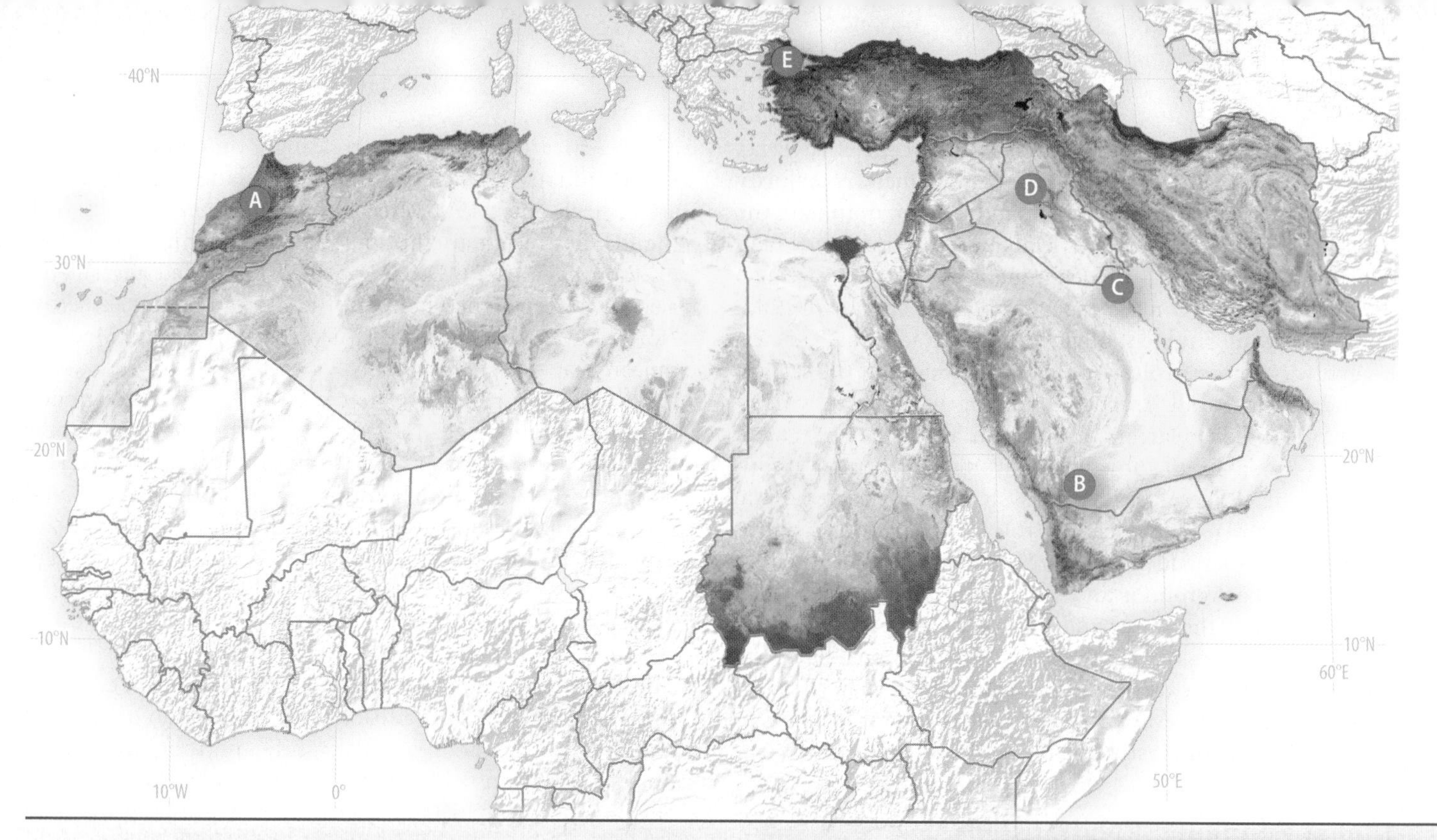

Physical Geography and Environmental Issues

Many nations within the region suffer from significant environmental challenges. Twentieth-century population growth across the region was dramatic, but it has been difficult and costly to expand the region's limited supplies of agricultural land and water resources. The results, apparent from the eroded soils of the Atlas Mountains to overworked garden plots along the Nile, are a classic illustration of the environmental price paid when population growth outstrips the ability of the land to support it.

7.1 If populations outstrip water supplies in North Africa's oasis settlements, how might residents adjust?

7.2 What are ways in which modern technology might address water shortages across the region? Are there limits or challenges to this approach?

Population and Settlement

The population geography of Southwest Asia and North Africa is strikingly uneven. Areas with higher rainfall or access to exotic water often have very high physiological population densities, whereas nearby arid zones remain almost empty of settlement.

7.3 Briefly describe the likely population density and land-use patterns you might see out the plane window on a flight between Riyadh (Saudi Arabia) and San'a (Yemen).

7.4 How might very low population densities impose special problems for maintaining effective political control across all portions of nations such as Saudi Arabia, Libya, or Algeria?

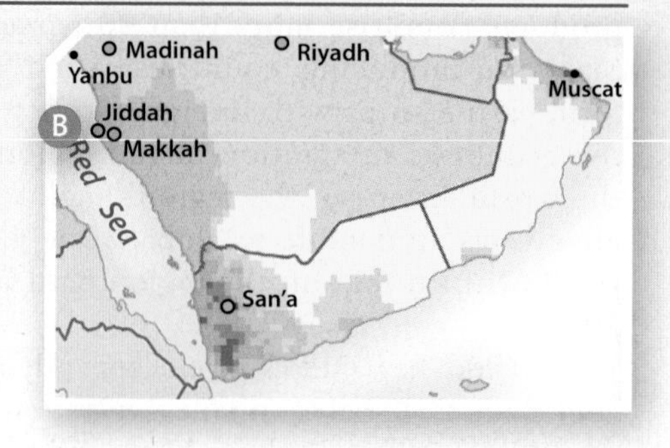

Key Terms

Arab League *(p. 334)*
Arab Spring *(p. 300)*
brain drain *(p. 345)*
choke point *(p. 308)*
culture hearth *(p. 300)*
domestication *(p. 311)*
exotic river *(p. 313)*
Fertile Crescent *(p. 311)*
fossil water *(p. 307)*
Hajj *(p. 322)*
hydropolitics *(p. 307)*
Islamic fundamentalism *(p. 300)*
Islamism *(p. 300)*
kibbutzim *(p. 313)*
Levant *(p. 301)*
Maghreb *(p. 301)*
medina *(p. 316)*
monotheism *(p. 321)*
Organization of the Petroleum Exporting Countries (OPEC) *(p. 300)*
Ottoman Empire *(p. 323)*
Palestinian Authority (PA) *(p. 332)*
pastoral nomadism *(p. 312)*
physiological density *(p. 310)*
protectorate *(p. 329)*
qanat system *(p. 306)*
Quran *(p. 322)*
sectarian violence *(p. 300)*
Shiites *(p. 322)*
Suez Canal *(p. 329)*
Sunnis *(p. 322)*
theocratic state *(p. 322)*
transhumance *(p. 312)*

In Review • Chapter 7
Southwest Asia and North Africa

Cultural Coherence and Diversity

Culturally, the region remains the hearth of Christianity, the spatial and spiritual core of Islam, and the political and territorial focus of modern Judaism. In addition, important sectarian divisions within religious traditions (especially the schism between Sunnis and Shiites), as well as long-standing linguistic differences, continue to shape the area's local cultural geographies and regional identities.

C

7.5 Why might it be said that Islam is both a powerful unifying and divisive cultural force within the region?

7.6 Why does Saudi Arabia remain such a pivotal part of the Islamic world?

Geopolitical Framework

Political conflicts have disrupted economic development across the region. Civil wars, sectarian violence, conflicts between states, and regional tensions have worked against initiatives for greater cooperation and trade. Perhaps most important, the region must deal with the conflict between modernity and more fundamentalist interpretations of Islam. One thing is certain: Future cultural change will be guided by complex responses to Western influences.

D

7.7 How likely is it that the cultural and religious divisions that run through Iraq, including Baghdad's varied neighborhoods, will be healed in the next 5–10 years? Why or why not?

7.8 Work with other students in the class to organize a debate on whether a renewed oil boom in the Iraqi economy might spur *greater* or *reduced* levels of sectarian violence within the country.

Economic and Social Development

Abundant reserves of oil and natural gas, coupled with the global economy's continuing reliance on fossil fuels, ensure that the region will remain prominent in world petroleum markets. Also likely are moves toward economic diversification and integration, which may gradually draw the region closer to Europe and other participants in the global economy.

E

7.9 What are likely to be the chief drivers of economic growth in settings such as Istanbul, Turkey, in the next 10–20 years?

7.10 Write an essay comparing and contrasting the challenges of producing sustained economic growth in Turkey and Saudi Arabia between 2015 and 2030.

Authors' Blogs

Scan to visit the author's blog for chapter updates

www.gad4blog.wordpress.com

Scan to visit the GeoCurrents blog

www.geocurrents.info

8 Europe

Physical Geography and Environmental Issues

Europe is a diverse region that ranges from the Mediterranean subtropics to the northern arctic lands. It is also one of the world's "greenest" regions, with strong regulations about recycling, pollution, and renewable energy.

Population and Settlement

Generally, Europe is a region of low natural growth and high rates of mobility and in-migration. Both are mixed blessings that resolve labor shortages in good times, but burden social welfare systems during economic downturns.

Alpine Europe. The picturesque village of Leogang, Austria, spreads out on the sunny side of the valley under Wildenkarkogel peak. For centuries, Leogang villagers eked out a subsistence from forest and dairy products. Then, in the late 20th century, tourists discovered this remote landscape, coming to ski in the winter and hike during the summer. Consequently, today's landscape is comprised of old farmhouses and new hotels, livestock pens and ski lifts, struggling farmers and rich tourists.

Cultural Coherence and Diversity

Europe has a long history of cultural tensions linked to internal differences in language and religion; however, today these tensions are primarily connected with immigration from other world regions.

Geopolitical Framework

Two world wars and a lengthy Cold War (1945–1989) divided 20th-century Europe into warring camps. Today, although largely an integrated and peaceful region, geopolitical tensions linked to micro-nationalism still persist.

Economic and Social Development

For half a century, the European Union (EU) has worked successfully to integrate Europe's diverse economies and political systems, making Europe into a global superpower. Today, however, serious internal economic and social development issues challenge the European region.

Europe is one of the most diverse regions in the world, encompassing a wide assortment of people and places in an area considerably smaller than North America. More than half a billion people reside in this region, living in 42 different countries that range in size from France, Spain, and Germany to microstates such as Andorra and Monaco (Figure 8.1).

The region's remarkable diversity produces a geographic mosaic of languages, religions, and landscapes. Commonly, in one day's journey you can find yourself hearing two or three languages, crossing several political borders, and possibly changing money as you travel about Europe.

Although visitors may revel in Europe's cultural and geographical diversity, these regional differences are also responsible for Europe's troubled past. In the 20th century alone, Europe was the principal battleground for two world wars, followed by the 45-year **Cold War** (1945–1990), which divided the continent (and the world) into two hostile and highly armed camps, pitting Europe and the United States against the former Soviet Union (now Russia) and its communist allies.

Europe is attempting to set aside nationalistic differences and, instead, work toward regional economic, political, and cultural integration through the European Union (EU).

Today, however, Europe is attempting to set aside nationalistic differences and, instead, work toward regional economic, political, and cultural integration through the **European Union (EU).** This supranational organization is made up of 28 countries, anchored by the larger western European states of Germany, France, Italy, and the United Kingdom (UK), and including smaller European countries from the Mediterranean to Scandinavia, as well as former Cold War Soviet satellite countries in the east. Although controversial, with some countries resisting membership, the EU has unquestionably emerged as an unparalleled regional organization that has largely eroded historical animosities (Figure 8.2).

Physical Geography and Environmental Issues: Human Transformation of a Diverse Landscape

Despite Europe's small size, its environmental diversity is extraordinary. A startling array of landscapes is found within its borders, from the arctic tundra of northern Scandinavia to the semiarid hillsides of the Mediterranean islands, with explosive volcanoes in southern Italy and glaciated seacoasts in Norway and Iceland.

Four factors explain this environmental diversity:

- The complex geology of this western extension of the Eurasian land mass has the newest, as well as the oldest, landscapes in the world.
- Europe's latitudinal extent, from the Arctic to the Mediterranean subtropics, affects climate, vegetation, and hydrology (Figure 8.3).
- Europe's high latitude is modified by the moderating influences of the Atlantic Ocean and its Gulf Stream, as well as the Baltic, Mediterranean, and Black seas.
- The long history of human settlement, spanning thousands of years, has transformed and modified Europe's landscapes in fundamental and diverse ways.

Landform Regions

Europe can be organized into four general topographic and landform regions: the European Lowland, forming an arc from southern France to the northeast plains of Poland, but also including southeastern England; the Alpine mountain system, extending from the Pyrenees in the west to the Balkan Mountains of southeastern Europe; the Central Uplands, positioned between the Alps and the European Lowland; and the Western Highlands, which include mountains in Spain and portions of the British Isles and the highlands of Scandinavia. Iceland, unquestionably a part of Europe, yet lying 900 miles (1500 km) west of Norway, has its own unique landforms, straddling as it does two different tectonic plates.

The European Lowland This lowland, also known as the North European Plain, is the unquestioned economic focus of western Europe, with its high population density, intensive agriculture, large cities, and major industrial regions.

LEARNING OBJECTIVES

After reading this chapter you should be able to:

- Describe, in general terms, the topography, climate, and hydrology of Europe.
- Identify the major environmental issues in Europe, as well as the pathways taken to resolve those problems.
- Provide examples of countries with different rates of natural growth.
- Describe the patterns of internal migration within Europe, as well as the geography of foreign migration to the region.
- Describe the major languages and religions of Europe.
- Summarize how the map of European states has changed in the last 100 years.
- Explain how Europe was divided during the Cold War and how it has changed since the Cold War's end in 1990.
- Describe Europe's economic and political integration as driven by the EU.
- Identify the major characteristics of Europe's current economic and social crisis.

▲ **Figure 8.1 Europe** Stretching from Iceland in the Atlantic to the Black Sea, Europe includes 42 countries, ranging in size from the large states, such as France and Germany, to the microstates of Liechtenstein, Andorra, San Marino, and Monaco. Currently, the population of the region is about 531 million.

▲ **Figure 8.2 European Union Controversy** The European Union (EU) was founded in the 1950s to facilitate greater cooperation among European countries. This has been achieved in many ways, yet it has also created controversy as the EU deliberately erodes the power of individual member countries. Here Germans protest the 2012 EU plan to bail out weak Mediterranean economies with subsidies from fiscally stronger EU countries.

Though not completely flat, most of this lowland lies below 500 feet (150 meters) in elevation. In places, however, the lowland is broken by rolling hills, plateaus, and even some uplands (such as in Brittany, France), where elevations exceed 1000 feet (300 meters). Many of Europe's major rivers (the Rhine, Loire, Thames, and Elbe) meander across this lowland and form broad estuaries before emptying into the Atlantic. Several of Europe's great ports, including London, Le Havre, Rotterdam, and Hamburg, are located on these lowland settings.

The Rhine River delta conveniently divides the unglaciated southern European Lowland from the glaciated plains to the north, which were covered by a Pleistocene ice sheet until about 15,000 years ago. Because of this continental glacier, the northern European Lowland, including the Netherlands, Germany, Denmark, and Poland, is far less fertile than the unglaciated portion in Belgium and France (Figure 8.4). Rocky clay materials in Scandinavia were eroded and transported south by glaciers. As the climate warmed and the glaciers retreated, piles of glacial debris were left on the plains of Germany and Poland. Elsewhere in the north, glacial meltwater created infertile outwash plains that have limited agricultural potential.

The Alpine Mountain System The Alpine mountain system forms the topographic spine of Europe and consists of a series of mountains running west to east from the Atlantic to the Black Sea and the southeastern Mediterranean. These mountain ranges carry distinct regional names, such as the Pyrenees, Alps, Apennines, Carpathians, Dinaric Alps, and Balkan Mountains, but they share geologic traits. All were created more recently (about 20 million years ago) than other upland areas of Europe, and all are constructed from a complex arrangement of rock types.

The Pyrenees form the political border between Spain and France and include the microstate of Andorra. This rugged range extends almost 300 miles (480 km) from the Atlantic to the Mediterranean. Within the mountain

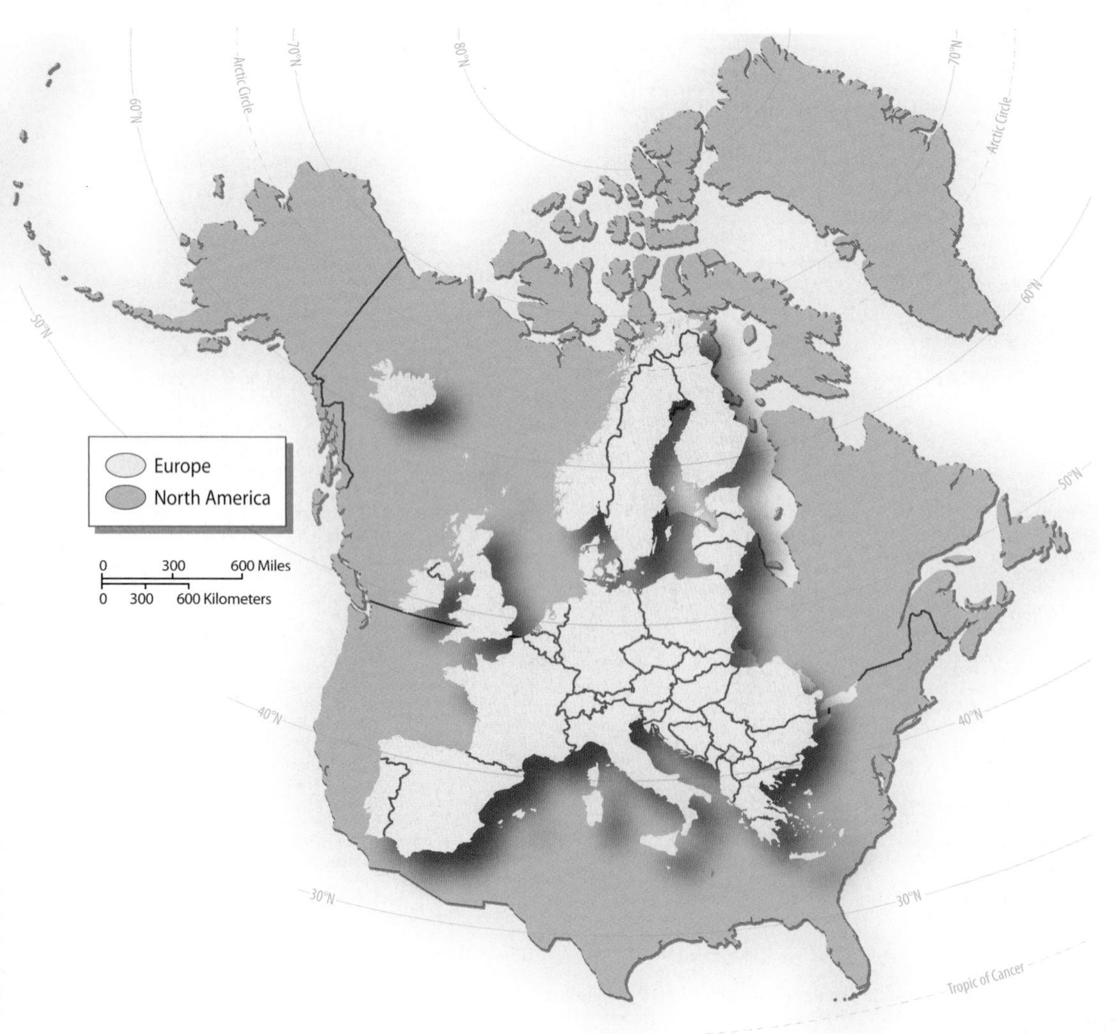

◀ **Figure 8.3 Europe's Size and Northerly Location** Europe is about two-thirds the size of North America, as shown in this cartographic comparison. Another important characteristic is the northerly location of the region, which affects its climate, vegetation, and agriculture. As depicted, much of Europe lies at the same latitude as Canada; even the Mediterranean lands are farther north than the United States–Mexico border. **Q: What parts of Europe are at the same latitude as your location in North America?**

▲ **Figure 8.4 The European Lowland** Also known as the North European Plain, this large lowland extends from southwestern France to the plains of northern Germany and into Poland. Numerous rivers drain interior Europe by crossing this region, giving rise to large port cities along the coastline. This photo is of the Normandy region of western France. The small island is Mont Saint-Michel, a historic fortified monastery.

range, glaciated peaks reaching to 11,000 feet (3350 meters) alternate with broad glacier-carved valleys. The Pyrenees are home to the Basque people in their western reaches, as well as to distinctive Catalan-speaking minorities in the east. Both groups have strong separatist traditions.

The centerpiece of Europe's geologic system is the prototypical mountain range, the Alps, running more than 500 miles (800 km) from France to eastern Austria. These impressive mountains are highest in the west, rising to more than 15,000 feet (4600 meters) in Mt. Blanc on the French-Italian border. In Austria, to the east, the Alps are much more subdued, and few peaks exceed 10,000 feet (3000 meters). Though easily crossed today by car or train through a system of long tunnels and valley-spanning bridges, these mountains have historically formed an important cultural divide between the Mediterranean lands to the south and central and western Europe in the north.

The Apennine Mountains are located south of the Alps; the two ranges, however, are physically connected by the hilly coastline of the French and Italian Riviera. Forming the mountainous spine of Italy, the Apennines are lower and lack the spectacular glaciated peaks and valleys of the true Alps. But farther to the south, the Apennines take on their own distinctive character with the explosive volcanoes of Mt. Vesuvius (just over 4000 feet, or 1200 meters) outside Naples and the much higher Mt. Etna (almost 11,000 feet, or 3350 meters) off Italy's toe on the island of Sicily. These two active volcanoes resulted from the meeting between the African and Eurasian tectonic plates and share geologic traits with faraway Iceland, on the tectonic boundary of the Eurasian and North American plates.

To the east, the Carpathian Mountains define the limits of the Alpine system in eastern Europe. They are a plow-shaped upland area that extends from eastern Austria to the Iron Gate gorge, which is a narrow passage for Danube River traffic where the borders of Romania and Serbia intersect. Although about the same length as the main Alpine chain, the Carpathians are not nearly as high. The highest summits in Slovakia and southern Poland are less than 9000 feet (2800 meters).

Central Uplands In western Europe, a much older highland region occupies an arc between the Alps and the European Lowland in France and Germany. These mountains are much lower in elevation than the Alpine system, with their highest peaks at 6000 feet (1800 meters). Formed about 100 million years ago, much of this upland region is characterized by rolling landscapes about 3000 feet (1000 meters) above sea level.

These uplands are important to western Europe because they contain the raw materials for Europe's industrial areas. In Germany and France, for example, they have provided the iron and coal central to each country's steel industry. To the east, mineral resources from the Bohemian highlands have also fueled major industrial areas in Germany, Poland, and the Czech Republic.

Western Highlands The Western Highlands define the western edge of the European subcontinent, extending from Portugal in the south, through portions of the British Isles in the northwest, to the highland backbone of Norway, Sweden, and Finland in the far north. These are Europe's oldest mountains, formed about 300 million years ago.

As with other upland areas that traverse many separate countries, specific names for these mountains differ from country to country. A portion of the Western Highlands forms the highland spine of England, Wales, and Scotland, where picturesque glaciated landscapes are found at modest elevations of 4000 feet (1200 meters) or less. These U-shaped glaciated valleys also appear in Norway's uplands, where they produce a spectacular coastline of **fjords**, or flooded valley inlets, similar to the coastlines of Alaska and New Zealand (Figure 8.5).

Though less elevated, the Fenno-Scandian Shield of Sweden and Finland is noteworthy because it is made up of some of the oldest rock formations in the world, dated conservatively at 600 million years. This landscape was

▼ **Figure 8.5 Fjord in Norway**
During the Pleistocene epoch, continental ice sheets and glaciers carved deep U-shaped valleys along what is now Norway's coastline. As the ice sheets melted and sea level rose, these valleys were flooded by Atlantic waters, creating spectacular fjords. Many fjord settlements are accessible only by boat, linked to the outside world by Norway's extensive ferry system.

eroded to bedrock by Pleistocene glaciers and, because of the cold climate and sparse vegetation, still has extremely thin soils that severely limit agricultural activity. Similar to other heavily glaciated areas, like the Canadian Shield of North America, numerous small lakes dot the countryside, giving evidence of the impressive erosional power of ice sheets.

Geologically, the far western edge of Europe is found in Iceland, which, as mentioned, is divided by the tectonic border of the Eurasian and North American plates. Like other tectonic boundaries, Iceland has many active volcanoes that occasionally spew ash into the atmosphere, causing serious problems for the airline traffic between Europe and North America.

Europe's Climate

Three principal climates characterize Europe (Figure 8.6). Along the Atlantic coast, a moderate and moist maritime climate dominates, modified by oceanic influences (see the climographs for Reykjavik, Belfast, London, and Paris). Farther inland, continental climates prevail, with hotter summers and colder winters (see the climographs for Oslo, Riga, Berlin, and Budapest). Finally, a dry-summer Mediterranean climate is found in southern Europe, from Spain to Greece (see the climographs for Barcelona, Rome, and Athens).

One of the most important climate controls is that of the Atlantic Ocean. Even though most of Europe is at a relatively high latitude (London, England, for example, is slightly farther north than Vancouver, British Columbia), the mild North Atlantic current, which is a continuation of the warm Atlantic Gulf Stream, moderates coastal temperatures from Iceland and Norway south to Portugal and inland to the western reaches of Germany. As a result, this maritime influence gives Europe a climate 5–10°F (3–6°C) warmer than regions at comparable latitudes, but lacking the effects of this warm ocean current. In the marine west coast climate region, no winter months average below freezing, even though cold rain, sleet, and an occasional blizzard are common winter events. Summers are often cloudy and overcast, with frequent drizzle and rain as moisture flows in from the ocean. Ireland, called the Emerald Isle, is an appropriate image for this maritime climate.

With increasing distance from the ocean (or where a mountain chain limits the maritime influence, as in Scandinavia), landmass heating and cooling becomes a strong climatic control, producing hotter summers and colder winters. Indeed, all continental climates average at least one month below freezing during the winter.

In Europe, the transition between maritime and continental climates takes place close to the Rhine River border of France and Germany. Farther north, although Sweden and other nearby countries are close to the moderating influence of the Baltic Sea, their higher latitude coupled with the blocking effect of the Norwegian mountains produces cold winter temperatures characteristic of true continental climates.

The Mediterranean climate is characterized by a distinct dry season during the summer, which results from the warm-season expansion of the Atlantic (or Azores) high-pressure area. This high pressure is produced by the global circulation of air warmed in the equatorial tropics. The warm air subsides (descends) between latitudes of 30 and 40 degrees, thus inhibiting summer rainfall. This same phenomenon also produces the Mediterranean climates of California, western Australia, parts of South Africa, and Chile. These rainless summers may attract tourists from northern Europe, but the seasonal drought can be problematic for agriculture. It is no coincidence that traditional Mediterranean cultures, such as the Arab, Moorish, Greek, and Roman, have been major innovators of irrigation technology.

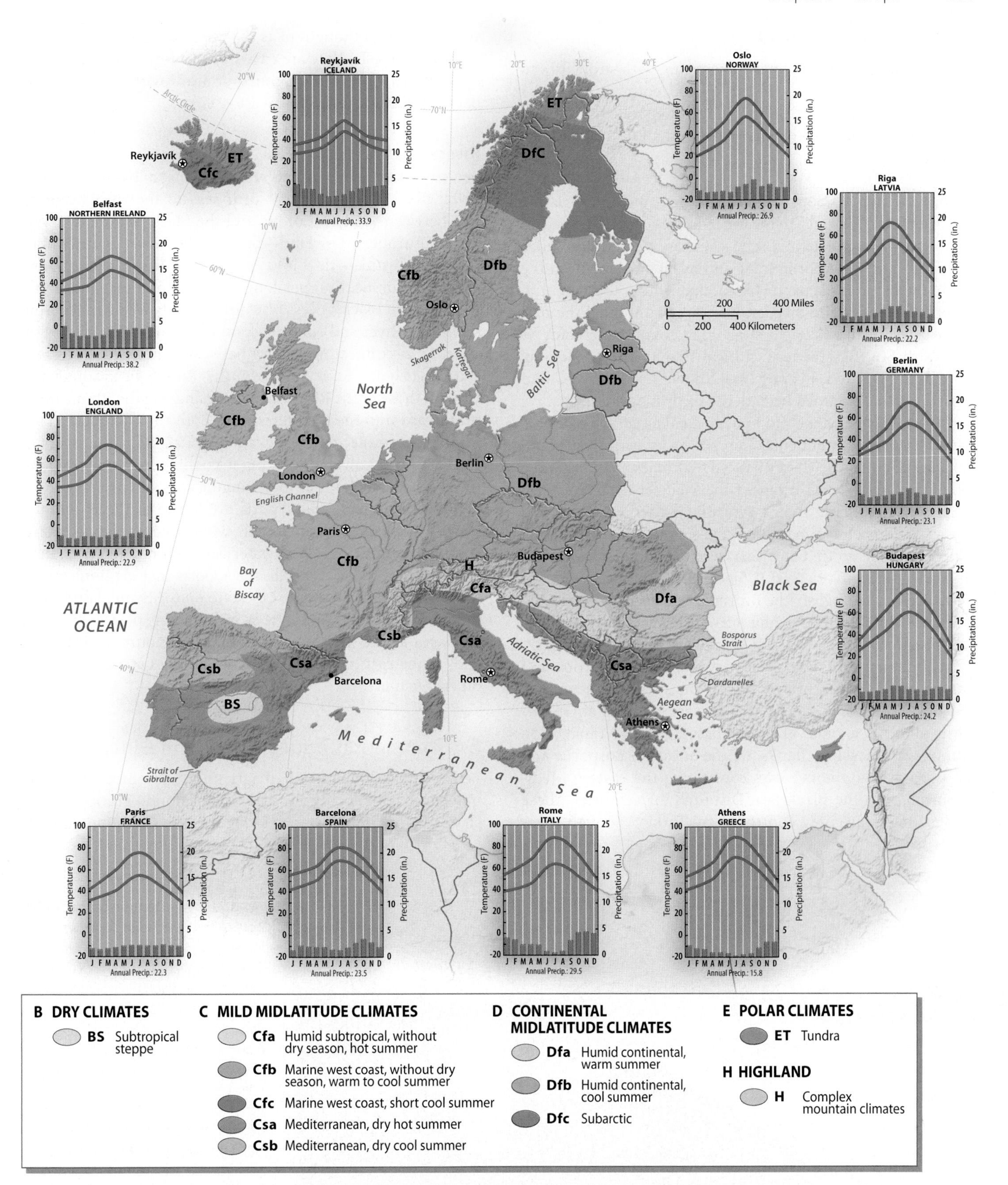

▲ **Figure 8.6 Climate of Europe** Three major climate zones dominate Europe. Close to the Atlantic Ocean, the marine west coast climate has cool seasons and steady rainfall throughout the year. Farther inland, continental climates have at least one month averaging below freezing, as well as hot summers, with a precipitation maximum occurring during the warm season. Southern Europe has a dry-summer Mediterranean climate. **Q: Where in Europe are there climates similar to the climate where you live in North America?**

Seas, Rivers, and Ports

In many ways, Europe is a maritime region with strong ties to its surrounding seas. Even landlocked countries such as Austria, Hungary, Serbia, and the Czech Republic have access to the ocean through an extensive network of navigable rivers and canals.

Europe's Ring of Seas Four major seas and the Atlantic Ocean encircle Europe; these water bodies are connected to each other through narrow straits with strategic importance for controlling waterborne trade and naval movement. In the north, the Baltic Sea separates Scandinavia from north-central Europe. Denmark and Sweden have long controlled the narrow Skagerrak and Kattegat straits that connect the Baltic to the North Sea. Besides its historical role as a major fishing ground, the North Sea is now well known for its rich oil and natural gas fields, mined from deep-sea drilling platforms.

The English Channel (in French, *La Manche*) separates the British Isles from continental Europe. At its narrowest point, the Dover Straits, the channel is only 20 miles (32 km) wide. Although England has regarded the channel as a protective moat, it has primarily been a symbolic barrier, for it deterred neither the French Normans from the continent nor the Viking raiders from the north. Only Nazi Germany found it a formidable barrier during World War II. Since 1993, and after decades of resistance by the English, the British Isles have been connected to France through the 31-mile (50-km) Eurotunnel, with its high-speed rail system carrying passengers, autos, and freight.

Gibraltar guards the narrow straits between Africa and Europe at the western entrance to the Mediterranean Sea, and Britain's stewardship of this passage remains an enduring symbol of a once great sea-based empire. Finally, on Europe's southeastern flanks are the Straits of Bosporus and the Dardanelles, the narrows connecting the eastern Mediterranean with the Black Sea. Disputed for centuries, these pivotal waters are now controlled by Turkey. Though these straits are often thought of as the physical boundary between Europe and Asia, they are easily bridged in several places to facilitate truck and train transportation within Turkey and between Europe and Southwest Asia.

Rivers and Ports Europe is a region of navigable rivers, connected by a system of canals and locks that allow inland barge travel from the Baltic and North seas to the Mediterranean and between western Europe and the Black Sea. Many rivers on the European Lowland—namely, the Loire, Seine, Rhine, Elbe, and Vistula—flow into Atlantic and Baltic waters. However, the Danube, Europe's longest river, flows east and south, rising in the Black Forest of Germany only a few miles from the Rhine River and running southeastward to the Black Sea, offering a connecting artery between central and eastern Europe (Figure 8.7). Similarly, the Rhône headwaters rise close to those of the Rhine in Switzerland, yet the Rhône flows southward into the Mediterranean. Both the Danube and the Rhône are connected by locks and canals with the rivers of the European Lowland, making it possible for barge traffic to travel between all of Europe's fringing seas and oceans.

As mentioned, major ports are found at the mouths of most western European rivers, where they serve as transshipment points for inland waterways and rail and truck networks. From south to north, these ports include Bordeaux at the mouth of the Garonne, Le Havre on the Seine, London on the Thames, Rotterdam (the world's largest port in terms of tonnage) at the mouth of the Rhine, Hamburg on the Elbe, and, to the east in Poland, Szczecin on the Oder and Gdansk on the Vistula.

Environmental Issues: Local and Global

Because of its long history of agriculture, resource extraction, industrial manufacturing, and urbanization, Europe has its share of environmental issues (Figure 8.8). Compounding the

◄ **Figure 8.7 Danube Barge Traffic** In 1992, the Danube River, Europe's longest, was connected by canal to the Rhine River, allowing commercial barge traffic to move throughout Europe and between the North and the Black seas. Because inland water traffic (IWT) is reportedly 80 percent cheaper than trucking and because industrialization is increasing in eastern Europe, barge traffic has grown considerably in the last decade. In response, several countries along the Danube are formulating plans to protect the Danube environment from increased IWT development. Here a tug pushes a barge on the Danube in Serbia.

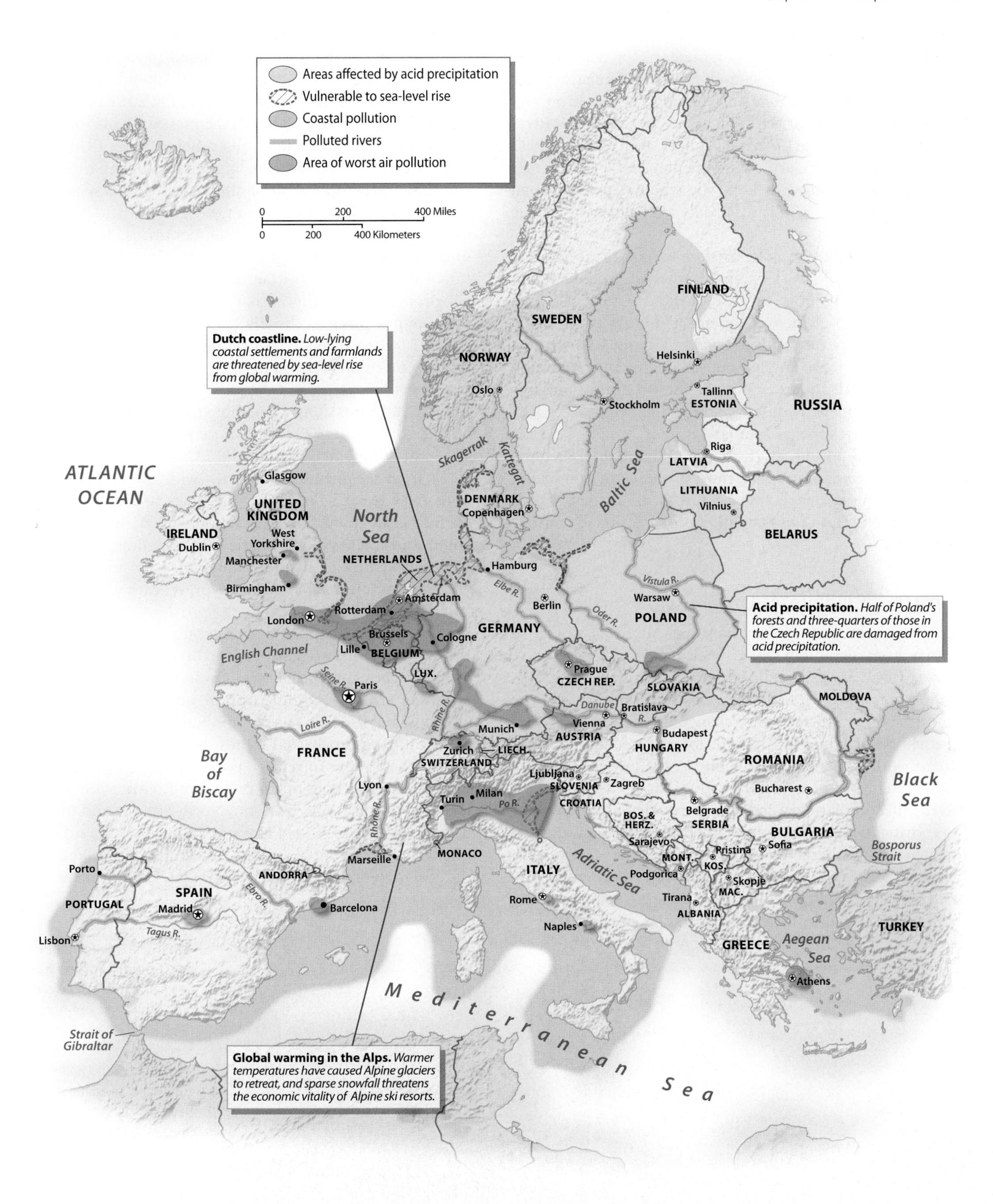

▲ **Figure 8.8 Environmental Issues in Europe** Although western Europe has worked energetically over the past 50 years to solve environmental problems such as air and water pollution, eastern Europe lags behind because environmental protection was not a high priority during the postwar communist period. Current efforts, however, show great promise.

situation is the fact that pollution rarely stays within political boundaries. Air pollution from England, for example, creates serious acid rain problems in Sweden. Similarly, water pollution from Swiss factories on the upper Rhine River creates major problems for the Netherlands, where Rhine River water is used for urban drinking supplies. As a result of these numerous trans-boundary environmental problems, the EU has taken the lead in addressing the region's environmental issues, and Europe is today probably the "greenest" of the major world regions.

Until recently, the countries of eastern Europe were plagued by far more serious environmental problems than their western neighbors because of their history as Soviet satellite countries, where economic planning emphasized short-term industrial output over environmental protection. In Poland, for example, industrial effluents reportedly had wiped out all aquatic life in 90 percent of the country's rivers, with damage from air pollution affecting over half of the country's forests. Similar legacies from the Soviet period were reported in the Czech Republic, Romania, and Bulgaria. Today, however, most of these environmental issues in eastern Europe have been resolved through funding from the EU and the strengthening of national environmental laws.

Climate Change in Europe

The fingerprints of global warming are everywhere in Europe, from dwindling sea ice, melting glaciers, and sparse snow cover in arctic Scandinavia to more frequent droughts in the water-starved Mediterranean area. Furthermore, the projections for future climate change are ominous: World-class ski resorts in the Alps are forecast to have warmer winters with less snow pack, while in the lowlands, higher summer temperatures could produce more frequent heat waves, affecting farmers and urban dwellers alike. In addition, rising sea levels from melting polar ice sheets will threaten the Netherlands, where much of the population lives in diked lands that are actually below sea level (Figure 8.9). Because of these threats, Europe has taken a strong stand in addressing climate change and, as a result, has numerous policies and programs to reduce greenhouse gas (GHG) emissions.

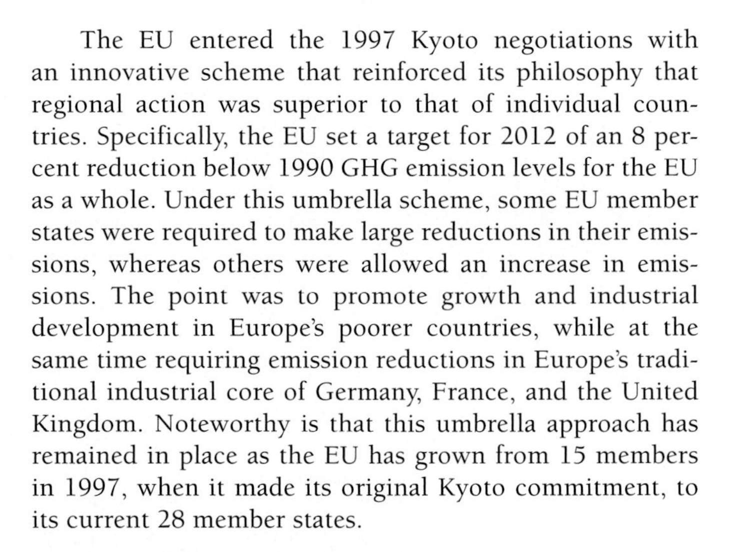

The EU entered the 1997 Kyoto negotiations with an innovative scheme that reinforced its philosophy that regional action was superior to that of individual countries. Specifically, the EU set a target for 2012 of an 8 percent reduction below 1990 GHG emission levels for the EU as a whole. Under this umbrella scheme, some EU member states were required to make large reductions in their emissions, whereas others were allowed an increase in emissions. The point was to promote growth and industrial development in Europe's poorer countries, while at the same time requiring emission reductions in Europe's traditional industrial core of Germany, France, and the United Kingdom. Noteworthy is that this umbrella approach has remained in place as the EU has grown from 15 members in 1997, when it made its original Kyoto commitment, to its current 28 member states.

Energy and Emissions Greenhouse gas emissions are closely linked to a country's energy mix, as well as to its population size. Not surprisingly, within the EU, the highest emissions come from the member countries with the largest populations, which burn the most fossil fuels. Germany, the largest European country by population with 82 million people, emits almost 800 million metric tons of pollutants each year. It is followed by the United Kingdom, Italy, and France, all with populations of about 60 million, yet each emitting about half as much as Germany.

As for Europe's fuel mix, generally speaking the region runs on fossil fuels—coal, gas, and oil (Table 8.1). Although Europe's early industrialization was based on coal, those resources are now running thin in western Europe, and, in fact, much of the EU's recent emission reductions have come from shutting down coal mines in England and Germany during the 1990s. To replace coal, Europe relies heavily on imported gas and oil, much of it from Russia, with the only local supplies coming from the North Sea gas and oil wells developed by the UK and Norway. Ironically, hydraulic fracturing (more commonly called "fracking") has driven down the cost of natural gas in the United States, resulting in many

◀ **Figure 8.9 Protecting Low Country Europe from Sea Level Rise** Conceived mtore than half a century ago, the Dutch Deltaworks was originally built to keep ocean storm surges and Rhine River flooding from southwestern Netherlands. But now, given the forecasts for sea level rise from global warming, the Deltaworks must be re-engineered and made higher to protect the 50 percent of the Netherlands that lies either below or within just 3.3 feet (1 meter) of the current sea level.

Table 8.1 European Energy Components

Country	Electricity from Fossil Fuels	Nuclear	Hydroelectric	Other Renewables
Germany	55%	23% (all nuclear shut down by 2020)	3%	13%
France	20.5%	53%	17.6%	5.3%
United Kingdom	75.4%	12.3%	1.9%	7.3%
Italy	65%	0%	18%	15.8%
Spain	48.5%	7.6%	13.7%	24.4%
Sweden	12.5%	25%	46.9%	15.2%
Norway	2.6%	0%	91.1%	2%
Iceland	4.7%	0%	72.9%	22.4%
Bulgaria	46.4%	20.3%	22.2%	1.9%
Serbia	66%	0%	26.6%	0%

Source: Data from *CIA World Factbook 2013.*

North American power plants switching from coal to gas. As a result, the price of coal has now dropped to the point where Europe is importing coal from North America because it is cheaper than imported natural gas from Russia. This strategy is injurious to the EU's emission reduction plans because coal emits far more CO_2 than does natural gas.

However, the increased use of coal may be only a short-term measure in Germany to compensate for the country's recent decision to shut down all its nuclear power plants by the year 2020 because of safety issues. Until this phaseout began, about a quarter of Germany's energy was generated by nuclear power. France apparently doesn't share its EU neighbor's concerns, since nuclear power generates fully 53 percent of that country's power.

Complementing EU emission reduction goals is a policy to increase the region's renewable energy resources to the point that the EU as a whole will be generating 20 percent of its power from hydropower, wind, solar, and biofuels by the year 2020. This goal seems well within reach given the existing hydropower facilities in the Alpine and Scandinavian countries, coupled with the expansion of wind and solar power in the last few years throughout the EU. Currently, Europe has over 25,000 wind farms, with this amount expected to double by 2015 (Figure 8.10; also see *Working Toward Sustainability: Germany's Energy Transformation*).

▼ **Figure 8.10 Wind Power in Northern Europe** Not only is Europe trying to reduce its carbon dioxide emissions, but also it is a world leader in generating renewable energy from wind, sun, and biofuels. This large wind farm is in Denmark.

Working Toward **Sustainability**

Germany's Energy Transformation

Germany is trying hard to become the world's first major industrial economy to be powered primarily by renewable energy, but the pathway forward is an uncertain one. In 2000, the country passed landmark legislation enabling a complete transformation of its energy industry so that solar, wind, and biomass energy would replace fossil fuel and nuclear power plants. The goals of this energy transformation are threefold: to reduce Germany's dependence on imported oil and gas; to reduce its global warming emissions; and to become a world leader in renewable energy technology. The timetable is to reach these goals by 2020—or 2030 at the latest.

If successful, Germany will provide the world with a workable model for sustainable energy. If it fails—and observers agree that serious technological and economic challenges lie ahead—the world's search for a clean-energy future could be irrevocably damaged.

Renewable Energy Incentives The renewable energy program was jumpstarted with an incentive to expand solar and wind power technology, requiring Germany's four major utility companies to buy all electricity generated by solar and wind generators. As a result, solar panels blossomed everywhere—on homes, on barns, in pastures, along highways, over parking lots—anywhere with space open to the southern sky, since there was money to be made selling solar energy to power utilities (**Figure 8.1.1**). Wind turbines, which are more expensive to install than rooftop solar panels, appeared on windy hilltops and along breezy coastlines, financed by companies, village cooperatives, and investment firms.

The early results were spectacular, with renewable energy currently making up over 20 percent of Germany's total energy picture. Even though solar panels are ubiquitous across the country's landscape, wind power currently produces most of the country's renewable energy (40 percent, contrasted to 8 percent for solar). Biomass fuel, often in the form of wood pellets for heating systems, is also a major contributor to the renewable energy mix.

Daunting Issues However, major challenges lie ahead. Solar and wind energy generates massive power surges only when the sun shines and the wind blows; thus, storage of this surplus is a major technological issue that needs to be resolved. Further, the country's power network needs to be completely rebuilt to efficiently distribute these new energy sources, which is an unwelcome and costly matter as the German economy stagnates. Until this new power grid is built, energy-intensive industries will continue to face service interruptions and increasing costs. Many fear that higher production costs may drive these industries out of Germany and into countries with a more reliable (and less expensive) energy infrastructure.

Accelerating the need for these changes is Germany's recent decision to phase out all nuclear power plants in the next decade because of safety concerns. Closing these plants, which formerly produced about 20 percent of Germany's power, has created a power shortfall that must be made up by expanding gas- and coal-fired utility plants. This will put more—not less—GHG emissions into the atmosphere. Despite these daunting issues, Germany remains committed to its plan with widespread popular and political support for a sustainable and clean energy future.

1. What other European countries draw heavily upon renewable energy?
2. Which European countries have the least amount of renewable energy? Why is that?

Google Earth Virtual Tour Video

http://goo.gl/U1nfJ7

◄ **Figure 8.1.1 Solar Panels in Germany** These solar panels covering the roof of a house near Tübingen, Germany, illustrate how homeowners are taking advantage of the country's feed-in tariff (FiT) law, which requires the four large utility companies to buy at a retail price all renewable energy produced. As a result, homeowners are not just utilizing solar energy for their own domestic uses, but also making a profit from the energy they generate.

Interesting to note is that on a per capita basis, Europeans generate only one-half to one-third the amount of CO_2 emissions as do Americans (5 to 9 per person in Europe compared to 18 tons per person in the United States). Higher fuel standards for cars and trucks, along with higher residential heating efficiency, generally explain this significant difference in per capita emissions.

As for emission reductions, despite its good intentions, the EU initially appeared to be falling far short of its 2012 target of an 8 percent reduction. Instead, in 2007 the emissions reduction was closer to just 1 percent. The main reason, the EU said, was the unanticipated growth in truck transport over the past decade, coupled with higher than anticipated emissions from industrial development in the newer EU countries of eastern Europe. However, in 2008 emissions dropped precipitously as economic activity throughout Europe slowed because of the EU financial crisis. As a result, the EU easily reached its 2012 Kyoto goal. More recently the EU agreed to a new higher target of a 20 percent emission reduction over its 1990 levels by the year 2020.

The EU's Emission Trading Scheme As part of its Kyoto emission reduction strategy, in 2005 the EU inaugurated the world's largest carbon trading scheme. Under this plan, specific yearly emission caps were set for the EU's largest GHG emitters. If these emitters exceeded those caps, they were to either purchase carbon emission equivalences from a factory or power plant below its own cap or, alternatively, buy credits from the EU carbon market. The goal of this cap-and-trade system was to make business more expensive for companies that pollute, while rewarding those staying under their carbon quota.

By early 2013, however, there were clear signs that the carbon trading scheme was not working well since the market price of permits had fallen so low that polluters were essentially ignoring the carbon market. An attempt to raise the price of carbon permits by restricting the number allocated failed to win necessary political support, due to fear that higher carbon permit prices would inhibit the EU's struggling economies. Although there are plans to revise the EU carbon trading scheme sometime soon, given the current economic situation, any changes that make carbon emissions more expensive are unlikely to gain the necessary political support.

REVIEW

8.1 Name and locate the major lowland and mountainous areas of Europe.

8.2 What are the three major climate regions of Europe?

8.3 How would inland barge traffic get from the mouth of the Rhine to the delta of the Danube?

8.4 Describe the major patterns of Europe's energy geography and how it is linked to the EU's agenda of global-warming emission reductions.

Population and Settlement: Slow Growth, Rapid Migration, and Widespread Urbanism

The major themes of Europe's population and settlement geography are its very low rates of natural growth; the complicated and often problematic patterns of internal and international migration; and a very high level of urbanization, particularly in the traditional industrial core area of western Europe. This densely settled core area includes England, northern France, Belgium, the Netherlands, Germany, and northern Italy (Figure 8.11).

Low (or No) Natural Growth

Probably the most striking characteristic of Europe's demography is the lack of natural growth (Table 8.2). That is, in most European countries the death rate exceeds the birth rate, resulting in either very slow or even no natural growth. Several countries, notably Germany and Italy, are actually experiencing negative natural growth, so that were it not for in-migration, they would actually decrease in size considerably over the next decades.

European population, like that of Japan and even the United States, is characterized by the fifth or postindustrial stage of the demographic transition (discussed in Chapter 1), where fertility falls below replacement levels. Germany is a good example, with its negative rate of natural increase (RNI) of –0.2. Similar patterns characterize England, France, and Italy, all of which have RNIs below population replacement levels. As a result, political concerns in these countries include future labor shortages, smaller internal markets, and declining tax revenues, which are needed to support social services (such as retirement pensions) for their aging populations (Figure 8.12, page 366).

Pro-Growth Policies Because many European governments consider low or no population growth an economic and social liability, programs and policies have been developed to promote population growth. These range from attempts to ban abortion and the sale of contraceptives to what are commonly called "family-friendly" policies. In many countries, family-friendly, pro-growth policies include full-pay maternity and paternity leaves for both parents, guarantees of continued employment once these leaves conclude, extensive child-care facilities for working parents, outright cash subsidies for having children, and free or low-cost public education and job training for their offspring. However, even with these family-friendly policies, no European country has a (TFR) above the replacement level of 2.1. It is only because of in-migration that any population growth takes place.

▲ **Figure 8.11 Population of Europe** The European region includes more than 531 million people, many of them clustered in large cities in both western and eastern Europe. As can be seen on this map, the most densely populated areas are in England, the Netherlands, Belgium, western Germany, northern France, and south across the Alps to northern Italy. **Q: What best explains the different population densities between eastern and western Europe?**

Table 8.2 Population Indicators

Country	Population (millions) 2013	Population Density (per square kilometer)	Rate of Natural Increase (RNI)	Total Fertility Rate	Percent Urban	Percent <15	Percent >65	Net Migration (Rate per 1000)
Western Europe								
Austria	8.5	101	0.0	1.4	67	14	18	5
Belgium	11.2	366	0.2	1.8	99	17	17	6
France	63.9	116	0.4	2.0	78	19	17	1
Germany	80.6	226	−0.2	1.4	73	13	21	5
Ireland	4.6	65	1.0	2.0	60	22	12	−7
Luxembourg	0.5	210	0.4	1.5	83	17	14	19
Netherlands	16.8	404	0.2	1.7	66	17	16	1
Switzerland	8.1	196	0.2	1.5	74	15	17	8
United Kingdom	64.1	264	0.4	2.0	80	18	16	2
Southern Europe								
Albania	2.8	96	0.5	1.8	54	20	12	−15
Bosnia & Herzegovina	3.8	75	−0.1	1.2	46	17	15	0
Croatia	4.3	75	−0.2	1.5	56	15	18	−1
Cyprus	1.1	123	0.6	1.5	62	17	13	14
Greece	11.1	84	−0.0	1.4	73	14	19	−1
Italy	59.8	199	−0.1	1.4	68	14	21	4
Kosovo	1.8	168	1.0	1.9	38	28	7	−2
Macedonia	2.1	80	0.2	1.5	65	17	12	0
Montenegro	0.6	45	0.3	1.7	64	19	13	−0
Portugal	10.5	114	−0.2	1.3	38	15	19	−4
Serbia	7.1	92	−0.5	1.4	59	14	17	1
Slovenia	2.1	102	0.1	1.5	50	14	17	0
Spain	46.6	92	0.1	1.3	77	15	18	−3
Northern Europe								
Denmark	5.6	130	0.1	1.7	87	17	18	4
Estonia	1.3	28	−0.1	1.6	69	16	18	−5
Finland	5.4	16	0.1	1.8	68	16	19	3
Iceland	0.3	3	0.8	2.0	95	21	13	−1
Latvia	2.0	31	−0.5	1.4	68	14	19	−2
Lithuania	3.0	45	−0.4	1.4	67	15	18	−7
Norway	5.1	13	0.4	1.8	80	18	16	9
Sweden	9.6	21	0.2	1.9	84	17	19	5
Eastern Europe								
Bulgaria	7.3	65	−0.6	1.5	73	14	19	−1
Czech Republic	10.5	133	0.0	1.5	74	15	16	1
Hungary	9.9	106	−0.4	1.3	69	15	17	1
Poland	38.5	123	−0.0	1.3	61	15	14	−0
Romania	21.3	89	−0.3	1.4	55	15	15	−0
Slovakia	5.4	110	0.1	1.3	54	15	13	1
Micro States								
Andorra	0.1	157	0.5	1.2	90	16	13	4
Liechtenstein	0.04	231	0.4	1.4	15	16	14	3
Malta	0.4	1,418	0.3	1.5	100	15	16	3
Monaco	0.04	237,172	0.0	1.4	100	13	24	13
San Marino	0.03	535	0.2	1.5	84	15	18	7
Vatican City	–	–	–	–	–	–	–	–

Source: Population Reference Bureau, *World Population Data Sheet, 2013.*

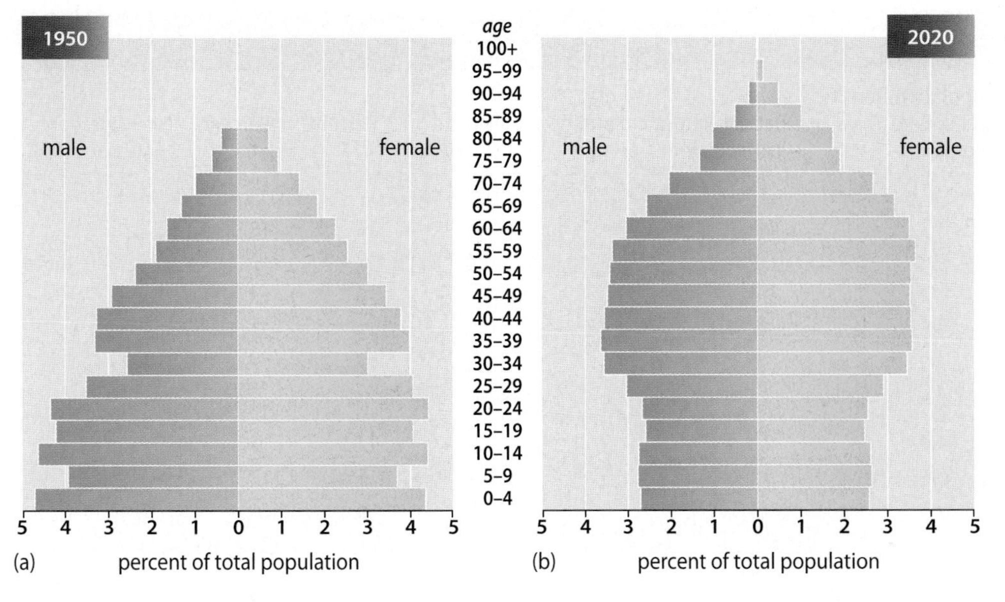

◀ **Figure 8.12 Population Pyramids of Germany** These two population pyramids illustrate the general aging of Europe by focusing on Germany, showing its relatively youthful population in 1950 (a), and contrasting that with Germany's projected population makeup in 2020 (b). A similar pattern would emerge for the other large Europe countries of France, Italy, Spain, and the United Kingdom.

Migration to and Within Europe

Since its origin in 1957, the EU has worked toward the goal of freer movement of both people and goods within its borders, and today residents of the 28 EU countries can pretty much move about as they please to work and study. In fact, these new freedoms have affected not just migration patterns, but also national population growth and loss. During the past decade, for example, some 16,000 Lithuanians (out of a total population of 3.3 million) moved to Ireland to take advantage of that country's booming economy while their home country languished in the economic doldrums. More recently, when the Irish boom burst, half of these EU migrants moved back to Lithuania. Similarly, a mass exodus of young people from Spain and Greece is presently taking place because of the dismal economic situation in their home countries, with England and Germany reportedly the beneficiaries of this internal European migration.

The Schengen Agreement Underlying this new European mobility is a formal political act that erodes Europe's historical national borders, the **Schengen Agreement**, named after the city in Luxembourg where it was signed in 1985. Before Schengen, crossing a European border usually involved showing passports and auto insurance papers, customs inspections, and so on at Europe's numerous border stations. Today, however, there are either no border stations or only cursory formalities when traveling between Schengen countries (Figure 8.13). To older Europeans and visitors who knew Europe before Schengen, it's a remarkable experience to cross a national border without even stopping (see *People on the Move: The Roma, Europe's Largest Ethnic Minority*).

Not all EU members are Schengen members. The United Kingdom, for example, is an EU member, but not a Schengen

▼ **Figure 8.13 Schengen Border** People stroll unhindered by document checks across the Poland–Germany border in the city of Görlitz. Earlier, before Poland became a member of the Schengen Agreement in late 2007, this border was a heavily policed "hard" border between Europe and the non-Schengen countries to the east, where passport and visa checks were rigorously enforced.

People on the Move

The Roma, Europe's Largest Ethnic Minority

The Romani people, also known as *Roma, Sinti, or Kale* (*Gypsies* is considered a pejorative term), are Europe's largest ethnic minority, with an estimated 10–12 million spread across Europe from the UK to the Balkans (**Figure 8.2.1**). Historically, the largest numbers of Roma have lived in eastern Europe; however, since the entry of those countries into the EU, increasing numbers of Romani have migrated to western Europe, with large populations today in Germany, France, and Spain.

However, this has not been a trouble-free migration; more bluntly, in many ways the Roma are the EU's nemesis. Although they embody the EU's agenda of free movement across national borders, the harsh reality is that no country really wants them because of their reputation (real or imagined) for crime, filth, and illiteracy.

A History of Discrimination Romani have endured centuries of discrimination since their arrival in Europe from northwestern India between 800 and 1300 CE. Despite the romantic legends of their music, freewheeling lifestyle, and craftsmanship, the Roma have been treated harshly. In 19th-century Romania, they were actually enslaved; during the 20th century, Nazi Germany targeted them for extinction as part of the Holocaust.

Today stories abound about their mistreatment. Thugs have attacked their camps in Italy and Hungary; Roma children are routinely denied schooling in Slovakia; France and Germany have deported thousands back to Romania as illegal migrants, despite legal questions about whether that is truly their country of origin. The Roma say they have no homeland, that their ethnic identity is not place-bound, but instead comes from their culture of mobility. As noted earlier, the contradiction with the EU's agenda of free movement within Europe is striking.

"Europeanization" of the Roma Problem Because no one country claims the Romani as its own or accepts responsibility for helping them socially and economically, the issue was taken on by the EU in 2005 with its "Decade of Roma Inclusion." With funds from the World Bank, the goal was simple: to improve living conditions for the Roma throughout Europe. This initiative has created education programs for Romani youth, job training for adults, and housing programs for those Roma wishing to settle down with a permanent home. But today, well more than halfway through the decade, the results are discouraging. Often local authorities do not cooperate with the EU for fear temporary Roma encampments may become permanent. Further, the Roma themselves have not always been cooperative: Ambitious youngsters are often held back by their intensely patriarchal and conservative families. Girls are married in their teens, and boys go to work at an early age, rather than attending school. Weary of the hostility they face from the outside world, Roma communities often cut themselves off from society and its laws. As a result, despite the best efforts of the EU, the Roma problem continues with no solution in sight.

▼ **Figure 8.2.1 Roma Family** Europe has about 10-12 million Romani people.

signatory because of the country's concerns about protecting its national borders. Consequently, entering the UK from another European country does involve passing through a customs check.

The Schengen Agreement, however, has become increasingly controversial in the last few years because of the increase in legal and illegal international migration to Europe. Once inside a Schengen country, such as Poland or Germany, an international migrant could move between countries freely just like a European resident or, for that matter, a tourist from the United States. But that fact has now caused some Schengen countries concern because of the large number of illegal migrants entering Europe. To counter this problem, several Schengen countries have discussed new border controls. Denmark, most notably, has reestablished formal border controls all along its southern boundary with Germany, and although these measures were challenged by the EU, Denmark's actions have been recently deemed legally permissible by an EU court.

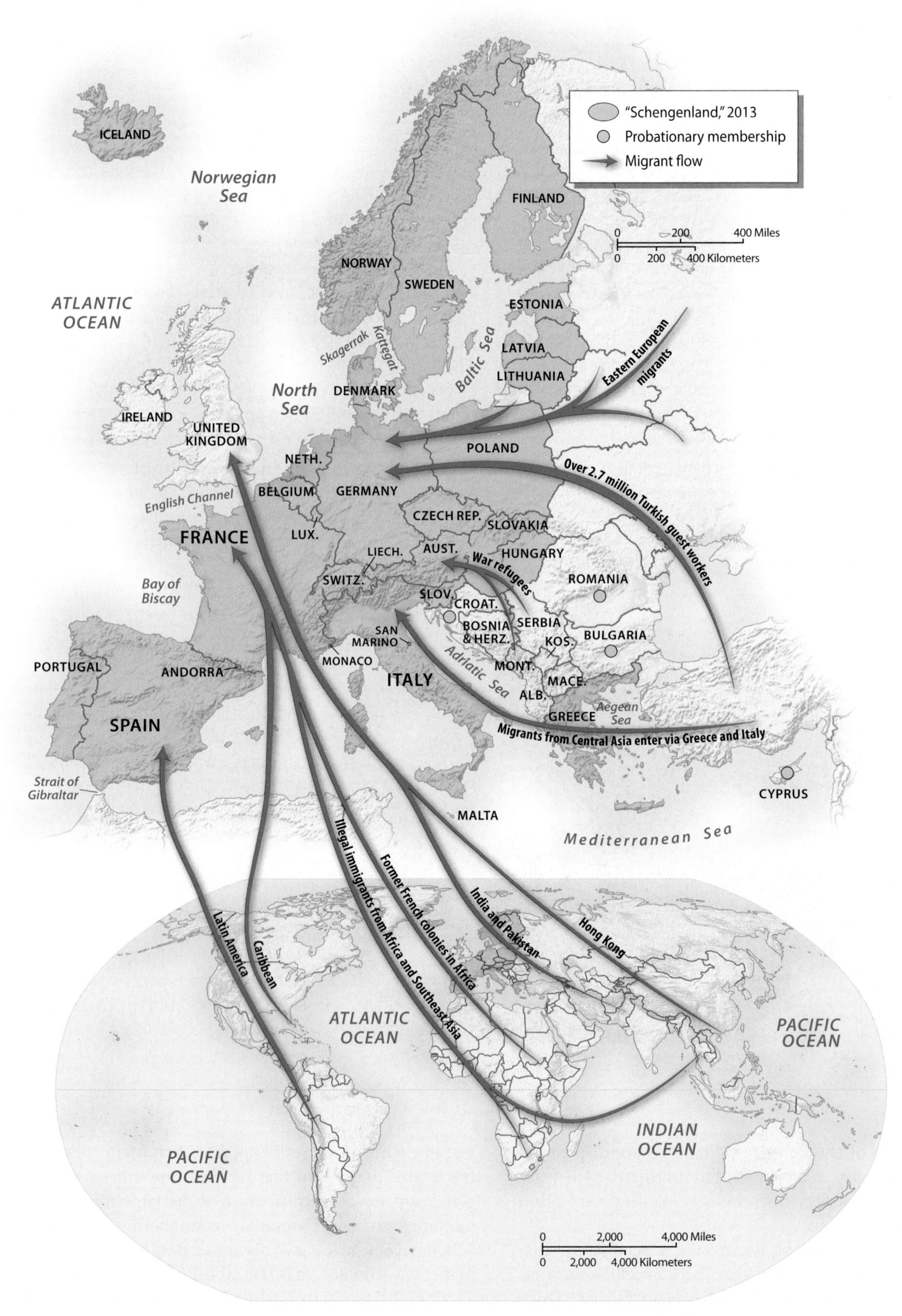

▲ **Figure 8.14 Migration into Europe** The routes and entry points for illegal migrants change frequently as countries such as Greece, Italy, and Spain increasingly police their borders. The porous border between Greece and Turkey, for example, has long been a favorite illegal entry point for migrants from Central Asia but has been hardened recently with high fences, watchtowers, and 24-hour guards.

▲ **Figure 8.15 Urban Landscapes** This aerial view of Grosseto, Italy, shows how the historic medieval city was encircled by the Renaissance–Baroque fortifications built to protect the settlement. Today parks and public buildings are located in place of the former walls and moat.

International Migration to Europe During western Europe's postwar rebuilding and recovery, its countries' economies were booming, and many looked to foreign migrant workers to ease their labor shortages. The former West Germany, for example, drew heavily on workers from Europe's rural and poorer periphery—Italy, the former Yugoslavia, Greece, and even Turkey—to fill industrial, construction, and service jobs. Similarly, France and England opened their doors to migrants from former colonies in Africa and Asia, resulting in thousands of workers arriving from India, Pakistan, Hong Kong, North Africa, and the Caribbean (Figure 8.14).

Later, with the collapse of Soviet border controls in 1990, emigrants from former Soviet satellite countries also poured into western Europe, seeking relief from the economic chaos in Russia and other eastern countries. This flight from the post-1990 economic and political chaos also included refugees from war-torn areas of the former Yugoslavia, particularly from Bosnia and Kosovo.

As a result of these different migration streams, foreigners now comprise about 10 percent of the population in Germany, France, and England, making many European countries ambivalent hosts to emerging multicultural societies. (For comparison, 11.7 percent of the U.S. population is comprised of people classified as foreigners.)

Leaky Borders and "Fortress Europe" With high unemployment and economic stress currently widespread in Europe, the region no longer needs to import workers internationally. In fact, given Europe's weak economy, many European countries fear that the influx of foreign migrants aggravates high domestic unemployment rates, as migrants work in an off-the-books "gray" economy.

As a result, Europe is working hard to stem the flow of illegal migration and also slow the amount of legal migration consisting of colonial residents or relatives of migrants already in Europe. Given the geography of Europe's long borders, however, this is a vexatious task. Both Spain and Italy, for example, have extensive coastal shores accessible to seafaring illegal migrants from Africa and the Near East. Greece has been accused of having a leaky border with Turkey, one that has become a favorite entry point for illegal migrants from Russia, Central Asia, and even South Asia. Bulgaria and Romania have similar problems that have actually prevented them from having full membership in the Schengen Agreement. Instead, they have probationary membership until they improve control over their Black Sea borders.

To help these perimeter countries police their borders, the EU has provided funds to strengthen their borders with border guards and, in some places, with physical border fences to inhibit illegal entry. For those with longer memories, these border fortifications are disturbingly reminiscent of the Cold War's Iron Curtain that divided Europe into west and east.

Today some observers describe Europe as being divided into a geographical system where its perimeter consists of hard borders—a "Fortress Europe," as critics (and anti-immigrant groups) call the plan—while its internal borders are deliberately soft and porous in the spirit of the Schengen Agreement. But, as noted, those soft internal Schengen borders are increasingly controversial, and the survival of Schengen, with its earlier lofty goals of a "Europe without borders," is now questionable.

Landscapes of Urban Europe

One of the major characteristics of Europe's settlement pattern is its high level of urbanization. All European countries except several small Balkan countries have more than half their populations in cities, with several western European countries more than 90 percent urbanized.

The Past in the Present North American visitors often find European cities far more interesting than our own because of the mosaic of both historical and modern landscapes, featuring medieval quarters and churches interspersed with high-rise buildings and modern department stores. Three historical periods dominate most European city landscapes. The medieval (900–1500), Renaissance–Baroque (1500–1800), and industrial (1800–present) periods have each left their characteristic traces on the European urban scene. Learning to recognize these stages of historical growth provides visitors to Europe's cities with fascinating insights into both past and present landscapes (Figure 8.15).

The **medieval landscape** is one of narrow, winding streets, crowded with three- or four-story masonry buildings

▲ **Figure 8.16 Urban Historic Preservation** Salzburg, Austria, was one of Europe's first cities to enact legal protection of its historic urban landscape, with a 1967 law. Since then, many other European cities have built upon the Salzburg approach to protect their own historic cityscapes. This view of Salzburg shows medieval dwellings and shops in the foreground and the medieval fortress on the hill, both bracketing the 17th-century Renaissance-Baroque cathedral.

with little setback from the street. This is a dense landscape with few open spaces, except around churches and town halls, where public squares or parks are clues to historical medieval open-air marketplaces.

As picturesque as we find medieval-era districts today, they nevertheless present challenges to contemporary inhabitants because of their narrow, congested streets and old housing. Modern plumbing and heating are often lacking, and rooms and hallways are small and cramped compared to present-day standards.

Many cities in Europe have enacted legislation to restore and protect their historic medieval landscapes. This movement began in the late 1960s and has become increasingly popular as cultures have worked to preserve the unique sense of place of their urban medieval sections. Because the costs of restoration are high, often these restoration projects lead to a demographic change where low- and fixed-income people are displaced by those able to pay higher rents. Further, historical areas often attract tourists, and with increased foot traffic the array of street-level shops also often changes from neighborhood-serving stores to those catering to tourists. Urban planners use the term *gentrification* to describe these changes to historical districts (Figure 8.16).

In contrast to the cramped and dense medieval landscape, those areas of the city built during the **Renaissance–Baroque period** are much more open and spacious, with expansive ceremonial buildings and squares, monuments, ornamental gardens, and wide boulevards lined with palatial residences. During this period (1500–1800), a new artistic sense of urban planning arose in Europe, resulting in the restructuring of many European cities, particularly the large capitals such as Paris and Vienna, where grandiose boulevards replaced older, more densely settled quarters. These changes were primarily for the benefit of the new urban elite—the royalty and rich merchants.

During the Renaissance–Baroque period, city fortifications limited the outward spread of these growing cities, thus aggravating crowding within. With the advent of assault artillery, European cities were forced to build an extensive system of defensive walls. Once encircled by these walls, the cities could not expand outward. Instead, as the demand for space increased within the cities, a common solution was to add several new stories to the medieval houses.

Industrialization dramatically altered the landscape of European cities. Historically, beginning in the early 19th century, factories clustered together in cities, drawn by their large markets and labor force and supplied by raw materials shipped by barge and railroad. Industrial districts of factories and worker tenements grew up around these transportation lines. In continental Europe, where many cities retained their defensive walls until the late 19th century, the new industrial districts were often located outside the former city walls, removed from the historic central city. In Paris, for example, when the railroad was constructed in the 1850s, it was not allowed to enter the city walls. Still today, even though the fortified walls of Paris are long gone, most train stations still lie beyond where the city's inner walls once stood.

Not to be overlooked are the post–World War II changes to European cities as they rebuilt from the war's destruction and adapted to the political and economic demands of the postwar era (see *Cityscapes: Dresden and Its Symbolic Landscape*). As in North American cities, suburban sprawl has become an issue in many European countries, as people seek lower-density housing in nearby rural environments. But unlike most North American cities, European urban areas generally have well-developed public transportation systems that offer attractive alternatives to commuting by car.

REVIEW

8.5 Which European countries have the highest and lowest rates of natural population increase?

8.6 Which European countries have the highest rates of out-migration? Of in-migration?

8.7 What is the Schengen Agreement?

8.8 Name three stages of historical urban development still commonly found in European urban landscapes.

Cityscapes

Dresden and Its Symbolic Landscape

Dresden, a major city in northeastern Germany that lies south of Berlin and close to the Czech border, is a testimonial to what, with only slight exaggeration, could be called the best and worst of Europe. As a result, Dresden's symbolic landscape is as important as its brick-and-mortar cityscape.

The Historic City Historically, 18th-century Dresden became known as the "Florence of the North" because of its stunning collection of Baroque churches and palaces clustered along the Elbe River. Like Florence itself, Dresden became a magnet for artists who captured the city's beauty in their paintings and sketches. But Dresden as a symbol of Europe's artistic sensibilities came to an abrupt end in February 1944 when American and British bombers destroyed the city in a series of controversial raids.

Ostensibly, the bombing aimed to destroy the railroad yards located just west of the historic old city in order to keep Nazi Germany from moving its troops from the eastern to the western front, where they would slow the advancing Allies. That goal was achieved, but another reality turned Dresden into a tragic symbol of war's unimaginable horrors. The same railroads that were moving army troops from east to west had also transported into Dresden thousands of refugees fleeing the war-torn east. Estimates are that several hundred thousand refugees were crowded into Dresden when the bombers attacked, dropping incendiaries that devastated the city in a nightmarish firestorm that killed an estimated 25,000 civilians. Controversy ensued immediately as to whether the firebombing was necessary, and it continues still today.

▼ **Figure 8.3.1 The Restored *Frauenkirche* in the Historic Old City** The reconstruction of this cathedral has come to symbolize post-war reconciliation between England, America, and Germany.

Rebuilding Dresden and Conflicts over Symbolic Space As a result of the bombing, rebuilding postwar Dresden became a symbol of reconciliation between former enemies as American and British citizens together donated millions of dollars for reconstructing the historic old city. A focus for these efforts was Dresden's magnificent landmark cathedral, the *Frauenkirche* (Figure 8.3.1). A good portion of the funds to restore the landmark church came from the people of Coventry, England, a city that similarly lost its historic cathedral to firebombing by Nazi Germany in 1940.

Sadly, Dresden as a symbol of reconciliation and peace was hijacked by Neo-Nazis. Peace activists have long gathered in Dresden on February 14 to memorialize Dresden's tragic past, while promoting peaceful reconciliation. But in 2005, thousands of Neo-Nazis and right-wing extremists poured into Dresden in a counterdemonstration to promote their fabrication that the Allied bombings somehow neutralized the horrors of the Nazi Holocaust. Joining the fray in following years were left-wing extremists and anarchists anxious to engage the Neo-Nazis in street battles. Most recently, in February 2013 peace activists kept the Neo-Nazis at bay by linking arms and encircling the complete inner city (Figure 8.3.2). At least temporarily, Dresden has regained its place as a symbol of peace and reconciliation.

1. London was also heavily damaged during the World War II. How was its cityscape rebuilt?.
2. List several other symbolic landscapes in European cities.

Google Earth Virtual Tour Video

http://goo.gl/sWDyi0

▼ **Figure 8.3.2 Anti-Nazi Demonstrations in Dresden** Peace activists march against extremists.

Cultural Coherence and Diversity: A Mosaic of Differences

The rich cultural geography of Europe demands our attention for several reasons. First, the highly varied mosaic of languages, customs, religions, ways of life, and landscapes that characterizes Europe has also strongly shaped regional identities that have all too often stoked the fires of conflict.

Second, European cultures have played leading roles in globalization as European colonialism brought about changes in languages, religion, economies, and social values in every corner of the globe. Examples include cricket games in Pakistan, high tea in India, Dutch architecture in South Africa, and the millions of French-speaking inhabitants of equatorial Africa.

Today, though, new waves of global culture are spreading into Europe (Figure 8.17). Although some European cultures embrace (or simply condone) these changes, other cultures actively resist. France, for example, struggles against both U.S.-dominated popular culture and the multicultural influences of its large African migrant population.

Geographies of Language

Language has always been an important component of nationalism and group identity in Europe (Figure 8.18). Today, although some small ethnic groups such as the Irish and the Bretons work hard to preserve their local language in order to reinforce their cultural identity, millions of Europeans are also busy learning multiple languages so they can better communicate across cultural and national boundaries. The EU itself, which is primarily an integrating force in Europe, celebrates its linguistic mosaic by recognizing more than 20 official languages.

▼ **Figure 8.17 Global Culture in Old Europe** A McDonald's fast-food outlet occupies a classic building within Belgrade, Serbia's historic central city.

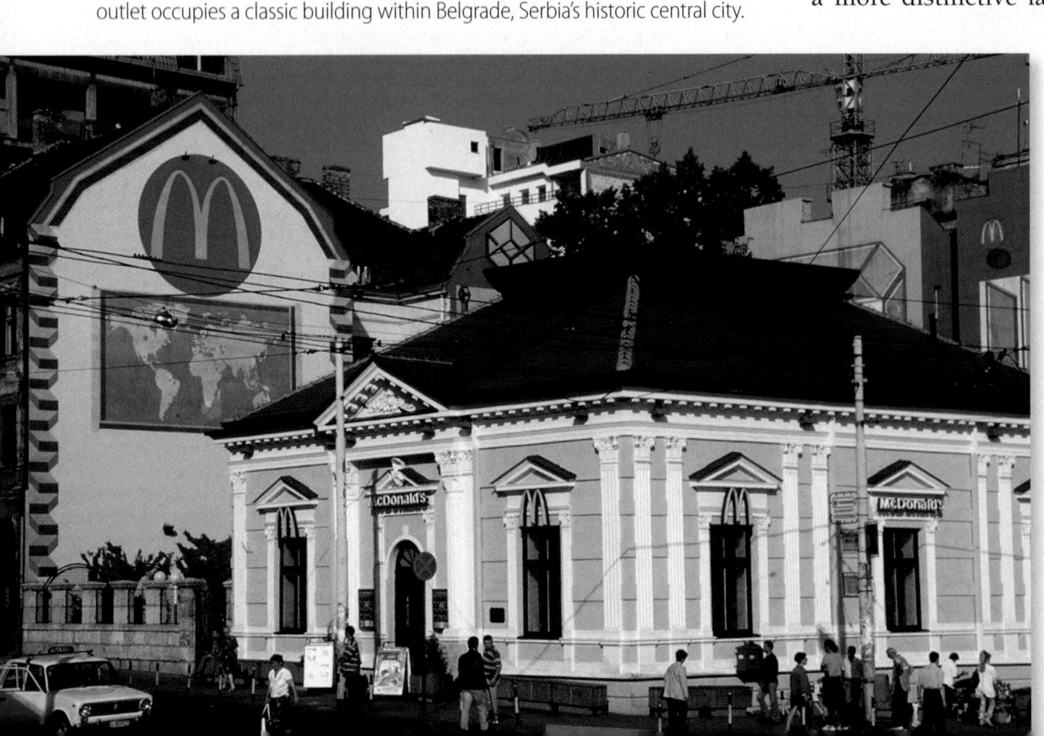

As their first language, 90 percent of Europe's population speaks Germanic, Romance, or Slavic languages, all of which are linguistic groups within the Indo-European family. Germanic and Romance speakers each number almost 200 million in the European region. Although Slavic languages are spoken by 400 million if you include Russia and its immediate neighbors, there are only 80 million Slavic speakers within Europe proper.

Germanic Languages Germanic languages dominate Europe north of the Alps. Today about 90 million people speak German as their first language, and it is the dominant language of Germany, Austria, Liechtenstein, Luxembourg, eastern Switzerland, and several small areas in Alpine Italy. Until recently, there were also large German-speaking minorities in Romania, Hungary, and Poland, but many of these people left eastern Europe and resettled in Germany when the Iron Curtain was lifted in 1990.

English is the second largest Germanic language, with about 60 million speakers using it as their first language. In addition, a large number of Europeans learn English as a second language, particularly in the Netherlands and Scandinavia, where many people are as fluent in English as are native speakers. Linguistically, English is closest to the Low German spoken along the coastline of the North Sea, which reinforces the theory that an early form of English evolved in the British Isles through contact with the coastal peoples of northern Europe. One of the distinctive traits of English that sets it apart from German, however, is that almost one-third of the English vocabulary is made up of Romance words brought to England during the Norman French conquest of the 11th century.

Elsewhere in this linguistic region, Dutch (in the Netherlands) and Flemish (in northern Belgium) together account for another 20 million people, with roughly the same number of Scandinavians speaking the closely related languages of Danish, Norwegian, and Swedish. Icelandic is a more distinctive language because of that country's geographic isolation from its Scandinavian roots.

Romance Languages Romance languages, including French, Spanish, and Italian, evolved from the vulgar (or everyday) Latin used within the Roman Empire. Today Italian is the most widely used of these Romance languages, with about 60 million Europeans speaking it as their first language. In addition to being spoken in Italy, Italian is an official language of Switzerland and is also spoken on the French island of Corsica.

French is spoken in France, western Switzerland, and southern Belgium (where it is known as *Walloon*).

▲ **Figure 8.18 Languages of Europe** Ninety percent of Europeans speak an Indo-European language. These languages can be grouped into the major categories of Germanic, Romance, and Slavic languages. Ninety million Europeans speak German as a first language, which places it ahead of the 60 million who list English as their native language. However, given the large number of Europeans who speak fluent English as a second language, you could make the case that English is the dominant language of modern Europe.

Today there are about 55 million native French speakers in Europe. As with other languages, French has very strong regional dialects. Linguists differentiate between two forms of French in France itself: that spoken in the north (the official form because of the dominance of Paris) and the language of the south, or *langue d'oc*. This linguistic divide expresses long-standing tensions between Paris and southern France.

Spanish also has very strong regional variations. About 25 million people speak Castilian Spanish, the country's official language, which dominates the interior and northern areas of that large country. However, the *Catalan* form, which some argue is a completely separate language, is found along the eastern coastal fringe, centered on Barcelona, Spain's second largest city. This distinct language reinforces a strong sense of cultural separateness that has led to the state of Catalonia being given autonomous status within Spain.

Portuguese is spoken by another 12 million speakers in that country and in the northwestern corner of Spain, although considerably more people speak the language in Brazil, a former Portuguese colony in Latin America. Finally, Romanian represents the most eastern extent of the Romance language family; it is spoken by 24 million people in Romania. Though unquestionably a Romance language, Romanian also contains many Slavic words.

▼ **Figure 8.19 Cyrillic Alphabet** A directional sign in downtown Sofia, Bulgaria uses both the Cyrillic and Roman alphabets to guide locals and visitors alike.

The Slavic Language Family Slavic is the largest European subfamily of the Indo-European languages. Slavic speakers are traditionally separated into northern and southern groups, divided by the non-Slavic speakers of Hungary and Romania.

To the north, Polish has 35 million speakers, with Czech and Slovakian totaling about 15 million. As noted earlier, these numbers pale in comparison with the number of northern Slav speakers in nearby Ukraine, Belarus, and Russia, which can easily reach more than 150 million. Southern Slav languages include three groups: 14 million Serbian and Croatian speakers (now considered separate languages because of the strong political and cultural differences between Serbs and Croats), 11 million Bulgarian and Macedonian speakers, and 2 million Slovenian speakers.

The use of two distinct alphabets further complicates the geography of Slavic languages (Figure 8.19). In countries with a strong Roman Catholic heritage, such as Poland and the Czech Republic, the Latin alphabet is used. In contrast, countries with close ties to the Orthodox church—Bulgaria, Montenegro, Macedonia, parts of Bosnia–Herzegovina, and Serbia—use the Greek-derived **Cyrillic alphabet**.

Geographies of Religion, Past and Present

Religion is an important component of the geography of cultural coherence and diversity in Europe because many of today's ethnic tensions result from historical religious events. To illustrate, significant cultural borders in the Balkans and eastern Europe are based upon the 11th-century split of Christianity into eastern and western churches, as well as the division between Christianity and Islam. In Northern Ireland, blood is still shed over the tensions resulting from the 17th-century split of Christianity into Catholicism and Protestantism. Also, much of the ethnic-cleansing terrorism in the former Yugoslavia during the 1990s resulted from the historical struggle between Christianity and Islam. In addition, considerable tension has arisen over the last decade regarding the large Muslim migrant populations residing in England, France, and Germany. Understanding these contemporary tensions involves taking a brief look at the historical geography of Europe's religions (Figure 8.20).

The Schism Between Western and Eastern Christianity In southeastern Europe, early Greek missionaries spread Christianity throughout the Balkans and into the lower reaches of the Danube. Progress was slower than in western Europe, perhaps because of continued invasions by peoples from the Asian steppes. Another factor is that Greek missionaries refused to accept the control of Roman Catholic bishops in western Europe.

This tension with western Christianity led to an official split of the eastern church from Rome in 1054 CE. This eastern church subsequently splintered into Orthodox sects closely linked to specific nations and states. Today, to illustrate, we find Greek Orthodox, Bulgarian Orthodox, and Russian Orthodox churches, all with different rites and rituals.

▲ **Figure 8.20 Religions of Europe** This map shows the divide in western Europe between the Protestant north and the Roman Catholic south. Historically, this distinction was much more important than it is today. Note the location of the former Jewish Pale, which was devastated by the Nazis during World War II. Today ethnic tensions with religious overtones are found primarily in the Balkans, where adherents of Roman Catholicism, Eastern Orthodoxy, and Islam are found in close proximity to one another. **Q: After comparing this map to the one of Europe's languages (Figure 8.18), list those areas where language families and religion appear to be related.**

Another factor that distinguished eastern Christianity from western was the Orthodox use of the Cyrillic alphabet. Because Greek missionaries were primarily responsible for the spread of early Christianity in southeastern Europe, it is not surprising that they used an alphabet based on Greek characters. More precisely, this alphabet is attributed to the missionary work of St. Cyril in the 9th century. As a result, the division between western and eastern churches, and between the two alphabets, remains one of the most prominent cultural boundaries in Europe.

The Protestant Revolt Besides the division between western and eastern churches, the other great split within Christianity occurred between Catholicism and Protestantism. This division arose in Europe during the 16th century and has divided the region ever since. However, with the exception of the Troubles in Northern Ireland, tensions today between these two major groups are far less problematic than in the past, when these religious differences led to several wars.

Conflicts with Islam Both the eastern and western Christian churches struggled with challenges from Islamic empires to Europe's south and east. Even though historical Islam was reasonably tolerant of Christianity in its conquered lands, Christian Europe was far less accepting of Muslim imperialism. The first crusade to reclaim Jerusalem from the Turks took place in 1095. After the Ottoman Turks conquered Constantinople in 1453 and gained control over the Bosporus strait and the Black Sea, they moved rapidly to spread a Muslim empire throughout the Balkans, arriving at the gates of Vienna in the middle of the 16th century. There Christian Europe stood firm militarily and stopped Islam from expanding into western Europe.

Ottoman control of southeastern Europe, however, lasted until the empire's demise in the early 20th century. This historical presence of Islam explains the current coexistence of religions in the Balkans, with intermixed areas of Muslims, Orthodox Christians, and Roman Catholics.

A Geography of Judaism Europe has long been a difficult home for Jews who were forced to leave Palestine during the Roman Empire. At that time, small Jewish settlements were located in cities throughout the Mediterranean. Later, by 900 CE, about 20 percent of the Jewish population was clustered in the Muslim lands of the Iberian Peninsula, where Islam showed greater tolerance for Judaism than had Christianity. Furthermore, Jews played an important role in trade activities both within and outside the Islamic lands. After the Christian reconquest of Iberia, however, Jews once more faced severe persecution and fled from Spain to more tolerant countries in western and central Europe.

One focus for this exodus was the area in eastern Europe that became known as the Jewish Pale. In the late Middle Ages, at the invitation of the Kingdom of Poland, Jews settled in cities and small villages in what is now eastern Poland, Belarus, western Ukraine, and northern Romania (see Figure 8.20). Jews collected in this region for several centuries in the hope of establishing a true European homeland.

Until emigration to North America began in the 1890s, 90 percent of the world's Jewish population lived in Europe, and most were clustered in the Pale region. Even though many emigrants to the United States and Canada came from this area, the Pale remained the largest population of Jews in Europe until World War II. Tragically, Nazi Germany devastated this ethnic cluster by focusing its extermination activities on the Pale.

In 1939, on the eve of World War II, 9.5 million Jews lived in Europe, or about 60 percent of the world's Jewish population. During the war, German Nazis murdered some 6 million Jews in the horror of the Holocaust. Today fewer than 2 million Jews live in Europe. However, since 1990 and the lifting of quotas on Jewish emigration from Russia, Belarus, and Ukraine, more than 100,000 Jews have emigrated to Germany, giving it the fastest-growing Jewish population outside Israel (Figure 8.21).

Patterns of Contemporary Religion Europe today has about 250 million Roman Catholics and fewer than 100 million Protestants. Generally, Catholics live in the southern half of the region, except for significant numbers in Ireland and Poland.

Protestantism is most widespread in northern Germany, the Scandinavian countries, and England, and it is intermixed

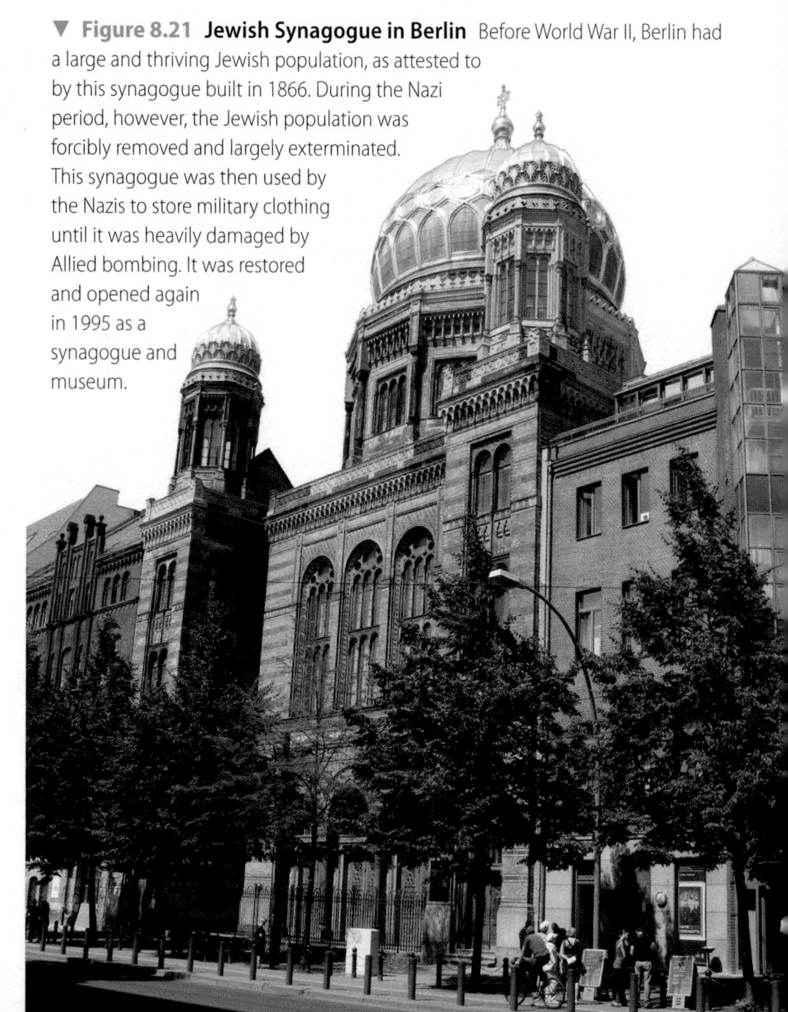

▼ **Figure 8.21 Jewish Synagogue in Berlin** Before World War II, Berlin had a large and thriving Jewish population, as attested to by this synagogue built in 1866. During the Nazi period, however, the Jewish population was forcibly removed and largely exterminated. This synagogue was then used by the Nazis to store military clothing until it was heavily damaged by Allied bombing. It was restored and opened again in 1995 as a synagogue and museum.

Everyday Globalization

International Students in Europe

Have you ever thought about spicing up your college career by studying at a European university? Great idea, and you'll be in good company with 1.3 million other international students, since Europe's colleges and universities are magnets for students from all over the world. England attracts the largest number of international students, but Germany, France, and the Netherlands are not far behind.

Language Issues The largest number of international students in Europe are from China, with second place going to India, followed by a handful of countries (including the United States) tying for third. Language skills, naturally, make a difference as to who goes where. South Asians and North Americans are obviously attracted to Britain because classes are taught in English. Similarly, the Netherlands is attractive for English speakers, since 75 percent of the classes are taught in English. Instruction in France is mainly in French, consequently, most international students in that country come from Francophone Africa. Although German universities offer some classes in English, mostly at the graduate level, undergraduate lectures are mainly in German. However, the German government provides stipends to international students who need to take intensive language classes before beginning their college studies, so the population of international students at major German universities is surprisingly high. At the University of Heidelberg, for example, 10 percent of the students are international (Figure 8.4.1 and 8.4.2).

▲ **Figure 8.4.1 Heidelberg** A full 10 percent of students at the University of Heidelberg are international.

Restructuring Higher Education in Europe Until recently, the degree structure of Europe's colleges and universities differed considerably between countries, but today there is movement toward having common degree programs. Earlier, British higher education was fairly similar to American universities, offering undergraduate B.A. or B.S. degrees, followed by graduate programs at the master's and doctoral levels. Germany and France (and most of continental Europe) had no undergraduate degree comparable to the B.A. or B.S.; instead, a student's first degree was closer to the American master's degree, both in academic focus and in time needed to earn the degree.

Today, however, most European universities are standardizing their degree programs so that they're similar to the U.S. and British model of a four-year B.A. in an undergraduate major and a two-year M.A. in a specialized field, with further study on a more specialized topic leading to the Ph.D. The goals of this standardization program, which is guided by the European Higher Education Area (EHEA), are, first, to have similar degree standards for all European colleges and universities and, second, to facilitate entry and college graduation for both European and international students by having similar degree requirements at all schools. The explosive growth in the number of international students now enrolled in Europe suggests that at least that component of the EHEA agenda is working well.

Google Earth
Virtual Tour Video

http://goo.gl/bukKHr

1. If you were to study in Europe, where would you go? Why?
2. Are there Europeans studying at your college? Interview them to find out about their experiences.

▼ **Figure 8.4.2 International Students in Heidelberg** The standardization of European degree programs with the U.S. and British has led to explosive growth of international students in Europe.

▲ **Figure 8.22 Muslim Men in a Berlin Mosque** After France, Europe's second-largest Muslim population is in Germany, a product of earlier guest worker policies that brought many Turks to the country and, more recently, Germany's liberal asylum policies towards refugees from civil conflicts throughout the world. A large Afghani population, for example, lives in Hamburg.

with Catholicism in the Netherlands, Belgium, and Switzerland. Because of Protestant reaction against the ornate cathedrals and statues of the Catholic Church, the landscape of Protestantism is much more sedate and subdued. Large cathedrals and religious monuments in Protestant countries are associated primarily with the Church of England because of its strong historical ties to Catholicism; St. Paul's Cathedral and Westminster Abbey in London are examples of such grandiose structures.

Not to be overlooked are the estimated 13 million Muslims in Europe. Most are migrants from Africa and southwestern Asia. France has the largest population of Muslims (4.7 million), with the second largest number (4.1 million) in Germany.

European Culture in a Global Context

Europe, like all other world regions, is currently caught up in a period of profound cultural change. In fact, many would argue that the pace of cultural change in Europe has been accelerated because of the complicated interactions between globalization and Europe's internal agenda of political and economic integration (see *Everyday Globalization: International Students in Europe*). Some pundits celebrate the "New Europe" of integration and unification, but other observers refer to a more tension-filled New Europe of foreign migrants and guest workers harassed by ethnic discrimination and racism.

Migrants and Culture Migration patterns are influencing the cultural mix in Europe. Historically, Europe spread its cultures worldwide through aggressive colonialization; today, however, the region is experiencing a reverse flow as millions of migrants move into Europe, bringing their own distinct cultures from the far-flung countries of Africa, Asia, and Latin America. Unfortunately, in some areas of Europe, the products of this cultural exchange are highly troubling.

Immigrant clustering, leading to the formation of distinctly ethnic neighborhoods and even ghettos, is now common in the cities and towns of western Europe (Figure 8.22). The high-density apartment buildings of suburban Paris, for example, are home to large numbers of French-speaking Africans and Arab Muslims caught in a web of high unemployment, poverty, and racial discrimination. As a result, cultural battles have emerged in many European countries. For example, in 2004, French leaders, unsettled by the country's large Muslim migrant population, attempted to speed assimilation of female high school students into French mainstream culture by banning a key symbol of conservative Muslim life: the head scarf (*hijab*). This rule triggered riots, demonstrations, and counterdemonstrations. As a result of these kinds of conflicts, the political landscape of many European countries now has far-right, nationalistic parties with thinly disguised platforms of excluding migrants from their countries.

Sports in Europe Soccer (which Europeans call football) is unquestionably Europe's national sport, played everywhere from sandlots to stadiums, by both women and men, at all levels from family picnics to multiple-level professional leagues (Figure 8.23). At the highest pro level, soccer teams draw crowds into NFL-sized stadiums holding 100,000 people. Smaller soccer stadiums seating 30,000–40,000 are common in every European town.

Like many sports throughout the world, soccer is irrevocably linked to globalized culture, with fanatical fans rooting for place-based teams comprised largely of international players who lack any local allegiance. But this contradiction doesn't keep soccer fans from taking their local allegiances across Europe's borders to rival towns and cities, where team loyalties sometimes turn violent. Soccer hooliganism, unfortunately, has become a common outlet for Europe's anti-migrant racism and xenophobia. It's also not uncommon for local soccer clubs to take on nationalistic, paramilitary behaviors far removed from the soccer pitch, with migrants often being their targets.

Aside from their homegrown sports like soccer and rugby, Europe has shown some interest in the North American sports of basketball, baseball, and American football.

▼ **Figure 8.23 Women's Sports in Europe** Women's soccer and basketball teams, at both the amateur and professional levels, are common throughout Europe. Here, soccer players from the German (in white) and French national teams do battle during the recent Women's World Cup competition.

▲ **Figure 8.24 Women in Europe's Business World** The employment of women in Europe's workforce differs significantly between different countries and regions. Scandinavia, for example, has the highest percentage of women in upper management positions, in contrast with the Mediterranean countries. This photo was taken at a business meeting in Berlin, Germany.

Basketball, though, is unquestionably Europe's favorite American sport, with hoops and courts increasingly common in the region's gyms and playgrounds. Pro leagues at all levels abound, both men's and women's, with most European cities supporting at least one pro team. In fact, it's now common for U.S. Women's National Basketball Association (WNBA) players to spend their off-season playing for a European pro team to augment their modest WNBA salaries. Also common is the increasing number of European basketball players (both male and female) who play for North American college and pro teams.

Baseball is fairly popular in Europe, having grown from seeds planted by postwar U.S. servicemen into several professional leagues. Today a handful of baseball academies have developed in Germany and France to train athletes who aspire to play Major League Baseball (MLB) in North America. Their goal is to break into MLB, just the way Latin Americans and Japanese did decades ago.

American football remains a novelty in most of Europe, even though the National Football League (NFL) usually schedules a preseason exhibition game in either England or Germany. Europeans applaud politely after a touchdown, but most agree that NFL football, with its frequent media timeouts, fails to capture the attention of crowds who thrive on soccer's nonstop action.

Gender Issues in Europe Despite the visibility of female political leaders and the fact that Europe is considered one of the most developed regions of the world, gender equity issues persist in government, business, and domestic life. For the EU countries as a whole, for example, male employment is 21 percent higher than for women, and women who work generally make 25 percent less than do males (Figure 8.24).

However, given the complexity of Europe, with its mixture of national, urban, rural, and migrant cultures, the nature and extent of gender issues differ widely among countries and regions. To illustrate, within the 28 EU countries about a quarter of the parliamentary offices are held by women. Sweden has the highest representation, with more than half of its ministers being female, whereas Cyprus has absolutely none. Similarly, in the business world, only 11 of Europe's largest companies have women in top management, yet women make up almost a third of top management in Norway, compared to just 1 percent in Luxembourg.

One interesting pattern is that female participation in the workforce is generally higher in the countries of eastern Europe and the Balkans. Two interrelated factors explain this. First, women were expected to work in the communist economies of these countries from 1945 to 1990. Second, families often needed two incomes to survive during the difficult economic transition that followed the collapse of the Soviet Union in 1990. Regardless of cause, the results are startling. Today Bulgaria has the highest ratio of female CEOs (21 percent) of any EU country, whereas Slovenia, formerly a part of socialist Yugoslavia, is the country with the least income disparity between men and women.

Women are highly represented in both government and business in the Scandinavian countries (Norway, Sweden, Denmark, Finland, and Iceland), but for very different reasons than in eastern Europe and the Balkans. It is generally agreed that the foundation of Scandinavia's gender equity comes from a combination of comprehensive child care; liberal maternity and paternity benefits that guarantee job security and career advancement after maternity and paternity leaves; and a tax code that does not punish dual-income families.

As a result, Norway and Sweden have the highest percentage of females in the workforce. Portugal has the third highest number at 71 percent; however, experts caution that working families in that country face significant differences, compared to Scandinavian families. In Portugal, because of its struggling economy, women usually work out of necessity rather than choice, with grandparents and other family members providing child care—unlike the government-sponsored child care common in Scandinavia.

Not to be overlooked are the extraordinarily complex gender issues within Europe's large migrant cultures and their host cultures. We mentioned earlier how France's national policies have become entangled with Muslim gender and cultural preferences. Other examples of these complexities can be found in Germany, as the state finds itself embroiled in cultural tensions. These range from women's freedom beyond the family household to prosecuting Turkish honor killings, where young women have paid with their lives for behaviors that are common to German culture, such as dating and marriage without parental consent, but that are unacceptable in traditional Turkish culture.

REVIEW

8.9 Describe the general location within Europe of the three major language groups: Germanic, Romance, and Slavic.

8.10 In general terms, describe the historic distribution within Europe of Catholicism, Protestantism, Judaism, and Islam.

8.11 Which countries have the highest numbers of Muslims? Why?

Geopolitical Framework: A Dynamic Map

One of Europe's unique characteristics is its dense fabric of 42 independent states within a relatively small area. Historically, the idea of democratic nation-states arose in Europe and, over time, replaced the fiefdoms and empires ruled by autocratic royalty. France, Italy, Germany, and the United Kingdom are major examples.

However, in many ways Europe's unique geopolitical landscape has been as much problem as promise. Twice in the past century, Europe shed blood to redraw its political borders, and within the past several decades nine new states have appeared—more than half through violent wars. Today, generally speaking, most of Europe's geopolitical hotspots are more about achieving regional autonomy than creating new nation-states (Figure 8.25).

Redrawing the Map of Europe Through War

Two world wars radically reshaped the geopolitical map of 20th-century Europe (Figure 8.26). Early in that century, Europe was divided into two opposing and highly armed camps that tested each other for a decade before the outbreak of World War I in 1914. France, Britain, and Russia were allied against Germany, the Austro-Hungarian Empire, and the remnant Ottoman Empire, which had formerly controlled an assortment of ethnic regions in the Balkans. Although World War I was referred to as the "war to end all wars," it fell far short of solving Europe's geopolitical problems. Instead, according to many experts, the peace treaty actually made another world war unavoidable.

When Germany and Austria–Hungary surrendered in 1918, the Treaty of Versailles peace process set about redrawing the map of Europe with two goals in mind: first, to punish the losers through loss of territory and severe financial reparations and, second, to recognize the nationalistic aspirations of unrepresented peoples by creating several new nation-states. As a result, the new states of Czechoslovakia and Yugoslavia were created. In addition, Poland was reestablished, as were the Baltic states of Finland, Estonia, Latvia, and Lithuania.

Though the goals of the treaty were admirable, few European states were satisfied with the resulting map. New states were resentful when their citizens were left outside the new borders. This created an epidemic of **irredentism**, or state policies for reclaiming lost territory and peoples. Examples were the large German population in the western portion of the newly created state of Czechoslovakia and the Hungarians stranded in western Romania by post–World War I border changes.

These imperfect geopolitical solutions were greatly aggravated by the global economic depression of the 1930s, which brought high unemployment, food shortages, and even more political unrest to Europe. Three competing ideologies promoted their own solutions to Europe's pressing problems: Western democracy (and capitalism); communism from the Soviet revolution to the east; and a fascist totalitarianism promoted by Mussolini in Italy and Hitler in Germany. With industrial unemployment at record rates in western Europe, public opinion fluctuated wildly between the extremist solutions of far-right fascism and leftist communism and socialism. In 1936, Italy and Germany joined forces through the Rome–Berlin "axis" agreement. As in World War I, this alignment was countered with mutual protection treaties among France, Britain, and the Soviet Union. When an imperialist Japan signed a pact with Germany, the scene was set for a second global war.

Nazi Germany tested Western resolve in 1938 by annexing Austria, the country of Hitler's birth, and then Czechoslovakia, under the pretense of providing protection for ethnic Germans located there. After signing a nonaggression pact with the Soviet Union, Hitler's armies invaded Poland on September 1, 1939. Two days later, France and Britain declared war on Germany. Within a month, the Soviet Union moved into eastern Poland, the Baltic states, and Finland to reclaim territories lost through the peace treaties of World War I. Nazi Germany then moved westward and occupied Denmark, the Netherlands, Belgium, and France, after which it began preparations to invade England.

In 1941, the war took several startling new turns. In June, Hitler broke the nonaggression pact with the Soviet Union and, catching the Red Army by surprise, took the Baltic states and then drove deep into Soviet territory. When Japan attacked the American naval fleet at Pearl Harbor, Hawaii, in December 1941, the United States entered the war in both the Pacific and Europe.

By early 1944, the Soviet army had recovered most of its territorial losses and moved against the Germans in eastern Europe, beginning the long communist domination in that region. By agreement with the Western powers, the Red Army stopped when it reached Berlin in April 1945. At that time, Allied forces crossed the Rhine River and began their occupation of Germany. Immediately after Hitler's suicide, Germany signed an unconditional surrender on May 8, 1945, ending the war in Europe. But with Soviet forces firmly entrenched in the eastern part of Europe, the military battles of World War II were immediately replaced by an ideological Cold War between communism and democracy that lasted 45 years, until 1990.

A Divided Europe, East and West

From 1945 until 1990, Europe was divided into two geopolitical and economic blocs, east and west, separated by the

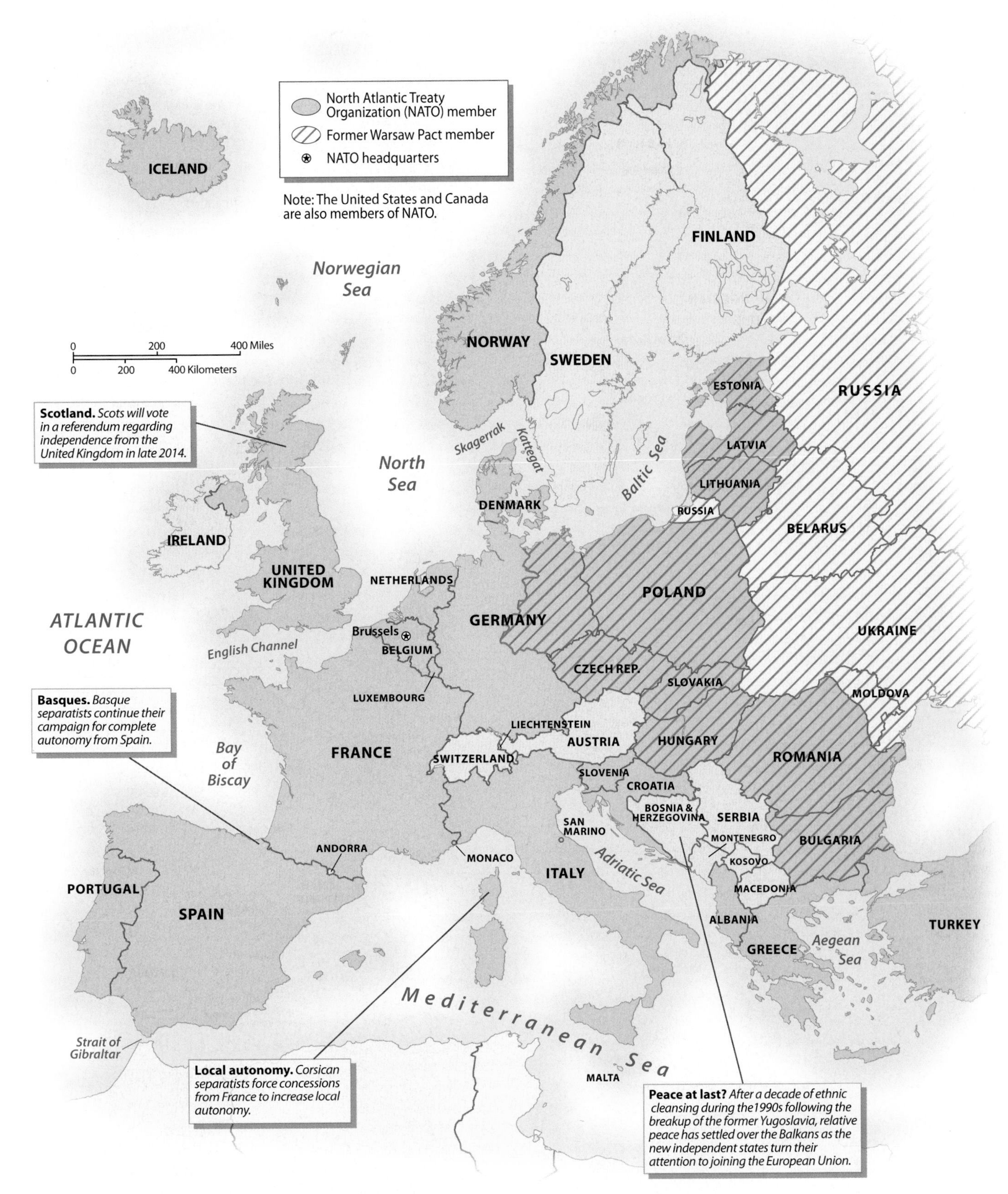

▲ **Figure 8.25 Geopolitical Issues in Europe** Although the major geopolitical issue of the early 21st century remains the integration of eastern and western Europe into the EU, numerous issues of micro- and ethnic nationalism also engender geopolitical fragmentation. In other parts of Europe, such as Spain, France, and Great Britain, questions of local ethnic autonomy within the nation-state structure challenge central governments.

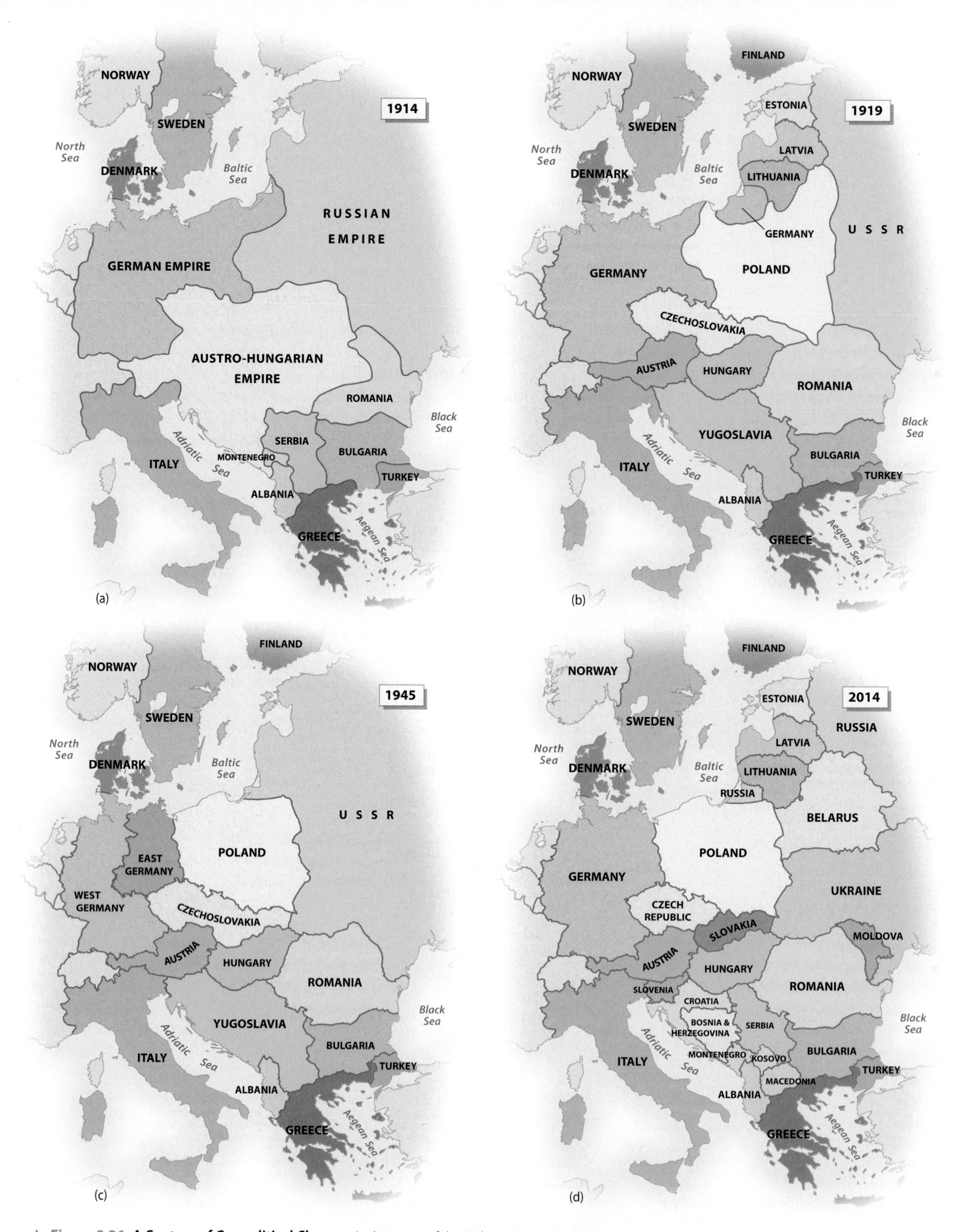

▲ **Figure 8.26 A Century of Geopolitical Change** At the outset of the 20th century, central Europe was dominated by the German, Austro-Hungarian (or Hapsburg), and Russian empires. Following World War I, these empires were largely replaced by a mosaic of nation-states. More border changes followed World War II, largely as a result of the Soviet Union's turning that area into a buffer zone between itself and western Europe. With the demise of Soviet hegemony in 1990, further political change took place. **Q: Where are the strongest relationships between political change and cultural factors such as language and religion?**

infamous **Iron Curtain**, which descended shortly after the peace agreement ending World War II (Figure 8.27). East of the Iron Curtain border, the Soviet Union imposed the heavy imprint of communism on all activities—political, economic, military, and cultural. To the west, as Europe rebuilt from the destruction of the war, new alliances and institutions were created to counter the Soviet presence in Europe.

Cold War Geography The seeds of the Cold War were planted at the Yalta Conference of February 1945, when the leaders of Britain, the Soviet Union, and the United States met to plan the shape of postwar Europe. Because the Red Army was already in eastern Europe and moving quickly on Berlin, Britain and the United States agreed that the Soviet Union would occupy eastern Europe and the Western allies would occupy parts of Germany.

The larger geopolitical issue, though, was the Soviet desire for a **buffer zone** between its own territory and western Europe. This buffer zone consisted of an extensive bloc of satellite countries, dominated politically and economically by the Soviet Union, that could cushion the Soviet heartland against possible attack from western Europe. In the east, the Soviet Union took control of the Baltic states, Poland, Czechoslovakia, Hungary, Bulgaria, Romania, Albania, and, briefly, Yugoslavia. Austria and Germany were divided into occupied sectors by the four (former) allied powers. In both cases, the Soviet Union dominated the eastern portion of the country, which contained the capital cities of Berlin and Vienna. Both capital cities, in turn, were divided into French, British, U.S., and Soviet sectors.

In 1955, with the creation of an independent and neutral Austria, the Soviets withdrew from their sector, effectively moving the Iron Curtain eastward to the Hungary–Austria border. Germany, however, quickly evolved into two separate states, West Germany and East Germany, which remained separate until 1990.

Along the border between east and west, two hostile military forces faced each other for almost half a century. Both sides prepared for and expected an invasion by the other across the barbed wire dividing Europe. In small military units from satellite countries, both the **North Atlantic Treaty Organization (NATO)** and the Warsaw Pact countries were armed with nuclear weapons, making Europe a tinderbox for a nightmarish third world war.

Berlin was the flashpoint that brought these forces close to a fighting war on two occasions. In the winter of 1948, the Soviets imposed a blockade on the city, denying Western powers access to Berlin across its East German military sector. This attempt to starve the city into submission by blocking food shipments from western Europe was thwarted by a nonstop airlift of food and coal by NATO. Then, in August 1961, the Soviets built the Berlin Wall to curb the flow of East Germans seeking political refuge in the West. The Wall became the concrete-and-mortar symbol of a firmly divided postwar Europe. For several days while the Wall was being built and the West agonized over destroying it, NATO and Warsaw Pact tanks and soldiers faced each other with loaded weapons at point-blank range. Though war was avoided, the Wall stood for 28 years, until November 1989.

The Cold War Thaw The symbolic end of the Cold War in Europe came on November 9, 1989, when East and West Berliners joined forces to rip apart the Berlin Wall with jackhammers and hand tools (Figure 8.28). By October 1990, East and West Germany were officially reunified into a single nation-state. During this period, all other Soviet satellite states, from the Baltic Sea to the Black Sea, also underwent major geopolitical changes that have resulted in a mixed bag of benefits and problems.

The Cold War's end came as much from a combination of problems within the Soviet Union (discussed in Chapter 9) as from rebellion in eastern Europe. By the mid-1980s, the Soviet leadership was advocating an internal economic restructuring and also recognizing the need for a more open dialogue with the West.

As a result of the Cold War thaw, the map of Europe began changing once again, with the unification of Germany, the peaceful "Velvet Divorce" of Czechs and Slovaks, and the reemergence of the Baltic states. More troublesome was the Balkan area, where the former Yugoslavia violently fractured into a handful of independent states.

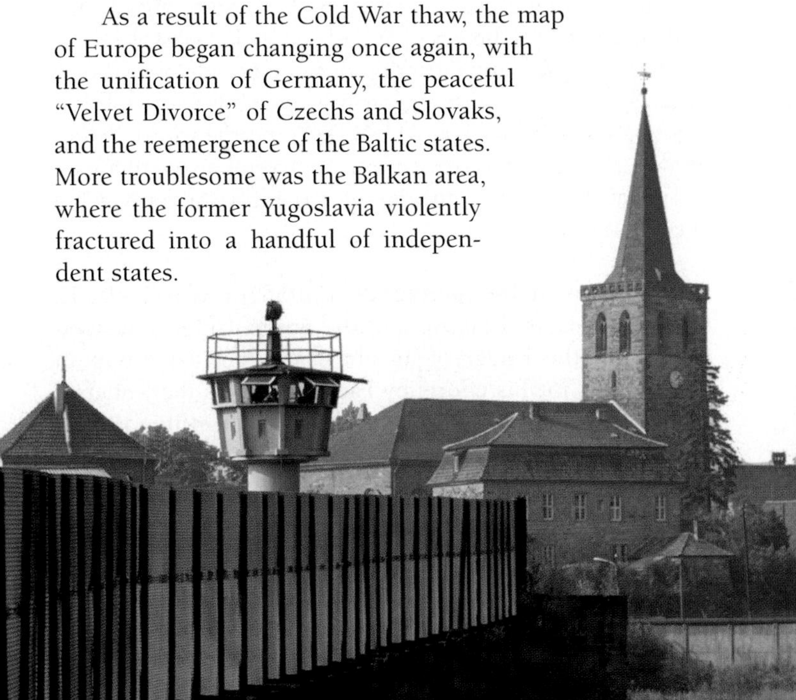

▶ **Figure 8.27 The Iron Curtain** From 1945 until 1989 Europe was divided politically and physically by the Iron Curtain, which separated the Soviet Union satellite countries of eastern Europe from Western Europe. This photo is of the border dividing the former East Germany from West Germany near Vaca, Germany. Besides the Iron Curtain itself, the border zone on the eastern side was commonly several miles wide and included military fortifications with severe restrictions on civilian movement.

(a)

(b)

▲ **Figure 8.28 The Berlin Wall** In August 1961, East Germany built a concrete and barbed wire structure along the border of East Berlin to stem the flow of refugees fleeing communist rule. The Wall was the most visible symbol of the Cold War division of East and West until November 1989, when the failing Soviet Union renounced its control over Eastern Europe. (a) The extent of the Wall zone at the Brandenburg Gate (East Berlin is to the left); (b) Berliners celebrate the end of the Wall with East German police who previously guarded the border zone with shoot-to-kill orders.

The Balkans: Waking from a Geopolitical Nightmare

The Balkans have long been a troubled area with their complex mixture of languages, religions, and ethnic allegiances. Throughout history, these allegiances have led to an often-changing geography of small countries (Figure 8.29). Indeed, the term **balkanization** is used to describe the geopolitical processes of small-scale independence movements based upon ethnic fault lines.

Following the fall of the Austro-Hungarian and Ottoman empires in the early 20th century, much of the region was unified under the political umbrella of the former Yugoslavian state. In the 1990s, however, Yugoslavia broke apart as ethnic factionalism and nationalism produced a decade of violence, turmoil, and wars of independence, creating a geopolitical nightmare for Europe, the EU, NATO, and the world. Today, though, despite lingering tensions in several areas, there are signs that the Balkan countries are moving toward a new era of peace and stability.

Balkan Wars of Independence Following World War II, Britain, the United States, and the Soviet Union rewarded Josip Tito, the leader of an anti-Nazi guerrilla group in the Balkans, for his efforts by backing him as the leader of a new socialist Yugoslavia. The three major ethnic groups in Yugoslavia—the Serbs, Croats, and Muslims—coexisted under Tito's strong leadership by subordinating their separatist agendas to the larger goal of a communist state independent of the Soviet Union. This relative peace continued for a decade after Tito's death in 1980. However, the region's temporary geopolitical stability completely unraveled in the 1990s, coincident with the lessening of Soviet control in other parts of eastern Europe.

In 1990, elections were held in Yugoslavia's different republics over the issue of secession from the mother state. Secessionist parties gained control in Slovenia and Croatia, but Serbian voters opted for continued Yugoslav unification, in what observers considered to be a government-controlled election. Nonetheless, Slovenia and Croatia declared independence in 1991, Macedonia in January 1992, and Bosnia and Herzegovina in April 1992. When the Yugoslavian army attacked Slovenia, the Balkan situation got Europe's full attention, and a negotiated settlement resulted in Slovenia's independence. In Bosnia–Herzegovina, however, Serb paramilitary units waged a ferocious war of ethnic cleansing against both Muslims and Croats in a war that lasted until 1995. At that time, a complex political arrangement created a Serb republic and a Muslim–Croat federation, both ruled by the same legislature and president. Croatia also fought a devastating, but successful war of independence against Serbian nationalists.

Kosovo, in the south of Serbia (as the former Yugoslavia was now called), was another trouble spot, with long-standing tensions between Serbs and Muslims. Although Kosovo had enjoyed differing degrees of autonomy within the former Yugoslavia, this autonomy was withdrawn by Belgrade in 1990 to protect the Serb minority population. Not surprisingly, the Muslim Kosovar rebels responded by proclaiming Kosovo's independence in 1991. This act was resisted vigorously by Serbia, which responded with a violent ethnic-cleansing program designed to oust the Muslims and make Kosovo a pure Serbian province (Figure 8.30). Diplomatic efforts failed to resolve this violence, and as warfare escalated in Kosovo, NATO (which included the United States) began bombing Belgrade in 1999 to force Serbia to accept a negotiated settlement. From 1999 until 2008, Kosovo was

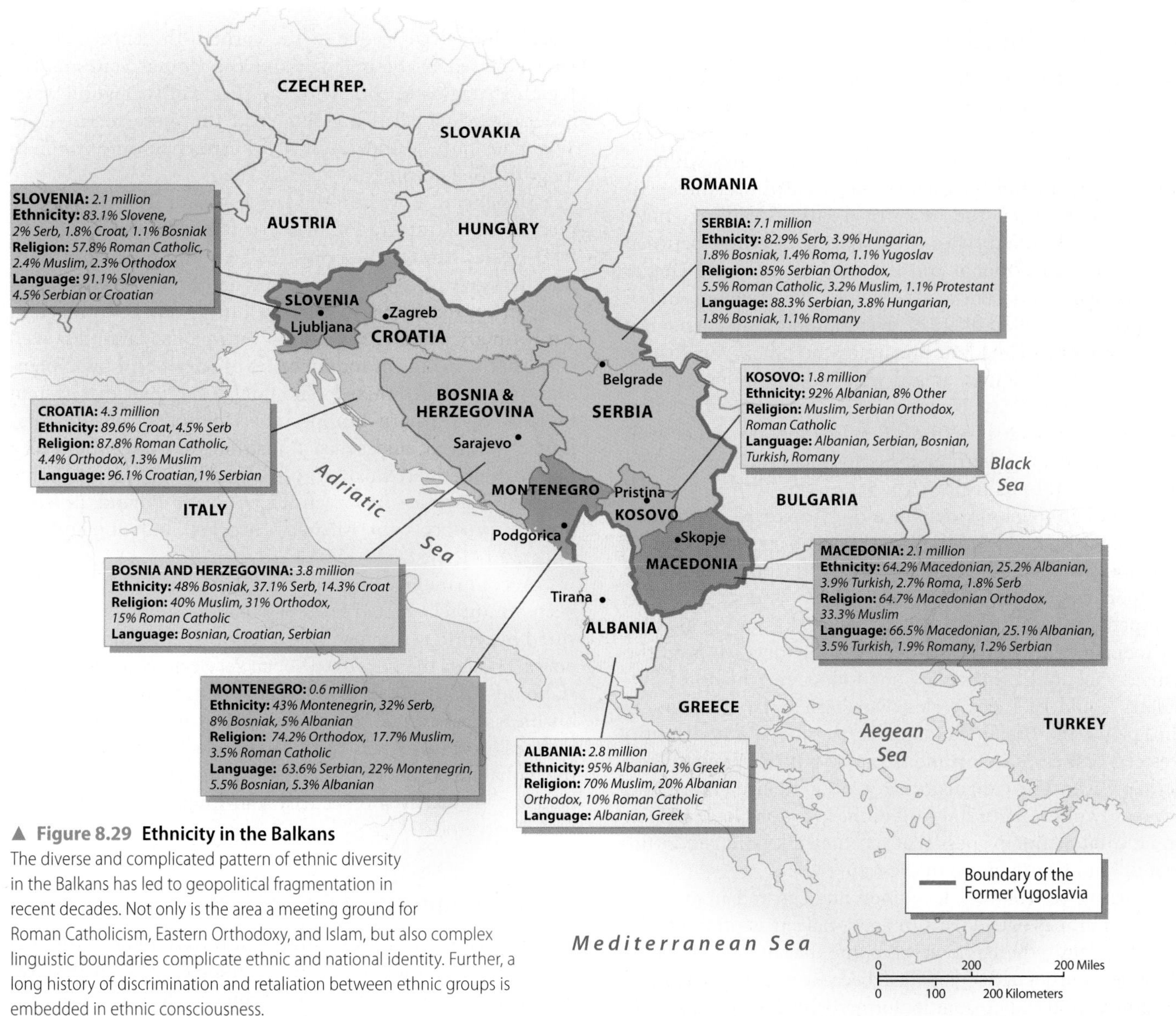

▲ **Figure 8.29 Ethnicity in the Balkans**
The diverse and complicated pattern of ethnic diversity in the Balkans has led to geopolitical fragmentation in recent decades. Not only is the area a meeting ground for Roman Catholicism, Eastern Orthodoxy, and Islam, but also complex linguistic boundaries complicate ethnic and national identity. Further, a long history of discrimination and retaliation between ethnic groups is embedded in ethnic consciousness.

administered by the United Nations (UN) as a protectorate, enforced by some 50,000 peacekeepers from 30 different countries.

Moving Toward Stability Although Serbia remains steadfast about reclaiming Kosovo and only grudgingly recognizes its independence, in most other matters a more moderate and less nationalistic government has led to Serbia's being reinstated in the UN and the Council of Europe and becoming an official candidate to the EU. Of the former Yugoslav republics, only Slovenia and Croatia have achieved membership in the EU, although all Balkan countries (including Albania, which was not a part of the former Yugoslavia) have begun the lengthy and complex EU membership process.

▼ **Figure 8.30 American Peacekeeping Troops in the Balkan Wars.** These U.S. soldiers were part of the United Nations force sent to Bosnia-Herzegovina to implement peace between different ethnic populations after Bosnia declared its independence from the former Yugoslavia.

A Europe of Small Regions?

As noted in Chapter 1, the concept of geopolitical **devolution** refers to a decentralization of power away from a central authority. In Europe, this takes many forms, ranging from an innocuous sharing of power with small states within a larger union to full-on calls for separatism and independence that threaten the unity of a larger political entity. Germany and France, with their sharing of power between the national government and the local districts and *Länder* (the German equivalent to U.S. states), illustrate one end of the spectrum. Scotland's 2014 referendum on independence from the United Kingdom illustrates the other. Somewhere in between lies Belgium, an awkward amalgamation of three distinct cultural areas (Flemish, Walloon, and German) that demonstrated in 2011–2012 it could function rather well without a federal government because so much power had been delegated to its regions.

Although regional demands for increased autonomy from larger states have long been a part of postwar Europe (Catalonia in Spain, Wales in the UK, and Corsica in France are good examples), all eyes are on Scotland as a clue to Europe's geopolitical future and fabric. If the Scots vote for independence and if they successfully devolve from the United Kingdom, others may soon follow, and the list of candidates could be long: Wales, Northern Ireland, Catalonia (in Spain), the Basque lands of northern Spain and southwestern France, Brittany and Corsica (both in France), and perhaps even the Languedoc area of southern France and Lombardy in northern Italy. All of these regions have a claim on cultural distinctiveness that has fueled separatism in the past or could easily do so in the future.

Clearly, geopolitical devolution and separatism are contrary to Europe's agenda of supranationalism as promoted by the EU. Many experts, however, opine that the EU has actually aided and abetted regionalism because it takes on many of the powers and responsibilities formerly held by national governments. Thus, the costs of regional independence are now lower than previously because microstates could exist comfortably under the EU's economic and political umbrella. Only time will tell, beginning with Scotland's 2014 referendum.

REVIEW

8.12 Describe briefly how the map of Europe changed with the Treaty of Versailles in 1918.

8.13 What European countries were considered Soviet satellites during the Cold War?

8.14 What countries made up former Yugoslavia?

Economic and Social Development: Integration and Transition

As the acknowledged birthplace of the Industrial Revolution, Europe in many ways invented the modern economic system of industrial capitalism. Though Europe was the world's industrial leader in the early 20th century, it was later eclipsed by both Japan and the United States as the region struggled to cope with the effects of two world wars, a decade of global depression, and the more recent Cold War. Currently, a drawn-out fiscal crisis continues to challenge Europe's economic structure.

In general, however, the last half-century of economic recovery and integration has been largely successful. In fact, western Europe's success at blending national economies has given the world a new model for regional cooperation, an approach that in the near future may be imitated in Latin America and Asia. Eastern Europe has fared less well because the results of four decades of Soviet economic planning were, at best, mixed. The total collapse of that system in 1990 cast eastern Europe into a period of chaotic economic, political, and social transition that has produced a highly differentiated pattern of rich and poor regions. Some countries, such as the Czech Republic and Poland, prosper, but the prospects for Albania, Hungary, and Romania are uncertain (Table 8.3).

Accompanying western Europe's economic boom has been an unprecedented level of social development as measured by worker benefits, health services, education, and literacy. Though the improved social services set an admirable standard for the world, today cost-cutting politicians and businesspeople argue that these services increase the cost of business so much that European goods cannot compete in the global marketplace. As a result, many of those traditional benefits, primarily job security and long vacation periods, have been eroded.

Europe's Industrial Revolution

Europe is the cradle of modern industrialism, with two fundamental innovations that enabled this **Industrial Revolution**: First, machines replaced human labor in many manufacturing processes, and, second, inanimate energy sources (water, steam, electricity, and petroleum) powered these new machines. England was the birthplace of this new system in the years between 1730 and 1850, but by the late 19th century this new industrialism had spread throughout Europe and, within decades, to the rest of the world.

Centers of Change England's textile industry, located on the flanks of the Pennine Mountains, was the center of the earliest industrial innovation. The county of Yorkshire, on the eastern side of the Pennines, had long been a hearth area for woolen textiles, drawing raw materials from the extensive sheep herds of that region and using the clean mountain waters to wash the wool before it was spun. Originally, waterwheels were used to power mechanized looms at the rapids and waterfalls of the Pennine streams, but by the 1790s the steam engine had become the preferred source of energy (Figure 8.31). Steam engines, however, needed fuel, and local wood supplies were quickly exhausted. Progress was stifled until the development

Table 8.3 Development Indicators

Country	GNI per capita, PPP 2011	GDP Average Annual %Growth 2000–11	Human Development Index (2011)[1]	Percent Population Living Below $2 a Day	Life Expectancy (2013)[2]	Under Age 5 Mortality Rate (1990)	Under Age 5 Mortality Rate (2011)	Adult Literacy (% ages 15 and older) Male/Female	Gender Inequality Index (2011)[3,1]
Western Europe									
Austria	42,030	1.8	.895	–	81	9	4	--/--	0.102
Belgium	39,150	1.5	.897	–	80	10	4	--/--	0.098
France	35,910	1.2	.893	–	82	9	4	--/--	0.083
Germany	40,190	1.1	.920	–	80	9	4	--/--	0.075
Ireland	33,520	2.4	.916	–	81	9	4	--/--	0.121
Luxembourg	64,110	2.9	.875	–	80	8	3	--/--	0.149
Netherlands	43,150	1.5	.921	–	81	8	4	--/--	0.045
Switzerland	52,530	1.9	.913	–	83	8	4	--/--	0.057
United Kingdom	35,950	1.7	.875	–	82	9	5	--/--	0.205
Southern Europe									
Albania	8,820	5.2	.749	<2	77	41	14	97/95	0.251
Bosnia & Herzegovina	9,190	4.2	.735	<2	76	19	8	99/96	–
Croatia	18,780	2.6	.805	<2	77	13	5	99/98	0.179
Cyprus	30,970	2.9	.848	–	78	11	3	99/97	0.134
Greece	25,110	1.8	.860	–	81	13	4	98/96	0.136
Italy	32,420	0.4	.881	–	82	10	4	99/99	0.094
Kosovo	–	5.2	–	–	69	–	–	--/--	–
Macedonia	11,370	3.3	.740	6.9	75	38	10	99/96	0.162
Montenegro	13,700	4.2	.791	<2	74	18	7	99/97	–
Portugal	24,620	0.5	.816	–	80	15	3	97/94	0.114
Serbia	11,550	3.7	.769	<2	74	29	7	99/97	–
Slovenia	26,500	2.9	.892	<2	80	10	3	100/100	0.080
Spain	31,440	2.1	.885	–	82	11	4	99/97	0.103
Northern Europe									
Denmark	41,920	0.8	.901	–	80	9	4	--/--	0.057
Estonia	20,850	3.9	.846	<2	76	20	4	100/100	0.158
Finland	37,660	1.9	.892	–	81	7	3	--/--	0.075
Iceland	31,020	2.7	.906	–	82	6	3	--/--	0.089
Latvia	19,090	4.1	.814	<2	74	21	8	100/100	0.216
Lithuania	20,760	4.7	.818	<2	74	17	6	100/100	0.157
Norway	61,450	1.6	.955	–	81	8	3	--/--	0.065
Sweden	42,210	2.3	.916	–	82	7	3	--/--	0.055
Eastern Europe									
Bulgaria	14,400	4.3	.782	<2	74	22	12	99/98	0.219
Czech Republic	24,490	3.7	.873	<2	78	14	4	--/--	0.122
Hungary	20,310	1.9	.831	<2	75	19	6	99/99	0.256
Poland	20,260	4.3	.821	<2	77	17	6	100/99	0.140
Romania	15,780	4.4	.786	<2	74	37	13	98/97	0.327
Slovakia	22,300	5.1	.840	<2	76	18	8	--/--	0.171
Micro States									
Andorra	–	5.9	.846	–	–	8	3	--/--	–
Liechtenstein	–	2.5	.883	–	82	10	2	--/--	–
Malta	24,480	1.8	.847	–	81	11	6	91/94	0.236
Monaco	–	4.3	–	–	–	8	4	--/--	–
San Marino	–	3.2	–	–	84	12	2	--/--	–
Vatican City	–	–	–	–		–	–	--/--	–

[1] United Nations, *Human Development Report, 2013*.

[2] Population Reference Bureau, *World Population Data Sheet, 2013*.

[3] Gender Inequality Index—A composite measure reflecting inequality in achievements between women and men in three dimensions: reproductive health, empowerment and the labor market that ranges between 0 and 1. The higher the number, the greater the inequality.

Source: World Bank, *World Development Indicators, 2013*.

▲ **Figure 8.31 Hearth Area of the Industrial Revolution** Europe's Industrial Revolution began on the flanks of England's Pennine Mountains, where swift-running streams were used to power mechanized looms to weave cotton and wool. Later, once the railroads were developed, many of these early factories switched to coal power.

of the railroad in the early 19th century, an innovation that allowed coal to be moved long distances at a reasonable cost.

Industrial Regions in Continental Europe The first industrial districts in continental Europe began appearing in the 1820s, located close to coalfields (Figure 8.32). The first area outside Britain was the Sambre-Meuse region, named for the two river valleys straddling the French-Belgian border. Like the English Midlands, it had a long history of cottage-based woolen textile manufacturing that quickly converted to the new technology of steam-powered mechanized looms.

By 1850, the dominant industrial area in all Europe (including England) was the Ruhr district in northwestern Germany, near the Rhine River. Rich coal deposits close to the surface powered the Ruhr's transformation from a small textile region to one of heavy industry, particularly of iron and steel manufacturing. Decades later, the Ruhr industrial region became synonymous with the industrial strength behind Nazi Germany's war machine, leading to its being bombed heavily in World War II.

Rebuilding Postwar Europe

As noted, Europe was the leader of the industrial world in 1900, when it produced 90 percent of the world's manufactured output. However, by 1945, after four decades of war and economic chaos, industrial Europe was in shambles, with many of its cities and industrial areas in ruins and much of the region's population dispirited, homeless, and hungry. Clearly, a new pathway for postwar Europe had to be forged to provide economic, political, and social security.

Evolution of the EU In 1950, the leaders of western Europe began discussing a new form of economic integration that would avoid a historical pattern of nationalistic independence leading to duplication of industrial effort. Robert Schuman, France's farsighted foreign minister, proposed the radical idea that instead of rebuilding separate iron and steel facilities in each country, Europe should share its natural resources. In May 1952, France, Germany, Italy, the Netherlands, Belgium, and Luxembourg ratified a treaty that joined them together in the European Coal and Steel Community (ECSC). Five years later, because of the resounding success of the ECSC, these six states agreed to work toward further integration by creating a larger European common market that would encourage the free movement of goods, labor, and capital. In March 1957, the Treaty of Rome was signed, establishing the European Economic Community (EEC).

In 1965, the EEC reinvented itself with the Brussels Treaty, which laid the groundwork for adding a political union to the already successful economic community. In this "Second Treaty of Rome," aspirations for more than economic integration were clearly envisioned, with the creation of an EEC council, court, parliament, and political commission. At that time, the EEC became the European Community (EC) and began expanding its membership beyond the original six member states. In 1991, the EC expanded its goals once again and became the European Union (EU), at which time the supranational organization moved even further into supranational affairs with discussions of common foreign policies and mutual security agreements.

In May 2004, the EU expanded beyond its core membership in western Europe by adding 10 new states, including a cluster of former Soviet-controlled communist satellites from eastern Europe. These new members—Latvia, Estonia, Lithuania, Poland, Slovakia, the Czech Republic, Hungary, Slovenia, Malta, and Cyprus—brought the total to 25. Bulgaria and Romania were admitted in January 2007 and Croatia in July 2013, resulting in the current 28 EU countries. As of this writing (late 2013), Iceland, Serbia, Montenegro, Macedonia, and Turkey are all formal candidates for EU membership (Figure 8.33).

Economic Disintegration and Transition in Eastern Europe

Eastern Europe has historically been less developed economically than its western counterpart. This is partially explained

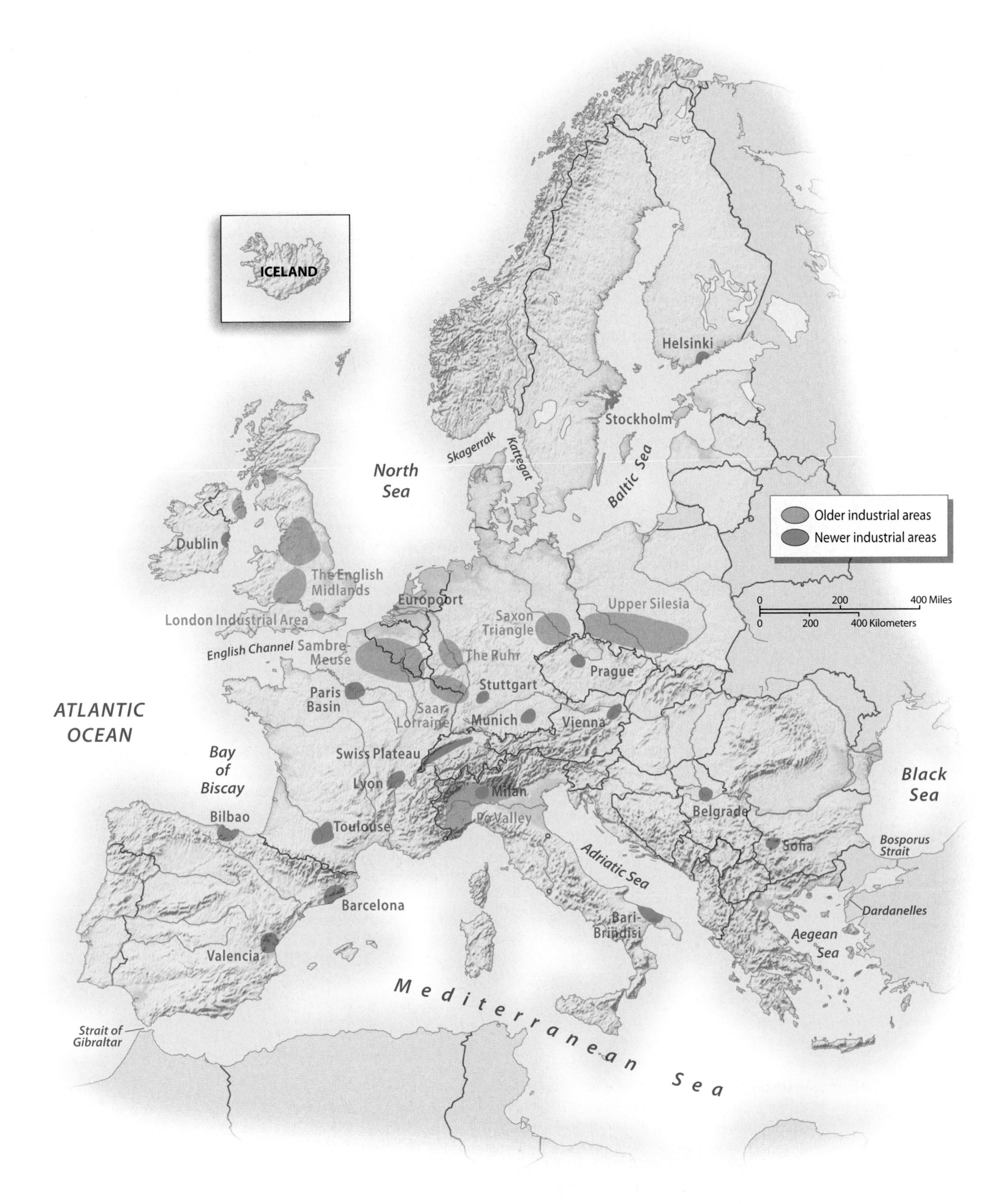

▲ **Figure 8.32 Industrial Regions of Europe** From England, the Industrial Revolution spread to continental Europe, starting with the Sambre-Meuse region on the French-Belgian border and then diffusing to the Ruhr area in Germany. Readily accessible surface coal deposits powered these new industrial areas. Early on, iron ore for steel manufacture came from local deposits, but later it was imported from Sweden and other areas in the shield country of Scandinavia. Most of the newer industrial areas are closely linked to urban areas.

▲ **Figure 8.33 The European Union** The driving force behind Europe's economic and political integration has been the EU, which was formed in the 1950s as an organization with six members focused solely on rebuilding the region's coal and steel industries. As of 2013, the EU has 28 members. Besides the official applicants of Turkey, Serbia, Macedonia, and Iceland, several Balkan countries are preparing applications to the EU. Note that Norway is not a member of the EU, primarily because membership would restrict that country's fishing industry. Those same fishing restrictions may also become problematic for Iceland's membership. **Q: Why is Switzerland not a number of the EU?**

▼ **Figure 8.34 Post-1990 Hardship in Eastern Europe** With the collapse of the Soviet Union in 1990, eastern Europe fell into a state of severe economic chaos that took its toll for several decades until countries could re-orient their economics (and political systems) to western Europe and the EU. This abandoned steelworks factory in Romania still stands as testimonial to those difficult times.

by the fact that this region is simply not as rich in natural resources as western Europe is. Also important is that in eastern Europe, these resources have long been exploited by outside interests, rather than being developed internally. This pattern began in the 19th century, when the Ottoman and Hapsburg empires dominated eastern Europe and the Balkans. Later, eastern Europe was exploited by Nazi Germany and, most recently, by the Soviet Union's centralized planning from 1945 to 1990.

Even though Soviet economic planning was ostensibly an attempt to develop eastern Europe's economy by coordinating resource usage, these efforts were, in fact, more to serve Soviet homeland interests. This system worked with mixed results for more than 40 years, but eastern European countries were plunged into economic and social chaos when the Soviet Union itself collapsed in 1991. Recovery and development since that time have been difficult, with some countries making the transition more rapidly and more fully than others, creating a geographic patchwork of both wealth and hardship throughout eastern Europe.

Change Since 1991 In place of Soviet coordination and subsidy came a painful period of economic transition that was outright chaotic in many eastern European countries. As the Soviet Union turned its attention to its own economic and political turmoil, it stopped exporting cheap natural gas and petroleum to eastern Europe. Instead, Russia sold these fuels on the open global market to gain hard currency. Without cheap energy, many eastern European industries were unable to operate and were forced to close, idling millions of workers (Figure 8.34). For example, in the first two years of the transition (1990–1992), industrial production fell 35 percent in Poland and 45 percent in Bulgaria. In addition, markets for eastern European products, guaranteed under Soviet planning, simply evaporated, further aggravating the collapse of eastern European economies.

To recover, eastern European countries began redirecting their economies toward western Europe. But to do this meant moving from a socialist-based economy of state ownership and control to a capitalist economy of private ownership and free markets. Without Soviet price supports and subsidies, a completely new economic system had to be constructed, one that could compete favorably in the new global marketplace. During this transitional period, financial security was elusive, with irregular paychecks for those with jobs and uncertain welfare benefits for those without. Cronyism prevailed, as those with strong social connections profited at the expense of others who were lost in the chaos. While some countries—Poland, the Czech Republic, Slovenia, and Slovakia—made the transition quickly, others—primarily the Balkan states—are taking longer (Figure 8.35). Of considerable help in this economic and social transition was EU membership, with a complex

▼ **Figure 8.35 Western Hypermarkets in Eastern Europe** A customer buys fish at a new Carrefour hypermarket on the outskirts of Sofia, Bulgaria. Consumer items, including food, were historically in short supply in eastern Europe during the postwar Soviet satellite period, with these shortages continuing through the difficult post-Soviet transition years. Today, however, there's no shortage as western super- and hypermarkets have populated eastern Europe. But whether local customers can actually afford these western goods remains unclear. Carrefour, a French firm, and the world's second largest retailer behind Walmart, is reportedly selling its eastern European stores in order to expand retailing efforts in China.

▲ **Figure 8.36 Development Issues in Europe: Economic Disparity** This map shows the economic differences in European countries based upon gross national income (GNI) per capita adjusted for purchasing power parity (PPP). Most striking are the economic disparities not only between western and eastern countries, but also between the richest (Norway, Luxembourg) and the poorest (Albania, Bosnia–Herzegovina). Unemployment rates, which can change rapidly, were included as indicators of economic vitality in mid-2013.

application process that was, nevertheless, accompanied by an infusion of vast amounts of funds to facilitate everything from environmental cleanup to development of small businesses. Important to remember is that this recovery and transition process in eastern Europe has not ended, but is still going on, with greater and lesser degrees of success.

Promise and Problems of the Eurozone

A common component of a country's sovereignty is the ability to control its own monetary system, which indeed was the case in Europe until a decade ago. At that time, people used marks in Germany, francs in France, lira in Italy, pounds in England, pesetas in Spain, and so on. Today, however, much of Europe has moved from individual state monetary systems to a common currency. This radical change began in 1999 when 11 of the then 15 EU member states joined together to form the European Monetary Union (EMU). At that time, cross-border business and trade transactions began taking place in a new monetary unit, called the euro. In 2002, euro coins and bills completely replaced traditional national currencies in the EMU countries, creating what is commonly called the **Eurozone**. Today it consists of 18 of the EU's 28 member states. In theory, all remaining EU members are obligated to join the Eurozone at some future date. Whether that happens or not remains to be seen.

By adopting a common currency, Eurozone members sought to increase the efficiency and competitiveness of both domestic and international business by eliminating costs associated with business payments made in different currencies. Germany, for example, exports two-thirds of its products to other EU members. With a common currency, this business is now essentially domestic trade, protected from the fluctuations of different currencies and not subject to transaction costs. Although many traditional economists had misgivings about this new European common currency system, the political goal of enhancing European political unity took precedence.

Some EU member countries, however—most notably, the United Kingdom—had (and still do have) considerable reservations about giving up control over their national monetary system and therefore continue to use their traditional currency. Indeed, the early controversy over Eurozone membership has been rekindled by the current economic crisis, bringing into question whether the EMU goal of having all EU members share a common currency will ever be achieved (Figure 8.36).

Europe's Current Economic Crisis Most analysts agree that two problems have caused Europe's present crisis. First, weak economies (namely, Greece, Spain, Ireland, and Portugal) were able to get cheap credit because of their EMU membership; with cheap money, governments and consumers spent too much, creating huge indebtedness. Second, before the EMU it was common for governments to manage debts by adjusting their currency by inflation or deflation; however, this is no longer allowed under the EMU. Instead, EMU members must seek help from more solvent EMU countries, which today means receiving bailouts funded by Germany, France, Luxembourg, and the Netherlands. Aggravating the crisis is the reality that the initial round of bailouts did not solve the problem, so Eurozone taxpayers are reluctant to fund further aid (Figure 8.37). Additionally, the list of weak

▼ **Figure 8.37 Euro Crisis and Protest** Greece suffered deeply because of the Euro crisis and its population has protested vehemently against the fiscal policies resulting from its bailout. Here, in Athens, Greece, labor unions protest against those policies. The term "troika" is originally a Russian word used to describe the three-part leadership of the Soviet era, and is used here in an unflattering reference to the three organizations administering the bailout funds to Greece: the EU, the European Central Bank, and the International Monetary Fund (IMF).

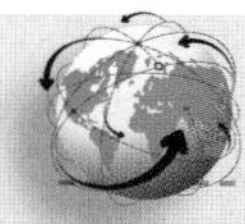

Exploring **Global Connections**

Gas, Money, and Politics in Europe's Southeast Corner

Europe's southeast corner is where East meets West, where cultures clash and geopolitical tensions fester. In the past, it was an arena of conflict between Christian and Muslim empires; today it's about offshore banking, money laundering, natural gas, and Middle East politics. One flashpoint for these geopolitical tensions is the divided island of Cyprus. (Figure 8.38)

Divided Cyprus Cyprus was fought over for centuries by Greeks and Turks—until UN peacekeepers separated the warring factions in 1964. They divided Cyprus between the majority Greek Cypriots in the south and the minority Turkish Cypriots to the north. In 1983, the Turkish Republic of Northern Cyprus (TRNC) declared itself an independent country; however, its independence is recognized only by Turkey (Figure 8.5.1).

In 2004, Cyprus entered the EU, with EU power and privileges extended only to "internationally recognized governments." Currently, this is only the Greek Cypriot half of the island. Then, in 2008, Cyprus joined the European Monetary Union, adopting the euro as its currency and setting the scene for the current crisis.

Money Troubles After the 1990 collapse of the Soviet Union, Cyprus found itself in the offshore banking/money laundering business, as newly rich Russian oligarchs stashed their money in Cypriot banks, far removed from the fickle Russian economy and safe (they thought) in European currency. Following their money, Russians also bought up island real estate, creating a cozy Russian community on this warm Mediterranean island, far from the cold Moscow winter (Figure 8.39).

However, because of their close economic and political ties to Greece, in late 2012 Cypriot banks found themselves in a serious fiscal crisis. Bank managers thought this crisis was only temporary because recent offshore oil and gas exploration had turned up a bonanza, that should bring riches in the near future. The government proposed a temporary measure to tide the banks over fiscally: a tax on Russian money held by the banks.

Gas Wars Now the plot thickens. Russia, of course, objected to the tax on its wealthy, but said it would gladly bail out Cypriot banks, provided Russia was included in any future natural gas deals. But Europe

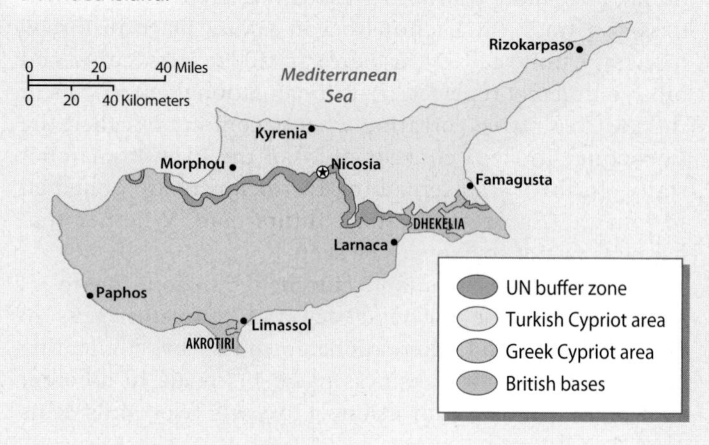

▼ **Figure 8.5.1 A Divided Cyprus** Geopolitical challenges keep Cyprus a divided island.

economies continues to grow; now on the endangered list is Italy, a country characterized as too big to fail, yet also too big to bail out. Billions of euros helped Greece, Slovenia, and Cyprus, but it will take trillions of euros, economists say, to save Italy.

Furthermore, there is controversy about how to move forward. Until recently, countries seeking EMU bailout money were required to adopt a policy of fiscal austerity, resulting in massive reductions of governmental spending. But that has led to stagnant economies and still higher unemployment in the affected countries. Other economists argue that bailout money should be used to stimulate growth, with massive government projects that create jobs and put money in the hands of consumers. More draconian scenarios are also discussed, such as whether weak countries should actually leave the Eurozone so they can manage their indebtedness by devaluing their currencies. However, to do so would have immense global implications, experts say, since this strategy would involve countries defaulting on their international indebtedness. Some say the negative effects on the United States would be far worse than those experienced during the 2007–2008 economic collapse, which is why the unresolved European financial crisis continues to unsettle U.S. and global stock markets.

REVIEW

8.15 What geographic factors are important in explaining the beginnings of the Industrial Revolution?

8.16 Describe the evolution of the European Union (EU).

8.17 What factors must be considered in describing eastern Europe's postwar economic geography?

8.18 Briefly describe the current pattern of regional differences in the economic geography of Europe.

said no, absolutely not, since Europe, which already depends on Russia for natural gas imports, does not want to see Russia involved in new Mediterranean gas fields.

Complicated? It gets worse: Now Turkey, Israel, Lebanon, Qatar, and the United States are involved (**Figure 8.5.2**). To start with, Turkey disputes Cyprus's ownership of the underwater gas fields and is angry that Cyprus granted leases to U.S. energy firms for resources Turkey claims are in Turkish territory. Israel is involved because it has territorial rights over a large adjacent offshore gas field; it needs to cut a deal with Turkey to get the gas onshore and into pipelines to Europe. Russia is courting both Israel and Turkey so it can expand its gas holdings. Qatar, the world's largest liquid natural gas exporter, doesn't want to lose its lucrative export business to Europe, so it wants a piece of the eastern Mediterranean action. Unquestionably, when natural gas, Russian money, and the Middle East meet, the action could be hot, to say the least.

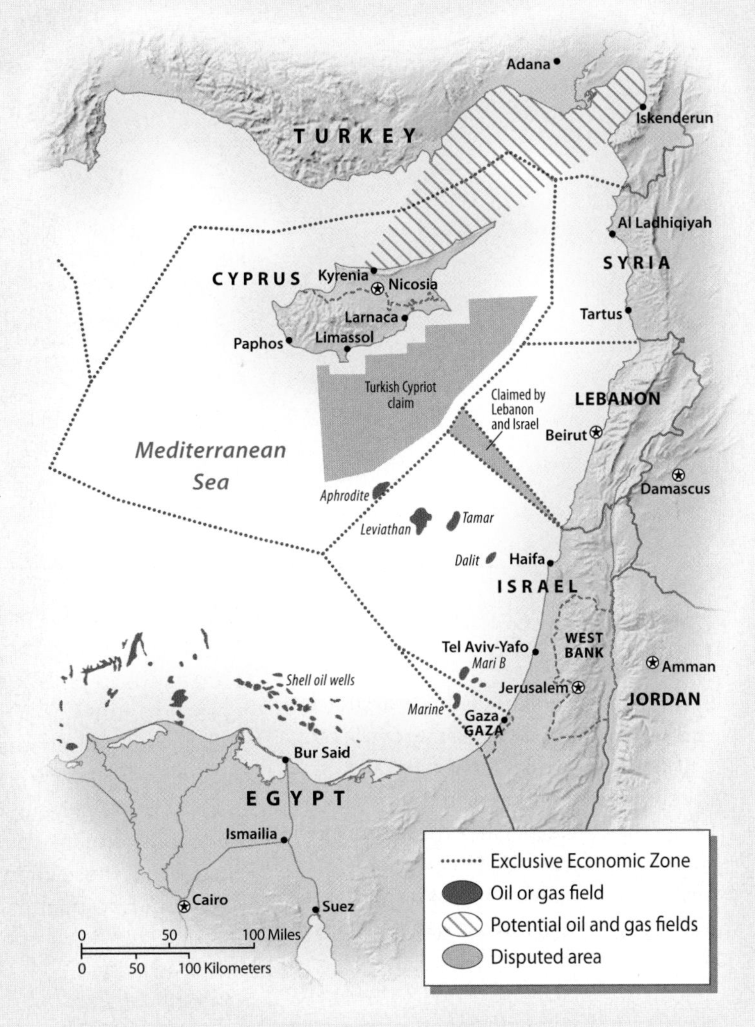

► **Figure 8.5.2 Eastern Mediterranean Offshore Gas Fields** A new geopolitical conflict is emerging as various states jockey for rights to offshore oil and gas exploration.

▼ **Figure 8.38 Cyprus** Located south of Turkey and northwest of Lebanon, the island is 9,250 square kilometers.

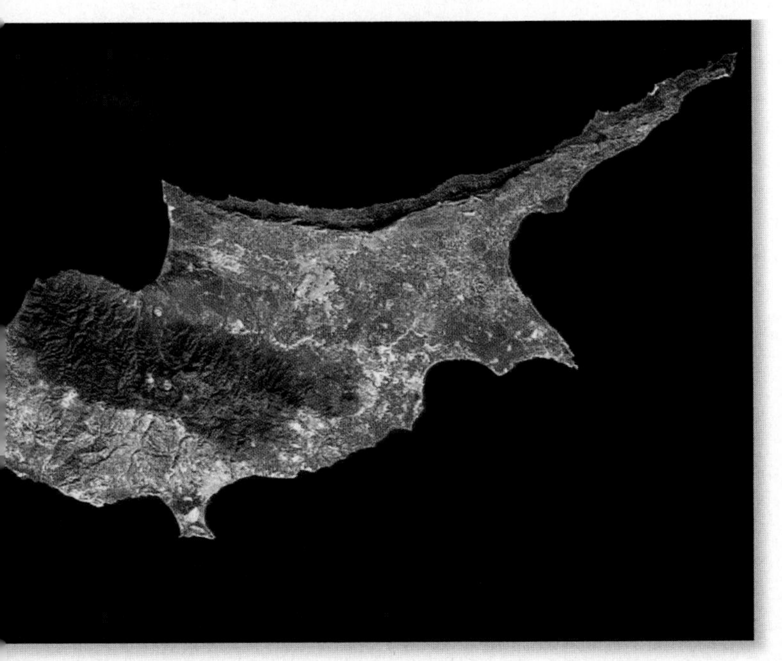

▼ **Figure 8.39 Russians in Cyprus** An influx of post-Soviet era Russian investments in the 1990's created significant Russian community and services- now challenged by severe economic crisis.

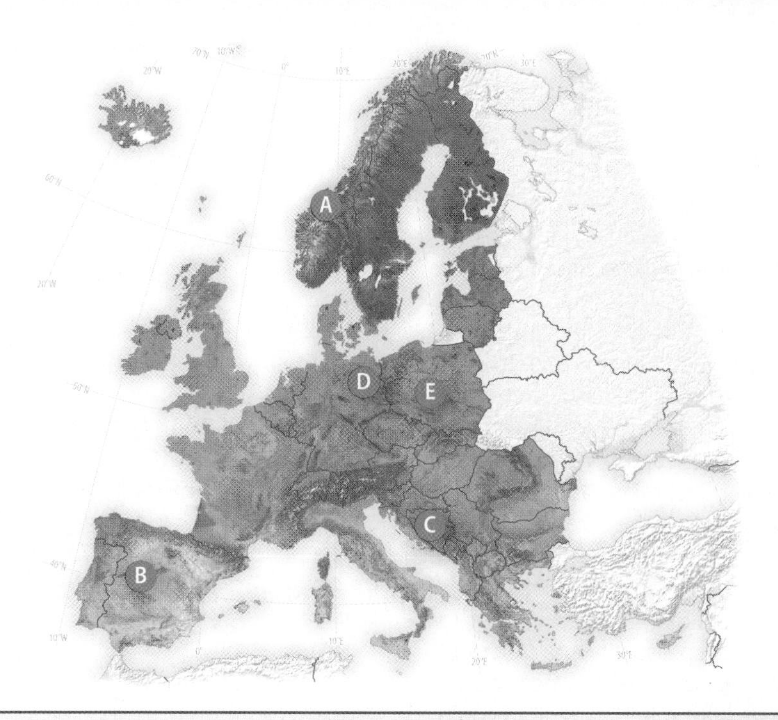

Physical Geography and Environmental Issues

Because of its immense latitudinal stretch, Europe includes a diverse array of climates and landscapes, from arctic tundra to dry summer Mediterranean shrublands. Under the EU's leadership, the region has successfully addressed trans-boundary environmental issues such as air and water pollution, but greenhouse gas emissions remains an unresolved issue.

8.1 What are the dominant landforms along the coast of Norway, and what geologic process created them?

8.2 Investigate whether fracking for natural gas will likely become common in Europe. Then link your findings to Europe's current dependence on imported fossil fuels. Might those imports increase or decrease? How might Russia's successful fracking change Europe's energy and emission picture?

Population and Settlement

With the notable exception of Ireland, all other European countries are below replacement levels in terms of natural growth. Unless this changes through family-friendly policies, future population growth (or decline) will be determined by in-migration. In terms of settlement patterns, Europe is a highly urbanized world region. Although many of Europe's cities were heavily damaged during wartime bombings, many have rebuilt and preserved their historic districts, demonstrating the importance of space and place to cultural identity.

8.3 Explain the reasons behind the differing population densities in Spain.

8.4 Create several plausible scenarios for Europe's economic vitality (or lack of it) in the year 2020. Then discuss how those scenarios might affect migration within as well as migration to Europe.

Key Terms

balkanization *(p. 384)*
buffer zone *(p. 383)*
Cold War *(p. 352)*
Cyrillic alphabet *(p. 374)*
devolution *(p. 386)*
European Union (EU) *(p. 352)*
Eurozone *(p. 393)*
fjord *(p. 355)*
Industrial Revolution *(p. 386)*
Iron Curtain *(p. 383)*
irredentism *(p. 380)*
medieval landscape *(p. 369)*
North Atlantic Treaty Organization (NATO) *(p. 383)*
Renaissance–Baroque period *(p. 370)*
Schengen Agreement *(p. 366)*

In Review • Chapter 8
Europe

Cultural Coherence and Diversity

Historically, Europe's diverse cultural geography was primarily a product of language and religion; today, however, it has become more complex because of diffuse global influences interacting with migrant cultures. The result is a complex mixture of traditional, global, and ethnic cultures. An example is how France tries to retain its traditional French language in the face of American English–speaking media and street-level African migrant speech.

C

8.5 What does the cultural landscape of Mostar, Bosnia-Herzegovina tell you about that country?

8.6 Find information on birth and death rates for a specific ethnic or migrant group within Europe. Compare these data to the national birth and death rates so that you can project how that specific group will grow (or not) compared to the larger national society.

Geopolitical Framework

Europe's borders changed often during the 20th century because of two world wars, the Cold War's end in 1990, and the devolution of the former Yugoslavia. Today there are different degrees of separatism, ranging from the autonomy given Basques and Catalonians in Spain to the very real possibility Scotland will separate from the United Kingdom.

D

8.7 This sign is in an icon of Berlin. Why?

8.8 If Scotland becomes independent, what are the implications in terms of its economic vitality, self-defense, monetary system, and all the other national housekeeping matters?

Economic and Social Development

After decades of postwar economic growth, Europe is now mired in a troublesome fiscal crisis that calls into question the EU's common currency program and, in many ways, aggravates the economic and social disparities between Europe's rich and not-so-rich countries.

E

8.9 After reviewing the data on Poland's economic condition, write a short description of how the activities along this street in Warsaw have changed since 1990.

8.10 Discuss the advantages and disadvantages of Greece, Spain, or Cyprus leaving the Eurozone as a way of resolving its fiscal crisis.

Authors' Blogs

Scan to visit the author's blog for chapter updates

www.gad4blog.wordpress.com

Scan to visit the GeoCurrents blog

www.geocurrents.info

9 The Russian Domain

Physical Geography and Environmental Issues

Many areas within the Russian domain suffered severe environmental damage during the Soviet era (1917–1991). Today air, water, toxic chemical, and nuclear pollution plague large portions of the region.

Population and Settlement

Urban landscapes within the Russian domain reflect a fascinating mix of imperial, socialist, and post-communist influences. Recently, many larger urban areas within the region have started to show similar trends toward sprawl and decentralization as those seen in North America and western Europe.

The Lena Pillars. Now protected as a UNESCO World Heritage Site, this spectacular Siberian setting south of Yakutsk in Russia's Sakha Republic is a reminder of the region's complex role as a storehouse of scenic treasures and natural resources.

Cultural Coherence and Diversity

Although Slavic cultural influences dominate the region, many non-Slavic minorities, including a variety of indigenous peoples in Siberia and a complex collection of ethnic groups in the Caucasus Mountains, shape the cultural and political geography of the domain.

Geopolitical Framework

The centralization of Russian political power has increased its dominance across the region, but democratic freedoms within the country have suffered amid crackdowns on the press and personal liberties.

Economic and Social Development

Russia's large supplies of oil and natural gas have made it a major player in the global economy, but future prosperity may increasingly hinge on unpredictable world prices for fossil fuels.

Travel slowly up the Lena River from the city of Yakutsk, the capital of the autonomous Sakha Republic. Be sure to choose summer for your journey: Yakutsk is one of the world's coldest cities, located deep in the heart of central Siberia. As you head upstream, admire the endless larch forests, part of the vast Siberian boreal ecosystem—millions of square miles of conifers—much of it underlain by permafrost. Less than a day's travel from Yakutsk, you see ragged pinnacles and gullied columns bordering the river's edge. Soon, the Lena Pillars fill your view, tall (500–1000 feet or 150–300 meters), naked, eroding cliffs (rich in fossils) of tan and yellow Cambrian rock. Long the traditional haunt of the reindeer-hunting Evenki people, the region was declared a UNESCO World Heritage Site in 2012, preserved for its scenic beauty and unique geological character. The Lena Pillars are an increasingly popular tourist destination for adventurous travelers, and they also symbolize for many regional residents the unparalleled riches and sheer scale of the Siberian landscape.

The Lena Pillars and their continental surroundings are an apt entryway into the huge and sprawling Russian domain, which occupies much of Earth's largest land mass. The region extends across the vast northern half of Eurasia and includes not only Russia itself, but also the nations of Ukraine, Belarus, Moldova, Georgia, and Armenia. The boundaries of the Russian domain have shifted over time. For decades, the regional definitions were relatively easy because the highly centralized Soviet Union (formally the Union of Soviet Socialist Republics or USSR) dominated the region's political geography. The country, born in a communist revolution in 1917, dwarfed all other states in the world in size and was powerfully united by the Soviet government and largely controlled by ethnic Russians. Most geographers agreed that the Soviet Union could thus be viewed as a single world region. Although the Soviet Union contained many cultural minorities, it wielded remarkable political and economic power from Leningrad (now St. Petersburg) on the Baltic Sea to Vladivostok on the Pacific Ocean. After World War II, some geographers even included much of Soviet-dominated eastern Europe within the region, in response to the country's expanded military role in nations such as East Germany, Poland, and Hungary.

The maps were suddenly redrawn late in 1991. The once-powerful Soviet state was officially dissolved, and in its place stood 15 former "republics" that had once been united under the Soviet Union. Now independent, each of these republics has tried to make its own way in a post-Soviet world. Some geographers initially treated the region as the "Former Soviet Union," but it quickly became clear that diverse cultural forces, economic trends, and political orientations were taking the republics in different directions. Even so, the Russian Republic remained dominant in size and influence and thus came to form the nucleus of a new Russian domain.

The Russian domain is rich with superlatives: Endless Siberian spaces, unlimited natural resources, legends of ruthless Cossack warriors, and tales of epic wars and revolutions are all part of the region's geographical and historical mythology.

The new regional definition reflects the changing political and cultural map since the breakup of the Soviet Union. The term *domain* suggests continuing Russian influence within the five other nations included in the region. Russia, Ukraine, and Belarus make up the core of the region. Enduring cultural and economic ties closely connect these three countries. Nearby Moldova and Armenia also broadly remain within Russia's geopolitical orbit, although forces within both countries are advocating closer ties with the European Union (EU). Relations between Russia and Georgia remain very strained, and recent political conflicts in Georgia brought Russian troops into that small country. Two significant areas that were once a part of the Soviet Union have been eliminated from the domain. The mostly Muslim republics of Central Asia and the Caucasus (Kazakhstan, Uzbekistan, Kyrgyzstan, Turkmenistan, Tajikistan, and Azerbaijan) have become aligned with a Central Asia world region (Chapter 10), while the Baltic republics (Estonia, Latvia, and Lithuania), all now in the EU and North Atlantic Treaty Organization (NATO), are best grouped with Europe (Chapter 8).

LEARNING OBJECTIVES

After reading this chapter you should be able to:

- Explain the close connection among latitude, regional climates, and agricultural production in Russia.
- Describe the major environmental issues affecting residents of the region.
- Identify the potential benefits and hazards of global warming within the region.
- Summarize major migration patterns, both in Soviet and in post-Soviet eras.
- Explain major urban land-use patterns in a large city such as Moscow.
- Describe the major phases of Russian expansion across Eurasia.
- Identify the key regional patterns of linguistic and religious diversity.
- Summarize the historical roots of the region's modern geopolitical system.
- Provide examples of how persistent cultural differences shape contemporary geopolitical tensions.
- Identify key ways in which natural resources, including energy, have shaped economic development in the region.
- Describe the key sectors of the Soviet-era economy and list major changes that have shaped the region's economy since the fall of the Soviet Union in 1991.

▲ **Figure 9.1 Cossacks** Natives of the Ukrainian and Russian steppe, the highly mobile Cossacks played a pivotal role in aiding Russian expansion into Siberia during the 16th century. Many modern descendants retain their skills of horsemanship and are proud of their distinctive ethnic heritage.

The Russian domain is rich with superlatives: Endless Siberian spaces, unlimited natural resources, legends of ruthless Cossack warriors, and tales of epic wars and revolutions are all part of the region's geographical and historical mythology (Figure 9.1). Indeed, the rise of Russian civilization remarkably parallels the story of the United States. Both cultures grew from small beginnings to become imperial powers that benefited from the fur trade, gold rushes, and transcontinental railroads during the 19th century. In addition, both countries experienced dramatic change resulting from industrialization in the 20th century.

Recently, however, the Russian domain has witnessed particularly breathtaking change. With the fall of the Soviet Union in 1991, new political and economic institutions reshaped everyday life. Economic collapse in the late 1990s produced steep declines in living standards throughout the region. Political instability grew between neighboring states, as well as within countries. After 2000, strong and increasingly centralized leadership within Russia set the region on a different course. With the help of higher energy prices (Russia is a major exporter of oil and natural gas), real economic improvements benefited most of the region, helping stabilize its economy and its larger political role in the world. Currently, however, many of the democratic reforms in Russia that were welcomed with the fall of communism in 1991 are being eroded by powerful governments committed to more direct state control. This new assertiveness on the part of the Russian government is increasing tensions within that country, as well as between Russia and its neighboring states.

Globalization is also shaping the Russian domain in complex ways. The region's relationship with the rest of the world shifted dramatically during the last 10 years of the 20th century. Until the end of 1991, all six countries belonged to the Soviet Union, the world's most powerful communist state. Under Soviet control, the region's 20th-century economy saw large increases in industrial output that made the nation a major global producer of steel, weaponry, and petroleum products. Its communist system offered a powerful set of ideas that promised economic prosperity and hope to residents within the region and around the world. Indeed, the political and military reach of the Soviet Union spanned the globe, making it a superpower on par with the United States. The Soviet presence dominated many eastern European countries, and nations from Cuba to Vietnam enjoyed close strategic ties with the country.

Suddenly, as the old communist order evaporated early in the 1990s, the now-independent republics of Russia, Ukraine, Belarus, Moldova, Georgia, and Armenia had to carve out new regional and global relationships. With the breakdown of Soviet control, the region also felt the growing presence of western European and American influences. Westernized popular culture combined with radical economic changes to transform the region. These new global relationships have not been easy. Social tensions have risen within the region as people grapple with fundamental economic changes. Political relationships with neighboring regions in Europe and Asia remain uncertain. The Russian domain has also been more fully exposed to both the opportunities and the competitive pressures of the global economy. The result is a world region that has seen its global linkages redefined in the recent past. Today fluctuating world oil markets, shifting patterns of foreign investment, new patterns of migration, and the shadowy flows of illegal drugs and Russian mafia money all demonstrate the unpredictable nature of the Russian domain's global connections.

Slavic Russia (population 140 million) dominates the region (Figure 9.2). Although only about three-quarters the size of the former Soviet Union, Russia's dimensions still make it the largest state on Earth. West of Moscow, the country's European front borders Finland and Poland, while far to the east, Mongolia and China share a sparsely populated boundary with sprawling Russian Siberia. Its area of 6.6 million square miles (17 million square kilometers) dwarfs even Canada, and its nine time zones are a reminder that when dawn arrives in Vladivostok on the Pacific Ocean, it is still only evening in Moscow.

With the demise of the Soviet Union, the Russians ended almost 75 years of Marxist rule. After a decade of political and economic instability (1991–2000), Russia has made

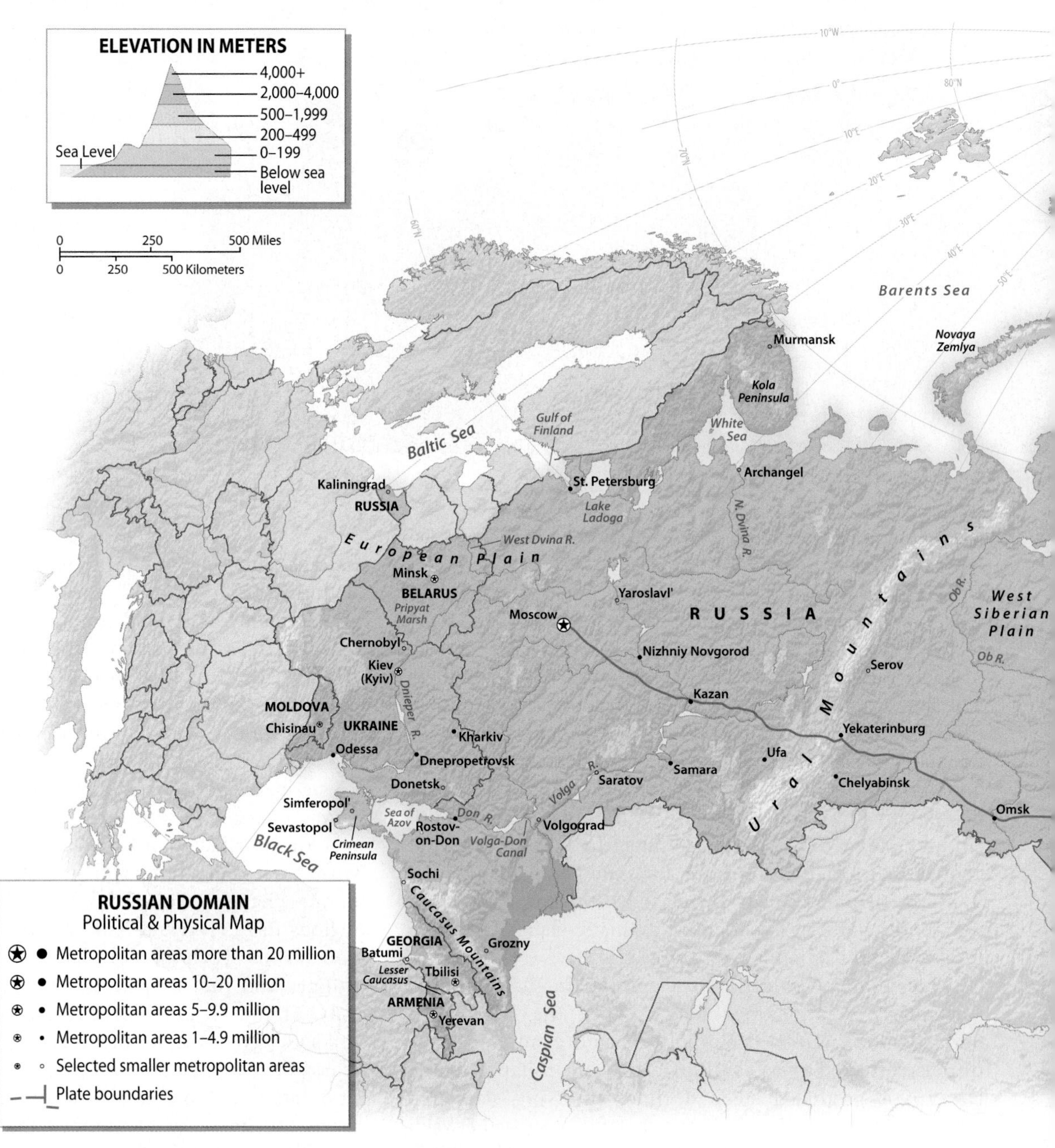

▲ **Figure 9.2 The Russian Domain** Russia and its neighboring states of Belarus, Ukraine, Moldova, Georgia, and Armenia make up a dynamic and unpredictable world region. Sprawling from the Baltic Sea to the Pacific Ocean, the region includes huge industrial centers, vast farmlands, and almost-empty stretches of tundra.

impressive progress in the new century. Russian President Vladimir Putin has built a reputation for strong leadership (2000–2008, 2012–), and he has stressed Russia's economic growth. Much of that growth has come to Russian cities, where expanding middle and professional classes are enjoying better living standards. Many rural areas, however, remain deeply mired in poverty. Another concern is that Putin's desire for wealth and power has been matched by his need for more-centralized political control inside the country. His close cooperation with the country's **oligarchs**, a small group of wealthy, very private businessmen who control (along with organized crime) important aspects of the Russian economy, also raises suspicions about Putin's actions. In addition, recent agreements on Russia's part to build closer economic ties with several neighboring nations (such as Belarus and Kazakhstan) suggest that the larger Russian

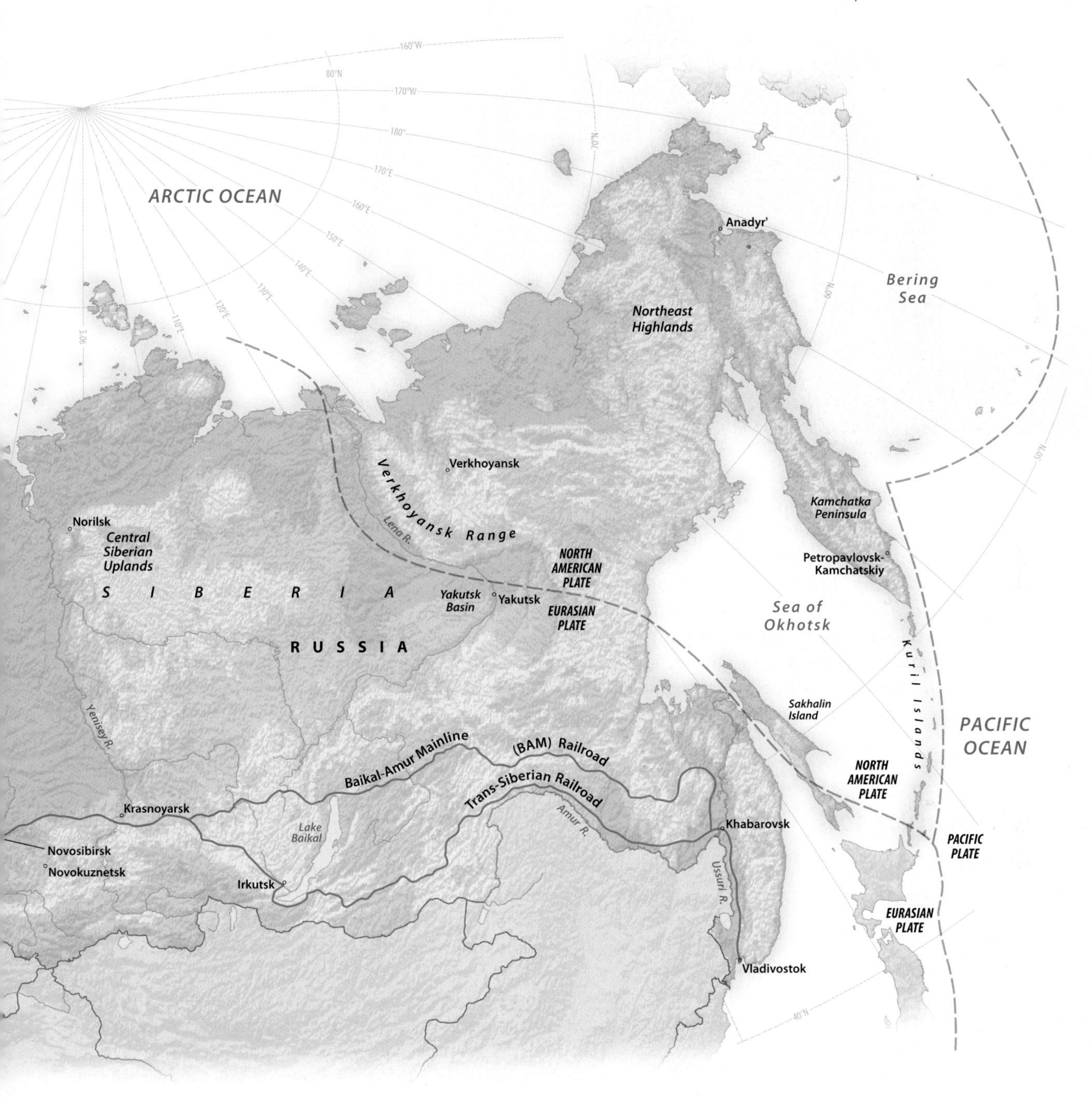

desire for more-centralized authority did not end with the demise of the Soviet Union. These agreements may also act as a counterbalance on the west to an expanding EU.

The bordering states of Ukraine, Belarus, Moldova, Georgia, and Armenia are inevitably linked to the evolution of their giant neighbor, even as they attempt to make their own way as independent nations. Emerging from the shadows of Soviet dominance has been difficult. Ukraine, in particular, has the size, population, and resource base to become a major European nation, but it has struggled to create real political and economic change since independence. With 45 million people and a rich storehouse of resources, Ukraine's size of 233,000 square miles (604,000 square kilometers) is similar to that of France. Nearby Belarus is smaller (80,000 square miles or 208,000 square kilometers), and its population of 10 million is likely to remain more closely tied economically and politically to Russia. Presently, its strikingly authoritarian and antiforeign leadership reflects many aspects of the old Soviet empire.

Moldova, with 4 million people, shares many cultural links with Romania, but its economic and political connections have kept it more closely tied to the Russian domain

◀ **Figure 9.3 Chisinau, Moldova** With a population of almost 1 million people, the Moldovan capital of Chisinau (called by its Russian name, *Kishinev*, during the Soviet period) is in the most economically affluent part of this small eastern European nation. Its landscape still reflects the influence of Soviet-era planning and urban design.

than to nearby portions of central Europe (Figure 9.3). South of Russia and beyond the bordering Caucasus Mountains, the Transcaucasian countries of Armenia and Georgia are similar in size to Moldova, but their populations differ culturally from that of their Slavic neighbor to the north. In addition, these two nations face significant political challenges: Armenia shares a hostile border with Azerbaijan (see Chapter 10), and Georgia's ethnic diversity and contentious relations with neighboring Russia threaten its political stability.

Physical Geography and Environmental Issues: A Vast and Challenging Land

The region's physical geography continues to shape its economic prospects in fundamental ways. For example, Russia's vast size poses special challenges, but its rich store of natural resources has offered unique economic benefits. At the same time, the Soviet period (1917–1991) also witnessed unparalleled, unrestrained economic development that damaged the region's environment in enduring ways. Today, more than two decades after the collapse of the Soviet Union, the environment still bears the scars of that legacy.

A Diverse Physical Setting

The region's northern latitudinal position is critical to understanding basic geographies of climate, vegetation, and agriculture (see Figure 9.4). Indeed, the Russian domain provides the world's largest example of a high-latitude continental climate, where seasonal temperature extremes and short growing seasons greatly limit opportunities for human settlement. In terms of latitude, Moscow is positioned as far north as Ketchikan, Alaska, and even the Ukrainian capital of Kiev (Kyiv) would sit north of the Great Lakes in Canada. Thus, apart from a small subtropical zone near the Black Sea, much of the region experiences a classic continental climate with hard, cold winters and marginal agricultural potential (see Figure 9.4).

The European West An airplane flight over the western portions of the Russian domain would reveal a vast, barely changing landscape below. European Russia, Belarus, and Ukraine cover the eastern portions of the vast North European Lowland, which runs from northern Germany to the Ural Mountains. One major geographical advantage of European Russia is that different river systems, all now linked by canals, flow into four separate drainage areas. The result is that trade goods can easily flow in many directions within and beyond the region. The Dnieper and Don rivers flow into the Black Sea; the West and North Dvina rivers drain into the Baltic and White seas, respectively; and the Volga River runs to the Caspian Sea (see Figure 9.2).

Most of European Russia experiences cold winters and cool summers by North American standards. Moscow, for example, is about as cold as Minneapolis in January, yet not nearly as warm in July. In Ukraine, Kiev is milder, however, and Simferopol', near the Black Sea, offers wintertime temperatures that average more than 20°F (11°C) warmer than those of Moscow.

Three distinctive environments shape agricultural potential in the European West (Figure 9.5). North of Moscow and St. Petersburg, poor soils and cold temperatures severely limit farming, and the region's boreal forests have been extensively logged (Figure 9.6). Belarus and central portions of European Russia possess longer growing seasons, but acidic **podzol soils**, typical of northern forest environments, limit output and the ability of the region to support a productive agricultural economy. Still, diversified agriculture includes grain (rye, oats, and wheat) and potato cultivation, swine and meat production, and dairying. South of 50° latitude, agricultural conditions improve across much of southern Russia and Ukraine. Forests gradually give way to steppe environments

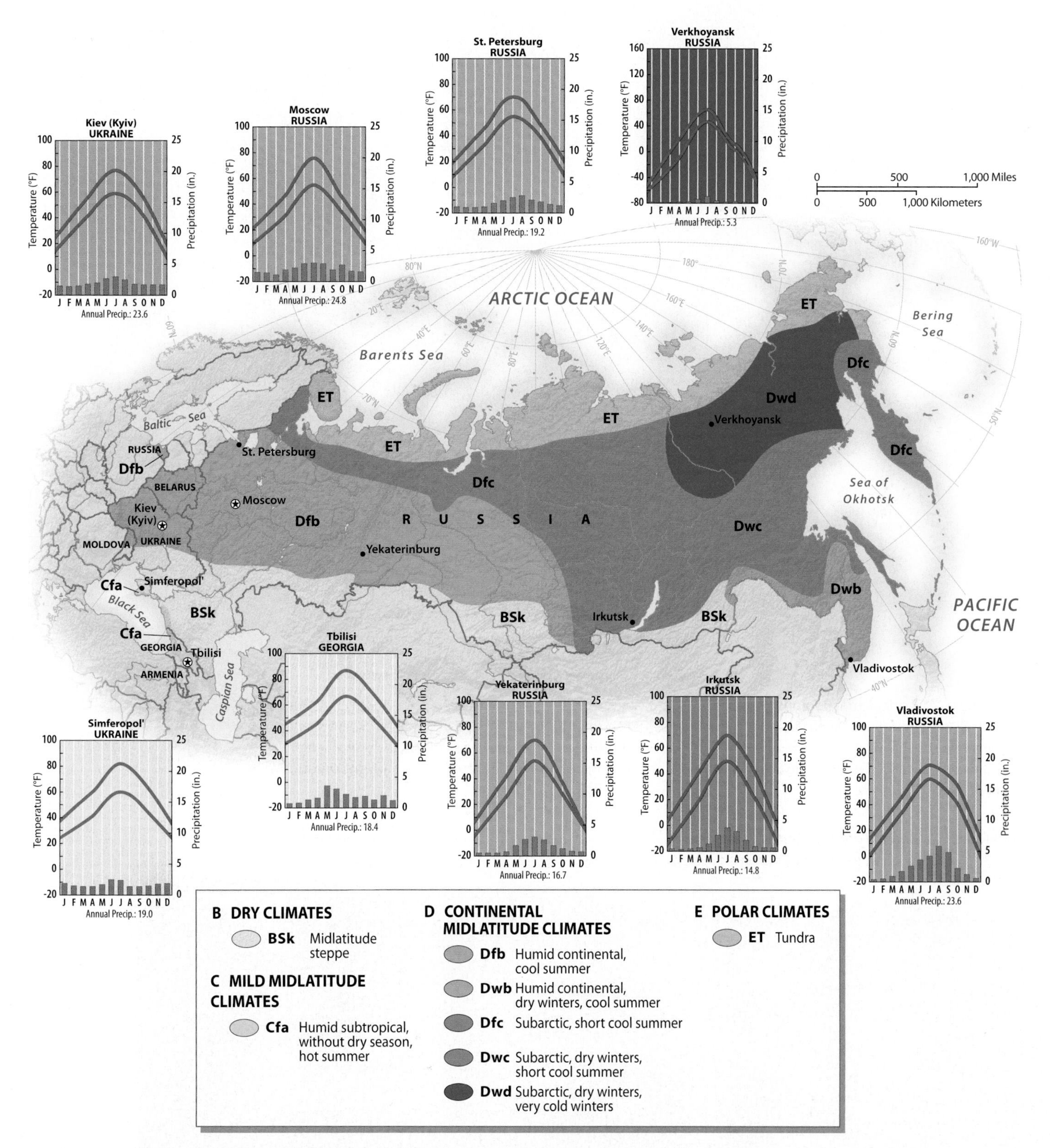

▲ **Figure 9.4 Climate of the Russian Domain** The region's northern latitude and large landmass produce dominant continental climates. Note the large seasonal ranges in temperature. Farming is limited by short growing seasons across the region. Aridity imposes limits elsewhere, especially across the southern Russian interior. Small zones of mild climate are found on the warming shores of the Black Sea in the far southwest corner of the region, producing subtropical conditions in Georgia.

▲ **Figure 9.5 Agricultural Regions** Harsh climate and poor soils combine to limit agriculture across much of the Russian domain. Better farmlands are found in Ukraine and in European Russia south of Moscow. Portions of southern Siberia support wheat production, but yield marginal results. In the Russian Far East, warmer climates and better soils translate into higher agricultural productivity. **Q: Describe the relationships between major agricultural zones and patterns on the climate map (Figure 9.4).**

▼ **Figure 9.6 Boreal Forest in Northwestern Russia** The once endless boreal forests of northern Russia are quickly being harvested for commercial lumber markets in nearby western Europe. Only a small percentage of roadless forest remains today.

▲ **Figure 9.7 Commercial Wheat Production** This Ukrainian wheat field typifies some of the region's most productive farmland. Longer growing seasons and good soils remain enduring assets in this portion of the Russian domain.

dominated by grasslands and by fertile "black earth" **chernozem soils**, which have proven valuable for commercial wheat, corn, and sugar beet cultivation and for commercial meat production (Figure 9.7).

The Ural Mountains and Siberia The Ural Mountains (see Figure 9.2) mark European Russia's eastern edge, separating it from Siberia. Despite their geographical significance as the traditional division between continents, the Urals are not a particularly impressive range; several of their southern passes are less than 1000 feet (300 meters) high. Still, its ancient rocks contain many valuable mineral resources, and it traditionally marked Russia's eastern cultural boundary. In early 2013, the southern Urals city of Chelyabinsk achieved momentary global notoriety when a large meteor exploded just above the city, injuring hundreds and damaging thousands of buildings (see *Exploring Global Connections: Russian Meteorite Fragments Go Global)*.

East of the Urals, Siberia unfolds across the landscape for thousands of miles. The great Arctic-bound Ob, Yenisey, and Lena rivers (see Figure 9.2) drain millions of square miles of northern country, including the flat West Siberian Plain, the hills and plateaus of the Central Siberian Uplands, and the rugged and isolated Northeast Highlands. Siberian vegetation and agriculture reflect the climatic setting. The northern portion of the region is too cold for tree growth and instead supports tundra vegetation, which is characterized by mosses, lichens, and a few ground-hugging flowering plants. Much of the tundra region is associated with **permafrost**, a cold-climate condition of unstable, seasonally frozen ground that limits the growth of vegetation and causes problems for railroad construction. South of the tundra, the Russian **taiga**, or coniferous forest zone, dominates a large portion of the Russian interior.

Along the Pacific, the Kamchatka Peninsula dangles dramatically into the waters of the North Pacific Ocean, and the region offers spectacular volcanic landscapes (Figure 9.8). Unlike much of the Russian domain, which has been poisoned by almost a century of reckless development and environmental exploitation, the Kamchatka region remains relatively untouched by the outside world.

▼ **Figure 9.8 Kamchatka Peninsula** The Kamchatka Peninsula offers many spectacular natural settings, as well as a habitat for a variety of Pacific salmon.

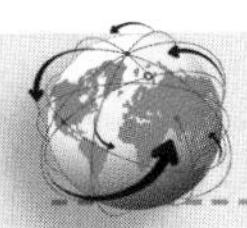

Exploring **Global Connections**

Russian Meteorite Fragments Go Global

It was literally the sound heard around the world. At 9:20 in the morning on February 15, 2013, the largest object to enter Earth's atmosphere in the last century (the Tunguska area of Siberia experienced a similar event in 1908) exploded above Chelyabinsk, Russia (**Figure 9.1.1**). The meteor, weighing between 7000 and 11,000 tons, made the loudest sound (as low-frequency pressure waves) ever recorded by the listening equipment that monitors compliance with the Nuclear Test Ban Treaty. Thousands of windows shattered near Chelyabinsk, and more than 1200 people were injured, mostly by flying glass. The fireball stunned local residents of the southern Urals city, and video footage of the event immediately went viral to every corner of the globe.

Treasures in the Snow In the days following the explosion, fragments of the meteorite, which were eagerly recovered by local residents, began their own worldwide journeys. The word was out: School children were digging through the snow, finding small black pebbles everywhere. One woman found a fist-sized stone in her woodshed that had come through the roof. Within a few days, strangers began showing up, their fat wallets filled with rubles and euros. They began quietly buying up the treasures (Russian authorities discourage the black market in meteorites) that had fallen from the sky. The woman with the large stone sold her find for $230, only to be offered over $1200 a few hours later. Quickly, eBay auctions were offering the fragments, an online meteorite sales site (star-bits.com) featured fresh finds from Chelyabinsk, and the International Meteorite Collectors Association (IMCA) Website was getting new hits from around the world (**Figure 9.1.2**).

▲ **Figure 9.1.1 Meteor in the Sky Above Chelyabinsk, 2013** Stunned residents of Chelyabinsk looked skyward on the morning of February 15, 2013, as a huge meteorite exploded above the city.

▲ **Figure 9.1.2 Meteorite Fragments Found Near Chelyabinsk** Many local residents of this southern Urals region cashed in on the lucrative global market for meteorite fragments.

A Long History of Collecting The Chelyabinsk phenomenon is nothing new. People have collected meteorites for thousands of years. Egyptian hieroglyphs refer to "iron from heaven," and some archaeologists believe ancient Egyptians made artifacts from iron- and nickel-rich meteorites that were found in the desert. In fact, dry, desert environments make for great meteorite hunting. In 2008, an Italian geologist browsing Google Earth thought he detected an unusual feature in the North African desert. Sure enough, it was a 150-foot-wide meteor crater, and he discovered thousands of meteorite fragments at the isolated site. But when he returned the following year, he found the site had been disturbed. Soon the Egyptian fragments were on sale at a collectors show in France.

By the time you are reading this essay, the fragments of the Chelyabinsk meteorite will have been redistributed around the world. The membership list of the IMCA reads like a roll call of the United Nations: There are collectors in Germany, New Zealand, Morocco, China, the Philippines, Brazil, Ireland, and, not surprisingly, Russia. What came out of the sky and fell to Earth (with utterly no warning from the scientific community) that February morning quickly acquired human meaning and monetary value and became in their own small and magical ways fragments of a much larger global economy.

1. What other examples of collectible artifacts and materials would have their own patterns of global movement and redistribution?
2. Given its relatively small population and isolated character, why is the interior of Russia so well known for its periodic encounters with Earth-bound meteors?

Google Earth Virtual Tour Video

http://goo.gl/BDsepG

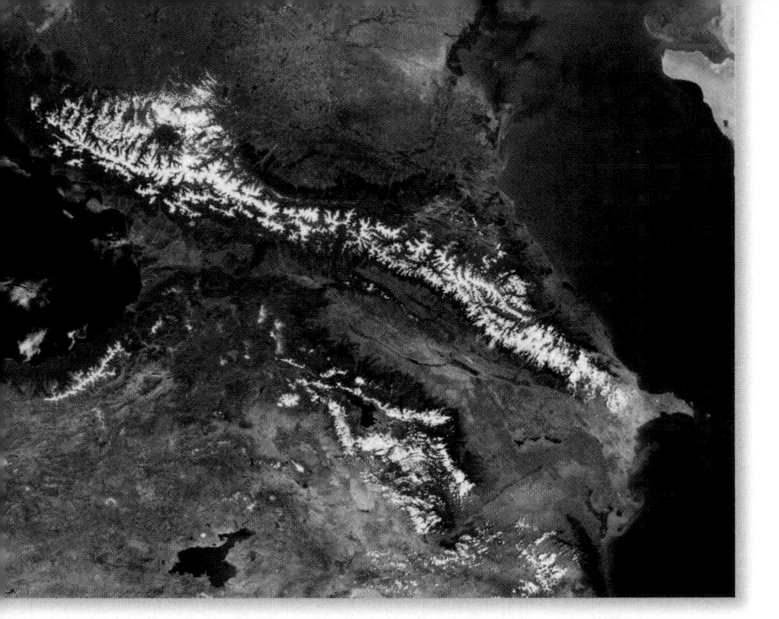

▲ Figure 9.9 **Satellite Image of the Caucasus Mountains** Aligned between the Black Sea (left) and the Caspian Sea (right), the rugged, snow-capped Caucasus Mountains prevent easy movement across the culturally diverse and politically contested borderlands between southern Russia and Georgia.

Six native species of Pacific salmon thrive in the region, spawning in the millions in the area's free-flowing rivers. The salmon are at the center of a complex environmental web: They provide food for the brown bears, seals, and Stellar's sea eagles that are abundant here; they also remain part of the subsistence diet of the Koryak and other native peoples of these coastal zones; and they offer ecotourism and sports fishing opportunities found nowhere else in the world. A concerted global effort is now under way to prevent development in seven sensitive spawning areas of Kamchatka wilderness. One Russian-supported plan calls for protecting an area nearly triple the size of Yellowstone National Park, and a recent consolidation of several national parks on the peninsula should make management of the region's natural resources more efficient and less costly.

The Russian Far East The Russian Far East is a distinctive subregion characterized by proximity to the Pacific Ocean, a more southerly latitude, and a pair of fertile river valleys. Located at about the same latitude as North America's New England, the region features longer growing seasons and milder climates than those found to the west or north. Here the continental climates of the Siberian interior meet the seasonal monsoon rains of East Asia. It is a fascinating zone of ecological mixing: Conifers of the taiga mingle with Asian hardwoods; and reindeer, Siberian tigers, and leopards also find common ground.

The Caucasus and Transcaucasia In European Russia's extreme south, flat terrain gives way first to hills and then to the Caucasus Mountains, a large range stretching between the Black and Caspian seas (Figure 9.9). Many delicate and relatively pristine mountain environments within the region have been recognized as extraordinarily fragile.

Farther south lies Transcaucasia and the distinctive natural setting of Georgia and Armenia. Patterns of both climate and terrain in the Caucasus and Transcaucasia are very complex. Rainfall is generally higher in the western zone, whereas the area's eastern valleys are semiarid. In areas of adequate rainfall or where irrigation is possible, agriculture can be quite productive. Georgia in particular has long been an important producer of subtropical fruits, vegetables, flowers, and wine (Figure 9.10).

► Figure 9.10 **Grape Harvest, Subtropical Georgia** The moderating influences of the Black Sea and a more southern latitude produce a small zone of humid subtropical agriculture in Georgia.

A Devastated Environment

The Russian domain has no shortage of fragile and endangered environments. The breakup of the Soviet Union and subsequent opening of the region to international public scrutiny revealed some of the world's most severe environmental degradation (Figure 9.11). Even official studies commissioned by the Russian government estimate that almost two-thirds of Russians live in an environment harmful to their health and that the country's environmental problems continue to worsen. The studies also suggest that nearly 65 million Russians live in areas of chronically poor air quality and that the drinking water is unsafe in half of the country.

The frenetic pace of seven decades of Soviet industrialization took its toll across the region. Even in some of the most remote reaches of Russia, careless mining and oil drilling, the spread of nuclear contamination, and rampant forest cutting have resulted in frightening environmental damage. New Russian environmental and antinuclear movements have protested these ecological disasters, but to date these movements remain a minor political voice in a region dominated by the desire for economic growth. Indeed, the magnitudes of many of these environmental challenges are so great that they have global implications and may affect world climate patterns, water quality, and nuclear safety. For example, since the 1980s, the global environmental costs of Siberian forests lost to lumbering and pollution may have exceeded those of the more widely publicized destruction of the Brazilian rainforest.

Air and Water Pollution Poor air quality plagues hundreds of cities and industrial complexes throughout the region (see Figure 9.11). The traditional Soviet practice of building large clusters of industrial processing and manufacturing plants in concentrated areas, often with minimal environmental controls, has produced an ongoing legacy of fouled air that stretches from Belarus to Siberia. A traditional reliance on abundant, but low-quality coal also contributes to pollution problems. The air quality in dozens of cities within the region typically fails to meet health standards,

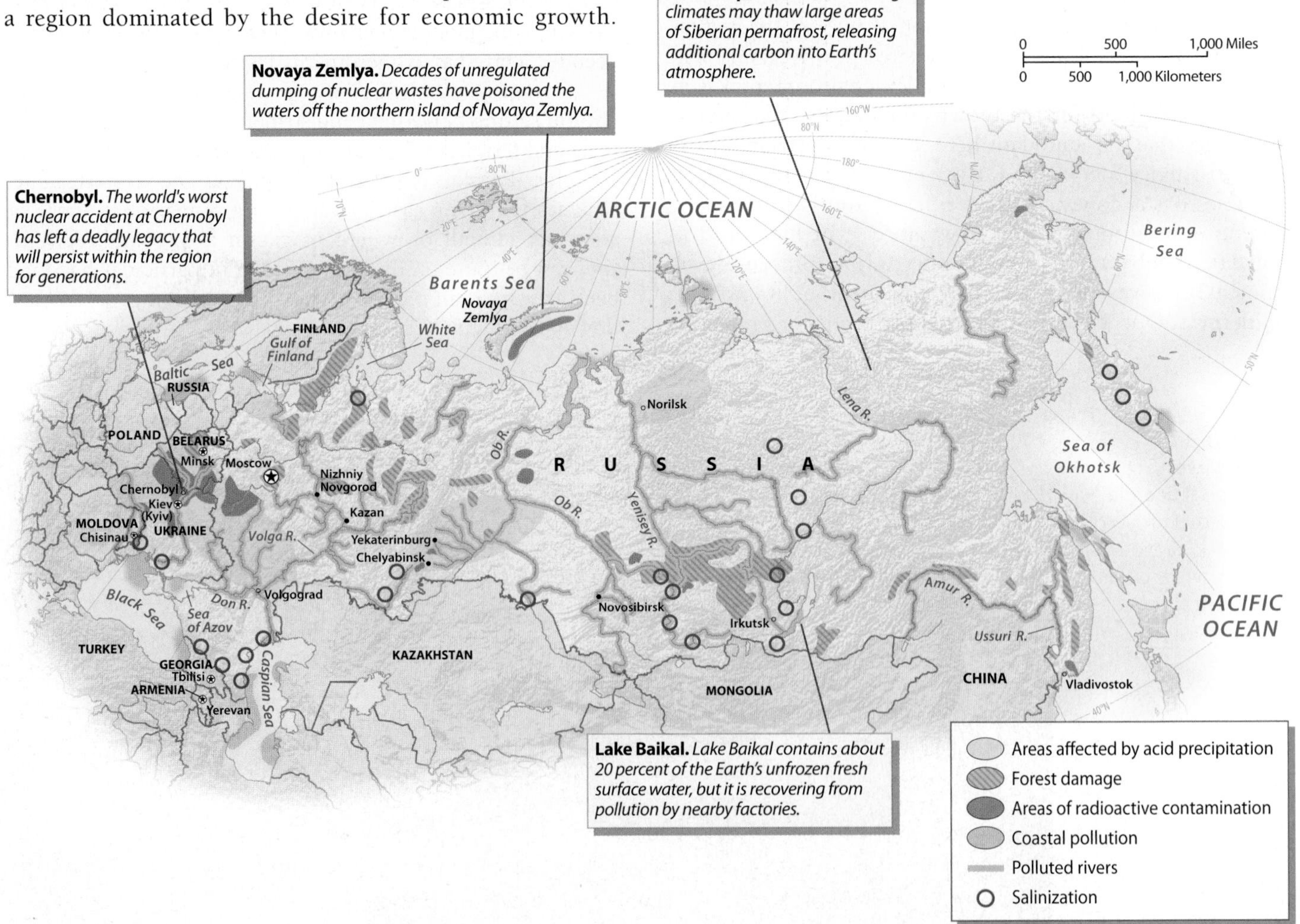

▲ **Figure 9.11 Environmental Issues in the Russian Domain** Varied environmental hazards have left a devastating legacy across the region. The landscape has been littered with nuclear waste, heavy metals, and air pollution. Fouled lakes and rivers pose additional problems in many localities. Present economic difficulties and political uncertainties only add to the costly challenge of improving the region's environmental quality in the 21st century.

► **Figure 9.12 Norilsk** The sprawling facilities of the Norilsk Nickel Plant dominate this portion of the city of Norilsk. Extensive air and water pollution are undesirable consequences of this industrial operation.

particularly in the winter when cold-air inversions trap the polluted atmosphere for days on end. Large numbers of urban residents across the region suffer from chronic respiratory problems.

Siberia's northern mining and smelting city of Norilsk is one of Russia's dirtiest urban areas, earning it the dubious distinction of appearing on the Blacksmith Institute's list of "ten most polluted places in the world" (Figure 9.12). In addition, a large swath of larch-dominated forest has died in a huge zone of contamination that stretches more than 75 miles (120 kilometers) east of the city. Norilsk Nickel (the major industrial polluter in the area) hopes to cut harmful sulfur dioxide emissions dramatically between 2015 and 2020. Elsewhere, however, growing rates of private car ownership have greatly increased automobile-related pollution. Today 90 percent of Moscow's air pollution has been linked to the city's growing automobile traffic.

Degraded water is another hazard that residents of the region must cope with daily (see Figure 9.11). Oil spills have harmed thousands of square miles in the tundra and taiga of the West Siberian Plain and along the Ob River. A 2011 study estimates that 5 million tons of oil (1 percent of Russia's annual oil production) are spilled every year, and the actual total could be higher.

Municipal water supplies are constantly vulnerable to industrial pollution, flows of raw sewage, and demands that increasingly exceed capacity. For example, the Baltic Sea near the city of St. Petersburg has reached a critical level of pollution that has killed fish and threatens to permanently damage the region's ecosystem. The biggest problem is that 30 percent of all the residential and industrial waste that enters the sea via the nearby Neva River is unfiltered raw sewage, a toxic mix of heavy metals and human waste that is rapidly killing nearby portions of the Baltic.

Elsewhere, the extensive industrialization and dam-building along Russia's Volga Valley have produced a corridor of degraded water that stretches for hundreds of miles. Water pollution has also affected much of the northern Black Sea, large portions of the Caspian Sea shoreline, and even Arctic Ocean waters off Russia's northern coast. Although Siberia's Lake Baikal has also suffered from industrial pollution, its future prospects appear brighter (see *Working Toward Sustainability: Lake Baikal's Success Story*).

The Nuclear Threat The nuclear era brought its own particularly deadly dangers to the region. The Soviet Union's aggressive nuclear weapons and nuclear energy programs expanded greatly after 1950, and issues of environmental safety were often ignored. Northeast Siberia's Sakha region, for example, suffered regular nuclear fallout in the era of above-ground nuclear testing. In other areas, nuclear explosions were widely utilized for seismic experiments, oil exploration, and dam-building projects. The once-pristine Russian Arctic has also been poisoned. During the Soviet era, the area around the northern island of Novaya Zemlya served as a huge and unregulated dumping ground for nuclear wastes. Nearby, dozens of atomic submarines have been abandoned to rust away among the fjords of the Kola Peninsula. Aging nuclear reactors also dot the region's landscape, often contaminating nearby rivers with plutonium leaks.

Nuclear pollution is particularly pronounced in northern Ukraine, where the Chernobyl nuclear power plant suffered a catastrophic meltdown in 1986. Large areas of nearby Belarus were also devastated in the Chernobyl disaster. Fallout contaminated fertile agricultural lands across much of the southern part of the country and rendered about 20 percent of the nation unhealthy to even live in. The reactor

Working Toward **Sustainability**

Lake Baikal's Success Story

Lake Baikal, located in southern Siberia (see **Figure 9.2**), has become one of Russia's most important settings for protecting its vast natural environment and for developing sustainable economic activities such as ecotourism. Remarkably, the lake contains about 20 percent of Earth's unfrozen fresh surface water. Not only is the lake almost 400 miles (644 kilometers) long, but it is also 5300 feet (1600 meters) deep, occupying a structural rift in the continental crust (**Figure 9.2.1**). It remains home to a large array of unique (or *endemic*) species found nowhere else on the planet, including the world's only freshwater seal.

A Threatened Treasure Lake Baikal suffered during the later Soviet period. Large pulp and paper mills were located along the lakeshore in the 1950s and 1960s because abundant forests were nearby and the lake's amazingly pure water was useful in producing high-quality wood fibers. Unfortunately, these industries discharged pollutants into the lake and into the surrounding atmosphere. With factory discharges, the lake's purity rapidly declined. However, things have improved since the early 1990s. Stricter regulations have reduced industrial pollution. Indeed, the lake has become *the* national "poster child" of the Russian environmental movement. In 1996, the lake became a UNESCO World Heritage Site, and three years later the Russian government formally created legislation designed to protect the lake.

Recently, the lake became the center of attention as Russia planned to expand a major Siberian oil pipeline linking Russian resources to East Asian markets. High oil prices have encouraged the Russians to make large new investments in their petroleum industry, but many environmentalists feared that these growing global demands for oil might have destructive local consequences for Lake Baikal. In 2006, major protests and petition-signing drives opposed the planned pipeline's close proximity to the north shore of the lake. The initiative caught the attention of Russian President Putin, who dramatically ordered that the pipeline be directed farther away from the lake's fragile ecosystem.

Opportunities for Ecotourism The lake has also been the setting for the development of ecotourism initiatives designed to maintain the region's high environmental quality, as well as provide jobs for local indigenous Siberian people. One project, begun in 2008 and coordinated with the help of the Global Nature Fund and other private donors, focused on creating a sustainable tourist infrastructure in Buryatia, a Russian republic along the lake's eastern shoreline. Thinly peopled by Altaic-speaking Buryats (many practice a form of Buddhism), the region is just now opening to tourists who are drawn there to enjoy the lake's natural features, as well as sample a bit of authentic Siberian culture (**Figure 9.2.2**). Buryats receive training on managing environmentally friendly tourist facilities, offering tours, and marketing local foods and crafts to visitors. The hope is that the effort will produce new jobs in an area of the country with chronic poverty and high unemployment, preserve the pristine quality of lake waters, and demonstrate for tourists Russia's 21st-century commitment to sustainable environmental practices.

▲ **Figure 9.2.1 Lake Baikal** Southern Siberia's Lake Baikal is one of the world's largest deep-water lakes. Industrialization devastated water quality after 1950 as pulp and paper factories poured wastes into the lake. Recent cleanup efforts have helped, but environmental threats remain.

1. Examine the maps of agricultural production (**Figure 9.5**), population (**Figure 9.14**) and industrial zones (**Figure 9.39**) in relation to Lake Baikal. How do these patterns help explain the relatively pristine character of the lake today?
2. Follow the Trans-Siberian Railroad in Google Earth along Lake Baikal's southern shoreline. What are the dominant features visible on your trip?

Google Earth Virtual Tour Video

http://goo.gl/JDxwFv

▼ **Figure 9.2.2 Buryats** Closely related to residents of Mongolia, Russia's Buryat population lives in the vicinity of Lake Baikal.

burned for 16 days, pouring smoke 2 miles into the sky and spreading nuclear contaminants from southern Russia to northern Norway. Thousands died directly from the accident, and millions of other residents across the region and in nearby affected portions of Europe suffered long-term health problems. It proved to be the world's worst nuclear accident and one of the greatest environmental disasters of the modern age, and it will continue to impact the ecological health of the region for decades to come.

The region's involvement with the nuclear age continues to take new forms. In 2001, Russia's Rostov Nuclear Energy Station went online, making it the region's first nuclear power plant since the Soviet era. Growing blackouts and energy shortages have produced a new government drive to revive Russia's nuclear industry. Recently, an ambitious plan to build 26 new reactors by 2020 produced a rash of protests across the country. At one gathering, protestors held signs simply declaring "One Chernobyl was plenty."

Post-Soviet Challenges The end of Soviet control has had mixed consequences for the region's environment. The demise of the Soviet Union initially brought about environmental improvement in some areas. Many factories shut down because they were no longer economically viable, which itself reduced pollution. The huge steel mills in the Russian city of Magnitogorsk, for example, produce less than half the raw steel they did 20 years ago, and global competition threatens to reduce demand further. Ironically, this decreased production has resulted in cleaner air. Although costly, advanced pollution control equipment is also beginning to be imported from western Europe. Elsewhere in Russia, nuclear warhead storage facilities have been consolidated, and government authorities are responsible for maintaining control over the nation's nuclear weapons. An environmental consciousness also is growing among young, educated Russians and Ukrainians, although public opinion polls across the region demonstrate that for most residents, environmental concerns remain very low priorities compared to issues such as economic opportunity and access to decent education and health care.

Russia has also depended heavily on its natural resources to finance its return to economic health, but this response to growing global demand has produced new environmental problems for the region. State-run petroleum and natural gas companies, for example, have tremendous power and freedom to produce profits as quickly as possible, regardless of the consequences. A similar mentality shapes the timber industry. With huge demand for lumber from nearby Japan and China, Siberian forests are disappearing at an alarming rate. These forests represent some of Earth's largest unlogged timber resources, and at present rates of cutting, much of this natural resource will vanish in the next 10 to 20 years. Thus, growing East Asian appetites for new homes and furniture are directly contributing to the destruction of Siberia's forests.

Russia also faces some interesting challenges related to wildlife. The old Soviet regime had some success in protecting both endangered species and sizable areas of natural habitat noted for their biological diversity. In the post-Soviet period, however, fewer regulations have opened the way for more uncontrolled hunting and trapping. It is questionable whether such animals as the Siberian tiger (only 350–450 still survive in the wild) and the Amur leopard (25–40 wild animals remain), both inhabiting the endangered forests of Russia's Far East, will survive the next few decades.

In contrast, Russian wolves are thriving, prompting Yegor Borisov, the governor of Siberian Yakutia, to declare a state of emergency in 2013. Far from revering wolves as charismatic symbols of environmental health, many Russians see them as dangerous predators. Pelt bounties ($660 per adult wolf pelt and $50 for the skin of a cub) can bring in a sizable income. Some communities offer bonuses such as free snowmobiles to hunters harvesting the most pelts. Following a cyclical collapse of the regional rabbit population (a key traditional food source), wolves turned to livestock, horses, and reindeer for food. Siberia's 3500 wolves killed about 16,000 domesticated reindeer in 2012. Even the World Wildlife Fund agrees that "there are too many wolves in Russia," suggesting that the furry predators may be in the crosshairs for quite a while.

Climate Change and the Russian Domain

Given its latitude and continental climates, the Russian domain is often cited as a world region that would benefit from a warmer global climate. But such an interpretation oversimplifies the complex natural and human responses to global climate change, some of which are already occurring across the region.

Potential Benefits Optimists point to some of the economic benefits that may result from warmer Eurasian climates. Some models predict, for example, that the northern limit of spring cereal cultivation in northwestern Russia will shift 60–90 miles (100–150 kilometers) poleward for every 1°C (1.8°F) of warming. Elsewhere, warmer springs in settings such as Belarus might allow for higher yields of maize and sunflowers. Areas once in tundra vegetation may be more suitable for boreal forest expansion.

Less severe winters may make energy and mineral development in arctic settings less costly. About 15 percent of the world's undiscovered oil reserves (and 30 percent of its undiscovered natural gas reserves) are probably located in these settings, and Russia has staked large claims to the region. In the Arctic Ocean and Barents Sea, warmer temperatures and less sea ice are translating into better commercial fishing, easier navigation, more high-latitude commerce, and more ice-free days in northern Russian ports. Indeed, by 2012, a record 46 commercial vessels negotiated the **northern sea route** along Siberia's northern coast. Northern ports such as Murmansk may reap the benefits of this arctic warming (Figure 9.13). In addition, less ice on northward-flowing Siberian rivers such as the Ob and Yenisey may make these waterways more valuable as corridors of commerce.

Many Russian scientists and political leaders have publicly dragged their feet at recent climate-change conferences. After all, they argue, warming could benefit the region. They also point out that the region, unlike China or India, is not witnessing a rapid increase in its greenhouse gas emissions. Indeed, Russian output of carbon emissions actually fell (because of declines in industrial production) more than 30 percent between 1990 and 2010.

Potential Hazards Even with such rosy scenarios within the Russian domain, might the long-term regional and global costs outweigh the benefits? Climate experts point to several areas of particular concern. First, hotter summers may increase the risk of wildfire. In what could be a sign of things to come, hundreds of wildfires broke out in Russia in the summer of 2010, mostly south and east of Moscow. The fires scorched more than 484,000 acres (196,000 hectares) of land, burning wheat fields and hundreds of structures and filling the skies of Moscow with smoke as visitors and residents experienced record heat.

Second, changes in ecologically sensitive arctic and subarctic ecosystems are already leading to major disruptions in wildlife and indigenous human populations in those settings of northern Russia. Take the example of the polar bear. The shrinking volume of arctic sea-ice habitat for the bears has meant that they are forced to widen their search for food. This has brought them into closer contact with arctic villages and also disrupted traditional hunting practices. Poachers have also profited, increasing their illegal harvests. Paradoxically, Russian officials and some wildlife management experts have argued that increasing legal and controlled hunting (versus illegal poaching) of the bears may be the best way to more effectively maintain their populations in this period of environmental flux.

Third, rising global sea levels will hit low-lying areas of the Black and Baltic seas particularly hard. Officials in St. Petersburg, Russia's second largest city, are already contemplating significant costs associated with controlling the Baltic's rising waters.

Finally, the largest potential change within the region, with truly global implications, relates to the thawing of the Siberian permafrost. Substantial areas of northern Russia are covered with permafrost that is already close to thawing. This same region of the world has witnessed some of the largest, most persistent global warming since 1950. Thus, even minor increases in temperature could have large and irreversible consequences for the region. Major changes in topography (mud flows, slumping, and erosion), drainage (lake coverage and rivers), and vegetation will be the result. Existing fish and wildlife populations will need to adjust in order to survive. Human infrastructure such as buildings, roads, and pipelines will also require substantial modification.

However, the greatest potential global impact may come with the huge release of carbon that is currently stored in existing permafrost environments. The soils frozen in permafrost contain large amounts of organic material that decomposes quickly when thawed. Most of the planet's permafrost could release its carbon reservoir within the next century, the equivalent of 80 years of burning fossil fuels. Such a contribution to the world's carbon budget, which would likely further warm Earth, is only beginning to be incorporated into models of global climate change. Thus, the survival of the Siberian permafrost may hold one of the keys to slowing or quickening further global warming.

REVIEW

9.1 Compare the climate, vegetation, and agricultural conditions of Russia's European West with those of Siberia and the Russian Far East.

9.2 Describe the high environmental costs of industrialization within the Russian domain.

◄ **Figure 9.13 Murmansk** The northern Russian port of Murmansk may see healthy growth and expansion of its harbor facilities as global warming brings more ice-free travel to arctic sea lanes.

Table 9.1 Population Indicators

Country	Population (millions) 2013	Population Density (per square kilometer)	Rate of Natural Increase (RNI)	Total Fertility Rate	Percent Urban	Percent < 15	Percent >65	Net Migration (Rate per 1000)
Armenia	3.0	102	0.4	1.6	63	17	10	0
Belarus	9.5	46	−0.1	1.6	76	15	14	1
Georgia	4.5	65	0.2	1.7	58	17	14	5
Moldova	4.1	122	−0.0	1.3	42	16	10	0
Russia	143.5	8	−0.0	1.7	74	16	13	2
Ukraine	45.5	75	−0.3	1.5	69	14	15	1

Source: Population Reference Bureau, *World Population Data Sheet, 2013.*

Population and Settlement: An Urban Domain

The six states of the Russian domain are home to about 200 million residents (Table 9.1). Although they are widely dispersed across a vast Eurasian land mass, most live in cities. The region's distinctive distributions of natural resources and changing migration patterns continue to shape its population geography. Government policies have encouraged migration into the eastern portions of the domain, but the population remains strongly concentrated in the traditional centers of the European West. Relatively low birth rates and higher death rates also remain a critical concern, as these trends threaten to shrink future regional populations.

Population Distribution

The favorable agricultural setting of the European West has offered a home to more people than live in the inhospitable conditions found across the larger areas of central and northern Siberia. Although Russian efforts over the past century have encouraged a wider dispersal of the population, it remains heavily concentrated in the west (Figure 9.14).

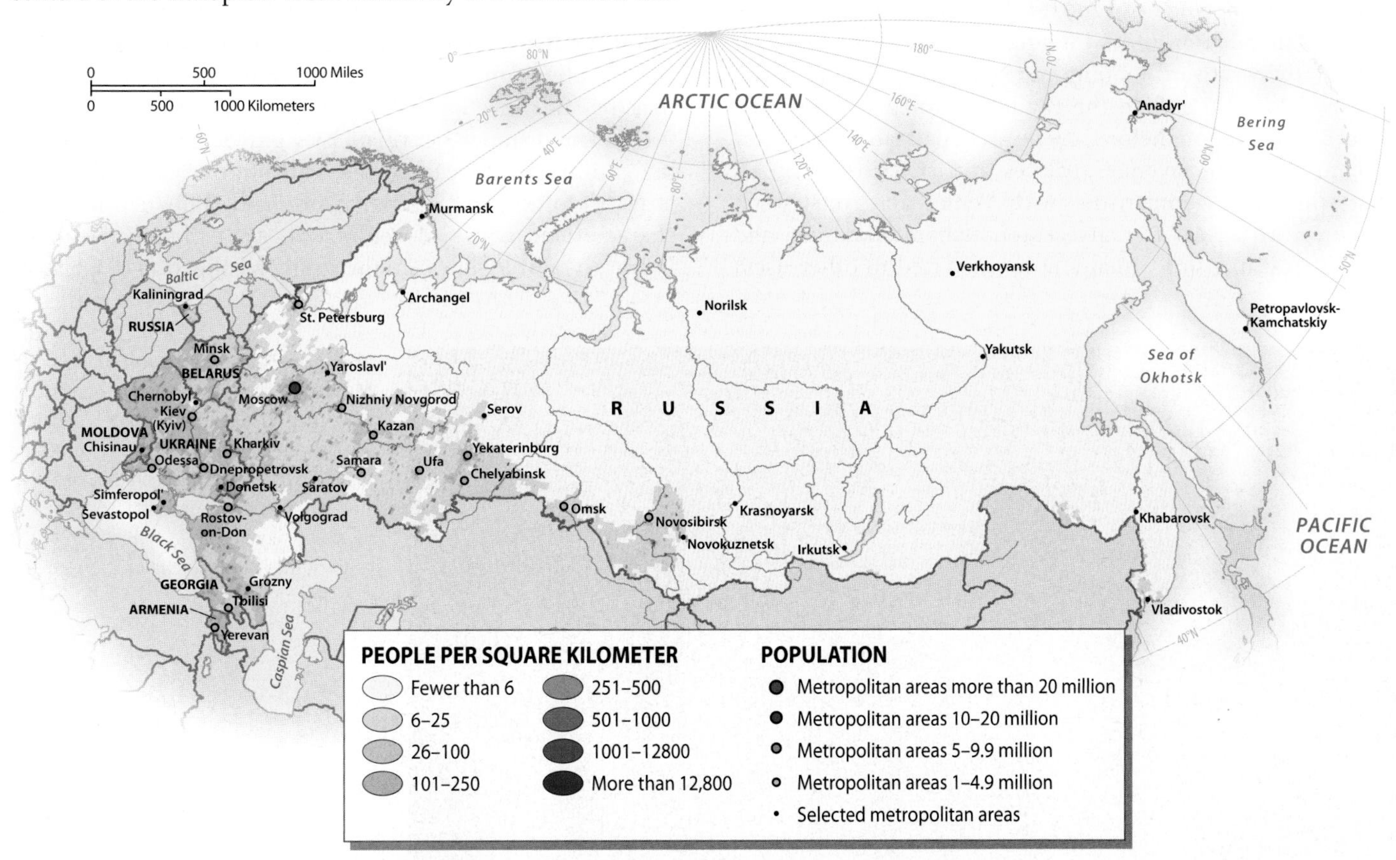

▲ **Figure 9.14 Population of the Russian Domain** Population within the region is strongly clustered west of the Ural Mountains. Dense agricultural settlements, extensive industrialization, and large urban centers are found in Ukraine, much of Belarus, and across western Russia south of St. Petersburg and Moscow. A narrower chain of settlements follows the better lands and transportation corridors of southern Siberia, but most of Russia east of the Urals remains sparsely settled.

European Russia is home to more than 100 million people, whereas Siberia, although far larger, holds only some 35 million. When you add in the 60 million inhabitants of Belarus, Moldova, and Ukraine, the imbalance between east and west becomes even more striking (see Table 9.1).

The European Core The region's largest cities, biggest industrial complexes, and most productive farms are located in the European Core, a subregion that includes Belarus, much of Ukraine, and Russia west of the Urals (see Figure 9.2). The sprawling city of Moscow and its nearby urbanized region clearly dominate the settlement landscape with a metropolitan area containing more than 11.5 million people (Figure 9.15). Unofficial estimates—which include the city's undocumented immigrants—are much higher, suggesting a metropolitan population between 15 and 17 million. Clearly Russia's primate city, Moscow has a population that is more than double the country's next largest metropolitan area (St. Petersburg), and it produces almost 20 percent of the entire nation's wealth. Looking ahead to 2020 and beyond, Moscow is projected to grow at a healthy pace, attracting domestic migrants from the country's less vibrant smaller cities and rural areas, as well as foreign immigrants from regions such as Central Asia.

On the shores of the Baltic, St. Petersburg (Leningrad in the Soviet period; 4.9 million people) has traditionally had a great deal of contact with western Europe. Literally rising from swamplands in 1703, St. Petersburg grew from the creative imagination of its founder, Tsar Peter the Great. It became the westward-looking tsarist capital until the Soviet takeover in 1917 (when the capital returned to Moscow). Peter's imagination helped create a Renaissance-style city built around grand avenues, palaces, and a network of canals that still draws comparisons with Venice and Amsterdam (Figure 9.16). Near the city center, elaborate baroque and neoclassical mansions (many now converted to other uses) and commercial buildings recall earlier days when Russians such as Tchaikovsky and Dostoyevsky walked the streets. Cutting through the heart of the city and bordered by the bright lights of shops, cafes, and theaters, Nevsky Prospekt ends near the Neva River, just two blocks from the world-famed Hermitage Museum, home to 2.8 million pieces of art.

Other urban clusters are oriented along the lower and middle stretches of the Volga River, including the cities of Kazan, Samara, and Volgograd. Industrialization within the region accelerated during World War II, as the region lay somewhat removed from German advances in the west. Today the highly commercialized river corridor, also containing important petroleum reserves, supports a diverse industrial base strategically located to serve the large populations of the European Core. Nearby, the resource-rich Ural Mountains include the gritty industrial landscapes of Yekaterinburg (1.4 million) and Chelyabinsk (1.1 million).

Beyond Russia, major population clusters within the European Core are also found in Belarus and Ukraine (see Table 9.1). The Belorussian capital of Minsk (1.8 million) is the dominant urban center in that country, and its landscape recalls the drab Soviet-style architecture of an earlier era. In nearby Ukraine, the capital of Kiev (Kyiv, 2.8 million) straddles the Dnieper River, and the city's old and beautiful buildings are a visual reminder of its historical role as a cultural and trading center within the European interior (see *Cityscapes: Kiev, Enduring Cultural Capital*).

Siberian Hinterland Leaving the southern Urals city of Yekaterinburg on a Siberia-bound train, you become aware that the land ahead is ever more sparsely settled (see Figure 9.14). The distance between cities grows, and the intervening countryside reveals a landscape shifting gradually from farms to forest. The Siberian hinterland is divided into two characteristic zones of settlement, each of which can be linked to railroad lines. To the south, a collection of isolated,

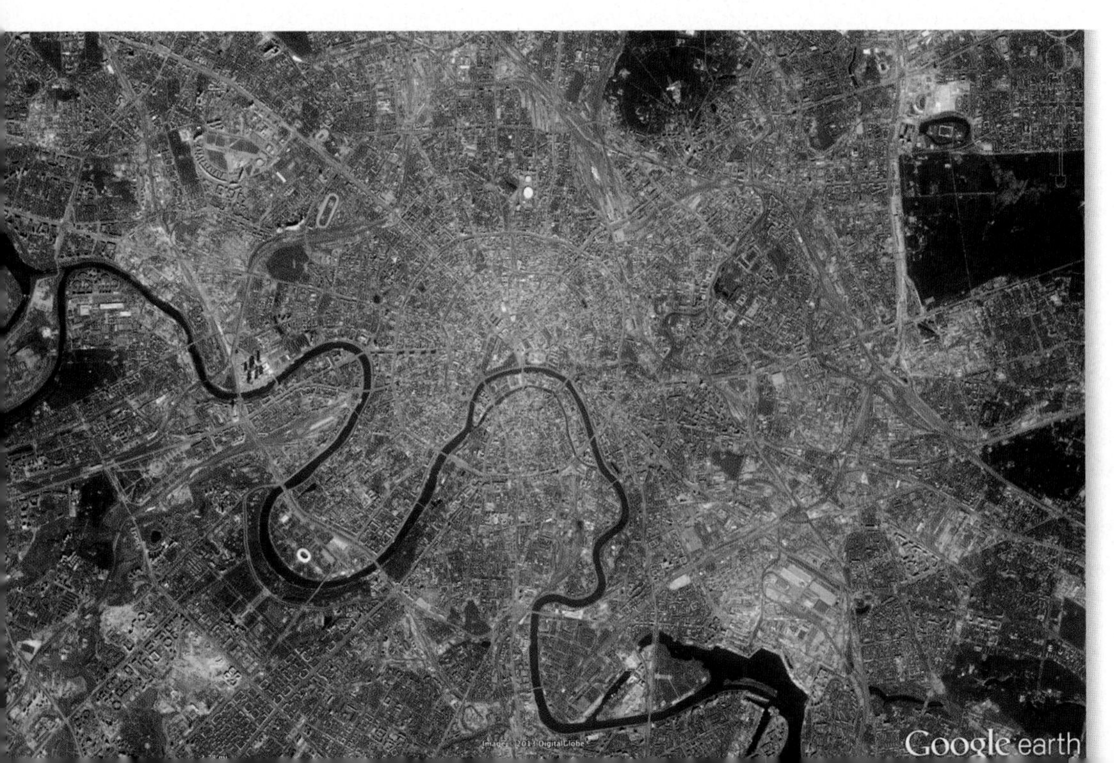

◄ **Figure 9.15 Metropolitan Moscow** Sprawling Moscow extends more than 50 miles (80 kilometers) beyond the city center. The city is home to more than 11.5 million people, and the relative strength of its urban economy continues to attract migrants from elsewhere in the country, thus putting more pressure on its infrastructure.

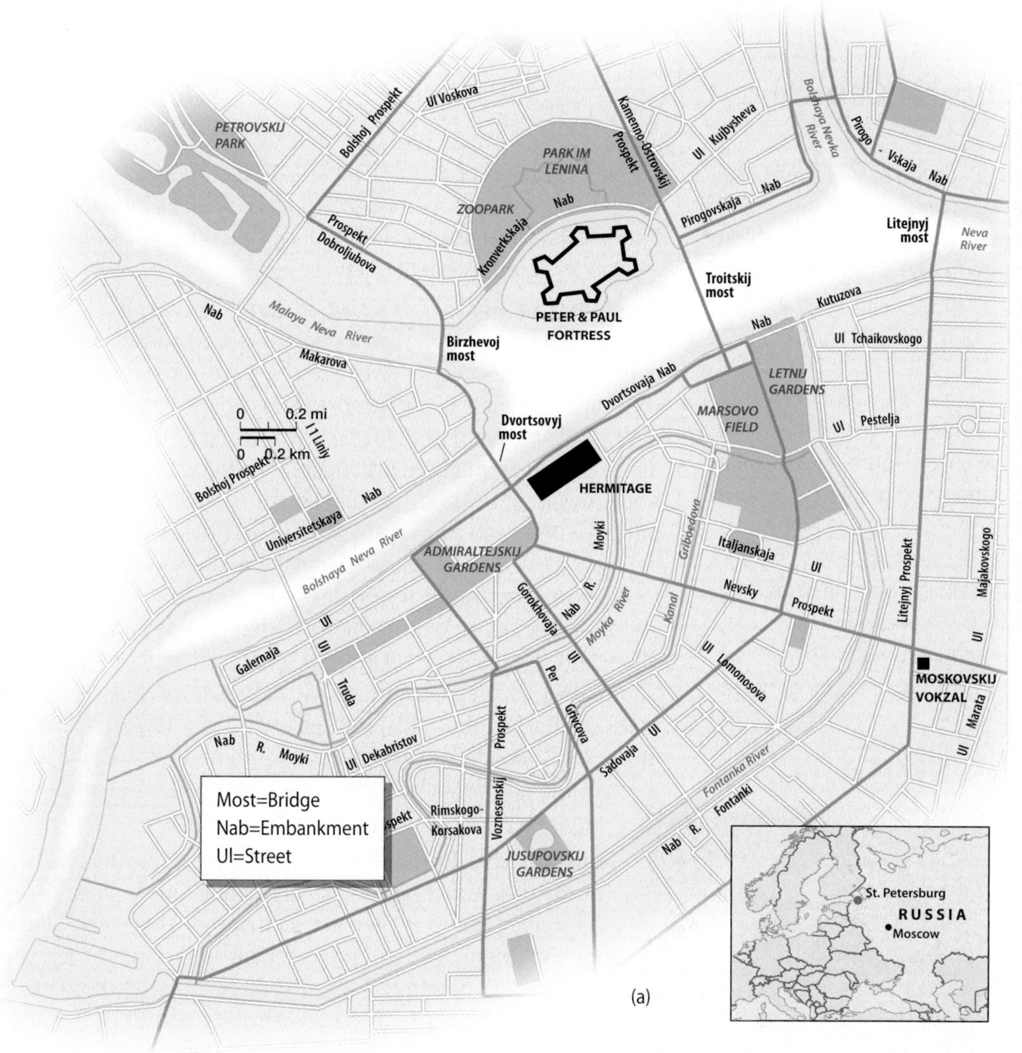

(b)

▲ **Figure 9.16 St. Petersburg** Often beloved as Russia's most beautiful city, St. Petersburg's urban design features a varied mix of gardens, open space, waterways, and bridges. The Hermitage (see map) is one of the largest and oldest museums in the world. Visible in the photo is the Church of the Savior on Spilled Blood, completed in 1907.

but sizable urban centers follows the **Trans-Siberian Railroad**, a key railroad passage to the Pacific completed in 1904 (see Figure 9.2). The eastbound traveler encounters Omsk (1.1 million) as the rail line crosses the Irtysh River, Novosibirsk (1.5 million) at its junction with the Ob River, and Irkutsk (600,000) near Lake Baikal. At the end of the line, the port city of Vladivostok (600,000) provides access to the Pacific. To the north, a thinner sprinkling of settlement appears along the more recently completed (1984) **Baikal–Amur Mainline (BAM) Railroad**, which parallels the older line, but runs north of Lake Baikal to the Amur River. From the BAM line to the Arctic, the almost empty spaces of central and northern Siberia dominate the scene, interrupted only rarely by small settlements, often oriented around points of natural resource extraction (Figure 9.17).

▼ **Figure 9.17 Abandoned Diamond Mine, Mirny, Siberia** The Siberian settlement of Mirny (population 37,000) is bordered by an abandoned diamond mine more than 1700 feet (525 meters) deep.

Cityscapes

Kiev, Enduring Cultural Capital

Like Pittsburgh and Paris, Kiev (sometimes spelled Kyiv) is a river city, blessed with bridges, shoreline walkways, and passing ships and barges (**Figure 9.3.1**). The city of 2.8 million people is the capital of Ukraine and the cultural and historical center of Ukrainian national identity. Kiev is one of eastern Europe's oldest cities, having functioned as an early and important center of Slavic culture, especially between 900 and 1200 CE.

Wander the "Right Bank" on the west side of the Dnieper River to explore the historical heart of the city and the location of many important business and governmental buildings (**Figure 9.3.2**). Much of the "Left Bank" (east) is residential land, incorporated into Kiev only within the last century.

Most of the modern Kievan urban landscape has taken shape in the past two centuries. By 1900, the city had become an important trade and administrative center within the Russian Empire. Later, as part of the USSR (1921–1991), Kiev acquired many Soviet-style buildings and a growing industrial infrastructure. The city survived a 2-year Nazi occupation (1941–1943) and the potentially catastrophic explosion of the nearby (60 miles or 100 kilometers north) Chernobyl nuclear reactor (1986). Since 1991, Kiev has served as the capital city of independent Ukraine.

West of the River Kiev's diverse urban geography reflects its varied natural and cultural environment. Set among rolling hills on either side of the Dnieper River, the city is blessed with many parks (Peremoha, or Victory Park, celebrates the Soviet defeat of the Nazis in World War II). West of the river, in the center of the city, Independence Square (formerly Soviet Square) celebrates Ukrainian freedom from Soviet control and has functioned repeatedly as a center of public protests and mass gatherings. One of the city's largest retailing malls (the Globe) is located beneath the square. A short distance to the northwest, Andriyivskyy Descent (or the Podil District) offers visitors to the city a rich collection of historical buildings and Ukrainian cultural experiences (**Figure 9.3.3**). Saint Andrew's Orthodox Church, with its green, baroque-style domes, dominates the neighborhood. Built in the mid-18th century, St. Andrews remains one of the city's architectural treasures and prime tourist destinations. Within a few blocks, many other historical homes and museums celebrate Ukrainian history and identity.

East of the River A different world sits on the east side of the river. Cross one of the city's numerous bridges or hop the underground (either the Red Line or the Green Line of the Kiev Metro system). The Left Bank landscape is filled with tall, massive, apartment-style buildings. Many of these newer neighborhoods were built between 1960 and 1985 during the Soviet period, and they feature mass-produced concrete-block construction. One city street looks like the next. Many of these high-rise complexes remain important residential areas today.

◄ **Figure 9.3.1 Kiev and the Dnieper River** The ornate domes of the Kiev Petchersk Lavra (one of the historical centers of Eastern Orthodox Christianity) frame this view of the city which includes many new buildings (in the distance) constructed on the east bank of the Dnieper River.

Regional Migration Patterns

Over the past 150 years, millions of people within the Russian domain have been on the move. These major migrations, both forced and voluntary, reveal sweeping examples of human mobility that rival the great movements from Europe and Africa or the transcontinental spread of settlement across North America.

Eastward Movement Just as settlers of European descent moved west across North America, exploiting natural resources and displacing native peoples, European Russians moved east across the vast Siberian frontier. Although the deeper historical roots of the movement extend back several centuries, the pace and volume of the eastward drift accelerated in the late 19th century as portions of the Trans-Siberian Railroad were being completed. Peasants were attracted to the region by its agricultural opportunities (in the south) and by greater political freedoms than they traditionally enjoyed under the **tsars** (or czars; Russian for "Caesar"), the authoritarian leaders who dominated politics during the pre-1917 Russian Empire. Almost 1 million Russian settlers moved into the Siberian hinterland between 1860 and 1914.

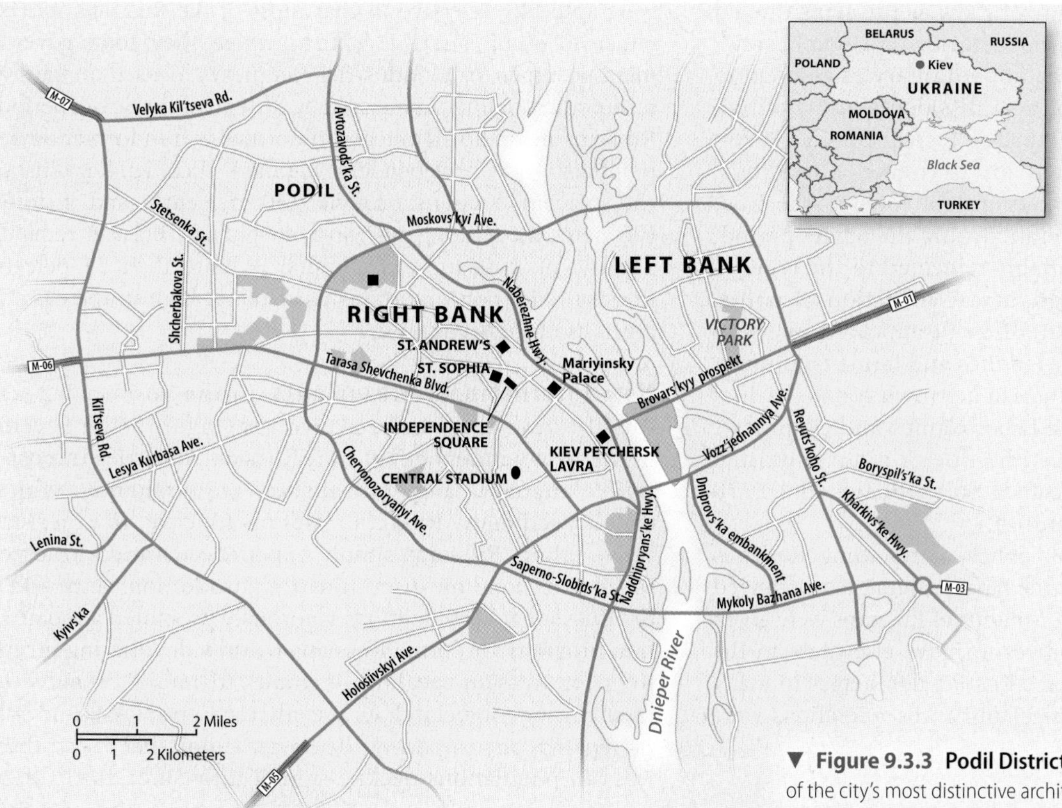

◄ **Figure 9.3.2 Kiev** The Ukrainian city of Kiev lines both sides of the Dnieper River. Many of the city's older neighborhoods are found west of the river (on the Right Bank), while an assortment of new office buildings and apartment houses are found east of the river (on the Left Bank).

▼ **Figure 9.3.3 Podil District** Kiev's Podil District features some of the city's most distinctive architecture, including the ornate green domes of St. Andrew's Orthodox Church (upper right).

More construction has accompanied Ukrainian independence, and many of the city's newer hotels and taller commercial buildings are also located on this side of the river, all seemingly a world away from the quiet hills and picturesque tree-lined streets of the Podil District.

1. Examine a map of Ukraine, and explain why Kiev residents are still thankful that a south wind was blowing across the city on April 26, 1986.
2. Follow the Dnieper River north and south of downtown Kiev. In broad terms, describe the changing land uses that are visible as you pass from the center of the city to its periphery.

Google Earth Virtual Tour Video

http://goo.gl/P2I5EA

The eastward migration continued during the Soviet period, once communist leaders consolidated power during the late 1920s and saw the economic advantages of developing the region's rich resource base. The German invasion of European Russia during World War II demonstrated that there were also strategic reasons for settling the eastern frontier, and this propelled further migrations during and following the war. Indeed, by the end of the communist era, 95 percent of Siberia's population was classified as Russian (including Ukrainians and other western immigrants). With completion of the BAM Railroad in the 1980s, another corridor of settlement opened in Siberia, prompting new migrations into a region once remote from the outside world.

Political Motives Political motives also have shaped migration patterns. Particularly in the case of Russia, leaders from both the imperial and the Soviet eras saw advantages in moving selective populations to new locations. The infilling of the southern Siberian hinterland had a political as well as an economic rationale. Both the tsars and the Soviet leaders saw their political power grow as Russians moved into the resource-rich Eurasian interior. For some, however, the move to Siberia was not voluntary as the region became a repository for political dissidents and troublemakers. The communist regime of Joseph Stalin (1928–1953) was particularly noted for its forced migrations, including the removal of thousands of Jews to the Russian Far East after 1928. Overall, during the Soviet period, uncounted millions were forcibly relocated to the region's infamous **Gulag Archipelago**, a vast collection of political prisons in which inmates often disappeared or spent years far removed from their families and home communities. (The term *Gulag* is a Russian acronym for the "Chief Administration of Corrective Labor Camps and Colonies"; the term *Gulag Archipelago* is the title of a novel written about these camps by Aleksandr Solzhenitsyn, the 1970 Nobel Prize winner in literature.)

Russification, the Soviet policy of resettling Russians into non-Russian portions of the Soviet Union, also changed the region's human geography. Millions of Russians were given economic and political incentives to move elsewhere in the Soviet Union in order to increase Russian dominance in many of the outlying portions of the country. The migrations were geographically selective in that most of the Russians moved either to administrative centers, where they took government positions, or to industrial complexes focused on natural resource extraction. As a result, by the end of the Soviet period, Russians made up significant minorities within former Soviet republics (now independent nations) such as Kazakhstan (30 percent Russian), Latvia (30 percent), and Estonia (26 percent). Among its Slavic neighbors, Belarus remains 11 percent Russian, and Ukraine is more than 24 percent Russian, with concentrations particularly high in the eastern portions of these countries.

New International Movements In the post-Soviet era, Russification has often been reversed (Figure 9.18). Several of the newly independent non-Russian countries imposed rigid language and citizenship requirements, which encouraged many Russian residents to leave. In other settings, ethnic Russians simply experienced varied forms of social and economic discrimination. In addition, since 2005, the Russian government has vigorously promoted a repatriation program for ethnic Russians worldwide, offering incentives for Russian-speaking migrants to return or move to their cultural homeland. As a result, the Central Asia and Baltic regions, once a part of the Soviet Union, have seen their Russian populations decline significantly since 1991, often

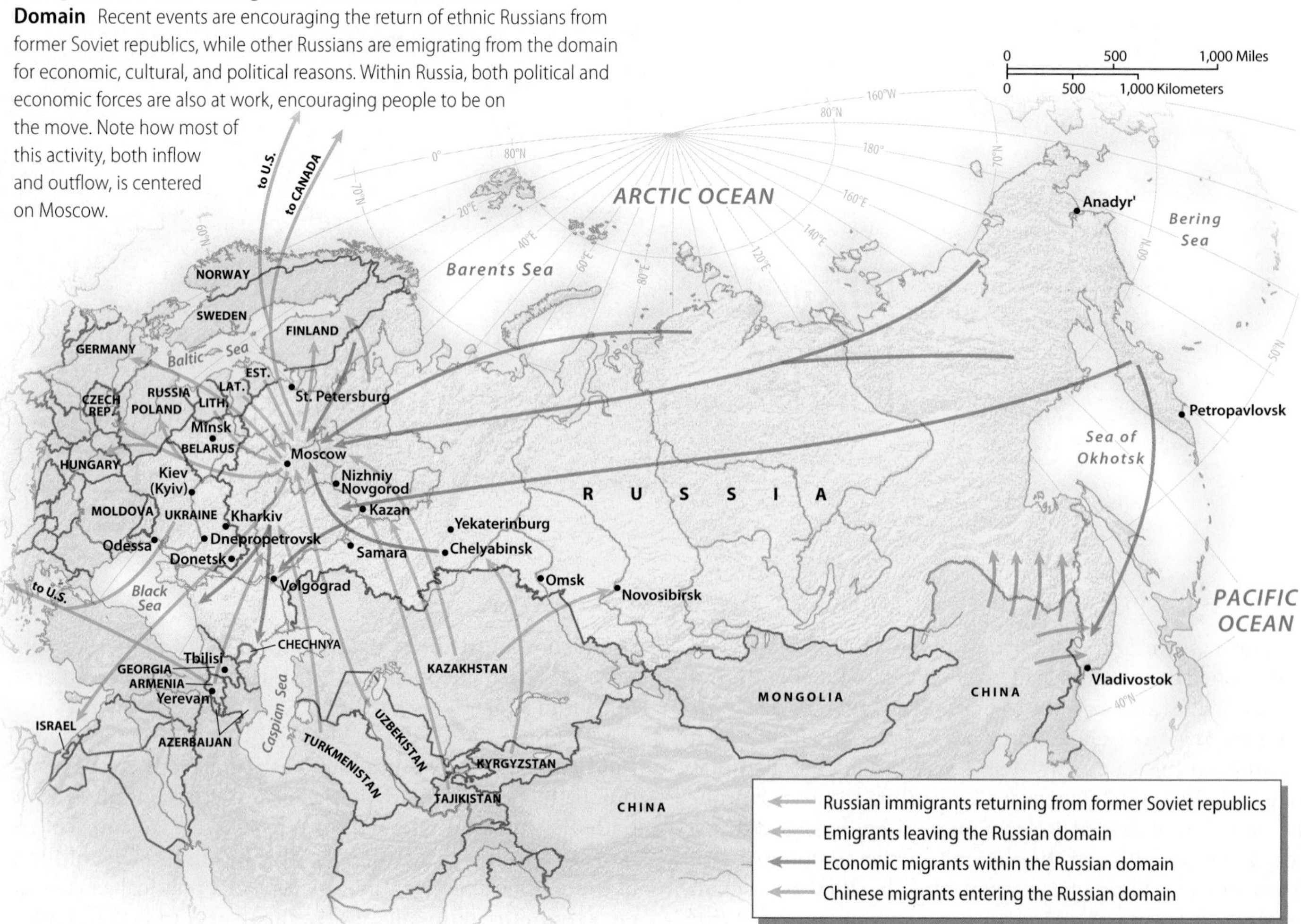

▼ **Figure 9.18 Recent Migration Flows in the Russian Domain** Recent events are encouraging the return of ethnic Russians from former Soviet republics, while other Russians are emigrating from the domain for economic, cultural, and political reasons. Within Russia, both political and economic forces are also at work, encouraging people to be on the move. Note how most of this activity, both inflow and outflow, is centered on Moscow.

by 20 to 35 percent. By 2010, more than 7 million Russians had left former Soviet republics and returned to their homeland.

Russia has also experienced a growing immigrant population, many of them undocumented migrants drawn to the region for work. Net in-migration to Russia totaled more than 320,000 people in 2011. The story has a familiar ring to it. More than 10 million undocumented immigrants are suspected in the country. Most are young, upwardly mobile people, predominantly male, who come for better-paying jobs. Many migrants send money they earn in Russia back home to their native lands. For example, one recent estimate suggests that about 20–30 percent of Tajikistan's economy is now based on funds sent to the country from émigrés living and working in Russia. Recently, the government has implemented tighter controls to restrict border crossings and enacted tougher penalties against businesses that hire illegal immigrants. There is a growing national debate concerning how many legal foreign workers should be allowed in the country and whether or not undocumented immigrants should be granted amnesty. Indeed, as a percentage of total population, there are more undocumented immigrants in Russia than in the United States.

The actual flows of undocumented immigrants are complex. While 80 percent of Russia's immigrants—legal and undocumented—come from portions of the former Soviet Union, a growing number of workers are from ethnically non-Slavic regions such as Central Asia. Street cleaners from Kyrgyzstan and shopkeepers from Uzbekistan are increasingly common sights in Moscow, where almost one-third of the country's undocumented immigrants may live (see *People on the Move: Moscow's Central Asian Communities*).

Finally, population movements in Russia's Far East, principally the growing numbers of arriving Chinese immigrants, are reshaping the economic and cultural geographies of that portion of the region (Figure 9.19). Somewhere between 5 and 7 million immigrants (many of them undocumented) may now live in the region. Much of the transformation has occurred since the early 1990s and is the result of a large, job-hungry Chinese population to the south, a shrinking supply of Russian workers to the north, and a regional economy that is increasingly being shaped by powerful forces of globalization. Walk through the Russian cities of Vladivostok and Khabarovsk, and you will see that street signs feature both Russian and Chinese lettering. Entire neighborhoods are dominated by immigrant populations. Chinese children are learning Russian in school, and Russians find themselves working for Chinese entrepreneurs. A significant nationalist backlash has occurred: Chinese have been attacked by gangs in cities, Chinese shopkeepers complain of being rousted by Russian police, and recent legislation has made it harder for Chinese to operate businesses in the region. On the other hand, many younger Russians have welcomed their Chinese counterparts, a growing number of joint Russian-Chinese companies have appeared in the region, and there is more intermarrying between the two groups as well.

▲ **Figure 9.19 Chinese Immigrants in the Russian Far East**
Many Chinese immigrants in the Russian Far East have become small-scale entrepreneurs in the region's commercial economy.

The domain's more open borders also have made it easier for other residents to leave the region (see Figure 9.18). Poor economic conditions and the region's unpredictable politics have encouraged many to emigrate. The "brain drain" of young, well-educated, upwardly mobile Russians has been considerable, and a World Economic Forum study in 2012 suggested many entrepreneurial young people feel uneasy living in a country where dependable legal institutions have been slow to develop and guarantee the rule of law. Sometimes, ethnic links also play a part in migration patterns. For example, many Russian-born ethnic Finns have moved to nearby Finland, much to the consternation of the Finnish government. Russia's Jewish population also continues to fall, a pattern begun late in the Soviet period. These emigrants have flocked mostly to Israel or the United States, where they locate in familiar Jewish neighborhoods such as Miami, Los Angeles, and New York City's Brighton Beach district.

The Urban Attraction Regional residents also have been bound for the cities. The Marxist philosophy embraced by Soviet planners encouraged urbanization. In 1917, the Russian Empire was still overwhelmingly rural and agrarian; 50 years later, the Soviet Union was primarily urban. Planners saw great economic and political advantages in efficiently clustering the population, and Soviet policies dedicated to large-scale industrialization obviously favored an urban orientation. Today Russian, Ukrainian, and Belorussian rates of urbanization are comparable to those of the industrialized capitalist countries (see Table 9.1).

Soviet cities grew according to strict governmental plans. Planners selected different cities for different purposes. Some were designed for specific industries, while others had primarily administrative roles. A system of internal passports prohibited people from moving freely from city to city. Instead, people generally went where the government assigned them jobs. Moscow, the country's

People on The Move

Moscow's Central Asian Communities

Stroll the corridors of a high-rise apartment house in Golyanovo, a suburb on Moscow's east side. You are liable to smell the *osh* cooking, a savory Central Asian blend of rice, lamb, carrots, peppers, onions, and garlic (Figure 9.4.1). You may pass a woman in the halls wearing a Muslim *hijab,* a veil that covers her head. The banter of children outside in the park may reflect their roots from Uzbekistan, Tajikistan, or Kyrgyzstan, although they are surely learning Russian in the local school.

A New Pattern of Immigrants Moscow is now home to a huge immigrant population (including perhaps 3–5 million undocumented residents), and many of them are from poor corners of the former Soviet Union, especially Central Asia (Figure 9.4.2). The majority are hard-working, young, and Muslim, trying to create new life opportunities for themselves and their families. They are fundamentally reworking the city's cultural geography. Many live in shabby trailers, cheap basement studios, shoddy boardinghouses, or inexpensive apartments. Some live and work near central Moscow, but many have located—often beyond the close scrutiny of authorities—in Moscow's sprawling suburbs. They frequently have jobs as construction workers, street vendors, cab drivers, and janitors, taking whatever low-paying work they can find.

▲ **Figure 9.4.2 Central Asian Immigrants in Moscow** These immigrants from Tajikistan work at an outdoor market near Moscow's Kiev Railway Station.

Moscow's Attraction In some ways, the story is an old one. Moscow has absorbed immigrants for centuries as Russia's influence expanded across Asia. This newest wave has grown tremendously since the collapse of the Soviet Union in 1991. Many Central Asian migrants, as citizens of former Soviet republics, found they did not need special visas to come to Russia. Once in the country, Moscow's bright lights, job prospects, and high wages (over 80 percent above the national norm) pulled them toward the national capital. Work permits for immigrants are legally required, but many have found ways to avoid the bureaucracy.

◀ **Figure 9.4.1 Osh** This savory Central Asian dish—a mix of rice, meat, vegetables, and spices—is an increasingly common sight in Moscow as immigrants from Uzbekistan, Tajikistan, and Kyrgyzstan arrive in large numbers.

Local Tensions Moscow's immigrant boom has also produced tension. With only a few mosques in the city (and perhaps 2 million Muslims), many immigrants have been pressing for more houses of worship, including a planned 13-story mega-mosque on the city's northwest side that might fit 40,000–50,000 people. Many native Russians have resisted, protesting the mosque project. Another new law—pushed by a group called the Movement Against Illegal Immigration—limits the number of non-Russians who can operate market stalls on the streets of the city. Growing violence against these non-Russian immigrants has also erupted in Moscow and elsewhere. As young Central Asian children flood Moscow schools (immigrant birth rates are far higher than Russian), many authorities struggle to keep up. But although *xenophobia* (fear or dislike of foreigners) may dominate the headlines, Moscow is also adjusting to, and even embracing, its Tajik, Uzbeki, and Kyrgyz residents. Many Russian churches have opened their doors and assisted new migrants, trendy Uzbeki-style eateries (featuring clay-oven bakeries) are popular among the city's young professionals, and most immigrants are able to find some kind of employment as they adjust to new lives in a far-off land.

leading administrative city, thrived under the Soviet regime. It formed the undisputed core of Soviet bureaucratic power, as well as the center of education, research, and the media. Specialized industrial cities also grew at a rapid pace. In the mining and metallurgical zone of the southern Urals, centers such as Yekaterinburg and Chelyabinsk mushroomed into major urban centers. Another cluster of specialized industrial cities, including Kharkiv and Donetsk, emerged in the coal districts of eastern Ukraine.

With the end of the Soviet Union, however, people gained basic freedoms of mobility. In addition, with a more open economy, especially in Russia, shifting urban employment opportunities increasingly reflect how effectively local economies could compete in the global market. This new economic reality has led to the depopulating of many older industrial areas, as they simply cannot produce raw materials or finished industrial goods at competitive global prices. Since 1989, for example, the Russian Northeast has seen its population decline as many workers have left harsh, unemployment-prone industrial centers for better opportunities elsewhere. In many cases, people are freely gravitating toward growth opportunities in locations of new foreign investment, principally in larger urban areas in western and southern Russia. Some also are receiving relocation assistance from a Russian government that admits many of these outlying settlements simply cannot be justified in a more market-oriented economy.

▲ **Figure 9.20 Downtown Moscow** Shoppers stroll pleasant Arbat Street in downtown Moscow. Upscale shops, restaurants, and entertainment venues attract both visitors and local residents.

Inside the Russian City

Today most people in the region live in a city, the product of a century of urban migration and growth (see Table 9.1). Large Russian cities possess a core area, or center, that features superior transportation connections; the best-stocked, upscale department stores and shops; the most desirable housing; and the most important offices (both governmental and private) (Figure 9.20). In the largest urban centers, cities such as Moscow and St. Petersburg also feature extensive public spaces and examples of monumental architecture at the city center. Within the city, there is usually a distinctive pattern of circular land-use zones, each of which was built at a later date moving outward from the center. Such a ringlike urban morphology is not unique. However, as a result of the extensive power of government planners during the Soviet period, this urban form is probably more highly developed here than in most other parts of the world.

At their very center, the cores of many older cities predate the Soviet Union. Pre-1900 stone buildings often dominate older city centers. Some of these are former private mansions that were turned into government offices or subdivided into apartments during the communist period, but are now being privatized again. Many of these older buildings, however, are being leveled in rapidly growing urban settings such as downtown Moscow. Urban preservation experts estimate that since 1992, several thousand historic buildings have been destroyed, including many structures that had supposedly received official protection. Retail malls replace many of these older structures in the city. Nearby, nightclubs and bars are filled with pleasure seekers as the city's professional elite mingles with foreign visitors and tourists. The Moscow International Business Center is also expanding into older industrial areas, offering new office space for Russian and international firms.

Farther out from the city centers are **mikrorayons,** large, Soviet-era housing projects of the 1970s and 1980s. Mikrorayons are typically composed of massed blocks of standardized apartment buildings, ranging from 9 to 24 stories in height. The largest of these supercomplexes contain up to 100,000 residents. Soviet planners hoped that mikrorayons would foster a sense of community, but most now serve as anonymous bedroom communities for larger metropolitan areas.

Some of Russia's most rapid urban growth has occurred on the metropolitan periphery, paralleling the North American experience. Moscow, for example, has seen its urban reach expand far beyond the city center. The surrounding administrative district (the Moscow Oblast) contains about 7 million residents and is home to more than 700 international companies. Land prices and tax rates are lower than in the central city; the bureaucracy is less onerous; and the transportation and telecommunications infrastructure is relatively new. New suburban shopping malls and housing districts featuring single-family homes, paralleling the North American model, are also popping up on the urban fringe,

▲ **Figure 9.21 New Single Family Housing, Moscow Suburbs** This newer upscale housing development in the Moscow suburbs was built by INKOM-Nedvizhimost. The large, neatly fenced lots and spacious homes build on many North American traditions of suburban taste and design.

allowing upscale residents to live and shop without having to visit the city center (Figure 9.21).

Elsewhere on Moscow's urban fringe, elite **dacha**, or cottage communities, appeal to many more well-to-do residents, particularly during the summer months. The tradition of rural retreats, dating back to the Russian Empire, also thrived during the Soviet era as Communist Party officials sought an escape from the dreary bureaucratic chores of the city. Today about 300 cottage settlements appear on the Moscow periphery, including many new elite privatized developments northwest and southeast of the city that cater to the country's business class. The older dacha belt to the west of the city along the Moskva River also offers summer homes, spa retreats, and holiday hotels in a zone dubbed the "Sub-Moscow Switzerland."

The Demographic Crisis

Since 2005, the Russian government has identified population loss as a key issue of national importance. Some government and United Nations (UN) estimates have predicted that Russia's population could fall by a startling 45 million by 2050. Similar conditions are affecting the other countries within the region (see Table 9.1). Beginning with World War II, large numbers of deaths combined with low birth rates to produce sizable population losses. Although population increases accelerated in the 1950s, growth slowed by 1970, and death rates began exceeding birth rates in the early 1990s. Two population pyramids tell the troubling tale (Figure 9.22). The first shows that, whereas adult-age populations are prominently represented (with the exception of older males and the birth crash of World War II), relatively few children are being added. Looking ahead to 2050, today's shrinking numbers take an even bigger toll as the pattern of small families is likely to continue.

In 2006, President Putin declared that demographic decline was Russia's "most acute problem." He and other Russian leaders have pushed for a long-term program aimed at raising birth rates and challenging the one-child family norm that has become widely accepted within the country. Under the plan, mothers of multiple children receive cash payments, extended maternity leave, and extensive daycare subsidies. Some Russian cities have even sponsored competitions, encouraging couples to have babies in the hopes of winning prizes such as automobiles. Recently, birth

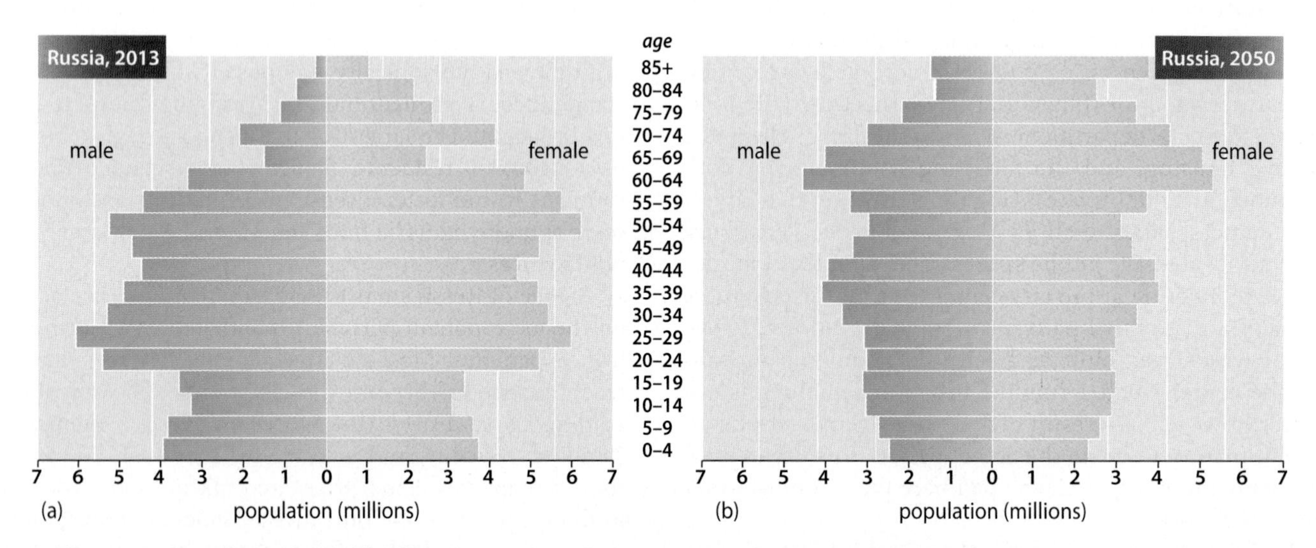

▲ **Figure 9.22 Population Pyramids of Russia** These two population pyramids provide (a) a recent glimpse (2013), as well as (b) predicted patterns (2050), of Russia's population structure. Present trends suggest that Russia's population will continue to age, with relatively fewer young people supporting a relatively large elderly population. **Q: Among Russian males, what is the evidence of the impact of earlier wars and higher death rates?**

rates have risen within Russia, perhaps a function of changing government policies. In particular, rates among ethnic non-Russians in the country (for example, in the Caucasus region and portions of Siberia) have been significantly above those of ethnic Russians. In 2009, the country reported its first year of natural population growth in more than 15 years. Indeed, both Russian and Ukrainian birth rates are now significantly higher than those of Germany, Italy, and Japan, and there are long waiting lists for urban kindergarten spots in Moscow. Employers are also offering more benefits to both mothers and fathers, making parenthood more appealing for some couples.

Still, many uncertainties remain concerning future demographic growth. Another severe and prolonged economic decline such as that experienced in the 1990s would likely depress birth rates. In addition, residents across the entire region still face significantly higher death rates than much of the developed world, and they also have daunting health-care challenges. The region's demographic structure also forecasts fewer women of child-bearing age in the near future, making the current improvements difficult to maintain. One other wild card remains: immigration. Some Russian population experts urge leaders to encourage more migration to the region, but this continues to unsettle others who see such inflows as a threat to Russian cultural dominance.

REVIEW

9.3. Discuss how major river and rail corridors have shaped the geography of population and economic development in the region. Provide specific examples.

9.4. Contrast Soviet and post-Soviet migration patterns within the Russian domain, and discuss the changing forces at work.

9.5. Describe some of the major land-use zones in the modern Russian city, and suggest why it is important to understand the impact of Soviet-era planning within such settings.

Cultural Coherence and Diversity: The Legacy of Slavic Dominance

The Russian domain remains at the heart of the Slavic world. For hundreds of years, Slavic peoples speaking the Russian language expanded their influence from an early homeland in central European Russia. Eventually, this Slavic cultural imprint spread north to the Arctic Sea, south to the Black Sea and Caucasus, west to the shores of the Baltic, and east to the Pacific Ocean. In this process of diffusion, Russian cultural patterns and social institutions spread widely, influencing many non-Russian ethnic groups that continued to live under the rule of the Russian Empire. The legacy of that Slavic expansion continues today. It offers Russians a rich historical identity and sense of nationhood. It also provides a meaningful context in which to understand how present-day Russians are dealing with forces of globalization and how non-Russian cultures have evolved within the region.

The Heritage of the Russian Empire

The expansion of the Russian Empire paralleled similar events in western Europe. As Spain, Portugal, France, and Britain carved out empires in the Americas, Africa, and Asia, the Russians expanded eastward and southward across Eurasia. Unlike other European empires, however, the Russian Empire formed one single territory, uninterrupted by oceans or seas. Only with the fall of the Soviet Union in 1991 did this transformed empire finally begin to dissolve.

Origins of the Russian State The origin of the Russian state lies in the early history of the **Slavic peoples**, defined linguistically as a distinctive northern branch of the Indo-European language family. The Slavs originated in or near the Pripyat marshes of modern Belarus. Some 2000 years ago they began to migrate to the east, reaching as far as modern Moscow by 200 CE. Slavic political power grew by 900 CE as Slavs intermarried with southward-moving warriors from Sweden known as *Varangians,* or *Rus*. Within a century, the state of Rus extended from Kiev (the capital) in modern Ukraine to Lake Ladoga near the Baltic Sea. The new Kiev–Rus state interacted with the rich and powerful Byzantine Empire of the Greeks, and this influence brought Christianity to the Russian realm by 1000 CE. Along with the new religion came many other aspects of Greek culture, including the Cyrillic alphabet. Even as groups such as the Russians and Serbs converted to **Eastern Orthodox Christianity**—a form of Christianity historically linked to eastern Europe and church leaders in Constantinople (modern Istanbul)—their Slavic neighbors to the west (the Poles, Czechs, Slovaks, Slovenians, and Croatians) accepted Catholicism. The resulting religious division split the Slavic-speaking world into two groups, one oriented to the west, the other to the east and south. This early Russian state soon faltered and split into several principalities that were then ruled by invading Mongols and Tatars (a group of Turkish-speaking peoples).

Growth of the Russian Empire By the 14th century, however, northern Slavic peoples overthrew Tatar rule and established a new and expanding Slavic state (Figure 9.23). The core of the new Russian Empire lay near the eastern fringe of the old state of Rus. The former center around Kiev was now a war-torn borderland (or "Ukraine" in Russian) contested by the Orthodox Russians, the Catholic Poles, and the Muslim Turks. Gradually, this area's language diverged from that spoken in the new Russian core, and *Ukrainians* and Russians developed into two separate peoples. A similar development took place among the northwestern Russians, who experienced several centuries of Polish rule and over time were transformed into a distinctive group known as the *Belorussians*.

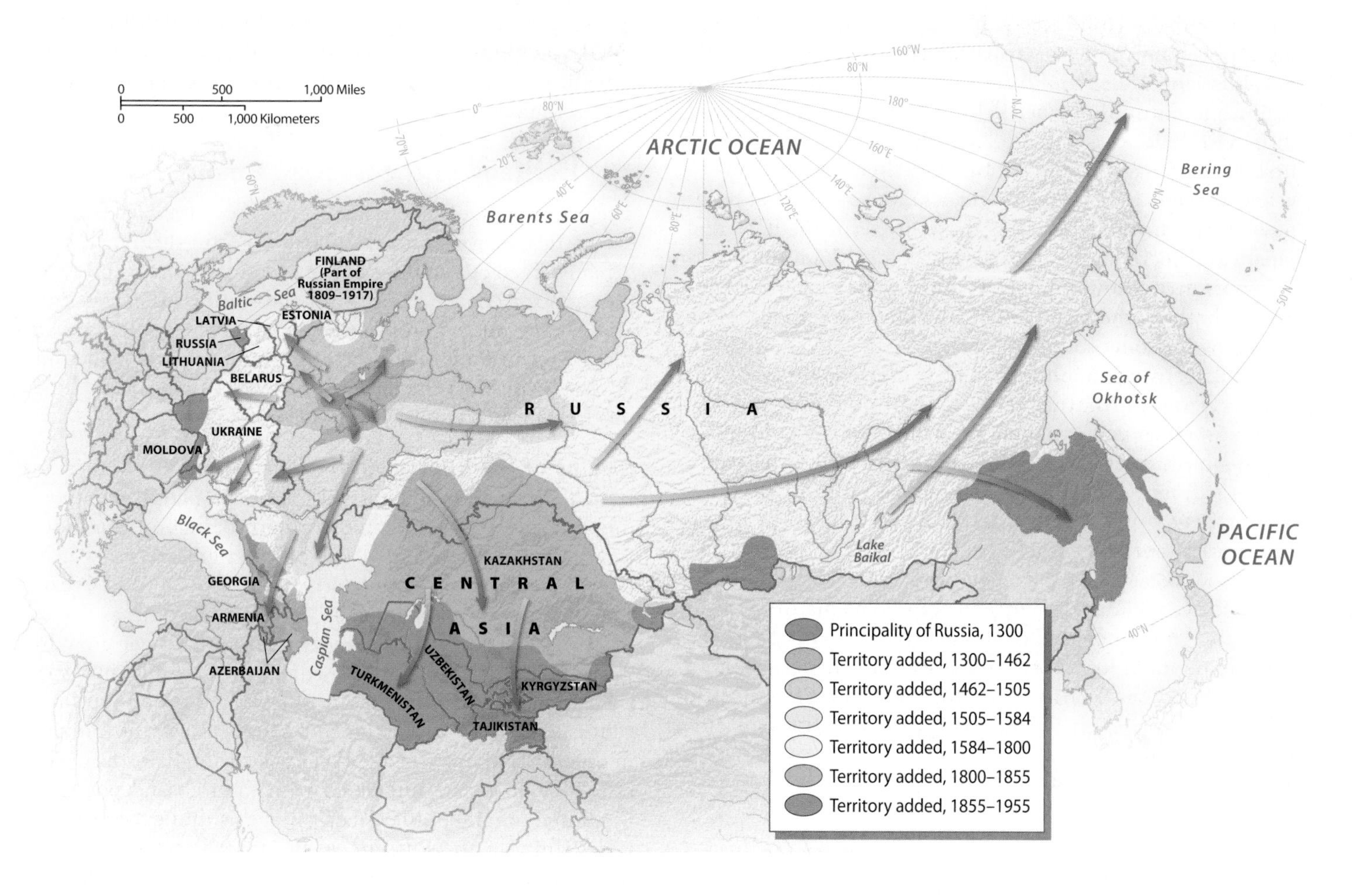

▲ **Figure 9.23 Growth of the Russian Empire** Beginning as a small principality in the vicinity of modern Moscow, the Russian Empire took shape between the 14th and 16th centuries. After 1600, Russian influence stretched from eastern Europe to the Pacific Ocean. Later, portions of the empire were added in the Far East, in Central Asia, and near the Baltic and Black seas.

The Russian Empire expanded remarkably in the 16th and 17th centuries (see Figure 9.23). Former Tatar territories in the Volga Valley (near Kazan) were incorporated into the Russian state in the mid-1500s. The Russians also allied with the seminomadic **Cossacks**, Slavic-speaking Christians who had earlier migrated to the region to seek freedom in the ungoverned steppes (see Figure 9.1). The Russian Empire granted them considerable privileges in exchange for their military service, an alliance that facilitated Russian expansion into Siberia during the 17th century. Premium furs were the chief lure of this immense northern territory. By the 1630s, Russian power was entrenched in central Siberia, and by the end of the century, it had reached the Pacific Ocean. Chinese resistance, however, delayed Russian occupation of the Far East region until 1858, and the imperial designs of the Japanese halted further expansion to the southeast when the Russians lost the Russo-Japanese War in 1905.

While the Russian Empire expanded to the east with great rapidity, its westward expansion was slow and halting. In the 1600s, Russia still faced formidable enemies in Sweden, Poland, and the Ottoman (Turkish) Empire. By the 1700s, however, all three of these states had weakened, allowing the Russian Empire to gain substantial territories. After defeating Sweden in the early 1700s, Tsar Peter the Great (1682–1725) obtained a foothold on the Baltic, where he built the

► **Figure 9.24 Peter the Great** Tsar Peter the Great ruled the Russian Empire between 1682 and 1725 and expanded imperial influence westward, reorienting Russia toward the Baltic Sea and its new capital of St. Petersburg.

new capital city of St. Petersburg (Figure 9.24). Later in the 18th century, Russia defeated both the Poles and the Turks and gained all of modern-day Belarus and Ukraine. Tsarina Catherine the Great (1762–1796) was particularly pivotal in colonizing Ukraine and bringing the Russian Empire to the warm-water shores of the Black Sea.

The 19th century witnessed the Russian Empire's final expansion. Large gains were made in Central Asia, where a group of once-powerful Muslim states was no longer able to resist the Russian army. The mountainous Caucasus region proved a greater challenge, as the peoples of this area had the advantage of rugged terrain in defending their lands. South of the Caucasus, however, the Christian Armenians and Georgians accepted Russian power with little struggle because they found it preferable to rule by the Persian or Ottoman empires.

The Legacy of Empire The expansion of the Russian people was one of the greatest human movements Earth has ever witnessed. By 1900, a traveler going from St. Petersburg on the Baltic to Vladivostok on the Sea of Japan would everywhere encounter Russian peoples speaking the same language, following the same religion, and living under the rule of the same government. Nowhere else in the world did such a tightly integrated cultural region cover such a vast space.

The history of the Russian Empire also reveals points of ongoing tension with the world beyond. One of these tensions centers on Russia's ambivalent relationship with western Europe. Russia shares with the West the historical legacy of Greek culture and Christianity. Since the time of Peter the Great, Russia has undergone several waves of intentional Westernization. At the same time, however, Russia has long been suspicious of—even hostile to—European culture and social institutions. Although elements of this debate were transformed during the Soviet period, the central tension remains, and it influences Russia to this day. As Europe further unifies through the expansion of the EU, this gap between East and West may take new forms, but it is unlikely to disappear anytime soon.

Geographies of Language

Slavic languages dominate the region (Figure 9.25). The distribution of Russian-speaking populations is complicated. Russian, Belorussian, and Ukrainian are closely related languages. Some linguists argue that they ought to be considered separate dialects of a single Russian language, as they are all mutually intelligible. Most Ukrainians, however, insist that Ukrainian is a distinct language in its own right, and there is a well-developed sense of national distinction between Russians and Ukrainians.

▼ **Figure 9.25 Languages of the Russian Domain** Slavic Russians dominate the region, although many linguistic minorities are present. Siberia's diverse native peoples add cultural variety in that area. To the southwest, the Caucasus Mountains and the lands beyond contain the region's most complex linguistic geography. Ukrainians and Belorussians, while sharing a Slavic heritage with their Russian neighbors, add further variety in the west.

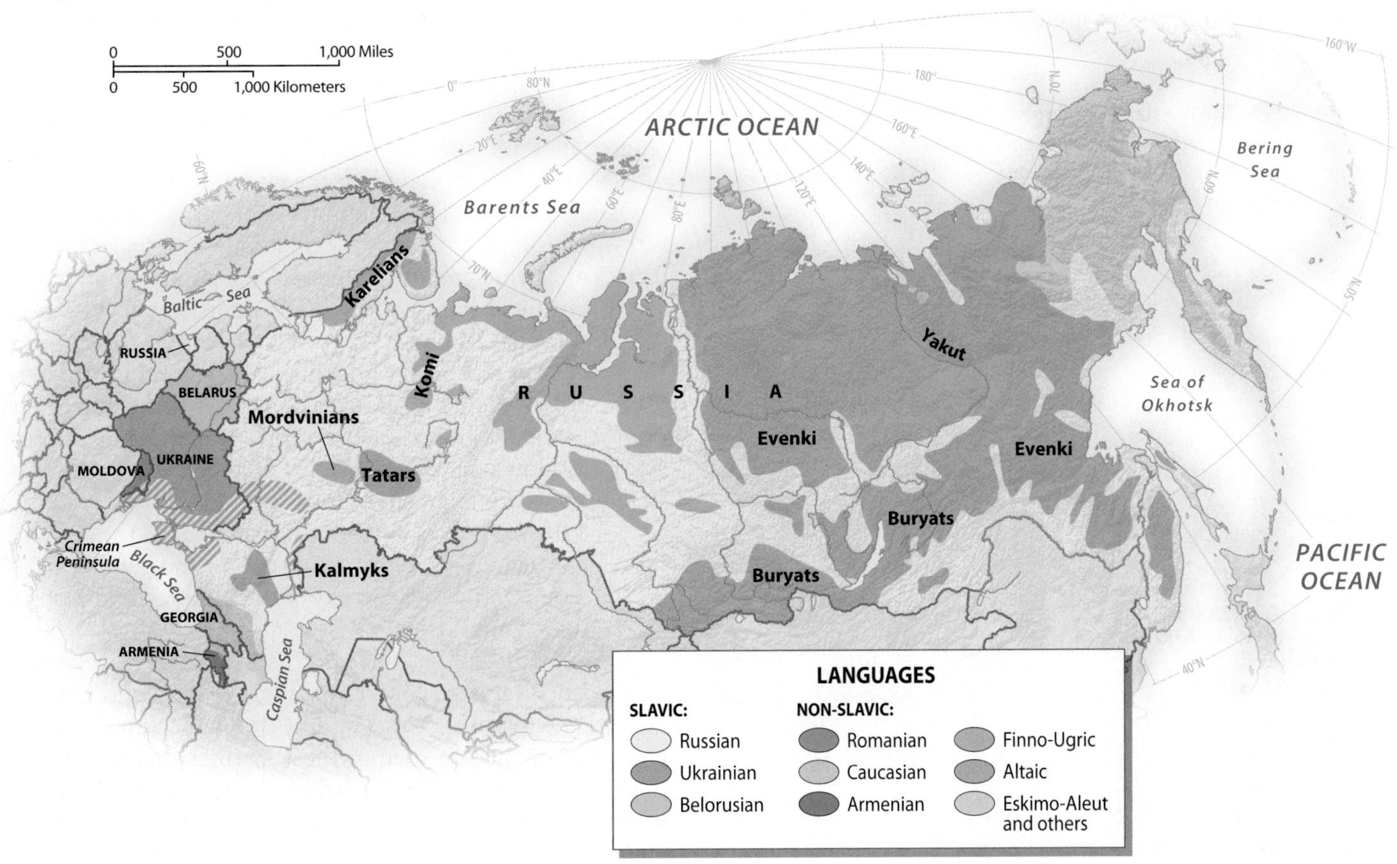

Belorussians, on the other hand, are more inclined to stress their close kinship with the Russians.

Patterns in Belarus, Ukraine, and Moldova The geographic pattern of the Belorussian people is relatively simple. The vast majority of Belorussians reside in Belarus, and most people in Belarus are Belorussians (see Figure 9.25). The country does, however, contain scattered Polish and Russian minorities. For the Russians, this presents few problems, since Russians and Belorussians can relatively easily assume each other's ethnic identity.

The situation in Ukraine, however, is more complex. Only about 67 percent of the population is Ukrainian. Russian speakers make up almost 25 percent of the population, but are strongly concentrated in eastern Ukraine, while they make up a much smaller portion of western Ukraine's population (Figure 9.26). Similarly, the Crimean Peninsula, now a part of Ukraine, has long ethnic and political connections to Russia. As a result, many Ukrainian citizens in the eastern and southern parts of the country rarely speak Ukrainian. Conversely, many Ukrainians born in the west never learn Russian. Kiev (Kyiv), the national capital, has been described as bilingual, "a Russian-speaking city whose people know how to speak Ukrainian."

Since 2004, Ukrainian speakers have increased their efforts to enforce Ukrainian cultural and linguistic traditions, a campaign that has provoked popular and political resistance in the eastern portion of the country. In 2012, the Ukrainian Parliament reaffirmed Ukrainian as the country's sole official language, angering many ethnic Russians. Even in Parliament, partisans struck blows, ripped clothes, and set off a brawl over the hotly contested legislation.

▼ **Figure 9.26 The Russian Language in Ukraine** Both eastern Ukraine and the Crimean Peninsula retain large numbers of Russian speakers, a function of long cultural and political ties that continue to complicate Ukraine's contemporary human geography.

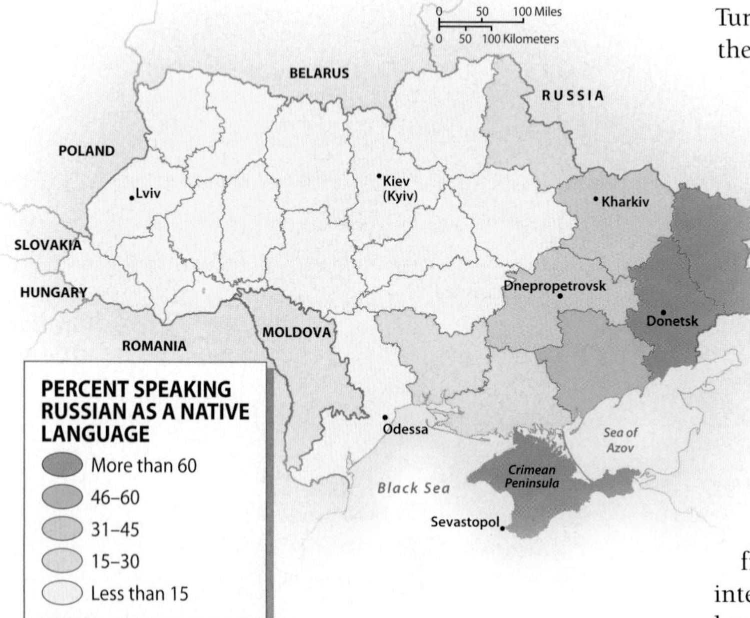

In nearby Moldova, Romanian (a Romance language) speakers are dominant, although ethnic Russians and Ukrainians each make up about 13 percent of the country's population. Use of the official Romanian language has been criticized by many Slavic speakers, particularly in the region east of the Dniester River (Transdniester) that borders Ukraine. During the Soviet period, such geographical complexities had little consequence because the distinctions among Russians, Ukrainians, and Moldavians were not viewed as important in official circles. Now that Russia, Ukraine, and Moldova are separate countries with a heightened sense of national distinction, this issue has emerged as a significant source of tension across the region.

Patterns Within Russia Approximately 80 percent of Russia's population claims a Russian linguistic identity. Russian speakers inhabit most of European Russia, but large enclaves of other peoples also occur. The Russian linguistic zone extends across southern Siberia to the Sea of Japan. In sparsely settled lands of central and northern Siberia, Russians are numerically dominant in many areas, but they share territory with varied indigenous peoples.

Finno-Ugric peoples, though small in number, dominate sizable portions of the non-Russian north. The Finno-Ugric (part of the Uralic language family) speakers make up an entirely different language family from the Indo-European Russians. Although many have been culturally Russified, distinct ethnic groups such as the Karelians, Komi, and Mordvinians remain a part of Russia's modern cultural geography. Karelians, in particular, identify strongly with their Finnish neighbors to the west.

Altaic speakers also complicate the country's linguistic geography. This language family includes the Volga Tatars, whose territory is centered on the city of Kazan in the middle Volga Valley. While retaining their ethnic identity, the Turkish-speaking Tatars have extensively intermarried with their Russian neighbors. Bilingual education is now common in the schools, with ethnic Tatars free to use their traditional language. Yakut peoples of northeast Siberia also represent Turkish speakers within the Altaic family. Other Altaic speakers in Russia belong to the Mongol group. In the west, examples include the Kalmyk-speaking peoples of the lower Volga Valley who migrated to the region in the 1600s. Far to the east, over 400,000 Buryats live in the vicinity of Lake Baikal and represent an indigenous Siberian group closely tied to the cultures and history of Central Asia.

The plight of many native peoples in central and northern Siberia parallels the situation in the United States, Canada, and Australia. Rural indigenous peoples in each of these settings remain distinct from dominant European cultures. Such groups also are internally diverse and often are divided into several unrelated linguistic clusters. One entire linguistic grouping of

▲ **Figure 9.27 Minority Evenki** Russia's indigenous Evenki population speaks an Altaic language. **Q: What social and economic problems might the Evenki share with many indigenous peoples in North America and Australia?**

Eskimo-Aleut speakers is limited to approximately 25,000 people, who are widely dispersed through northeast Siberia. Another indigenous Siberian group is the Altaic-speaking Evenki, whose traditional territory covers a large portion of central and eastern Siberia (Figure 9.27). At home in the taiga, traditional Evenki hunted animals (such as elk and wild reindeer) and raised livestock (horses and reindeer). Many of these Siberian peoples have seen their traditional ways challenged by the pressures of Russification, just as indigenous peoples elsewhere in the world have been subjected to similar pressures of cultural and political assimilation. Unfortunately, other common traits seen within such settings are low levels of education, high rates of alcoholism, and widespread poverty.

Transcaucasian Languages Although small in size, Transcaucasia offers a bewildering variety of languages (Figure 9.28). From Russia, along the north slopes of the Caucasus, to Georgia and Armenia east of the Black Sea, a complex history and a fractured physical setting have combined to produce some of the most complicated language patterns in the world. No fewer than three language families (Caucasian, Altaic, and Indo-European)

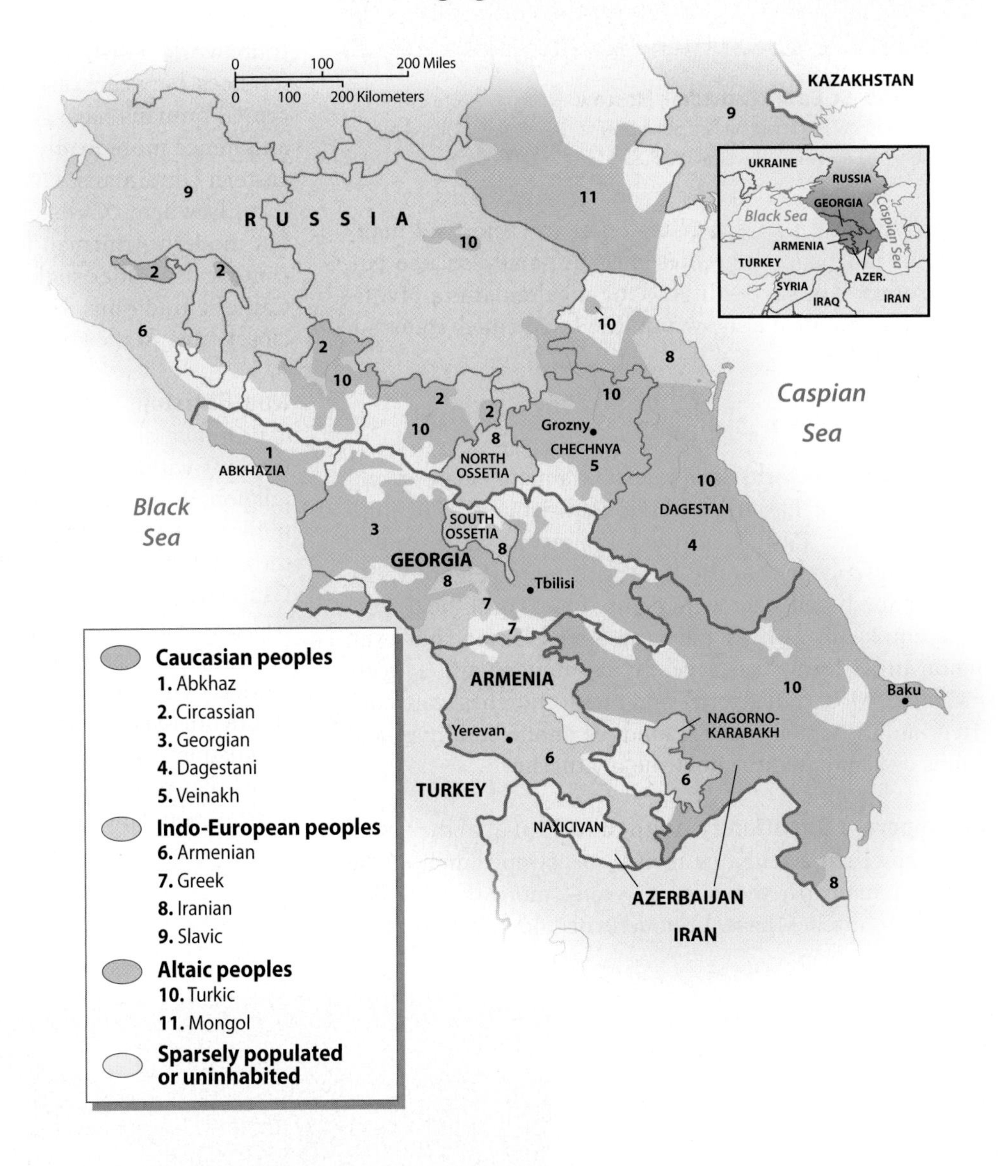

► **Figure 9.28 Languages of the Caucasus Region** A complicated mosaic of Caucasian, Indo-European, and Altaic languages characterizes the Caucasus region of southern Russia and nearby Georgia and Armenia. Persistent political problems have erupted periodically in the region as local populations struggle for more autonomy. Recent examples include independence movements in Chechnya and in nearby Dagestan.

▲ **Figure 9.29 St. Basil's Cathedral, Moscow** Built in the 16th century, this famous Moscow landmark on Red Square remains a symbol of Russian culture and of the enduring power of the Russian Orthodox Church.

are spoken within a region smaller than Ohio, and many individual languages are represented by small, isolated cultural groups. Not surprisingly, language remains a pivotal cultural and political issue within the fragmented Transcaucasian subregion.

Geographies of Religion

Most Russians, Belorussians, and Ukrainians share a religious heritage of Eastern Orthodox Christianity. For hundreds of years, Eastern Orthodoxy served as a central cultural presence within the Russian Empire (Figure 9.29). Indeed, church and state were tightly fused until the demise of the empire in 1917. Under the Soviet Union, however, religion in all forms was severely discouraged and actively persecuted. Most monasteries and many churches were converted into museums or other kinds of public buildings, and schools disseminated the doctrine of atheism.

Contemporary Christianity With the downfall of the Soviet Union, however, a religious revival has swept much of the Russian domain. In the past 20 years, more than 12,000 Orthodox churches have been returned to religious uses. Now, an estimated 75 million Russians are members of the Orthodox Church, including almost 500 monastic orders dispersed across the country. Within Russia, the Orthodox Church appears headed toward a return to its former role as official state church. The government increasingly is using church officials to sanction various state activities, which is particularly ironic because many of Russia's current leaders played roles in the earlier Soviet period when religious observances were banned. For example, when he became president in 2000, former Soviet-era KGB agent Vladimir Putin professed his Orthodox beliefs and claimed that his mother baptized him during the height of Soviet repression. In turn, the Orthodox Church has supported recent Russian political campaigns, both domestically and internationally. In addition, various forms of evangelical Protestantism (especially from the United States) have been on the rise in Russia since the disintegration of the Soviet Union.

Other forms of Western Christianity are also present in the region. For example, the people of western Ukraine, who experienced several hundred years of Polish rule, eventually joined the Catholic Church. Eastern Ukraine, on the other hand, remained fully within the Orthodox framework. This religious split reinforces the cultural differences between eastern and western Ukrainians. Western Ukrainians have generally been far more nationalistic, and hence more firmly opposed to Russian influence, than eastern Ukrainians.

Elsewhere, Christianity came early to the Caucasus, but modern Armenian forms—their roots dating to the 4th century CE—differ slightly from both Eastern Orthodox and Catholic traditions. Georgian Christianity, however, is more closely tied to the Orthodox faith.

Non-Christian Traditions Non-Christian religions also appear and, along with language, shape ethnic identities and tensions within the region. Islam is the largest non-Christian religion. Russia has some 7000 mosques and approximately 20 million adherents (Figure 9.30). Most are Sunni Muslims, and they include peoples in the North Caucasus, the Volga Tatars, and Central Asian peoples near the Kazakhstan border. Growth rates among Russia's Muslim population are three times that of the non-Muslim population.

► **Figure 9.30 Kazan Mosque** Kazan, the capital of Russia's Tatarstan Republic, has long been a focus of Muslim populations in the country.

Islamic fundamentalism also has been on the rise, particularly among Muslim populations in the Caucasus region, who increasingly resist what they see as strong-arm tactics and repressive actions on the part of the Russian government. In 2012, a growing number of young, conservative Muslim Salafists living in Russian Tatarstan also drew sharp distinctions between themselves and traditional Sunnis in the region, raising ethnic and religious tensions. Local authorities, as well as Moscow leaders, have drafted laws designed to limit the rising influence of the Salafists, but many observers feel this will only add to their appeal among the region's disaffected youth.

Russia, Belarus, and Ukraine are also home to more than 1 million Jews, who are especially numerous in the larger cities of the European West. Jews suffered severe persecution under both the tsars and the communists. Recent out-migrations, prompted by new political freedoms, have further reduced their numbers in Russia, Belarus, and Ukraine. Buddhists also are represented in the region, associated with the Kalmyk and Buryat peoples of the Russian interior. Indeed, Buddhism has witnessed a recent renaissance and now claims approximately 1 million practitioners, mostly in Asiatic Russia.

Russian Culture in Global Context

Russian culture has enriched the world in many ways. Russian cultural norms have for centuries embodied both an inward orientation toward traditional forms of expression (and nationalism) and an outward orientation directed primarily to western Europe. By the 19th century, even as Russian peasants interacted rarely with the outside world, Russian high culture had become thoroughly Westernized, and Russian composers, novelists, and playwrights gained considerable fame in Europe and the United States. Composers such as Tchaikovsky, Rachmaninoff, and Stravinsky stand as hugely important figures in the world of classical music. In literature, high school and college students around the world still bury themselves in the enduring stories told by legendary writers such as Leo Tolstoy (*War and Peace, Anna Karenina*) and Fyodor Dostoyevsky (*Crime and Punishment, The Brothers Karamazov*). Even in Russia today, there is great reverence and appreciation for the arts (the Bolshoi Ballet remains one of the world's leading dance companies) and for the ornate, baroque-style buildings that adorn older city centers such as St. Petersburg (see Figure 9.16).

Soviet Days During the Soviet period, a new mixture of cultural relationships unfolded within the socialist state. Initially, European-style modern art flourished in the Soviet Union, encouraged by the radical ideas of the new rulers. By the late 1920s, however, Soviet leaders turned against modernism, which they viewed as the decadent expression of a declining capitalist world. Many Soviet artists fled to the West, and others were exiled to Siberian labor camps. Increasingly, state-sponsored Soviet artistic productions centered on **socialist realism**, a style devoted to the realistic depiction of workers harnessing the forces of nature or struggling against capitalism. Still, traditional high arts, such as classical music and ballet, continued to receive lavish state subsidies, and to this day, Russian artists regularly achieve worldwide fame.

Turn to the West By the 1980s, it was clear that the attempt to fashion a new Soviet culture based on socialist realism and working-class solidarity had failed. The younger generation instead adopted a rebellious stance, turning for inspiration to fashion and rock music from the West. The mass-consumer culture of the United States proved immensely popular, symbolized above all by brand-name chewing gum, jeans, and cigarettes. The Soviet government attempted to ward off this perceived cultural onslaught, but with little success. Soviet officials could more easily censor books and other forms of written expression than Western countries, but even here they were increasingly frustrated. By the end of the Soviet period, the persistent secrecy that separated the Soviet Union from the West had broken down, thus enabling more people and influence to flow into the region.

After the fall of the Soviet Union in 1991, basic freedoms brought an inrush of global cultural influences, particularly to the region's larger urban areas such as Moscow. Shops were quickly flooded with Western books and magazines; people pondered the financial mysteries of home mortgages and condominium purchases; and they reveled in the new-found pleasures of fake Chanel handbags and McDonald's hamburgers. English-language classes became even more popular in cities such as Moscow, where Russians hurried to embrace the world their former leaders had warned them about for generations. Cultural influences streaming into the country were not all Western in inspiration. Films from Hong Kong and Mumbai (Bombay), as well as the televised romance novels (*telenovelas*) of Latin America, for example, proved far more popular in the Russian domain than in the United States. This onrush of global cultural influences, however, has not equally spread throughout the region. While urban residents have access and money to explore such options, rural life across the Russian domain remains far more wedded to traditional cultural institutions and values.

The Music Scene Younger residents of the region also have embraced the world of popular music, and their enthusiasm for American and European performers, as well as their support of a budding home-grown music industry, symbolizes the changing values of an increasingly post-Soviet generation. MTV Russia went on the air at midnight on September 26, 1998. A year later, the network sponsored a huge open-air concert on Red Square. Headlined by the Red Hot Chili Peppers, the event was a far cry from Soviet-era tank parades in front of Communist Party officials. Today Russian MTV reaches most of the nation's younger viewers.

Major global media companies such as Sony Music Entertainment established Russian operations in the post-Soviet era. Several of the world's other major recording companies have signed multiple Russian artists for domestic markets, helping boost a home-grown pop-music culture. Universal also has opened the way for Russian performers to go global.

For example, it sponsored the English-language debut of Russian pop singer Alsou in 2001. Three years later, Ukrainian singer Ruslana won the coveted Eurovision 2004 award for her song "Wild Dance." In a superb example of cultural globalization, the high-tech performance of the song at the awards show in Istanbul, Turkey, was seen by more than 100 million television viewers worldwide, and it featured a mix of Ukrainian folk music traditions, leather-clad female dancers, and on-stage acrobatics. More recently, Moscow hosted the 2009 Eurovision awards, which featured a strong performance from Ukrainian singer, composer, and television personality Svetlana Loboda (Figure 9.31). She used her performance of "Be My Valentine (Anti-Crisis Girl)" and her popularity in both Russia and Ukraine as a platform to speak out against domestic violence against women, a widespread problem in her home country. In 2012, Loboda, often referred to as the Ukrainian Lady Gaga, traveled to Los Angeles to produce a new album that soared to the top of the charts in her native land and brought her more fame on the global stage.

A Revival of Russian Nationalism Even as the Russian domain finds itself increasingly swept up in external cultural influences, countertrends that emphasize Russian nationalism persist. Extremely conservative nationalists resist foreign influences such as Western-style music, and they have often been supported by more traditional elements of the Russian Orthodox Church. Even President Putin and the Russian government have carefully resurrected numerous symbols from the Russian past in order to cultivate a renewed sense of Russian identity within the region. Recently, Russian leaders proposed a return to the tsarist-era coat of arms (the double-headed eagle) to symbolize the ongoing connections between Russian society today and the earlier culture of Dostoyevsky and Tolstoy. The glory of Soviet communist days (1917–1991) also has been revived, with a strong Russian tone. The Soviet-era national anthem was reintroduced (with new lyrics) in 2001, and several Soviet holidays, celebrations, and national awards have been reinstated. Recent Russian textbooks have even resurrected the virtues of Joseph Stalin. Once discredited as a brutal, cruel leader, Stalin is now seen as an "effective manager" of Soviet affairs in an earlier era.

▼ **Figure 9.31 Svetlana Loboda** Ukrainian singing star Svetlana Loboda was one of the top performers at Eurovision 2009, held in Moscow. In addition to a successful career in music and television, Loboda has championed the plight of battered women throughout the region.

Sports and National Identity The Soviet era also revealed a close and powerful connection between popular sports and Soviet nationalism. Russia's famed love affair with chess, kindled during tsarist times, received renewed attention during the Soviet period. Lenin's army commander opened chess schools and promoted the game after 1921. Stalin played often, believing that the game embodied the Soviet knack for intellect and strategy. Indeed, the board game's popularity outlived the Soviet era, and 21st-century Russians still embrace the game.

Both the Summer and the Winter Olympic Games have also been signature events to display Soviet superiority over their American and western European rivals. Indeed, from figure skating to gymnastics, Soviet-era athletes often dominated particular events for years. American fans were occasionally on the winning side as well: The miracle victory over the Soviet team in ice hockey at the 1980 Winter Olympics (in Lake Placid, New York) remains one of the most memorable moments in modern global sports history.

In the post-Soviet era, popular sports remain hugely important throughout the region, and they serve as signatures of national pride and identity. Whether it is Moldovan Trânta (a form of wrestling), Ukrainian soccer, or Russian ice hockey, enthusiasm for winning teams and leading players is equal to anything an American basketball or football league could hope for. One of the reasons the Russians (including President Putin) prepared so carefully for the 2014 Winter Olympics in Sochi (in southern Russia) was to demonstrate the country's ability to once again host that important global sports event. Coined the "greenest Olympics in history" by its promoters because of their use of environmentally friendly technologies, the Sochi games boasted an Olympic-sized price tag of more than $50 billion (much of it paid by the country's wealthy oligarchs). Regardless of how many gold medals Russian athletes garnered, the Sochi games were designed to be a monument to President Putin's economic miracle and a symbol of Russia's renewed presence on the world stage (Figure 9.32).

REVIEW

9.6 What were the key phases of colonial expansion during the rise of the Russian Empire, and how did each enlarge the reach of the Russian state?

9.7 What are some of the key ethnic minority groups (as defined by language and religion) within Russia and neighboring states?

Geopolitical Framework: Resurgent Global Superpower

The geopolitical legacy of the former Soviet Union still weighs profoundly upon the Russian domain. After all, the bold lettering of the "Union of Soviet Socialist Republics" dominated the Eurasian map for much of the 20th century, and the country's global political reach left no corner of the world untouched. Former Soviet republics still struggle to define new geopolitical identities for themselves. Neighboring states continue to view the region with nervousness left over from the days of Soviet political and military power. Present demands for more local and regional political control within countries such as Russia, Ukraine, Moldova, and Georgia can still be understood in the context of the Soviet era, when a highly centralized government gave little voice to regional dissent.

At the same time, Russia's renewed global visibility and its desire to recentralize authority in Moscow have raised concerns elsewhere within the domain, in Europe, and in the United States. President Putin's reelection in 2012 (he captured 64 percent of the vote) was a reminder that the era of strong centralized rule in the Russian domain is far from over, although growing undercurrents of discontent may make Putin's newly extended six-year term in office (2012–2018) seem endless, both for the leader and for those who voted for him (Figure 9.33).

▲ **Figure 9.33 Vladimir Putin** Since 2000, Vladimir Putin's influence across the Russian domain has been immense. Within Russia, Putin (as both president and prime minister) has managed an impressive economic recovery, while limiting civil liberties. Beyond Russia, Putin has reestablished the region's geopolitical presence on the global stage.

Geopolitical Structure of the Former Soviet Union

The Soviet Union rose from the ashes of the Russian Empire, which collapsed abruptly in 1917. The ultraconservative policies of the Russian tsar generated opposition among businesspeople and workers, while the peasants, who formed the majority of the population, had always resented the powerful landowning aristocracy. After the fall of the tsar and the aristocracy, a broad-based coalition government assumed authority. Several months later, however, the **Bolsheviks**, a faction of radical Russians representing the interests of the industrial workers, seized power within the country. They espoused the doctrine of **communism**, a belief that was based on the writings of Karl Marx and that promoted the overthrow of capitalism by the workers, the large-scale elimination of private property, state ownership and central planning of major sectors of the economy (both agricultural and industrial), and one-party authoritarian rule. The leader of these Russian communists was Vladimir Ilyich Ulyanov, usually

▼ **Figure 9.32 Winter Olympics Venue, Sochi** President Putin authorized large new investments in the southern Russian city of Sochi to guarantee the success of the 2014 Winter Olympics.

▲ **Figure 9.34 Soviet Geopolitical System** During the Soviet period, the boundaries of the country's 15 internal republics often reflected major ethnic divisions. Ultimately, however, many of the ethnically non-Russian republics pressured the Soviet government for more political power. As the Soviet Union disintegrated, the former republics became politically independent states and now form an uneasy ring of satellite nations around Russia.

known as Lenin, his self-selected name. Lenin was a key architect of the Soviet Union, which fully emerged between 1917 and 1922. It became the Russian Empire's successor state.

The new **socialist state**, in which most key economic sectors were controlled by the government, reconfigured Eurasian political geography. Although it resembled the territorial contours of the Russian Empire and centralized authority continued to be concentrated in European Russia, the political and economic structure of the country was radically transformed. Lenin and the other communist leaders were aware that they faced a major challenge in organizing the new state.

The Soviet Republics and Autonomous Areas Soviet leaders designed a geopolitical solution that maintained their country's territorial boundaries and acknowledged, at least theoretically, the rights of its non-Russian citizens. Each major nationality was to receive its own "union republic," provided it was situated on one of the nation's external borders (Figure 9.34). Eventually, 15 such republics were established, creating the Soviet Union. Some of these republics were quite small, while the massive Russian Republic sprawled over roughly three-quarters of the Soviet terrain. Each republic was to have considerable political autonomy and was vested with the right to withdraw from the union if it so desired. In practice, however, the Soviet Union remained a centralized state, with important decisions made in the capital of Moscow.

The Soviets came up with different geopolitical solutions to acknowledge smaller ethnic groups and nationalities that were not situated on the country's external borders. Indeed, dozens of significant minority groups pressed for recognition. One solution to this problem was the creation of **autonomous areas** of varying sizes that recognized special ethnic homelands, but did so within the structure of existing republics. Thus, within the Russian Republic, the larger nationalities, such as the Yakut and the Volga Tatars, were granted their own "autonomous republics" (not to be confused with the 15 higher-level union republics, such as Russia or Ukraine). The system's main deficiency was that the autonomy it was designed to provide proved to be more of a charade than a reality.

Centralization and Expansion of the Soviet State In the early Soviet era, it appeared that the framework of separate republics and autonomous areas might allow non-Russian peoples to protect their own cultures and establish their own social and economic policies (provided, of course, that such policies embodied Marxist principles). From the beginning, however, it was clear that such self-determination would be a temporary measure. According to the official ideology of Marxist beliefs, the gradual development of a communist society would see the withering away of all significant ethnic differences and the disappearance of religion. In the future, a new classless Soviet society was supposed to emerge.

By the 1930s, it was clear that national autonomy would not have any real significance within an increasingly centralized Soviet state. The chief architect of this political consolidation was Soviet leader Joseph Stalin, who did everything he could to centralize power in Moscow and to assert Russian authority. Stalin launched a ruthless plan of state-controlled agricultural production and industrialization. Although many people initially resisted his policies, Stalin never hesitated to use force to bring about his vision of a purer socialist revolution within the Soviet Union. Stalin also utilized more-subtle methods of control in his quest to produce a new communist society. The Soviet landscape itself came to reflect communist ideas. Soviet-style architecture, paintings, cinema, and many other visual elements in the everyday world came to symbolize the new world view.

The Stalin period also saw the geographical enlargement of the Soviet Union. As a victorious power in World War II, the country acquired southern Sakhalin and the Kuril Islands from Japan. It regained the Baltic republics (Lithuania, Latvia, and Estonia—independent between 1917 and 1940), as well as substantial territories formerly belonging to Poland, Romania, and Czechoslovakia.

After World War II, the Soviet Union also gained significant authority, although not actual sovereignty, over a broad swath of eastern Europe. As they pushed the German army west toward the end of the war, Soviet troops advanced across much of the region, actively working to establish communist regimes thereafter. In the words of British leader Winston Churchill, the Soviets extended an "**Iron Curtain**" between their eastern European allies and the more democratic nations of western Europe by restricting the free flow of people and information.

As eastern Europe disappeared behind the Iron Curtain, the Soviet Union and the United States, although allies during World War II, became antagonists in a global **Cold War** of escalating military and economic competition that lasted from 1948 to 1991. In its post–World War II heyday, the Soviet Union became a global superpower, one of only two countries equipped with enough nuclear weapons to ensure global destruction. At the height of its power in the late 1970s, it enjoyed close military and economic alliances not only with eastern Europe, but also with certain countries in Asia (Mongolia, North Korea, Vietnam, Laos, and Cambodia), the Caribbean region (Cuba and Nicaragua), and Africa (Angola, Somalia, and several others).

End of the Soviet System Ironically, Lenin's system of culturally defined republics helped sow the seeds of the Soviet Union's downfall. Even though the nationally based republics and autonomous areas of the Soviet Union were never allowed real freedom, they did provide a lasting political framework for the maintenance of distinct cultural identities. Indeed, contrary to the expectations of Soviet leaders, ethnic nationalism intensified in the post–World War II era as the Soviet system grew less repressive. When Soviet president Mikhail Gorbachev initiated his policy of **glasnost**, or greater openness, during the 1980s, several republics—most notably the Baltic states of Lithuania, Latvia, and Estonia—demanded outright independence.

Other forces also worked toward the political end of the Soviet regime. A failed war in Afghanistan in the early 1980s frustrated both the Soviet leaders and the Soviet population. In eastern Europe, multiple protests over Soviet dominance emerged after World War II from Hungary, Czechoslovakia, and Poland. Worsening domestic economic conditions, increasing food shortages, and a declining quality of life led to fundamental questions concerning the value of centralized planning within the country. In response, President Gorbachev introduced **perestroika**, or planned economic restructuring, aimed at making production more efficient and more responsive to the needs of Soviet citizens. In 1991, however, Gorbachev saw his authority slip away amid rising pressures for political decentralization and more-dramatic economic reforms. During the summer, Gorbachev's regime was further imperiled by the popular election of reform-minded Boris Yeltsin as the head of the Russian Republic and by a failed military coup by communist hard-liners. By late December, all of the country's 15 constituent republics had become independent states, and the Soviet Union ceased to exist.

Current Geopolitical Setting

Post-Soviet Russia and the nearby independent republics have radically rearranged their political relationships since the collapse of the Soviet Union in 1991. All of the former republics still struggle to establish stable political relations with their neighbors, and tensions within Russia continue as President Putin has pushed for more centralized control and more limits on media independence and personal liberties. For a time, it seemed that a looser political union of most of the former republics, called the **Commonwealth of Independent States (CIS)**, would emerge from the ruins of the Soviet Union. All the former republics, with the exception of the three Baltic states, joined the CIS soon after the dismemberment of the old union (Figure 9.35). By the early 21st century, however, the CIS had developed into little more than a forum for discussion, without real economic or political power. In addition, Georgia withdrew from participation in 2009 amid its sharpening disagreements with its Russian neighbor.

Russia continues to maintain a military presence in many of the former Soviet republics across the region.

Denuclearization, the return of nuclear weapons from outlying republics to Russian control and their partial dismantling, was initiated during the 1990s. Soviet-era nuclear arsenals in Kazakhstan, Ukraine, and Belarus were removed in the process. Elsewhere, Tajikistan invited the Russian army in during the early 1990s to quell its own internal ethnic struggles, and Armenia has maintained close military ties with Moscow. Russia also maintains naval bases in Ukraine's Crimea region. The Soviet-era Black Sea fleet was traditionally based in the Crimean port of Sevastopol, now in Ukraine. After lengthy negotiations, Russia and Ukraine agreed in 1997 to share the naval base, and in 2010, the lease agreement was extended until at least 2042. Russia retains some 80 vessels and 15,000 service members in Ukraine. But the longer-term political sovereignty of the entire Crimean Peninsula remains in doubt. The peninsula has been contested real estate within the Russian domain for centuries. Some Ukrainians would like nothing better than to see the Russians leave. Many ethnic Russians living in the area, however, see the continuing Russian military presence as protecting their long-term interests.

Geopolitics in the South and West

Russia shares many cultural and political ties with Belarus and Ukraine, and these connections have grown stronger since 2010. In Belarus, leaders have been slow to open to political and economic opportunities in western and central Europe. The country remains firmly within Russia's political orbit. In 2010, the two countries pledged to move toward a "union state" and expanded their common military maneuvers. Specifically, Russia and Belarus have protested the growth of NATO influence in central Europe, seeing that expansion as a potential military threat. As part of recent agreements, Russia also affirmed its right to use its military within Belarus to defend against these regional threats. In 2012, Belorussian President Alyaksandr Lukashenka (often called Europe's last dictator) adopted one of the world's most restrictive policies on free use of the Internet within the nation, strongly discouraging the use of any "foreign" Websites. Many political dissidents inside the country remain imprisoned.

Russia's relationships with Ukraine have been less predictable. Ukraine remains highly dependent on Russian

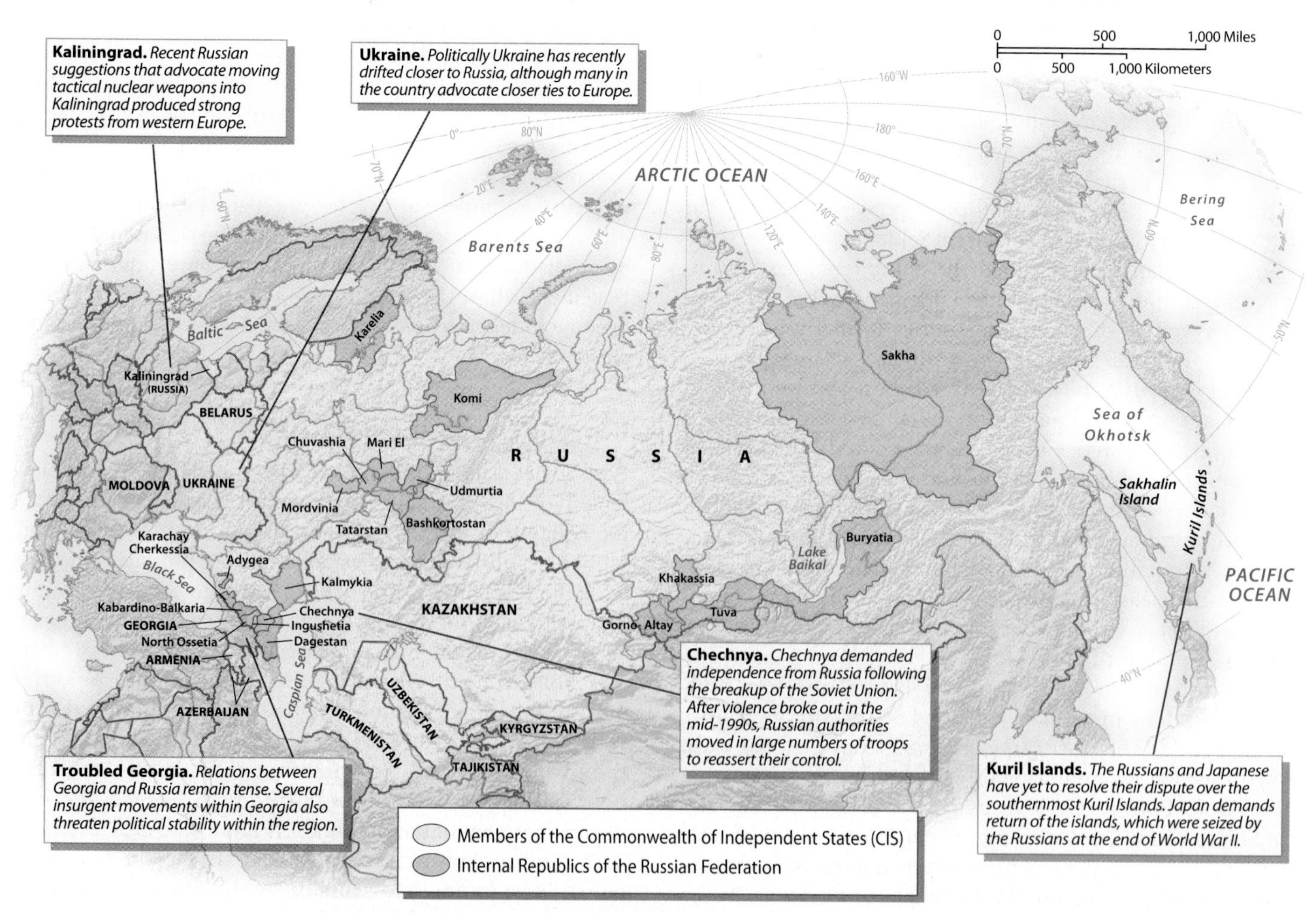

▲ **Figure 9.35 Geopolitical Issues in the Russian Domain** The Russian Federation Treaty of 1992 created a new internal political framework that acknowledged many of the country's ethnic minorities. Recently, however, Russian authorities have moved to centralize power and limit regional dissent. Russia's relations with several nearby states remain strained.
Q: Cite some similarities between Russia's internal republics shown here and the regional map of languages (Figure 9.25).

energy supplies. In addition, major Russian pipelines to central and western Europe pass through Ukraine. The result is that the two nations have often wrangled over energy prices and the availability of oil and natural gas supplies. Early in 2009, for example, Russia briefly shut off all natural gas supplies to Ukraine amid pricing and payment disputes between Ukraine and Gazprom, the Russian energy company.

Ukraine's internal politics have also riled Russia. In 2004, Viktor Yushchenko, a reformist leader and Ukrainian nationalist, came to power after a controversial election that included an attempt to poison him. Yushchenko often criticized Russian leaders and excessive Russian interference with Ukraine's affairs. He also explored Ukraine's admission into the EU and NATO. However, Viktor Yanukovich, one of Yushchenko's rivals, argued for closer relations with Moscow. Steep declines in Ukraine's economy in 2009 prompted widespread public protests in Kiev and calls for early elections, challenging Yushchenko. Elections in 2010 brought Yanukovich into power. Voting patterns revealed Yanukovich's popularity among Russian speakers in the country (see Figure 9.26), but he also attracted disgruntled Ukrainians weary of the nation's battered economy. These results appeared to signal a tilt toward Russia. A 2012 free-trade agreement between the two countries also suggests closer ties, and Russia is pressing its southern neighbor for an even stronger customs union that will further wed their political and economic systems. However political winds blow in Ukraine, Russia will continue to influence Ukrainian politics in any way it can to serve its own political and economic interests.

Tiny Moldova has also witnessed political tensions in the post-Soviet era. Conflict has repeatedly flared in the Transdniester region in the eastern part of the country, where Russian troops remain and where Slavic separatists have pushed for independence from a central government dominated by Romanian-speaking Moldovans. Russia has encouraged the case for "independence," seeing it as a way to increase its influence in the region. On the other hand, a growing number of Romanians have suggested that Moldovans join with Romania (which is now in NATO and the EU) and dump troublesome Transdniester in the process.

Transcaucasia also remains unstable. Since 2003, the Georgian government has moved toward closer ties with the United States, even suggesting that it might join NATO. At the same time, Georgia's own internal politics recently provoked an invasion from neighboring Russia. In 2008, Georgia attempted to reassert its control over Abkhazia and South Ossetia, two breakaway regions that border Russia and that have wide support in that nearby country. The Russians responded with tanks, for a time occupying large sections of Georgia's territory. Almost 1000 people died in the conflict, and more than 30,000 people were displaced from their homes. The situation remains tense today, with Georgia still claiming control over Abkhazia and South Ossetia and with Russia formally recognizing the "independence" of these microstates.

In nearby Armenia, the territories of the Christian Armenians and the Muslim Azeris interpenetrate one other in a complex fashion. The far southwestern portion of Azerbaijan (Naxicivan) is actually separated from the rest of the country by Armenia, while the important Armenian-speaking district of Nagorno-Karabakh is officially an autonomous portion of Azerbaijan. After Armenia successfully occupied much of Nagorno-Karabakh in 1994, fighting between the countries diminished. No final peace treaty has been signed, however, and Azerbaijan demands the return of the territory. Meanwhile, Armenia's traditionally close connections with Russia are increasingly counterbalanced with the country's interest in building ties with the United States and the EU. Elections held early in 2013 suggested this search for balance will continue in the future.

Geopolitics Within Russia Within Russia, further pressures for devolution, or more localized political control, produced the March 1992 signing of the Russian Federation Treaty. The treaty granted Russia's internal autonomous republics and its lesser administrative units greater political, economic, and cultural freedoms, including more control of their natural resources and foreign trade. Conversely, it weakened Moscow's centralized authority to collect taxes and to shape policies within its varied hinterlands. Defined essentially along ethnic lines, 21 regions possess status as republics within the federation and now have constitutions that often run counter to national mandates. Scattered from central Siberia to the north-facing slopes of the Caucasus, these republics reflect much of the linguistic and religious diversity of the nation (see Figure 9.35).

Since 2000, Russian leaders, especially Vladimir Putin, have pushed for more centralized control in the country. Putin's prominence has been enduring: He served two terms as president (2000–2008) and one term as prime minister (2008–2012); then he was reelected as Russia's president in March 2012. Putin, a former internal security agent with the KGB during the Soviet period, has steadily consolidated power in the country, pushed for strong economic growth oriented around Russia's energy economy, and reasserted Russia's political and military role, both on the world stage and as a dominant regional power.

However, cracks have appeared within Russia's more centralized political power structure. The Caucasus region and several of its internal republics (specifically Ingushetia, Chechnya, and Dagestan) remain very unstable (see Figure 9.35). Although Russia officially claimed an end to hostilities within Chechnya in 2009 (where Muslim rebels have pushed for independence since 1994), the situation on the ground remains unpredictable, particularly beyond major urban areas. Many Chechen Muslims, frustrated from their failed attempt at independence from Russia in the 1990s, would like nothing more than freedom from distant autocrats in Moscow. Even as Grozny (Chechnya's capital) (Figure 9.36) rebuilds from the war years (thanks to Russian aid), antipathy toward Russia continues to simmer beneath the surface.

In adjacent Ingushetia, observers describe a widespread lack of civil order, and rebel groups have been increasingly bold in their attacks against government officials. Instability in the Caucasus also has wider implications: Rebel groups from the region claimed responsibility for a deadly explosion on an express train between Moscow and St. Petersburg early in 2010, and more generally, the region's instability challenges the overall authority of the central government in Moscow.

Russian Challenge to Civil Liberties Growing public protests since 2009 have challenged Putin's authority. After his 2012 election, thousands of protesters marched in the streets of Moscow and St. Petersburg, and hundreds were arrested. Disgruntled members of the urban middle class (he received less than 50 percent of the vote in Moscow), human rights groups, and opposition political parties resent Putin's grip on power. Protestors demand a freer press, more democracy, more open elections, and a broader commitment to economic growth. They have criticized Putin's strong-arm leadership style. They also note Putin's increasingly close ties to the **siloviki**, members of the nation's military and security forces.

Protests against the Russian central government are in part a response to the government's crackdown on civil liberties. Immediately after the fall of the Soviet Union, Russia enjoyed a genuine flowering of democratic freedoms. A multiparty political system, independent media, and a growing array of locally and regionally elected political officials signaled real change from the authoritarian legacy of the Soviet period.

Since 2002, however, many hard-won civil liberties have slipped away, victims of Putin's campaign to consolidate political power, increase the authority of the central government, limit press freedoms, and silence critics who disagreed with his policies. For example, the Russian president now has more direct control of nominating candidates for dozens of Russian governorships and mayoral positions. Many of the country's media outlets have also lost their autonomy and are now under more direct government ownership and influence. Even more disturbingly, many outspoken journalists critical of the government (such as Anna Politkovskaya) have died under suspicious circumstances or been murdered. Recently, Russian officials have also increased their surveillance and regulation of the Internet, another move to silence opposition. In 2013, even former Soviet-era leader Mikhail Gorbachev proclaimed that new laws against public protests and Internet freedom "attack the rights of citizens," and he suggested that Putin was afraid of his own people.

In 2012, Pussy Riot, a Russian female punk rock band, made global headlines with its performance of a song critical of President Putin in Moscow's Christ Savior Cathedral, one of Russia's largest houses of worship (Figure 9.37). Band members—who were decked out in fluorescent stockings as they sang "Mother of God, blessed Virgin, drive out Putin!"—were accused of "hooliganism" and jailed. Although the women claimed their actions were a political protest, they were given multiple-year sentences (one was later released), igniting a flurry of public reactions against the government and suggesting that Putin may face political challenges from unorthodox places in the years ahead.

▼ **Figure 9.36 Grozny in Russian Chechnya** Large government-managed investments have helped rebuild the war-torn Chechen capital, even though anti-Russian sentiment still runs high in the countryside.

The Shifting Global Setting

Since 1991, regional political tensions have continued to challenge the Russians, in both the east and the west. In East Asia, the boundary between Russia and China imposed by the Russian Empire in 1858 has never been fully accepted by Beijing. Relations between the two nations improved after a 2004 agreement clarified territorial claims along the Amur and Ussuri rivers, but the potential for renewed conflict remains. In addition, Russia has played an important role in containing North Korea's nuclear ambitions in the region, pressuring its Far East neighbor to limit uranium enrichment and weapons development projects. Territorial disagreements—specifically, a dispute over the Kuril Islands—also complicate Russia's relationship with Japan.

To the west, Russia worries about the expansion of NATO. Although most Russian leaders accepted the inevitable inclusion of Poland, Hungary, and the Czech Republic in NATO, they strongly opposed the addition of the Baltic republics (Estonia, Latvia, and Lithuania) to the increasingly powerful organization. In fact, the move provoked some Russian military leaders to recommend new troop and weapon deployments on the country's western borders. The plight of Russian Kaliningrad is a related sore point (Figure 9.38). After World War II, the former Soviet Union added a small, but strategic territory on the Baltic Sea.

▲ **Figure 9.37 Pussy Riot** This Russian punk rock band went to jail for hooliganism when they appeared in a prominent Russian Orthodox Church and sang a song protesting President Putin's rule.

It was the northern portion of East Prussia (now the port of Kaliningrad), previously part of Germany. It still forms a small, but highly strategic Russian **exclave**, which is defined as a portion of a country's territory that lies outside its contiguous land area. Kaliningrad (population 950,000) remains a part of the Russian Federation; but since 2004, neighboring Lithuania and Poland have both joined NATO and the EU. Russian officials fear that being in the midst of NATO and EU nations may encourage some residents of Kaliningrad to think about independence. In response, the Russians have beefed up their military presence in the exclave, making it clear that they have no intention of giving up one of their valuable port cities on the Baltic Sea.

Today Russian leaders are reasserting their nation's global political status. Russia retains a permanent seat on the UN Security Council, and its inclusion in the G8 economic meetings signifies its growing international clout. The country's nuclear arsenal, while reduced in size, remains a powerful counterpoint to American, European, and Chinese interests. Russia no longer directly challenges the United States as it did in Soviet days, but it still acts as a partial counterweight to the United States in international maneuverings. Since his reelection to the presidency in 2012, Putin has also been taking an increasingly anti-American slant in his foreign policy, hoping such tactics increase his popularity on the home front. Recent decisions to bar USAID (a U.S.-based development and humanitarian aid agency) operations in the country and to legally ban American parents from adopting Russian children illustrate the point.

Despite their nominal independence from direct Russian control, the other nations within the Russian domain clearly feel the presence of their mammoth neighbor. Indeed, the long geopolitical history of the region suggests that Russia's recent reemergence as a more powerful, centralized state upon the global stage is a sign of things to come.

REVIEW

9.8 How do current geopolitical conflicts reflect long-standing cultural differences within the region?

9.9 Describe how Vladimir Putin has played a key role in consolidating Russia's power since 2000, both within the country and beyond.

► **Figure 9.38 Kaliningrad** A Russian exclave on the Baltic Sea, is Kaliningrad now surrounded by Poland and Lithuania, both EU- and NATO-member states. The multistory Hotel Kaliningrad (photo, right) borders the Lenin Prospekt in this downtown view of the Russian city.

Economic and Social Development: The Key Role of Energy

Since the disintegration of the Soviet Union, the economy of the Russian domain has fluctuated widely, alternating between significant declines and cyclical recoveries. Economic declines devastated the region for much of the 1990s. Between 2002 and 2008, especially for Russia, higher oil and gas prices brought significant, but selective economic improvement. The entire region was hit hard in the 2008–2010 economic downturn. Recently, however, economic growth has picked up, particularly in Russia, where higher energy prices once again have bolstered the economy (Table 9.2).

The true economic potential of the Russian domain has always been difficult to gauge. Optimists point to the vast size, abundant natural resources, and well-educated, urbanized populations of the region as significant assets. Indeed, in its heyday, the Soviet Union rose to become one of the great industrial powers in the world, and it did so in a remarkably short period of time.

Skeptics note that size also brings its disadvantages, particularly by raising transportation costs within the region's economy. They also point out that Russia's northern location always has made food production problematic. Most notably, since the breakup of the Soviet Union, most economies within the region have struggled to evolve in a stable and predictable fashion toward greater productivity and output. Thus, the region's future path to prosperity remains clouded as it spends the early years of the 21st century reconfiguring an economy that was created largely during the Soviet period.

The Legacy of the Soviet Economy

The creation of the Soviet Union in 1917 initiated a radical change within the region's economy. Under the Russian Empire, most people were peasant farmers. Following the revolution, however, the Soviet Union quickly emerged to rival, and even surpass, many of the most powerful industrial economies on Earth. During that era of unmatched growth, much of the region's present economic infrastructure was established, including new urban centers and industrial developments, as well as a modern network of transportation and communication linkages.

As communist leaders such as Stalin consolidated power in the 1920s and 1930s, they nationalized Russian industries and agriculture, creating a system of **centralized economic planning** in which the state controlled production targets and industrial output. The Soviets emphasized heavy, basic industries (steel, machinery, chemicals, and electricity generation), postponing demand for consumer goods to the future. By the late 1920s, Stalin also shifted agricultural land into large-scale collectives and state-controlled farms.

Much of the Russian domain's basic infrastructure—its roads, rail lines, canals, dams, and communications networks—originated during the Soviet period (Figure 9.39). Dam and canal construction, for example, turned the main rivers of European Russia into a virtual network of interconnected reservoirs. Invaluable links such as the Volga–Don Canal (completed in 1952), which connected those two key river systems, have greatly eased the movement of industrial raw materials and manufactured goods within the country (Figure 9.40). The Soviets also added thousands of miles of new railroad tracks in the European west, and the Trans-Siberian Railroad was modernized and complemented by the addition of the BAM link across central Siberia. Farther north, the Siberian Gas Pipeline was built to link the energy-rich fields of the Soviet Arctic with growing demand in Europe. Overall, the postwar period produced real economic and social improvements for the Soviet people.

Despite these successes, problems increased during the 1970s and 1980s. Soviet agriculture remained inefficient, and grain imports grew. Manufacturing efficiency and quality failed to match the standards of the West, particularly in regard to consumer goods. Equally troubling was the fact

Table 9.2 Development Indicators

Country	GNI per capita, PPP 2011	GDP Average Annual % Growth 2000–11	Human Development Index (2011)[1]	Percent Population Living Below $2 a Day	Life Expectancy (2013)[2]	Under Age 5 Mortality Rate (1990)	Under Age 5 Mortality Rate (2011)	Adult Literacy (% ages 15 and older) Male/Female	Gender Inequality Index (2011)[3,1]
Armenia	6,100	8.2	0.729	19.9	74	47	18	100/99	0.340
Belarus	14,460	7.8	0.793	<2	72	17	6	100/99	–
Georgia	5,350	6.6	0.745	35.6	75	47	21	100/100	0.438
Moldova	3,640	5.1	0.660	4.4	71	35	16	99/98	0.303
Russia	20,410	5.1	0.788	<2	70	27	12	100/99	0.312
Ukraine	7,040	4.3	0.740	<2	71	19	10	100/100	0.338

[1]United Nations, *Human Development Report, 2013.*

[2]Population Reference Bureau, *World Population Data Sheet, 2013.*

[3]Gender Inequality Index—A composite measure reflecting inequality in achievements between women and men in three dimensions: reproductive health, empowerment and the labor market that ranges between 0 and 1. The higher the number, the greater the inequality.

Source: World Bank, *World Development Indicators, 2013.*

▼ **Figure 9.39 Major Natural Resources and Industrial Zones** The region's varied natural resources and chief industrial zones are widely distributed. In southern Siberia, rail corridors offer access to many mineral resources. In the mineral-rich Urals and eastern Ukraine, proximity to natural resources sparked industrial expansion, while Moscow's industrial power is related to its proximity to markets and capital. **Q: Looking at the map, why might it be argued that Russia's size is both a blessing and a curse?**

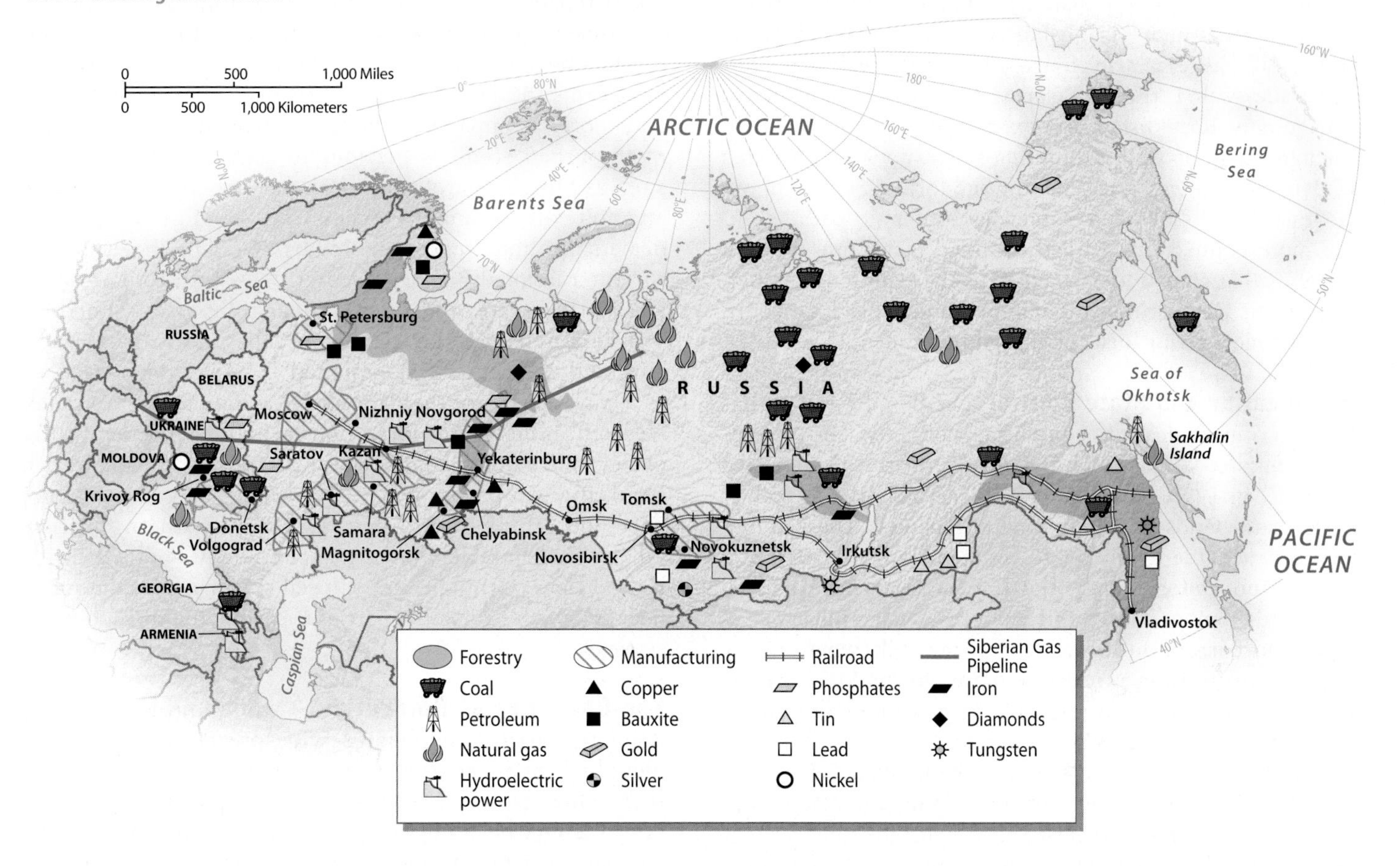

that the Soviet Union was failing to participate fully in the technological revolutions that were transforming the United States, Europe, and Japan. Disparities also visibly grew between the Soviet elite and a national population that still enjoyed few personal freedoms. By the late 1980s, the Soviet Union had reached both an economic and a political impasse.

The Post-Soviet Economy

Fundamental economic changes have shaped the Russian domain since 1991. Particularly within Russia itself, much of the highly centralized state-controlled economy has been replaced by a mixed economy of state-run operations and private enterprise. The collapse of the communist state also meant that economic relationships between the former Soviet republics were no longer controlled by a single, centralized government. Fundamental problems of unstable currencies, corruption, and changing government policies plagued the system for much of the 1990s. Higher oil and natural gas prices have led to growth in the Russian economy, especially between 2002 and 2007 and since 2010.

▼ **Figure 9.40 Volga–Don Canal** This view near Volgograd suggests the enduring economic importance of the Volga–Don Canal. Built during the Soviet era, the canal remains a key commercial link that facilitates the economic integration of southern Russia.

Redefining Regional Economic Ties Many economic ties still link the region. Russia enjoys especially close economic ties with both Belarus and Ukraine. Russian dominance is likely to continue, as suggested by that country's leading role in the creation of a new customs union in 2010 with neighboring Belarus and Kazakhstan (and potentially Ukraine). In late 2011, Russia also signed a free-trade deal to encourage trade with Ukraine, Belarus, Moldova, and Armenia. Intriguingly, Kazakhstan, Kyrgyzstan, and Tajikistan joined the agreement as well, prompting suggestions that Russian leaders were quietly reassembling the former Soviet Union. Georgia has followed a somewhat different path, seeking more connections with the United States and western Europe. Georgian leaders have actively explored membership in the EU. Batumi, the nation's largest Black Sea port, has become a major focus of foreign investment.

Privatization and State Control The post-Soviet era has brought a great deal of economic uncertainty. The government initiated a massive program to privatize the Russian economy in 1993. Millions of Russians had options to buy agricultural lands and industrial companies. These initiatives opened the economy to more private initiative and investment. Unfortunately, the lack of legal and financial safeguards invited abuses and often resulted in mismanagement and corruption in the new system. Elsewhere in the region, privatization proceeded more slowly. Much of the Belorussian economy, for example, remains mired in inflexible, corrupt state-controlled companies.

In Russia, almost 90 percent of the farmland was privatized by 2003, with many farmers forming voluntary cooperatives or joint-stock associations to work the same acreage as under the Soviet system. Although crop prices have risen, costs have gone up even faster, and many farmers are not skilled in dealing with the uncertainties of a market-driven economy. Overall, agriculture employs about 10 percent of Russia's workforce, but the basic distribution of crops remains little changed from the Soviet era. Short growing seasons, poor soil, and moisture deficiencies still pose challenges.

Russia encouraged more privatization in the service sector. Thousands of retailing establishments have appeared, and they now dominate that portion of the economy. In addition, the long-established "informal economy" continues to flourish. Even during the Soviet era, millions of citizens earned extra money by informally selling Western consumer goods, manufacturing food and vodka, and providing skilled services such as computer and automobile repairs. Today these barter transactions and informal cash deals form a huge part of the economy and are never reported to government authorities.

The natural resource and heavy industry sectors of the economy were initially privatized in Russia, but in recent years, under Putin's management, state-run enterprises took back more control of the nation's energy assets and infrastructure. Gazprom, the huge Russian natural gas company, was privatized in 1994, but since 2005, its activities have increasingly been controlled by the state. This company is seen as a critical part of a newly emerging "state industrial policy," in which the central government is playing a more direct role in the economy. Nicknamed "Russia, Inc.," Gazprom remains one of the world's largest companies, employing more than 300,000 people and controlling huge natural gas reserves.

Especially in Russia—and particularly in that country's cities—the successes of the new economy are increasingly visible on the landscape. Luxury malls, office buildings, and more-fashionable housing subdivisions are now part of the urban scene as the middle class grows in settings such as Moscow. Stroll the area near that city's Red Square and you will encounter Planet Sushi, the Moscow Maserati dealership, and trendy nightclubs. West of the Kremlin, the International Business Center plays host to a growing array of domestic and international companies. Skolkovo Village, Russia's answer to Silicon Valley, is also near Moscow, and the center (championed by government leaders as a way to harness the nation's scientific talent) promotes research in biotech, information technologies, energy efficiency, and nuclear and space technologies. On the other hand, the gap has grown between increasing urban affluence and grinding rural poverty. Many southern Siberian villages have no telephones, jobs, or money. Vodka is easier to find than running water. Closed shops and a continued lack of services remain a part of everyday life in such rural settings.

The Challenge of Corruption Throughout the Russian domain, corruption remains widespread. A 2011 study identified Russia as one of the world's most corrupt countries, where bribery was especially widespread. Doing business often means lining the pockets of government officials, company insiders, or trade union representatives. Organized crime remains pervasive in Russia. Many ties also remain between organized crime and Russian intelligence agencies. Much of the country's real wealth has been exported to foreign bank accounts. Various local and regional crime organizations divide up much of the economy. Occasionally, violence and gangland-style murders still unfold on the streets of Moscow, much to the embarrassment of government officials. The Russian mafia has also gone global; it has been implicated in huge money-laundering schemes involving Russian, British, and U.S. banks, as well as the flow of International Monetary Fund investments into the region.

Problems of Health Care and Alcoholism Health care is another major social problem within the Russian domain. Annual health-care expenditures remain only a fraction of what they were during the Soviet period (Figure 9.41). To put the problem in global perspective, Ukrainians spend only about 16 percent of what Japanese spend annually on health care ($495 vs. $3045), and Russians typically survive on less than 13 percent of what most Americans spend annually on health care ($1043 vs. $7960). Mortality rates for Russian men are especially grim. One in three Russian men dies before retirement. Cardiovascular disease, often

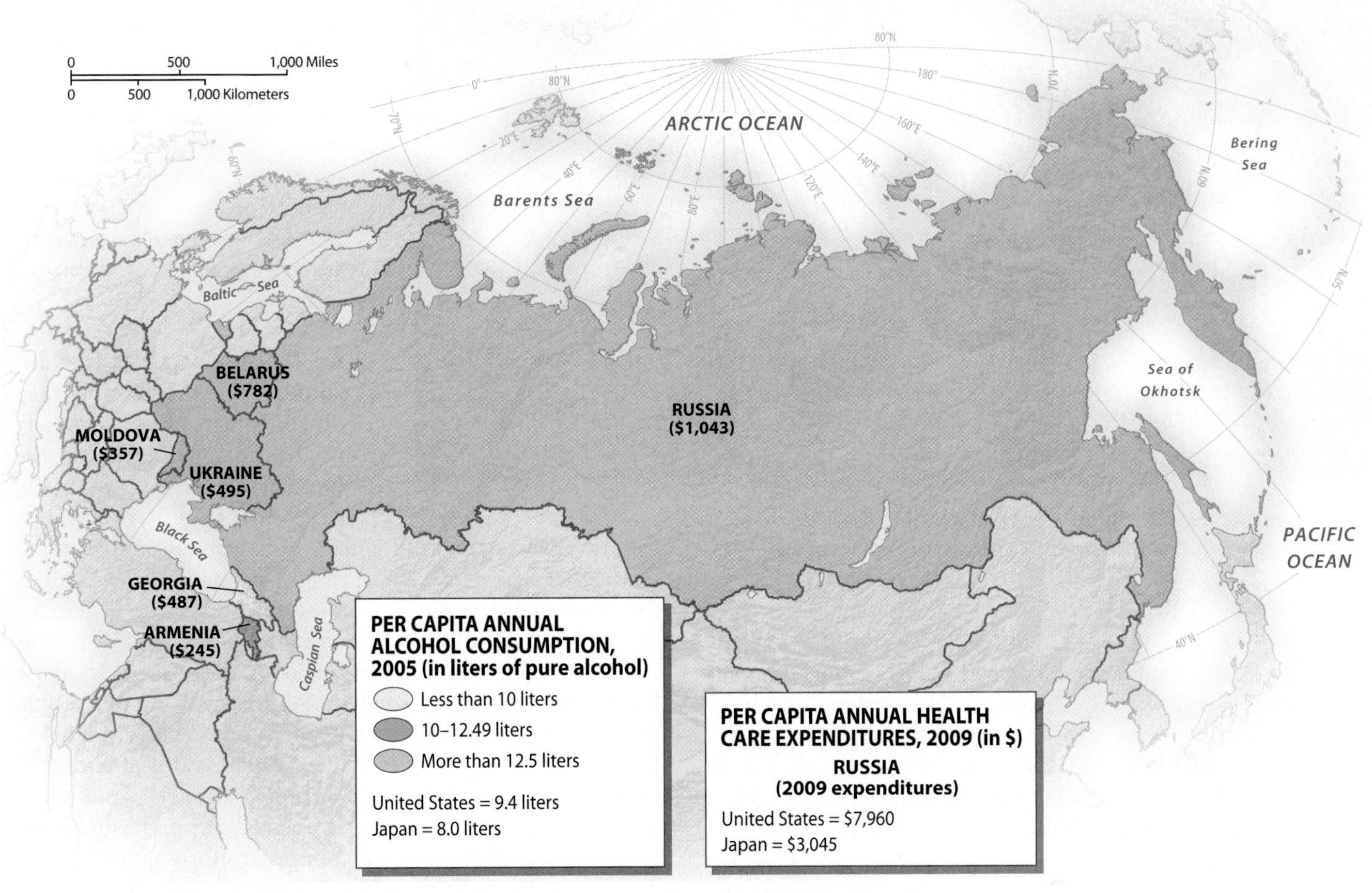

▲ **Figure 9.41 Development Issues in the Russian Domain: Health Care and Alcohol** People in some portions of the Russian domain survive on less than $500 per year for health care, compared with much higher expenditures in much of the rest of the developed world. Alcohol consumption remains highest in Russia, where it has been targeted as a key social problem by politicians and health-care experts. *(United Nations, WHO statistics, 2012)*

related to high-fat diets and physical inactivity, is a key contributor to these elevated death rates. Smoking also remains widely popular (54 percent of physicians smoke), with Russian men claiming the dubious distinction of having the highest smoking rates on Earth. HIV-AIDS is also a major problem. In Russia, it is estimated that about 2.5 million people are infected and that over 8 percent of deaths can be attributed to the virus (with rates of infection peaking around 2015).

Alcohol use in Russia (15.73 liters of pure alcohol per person annually) remains far above the global average (6.13 liters) (see Figure 9.41). Some estimates that include illegal black market production put the Russian figure at more than 18 liters per person. For comparison, the average alcohol consumption in Japan (8 liters per year) is typical of many other developed-world settings, while India (2.64 liters per year) and Egypt (0.4 liters per year) exemplify settings where religious and cultural traditions discourage high rates of alcohol consumption. Furthermore, Russians imbibe mainly hard spirits, especially vodka. Russian leaders initiated an antidrinking campaign in early 2010, calling their country's plight (they rank among the top five nations in the world in alcohol consumption) "a national disaster." Prices for vodka were raised, although bootleg liquor is cheap and widely available. An ambitious goal was set to reduce alcohol consumption by 50 percent within the next decade. Meanwhile, however, binge drinking and chronic high levels of alcohol consumption continue to be life-threatening problems for millions of people throughout the region. In many settings, alcoholism is the leading cause of death for people (especially for men) between 15 and 64 years of age.

Challenges for Women Women, while often better educated than men, usually earn substantially less money performing the same work. Women are also underrepresented in positions of corporate or political power, often faring worse than their western European or American counterparts. In many settings in the Russian domain, women are also abused. Violence against women has been widely reported in the post-Soviet era. Beatings and rapes are common. A survey in Moscow suggested that one-third of divorced women had experienced domestic violence, and a women's rights group in Ukraine

▲ **Figure 9.42 FEMEN Protest** This group of Ukrainian feminists has vigorously protested the exploitation of women in their own country and around the world.

reported that rape was common in many villages. International organizations have sharply criticized government authorities for doing little to change the situation.

In addition, **human trafficking** (a practice in which women are lured or abducted into prostitution) is a widespread problem. Armenia, Ukraine, Moldova, and rural districts of Russia are major sources of young women who are forced into prostitution in Europe and the Middle East. It is a multi-billion-dollar business, involving the large-scale participation of organized crime and hundreds of thousands of young women. Some estimates suggest that in addition to Ukraine's large domestic sex tourism industry, more than 500,000 women have emigrated as sex workers elsewhere, mostly in western Europe. Since 2008, the Ukrainian feminist organization known as FEMEN has made headlines around the world with its members' high-profile topless protests of that country's sex industry (their slogan is "Ukraine is not a brothel") (Figure 9.42). The region is also a major global source for Internet brides, a practice that invites additional violence against women.

Growing Economic Globalization

The relationship between the Russian domain and the world beyond has shifted greatly since the end of communism. During much of the Soviet era, the region was quite isolated from the world economic system. By the 1970s, however, the Soviet Union had begun to export large quantities of fossil fuels to the West, while importing more food products. Although connections with the global economy have grown with the downfall of the Soviet Union, the region remains one of the least globalized parts of the more developed world. That is changing, however, particularly with Russia's 2012 admission into the World Trade Organization (WTO), which should open many new global trade possibilities by 2020. Optimists suggest that Russia's new status in the organization will encourage the country to modernize its regulatory environment and to reduce restrictions on foreign imports, giving Russian consumers better access to world markets.

A More Globalized Consumer Most visibly, a barrage of consumer imports already reaches many residents of the Russian domain, particularly those living in its larger cities. All of the symbols of global capitalism are visible in the heart of Moscow and, increasingly, in many other settings throughout the region. According to the fast-food industry, by 2015 Russia will be home to more than 300 KFC restaurants and more than 500 McDonald's. Luxury goods from the West have also found a small, but enthusiastic, highly visible market among the Russian elite, a group noted for its devotion to BMW automobiles, Rolex watches, and other status emblems. Most Russians find such luxuries beyond their limited budgets, but they are interested in purchasing basic foods, cheap technology, and popular clothing and media from their Western neighbors or from eastern and southern Asia. Russia's largest sources of imports are Germany, China, and Japan, but many American consumer products are also in high demand. It is also a two-way street: The region's software engineers and video game developers have had a large impact on these industries around the world (see *Everyday Globalization: How the Russian Domain Shapes the Virtual World*).

Attracting Foreign Investment Despite all of the political and economic uncertainties, most countries of the Russian domain are attracting some foreign investment, and Russia's WTO membership may encourage even more in the years ahead. As measured by total foreign investment, the strongest global ties by far have been with the United States, Japan, and western Europe, particularly Germany and Great Britain. The success of Russian equity markets and the relative stability of its financial sector have been encouraging. Still, many potential investors are put off by continuing uncertainties with Russian legal frameworks, lingering problems with slow bureaucracies and red tape, and growing concerns about the ultimate political aims of Russian leaders. In a World Bank survey measuring the business climate for foreign investors, Russia ranked in 120th place (out of 181 nations surveyed), behind Nigeria. Some of the largest outside investments have been made in Russia's oil and gas economy, but recently some of those opportunities have cooled amid the country's desire to

Everyday **Globalization**

How the Russian Domain Shapes the Virtual World

It's a tough virtual world out there, especially when tanks, pirates, spacecraft, and battleships threaten us on every front.

The Russian Connection As every American college student knows, the video game and online gaming landscape has changed dramatically in the last 30 years, since Russian Alexie Pajitnov invented Tetris at the Soviet Academy of Sciences in 1984 (it has sold over 125 million copies on 30 different platforms). Less apparent, however, is the enduring connection between the Russian domain and the multi-billion-dollar video-gaming industry. Russian (1C Company), Belorussian (Wargaming.net), and Ukrainian (Persha Studia) software companies have all played pivotal roles in shaping the world's virtual landscapes. This includes outmaneuvering pirates in the Caribbean (*Age of Pirates*), organizing massive online tank-based clan wars (with more than 190,000 players online simultaneously) somewhere on the world map (*World of Tanks*), and completing house-to-house searches on the European battlefield during World War II (*Men of War*) (Figure 9.5.1).

Regional Advantages How did this region become so central in creating the virtual worlds shared today by hundreds of millions of gamers? Part of the answer is no doubt Soviet-era investments in pioneering computer technology and software development, much of it linked with the Cold War. Add to this a generation of sophisticated, technically trained computer geeks such as Pajitnov (who now lives in Washington State and works for an American software gaming company), who were well positioned to master the programming challenges of the budding industry. The Russian domain also offers a less expensive and less regulated environment where programmers have enjoyed considerable intellectual freedom beneath the radar of the bureaucracy.

Russia's Microsoft Although Belarus and Ukraine software developers certainly participate in the industry, Russia dominates the game. Boris Nuraliev was one of the corporate founders of the movement in the early 1990s. He created the 1C Company—often called Russia's Microsoft—which moved from the rather ordinary world of business software into the more extraordinary world of gaming (*Theater of War, Kings Bounty: The Legend, Pacific Fighters*, etc.). Today the company, based in Moscow, employs almost 1000 people (including 250 internal game developers) and is the largest game publisher and developer in the region. The company is also the most visible participant in the annual KRI (Russian Game Developers) Conference (begun in 2003), which is *the* place to be if you want to know the latest about online gaming, anti-piracy initiatives, or virtual worlds coming soon to a screen near you.

You might ponder the cultural significance of this massive Russian participation in the creation of our virtual worlds (Figure 9.5.2). Think about the landscapes we navigate, the strategic challenges we face, and the fascinating mix of fact (historical tanks and weapons, battle settings, and costumes are meticulously and accurately displayed) and fancy that makes up that world. Not surprisingly, a little bit of traditional Russian culture also gets passed our way. Just listen to the melody most associated with Tetris: Nikolai Nekrasov wrote the poem titled *Korobeiniki* in 1861, and the verse later became a Russian folk song that all Tetris enthusiasts have heard endless times as they skillfully maneuver their tetrominoes into place.

▲ **Figure 9.5.1 Belorussian Tanks Roll into California** The World of Tanks, a popular multiplayer online game, is being promoted in this publicity shot at the 2013 E3 Video Game Expo in Los Angeles. The game is produced by Wargaming.net, one of many Belorussian companies specializing in this global industry.

▲ **Figure 9.5.2 Russia's Blossoming Virtual World** These Russian youngsters eagerly explore the gaming cyberworld at the 2009 GameWorld interactive entertainment exhibition in Moscow.

limit foreign ownership of its energy resources and infrastructure.

Elsewhere in the region, connections with the global economy vary. In Belarus, a lack of economic reforms continues to slow new investment. About 80 percent of the nation's industrial sector is controlled by the state, and the political climate seems opposed to foreign investment. Neighboring Ukraine has succeeded in attracting more capital since 2004, but it is still early in the process of fully opening its economy to outside investment. For example, a contract signed in 2013 with Royal Dutch Shell will initiate a major drilling program to develop the country's natural gas resources. Many Ukrainians are particularly keen on producing more domestic energy as one way to decrease their reliance on Russia. Moldova and Armenia have both made some progress in economic reforms that help reduce barriers to foreign investment, but the ultimate economic success of these nations remains closely linked to policies and conditions that unfold within Russia.

Globalization and Russia's Petroleum Economy

Russia's oil and gas industry remains one of the strongest economic links between the region and the global economy, and the diverse international connections it has forged suggest the increasing importance of this sector to the region's future. The statistics are impressive: Russia's energy production makes up more than one-quarter of its economic output and two-thirds of its exports. Russia has 26 percent of the world's natural gas reserves (mostly in Siberia) and is the world's largest gas exporter. As for oil, Russia is by far the world's largest non-OPEC producer and the second largest oil exporter in the world (behind Saudi Arabia). It far outpaces the United States in annual output (major oilfields are in Siberia, the Volga Valley, the Far East, and the Caspian Sea region), and it possesses more than 75 billion barrels of proven reserves.

The dynamic nature of the Russian oil and gas business exemplifies how globalization is changing the region's economy. In the Soviet era, about half of Russia's oil and gas exports went to other Soviet republics, such as Ukraine and Belarus. These two nations still depend on Russian supplies, but the primary destination for Russian petroleum products has overwhelmingly shifted to western Europe. Russia now supplies that region with more than 25 percent of its natural gas and 16 percent of its crude oil. An agreement between Russia and the EU in 2000 aimed at the rapid expansion of these East–West linkages. The Siberian Gas Pipeline already connects distant Asian fields with western Europe via Ukraine (see Figure 9.39). Those connections are supplemented by lines through Belarus (the Yamal–Europe Pipeline) and Turkey (the Blue Stream Pipeline). Underwater pipelines beneath the Baltic Sea (Nord Stream—completed in 2012) and Black Sea (South Stream—to be completed in 2015) will deliver even greater volumes of gas to northern and southern Europe. Far to the east, Russian energy companies are adding to their pipeline connections with energy-rich Sakhalin Island, and they also will construct a large new gas pipeline from Yakutia (in northern Siberia) to Vladivostok.

Expansions are also refashioning the geography of oil exports. An expanding oil port facility at Primorsk (near St. Petersburg) serves western Europe and other global markets. To the south, a large export terminal opened at Novorossiysk (on the Black Sea) in 2001, delivering Caspian Sea oil supplies to the world market via a pipeline passing through troubled Chechnya (Figure 9.43).

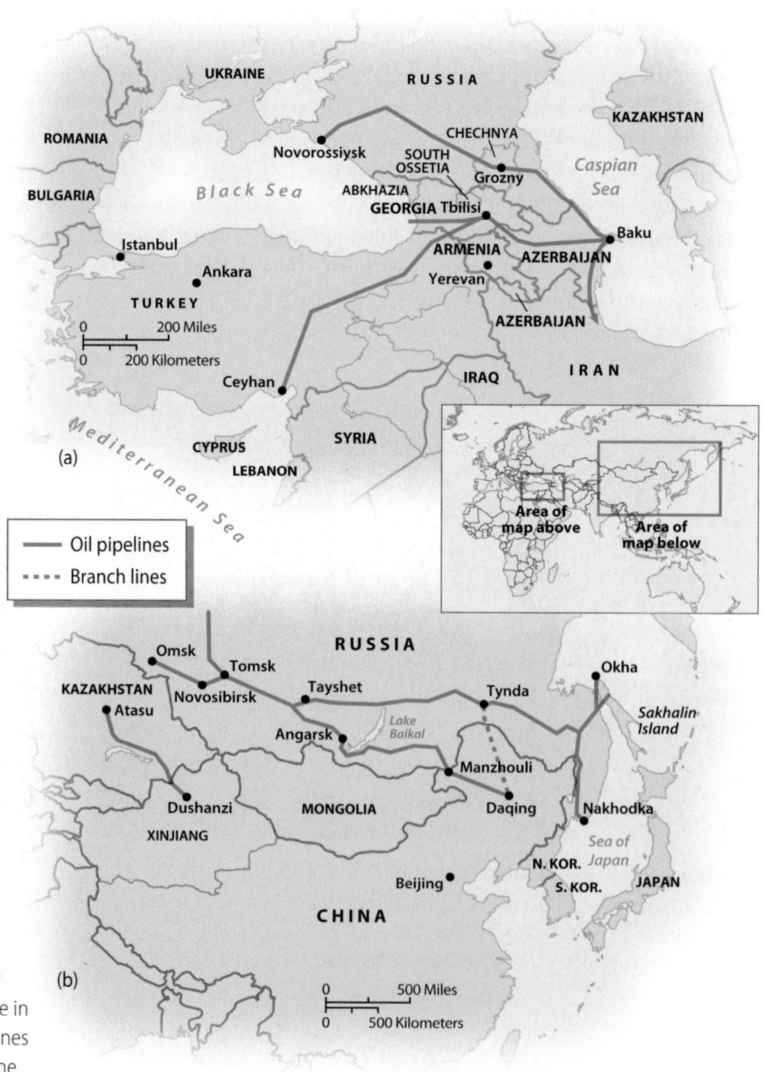

▶ **Figure 9.43 Russia's Expanding Pipelines** These two maps show new and planned oil pipelines that are designed to expand Russia's presence in the global petroleum economy. (a) Projects near the Caspian Sea take pipelines through politically unstable portions of the region, while (b) different pipeline projects in the Far East would benefit nearby China or Japan.

Nearby, oil pipelines between Baku (on the Caspian Sea) and the Black and Mediterranean seas cross Azerbaijan and Georgia. In the Russian Far East, both China and Japan are lobbying hard for more pipeline projects. Russians are building a large new Siberian Pacific Pipeline to link the Siberian fields to Asian markets. The Chinese want Russian oil to flow to Daqing, where it could be refined for national and regional markets. Japan prefers a large new facility at the Pacific port of Nakhodka, well positioned to supply Japan and offering Russia easy access to global markets via the Pacific Ocean. Other links connect the system with developments on Sakhalin Island, where several major energy projects are currently being completed (see Figure 9.43). Recently, Chevron Corporation entered into negotiations to help develop Russia's large arctic oil and gas resources, and other global energy companies may participate in the venture in coming years as the region is opened to development (facilitated by warmer, ice-free conditions).

Although many global companies have participated in the development of Russia's energy infrastructure, recent events suggest that state-controlled Russian companies may play a larger role in the future. Some private companies that were based in Russia, but that also utilized foreign capital, have simply been nationalized. Other North American, European, and Japanese energy companies have seen their role in Russian projects limited or eliminated. On Sakhalin Island, for example, Gazprom, the state-controlled Russian gas company, has taken control of several ventures originally dominated by companies such as Royal Dutch Shell. Ironically, Russian authorities have stripped control of these projects from foreign companies by suggesting that they failed to follow environmental regulations. Once these projects were in Russian hands, environmental concerns faded away.

Local Impacts of Globalization As the discussion of the petroleum industry suggests, local impacts of globalization are highly selective. Obviously, portions of the region that are close to oil and gas wells, pipeline and refinery infrastructure, and key petroleum shipping points are greatly affected (both economically and environmentally) by the changing global energy economy. The same is true more broadly: Globalization has affected different locations in the Russian domain in very distinctive ways. In Russia, capitalism has brought its most dramatic, though selective, benefits to Moscow. Indeed, much of the country's foreign investment has flowed into the Moscow area. Beyond Moscow, St. Petersburg and the Siberian cities of Omsk and Novosibirsk (Figure 9.44) have also seen growing global investment. New oil and gas prospects in Siberia and the Russian Far East (Sakhalin Island) and near the Caspian Sea have attracted other investments. In addition, port cities such as Vladivostok are well positioned to take advantage of their accessibility to nearby markets.

Elsewhere, globalization imposes penalties. Older, less competitive locales have been hit hard. Aging steel plants, for example, no longer have the guaranteed markets for their high-cost, low-quality products that they did in the days of the planned Soviet economy. Instead, they must compete on the global market, a market that is increasingly prone to lower prices and weakening demand for many of the industrial goods that the region produces. Russia's recent membership in the WTO will only make matters worse for many domestic companies because the new agreements will make it easier for highly competitive foreign corporations to enter Russian markets.

REVIEW

9.10 Describe how centralized planning created a new economic geography across the former Soviet Union. What is its lasting impact?

9.11 Briefly summarize the key strengths and weaknesses of the post-Soviet Russian economy, and suggest how globalization has shaped its evolution.

► **Figure 9.44 Novosibirsk** The city of Novosibirsk, the third largest city in Russia, has become a major focus of investment and urban growth in Siberia since the disintegration of the Soviet Union.

Physical Geography and Environmental Issues

Huge environmental challenges remain for the Russian domain. The legacy of the Soviet era includes polluted rivers and coastlines, poor urban air quality, and a frightening array of toxic wastes and nuclear hazards.

9.1 Why is the Volga River often referred to as Russia's version of the Mississippi?

9.2 Join with a group of students to debate another student group on the question of whether Russia's natural environment is one of its greatest assets or one of its greatest liabilities.

Population and Settlement

Declining and aging populations are part of the sobering reality for much of the region. Although some localities see modest population growth related to in-migration (mostly toward expanding urban areas), many rural areas and less competitive industrial zones are likely to see continued outflows of people and very low birth rates.

9.3 Traditionally, why is the large area of Russia located south of Volgograd so sparsely populated?

9.4 Given recent economic developments near the Caspian Sea, why might population in this area increase in the future?

Key Terms

autonomous area *(p. 434)*
Baikal–Amur Mainline (BAM) Railroad *(p. 417)*
Bolsheviks *(p. 433)*
centralized economic planning *(p. 440)*
chernozem soils *(p. 407)*
Cold War *(p. 435)*
Commonwealth of Independent States (CIS) *(p. 435)*
communism *(p. 433)*
Cossacks *(p. 426)*
dacha *(p. 424)*
denuclearization *(p. 436)*
Eastern Orthodox Christianity *(p. 425)*
exclave *(p. 439)*
glasnost *(p. 435)*
Gulag Archipelago *(p. 420)*
human trafficking *(p. 444)*
Iron Curtain *(p. 444)*
mikrorayon *(p. 423)*
northern sea route *(p. 413)*
oligarch *(p. 402)*
perestroika *(p. 435)*
permafrost *(p. 407)*
podzol soils *(p. 404)*
Russification *(p. 420)*
siloviki *(p. 438)*
Slavic peoples *(p. 425)*
socialist realism *(p. 431)*
socialist state *(p. 434)*
taiga *(p. 407)*
Trans-Siberian Railroad *(p. 417)*
tsar *(p. 418)*

In Review • Chapter 9
Russian Domain

Cultural Coherence and Diversity

Much of the region's underlying cultural geography was formed centuries ago from the complex mix of Slavic languages, Orthodox Christianity, and numerous ethnic minorities that continue to complicate the scene today. Further changing the country are new global influences—a set of products, technologies, and attitudes that often clash with traditional cultural values.

9.5 Where in the Russian domain would you be most likely to encounter Yakut-speaking peoples?

9.6 Cite some key similarities and differences you might observe in comparing the lifestyles of the Yakut with those of native North American populations.

Geopolitical Framework

Much of the region's political legacy is rooted in the Russian Empire, a land-based system of colonial expansion that greatly enlarged Russian influence after 1600 and then reappeared as the Soviet Union expanded its influence. Only large remnants of that empire survive on the modern map, yet it has stamped the geopolitical character of the region in lasting ways.

9.7 Why is this area of South Ossetia troublesome for Georgia's government?

9.8 Why is Russia one of the few nations in the world to recognize South Ossetia as an independent nation?

Economic and Social Development

The region's future economic geography, particularly in Russia, remains tied to the fortunes of the unpredictable global energy economy. Russia's recent admission into the WTO signals its growing integration with the global economy.

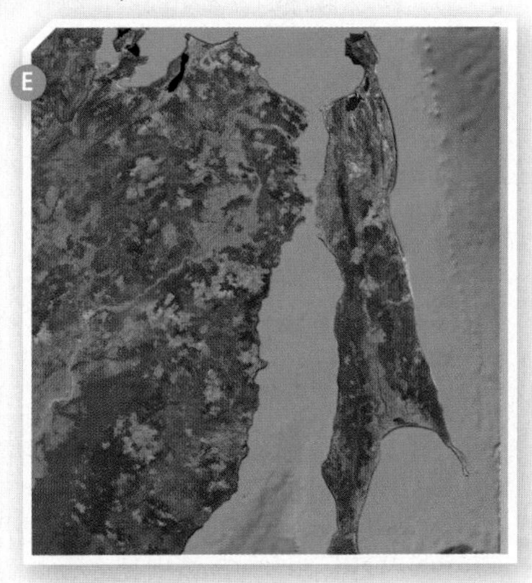

9.9 What global energy markets are most likely to be served by oil and natural gas produced on Russia's Sakhalin Island?

9.10 What are some of the key environmental and cultural challenges that face zones of rapid energy development such as Sakhalin?

MasteringGeography™

Looking for additional review and test prep materials? Visit the Study Area in **MasteringGeography™** to enhance your geographic literacy, spatial reasoning skills, and understanding of this chapter's content by accessing a variety of resources, including **MapMaster™** interactive maps, videos, RSS feeds, flashcards, web links, self-study quizzes, and an eText version of *Diversity Amid Globalization*.

Authors' Blogs

Scan to visit the author's blog for chapter updates

www.gad4blog.wordpress.com

Scan to visit the GeoCurrents blog

www.geocurrents.info

10 Central Asia

Physical Geography and Environmental Issues

Central Asia includes some of the world's most extreme deserts, as well as some of its highest mountains. Intensive agriculture along the rivers that flow out of the highlands and into the deserts has resulted in water shortages, leading to the desiccation of many of the region's lakes and wetlands.

Population and Settlement

Pastoral nomadism, the traditional way of life across much of Central Asia, is gradually disappearing as people settle in towns and cities.

Modern Astana. The capital of Kazakhstan since 1997, Astana is a rapidly growing city with a population approaching 800,000. The city center is noted for its lavish modern architecture made possible by Kazakhstan's oil wealth.

Cultural Coherence and Diversity

In much of eastern Central Asia, the growing Han Chinese population threatens the indigenous Tibetan and Uyghur cultures. In the west, the role of Islam in social and political life remains a major issue.

Geopolitical Framework

Afghanistan and its neighbors to the north are frontline states in the struggle between radical Islamic fundamentalism and secular governments.

Economic and Social Development

Despite its abundant resources, Central Asia remains a relatively poor region, although much of it does enjoy relatively high levels of social development.

Visitors to the sprawling, oil-rich country of Kazakhstan are often surprised by the extensive, highly modern architecture of the capital city of Astana. Located in the semiarid grasslands in the northern part of the country, Astana was a relatively minor city before Kazakhstan gained its independence from the Soviet Union in 1991. Several years later, the new country's leaders decided to relocate their capital city from Almaty in the south to the more centrally located Astana. The move was opposed by many Kazakhstanis, as the new capital is windswept and frigid during the long winter; with an average January temperature of 6.4°F (–14.2 °C), it is the world's second coldest capital, after Mongolia's Ulaanbaatar. Kazakhstan's leaders, however, wanted to move their seat of government farther to the north in part to fend off the possibility that northern Kazakhstan, heavily populated by ethnic Russians, might want to secede.

By most measures, the development of Astana has been a success. The city's population has expanded rapidly, growing from 281,000 in 1999 to 708,000 in 2010. Its demographical characteristics have also changed; formerly a Russian-majority city, its population is now roughly 60 percent ethnic Kazakh. Young professionals have been moving into the city, as have poorer workers from neighboring countries. Owing to Kazakhstan's oil wealth, several multinational corporations have established offices in the city, and large hotel chains have moved in as well. Kazakhstan's leadership has paid close attention to Astana's architecture, wanting the city to be a showcase of modernity. Acclaimed international architects were hired from such countries as Japan and the United Kingdom. As new buildings went up, old ones came down, especially those built during the Soviet period. Astana is designed to be a symbol of the new, not a reminder of the past.

Tremendous change over the past quarter-century has characterized not just Kazakhstan, but most other parts of Central Asia as well. Until 1991, most of this lightly populated and relatively arid region was controlled by the Soviet Union, and as result, it did not appear separately in world geography texts. When the Soviet Union collapsed, the position of Central Asia began to be reconsidered. Suddenly, several new countries appeared on the international scene, prompting scholars to reexamine the manner in which they divided the world. In some textbooks, the former Soviet Central Asia is grouped with Southwest Asia and North Africa, whereas in others it continues to be placed with a region centered on Russia. Elsewhere, Central Asia is deemed a world region in its own right, a usage we follow here.

> Although it was once a power center as well as one of the main conduits of global trade, the landlocked region of Central Asia has been relatively isolated for most of the past 200 years, and politically dominated by countries in other world regions.

Even those writers who do distinguish a Central Asian world region tend to define it in different ways. Most authorities agree that it includes five newly independent (former Soviet) republics: Kazakhstan, Kyrgyzstan, Uzbekistan, Tajikistan, and Turkmenistan. This chapter also includes other territories—Mongolia, Afghanistan, Azerbaijan, and the autonomous regions of western China (Tibet, Xinjiang, and Inner Mongolia [Nei Mongol]) (Figure 10.2). The inclusion of these additional territories within Central Asia is controversial. Azerbaijan is often classified with its neighbors in the Caucasus region (Georgia and Armenia); western China is obviously part of East Asia by political criteria; and Mongolia is also often placed within East Asia because of both its location and its historical connections with China. Afghanistan can alternatively be located within either South Asia or Southwest Asia.

However, considering Central Asia's historical unity, its common environmental circumstances, and its reentry onto

LEARNING OBJECTIVES

After reading this chapter you should be able to:

- Explain how the key environmental differences among Central Asia's desert areas, its mountain and plateau zone, and its steppe (grassland) belt influence human settlement and economic development.
- Identify the main reasons for the disappearance of the Aral Sea and outline the economic and environmental consequences of the loss of this once-massive lake.
- Summarize the reasons why water resources are of such great importance in Central Asia and describe the ways in which people are responding to water shortages.
- Explain why Central Asia's population is so unevenly distributed, with some areas densely settled and other essentially uninhabited.
- Describe the differences between Central Asia's historical cities and those that have been established within the past 100 years.
- Outline the ways in which religion divides Central Asia and describe how religious diversity has influenced the history of the region.
- Identify distinct ways in which cultural globalization has impacted different parts of Central Asia and explain why cultural globalization is controversial in much of the region.
- Describe the geopolitical roles played in Central Asia by Russia, China, and the United States and explain why the region has been the site of pronounced geopolitical tension over the past several decades.
- Describe the ways in which ethnic conflict has contributed to instability in Afghanistan and assess the potential of ethnic tension to destabilize the rest of the region.
- Explain the role of oil and natural gas production in generating extremely uneven levels of economic and social development across Central Asia.

▶ **Figure 10.1 Tibetan Plateau** Alpine grasslands and tundra, interspersed with rugged mountains and saline lakes, dominate the Tibetan Plateau. In summer, the sparse vegetation offers forage for the herds of yaks tended by nomadic Tibetan pastoralists. Much of the northern part of Tibet, however, is too high to support pastoralism and is therefore uninhabited.

the stage of global geopolitics, we think that it deserves consideration in its own right. It also makes sense to define its limits rather broadly. For example, Azerbaijan is linked both culturally (language and religion) and economically (oil) more to Central Asia than to Armenia and Georgia.

At the same time, any unity that Central Asia possesses is far from stable. Continuing Chinese political control over southeastern Central Asia and Han Chinese (meaning people of Chinese cultural heritage) migration into that region threaten any claims for regional coherence. Moreover, Central Asia itself remains deeply divided along cultural lines. Most of the region is Muslim in religious orientation and Turkic in language, but both the northeastern and the southeastern sections (Mongolia and Tibet) are traditionally Buddhist. Only time will tell whether Central Asia will continue to merit recognition as a distinct world region in its own right.

One reason for Central Asia's former lack of prominence is that it was poorly integrated into international trade networks. This situation began to change in the 1990s as large oil and gas reserves were found, especially in Kazakhstan, Turkmenistan, and Azerbaijan. Moreover, some external countries are seeking to exert geopolitical influence over Central Asia, including Iran, Pakistan, India, the United States, and Russia. Controversies surrounding China's strict control of its Central Asian lands also highlight the significance of the region.

Physical Geography and Environmental Issues: Steppes, Deserts, and Threatened Lakes

Central Asia forms a large, compact region in the center of the Eurasian landmass. Alone among the world regions, it lacks ocean access. Owing to its continental position in the center of the world's largest landmass, Central Asia is noted for its harsh climate. High mountains, deep basins, and extensive plateaus magnify its climatic extremes. The aridity of the region, as we shall see, has also contributed to some of the most severe environmental problems in the world. Other parts of the region, however, have relatively few, minor environmental problems. To understand the variation in Central Asia's environmental circumstances, it is necessary to examine its distinctive physical regions.

Central Asia's Physical Regions and Climate

In general terms, Central Asia is dominated by grassland plains (or **steppes**) in the northern area, desert basins in the southwestern and central areas, and high plateaus and mountains in the south-central and southeastern areas. Several mountain ranges also extend into the heart of the region, dividing the desert zone into a series of separate basins and giving rise to the rivers that flow first into the deserts and then into the imperiled lakes.

The Central Asian Highlands The highlands of Central Asia originated in one of the great tectonic events of Earth's history: the collision of the Indian subcontinent into the Asian mainland. This ongoing impact has created the highest mountain range in the world, the Himalayas, located along the boundary of South Asia and Central Asia.

Yet the Himalayas are merely one portion of a much larger network of high mountains and plateaus. In the northwest portion of the range, the Himalayas merge with the Karakoram Range and then the Pamir Mountains. From the so-called Pamir Knot—a complex tangle of mountains situated where Pakistan, Afghanistan, China, and Tajikistan converge—other towering ranges radiate outward in several directions. The Hindu Kush sweeps to the southwest through central Afghanistan; the Kunlun Shan extends to the east (along the northern border of the Tibetan Plateau); and the Tien Shan swings out to the northeast into China's Xinjiang Province. All of these ranges have peaks higher than 20,000 feet (6000 meters) in elevation. Much lower, but still significant ranges are found along Turkmenistan's

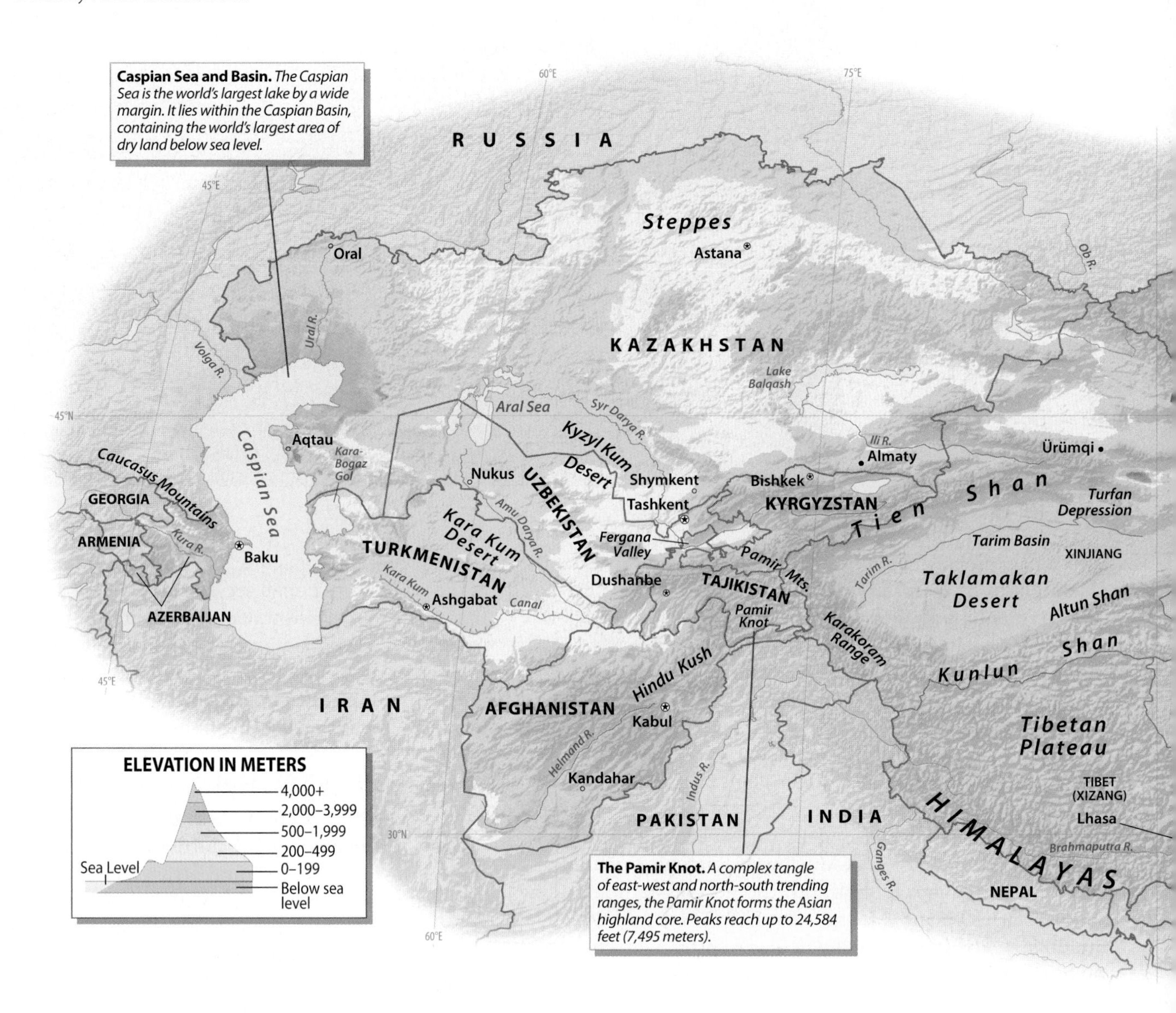

▲ **Figure 10.2 Central Asia** A vast, sprawling region in the center of the Eurasian continent, Central Asia, a vast, sprawling region in the center of the Eurasian continent, is dominated by arid plains and basins in the west and by lofty mountain ranges and plateaus in the east. Eight independent countries—Kazakhstan, Turkmenistan, Uzbekistan, Kyrgyzstan, Tajikistan, Azerbaijan, Afghanistan, and Mongolia—form Central Asia's core. China's lightly populated far west and north are often placed within Central Asia as well, due to patterns of cultural and physical geography.

boundary with Iran and Azerbaijan's boundaries with Russia, Armenia, and Iran.

More extensive than these mountain ranges, however, is the Tibetan Plateau (Figure 10.1). This massive upland extends some 1250 miles (2000 km) from east to west and 750 miles (1200 km) from north to south. More remarkable than its size is its elevation; virtually the entire area is higher than 12,000 feet (3700 meters) above sea level. Most of the large rivers of South, Southeast, and East Asia originate on the Tibetan Plateau and adjoining mountains, including the Indus, Ganges, Brahmaputra, Mekong, Yangtze, and Huang He.

The average elevation of the Tibetan Plateau is 14,800 feet (4500 meters), which is near the maximum height at which human life can exist. Although most of the plateau is lightly populated by nomadic herders, a large zone in the northwest is too high for human habitation. Rather than forming a flat, tablelike surface, the plateau is punctuated with east–west-running ranges alternating with basins that have no river drainage to the sea. Although the southeastern sections of the plateau receive ample precipitation, most of Tibet is arid. Cut off from any source of moisture by high ranges, large areas of the plateau receive only a few inches of rain a year. Winters on the Tibetan Plateau are cold, and although summer afternoons can be warm, summer nights remain chilly.

The Plains and Basins Although the mountains of Central Asia are higher and more extensive than those found anywhere else in the world, most of the region is characterized

The gorge country of eastern Tibet. *Several extremely steep canyons alternate with lofty ridges, making eastern Tibet one of the most topographically forbidding places in the world.*

by plains and basins of low and intermediate elevation. This lower-lying zone can be divided into two main areas: a central belt of deserts punctuated by lush river valleys and a northern swath of semiarid steppe.

Central Asia's desert belt is itself divided into two segments by the Tien Shan and Pamir mountains (Figure 10.2). To the west lie the arid plains of the Caspian and Aral sea basins, located primarily in Turkmenistan, Uzbekistan, and southern Kazakhstan (Figure 10.4, page 457). Most of this area is relatively flat and very low; the surface of the Caspian Sea lies 92 feet (28 meters) below sea level. The climate of this region is characterized by pronounced seasonal differences; summers are dry and hot, whereas winter temperatures average well below freezing (see the climographs for Tashkent and Almaty in Figure 10.3). Central Asia's eastern desert region extends for almost 2000 miles (3200 km) from the extreme west of China to the southeastern edge of Inner Mongolia. It is conventionally divided into two deserts: the Taklamakan, found in the Tarim Basin of Xinjiang, and the Gobi, which runs along the border between Mongolia proper and the Chinese region of Inner Mongolia. The Gobi, lying in the rain shadow of the Tibetan Plateau, is usually considered to be Asia's largest desert.

The environment of western Central Asia is distinguished from that of eastern Central Asia in part by its larger rivers. More snow falls on the western than the eastern slopes of the Pamir Mountains, giving rise to more abundant runoff. The largest of these rivers historically flowed into sizable lakes, although, as we shall see, several of these water bodies have almost disappeared over the past several decades. Other rivers, such as the Helmand of Afghanistan, terminate in shallow lakes or extensive marshes and salt flats. In Xinjiang, one substantial river, the Tarim, flows out of the highlands onto the basin floor, where it frequently shifts course across the sandy lowlands. Before the completion of irrigation projects in the 1960s, it terminated in a salty lake called Lop Nor. Subsequently, the newly dried-out Lop Nor salt flat was used periodically by China for testing nuclear weapons.

North of the desert zone, rainfall gradually increases, and desert eventually gives way to the great grasslands, or steppe, of northern Central Asia. Near the region's northern boundary, trees begin to appear in favored locales, outliers of the great Siberian taiga (coniferous forest) of the north. A nearly continuous swath of grasslands extends some 4000 miles (6400 km) east to west across the entire region, only partially broken by Mongolia's Altai Mountains. Summers on the northern steppe are usually pleasant, but winters can be brutally cold (see the climographs for Urumqi and Ulaanbaatar in Figure 10.3).

Central Asia's Environmental Challenges

Much of Central Asia has a fairly clean environment, owing mainly to low population density. In fact, some parts of Central Asia, such as northwestern Tibet, remain almost pristine, with little human impact of any kind. Industrial pollution, however, is a serious problem in the larger cities, such as Uzbekistan's Tashkent and Azerbaijan's Baku. Mongolia's capital, Ulaanbataar, now has one of the world's worst air pollution problems, owing to its rapid growth, use of coal for heating, and stagnant wintertime air masses. Elsewhere, the typical environmental dilemmas of arid environments plague the region: **desertification** (the spread of deserts); **salinization** (the accumulation of salt in the soil); and **desiccation** (the drying up of lakes and wetlands) (Figure 10.5, page 458). The destruction of the Aral Sea, located on the boundary of Kazakhstan and Uzbekistan in western Central Asia, has been particularly tragic.

The Shrinking Aral Sea In April 2010, United Nations (UN) Secretary General Ban-Ki Moon examined the remains of the Aral Sea by helicopter. He later told reporters that he had been shocked by what he saw. "It is clearly one of the

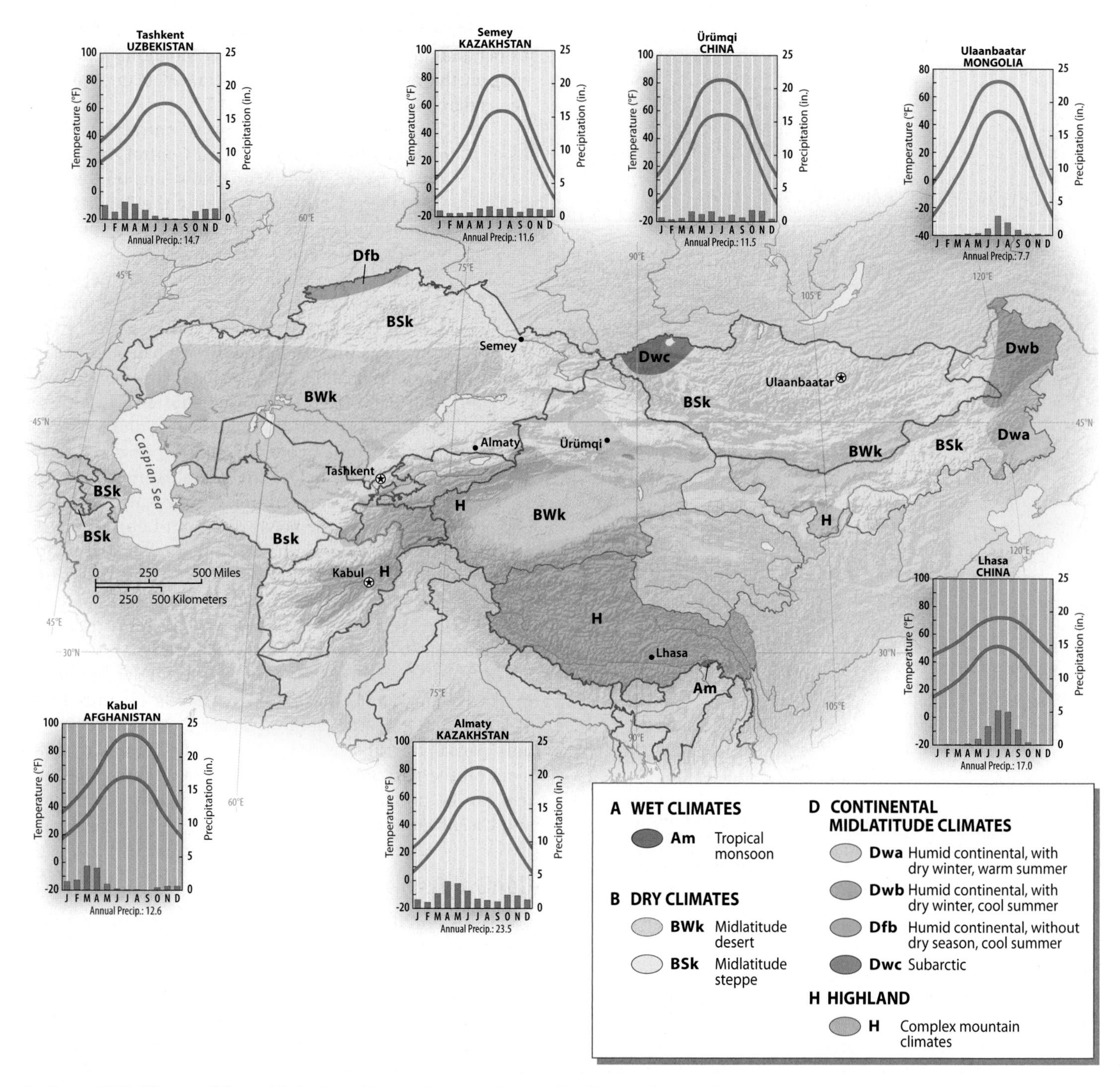

▲ **Figure 10.3 Climate of Central Asia** Central Asia is a dry region dominated by desert and steppe climates. Even in most of Central Asia's highlands, marked "H" on this map, arid conditions predominate. Truly humid areas in Central Asia are limited to the far north and extreme southeast. As a midlatitude region located in the interior of a vast continent, Central Asia is marked by pronounced continentality, experiencing profound differences between winter and summer temperatures.

worst environmental disasters of the world," he said. "It really left with me a profound impression, one of sadness that such a mighty sea has disappeared" (Figure 10.6).

Until the late 20th century, the Aral Sea was the world's fourth largest lake. Its only significant sources of water are the Amu Darya and Syr Darya rivers, which flow out of the Pamir Mountains, some 600 miles (960 km) to the southeast. Over thousands of years, water diversions for irrigation have lowered flows in both of these rivers, but after 1950 the scale of diversion vastly expanded. The valleys of the two rivers formed the southernmost farming districts of the Soviet Union, and the rivers became vital suppliers of water for warm-season crops, especially cotton. The largest of these projects was the Kara Kum Canal, which carries water from the Amu Darya across the deserts of southern Turkmenistan.

Unfortunately, the more river water was diverted for crop production, the less freshwater was available for the Aral Sea. The Aral proved to be particularly sensitive because it is relatively shallow. By the 1970s, the shoreline was retreating at an unprecedented rate; eventually, several "seaside" villages found themselves stranded more than 40 miles (64 km) inland. As salinity levels increased, most fish species disappeared. New islands began to emerge, and by the 1990s the Aral Sea consisted of two separate lakes.

The destruction of the Aral Sea resulted in economic and cultural damage, as well as ecological devastation. Fisheries were wiped out, and even local agriculture suffered. The retreating lake left large salt flats on its exposed beds; windstorms picked up the salt, along with the agricultural chemicals that had accumulated in the lake's shallows, and deposited it in nearby fields. Crop yields have thus declined; desertification has accelerated; and public health has been undermined. Particularly hard hit by this destruction are the Karakalpak people, members of a relatively powerless ethnic group who inhabit the formerly rich delta of the Amu Darya River.

In 2001, the oil-rich government of Kazakhstan decided to save what was left of its portion of the Aral Sea. By reconstructing canals, sluices, and other waterworks along the Syr Darya River, it doubled the flow of water reaching the northern Aral, which correspondingly began to rise. A large dam built across the lakebed prevented the extra water from flowing south into the salty waters and onto the extensive salt flats of the southern Aral Sea. By 2010, the area of the northern lake had increased by 50 percent, and its salt content had dropped low enough to allow the return of indigenous fish species. Currently, the World Bank and the government of Kazakhstan are discussing an extension of the project, hoping to extend the lake as far as the former port city of Aralsk.

The southern Aral Sea, meanwhile, continues to shrink, and its water quality continues to deteriorate. It, too, has been divided into two water bodies, and its eastern basin now dries up completely during the summer. Thus far, the oil-poor, agriculture-dependent government of Uzbekistan has done little to salvage what remains of its portion of the Aral Sea. It is worried, however, that Tajikistan will further reduce the flow of local rivers by its own water development projects, resulting in significant tension between the two countries.

Fluctuating Lakes The Aral Sea is not the only large lake-basin in Central Asia. Despite its general aridity, the region contains large lakes because it forms a low-lying basin, without drainage to the ocean, that is virtually surrounded by mountains and other more humid areas. The world's largest lake, by a huge margin, is the Caspian Sea, located along the region's western boundary; and the 15th largest is Lake Balkhash in eastern Kazakhstan. Several other large lakes, such as Kyrgyzstan's spectacular Issyk Kul, the world's second largest alpine lake, are also located in the region.

The Aral and the Caspian are not true seas because they are not connected with the ocean. The Caspian is less salty than the ocean (particularly in the north), while until the 1970s the Aral was only slightly brackish (salty). Lake Balkhash is almost fresh in the western part, but its long eastern extension is quite salty. Because none of these lakes is drained by rivers, all naturally fluctuate in size, depending on how much precipitation falls in their drainage basin in a given year. Like the Aral Sea, some of Central Asia's lakes have suffered from reduced water flow and hence increasing salinity; Balkhash, for example, has recently shrunk by some 770 square miles (2000 square km).

The story of the Caspian is more complicated. The Caspian Sea receives most of its water from the large rivers of the north—the Ural and the Volga—which drain much of European Russia. Owing to the development of extensive irrigation facilities in the lower Volga Basin, the volume of

▶ **Figure 10.4 Central Asian Desert** Much of Central Asia is dominated by deserts and other arid lands. The Karakum Desert of Turkmenistan is especially dry, supporting little vegetation.

► **Figure 10.5 Environmental Issues in Central Asia** Central Asia has experienced some of the world's most severe desertification problems. Soil erosion and overgrazing have led to the advance of desertlike conditions in much of western China and Kazakhstan. In western Central Asia, the most serious environmental problems are associated with the diversion of rivers for irrigation and the corresponding desiccation of lakes. Oil pollution is a particularly serious issue in the Caspian Sea area.

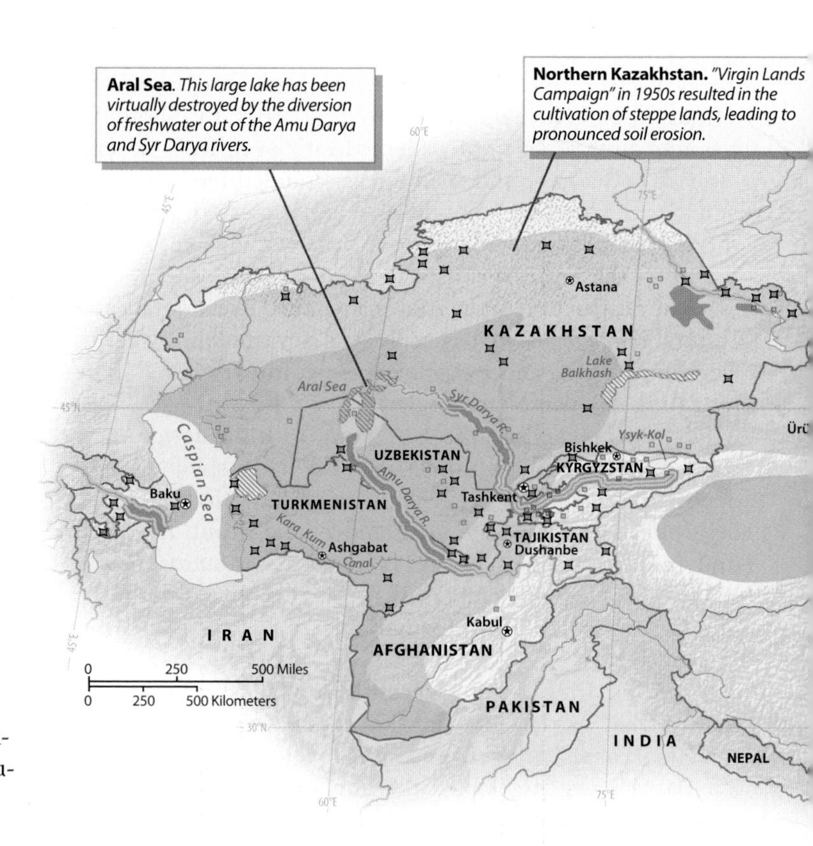

freshwater reaching the Caspian began to decline in the second half of the 20th century. With a reduced influx of water, the level of the great lake dropped, exposing as much as 15,000 square miles (39,000 square km) of former lake bed. A reduced volume of water resulted in increased salinity levels, undermining fisheries. After reaching a low point in the late 1970s, the Caspian began to rise, presumably because of higher than normal precipitation in its drainage basin. By the mid-1990s, it had risen some 8.2 feet (2.5 meters). This enlargement, too, has caused problems, inundating, for example, some of the newly reclaimed farmlands in the Volga Delta. Since 1995, the level of the lake has generally stabilized, experiencing only minor fluctuations. The most serious current environmental threat to the Caspian is pollution from the oil and natural gas industry, rather than fluctuation in size.

▼ **Figure 10.6 The Shrinking Aral Sea** These satellite images show the steady shrinkage of the Aral Sea, which was a massive lake as recently as the 1970s, and is now divided into several much smaller lakes. **Q: Why has the northern portion of the Aral Sea retained its water volume, unlike the rest of the lake, and why has the southeastern portion disappeared altogether?**

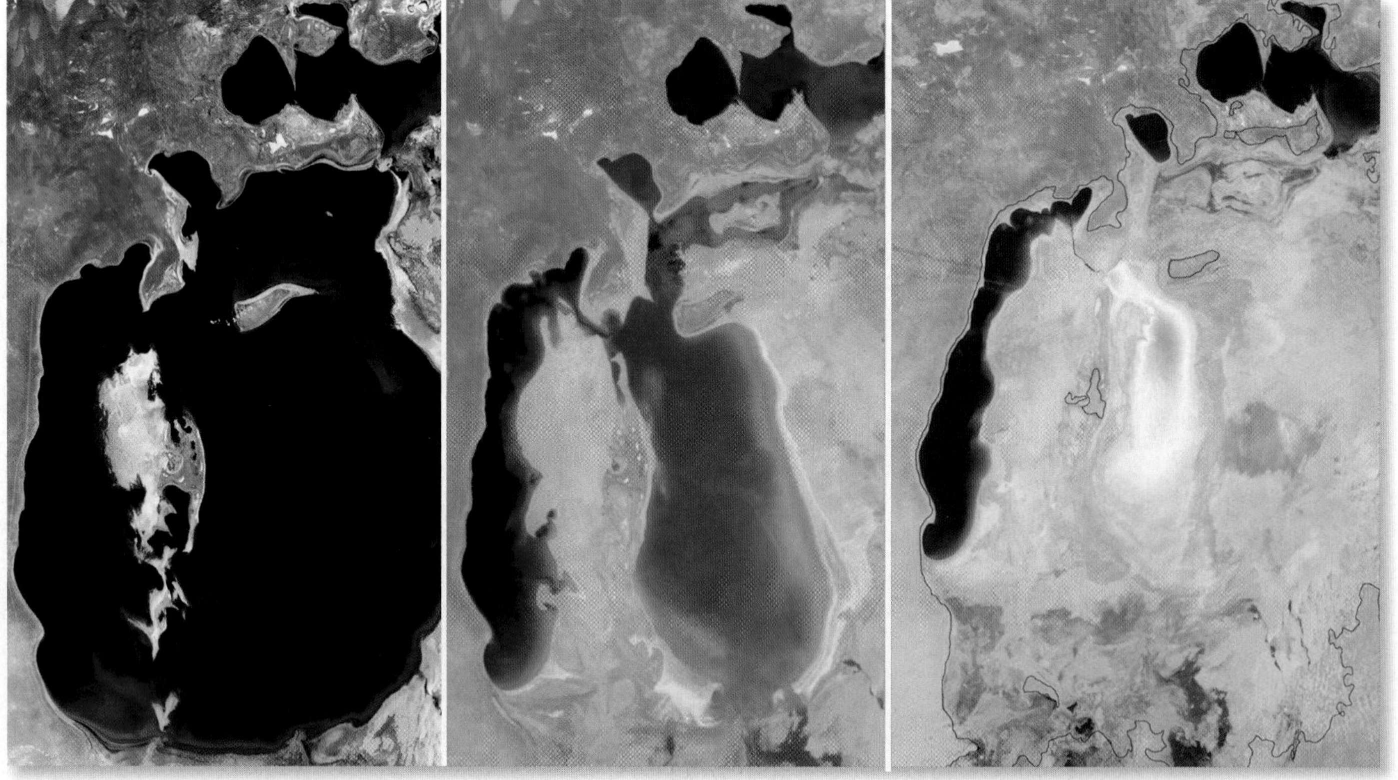

(a) July 1989 (b) August 2003 (c) August 2009

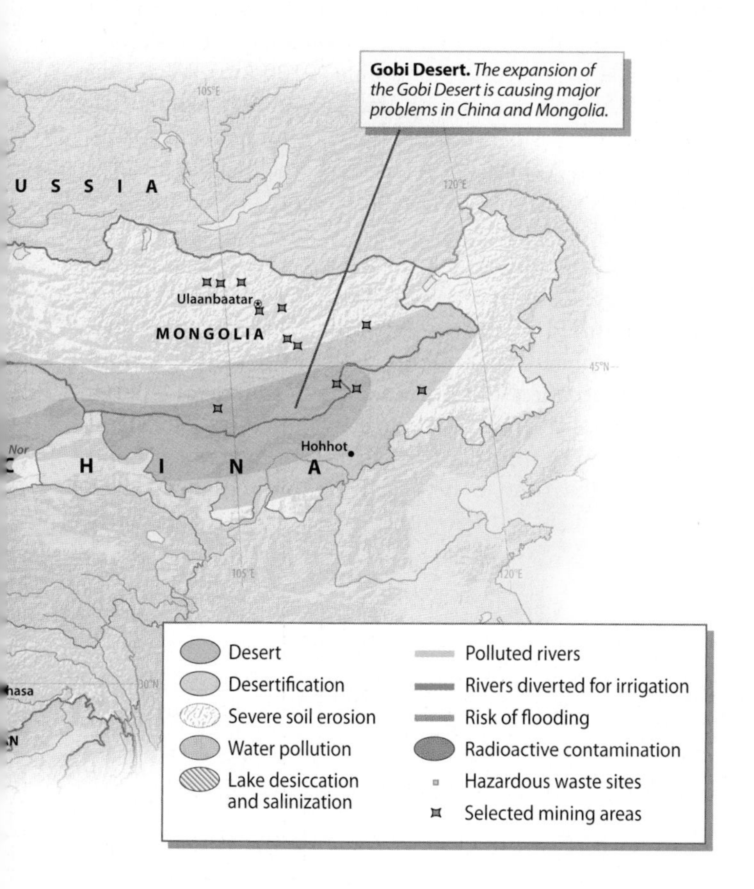

Water development projects have also begun to impact the Tibetan Plateau. Concerned about water shortages in the North China Plain, China is now diverting part of the flow of the headwaters of the Yangtze River into the headwaters of the Huang He (or Yellow River). This and other transfer schemes now threaten several of the Tibetan Plateau's large wetlands, havens for many rare species of migrating waterfowl.

Desertification and Deforestation Desertification is another major concern in Central Asia. In the eastern part of the region, the Gobi and Taklamakan deserts have gradually spread southward, encroaching on settled lands in northeastern China. Farmers sometimes have to relocate their houses three or four times over the course of their lives in order to avoid the sands. The Chinese government has tried to stabilize dune fields with massive tree- and grass-planting campaigns, but such efforts have been only partially successful. As a result, sand and dust storms have increased in frequency and now often affect much of northern China (Figure 10.7). In April 2010, sandstorms in the area were so severe that they were clearly visible on satellite images.

The former Soviet Central Asia has also seen extensive desertification. Northern Kazakhstan was one of the main sites of the ambitious Soviet "Virgin Lands Campaign" of the 1950s, in which semiarid grasslands were plowed and planted with wheat. Many of these lands have since returned to native grasses, but not before erosion stripped away much of their productivity. Some reports claim that up to 50 percent of Kazakhstan's farmland has been abandoned since 1990, due in part to desertification. In the irrigated lands of Uzbekistan, salinization has been a major problem, forcing the abandonment of some agricultural areas to the desert. Throughout the region, however, projects are being initiated to address such problems. In early 2013, for example, Turkmenistan initiated a massive tree-planting campaign aimed at stopping the spread of desert sands. Local officials were ordered to plant some 3 million seedlings and to make sure that they received adequate water from irrigation canals.

Deforestation has also harmed the region, resulting in wood shortages and reduced dry-season water supplies. Although most of Central Asia is too dry to support forests,

▶ **Figure 10.7 Dust Storm in the Gobi Desert** Straddling the border between Mongolia and China, the Gobi Desert is noted for its harsh climate, with frequent droughts and dust storms. The region's indigenous Mongolian inhabitants raise livestock and live in felt-covered tents, as can be seen in this photograph.

many of its mountains were once well wooded. Today extensive forests can be found only in the wild gorge country of the eastern Tibetan Plateau; in some of the more remote west- and north-facing slopes of the Tien Shan, Altai, and Pamir mountains; and in the mountains of northern Mongolia. Some of Central Asia's woodlands, particularly the walnut forests of Kyrgyzstan, are noted for their high biodiversity and hence form specific target areas for conservation groups (see *Everyday Globalization: Kazakhstan, Home of the Apple*).

Energy Issues in Central Asia Central Asia is home to many vast oil and natural gas fields. Combined with the region's low population density, these resources make petrochemical export a key building block of local economies (Figure 10.8). China's Xinjiang Province contains most of the country's oil and natural gas reserves, and several large pipelines have been built since 2002 to bring fossil fuel to the major Chinese population centers along the coast. The West–East Gas Pipeline, opened for full operation in 2005, stretches 2,485 miles (4000 km) from the Tarim Basin to Shanghai. It has already been upstaged by a second 5,650-mile (9100-km) West–East Gas Pipeline that connects with the Central Asia–China Gas Pipeline, moving gas overland all the way from Turkmenistan to eastern China. A third West–East Gas Pipeline, planned for completion in 2015, will connect Xinjiang to Fujian Province in southern China.

The western side of Central Asia is no less dependent on energy resources than the east. In 2011, Azerbaijan's exports amounted to nearly $25 billion, of which almost $24 billion came from mineral fuels, oil, and associated products. Azerbaijan's importance as a source of energy dates back far into the past, with evidence of a local oil industry spanning thousands of years. Oil fields near Baku were among the first to be heavily exploited in modern times, the earliest derrick going up in 1871. This early

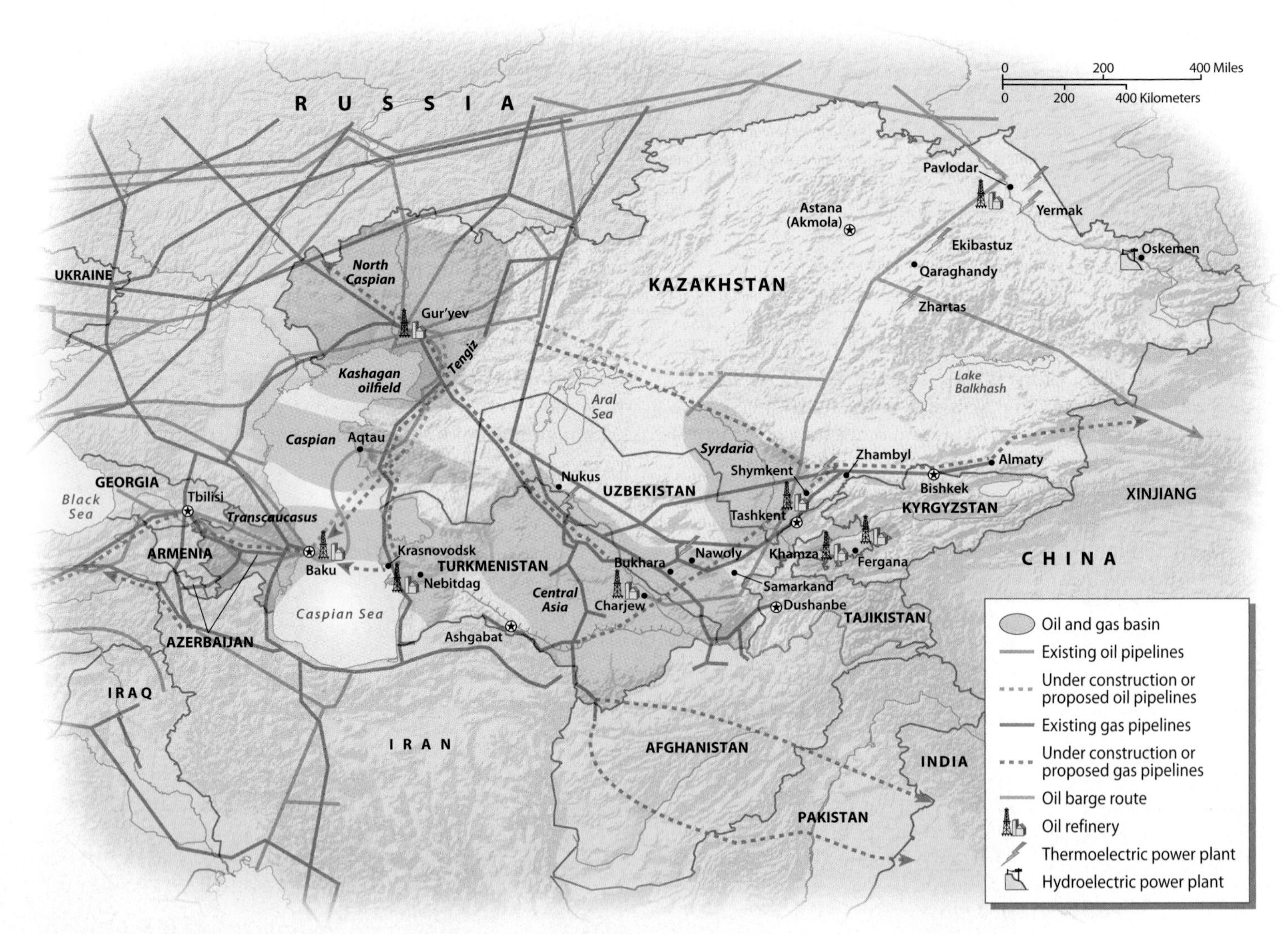

▲ **Figure 10.8 Oil and Gas Pipelines** Central Asia has some of the world's largest oil and gas deposits and recently has emerged as a major center for drilling and exploration. Because of its landlocked location, Central Asia cannot easily export its petroleum products. Pipelines have been built to solve this problem, and others are currently being planned. Pipeline construction is a contentious issue, however, as several of the potential pathways lie across Iran, a country that remains under U.S. sanctions.

Everyday Globalization

Kazakhstan, Home of the Apple

The average American eats around 65 apples a year, making the apple one of the world's most popular fruits. Grown over much of the world today, apples evolved in the forests of Central Asia, most likely in Kazakhstan's Tien Shan Mountains. Kazakhstan's forests feature at least 56 types of rare wild apples that can still be found in abundance, far more than can be found in any other part of the world (**Figure 10.1.1**). The Silk Road traders who crisscrossed the area since antiquity, trading European goods for silk and other goods of the Far East, took apples and their seeds east and west from the area. Over time, they created the proliferation of domesticated varieties we enjoy today. *Almaty,* the name of Kazakhstan's largest city, translates to "father of apples." The name is telling, since it long predates the genetic evidence indicating that apples evolved in the vicinity.

Genetic Diversity The fragmented and isolated forests of the Tien Shan Mountains, which act to prevent the mixing of genes among groups of trees, provide an ideal natural laboratory for the evolution of different types of apples. Wild apple trees in the area are accordingly quite diverse in their genetic makeup, making the area a key destination for scientists looking to create more robust apple varieties. The vast majority of apples eaten today either are of the classic Red Delicious and Golden Delicious varieties or derived the vast majority of their genes from one of those two varieties (like Fuji and Gala apples). Such homogeneity in the world's apple supply leaves orchards the world over vulnerable to pests and diseases.

By incorporating new genes from other types of apples, researchers hope to inhibit the spread of and devastation caused by these afflictions. In addition, the severe winters of Central Asia's mountains require greater tolerance to cold than today's domestic apples can endure. Exploiting the innate toughness of wild apples could potentially expand the range of climates where apple cultivation is possible. The wild apples of Central Asia are so diverse, however, that attempts to hybridize them with domestic apples usually produce inconsistent results.

▼ **Figure 10.1.1 Wild Apple Forest in Kazakhstan** The domesticated apple probably originated in the mountains of southern Kazakhstan. Many varieties of wild apple still grow in this area.

Threats to the Apple Forests Though many of Kazakhstan's apple forests now enjoy legal protection from development and destruction, they retain only a small fraction of their former range. According to the Kazakh Commission of the Environment, only 30 percent of wild apple stands remain, and some of them are being cleared to make way for urban growth. Researchers consider their work a race against time, since any decrease in the diversity of wild apples represents an irreparable loss.

1. Why do plant breeders care so deeply about the area in which a particular species was first domesticated?
2. Where are the main apple-growing areas of the world located today?

Google Earth Virtual Tour Video

http://goo.gl/aT13LK

abundance of cheap energy helped to make Baku the major city it is today. Kazakhstan is also a major oil exporter, and Turkmenistan has the third largest reserves of natural gas in the world.

Several Central Asian countries, including Tajikistan and Afghanistan, lack significant fossil fuel reserves and must seek out other sources of energy. Tajikistan relies on hydropower for over 90 percent of its electricity needs, taking advantage of the Vakhsh and Panj rivers, both major tributaries of the Amu Darya. The Nurek Dam, standing athwart the Vakhsh, was completed in 1980 under the auspices of the Soviet Union and still stands as the world's tallest dam at 997 feet (304 meters). The country's abundance of dams allows it to export excess electricity to Uzbekistan, from which it also imports a great deal of oil and natural gas. Uzbekistan strongly objects, however, to Tajikistan's dam-building program, fearing that it will result in the diminished flow of water into its arid agricultural regions.

▲ **Figure 10.9 Shrinking Glacier in Tibet** Due to global climate change, many of the glaciers on the Tibetan Plateau are retreating. As the glacier visible in this photo pulls back, the lake in front of it expands.

Low energy efficiency remains a major problem for Central Asian economies, which tend to exceed both low-income and wealthy countries in energy intensity of gross domestic product (GDP). Energy consumption is set to continue rising in the future as the region grows economically. Central Asia's more energy-rich countries, such as Uzbekistan, Kazakhstan, and Turkmenistan, are generally also those with the most voracious energy consumption habits.

Climate Change in Central Asia

Most climate change experts predict that Central Asia will be hard hit by global warming, largely because the region depends so heavily on snow-fed rivers. The Tibetan Plateau has already seen marked increases in temperature, resulting in the reduction of permafrost and the rapid drawback of mountain glaciers. Eighty percent of Tibet's glaciers are presently retreating, some at rates of up to 7 percent a year, leading to controversial predictions that the region could lose up to half of its ice fields within the next 50 years. It has also been predicted that Tajikistan could lose 20 percent of its ice cover by 2050. Tree-ring data from Mongolia indicate that the last century has been the warmest in more than a thousand years. Some researchers claim that Mongolia has experienced more warming over the past half-century, about 4°F (2°C), than any other part of the world, resulting in the spread of deserts and the loss of pasturelands.

The retreat of Central Asia's glaciers is especially worrisome because of their role in providing dependable flows of water for local rivers (Figure 10.9). A 2012 study found a 27 percent increase in the area covered by glacial lakes in the Tibetan Plateau, a good indicator of melting glaciers. In some areas, the melting of ice has resulted in the temporary flooding of lowland basins, but the long-term result will be to further

▼ **Figure 10.10 Population of Central Asia** Central Asia as a whole remains one of the world's least densely populated regions, although it does contain distinct clusters of higher population density. Most of Central Asia's large cities are located near the region's periphery or in its major river valleys.

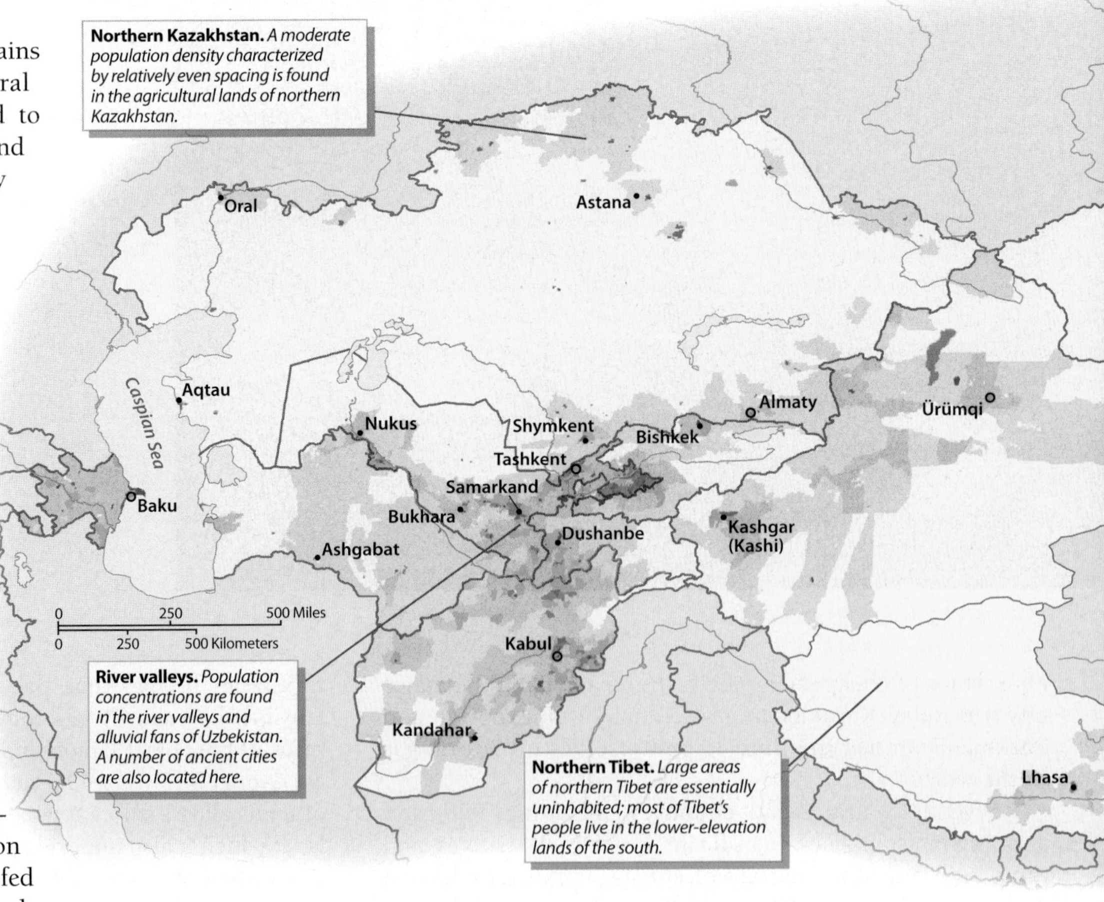

reduce freshwater resources in a region already burdened by aridity. Climate change from global warming could also reduce precipitation in the arid lowlands of western Central Asia, compounding the problem. Prolonged and devastating droughts have recently struck Afghanistan and several of its neighbors, indicating a possible shift to a drier climate. As a result, the UN Intergovernmental Panel on Climate Change (IPCC) has predicted a 30 percent crop decline for Central Asia as a whole by mid-century.

However, as is true elsewhere in the world, global warming will not affect all parts of Central Asia in a uniform manner. Some areas, including the Gobi Desert and parts of the Tibetan Plateau, could possibly see a long-term increase in precipitation. Such forecasts are scant consolation, however, in a region that is experiencing rampant desertification. Some global climate models also indicate increased precipitation in the boreal forest zone of Siberia that extends into the mountains of northern Mongolia. But as the rivers coming out of these mountains generally flow north, few benefits would be reaped by the rest of Central Asia.

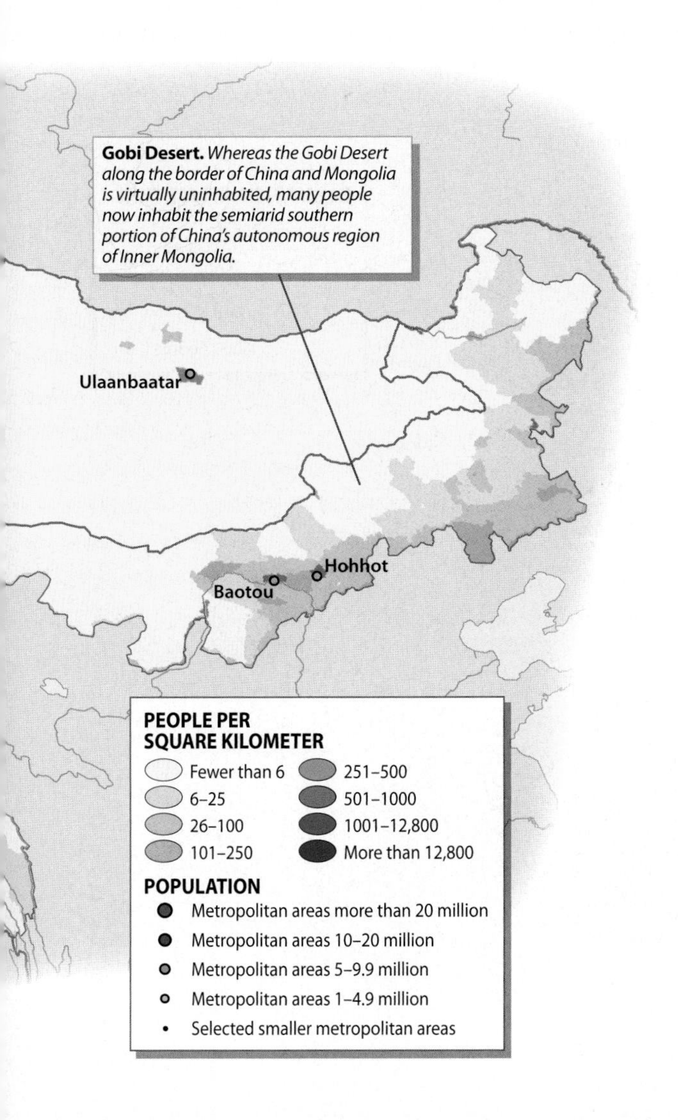

REVIEW

10.1 Why does Central Asia have such large lakes, and why are many of these lakes so deeply threatened?

10.2 How does the location of Central Asia, near the center of the world's largest landmass and at the junction of several large tectonic plates, influence the region's climate and landforms?

10.3 Why does Central Asia figure so prominently in many discussion of climate change?

Population and Settlement: Densely Settled Oases Amid Vacant Lands

Most of Central Asia is sparsely populated (Figure 10.10). Large areas are essentially uninhabited, either too arid or too high to support human life. Other vast expanses are populated by widely scattered groups of nomadic **pastoralists** (people who raise livestock for subsistence purposes). Mongolia, which is more than twice the size of Texas, has only 3.2 million inhabitants. But as is common in arid environments, those few lowland areas with fertile soil and dependable water supplies are thickly settled. Despite its overall aridity, Central Asia is well endowed with perennial rivers and productive oases. The nomadic pastoralists of the steppe and desert zones have dominated the history of Central Asia, but the sedentary peoples of the river valleys have always been more numerous.

Highland Population and Subsistence Patterns

The environment of the Tibetan Plateau is particularly harsh. Not only is the climate cold and water often scarce or brackish, but also ultraviolet radiation in this high-elevation area is always pronounced. Only sparse grasses and herbaceous plants—so-called mountain tundra—can survive, and human subsistence is obviously difficult under such conditions. The only feasible way of life over much of the Tibetan Plateau is nomadic pastoralism based on the yak, an altitude-adapted relative of the cow. Several hundred thousand people manage to make a living in such a manner, roaming with their herds over vast distances.

Farming in Tibet is possible only in a few favorable locations, generally those that are *relatively* low in elevation and that have good soils and either adequate rain or a dependable irrigation system. The main zone of agricultural settlement lies in the far south, where protected valleys offer favorable conditions. The population of Tibet proper (the Chinese autonomous region of Xizang) is only 3 million, a small number indeed considering the vast size of the area.

Population densities are also low in the other highland areas of Central Asia, although densely settled agricultural communities can be found in protected valleys.

The complex topography of the Pamir Range in particular offers a large array of small and nearly isolated valleys that are suitable for agriculture and intensive human settlement. Not surprisingly, this area is marked by great cultural and linguistic diversity. Many villages here are noted for their agricultural terraces and well-tended orchards.

Central Asia's mountains are vitally important for people living in the adjacent lowlands, whether they are settled farmers or migratory pastoralists. Many herders use the highlands for summer pasture; when the lowlands are parched, the high meadows provide rich grazing. The Kyrgyz (of Kyrgyzstan) are noted for their traditional economy based on **transhumance**, moving their flocks from lowland pastures in the winter to highland meadows in the summer. The farmers of Central Asia rely on the highlands for their wood supplies and, more important, for their water. Settled agricultural life in most of Central Asia is possible only because of the rivers and streams flowing out of the region's mountains.

Pastoralism and Farming in the Lowlands

Most of the inhabitants of Central Asia's deserts live in the narrow belt where the mountains meet the basins and plains. Here water supplies are adequate, and soils are not contaminated with salt or alkali, as is often the case in the basin interiors. As a result, the population distribution pattern of China's Tarim Basin forms a ringlike structure (Figure 10.11). Streams flowing out of the mountains are diverted to irrigate fields and orchards in the narrow band of fertile cropland situated between the steep slopes of the mountains and the nearly empty deserts of the central basin.

The population west of the Pamir Range, in the former Soviet Central Asia, is also concentrated in the transitional zone nestled between the highlands and the plains. A series of **alluvial fans**—fan-shaped deposits of sediments dropped by streams flowing out of the mountains—has long been devoted to intensive cultivation. Fertile **loess**, a silty soil deposited by the wind, is widespread, and in a few favored areas winter precipitation is high enough to allow rain-fed agriculture. Several large valleys in this area, such as the Fergana Valley of the upper Syr Darya River, offer fertile and easily irrigated farmland. In the far west of the region, Azerbaijan's Kura River Basin also supports intensive agriculture and concentrated settlement.

Unlike the other deserts of the region, the Gobi has few sources of permanent water. Rivers draining Mongolia's highlands flow to the north or terminate in interior basins, whereas only a few of the larger streams from the Tibetan Plateau reach the Gobi proper. (The Huang He, however, does swing north to reach the desert edge before turning

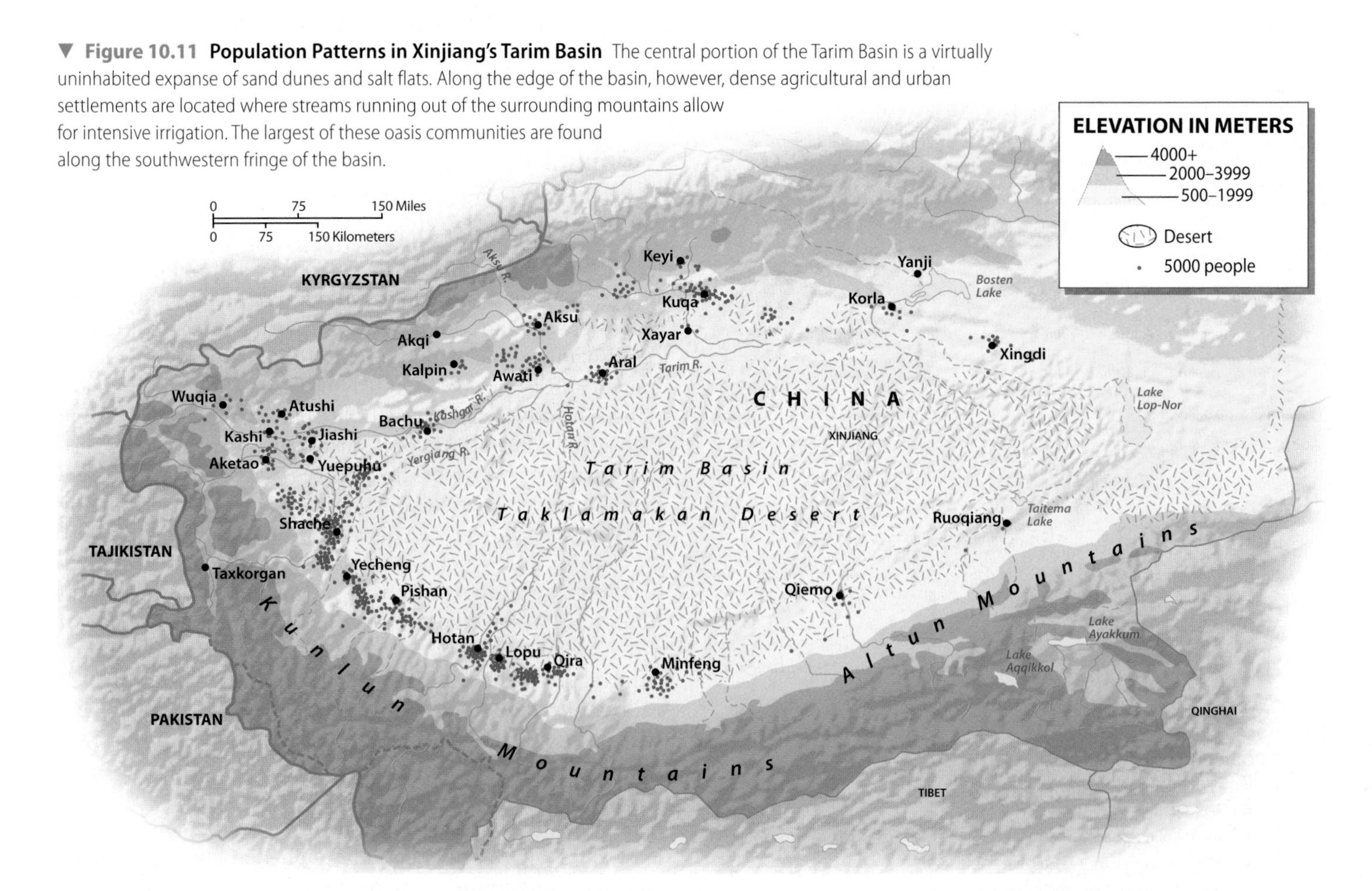

▼ **Figure 10.11 Population Patterns in Xinjiang's Tarim Basin** The central portion of the Tarim Basin is a virtually uninhabited expanse of sand dunes and salt flats. Along the edge of the basin, however, dense agricultural and urban settlements are located where streams running out of the surrounding mountains allow for intensive irrigation. The largest of these oasis communities are found along the southwestern fringe of the basin.

▶ **Figure 10.12 Steppe Pastoralism** The steppes of northern and central Mongolia offer lush pastures during the summer. Mongolians, some of the world's most skilled horse riders, have traditionally followed their herds of sheep and cattle, living in collapsible, felt-covered yurts. Many Mongolians still follow this way of life. **Q: Why is nomadic pastorialism so much more widespread in Mongolia than in most other countries?**

south to flow through the Loess Plateau.) Owing to this scarcity of **exotic rivers** (those originating in more humid areas) and to its own aridity, the Gobi remains one of Asia's least-populated areas.

The steppes of northern Central Asia are the classical land of nomadic pastoralism. Until the 1900s, virtually none of this area had ever been plowed and farmed. To this day, pastoralism remains a common way of life across the grasslands. In northwestern China and the former Soviet republics, however, most pastoral peoples have been forced to adopt sedentary lifestyles. National governments in the region, like those in most other parts of the world, find migratory people hard to control and difficult to provide with social services. Migratory herders are also highly vulnerable to natural disasters. Severe cold in the winter of 2010, for example, killed a quarter of the domesticated animals in Mongolia, resulting in a subsistence crisis over much of the country. Although roughly one-third of Mongolians are still nomadic pastoralists (Figure 10.12), the country's prime minister recently announced that he expected this way of life to diminish over the next few decades (see *Working Toward Sustainability: Pastoralism in Mongolia*).

In northern Kazakhstan, the Soviet regime converted the most productive pastures into farmland in the mid-1900s in order to increase the country's supply of grain. Some of these lands have since reverted to steppe, but large areas remain under the plow; Kazakhstan is the world's sixth largest exporter of wheat. Consequently, northern Kazakhstan has the highest population density of the steppe belt.

Population Issues

Some portions of Central Asia are growing at a moderately rapid pace. The demographic data on the region show overall fertility rates near the global middle, while exhibiting pronounced variations between different areas (Table 10.1). Afghanistan, the least-developed and most male-dominated

Table 10.1 Population Indicators

Country	Population (millions) 2013	Population Density (per square kilometer)	Rate of Natural Increase (RNI)	Total Fertility Rate	Percent Urban	Percent <15	Percent >65	Net Migration (Rate per 1,000)
Afghanistan	30.6	47	2.8	5.4	24	49	2	–5
Azerbaijan	9.4	109	1.3	2.3	53	22	6	0
Kazakhstan	17.0	6	1.4	2.6	55	25	7	–0
Kyrgyzstan	5.7	28	2.1	3.1	34	31	4	–7
Mongolia	2.8	2	2.1	2.8	63	27	4	–1
Tajikistan	8.1	56	2.5	3.7	26	36	3	–1
Turkmenistan	5.2	11	1.4	2.5	47	32	5	–2
Uzbekistan	30.2	68	1.6	2.3	51	29	4	–2

Source: Population Reference Bureau, *World Population Data Sheet, 2013*.

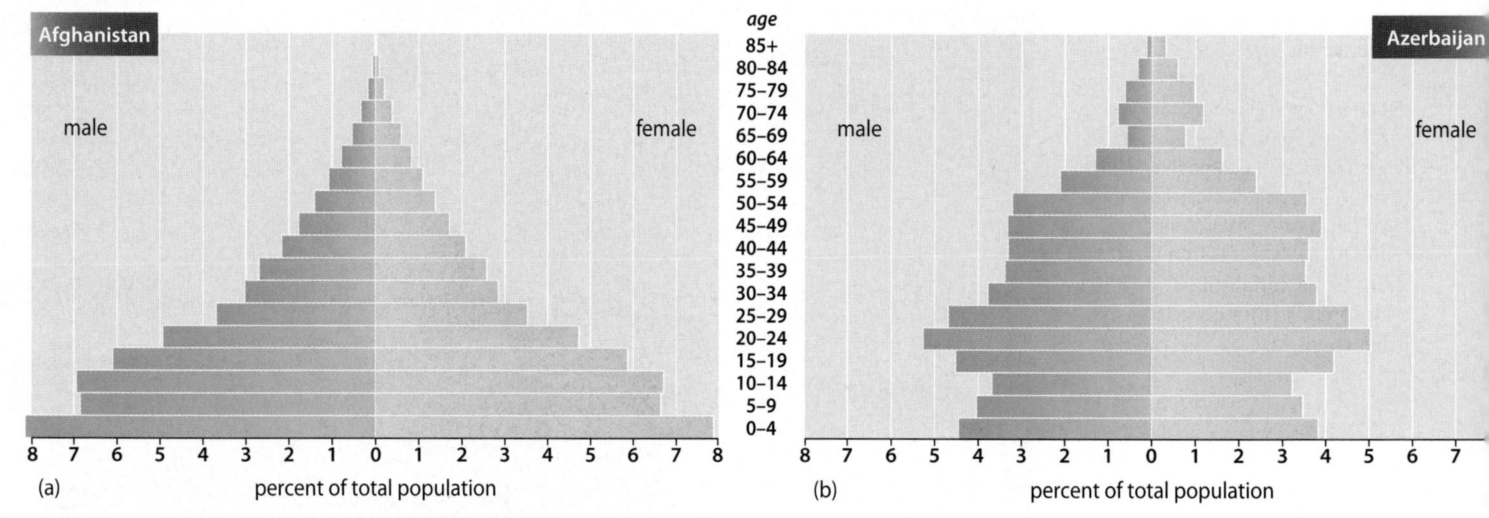

▲ **Figure 10.13 Population Pyramids of Afghanistan and Azerbaijan** (a) In Afghanistan, high birth rates coupled with high death rates have created a triangular population pyramid. In this poor, male-dominated country, elderly men and elderly women are equally few in numbers. (b) In contrast, Azerbaijan has a relatively balanced population pyramid, reflecting the fact that it is close to population stability. Azerbaijan's birth rate, however, has increased in recent years, as is indicated by the wider bar for ages "0–4" than for "10–14." Note also that women outlive men in Azerbaijan, which is the normal pattern.
Q: In Afghanistan, why are so many more people in the "0–4" age bracket than in the "5–9" age bracket?

country of the region, has the highest birth rate by a substantial margin. Tajikistan, also noted for its low levels of economic development, also has an elevated birth rate. As you can see in Figure 10.13, the fact that Azerbaijan has a much lower birth rate than Afghanistan results in markedly different population pyramids for the two countries.

Low fertility, however, can be counteracted by immigration, as is the case in much of western China. Here substantial population growth over the past 30 years has stemmed from the migration of Han Chinese into the area. Many of the indigenous inhabitants resent this influx, but the Chinese government claims that it is necessary for economic development.

Migration has also complicated demographic patterns in the former Soviet zone. After the breakup of the Soviet Union, millions of Russians and Ukrainians left Central Asia to return to their homelands. More recently, Russia's economic boom has led many Azerbaijanis, Tajiks, Uzbeks, and other Central Asians to seek employment in Moscow and St. Petersburg, even though they face substantial discrimination and occasional violence. Russian youth gangs often attack foreigners, particularly those from Central Asia. After the Russian government severely reduced the quota of immigrant workers in 2011, extreme nationalists began to form paramilitary groups to harass suspected undocumented laborers, a move supported by many Russian citizens.

Urbanization in Central Asia

Although the steppes of northern Central Asia had no real cities before the modern age, the river valleys and oases have been partially urbanized for thousands of years. Cities such as Samarkand and Bukhara in Uzbekistan were famous even in medieval Europe for their riches and their lavish architecture. This early urban development was built upon the region's economic and political prominence. The Amu Darya and Syr Darya valleys lay near the midpoint of the overland trans-Eurasian silk route, and they formed the core of several ancient empires based on the cavalry forces of the steppe. The modern era of steamships and oceanic trade undermined this system, bringing hardship to the cities of Central Asia.

The conquest of Central Asia by the Russian and Chinese empires ushered in a new wave of urban formation. Cities slowly began to appear on the Kazakh steppe, where none had previously existed. The Manchu and Chinese conquerors of Inner Mongolia and Xinjiang built new administrative and garrison cities, often placing them only a few miles from indigenous urban sites. The old indigenous cities of the region are characterized by complex and almost mazelike networks of streets and alleyways, whereas the neighboring Chinese cities were constructed according to a strict geometrical order. The continuing growth of urban populations, along with the establishment of new urban policies, is now obscuring this old dualism. In 2009, Beijing announced that the old city of Kashgar (Kashi) in China's far west would be largely demolished, due mainly to concerns about earthquake safety. Although China promised to save many of Kashgar's most historic buildings and to build new residences for its inhabitants, some local people view the plan as an assault on their cultural identity.

Although it is possible to distinguish Russian/Soviet cities from indigenous cities in the former Soviet zone, this dichotomy is not clear-cut. In Uzbekistan, for example,

Working Toward Sustainability

Pastoralism in Mongolia

For thousands of years, animal husbandry has been the key economic and cultural activity across Mongolia. Even today livestock herding employs roughly 40 percent of Mongolia's workers and accounts for 20 percent of its economic output. Eighty-two percent of Mongolia's land area is pastureland, supporting over 40 million livestock. However, according to a study by the Swiss Agency for Cooperation and Development (SDC), increasing degradation of pastureland threatens pastoralism in the country, potentially compromising the livelihoods of many of Mongolia's poorest citizens (Figure 10.2.1).

Problems with Pastures One major cause of pasture degradation in Mongolia is thought to be the de facto open-access system of pasture management that has prevailed since the end of communist government during the 1990s. In the absence of effective pasture management, herding families are often reluctant to move their animals to new pastures for fear of losing access to their current grazing area, and locations convenient to water sources tend to be fiercely contested. An influx of poor job seekers from urban areas, accompanying the post-communist era of high unemployment in the 1990s, undermined incentives to conserve pastureland. The result is that many are now barely able to make a living. Even as Mongolia experiences strong overall economic growth, the poverty rate in rural areas is nearly 50 percent.

Improved Management To combat the problem, the SDC has helped introduce so-called Pasture-User Groups (PUGs), territorially defined autonomous organizations supported by local government that include all herders in a given area. Proponents hope that PUGs, by providing a social and legal framework for coordination of herding activities, can prevent conflicts between herders and eliminate the individual incentives for overgrazing that plague Mongolia's pastureland. So far, the PUG system has shown itself to be effective on a limited scale, although relationships with nonmembers have proven to be a stumbling block and PUGs do not seem to work well in the more arid zones.

▼ **Figure 10.2.1 Degraded Pasture in Mongolia** Many of Mongolia's pasturelands have been overgrazed, leading to the spread of desert conditions into grassland areas.

The move toward more formal management of grazing areas is in many ways a return to Mongolia's past. As attested by a legal code issued in 1229, specific groups of herders have long been linked directly to pastureland, which was usually managed by tribal leaders for the perceived benefit of the group. The Soviet era saw the elimination of clan ownership of livestock, with herders instead raising government-owned animals as state employees. The future economic success or failure of Mongolia's rural areas will depend to a great extent on the effectiveness of PUGs or other methods of pastureland management.

Google Earth Virtual Tour Video

http://goo.gl/ps0MDc

1. Why is the condition of pastureland such an important issue in Mongolia?
2. How have political and economic changes in Mongolia since 1991 affected the country's pastoral economy?

Tashkent is largely a Soviet creation, while many parts of Bukhara and Samarkand still reflect the older urban patterns and architectural forms (Figure 10.14). Several major cities, such as Kazakhstan's former capital of Almaty, did not exist before Russian colonization. In Azerbaijan, Baku emerged as a major city in the early 20th century as the first Caspian oil fields began to be intensively exploited. Almost everywhere in the former Soviet zone, the effects of centralized urban planning and design are evident.

Urbanization is gradually, but unevenly spreading across Central Asia. Afghanistan has recently seen a huge expansion of its capital and major city, Kabul (see *Cityscapes: Kabul, A Precariously Booming City*). Even Mongolia, long a land virtually without permanent settlements, now has more people living in cities than in the countryside. More than 1.2 million people, more than a third of Mongolia's total population, now live in Ulaanbataar, the capital city. Ulaanbataar has grown so rapidly that many of its residents

▲ **Figure 10.14 Traditional Architecture in Samarkand** Samarkand, Uzbekistan, is famous for its lavish Islamic architecture, some of it dating back to the 1400s. The city owes its rich architectural heritage in part to the fact that it was the capital of the empire created by the great medieval conqueror Tamerlane.

still live in traditional felt-covered tents, with few urban services. In some parts of the region, however, cities remain relatively few and far between. Only 26 percent of the people of Tajikistan, for example, are urban residents. Tibet similarly remains a predominantly rural society, although the influx of Han Chinese into the region is creating a larger urban system. To understand this movement and its broader ramifications, we must examine Central Asia's patterns of cultural geography.

REVIEW

10.4 Why are large parts of Central Asia so sparsely settled, yet others have dense populations?

10.5 Why is the urban environment of Central Asia changing so rapidly?

10.6 Why is nomadic pastoralism so historically important in Central Asia, and how is this adaptation currently changing?

Cultural Coherence and Diversity: A Meeting Ground of Different Traditions

Although Central Asia has a certain environmental unity, its cultural coherence is more questionable. The western half of the region is largely Muslim and is often classified as part of Southwest Asia. On the other side of Central Asia, Mongolia and Tibet are characterized by a distinctive form of Tibetan Buddhism. Tibet is culturally linked to both South and East Asia, and Mongolia is intimately associated with China, but neither fits easily within any world region. Such complexity can best be understood by examining the region's historical geography.

Historical Overview: Steppe Nomads and Silk Road Traders

The river valleys and oases of Central Asia were early sites of agricultural communities. Archaeologists have discovered abundant evidence of farming villages dating back to the Neolithic period (beginning circa 8000 BCE) in the Amu Darya and Syr Darya valleys and along the rim of the Tarim Basin. After the domestication of the horse around 4000 BCE, nomadic pastoralism emerged in the steppe belt as a new human adaptation. Eventually, pastoral peoples gained power over the entire region, transforming not only the history of Central Asia, but also that of virtually all of Eurasia. In the premodern period, highly mobile pastoral nomads enjoyed major military advantages over sedentary societies. Not until the age of gunpowder were the benefits of pastoralism offset by the demographic and economic advantages held by the more populous, agriculturally based states.

Before the modern era of steamships, Central Asia was also a crucial location for long-distance trade. Much of the economic exchange between East Asia and both Europe and Southwest Asia passed along several parallel routes traversing the region, known as the **Silk Road.** Religious practices and other cultural forms also moved along these trade routes. Buddhism spread from India to China, for example, by way of Central Asia.

The linguistic geography of Central Asia has undergone major changes over the ages. Into the first millennium CE, the inhabitants of the region spoke Indo-European languages closely related to Persian. More than a thousand years ago, Indo-European languages began to be replaced on the steppe by languages in the Altaic family, which include Mongolian and Turkic. By the second century BCE, a powerful nomadic empire of Turkic-speaking peoples arose in what today is Mongolia. As Turkic power spread through most of Central Asia, Turkic languages gradually began to replace Indo-European tongues in the agricultural communities. This process was never completed, however, and southwestern Central Asia remains a meeting ground of the Persian and Turkic languages. Uzbek, for example, is classified in the Turkic language family, but much of its basic vocabulary is of Persian origin.

Cityscapes

Kabul, A Precariously Booming City

Located in a fertile river valley astride the ancient Silk Road, Kabul is one of Central Asia's oldest and largest cities. Though convulsed by war several times over the past several decades, Kabul was home to more than 3.5 million people in 2010 and continues to grow at the astounding pace of nearly 5 percent a year. It follows that Kabul is a city of rapid change and considerable uncertainty. The rapid expansion of the urban area and a lack of both greenery and pavement contribute to the formation of dust clouds that tend to hang over the city. The influx of so many new residents, moreover, has strained the city's infrastructure almost to its breaking point.

Poor Neighborhoods Kabul's suburban areas are home to some 20 percent of the city's population—and a disproportionate number of its poorest residents. Many of these neighborhoods, which stretch far into the hills surrounding Kabul Valley, lack running water, sanitation, and paved streets. The vistas are stunning, but the future of these informal settlements—which almost universally conflict with Kabul's zoning codes—remains clouded. The expected drawdown in construction work following a U.S. military withdrawal from Afghanistan is also expected to sharply cut one of the area's main sources of income.

Upscale Neighborhoods Despite the violence that has gripped the city in recent years, many parts of Kabul increasingly resemble the upscale areas of other world cities. The Khair Khāna neighborhood, located on the northwest side of the city, is the scene of a recent high-rise construction boom, and residents flock to the Lycee Maryam, one of Kabul's most conspicuous shopping streets (Figure 10.3.1). Kabul's downtown area features many new developments, including five-star hotels and Kabul City Center, a nine-story shopping mall. The mall, which opened in 2005, is the first of its kind in Afghanistan. Like many other major attractions in the city, it features explosive-resistant windows and metal detector screenings at entrances to help deter the ever-present threat of suicide bombings.

One of Afghanistan's most ambitious current undertakings is the so-called Kabul New City project. This project aims to provide new housing, workspace, and parks together with quality infrastructure to accommodate the region's growing population. It also seeks to provide an alternative to the shantytowns sprouting up on surrounding hillsides. The project aims to eventually house 3 million inhabitants, which would constitute more than half of Kabul's current population.

▼ **Figure 10.3.1 Khair Khāna Neighborhood in Kabul** Over the past several years, the Khair Khāna area has been transformed into one of Kabul's most up-scale neighborhoods.

1. Why has Kabul grown so much more rapidly than any other city in Afghanistan?
2. How does the climate of Kabul contribute to the city's environmental problems?

Google Earth Virtual Tour Video

http://goo.gl/UKOSPD

The Turks were eventually replaced on the eastern steppes by another group of Altaic speakers, the Mongols. In the late 1100s, the Mongols, under the leadership of Genghis Khan, united the pastoral peoples of Central Asia and used the resulting force to conquer nearby sedentary societies. By the late 1200s, the Mongol Empire had grown into the largest contiguous empire the world had ever seen, stretching from Korea and southern China in the east to the Carpathian Mountains and the Euphrates River in the west (Figure 10.15). In today's Mongolia, Genghis Khan has reemerged as the country's national hero. China is also trying to claim the same legacy, having recently built a Genghis Khan theme park in Inner Mongolia.

Protected by mountain barriers and by the rigorous conditions of the plateau, Tibet has taken a different course from the rest of Central Asia. Tibet emerged as a strong, unified kingdom around 700 CE. Tibetan unity and power did not persist, however, and the region reverted to its former state of semi-isolation. Tibet was incorporated for a short period in the 1200s into the Mongol Empire, and in later

▲ **Figure 10.15 The Mongol Empire of the 1200s** In the 1200s, the Mongols carved out the largest land-based empire the world has ever seen. From a core area in modern-day Mongolia, the Mongol conquests extended as far as southern China in the southeast, Ukraine in the west, and Iraq in the southwest. Although the empire did not long remain unified, it did have profound effects on the subsequent political and economic history of Eurasia.

centuries other Mongol states occasionally enjoyed limited power over the Tibetans. These interactions resulted in the establishment of Mongolian communities in the northeastern portion of the plateau and in the eventual conversion of the Mongolian people to Tibetan Buddhism.

Contemporary Linguistic and Ethnic Geography

Today most of Central Asia is inhabited by peoples speaking Altaic languages, but patterns of linguistic geography remain complex (Figure 10.16). A few indigenous Indo-European languages are confined to the southwest, and Tibetan remains the main language of the Tibetan Plateau. Russian is also widely spoken in the west, and in most former Soviet countries Russian retains an official status. Similarly, Mandarin Chinese is increasingly important in the east. In both Tibet and Xinjiang, Mandarin Chinese is now the basic language of higher education. In Tibet, most urban merchants are now Chinese-speaking immigrants, causing much concern among the indigenous people.

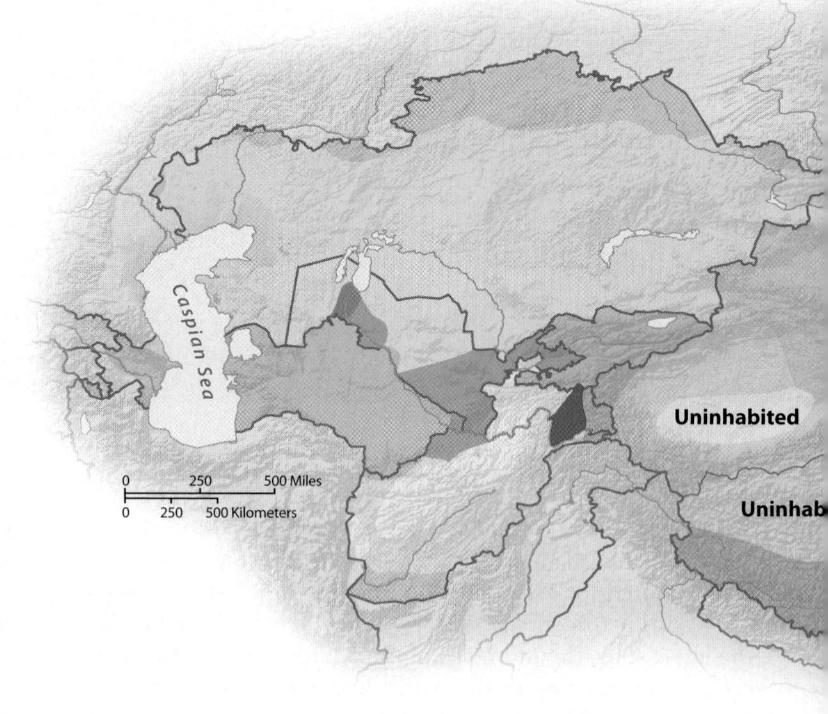

Tibetan Tibetan is divided into several distinct dialects that are spoken over almost the entire inhabited portion of the Tibetan Plateau. Approximately 6 million people speak Tibetan; of these, roughly 2.5 million live in Tibet proper, while most of the rest reside in China's provinces of Qinghai and Sichuan. Tibetan is usually placed in the Sino-Tibetan family, implying a shared linguistic ancestry between the Chinese and the Tibetan peoples, but some scholars argue that no definite relationship between the two has been established. Tibetan has an extensive literature written in its own script, most of which is devoted to religious topics.

Mongolian Mongolian is comprised of a cluster of closely related dialects spoken by approximately 5 million people. The standard Mongolian of both the independent country of Mongolia and China's Inner Mongolia is called Khalkha; other Mongolian dialects include Buryat (found in southern Siberia) and Kalmyk (found to the northwest of the Caspian Sea). Mongolian has its own distinctive script, which dates back some 800 years, but Mongolia itself adopted the Cyrillic alphabet of Russia in 1941. In China's region of Inner Mongolia, however, the old script is still widely used.

▼ **Figure 10.16 Languages of Central Asia** Most of Central Asia is dominated by languages in the Altaic family, which includes both the Turkic languages (found through most of the center and the west of the region) and Mongolian (found in Central Asia's northeast). Several Indo-European languages, however, are located in both the far northwestern and the south-central regions, while Tibetan, a Sino-Tibetan language, covers most of the Tibetan Plateau in southeastern Central Asia.

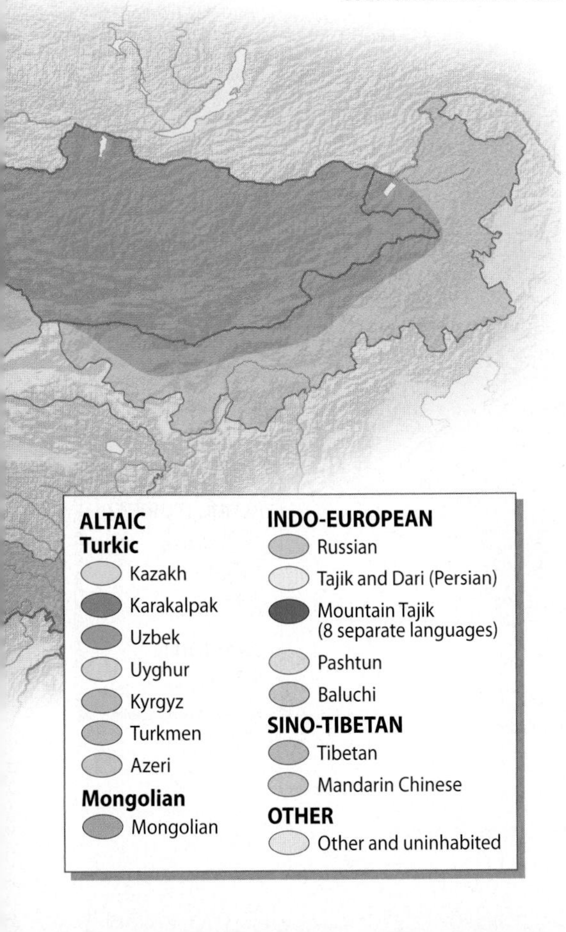

Although Mongolian speakers form about 90 percent of the population of Mongolia, in China's Inner Mongolian Autonomous Province they have been partly submerged by a wave of Han Chinese migrants over the past 50 years. Today only about 4 million out of the 25 million residents of Inner Mongolia identify themselves as Mongols. Even so, more Mongols live in China than in Mongolia, as Mongolia's total population is only around 3 million.

Turkic Languages Far more Central Asians speak Turkic languages than Mongolian and Tibetan combined. Central Asia's Turkic linguistic sphere extends from Azerbaijan in the west through China's Xinjiang Province in the east. The various Turkic languages are not as closely related to each other as are the dialects of Mongolian, but they are still obviously kindred tongues. Six main Turkic languages are found in Central Asia; five are associated with newly independent (former Soviet) republics of the west, and the sixth, Uyghur, is the main indigenous language of northwestern China.

Uyghur is an old and important language. The Uyghur number about 8 million, almost all of whom live in Xinjiang. As recently as 1949, the Uyghur formed about 90 percent of the population of Xinjiang; now, because of Han Chinese immigration, that figure has dropped to 44 percent. The eastern and northern areas of Xinjiang are now largely Chinese-speaking. There are also about 1.8 million Kazakh speakers in far northern Xinjiang.

Of the six countries of the former Soviet Central Asia, five—Kazakhstan, Uzbekistan, Turkmenistan, Kyrgyzstan, and Azerbaijan—are named after the Turkic languages of their dominant populations. In three of these countries, the indigenous people form a substantial majority. Over 90 percent of the people of Azerbaijan are classified as Azeri (there are more Azeris in northern Iran, however, than there are in Azerbaijan); some 80 percent of the people of Uzbekistan are classified as Uzbek; and between 80 and 90 percent of the people in Turkmenistan are classified as Turkmen (figures here vary widely). Most of these people speak their indigenous languages at home, although many Kazakhs are more comfortable speaking Russian than Kazakh.

With more than 26 million speakers, Uzbek is the most widely spoken Central Asian language. In the Amu Darya Delta in the far north of Uzbekistan, however, many people speak a Turkic language more closely related to Kazakh called Karakalpak. Kazakh speakers are widely scattered across Uzbekistan's sparsely populated deserts. In Uzbekistan's older cities, such as Samarkand, part of the residents still speak Tajik (or Persian). Tajik activists, moreover, claim that the number of Tajik speakers in Uzbekistan is much greater than what the official statistics indicate, which peg the figure at about 5 percent, and that many local Tajiks have been pressured into assuming an Uzbek identity.

In the two other Turkic republics, the nationality for which the country is named is not so dominant. At the time of independence in 1991, the Kyrgyz formed only a bare minority in Kyrgyzstan. Due to Russian emigration and differential birth rates, however, over 70 percent of the country's inhabitants now identify themselves as Kyrgyz. The situation is similar in Kazakhstan, where Kazakhs and Russians each constituted about 40 percent of the national population in 1991. Today the figures are 63 percent and 23 percent, respectively. The other residents of these countries identify themselves as Uzbeks, Ukrainians, and members of a wide assortment of other groups. In general, the Kazakhs and other Turkic peoples live in the center and south of Kazakhstan, while the people of European descent live in the agricultural districts of the north and in the cities of the southeast. Roughly a quarter million Central Asians are of ethnic Korean background, a legacy of the Stalinist period in the Soviet Union, when Koreans were deported from the Russian Far East to Uzbekistan, Kazakhstan, and Kyrgyzstan.

Linguistic Complexity in Tajikistan The sixth republic of the former Soviet Central Asia, Tajikistan, is dominated by people who speak an Indo-European rather than a Turkic language. Tajik is so closely related to Persian that it is considered to be a Persian (or Farsi) dialect. Roughly 6.3 million people in Tajikistan, about 84 percent of the total population, identify themselves as Tajiks. The remote mountains of eastern Tajikistan are populated by peoples speaking a variety of distinctive Indo-European languages, sometimes collectively referred to as "Mountain Tajik."

Tajikistan, like much of the rest of the former Soviet portion of Central Asia, is noted for its complex mixture of language and ethnic groups. About 14 percent of its people, for example, are Uzbeks. Today ethnic suspicion pervades much of the region, in part because the governments of Tajikistan and Uzbekistan have poor relations. Before the Soviet period, however, mode of life was as important as language in determining identity; settled farmers and city dwellers were usually classified as "Sarts," regardless of whether they spoke Tajik or a local Turkic language. Soviet planners essentially created the modern nationalities of Tajik and Uzbek, and they tried to force people into one category or the other. Because the two groups were generally mixed together, the Soviet Union also had to draw extremely convoluted boundaries around the republics of Uzbekistan and Tajikistan. The resulting political geography causes problems today for both countries.

Language and Ethnicity in Afghanistan The linguistic geography of Afghanistan is even more complex than that of Tajikistan (Figure 10.17). Afghanistan was never colonized by outside powers, and it is one of the few countries of the world to have inherited the boundaries of a premodern, indigenous kingdom. This kingdom emerged in the 1700s on traditional dynastic lines that did not reflect ethnic or linguistic divisions (dynastic linkages are those based on the family of the monarch). The modern nation-state ideal—that each country should be identified with a particular national group—has been slow to develop in Afghanistan.

The 18th-century creators of Afghanistan were mostly members of the Pashtun ethnic group, but they did not attempt to build a nation-state around Pashtun identity.

▶ **Figure 10.17 Afghanistan's Ethnolinguistic Patchwork** Afghanistan is one of the world's more ethnically complex countries. Its largest ethnic group is that of the Pashtuns, a people who inhabit most of the southern portion of the country, as well as the adjoining borderlands of Pakistan. Northern Afghanistan is inhabited mostly by Uzbeks, Tajiks, and Turkmens, whose main population centers are located in Uzbekistan, Tajikistan, and Turkmenistan, respectively. The Hazaras of Afghanistan's central mountains, like the Tajiks, speak a form of Persian, but are considered to be a separate ethnic group, in part because they, unlike other Afghans, follow Shiite rather than Sunni Islam. Smaller groups are found elsewhere in the country.

Indeed, more Pashtuns live in Pakistan than in Afghanistan. In Afghanistan, estimates of the proportion of the populace speaking the Pashto language of the Pashtuns vary from 38 to 50 percent. Most of them live to the south of the Hindu Kush Mountains.

Approximately half of the people of Afghanistan speak Dari, Afghanistan's variant of Persian, as their first language. Dari speakers are concentrated in the north. Three separate ethnicities are ascribed to the Dari-speaking people: the settled farmers and townspeople in the west and north are Tajiks; the traditionally seminomadic people of the west-center are Aimaks; and the East Asian–appearing villagers of the central mountains are Hazaras, reputed to be descendants of medieval Mongol conquerors. Although traditionally forming something of an underclass, Afghanistan's Hazaras are now showing more dedication to modern education than the country's other ethnic groups and as a result are beginning to prosper. The Tajiks are the most important non-Pashtun group, comprising between a quarter and a third of the total Afghan population.

Although Pashto and Dari are Afghanistan's official languages, a variety of other languages are also spoken in the country, creating an especially complex ethnic patchwork. Approximately 10 percent of the people of Afghanistan speak Turkic languages, including both Uzbek and Turkmen. Other ethnolinguistic groups in Afghanistan include the Baluchis, the Nuristanis (who speak five separate languages), and the Pashai, all of whose languages are Indo-European, and the Brahui, who speak a Dravidian language related to the tongues of southern India. Such complex ethnic divisions give Afghanistan a weak national foundation, presenting serious challenges for the country's government.

Geography of Religion

An assortment of religions characterized ancient and medieval Central Asia. The major overland trading routes of premodern Eurasia crossed the region, giving easy access to both merchants and missionaries. Several varieties of Buddhism, Islam, Christianity, and Judaism, as well as several minor religions, have all thrived at various times and places within the region. Eventually, however, religious lines hardened, and Central Asia was divided into two opposed spiritual camps: Islam triumphed in the west and center (Figure 10.18), while Tibetan Buddhism prevailed in Tibet and Mongolia.

Islam in Central Asia As is true elsewhere in the Muslim world, different Central Asian peoples are known for their different interpretations of Islamic orthodoxy. The Pashtuns of Afghanistan are noted for their strict Islamic ideals—although critics contend that Pashtun religious practices, such as forbidding women's faces to be seen in public, are based mostly on their own customs. By contrast, the traditionally nomadic groups of the northern steppes, such as the Kazakhs, have historically been considered lax in their religious practices. Although most of the region's Muslims are Sunnis, Shiism is dominant among the Hazaras of central Afghanistan, the Azeris of Azerbaijan, and the mountain dwellers of eastern Tajikistan.

Under the communist rule of China, the Soviet Union, and Mongolia, all forms of religion were discouraged. Chinese authorities and student radicals suppressed Islam in Xinjiang during the Cultural Revolution of the late 1960s and early 1970s. Mosques were destroyed or converted to museums, and religious schools were closed. Periodic persecution of Islam also occurred in Soviet Central Asia, and until the 1970s many observers thought that Islam was slowly diminishing throughout the region.

Religious expression was not so easily discouraged, however, and interest in Islam began to grow again in the former Soviet Central Asia in the 1980s. In the post-Soviet period, Islam continues to revive as people reassert their indigenous heritage and identity. Thus far, signs of Islamic political fundamentalism are few in most areas. Still, most Central Asian leaders remain wary of any signs of radical Islamic practices. In 2009, for example, Tajikistan prohibited teachers under the age of 50 from growing beards, commonly regarded as

▼ **Figure 10.18 Entrance to the Poi-Kalyan Mosque, Bukhara, Uzbekistan** With the exceptions of Tibet and Mongolia, religion in most of Central Asia is dominated by Islam. The mosques of Central Asia's older cities, such as Bukhara in Uzbekistan, are noted for their striking architecture.

a symbol of fundamentalism, and it heightened its campaign to shut down unregistered mosques. In response to widespread religious criticism, Tajikistan's government also banned revealing Western clothing styles—as well as graduation parties and the use of mobile phones in the country's secondary schools.

In China, Muslims now enjoy basic freedom of worship, but the state still closely monitors religious expression out of fear that it might become entangled with political separatism. Although some evidence suggests that Islam has served as a focal point of an anti-Chinese movement among the Uyghur people, most Uyghur leaders insist that their program is not radically fundamentalist. Still, ethnic relations in northwestern China remain very tense. In 2009, ethnic rioting pitting Uyghurs against Han Chinese took as many as 200 lives in Ürümqi, the capital city of Xinjiang Province. Although the region has quieted down, Uyghur activists complain of harassment by Chinese officials. According to a recent report in *Eurasia Review*, one Uyghur farmer had his house raided 15 times in 2012.

In Afghanistan, Islamism (or politically radical Islamic fundamentalism) did emerge as a powerful political movement. From the mid-1990s until 2001, most of the country was controlled by the **Taliban,** an extremist organization that insisted that all aspects of society conform to its own harsh version of Islamic orthodoxy. This commitment was demonstrated in early 2001 when Afghanistan's religious authorities oversaw the destruction of Buddhist statues in the country—including some of the world's largest and most magnificent works of art—as symbols of a non-Islamic past. Although the Taliban regime fell in late 2001, Taliban insurgents continue to fight, hoping to regain power. Islamism remains a powerful force in Afghanistan, particularly among the Pashtun people of the south.

Islam is not the only religion represented in western Central Asia. Many Russians belong to the Russian Orthodox Church, and Uzbekistan has a small Jewish population. Kazakhstan also counts some 100,000 Catholics, mostly people of Polish or German descent.

Tibetan Buddhism Mongolia and Tibet stand apart from the rest of Central Asia in their people's practice of Tibetan Buddhism. Buddhism entered Tibet from India many centuries ago, where it merged with the indigenous religion of the area, called Bon. The resulting faith is more oriented toward mysticism than are other forms of Buddhism, and it is more hierarchically organized. Standing at the apex of Tibetan Buddhism is the **Dalai Lama** (a term derived from the words for "Great Teacher"), considered to be a reincarnation of the **Bodhisattva** of Compassion (a Bodhisattva is a spiritual being who helps others attain enlightenment). Ranking below him is the Panchen Lama, followed by other religious officials. Until the Chinese conquest, Tibet was essentially a **theocracy,** or religious state, with the Dalai Lama enjoying political as well as religious authority. A substantial proportion of Tibet's male population has traditionally become monks, and monasteries once wielded both economic and political power (Figure 10.19).

Tibetan Buddhists suffered severe persecution in the 1960s. The Dalai Lama fled to India with many of his followers after China unleashed a military crackdown on Tibet in

▼ **Figure 10.19 Tibetan Buddhist Monastery** Tibet is well known for its large Buddhist monasteries, which at one time served as seats of political as well as religious authority. The Ganden Monastery, pictured here, is one of the three great "university monasteries" of Tibet.

1959 and has since been a powerful advocate for the Tibetan cause in international circles. During the 1960s and 1970s, an estimated 6000 Tibetan Buddhist monasteries were destroyed, and thousands of monks were killed. Although China later allowed many monasteries to reopen, the number of active monks today is only about 5 percent of what it was before the Chinese occupation. Still, the Buddhist faith continues to form the basis of Tibetan identity and, in so doing, helps keep alive the dream of independence, or at least of real autonomy, within Tibet. In 2008, Tibetan monks led a series of protests against the Chinese government, and as a result, ethnic riots between Tibetans and Han Chinese broke out across the Tibetan Plateau. The Chinese government reacted harshly and later blamed the Dalai Lama for inciting the unrest. More recently, Tibetan monks and nuns have been burning themselves to death to protest Chinese policies. As of March 2013, over 100 Tibetans have engaged in such acts of self-immolation.

In Mongolia, the downfall of communism has allowed the Tibetan Buddhist faith to experience a renaissance. Several monasteries have been refurbished, and many people are returning to their national religion. The intensity of Buddhist belief, however, is not as strong in Mongolia as it is in Tibet. It has been estimated that about half of the people of Mongolia actively follow the Buddhist religion today, whereas about 40 percent identify themselves as nonreligious.

Central Asian Culture in International Context

Although Central Asia is remote and poorly integrated into global cultural circuits, it is hardly immune to the forces of globalization. Certainly, the elaborate hyper-modern buildings appearing in Astana (in Kazakhstan) and Baku (in Azerbaijan) show strong global influences (Figure 10.20). Even the tensions existing throughout the region between religious and secular orientations and between ethnic nationalism and multiethnic inclusion are aspects of the current global condition. So, too, the increased usage of English throughout Central Asia shows that this part of the world is not cut off from global culture. Such influences are especially marked in the oil cities of the Caspian Basin, such as Azerbaijan's Baku. In the former Soviet sphere, however, Russian is much more important than English.

The Russian Language Issue During the Soviet period, the Russian language spread widely through western Central Asia. Russian served both as a common language and as a means of instruction in higher education. People had to be fluent in Russian in order to reach any position of responsibility. The Cyrillic (or Russian) script, moreover, replaced the modified Arabic script that was previously used for the indigenous languages. After the fall of the Soviet Union, however, most Russians migrated back to Russia, especially from the region's poorer countries. As a result, the use of Russian in education, government, business, and the media has declined in favor of local languages, and the Cyrillic alphabet has been gradually replaced by the Latin one.

Despite its recent decline, Russian remains the common language of the former Soviet Central Asia. Even some of the leaders of these countries speak their own languages imperfectly, and Russian is often the only way to communicate across Central Asia's national boundaries. The continuing prominence of Russian, however, irritates many people. Kazakh nationalists, for example, are upset that Russian remains the country's "official language of communication between different ethnic groups." In early 2010, hundreds of Kazakh students protested their country's language laws, demanding that government meetings, sessions of parliament, and presidential speeches be conducted only in Kazakh. In 2007, Tajikistan decided to ban the use of Russian in advertising, business communication, and governmental documents, but in 2011 it dropped the policy, making Russian a "permissible" language for interethnic communication and government documents. Up to a third of

▼ **Figure 10.20 Baku, Azerbaijan** Baku is a vibrant and increasingly cosmopolitan city closely linked to global economic and cultural networks. Baku is now well known for its new buildings and modern architecture.

Tajikistan's workers reside on a temporary basis in Russia, so Russian language skills are recognized by many to be more important than ever.

Globalization and Sports in Central Asia In many ways, Central Asia has been slower to embrace the globalization of sports than other regions. Mongolia's sporting culture is perhaps the most distinctive, where traditional forms of wrestling and horseback riding remain the marquee sporting events. Every summer, the three-day Naadam festival brings thousands of competitors to Ulaanbaatar to engage in wrestling, horse racing, and archery (Figure 10.21). Alhough the Ulaanbaatar Naadam dominates Mongolia's sporting calendar, smaller festivals featuring the same events take place frequently in many places.

Mongolian horse racing is deeply connected to the country's nomadic past. In previous centuries, Mongolian horses were prized in warfare for their small stature and superior endurance, allowing the Mongolians to outmaneuver their enemies and build a continent-spanning empire. Today endurance is still the most prized trait among Mongolian horses, and most races are cross-country events that span over 10 miles, quite in contrast to the short-distance horse racing that dominates in most of the rest of the world. In providing a living link to the past, equestrian sports—together with archery and wrestling—help provide a solid foundation for Mongolian culture and unity.

Although soccer is the most popular sport in most Central Asian countries, its supremacy is far from complete. In Afghanistan, cricket is a key point of national pride. The country's national team, although certainly not a juggernaut, has shown steady improvement and has become one of the world's better squads. Each province has a team that plays domestically, and several new stadiums are under construction or have been built recently. Kazakhstan and Uzbekistan are home to many successful individual-sport athletes, producing world-class weightlifters, wrestlers, and cyclists.

▼ **Figure 10.21 Mongolian Archery Contest.** Traditional Mongolian sports are making a comeback as the country seeks to maintain its culture. Women as well as men now participate in such athletic events as archery. **Q: What historical factors might help explain the popularity of horse racing, archery, and wrestling in Mongolia?**

REVIEW

10.7 How do patterns of religious affiliation divide Central Asia into distinct regions, and why is religion in the region the source of increasing tensions?

10.8 How have the patterns of linguistic geography in Central Asia been transformed over the past two decades?

10.9 What was the historical significance of the Silk Road for Central Asia?

Geopolitical Framework: Political Reawakening in a Power Void

Central Asia has played a minor role in global political affairs for the past 300 years. Before 1991, the entire region, except Mongolia and Afghanistan, lay under direct Soviet or Chinese control. Mongolia, moreover, was a Soviet satellite, and even Afghanistan came under Soviet domination in the late 1970s. The breakup of the Soviet Union, however, saw the emergence of six new countries, helping to reestablish Central Asia as a world region (Figure 10.22).

Partitioning of the Steppes

Before the 1700s, Central Asia had been a power center whose mobile armies threatened the far more populous, sedentary peoples of Asia and Europe. However, military developments, particularly the development of advanced firearms, changed the balance of power, allowing the wealthier agricultural states to defeat the nomadic pastoralists and take over their lands by the late 1700s. The winners in this struggle were the two largest states bordering the steppes: Russia and China.

The Manchu conquest of China in 1644 undercut the autonomy of the peoples of the eastern steppe. The Manchus came from Manchuria, the eastern borderlands of Central Asia, and were themselves skilled in the arts of cavalry warfare. By the late 1700s, China stood at its greatest territorial extent. Within its grasp lay not only Mongolia and Xinjiang, but also Tibet and a slice of modern Kazakhstan. Although Manchu-ruled China declined rapidly after 1800, it was still able to retain most of its Central Asian territories. But by the early 1900s, Chinese authority had begun to diminish in Central Asia as well. When the Manchu (also called Ch'ing or Qing) Dynasty fell in 1912, Mongolia became independent, although China did keep the extensive borderlands of Inner Mongolia (Nei Mongol). Tibet had earlier gained de facto independence,

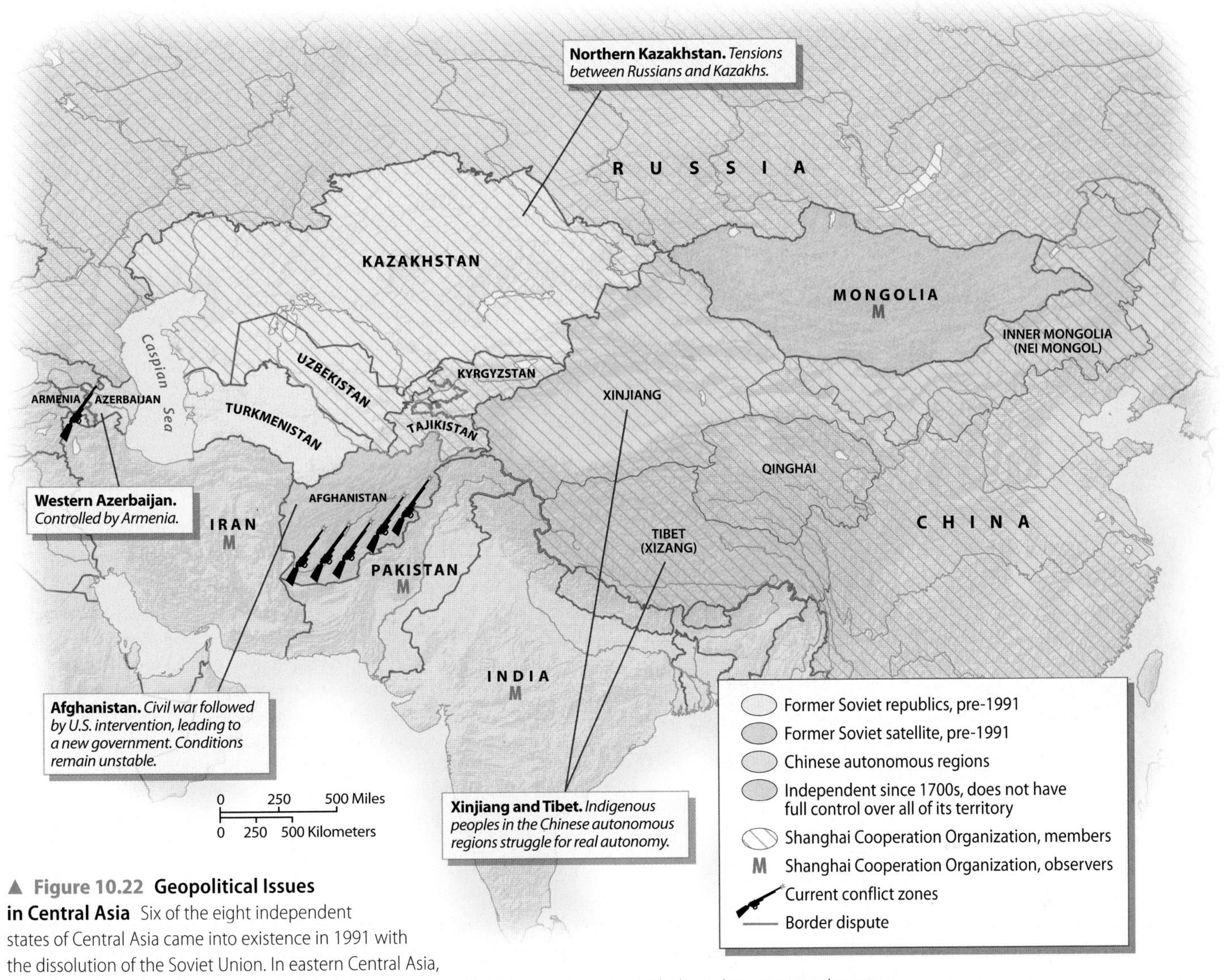

▲ **Figure 10.22 Geopolitical Issues in Central Asia** Six of the eight independent states of Central Asia came into existence in 1991 with the dissolution of the Soviet Union. In eastern Central Asia, the most serious difficulties stem from China's maintenance of control over areas in which the indigenous peoples are not Chinese. Afghanistan, scene of a prolonged and brutal civil war, has experienced the most extreme forms of geopolitical tension in the region.

and even Xinjiang lay beyond the reach of effective Chinese authority during the 1920s.

Russia began to advance into Central Asia at roughly the same time as China. In the 1700s, the Russian Empire undertook the systematic conquest of the Kazakh steppe. Expansion farther to the south, however, was blocked by the agricultural states of Uzbekistan. Only in the late 1800s, when European military techniques and materials raced ahead of those of Asia, was Russia able to conquer the Amu Darya and Syr Darya valleys. Its subjugation of this area was not completed until the early 1900s, just before the Soviet Union replaced the Russian Empire. The Russians advanced into Central Asia in part because of their concern over possible British influence in the area. Britain attempted to conquer Afghanistan, but was rebuffed by Afghan forces—and by the country's forbidding terrain. Subsequently, Afghanistan's position as an independent "buffer state" between the Russian Empire (later the Soviet Union) and the British Empire in South Asia remained secure.

Central Asia Under Communist Rule

Western Central Asia came under communist rule soon after the founding of the Soviet Union in 1917, with Mongolia following in 1924. After the Chinese revolution of 1949, the communist system was imposed on Xinjiang, Tibet, and Inner Mongolia. In all of these areas, major changes in the geopolitical order soon followed.

Soviet Central Asia Although the Soviet Union inherited the Russian imperial domain in Central Asia virtually intact, it enacted new policies. The Soviet regime sought to create a socialist economy and to build a new society that would eventually knit together all of the massive territories of the country. Central Asia's leaders were

Figure 10.23 Political Boundaries Around the Fergana Valley Some of the world's most convoluted political boundaries can be found in the vicinity of the Fergana Valley. The central portion of the valley belongs to Uzbekistan, which is otherwise separated from it by high mountains. The lower valley, on the other hand, is part of Tajikistan, the core area of which is likewise separated from the valley by highlands. The Fergana's upper periphery belongs to Kyrgyzstan. Note also the small exclaves of Uzbekistan within Kyrgyzstan.

replaced by Communist Party officials loyal to the new state; Russian immigration was encouraged; and local languages had to be written in the Cyrillic (Russian) rather than the Arabic script.

Although Soviet authorities foresaw the emergence of a single Soviet nationality, they realized that local ethnic diversity would not disappear overnight. Early Soviet leaders such as Vladimir Lenin also claimed that they wanted to insulate non-Russian peoples from Russian domination. The new government therefore divided the Soviet Union into a series of nationally defined "union republics" in which a certain degree of cultural autonomy would be allowed. It was uncertain, however, what the relevant units in Central Asia should be. Was there a single Turkic-speaking nationality, or were the Turkic peoples divided into several separate nationalities, with the Tajiks forming yet another? For several years, boundaries shifted as new "republics" appeared on the map. Finally, in the 1920s the modern republics of Kazakhstan, Kyrgyzstan, Tajikistan, Uzbekistan, Turkmenistan, and Azerbaijan assumed their present configurations.

In certain areas of the former Soviet portion of Central Asia, such as the fertile Fergana Valley, the political boundaries drawn between the republics were extremely complex (Figure 10.23). These intricate borders continue to cause problems in the region. In early 2013, for example, Kyrgyzstan temporarily cut off access to several **exclaves** (an exclave is the territory of one nation surrounded by the territory of another nation) of Uzbekistan surrounded by Kyrgyz territory, infuriating the Uzbek government. At roughly the same time, Uzbekistan announced plans to build a railway line into its portion of the valley, bypassing Tajikistan's territory. This move angered Tajikistan, as it has profited from transit fees on the old transnational railway.

Some scholars argue that the Soviet nationalities policy backfired severely. Rather than forming a transitional step on the way to a Soviet identity, the constituent republics of the Soviet Union nurtured local nationalisms that ultimately undermined the Soviet system. Identities such as Turkmen, Uzbek, and Tajik that had been vague in the pre-Soviet period were now given real political significance. Another problem undercutting Soviet unity was the fact that the cultural and economic gaps separating Central Asians from Russians did not diminish as much as planned.

The Chinese Geopolitical Order After China reemerged as a united country in 1949, it, too, was able to reclaim most of its old Central Asian territories. China's communist leaders promised the non-Chinese peoples a significant degree of political self-determination and cultural autonomy—and thus found much local support in Xinjiang. Tibet, isolated behind its mountain walls and virtually independent for many years, presented a greater obstacle. China occupied Tibet in 1950, but the Tibetans launched a rebellion in 1959. When the rebellion was crushed, the Dalai Lama and some 100,000 followers found refuge in India, where they maintain the "Tibetan Government in Exile." The current Dalai Lama, however, retired as the "head of state" of this organization in 2011 and hinted that the office of the Dalai Lama could be occupied by a woman in the future or might even be abolished altogether. The Chinese government objected to these statements, insisting that it has the right to appoint the religious leaders of Tibetan Buddhism.

Loosely following the Soviet nationalities model, China established autonomous regions in areas occupied primarily by non-Han Chinese peoples, including Xinjiang, Tibet proper (called Xizang in Chinese), and Inner Mongolia. Such autonomy, however, often turned out to be more symbolic than real, and it did not prevent the massive immigration of Han Chinese into these areas. Nor were all parts of Chinese Central Asia granted autonomous status. The large and historically Tibetan and Mongolian province of Qinghai, for example, remained an ordinary Chinese province.

Current Geopolitical Tensions

Although the former Soviet portion of Central Asia weathered the post-1991 transition to independence relatively smoothly, the region presently suffers from several ethnic conflicts, as well as from the struggle between radical Islam and secular governments. Much of China's Central Asian territory is troubled, but China retains a firm grip. Such troubles are insignificant, however, compared to the situation in Afghanistan, where brutal warfare has been going on for decades.

Independence in Former Soviet Lands After 1991, the six newly independent countries of the former Soviet Union did not find it easy to chart their own political courses. They still had to cooperate with Russia on security issues, and all initially opted to join the Commonwealth of Independent States, the rather hollow successor of the Soviet Union. In most cases, authoritarian rulers, rooted in the old order, retained power and sought to undermine opposition groups. All told, democracy made less progress in Central Asia than in other parts of the former Soviet Union. Kyrgyzstan forms a partial exception, as it generally runs open elections. Such a policy, however, has not generated stability. In April 2010, demonstrators brought down the government, forcing Kyrgyzstan's president to flee. Three months later, massive ethnic riots broke out in southern Kyrgyzstan as supporters of the country's ousted president attacked ethnic Uzbeks. As many as 2000 people were killed in the rioting, and up to 300,000 were displaced (Figure 10.24).

In Tajikistan, war broke out almost immediately after independence in 1991. Many members of the smaller ethnic groups of the mountainous east resented the authority of the Tajiks and rebelled. They were joined by Islamist groups seeking to overthrow the secular state. The civil war was officially ended in 1997 when the Islamists agreed to work through normal political channels. However, continuing instability led Tajikistan's government to increase its pressure on the Islamic Renaissance Party, the only legal Islamist party in the former Soviet Central Asia. The Islamic Renaissance Party responded by modernizing its image to appeal to educated people, youth, and women, but in the 2010 elections it received only 8 percent of the vote.

Azerbaijan also experienced strife following the breakup of the Soviet Union. Armenia invaded and occupied a portion of far western Azerbaijan, allowing the Armenian-speaking highlands to form the "breakaway republic" of Nagorno-Karabakh. Hostile relations with Armenia make it difficult for Azerbaijan to govern its exclave of Nakhchivan, a piece of Azerbaijani territory separated from the rest of the country by Armenia and Iran. Although the Nagorno-Karabakh issue is often considered inactive, forming a so-called frozen conflict, relations remain tense. In early 2013, for example, Azerbaijan accused Armenian-backed troops of killing one

▼ **Figure 10.24 Ethnic Tension in Kyrgyzstan** Conflicts between Kyrgyz and Uzbek people in the Fergana Valley area have resulted in much tension in recent year. Shown here are young men marching under the flag of Kyrgyzstan toward a private university in an Uzbek neighborhood in the city of Jalal-Abad in Kyrgyzstan.

of its soldiers near the de facto border, leading to warnings about the possibility of renewed fighting.

Although Uzbekistan and Turkmenistan had relatively easy transitions out of the Soviet Union, both countries are ruled by repressive governments that allow little room for personal freedom or opposition movements. In 2012, Uzbekistan increased its censorship of the Internet and continued to crack down on human rights. Until his death in late 2006, Turkmenistan's authoritarian president, Saparmurat Niyazov—who called himself simply "Turkmenbashi" ("father of the Turkmens")—maintained a lavish personality cult, forcing all students in his country to read and honor his lengthy book on Turkmen history, philosophy, and mythology. Since Niyazov's death, Turkmenistan has reduced some of its restrictions on daily life, but it remains one of the world's most repressive countries. Meanwhile, its current president, Gurbanguly Berdymukhammedov, is busy constructing his own personality cult. After he played a song on national television in 2011, for example, the guitar that he had strummed was immediately sent to a museum to be cherished as a "great treasure."

Strife in Western China Local opposition to Chinese rule is pronounced in many of the region's indigenous communities. Protests by Tibetans have not been successful, but they have brought the attention of the world to their struggle. China maintains several hundred thousand troops in a region that has only some 3 million civilian inhabitants. Such an overwhelming military presence is considered necessary because of both Tibetan resistance and the strategic importance of the region. Massive Tibetan protests in 2008 were handled harshly by the Chinese government. Tibetan activists, as noted earlier, have taken to burning themselves to protest Chinese policies. China has again reacted with force. In February 2013, for example, it arrested 70 Tibetan leaders accused of being active in the protest movement.

China's control of Xinjiang is more secure than its hold on Tibet. Although fragmented by deserts and mountains, Xinjiang's terrain is less forbidding than that of Tibet, and it is now the home of millions of Han Chinese immigrants. Xinjiang is also an economically vital part of China. It contains a variety of mineral deposits (including oil) essential for Chinese industry, and it has been the site of nuclear weapons tests. Many Uyghurs, not surprisingly, oppose such uses of their homeland, and they resent the periodic suppression of their religion by the communist regime. Ethnic strife between Han Chinese and Uyghurs resulted in massive unrest in 2009, leading to a number of deaths and large-scale destruction of property (Figure 10.25). In 2011, 18 people were killed in what government officials described as a terrorist attack, but the World Uyghur Congress described as a police attack on unarmed protestors.

China's position is that all of its Central Asian lands are integral portions of its national territory. Those who advocate independence are viewed as traitors and are sometimes considered to be in league with Western political forces that have sought to keep China weak and divided for the past 200 years.

War in Afghanistan None of the conflicts elsewhere in Central Asia compares in intensity to the situation in Afghanistan. Afghanistan's troubles began in 1978 when a Soviet-supported military "revolutionary council" seized power. The new Marxist-oriented government suppressed religion, which resulted in rebellion. When the government was about to collapse, the Soviet Union launched a massive invasion. Despite its power, the Soviet military was never able to gain control of the more rugged parts of the country. Moreover, Pakistan, Saudi Arabia, and the United States ensured that the anti-Soviet forces remained well armed and financed.

The exhausted Soviets finally withdrew their troops in 1989. The puppet government that they installed remained

▼ **Figure 10.25 Ethnic Tension in Xinjiang** In 2009, ethnic rioting pitting the indigenous Uyghurs against Han Chinese immigrants broke out in several cities in Xinjiang. Chinese security forces reacted harshly to Uyghur protesters.

▲ **Figure 10.26 War in Afghanistan** Warfare continues in Afghanistan as troops from the United States and other countries attempt to root out remnants of the Taliban and maintain order. Tense encounters with local residents often result when foreign soldiers search for weapons and insurgent fighters.

to face the insurgents alone. It managed to hold the country's core around the city of Kabul for a few years, largely because the opposition forces were themselves divided. Local warlords grabbed power in the countryside, destroying any semblance of central authority. Warlords and their forces committed many atrocities, including the shelling of civilian areas, the abduction of noncombatants, and the mass raping of women from enemy communities.

In 1995–96, the Taliban arrived on the Afghan scene. The Taliban was founded by young Muslim religious students disgusted with the anarchy that was consuming their country. They were convinced that only the firm imposition of Islamic law could end corruption and quell the disputes among the country's different ethnic groups. With large numbers of soldiers flocking to their standard and with military support from Pakistan, the Taliban won control of most of Afghanistan by the late 1990s. As its own power grew more secure, the Taliban offered refuge in Afghanistan to the radical Islamist organization Al Qaeda, led by Osama bin Laden.

The Taliban acquired power not only from its religious nature, but also from the ethnic divisions of Afghanistan. It gained most of its strength from the Pashtuns of the southern half of the country. The Pashtuns have a reputation for militarism, as well as good connections for military supplies among their relatives across the Pakistan border. The main opposition to the Taliban came from the country's other ethnic groups, especially the Uzbeks and Tajiks of the north and the Shiite Hazaras of the central mountains.

By the end of the century, most of the people of Afghanistan lost faith in the Taliban, largely because of the severe restrictions imposed on daily life. Such constraints were most pronounced for women, but even men were compelled to obey the Taliban's numerous decrees. Most forms of recreation were outlawed, including television, films, music, and even kite flying. Little escaped the watchful eyes of the Taliban's religious police.

The aftermath of September 11, 2001, completely changed the balance of power in the region. The United States and Britain, working closely with anti-Taliban groups concentrated in northern Afghanistan, launched a war against Al Qaeda and the Taliban government. Within a few months, formal Taliban power collapsed, allowing the emphasis to shift to rebuilding Afghanistan's institutions and establishing an effective central government. Such projects were not very successful, however, as governmental corruption and ethnic animosity have prevented Afghanistan from becoming a stable country. By 2004, the Taliban had regrouped, operating from safe havens in Pakistani tribal areas (see Chapter 12). Pakistan's government worries that the current Afghan administration is too friendly with India, and it is irritated by the fact that Kabul does not accept the legitimacy of the boundary separating the two countries. As a result, some experts believe, Pakistan would like the Taliban to regain power—or at least keep Afghanistan in a weak position.

Afghanistan's new government has had to rely on the military power of the International Security Assistance Force (ISAF), led by the North Atlantic Treaty Organization (NATO). But despite its more than 58,000 troops, the ISAF has been unable to stop the Taliban resurgence (Figure 10.26). By 2008, many observers feared that Afghanistan was on the verge of becoming a failed state, and in 2009, the new administration of the United States responded by sending an additional 17,000 troops to the country. Early 2010 saw major offensives against Taliban strongholds in the Kandahar area. The U.S. forces have also targeted high-level Taliban leaders, often by bombing their compounds. This strategy has weakened the Taliban command structure, but has also resulted in large numbers of civilian casualties, reducing local support for the war effort.

In 2012, the United States and its coalition partners endorsed an exit strategy from the Afghan war, agreeing to begin withdrawing troops and to turn over command of military operations by mid-2013. Continued Taliban strikes soon put this strategy in jeopardy, but the United States has still insisted that it will withdraw all of its combat troops by the end of 2014, although more-specialized military personnel could remain in the country for another decade. The continuing presence of American trainers, advisors, and counter-terrorism experts after that date, however, remains highly controversial. In March 2013, Afghan president Hamid Karzai infuriated officials in Washington by charging that the U.S. forces were actually working with the Taliban in order to destabilize Afghanistan, thereby maintaining an excuse to keep foreign troops in the country.

Global Dimensions of Central Asian Geopolitics

As the previous discussion indicates, Central Asia has recently emerged as a key arena of geopolitical tension. Vying for power and influence in the region are several important countries, including China, Russia, Pakistan, India, Iran, Turkey, and the United States. In addition, the revival of Islamic fundamentalism has generated international geopolitical repercussions.

A Continuing U.S. Role Afghanistan is not the only Central Asian country to have experienced a substantial U.S. armed presence. After 9/11, the United States established military operations in Uzbekistan, Tajikistan, and Kyrgyzstan. These countries were eager to host the U.S. military because they feared that Islamic fundamentalist movements could overthrow their governments. Already by the late 1990s, the Islamic Movement in Uzbekistan (IMU), a group dedicated initially to the creation of an Islamic state in the Fergana Valley, was engaging the armies of Uzbekistan, Tajikistan, and Kyrgyzstan. As a result, military budgets increased throughout the region, straining governmental finances, but also allowing Central Asian governments to get an upper hand on the IMU. By 2010, experts had concluded that the IMU was no longer operating in the former Soviet Central Asia.

Since 2005, the influence of the United States has declined across much of the former Soviet portion of Central Asia. In that year, Uzbekistan shut down an American military base in the country after the United States criticized the Uzbek government for human rights abuses. In 2012, moreover, Kyrgyzstan announced that it would close down a U.S. air base in its territory, the so-called Transit Center at Manas, by the end of 2014. The potential for a Taliban resurgence after the planned withdrawal of U.S. troops from Afghanistan, however, has resulted in a certain degree of reassessment. In 2012, for example, Uzbekistan announced that it would consider allowing the reestablishment of a U.S. military base in the country.

The United States' relations with Mongolia, however, remain relatively close, and have strengthened in recent years. Mongolia seeks close ties with the United States, along with South Korea and Japan, due largely to its concerns about falling under the influence of China or Russia.

Relations with China and Russia After the breakup of the Soviet Union, relations between the newly independent countries of Central Asia and China remained tense for several years. China objected to the boundary lines separating it from Tajikistan and Kyrgyzstan and asked for adjustments. These boundary disputes were peacefully settled in the early years of the new millennium, with Kyrgyzstan ceding 222,400 acres (90,000 hectares) of disputed land to China. China subsequently initiated a series of diplomatic maneuvers designed to quell tensions, fight Islamic separatism, and gain greater access to the natural resources of Central Asia.

After the fall of the Soviet Union, Russia continued to regard all of the former Soviet territories of Central Asia as lying within its sphere of influence and, as such, has resented U.S. military initiatives. Russia has, moreover, retained or built military bases of its own in both Tajikistan and Kyrgyzstan. Economics and infrastructure also tie Russia to the former Soviet Central Asian sphere. The main transportation line linking European Russia to central Siberia, for example, cuts across northern Kazakhstan, and the former Soviet Central Asia's rail and pipeline links to the outside world are still largely oriented toward Russia. Kazakhstan also contains the huge Baikonur Cosmodrome, Russia's main site for rocket launches (Figure 10.27).

In the early years of the new millennium, most Central Asian leaders concluded that the potential economic and political advantages to be gained by cooperation with Russia and China outweighed the disadvantages. The major consequence of this attitude was the formation of the **Shanghai**

◄ **Figure 10.27 Baikonur Cosmodrome** The world's largest operational space-launch facility, the Baikonur Cosmodrome, is situated in Kazakhstan, but is leased and operated by Russia. **Q: Why does Russia maintain its most important space facility in Kazakhstan, rather than in its own territory?**

People on the Move

Afghan Refugees in Iran

For much of the latter half of the 20th century, Iran was a key destination for Afghan refugees looking for work and ways to escape domestic strife. Immigration began in earnest after a drought-fueled famine killed some 100,000 Afghans during the early 1970s, and it intensified dramatically with the 1979 Soviet invasion of Afghanistan. The number of Afghans living in Iran climbed to roughly 3 million by 1992, but in that year some 1.4 million Afghans returned to their homeland. However, civil war in Afghanistan followed shortly, and before long thousands again made their way out of the country and into Iran. Today 6.5 million Afghans live outside the country, and about 1.5 million of these live in Iran.

Growing Resentment in Iran Since the 1990s, Afghans have found themselves increasingly unwelcome in Iran. Many Iranians blame Afghan migrants, who are usually willing to work for exceptionally low pay, for unemployment and economic hardship in their country. Usually, immigrants find themselves working in construction or performing various other types of physical labor. Conditions for immigrants have been especially harsh since 2007, when a major crackdown by the Iranian government led to the deportation of roughly half a million refugees (**Figure 10.4.1**). The precarious immigration status of dislocated families often puts them at the mercy of Iranian employers and moneylenders, who sometimes withhold or demand money in exchange for not reporting refugees to immigration officials.

For cultural, economic, and religious reasons, Afghan immigrants tend to have very high birth rates, giving rise to a large number of second-generation Afghan children growing up in Iran. In 2006, 82 percent of Afghans in Iran under age 15 and 42 percent of those between the ages of 15 and 29 were born in Iran. Second-generation Afghans in Iran usually work long hours from a young age to help support their families; most are also barred from attending school, and families that want to see their children educated are forced to pay high fees to private teachers. Iran outlawed marriage between Afghan refugees and Iranian citizens in 2001, which prevents children from obtaining birth certificates and puts further strain on relationships.

Return Home? The Afghan government actively encourages the country's emigrants to return, especially since 2007. Nevertheless, most who make their way back to Afghanistan find it difficult to locate jobs and living space. The most common destination is the capital city of Kabul, where millions now crowd into peripheral suburbs that lack basic services like plumbing and electricity, helping to make Kabul one of the world's fastest-growing cities (see *Cityscapes: Kabul, A Precariously Booming City*).

▼ **Figure 10.4.1 Afghan Refugees in Iran** Hundreds of thousands of Afghan refugees currently reside in Iran. Most live in grim camps like the one shown in this photograph.

Cooperation Organization (SCO), composed of China, Russia, Kazakhstan, Kyrgyzstan, Tajikistan, and Uzbekistan (and commonly referred to as the "Shanghai Six"). The SCO seeks cooperation on such security issues as terrorism and separatism, aims to enhance trade, and serves as a counterbalance against the United States. Although not members, Mongolia and Afghanistan have "observer" status in the SCO, as do Iran, Pakistan, and India. The United States applied for such a status in 2006, but its bid was rejected. Turkmenistan, on the other hand, has refused to join. By 2010, some observers were arguing that economic competition between Russia and China would weaken the SCO, reducing Russian influence in the region. In early 2013, news reports stated that Russia was growing increasingly suspicious of the organization and that China was seeking to strengthen it. Russia is instead emphasizing the Collective Security Treaty Organization, a weak military alliance that includes, in addition to Russia itself, Kazakhstan, Kyrgyzstan, Tajikistan, Armenia, and Belarus.

The Roles of Iran, Pakistan, and Turkey Iran is also interested in the new republics of Central Asia. Iran is a major trading partner and offers a good potential route to the ocean. Since the completion of a rail link between Iran and Turkmenistan in the late 1990s, some of Central Asia's global trade has been reoriented toward Iran's ports. Iran's cultural links with the region are old and deep, particularly in Tajikistan and northern Afghanistan. The Iranian government has gained some influence in Afghanistan, especially in the western region of Herat and among the Shiite Hazaras, but problems with Afghan refugees in Iran have created tensions between the two countries (see *People on the Move: Afghan Refugees in Iran*). Iran's relations with Azerbaijan, on the other hand, remain tense, in part because many Azerbaijanis refer to northwestern Iran—an Azeri-speaking region—as "Southern Azerbaijan."

Pakistan has also strived to gain influence in Central Asia, hoping that pipelines will eventually carry Central Asian oil and natural gas to its new deep-water port at Gwadar. Any such pipelines, however, would have to pass through the rugged topography and perilous political environment of Afghanistan. During the Taliban period, Pakistan enjoyed close relations with Afghanistan's government. The aftermath of 9/11 thus put Pakistan in a serious bind. In response to both pressure and promises from the United States, Pakistan joined the anti-Taliban coalition and supplied the U.S. military with valuable intelligence. The subsequent retreat of the Taliban into northwestern Pakistan, however, has resulted in extremely tense relations between the two countries. Also complicating relations is the fact that Afghanistan refuses to recognize the legitimacy of the border that separates them, claiming that it was a British imperial imposition.

Turkey's connections with Central Asia are potentially close. Most Central Asians speak Turkic languages, and Turkey also offers itself as the model—contrary to those of Iran and Pakistan—of the modern, secular state of Muslim heritage. It is unclear, however, whether the Turkish system can be exported easily, even to other Turkic-speaking states. In early 2013, Turkish president Recep Erdogan surprised the world by announcing that Turkey might opt to join the SCO, especially if its bid to join the European Union continued to be forestalled.

REVIEW

10.10 How did the collapse of the Soviet Union in 1991 change the geopolitical structure of Central Asia, and how has Russia attempted to maintain influence in the region?

10.11 How does the geopolitical situation of Afghanistan differ from those of the other countries of the region, and why is Afghanistan's political history so different from those of its neighbors?

10.12 What role do international treaty organizations play in the geopolitical situation of Central Asia?

Economic and Social Development: Abundant Resources, Struggling Economies

By most measures, Central Asia is one of the poorer regions of the world. Afghanistan in particular is near the bottom of almost every list of economic and social indicators. Other Central Asian countries, however, have enjoyed relatively high levels of health and education, a legacy of the social programs enacted by their former communist regimes. Unfortunately, these same governments also built inefficient economic systems, and after the fall of the Soviet Union, western Central Asia experienced a spectacular economic decline. Although much of the region is now economically expanding, growth has been largely based on the extraction of natural resources, particularly oil and natural gas. Central Asia's landlocked position, however, poses particular challenges to all forms of economic development, as it greatly increases transportation costs. Even fossil fuel exploitation here requires the construction of extremely long and expensive pipelines.

Economic Development in Central Asia

Soviet economic planners sought to spread the benefits of economic development widely across their country. This required building large factories, even in remote areas such as Tajikistan, regardless of the costs involved. Such Central Asian industries relied heavily on subsidies from the Soviet central government. When those subsidies ended, the industrial base of the region began to collapse, leading to plummeting living standards. But as is true elsewhere in the former Soviet Union, certain well-connected individuals have grown very wealthy since the fall of communism. The growing gap between the rich and the poor remains a major concern across most of Central Asia.

The Post-Communist Economies As Table 10.2 shows, no Central Asian country or region could be considered prosperous by global standards. Kazakhstan generally stands as the most developed, and it probably has the best prospects. Kazakhstan has two of the world's largest underutilized deposits of oil and natural gas—the vast Tengiz and Kashagan fields in the northeastern Caspian Basin (Figure 10.28)—as well as sizable deposits of other minerals. Kazakhstan has signed agreements with Western oil companies to exploit its oil reserves and has received major investments from China. Between 2001 and 2008, Kazakhstan's oil and gas wealth made it one of the world's fastest-growing economies, with annual growth averaging more than 8 percent. Growth subsequently slowed, but the Kazakh economy remains healthy. The country's impending accession into the World Trade Organization (WTO), moreover, has recently boosted its prospects.

Owing to its sizable population, Uzbekistan has the second largest economy in the region. Uzbekistan did not

Table 10.2 Development Indicators

Country	GNI per capita, PPP 2011	GDP Average Annual %Growth 2000–11	Human Development Index (2011)[1]	Percent Population Living Below $2 a Day	Life Expectancy (2013)[2]	Under Age 5 Mortality Rate (1990)	Under Age 5 Mortality Rate (2011)	Adult Literacy (% ages 15 and older) Male/ Female	Gender Inequality Index (2011)[3,1]
Afghanistan	1,140	–	0.374	–	60	192	101	--/--	0.712
Azerbaijan	8,950	16.0	0.734	2.8	74	95	45	100/100	0.323
Kazakhstan	11,250	8.0	0.754	<2	69	57	28	100/100	0.312
Kyrgyzstan	2,290	4.4	0.622	22.9	70	70	31	100/99	0.357
Mongolia	4,290	7.4	0.675	–	69	107	31	97/98	0.328
Tajikistan	2,300	8.0	0.622	27.7	67	114	63	100/100	0.338
Turkmenistan	8,690	8.7	0.698	49.7	65	94	53	100/99	–
Uzbekistan	3,420	7.3	0.654	–	68	75	49	100/99	–

[1]United Nations, *Human Development Report, 2013.*

[2]Population Reference Bureau, *World Population Data Sheet, 2013.*

[3]Gender Inequality Index—A composite measure reflecting inequality in achievements between women and men in three dimensions: reproductive health, empowerment and the labor market that ranges between 0 and 1. The higher the number, the greater the inequality.

Source: World Bank, *World Development Indicators, 2013.*

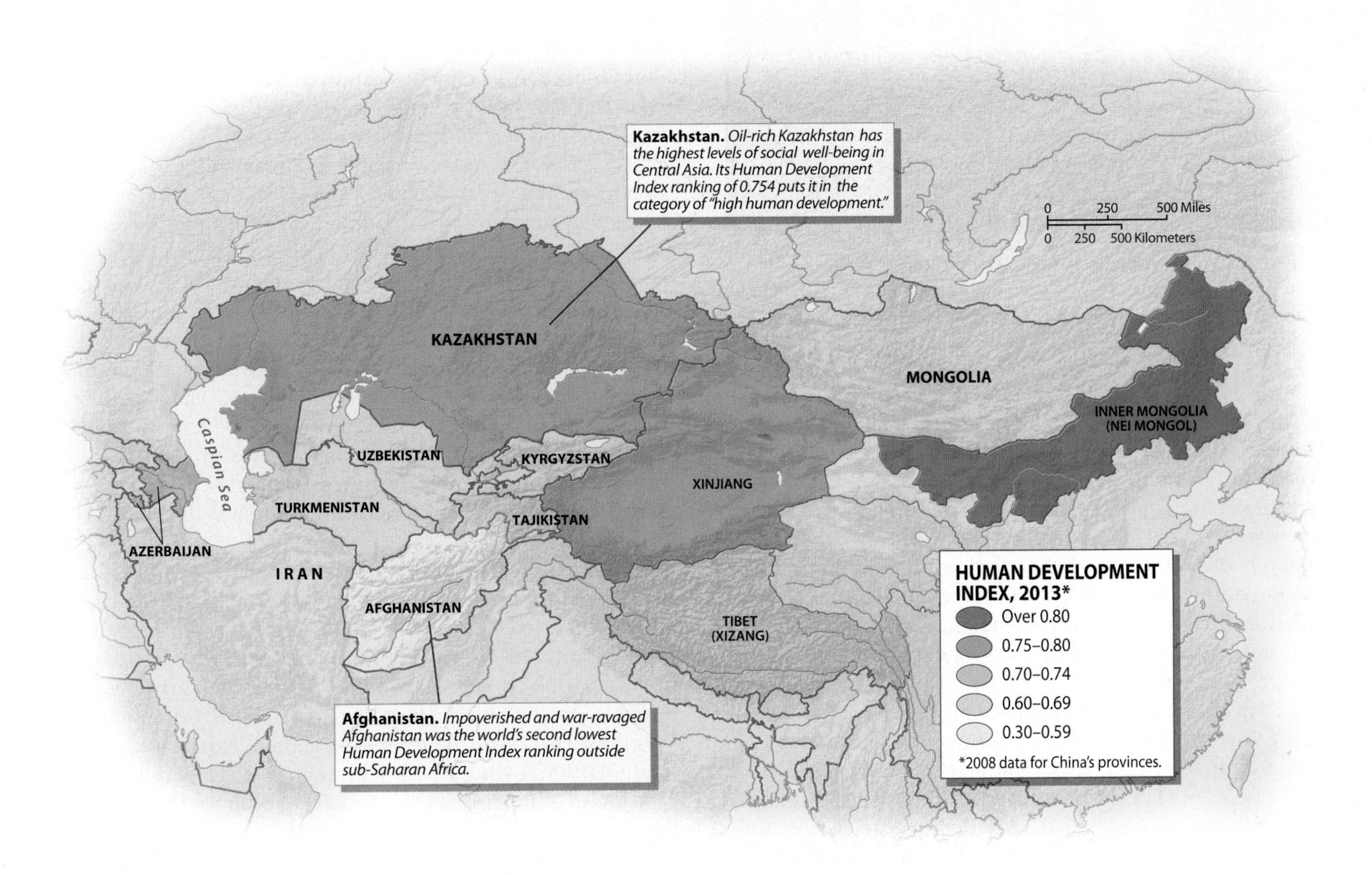

▲ **Figure 10.28 Development Issues in Central Asia: Human Development** Levels of social development vary significantly across Central Asia. Afghanistan ranks as one of the least-developed countries in the world, whereas both Azerbaijan and Kazakhstan rank relatively high.

▲ **Figure 10.29 Cotton Growing in Uzbekistan** Uzbekistan is one of the world's main cotton-growing and -exporting countries. Although the cotton crop is essential for the country's economy, it also results in extensive environmental degradation.

decline as sharply as its neighbors after the fall of the Soviet Union, largely because it retained many aspects of the old command economy (that is, an economy run by governmental planners, rather than by private firms responding to the market). Uzbekistan remains the world's second largest exporter of cotton, and it has significant gold and natural gas deposits (Figure 10.29). However, environmental degradation threatens cotton production, and both inefficient state management and widespread corruption continue to hamper production.

Kyrgyzstan, in contrast, moved aggressively after the fall of the Soviet Union to privatize former state-run industries. Along with Mongolia, it is Central Asia's only member of the WTO as of March 2013. However, Kyrgyzstan's economy is heavily agricultural, few of its industries are competitive, and political strife has discouraged foreign investment. On the bright side, Kyrgyzstan does enjoy the largest supply of freshwater in Central Asia, which will likely become increasingly valuable in years to come, and its mineral reserves are substantial. Gold exports are particularly important; when gold exports slump, as they did in 2009, Kyrgyzstan experiences economic hardship. Political instability in early 2010 also hurt the country's economy.

Turkmenistan also has a substantial agricultural base, due mainly to Soviet irrigation projects, and it remains a major cotton exporter. It retains a state-run economy and has resisted pressure for economic liberalization. A gas pipeline to China that became operational in 2010, moreover, has given the country an economic boost. Turkmenistan wants to encourage foreign investment, but corruption and bureaucratic obstacles discourage most potential investors.

Central Asia's oldest fossil fuel industry is located in Azerbaijan. Azerbaijan has attracted a great deal of international interest and investment, promising to revitalize its oil industry. Through the 1990s, its economy responded slowly, and Azerbaijan remains a poor country. In 2006, however, its oil-fueled economy reportedly grew by 34 percent, the fastest rate of expansion in the world. Since then, its economy has been highly volatile, undergoing rapid shifts from expansion to stagnation. Critics contend that the country has not enacted adequate market reforms and remains far too dependent on its oil industry.

The most economically troubled of the former Soviet republics is Tajikistan. With a per capita gross national income (GNI) of only some $2200, Tajikistan is poor indeed. It is burdened by its remote location far from the region's main roads and railways, its rugged topography, its lack of natural resources, and its political strife. Almost half of Tajikistan's labor force works abroad, mostly in Russia and Kazakhstan, supporting the local economy through their remittances. Tajikistan's government hopes to revive its fortunes by opening new transportation routes to China and by further developing its generation of hydroelectricity (Figure 10.30). A large new dam was finished in 2009, another was completed in 2011, and others are under construction. Enhanced electricity supplies are allowing Tajikistan to expand its aluminum smelting, one of its few competitive industries. The biggest of these projects is the Rogun Dam, which will be the tallest dam in the world if and when it is ever completed. Although construction continues, Tajikistan has not yet gained the financing necessary to complete this massive project.

Mongolia, although never part of the Soviet Union, was a close Soviet ally run by a communist party. It, too, suffered an economic collapse in the 1990s when Soviet subsidies came to an end. Mongolia thus emerged in the post-communist period as a poor country. During the early 2000s, however, major mining investments by both Chinese and Western firms resulted in rapid economic expansion, with total GDP increasing by 17.5 percent in 2011 and 12 percent in 2012. In early 2010, Mongolia began to develop the massive Oyu Tolgoi copper–gold project, one of the largest mining operations the world has ever seen. In early 2013, however, a planned expansion of the mine was delayed when the World Bank refused to release funding due to social and environmental concerns. The Mongolian mining boom has generated many such problems. Many Mongolians

▲ **Figure 10.30 Nurek Dam, Tajikistan** Built during the Soviet period, Nurek is the tallest dam in the world. The electricity that it provides is essential to the economy of Tajikistan.

fear that their Chinese mining companies have too much power. Small-scale, unlicensed "ninja miners," moreover, have caused a significant amount of local environmental degradation.

The Economy of Western China The Chinese portions of Central Asia did not suffer the same economic crash that visited other parts of the region with the fall of the Soviet Union. China has one of the world's fastest-growing economies, although its centers of dynamism are located in the distant coastal zone. Still, even the more remote parts of the country have experienced a significant degree of development in recent years. From 2002 to 2012, Inner Mongolia's economy expanded at a scorching annual rate of around 17 percent, owing in large part to its extensive deposits of rare earth minerals and other valuable mineral resources.

Tibet in particular remains burdened by poverty. Most of the plateau is relatively cut off from the Chinese economy and is even more isolated from the global economy. Hoping to reduce Tibetan separatism, China has been investing large amounts of money in infrastructural projects. In 2006, it inaugurated the monumentally expensive Qinghai–Tibet Railway, linking Tibet's capital city of Lhasa to the rest of the country. Reaching heights of over 16,400 feet (5000 meters), the trains on this railway have to be equipped with supplemental oxygen (Figure 10.31). As a result of this and other transportation projects, Tibet's tourism economy is booming. Many Tibetans, however, argue that most of the benefits are flowing to Han Chinese

▼ **Figure 10.31 Qinghai–Tibet Railway** China is engaged in a massive program to extend roads and railroads into the western half of the country. The main connection with the Tibetan Plateau is the recently completed Qinghai–Tibet Railway. This railway has resulted in a large increase in tourism in Tibet. **Q: Why have the railroad tracks been elevated so high above the valley floor?**

Exploring Global Connections

The Heroin and Opium Trade from Afghanistan

No country depends more on illicit activity for its economic livelihood than Afghanistan. Opium, the final product of which is usually heroin, amounts to as much as 15 percent of Afghanistan's economy, providing billions of dollars worth of annual exports in a country with little else to offer world markets. Ninety percent of the world's opium supply comes from Afghanistan, usually making its way through Central Asia and into Europe.

The Economics of Opium For Afghans, the decision to grow poppies (from which opium is derived) is often a simple acknowledgment of economic reality. A given tract of farmland can earn its owner up to 11 times more money producing poppies than growing wheat or a similar crop, a premium that is sometimes necessary just to achieve subsistence (**Figure 10.5.1**). Operating in the black market has its drawbacks, however, and farmers unable to pay off debts risk seeing their daughters kidnapped by smugglers.

The ubiquity of poppies in Afghanistan also contributes to wider social problems that reach beyond farming communities. One out of every 30 Afghans is estimated to be a heroin or opium addict, a proportion that is the largest in the world. This figure also appears to be rising, due in part to increasing addiction among women. In addition to damaging health and depressing productivity, drug addiction represents an enormous financial burden to many Afghans who can ill afford the $4 a day needed to feed their habit. Opium production is also a major source of funds for the Taliban and other anti-government organizations in the country. In 2009, the UN estimated that the Taliban alone trafficked 275 tons of heroin, charging a 10 percent tax to farmers.

Trade Routes and Repercussions Opium leaves Afghanistan along three major routes: one heading to South Asia and East Asia via Pakistan; another moving through Iran to Turkey and Europe; and another running through Central Asia to Russia. The ramifications of this trade are enormous for all of the countries involved. Heroin addiction in the countries bordering Afghanistan is highly elevated, and the product provides a revenue stream for criminal organizations everywhere from Central Asia to North America.

Enforcing the legal prohibition against poppy farming and opium production in Afghanistan presents government officials and their U.S. allies with a difficult dilemma. Destroying crops can be a devastating blow to individual families, helping to radicalize them and breed hostility toward the central government. Allowing the industry to flourish, however, is hardly an option for the reasons mentioned above. Enforcement often resembles a game of Whack-a-mole. Despite destroying 48 heroin laboratories, seizing 2188 kilograms of heroin, and destroying countless poppy farms in 2009, poppy farming, opium production, and drug consumption continue to grow.

▼ **Figure 10.5.1 Afghan Farmer in His Poppy Field** Most of the world's opium is produced in southern and eastern Afghanistan. Opium-poppy farmers can earn an adequate income from a small plot of land, such as the one visible in this photograph.

immigrants, rather than to the indigenous peoples of the region. Still, the Chinese government claims that as of 2011 the Tibetan economy was growing by more than 12 percent a year, and in 2012 it announced that new mining developments in the region would further enhance economic expansion.

Most observers agree that Xinjiang has tremendous economic potential. It boasts significant mineral wealth, including most of China's oil and natural gas reserves, and it has benefited from infrastructure development projects over the past several decades. The recent completion of oil and natural gas pipelines from Kazakhstan and Turkmenistan has also bolstered the region's petrochemical industries. In 2012, the Chinese government announced that Xinjiang's economy was growing by more than 10 percent a year, a rate slightly higher than the national average. But as is the case in Tibet, many of the indigenous peoples of Xinjiang believe that the wealth of their region is being monopolized by the

Chinese state and the Han Chinese immigrants. Some point to the Xinjiang Production and Construction Corps, a quasi-military firm that runs hundreds of factories, farms, schools, and hospitals. This huge company was initially formed by the army that rejoined Xinjiang to China in 1949, and it is still run almost entirely by Han Chinese.

Economic Misery in Afghanistan Afghanistan is Central Asia's poorest country by a substantial margin. Since the late 1970s, it has suffered nearly continuous war, undermining virtually all economic endeavors. Even before the war, it had experienced little industrial or commercial development. Its only significant legitimate exports are animal products, handwoven carpets, and a few fruits, nuts, and semiprecious gemstones. An ineffective government, poor infrastructure, and endemic corruption all plague economic development. Yet Afghanistan does remain economically competitive in one highly problematic area: the production of illicit drugs for the global market (see *Exploring Global Connections: The Heroin and Opium Trade from Afghanistan*).

Since 2003, the economy of Afghanistan has experienced some growth, although much of the gains have come through foreign assistance. International donors pledged some US$67 billion in assistance funds between 2003 and 2010, and in 2012 another US$67 billion were promised though 2016. Thus far, such aid has not been able to jump-start the Afghan economy. Afghanistan does have vast mineral deposits, which would quickly be developed if security could be guaranteed. Despite the seemingly endless war, China decided to invest $2.8 billion in a massive copper-mining scheme in 2008, and in 2013 the United Kingdom agreed to invest in oil and natural gas ventures in the country. The prospects for sustained economic development in Afghanistan, however, still seem unlikely.

Central Asian Economies in Global Context Even Afghanistan—despite its poverty and relative isolation—is thoroughly embedded in the global economy, albeit through illicit products. But with the exceptions of the drug trade, the oil and natural gas industries, and other mining ventures, the overall extent of globalization in Central Asia remains low.

In the former Soviet area, the most important international connections remain with Russia, although economic ties with China are rapidly expanding. In the early post-Soviet period, both Russia and the newly independent Central Asian countries were ambivalent about their continuing economic relationships. Disentangling Central Asian economies from that of Russia, however, proved difficult, largely because of infrastructural linkages inherited from the Soviet period. In 2007, two-thirds of Turkmenistan's natural gas exports flowed through Russia's state-owned Gazprom pipelines. Since the creation of the SCO, moreover, Uzbekistan, Kazakhstan, Kyrgyzstan, and Tajikistan have eagerly sought closer linkages with both Russia and China. Connections with China have been greatly enhanced by the recent completion of several important gas pipelines.

The United States and other Western countries, as well as India, are also drawn to the area by its oil and natural gas deposits. Many large oil companies operate in Azerbaijan and Kazakhstan. The delivery of Central Asian oil to the west was enhanced in 2006 with the opening of the extraordinarily expensive Baku–Tbilisi–Ceyhan Pipeline—the second longest in the world—which runs from the Caspian Sea to the Mediterranean at the Turkish port of Ceyhan (Figure 10.32). If the planned, but controversial Nabucco Pipeline is built, Azerbaijan will be exporting large quantities of natural gas to Europe by 2017.

Central Asia's spectacular natural scenery and generally low population density present opportunities for the development of international ecotourism. Kyrgyzstan, for example, is working hard to build a sustainable tourism sector focused on Issyk Kul, the world's second largest alpine lake. Ecotourism companies in Mongolia offer tourists from wealthy countries the opportunity to experience the nomadic lifestyle of traditional pastoral people, sleeping in felt tents and riding horses over the great expanses of the Mongolian grasslands.

▼ **Figure 10.32 Baku–Tbilisi–Ceyhan Pipeline** This pipeline, which transports oil from Azerbaijan to the Mediterranean coast of Turkey, is one of the world's most important. It began operation in 2006.

Social Development in Central Asia

Social conditions in Central Asia vary more than economic conditions. In the former Soviet territories, levels of health and education remain fairly high, but are generally declining. Not surprisingly, Afghanistan is at the bottom of the scale by every measure.

Social Conditions and the Status of Women in Afghanistan Social conditions in Afghanistan are no better than economic circumstances, although reliable information is difficult to obtain. As indicated in Table 10.2, the average life expectancy in the country is a mere 60 years, one of the lowest figures in the world. Infant and childhood mortality levels remain extremely high. Not only does Afghanistan suffer from constant warfare, but also its rugged topography hinders the provision of basic social and medical services. Illiteracy is commonplace and is notably gender-biased. Afghanistan's adult female literacy level is estimated by some experts at only around 12 percent, which would make it one of the lowest figures in the world.

Despite these dismal figures, some progress has occurred. As recently as 2004, Afghanistan's average life expectancy was a mere 44 years. Education levels are also improving, for girls as well as boys, and today an estimated 8 million Afghan children regularly attend school. Such gains have been especially pronounced in the Hazara area in the central part of the country and, to a lesser extent, in the Tajik-dominated areas of the north. Such gains, however, must be regarded as fragile. The Taliban opposes female education and periodically attacks schools for girls. If the Taliban were to regain power, Afghanistan social development would likely plummet again.

Women in most parts of Afghanistan—and especially in Pashtun areas—continue to lead highly constrained lives. In many areas, they must completely conceal their bodies, including their faces, when venturing into public. Such restrictions intensified in the 1990s. Dress controls were strictly enforced where the Taliban gained authority. In most areas, Taliban forces prevented women from working, attending school, and often even obtaining medical care. The assault on women by the Taliban did not go uncontested in Afghanistan. Many women risked their lives to document the horrific abuses that Afghan females were suffering. Although Taliban leaders argued that they were upholding Islamic orthodoxy, their opponents contended that they were actually enforcing an extreme version of Pashtun customary law.

The fall of the Taliban brought temporary joy to most of the women of Afghanistan. Many promptly uncovered their faces and began working or seeking work. The new Afghan constitution proclaimed that henceforth men and women would enjoy equal rights. However, in most parts of the country, little actually changed with the coming of the new regime. Social customs continued to force most women into arranged marriages, followed by domestic seclusion. Indeed, in some areas of the country the upsurge in both crime and political violence has made the position of women increasingly insecure. According to some Afghan women's groups, over 70 percent of women in the country face forced marriages, and roughly one-third suffer physical, psychological, or sexual violence.

Social Conditions Elsewhere in Central Asia The figures in Table 10.2 reveal relatively favorable levels of social welfare overall for the former Soviet zone. This is especially notable when you consider the poorly developed economies of such countries as Tajikistan and Uzbekistan or contrast conditions with those of neighboring Afghanistan. Tajikistan, with a per capita GNI of only $2300, has a female life expectancy of 69 years, as well as almost universal adult literacy. The social successes of Tajikistan and its northern and western neighbors reflect investments made during the Soviet period. Health and educational facilities, however, have declined in the face of economic and political turmoil. Certainly, the region's relatively high levels of infant and childhood mortality do not bode well. Whereas Tajikistan might hope that its educated workforce will attract foreign investment, its remote location and political tensions make such a scenario unlikely.

The situation is less clear in the Central Asian portions of China. Official figures indicate that social development is more advanced in Inner Mongolia than in any other part of Central Asia and that Xinjiang has relatively high social development levels as well (see Figure 10.28). But although all parts of China have made significant progress in health and education, many reports suggest that the indigenous peoples of Tibet and Xinjiang have been left behind. Some sources claim that more than half of the non-Han people of Tibet remain illiterate. Chinese sources, on the other hand, maintain that illiteracy has been declining rapidly. In 2008, China's official media outlets reported that over 95 percent of the young people of Tibet had learned to read and write. Tibetan activists complain, however, that most such education is conducted in Mandarin Chinese, rather than in the Tibetan language.

China's minority peoples have been granted widespread exemptions from the country's strict population control measures. But there are many reports from Xinjiang and elsewhere of forced sterilizations and abortions, adding to the local political tensions.

Central Asian Gender Issues Despite widespread poverty and a political scene dominated by men, the status of women in Central Asia, with the notable exception of Afghanistan, compares favorably to their status in other regions at similar levels of income. According to the World Economic Forum's *Global Gender Gap Report 2012*, female professional workers and enrollees in tertiary education outnumber men in Kazakhstan, Kyrgyzstan, and Mongolia, and Mongolian women lead the world in overall economic participation and

▶ **Figure 10.33 Afghan Women in Parliament** Women in Afghanistan generally have a low social position, and they suffer from many restrictions, especially in the rural parts of the country. They do have a number of reserved seats in the country's parliament, although critics contend that female legislators have little real power.

opportunity relative to men. Women in the region generally receive similar pay for similar work and are expected to continue their educations past primary school. Much of this achievement can be attributed to investment in education during the Soviet period, though doubts have arisen as to whether the region's relative equality can persist into the future. The situation of women in much of Afghanistan, moreover, remains dire. But it is also true that women now have constitutionally protected rights in Afghanistan, and 64 seats are reserved in the country's parliament for female delegates (Figure 10.33).

Success in the education and economic spheres has thus far failed to fully translate into political influence, as men continue to significantly outnumber women in all Central Asian parliaments. Roza Otunbayeva, who served as President of Kyrgyzstan from mid-2010 through 2011, was the first female head of state in Central Asia since Mongolia's Sükhbaataryn Yanjmaa left office in 1954. Some evidence suggests, moreover, that the position of women is in jeopardy in some parts of the region and has even begun to decline. In Kazakhstan, leaders have blamed feminism for the country's low birth rate, and in much of the region newly rich men are increasingly practicing polygamy. As education becomes more expensive, moreover, girls are increasingly staying at home. Women are also trafficked from the poorer countries of the region, particularly Uzbekistan and Tajikistan, into prostitution in Russia, the United Arab Emirates, and elsewhere.

Another troubling gender issue in Central Asia is that of "bride kidnapping." In Kyrgyzstan, Kazakhstan, and parts of Uzbekistan, men often abduct women whom they want to marry. In some cases, the woman in question has already agreed to go along, but in many cases she is taken against her will. This practice is contrary to both Islamic law and national legal codes, yet it has been very persistent. In January 2013, however, Kyrgyzstan increased the penalty for bride kidnapping to a maximum of 10 years in prison, hoping to eliminate the practice.

REVIEW

10.13 How does the distribution of fossil fuel deposits influence economic and social development in Central Asia?

10.14 Why does the social position of women vary so much across Central Asia, and why is Afghanistan particularly problematic in this regard?

10.15 How is the Soviet legacy reflected in the social and economic development of western Central Asia?

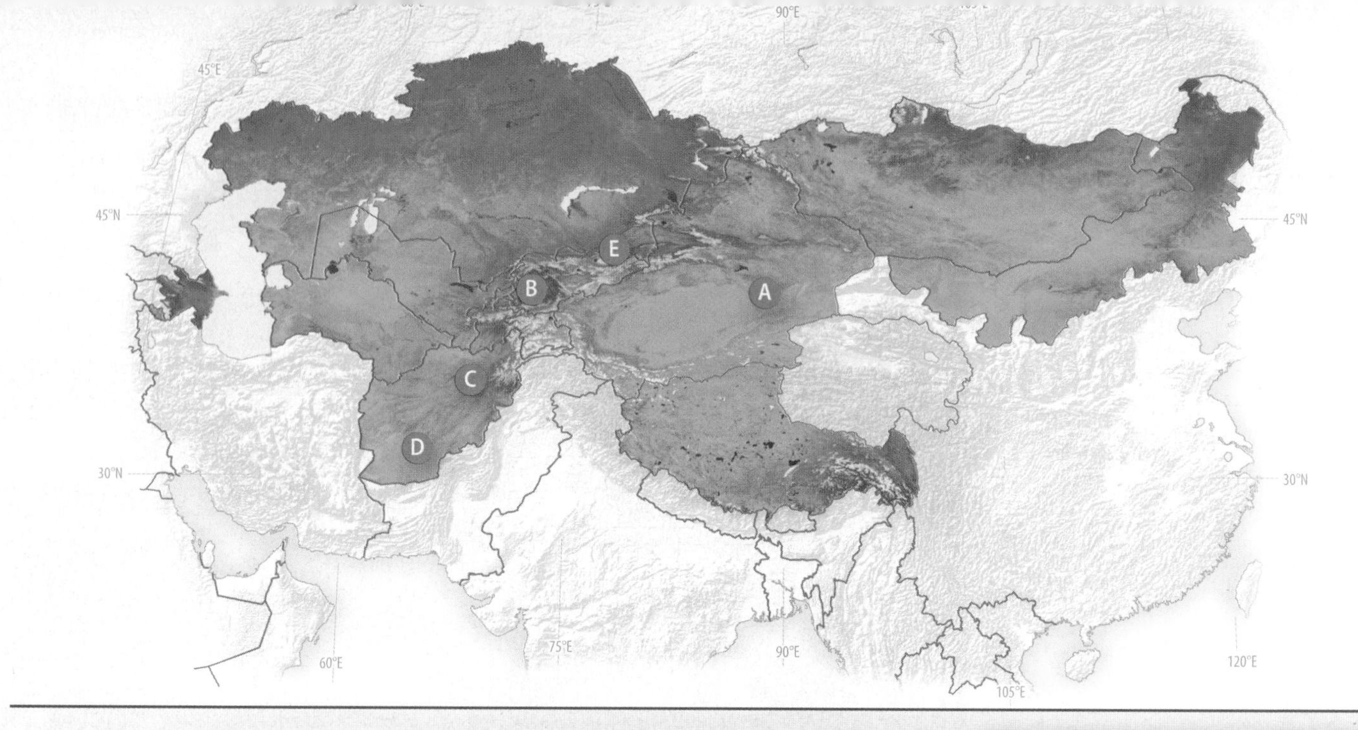

Physical Geography and Environmental Issues

Environmental problems, among others, have brought the region to global attention. The destruction of the Aral Sea is one of the worst environmental disasters the world has seen, and many other lakes in the region have experienced similar problems. Desertification has devastated many areas, and the booming oil and gas industry has created its own environmental disasters.

10.1 Lop Nur in China's Xinjiang autonomous region was once a lake, but now it is a dried-out salt flat that has been used for nuclear tests and that contains the world's largest potash fertilizer plant. Why did the lake disappear, and why has China used the desiccated lakebed for nuclear tests and potash processing?

10.2 What strategies might be used to prevent the desiccation of other lakes in Central Asia?

Population and Settlement

Large movements of people in Central Asia have attracted global attention. The migration of Han Chinese into Tibet and Xinjiang is steadily turning the indigenous peoples of these areas into minority groups. The migration of Russian speakers out of the former Soviet areas of Central Asia has also resulted in major transformations. People are currently moving from the poorer countries of the former Soviet zone to oil-rich Kazakhstan. In Afghanistan, war and continuing chaos have generated large refugee populations.

10.3 An area of very high population density is found near the center of this map. Why is this particular region so crowded, and why is it surrounded by areas of much lower population density?

10.4 How might such steep population gradients influence political tension and economic development in the region?

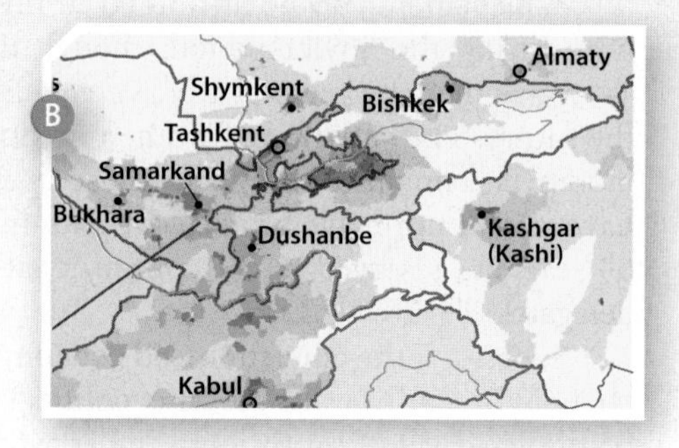

Key Terms

alluvial fan *(p. 464)*
Bodhisattva *(p. 474)*
Dalai Lama *(p. 474)*
deforestation *(p. 459)*
desertification *(p. 455)*
desiccation *(p. 455)*
exclave *(p. 478)*
exotic river *(p. 465)*
loess *(p. 464)*
pastoralist *(p. 463)*
salinization *(p. 455)*
Shanghai Cooperation Organization (SCO; "Shanghai Six") *(p. 482)*
Silk Road *(p. 468)*
steppe *(p. 453)*
Taliban *(p. 474)*
theocracy *(p. 474)*
transhumance *(p. 464)*

In Review • Chapter 10
Central Asia

Cultural Coherence and Diversity

Religious tension has recently emerged as a major cultural issue throughout much of western Central Asia. Radical Islamic fundamentalism remains a potent force in southern and western Afghanistan and, to a somewhat lesser extent, in the Fergana Valley of Uzbekistan, Tajikistan, and Kyrgyzstan. Although moderate forms of Islam are prevalent in most of the region, Central Asian leaders have used the fear of religiously inspired violence to maintain repressive and antidemocratic policies.

10.5 **The giant statues of the Buddha in what is now central Afghanistan were constructed in the early 6th century and dynamited in 2001 by the Taliban. Why would the Taliban government of Afghanistan have decided to destroy these world-famous statues?**

10.6 Controversial efforts are now under way to restore the statues. Is such a project worthwhile, and why has it proved to be so controversial?

Geopolitical Framework

China maintains a firm grip on Tibet and Xinjiang, but the rest of Central Asia has become a key area of geopolitical competition. Russia, the United States, China, Iran, Turkey, Pakistan, and India all contend for influence. Central Asian countries have attempted to play these powers against each other in order to maintain or increase their world status. Overall, political regimes throughout the region remain largely authoritarian, although popular uprisings have brought down two governments in Kyrgyzstan.

10.7 **The Soviet invasion and subsequent occupation of Afghanistan (1979–1989) proved very costly to the Soviet Union. Why did Soviet leaders feel compelled to invade Afghanistan, and why did their occupation of the country ultimately prove unsuccessful?**

10.8 Is it reasonable to draw connections between the Soviet and the U.S. experiences in Afghanistan? In what ways have they been similar and different?

Economic and Social Development

The economies of Central Asia are gradually opening up to global connections, largely because of their substantial fossil fuel reserves. However, Central Asia is likely to face serious economic difficulties for some time, especially since the region is not a significant participant in global trade and has attracted little foreign investment outside the fossil fuel sector. The growing of opium and the manufacture of heroin remain a serious problem in Afghanistan. Drug trafficking, moreover, has weakened legitimate economic activities in other countries of the region.

10.9 **The Canadian-owned Kumtor Gold Mine in Kyrgyzstan has been very important for the country's economy, yet it is increasingly controversial. Why do so many people in Kyrgyzstan oppose this particular kind of economic development?**

10.10 What are some of the potential disadvantages faced by a country that focuses much of its economic development plans on the exploitation of a particular natural resource such as gold?

MasteringGeography™

Looking for additional review and test prep materials? Visit the Study Area in **MasteringGeography™** to enhance your geographic literacy, spatial reasoning skills, and understanding of this chapter's content by accessing a variety of resources, including **MapMaster™** interactive maps, videos, RSS feeds, flashcards, web links, self-study quizzes, and an eText version of *Diversity Amid Globalization*.

Authors' Blogs

Scan to visit the author's blog for chapter updates

www.gad4blog.wordpress.com

Scan to visit the GeoCurrents blog

www.geocurrents.info

11 East Asia

Environmental Geography

China has long experienced severe deforestation and soil erosion, and its current economic boom is generating the worst pollution problems in the world. Japan, South Korea, and Taiwan all maintain extensive forests and relatively clean environments.

Population and Settlement

China is currently undergoing a major transformation as tens of millions of peasants move from poor villages in the interior to booming cities along the coastal region. Birth rates are low, and populations are aging throughout the region.

Gangnam District. Seoul is the capital city, economic hub, and cultural center of South Korea. The Gangnam District of Seoul, made famous by the South Korean musician Psy, is noted for its expensive shops and luxury apartments.

Cultural Coherence and Diversity

Despite the presence of several unifying cultural features, East Asia in general and China in particular are divided along several striking cultural lines. Historically, however, the region has been tied together by Mahayana Buddhism, Confucianism, and the Chinese writing system.

Geopolitical Framework

The growing military power of China is generating tension with other East Asian countries, while Korea remains a divided nation. Japan, South Korea, and Taiwan have responded to China's growth by strengthening ties with the United States.

Economic and Social Development

Over the past several decades, East Asia has emerged as a core area of the world economy, with China experiencing one of the most rapid economic expansions the world has ever seen. Japan and South Korea also have strong economies, but North Korea remains desperately poor.

K-pop, or Korean popular music, gained a huge base of fans in Asia and across much of the rest of the world in the early 2000s. Its influence in the United States and Europe, however, was less pronounced until 2012. In that year, the South Korea pop star PSY released the mega-hit "Gangnam Style," which climbed to the top of the music charts in more than 30 countries. By January 15, 2013, the "Gangnam Style" YouTube video had been viewed more than 1.2 billion times, smashing all records on the Website. Korean popular music is now a fully global phenomenon.

PSY's hit song refers to the Gangnam District, which is the wealthiest area of Seoul, South Korea's capital and largest city. In the 1970s, the district was the least developed part of the city, and South Korea itself was a rapidly growing, but still relatively poor country. Today South Korea is a global technological powerhouse, and its prosperity is showcased in Gangnam. The district's new wealth, however, has resulted in a degree of extravagance, and its inhabitants are often accused of acting in a trendy manner and showing off their status. PSY's song and video can be seen as mocking the lavish lifestyles and social pretensions of Gangnam residents, although the artist himself maintains that he is actually poking fun at people who claim to follow the "Gangnam style" even though they do not live in the exclusive district.

Despite the presence of several unifying cultural features, East Asia in general and China in particular are divided along several striking cultural lines.

The Gangnam style phenomenon says a lot about economic development, cultural change, and globalization in East Asia, a region composed of China, Japan, South Korea, North Korea, and Taiwan (or, in formal terminology, the People's Republic of China, the State of Japan, the Republic of Korea, the Democratic People's Republic of Korea, and the Republic of China, respectively; Figure 11.1). Massive cities have mushroomed over the past several decades as East Asia has emerged as a core zone of the global system, noted for its economic, technological, and cultural power. But not all parts of the region have been part of this story. Seoul itself sits only about 31 miles (50 km) from the North Korean

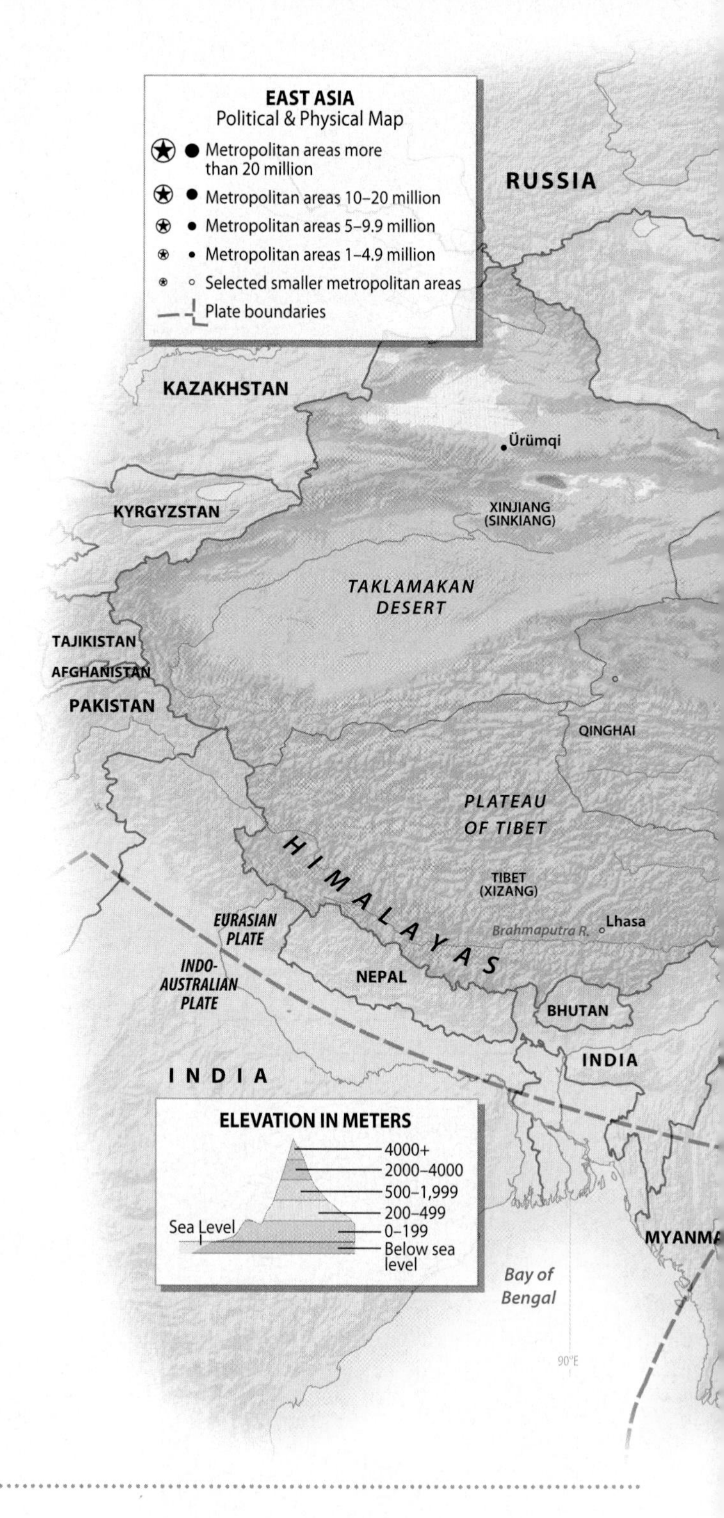

LEARNING OBJECTIVES

After reading this chapter you should be able to:

- Identify the key environmental differences between the island portions of East Asia (Japan and Taiwan) and the mainland.
- Describe the main environmental problems China faces today and compare them with the environmental challenges faced by Japan, South Korea, and Taiwan.
- Summarize the relationships among topography, climate, rice cultivation, and population density across East Asia.
- Explain why China's population is so unevenly distributed, with some areas densely settled and other almost uninhabited.
- Outline the distribution of major urban areas across East Asia and explain why the continued expansion of the region's largest cities is often viewed as a problem.
- Describe the ways in which religion and other systems of belief both unify and divide East Asia.
- Explain the distinction between the Han Chinese and other ethnic groups of China, paying particular attention to language.
- Describe the geopolitical division of East Asia during the Cold War period and explain how the division of that period still influences East Asian geopolitical relations.
- Identify the main reasons behind East Asia's rapid economic growth in recent decades and discuss any possible limitations to continued expansion at such a rate.
- Describe the differences in economic and social development found across China and, more generally, across East Asia as a whole.

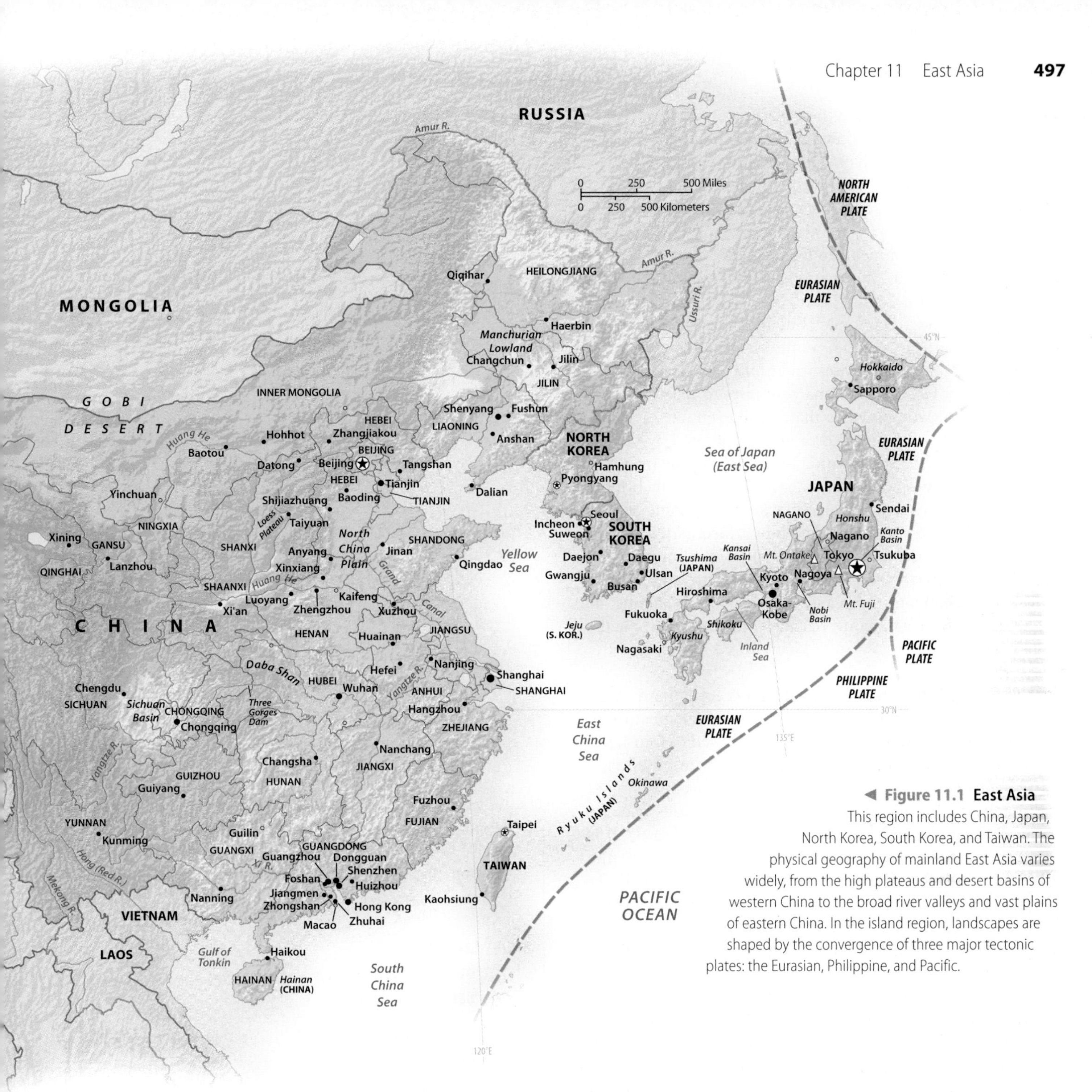

Figure 11.1 East Asia This region includes China, Japan, North Korea, South Korea, and Taiwan. The physical geography of mainland East Asia varies widely, from the high plateaus and desert basins of western China to the broad river valleys and vast plains of eastern China. In the island region, landscapes are shaped by the convergence of three major tectonic plates: the Eurasian, Philippine, and Pacific.

border, beyond which you find very different conditions. In stark contrast to the south, North Korea is noted for its poverty, economic stagnation, and resistance to globalization. Such disparities are not so extreme elsewhere in East Asia, but vast differences still divide the region's dynamic, globalized cities from its more inward-looking rural areas, especially in China.

East Asian Unity and Divisions Although it contains few countries, East Asia is the most populous region of the world. China alone is inhabited by more than 1.3 billion people—more than live in any other region of the world except South Asia. "Vast differences still divide the region's dynamic, globalized cities from its more inward-looking rural areas." Although East Asia is historically unified by cultural features, in the second half of the 20th century it was divided ideologically and politically, with the capitalist economies of Japan, South Korea, Taiwan, and Hong Kong separated from the communist bloc of China and North Korea. As Japan reached the pinnacle of the global economy in the 1980s, much of China remained mired in extreme poverty. Since the 1990s, however, economic disparities within East Asia have decreased significantly. Although China is still governed by the Communist Party, it has embarked on a path of capitalist development that has resulted in extremely rapid economic growth, as well as extremely severe environmental problems. (Briefly, **communism** is an economic and political system in which the government, as representative of the people, owns all land, business, and industry and in which one political party controls the state; in **capitalism**, land, business, and industry are owned by private companies or individuals for profit.)

Economic development and technological progress have resulted in increasingly close business connections among China, Japan, South Korea, and Taiwan. Tightening economic linkages, however, have not solved the region's geopolitical problems. Relations between North Korea and South Korea remain extremely tense, and China continues to regard Taiwan as a renegade province that it will eventually reclaim. Japan's relations with South Korea are complicated by memories of World War II, along with a territorial dispute over several tiny islands. Disputed islands have generated much more serious tension between China and Japan, which some experts think have the potential to escalate into armed conflict. More generally, China's economic rise has been accompanied by a much more assertive foreign policy and a growing military budget, developments that cause much concern among its neighbors, as well as among distant countries, such as the United States.

East Asia as a World Region East Asia is easily marked off on the world map by the territorial extent of its constituent countries. Japan is an island country composed of four main islands and a chain of much smaller islands (the Ryukyus) that extends almost to Taiwan. Taiwan is basically a single-island country, but its political status is ambiguous because China claims it as part of its own territory. China itself is a vast continental state—the third most extensive in the world—that reaches well into Central Asia. It also includes the large island of Hainan in the South China Sea. Korea constitutes a compact peninsula between northern China and Japan, but it is divided into two sovereign states: North Korea and South Korea.

In political terms, such a straightforward definition of East Asia is appropriate. If you turn to cultural considerations, however, the issue becomes more complicated. The main cultural issue concerns the western half of China. This is a huge, but lightly populated area; only about 5 percent of the residents of China live in the west. By some criteria, only the populous eastern half of China (sometimes called **China proper**) fits into the East Asian world region. The indigenous inhabitants of western China are not Chinese by culture and language, and they have never accepted the religious and philosophical beliefs that give historical unity to East Asian civilization. In the northwestern quarter of China, called *Xinjiang* in Chinese, most indigenous inhabitants speak Turkic languages and are Muslim in religion. Tibet, in southwestern China, has a highly distinctive culture, and the Tibetans in general resent Chinese authority. Xinjiang and Tibet can thus be classified within both East Asia and Central Asia (covered in Chapter 10). In this chapter, we examine western China largely to the extent that it is politically part of China.

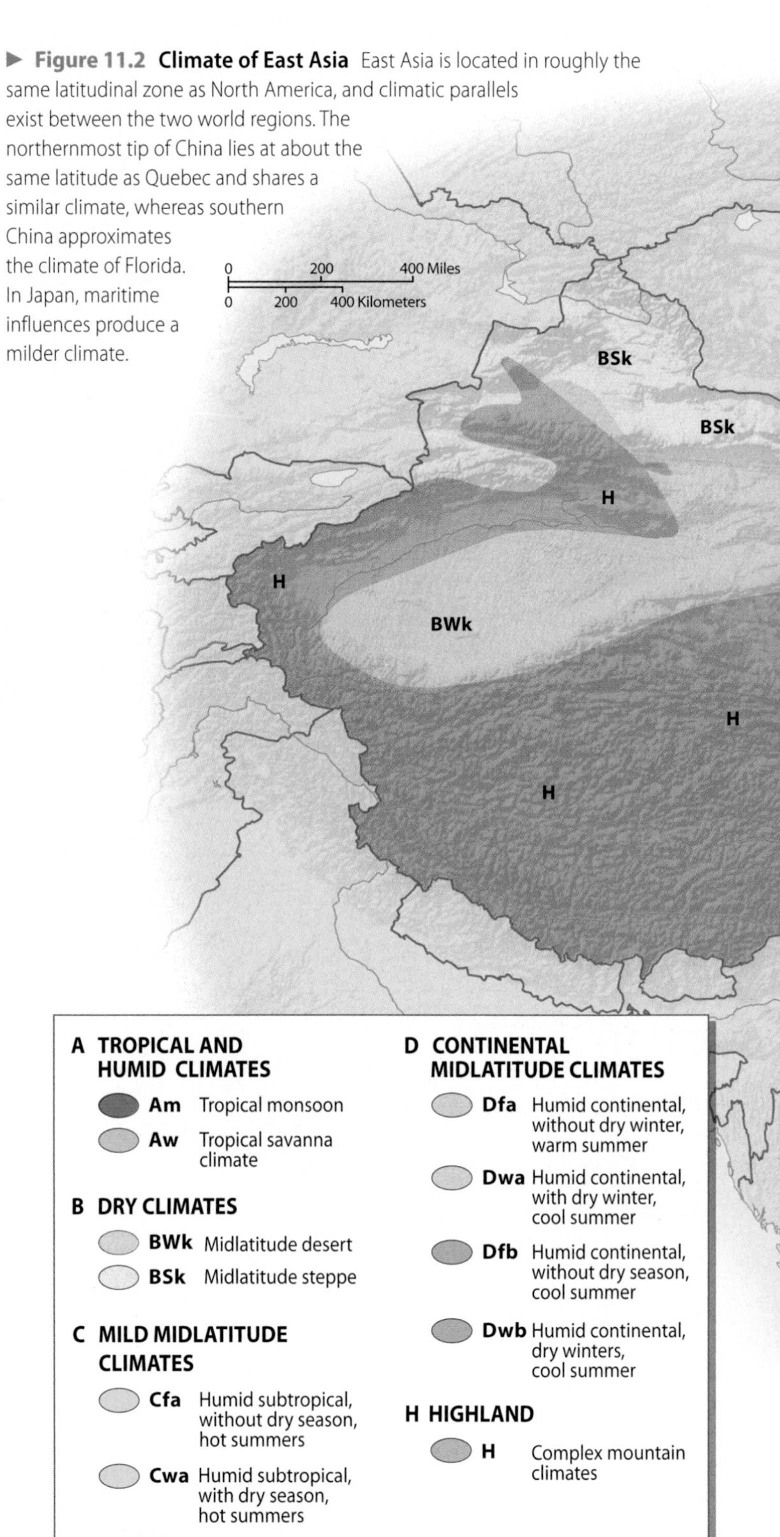

▶ **Figure 11.2 Climate of East Asia** East Asia is located in roughly the same latitudinal zone as North America, and climatic parallels exist between the two world regions. The northernmost tip of China lies at about the same latitude as Quebec and shares a similar climate, whereas southern China approximates the climate of Florida. In Japan, maritime influences produce a milder climate.

Physical Geography and Environmental Issues:

Resource Pressures in a Crowded Land

Environmental problems in East Asia are particularly severe owing to a combination of the region's large population, its massive industrial development, and its physical geography. Steep slopes and heavy rainfall make many areas vulnerable

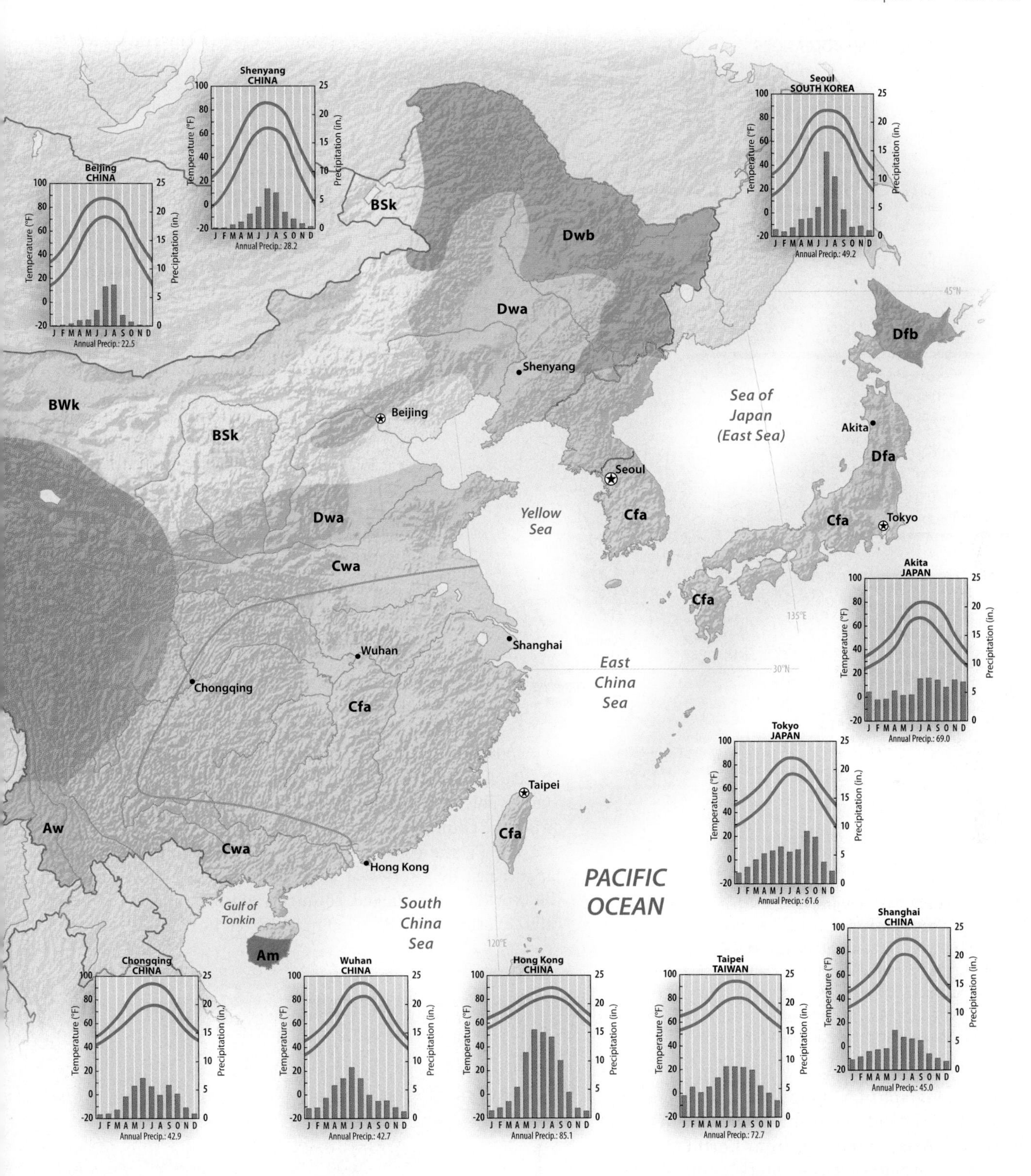

to soil erosion and mudslides, and a seismically active environment generates threats from earthquakes and **tsunamis**. (A tsunami is a large ocean wave, usually caused by an earthquake, that can be very destructive when reaching the coast.) Owing to its rapid economic growth and lax regulations, China probably suffers from the world's most severe air and water pollution. Japan, South Korea, and Taiwan, on the other hand, have invested heavily in environmental protection, resulting in much cleaner natural environments.

East Asia's Physical Geography

East Asia is situated in the same general latitudinal range as the United States, although it extends considerably farther north and south. The northernmost tip of China lies as far north as central Quebec, and China's southernmost point is at the same latitude as Mexico City. Thus, the climate of southern China is roughly comparable to that of southern Florida and the Caribbean, whereas the climate of northern China is more similar to that of south-central Canada (Figure 11.2; note

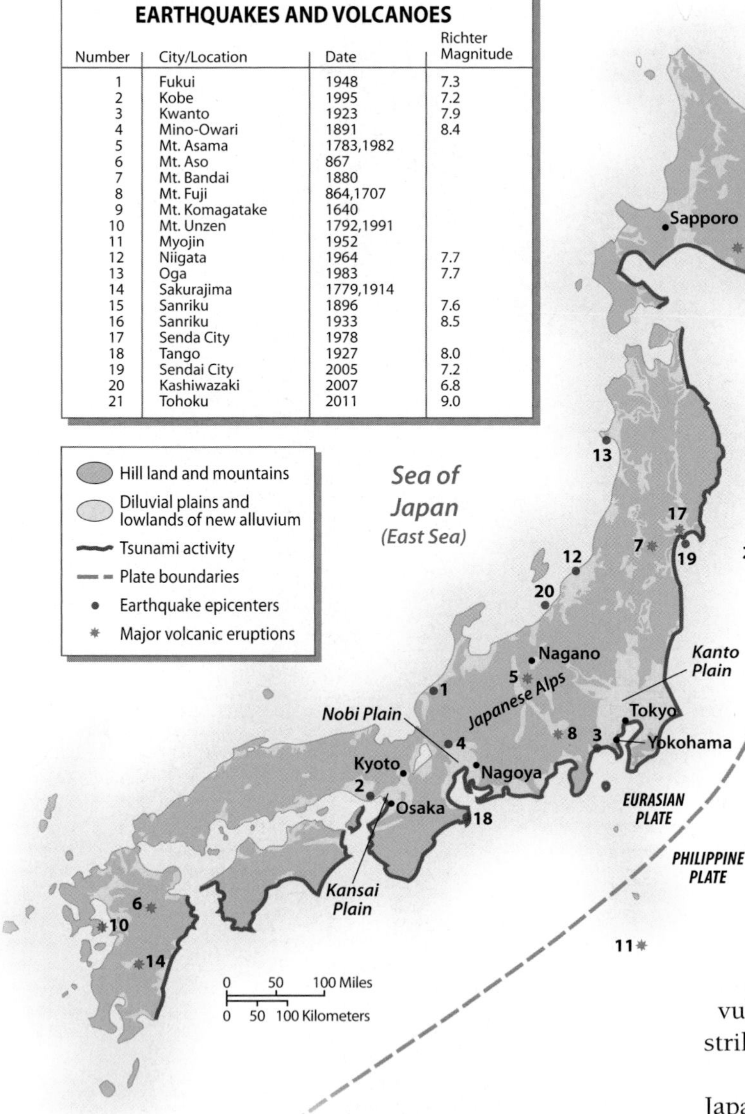

EARTHQUAKES AND VOLCANOES

Number	City/Location	Date	Richter Magnitude
1	Fukui	1948	7.3
2	Kobe	1995	7.2
3	Kwanto	1923	7.9
4	Mino-Owari	1891	8.4
5	Mt. Asama	1783,1982	
6	Mt. Aso	867	
7	Mt. Bandai	1880	
8	Mt. Fuji	864,1707	
9	Mt. Komagatake	1640	
10	Mt. Unzen	1792,1991	
11	Myojin	1952	
12	Niigata	1964	7.7
13	Oga	1983	7.7
14	Sakurajima	1779,1914	
15	Sanriku	1896	7.6
16	Sanriku	1933	8.5
17	Senda City	1978	
18	Tango	1927	8.0
19	Sendai City	2005	7.2
20	Kashiwazaki	2007	6.8
21	Tohoku	2011	9.0

▲ **Figure 11.3 Japan's Physical Geography** Japan has several sizable lowland plains, primarily along the coastline, but they are interspersed among rugged mountains and uplands. Because of its location at the convergence of three major tectonic plates, Japan experiences numerous earthquakes. Volcanic eruptions can also be hazardous and are linked directly to Japan's location on tectonic plate boundaries. Additionally, much of Japan's coast is vulnerable to devastating tsunamis caused by earthquakes in the Pacific Basin.

particularly the climographs for Hong Kong and Chongqing in the south and Beijing and Shenyang farther north). The island belt of East Asia, extending from northern Japan through Taiwan, is situated at the intersection of three tectonic plates (the basic building blocks of Earth's crust): the Eurasian, the Pacific, and the Philippine plates. This area, particularly Japan, is therefore geologically active, experiencing numerous earthquakes and being dotted with volcanoes (Figure 11.3).

Japan's Physical Environment Although slightly smaller than California, Japan extends farther north and south. As a result, Japan's extreme south, in southern Kyushu and the Ryukyu Archipelago, is subtropical, whereas northern Hokkaido is almost subarctic. Most of the country, however, including the main island of Honshu, is temperate. The climate of Tokyo is not unlike that of Washington, DC, although Tokyo is distinctly rainier.

Japan's climate varies not only from north to south, but also from southeast to northwest, across the main axis of the archipelago. In winter, the area facing the Sea of Japan receives much more snow than the Pacific Ocean coastline. During this time of the year, cold winds from the Asian mainland blow across the relatively warm waters of the Sea of Japan. The air picks up moisture over the sea and deposits it, usually as snow, when it hits land (Figure 11.4). The Pacific coast of Japan, on the other hand, is far more vulnerable to typhoons (hurricanes), which frequently strike the country.

The Pacific coast of Japan is separated from the Sea of Japan coast by a series of mountain ranges. Japan is one of the world's most rugged countries, with mountainous terrain covering some 85 percent of its territory. Most of these

▶ **Figure 11.4 Heavy Snow in Japan's Mountains** Cold, moist air moving off the Sea of Japan produces heavy snows in northwestern Japan and along the country's mountainous spine. Numerous major ski areas dot the Japanese Alps, several of which have hosted world-class sports competitions, including the 1998 Winter Olympics.

► **Figure 11.5 2011 Tohoku Earthquake and Tsunami** The March 11, 2011, Tohoku earthquake and tsunami were one of the most devastating natural disasters of recent times. At its extreme, the tsunami wave was over 133 feet (40 meters) high, giving it enough power to wash away entire villages. More than 15,000 people died as a result. **Q: Why did the 2011 Tohoku earthquake and tsunami generate such profound consequences for the Japanese economy and political system?**

uplands are thickly wooded, making Japan also one of the world's most heavily forested countries. Japan owes its lush forests both to its mild, rainy climate and to its long history of forest conservation. For hundreds of years, both the Japanese state and its village communities have enforced strict conservation rules, ensuring that timber and firewood extraction is balanced by tree growth.

Along Japan's coastline and interspersed among its mountains are limited areas of alluvial plains. Although these lowlands were once covered by forests and wetlands, they have long since been cleared and drained for intensive agriculture. The largest Japanese lowland is the Kanto Plain to the north of Tokyo, but it is only some 80 miles wide and 100 miles long (130 by 160 km). The country's other main lowland basins are the Kansai, located around Osaka, and the Nobi, centered on Nagoya. In the mountainous province of Nagano, smaller basins are sandwiched between the imposing peaks of the Japanese Alps.

The Tsunami Threat Japan's location makes it particularly vulnerable to earthquakes, just as its extensive coastlines make it susceptible to tsunamis. The power of such events was demonstrated March 11, 2011, when northeastern Japan was devastated by one of the largest and most destructive natural events of modern times, the 2011 Tohoku earthquake and tsunami (Figure 11.5). Over 15,000 lives were lost. Beyond the loss of lives and homes, the local economy was devastated. In the fishing town of Miyako, some 900 out of 960 fishing boats were ruined, along with the fish market, seafood-processing factories, and refueling facilities and cranes.

The repercussions of the earthquake and tsunami were enormous across Japan. The Japanese government had invested billions of dollars in anti-tsunami seawalls, which were widely believed to provide adequate security, along almost 40 percent of its coastline. The surge of water from the Tohoku tsunami, however, simply washed over the tops of the barriers, collapsing many in the process. Japan, an energy-poor country, had also invested heavily in nuclear power, which was widely regarded as relatively safe and secure. The earthquake and tsunami, however, caused severe damage to the Fukushima Daiichi nuclear power plant, releasing large quantities of radiation and forcing the evacuation of over 200,000 people. The crisis led to widespread loss of confidence in both the Japanese government and nuclear power.

Taiwan's Environment Taiwan, an island about the size of Maryland, sits at the edge of the continental land mass. To the west, the Taiwan Strait is only about 200 feet (60 meters) deep; to the east, ocean depths of many thousands of feet are found 10 to 20 miles (16 to 32 km) offshore.

Taiwan itself forms a large tilted block. Its central and eastern regions are rugged and mountainous, whereas the west is dominated by an alluvial plain. Bisected by the Tropic of Cancer, Taiwan has a mild winter climate, but it is sometimes battered by typhoons in the early autumn. Unlike nearby areas of China proper, Taiwan still has extensive forests, concentrated in its eastern upland areas.

Chinese Environments Even if you exclude the Central Asian provinces of Tibet and Xinjiang, China is a vast country with diverse environmental regions. For the sake of convenience, we can divide China proper into two main areas: Northern China lies to the north of the Yangtze River Valley, and southern China includes the Yangtze and all areas to the south.

Southern China is a land of rugged mountains and hills interspersed with lowland basins. The lowlands of southern China are far larger than those of Japan. One of the most distinctive regions is the former lake bed of central Sichuan (see Figure 11.1). Protected by imposing mountains, Sichuan is noted for its mild winter climate. Eastward from Sichuan, the Yangtze River passes through several other broad basins (in the Middle Yangtze region), partially separated from each other by hills and low mountains, before flowing into a large delta near Shanghai.

South of the Yangtze Valley, the mountains are higher (up to 7000 feet, or 2150 meters, in elevation), but they are still interspersed with alluvial lowlands. Sizable valleys are found in the far south, such as the Xi Basin in Guangdong Province. Here the climate is truly tropical, free of frost. West of Guangdong lie the moderate-elevation plateaus of Yunnan and Guizhou, the former noted for its perennial springlike weather. Finally, to the northeast of Guangdong lies the rugged coastal province of Fujian (Figure 11.6). With its narrow coastal plain and deeply indented coastline, Fujian's landscape does not support a productive agricultural economy; as a result, the Fujianese people have often sought a maritime way of life, many of them working in the fishing and shipping industries.

North of the Yangtze Valley, the climate is both colder and drier than it is to the south. Summer rainfall is generally abundant except along the edge of the Gobi Desert, but the other seasons are dry. Desertification is a major threat in parts of the North China Plain, which have experienced prolonged droughts in recent years. With the exception of a few low mountains in Shandong Province, the entire area is a virtually flat plain. Seasonal water shortages in this area are growing increasingly severe as water withdrawals for irrigation and industry increase, leading to concerns about an impending crisis.

▲ **Figure 11.6 The Fujian Coast of China** The rugged coastal province of Fujian lies in southeastern China. Here the coastal plain is narrow and the shoreline deeply indented, producing a picturesque landscape. Because of limited agricultural opportunities along this rugged coastline, many Fujianese people work in maritime activities.

West of the North China Plain sits the Loess Plateau. This is a fairly rough upland of moderate elevation and uncertain precipitation. It does, however, have fertile soil—as well as huge coal deposits. Farther west are the semiarid plains and uplands of Gansu Province, situated at the foot of the great Tibetan Plateau.

China's far northeastern region is called *Dongbei* in Chinese and Manchuria in English. Manchuria is dominated by a broad, fertile lowland basin sandwiched between mountains stretching along China's borders with North Korea, Russia, and Mongolia. Although winters here can be brutal, summers are usually warm and moist. Manchuria's peripheral uplands have some of China's best-preserved forests and wildlife refuges.

◀ **Figure 11.7 The Mountains and Lowlands of Korea** Like Japan, Korea has extensive uplands interspersed with lowland plains. The country's highest mountains are in the north, whereas its most extensive alluvial plains are in the south. South Korea's provinces, shown here, are culturally as well as physically distinctive.

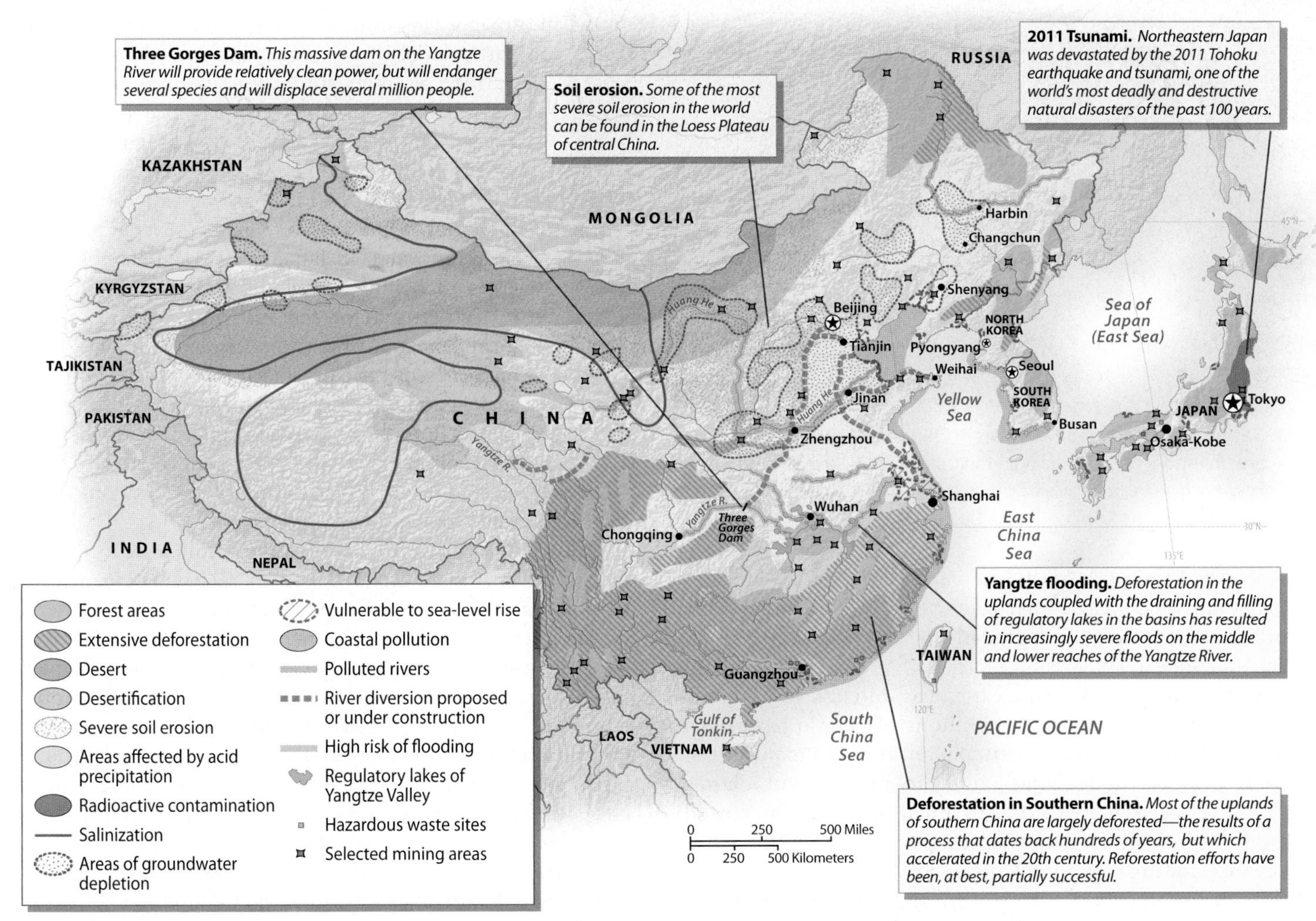

▲ **Figure 11.8 Environmental Issues in East Asia** This vast world region has been almost completely transformed from its natural state and continues to have serious environmental problems. In China, some of the more pressing environmental issues involve deforestation, flooding, water control, and soil erosion.

Korean Landscapes Korea forms a well-demarcated peninsula, partially cut off from Manchuria by rugged mountains and sizable rivers (Figure 11.7). Like Japan, its latitudinal range is pronounced. The far north, which just touches Russia's Far East, has a climate not unlike that of Maine, whereas the southern tip is more reminiscent of the Carolinas. Korea is a mountainous country with scattered alluvial basins, a landscape that has, as in many other parts of East Asia, deeply influenced its demographic and agricultural development. The lowlands of the southern portion of the peninsula are more extensive than those of the north, giving South Korea a distinct agricultural advantage over North Korea. However, North Korea has more abundant mineral deposits and hydroelectric resources.

In the early 20th century, the northern half of the Korean Peninsula also had much more extensive forests than the south. Over the past several decades, South Korea has made substantial progress in reforesting its extensive mountain areas. In the same period, impoverished North Korea has experienced an increase in deforestation, resulting in a stark landscape contrast on either side of the Demilitarized Zone (DMZ) that separates the two states.

East Asia's Environmental Challenges

Deforestation is a major problem not just in North Korea, but in much of eastern China as well. Pollution and the reduction of biological diversity are also major issues in China and other parts of East Asia. The region as a whole, moreover, has a serious shortage of energy resources (Figure 11.8).

Deforestation and Desertification Many of the upland regions of China and North Korea support only grass, meager scrub, and stunted trees (Figure 11.9). China lacks the historical tradition of forest conservation that characterizes Japan. During earlier periods of Chinese history, hillsides were often cleared of wood for fuel, and in some instances entire forests were burned for ash that could be used as fertilizer. In much of southern China, sweet potatoes, maize, and other

▲ **Figure 11.9 Denuded Hillslopes in China** Because of the need to clear forests for wood products and agricultural lands, China's mountain slopes have long been deforested. Without forest cover, soil erosion is a serious issue.

crops have been grown on steep and easily eroded hillsides for hundreds of years. After centuries of exploitation, many hillslopes are now so degraded that they cannot easily regenerate forests.

Over the past 50 years, China has initiated several successful reforestation efforts. Millions of people have been mobilized by the government to plant trees. In many areas, such projects have failed, but the government claims that overall the area covered by forests has increased from 9 percent in 1949 to 13 percent in 2008. After severe flooding along the Yangtze River in 1998, reforestation efforts in central China were significantly enhanced. Still, it will be many years before such new forests can be logged. At present, substantial timber reserves are found only in China's far northeast, where a cold climate limits tree growth, and along the eastern slopes of the Tibetan Plateau, where rugged terrain restricts commercial forestry. As a result, China suffers a severe shortage of forest resources.

Desertification is also a major problem for China. In many areas, sand dunes have pushed south from the Gobi Desert, smothering farmlands. Some reports claim that more desertification occurs in China than in any other country. The Chinese government has responded with massive dune-stabilization schemes, which usually involve planting grasses to stop sand from blowing away (Figure 11.10). The most ambitious of such projects involved planting more than 300 million trees along a 3000-mile (4800-km) swath of northern China. This "great green wall" was designed both to stop the southward expansion of the desert and to reduce the dust storms that often hit northern China. Overall, the rate of desertification in China is reported to have declined from 1300 square miles (3400 square km) per year in the 1990s to 500 square miles (1300 square km) in the early 2000s.

Mounting Pollution As China's industrial base expands, other environmental problems, such as water pollution and toxic-waste dumping, are growing more acute, particularly in the booming coastal areas. The burning of high-sulfur coal has resulted in severe air pollution, a problem aggravated by an increasing number of automobiles (Figure 11.11). According to a World Health Organization (WHO) report, air pollution kills 656,000 Chinese citizens each year, while another 95,000 die from polluted drinking water. In January 2013, an air pollution crisis in Beijing forced the Chinese government to restrict outdoor activities and suspend certain forms of industrial production. Although Chinese citizens have organized thousands of protests against pollution and other forms of environmental destruction, the government has generally insisted that economic growth remain the top priority. China's government does, however, support the development of clean, renewable sources of energy.

Considering Japan's large population and intensive industrialization, its environment is relatively clean. The very density of its population gives certain environmental advantages, such as allowing a highly efficient public transportation system. In the 1950s and 1960s, Japan did suffer from some of the world's worst pollution. Soon afterward the Japanese government passed stringent environmental laws covering both air and water pollution. Japan has continued to make progress on these fronts (see *Working Toward Sustainability: Japan's Smart City Movement*).

Japan's cleanup was aided by its insular location because winds usually carry smog-forming chemicals out to sea. Equally important has been the phenomenon of **pollution exporting**. Because of Japan's high cost of production and its strict environmental laws, many Japanese companies have moved their dirtier factories overseas. In effect, Japan's pollution has been partially displaced to poorer countries. The same argument can also be made about the United States and western Europe, but to a somewhat lesser extent. Taiwan and South Korea have also followed Japan's footsteps by setting up new factories in countries—including China—that have less strict environmental standards.

Endangered Species The growing number of endangered species in East Asia has been linked to the region's economic rise. Many forms of traditional Chinese medicine are based on products derived from rare and exotic animals. Deer antlers, bear gallbladders, snake blood, tiger penises, and rhinoceros horns are believed to have medical effectiveness, and certain individuals will pay fantastic sums of money to purchase them. As wealth has accumulated, trade in such substances has expanded. China itself has relatively little remaining wildlife, but other areas of the world help supply its demand for wildlife products.

China has, however, moved strongly to protect some of its remaining areas of habitat. Among the most important of these are the high-altitude forests and bamboo thickets of

◄ **Figure 11.10 Dune Stabilization in China** The spread of sand dunes is a significant hazard in dry areas of northern China. By spreading dried vegetation and then planting grass seeds, the dunes can be stabilized.

▼ **Figure 11.11 Air Pollution in China** Major Chinese cities suffer from some of the worst air pollution in the world. As evidenced in this photo of Shanghai, the resulting haze can make it difficult to see for more than a few city blocks.

Working Toward Sustainability

Japan's Smart City Movement

Japan's historical encouragement of dense settlement has made the country a leader in energy efficiency almost by default. Nevertheless, Japan is embarking on several joint public-private endeavors to make its communities more sustainable. Four cities have joined the Next-Generation Energy and Social Systems Verification Experiment, a project aimed at establishing working models of urban sustainability that can later be adopted more easily at a large scale. A key theme in the experiment is demand-side energy management. Demand for electricity usually peaks during the evening, but since solar electricity production can take place only when the Sun shines, converting a normal city to run on solar power requires either extremely costly batteries or backup power plants, which generate additional pollution. The aim of this experiment is to establish the technologies and practices required to shift electricity demand from traditional peak times to times when the Sun is shining brightly.

The Yokohama Smart City Project, a joint venture between the city government and the Toshiba Corporation, is perhaps the most ambitious of these undertakings, involving thousands of residential units and several large industrial and commercial buildings. A key goal of the undertaking is to encourage the adoption of electric vehicles and ensure that many or most are charged during times of peak solar power. The project also aims to give power consumers, including businesses and factories, more information on their electricity consumption habits so that they can use the information to lower both their own electric bills and the social cost of electricity.

Fujisawa Sustainable Smart Town (SST) is another public-private project intended to create an energy-efficient community of 1000 households from scratch (Figure 11.1.1). The town aims to satisfy 30 percent of its energy needs with solar panels arrayed on rooftops and in public areas and to reduce water consumption by 30 percent and carbon dioxide emissions by 70 percent relative to similar towns. Fujisawa SST plans to emphasize car sharing, electric vehicles, and bicycles. It will also introduce a Fujisawa Smart Points program to offer residents rewards for actions taken that reduce pollution and energy use.

▲ **Figure 11.1.1 Fujisawa Sustainable Smart Town** The president of Panasonic Corporation, the mayor of Fujisawa City, and other dignitaries are shown here at a new conference focused on the Fujisawa Sustainable Smart Town initiative.

western Sichuan Province, home of the panda bear. Efforts are also being made to preserve wilderness in northern Manchuria, where evidence of a surviving population of Siberian tigers was discovered in 1997, and in the canyon lands of northwestern Yunnan, which were declared off-limits to loggers in 1999.

Wildlife is also scarce in Korea. Ironically, however, the so-called Demilitarized Zone that separates North from South Korea functions as an unintentional wildlife sanctuary, supporting populations of several endangered species, including the Asiatic black bear.

Energy in East Asia East Asia's rapid industrialization has forced the region to expand its energy generation at breakneck speed. China is now the world's largest energy consumer and derives 70 percent of that energy from coal (Figure 11.12). China also consumes more oil than any other country except the United States. The extreme air pollution that plagues China's urban centers has led the country to seek alternative sources of energy. Hydropower is a significant source of electricity in China, accounting for 22 percent of its installed capacity in 2009. Roughly 3 percent of China's electricity comes from the large—and controversial—Three Gorges Dam. China is currently building several large dams in the rugged lands of Tibet and Yunnan, causing concern in the downstream countries of South and Southeast Asia. China has also become an investment leader in other forms of renewable energy, with plans to spend $473 billion on wind, solar, and hydroelectric power between 2011 and 2015

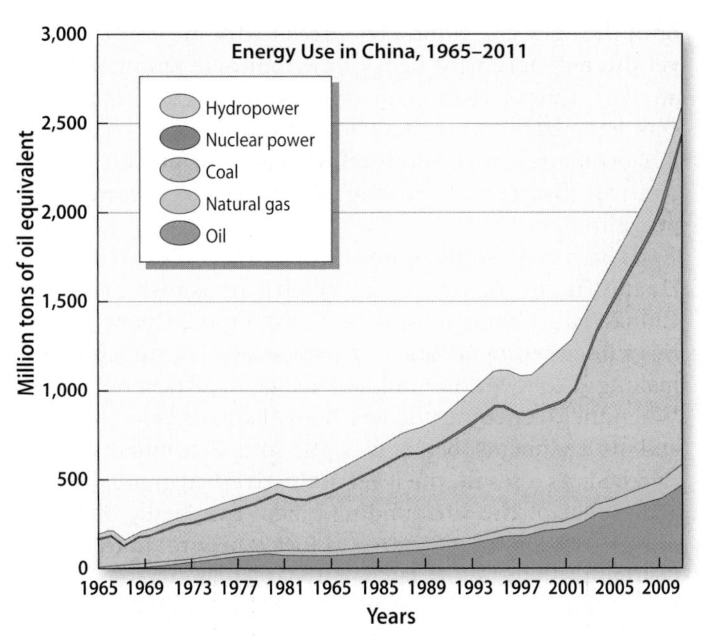

▲ **Figure 11.12 Energy Use in China** Most of China's surging energy demand is being met by burning coal and oil.

as part of the country's latest Five-Year Plan. China is already the world's largest producer of hydroelectric power and the second largest producer of wind power (Figure 11.13).

Throughout its industrialized past, Japan has strained to deal with a relative scarcity of natural energy resources. The country lacks significant coal, oil, and natural gas reserves and depends almost entirely on imported fuel. Oil, coal, and natural gas together make up 82 percent of Japan's total energy use, with oil leading the way at 42 percent. Hydroelectric power in Japan is an important source of domestic energy, contributing 7 percent of the country's electricity and 3 percent of its total energy.

Nuclear energy accounted for 13 percent of Japan's energy consumption until the 2011 earthquake and tsunami. The disaster resulted in the eventual shutdown of all Japan's nuclear generation capacity, though most of Japan's reactors are set to resume electricity generation in the near future. During the interim, most of the demand has been filled by increased consumption of natural gas. Current political opposition to nuclear power in Japan likely means that the country will need to rely increasingly on other power sources, such as natural gas or renewables in the medium to long term. South Korea, which faces energy challenges similar to those of Japan, reaffirmed its commitment to nuclear energy in the wake of Fukushima. South Korea's scheduled goal is to generate almost half its electricity from nuclear reactors by 2024.

Dams, Flooding, and Soil Erosion in China

Most of China's severe environmental problems also characterize other densely populated, rapidly industrializing countries. Other issues, however, are unique to the China, linked by the country's specific environmental conditions. Many of these problems are focused on dams, flooding, and soil erosion. The building of the Three Gorges Dam on the Yangtze River in particular generated a global environmental controversy.

The Three Gorges Dam Controversy The Yangtze River (also called the Chang Jiang) is one of the most important physical features of East Asia. This river, the third largest (by volume) in the world, emerges from the Tibetan highlands onto the rolling lands of the Sichuan Basin and passes through what used to be a magnificent canyon in the Three Gorges area (Figure 11.14). It then meanders across the lowlands of central China before entering the sea near the city of Shanghai. The Yangtze has historically been the main avenue of entry into the interior of China and is celebrated in Chinese literature for its beauty and power. Since the 1990s, however, it has become the focal point of an environmental controversy of global proportions.

Finished in 2006, the Three Gorges Dam—600 feet (180 meters) high and 1.45 miles (2.3 km) long—is the largest hydroelectric dam in the world, requiring $26 billion to build. The reservoir it created is 350 miles (563 km) long,

▼ **Figure 11.13 Chinese Windfarm** China has a large potential for wind energy, and has been building windfarms at a rapid rate. This photograph shows newly constructed windmills in China's Inner Mongolia Autonomous Region.

has displaced an estimated 1.2 million people, and has inundated a major scenic attraction (the Three Gorges of the Yangtze, for which the dam is named). The dam also traps sediment and pollutants in the reservoir and disrupts habitat for several endangered species, including the Yangtze River dolphin. The ecological and human rights consequences of the dam are so negative that the World Bank withdrew its support before the project was completed. The Chinese government, however, argues that the clean power generated by the dam offsets its environmental problems.

Chinese planners also praise the water-control benefits provided by dams. The lower and middle stretches of the Yangtze periodically suffer from devastating floods, and planners hope that the Three Gorges Dam will store enough water to help protect 15 million people and over 3 million acres (1.2 million hectares) of farmland on the lower Yangtze floodplains. However, flooding has been exacerbated over the past several decades by the declining area of lakes on the Yangtze floodplain, which function as overflow areas during floods. East of Sichuan, the Yangtze passes through several basins (in Hunan, Hubei, and Jiangxi provinces), each containing a group of lakes. During periods of high water, river flows are diverted into these **regulatory lakes,** thus reducing the flow downstream. After the flood season is over, water drains from the lakes and therefore helps maintain the water level in the river downstream. In the 1950s and 1960s, however, the Chinese government drained many of these lakes in order to convert the lake beds to farmland. As a result, flooding downstream intensified.

Flooding in Northern China The North China Plain, which has been deforested for thousands of years, is plagued by both drought and flood. This area is dry most of the year, yet often experiences heavy downpours in summer. Since ancient times, large-scale hydraulic engineering projects have both controlled floods and allowed irrigation. But no matter how much effort has been put into water control, disastrous flooding has never been completely prevented.

The worst floods in northern China are caused by the Huang He, or Yellow River, which cuts across the North China Plain. Owing to upstream erosion, the Huang He carries a huge **sediment load,** or suspended clay, silt, and sand, making it the world's muddiest major river (Figure 11.15). When the river enters the low-lying plain, its velocity slows, and its sediments begin to settle and accumulate in the riverbed. As a result, the level of the riverbed gradually rises above that of the surrounding lands. Eventually, the river must break free of its course to find a new route to the sea over lower-lying ground. Twenty-six such course changes have been recorded for the Huang He throughout Chinese history.

Through the process of sediment deposition and periodic course changes, the Huang He has actually created the vast North China Plain. In prehistoric times, the Yellow Sea extended far inland. Even today the sea is retreating as the Huang He's delta expands; one study revealed a 6-mile (9.6 km) advance of the land in a three-year period. Such a process occurs in other alluvial plains, but nowhere else in the world is it so pronounced. Nowhere else, moreover, is it so destructive. The world's two deadliest recorded natural disasters, each of which killed between 1 and 2 million people, involved flooding of the Yellow River (in 1887 and 1931). Since ancient times, the Chinese have attempted to keep the river within its banks by building progressively larger dikes. Eventually, however, the riverbed rises so high

▼ **Figure 11.14 The Three Gorges of the Yangtze** (a) The Three Gorges on the Yangtze River was once one of China's most scenic natural attractions. (b) It is now flooded by a reservoir created by the Three Gorges Dam, which produces more hydroelectricity than any other facility in the world.

(a)

(b)

▲ **Figure 11.15 Massive Discharge on China's Yellow River** The Yellow River (or Huang He) is considered to be the world's muddiest major river. Silt deposits increase the risk of flooding and clog up reservoirs. Here visitors view a massive gush of water released from the Xiaolangdi Reservoir in order to clear out the accumulated silt deposits.

that the flow can no longer be contained and catastrophic flooding results. The river has not changed its course since the 1930s, but most geographers think that another course correction is inevitable.

Although flooding is occasionally catastrophic along the Huang He, the more common problem is a lack of water. The lower course of the river sometimes dries out completely, resulting in serious water shortages (Figure 11.16). As a result, China is currently working on several major projects designed to transfer water from the Yangtze River to the Huang He. Diversion canals designed to link the two rivers are planned in both the headwaters area on the Tibetan Plateau and the North China Plain. Current cost estimates put these projects at roughly US$62 billion, a figure more than twice that required for the Three Gorges Dam. China is also developing desalinization plants in the region, but the resulting water is very expensive.

Erosion on the Loess Plateau The Huang He's sediment burden is derived from the eroding soils of the Loess Plateau, located to the west of the North China Plain. **Loess** consists of fine, windblown sediment that was deposited on this upland area during the last ice age, accumulating in some places to depths of several hundred feet.

Loess forms fertile soil, but it washes away easily when exposed to running water. At the dawn of Chinese civilization, the semiarid Loess Plateau was covered with tough grasses and scrubby forests that retained the soil. Chinese farmers, however, began to clear the land for the abundant crops that it yields when rainfall is adequate. Cultivation required plowing, which, by exposing the soil to water and wind, exacerbated erosion. As the population of the region gradually increased, the remaining areas of woodland diminished, leading to faster rates of soil loss. As the erosion process continued, great gullies cut across the plateau, steadily reducing the extent of arable land.

Today the Loess Plateau is one of the poorest parts of China. Population is only moderately dense by Chinese standards, but good farmland is limited, and drought is common. The Chinese government has long encouraged the construction of terraces to conserve the soil, but such efforts have not been effective everywhere. More recently, the World Bank and the government of China have pushed through the Loess Plateau Watershed Rehabilitation Project, which has achieved considerable success in some areas (Figure 11.17). An estimated 51 square miles (133 square km) of degraded land have been fully reforested.

▼ **Figure 11.16 Drought in Northern China** The Yellow River (or Huang He) is sometimes called the "cradle of Chinese civilization" owing to its historical importance. However, due to the increasing extraction of water for agriculture and industry, the river now often runs dry in its lower reaches. In this photo, two children are playing on a boat in the dried-up riverbed in China's Henan Province.

▲ **Figure 11.17 Reforestation on the Loess Plateau** Large-scale eco-development projects funded by the World Bank and other organizations have finally begun to reduce and even reverse the environmental degradation of China's vulnerable Loess Plateau. Terracing and tree-planting efforts have been especially successful. **Q: Why have such large ridges been constructed around every tree planted on this slope, and why are they shaped this way?**

Climate Change in East Asia

East Asia has come to occupy a central position in global warming debates, largely because of China's rapid increases in carbon emissions. From a level only about half that of the United States in 2000, China's total production of greenhouse gases (GHGs) surpassed that of the United States in 2007 (Figure 11.18). This staggering rise has been caused both by China's explosive economic growth and by the fact that it relies on burning coal to meet most of its energy needs.

The potential effects of global warming in China have serious implications for human populations. According to the 2007 Chinese National Climate Change Assessment, the country's production of wheat, corn, and rice could fall by as much as 37 percent if average temperatures increase by 3.5 to 5.5°F (2 to 3°C) over the next 50–80 years. Increased evaporation rates, coupled with the melting of glaciers on the Tibetan Plateau, could greatly intensify the water shortages that already plague much of northern China. In the wet zones of southeastern China, global warming concerns center on the possibility of more intense storms in this already flood-prone region.

For many years, the Chinese government insisted that economic growth take priority over reducing GHG emissions. Officials argued that China's per capita emissions were still far below those of the United States and that wealthier countries should take the lead in developing climate-friendly technologies. Such attitudes, however, began to change significantly around 2008, when the Chinese government began to push hard for energy efficiency and to invest massively in renewable energy. These steps have not yet had a large impact on the country. Today, as China's industrial output and standard of living continue to surge ahead, so, too, does its production of GHGs. Chinese CO_2 emissions are not predicted to peak until around 2030.

China's contribution to the causes of global warming has overshadowed those of other East Asian countries. Japan, South Korea, and Taiwan are all major emitters of GHGs, but all also have energy-efficient economies, releasing far less carbon than the United States on a per capita basis. Overall, Japan has been a strong proponent of international treaties designed to force reductions in GHG emissions. In 2009, for example, its government pledged to reduce its carbon footprint by 25 percent by 2020. The earthquake and tsunami catastrophe of 2011, however, undermined all such plans. After shutting down its nuclear power plants in the wake of the disaster, Japan was forced to burn more fossil fuels to generate electricity, increasing its GHG emissions by roughly 4 percent. Japan is now engaged in a divisive national debate about the future of nuclear power and the possibilities of developing alternative sources of energy.

▼ **Figure 11.18 Annual Carbon Dioxide Emissions from China and Other Countries** As can be seen in this graph, China's carbon dioxide emissions have surged ahead, overtaking those of other countries.

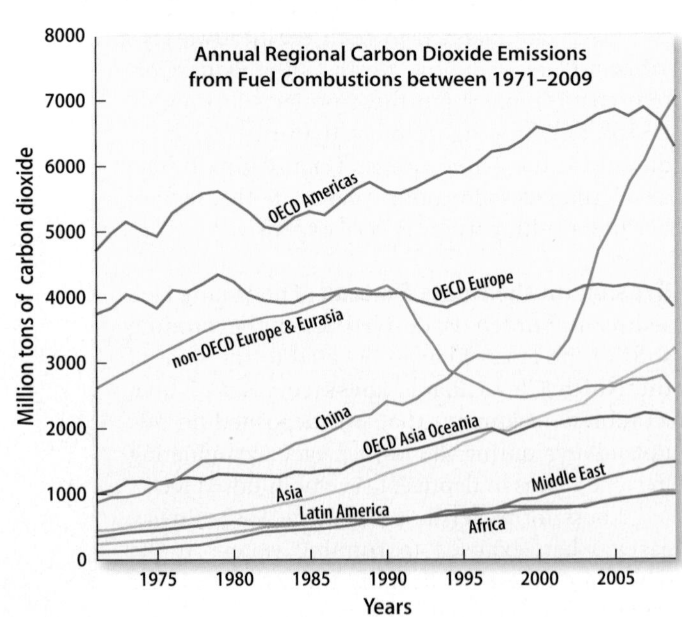

REVIEW

11.1 Why has China become the world's largest emitter of GHGs, and what is its government doing about this problem?

11.2 Why is Japan so much more heavily forested than China?

11.3 How are the basic patterns of physical geography in China linked to the country's historical problems with floods and droughts?

Population and Settlement:

A Realm of Crowded Lowland Basins

East Asia, along with South Asia, is the most densely populated region of the world (Table 11.1). The lowlands of Japan, Korea, and China are among the most intensely used portions of Earth, containing East Asia's major cities and most of its agricultural lands (Figure 11.19). Although the population density of East Asia is extremely high, the region's demographic growth rate has plummeted since the 1970s. In Japan especially, the current concern is population loss, leading to an aging population that will need to be supported by a shrinking number of younger workers. China's population is much younger than that of Japan (Figure 11.20), but it is heading in the same direction.

Agriculture and Settlement in Japan

Japan is a highly urbanized country, supporting two of the largest urban agglomerations in the world: Tokyo and Osaka. Yet it is also one of the world's most mountainous countries, with lightly inhabited uplands. Agriculture must therefore share the limited lowlands with cities and suburbs, resulting in extremely intensive farming practices.

Japan's Agricultural Lands Japanese agriculture is largely limited to the country's coastal plains and interior basins. Japanese rice farming has long been one of the most productive forms of agriculture in the world, helping support a population of 127 million people on a relatively small and rugged land. Although rice is grown in almost all Japanese lowlands, the country's premier rice-growing districts lie along the Sea of Japan coast in Honshu. Vegetables and even rice crops are cultivated intensively on tiny patches within suburban and urban neighborhoods (Figure 11.21). The valleys of central and northern Honshu are famous for their temperate-climate fruit, while citrus comes from the milder southwestern reaches of the country. Crops that thrive in a cooler climate, such as potatoes, are produced mainly in Hokkaido and northern Honshu.

Settlement Patterns All Japanese cities—and the vast majority of the Japanese people—are located in the same lowlands that support the country's agriculture. Not surprisingly, the three largest metropolitan areas—Tokyo, Osaka, and Nagoya—sit near the centers of the three largest plains.

The fact that Japan's settlements are largely restricted to roughly 15 percent of its land area means that the country's effective population density—the actual crowding it experiences—is one of the highest in the world. This is especially notable in the main industrial belt, which extends from Tokyo through Nagoya and Osaka and then along the Inland Sea (the maritime region sandwiched between Shikoku, western Honshu, and Kyushu) to the northern coast of Kyushu. This area is perhaps the most intensively used part of the world for human habitation, industry, and agriculture.

Japan's Urban-Agricultural Dilemma Due to such space limitations, all Japanese cities are characterized by dense settlement patterns. In the major urban areas, the amount of available living space is highly restricted for all but the most affluent families. Critics have long argued that Japan should allow its cities and suburbs to expand into nearby rural areas. However, because most uplands are too steep for residential use, such expansion would have to come at the expense of farmland.

As it now stands, the wholesale conversion of farmland to neighborhoods would be difficult. Although most farms are extremely small (the average is only several acres) and only marginally viable, farmers are politically powerful, and

Table 11.1 Population Indicators

Country	Population (millions) 2013	Population Density (per square kilometer)	Rate of Natural Increase (RNI)	Total Fertility Rate	Percent Urban	Percent <15	Percent >65	Net Migration (Rate per 1000)
China	1,357.4	142	0.5	1.5	53	16	9	−0
Hong Kong	7.2	6,556	0.7	1.3	100	11	14	2
Japan	127.3	337	−0.2	1.4	91	13	25	1
North Korea	24.7	205	0.5	2.0	60	22	9	−0
South Korea	50.2	505	0.4	1.3	82	16	11	1
Taiwan	23.4	649	0.3	1.3	78	15	11	1

Source: Population Reference Bureau, *World Population Data Sheet, 2013.*

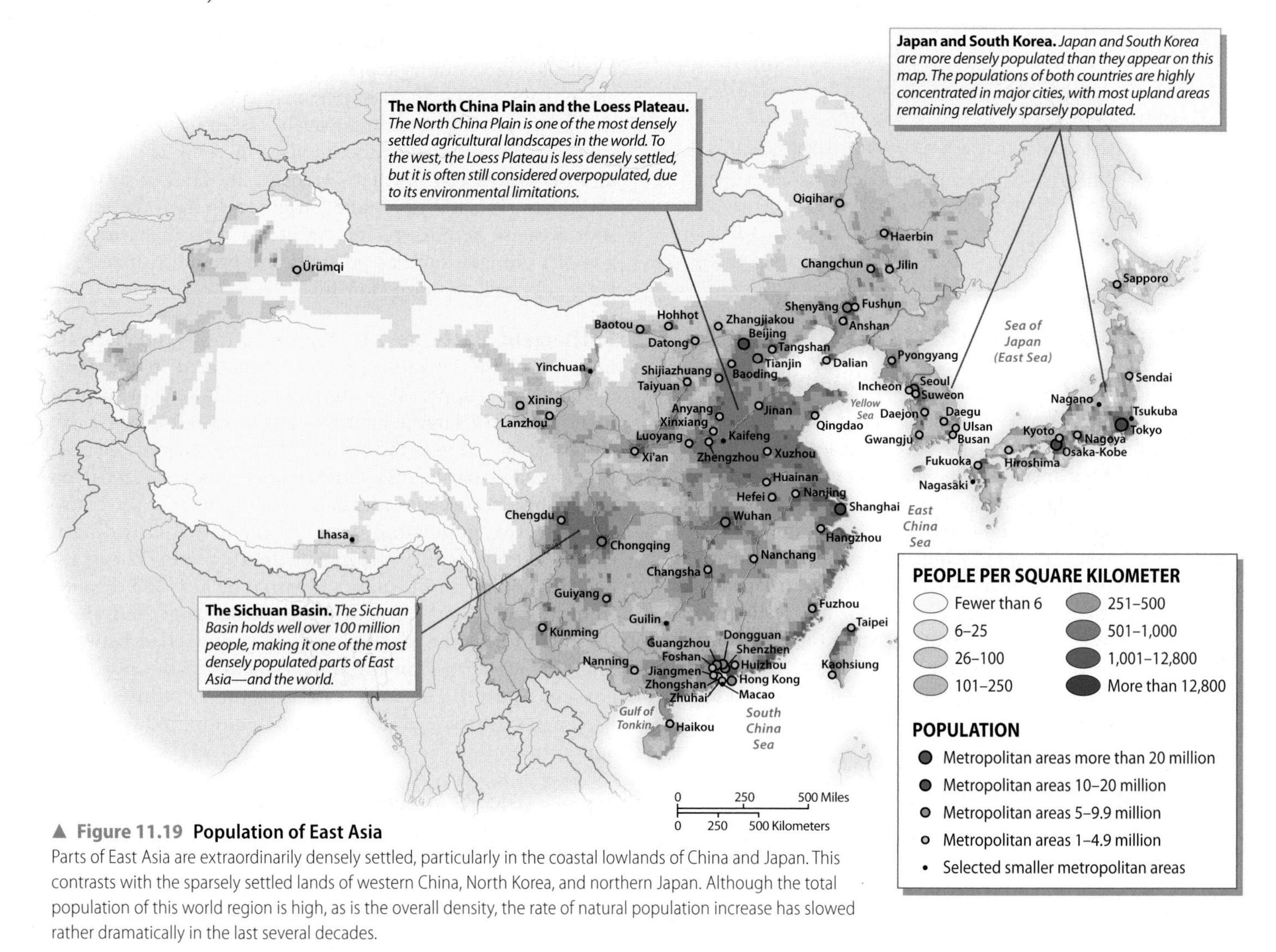

▲ **Figure 11.19 Population of East Asia**
Parts of East Asia are extraordinarily densely settled, particularly in the coastal lowlands of China and Japan. This contrasts with the sparsely settled lands of western China, North Korea, and northern Japan. Although the total population of this world region is high, as is the overall density, the rate of natural population increase has slowed rather dramatically in the last several decades.

▼ **Figure 11.20 Population Pyramids of China and Japan** (a) China has a balanced demographic profile, but it also has a low birth rate, which could result in problems similar to Japan's in another 10 to 20 years. (b) Japan has one of the world's oldest and most rapidly aging populations, due to both its low birth rate and its high life expectancy. The large Japanese population in the over-60 category places a heavy burden on the country's economy.

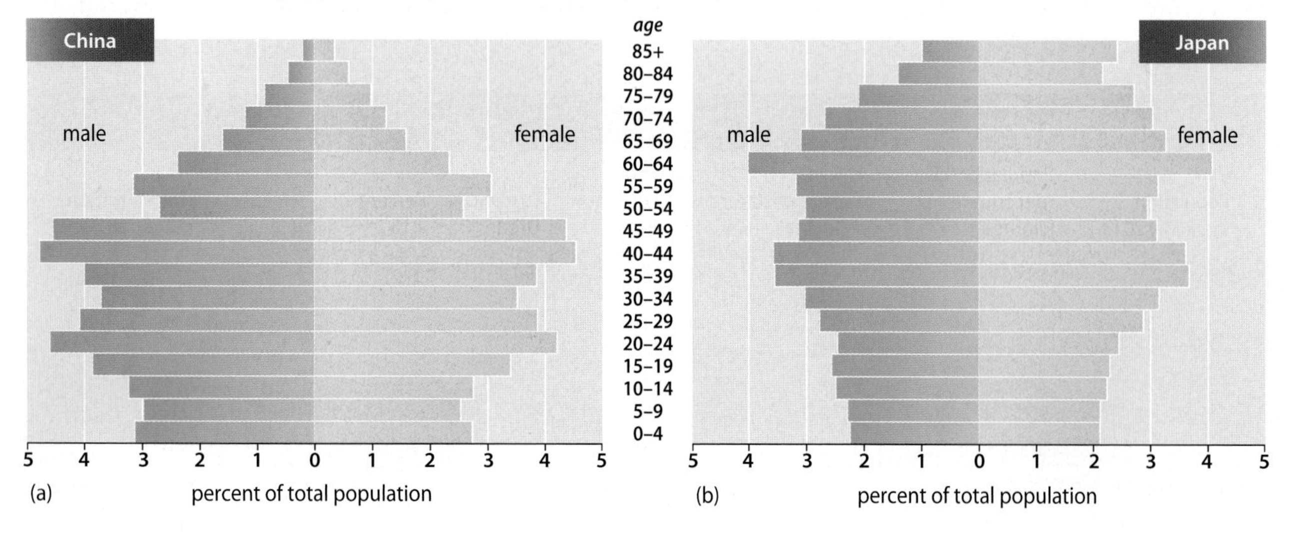

▲ **Figure 11.21 Japanese Urban Farm** Japanese landscapes often combine dense urban settlement with small patches of intensively farmed land. Here a small farm is immediately adjacent to an urban neighborhood.

croplands are often protected by the tax code. Moreover, most Japanese citizens believe that it is vitally important for their country to remain self-sufficient, particularly in rice. Rice imports are therefore highly restricted.

Economic interests in the United States, along with some Japanese consumer advocates, stress the fact that the present system forces Japanese consumers to pay several times the world market price for rice, their staple food. Critics also point out that Japan relies on imports for more than half of its overall food needs, as well as for the energy needed to grow rice. But whatever arguments are made, change will come only gradually, if at all. For the moment, Japan must live with the tensions resulting from the fact that its affluent population is forced to live in tight proximity and to pay high prices for basic staples.

Agriculture and Settlement in China, Korea, and Taiwan

Like Japan, Taiwan and Korea are urban societies. China has recently passed the 50 percent urbanization level, and its government predicts that 70 percent of Chinese citizens will live in cities by 2035. Chinese cities are rather evenly distributed across the plains and valleys of eastern China. As a result, the overall pattern of population distribution in China closely follows the geography of agricultural productivity.

China's Agricultural Regions A line drawn just to the north of the Yangtze Valley divides China into two main agricultural regions. To the south, rice is the dominant crop. To the north, wheat, millet, and sorghum are the most common.

In southern and central China, population is highly concentrated in the broad lowlands, which are famous for their fertile soil and intensive agriculture. More than 100 million people live in the Sichuan Basin, while more than 70 million reside in and near the Yangtze River Delta in Jiangsu, a province smaller than Ohio. Growing crops occurs year-round in most of southern and central China; summer rice alternates with winter barley or vegetables in the north, whereas two rice crops can be harvested in the far south. Southern China also produces a wide variety of tropical and subtropical crops, and moderate slopes throughout the area supply sweet potatoes, corn, and other upland produce.

The North China Plain is one of the world's most thoroughly **anthropogenic landscapes;** that is, it has been heavily transformed by human activities. Virtually its entire extent is either cultivated or occupied by houses, factories, and other structures of human society. Too dry in most areas for rice, the North China Plain supports the main crops of wheat, millet, and sorghum.

Northeastern China, which is usually called Manchuria in English, was a lightly populated frontier zone as recently as the mid-1800s. Today, with a population of more than 100 million, its central plain is thoroughly settled. Still, Manchuria remains less crowded than many other parts of China, and it is one of the few parts of China to produce a consistent food surplus.

The Loess Plateau is more thinly settled yet, supporting only some 70 million inhabitants. However, considering the area's aridity and widespread soil erosion, this is a high figure.

Like other portions of northern China, the Loess Plateau produces wheat and millet. One of its unique settlement features is subterranean housing. Although loess erodes quickly under the impact of running water, it holds together well in other circumstances. For millennia, villagers have excavated pits on the surface of the plateau, from which they have tunneled into the earth to form underground houses (Figure 11.22). These subterranean dwellings are cool in the summer and warm in the winter. Unfortunately, they tend to collapse during earthquakes. One quake in 1920 killed an estimated 100,000 persons; another in 1932 killed some 70,000.

Patterns in Korea and Taiwan Like China and Japan, Korea is a densely populated country. It contains some 75 million people (25 million in the north and 50 million in the south) in an area smaller than Minnesota. South Korea's population density is approximately 1200 per square mile (490 per square km), a density higher than Japan's. Most of South Korea's people are crowded into the alluvial plains and basins of the west and south. South Korean agriculture is dominated by rice. North Korea, in contrast, relies heavily on corn and other upland crops that do not require irrigation.

Taiwan is the most densely populated state in East Asia. Roughly the size of the Netherlands, it contains more than 22 million inhabitants. Its overall population density is more than 1500 people per square mile (645 per square km), one of the highest figures in the world. Because mountains cover most of central and eastern Taiwan, virtually the entire population is concentrated in the narrow lowland belt in the north and west. Large cities and numerous factories are scattered here amid lush farmlands. Despite its highly productive agriculture, Taiwan, like other East Asian countries, is forced to obtain much of its food from abroad.

▼ **Figure 11.22 Loess Settlement** In the Loess Plateau of central China, houses are often constructed underground. Although these dwellings are cool in the summer and warm in the winter, they tend to collapse easily during earthquakes.

Agriculture and Resources in Global Context

Japan, South Korea, and Taiwan are major food importers, and China has recently moved in the same direction. Other resources also are being drawn in from all quarters of the world by the powerful economies of East Asia.

Global Dimensions of Japanese Agriculture and Forestry Japan may be mostly self-sufficient in rice, but it is still one of the world's largest food importers. As the Japanese have grown more prosperous over the past 60 years, their diet has grown more diverse. That diversity is made possible largely by importation of food.

Japan imports food from a wide array of other countries. It procures both meat and the feed used in its domestic livestock industry from the United States, Canada, and Australia. These same countries supply wheat needed to produce bread and noodles. Japan is now the world's second largest wheat importer. Even soybeans, long a staple of the Japanese diet, must be purchased from Brazil, the United States, and other countries. Japan has one of the highest rates of fish consumption in the world, and the Japanese fishing fleet scours the world's oceans to supply the demand (Figure 11.23). Japan also purchases prawns and other seafood, much of it farm-raised in former mangrove swamps, from Southeast Asia and Latin America.

Japan also depends on imports to supply its demand for forest resources. Although its own forests produce high-quality cedar and cypress logs, it obtains most of its construction lumber and pulp (for papermaking) from western North America and Southeast Asia. As the rainforests of Malaysia, Indonesia, and the Philippines diminish, Japanese interests are beginning to turn to Latin America and Africa as sources of tropical hardwoods. Japanese and South Korean firms are also extracting resources from eastern Russia, a nearby and previously little-exploited forest zone.

Japan is able to support its large and prosperous population on such a restricted land base because it can purchase resources from abroad. Almost all of the oil, coal, and other minerals that it consumes are imported. Although geographers once looked at resource endowments as a main support of each country's economy, such a view is no longer supportable. As long as a country can export products of value, it can obtain whatever imports it requires. This also means, however, that the environmental degradation generated by a successful economy, such as that of Japan, becomes globalized as well.

Global Dimensions of Chinese Agriculture Until the late 1990s, China was self-sufficient in food, despite its huge population and crowded lands. But rapid economic growth has resulted in an increased consumption of meat, which requires large amounts of feed grain. Growth also has brought about the loss of agricultural lands to residential and industrial development. China is now the world's largest importer of soybeans and vegetable oil, and it imports significant quantities of corn (maize) and other grains in most years. In 2012, imports accounted for roughly 12 percent of China's food needs. Its government aims to be 95 percent self-sufficient in food, but planners are increasingly skeptical that this figure can be reached.

Although China is a net food importer when it comes to total calories, the situation is not so clear in regard to total value. China exports many high-value specialty crops and processed foods, such as garlic, apples, farm-raised fish, and many vegetables. Farmers in both wealthy countries such as the United States and poor countries such as the Philippines often complain that China sells its fruit and vegetables abroad at below-market rates, seeking to build export dominance. In addition, consumers in both China and foreign markets are concerned about the safety of Chinese food products. In 2009, 300,000 Chinese infants became ill and 6 died after drinking milk laced with the industrial chemical melamine; in response, China executed two executives in the milk industry. Chinese consumers are still concerned, however, and in early 2013 Australian stores reported that they were running out of infant formula, allegedly due to Chinese tourists buying safe Australian products in bulk.

Korean Agriculture in a Global Context South Korea has also made the transition, like Japan and Taiwan, to a global food and resource procurement pattern. Although its government aims to be self-sufficient in rice and vegetables, it is a major importer of feed grains, wheat, and soybeans. South Korea also imports large quantities of beef, particularly from Australia. In 2003, it banned beef imports from the United States due to fears about mad cow disease. In 2008, U.S. beef was again allowed into the country, but the decision prompted massive protests by South Korean farmers and others concerned about food safety. South Korea is now the third largest importer of U.S.-produced beef, trailing only Mexico and Canada.

North Korea, on the other hand, has pursued a goal of self-sufficiency. This policy was relatively successful for several years, but in the mid-1990s a series of floods, followed by drought, destroyed most of the country's rice and corn crops, leading to widespread famine. Since then, North Korea has relied heavily on international aid—much of it coming from South Korea—to feed its people, but malnutrition remains widespread. International food aid has varied widely in recent years, dropping from 1508 thousand tons in 2001 to 47 thousand tons in 2011.

Urbanization in East Asia

China has one of the world's oldest urban foundations, dating back more than 3500 years. In medieval and early modern times, East Asia supported a well-developed system of cities. In the early 1700s, Tokyo, then called Edo, probably overshadowed all other cities, with a population of more than 1 million.

Despite this early start, East Asia was largely rural at the end of World War II. Some 90 percent of China's people then lived in the countryside, and even Japan was only about 50 percent urbanized. However, as the region's economy began to grow after the war, so did its cities. Japan, Taiwan, and South Korea are now between 78 and 86 percent urban, which is typical for advanced industrial countries. Although almost half of China's population still lives in rural areas, the country is urbanizing very rapidly.

Chinese Cities Traditional Chinese cities were clearly separated from the countryside by defensive walls. Most were planned in accordance with strict geometrical principles that were thought to reflect the cosmic order. The old-style Chinese city was dominated by low buildings and characterized by straight streets. Houses were typically built around courtyards, and narrow alleyways served both commercial and residential functions.

China's urban fabric began to change during the colonial period. A group of port cities was taken over by European interests, which proceeded to build Western-style buildings and modern business districts. By far the most important of these semicolonial cities was Shanghai, built near the mouth of the Yangtze River, the main gateway to interior China. When the communists came to power in 1949, Shanghai, with a

▼ **Figure 11.23 Japanese Fishing Boat** Japan has one of the world's largest fishing fleets, and its fishing vessels operate in international waters across the world.

population of more than 10 million, was the second largest city in the world. The new authorities, however, viewed it as a decadent, foreign creation. They therefore milked it for taxes, which they invested elsewhere. As a result, much of the city began to decay.

Since the late 1980s, Shanghai has experienced a major revival and is again in many respects China's premier city. Migrants are now pouring into the city, even though the state still tries to restrict the flow, and building cranes crowd the skyline. Official statistics now put the population of the municipality at more than 23 million, and planners suggest that it will reach 30 million by 2020. The new Shanghai is a city of massive high-rise apartments and concentrated industrial developments (Figure 11.24). In 2010, Shanghai surpassed Singapore to become the busiest container port in the world. However, despite Shanghai's revived economic fortunes, the city remains politically secondary to Beijing, China's capital.

Beijing was China's capital during the Manchu period (1644–1912), a status it regained in 1949. Under communist rule, Beijing was radically transformed; old buildings were razed, and broad avenues were plowed through neighborhoods. Crowded residential districts gave way to large blocks of apartment buildings and massive government offices to accommodate a population that now stands at roughly 19 million. Some historically significant structures were saved; the buildings of the Forbidden City, for example, where the Manchu rulers once lived, survived as a complex of museums. The area immediately in front of the Palace Museum, however, was cleared. The resulting plaza, Tiananmen Square, is reputed to be the largest open square in any city of the world (Figure 11.25).

The Chinese Urban System China's urban system as a whole is fairly well balanced, with sizable cities relatively evenly spaced across the landscape and with no city overshadowing all others. This balance stems from China's heritage of urbanism, its vast size and distinctive physical-geographical regions, and its legacy of socialist planning. In the early 1960s, noted China scholar G. William Skinner argued that the distribution of Chinese cities is best explained through the use of **central place theory**. Central place theory holds that an evenly distributed rural population will give rise to a regular hierarchy of urban places, with uniformly spaced larger cities surrounded by constellations of smaller cities, each of which, in turn, will be surrounded by smaller towns.

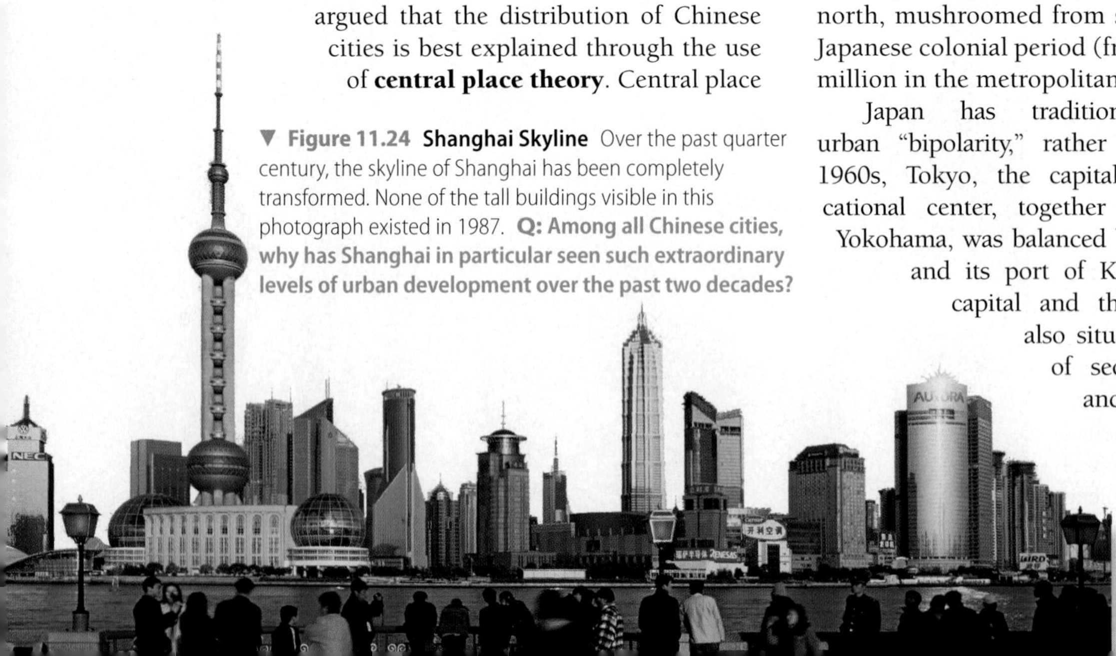

▼ **Figure 11.24 Shanghai Skyline** Over the past quarter century, the skyline of Shanghai has been completely transformed. None of the tall buildings visible in this photograph existed in 1987. **Q: Among all Chinese cities, why has Shanghai in particular seen such extraordinary levels of urban development over the past two decades?**

In the 1990s, Beijing and Shanghai vied for the first position among Chinese cities, with Tianjin, serving as Beijing's port, coming in third. All three of these cities, along with Chongqing in the Sichuan Basin, have been removed from the regular provincial structure of the country and granted their own metropolitan governments. In 1997, another major city, Hong Kong, passed from British to Chinese control. Rather than becoming an ordinary part of China, Hong Kong was granted a unique status as an autonomous Special Administrative Region (SAR) of China and allowed to largely manage its own affairs (see *Cityscapes: Hong Kong, the Vertical City*). Although not as populous as Beijing or Shanghai, Hong Kong is far wealthier. The greater metropolitan area of the Xi Delta, composed of Hong Kong, Shenzhen, and Guangzhou (called Canton in the West), is certainly one of China's premier urban areas. Many other Chinese cities scattered across the country are growing rapidly. By 2013, almost 30 Chinese cities had populations of over 3 million.

Urban Patterns in South Korea, Taiwan, and Japan Unlike China, South Korea and Taiwan are noted for their **urban primacy**, with their urban population concentrated in a single city. Seoul, the capital of South Korea, overwhelms all other cities in the country. Seoul itself is home to more than 10 million people, and its metropolitan area contains some 24 million, almost half of South Korea's population. All of South Korea's major governmental, economic, and cultural institutions are concentrated there. Seoul's explosive and generally unplanned growth has resulted in severe congestion. The South Korean government is promoting industrial growth in other cities, but none of its actions has yet challenged the primacy of the capital.

Taiwan is similarly characterized by a high degree of urban primacy. The capital city of Taipei, located in the far north, mushroomed from some 300,000 people during the Japanese colonial period (from 1895 to 1945) to more than 7 million in the metropolitan area today.

Japan has traditionally been characterized by urban "bipolarity," rather than urban primacy. Until the 1960s, Tokyo, the capital and main business and educational center, together with the neighboring port of Yokohama, was balanced by the mercantile center of Osaka and its port of Kobe. Kyoto, the former imperial capital and the traditional cultural center, is also situated in the Osaka region. A host of secondary and tertiary cities balance Japan's urban structure. Nagoya, with a metropolitan area of almost 9 million people, remains the center of the automobile industry. As Japan's economy

▲ **Figure 11.25 Tiananmen Square** This open area in Beijing, one of the world's largest city squares, is of great historical and political significance. Massive public protests here in 1989 resulted in a harsh crackdown and the temporary imposition of martial law

boomed in the 1960s, 1970s, and 1980s, so did Tokyo. The capital city then outpaced all other urban areas in almost every urban function. Tokyo itself has about 13 million inhabitants, but the Greater Tokyo area now contains more than roughly 35 million people, making it the world's most populous urbanized area. The Osaka–Kobe–Kyoto metropolitan area stands second, with almost 19 million inhabitants. Concerned about the increasing primacy of Tokyo, the Japanese government has been trying to steer new developments to other parts of the country.

Most of Japan's other main cities have seen modest population gains over the past several decades. Urban growth, in general, has been supported by rural depopulation. Metropolitan expansion has been particularly pronounced in the cities linking Tokyo to Osaka, an area known as the Tokkaido corridor. Transportation connections are superb along this route, and proximity to Tokyo, Osaka, and Nagoya encourages development. The result has been the creation of a **superconurbation**, sometimes called a megalopolis: a huge zone of coalesced metropolitan areas (see Figure 11.26). You can travel from Tokyo to Osaka on the main rail line, a distance of almost 300 miles (480 km), and never leave the urbanized area. By some accounts, 65 percent of Japan's people are crowded into this narrow supercity.

Japanese cities sometimes strike foreign visitors as rather gray and monotonous places, lacking historical interest (Figure 11.27). Little of the country's older architecture remains intact. Traditional Japanese buildings were made of wood, which survives earthquakes much better than stone or brick. Fires have therefore been a long-standing hazard, and in World War II the U.S. Air Force firebombed most Japanese cities, virtually obliterating them. In addition, Hiroshima and Nagasaki were completely destroyed by atomic bombs. The one exception was Kyoto, the old imperial capital, which was spared devastation. As a result, Kyoto is famous for its beautiful (wooden) Buddhist monasteries and

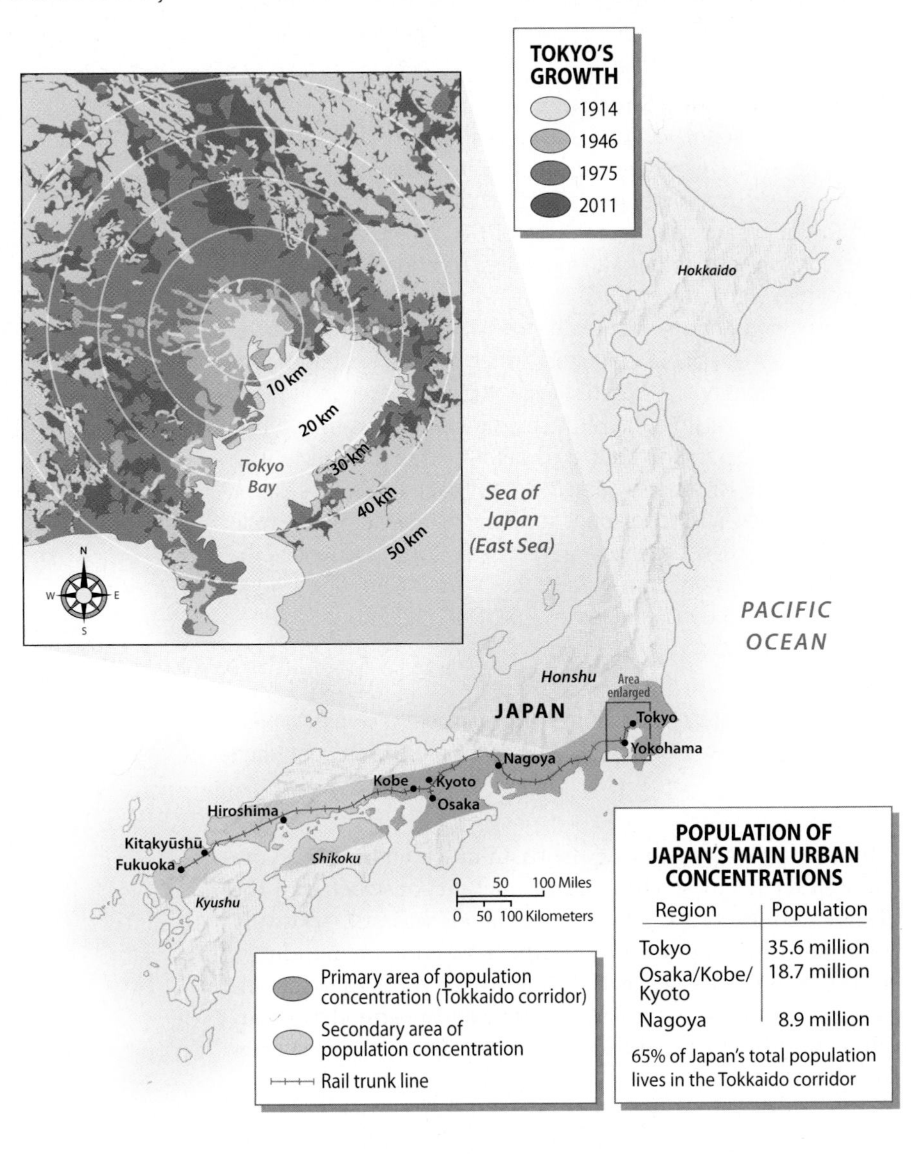

Region	Population
Tokyo	35.6 million
Osaka/Kobe/Kyoto	18.7 million
Nagoya	8.9 million

► **Figure 11.26 Urban Concentration in Japan** The inset map shows the rapid expansion of Tokyo in the postwar decades. Today the Greater Tokyo metropolitan area is home to roughly 35 million people. The larger map shows the cluster of urban settlements along Japan's southeastern coast. The major area of urban concentration is between Tokyo and Osaka, a distance of some 300 miles (482 km), known as the Tokkaido corridor. By some accounts, 65 percent of Japan's population lives in this area.

Cityscapes

Hong Kong, the Vertical City

If there is one characteristic that defines the built environment of Hong Kong, it is the city's verticality. Of Hong Kong's 8096 buildings, 1308 are skyscrapers of at least 40 stories, and 6588 are high-rise buildings of at least 12 stories. No other city in the world comes close. Such architectural extremes are necessary, since despite having a population of over 7 million on a mere 426 square miles (1104 square km) of land area, only 17 percent of Hong Kong is built up. Most of the rest of the SAR is too steep to build on, creating stunning vistas when juxtaposed with adjacent skyscrapers.

The historic core of Hong Kong centers on the northern part of Hong Kong Island, where the deep waters of Victoria Harbor made the area a key entrepôt for international shipping. Hong Kong Island is also home to some of the world's most expensive real estate. Land on Victoria Peak, majestically overlooking the skyscrapers below, sells for as much as $78,200 per square meter. Hong Kong's booming real estate market, however, is not without a downside. Many of Hong Kong's poorer residents find themselves increasingly unable to afford rents, and some pay more than $167 per month to live in small, bedbug-infested steel cages.

Transportation and Shopping

Despite Hong Kong's vertical density, most of the city offers an intimate pedestrian experience at street level, allowing residents and visitors to fully soak in the city's hybrid culture. Such a feat is possible due to the overwhelming popularity of Hong Kong's public transportation, which, together with foot travel, accounts for over 90 percent of trips within the city.

Most Hong Kong residents do their shopping at small convenience stores, which are ubiquitous in the city. Hong Kong boasts the highest density of 7-Eleven stores in the world, even though the local convenience store market is split between 7-Eleven and rival Circle-K. Walking around the city, it is not hard to find blocks with several convenience stores packed closely together. Perhaps unsurprisingly, 7-Eleven is Hong Kong's most popular brand. Shopping in Hong Kong is very convenient, as nearly all stores—as well as the city's public transportation—accept the Octopus Card, which works something like a debit card. As a result, Hong Kong is quickly becoming a cashless society.

New Towns

Much of Hong Kong's population growth in recent years has come in so-called New Towns located outside the main urban areas of Hong Kong Island and adjacent Kowloon (**Figure 11.2.1**). Tsuen Wan was Hong Kong's first New Town, beginning construction in 1959. Well over 2 million people now live in such developments, which have become dense agglomerations in their own right. With the exception of Tsuen Wan, which is very close to Kowloon, most New Towns are detached from the main city's street fabric and connected only by long, winding highways and rail lines.

1. How have geopolitical, economic, and physical geographical features together created such a "vertical city" in Hong Kong?
2. As Hong Kong is part of the People's Republic of China, why is it so different from other Chinese cities?

Google Earth Virtual Tour Video

http://goo.gl/X0b2MZ

▲ **Figure 11.2.1 Hong Kong** The core of Hong Kong consists of the densely packed neighborhoods of Victoria and Kowloon, but the territory of the Special Administrative Region of Hong Kong is much larger. The Sai Kung Peninsula in the east, for example, is a sparsely populated area devoted mostly to parks.

Shinto temples, which ring the basin in which central Kyoto lies. Other Japanese cities were largely reconstructed in the 1950s and 1960s, a period when Japan was still relatively poor and could afford only inexpensive concrete buildings. In the boom years of the 1980s, however, modern skyscrapers rose in many of the larger cities, and postmodernist architecture began to add variety, especially to the wealthier neighborhoods.

REVIEW

11.4 Why does East Asia get so much of its food and so many of its natural resources from other parts of the world?

11.5 How is the urban landscape of East Asia, and particularly that of China, currently changing?

11.6 How have East Asia's patterns of population density influenced the region's agricultural and settlement systems?

▲ **Figure 11.27 Tokyo Apartments** The extremely high population density of Tokyo and other large Japanese cities forces most people to live in crowded apartment blocks. The photo depicts Tokyo's Danchi high-rise apartment complex.

Cultural Coherence and Diversity: The Historical Influence of Confucianism and Buddhism

East Asia is in some respects one of the world's most unified cultural regions. Although different parts of East Asia have their own unique cultural features, the entire region shares certain historically rooted ways of life and systems of ideas.

Most of these East Asian commonalities can be traced back to ancient Chinese civilization. This society emerged roughly 4000 years ago, largely in isolation from the Eastern Hemisphere's other early centers of civilization in the valleys of the Indus, Tigris-Euphrates, and Nile rivers. As a result, East Asian civilization developed along several unique lines.

Unifying Cultural Characteristics

The most important unifying cultural characteristics of East Asia are related to religious and philosophical beliefs. Throughout the region, Buddhism and Confucianism have shaped not only individual beliefs, but also social and political structures. Although the role of traditional belief systems has been seriously challenged over the past 50 years, especially in China, historically rooted cultural patterns have not disappeared.

Writing Systems The clearest distinction between East Asia and the world's other cultural regions is found in written language. Existing writing systems elsewhere in the world are based on the alphabetic principle, in which each symbol represents a distinct sound. East Asia, in contrast, evolved an entirely different system, known as **ideographic writing**. In ideographic writing, each symbol (commonly called a character) primarily represents an idea, rather than a sound (although the symbols can denote sounds in certain circumstances). As a result, ideographic writing requires the use of a large number of distinct symbols.

The Chinese ideographic writing system has one major disadvantage and one major advantage when compared with alphabetic systems, both of which stem from the fact that it is largely divorced from spoken language. The disadvantage is that it is difficult to learn; to be literate, a person must memorize thousands of characters. The advantage is that two literate people do not have to speak the same language to be able to communicate because the written symbols that they use to express their ideas are the same.

In Korea, Chinese characters were adopted at an early date and were used exclusively for hundreds of years. In the 1400s, however, Korean officials decided that Korea needed its own alphabet. They wanted to allow more widespread literacy, hoping also to more clearly differentiate Korean culture from that of China. The use of the new script spread quickly throughout the country. Korean scholars and officials, however, continued to use Chinese characters, regarding their own script as suitable only for popular writings. Today the Korean script is used for almost all purposes, but scholarly works still contain scattered Chinese characters.

The writing system of Japan is even more complex. Initially, the Japanese simply borrowed Chinese characters, referred to in Japanese as kanji. Owing to the grammatical differences between the Japanese and Chinese languages, the exclusive use of kanji resulted in awkward sentence construction. The Japanese solved this quandary by developing a quasi-alphabet known as hiragana that allowed the expression of words and parts of speech not easily represented by Chinese characters. In hiragana, each symbol represents a distinct syllable, or combination of a consonant and a vowel sound. (A different, but essentially parallel system, called katakana, is used in Japan for spelling words of foreign origin.) Eventually, the two styles of writing merged, and written Japanese came to employ a complex mixture of symbols. Japanese kanji today differ slightly from Chinese characters.

The Confucian Legacy Just as the use of a common writing system helped forge cultural linkages throughout East Asia, so, too, **Confucianism** (the philosophy developed by Confucius) came to occupy a significant position in all of the societies of the region. So strong is the heritage of Confucius that some writers refer to East Asia as the "Confucian world." The premier philosopher of Chinese history, Confucius (or *Kongfuzi,* in Mandarin Chinese) lived during the 6th century BCE, a period of marked political instability. Confucius's goal was to create a philosophy that could generate stability. Although Confucianism is often considered to be a religion, Confucius himself was mostly interested in the "here and now," focusing his attention on how to lead a correct life and organize a proper society (Figure 11.28).

Confucius stressed deference to the proper authority figures, but he also thought that authority has a responsibility to act in a benevolent manner. The most basic level of the traditional Confucian moral order is the family unit, considered the bedrock of society. The ideal family structure is patriarchal, and children are told to obey and respect their parents—especially their fathers—as well as their elder brothers.

Confucian philosophy also stresses the need for a well-rounded and broadly humanistic education. To a certain extent, Confucianism advocates a kind of meritocracy, holding that an individual should be judged on the basis of behavior and education, rather than on family background. The high officials of Imperial China (pre-1912)—the powerful **Mandarins**—were thus selected by competitive examinations. Only wealthy families, however, could afford to give their sons the education needed for success on those grueling tests.

In Japan, Confucianism was never as important as it was on the mainland. Japanese officials were actually able to exclude certain Confucian beliefs that they considered dangerous. The most important of these was the revocable "mandate of heaven." According to this notion, the emperor of China derived his authority from the principle of cosmic harmony, but such a mandate could be withdrawn if he failed to fulfill his duties. This idea was used both to explain and to legitimize the rebellions that occasionally resulted in a change of China's ruling dynasty. In Japan, on the other hand, a single imperial dynasty has persisted throughout the entire period of written history. Although the emperor of Japan has had little real power for more than a thousand years, the sanctity of his family lineage continues to form a basic principle of Japanese society.

▼ **Figure 11.28 Confucian Temple in China** Confucianism is usually considered to be a philosophy rather than a religion, but it does have religious aspects. In Confucian temples, which are found through many parts of East Asia, the spirit and philosophy of Confucius are honored and often worshipped. **Q: To what extent can Confucianism be said to unify East Asia, both in the past and in the contemporary period?**

The Modern Role of Confucian Ideology The significance of Confucianism in East Asian development has been hotly debated for the past 100 years. In the early 1900s, many observers believed that the conservatism of the philosophy, derived from its respect for tradition and authority, was responsible for the economically backward position of China and Korea. However, because East Asia has enjoyed the world's fastest rates of economic growth over the past several decades, such a position is no longer supportable. Some scholars now argue that Confucianism's respect for education and the social stability that it generates give East Asia an advantage in international competition.

Over the past 100 years, Confucianism has lost much of the hold that it once had on public morality throughout East Asia, especially in China. After the communist revolution, China's rulers sought for decades to discourage, if not eliminate, Confucian thought. Currently, however, Chinese officials are pushing for a revival of the philosophy, hoping that it will lead to enhanced social stability. Confucian schools are multiplying, and recent books, films, and television shows about Confucius have been popular. The Chinese government also maintains 322 Confucius Institutes in more than 90 countries (as of 2010), designed to promote Chinese culture and language. Some critics, however, have accused these institutes as acting as propaganda arms of the Chinese government.

Religious Unity and Diversity

Certain religious beliefs have worked alongside Confucianism to cement together cultures of the East Asian region. The most important culturally unifying beliefs are associated with Mahayana Buddhism. Other religious practices, however, have had a more divisive role.

Mahayana Buddhism Originating in India in the 6th century BCE, Buddhism stresses escape from an endless cycle of rebirths to reach union with the divine cosmic principle (or nirvana). By the 2nd century CE, Buddhism was spreading throughout China, and within a few hundred years it had expanded into East Asia. Today Buddhism remains widespread everywhere in the region, although it is far less significant here than it is in mainland Southeast Asia and Sri Lanka.

The variety of Buddhism practiced in East Asia—Mahayana, or Greater Vehicle—is distinct from the Theravada Buddhism of Sri Lanka and Southeast Asia (Figure 11.29). Most important, Mahayana Buddhism simplifies the quest for nirvana, in part by suggesting that entities (boddhisatvas) exist who refuse divine union for themselves in order to help others spiritually. Mahayana Buddhism is also nonexclusive; in other words, one may follow it while simultaneously practicing other faiths. Thus, many Chinese consider themselves to be both Buddhists and Taoists, whereas most Japanese are at some level both Buddhists and followers of Shinto.

As Mahayana Buddhism spread through East Asia in the medieval period, many different sects emerged. Probably the best known is Zen, which demands that its followers engage in the rigorous practice of "mind emptying." At one time, Buddhist monasteries associated with Zen and other sects were rich and powerful. In all East Asian countries, however, periodic reactions against Buddhism resulted in the persecution of monks and the suppression of monasteries. Despite such hardships, East Asian Buddhism was never extinguished. However, it also never became the focal point of society that it did in mainland Southeast Asia. In Japan, that position was partially captured by a different religion, Shinto.

▲ **Figure 11.29 The Buddhist Landscape** Mahayana Buddhism has been traditionally practiced throughout East Asia. This Golden Buddha statue is located in Baomo Park in Chi Lei Village, Guangdong Province, China. **Q: What is the symbolic significance of the flowers growing in the water around the statue of the Buddha?**

Shinto The religious practice of Shinto is so closely bound to the idea of Japanese nationality that it is questionable whether a non-Japanese person can follow it. Shinto began as the animistic worship of nature spirits, but it was gradually refined into a subtle set of beliefs about the harmony of nature and its connections with human existence. Until the late 1800s, Buddhism and Shinto were complexly intertwined. Subsequently, the Japanese government began to disentangle the two faiths, while elevating Shinto into a nationalistic cult focused on the divinity of the Japanese imperial family. After World War II, the more excessive aspects of nationalism were removed from the religion.

Shinto is still a place- and nature-centered religion. Certain mountains, particularly the volcanic Mount Fuji, are considered sacred and are thus climbed by large numbers of people. Major Shinto shrines, often located in scenic places, attract numerous pilgrims; most notable is the Ise Shrine south of Nagoya, site of the cult of the emperor. Local Shinto shrines, as well as Buddhist temples, offer leafy oases in otherwise largely treeless Japanese urban neighborhoods.

Taoism and Other Chinese Belief Systems The Chinese religion of Taoism (or Daoism) is similarly rooted in nature worship. Like Shinto, it stresses the acquisition of spiritual harmony and the pursuit of a balanced life. Taoism is indirectly associated with feng shui, also called **geomancy**, the Chinese and Korean practice of designing buildings in accordance with the spiritual powers that supposedly course through the local topography (Figure 11.30). Even in hypermodern Hong Kong, skyscrapers worth millions of dollars have occasionally gone unoccupied because people believed that their construction failed to accord with geomantic principles. The government of China officially recognizes Taoism as a religion and seeks to regulate it through

◄ **Figure 11.30 Feng Shui Compass** An ancient philosophical system designed to harmonize human existence with nature, feng shui is still widely practiced in China. The feng shui compass, shown here, helps practitioners align buildings with energy flows that they believe course through the natural landscape.

the state-run China Taoist Association. In Taiwan, government statistics state that roughly a third of the population currently follows Taoism.

Despite the fact that both Taoism and Buddhism were historically followed throughout the country, traditional religious practice in China has always embraced local particularism; in other words, it has focused on the unique attributes of particular places. Many minor gods were traditionally associated with single cities or other specific areas. In modern-day rural China, village gods are often still honored.

Minority Religions Followers of virtually all world religions can be found in the increasingly cosmopolitan cities of East Asia. Christianity is well represented. Over a million Japanese belong to Christian churches, whereas South Korea is roughly 30 percent Christian (mostly Protestant). South Korea reportedly sends more missionaries abroad than any country except the United States. Some reports indicate that Christianity is growing rapidly in China—despite state discouragement—but reliable information is scarce. Officially, China acknowledges a Christian population of only around 15 million, but outside experts estimate that some 40–80 million Chinese citizens attend unofficial "house churches," most of which are Protestant.

China also has a long-established Muslim community, estimated as having between 20 and 60 million adherents. Most Chinese Muslims are members of minority ethnic groups living in the far west. The roughly 10 million Chinese-speaking Muslims, called *Hui*, are concentrated in Gansu and Ningxia in the northwest and in Yunnan Province in the south. Smaller clusters of Hui, often segregated in their own villages, live in almost every province of China. The only Muslim congregations in Japan and South Korea, on the other hand, are associated with recent and often temporary immigrants from South, Southeast, and Southwest Asia.

Secularism in East Asia For all of these varied forms of religious expression, East Asia is one of the most secular regions of the world. In South Korea, for example, roughly half of the population is not religious. In Japan, most people occasionally observe Shinto or Buddhist rituals and maintain a small shrine for their ancestors, but only a small segment of the population is deeply religious. As a result, statistics on religious affiliation in Japan vary tremendously, with estimates of the followers of Shinto ranging between 4 million and 119 million! Japan also has several "new religions," sometimes called cults, a few of which are noted for their fanaticism. But for Japanese society as a whole, religion is not tremendously important.

After the communist regime took power over China in 1949, all forms of religion and traditional philosophy were discouraged and sometimes severely repressed. Under the new regime, the atheistic philosophy of **Marxism**—the communistic belief system developed by Karl Marx—became the official ideology. In the 1960s, many observers thought that traditional forms of Chinese religion would survive only in overseas Chinese communities. With the easing of Marxist orthodoxy during the 1980s and 1990s, however, many forms of spiritual expression began to return.

Like China, North Korea adopted communist ideology when it became independent after World War II. As a result, religious beliefs were severely repressed. North Korean leaders also supplemented Marxism with a unique official ideology called *juche*, or "self-reliance." Ironically, juche demands absolute loyalty to North Korea's political leaders. It also involves intense nationalism based on the idea of Korean racial purity. Over time, Marxist beliefs have gradually been supplanted by those of juche; in North Korea's revised 2009 constitution, all references to communism were removed. North Korea has also deviated from Marxism by instituting the principle of **dynastic succession**, with power flowing from father to son. When the previous leader Kim Jong-il died in late 2011, he was immediately replaced by his untested 28-year-old son, Kim Jong-un (Figure 11.31).

Linguistic and Ethnic Diversity

Japanese and Mandarin Chinese may partially share a system of writing, but the two languages bear no direct relationship (Figure 11.32). In their grammatical structures, Chinese and Japanese are more different from each other than are Chinese and English. Japanese, however, has adopted many words of Chinese origin.

Language and National Identity in Japan According to most scholars, Japanese is not related to any other language. Korean is also usually classified as the only member of its

▲ **Figure 11.31 Funeral of North Korean Leader Kim Jong-il** North Korea is a highly repressive state that has cultivated "personality cults" around its leaders. The funeral of Kim Jong-il in December 2011 thus entailed a massive public ceremony.

language family. Many linguists, however, think that Japanese and Korean should be grouped together because they share many grammatical features. Only the mutual distrust between the Japanese and the Koreans, some suggest, has prevented this linguistic relationship from being acknowledged.

From several perspectives, the Japanese form one of the world's most homogeneous peoples, and they tend to regard themselves in such a manner. To be sure, minor cultural and linguistic distinctions are noted between the people of western Japan (centered on Osaka) and eastern Japan (centered on Tokyo), and many individual regions have distinctive customs and lifestyles. Overall, however, such differences are of little significance. Only in the Ryukyu Islands can you find variants of Japanese so distinct that they can be considered separate languages. Ethnically, this region of Japan is also distinct. In fact, many Ryukyu people believe that they have not been treated as full members of the Japanese nation and that they have suffered from discrimination.

Minority Groups in Japan In earlier centuries, the Japanese archipelago was divided between two very different peoples: the Japanese living to the south and the Ainu inhabiting the north. The Ainu are completely distinct from the Japanese. They possess their own language and have a recognizable physical appearance (Figure 11.33). Owing to their facial features, the Ainu were once categorized as members of the "Caucasian race." However, few scholars now believe that humankind is divided into separate races, and Ainu are not related to Europeans by genetic criteria..

For centuries, the Japanese and the Ainu competed for land, and by the 10th century CE the Ainu had been largely driven off the main island of Honshu. Until the 1800s, however, Hokkaido largely remained Ainu territory. The Japanese people subsequently began to colonize Hokkaido, putting renewed pressure on the Ainu. Today, according to official statistics, around 25,000 Ainu remain, although some sources put the actual number as high as 200,000. By the 1990s, only about a dozen people still spoke the Ainu language, although efforts are now under way to revive its use.

The approximately 800,000 people of Korean descent living in Japan today also have felt discrimination. Many of them were born in Japan (their parents and grandparents having left Korea early in the 1900s) and speak Japanese, rather than Korean, as their primary language. But despite their deep bonds to the country, only around 300,000 people of Korean background have been able to gain Japanese citizenship, the rest being treated as resident aliens. Perhaps as a result of such treatment, many Japanese Koreans hold radical political views and support North Korea.

Starting in the 1980s, other immigrants began to arrive in Japan, mostly from the poorer countries of Asia. Most do not have legal status. Men from China and southern Asia typically work in the construction industry and in other dirty and dangerous jobs; women from Thailand and the Philippines often work as entertainers and sometimes as prostitutes. Roughly 200,000 Brazilians of Japanese ancestry have returned to Japan for the relatively high wages they can earn. However, immigration is less pronounced in Japan than in most other wealthy countries, and relatively few migrants acquire permanent residency, let alone citizenship. Because of Japan's impending population decline, however, some scholars estimate that it may need to import several million foreign workers over the next decade.

The most victimized people in Japan could be the **Burakumin**, or *Eta*, an outcast group of Japanese whose ancestors worked in "polluting" industries such as leathercraft. Discrimination is now illegal, but the Burakumin are still among the poorest and least-educated people in Japan. Private detective agencies do a brisk business checking prospective marriage partners and employees for possible Burakumin ancestries, which might prevent the marriage or job offer. The Burakumin, however, have banded together to demand their rights; the Buraku Liberation League is politically powerful and is reputed to have close connections with the Yakuza (the Japanese criminal organization). The Burakumin usually live in separate neighborhoods and are concentrated in the Osaka region of western Japan.

▶ **Figure 11.32 Languages of East Asia** The linguistic geography of Korea and Japan is very straightforward, as the vast majority of people in those countries speak Korean and Japanese, respectively. In China, the dominant Han Chinese speak a variety of closely related Sinitic languages, the most important of which is Mandarin Chinese. In the peripheral regions of China, a large number of languages—belonging to several different linguistic families—can be found.

Language and Identity in Korea The Koreans, like the Japanese, are a relatively homogeneous people. The vast majority of people in both North and South Korea speak Korean and unquestioningly consider themselves to be members of the Korean nation.

South Korea does, however, have a strong sense of regional identity, which can be traced back to the medieval period when the peninsula was divided into three separate kingdoms. The people of southwestern South Korea, centered on the city of Gwangju, tend to be viewed as distinctive, and many southwesterners believe that they have suffered periodic discrimination. In contrast, the Kyongsang region of southeastern South Korea has supplied most of the country's political leaders and has received more than its share of development funds.

Korean identity also has an international component. Several million Korean speakers reside directly across the border in northern China. Because of deportations ordered by the Soviet Union in the mid-20th century, substantial Korean communities also can be found in Kazakhstan in Central Asia. Over the past several decades, a Korean **diaspora**—the scattering of a particular group of people over a vast geographical area—has brought hundreds of thousands of Koreans to the United States, Canada, Australia, and New Zealand. South Korea's prosperity has also attracted hundreds of thousands of immigrants into the country. Particularly notable are the almost 100,000 Vietnamese people living in the country, many of whom are women who have married Korean farmers who have a difficult time attracting local wives.

▼ **Figure 11.33 Ainu Men** The indigenous Ainu people of northern Japan are much reduced in population, but they still maintain some of their cultural traditions. Here Ainu men participate in the Marimo Festival on the northern Japanese island of Hokkaido.

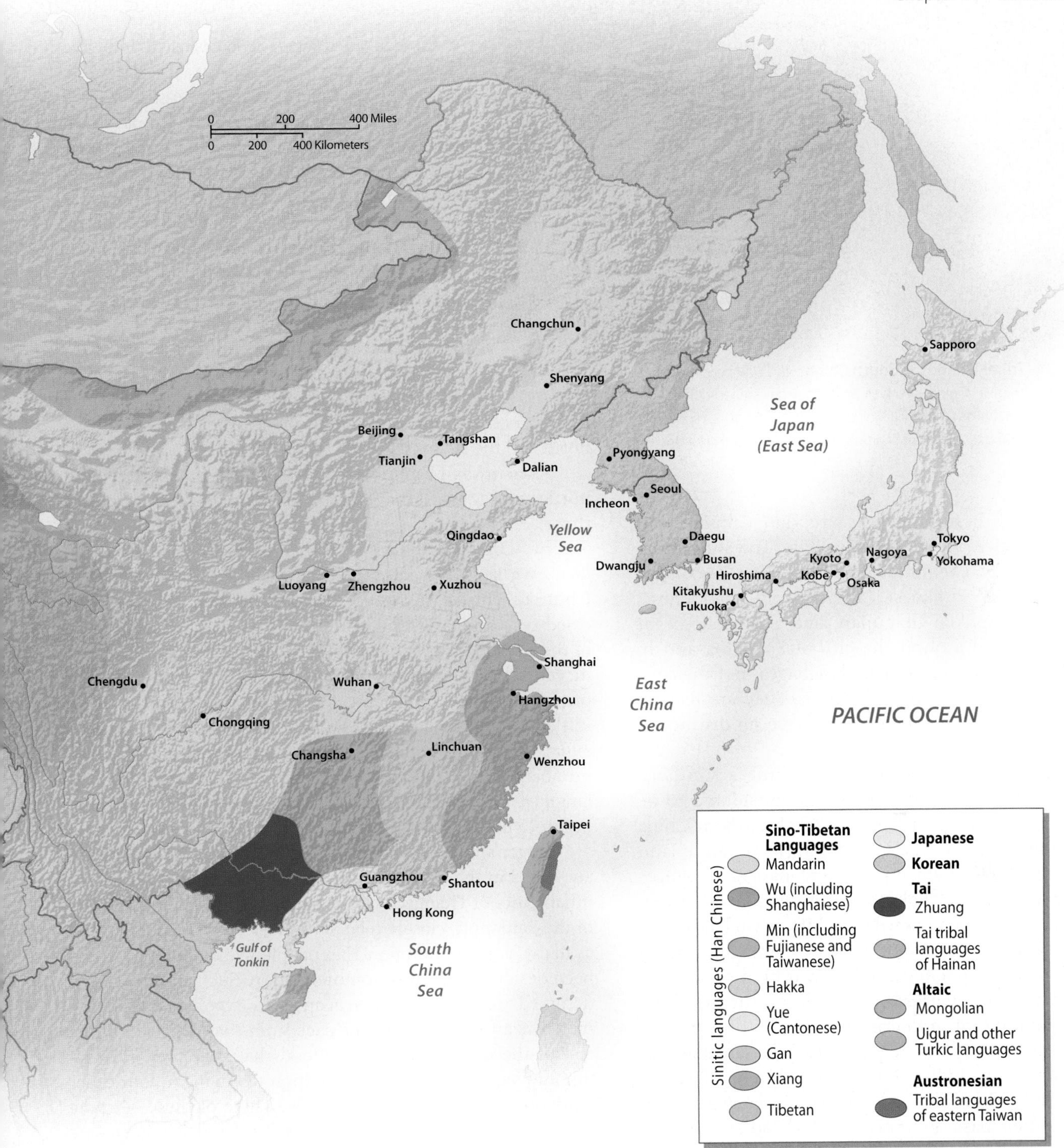

Language and Ethnicity Among the Han Chinese The geography of language and ethnicity in China is far more complex than that of Korea or Japan. This is true even if we consider only the eastern half of the country, so-called China proper. The most important distinction is that separating the Han Chinese from the non-Han peoples. The Han, who form the vast majority, are those people who have long been incorporated into the Chinese cultural and political systems and whose languages are expressed through Chinese writing. The Han do not, however, all speak the same language.

The northern, central, and southwestern regions of China—a vast area extending from Manchuria through the middle and upper Yangtze Valley to the valleys and plateaus of Yunnan in the far south—constitute a single linguistic zone. The spoken language here is generally called *Mandarin Chinese*. Mandarin is divided into several dialects, but the standard dialect of the Beijing area is gradually spreading. Standard Mandarin—called *Pǔtōnghuà* locally—is China's official national language.

In southeastern China, from the Yangtze Delta to China's border with Vietnam, several separate languages are spoken.

▲ **Figure 11.34 Tribal Villages in South China** Non-Han people are usually classified as "tribal" in China, which assumes they have a traditional social order based upon autonomous village communities. The photo shows a Miao village in the Xiangxi Tujia and Miao Autonomous Prefecture, Hunan Province.

Peoples speaking these languages are Han Chinese, but they are not native Mandarin speakers. Traveling from south to north, we encounter Cantonese (or Yue), spoken in Guangdong; Fujianese (alternatively Hokkienese, or Min, locally), spoken in Fujian; and Shanghaiese (or Wu), spoken in and around the city of Shanghai and in Zhejiang Province. These are true languages, not dialects, since they are not mutually intelligible. They are usually called dialects, however, because they have no distinctive written form.

The Hakka, a group of people speaking a southern Chinese language, are occasionally not considered to be Han Chinese. Evidently, their ancestors fled northern China roughly a thousand years ago to settle in the rough upland area where the Guangdong, Fujian, and Jiangxi provinces meet. Later migrations took them throughout southern China, where they typically settled in hilly wastelands. The Hakka traditionally made their living by growing upland crops such as sweet potatoes and by working as loggers, stonecutters, and metalworkers. Today they form one of the poorest communities of southern China.

Despite their many differences, all of the languages of the Han Chinese (including Hakka) are closely related to each other and belong to the same Sinitic language subfamily. Since their basic grammars and sound systems are similar, a person speaking one of these languages can learn another relatively easily. It is usually difficult, however, for speakers of European languages—or of Japanese or Korean—to gain fluency in these tongues. All Sinitic languages are **tonal** and monosyllabic; their words are all composed of a single syllable (although compound words can be formed from several syllables), and the meaning of each basic syllable changes according to the pitch in which it is uttered.

The Non-Han Peoples Many of the more remote upland districts of China proper are inhabited by groups of non-Han peoples speaking non-Sinitic languages. According to the Chinese government, the country contains 55 such ethnic groups. Most of these peoples are classified as tribal, implying that they have a traditional social order based on self-governing village communities (Figure 11.34). Such a view is not entirely accurate, however, because some of these groups once had their own kingdoms. What they do have in common is a heritage of cultural and sometimes political tension with the Han Chinese.

Over the course of many centuries, the territory occupied by these non-Han communities has steadily declined in size, caused by the continued expansion of the Han, as well as by non-Han emigration. Acculturation into Chinese society and intermarriage with the Han also have reduced many non-Han groups. Their main concentrations today are in the rougher lands of the far north and the far south.

As many as 11 million Manchus live in the more remote portions of Manchuria. The Manchu language is related to those of the tribal peoples of central and southeastern Siberia. Only a handful of Manchus, however, still speak their own language, the rest having abandoned it for Mandarin Chinese. This is an ironic situation because the Manchus ruled the Chinese Empire from 1644 to 1912. Until the end of this period, the Manchus prevented the Han from settling in central and northern Manchuria, which they hoped to preserve as their homeland. Once Chinese were allowed to settle in Manchuria in the 1800s—in part to prevent Russian expansion—the Manchus soon found themselves vastly outnumbered. As they began to intermarry with and adopt the language and customs of the newcomers, their own culture began to disappear.

Larger and more secure communities of non-Han peoples are found in the far south, especially in Guangxi. Most of the inhabitants of Guangxi's more remote areas speak languages of the Tai family, closely related to those of Thailand. Because as many as 18 million non-Han people live in Guangxi, it has been designated an **autonomous region**. Such autonomy was designed to allow non-Han peoples to experience "socialist modernization" at a different pace from that expected of the rest of the country. Critics contend that very little real autonomy has ever existed. (In addition to Guangxi, four other autonomous regions exist in China. Three of these—Xizang [Tibet], Nei Monggol [Inner Mongolia], and Xinjiang—are located in Central Asia and are discussed in Chapter 10. The final autonomous region, Ningxia, located in northwestern China, is distinguished by its large concentration of Hui [Mandarin-speaking Muslims].)

Other areas with sizable numbers of non-Han peoples are Yunnan and Guizhou, in southwestern China, and western Sichuan. Most tribal peoples here practice swidden agriculture (also called "slash and burn") on rough slopes; flatter lands are generally occupied by rice-growing Han Chinese. A wide variety of separate languages, falling into several linguistic families, is found among the ethnic groups living in the uplands. Figure 11.35 shows that in Yunnan, the resulting ethnic mosaic is staggeringly complex.

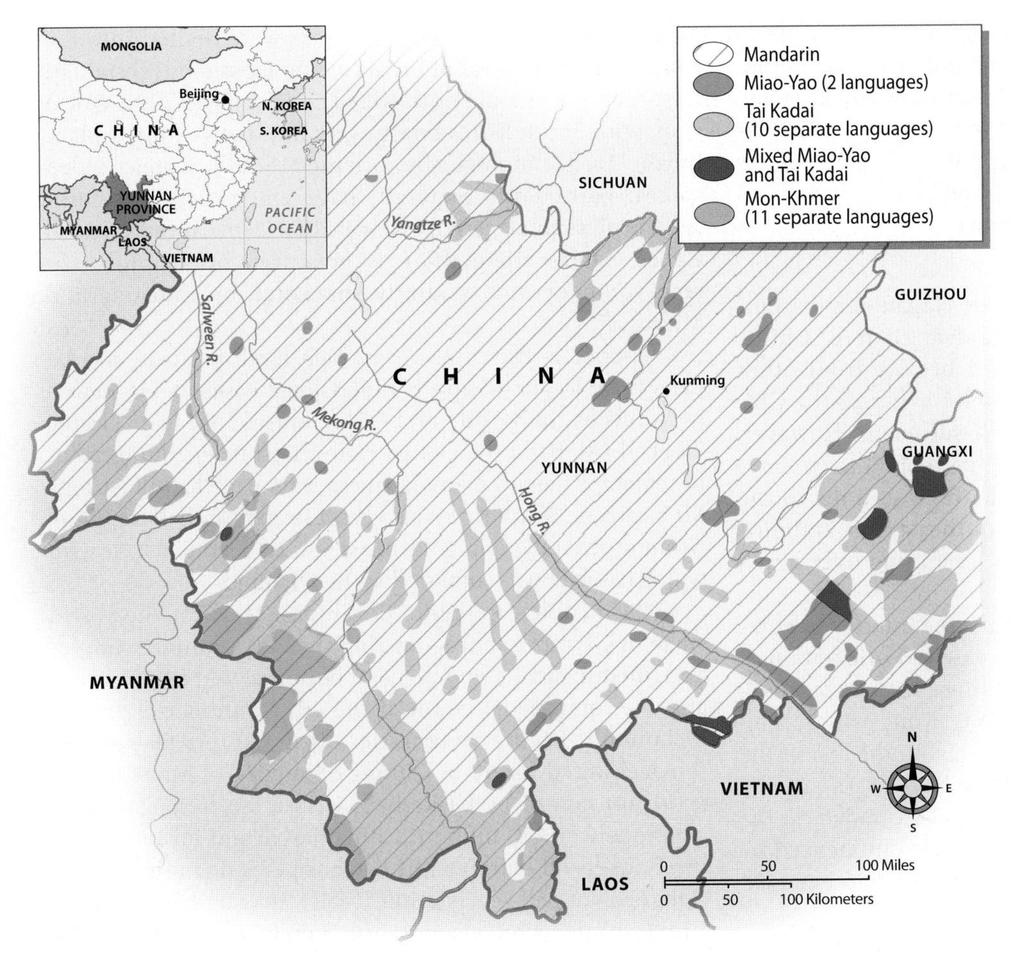

▲ **Figure 11.35 Language Groups in Yunnan** China's Yunnan Province is the most linguistically complex area in East Asia. In Yunnan's broad valleys, its relatively level plateau areas, and its cities, most people speak Mandarin Chinese. In the hills, mountains, and steep-sided valleys, however, people speak a wide variety of tribal languages, falling into several linguistic families. In certain areas, several different languages can be found in very close proximity.

Language and Ethnicity in Taiwan Taiwan is also noted for its linguistic and ethnic complexity. In the island's mountainous eastern region, several small groups of "tribal" peoples speak languages related to those of Indonesia (belonging to the Austronesian language family). These peoples resided throughout Taiwan before the 16th century. At that time, however, Han migrants began to arrive in large numbers. Most of the newcomers spoke the Fujianese (or Hokkien) dialect, which evolved into the distinctive language of Taiwanese.

Taiwan was transformed almost overnight in 1949, when China's nationalist forces, defeated by the communists, sought refuge on the island. Most of the nationalist leaders spoke Mandarin, which they made the official language. Taiwan's new leadership discouraged Taiwanese, viewing it as a mere local dialect. As a result, tension developed between the Taiwanese and the Mandarin communities. Only in the 1990s did Taiwanese speakers begin to reassert their language rights, resulting in the use of Taiwanese in schools alongside Mandarin Chinese. Today most Taiwanese citizens are fluent in both languages. In regard to Taiwan's television industry, dramas and variety shows tend to be in Taiwanese, whereas game shows and documentaries are more often done in Mandarin.

East Asian Cultures in Global Context

East Asia has long exhibited tensions between an internal orientation and tendencies toward cosmopolitanism. This dichotomy has both a cultural and an economic dimension. Until the mid-1800s, East Asian countries attempted to insulate themselves from Western cultural influences. Japan subsequently opened its doors, but remained ambivalent about foreign ideas. Only after its defeat in 1945 did Japan really opt for a globalist orientation. It was followed in this regard by South Korea, Taiwan, and Hong Kong (then a British colony). The Chinese and North Korean governments, to the contrary, decided during the early **Cold War** decades of the 1950s and 1960s to isolate themselves as much as possible from Western and global culture.

The Cosmopolitan Fringe Japan, South Korea, Taiwan, and Hong Kong are characterized by a vibrant internationalism, which coexists with strong national and local cultural identity. Virtually all Japanese, for example, study English for 6 to 10 years, and, although relatively few learn to speak it fluently, most can read and understand a good deal. Business meetings among Japanese, Chinese, and Korean firms are often conducted in English. Relatively large numbers of advanced students, especially from Taiwan, study in the United States and other English-speaking countries and thus bring home a kind of cultural bilingualism.

The current cultural flow is not merely from a globalist West to a previously isolated East Asia. Instead, the exchange is growing more reciprocal. Hong Kong's action films are popular throughout most of the world and have influenced filmmaking techniques in Hollywood. Over the past few years, South Korean popular culture has spread throughout East Asia and in much of the rest of the world. Japan almost dominates the world market in video games, and its ubiquitous comic-book culture and animation techniques are now following its karaoke bars in their overseas march.

Cultural globalization is, of course, as controversial in Japan as it is elsewhere. Japanese ultranationalists are few, but vocal, calling their fellow citizens to resist the decadence of the West and to return to the military and cultural traditions of the **samurai**, the warrior class of feudal Japan. Many other Japanese people, however, contend that their country is too insular, worrying that they do not possess the English-language and global cultural skills necessary to operate effectively in the world economy.

The Chinese Heartland In one sense, Japan is more culturally predisposed to cosmopolitanism than is China. The Japanese have always borrowed heavily from other cultures (particularly from China itself), whereas the Chinese have historically been more self-sufficient. However, the southern coastal Chinese have long tended to be more international in their orientation, with close links to the Chinese diaspora communities of Southeast Asia and ultimately to maritime trading circuits extending over much of the globe.

In most periods of Chinese history, the interior orientation of the center prevailed over the external orientation of the southern coast. After the communist victory of 1949, only the small British enclave of Hong Kong was able to pursue international cultural connections. In the rest of the country, a dour and puritanical cultural order was rigidly enforced. Although this culture was largely founded on the norms of Chinese peasant society, it was also influenced by the communist-oriented cultural system that had emerged in the Soviet Union.

After China began to liberalize its economy and open its doors to foreign influences in the late 1970s and early 1980s, the southern coastal region suddenly assumed a new prominence. Through this gateway, global cultural patterns began to penetrate the rest of the country. The result has been the emergence of a vibrant, but somewhat showy urban popular culture throughout China, replete with such global features as nightclubs, karaoke bars, fast-food franchises, and theme parks.

Globalization and Sports in East Asia Although the sporting culture of East Asia is as varied as the region's people, the last half-century has seen the area increasingly dominated by popular international sports. Japan's national sport of Sumo wrestling dates back to at least the 8th century and holds an important place in the country's psyche (Figure 11.36). The country's top spectator sports—baseball and soccer (association football)—trace their Japanese roots back to the Meiji period (1868–1912), when foreign advisors were invited into the country with the goal of modernizing Japanese technology. In soccer, Japan and South Korea cohosted the 2002 World Cup, with South Korea boasting an impressive fourth-place finish. In technologically oriented South Korea, video games, particularly Starcraft, have emerged as a significant professional activity since 2000. The annual World Cyber Games competition is organized by a South Korean company and is jointly sponsored by Samsung and Microsoft.

Baseball is played at a very high level in Japan, and Japan's main league—Nippon Professional Baseball—is viewed by scouts as the world's best league outside of Major League Baseball (MLB) in the United States. Despite encouragement to remain in Japan, many Japanese stars have had illustrious MLB careers. South Korea also has a strong baseball tradition, winning the Olympic gold medal in 2008, the last year in which baseball was contested at the Olympics.

The People's Republic of China excels on the world stage in Olympic sports, hosting the 2008 Summer Games and finishing in second place in both the gold medal count and the total medal count at the 2012 Summer Olympics in London. Diving, gymnastics, swimming, and badminton accounted for 40 of China's 88 medals. China's largest sport appears to be basketball, with a fan base estimated generously at 450 million. China native and former Houston Rockets all-star Yao Ming, whose U.S. debut in the National Basketball Association (NBA) attracted 200 million Chinese television viewers, did much to popularize the game in the early 2000s. American NBA stars like Kobe Bryant and LeBron James are among China's most well-known celebrities.

REVIEW

11.7 What features mark the Han Chinese as an ethnic group?

11.8 How has the geography of religion changed in East Asia since the end of World War II?

11.9 What have been the main historical factors in the creation of an East Asian world region?

▼ **Figure 11.36 Sumo Wrestlers** One of the most popular sports in Japan is Sumo wrestling. This ancient sport still has many ritual aspects, as can be seen in this photograph.

Geopolitical Framework: Struggles for Regional Dominance

Much of the political history of East Asia revolves around the centrality of China and the ability of Japan to remain outside China's grasp. The traditional Chinese conception of geopolitics was based on the idea of a universal empire: All territories either were a part of the Chinese Empire, paying tribute to it and acknowledging its supremacy, or stood outside the system altogether. Until the 1800s, the Chinese government would not recognize any other as its diplomatic equal. When China could no longer maintain its power in the face of European aggression, the East Asian political system fell into disarray. As European power declined in the 1900s, China and Japan contended for regional leadership. After World War II, East Asia was split by larger Cold War rivalries (Figure 11.37).

The Evolution of China

The original core of Chinese civilization was the North China Plain and the Loess Plateau. For many centuries, periods of unification alternated with times of division into

▼ **Figure 11.37 Geopolitical Issues in East Asia** This region remains one of the world's geopolitical hot spots. Tensions are particularly severe between capitalist, democratic South Korea and the isolated communist regime of North Korea and also between China and Taiwan. China has had several border disputes, whereas Japan and Russia have not been able to resolve their quarrel over the southern Kuril Islands.

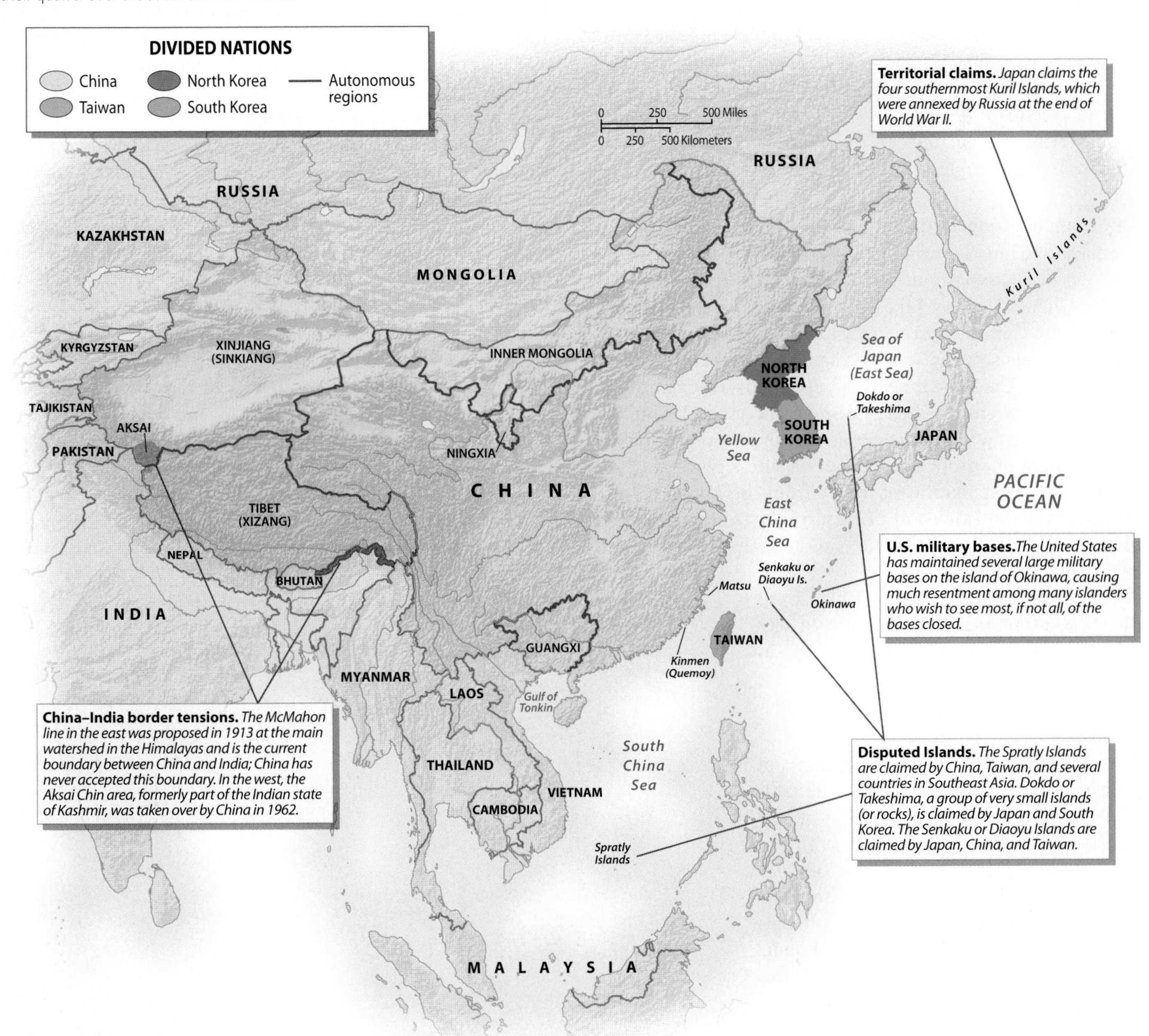

competing states. The most important episode of unification occurred in the 3rd century BCE. Once political unity was achieved, the Chinese Empire began to expand vigorously to the south of the Yangtze Valley. Subsequently, the ideal of the imperial unity of China triumphed, with periods of division seen as indicating disorder. This ideology helped cement the Han Chinese into a single people.

Several Chinese dynasties rose and fell between 219 BCE and 1912, most of them controlling roughly the same territory. The core of the Chinese Empire remained China proper, excluding Manchuria. Other lands, however, were sometimes ruled as well. Until around 1800, the Chinese Empire was the world's wealthiest and most powerful state. Its only real threat came from the pastoral peoples of Mongolia and Manchuria. Although vastly outnumbered by China, these societies were organized on a highly effective military basis. Usually, the Chinese and the Mongols enjoyed a mutually beneficial trading relationship. Periodically, however, they waged war, and on several occasions the northern nomads conquered China (see Chapter 10). The Great Wall along China's border did not, in other words, provide adequate defense. In time, the conquering armies adopted Chinese customs in order to govern the far more numerous Han people.

The Manchu Qing Dynasty The final and most significant conquest of China occurred in 1644, when the Manchus toppled the Ming Dynasty and replaced it with the Qing (also spelled Ch'ing) Dynasty. Like earlier conquerors, the Manchus retained the Chinese bureaucracy and made few institutional changes. Their strategy was to adapt themselves to Chinese culture, yet at the same time to preserve their own identity as an elite military group. Their system functioned well until the mid-19th century, when the empire began to crumble before the onslaught of European and, later, Japanese power.

China's most significant legacy from the Manchu Qing Dynasty was the extension of its territory to include much of Central Asia. The Manchus subdued the Mongols and eventually established control over eastern Central Asia, including Tibet. Even the states of mainland Southeast Asia sent tribute and acknowledged Chinese supremacy. Never before had the Chinese Empire been so extensive or so powerful.

The Modern Era From its height of power and extent in the 1700s, the Chinese Empire descended rapidly in the 1800s as it failed to keep pace with the technological progress of Europe. Threats to the empire had always come from the north, and Chinese and Manchu officials saw little peril from European merchants operating along their coastline. The Europeans were distressed by the amount of silver needed to obtain Chinese silk, tea, and other products and by the fact that the Chinese disdained their manufactured goods. In response, the British began to sell opium, which Chinese authorities viewed as a threat. When the imperial government tried to suppress the opium trade in the 1839, the British attacked and quickly prevailed (Figure 11.38).

This first "opium war" ushered in a century of political and economic chaos in China. The British demanded free trade in selected Chinese ports and in the process overturned the traditional policy of managed trade based on the acknowledgment of Chinese supremacy. As European enterprises penetrated China and undermined local economic interests, anti-Manchu rebellions began to break out. In the 1800s, all such uprisings were crushed, but not before causing tremendous destruction. Meanwhile, European power continued to advance. In 1858, Russia annexed the northernmost reaches of Manchuria, and by 1900 China had been divided into separate **spheres of influence** where the interests of different European countries prevailed (Figure 11.39). (In a sphere of influence, the colonial power had no formal political authority, but did enjoy informal influence and tremendous economic clout.)

A successful rebellion launched in 1911 finally toppled the Manchus, but subsequent efforts to establish a unified Chinese Republic were not successful. In many parts of the country, local military leaders, or "warlords," grabbed power for themselves. By the 1920s, it appeared that China might be completely dismembered. The Tibetans had regained autonomy; Xinjiang was under Russian influence; and in China proper Europeans and local warlords vied with the weak Chinese Republic for power. In addition, Japan was increasing its demands and seeking to expand its territory.

▼ **Figure 11.38 Opium War** Great Britain humiliated China in two "opium wars" in the early 1800s, forcing the much larger country to open its economy to foreign trade and to grant Europeans extraordinary privileges. The East India Company steamer *Nemesis* destroyed Chinese war junks in January 1841.

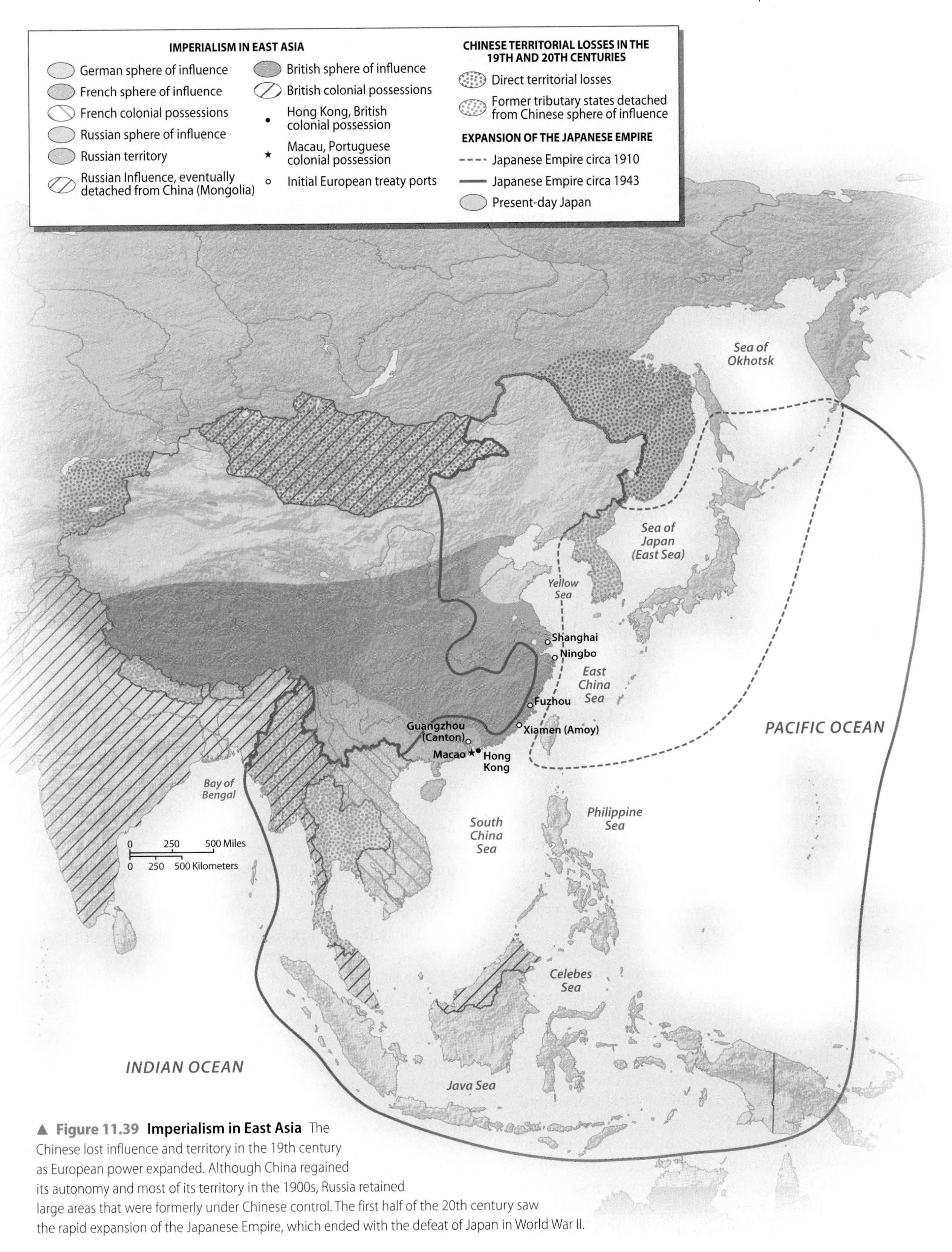

▲ **Figure 11.39 Imperialism in East Asia** The Chinese lost influence and territory in the 19th century as European power expanded. Although China regained its autonomy and most of its territory in the 1900s, Russia retained large areas that were formerly under Chinese control. The first half of the 20th century saw the rapid expansion of the Japanese Empire, which ended with the defeat of Japan in World War II.

The Rise of Japan

Japan did not emerge as a strong, unified state until the 7th century. From its earliest days, Japan looked to China (and, at first, to Korea as well) for intellectual and political models. Its offshore location insulated Japan from the threat of rule by the Chinese Empire. At the same time, the Japanese conceptualized their islands as a separate empire, equivalent in certain respects to China. Between 1000 and 1580, however, Japan was usually divided into several mutually hostile feudal realms. Around 1600, Japan was reunited by the armies of the Tokugawa **Shogunate** (a shogun was a supreme military leader who theoretically remained under the emperor). Shortly afterward, Japan attempted to isolate itself from the rest of the world. Until the 1850s, Japan allowed little trade with outsiders.

Japan remained largely closed to foreign commerce and influence until U.S. gunboats sailed into Tokyo Bay to demand access in 1853. Aware that they could no longer keep the Westerners out, Japanese leaders set about modernizing their economic, administrative, and military systems. This effort accelerated when the Tokugawa Shogunate was toppled in 1868 by the Meiji Restoration. (It is called a restoration because it was carried out in the emperor's name, but it did not give the emperor any real power.) Unlike China, Japan successfully accomplished most of its reform efforts.

The Japanese Empire Japan's new rulers realized that their country remained threatened by European imperial power. They decided that the only way to meet the European challenge was to industrialize and become expansionistic themselves. Japan therefore took control over Hokkaido and began to move farther north into the Kuril Islands and Sakhalin. In 1895, the Japanese government tested its newly modernized army against China, winning a quick and profitable victory that gave it authority over Taiwan. Tensions then mounted with Russia as the two countries vied for power in Manchuria and Korea. The Japanese defeated the Russians in 1905, giving them considerable influence in northern China. With no strong rival in the area, Japan annexed Korea in 1910. Alliance with Britain, France, and the United States during World War I brought further gains, as Japan was awarded Germany's island colonies in Micronesia.

The 1930s brought a global depression, putting a resource-dependent Japan in a difficult situation. The country's leaders sought a military solution, and in 1931 Japan conquered Manchuria. In 1937, Japanese armies moved south, occupying the North China Plain and the coastal cities of southern China. The Chinese government withdrew to the relatively inaccessible Sichuan Basin to continue the struggle. During this period, Japan's relations with the United States steadily deteriorated. In 1941, Japan's leaders decided to destroy the American Pacific fleet in order to clear the way for the conquest of resource-rich Southeast Asia. Their grand strategy was to unite East and Southeast Asia into a "Greater East Asia Co-Prosperity Sphere." This "sphere" was to be ruled by Japan, however, and was designed to keep the Americans and Europeans out. Japanese forces elsewhere in East Asia sometimes engaged in brutal acts. As a result, serious tensions emerged between the Japanese and the other East Asians—tensions that persist to this day.

Postwar Geopolitics

With the defeat of Japan at the end of World War II, East Asia became an arena of rivalry between the United States and the Soviet Union. Initially, American interests prevailed in the maritime fringe, whereas Soviet influence advanced on the mainland. In time, however, East Asia began to experience its own revival.

Japan's Revival Japan lost its colonial empire when it lost World War II. Its territory was reduced to the four main islands plus the Ryukyu Archipelago and a few minor outliers. In general, the Japanese government accepted this loss of land. The only remaining territorial conflict concerns the four southernmost islands of the Kuril chain, which were taken by the Soviet Union in 1945. Although Japan still claims these islands, Russia refuses even to discuss relinquishing control, straining Russo-Japanese relations.

After losing its overseas possessions, Japan had to turn to trade to obtain the resources needed for its economy. Japan's military power was strictly limited by the constitution imposed on it by the United States, forcing Japan to rely on the U.S. military for much of its defense needs. The U.S. Navy continues to patrol many of Japan's vital sealanes, and U.S. armed forces maintain several bases within the country. This U.S. military presence, however, is controversial. Particularly contentious are the U.S. bases on Okinawa, which take up much of the island's territory. In 2008, the United States agreed to relocate many of the U.S. Marines stationed on Okinawa to Guam and other locations, a process that will not be completed until 2015.

Slowly, but steadily, Japan's own military has emerged as a strong regional force despite the constitutional limits imposed on it. In recent years, fears about North Korea's nuclear program and China's military growth have led the Japanese government to reconsider its constitutionally imposed limitations on military spending. In 2010, Japan announced that it would change the focus of its military away from the old Cold War struggle with Russia and toward the rising threat posed by China. As tensions mounted with China in late 2012 and early 2013, the Japanese government announced its intentions to seek closer military ties with the United States.

Other East Asian countries are concerned about the potential threat posed by a remilitarized Japan. Such a perception was strengthened in the early years of the new millennium when Japan's former prime minister visited the Yasukuni Shrine, which contains a military cemetery in which several war criminals from World War II are buried. Anti-Japanese sentiments in China—which have occasionally boiled over into huge public protests—have also been encouraged by the publication of Japanese textbooks that minimize the country's atrocities during the war.

The Division of Korea The aftermath of World War II brought much greater changes to Korea than to Japan. As the end of the war approached, the Soviet Union and the United States agreed to divide Korea at the 38th parallel. Soon two separate regimes were established, the northern one allied with the Soviet Union, the southern one with the United States. In 1950, North Korea invaded South Korea, seeking to reunify the country. The United States, with support from the United Nations (UN), supported the south, while China aided the north. The war ended in a stalemate, and Korea has remained a divided country, its two governments locked in a continuing Cold War (Figure 11.40).

Large numbers of U.S. troops remained in the south after the war. South Korea in the 1960s was a poor, agrarian country that could not defend itself. Over the past 30 years, however, the south has emerged as a wealthy trading nation, while the fortunes of the north have plummeted. The southerners' fear of the north gradually lessened, especially among the members of the younger generation. Many South Korean students have resented the presence of U.S. forces, seeking instead to build friendly relations with North Korea. Changing military priorities, meanwhile, led the United States to reduce its presence in Korea, dropping its troop strength in 2004 from roughly 37,000 to 28,500.

In the late 1990s, the South Korean government began to pursue better relations with North Korea, giving it large amounts of food aid and industrial investment in exchange for reduced hostility. In 2008, however, a new South Korean government concluded that this so-called Sunshine Policy had not restrained North Korean belligerency. North Korean nuclear testing further strained relations, as did the March 2010 sinking of a South Korean naval vessel, an action denied by North Korea. In the same year, the North Korean military fired artillery shells at a South Korean island, killing several people. As tensions between the south and north mounted, the South Korean government sought closer military ties with the United States. In early 2013, North Korea warned that "dark clouds of war" were gathering as the United States and South Korea conducted joint military training operations.

The Division of China World War II brought tremendous destruction and loss of life to China. Even before the war began, China had been engaged in a civil conflict between nationalists (who favored an authoritarian capitalist economy) and communists. The communists had originally been based in the middle Yangtze region, but in 1934 nationalist pressure forced them out. Under the leadership of Mao Zedong, they retreated in the "Long March," which took them to the Loess Plateau, an area close to both the traditional power center of northern China and the industrialized zones of Manchuria. After the Japanese invaded China proper in 1937, the two camps cooperated, but as soon as Japan was defeated, China was again embroiled in civil war. In 1949, the communists proved victorious, forcing the nationalists to retreat to Taiwan.

A latent state of war has persisted ever since between China and Taiwan. Although no battles have been fought, gunfire has periodically been exchanged over the Matsu Islands, several small Taiwanese islands just off the mainland. The Beijing government still claims Taiwan as an integral part of China and vows eventually to reclaim it. The nationalists who gained power in Taiwan long maintained that they represented the true government of China.

The idea of the intrinsic unity of China continues to be influential. In the 1950s and 1960s, the United States recognized Taiwan as the only legitimate government of China, but its policy changed in the 1970s after U.S. leaders decided that it would be more useful and realistic to recognize Mainland China. Soon China entered the United Nations, and Taiwan found itself diplomatically isolated. As of 2013, only 22 members of the United Nations, mostly small African, Pacific, and Caribbean states, continue to recognize Taiwan and, in return, to receive Taiwanese aid. In reality, however, Taiwan is a fully separate country. In 2001, both China and Taiwan entered the World Trade Organization (WTO), but Taiwan was forced to join under the awkward name of "Chinese Taipei," to avoid the suggestion that it is a sovereign country.

▼ **Figure 11.40 The Demilitarized Zone in Korea** North and South Korea were divided along the 38th parallel after World War II. Today, even after the conflict of the early 1950s, these states are separated by the demilitarized zone, or DMZ, which runs near the parallel and which is actively patrolled by armed forces.

In the early years of the new century, Taiwan seemed to be moving toward formal separation from China. In the 2008 Taiwanese election, however, the anti-independence Kuomingtang Party returned to power. The new president Ma Ying-Jeou made it clear that he regarded Taiwan and China as two separately governed areas of a single country. The election of Ma greatly reduced tensions between China and Taiwan. Trade between the island and the mainland is booming, as is tourism. But the de facto independence of Taiwan continues to irritate Beijing. In 2010, China harshly criticized the United States for agreeing to sell $6.4 billion worth of military equipment to the Taiwanese government.

Chinese Territorial Issues Despite the fact that it has been unable to regain Taiwan, China has successfully retained the Manchu territorial legacy. In the case of Tibet, this has required considerable force; resistance by the Tibetans compelled China to launch a full-scale invasion in 1959. The Tibetans, however, have continued to struggle for real autonomy if not actual independence, as they fear that the Han Chinese now moving to Tibet will eventually outnumber them (this controversial issue is covered in more detail in Chapter 10).

China claims several other areas that it does not control. It contends that former Chinese territories in the Himalayas were illegally annexed by Britain when it controlled South Asia, resulting in a border dispute with India. The two countries went to war in 1962 when China occupied an uninhabited highland district in northeastern Kashmir. Tensions between India and China have eased over the past decade, but their territorial disagreement remains unresolved.

One territorial issue was finally resolved in 1997 when China reclaimed Hong Kong. In the 1950s, 1960s, and 1970s, Hong Kong acted as China's window on the outside world, and it grew prosperous as a capitalist enclave. As Chinese relations with the outer world opened in the 1980s, Britain decided to honor its treaty provisions and return Hong Kong to China. China, in turn, promised that Hong Kong would retain its fully capitalist economic system and its partially democratic political system for at least 50 years. Civil liberties not enjoyed in China itself were to remain protected in Hong Kong.

Hong Kong's autonomy, however, remains insecure. Over the past several years, China has thwarted efforts by its people to install a more representative government. As a result of such incidents, hundreds of thousands of people have taken to the streets of Hong Kong in protest. A massive demonstration on New Year's Day 2013, for example, focused on accusations that Hong Kong's chief executive had both mislead the public on controversial real estate developments and has acted as a puppet of the Beijing government.

In 1999, Macau, the last colonial territory in East Asia, was returned to China (Figure 11.41). Like Hong Kong, it is classified as a SAR, under its own autonomous government and subject to its own laws. This small former Portuguese enclave, located across the estuary from Hong Kong, has functioned largely as a gambling refuge (see *Exploring Global Connections: China Courts the Portuguese-Speaking World*). Now the largest betting center in the world, Macau derives 40 percent of its gross domestic product from gambling.

▲ **Figure 11.41 Macau Casino** Macau, reclaimed by China from Portugal in 1999, retains a unique mixture of Chinese and Portuguese cultural influences. Its economic mainstay remains gambling, which is prohibited in the rest of China. **Q: What is the significance of the name of the casino and the fact that two different writing systems are employed in the sign?**

The Senkaku/Diaoyu Crisis The most serious territorial dispute in East Asia concerns a group of tiny islands to the northeast of Taiwan that are called the Senkaku Islands in Japanese and the Diaoyu Islands in Chinese (Figure 11.42). Although the five uninhabited islands total only 2.7 square miles (7 square km), they are probably situated over substantial oil and natural gas reserves. Japan has controlled the islands since 1895, except for the period after World War II when they were under U.S. authority, but China maintains that it has better historical grounds for claiming them than does Japan. Tensions heated up in 2012, when the Japanese government purchased three of the islands from a private Japanese family that had previously owned them. China objected strenuously, viewing this act as an infringement on its sovereignty. By early 2013, increased naval and air-force patrols in the vicinity by both countries led some observers to fear that an actual military conflict could break out. But despite their tense relations, Japan and China are economically integrated to a substantial degree, and neither country wants to engage in a battle that would have severe economic and international consequences.

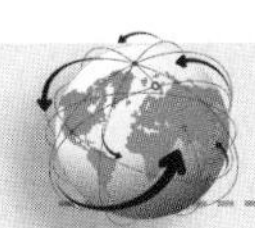

Exploring **Global Connections**

China Courts the Portuguese-Speaking World

In 2012, representatives from China and seven Portuguese-speaking (or Lusophone) countries—Angola, Brazil, Cape Verde, Guinea Bissau, Mozambique, Portugal and East Timor—met in Hohhot, the capital of the Chinese autonomous region of Inner Mongolia, to discuss trade and investment opportunities. The meeting took place under the auspices of the so-called Macau Forum (officially designated as the Forum for Economic and Trade Cooperation Between China and Portuguese-Speaking Countries), which has been meeting periodically since 2003. The organization is widely regarded as a Chinese diplomatic success that has paid substantial economic dividends. From 2003 to 2011, trade between China and the Lusophone countries grew from US$10 billion to US$117 billion. In 2009, Brazil, the world's seventh largest economy, became China's largest trading partner in the Southern Hemisphere, with bilateral trade reaching US$42 billion.

The Portuguese Language in China The connection between China and the Portuguese-speaking world is enhanced by the fact that China is itself, to a very small extent, a Lusophone country. Macau, a Portuguese colony until 1999, is now a SAR; although it has its own laws, political system, and currency, it is nonetheless fully under Chinese sovereignty. Although less than 1 percent of the half-million residents of gambling-focused Macau speak Portuguese at home, the language has official status in the SAR, along with Mandarin Chinese. Portuguese cultural connections, moreover, are carefully maintained. Macau's chief executive, Chui Sai On, recently stated that the territory has "vast potential for cooperation that need[s] to be explored with Portugal," noting as well that "Portuguese characteristics contributed to Macau's development of its role as a trade and services platform between China and the Portuguese-speaking countries."

The African Connection Chinese investments in Lusophone Africa have mostly targeted oil-rich Angola. One large project is the $1.5 billion Benguela Railway, financed and engineered by a Chinese firm and largely built by Chinese labor. More problematic is Nova Cidade de Kilamba, a massive residential and business development located near the capital city of Luanda (**Figure 11.3.1**). According to a recent *Business Insider* report, "The $3.5 billion development covers 12,355 acres and was built to house about 500,000 people, and this is one of several satellite cities being constructed by Chinese firms around Angola." The same story, however, claims that the new city is a virtual ghost town, as few Angolans can afford to live there. Such underutilized instant cities are relatively common in China and are often noted by those who think that the Chinese economy is headed for trouble.

Chinese investments in Mozambique are also rapidly rising. Iron ore and other minerals are the focus of most attention, but infrastructural projects are also entailed. China's Exim Bank, for example, recently extended a $682 million loan to Mozambique to build a suspension bridge in the capital. Portugal had been expected to play this role, but was forced to pull out due to the European debt crisis.

1. To what extent does Macau's historical relationship with Portugal create connections between China and the Portuguese-speaking world
2. Why are Chinese investors particularly interested in business opportunities in Angola?

Google Earth Virtual Tour Video

http://goo.gl/C6D01b

▼ **Figure 11.3.1 Nova Cidade de Kilamba** Most of the new buildings in Angola's city of Nova Cidade de Kilamba have not yet been occupied.

Global Dimensions of East Asian Geopolitics

In the early 1950s, East Asia was divided into two hostile Cold War camps: China and North Korea were allied with the Soviet Union, while Japan, Taiwan, and South Korea were linked to the United States. The Chinese-Soviet alliance soon deteriorated into mutual hostility, and in the 1970s China and the United States found that they could accommodate each other, sharing an enemy in the Soviet Union. In contrast, North Korea's relations with the United States and many other countries have only grown more heated during the same period. As China's economy and military have steadily expanded, however, international geopolitical tensions across the region have again intensified.

The Ongoing North Korean Crisis In 1994, North Korea refused to allow international inspections of its nuclear power facilities—which many countries believed were

◀ **Figure 11.42 Senkaku/Diaoyu Islands** These rugged and uninhabited islands in the East China Sea are administered by Japan but claimed by both China and Taiwan. As China has increasingly emphasized its claims to these rocky islands, tensions with Japan have mounted. **Q: Why have these small and relatively insignificant islands generated such potentially dangerous conflict over the past several years?**

being used for weapons production—provoking an international crisis. It eventually relented, however, in exchange for energy assistance from the United States. But by 2002, the agreement had fallen apart, as evidence mounted that North Korea was continuing to pursue nuclear weapons. Other complaints by the international community focused on North Korea's trafficking in illegal drugs and the fact that it holds hundreds of thousands of its own citizens in brutal labor-camp prisons.

In response to North Korea's nuclear ambitions, the United States has advocated multilateral negotiations involving South Korea, China, Japan, and Russia. North Korea, however, insisted for years that it would negotiate only with the United States. Owing in part to pressure from China, its main trading partner, North Korea finally agreed to the broader framework, participating in five rounds of talks between 2003 and 2007. Little progress was made, however, and in 2009 North Korea pulled out of the negotiations, expelled all nuclear inspectors, and resumed nuclear enrichment. In the same year, it detonated a nuclear weapon, launched a partially successful space booster rocket, and tested a series of ballistic missiles. Additional missile tests occurred over the next few years, and in early 2013 North Korea detonated another nuclear weapon. Such actions have put China in a difficult diplomatic situation. China has officially objected to North Korea's nuclear tests, and it agreed to limited UN sanctions on the country, but it has continued to support the Pyongyang government in other ways. Many experts think that China is concerned that North Korea could collapse, leading to reunification with South Korea, which would result in a refugee crisis, as well as a shared border with a U.S. ally.

China on the Global Stage The end of the Cold War, coupled with the rapid economic growth of China, reconfigured the balance of power in East Asia. The U.S. military became increasingly worried about the growing power of the Chinese armed forces, as did China's neighbors. China now has the largest army in the world, as well as nuclear capability and sophisticated missile technology. Through the early years of the new century, China's military budget grew at an average annual rate of about 10 percent, creating one of the world's most powerful armed forces. China is also rapidly building up its navy, buying advanced ships from Russia and beginning to construct its own.

China is thus coming of age as a major force in global politics. Whether it is a force to be feared by other countries is a matter of considerable debate. Chinese leaders insist that they have no expansionist designs and no intention of interfering in the internal affairs of other countries. They regard concerns expressed by the United States and other countries about their human rights record, as well as their activities in Tibet, as undue meddling in their own internal affairs.

Current opinions on China in the United States vary tremendously. Many U.S. leaders, particularly those in the business community, contend that the two countries should ignore their political differences and develop closer economic and cultural ties. Some East Asian experts similarly argue that the United States must respect China's sovereignty more carefully—or risk a future conflict in the region. Critics, on the other hand, think that China's trade practices are unfair; its labor, human rights, and environmental records appalling; and its actions in Tibet unsupportable.

REVIEW

11.10 How have geopolitical issues in East Asia changed since the end of the Cold War?

11.11 How did the decline of China during the 1800s affect the geopolitical structure of East Asia?

11.12 To what extent do conflicts over natural resources influence geopolitical tensions across East Asia?

Table 11.2 Development Indicators

Country	GNI per capita, PPP 2011	GDP Average Annual % Growth 2000–11	Human Development Index (2011)[1]	Percent Population Living Below $2 a Day	Life Expectancy (2013)[2]	Under Age 5 Mortality Rate (1990)	Under Age 5 Mortality Rate (2011)	Adult Literacy (% ages 15 and older) Male/Female	Gender Inequality Index (2011)[3,1]
China	8,390	10.8	.699	27.2	75	49	15	97/91	0.213
Hong Kong	52,350	4.6	.906	–	83	–	–	--/--	–
Japan	35,330	0.7	.912	–	83	6	3	--/--	0.131
North Korea	1,800*	–	–	–	69	45	33	100/100	–
South Korea	30,370	4.0	.909	–	81	8	5	--/--	0.153
Taiwan	39,400*	5.4*	–	–	79	–	–	96*	–

[1]United Nations, *Human Development Report, 2013.*

[2]Population Reference Bureau, *World Population Data Sheet, 2013.*

[3]Gender Inequality Index—A composite measure reflecting inequality in achievements between women and men in three dimensions: reproductive health, empowerment and the labor market that ranges between 0 and 1. The higher the number, the greater the inequality.

Source: World Bank, *World Development Indicators,* 2013.

*Additional data from the CIA Factbook 2013

Economic And Social Development: A Core Region of the Global Economy

East Asia exhibits a wide range of economic and social development. Japan's urban belt contains one of the world's greatest concentrations of wealth, whereas North Korea is one of the world's most impoverished societies. Overall, however, East Asia has experienced rapid economic growth since the 1970s, with two of its economies, those of Taiwan and South Korea, jumping from the ranks of underdeveloped to developed (Table 11.2). As the economic development of Taiwan and South Korea converged with that of Japan, some geographers began to argue that these three countries should be grouped together as the "Jakota Triangle." Over the past two decades, however, China has experienced the most rapid economic expansion in the region, turning it into the world's second largest economy, after the United States.

Japan's Economy and Society

Japan was the pacesetter of the world economy in the 1960s, 1970s, and 1980s. In the early 1990s, however, the Japanese economy experienced a major setback, and growth has been slow ever since. But despite its recent problems, Japan is still one of the world's largest economic powers.

Japan's Boom and Bust Although Japan's heavy industrialization began in the late 1800s, most of its people remained poor. The 1950s, however, saw the beginnings of the Japanese "economic miracle." Shorn of its empire, Japan was forced to export manufactured products. Beginning with inexpensive consumer goods, Japanese industry moved to more sophisticated materials, including automobiles, cameras, electronics, machine tools, and computer equipment. By the 1980s, it was the leader in many segments of the global high-tech economy.

In the early 1990s, Japan's inflated real estate market collapsed, leading to a banking crisis. At the same time, many Japanese companies discovered that producing labor-intensive goods at home had become too expensive. They therefore began to relocate factories to Southeast Asia and China. Because of these and related difficulties, Japan's economy slumped through the 1990s and into the new millennium. The Japanese government made several attempts to revitalize the economy through state spending, resulting in a huge government debt. Although the Japanese economy began to expand again in 2009 and 2010, the earthquake and tsunami of 2011 dealt a severe economic blow. In early 2013, the Japanese government undertook a series of measures to weaken the value of its currency, the yen, in order to boost exports and thus jump-start economic growth.

Despite its problems, Japan remains a core country of the global economic system. Its economy spans the globe as Japanese multinational firms invest heavily in production facilities in North America and Europe, as well as in poorer countries. Japan is a world leader in a large array of high-tech fields, including robotics, optics, and machine tools for the semiconductor industry (Figure 11.43). It is also one of the world's largest creditor nations, despite its own debts, owning a large percentage of U.S. government bonds. Japanese firms, however, are sometimes regarded as having become less innovative than their Chinese and especially their South Korean rivals. Critics point out that most Japanese companies remain heavily reliant on fax machines, a technology that has been largely bypassed in most other advanced economies.

Living Standards in Japan Despite its affluence, living standards in Japan remain somewhat lower than those of the United States. Housing, food, transportation, and services are particularly expensive in Japan. However, although the Japanese live in cramped quarters and pay high prices for

▲ **Figure 11.43 Japanese Robot** Japan is a world leader in robotics. Japanese researchers are now working hard to create useful humanoid robots, like the one seen in the photograph here.

basic products, they also enjoy many benefits unknown in the United States. Unemployment remains lower than in the United States; health care is provided by the government; and crime rates are extremely low. By such social measures as literacy, infant mortality, and average longevity, Japan surpasses the United States. Japan also lacks the extreme poverty found in certain pockets of American society.

Japan, of course, has its share of social problems. Koreans and alien residents from other Asian countries suffer discrimination, as do members of the indigenous Japanese underclass, the Burakumin. Japan's more remote rural areas have few jobs, and many have seen population decreases. In many small villages, most of the remaining people are elderly. Farming is an increasingly marginal occupation, and many farm families survive only because one family member works in a factory or office. Professional and managerial occupations in Japan's cities are notable for their long hours and high levels of stress.

The Newly Industrialized Countries

In the 1960s, 1970s, and 1980s, the Japanese path to development was successfully followed by its former colonies, South Korea and Taiwan. Hong Kong also emerged as a newly industrialized economy in this period, although its economic and political systems remained distinctive.

The Rise of South Korea The postwar rise of South Korea was even more remarkable than that of Japan. During the period of Japanese occupation, Korean industrial development was concentrated in the north, which is rich in natural resources. The south, in contrast, remained a densely populated, poor, agrarian region. South Korea emerged from the bloody Korean War as one of the world's least-developed countries. In the 1960s, the South Korean government initiated a program of export-led economic growth. It guided the economy with a heavy hand and denied basic political freedom to the Korean people. By the 1970s, such policies had proved highly successful in the economic realm. Huge Korean industrial conglomerates, known as *chaebol,* moved from manufacturing inexpensive consumer goods to heavy industrial products and then to high-tech equipment.

By the 1990s, South Korea emerged as one of the world's main producers of semiconductors, as well as the world's largest shipbuilder. South Korean wages also rose at a rapid clip. The country has invested heavily in education (by some measures, it has the world's most intensive educational system), which has served it well in the global high-tech economy. Large South Korean companies are themselves now strongly multinational, building new factories in the low-wage countries of Southeast Asia and Latin America, as well as in the United States and Europe. South Korea is also noted for its ready embrace of cutting-edge technology. By 2010, it was widely regarded as being the world leader in Internet connectivity and speed. At the same time, many of its high-tech firms were surpassing their Japanese rivals (see *Everyday Globalization: The East Asian Smartphone Connection*).

Contemporary Korea The political and social development of South Korea has not been nearly as smooth as its economic progress. Throughout the 1960s and 1970s, student-led protests against the dictatorial government were brutally repressed. As the South Korean middle class expanded and prospered, pressure for democratization grew, and by the late 1980s it could no longer be denied. But even though democratization has been successful, political tension has not disappeared. Critics contend that the South Korean economy needs substantial reforms—in particular, the breaking up of the large conglomerates. In recent years, however, the South Korean economy has been characterized by steady growth, moderate inflation, low unemployment, and large export surpluses.

Between 1998 and 2008, South Korea attempted under the Sunshine Policy to use its own economic success to help North Korea, encouraging South Korean companies to invest in jointly operated factories in North Korea. During the same period, North Korea allowed the limited emergence of public markets. Joint ventures with South Korea, however, were never particularly successful, and after the Korean geopolitical crisis of 2010 such economic ventures were curtailed. During the same period, the Pyongyang government also moved to restrict its small private markets. By 2013, however, North Korea's underground market economy was again growing, bringing a certain measure of prosperity to a small class of entrepreneurs, particularly those with good political connections. The North Korean government has also increasingly shunted development funds into Pyongyang. But although the capital city has experienced improved condi-

Everyday Globalization

The East Asian Smartphone Connection

Over the last several years, smartphones have gone from a niche curiosity to a must-have everyday tool. The total number of U.S. smartphone users reached 115.8 million in 2012, and the total number of smartphones shipped worldwide came in at 712 million in the same year. To build their products, the two leading smartphone makers, Apple and Samsung, depend on complex and far-reaching supply chains that rely critically on East Asian manufacturing and materials. Samsung itself is an old and successful South Korean company whose output accounts for roughly one-fifth of the country's economy.

Beginning in 1938 as a small trading company focused primarily on foodstuffs, Samsung gradually grew to become a major heavy-industry and insurance firm. In 1969, the company entered the electronics market, producing its first television in 1970. Today Samsung's Galaxy smartphones, which use the Google Android operating system, have taken the world by storm, vying with California-based Apple's iPhone for popularity.

The Rare Earth Issue Smartphone production—regardless of the model—begins with the mining of various metals in locations all over the world. In addition to relatively common metals like tin, which are found in many countries, smartphones require several so-called rare earth elements to function, many of which come from China. As of 2012, China mined 90 percent of the world's rare earth elements, though it has only 23 percent of the world's known reserves. Nearly all of China's rare earth production comes from the province of Inner Mongolia—and specifically from the city of Baotou. Home to about 2 million people, Baotou is infamous for the dangerous pollution and environmental degradation that accompany rare-earth mining.

The Labor Issue China is also home to the majority of manufacturing firms—numbering in the hundreds—that build smartphones and other electronics under contract for Apple and Samsung. China and other Asian countries have long been havens for manufacturing due to low wages and relatively loose labor and environmental regulations. Both Apple and Samsung have come under fire in recent years over the working conditions at Chinese contractors. Foxconn, a Taiwanese company that has grown into China's largest private-sector employer, manufactures many Apple products, including the iPhone (Figure 11.4.1). The firm has been heavily criticized since a spate of suicides came to light in 2010. Both Apple and Foxconn have since taken steps to improve working conditions, though some observers remain unsatisfied.

In 2012, Apple concluded an audit of 393 Chinese suppliers, finding 11 facilities employing illegal child labor. This led the company to cut ties with circuit board manufacturer Guangdong Real Faith Pingzhou Electronics. In the fall of 2012, Samsung also confronted accusations of child labor violations at its Chinese suppliers and has since denied the reports following an audit, though the company says it uncovered other violations and will be more closely monitoring 249 of its Chinese contractors.

1. What in particular about smartphones makes them such a highly globalized product?
2. Why are many countries, and companies, concerned about China's control over most of the world's rare earth elements?

Google Earth Virtual Tour Video

http://goo.gl/xYRjl1

▲ **Figure 11.4.1 Foxconn Factory in China** The Taiwanese firm Foxconn operates many factories in China, several of which make cell phones for multinational corporations.

tions as a result, smaller cities and especially the countryside remain as impoverished as ever (see *People on the Move: North Korea's Hidden Migration Streams*).

Geographical Tools: Google Earth and North Korea Getting accurate economic as well as political information from North Korea is almost impossible, due mainly to the secretive and often seemingly paranoid behavior of the country's leadership. But in recent years, Google Earth has become a prime source of data. Led by Curtis Melvin, who blogs at North Korea Economy Watch, a group of scholars and amateur geographers has learned how to ferret out detailed information on economic development, political repression, and the armaments industry from satellite images. In early 2013, for example, new Google Earth images showed a significant expansion of the Kwan-il-so Labor Camp, noted for its brutal and often deadly treatment of political prisoners (Figure 11.44). Melvin and his colleagues have also demonstrated the lavish lifestyles

◀ **Figure 11.44 Kwan-il-so Labor Camp** Many North Korean political prisoners live in dismal labor camps, suffering through very challenging conditions. **Q: Why do researchers rely so extensively on satellite images for investigating labor camps and other political prisons in North Korea?**

enjoyed by North Korea's narrow elite class, showcasing such indulgences as yachts and a gigantic private waterslide. In 2012, members of the same team released the DPRK Digital Atlas, based on a searchable Google Earth platform. This innovative atlas allows members of the general pubic to easily explore inaccessible areas in what is probably the world's most guarded country.

Taiwan and Hong Kong Like South Korea, Taiwan and Hong Kong have experienced rapid economic growth since the 1960s. The Taiwanese government, like those of South Korea and Japan, has guided the economic development of the country. Hong Kong, unlike its neighbors, has been characterized by one of the most **laissez-faire** economic systems in the world, one with a great amount of market freedom from government interference (*laissez-faire* is a French term meaning "let it be"). Hong Kong traditionally functioned as a trading center, but in the 1960s and 1970s it emerged as a major producer of textiles, toys, and other consumer goods. By the 1980s, however, such cheap products could no longer be made profitably in such an expensive city. Hong Kong's industrialists subsequently began to move their plants to nearby areas in southern China, while Hong Kong itself increasingly specialized in business services, banking, telecommunications, and entertainment. By 2012, however, surging property valuation and an extremely high cost of living were linked by some experts to an economic slowdown, leading to some concern about the city's future.

Both Taiwan and Hong Kong have close overseas economic connections. Linkages are particularly tight with Chinese-owned firms located in Southeast Asia. Taiwan's high-technology businesses are also intertwined with those of the United States; a constant back-and-forth flow of talent, technology, and money occurs between Taipei and Silicon Valley. Hong Kong's economy also is closely bound with that of the United States (as well as those of Canada and Britain), but its closest connections are with the rest of China. Taiwan is moving in the same direction; China is now its largest export market and its second largest source of imports after Japan.

Chinese Development

China dwarfs all the rest of East Asia in both physical size and population. Its economic takeoff is thus reconfiguring the economy of the entire region and, to some extent, that of the world as a whole. China now has a vast middle class, some 350 million strong, that is able to afford a broad array of expensive consumer goods. As that class is expected to grow to 600 million by 2020, both the Chinese economy and that of the world as a whole are being rapidly transformed. However, despite its recent success, China's economy has several weaknesses. The future of the Chinese economy is one of the biggest uncertainties facing both the East Asian and the global economies.

China Under Communism More than a century of war, invasion, and near-chaos in China ended in 1949 when the communist forces led by Mao Zedong seized power. The new government, inheriting a weak economy, set about nationalizing private firms and building heavy industries. Some successes were realized, especially in Manchuria, where a large amount of heavy industrial equipment was inherited from the Japanese colonial regime.

In the late 1950s, however, China experienced an economic disaster ironically called the "Great Leap Forward." One of the main ideas behind this scheme was that small-scale village workshops could produce the large quantities of iron needed for sustained industrial growth. Communist Party officials demanded that these inefficient workshops meet unreasonably high production quotas. In some cases, the only way they could do so was to melt peasants' agricultural tools. Peasants also were forced to contribute such a large percentage of their crops to the state that many went hungry. The result was a horrific famine that may have killed 20 million people.

The early 1960s saw a return to more pragmatic policies, but toward the end of the decade a new wave of radicalism swept through China. This "Cultural Revolution" was aimed at mobilizing young people to stamp out the remaining vestiges of capitalism. Thousands of experienced industrial managers and college professors were expelled from their positions. Many were sent to villages to be "reeducated" through hard physical labor; others were simply killed. The economic consequences of such policies were devastating.

People on the Move

North Korea's Hidden Migration Streams

Most of the migration in East Asia during recent years has been due to *pull factors,* such as the lure of jobs and higher wages in urban areas. However, emigration from North Korea and within North Korean provinces is more often the result of famine and destitution.

Escaping Famine and Poverty Although population movement between Korea and northeastern China has a long history, migration to China from North Korea reached new heights of intensity and international notoriety when famine gripped the country during the late 1990s. Estimates of the famine's death toll range from around 200,000 to 3 million people. Poverty and malnutrition continue to plague the North Korean countryside, keeping strong pressure on residents to leave the country. Some 200,000 North Korean refugees are estimated to live in hiding in China, with another 24,000 in South Korea (**Figure 11.5.1**). The number of North Koreans arriving in South Korea annually has fallen in recent years, however, dropping from around 2500 in the mid-2000s to 1500 in 2012.

The absolute number of North Korean refugees fleeing the country over the last two decades is not especially large compared to other population movements in East Asia, but the difficulty of the journey and of life as a refugee hints at the latent demand for emigration. North Korea maintains tight control of its borders, using lethal force or arrest to bar its people from leaving. Most who manage to escape do so by paying bribes to smugglers, but a recent crackdown by the North Korean government on such practices has made leaving the country more difficult than ever.

Famine has also caused considerable migration within North Korea. Although the government maintains that such movements are negligible, outside estimates point to a migration rate of 18.7 percent, with more than 30 percent of such migrants citing the "search for food" as their primary reason for moving. Migration within North Korea is heavily restricted, making most internal population movements illegal and "irregular." A dearth of reliable census data on the subject makes it difficult to examine in more detail.

Refugees in China Within China, North Korean refugees are treated as illegal economic immigrants and face deportation if discovered by authorities. Around 5000 refugees are sent back to North Korea from China annually, a figure that now likely exceeds the number of North Koreans entering China. Some North Korean refugees in China try to relocate to other countries, often in Southeast Asia.

Why do you think that the government of China treats North Korean refugees in such a harsh manner?

▲ **Figure 11.5.1 North Korean Refugees in China** North Koreans who flee to China are often forced to to do menial work on small Chinese farms.

Toward a Post-communist Economy When Mao Zedong died in 1976, China faced a crucial turning point. Its economy was nearly stagnant, and its people were desperately poor. A political struggle ensued between pragmatists hoping for change and dedicated communists. The pragmatists emerged victorious, and by the late 1970s it was clear that China would embark on a different economic path. The new China would seek closer connections with the world economy and take a modified capitalist road to development. China did not, however, transform itself into a fully capitalist country. The state continued to run most heavy industries, and the Communist Party retained a monopoly on political power. Instead of suddenly abandoning the communist model, as the former Soviet Union later did, China allowed cracks to appear in which capitalist ventures could take root and thrive. In 1995, almost 80 percent of the Chinese economy was controlled by the government, but by 2009 over 70 percent was controlled by private enterprise.

One of China's first capitalist openings was in agriculture, which had previously been dominated by large-scale communal farms. Individuals were suddenly allowed to act as agricultural entrepreneurs, selling produce in the open market. Owing to this change, the income of many farmers rose dramatically. By the late 1980s, however, the focus of growth had shifted to the urban-industrial sector. As the government became concerned about inflation, it placed price caps on agricultural products and increased taxes on farmers. By the early years of the new millennium, many rural areas in China's interior provinces were experiencing economic distress, prompting many people to relocate to China's coastal cities or to foreign countries.

The Era of Rapid Growth One of China's most important early industrial reforms involved opening **Special Economic Zones (SEZs)**, in which foreign investment was welcome and state interference minimal. The Shenzhen SEZ, adjacent to Hong Kong, proved particularly successful after Hong Kong manufacturers found it a convenient source of cheap land and labor (Figure 11.45). Additional SEZs were soon opened, mostly in the coastal region. The basic strategy was to attract foreign investment that could generate exports, the income from which would supply China with the capital it needed to build its infrastructure and thus achieve conditions for sustained economic growth.

Other market-oriented reforms followed. From 1980 to 2010, the Chinese economy grew at an average rate of roughly 10 percent a year, perhaps the fastest rate of expansion the world has ever seen. China emerged as a major trading nation, amassing large trade surpluses and foreign reserves. As of 2012, China's foreign exchange reserves stood at $3311 billion, dwarfing second-place Japan's figure of $1268 billion. In 2012, however, mounting inflation and other problems resulted in a decrease in growth. Still, the 8 percent economic expansion recorded for the year was still one of the highest in the world.

China's economic growth has increased tensions with the United States. China exports far more to the United States than it imports, leading some U.S. politicians to demand that China allow its currency, the renminbi (the primary unit of which is the yuan), to rise against the dollar. In 2005, China officially unhooked its currency from the U.S. dollar, and within a few years it had appreciated 21 percent. Critics in the United States accuse China of unfairly keeping its currency undervalued in order to enhance its exports. Concern about inflation, however, did lead China to allow its currency to rise slightly in late 2012 and early 2013.

▼ **Figure 11.45 Shenzhen** The city of Shenzhen, adjacent to Hong Kong, was one of China's first Special Economic Zones. It has recently emerged as a major city in its own right.

Economists are divided as to whether China will be able to maintain rapid rates of economic growth over the next few decades. Wages in Chinese factories are rapidly increasing, convincing some foreign firms to return production to their homelands or to relocate to poorer countries, such as Vietnam or Bangladesh. Inflation, social unrest, and environmental degradation are also increasing concerns. Some experts maintain that China has invested unwisely in some areas, building new housing developments and even instant cities that have failed to attract residents. However, it is also true that critics have been predicting a Chinese economic slowdown or even collapse for years, yet the Chinese economy continues to be the marvel of the world.

Social and Regional Differentiation The Chinese economic surge unleashed by the reforms of the late 1970s and 1980s resulted in growing **social and regional differentiation**. In other words, certain groups of people—and certain portions of the country—prospered far more than others. Despite its official adherence to socialism, the Chinese state encouraged the formation of an economic elite, concluding that wealthy individuals are necessary to transform the economy. The least-fortunate Chinese citizens were sometimes left without work, and many millions have migrated to the booming coastal cities, generating one of the largest mass migrations the world has ever seen. The government has tried to control the transfer of population, but with only partial success. As China's economic boom accelerated, economic disparities mounted. The rapid growth of the elite population made China the world's fastest-growing market for luxury automobiles and golf-course developments. At the same time, vulnerable state-owned enterprises have increasingly abandoned their provision of housing, medical care, and other social services. By early 2013, the gap between the rich and the poor had become so large that the government pledged that it would undertake efforts to reduce such disparities.

Economic disparities in China, as in other countries, are geographically structured. Before the reform period, the communist government attempted to equalize the fortunes of the different regions, giving special privileges to individuals from poor places. Such efforts were not wholly successful, and some provinces continued to be deprived. Since the coming of market reforms, moreover, the process of regional economic differentiation has accelerated (Figure 11.46). Urban areas are now much better off than rural zones, with the average city-dweller having three times as much disposable income as the average person living in a vil-

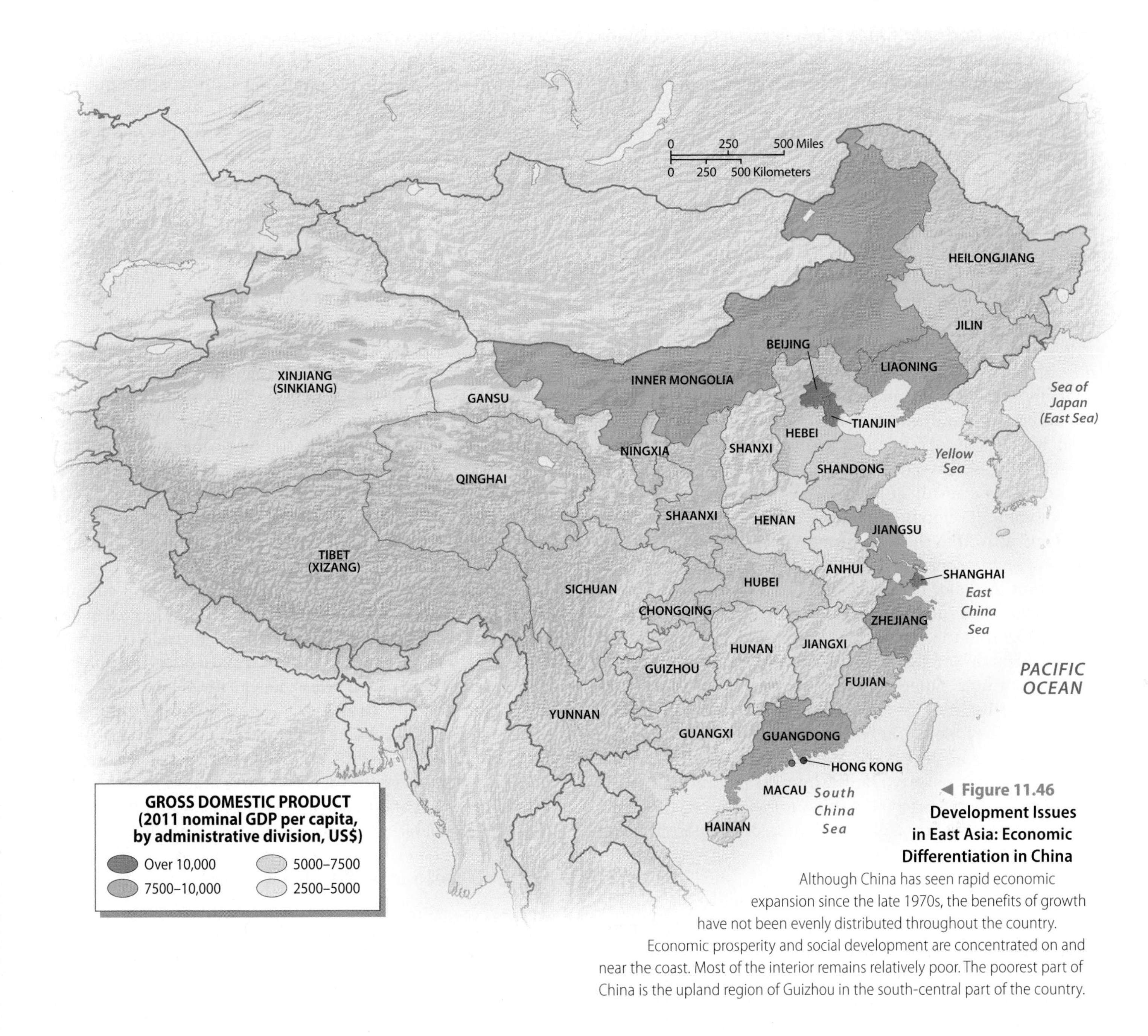

Figure 11.46 Development Issues in East Asia: Economic Differentiation in China Although China has seen rapid economic expansion since the late 1970s, the benefits of growth have not been evenly distributed throughout the country. Economic prosperity and social development are concentrated on and near the coast. Most of the interior remains relatively poor. The poorest part of China is the upland region of Guizhou in the south-central part of the country.

lage. Those living in the country's booming coastal cities have been particularly fortunate over the past few decades.

The Booming Coastal Region Most of the benefits from China's economic transformation have flowed to the coastal region and to the capital city of Beijing. The first beneficiaries were the southern provinces of Guangdong and Fujian. This region was perhaps predisposed to the new economy because the southern Chinese have long been noted for engaging in overseas trade. Guangdong and Fujian have also benefited from their close connections with the overseas Chinese communities of Southeast Asia and North America. Proximity to Taiwan and especially Hong Kong also proved helpful. Vast amounts of capital have flowed to the south coastal region from foreign (and Hong Kong–based) Chinese business networks.

By the early 2000s, the Yangtze Delta, centered on the city of Shanghai, reemerged as the most dynamic region of China. The delta was the traditional economic (and intellectual) core of China, and before the communist takeover Shanghai was its premier industrial and financial center. The Chinese government, moreover, has encouraged the development of huge industrial, commercial, and residential complexes, hoping to take advantage of the region's dynamism. In 2013, Shanghai officials unveiled a new five-year plan focused on luring more multinational firms to the city, with the goal of doubling municipal economic production from 2010 to 2020.

Interior and Northern China The interior regions of China have seen much less economic expansion. The central and northern areas of Manchuria were formerly quite prosperous, owing to fertile soils and early industrialization, but have not participated much in the recent boom. Many of the state-owned heavy industries of the Manchurian **rust belt**, or zone of decaying factories, are not efficient. In 2012, six cities in Liaoning Province announced that they were going to convert several heavily polluted industrial sites to farmland. Manchuria's once-productive oil wells, moreover, are largely exhausted. China and Russia have, however, recently constructed an oil pipeline from Siberia to Daqing in Manchuria, hoping to revive the economy of this decaying oil town.

Most of the interior provinces of China also largely missed the initial wave of growth in the 1980s and 1990s. In many areas, rural populations continued to grow while the natural environment deteriorated. One consequence has been high levels of underemployment and out-migration. By most measures, poverty increases with distance from the coast. As a result of such discrepancies, China is now encouraging development in the west (including Tibet and Xinjiang, as well as the western provinces of China proper), focusing on transportation improvement and natural resource extraction. As labor shortages begin to appear in the booming coastal areas, industrialists are responding by building new factories in the interior portions of the country.

Scholars debate the conditions found in the poorer parts of interior China. Some believe that China's official statistics are too positive, hiding significant poverty. Others think that the country's economic boom has substantially raised living standards even in the poorest districts. According to official statistics, China's poverty rate fell from 53 percent in 1981 to less than 3 percent in 2009. Critics note, however, that as of 2008, almost 30 percent of the Chinese population lived on less than $2 a day.

Rising Tensions China's explosive, but uneven economic growth has generated some problems. Inflation has made planning difficult and created hardship for those on fixed incomes. Corruption by state officials is by some accounts rampant; success often seems to depend on knowing the right people and having the right connections. In 2009 alone, more than 100,000 public officials were convicted on corruption charges. Not only do poorer people have to pay frequent bribes, but also many have been evicted from their homes to make way for property developments. Widespread public anger over such matters has led to periodic governmental crackdowns on corruption; one such drive in early 2013 uncovered a provincial official supporting 47 mistresses. Critics complain, however, that such efforts merely target a few particularly blatant instances of corruption.

An equally significant issue has been the struggle for free expression and political openness. In 1989, as the Cold War was coming to an end, the state crushed a popular student-led protest movement for government accountability and democratic reform. After as many as 1 million citizens had gathered in Beijing's Tiananmen Square to demand change, the government sent in troops, declared martial law, and arrested hundreds of student leaders. Officials who were thought to be sympathetic to the protests were removed from office, and the opposition was forced to go underground.

In the years after the Tiananmen event, anti-government sentiments declined as most people in China concentrated on economic development. However, in the early 2000s, the local protest movements gained strength, facilitated by tech-savvy activists using social media to organize their actions. Experts estimate that more than 100,000 organized protests occur each year. Although most such demonstrations focus on local issues, pro-democracy protests did break out in several Chinese cities in 2011.

China has responded to such protest movements in part by increasing its restrictions on free speech. Internet censorship has become a particularly contentious issue. The Chinese government is currently estimated to employ an "Internet police force" roughly 30,000 strong. Many Chinese Internet users, however, have learned to use sophisticated techniques to avoid official censorship. Others have taken to the streets: In early 2013, free-speech protesters in the important southern city of Guangzhou clashed with Communist Party loyalists holding posters of Chairman Mao.

China's political and human rights policies have complicated its international relations. Tension with the United States and other wealthy countries also stems from economic issues. China's large and growing trade surplus and its reluctance to enforce copyright and patent laws irritate many of its trading partners. Numerous U.S. firms have accused Chinese businesses of pirating music, software, and brand names. China passed a new trademark law in 2011, but enforcement has been spotty. As the global market grows and as popular culture becomes globalized, copyright and trademark infringement is becoming increasingly profitable and hence increasingly difficult to control.

Social Development in East Asia

Levels of social development in East Asia, as you might expect, vary significantly across the region. Health and education standards are very high in the more economically advanced parts of the region and lag behind in the poorer parts of China and in North Korea. But despite its poverty and political repression, even North Korea posts average life expectancy figures slightly above the world average. The greatest advances in social development over the past few decades have occurred in China.

Social Conditions in China Since coming to power in 1949, China's communist government has made large investments in medical care and education, and today China boasts fairly impressive health and longevity figures. However, as China has moved to a market-based economy, its formerly extensive system of rural medical clinics has deteriorated. Some observers fear that levels of both health and education may be declining for the poorest segments of China's population.

Human well-being in China is also geographically structured. The literacy rate, for example, remains lower in many of the poorer parts of China, including the uplands of Yunnan and

▲ **Figure 11.47 Undocumented Migrants in Chinese Cities** As China's economy grows, tens of millions of people are moving from the countryside to urban areas. Many of these people, however, do not have official permission to live in cities, and are thus undocumented migrants in their own country.

Guizhou and the interior portions of the North China Plain, than in the most prosperous coastal regions of the country. Increasingly, the largest gap separates the booming cities from the languishing rural areas. People from poor, rural areas often move to the thriving cities, but the government has put strict limits on such movement, and as a result many urban migrants have no official residency status. As a result, they essentially live as "undocumented migrants" in their own country, and as a result have a difficult time getting access to governmental services, including education for their children (Figure 11.47).

Population policy also remains an unsettling issue for China. With 1.3 billion people highly concentrated in less than half of its territory, China has one of the world's highest effective population densities. By the 1980s, its government had become so concerned that it instituted the famous one-child policy. Under this plan, couples in normal circumstances are expected to have only a single offspring and can suffer financial and other penalties if they do not comply. However, there have always been numerous exceptions to the policy, especially in rural areas and minority regions, and as a result the one-child rule was applied to only about 35 percent of China's total population as of 2007. Due both to the one-child policy and to general urban and economic development, China's total fertility rate has plummeted and is now at 1.6, well below the replacement level. Even so, China's population will likely reach roughly 1.5 billion persons before it stabilizes.

The decline in Chinese fertility has brought about problems of its own. China is now worried about impending labor shortages, as well as its aging population. China's population policy has also generated social tensions and human rights abuses. Particularly troubling is the growing gender imbalance in the Chinese population. In 2009, 119 boys were born for every 100 girls, and in the province of Jiangxi the figure was 143. This asymmetry reflects the practice of honoring one's ancestors; because family lines are traced through male offspring, a family must produce a male heir to maintain its lineage. Many couples are therefore desperate to produce a son. One option is gender-selective abortion; if ultrasound reveals a female fetus, the pregnancy is sometimes terminated. Baby girls also are commonly abandoned, and well-substantiated rumors of female infanticide circulate. Baby boys, in contrast, are sometime kidnapped and sold to young couples desperate to raise a son.

Gender Issues in East Asia Gender roles throughout East Asia have undergone significant changes during the last century as industrialization and changing social values have transformed the societies of the region. Nevertheless, East Asian countries generally lag behind other industrialized countries in gender equality. According to the 2012 Global Gender Gap Report, China ranks 69th out of 135 countries, Japan ranks 101st, and South Korea ranks 108th (North Korea and Taiwan were not ranked by the report).

In Japan, the first East Asian country to industrialize, women were denied most political and legal rights under the 1889 Meiji Constitution and in the Meiji Civil Code of 1898. At the same time, a growing urban economy opened up new opportunities for work outside the home. Since the end of World War II, women in Japan have enjoyed a full slate of political and legal rights, though their opportunities for career advancement lag behind those of men. The traditional expectation that women take the leading role in caring for children forces many to choose between advancing their careers and starting families. Though this is a common dilemma in the industrialized world, the theme is particularly pronounced in Japan, as evidenced by the country's exceptionally low birth rate and declining population. The status of women in South Korea and Taiwan is similar to that of women in Japan. Though fully equal to males in the eyes of the law, few women reach positions of power in business or government.

The legacy of Confucianism throughout East Asia, particularly in China, emphasizes the desirability of obedience to a male patriarch within the home. The ideology is both a reflection and a cause of the difficulties women have faced historically in Chinese society. In the 1800s, the widespread practice of foot binding—involving the breaking and constriction of women's feet so as to achieve a dainty aesthetic—grew to involve nearly half the women in China before the decline of the custom in the earltty 20th century. Economic growth and the relaxation of traditional cultural norms have clearly improved the position of women in China. Women now form over 42 percent of the Chinese workforce and made up 23 percent of the 2012 National Congress of the Communist Party of China. Nevertheless, the upper echelons of Chinese politics remain largely the purview of men, and women's reproductive freedom is challenged by China's population policies.

REVIEW

11.13 How has the economic development of Japan, South Korea, Taiwan, and China been similar since the end of World War II, and how has it differed between these countries?

11.14 Why do levels of social and economic development vary so extensively from the coastal region of China to the interior portions of the country?

11.15 Do the geographical patterns of social development in East Asia differ from those of economic development? If so, how?

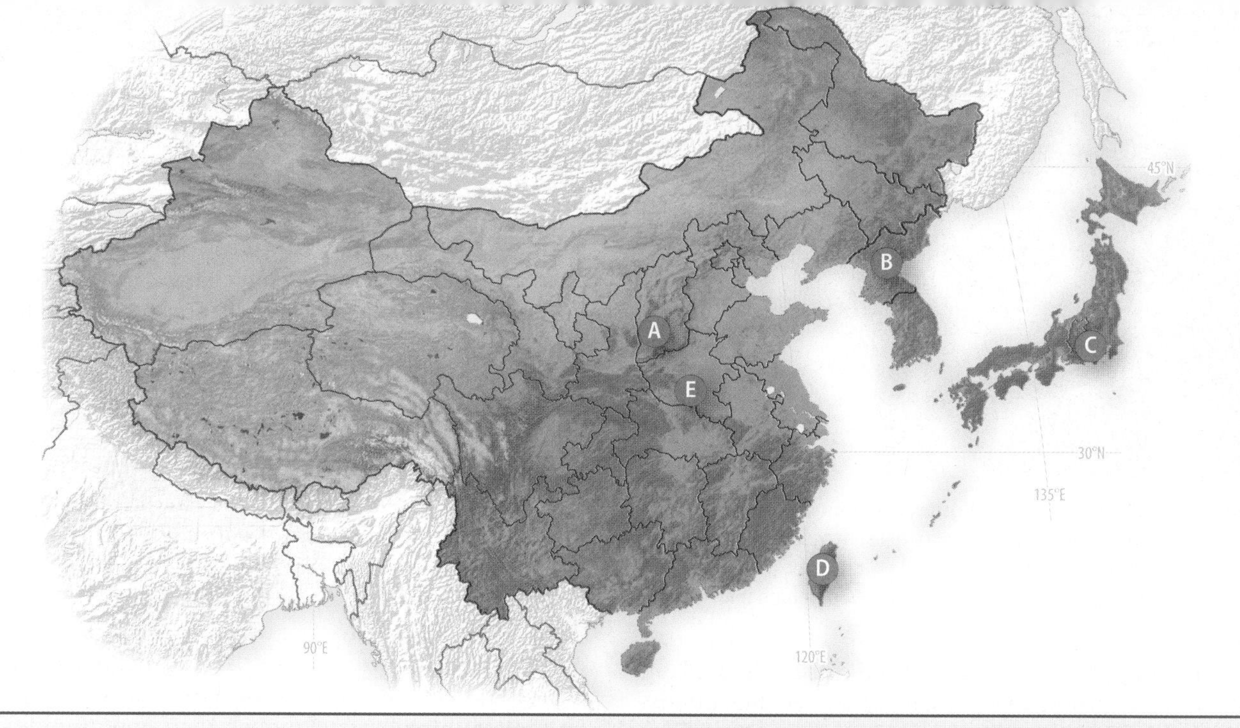

Physical Geography and Environmental Issues

The economic success of East Asia has been accompanied by severe environmental degradation. Japan, South Korea, and Taiwan have responded by enacting strict environmental laws and by moving many of their most polluting industries overseas. The major environmental issue in the region today concerns the rapid growth of the Chinese economy. Pollution in Chinese cities is so serious that it has major negative effects on human health, and much of the Chinese countryside suffers from environmental problems such as soil erosion and desertification. China is responding, however, with a major program for renewable energy.

11.1 This map detail shows an area characterized by severe soil erosion, a strong risk of flooding, and a new water transfer project. How are these phenomena related,?

11.2 What are the potential advantages and disadvantages of diverting water flows from one river basin to another?

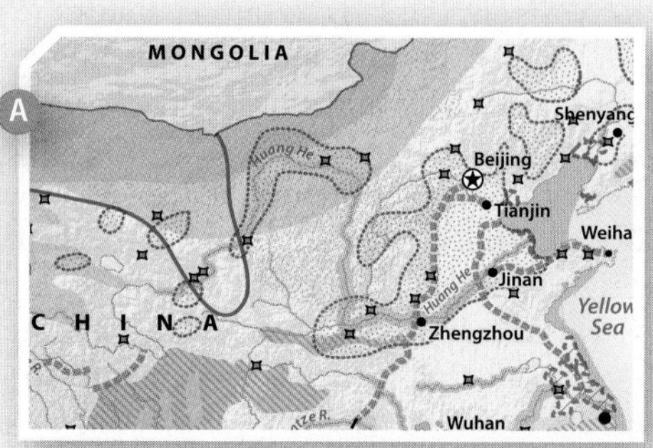

Population and Settlement

East Asia is a densely populated region, but its birth rates have plummeted in recent decades. Japan is presently experiencing population decline, which will put pressure on the Japanese economy. In China, the biggest demographic challenge results from the massive movement of people from the interior to the coast and from rural villages to the rapidly expanding cities. China has been trying to redirect development toward the interior, but thus far has had little success.

11.3 Why is animal power used so often in North Korean agriculture, and what implications does this have for the North Korean population?

11.4 What kind of consequences would a country be likely to face in pursuing a policy of food self-sufficiency?

Key Terms

anthropogenic landscape *(p. 513)*
autonomous region *(p. 526)*
Burakumin *(p. 523)*
capitalism *(p. 498)*
central place theory *(p. 516)*
China proper *(p. 498)*
Cold War *(p. 527)*
communism *(p. 497)*
Confucianism *(p. 520)*
diaspora *(p. 524)*
dynastic succession *(p. 522)*
geomancy *(p. 521)*
ideographic writing *(p. 519)*
laissez-faire *(p. 540)*
loess *(p. 509)*
Mandarin *(p. 520)*
Marxism *(p. 522)*
pollution exporting *(p. 504)*
regulatory lake *(p. 508)*
rust belt *(p. 544)*
samurai *(p. 528)*
sediment load *(p. 508)*
Shogunate *(p. 532)*
social and regional differentiation *(p. 542)*
Special Economic Zone (SEZ) *(p. 542)*
sphere of influence *(p. 530)*
superconurbation *(p. 517)*
tonal language *(p. 526)*
tsunami *(p. 499)*
urban primacy *(p. 516)*

In Review • Chapter 11
East Asia

Cultural Coherence and Diversity

East Asia is unified by deep cultural and historical bonds. China has had the largest influence on East Asia because, at one time or another, it ruled most of the region. Although Japan has never been under Chinese rule, it still has profound historical connections to Chinese civilization. The more prosperous parts of East Asia have welcomed cultural globalization over the past several decades.

11.5 Mt. Fuji is often viewed as a national symbol of Japan. How does the Japanese religion of Shinto contribute to the cultural significance of this mountain?

11.6 How does the natural world figure into beliefs and practices found among other religions around the world?

Geopolitical Framework

Geopolitically, East Asia has been characterized by much strife since the end of World War II. China and Korea are still suspicious of Japan, and they worry that it might rebuild a strong military force. Japan is concerned about the growing military power of China and about the nuclear arms and missiles of North Korea. Territorial disputes over islands complicate relations between China and Japan and also between Japan and South Korea. Relations between North and South Korea improved between 2000 and 2008, but subsequently deteriorated as North Korea continued to test nuclear weapons.

11.7 If Taiwan is already an independent country, why are the Taiwanese protesters in this photo demanding Taiwanese independence?

11.8 Should the international community consider acknowledging the independent existence of Taiwan, or is the "one China policy" necessary for regional and global stability?

Economic and Social Development

With the notable exception of North Korea, all East Asian countries have experienced rapid economic growth since the end of World War II. In the 2000s, the most important story was the rise of China. China's economic expansion has reduced poverty nationwide, but has also generated serious tensions between the wealthier, more globally oriented coastal regions and the less-prosperous interior provinces. The rise of China also has global implications, as many countries in all regions of the world have profited by exporting the raw materials needed by China's booming industries.

11.9 China has recently been rapidly expanding its urban system, yet in some areas huge new apartment blocks have gone unoccupied. Why have such projects been built, and what consequences might they have for China's economy?

11.10 What are some of the advantages and disadvantages of massive planned residential developments?

MasteringGeography™

Looking for additional review and test prep materials? Visit the Study Area in **MasteringGeography**™ to enhance your geographic literacy, spatial reasoning skills, and understanding of this chapter's content by accessing a variety of resources, including **MapMaster** interactive maps, videos, RSS feeds, flashcards, web links, self-study quizzes, and an eText version of *Diversity Amid Globalization*

Authors' Blogs

Scan to visit the author's blog for chapter updates

www.gad4blog.wordpress.com

Scan to visit the GeoCurrents blog

www.geocurrents.info

12 South Asia

The New Face of Corporate India. Trainees walk past the newly constructed Multiplex Dome on the Infosys corporate campus in Mysore, India.

Physical Geography and Environmental Issues

The arid parts of South Asia suffer from water shortages and salinization of the soil, whereas the humid areas often experience devastating floods. Pollution levels in urban areas are often extremely high.

Population and Settlement

South Asia will soon become the most populous region in the world. Birth rates in much of the region, however, have dropped sharply, falling below the replacement level in most of southern and western India. Fertility rates remain much higher, however, in northern India and Pakistan.

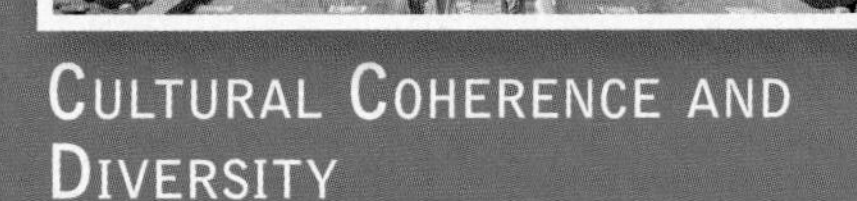

Cultural Coherence and Diversity

South Asia is one of the most culturally diverse regions of the world, with India alone having more than a dozen official languages, as well as numerous followers of most major religions. Religious and ethnic tensions have increased in many areas over the past several decades, especially in Pakistan.

Geopolitical Framework

South Asia is burdened not only by a large number of violent secession movements, but also by the struggle between the nuclear-armed countries of India and Pakistan, which threatens both regional and global stability. The relationship between Pakistan and the United States is tense, as is that between India and China.

Economic and Social Development

Although South Asia is one of the poorest world regions, parts of India are experiencing rapid development based on the high-tech skills of its educated people. Although extremely poor, Bangladesh has also made significant social and economic progress, whereas Pakistan has lagged a bit behind.

When people from Europe or North America imagine Indian cities, they generally picture crowded and chaotic streets, impoverished slums, or industrial blight. Few think of leafy, spacious corporate campuses with modern buildings and world-class infrastructure. Yet this form of development is increasing in many Indian cities. As you can see from the photograph, the Infosys campus in the southern city of Hyderabad would not look out of place in California's Silicon Valley. Hyderabad, like other South Asian cities, still has its share of slums and overcrowded streets. But it is important to complement the traditional view of South Asian poverty with one of modernity and development. Both visions capture important realities of the region in the early 21st century.

Although South Asia as a whole is one of the poorest regions of the world, parts of India are experiencing rapid economic development based on the high-tech skills of the educated segment of its population.

Hyderabad, India, has a metropolitan population of almost 8 million, making it one of India's largest cities. It has recently emerged as a major high-tech center, although it is often overshadowed in this regard by Bangalore (or Bengaluru), another southern Indian city. Infosys, which is headquartered in Bangalore, is India's third largest information technology (IT) company, with annual revenues of nearly US$7.5 billion. It specializes in software, outsourcing, and business consulting. Infosys is also highly connected with the high-tech economy of the United States, particularly that of California's Bay Area, and its operations in the United States are extensive. In 2012, Infosys was the second largest employer of professionals with H-1B visas in the United States. (H-1B visas allow firms in the United States to employ foreign workers with specialized skills.)

However, although the shiny, new Infosys corporate campus says something important about recent developments in South Asia, it is also essential to remember that this region is extremely diverse and is still one of the most impoverished parts of the world. To understand the complex realities of contemporary South Asia, it is first necessary to examine how the region has been defined and separated from the rest of the world.

South Asia as a World Region

South Asia is easily defined in terms of physical geography. The bulk of the region forms a subcontinent—often called the **Indian Subcontinent**—separated from the rest of Asia by formidable mountain ranges. Located here are the region's major countries of India, Pakistan, and Bangladesh, as well as the smaller mountainous states of Nepal and Bhutan. Also placed within South Asia are several islands in the Indian Ocean, including the countries of Sri Lanka and the Maldives, as well as the Indian territories of Lakshadweep and the Andaman and Nicobar islands.

India is by far the largest South Asian country, both in size and in population. Covering more than 1 million square miles (2,590,000 square km), India is the world's seventh largest country in area and, with more than 1.2 billion inhabitants, second only to China in population. Pakistan and Bangladesh are next largest in size and population, with more than 180 and 150 million inhabitants, respectively. A compact country about the size of Wisconsin, Bangladesh is one of the most densely populated places in the world. Bangladesh has a short border with Burma (Myanmar), but it is otherwise virtually surrounded by India, which wraps around the country to the north and the northeast.

South Asia is historically united by deep cultural commonalities. Religious ideas associated with Hinduism and Buddhism were once found throughout the region. For the past thousand years, however, Islam has also played a major role, and South Asia has some of the largest Muslim

LEARNING OBJECTIVES

After reading this chapter you should be able to:

- Explain how the monsoon is generated and describe its importance for South Asia.
- Describe the geological relationship between the Himalayas and other high mountains of northern South Asia and the flat, fertile plains of the Indus and Ganges river valleys.
- Outline the ways in which the patterns of human population growth in South Asia have changed over the past several decades and explain why they vary so strikingly from one part of the region to another.
- Identify the causes of the explosive growth of South Asia's major cities and describe both the benefits and the problems that result from the emergence of such large cities.
- Compare and contrast the ways in which India and Pakistan have dealt with the problems of building national cohesion, considering the fact that both countries contain numerous distinctive language groups.
- Summarize the historical relationship between Hinduism and Islam in South Asia and explain why so much tension exists between the two religious communities today.
- Explain why South Asia was politically partitioned at the end of the period of British rule and show how the legacies of partition have continued to generate political and economic difficulties in the region.
- Describe the various challenges that India, Pakistan, and Sri Lanka have faced from insurgency movements that seek to carve out new independent states from their territories.
- Explain why European merchants were so eager to trade in South Asia in the 16th, 17th, and 18th centuries and describe how their activities influenced the region's later economic development.
- Summarize the ways in which economic and social development varies across the different regions of South Asia and explain why such variability is so pronounced.

populations in the world. Religious tensions in the region have influenced geopolitical conflicts, which are a major concern. Although many observers are optimistic about South Asia's future, others focus on the region's huge challenges.

South Asia's Geopolitical Challenges

South Asia has experienced intense political conflict for decades. Since independence from Britain in 1947, India and Pakistan, South Asia's two largest countries, have fought several wars and remain locked in a bitter dispute concerning possession of the disputed territory of Kashmir in the northern reaches of the region. Religious divisions are linked to this geopolitical turmoil, for India is primarily a Hindu country with a large Muslim minority, while Pakistan is almost entirely Muslim. Religious and ethnic tensions also abound within both countries and in the neighboring countries of Bangladesh, Nepal, and Sri Lanka as well.

Parallel to these geopolitical tensions are demographic and economic concerns. Although fertility levels have dropped dramatically in recent years across much of the region, they remain elevated in some areas, particularly north-central India and Pakistan. As a result, some experts are concerned about how South Asia will be able to sustain its population. Although agricultural production has kept pace with population growth over the past four decades, many South Asian environments are experiencing pronounced stress. Poverty compounds such problems. Along with Sub-Saharan Africa, South Asia is the poorest part of the world. Roughly one-third of India's people subsist on less than $1 a day.

To understand South Asia's continuing challenges, it is first necessary to examine the region's physical and environmental geography in some detail.

Physical Geography and Environmental Issues: From Tropical Islands to Mountain Rim

South Asia's environmental geography covers a wide spectrum that ranges from the highest mountains in the world to densely populated islands barely above sea level; from one of the wettest places on Earth to dry, scorching deserts; from tropical rainforests to degraded scrublands to coral reefs (Figure 12.1). All of these ecological zones have their own distinct and complex environmental problems.

The Four Physical Subregions of South Asia

South Asia is separated from the rest of the Eurasian continent by a series of sweeping mountain ranges, including the Himalayas—the highest mountains in the world. To better understand environmental conditions in this vast region, the Indian subcontinent can be broken down into four physical subregions, starting with the high mountain ranges of its northern fringe and extending to the tropical islands of the far south.

Mountains of the North South Asia's northern rim of mountains is dominated by the great Himalayan Range, forming the northern borders of India, Nepal, and Bhutan. These mountains are linked to the equally high Karakoram Range to the west, extending through northern Pakistan. More than two dozen peaks exceed 25,000 feet (7600 meters), including the world's highest mountain, Everest, on the Nepal–China (Tibet) border at 29,028 feet (8848 meters). To the east are the lower Arakan Yoma Mountains, forming the border between India and Burma (Myanmar) and separating South Asia from Southeast Asia.

These formidable mountain ranges were produced by tectonic activity caused by peninsular India pushing northward into the larger Eurasia Continental Plate. As a result of the collision between these two tectonic plates, great mountain ranges were folded and upthrust. The entire region is seismically active, putting all of northern South Asia in serious earthquake danger (see Figure 12.1). A massive earthquake in the Pakistani-controlled section of Kashmir on October 8, 2005, for example, resulted in roughly 80,000 deaths and left more than 3 million people homeless.

Indus–Ganges–Brahmaputra Lowlands South of the northern highlands lie vast lowlands created by three major river systems that have carried sediments eroded off the mountains through millions of years, building alluvial plains of fertile and easily farmed soils. These lowlands are densely settled and constitute the core population areas of Pakistan, India, and Bangladesh.

Of these three rivers, the Indus is the longest, covering more than 1800 miles (2880 km) as it flows southward from the Himalayas through Pakistan to the Arabian Sea, providing much-needed irrigation waters to Pakistan's desert areas. The broad band of cultivated land in the desert zone of central and southern Pakistan that is made possible by the Indus is clearly visible on satellite images (Figure 12.2). Pakistan is currently concerned about India's dam-building projects on several tributaries of the Indus, which could reduce the flow of this all-important river.

Even more densely settled is the vast lowland of the Ganges, which, after flowing out of the Himalayas, travels southeasterly some 1500 miles (2400 km) to empty into the Bay of Bengal. The Ganges not only provides the fertile alluvial soil that has made northern India a densely settled area for thousands of years, but also has long served as a vital transportation corridor. Given the central role of this river in South Asia's past and present, it is understandable that Hindus consider the Ganges sacred. But while the waters of the Ganges are reputed to have healing powers, the river is considered to be one of the world's most polluted watercourses.

Although this large South Asian lowland often is referred to as the Indus–Ganges Plain, this term neglects the Brahmaputra River. This river rises on the Tibetan Plateau and flows easterly, then southward, and then westerly over 1700 miles (2700 km), joining the Ganges in central Bangladesh and spreading out over the vast delta, the largest in the world. Unlike the sparsely populated Indus Delta in Pakistan,

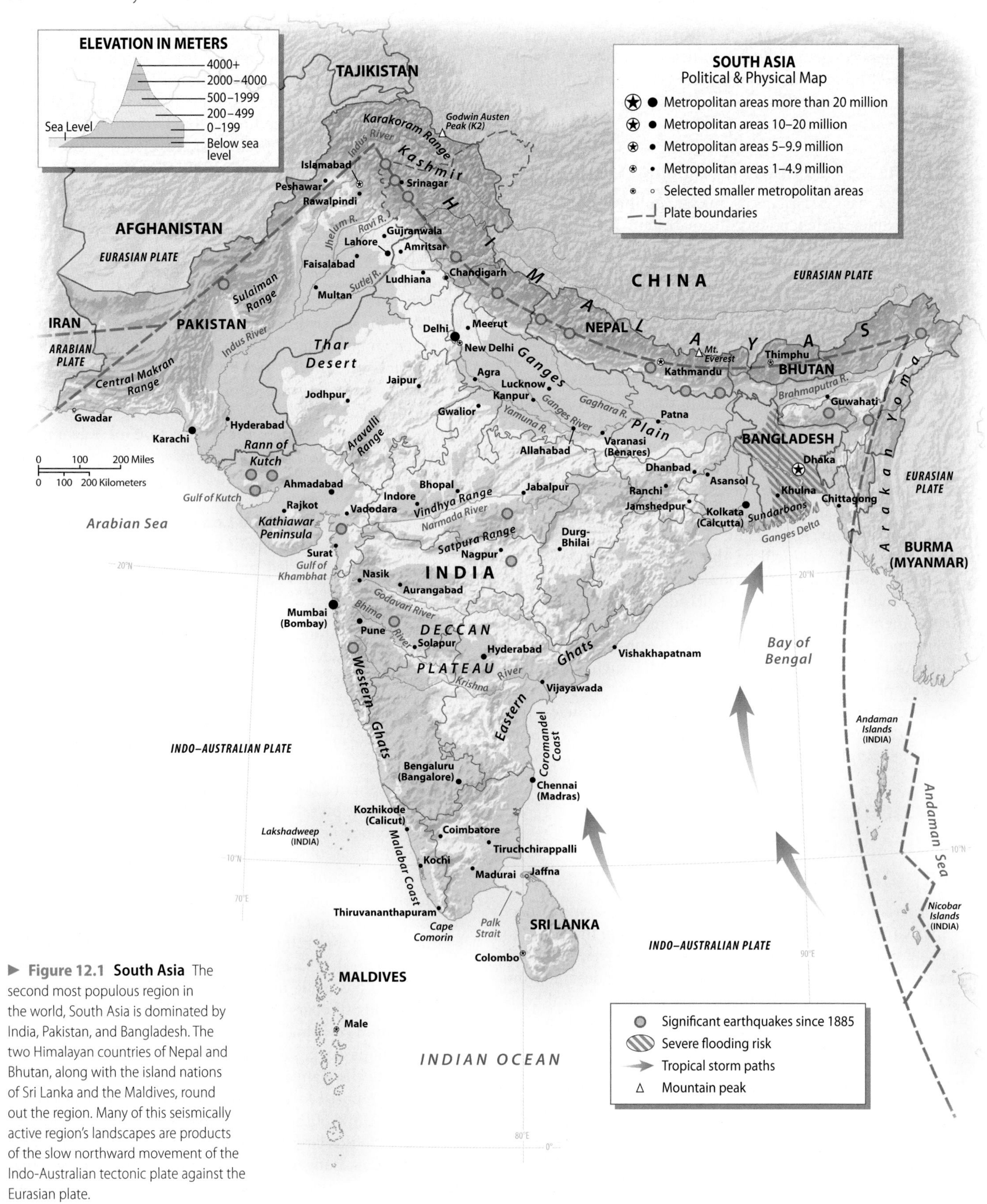

▶ **Figure 12.1 South Asia** The second most populous region in the world, South Asia is dominated by India, Pakistan, and Bangladesh. The two Himalayan countries of Nepal and Bhutan, along with the island nations of Sri Lanka and the Maldives, round out the region. Many of this seismically active region's landscapes are products of the slow northward movement of the Indo-Australian tectonic plate against the Eurasian plate.

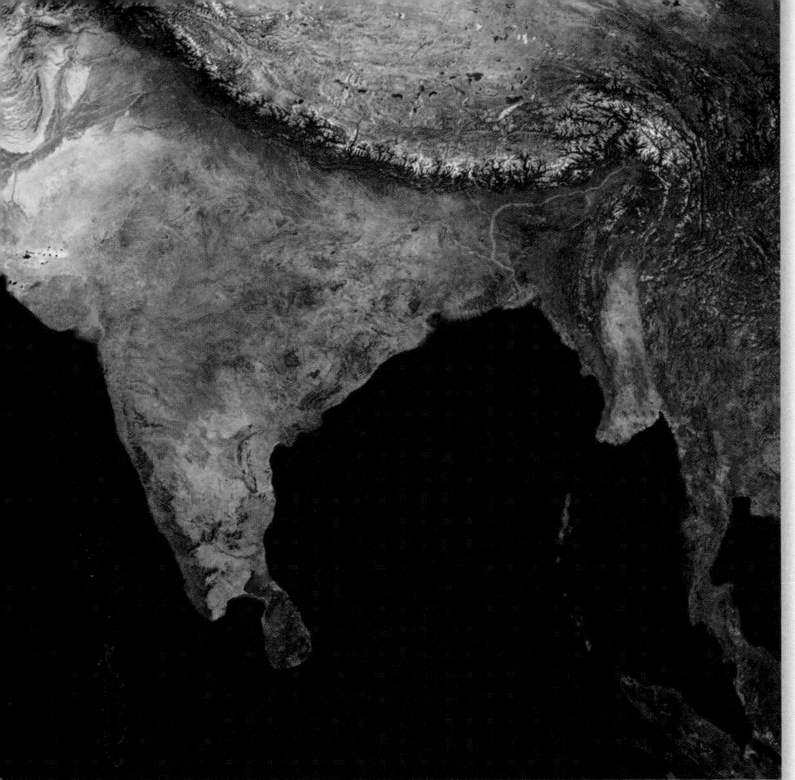

▲ **Figure 12.2 South Asia from Space** This satellite photo shows the four physical subregions of South Asia, from the snow-clad Himalayas in the north to the islands of the south. The irrigated lands of the Indus River Valley in Pakistan are clearly visible in the upper left.

the Ganges–Brahmaputra Delta is one of the most densely settled areas in the world, containing more than 3000 people per square mile (1200 per square km).

Peninsular India Jutting southward from the Indus–Ganges Plain is the familiar shape of peninsular India, made up primarily of the Deccan Plateau. This plateau zone is bordered on each coast by narrow coastal plains backed by elongated north–south mountain ranges. On the west are the higher Western Ghats, which are generally about 5000 feet (1500 meters) in elevation, but reach more than 8000 feet (2400 meters) near the peninsula's southern tip. To the east, the Eastern Ghats are lower and discontinuous, thus forming less of a transportation barrier to the broader eastern coastal plains. On both coastal plains, fertile soils and adequate water supplies support population densities comparable to those of the Ganges lowlands.

Soils are poor or average over much of the Deccan Plateau, but in Maharashtra state basaltic lava flows have produced fertile black soils. A reliable water supply for agriculture is a major problem in most areas. Much of the western plateau lies in the rain shadow of the Western Ghats, giving it a semiarid climate. Small reservoirs or tanks have been the traditional method for collecting monsoon rainfall for use during the dry season. More recently, deep wells and powerful pumps have provided groundwater to support irrigated crops and village water needs.

Partly because of the overuse of these aquifers, the Indian government is building a series of large dams to generate electricity and allow more extensive irrigation. These projects are highly controversial, largely because the reservoirs have displaced hundreds of thousands of rural residents and will continue to displace many more. The Sardar Sarovar Dam on the Narmada River in the state of Gujarat, commissioned in 2006, has been particularly controversial (Figure 12.3). This dam is part of the much larger Narmada Project, which involves the construction of 30 major dams, as well as extensive canal systems. Proponents claim that the scheme will allow India to feed an additional 20–30 million people, while opponents claim that the human costs are too high, pointing out that some 200,000 displaced people had not yet received compensation by the Indian government.

The Southern Islands At the southern tip of peninsular India lies the island country of Sri Lanka, which is almost linked to India by a series of small islands called Adam's Bridge. Sri Lanka is ringed by extensive coastal plains and low hills, but mountains reaching more than 8000 feet (2400 meters) cover the southern interior, providing a cool, moist climate. Because the main winds arrive from the southwest, that portion of the island is much wetter than the rain-shadow area of the north and east.

Forming a separate country are the Maldives, a chain of more than 1200 islands stretching south to the equator some 400 miles (640 km) off the southwestern tip of India. The combined land area of these islands is only about 116 square miles (290 square km), and only a quarter of the islands are

▼ **Figure 12.3 Sardar Sarovar Dam** The Sardar Sarovar Dam on the Narmada River in the Indian state of Gujarat was completed in 2008. **Q: Why is this dam relatively popular in Gujarat but highly unpopular in the neighboring state of Madhya Pradesh?**

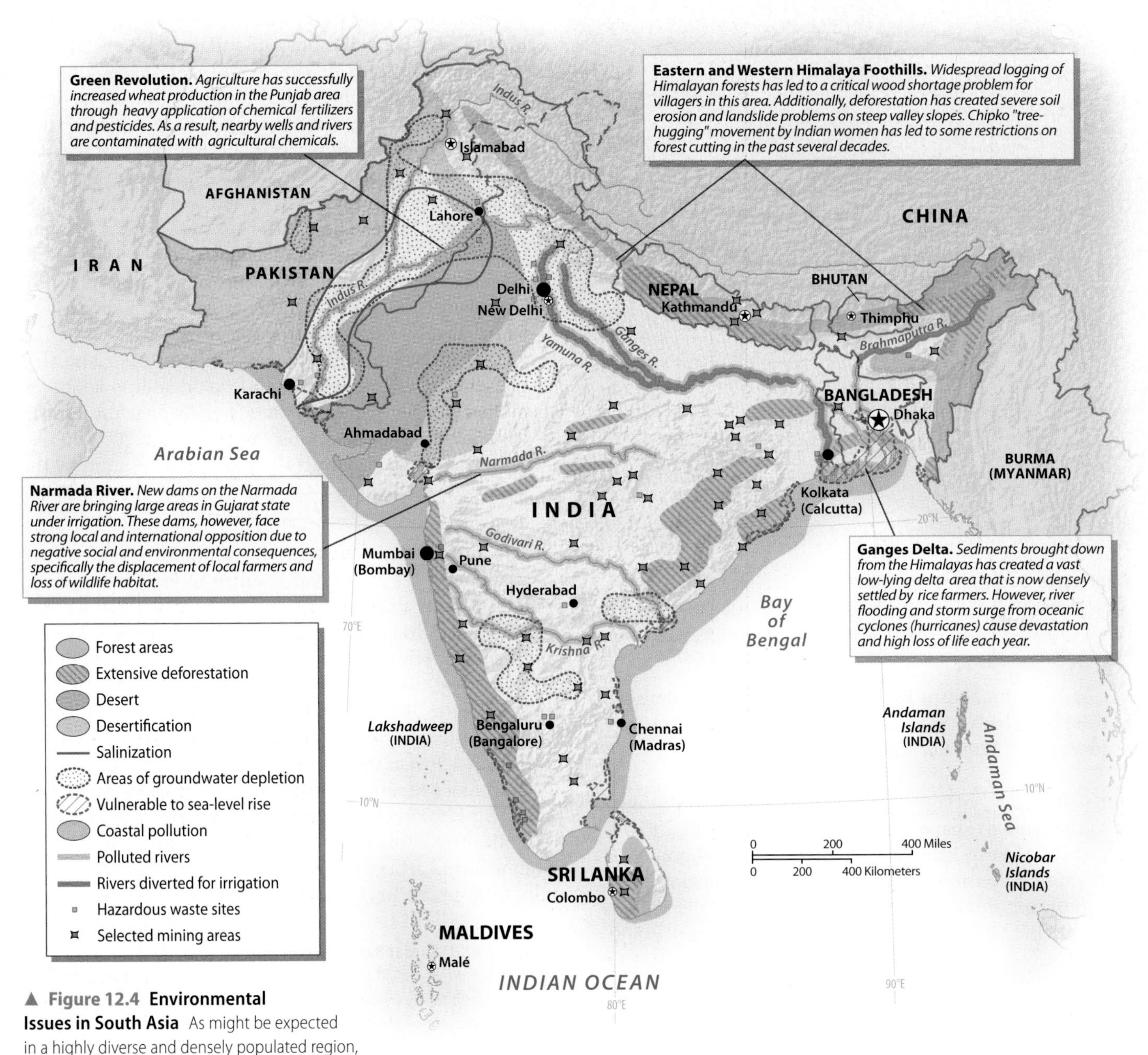

▲ **Figure 12.4 Environmental Issues in South Asia** As might be expected in a highly diverse and densely populated region, South Asia suffers from some major environmental problems. These range from salinization of irrigated lands in the dry areas of Pakistan and western India to groundwater pollution from Green Revolution fertilizers and pesticides. Deforestation and erosion are widespread in upland areas.

actually inhabited. The islands of the Maldives are flat, low coral atolls, with a maximum elevation of just over 6 feet (2 meters) above sea level.

Environmental Issues in South Asia

As is true in other poor and densely settled regions of the world, a raft of serious ecological issues plagues South Asia. The region also suffers from the usual environmental problems of water and air pollution that accompany early industrialization and manufacturing (Figure 12.4). Compounding all of these problems are the immense numbers of new people added each year through natural population growth. In all countries in the region, however, major efforts are under way to address South Asia's environmental problems (see *Working Toward Sustainability: Green Schools and Eco-Tourism in Bhutan*).

The Precarious Situation of Bangladesh The link between population pressure and environmental problems is nowhere clearer than in the delta area of Bangladesh, where the search for fertile land has driven people into hazardous areas, putting millions at risk from seasonal flooding, as well as from the powerful **cyclones** that form over the

Working Toward **Sustainability**

Green Schools and Eco-Tourism in Bhutan

Since 1971, Bhutan has raised eyebrows worldwide with its "gross national happiness" approach to measuring output, which includes several psychological as well as environmental indices. Over time, Bhutan has increasingly tried to incorporate these principles into daily life, putting particular focus on environmental well-being. Bhutan's Green Schools for Green Bhutan program, begun in 2009, attempts to imbue Bhutan's younger generations with a deeply ingrained understanding of conservation and sustainability principles (Figure 12.1.1).

Guiding Principles Green schools aim to fulfill five major criteria: (1) Student learning should be organized around environmental concepts; (2) school grounds themselves should be utilized as an educational tool; (3) community enhancement projects should be incorporated into the curriculum; (4) schools should deal with waste, energy, and water in sustainable ways; and (5) school administration should continue to search for and promote new kinds of green activities. Proponents of this kind of educational experience hope that students will become well versed in the environmental challenges facing the country and develop a general respect for the natural world. Though the ultimate goals are rather profound, many of the actionable steps are mundane and achievable, such as encouraging students to walk to school.

The Tourism Challenge One way that Bhutan strives to improve its gross national happiness is by leveraging natural beauty for tourist dollars, which also form one of its few links to the broader world economy. Bhutan was closed completely to foreigners until 1974, and since then it has continued to limit tourism, preferring to minimize the impact of visitors by attracting a smaller number of wealthier tourists. Although tourism accounts for only around 1 percent of Bhutan's economic output, it is an important source of foreign exchange earnings, bringing in around $40 million per year. Ecotourism also has the potential to bring money to Bhutan's poorer rural areas, where around 60 percent of the population depends on farming or forestry.

The high-end tourism that Bhutan emphasizes demands a clean natural environment, but as the country gradually develops, pollution and other environmental problems are mounting. Bhutan has addressed the issue of wood smoke pollution by electrification, which, in turn, relies on the extensive building of dams to generate hydropower. Dam-building, however, is itself environmentally controversial. The lack of adequate waste disposal facilities is another growing issue. A more intractable problem is the wafting of air pollution from neighboring India into the highlands of Bhutan.

1. Why do you think that Bhutan in particular has focused so much attention on sustainable development?
2. Do you think that "Gross National Happiness" is an appropriate way to conceptualize national development? If so, how might this concept be adequately measured?

Google Earth Virtual Tour Video

http://goo.gl/qMo0hM

▲ **Figure 12.1.1 Rural School in Bhutan** Children relax outside of one of Bhutan's Green Schools.

Bay of Bengal. For millennia, drenching monsoon rains have eroded and transported huge quantities of sediment from the Himalayan slopes to the Bay of Bengal by the Ganges and Brahmaputra rivers, gradually building this low-lying and fertile deltaic environment. With continual population growth, people have gradually moved into the swamps to transform them into highly productive rice fields. This agricultural activity has supported Bangladesh's vast population, but it has intensified the effects of the area's natural hazards.

Although periodic floods are a natural, even beneficial, phenomenon that enlarges deltas by depositing fertile riverborne sediment, flooding has become a serious problem for people inhabiting these low-lying areas. In September 1998, for example, more than 22 million Bangladeshis were made homeless when water covered two-thirds of the country (Figure 12.5). In June 2010, fairly typical monsoon downpours killed dozens of people and marooned more than 150,000.

With the populations of both Bangladesh and northern India continuing to grow, there is a strong possibility that flooding will take even higher tolls in the coming decade, as desperate farmers relocate into the hazardous lower floodplains. Deforestation of the Ganges and Brahmaputra headwaters magnifies the problem. When forest cover and ground vegetation are removed, rainfall does not soak into the ground to slow runoff and replenish groundwater supplies. Thus, deforestation in river headwaters results in increased flooding in the wet season, as well as a lower water level during the dry season, when

▲ **Figure 12.5 Flooding in Bangladesh** Devastating floods are common in the low-lying delta lands of Bangladesh. Heavy rains come with the southwest monsoon, especially to the Himalayas, and powerful cyclones often develop over the Bay of Bengal.

river flows are supplemented by groundwater. Bangladesh's water problems are further magnified by the fact that many of its aquifers are contaminated with arsenic from natural sources, threatening the health of as many as 80 million people.

Forests and Deforestation Forests and woodlands once covered most of South Asia, except for the desert areas in the northeast. However, in most places, tree cover has vanished as a result of human activities. The Ganges Valley and coastal plains of India, for example, were largely deforested thousands of years ago to make space for agriculture. Elsewhere, forests were cleared more gradually for agricultural, urban, and industrial expansion. Railroad construction in the 19th century was especially destructive. More recently, hill slopes in the Himalayas and in the remote lands of eastern India have been logged for commercial purposes.

Beginning in the 1970s, India embarked on several reforestation projects. Some of these efforts have been successful; according to some sources, Indian forest coverage actually increased by almost 6 percent between 1990 and 2005. As is true in many parts of the world, however, reforested areas are often covered by nonnative trees, such as eucalyptus, that support little wildlife. According to a 2010 report, India's single-species tree plantations are expanding by up to 7000 square miles (18,000 square km) every year (Figure 12.6). But in many of India's standing natural forests, wood removal is estimated to exceed growth by a substantial margin.

As a result of deforestation, many villages of South Asia suffer from a shortage of fuelwood for household cooking, forcing people to burn dung cakes from cattle. Although this low-grade fuel provides adequate heat, it diverts nutrients that could be used as fertilizers to household fires. It also results in high levels of air pollution, both indoors and outside. Where wood is available, collecting it may involve many hours of female labor because the remaining sources of wood are often far from the villages.

Villagers living in India's forest areas have suffered extensively from deforestation. As early as the 1970s, social movements designed to protect remaining groves gained prominence in some areas. The most noted of these movements, called Chipko, mobilized women in northern India to engage in "tree-hugging" campaigns to stop logging. As a result of such social pressure, India began to involve local residents in forestry and conservation projects. Increasingly, villagers have been demanding their own rights to land and resources in wooded areas. In the landmark 2006 Forest Rights Act, the Indian government granted extensive rights to members of forest-dwelling communities. Many urban environmentalists, however, argue that this legislation could actually increase deforestation, as expanding villages often seek to convert wooded lands to agricultural fields. Some also fear that it could harm the efforts to preserve tigers and other large animal species.

Wildlife: Extinction and Protection Although the overall environmental situation in South Asia is worrisome, wildlife protection inspires some optimism. The region has managed to retain a diverse assemblage of wildlife despite population

▼ **Figure 12.6 Logging in India** A stand of small trees is being logged off in the southern Indian state of Kerala. Such forestry practices leave little habitat for wildlife.

pressure and intense poverty. The only remaining Asiatic lions live in India's Gujarat state, and even Bangladesh retains a viable population of tigers in the Sundarbans, the mangrove forests of the Ganges Delta. Wild elephants still roam several large reserves in India, Sri Lanka, and Nepal.

The protection of wildlife in India far exceeds that in most other parts of Asia. The official Project Tiger, which currently operates more than 53 preserves, has been credited with increasing the country's tiger population from 1200 in the 1970s to roughly 3500 by the 1990s. Continued forest clearance and population increase, however, led to a drop in the tiger population to around 1700 by 2010 (Figure 12.7). As a result, the government has set up a Tiger Protection Force to fight against poaching and to relocate villagers from crucial habitat areas. In 2013, officials began to discuss the possibility of removing dogs from some tiger areas, due to the threats posed by the canine distemper virus.

As the situation with tigers demonstrates, wildlife conservation in a country as poor and crowded as India is not easy. In many areas, pressure is mounting to convert remaining wildlands to farmlands. The best remaining zone of extensive wildlife habitat is in India's far northeast, an area subjected to rapid immigration and political unrest. Moreover, wild animals, particularly tigers and elephants, threaten crops, livestock, and even people living near the reserves. When a rogue elephant herd ruins a crop or a tiger kills livestock, government agents are usually forced to destroy the animal.

South Asia's Monsoon Climates

The dominant climatic factor for most of South Asia is the **monsoon**, the distinct seasonal change of wind direction that corresponds to wet and dry periods. Most of South Asia has three distinct seasons. First is the warm and rainy season of the southwest monsoon from June through October (the summer monsoon). This is followed by a relatively cool and dry season, extending from November until February, when the dominant winds are from the northeast (the winter monsoon). Only a few areas in far northwestern and southeastern South Asia get substantial rainfall during this period. Next comes the hot season from March to early June, during which great heat and humidity build up until the monsoon's much anticipated and rather sudden "burst."

This monsoon pattern is caused by large-scale meteorological processes that affect much of Asia (Figure 12.8). During the Northern Hemisphere's winter, a large high-pressure system forms over the cold Asian landmass. Cold, dry winds flow outward from the interior of this high-pressure cell, over the Himalayas and down across South Asia. As winter turns to spring, these winds diminish, resulting in the hot, dry season. Eventually, this buildup of heat over South and Southwest Asia produces a large thermal low-pressure cell. By early June, this low-pressure cell is strong enough to draw in warm, moist air from the Indian Ocean.

Heavy **orographic rainfall** results from the uplifting and cooling of moist monsoon winds over the Western Ghats. As a result, some stations receive more than 200 inches (508 cm) of rain during the four-month wet season. On the climate map (Figure 12.9), these are the areas of Am, or tropical monsoon, climate. Inland, however, a strong rain-shadow effect dramatically reduces rainfall on the Deccan Plateau. Farther north, as the monsoon winds are forced up and over the mountains, copious amounts of rainfall are characteristic. Cherrapunji, India, at an elevation of 4000 feet (1200 meters), is a strong contender for the title of world's wettest place, with an average rainfall of 450 inches (1150 cm); see Figure 12.9.

Not all of South Asia receives substantial rainfall from the southwest monsoon. In much of Pakistan and the Indian state of Rajasthan, precipitation is low and variable, resulting in steppe and desert climates. In Karachi, the annual total is less than 10 inches (25 cm); see Figure 12.9. But even in the drier parts of South Asia, heavy monsoonal rain sometimes hits. In the summer of 2010, for example, prolonged downpours resulted in devastating flooding across much of Pakistan. With damages estimated at US$43 billion, more than 2000 people dead, and more than 20 million people affected, the 2010 Pakistan flood is commonly regarded as the worst natural disaster that the country has ever experienced.

Regardless of whether rainfall is heavy or light, the monsoon rhythm affects all of South Asia in many different ways, from the delivery of much-needed water for crops and

▼ **Figure 12.7 Human–Tiger Interactions** Tourists photograph a tiger in Ranthambhore National Park in the Indian state of Rajasthan. Ranthambhore is one of the best place in South Asia to see wild tigers.

▼ **Figure 12.8 The Summer and Winter Monsoons** (a) Low pressure centered over South and Southwest Asia draws in warm, moist air masses during the summer, bringing heavy monsoon rains to most of the region. Usually, these rains begin in June and last for several months. (b) During the winter, high pressure forms over northern Asia. As a result, winds are reversed from those of the summer. During this season, only a few coastal locations along India's east coast and in eastern Sri Lanka receive substantial rain.

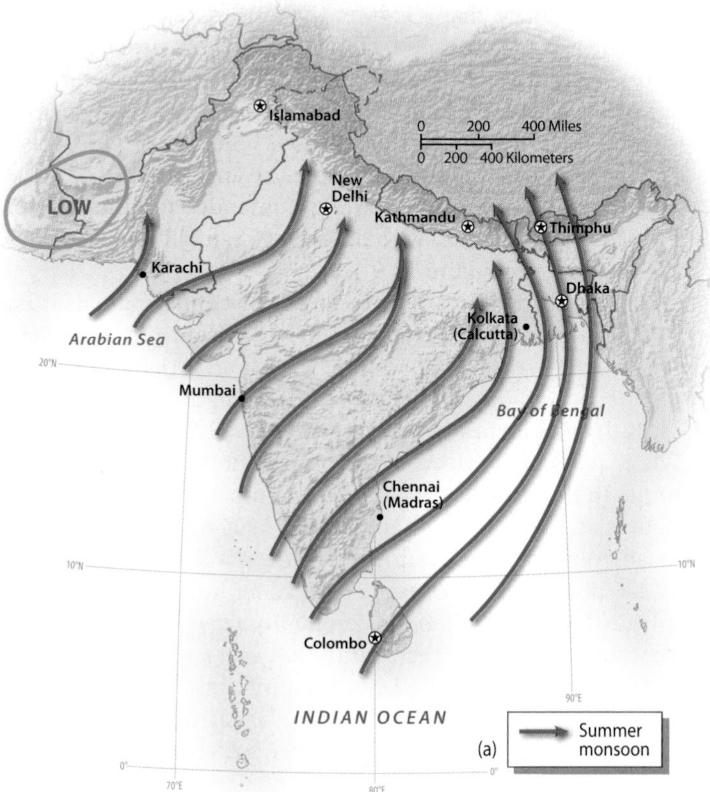

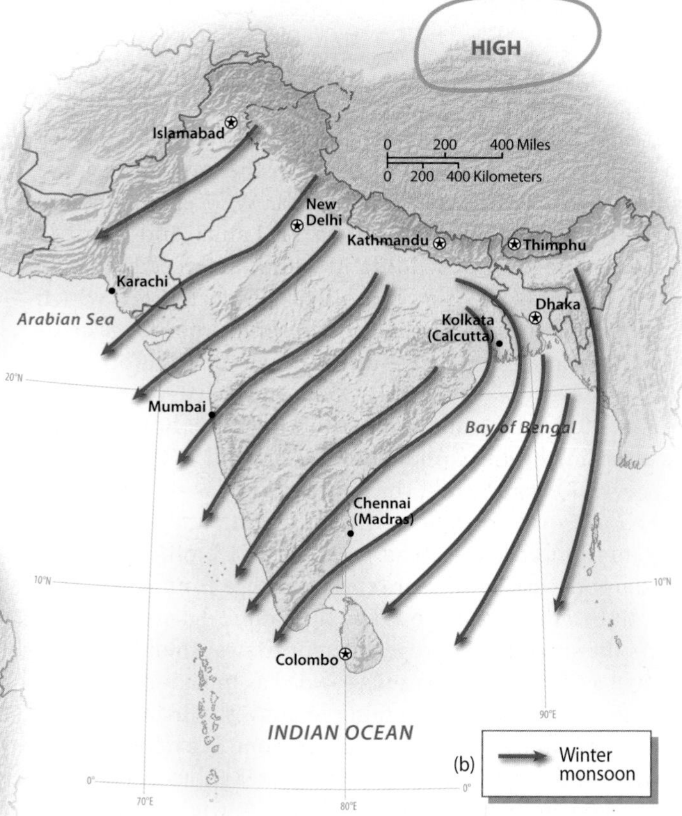

villages to the mood of millions of people as they eagerly await relief from oppressive heat (Figure 12.10). Some years the monsoon delivers its promise with abundant moisture; in other years, though, it brings scant rainfall, resulting in crop failure and hardship. In 2012, the monsoon rains in northern India were much delayed, harming crops, and when they finally hit in late August, flash flooding caused additional damage.

Climate Change in South Asia

Many areas of South Asia are particularly vulnerable to the effects of global warming. Even a minor rise in sea level will inundate large areas of the Ganges–Brahmaputra Delta in Bangladesh. Already, over 18,500 acres (7500 hectares) of swampland in the Sundarbans have been submerged. Several small islands in this area have recently disappeared due to a combination of sea-level rise and subsidence. A 2007 report from the Indian government suggests that up to 7 million people could be displaced by the end of the century due to a predicted 3.3-foot (1-meter) rise in sea level. If the most severe sea-level forecasts come to pass, the atoll nation of the Maldives will simply vanish beneath the waves.

South Asian agriculture is likely to suffer from several problems linked to global climate change. Many Himalayan glaciers are rapidly retreating, threatening the dry-season water supplies of the Indus–Ganges Plain, an area that already suffers from overuse of groundwater resources. Winter temperature increases of up to 6.4°F (3°C), along with decreased rainfall, could threaten the vital wheat crop of Pakistan and northwestern India, undermining the food security of both countries. In parts of South Asia, however, global warming could result in increased rainfall due to an intensification of the summer monsoon. Unfortunately, such a new precipitation regime will likely be characterized by more intense cloudbursts coupled with fewer episodes of gentle, prolonged rain. A 2013 World Bank report warned that "huge disruptive" monsoon events, which currently strike once a century on average, could occur once in every decade.

Responses to the Climate Challenge India signed the Kyoto Protocol in 2002, but as a developing country, it has been exempt from the main provisions of the treaty. With its poor and still largely nonindustrial economies, South Asia still has a low per capita output of greenhouse gases. But India's economy in particular is not only growing rapidly,

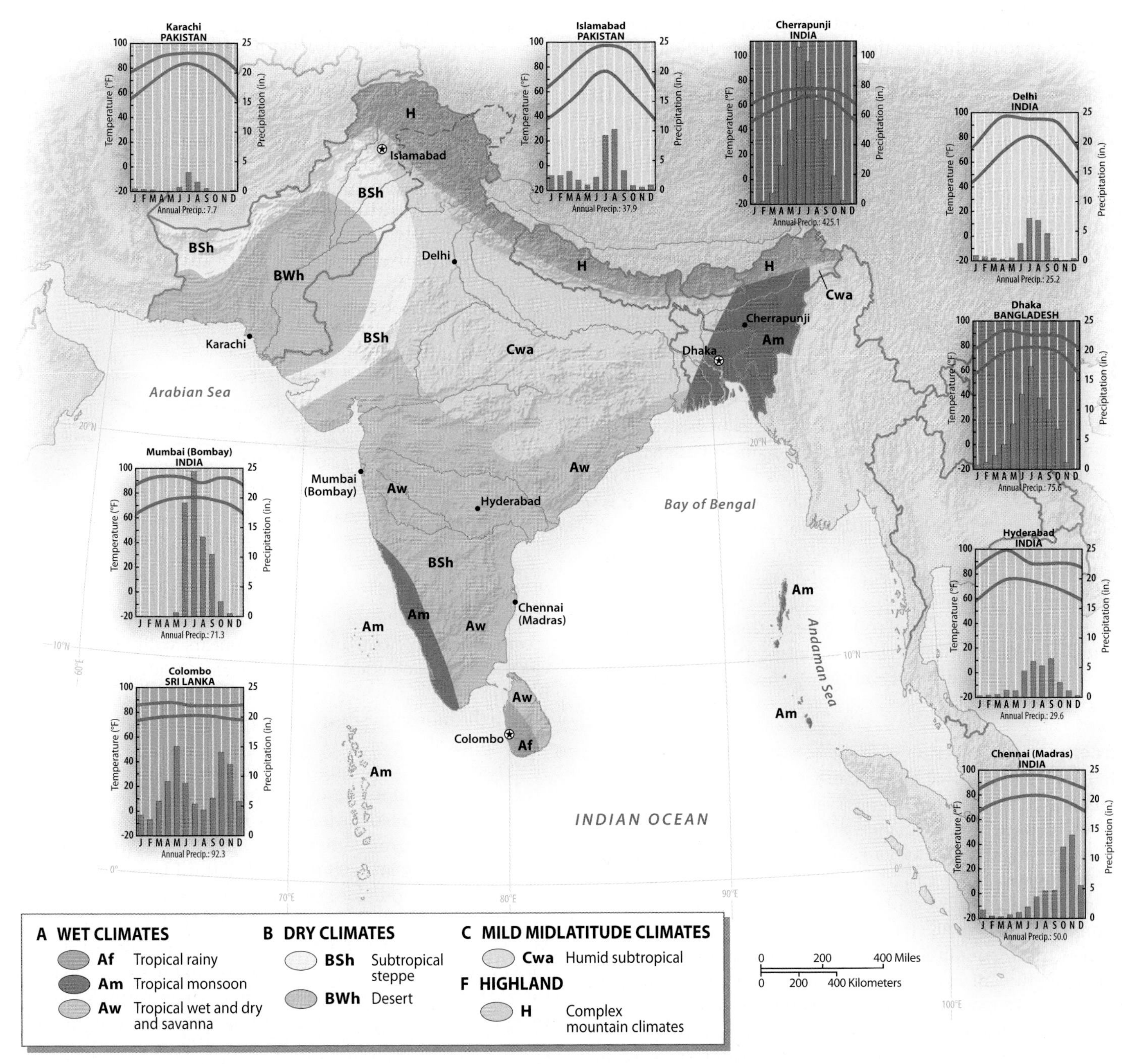

▲ **Figure 12.9 Climate of South Asia** Except for the extensive Himalayas, South Asia is dominated by tropical and subtropical climates. Many of these climates show a distinct summer rainfall season associated with the southwest monsoon. The climographs for Mumbai (Bombay) and Delhi are excellent illustrations. However, the climographs for locations on the east coast, such as Madras, India, and Colombo, Sri Lanka, show how some locations also receive rain from the northeast monsoon of the winter.

► **Figure 12.10 Monsoon Rain** During the summer monsoon, some Indian cities such as Mumbai (Bombay) receive more than 70 inches (180 cm) of rain in just three months. These daily torrents cause floods, power outages, and daily inconvenience—as well as joy. The monsoon rains are crucial to India's agriculture. If the rains are late or abnormally weak, crop failure often results.

but also heavily dependent on burning coal to generate electricity. Currently, India is the world's sixth largest emitter of carbon dioxide. In 2012, its government said that it would not even consider reducing greenhouse gas emissions until 2020 and that even then it would do so only in exchange for foreign financial assistance.

Climate change may put Pakistan in a particular bind. As much as 90 percent of the irrigation water for this mostly desert country comes from the mountains of Kashmir, a contested region divided between Pakistan and India. According to estimates of the Intergovernmental Panel on Climate Change, the glaciers of Kashmir that feed Pakistan's rivers will be much reduced by the middle of the century. Because Pakistan already experiences periodic water shortages, many experts believe that it must cooperate with India to develop and conserve the water resources of the Kashmir region. However, considering the geopolitical tension between the two countries, which is focused on Kashmir, any such agreement seems unlikely.

Efforts are being made to prepare South Asia for possible climate change. In 2011, the global organization called CGIAR instituted a program aimed at creating "climate-smart villages" across the region. The program encourages tree planting, rainwater harvesting, careful water management, and soil conservation. It also facilitates the sharing of information on weather and climate, as well as on market conditions, through smartphone networks.

Energy in South Asia Adequately responding to the challenge of climate change in South Asia will require major changes to the region's energy systems. The primitive state of the electric grid in many parts of South Asia also presents an enormous challenge for regional development. The problem is particularly acute in Bangladesh, where in 2009 just slightly over 40 percent of the country's population had access to electricity. Even in areas with electricity access, power is often intermittent and unreliable, creating a significant barrier to capital investment. The dearth of power transmission lines in rural areas, paired with reliably intense sunlight, makes both Bangladesh and the wider region excellent candidates for greater small-scale solar electricity generation. Still, the lion's share of recent investment has consisted of gas-, coal-, and oil-based power plants.

International cooperation in power transmission plays an increasingly central role in energy policy for South Asian countries, holding the potential both to bring power to millions and to improve the formerly icy international relations between the region's neighboring countries. In June 2013, Pakistan, which also suffers from devastating electricity shortages, indicated a desire to purchase 500 to 1000 megawatts of electricity from India. Pakistan is also attempting to import electricity from other nearby countries, along with natural gas and gasoline from India. Though cooperation between Pakistan and India in the energy sector appears to be gaining momentum, political rivalry may yet put these recent plans on hold. India also appears willing to export electricity to Bangladesh, and it currently trades power with Bhutan, which provides India with excess hydroelectricity during the monsoon season and imports electricity from it at other times of the year.

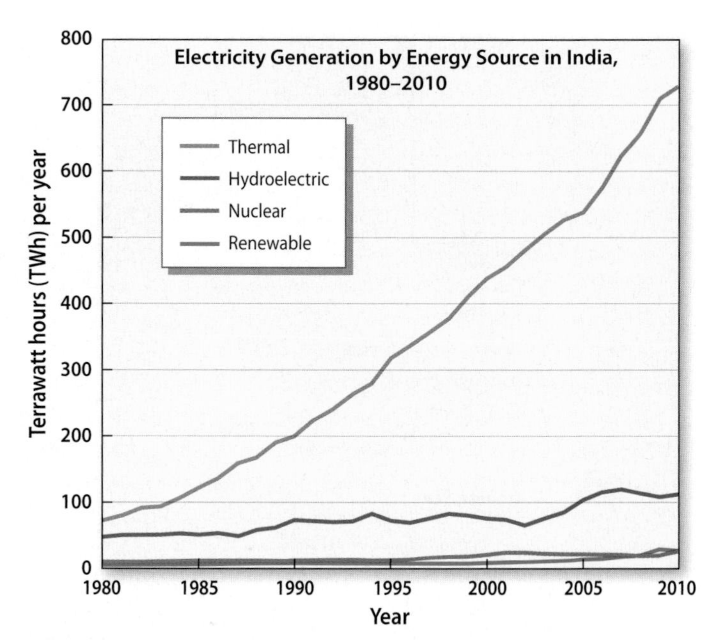

▲ **Figure 12.11 Energy Use in India** Most of India's surging electricity demand is met by burning coal and other fossil fuels.

Despite the booming investment in fossil fuel–based power generation in the region, South Asia lacks adequate supplies of domestic energy resources and imports staggering amounts from other parts of the world. India is the region's energy giant and the world's third largest coal producer, yet the country still imported 16.77 million tons of coal in May 2013 alone, a 43 percent increase from May 2012. As you can see in Figure 12.11, almost all of India's rapidly growing electricity demand has been met by thermal sources, which usually entail the burning of coal. Insignificant oil reserves force India to import roughly 80 percent of its oil, which comes mainly from Iran, Venezuela, Iraq, and Oman. By virtue of the region's large and growing population, the speed and manner in which energy access in South Asia proceeds over the next few decades promise to be one of the most significant areas of global concern as the desire of millions to lead more comfortable modern lives clashes with the global environmental consequences of that trajectory.

REVIEW

12.1 Why is the monsoon so crucial to life in South Asia?

12.2 Why is flooding such an important environmental issue in Bangladesh and adjacent areas of northeastern India?

12.3 What particular physical features of South Asia make the region especially vulnerable to climate change?

Table 12.1 Population Indicators

Country	Population (millions) 2013	Population Density (per square kilometer)	Rate of Natural Increase (RNI)	Total Fertility Rate	Percent Urban	Percent <15	Percent >65	Net Migration (Rate per 1000)
Bangladesh	156.6	1,087	1.5	2.3	26	31	5	−3
Bhutan	0.7	16	1.5	2.6	36	30	5	0
India	1276.5	388	1.5	2.4	31	30	6	−0
Maldives	0.4	1,208	1.9	2.3	35	27	5	−0
Nepal	26.8	182	1.7	2.6	17	35	5	−5
Pakistan	190.7	230	2.3	3.8	35	37	4	−2
Sri Lanka	20.5	312	1.2	2.1	15	26	8	−2

Source: Population Reference Bureau, *World Population Data Sheet, 2013.*

Population and Settlement: The Demographic Dilemma

South Asia soon will surpass East Asia as the world's most populous region (Table 12.1). Overall, fertility levels have dropped markedly in recent years, but population continues to grow rapidly in many areas, particularly in north-central India and Pakistan. And although South Asia has made remarkable agricultural gains since the 1960s, there is still widespread concern about the region's ability to feed itself. The threat of crop failure remains, in part because much South Asian farming is vulnerable to the unpredictable monsoon rains.

The Geography of Population Expansion

South Asia's recent decline in human fertility shows distinct geographical patterns. All the states of southern and western India, along with Sri Lanka, now have fertility rates at or below the replacement level and hence should see population stabilization within a few decades. In north-central India and in Pakistan, however, the birth rate remains elevated (Figure 12.12). All South Asian countries have established family-planning programs, but the commitment to these policies varies widely from place to place.

India Widespread concern over India's population growth began in the 1960s. To a significant extent, family-planning measures, along with general economic and social development, have been successful; the total fertility rate (TFR) dropped from 6 in the 1950s to the current rate of 2.4. Fertility rates vary widely within India, from below 2.0 in the states of Goa, Kerala, Tamil Nadu, Andhra Pradesh, and Himachal Pradesh, to a problematic high of 3.4 in Uttar Pradesh and 3.7 in Bihar. (Uttar Pradesh, with some 200 million people, would be the sixth largest country in the world if it were independent.) A strong relationship is evident between women's education and family planning; where female literacy has increased most dramatically, fertility levels have rapidly declined.

A distinct cultural preference for male children is found in most of South Asia, a tradition that complicates family planning. Abortion rates of female fetuses are much higher than those of males, even though the practice of sex-selective abortion is illegal. In much of northern India, a lower fertility rate seems to be accompanied by an even lower ratio of female to male infants. India's prosperous

▼ **Figure 12.12 Population Pyramids of Sri Lanka and Pakistan** (a) Sri Lanka has had a relatively low birth rate for decades, as well as relatively high longevity figures, and thus has a well-balanced population pyramid. (b) Pakistan, in contrast, has a much higher birth rate, as well as a lower average lifespan, and thus has a bottom-heavy pyramid.

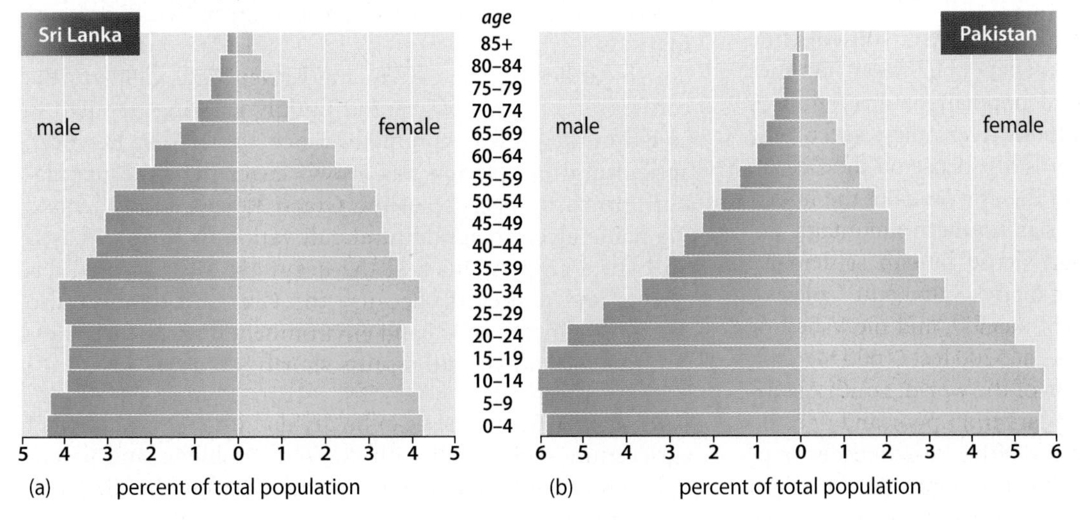

northwestern state of Haryana has recently seen its TFR drop to just below the replacement level, but at the same time its population has grown increasingly male-dominated. In 2011, Haryana's sex ratio at birth was only 830 girls for every 1000 boys. Recent reports, however, indicate that this situation is beginning to change. In 2001, the northwestern Indian state of Punjab reported a sex ratio of 874 females per 1000 males, but by 2011 it reported 893 females per 1000 males.

Pakistan and Bangladesh Pakistan, with a population of more than 187 million, has long had a somewhat ambivalent attitude toward family planning. Although the government's official position is that the birth rate is excessive, the country was slow to develop an effective, coordinated family-planning program. As a result, the TFR long remained elevated. Pakistan's concerned government then launched the National Population Policy 2010 program, which aims to reduce the fertility rate to 3.0 by 2015. The program aims to make family planning more accessible and seeks to recruit religious leaders to support the program. Such measures seem to have been effective, as the country's fertility rate has begun to drop more rapidly, and it now stands at 3.3. Yet even at this birth rate, Pakistan's population would expand to 380 million by 2050.

Densely settled Bangladesh, with a population about half that of the United States packed into an area smaller than the state of Wisconsin, has made huge strides in population stabilization. As recently as 1975, its TFR was 6.3, but it dropped to 2.2 by 2011. Its family-planning success can be attributed to strong support from the Bangladesh government for advertising through radio and billboards. Also important are more than 35,000 women fieldworkers who take information about family planning into every village in the country. In 2010, the president of Bangladesh urged the entire country to adopt the slogan "Not more than two children, one is better."

Migration and the Settlement Landscape

The most densely settled areas of South Asia still coincide with zones of fertile soils and dependable water supplies (Figure 12.13). The largest rural populations are found in the core area of the Ganges and Indus river valleys and on the coastal plains of India. Settlement is less dense on the Deccan Plateau and is relatively sparse in the arid lands of the northwest. Although most of South Asia's northern mountains are too rugged and high to support dense human settlement, major population clusters occur in the Katmandu Valley of Nepal, situated at 4400 feet (1300 meters), and the Valley, or Vale, of Kashmir in northern India, at 5200 feet (1600 meters).

As is true in most other parts of the world, South Asians have migrated for hundreds of years from poor and densely populated areas to places that are either less densely populated or wealthier. This process is ubiquitous in South Asia today, but several areas stand out as zones of intensive out-migration: Bangladesh; the northern Indian states of Bihar, Punjab, and Rajasthan; and the northern portion of India's Andhra Pradesh (Figure 12.14). Migrants from Bangladesh are settling in large numbers in rural portions of adjacent Indian states, exacerbating ethnic and religious tensions. In Nepal, migrants have been moving over the past 50 years from crowded mountain valleys to formerly malaria-infested lowlands along the Indian border, again generating ethnic strife. Sometimes migrants are forced out by war; over the past 20 years, sizable streams of people from northern Sri Lanka and from Kashmir have sought security away from their battle-scarred homelands.

South Asia is one of the least urbanized regions in the world, with only about 35 percent of its population living in cities. Most South Asians still reside in compact rural villages, but large numbers are streaming into the region's rapidly growing cities. As a result, India's urbanization rate is expected to exceed 40 percent by 2030. Such urbanization often stems more from desperate conditions in the countryside than from the attractions of city life. As farms begin to mechanize, farm laborers often have no choice but to migrate to urban areas. Many small farmers, moreover, have difficulty competing with their wealthier neighbors, who can more easily afford modern fertilizers and other agricultural inputs. The agricultural situation over much of India has become so desperate that the country is experiencing an epidemic of farmer suicides. Official statistics indicate that at least 270,940 Indian famers committed suicide between 1995 and 2013.

Agricultural Regions and Activities

South Asian agriculture has historically been less productive than that of East Asia. Even today, rice yields per unit of land in India are about half the level found in China. Although the reasons behind such poor yields are complex, many experts cite the relatively low social status of most cultivators and the fact that much farmland has long been controlled by wealthy landlords, who hire others to work their plots. Some scholars also blame British colonialism, which emphasized export crops for European markets.

Regardless of the causes, low agricultural yields in the context of a huge, hungry, and rapidly growing population constitute a pressing problem. Since the 1970s, however, agricultural production has grown faster than the population, primarily because of the **Green Revolution**, which is the name given to agricultural cultivation techniques based on hybrid crop strains and the heavy use of industrial fertilizers and pesticides. Because the Green Revolution also carries significant social and environmental costs, it remains controversial, as we will discuss shortly.

Crop Zones South Asia can be divided into several distinct agricultural regions, all with different problems and potentials. The most fundamental division is among the three primary subsistence crops of rice, wheat, and millet.

Rice is the main crop and foodstuff in the lower Ganges Valley, along the lowlands of India's eastern and western coasts, in the delta lands of Bangladesh, along Pakistan's lower Indus Valley, and in Sri Lanka. This distribution reflects the large volume of irrigation water needed to grow rice (Figure 12.15). The amount of rice grown in South Asia is impressive: India ranks behind only China in world rice production, and Bangladesh is the fourth largest producer.

Wheat is the principal crop in the northern Indus Valley and in the western half of India's Ganges Valley. South Asia's "breadbasket" lies in the northwestern Indian states of Punjab and Haryana, along with adjacent areas in Pakistan.

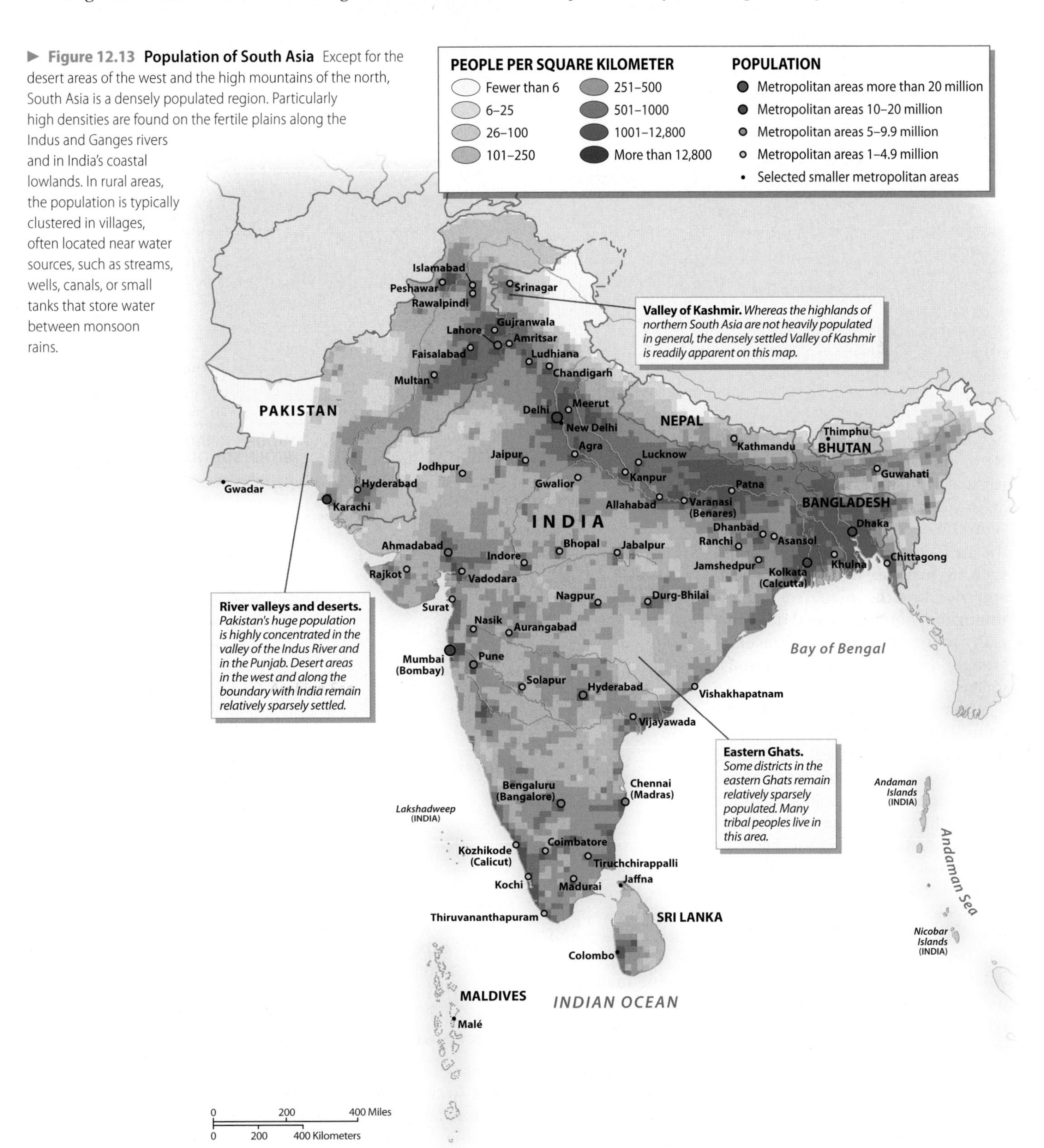

► **Figure 12.13 Population of South Asia** Except for the desert areas of the west and the high mountains of the north, South Asia is a densely populated region. Particularly high densities are found on the fertile plains along the Indus and Ganges rivers and in India's coastal lowlands. In rural areas, the population is typically clustered in villages, often located near water sources, such as streams, wells, canals, or small tanks that store water between monsoon rains.

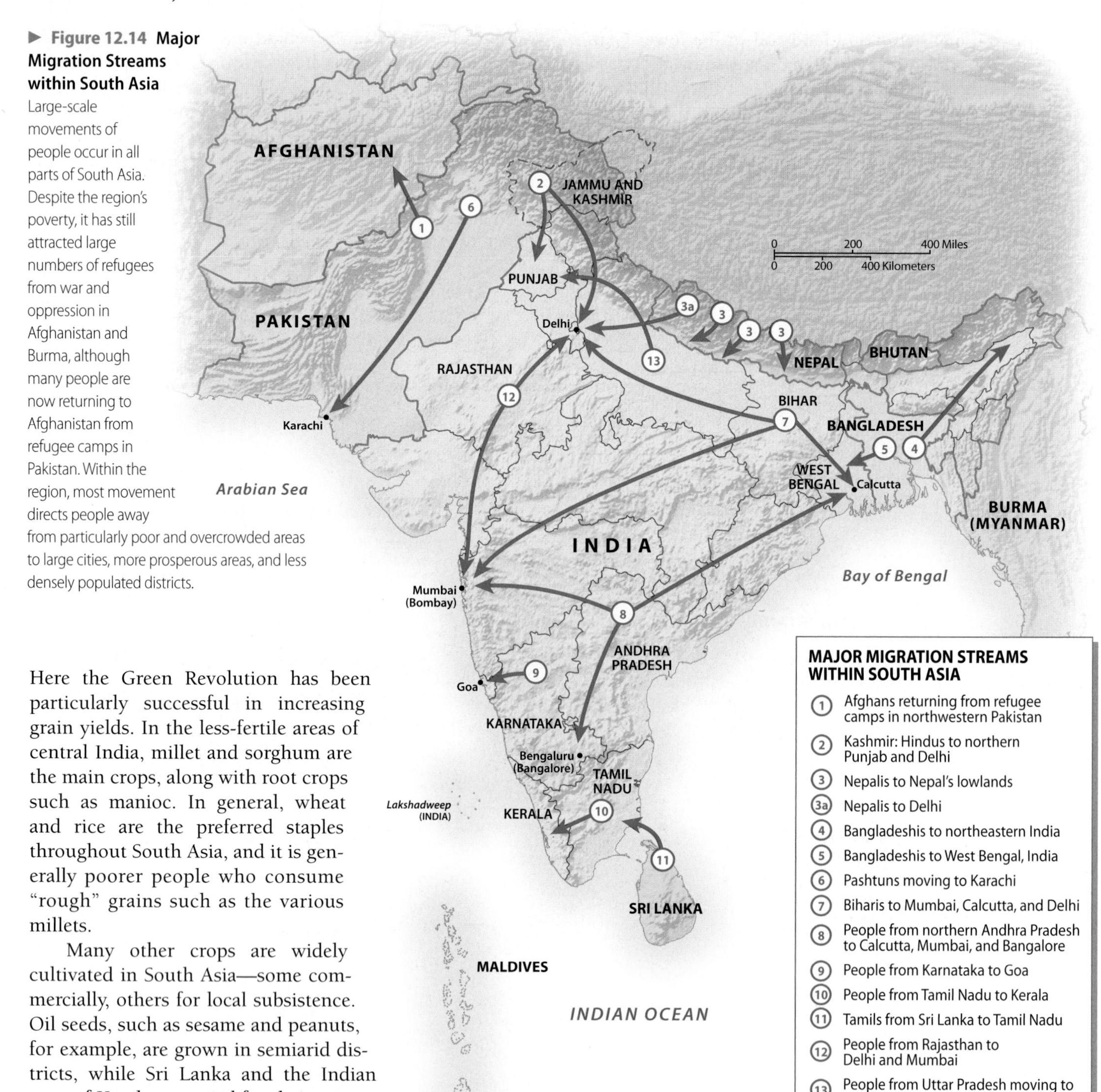

▶ **Figure 12.14 Major Migration Streams within South Asia** Large-scale movements of people occur in all parts of South Asia. Despite the region's poverty, it has still attracted large numbers of refugees from war and oppression in Afghanistan and Burma, although many people are now returning to Afghanistan from refugee camps in Pakistan. Within the region, most movement directs people away from particularly poor and overcrowded areas to large cities, more prosperous areas, and less densely populated districts.

Here the Green Revolution has been particularly successful in increasing grain yields. In the less-fertile areas of central India, millet and sorghum are the main crops, along with root crops such as manioc. In general, wheat and rice are the preferred staples throughout South Asia, and it is generally poorer people who consume "rough" grains such as the various millets.

Many other crops are widely cultivated in South Asia—some commercially, others for local subsistence. Oil seeds, such as sesame and peanuts, for example, are grown in semiarid districts, while Sri Lanka and the Indian state of Kerala are noted for their coconut groves, spice gardens, and tea plantations. In both Pakistan and west central India, cotton is widely cultivated, and Bangladesh has long supplied most of the world's jute, a tough fiber used in the manufacture of rope.

Many, if not most, South Asians receive inadequate protein, as meat consumption is extremely low. This partly reflects the region's poverty because meat is expensive to produce. In India, religion is equally important, as most Hindus are vegetarians. Despite this prohibition against eating meat, animal husbandry is vitally important throughout South Asia. India has the world's largest cattle population—in part because cattle are sacred in Hinduism, but also because milk is one of South Asia's main sources of protein. Although the cattle of India have traditionally yielded little milk, a so-called white revolution has increased dairy efficiency. In 2010, India produced more milk than any other country by a substantial margin, while Pakistan ranked fourth, trailing only India, the United States, and China.

The Green Revolution The main reason South Asian agriculture has kept up with population growth is the Green Revolution, which originated during the 1960s in agricultural

▶ Figure 12.15 **Rice Cultivation** A large amount of irrigation water is needed to grow rice, as is apparent from this photo from Sri Lanka. Rice also is the main crop in the lower Ganges Valley and Delta, along the lower Indus River of Pakistan, and on India's coastal plains.

research stations established by international development agencies. One of the major problems researchers faced was the fact that they could not attain higher yields by simply fertilizing local seed strains because the plants would grow taller and then fall to the ground before the grain could mature. The solution was to cross-breed new "dwarf" crop strains that would respond to heavy chemical fertilization by producing more grain rather than longer stems.

By the 1970s, it was clear that these efforts had succeeded in reaching their initial goals. The more prosperous farmers of the Punjab quickly adopted the new "miracle wheat" varieties, solidifying the Punjab's position as the region's breadbasket. Green Revolution rice strains also were adopted in the more humid areas. As a result, South Asia was transformed from a region of chronic food shortages to one of self-sufficiency. India more than doubled its annual grain production between 1970 and the mid-1990s, from 80 to 191 million tons.

Although the Green Revolution was clearly an agricultural success, many argue that it has been an ecological and social disaster. Serious environmental problems result from the chemical dependency of the new crop strains. Not only do they typically need large quantities of industrial fertilizer, which is both expensive and polluting, but also they require frequent pesticide applications because they often lack natural resistance to local plant diseases and insects. A 2009 study showed that 20 percent of the wells in India's Punjab are seriously contaminated by nitrates, derived largely from artificial fertilizers.

Social problems have also followed the Green Revolution. In many areas, only the more prosperous farmers are able to afford the new seed strains, along with the irrigation equipment, farm machinery, fertilizers, and pesticides necessary to support this new high-technology agriculture (Figure 12.16). As a result, poorer farmers often go deeply into debt and are then forced off the land when they fail to repay their loans.

▼ Figure 12.16 **Green Revolution Farming** Because of "miracle" wheat strains that have increased yields in the Punjab area, this region has become the breadbasket of South Asia. Similarly intensive methods are used in the Punjab for other crops, such as the tomatoes being grown under plastic covers. Increased production, however, has led to both social and environmental problems.

Future Food Supply The Green Revolution has fed South Asia's expanding population over the past several decades, but whether it will be able to continue doing so remains unclear. Many of the crop improvements have seemingly exhausted their potential, and much of South Asia's agricultural economy is currently in a state of crisis as farmers fail to make ends meet.

Optimists believe, however, that South Asia's food production could be substantially increased. Further improvements in highways and railroads, for example, would result in less wastage, as well as higher profit margins for struggling farmers. Higher yields would also result if the techniques pioneered in northeastern India were applied throughout the country. In Punjab, for example, the average farm produces 3130 kilograms of rice per acre, whereas the corresponding figure in Bihar is only 1370 kilograms per acre. Green Revolution techniques, moreover, might be profitably applied to secondary grain crops, and genetic engineering might provide another breakthrough, ushering in a second wave of increased crop yields. Most environmentalists, however, see serious dangers in this new technology.

Another option is expanded water delivery because many fields remain unirrigated in South Asia's semiarid areas. Even in the humid zone, dry-season fallow is usually the norm.

Irrigation, however, brings its own problems. In much of Pakistan and northwestern India, where irrigation has been practiced for generations, soil **salinization** (see Figure 12.4), or the buildup of salt in agricultural fields, is already a major constraint. An estimated 27 million acres (11 million hectares) of land in Pakistan are now too salty for normal farming. Additionally, groundwater levels are falling in Punjab, India's breadbasket, because double-cropping has pushed water use beyond the sustainable yield of the underlying aquifers.

Urban South Asia

Although South Asia is one of the least urbanized regions of the world (see Table 12.1), the cities it does have are some of the world's largest urban agglomerations. India alone lists some 46 cities with populations greater than a million, most of which are growing rapidly. Mumbai's (Bombay) population is now roughly 14 million, with up to 22 million living in the city's metropolitan area.

Because of this rapid growth, most South Asian cities have staggering problems with homelessness, poverty, congestion, water shortages, air pollution, and sewage disposal. Kolkata's (Calcutta) homeless are legendary, with perhaps half a million people sleeping on the streets each night. In that city and others, sprawling squatter settlements, or bustees, mushroom in and around urban areas, providing temporary shelter for many urban migrants (Figure 12.17). A brief survey of the region's major cities indicates the problems and prospects of the region's urban areas (see also *Cityscapes: Karachi, Pakistan's Sprawling Megacity*).

Mumbai (Bombay) The largest city in South Asia, Mumbai (often called by its former name, Bombay) is India's financial, industrial, and commercial center. Mumbai is responsible for much of India's foreign trade, has long been a manufacturing center, and is the focus of India's film industry—the world's largest. Mumbai's economic vitality draws people from all over India, resulting in simmering ethnic tensions.

Because of the city's restricted space, most of Mumbai's growth has taken place to the north and east of the historic city. Building restrictions in the downtown area have resulted in skyrocketing commercial and residential rents, which are some of the highest in the world. Even members of the city's thriving middle class have difficulty finding adequate housing. Hundreds of thousands of less-fortunate immigrants live in "hutments," crude shelters built on formerly busy sidewalks. The least fortunate sleep on the street or in simple plastic tents, often placed along busy roadways.

Mumbai's notorious road congestion eased somewhat in 2009, after the completion of a massive eight-lane, $340 million bridge linking the central city to its northern suburbs. The Mumbai Metro, an ambitious rapid transit system, is scheduled to be completed in 2021, promising further improvements.

▼ **Figure 12.17 Mumbai (Bombay) Hutments** Hundreds of thousands of people in Mumbai (Bombay) live in crude hutments, with no sanitary facilities, built on formerly busy sidewalks. Hutment construction is forbidden in many areas, but wherever it is allowed, sidewalks quickly disappear.

Delhi The sprawling capital of India, Delhi has roughly 22 million people in its metropolitan area. It consists of two contrasting landscapes expressing its past: Delhi (or old Delhi), a former Muslim capital, is a congested town of tight neighborhoods; New Delhi, in contrast, is a city of wide boulevards, monuments, parks, and expensive residential areas. It was born as a planned city when the British decided to move the colonial capital from Calcutta in 1911. Located here are the embassies, luxury hotels, government office buildings, and airline offices necessary for a vibrant political capital. South of the government area are more expensive residential areas, some of which are focused on large parks and ornamental gardens. Rapid growth, along with the government's inability to control auto and industrial emissions, made Delhi one of the world's ten most polluted cities in the late 1990s. As a result, the government began to shut down many of the city's small-scale industries, although local residents have banded together to protect their livelihoods. Overall, significant environmental progress has been made.

Kolkata (Calcutta) To many, Kolkata is emblematic of the problems faced by rapidly growing cities in developing countries. Not only is homelessness widespread, but also this metropolitan area of some 15 million falls far short of supplying its residents with water, power, or sewage treatment. Electrical power is so inadequate that every hotel, restaurant, shop, and small business has to have some sort of standby power system. During the wet season, many streets are routinely flooded.

Kolkata faces continued rapid growth as migrants drain in from the countryside, a mixed Hindu-Muslim population that generates ethnic rivalry, a troubled economic base,

Cityscapes

Karachi, Pakistan's Sprawling Megacity

Karachi is Pakistan's largest city—the third largest in the world by some measurements—with 13 million people living in the city proper and as many as 23.5 million in the metropolitan area (**Figure 12.2.1**). Twenty percent of Pakistan's gross domestic product (GDP) is produced within the city's borders, and 90 percent of Pakistan's financial and multinational companies are headquartered here. Karachi was Pakistan's capital before the decision was made in 1960 to relocate the government to a new city—Islamabad—located near the contested region of Kashmir. Karachi lies adjacent to the Arabian Sea, and its excellent harbors have made it an important trading center for hundreds of years. Today 95 percent of Pakistan's foreign trade goes through Karachi's two main ports. Karachi's economic centrality in modern Pakistan makes the city's increasing ethnic and religious tensions a major concern for Pakistan's leaders.

The Urban Core The traditional center of Karachi is the area in and around Saddar Town, located in the southern part of the city near the mangrove swamps and beaches of the coast. Saddar (the central business district of Saddar Town) hosts most of Karachi's largest businesses, universities, and shopping areas, as well as its densest residential areas. The Empress Market, named in honor of Queen Victoria and completed in 1889, is a famous example of Victorian architecture and one of the few historical buildings remaining in Karachi. The building continues to serve as Karachi's primary shopping hub, and traditional vendors selling produce and copious arrays of spices still dominate the area. Another important downtown asset is Zaibunnisa Street, also a well-traveled business area since colonial times.

Changing Karachi As in many of the dense and quickly growing cities of South Asia, encroachment is a widespread grievance of Karachi residents. High levels of all kinds of traffic on city streets often slow movement to a crawl and make it tempting for vendors to set up shop along street edges, slowing movement even further. Eventually, permanent buildings arrive, making it very difficult to restore a lost right-of-way. Karachi has attempted to combat this problem through near-constant antiencroachment programs, as well as the construction of new areas of settlement further afield. Several new skyscrapers have opened in Karachi recently, and although they are modest by the standards of other world cities, they remain somewhat of an aberration in South Asia. Two of the city's three largest buildings—Ocean Towers and Centre Point Tower—are dedicated to commercial space, as opposed to housing or offices.

Rising Tensions Unfortunately, Karachi suffers from serious political and ethnic tensions that have turned parts of the city into armed camps. Ethnic conflicts have generally pitted Sindhis, the region's indigenous inhabitants, against the Muhajirs, Muslim refugees from India who settled in and around the city after independence in 1947. More recently, the migration of Pashtuns from northwest Pakistan has further destabilized the situation, as have clashes between Sunni and Shiite Muslims. In 2012, almost 2000 people lost their lives due to ethnic and sectarian violence in the city. Several radical Islamist groups, including the Taliban and Al Qaeda, are also reputed to have important bases in the city.

▼ **Figure 12.2.1 Karachi** With a population of some 13 million, Karachi is by far the largest and most cosmopolitan city in Pakistan. This strife-torn port city was also the capital of Pakistan from 1947 until the early 1960s, when the seat of government was gradually transferred to Islamabad.

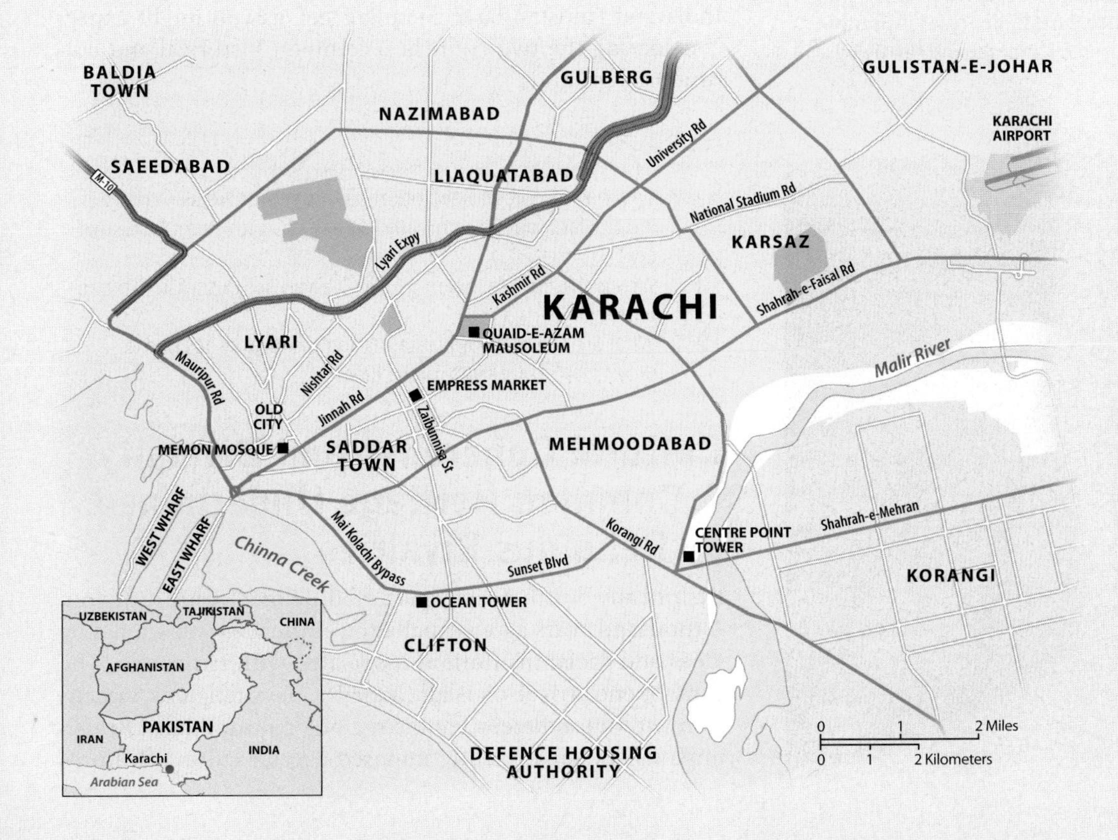

1. In what ways can you see the legacy of British colonialism in contemporary Karachi?
2. Why has Karachi seen so much more ethnic tension than Pakistan's other major cities?

Google Earth Virtual Tour Video

http://goo.gl/ibO8xv

▲ **Figure 12.18 Dhaka City View** Dhaka, the capital of Bangladesh, has emerged as a vibrant metropolis in recent decades. Although its slums are extensive, its central commercial district is relatively orderly and prosperous.

and an overloaded infrastructure. Clearly, Kolkata's future will have problems. Yet it remains a culturally vibrant city noted for its fine educational institutions, theaters, and publishing firms.

Dhaka As the capital and major city of Bangladesh, Dhaka (also spelled *Dacca*) has experienced rapid growth due to migration from the surrounding countryside. In 1971, when the country gained independence from Pakistan, Dhaka had about 1 million inhabitants; today its metropolitan area numbers almost 15 million (Figure 12.18). Dhaka's economic vitality has increased since independence because it combines the administrative functions of government with the largest industrial concentration in Bangladesh. Cheap and abundant labor in Dhaka has made the city a global center for clothing, shoe, and sports equipment manufacturing. North American shoppers are readily reminded of this new role when looking through clothing tags at any department store.

▼ **Figure 12.19 Islamabad Landscape** Islamabad has been the capital of Pakistan since the 1960s. **Q: Unlike most major cities in South Asia, Islamabad is characterized by straight and relatively wide streets. Why is this so?**

Islamabad Upon independence, Pakistan's leaders determined that Karachi was too far from the center of the country and that an entirely new capital was necessary. This planned city would make a statement through its name—Islamabad—about the religious foundation of Pakistan. Located close to the contested region of Kashmir, it would also make a geopolitical statement. In geographic terms, such a city is referred to as a **forward capital** because it signals—both symbolically and geographically—the intentions of the country. By building its new capital in the north, Pakistan sent a clear message that it would not abandon its claims to the portion of Kashmir controlled by India. Islamabad is closely linked to the historical city of Rawalpindi, once a major British encampment, a few miles away. These two cities form a single metropolitan region of about 4.5 million people, but they are completely different in appearance and character. To avoid congestion, planners designed Islamabad around self-sufficient sectors, each with its own government buildings, residences, and shops (Figure 12.19).

The closest parallel in India to Islamabad is Chandigarh, the modern planned city that serves as the capital of two Indian states: Punjab and Haryana. All told, the cities of India and Pakistan have a similar feel, as you might expect considering the two countries' common historical and cultural backgrounds.

REVIEW

12.4 Why has the Green Revolution been so controversial, considering the fact that it has greatly increased South Asia's food supply?

12.5 Why is the Punjab region usually viewed as South Asia's "breadbasket"?

12.6 What are the major advantages and disadvantages of the growth of South Asia's megacities?

Cultural Coherence and Diversity: A Common Heritage Undermined by Religious Rivalries

Historically, South Asia forms a well-defined cultural region. A thousand years ago, virtually the entire area was united by ideas and social institutions associated with Hinduism. The subsequent arrival of Islam added a new religious dimension without undercutting the region's cultural unity. British imperialism subsequently imposed several cultural features

▲ **Figure 12.20 Religious Tension and Cooperation** The dismantling of the Babri Mosque in Ayodhya by Hindu nationalists in 1992 caused intense religious conflict in many parts of India. More recently, Hindu-Muslim brotherhood groups (Bhai-Bhai) have emerged to try to foster understanding and mutual respect. A rally of one such group is shown here.

over the entire region, from the widespread use of English to a common passion for cricket. Since the mid-20th century, however, religious and political strife has intensified, leading some to question whether South Asia can still be conceptualized as a culturally coherent world region.

India has been a secular country since its inception, with the Congress Party, its early guiding political organization, struggling to keep politics and religion separate. In doing so, it relied heavily on support from Muslims. Since the 1980s, this secular political tradition has come under increasing pressure from **Hindu nationalism**. Hindu nationalists promote the religious values of Hinduism as the essential fabric of Indian society.

Hindu nationalists gained considerable political power both at the federal level and in many Indian states through the Bharatiya Janata Party (BJP), leading to widespread agitation against the country's Muslim minority in the mid-1990s. In several high-profile instances, Hindu mobs demolished Muslim mosques that were allegedly built on the sites of ancient Hindu temples. The destruction of a mosque in Ayodhya in the Ganges Valley galvanized the nationalistic BJP membership in 1992. In 2002, more than 2000 people—mostly Muslims—were killed during religious riots in India's state of Gujarat. Two years later, however, India's Hindu nationalist movement suffered a huge electoral defeat, perhaps indicating that the wave of religious strife had peaked. Relations between Hindus and Muslims are now generally peaceful, although violent incidents still occur (Figure 12.20).

In predominantly Muslim Pakistan, rising Islamic fundamentalism has generated severe conflict. Radical fundamentalist leaders want to make Pakistan a fully religious state under Islamic law, a plan rejected by the country's secular intellectuals and international businesspeople. The government has attempted to intercede between the two groups, but is often viewed as biased toward the Islamists. Anti-blasphemy laws, for example, have been used to persecute members of the country's small Hindu and Christian communities, as well as liberal Muslim writers. Large amounts of money from Saudi Arabia, moreover, have gone into the development of fundamentalist religious schools that are reputed to encourage extremism.

Religious and political extremists in South Asia argue that the current struggles reflect deeply rooted historical divisions. Most scholars disagree, regarding religious conflict as a recent development. To weigh these contrasting claims, it is necessary to examine the cultural history of South Asian civilization.

Origins of South Asian Civilizations

Many scholars think that the roots of South Asian culture extend back to the Indus Valley civilization, one of the world's first urban-oriented cultures, which flourished from 3300 to 1300 BCE in what is now Pakistan and northwestern India. This remarkable society is poorly understood, as its script has not been deciphered. It may have been characterized by tight political centralization, however, as many of its artifacts are extraordinarily uniform. Despite its long period of power, the Indus Valley civilization vanished almost entirely in the second millennium BCE, after which the record grows dim. By 800 BCE, however, a new urban focus had emerged in the middle Ganges Valley (Figure 12.21). The social, religious, and intellectual customs associated with this civilization eventually spread throughout the lowlands of South Asia.

Hindu Civilization The religious complex of this early South Asian civilization was an early form of Hinduism, a complicated faith that incorporates diverse forms of worship and that lacks any standard system of beliefs (Figure 12.22). Certain deities are recognized, however, by all believers, as is the notion that these various gods are all manifestations of a single divine entity. All Hindus, moreover, share a common set of epic stories, usually written in the sacred language of Sanskrit. Hinduism is noted for its mystical tendencies, which have long inspired many men (and a few women) to seek an ascetic lifestyle, renouncing property and sometimes all regular human relations. One of its hallmarks is a belief in the transmigration of souls from being to being through reincarnation, wherein the nature of one's acts in the physical world influences the course of these future lives.

Scholars once confidently argued that Hinduism originated from the fusion of two distinct religious traditions: the mystical beliefs of the subcontinent's indigenous inhabitants (including the people of the ill-fated Indus Valley civilization) and the sky-god religion of the Indo-European invaders who swept into the region from Central Asia sometime in the second millennium BCE. Such a scenario also proved convenient for explaining India's **caste system**, the strict division of society into different hierarchically ranked hereditary groups. The elite invaders, according to this theory, wished to remain separate from the people they had defeated, resulting in an elaborate system of social division. Recent research, however, indicates that the caste system, like Hinduism itself, emerged through more gradual processes of social and cultural evolution. Critics also note that the British imperial authorities tended to harden

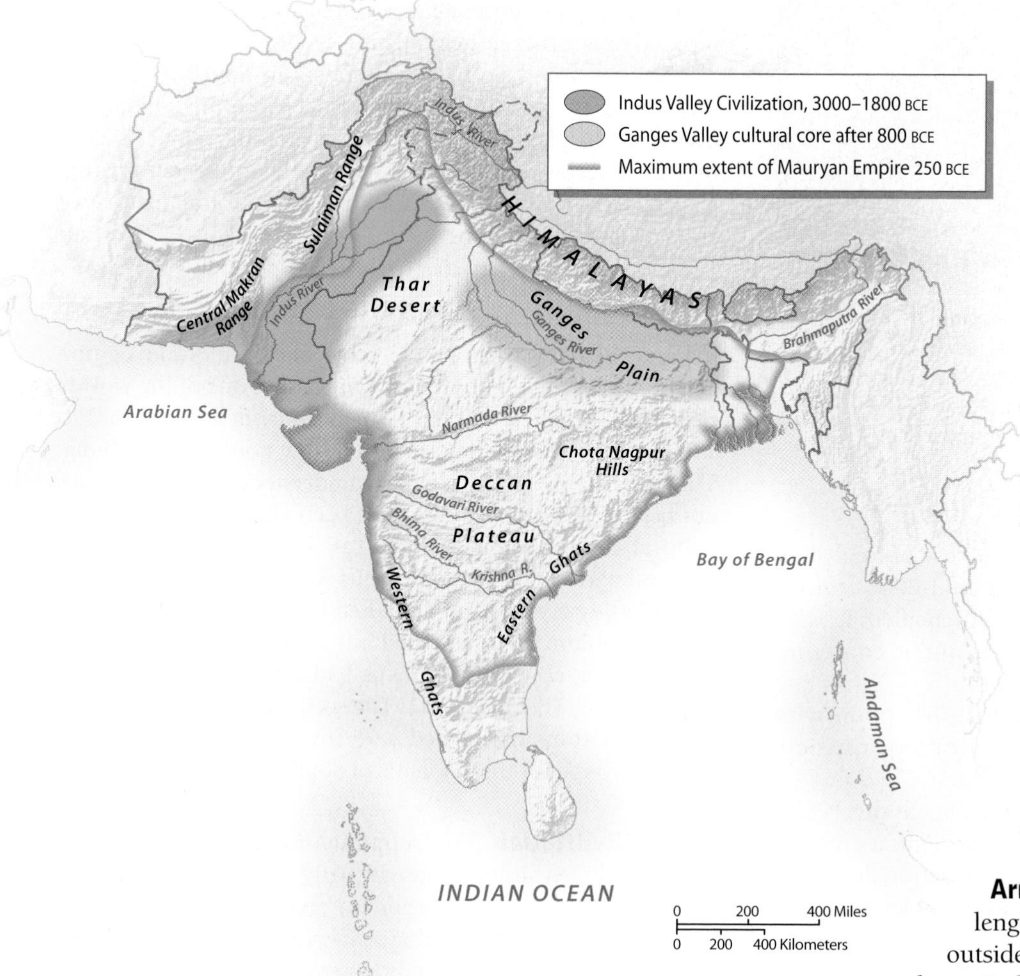

◀ **Figure 12.21 Early South Asian Civilizations** The roots of South Asian culture may extend back 5000 years to an Indus Valley civilization based on irrigated agriculture and vibrant urban centers. What happened to that civilization remains a topic of conjecture because the archaeological record grows dim by 1800 BCE. Later, a new urban focus emerged in the Ganges Valley, from which social, religious, and intellectual influences spread throughout lowland South Asia.

caste divisions, as doing so helped them maintain power over the large Indian population.

Buddhism Although a caste system of some sort seems to have existed in the early Ganges Valley civilization, it was soon challenged from within by Buddhism. Siddhartha Gautama, the Buddha, was born in 563 BCE in an elite caste. He rejected the life of wealth and power that was laid out before him, however, and sought instead to attain enlightenment, or mystical union with the cosmos. He preached that the path to such "nirvana" was open to all, regardless of social position. His followers eventually established Buddhism as a new religion. Buddhism spread through most of South Asia, becoming something of an official faith under the Mauryan Empire, which ruled much of the subcontinent in the 3rd century BCE. Later centuries saw Buddhism expand through most of East, Southeast, and Central Asia.

But for all of its successes abroad, Buddhism never replaced Hinduism in India. It remained focused on monasteries rather than spreading throughout the wider society. Many Hindu priests, moreover, struggled against the new faith. One of their techniques was to embrace many of Buddhism's philosophical ideas and enfold them within the intellectual system of Hinduism. By 500 CE, Buddhism was on the retreat throughout South Asia, and within another 500 years it had virtually disappeared from the region. The only major exceptions were the island of Sri Lanka and the high Himalayas, both of which remain mostly Buddhist to this day.

Arrival of Islam The next major challenge to Hindu society—Islam—came from outside the region. Arab armies conquered the lower Indus Valley around 700 CE, but advanced no farther. Then around the year 1000, Turkish-speaking Muslims began to invade from Central Asia. At first, they merely raided, but eventually they began to settle and rule on a permanent basis. By the 1300s, most of South Asia lay under Muslim power, although Hindu kingdoms persisted in southern India and in the arid lands of Rajasthan in northwest India. Later, during the 16th and 17th centuries, the **Mughal Empire** (also spelled *Mogul*) dominated much of the

▼ **Figure 12.22 Hindu Temple** Although India has long been noted for its ancient temples, lavish new Hindu religious complexes continue to be constructed. The Akshardham Temple here was inaugurated in New Delhi in 2005. Expected to be a major tourist attraction, Akshardham also serves as an educational center focused on the culture of India.

◄ **Figure 12.23 The Red Fort** The Red Fort of Delhi, completed in 1648, was the military center of the Moghul Empire. Today this massive fortification, one of the largest in the world, is a major tourist destination.

region from its power center in the upper Indus–Ganges Basin (Figure 12.23).

At first, Muslims formed a small ruling elite, but over time, increasing numbers of Hindus converted to the new religion, particularly those from lower castes, who sought freedom from the rigid social order. Conversions were most pronounced in the northwest and northeast, and eventually the areas now known as Pakistan and Bangladesh became predominantly Muslim.

At first glance, Islam and Hinduism are strikingly divergent faiths. Islam is resolutely monotheistic, austere in its ceremonies, and spiritually egalitarian (all believers stand in the same relationship to God). Hinduism, by contrast, is polytheistic (at least on the surface), lavish in its rituals, and caste-structured. Because of these profound differences, Hindu and Muslim communities in South Asia are sometimes viewed as utterly distinct, living in the same region, but not sharing the same culture or civilization. Increasingly, such a view is expressed in South Asia itself. Many residents of modern Pakistan stress the Islamic nature of their country and its complete separation from India.

However, overemphasizing the separation of Hindu and Muslim communities risks missing much of what is historically distinctive about South Asia. Until the 20th century, Hindus and Muslims usually coexisted on friendly terms; the two faiths stood side by side for hundreds of years, during which time they came to influence each other in many ways. Moreover, aspects of caste organization have persisted among South Asian Muslims, just as they have among India's Christians.

The Caste System

Although caste is one of the historically unifying features of South Asia, the system is not uniformly distributed across the subcontinent. It has never been significant in India's tribal areas; in modern Pakistan and Bangladesh its role is fading; and in the Buddhist society of Sri Lanka its influence has long been somewhat marginal. Even in India, caste is now deemphasized, especially among the more urban and educated segments of society. But with all that said, caste continues to structure day-to-day social existence for hundreds of millions of Indians. It is estimated that roughly 1000 young Indians are murdered every year by relatives who will not tolerate their marriages across caste lines.

Caste is actually a rather clumsy term for denoting the complex social order of the Hindu world. The word itself, of Portuguese origin, combines two distinct local concepts: *varna* and *jati*. *Varna* refers to the ancient fourfold social hierarchy of the Hindu world, whereas *jati* refers to the hundreds of local endogamous ("marrying within") groups that exist at each varna level (different jati groups are thus usually called *subcastes*). Jati, like varna, are hierarchically arranged, although the exact order of precedence is not so clear-cut.

It often has been argued that the essence of the caste system is the notion of social pollution. The lower one's position in the hierarchy is, the more potentially polluting one's body supposedly is. Members of higher castes were traditionally not supposed to eat or drink with, or even use the same utensils as, members of lower castes.

The Main Caste Groups Three varna groups constitute the traditional elite of Hindu society. At the apex sit the Brahmins, members of the traditional priestly caste. Brahmins perform the high rituals of Hinduism, and they form the traditional intellectual elite of India. Most Brahmins value education highly, and today they are disproportionately represented among India's professional classes. Below the Brahmins in the traditional hierarchy are the Kshatriyas, members of the warrior or princely caste. In premodern India, this group actually had more power and wealth than did the Brahmins; it was they who ruled the old Hindu kingdoms. Next stand the Vaishyas, members of the traditional merchant caste. In earlier centuries, a near monopolization of long-distance trade and money lending in northern India gave many Vaishyas ample opportunities to accumulate wealth. The precepts of vegetarianism and nonviolence are particularly strong among certain merchant subcastes of western India. One prominent representative of this tradition was Mohandas Gandhi, the founder of modern India and one of the 20th century's greatest leaders.

The majority of India's population fits into the fourth varna category, that of the Sudras. The Sudra caste is composed of an especially large array of subcastes (jati), most of which originally reflected occupational groupings. Most Sudra subcastes were traditionally associated with peasant farming, but others were based on craft occupations, including those of barbers, smiths, and potters.

The Brahmins, Kshatriyas, Vaishyas, and Sudras form the basic fourfold scheme of caste society, but another sizable group stands outside the varna system altogether. These are the so-called *untouchables,* or **dalits**, as they are now preferably called. Dalits were not traditionally allowed to enter Hindu temples. Such low-status positions were derived historically from "unclean" occupations, such as those of leather workers (who dispose of dead animals), scavengers, latrine cleaners, and swine herders. In the southern Indian state of Kerala until the 19th century, some groups were actually considered "unseeable"; these unfortunate people had to hide in bushes and avoid walking on the main roads to protect members of higher castes from "visual pollution." Not surprisingly, many Indian dalits have converted to Islam, Christianity, and Buddhism in an attempt to escape from the caste system. Even so, they continue to suffer discrimination. But some dalits have managed to reach high positions. In fact, the chief architect of India's constitution, B. R. Ambedkar, was from an "untouchable" caste (Figure 12.24).

The Changing Caste System The caste system is clearly in a state of flux in India today. Its original occupational structure has long been undermined by the necessities of a modern economy, and various social reforms have chipped away at the discrimination that it embodies. The dalit community itself has produced several notable national leaders who have waged partially successful political struggles. Owing to such efforts, the very concept of "untouchability" is now technically illegal in India. The Indian central government also insists that a significant percentage of university seats and governmental jobs be reserved for students from low-caste backgrounds, while a number of Indian states have set higher quotas, both for education and for governmental employment. Some states also reserve positions in universities for Muslims and Christians, as members of religious minorities. Such "reservations," as they are called, are controversial, as many people think that they unfairly penalize people of higher-caste background.

▼ **Figure 12.24 B. R. Ambedkar** The main architect of India's constitution was B.R. Ambedkar, who was also noted as a philosopher, historian, and economist. Ambedkar was born to a poor Dalit ('Untouchable") family, and faced significant discrimination as a child and young man.

Contemporary Geographies of Religion

South Asia, as we have seen, has a predominantly Hindu heritage overlain by a substantial Muslim imprint. Such a picture fails, however, to capture the enormous diversity of modern religious expression in contemporary South Asia. The following discussion looks specifically at the geographical patterns of the region's main faiths (Figure 12.25).

Hinduism Less than 1 percent of the people of Pakistan are Hindu, and in Bangladesh and Sri Lanka, Hinduism is a distinctly minority religion. Almost everywhere in India, however—and in Nepal, as well—Hinduism is very much the faith of the majority. In east-central India, more than 95 percent of the population is Hindu. Hinduism is itself a geographically complicated religion, with aspects of faith varying in different parts of India. Even within a given Indian state, forms of worship often differ from region to region and from caste to caste.

Islam Islam may be considered a "minority" religion for the region as a whole, but such a designation obscures the tremendous importance of this religion in South Asia. With more than 400 million members, the South Asian Muslim community is the largest in the world. Bangladesh and especially Pakistan are overwhelmingly Muslim. India's Islamic community, although constituting only some 13–15 percent of the country's population, is still roughly 160 million strong, making it the world's third largest Muslim group.

Although Muslims live in almost every part of India, they are concentrated in four main areas: in most of India's cities; in Kashmir, particularly in the densely populated Vale of Kashmir; in the upper and central Ganges Plain; and in the southwestern state of Kerala. In northern India especially, Muslims tend to be poorer and less educated than their Hindu neighbors, leading some observers to call them the country's "new dalits."

Interestingly, the state of Kerala is about 25 percent Muslim even though it was one of the few parts of India that never experienced prolonged Muslim rule. Islam in Kerala is historically connected not to Central Asia, but rather to trade across the Arabian Sea. Kerala's Malabar Coast long supplied spices and other luxury products to Southwest Asia, enticing many Arabian traders to settle in this part of India. Gradually, many of Kerala's native inhabitants converted to

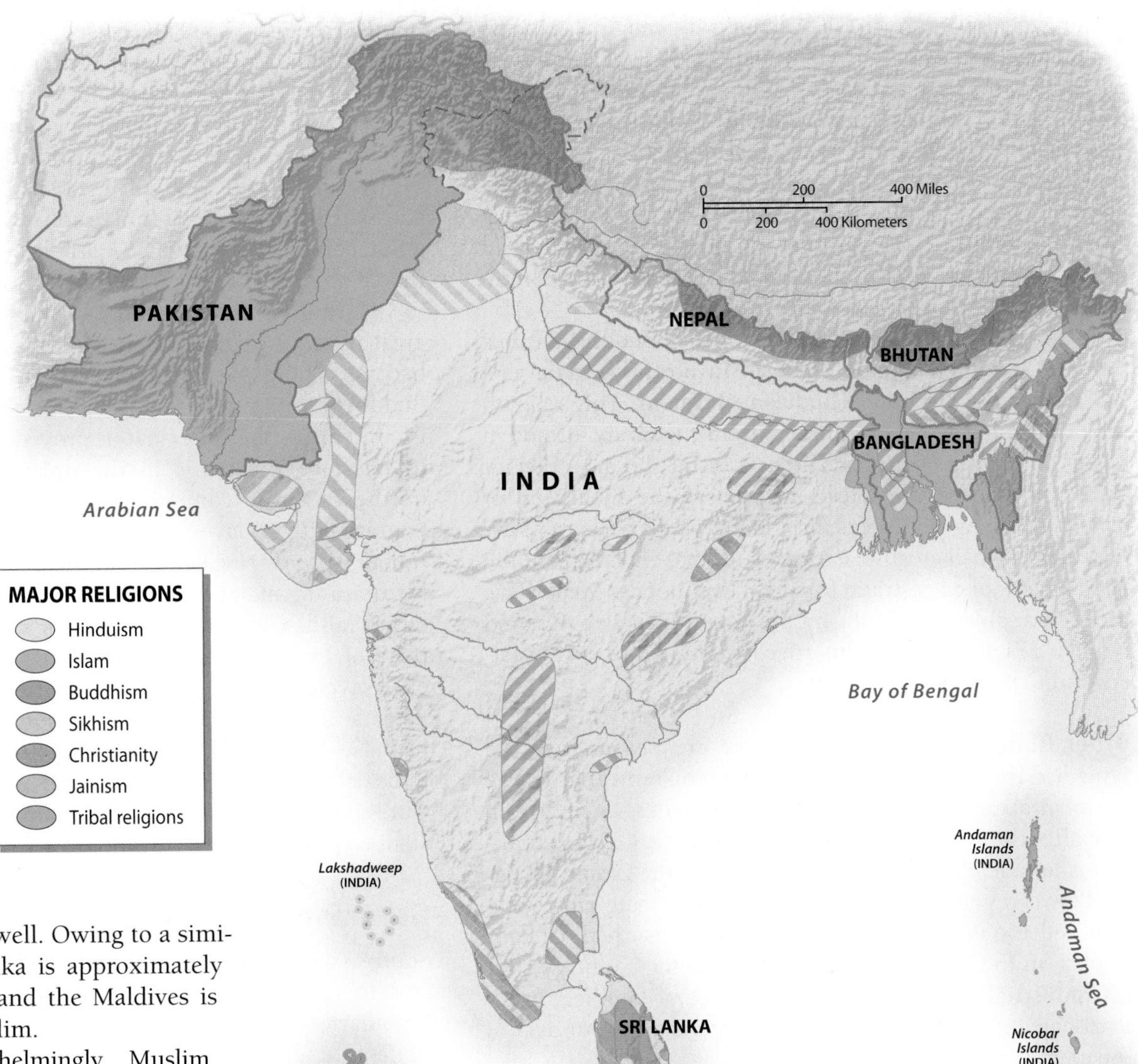

▶ **Figure 12.25 Religions of South Asia** Hindu-dominated India is bracketed by the two important Muslim countries of Pakistan and Bangladesh. Some 150 million Muslims, however, live within India, constituting roughly 15 percent of the total population. Of particular note are the Muslims in northwestern Kashmir and in the Ganges Valley. Sikhs form the majority population in India's state of Punjab. Also note the Buddhist populations in Sri Lanka, Bhutan, and northern Nepal; the areas of tribal religion in the east; and the centers of Christianity in the southwest.

the new religion as well. Owing to a similar process, Sri Lanka is approximately 9 percent Muslim, and the Maldives is almost entirely Muslim.

As an overwhelmingly Muslim country, Pakistan officially calls itself an Islamic Republic. As such, Islamic law is technically supposed to override the country's secular laws, although in actuality the situation remains ambiguous. In Pakistan's two most conservative provinces, Balochistan and Khyber Pakhtunkhwa (previously known as the North-West Frontier Province), Islamic law tends to be much more strictly enforced.

Sikhism The tension between Hinduism and Islam in medieval northern South Asia helped give rise to a new religion called **Sikhism**. Sikhism originated in the late 1400s in Punjab, near the modern boundary between India and Pakistan. Punjab was the site of religious fervor at the time; Islam was gaining converts, and Hinduism was increasingly on the defensive. The new faith combined elements of both religions and thus appealed to many who felt trapped between their competing claims. Many orthodox Muslims, however, viewed Sikhism as a heresy precisely because it incorporated elements of their own religion. Periodic bouts of persecution led the Sikhs to adopt a militantly defensive stance, and in the political chaos of the early 1800s they were able to carve out a powerful kingdom for themselves. Even today Sikh men are noted for their work as soldiers and bodyguards.

At present, the Indian state of Punjab is approximately 60 percent Sikh. Small, but often influential groups of Sikhs are scattered across the rest of India. Devout Sikh men are immediately visible because they do not cut their hair or their beards. Instead, they wear their hair wrapped in a turban and often tie their beards close to their faces.

Buddhism and Jainism Although Buddhism virtually disappeared from India in medieval times, it persisted in Sri Lanka. Among the island's dominant Sinhalese people, Theravada Buddhism developed into a virtual national religion, fostering close connections between Sri Lanka and mainland Southeast Asia. In the high valleys of the Himalayas,

Buddhism also survived as the majority religion. Tibetan Buddhism, with its esoteric beliefs and huge monasteries, has been better preserved in Bhutan and in the Ladakh region of northeastern Kashmir than in Chinese-controlled Tibet itself. The small city of Dharamsala in the northern Indian state of Himachal Pradesh is the seat of Tibet's government-in-exile and of its spiritual leader, the Dalai Lama, who fled Tibet in 1959. Another group of Buddhists lives in central India, where 20th-century efforts to convert Hindu dalits to the faith were particularly successful.

At roughly the same time as the birth of Buddhism (circa 500 BCE), another religion emerged in northern India as a protest against orthodox Hinduism: **Jainism**. This religion took nonviolence to its ultimate extreme. Jains are forbidden to kill any living creatures, and as a result the most devout adherents wear gauze masks to prevent the inhalation of small insects. Agriculture is forbidden to Jains because plowing can kill small creatures. As a result, most members of the faith have looked to trade for their livelihoods. Many have prospered, aided no doubt by the frugal lifestyles required by their religion. Today Jains are concentrated in northwestern India, particularly Gujarat.

Other Religious Groups Even more prosperous than the Jains are the Parsis, or Zoroastrians, concentrated in the Mumbai area. The Parsis arrived as refugees, fleeing from Iran after the arrival of Islam in the 7th century. Zoroastrianism is an ancient religion that focuses on the cosmic struggle between good and evil. Although numbering only a few hundred thousand, the Parsi community has had a major impact on the Indian economy. Several of the country's largest industrial firms were founded by Parsi families. Intermarriage and low fertility, however, threaten the survival of the community (Figure 12.26).

Indian Christians are more numerous than either Parsis or Jains. Their religion arrived some 1800 years ago; early contact between the Malabar Coast and Southwest Asia brought Christian as well as Muslim traders. A Jewish population also established itself, but later declined; today it numbers only a few hundred. Kerala's Christians, by contrast, are counted in the millions, constituting some 20 percent of the state's population. Several Christian sects are represented, but the largest are historically affiliated with the Syrian Church of Southwest Asia. Another stronghold of Christianity is the small Indian state of Goa, a former Portuguese colony, where Roman Catholics make up a little more than a quarter of the population.

During the colonial period, British missionaries went to great efforts to convert South Asians to Christianity. They had very little success, however, in Hindu, Muslim, and Buddhist communities. The remote tribal districts of British India, on the other hand, proved to be more receptive to missionary activity. In the uplands of India's extreme northeast, entire communities abandoned their traditional animist faith in favor of Protestant Christianity. Today the Indian state of Nagaland has a Baptist majority, whereas in Mizoram the majority of the population follows the Presbyterian faith. Christian missionaries are still active in many parts of India, but during the 1990s they began to experience severe pressure in many areas from Hindu nationalists. Pakistan's Christian community, some 2.8 million strong, has also been under pressure—in this case, from Muslim radicals. In March 2013, a mob of people burned down a Christian neighborhood of some 200 buildings in the Pakistani city of Lahore in response to allegations of blasphemy.

Geographies of Language

South Asia's linguistic diversity matches its religious diversity. In fact, one of the world's most important linguistic boundaries runs directly across India (Figure 12.27). North of the line, almost all languages belong to the Indo-European group, the world's largest linguistic family. The languages of southern India belong to the **Dravidian** language family, a linguistic group unique to South Asia. Along the mountainous northern rim of the region, a third linguistic fam-

◀ **Figure 12.26 Zoroastrianism in India** India's small, but influential Parsi community follows the ancient Iranian religion of Zoroastrianism. These decorations in the style of ancient Iran are on a Parsi temple in Mumbai, India. **Q: Why is this Parsi temple in India decorated in a style that derives from ancient Iran rather than ancient India?**

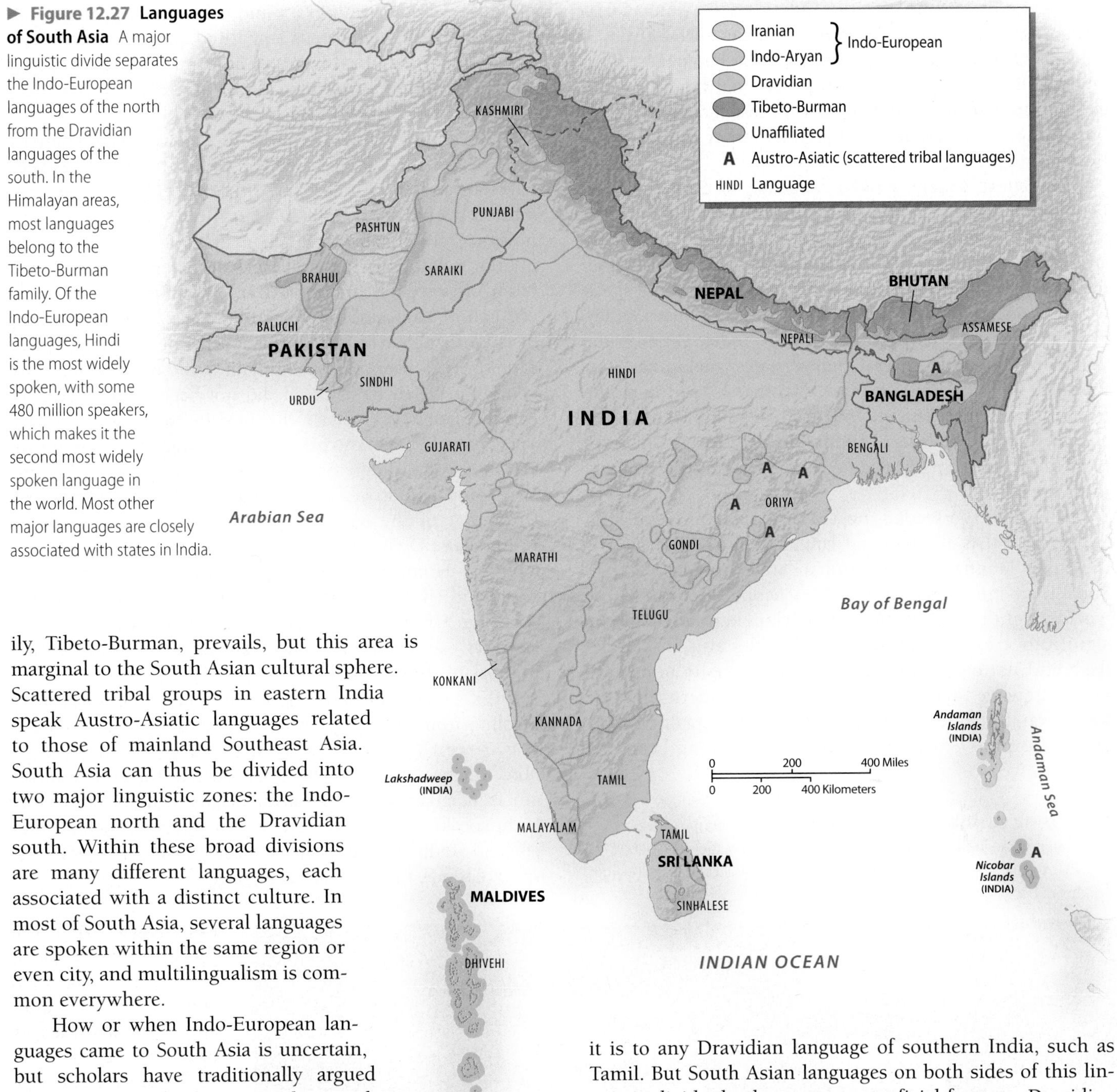

▶ **Figure 12.27 Languages of South Asia** A major linguistic divide separates the Indo-European languages of the north from the Dravidian languages of the south. In the Himalayan areas, most languages belong to the Tibeto-Burman family. Of the Indo-European languages, Hindi is the most widely spoken, with some 480 million speakers, which makes it the second most widely spoken language in the world. Most other major languages are closely associated with states in India.

ily, Tibeto-Burman, prevails, but this area is marginal to the South Asian cultural sphere. Scattered tribal groups in eastern India speak Austro-Asiatic languages related to those of mainland Southeast Asia. South Asia can thus be divided into two major linguistic zones: the Indo-European north and the Dravidian south. Within these broad divisions are many different languages, each associated with a distinct culture. In most of South Asia, several languages are spoken within the same region or even city, and multilingualism is common everywhere.

How or when Indo-European languages came to South Asia is uncertain, but scholars have traditionally argued that they arrived when nomadic peoples from Central Asia invaded the subcontinent in the second millennium BCE, largely supplanting the indigenous Dravidian peoples. According to this hypothesis, offshoots of the same original cattle-herding people also swept across both Iran and Europe, bringing their language to all three places. Supposedly, the ancestral Indo-European tongue introduced to India was similar to Sanskrit, the sacred language of Hinduism. This rather simplistic scenario, however, is now regarded with some suspicion, and many scholars argue for a more gradual infiltration of Indo-European speakers from the northwest.

Any modern Indo-European language of India, such as Hindi or Bengali, is more closely related to English than it is to any Dravidian language of southern India, such as Tamil. But South Asian languages on both sides of this linguistic divide do share some superficial features. Dravidian languages, for example, have borrowed many words from Sanskrit, particularly those associated with religion and scholarship, whereas the Indo-European languages have borrowed many sounds from the Dravidian languages.

The Indo-European North South Asia's Indo-European languages are themselves divided into two subfamilies: Iranian and Indo-Aryan. Iranian languages, such as Baluchi and Pashto, are found in western Pakistan, near the border with Iran and Afghanistan. Languages of the strictly South Asian Indo-Aryan groups are closely related to each other, but are still quite distinctive, often being written in different scripts. Each of the major languages of India is associated with one or more Indian states. Thus, Gujarati is found in Gujarat, Marathi in Maharashtra, Oriya in Odisha (Orissa), and so on. Two of these languages, Punjabi and Bengali, span India's international boundaries to extend into Pakistan and Bangladesh, respectively, as these borders were established on religious rather than linguistic lines.

The most widely spoken language of South Asia is Hindi—not to be confused with the Hindu religion. With almost 500 million speakers, Hindi is the second most widely spoken language in the world. It occupies a prominent role in contemporary India, both because so many people speak it and because it is the main language of the Ganges Valley, the historical and demographic core of India. Hindi is the dominant tongue of several Indian states, including Uttar Pradesh, Madhya Pradesh, and Haryana. In addition, the main forms of speech found in Rajasthan and Bihar are often considered to be dialects of Hindi. Most students from other parts of the country learn some Hindi, often as their second or third language.

Bengali is the second most widely spoken language in South Asia. It is the national language of Bangladesh and the main language of the Indian state of West Bengal. Spoken by more than 200 million people, Bengali is the world's eighth or ninth most widely spoken language. Its significance extends beyond its official status in Bangladesh and its total numerical strength. Equally important is its extensive literature, highlighted in modern times by Rabindranath Tagore (1861–1941). West Bengal, particularly its capital of Kolkata (Calcutta), has long been one of South Asia's main literary and intellectual centers. Kolkata may be noted for its appalling poverty, but it also has one of the highest levels of cultural production in the world, as measured by the output of drama, poetry, novels, and film (Figure 12.28).

The Punjabi-speaking zone in the west was similarly split at the time of independence—in this case, between Pakistan and the Indian state of Punjab. An estimated 90 million people speak Punjabi, but it does not have the significance of Bengali. Although Punjabi is the main vehicle of Sikh religious writings, it lacks an extensive literary tradition. In recent years, moreover, the people of the southern half of Pakistan's province of Punjab have successfully insisted that their dialect forms a separate language, called Saraiki. Also, even though Punjabi is the most widely spoken language in Pakistan, it did not become the country's national language. Instead, that position was given to Urdu.

Urdu, like Hindi, originated on the plains of northern India. The difference between the two was largely one of religion: Hindi was the language of the Hindu majority, Urdu that of the Muslim minority, including the former ruling class. Owing to this distinction, they are written differently—Hindi in the Devanagari script (derived from Sanskrit) and Urdu in a modified version of the Arabic script. Although Urdu contains many words borrowed from Persian, its basic grammar and vocabulary are almost identical to those of Hindi.

With independence in 1947, millions of Urdu-speaking Muslims from the Ganges Valley fled to the new state of Pakistan. Since Urdu had a higher status than Pakistan's indigenous tongues, it was quickly established as the new country's official language. Karachi, Pakistan's largest city, is now mostly Urdu-speaking, but elsewhere other languages, such as Punjabi and Sindhi, remain primary. Most Pakistanis, however, do speak Urdu as their common language.

▼ **Figure 12.28 Kolkata (Calcutta) Bookstore** Although Kolkata (Calcutta) is noted in the West mostly for its abject poverty, the city is also known in India for its vibrant cultural and intellectual life, illustrated by its large number of bookstores, theaters, and publishing firms.

Languages of the South Four thousand years ago, Dravidian languages were probably spoken across most of South Asia. As Figure 12.27, page 575 indicates, a Dravidian tongue called *Brahui* is still found in the uplands of western Pakistan. The four main Dravidian languages, however, are confined to the south. As in the north, each language is closely associated with an Indian state: Kannada in Karnataka, Malayalam in Kerala, Telugu in Andhra

Pradesh, and Tamil in Tamil Nadu. Tamil is usually considered the most important member of the family because it has the longest history and the largest literature. Tamil poetry dates back to the first century CE, making it one of the world's oldest written languages.

Although Tamil is spoken in northern Sri Lanka, the country's majority population, the Sinhalese, speak an Indo-European language. Apparently, the Sinhalese migrated from northern South Asia several thousand years ago. Although this movement is lost to history, the migrants evidently settled on the island's fertile and moist southwestern coastal and central highland areas, which formed the core of a succession of Sinhalese kingdoms. These same people also migrated to the Maldives, where the national language, Divehi, is essentially a Sinhalese dialect. The drier north and east coasts of Sri Lanka, in contrast, were settled long ago by Tamils from southern India. In the 1800s, British landowners imported Tamil peasants from the mainland to work on their tea plantations in the central highlands, giving rise to a second population of Tamil speakers in Sri Lanka.

Linguistic Dilemmas The multilingual countries Sri Lanka, Pakistan, and India have all experienced linguistic conflicts. Such problems are most complex in India, simply because India is so large and has so many different languages. India's linguistic environment is changing in complicated ways, pushed along by modern economic and political forces.

Indian nationalists have long dreamed of a national language that could help forge the different communities of the country into a more unified nation. But this **linguistic nationalism**, or the linking of a specific language with nationalistic goals, faces the stiff resistance of provincial loyalty, which itself is intertwined with local languages. The obvious choice for a national language would be Hindi, and Hindi was indeed declared as such in 1947. Raising Hindi to this position, however, alienated speakers of important northern languages such as Bengali and Marathi, and even more so the speakers of the Dravidian tongues, especially Tamil. As a result of this cultural tension, in the 1950 Indian constitution, Hindi was demoted to sharing the position of "official language" of India with 14 other languages. Over time, additional languages were added to this list, such that India now has 23 separate languages with an official status.

Regardless of opposition, the role of Hindi is expanding, especially in the Indo-European-speaking north. Here local languages are closely related to Hindi, which can therefore be learned without too much difficulty. Hindi is spreading through education, but even more significantly through popular media, especially television and motion pictures. Films and television programs are made in several Indian languages, but Hindi remains the primary vehicle. The most popular movies coming out of Mumbai's "Bollywood" film industry actually tend to be delivered in a neutral dialect that is as close to Urdu as to Hindi—and with a good deal of English thrown in as well.

The Role of English Even if Hindi is spreading, it still cannot be considered anything like a common Indian language. In the Dravidian south, more importantly, its role remains secondary. National-level political, journalistic, and academic communication thus cannot be conducted in Hindi or in any other indigenous language. Only English, an "associate official language" of contemporary India, serves this function.

Before independence, many educated South Asians learned English for its political and economic benefits under British colonialism. It therefore emerged as the de facto common tongue of the upper and middle classes. Today a few extreme nationalists want to deemphasize English, but most Indians, and particularly those of the south, advocate it as a neutral national language because all parts of the country have an equal stake in it. Furthermore, English confers substantial international benefits.

English is thus the main integrating language of India, and it remains widely used elsewhere in South Asia. Roughly a third of the people of India are able to carry on a conversation in English. English-medium schools abound in all parts of the region, and many children of the elite learn this global language well before they begin their schooling.

The English spoken in South Asia has forms and vocabulary elements that sometimes make cross-cultural communication difficult. As a result, IT companies in India, particularly those that run international call centers, sometimes ask their employees to watch reruns of popular television shows from the United States in order to gain fluency in American pronunciation and slang.

South Asians in a Global Cultural Context

The widespread use of English in South Asia has not only facilitated the spread of global culture into the region, but also helped South Asian cultural production reach a global audience, in addition to having many economic benefits (see *Everyday Globalization: Customer Support Call Centers in India*). The global spread of South Asian literature, however, is nothing new. As early as the turn of the 20th century, Rabindranath Tagore gained international acclaim for his poetry and fiction, earning the Nobel Prize for Literature in 1913. In the 1980s and 1990s, such Indian novelists as Salman Rushdie and Vikram Seth became major literary figures in Europe and North America. The films of Bollywood, moreover, are finding increasingly large audiences abroad. In 2009, Indian films sold 3.6 billion tickets worldwide, whereas those of Hollywood sold only 2.6 billion.

The spread of South Asian culture abroad has been accompanied by the spread of South Asians themselves. Migration from South Asia during the time of the British Empire led to the establishment of large communities in such far-flung places as eastern Africa, Fiji, and the southern Caribbean (Figure 12.29). Subsequent migration targeted the developed world; there are now several million people of South Asian descent living in Britain and a similar number in North America. Many contemporary migrants to the United States are doctors, software engineers, and other

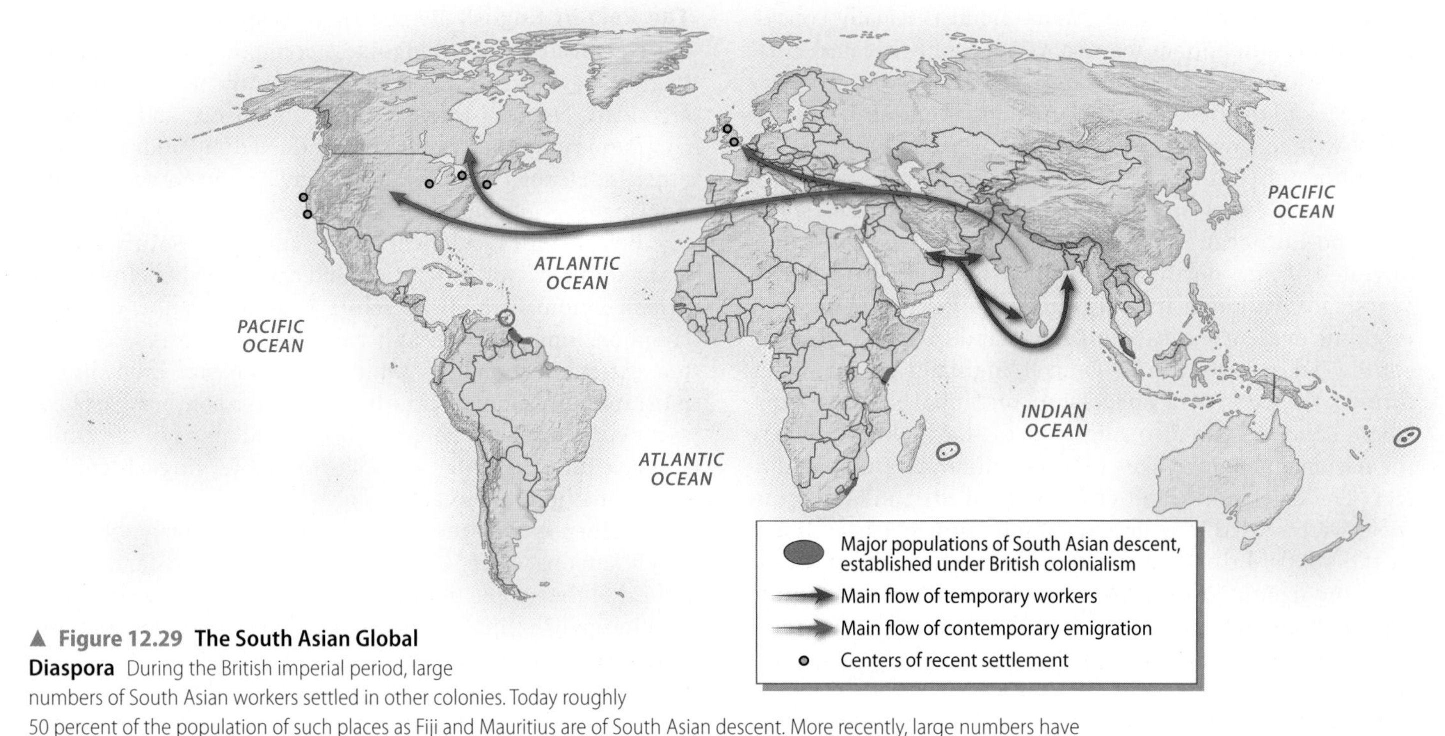

▲ **Figure 12.29 The South Asian Global Diaspora** During the British imperial period, large numbers of South Asian workers settled in other colonies. Today roughly 50 percent of the population of such places as Fiji and Mauritius are of South Asian descent. More recently, large numbers have settled, and are still settling, in Europe (particularly Britain) and North America. Large numbers of temporary workers, both laborers and professionals, are employed in the wealthy oil-producing countries of the Persian Gulf.

professionals, making Indian-Americans one of the country's wealthiest ethnic groups.

Globalization and Cultural Tensions In South Asia itself, the globalization of culture has brought tensions as severe as those felt anywhere in the world. Traditional Hindu and Muslim religious customs frown on any overt display of sexuality—a staple feature of global popular culture. Even though romance is a recurrent theme in the often melodramatic Bollywood films of Mumbai, kissing is considered risqué. Religious leaders thus often criticize Western films and television shows as being immoral. Although India is a relatively free country, both the national government and the state governments periodically ban films and books that are considered too sexual. In 2012, for example, the film *The Girl with the Dragon Tattoo* was banned because of its adult scenes.

In the tourism-oriented Indian state of Goa, such cultural tensions are on full display. There German and British sun-worshipers often wear nothing but thong bikini bottoms, whereas Indian women tourists go into the ocean fully clad. Young Indian men, for their part, often simply walk the beach and gawk, good-naturedly, at the outlandish foreigners.

Globalization and Sports The athletic culture of South Asia features an eclectic mix of immensely popular regional sports, increasingly popular global sports, and locally significant styles of martial arts. Rapid economic expansion and population growth provide an ever-larger audience for competition of all kinds. Yet despite the myriad changes convulsing the region, South Asia's most popular sports share their origins in the region's colonial past.

Cricket was introduced to India by sailors from the British East India Company at some point during the early 18th century, with historically attested matches taking place after 1721. The sport steadily grew in popularity over the succeeding decades, and well before 1900 India had become one of the world's elite cricket destinations. Currently, only ten countries can claim full membership in the sport's main international governing body, the International Cricket Council, and four of those—India, Pakistan, Bangladesh, and Sri Lanka—lie in South Asia. In recent years, the public popularity of top cricket stars has come to exceed that of Bollywood film stars.

Two other popular South Asian sports that link the region to British culture are field hockey and badminton. India's national field hockey team leads the world in Olympic gold medal victories with eight, earning its first in 1928. The first set of official badminton rules was laid out in the Indian city of Pune in 1873, representing an evolution of the traditional English game of battledore and shuttlecock and the native Indian game now known as ball badminton. Although elite badminton competitions now tend to be dominated by Chinese, South Korean, and Southeast Asian players, India remains competitive, and the sport continues to be one of the country's most popular pastimes.

Many of the world's most popular sports, such as soccer and basketball, have been slow to make inroads in South Asia. Despite a vast population from which to draw players, India's national soccer team has only "qualified" for the World Cup once—in 1950, all of India's qualifying rivals withdrew from competition, making the Indian team eligible by default. The Indian team then declined to participate in the World Cup, citing

Everyday Globalization

Customer Support Call Centers in India

Beginning in the 1990s and continuing in earnest into the 2000s, consumers throughout the English-speaking world were introduced to hundreds of thousands of Indian call center workers. As the high-tech economy expanded, customers increasingly had to request advice and assistance by telephone, turning to such call centers to do so. Western companies eager to reduce their labor costs shifted many of their call centers to India to take advantage of low wages, and Indian companies specializing in such operations emerged to do business for other clients (Figure 12.3.1). The main focus of the Indian call center industry is the city of Bangalore, where a growing concentration of technology firms has lead to the designation "Silicon Valley of India" or "Silicon Plateau."

▼ **Figure 12.3.1 Indian Call Center** With its considerable technical and English-language skills, India has emerged as a major provider of international telephone-based customer services. The call center shown here is run by the Indian firm ITC Infotech in the city of Bangalore.

Challenges to the Business Roughly 350,000 Indians now work in call centers across the country. After years of rapid growth, the number has leveled off as companies are looking elsewhere to fill their customer service needs. The most popular new destination is now the Philippines, where a labor force speaking less-heavily accented English and more closely immersed in Western culture makes the area a more attractive location for call centers, despite marginally higher wages. Still, call centers represent an important and relatively attractive source of employment in India.

The drawbacks of working in Indian call centers are well known. Working nights and sleeping during the daytime to match the schedules of Americans takes its toll on workers. According to one study, 81 percent of workers show signs of depression, and a quarter use illegal drugs. Nevertheless, call center jobs are heavily sought after, especially by young Indian men. For them, a job at a call center represents independence and a path from poverty to a comfortable lifestyle. For many, call center jobs are almost a prerequisite for marriage and life away from one's parents.

Cultural Translation Because of the cultural gap that exists between Indians and Americans, many Indians attend specialized schools and courses geared toward call center employment. Most training aims to help students mimic American accents and teach them how to handle various call situations adroitly. Many prospective employees take expensive "accent neutralization classes" designed to eliminate "mother-tongue influences" in speaking English. Such students are often forced to repeat simple syllables, such as "pa," for up to a half hour at a time to help them lose their Indian accents. Through such educational efforts, as well as through the day-to-day phone conversations that workers must engage in, call centers help stimulate globalization in personal as well as corporate ways.

travel expenses. Bangladesh, Pakistan, and Sri Lanka have never qualified, though the Pakistani team has traditionally performed well at the South Asian games. All signs indicate that interest in soccer among South Asians is currently growing.

Contemporary popular culture in South Asia thus reveals global linkages as well as divisions. The same tensions can be seen, and in much stronger form, in the region's geopolitical framework.

REVIEW

12.7 Why has religion become such a contentious issue in South Asia over the past several decades?

12.8 How have India and Pakistan, the two largest countries of South Asia, tried to foster national unity in the face of ethnic and linguistic fragmentation?

12.9 What cultural features were spread by British imperialism in South Asia?

Geopolitical Framework: A Deeply Divided Region

Before the 1800s, South Asia had never been politically united. Although a few empires covered most of the subcontinent at various times, none spanned its entire extent. Whatever unity the region had was cultural, not political. The British, however, brought the region into a single political system by the middle of the 19th century. Independence in 1947 witnessed the traumatic separation of Pakistan from India; in 1971, Pakistan itself was divided, with the independence of Bangladesh, formerly East Pakistan. Serious internal tensions, moreover, began to rise in several South Asian countries in the late twentieth century, although tensions eased in many areas in the new millennium (Figure 12.30). The most important geopolitical issue is the continuing tension between Pakistan and India, both of which are nuclear powers.

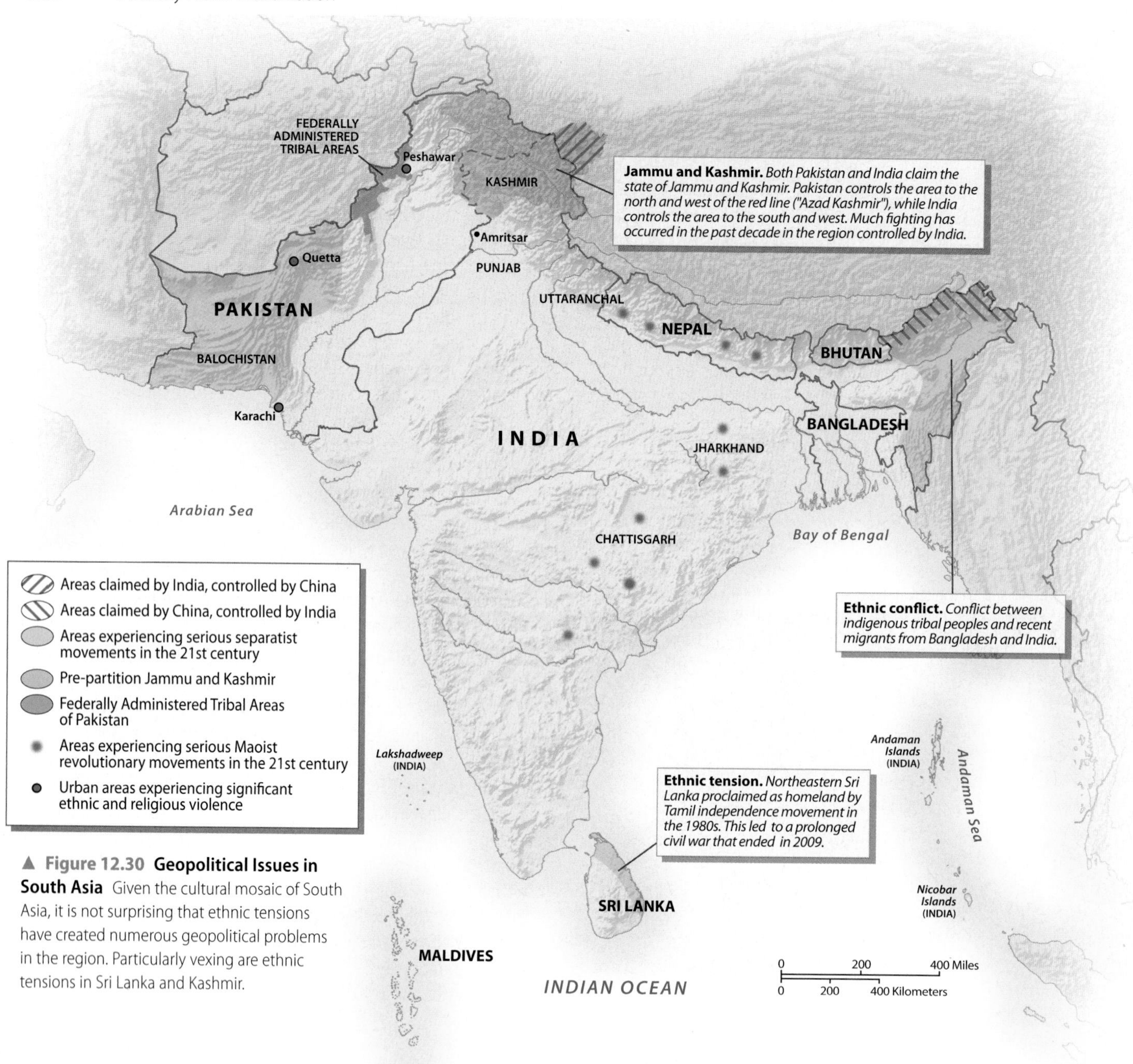

▲ **Figure 12.30 Geopolitical Issues in South Asia** Given the cultural mosaic of South Asia, it is not surprising that ethnic tensions have created numerous geopolitical problems in the region. Particularly vexing are ethnic tensions in Sri Lanka and Kashmir.

South Asia Before and After Independence in 1947

During the 1500s, when Europeans began to arrive on the coasts of South Asia, most of the northern subcontinent came under the power of the Mughal Empire, a powerful Muslim state ruled by people of Central Asian descent (Figure 12.31). Southern India remained under the control of a Hindu kingdom called Vijayanagara. European merchants, keen to obtain spices, textiles, and other Indian products, established a series of coastal trading posts. The Mughals and other South Asian rulers were little concerned with the growing European naval power, as their own focus was the control of land. The Portuguese carved out an enclave in Goa on the west coast, while the Dutch gained control over much of Sri Lanka in the 1600s, but neither was a threat to the Mughals.

The Mughal Empire grew stronger in the 1600s, whereas Hindu power declined until it was limited to the peninsula's extreme south. In the early 1700s, however, the Mughal Empire weakened rapidly. Several contending states—some ruled by Muslims, others by Hindus, and one by Sikhs—emerged in former Mughal territories. As a result, the 18th century in South Asia was a time of political and military turmoil.

The British Conquest These unsettled conditions provided an opening for European imperialism. The British and French, having largely displaced the Dutch and Portuguese, competed for trading posts. Because Indian cotton textiles were the best in the world prior to the Industrial Revolution, British and French merchants wanted to obtain huge quantities for their global trading networks. With Britain's overwhelming victory over France in the Seven Years' War (1756–1763), France was reduced to a few marginal coastal possessions in southern

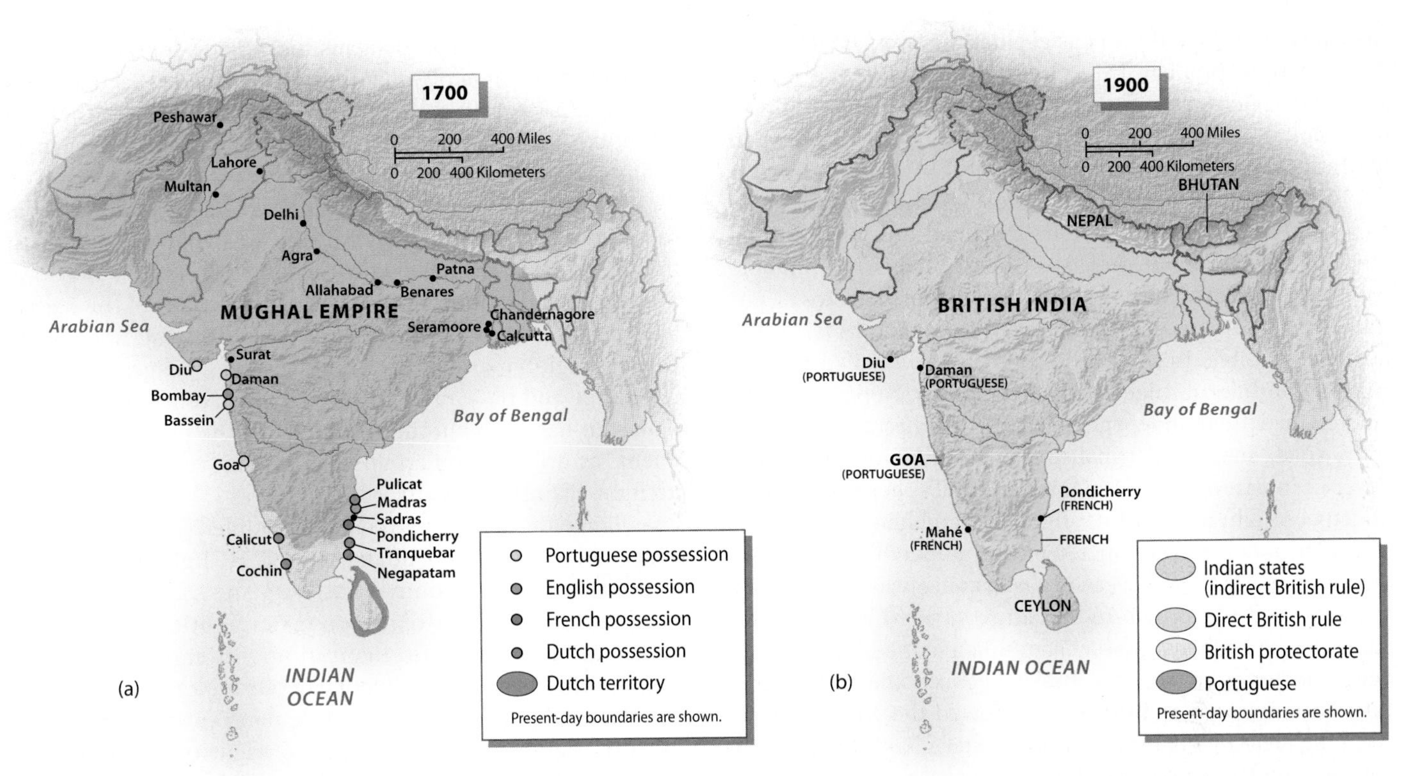

▲ **Figure 12.31 Geopolitical Change** (a) At the onset of European colonialism before 1700, much of South Asia was dominated by the powerful Mughal Empire. (b) Under Britain, the wealthiest parts of the region were ruled directly, but other lands remained under the partial authority of indigenous rulers. Independence for the region came after 1947, when the British abandoned their extensive colonial territory. (c) Bangladesh, formerly East Pakistan, gained its independence at the end of 1971 after a short struggle against centralized Pakistani rule from the west.

India. Britain, or more specifically the **British East India Company**, the private firm that acted as an arm of the British government, now monopolized overseas trade in the area and began to stake out a South Asian empire of its own.

The company's usual method was to make strategic alliances with Indian states to defeat the latter's enemies, most of whose territories it would then grab for itself. As time passed, its army, largely composed of South Asian mercenaries, grew ever more powerful. Several Indian states put up strong resistance, but none could ultimately resist the immense resources of the East India Company. Valuable local allies, as well as a few former enemies, were allowed to remain in power, provided that they no longer threatened British interests. The territories of these indigenous states, however, were gradually whittled back, and British advisors increasingly dictated their policies.

From Company Control to British Colony The continuing reduction in size of the Indian states, coupled with the growing arrogance of British officials, led to a rebellion in 1857 across much of South Asia. When this uprising (called the *Sepoy Mutiny* by the British) was finally crushed, a new political order was implemented. South Asia was now to be ruled by the British government, with the monarch of England serving as its head of state.

Until 1947, the British government maintained direct control over South Asia's most productive and densely populated areas, including virtually the entire Indus–Ganges Valley and most of the coastal plains. The British also ruled Sri Lanka, having supplanted the Dutch in the 1700s. The major areas of indirect rule, where Indian rulers retained their princely states under British advisors, were in Rajasthan, the uplands of central India, and southern Kerala and along the frontiers (see Figure 12.31). The British administered this vast empire through three coastal cities that were largely their own creation: Bombay (Mumbai), Madras (Chennai), and, above all, Calcutta (Kolkata). In 1911, they began building a new capital in New Delhi, near the strategic divide between the Indus and Ganges drainage systems.

While the political geography of British India stabilized after 1857, the empire's frontiers remained unsettled. British officials worried about threats to their immensely profitable colony, particularly from the Russians advancing across Central Asia. In response, they attempted to expand as far to the north as possible. In some cases this merely entailed making alliances with local rulers. In such a manner, Nepal and Bhutan retained their independence. In the extreme northeast, some small states and tribal territories, most of which had never been part of the South Asian cultural sphere, were more directly brought into the British Empire, and hence into India. A similar policy was conducted on the vulnerable northwestern frontier. Here, however, local resistance was much more effective, and the British-Indian army suffered defeat at the hands of the Afghans. Afghanistan thus retained its independence, forming an effective buffer between the British and Russian empires. The British also allowed the tribal Pashto-speaking areas of what is now northwestern Pakistan to retain almost complete autonomy, thus forming a secondary buffer.

Independence and Partition The framework of British India began to unravel in the early 20th century as the people of South Asia increasingly demanded independence. The British, however, were equally determined to stay, and by the 1920s the region was embroiled in massive political protests.

The rising nationalist movement's leaders faced a major dilemma in attempting to organize a potentially independent regime. Many leaders—including Mohandas Gandhi, the main figure of Indian independence—favored a unified state that would encompass all British territories in mainland South Asia. However, most Muslim leaders feared that a unified India would leave their people in a vulnerable position. They therefore argued for the division of British India into two new countries: a Hindu-majority India and a Muslim-majority Pakistan. One problem with this idea was that in several parts of northern South Asia, Muslims and Hindus were settled in roughly equal proportions. A more significant obstacle was the fact that the areas of clear Muslim majority were located on opposite sides of the subcontinent, in present-day Pakistan and Bangladesh.

No longer able to maintain their world empire after World War II, the British withdrew from South Asia in 1947. As this occurred, the region was indeed divided into two countries: India and Pakistan. Partition itself was a horrific event, with fighting between Hindus and Muslims resulting in the death of hundreds of thousands of people. Roughly 7 million Hindus and Sikhs fled Pakistan, to be replaced by roughly 7 million Muslims fleeing India.

The Pakistan that emerged from partition was for several decades a clumsy two-part country, with its western section in the Indus Valley and its eastern portion in the Ganges Delta. The Bengalis, occupying the poorer eastern section, complained that they were treated as second-class citizens. In 1971, they launched a rebellion and, with the help of India, quickly prevailed. Bangladesh then emerged as a new country.

This second partition did not solve Pakistan's problems, however, as it remained politically unstable and prone to military rule. Pakistan retained the British policy of allowing almost full autonomy to the Pashtun tribes of the Federally Administered Tribal Areas of the northwest, a relatively lawless region marked by clan fighting and vengeance feuds. The large and poor province of Balochistan in southwestern Pakistan has been another problem for the national government, as a long-simmering separatist movement continues to fight against Pakistan's military forces.

Bangladesh has also had a troubled political career since achieving independence in 1971. High levels of corruption, growing Islamic radicalism, and street-level fighting between members of its two major political parties have damaged its democratic institutions. In early 2013, hundreds of thousands of radical Islamists took to the streets to challenge the government and demand laws against blasphemy (Figure 12.32). In response, security forces cracked down on the Islamist group Hefajat-e-Islam, killing dozens of its supporters in an episode described by some as a massacre.

Geopolitical Structure of India With independence in 1947, the leaders of India, committed to democracy, faced a major challenge in organizing such a large and culturally diverse country. They decided to chart a middle ground between centralization and local autonomy. India itself was thus organized as a **federal state**, with a significant amount of power given to its individual states. The national government, however, retained full control over foreign affairs and a large degree of economic authority.

Following independence, India's constituent states were reorganized to match the country's linguistic geography. The idea was that each major language group should have its own

▼ **Figure 12.32 Anti-blasphemy Protests in Bangladesh** Muslim activists in Bangladesh want the country to enact laws against blasphemy, similar to those in place in Pakistan. Protestors demanding such a law are shown here blocking a main road in the capital city of Dhaka in May, 2013

state (with the massive Hindi-speaking population having several), hence a degree of political and cultural autonomy. Yet only the largest groups received their own territories, which has led to recurring demands from smaller groups that have felt politically excluded. Over time, several new states have been added to the map. Goa, after the Portuguese were forced out in 1961, became a separate state in 1987, over the objection of its large northern neighbor of Maharashtra. In 2000, three new states were added (Jharkand, Uttaranchal, and Chhattisgarh), and in 2013 the government agreed that a state called Telangana would be carved out of the northwestern districts of Andhra Pradesh. More than a dozen additional new states have been proposed elsewhere in India.

Ethnic Conflicts and Tensions in South Asia

The movement for new states in India has been largely rooted in ethnic tensions. Unfortunately, violent ethnic conflicts persist in many parts of South Asia. Of these conflicts, the most complex—and perilous—is in Kashmir.

Kashmir Relations between India and Pakistan were hostile from the start, and the situation in Kashmir has kept the conflict burning (Figure 12.33). During the British period, Kashmir was a large state with a primarily Muslim core joined to a Hindu district in the south (Jammu) and a Tibetan Buddhist district in the far northeast (Ladakh). Kashmir was then ruled by a Hindu **maharaja**, or a king subject to British advisors. During partition, Kashmir came under severe pressure from both India and Pakistan. Troops linked to Pakistan gained control of western and much of northern Kashmir, at which point the maharaja opted for union with India. India thus retained the core areas of the state, but neither country would accept the other's control over any portion of Kashmir. As a result, India and Pakistan have fought several inconclusive wars over the issue.

The Indo-Pakistani boundary has remained fixed, but the struggle in Kashmir has intensified, flaming out into an open insurgency in 1989. Although some Kashmiris would like to join their homeland to Pakistan, a 2007 poll indicated that 87 percent of the people of Kashmir Valley preferred independence. Opposition to Kashmir's independence is one of the few things on which India and Pakistan can agree. India accuses Pakistan of supporting training camps for Islamist militants and of helping them sneak across the border. As a result, India began the construction of a fortified fence along the so-called line of control that divides Kashmir. The result of all of this activity has been a low-level, but periodically brutal war that has claimed over 40,000 lives and displaced one out of every six inhabitants of Kashmir.

The situation in Kashmir began to improve in 2004 when India and Pakistan initiated a serious round of negotiations. India agreed that Pakistan must play a role in any Kashmir peace settlement, whereas Pakistan's government reduced its support of Muslim militants. By 2012, even tourism showed some signs of revival in the Vale of Kashmir, noted for its lush gardens and orchards nestled among some of

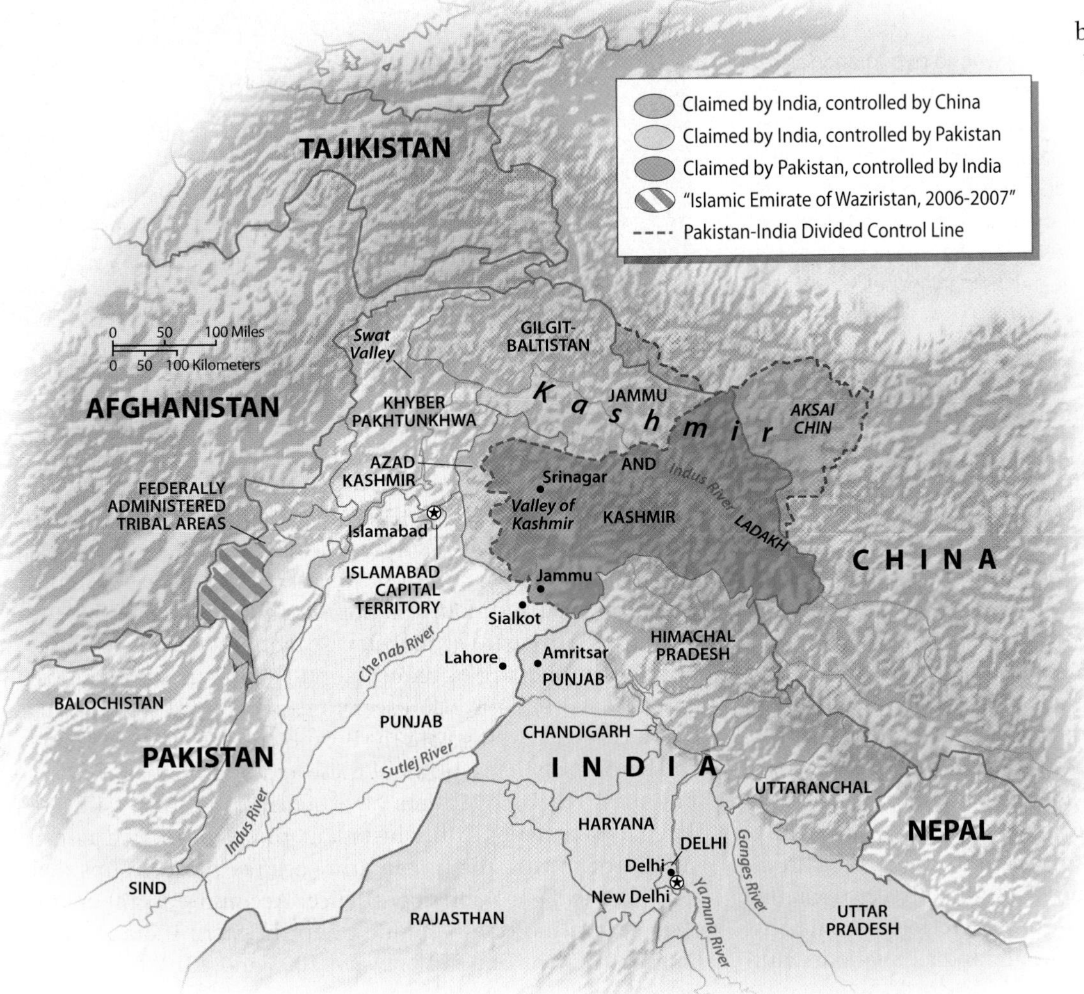

◀ **Figure 12.33 Areas of Conflict in Northern South Asia** Unrest in Kashmir inflames the continually hostile relationships between the two nuclear powers of India and Pakistan. Today many Kashmiris wish to join Pakistan, while many others argue for an independent state. Note also Pakistan's Federally Administered Tribal Areas has also experienced much recent fighting.

People on the Move

Migration from Bangladesh to Assam, India

Migration from Bangladesh to the eastern Indian state of Assam constitutes one of the most controversial political issues facing South Asia. Clear statistics on Bangladeshi migration to Assam are lacking, since much of it is illegal. However, experts agree that hundreds of thousands and probably millions of Bangladeshis currently reside in this troubled Indian state. Cultural differences between Assamese and Bangladeshi immigrants generate misunderstanding and conflict. Most of the Bangladeshi immigrants are Muslim, in contrast to the local Hindu majority, and although the Assamese and Bengali languages are closely related, they are different enough that communication is rendered difficult.

Historical Background and Developments At the heart of the matter is persistent violence in Assam that many Assamese blame on immigration—an accusation that Bangladeshis say is unfair. "Assam Agitation" began in the late 1970s, initially led by the All Assam Students Union. This movement eventually precipitated the 1983 Nellie massacre, in which at least 2191 people, mostly Bangladeshi migrants, lost their lives. In order to protect Assam from "adverse effects upon the political, social, cultural and economic life of the state," India passed the Assam Accords, a memorandum of settlement with certain anti-immigrant groups, in 1981. The accord disenfranchised any illegal immigrants who arrived after 1966 and mandated the expulsion of those who arrived after March 24, 1971.

The logistical difficulties of determining who to deport and how to proceed proved overwhelming for authorities. As a result, relatively little has been done to enforce the accord. Over the years, the issue continued to fester, but by 2012 the government could claim that immigration from Bangladesh had slowed down and that violence between immigrants and local ethnic groups was at an all-time low. Unfortunately, those claims were soon made obsolete, as a fresh wave of violence claimed 77 lives in Assam and caused other riots across India.

The Bodoland Issue The influx of immigrants from Bangladesh is not the only ethnic problem facing Assam. The state contains several indigenous ethnic groups in addition to the majority Assamese, and they, too, have been clamoring for their own rights and recognition. The Bodo people in particular, who speak a language related to Tibetan, have demanded the creation of their own state, which would be called Bodoland (Figure 12.4.1). The Bodos complain that they have been denied educational opportunities and that their lands have seen little infrastructure investment. Like other indigenous groups of Assam, they are particularly concerned about immigration into their homeland. In 2012, as many as 200,000 Bengali-speaking Muslims were forced to flee their homes as a result of attacks by Bodo militias and mobs.

▼ **Figure 12.4.1 Rally for Bodoland** A major source of political controversy in India is the drive to create new states. The student protestors seen in this photograph want to break up the northeastern state of Assam to create a new state that would provide a political unit for the Bodo ethnic group.

the world's most spectacular mountains. Most Kashmiris, however, resent the Indian military presence, while India believes that it must maintain a firm grip on the region to prevent the infiltration of militants from Pakistan. In July 2013, for example, the Indian military claimed that it stopped a major incursion of heavily armed fighters from across the border, killing nine in the process.

India's Northeastern Fringe A more complicated ethnic conflict emerged in the late 1900s in the uplands of India's extreme northeast, particularly in the states of Arunachal Pradesh, Nagaland, Manipur, and portions of Assam. One underlying problem stems from demographic change and cultural collision. Much of this area is still relatively lightly populated, and as a result it has attracted many migrants from Bangladesh and adjacent provinces of India (see *People on the Move: Migration from Bangladesh to Assam, India*). Tensions in the northeast have thus complicated India's relations with Bangladesh. India accuses Bangladesh of allowing separatists sanctuary on its side of the border and objects as well to continuing Bangladeshi emigration. As a result, India is currently building a 2500-mile (4000-km), $1.2 billion fence along the border between the two countries (Figure 12.34).

Through much of the northeast, insurgent groups continue to seek autonomy, if not statehood. After 2000, the Indian government began to invest more money in this troubled region, hoping to reduce popular support for separatist movements. India is also eager to expand trade with Burma and has been working with the Burmese government to secure the border zone. As a result of these and other initiatives, several local rebel movements have signed cease-fires with the Indian government. Other insurgent groups have even agreed to cooperate with the Indian army against those groups that continue to fight for independence. According to the South Asia Terrorism Portal, fighting in northeastern India resulted in roughly 5000 fatalities between 2005 and 2013.

▲ **Figure 12.34 India–Bangladesh Border Barrier** India began building a fence between its territory and that of Bangladesh in 2003 in order to reduce illegal immigration and stop the influx of militants. Here members of the Indian Border Security Force are patrolling a segment of the border barrier. **Q: Why is India so much more concerned about its border with Bangladesh than it is about its borders with Nepal, Bhutan, and Burma?**

Sri Lanka Until recently, interethnic violence in Sri Lanka was also severe. Here the conflict has roots in both religious and linguistic differences. Northern Sri Lanka and parts of its eastern coast are dominated by Hindu Tamils, while the island's majority group is Buddhist in religion and Sinhalese in language (Figure 12.35). Relations between the two communities have historically been fairly good, but tensions mounted soon after independence.

The basic problem is that Sinhalese nationalists favor a unitary government, some of them going so far as to argue that Sri Lanka ought to be defined as a Buddhist state. Most Tamils, on the other hand, support political and cultural autonomy, and they have accused the government of discriminating against them. Overall, levels of education are higher among the Tamils, but the government has favored the Sinhalese majority. In 1983, war erupted when the rebel force known as the **Tamil Tigers** (or the *Liberation Tigers of Tamil Eelam*) attacked the Sri Lankan army. Both extreme Tamil and extreme Sinhalese nationalists remained unwilling to compromise, prolonging the war. Several times Norwegian-backed efforts to reach a peace settlement seemed promising, but all ended in failure

The Sri Lankan conflict intensified in March 2007, when the Tamil Tigers cobbled together a rudimentary air force and bombed Sri Lanka's main airbase. The government subsequently abandoned negotiations, launching instead an all-out offensive. In May 2009, its military crushingly defeated the Tamil Tiger army, killing the organization's leaders. The defeat of the Tamil Tigers also brought about a humanitarian disaster, as many civilians were killed and over 300,000 were displaced. Sri Lanka is now finally at peace, but ethnic tensions are still pronounced, and allegations of discrimination against Tamils persist. In March 2013, the United Nations Human Rights Council passed a critical resolution urging Sri Lanka to conduct investigations into its own alleged war crimes during the insurgency. Other organizations have accused the Sri Lankan government of prohibiting dissent, especially in the Tamil-speaking areas.

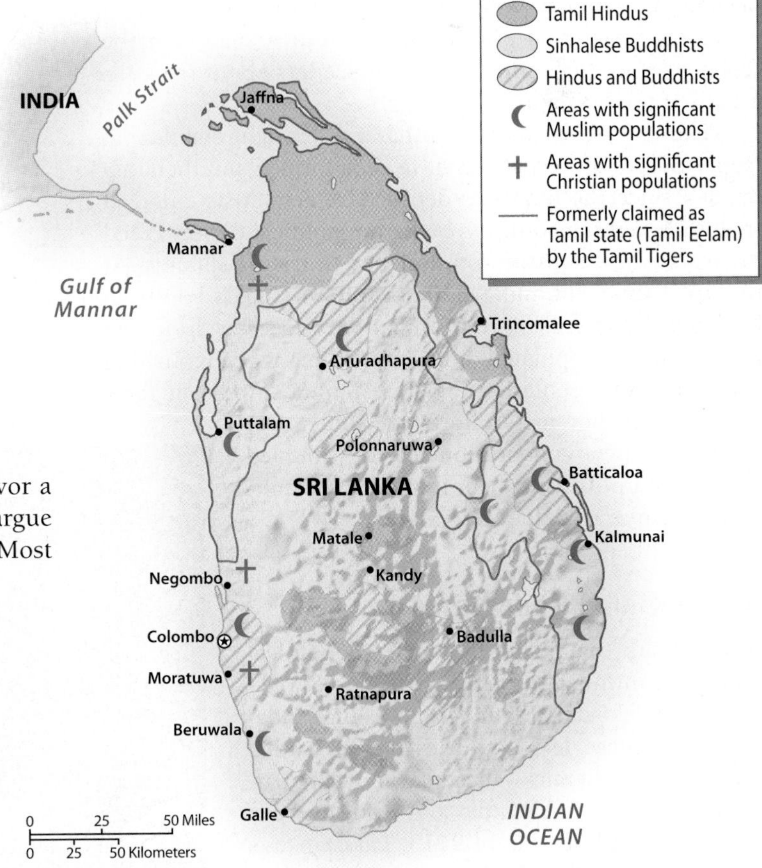

► **Figure 12.35 Ethnic Tensions in Sri Lanka** The majority of Sri Lankans are Sinhalese Buddhist, many of whom maintain that their country should be a Buddhist state. A Tamil-speaking Hindu minority in the northeast strenuously resists this idea. Tamil militants, who waged war against the Sri Lankan government for several decades until defeated in 2009, hoped to create an independent country in their northern homeland. Separate Christian and Muslim populations make for a complex social environment in Sri Lanka.

The Maoist Challenge

Not all of South Asia's current conflicts are rooted in ethnic or religious differences. Poverty and inequality in east-central India, for example, have generated a persistent revolutionary movement that finds inspiration in the former Chinese communist leader Mao Zedong. Tribal people whose lands are experiencing resource exploitation by outsiders seem to form the bulk of the Maoist fighters. Violence associated with this movement resulted in over 1100 deaths in 2009 alone. In May 2013, India was shocked when a convoy of politicians returning from a rally in Chhattisgarh state was attacked by several hundred Maoist insurgents, resulting in at least 29 civilian deaths. Such violence has prevented investment in some of India's least-developed areas, intensifying the underlying economic and social problems that gave rise to the insurgency in the first place.

India's Maoist rebellion is too small to effectively challenge the state, but the same cannot be said in regard to Nepal. Nepalese Maoists, frustrated by the lack of development in rural areas, emerged as a significant force in the 1990s. In 2002, Nepal's king, citing the communist threat, dissolved parliament and took over total control of the country's government. This move only intensified the struggle, however, and within a few years the rebels had gained control of over 70 percent of the country. Both sides committed atrocities, resulting in more than 13,000 deaths. By 2005, Nepal's urban population also turned against the monarchy, launching massive protests in Katmandu. In 2008, the king stepped down, and Nepal became a republic, with the leader of the former Maoist rebels serving as prime minister. But a year later, he resigned after quarreling with army leaders about the still-armed Maoist militias.

The end of the monarchy has not brought stability to Nepal. Several governments have been formed and then disbanded since the king was deposed. A new crisis emerged in 2012 when the country's bickering political parties failed to create a new constitution designed to restore stability. At the same time, the indigenous people of Nepal's southern lowlands, distressed by the migration of settlers from the more densely populated hill country, have been pushing for greater representation and threatening to rebel if their demands are not met. Efforts are under way to create a new constitution for the troubled country, but political turmoil and ethnic conflict continue to delay the process.

International Geopolitics

South Asia's major international geopolitical problem continues to be the struggle between India and Pakistan (Figure 12.36). Since independence, these two countries have regarded each other as enemies. The stakes now are particularly high as both India and Pakistan have nuclear capabilities. As of 2013, Pakistan was estimated to have between 110 and 120 warheads, whereas India is estimated to have between 90 and 110. Relations between the two countries sharply deteriorated in November 2008 when terrorists operating from Pakistan launched a series of coordinated attacks on tourist facilities and public places in Mumbai, killing 173 people. Pakistan responded by investigating the event and arresting several alleged plotters, but many Indians suspect that the attack had the support of certain elements of Pakistan's government and military. Both Indian and Pakistani leaders, however, decided that it would be in their own best interests to reduce tensions. Talks between the two countries resumed in 2011, and in 2012 India agreed to remove restrictions on investments from Pakistan in Indian companies.

Relations with China and the United States During the global Cold War, Pakistan allied itself with the United States, and India remained neutral, while leaning slightly toward the Soviet Union. Such entanglements fell apart with the end of the superpower conflict in the early 1990s. Subsequently, Pakistan forged an informal alliance with China, from which Pakistan obtained sophisticated military equipment. India, on the other hand, has strengthened its ties with the United States during the same period.

China's military connection with Pakistan is rooted in its own animosity toward India. In 1962, China defeated India in a brief war, gaining control over the virtually uninhabited territory of Aksai Chin in northern Kashmir. Growing trade has brought the two countries closer together in some respects, but the fact that China continues to control Aksai Chin and to claim the entire northeastern Indian state of Arunachal Pradesh as its own territory ensures that relations between the two massive countries remain tense. In 2013, India accused China of sending troops over the "line of control" established after the 1962 war in the Himalayas, resulting in a three-week standoff. In the end, China agreed

▼ **Figure 12.36 India–Pakistan Tensions** An Indian officer looks through binoculars in war-torn Kashmir. Relationships between India and Pakistan have remained extremely tense since independence in 1947. Moreover, with both countries now nuclear powers, the fear that border hostilities will escalate into wider warfare has become a nightmarish possibility.

to withdraw its forces in exchange for India demolishing several bunkers in another disputed area.

Pakistan's Complex Geopolitics Pakistan's geopolitical situation became more complex in the aftermath of the attacks in the United States on September 11, 2001. Until that time, Pakistan had strongly supported Afghanistan's Taliban regime (see Chapter 10). After the attack on the World Trade Center and Pentagon (orchestrated from Afghanistan by Osama bin Laden's Al Qaeda terrorist organization), the United States gave Pakistan a stark choice: Either it would assist the United States in its fight against the Taliban and receive in return debt reductions and other forms of aid, or it would lose favor. Pakistan quickly agreed to help, and Pakistan offered both military bases and valuable intelligence to the U.S. military.

Pakistan's decision to help the United States came with large risks. Both Al Qaeda and the Taliban enjoy substantial support among the Pashtun people of northwestern Pakistan. After suffering several military reversals, Pakistan decided to negotiate with these radical Islamists, and on several occasions it gave them virtual control over sizable areas. From these bases, militants have launched numerous attacks on U.S. forces in Afghanistan and have attempted to gain control over broader swaths of Pakistan's territory. The United States has responded by using robotic drone aircraft to attack insurgent leaders, a tactic that has resulted in large numbers of civilian casualties, infuriating most Pakistanis and generating pronounced anti-American sentiment throughout the country. Relations with the United States further deteriorated in May 2011 after U.S. forces launched a raid deep into Pakistan's territory that resulted in the death of Osama bin Laden. An official Pakistani report released in July 2013 claimed that the raid that killed bin Laden was an "American act of war against Pakistan" illustrating "contemptuous disregard of Pakistan's sovereignty, independence and territorial integrity in the arrogant certainty of its unmatched military might." But despite the deep tensions between Pakistan and the United States, the latter continues to support the Pakistani military with weapons and financial aid.

The security crisis in Pakistan has destabilized the country's internal politics as well. In November 2008, Pakistan's president declared a state of emergency, and in the following month, Benazir Bhutto, Pakistan's former prime minister, was assassinated as she campaigned for the 2008 election. After the election, hard-line Islamists expanded their control over remote areas. In April 2009, Pakistan's army launched a major invasion of the Swat Valley, a former vacation zone 100 miles (160 km) northwest of the capital city of Islamabad that had been taken by Islamic militants a few months earlier. As the military advanced, hundreds of thousands of civilians fled, generating a humanitarian catastrophe. Although Pakistan is now a functioning democracy that pulled off a successful general election in 2013, animosity between different political parties, ethnic groups, and religious communities remains pronounced. As a result, a renewal of military rule seems likely to some observers. But many skeptical outsiders fear that the Pakistani military, and particularly its extremely powerful Directorate for Inter-Services Intelligence (ISI), has been infiltrated by radical Islamist elements.

REVIEW

12.10 How have relations between India and Pakistan influenced South Asian geopolitical developments over the past several decades?

12.11 Why has South Asia experienced numerous insurgencies and other political conflicts since the end of British rule in 1947?

12.12 How has the rise of China influenced geopolitical developments in South Asia over the past several decades?

Economic and Social Development: Rapid Growth and Rampant Poverty

South Asia is a land of developmental paradoxes. It is, along with Sub-Saharan Africa, the poorest world region, yet it is also the site of some immense fortunes. South Asia has achieved many world-class scientific and technological accomplishments, but it also has some of the world's highest illiteracy rates. Although South Asia's high-tech businesses are closely integrated with centers of the global information economy, the South Asian economy as a whole was long one of the world's most self-contained and inward looking.

It is difficult to exaggerate South Asia's poverty. Roughly 300 million Indians live below their country's official poverty line, which is set at a very meager level (Figure 12.37).

▼ **Figure 12.37 Poverty in India** India's rampant poverty results in a significant amount of child labor. This 10-year-old boy is moving a large burden of plastic waste by bicycle.

Approximately 20 percent of India's citizens are seriously undernourished, as are 30 percent of the people of Bangladesh. By measures such as infant mortality and average longevity, Nepal is in even worse condition. In urban slums throughout South Asia, rapidly growing populations have little chance of finding housing or basic social services. Estimates indicate that up to half a million South Asian children work as virtual slaves in carpet-weaving workshops and other small-scale factories.

Despite such deep and widespread poverty, South Asia should not be regarded as a zone of uniform misery. India especially has a large and growing middle class, as well as a small, but wealthy upper class. Roughly 250 million Indians are able to purchase such modern consumer items as televisions, motor scooters, and washing machines. India's economy grew from the 1950s to the 1990s at a moderate, but accelerating pace, and by the new millennium it was booming. Even in the global recession years of 2008 and 2009, the Indian economy expanded at a brisk annual rate of roughly 7 percent.

In the summer of 2013, however, a sharp fall in the value of the Indian currency coupled with a decline in consumption resulted in an economic crisis. As a result, many experts think that India will need to undertake substantial economic reforms if it is to continue to experience sustained economic growth. Prospects also vary significantly across the country, as several Indian states have shown marked economic vitality (Figure 12.38) whereas others have seen much less growth and development.

Geographies of Economic Development

After independence, the governments of South Asia attempted to create new economic systems that would benefit

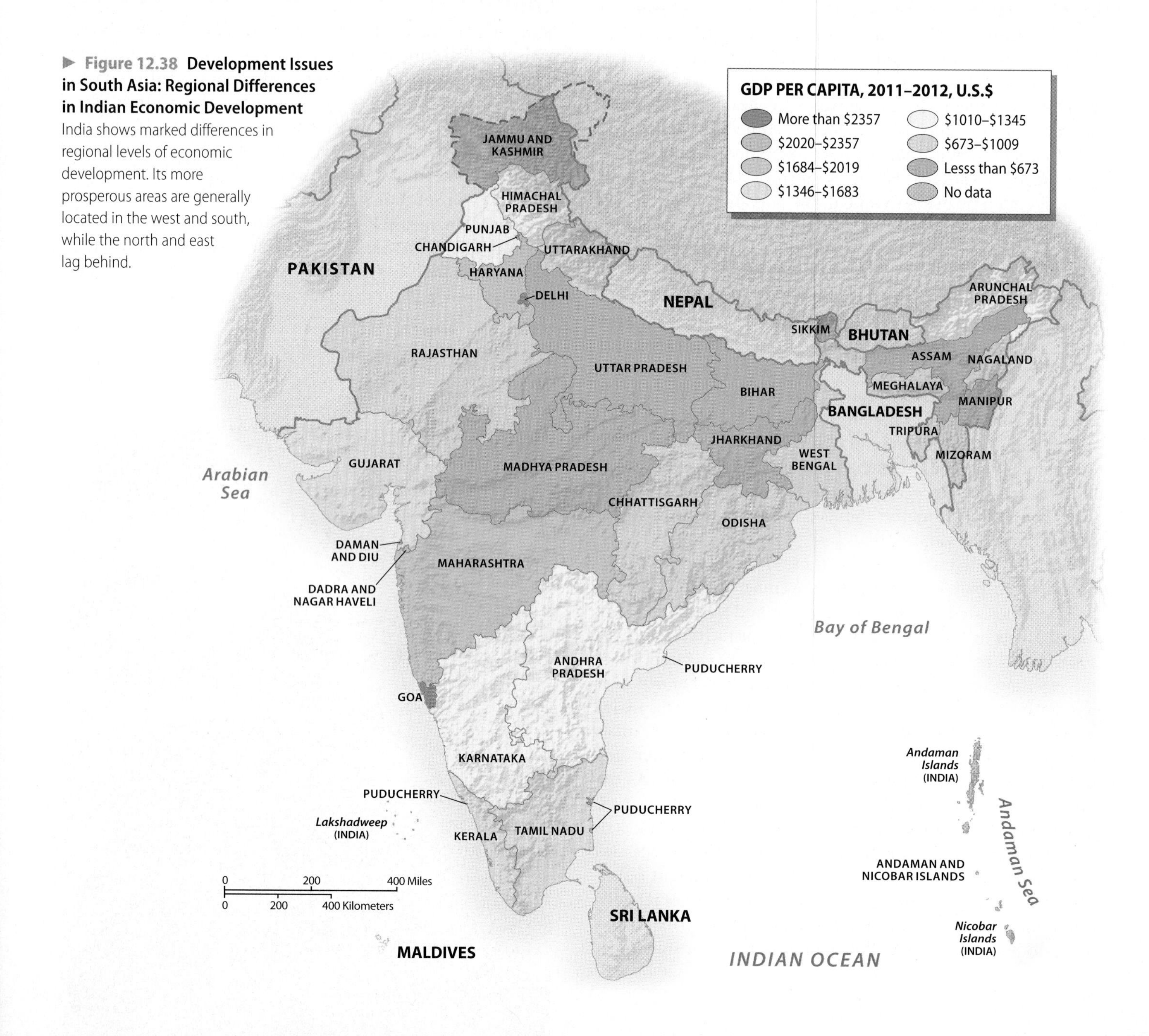

▶ **Figure 12.38 Development Issues in South Asia: Regional Differences in Indian Economic Development** India shows marked differences in regional levels of economic development. Its more prosperous areas are generally located in the west and south, while the north and east lag behind.

Table 12.2 Development Indicators

Country	GNI per capita, PPP 2011	GDP Average Annual % Growth 2000–11	Human Development Index (2011)[1]	Percent Population Living Below $2 a Day	Life Expectancy (2013)[2]	Under Age 5 Mortality Rate (1990)	Under Age 5 Mortality Rate (2011)	Adult Literacy (% ages 15 and older) Male/Female	Gender Inequality Index (2011)[3,1]
Bangladesh	1,940	6.0	.515	76.5	70	139	46	61/52	0.518
Bhutan	5,570	8.5	.538	29.8	67	138	54	65/39	0.464
India	3,640	7.8	.554	68.7	66	114	61	75/51	0.610
Maldives	7,430	7.2	.688	12.2	74	105	11	98/98	0.357
Nepal	1,260	3.9	.463	57.3	68	135	48	73/48	0.485
Pakistan	2,870	4.9	.515	60.2	66	122	72	69/40	0.567
Sri Lanka	5,520	5.8	.715	23.9	74	29	12	93/90	0.402

[1]1United Nations, *Human Development Report, 2013.*

[2]Population Reference Bureau, *World Population Data Sheet, 2013.*

[3]Gender Inequality Index—A composite measure reflecting inequality in achievements between women and men in three dimensions: reproductive health, empowerment and the labor market that ranges between 0 and 1. The higher the number, the greater the inequality.

Source: World Bank, *World Development Indicators, 2013.*

their own people rather than foreign countries or corporations. As in most other parts of the world, planners stressed heavy industry and economic autonomy. Some major gains were realized, but the overall pace of development remained slow. Since the 1990s, governments in the region, and especially that of India, have gradually opened their economies to the global economic system. In the process, core areas of economic development have emerged, surrounded by peripheral areas that have lagged behind, creating landscapes of striking economic disparity (Table 12.2).

The Himalayan Countries Both Nepal and Bhutan are disadvantaged by their rugged terrain and remote locations, as well as by the fact that they remain relatively isolated from modern technology and infrastructure. But such factors are somewhat misleading, especially for Bhutan, because they fail to take into account the fact that many areas in the Himalayas are still subsistence-oriented.

Bhutan has purposely remained somewhat disconnected from the modern world economy, its small population living in a relatively pristine natural environment. Bhutan is so isolationist that it has only recently allowed tourists to enter—provided that they agree to spend substantial amounts of money while in the country. Its government has made the unusual move of downplaying conventional measures of economic development, attempting to substitute "gross national happiness" for "gross national product."

Bhutan is not, however, completely cut off from the rest of the world. It exports substantial amounts of hydroelectric power to India, helping its economy grow by a staggering rate of 19 percent in 2008 and 9.9 percent in 2012. Such growth has brought immigrants; approximately 100,000 Indian laborers are now working on roads and other infrastructure projects in Bhutan. Local resentment against these and other newcomers from India and especially Nepal has resulted in ethnic tensions, forcing up to 60,000 people to flee from Bhutan to refugee camps in Nepal.

Nepal is more heavily populated and suffers much more severe environmental degradation than Bhutan. It also is more closely integrated with the Indian, and ultimately the world, economy, although three-quarters of its people still depend on small-scale agriculture for their livelihoods. Nepal's economy relies heavily on international tourism (Figure 12.39). Tourism has brought some prosperity to a

▼ **Figure 12.39 Tourism in Nepal** Nepal has long been one of the world's main destinations for adventure tourism, although business has suffered in recent years due to the county's political instability. Many tourists in Nepal stay in rustic lodges, several of which have posted advertisements visible here.

few favored locations, but often at the cost of heightened ecological damage. Tourism in Nepal began to suffer, moreover, after the country entered a period of political turmoil in 2002. Remittances from Nepalese workers living abroad currently help sustain the country's fragile economy.

Bangladesh The economic figures for Bangladesh are not as low as those of the Himalayan countries, but they are more indicative of widespread hardship because most people there require cash to meet their basic needs. Partly because of the country's massive population, poverty is extreme and widespread.

Environmental degradation has contributed to Bangladesh's impoverishment, as did the partition of 1947. Most of prepartition Bengal's businesses were located in the western area, which went to India. The division of Bengal tore apart an integrated economic region, much to the detriment of the poorer and mainly rural eastern section. Bangladesh also has suffered because of its agricultural emphasis on jute, a plant that yields tough fibers useful for making ropes and burlap bags. When synthetic materials began to undercut the global jute market, Bangladesh failed to discover any major alternative export crops.

Not all of the economic news coming from Bangladesh is negative. Low-interest credit provided by the internationally noted Grameen Bank has given hope to many poor women in Bangladesh, allowing the emergence of some vibrant small-scale enterprises (Figure 12.40). Such **micro-credit** operations were pioneered in Bangladesh and have more recently spread across many of the underdeveloped parts of the world. By 2000, with the country's birth rate steadily falling, the Bangladeshi economy was finally beginning to grow substantially faster than its population.

The country is internationally competitive in textile and clothing manufacture; in 2009, garment exports earned US$12.3 billion. Bangladesh's clothing industry, however, has a negative side, as wages are low and working conditions are often brutal and extremely unsafe. Such problems were brought to the attention of the world in April 2013 when the eight-story Rana Plaza building, which contained several clothing factories, collapsed, killing 1127 people in one of the most severe industrial catastrophes that the world has ever seen. Since that event, several global clothing retailers have promised to work with the government of Bangladesh to improve working conditions in these factories.

▼ **Figure 12.40 Grameen Bank** The internationally famous Grameen Bank supplies low-interest micro-loans to the women of Bangladesh. This photo shows funds being dispensed at a Grameen branch meeting. **Q: Why is the Grameen Bank focused on providing micro-loans to women rather than to men?**

Pakistan Pakistan also suffered the effects of partition in 1947. But unlike Bangladesh, Pakistan at least inherited a reasonably well-developed urban infrastructure. It also has a productive agricultural sector, as it shares the fertile Punjab with India. In addition, Pakistan boasts a large textile industry, based in part on its huge cotton crop. Several Pakistani cities, moreover, have developed important niches in the global economy; Sialkot, for example, is famous for its high-end sporting goods and surgical implements.

Yet in recent years, Pakistan's economy has been faltering, hampered by a high inflation rate and slow growth. Pakistan is burdened by extremely high levels of military spending, yet at the same time it has experienced growing internal strife. Its electricity supply is woefully inadequate, resulting in long brownouts that often force factories to shut down their production lines. Additionally, a small, but powerful landlord class that pays virtually no taxes to the central government controls many of its best agricultural lands. Unlike India, moreover, Pakistan has not been able to develop a successful IT industry.

Like India, Pakistan has inadequate energy supplies, which has led it to look to both Central Asia and Southwest Asia for fossil fuels. One consequence of this policy has been the construction of a huge new deep-water port at Gwadar near the border with Iran, adjacent to some of the world's most important oil-tanker routes. Built with massive Chinese engineering and financial assistance, the port of Gwadar is managed by a state-owned Chinese firm.

Sri Lanka and the Maldives As you can see in Table 12.2, Sri Lanka's economy is the second most highly developed in South Asia by conventional criteria. Its exports are concentrated in textiles and agricultural products such as rubber and tea. By global standards, however, it is still a poor country. Its progress, moreover, was long undercut by its civil war. In 2009, when the war finally ended, the Sri Lankan stock market posted gains of over 100 percent, reflecting renewed confidence in the country's economy. Sri Lanka hopes to benefit more from the prime location of its port of Colombo, from its high levels of education, and from its great tourism potential. To do so, however, it needs to attract large quantities of foreign investment. Sri Lanka currently depends heavily on the remittances sent home by its workers living abroad, estimated to total some $3 billion a year.

The Maldives is the most prosperous South Asian country based on per capita GNI, but its total economy, like its population, is tiny. Most of its revenues are gained from fishing and international tourism (see *Exploring Global Connections: The Troubled Tourism Industry in the Maldives*). The benefits from the tourist economy, however, flow mainly to the country's small elite population, resulting in large-scale public discontent and political repression. Tourism is also highly vulnerable to international recessions, resulting in an economic crisis in 2008–2009. The Maldivian economy has more recently recovered, but growth remains slow.

India's Less-Developed Areas India's economy, like its population, dwarfs those of the other South Asian countries. Although India's per capita gross national income (GNI) is roughly comparable to that of Pakistan, its total economy is many times larger. As the region's largest country, India also exhibits far more internal variation in economic development than its neighbors. The most basic economic division is between India's more prosperous west and south and its poorer districts in the north and east.

The remote states of India's northeastern fringe rank relatively low on the economic ladder, as measured by per capita GNI, but the prevalence of subsistence economies makes such statistics misleading. More extreme deprivation is found in the densely populated lower and middle Ganges Valley, where cash economies generally prevail. Bihar is India's poorest state by virtually all economic indicators, and neighboring Uttar Pradesh, India's most populous state, also is poverty-stricken. The other states of north-central India, such as Madhya Pradesh, Jharkhand, and Chhattisgarh, have also experienced relatively little economic development. Both Bihar and Uttar Pradesh have fertile soils, but their agricultural systems have not profited as much from the Green Revolution as have those of Punjab. In both states, the caste system is deeply entrenched, tensions between Hindus and Muslims are bitter, and opportunities for most peasants are limited. Ironically, South Asia's wealth was historically concentrated in the fertile lowlands of the Ganges Valley, yet today the area ranks among the poorest parts of an impoverished world region.

Despite its deeply entrenched poverty, north-central India experienced a surprising resurgence beginning around 2008. Since then, Bihar in particular has posted double-digit economic growth in most years, a turnaround partly attributable to reduced corruption. Several other states in the region have also done well, and in 2013 Madhya Pradesh surpassed Bihar as India's economically fastest-growing state. Massive Uttar Pradesh has also been doing relatively well in recent years, with much of its growth attributable to a state-led highway-building boom.

Other states in eastern India such as Odisha (formerly Orissa) and Assam are also quite poor, but the large and important state of West Bengal ranks about average for India as a whole. Some of the worst slums in the world are located in West Bengal's Kolkata (Calcutta). But Kolkata also supports a substantial and well-educated middle class, and it is the site of a sizable industrial complex. For most of the period of Indian independence, West Bengal has been governed by a leftist party that has fostered, with little success, heavy, state-led industry. In a dramatic turnaround in the 1990s, West Bengal's Marxist leaders began to advocate internationalization, encouraging large multinational firms to build new factories in "special economic zones" subjected to low levels of taxation. Such programs have generated substantial opposition that occasionally becomes violent.

Although western India is in general much more prosperous than eastern India, the large western state of Rajasthan ranks well below average. Rajasthan suffers from an arid climate; nowhere else in the world are deserts and semideserts so densely populated. It also is noted for its social conservatism. During the British period, almost all of this large state remained outside the sphere of direct imperial power. Here, in the courts of maharajas, the military and political traditions of Hindu India persisted up until recent times. Rajasthan's rulers not only maintained elaborate courts and fortifications, but also supported many traditional Indian arts. Because of this political and cultural legacy, Rajasthan is one of India's most important destinations for international tourists.

India's Centers of Economic Growth North of Rajasthan lie the Indian states of Punjab and Haryana, showcases of the Green Revolution. Their economies have relied largely on agriculture, but recent investments in food processing and other industries have been substantial. Punjab has the lowest levels of malnutrition in India and the most highly developed infrastructure. Despite its relative prosperity, this region has seen rural unrest in recent years. On Haryana's eastern border lies the capital district of New Delhi, where India's political power and much of its wealth are concentrated.

India's west-central states of Gujarat and Maharashtra are noted for their industrial and financial clout, as well as for their agricultural productivity. Gujarat was one of the first parts of South Asia to experience substantial industrialization, and its textile mills are still among the most productive in the region (Figure 12.41). Gujaratis have long been famed as merchants and overseas traders, and they are disproportionately represented in the **Indian diaspora**, the migration of large numbers of Indians to foreign countries. As a result, cash remittances from these emigrants help to bolster the state's economy. In recent years, Gujarat's government has been deeply devoted to economic development, resulting in extremely rapid economic growth. Critics contend, however, that social development has lagged behind, due in part to Gujarat's focus on business.

The large state of Maharashtra is usually viewed as India's economic pacesetter. The huge city of Mumbai (Bombay) has long been the financial center, media capital, and manufacturing powerhouse of India. According to official figures, the Mumbai metropolitan area accounts for 25 percent of India's industrial output, 40 percent of its maritime trade, and 70 percent of its major financial transactions. Large industrial zones are located in several other cities of Maharashtra, especially Pune and Nagpur. In recent years, Maharashtra's

Exploring Global Connections

The Troubled Tourism Industry in the Maldives

The Maldives, 80 percent of which rises less than one meter above sea level, is a country defined by its relationship to the sea and the tourist dollars generated by its pristine beaches and resorts. Roughly 1 million tourists visit the Maldives each year, generating 30 percent of Maldivian GDP and the overwhelming majority of its tax revenue. Entire Maldivian islands have been turned over entirely to the tourism industry, generally supporting a single resort hotel. Such small islands, averaging about 40 acres (16 hectares) in size, typically have their own coral reefs and shallow, protected lagoons for swimming and snorkeling (Figure 12.5.1). Workers live on other islands, boating in for their shifts.

Political Troubles Lacking other industries and forced to confront poverty, the Maldives relies on tourism as an essential resource. However, tourism is an industry that tends to be very sensitive to political and human rights considerations, and recent events spotlighting the Maldives' poor record on this front pose a challenge to the country's sandy economic engine.

The Maldives functioned as an autocracy until elections began in 2008, and democratic institutions there remain fragile. Former President Mohamed Nasheed, generally admired internationally, was driven from power in 2012 in what he calls a coup, and upon leaving office he urged tourists to "be more aware of what is going on" in Malé, the Maldivian capital. The turmoil surrounding his exit brought with it several calls for tourism boycotts, damaging travel advisories, and as much as $100 million in losses. The Maldives hopes to replace some of the European visitors it appears to be losing with Chinese vacationers, who it perceives as more tolerant of the country's institutions.

Cultural and Environmental Problems Beyond high politics, international outcry focuses particularly on the country's judicial system and its often-draconian laws. A recent case involving a teenage rape victim who was sentenced to 100 lashes for premarital sex generated a petition calling for a boycott of the Maldivian tourist industry until more steps are taken to protect the innocent. As of April 2013, over 1.87 million people had signed the petition. Attention has also been given to the country's poor environmental record. Though perhaps to be expected, the Maldives struggles to safely dispose of waste generated by residents and tourists and has dealt with the situation by turning one of its islands near the densely packed capital of Malé —Thilafushi—into what amounts to a burning hellscape of toxic waste. To top it all off, the rising sea level associated with global warming could result in the disappearance of the Maldivian tourist islands by the end of the century—and much of the rest of the country as well.

Google Earth
Virtual Tour Video

http://goo.gl/Fy8E5M

1. What are some of the reasons why most tourism in the Maldives takes place on tiny islands, such as the one shown in this photograph?
2. Why do you think the government of the Maldives focuses so much on tourism as a source of foreign revenue?

▼ **Figure 12.5.1 Tourist Island in the Maldives** Like most tourism facilities in the Maldives, Cocoa Island Resort, shown in this photograph, caters to well-off foreign visitors. The vacation experience in the Maldives usually focuses more on the sea, and the beach, than on the land, as most tourist islands are extremely small.

▼ **Figure 12.41 Textile Factory in Gujarat** The western Indian state of Gujarat is one of India's main manufacturing centers. This modern cotton mill is in the city of Ahmedabad.

economy has grown more quickly than those of most other Indian states, reinforcing its primacy. Its per capita level of economic production is now roughly 50 percent greater than that of India as a whole.

The center of India's fast-growing high-technology sector lies farther to the south, especially in Karnataka's capital of Bangalore, which in 2007 changed its name to Bengaluru. The Indian government selected the upland Bengaluru area for its fledgling aviation industry in the 1950s. Other technologically sophisticated ventures soon followed. In the 1980s and 1990s, a quickly growing computer software industry emerged, earning Bengaluru the label of "Silicon Plateau." In the 1980s, growth was spurred by the investments of U.S. and other foreign corporations eager to hire relatively inexpensive Indian technical talent. Since the 1990s, these multinational companies have been joined by a rapidly expanding group of locally owned firms. Biotechnology is also thriving. Unfortunately, Bengaluru's rapid growth has stretched the city's infrastructure to the breaking point. Roads are commonly jammed, electricity supplies are inadequate, and many parts of the city can count on only three hours of running water a day.

Partly because of Bengaluru's problems, other cities in southern India have recently emerged as rival high-tech centers. Hyderabad in Andhra Pradesh, often called "Cyberabad," is well known for its IT and pharmaceutical firms, as well as its film industry, India's second largest. Chennai (Madras) in Tamil Nadu, recently voted as having the highest quality of life among India's major cities, is noted for its software production, as well as its financial services and automobile industries.

India has proved especially competitive in software because software development does not require a sophisticated infrastructure; computer code can be exported via wireless telecommunication systems without the use of modern roads or port facilities. What is necessary, of course, is technical talent, and this India has in abundance. Many Indian social groups are highly committed to education, and India has been a scientific power for decades. With the growth of the software industry, India's brainpower has finally begun to translate into economic gains. Whether such developments can spread benefits throughout the country remains to be seen. Most of India's rural areas are not prospering, and malnutrition remains widespread. What is certain, however, is that IT has tightly linked certain parts of India to the global economy.

▼ **Figure 12.42 South Asia Global Linkages: Exports, Direct Foreign Investment, Internet Uses, Tourism** Despite South Asia's growing global connections, the region as a whole is still relatively self-contained, especially in regard to finance. Internet use remains low, especially in Bangladesh and Nepal.

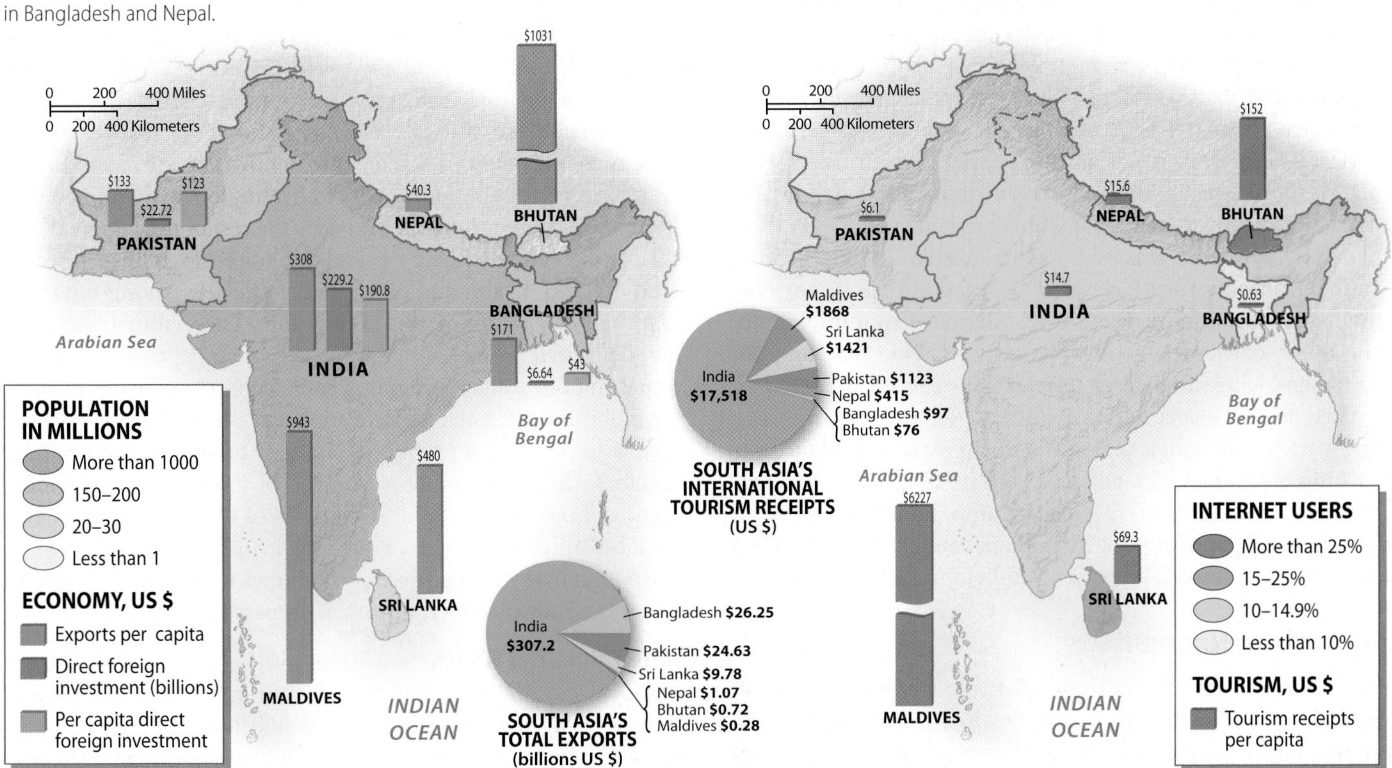

Globalization and South Asia's Economic Future

South Asia is not one of the world's most globalized regions by conventional economic criteria. The volume of foreign trade is relatively small, foreign direct investment is still modest, and (with the exception of the Maldives) international tourists are few (Figure 12.42). But globalization in South Asia is advancing rapidly.

To understand the low globalization indicators for the region, it is necessary to look at its recent economic history. India's postindependence economic policy, like those of other South Asian countries, was based on widespread private ownership combined with high-tariff barriers and governmental control of planning, resource allocation, and certain heavy industrial sectors. This mixed socialist-capitalist system initially brought a fairly rapid development of heavy industry and allowed India to become virtually self-sufficient.

By the 1980s, however, problems with India's economic model were becoming apparent, and frustration was mounting among the business and political elite. Although growth was continual, it remained in most years only a percentage point or two above the rate of population expansion. The percentage of Indians living below the poverty line, moreover, remained virtually constant. At the same time, countries such as China and Thailand were experiencing rapid development after opening their economies to global forces. Many Indian businesspeople were irritated by the governmental regulations that undermined their ability to expand. In the 1980s, foreign indebtedness began to mushroom, putting further pressure on the economy.

In response to these difficulties, the Indian government began to liberalize its economy in 1991. Many regulations were modified and some were eliminated, and the economy was gradually opened to imports and multinational businesses.

Firms in the advanced industrial countries increasingly turned to **outsourcing** many of their labor-intensive technical tasks, contracting them out to Indian companies. Other South Asian countries have followed a somewhat similar path, although generally with less success than India. Pakistan, for example, began to privatize many of its state-owned industries in 1994, and in 2000 it began to turn over the banking sector to private enterprise.

The gradual internationalization and deregulation of the Indian economy has generated substantial opposition. As early as 1984, opposition to foreign investment mounted when a gas leak from a poorly run Union Carbide pesticide factory in Bhopal, India, killed almost 4000 people and injured as many as 500,000. More recently, foreign competitors began to challenge many domestic firms. Cheap manufactured goods from China are seen as an especially serious threat. India has a strong heritage of economic nationalism, stemming from the colonial exploitation it long suffered. Agricultural liberalization, a central agenda item of the World Trade Organization, became an especially contentious issue by the early 2000s because most of India's huge farming sector is not globally competitive. In 2006, mounting opposition led India's government to stall further plans for the privatization of state-controlled economic activities.

Opposition has also grown in the United States and Europe to the outsourcing of jobs to India. Due both to such resistance and to rising wages in India's high-tech centers, Indian companies are now beginning to turn the tables by relocating some of their operations back to North America. Mumbai-based Aegis Communications, for example, recently opened a call center in New York, which focuses on providing medical information.

Although economic globalization is rapidly advancing in India, the country has not seen the levels of direct foreign investment that have recently transformed the Chinese economy. India also devotes a far smaller percentage of its national budget to infrastructure improvements than does China. Because of such investment shortages, India, like the rest of South Asia, does not have an transportation system adequate to meet its economic needs, and its supply of electricity is woefully short. As a result, many observers doubt that it will be able to match the growth rates that China has achieved.

Although India gets most of the media attention, other South Asian countries have also experienced significant economic globalization in recent years. Besides exporting textiles and other consumer goods, Bangladesh, Pakistan, and Sri Lanka send large numbers of their citizens to work abroad, particularly in the Persian Gulf countries. Remittances from foreign workers are Bangladesh's second largest source of income, reaching more than $11 billion a year in 2012. On a per capita basis, remittances are even more important for Sri Lanka. Out of a total population of 21 million, roughly 1.5 million Sri Lankans work abroad, 90 percent of them in the Middle East.

Social Development

South Asia's social indices show relatively low levels of health and education, which is hardly surprising considering the region's poverty. Levels of social well-being vary greatly across the region. As might be expected, people in the more prosperous areas of western India are healthier, live longer, and are better educated, on average, than people in the poorer areas, such as the lower Ganges Valley. Bihar thus stands at the bottom of most social as well as economic measurements, whereas Punjab, Gujarat, and Maharashtra stand near the top. In much of Bihar, where almost half of the population lives below the poverty line, the situation is dire; considering the fact that over a third of the state's teachers are absent on any given day, progress has not been easy. Bihar's recent economic turnaround, however, has prompted guarded optimism about its social development among informed observers.

Several key measures of social welfare are higher in India than in Pakistan. Pakistan, with a literacy rate of only about 56 percent, has done a particularly poor job of educating its people. Still, it is important to realize how much progress has been made. The province of Balochistan, for example, saw its literacy rate increase from 10 percent in 1981 to over 40 percent in 2012. Such gains have not been realized in the Federally Administered Tribal Areas (FATA), where the literacy rate is still below 20 percent overall and under 5 percent for women. FATA's lack of formal education has provided an opening for radical Islamist organizations, which often offer the only options for schooling. In several other areas of social development, Pakistan is still ahead of India. Pakistan probably suffers less malnutrition, and the average Pakistani lives a few more years than the average Indian. Pakistan also has fewer beggars and people living on the streets than does India, partly due to its widespread Islamic charity system.

Several discrepancies stand out when you compare South Asia's map of economic development with its map of social well-being. Portions of India's extreme northeast, for example, have high literacy rates despite their general lack of economic development, owing to the educational efforts of Christian missionaries. In Mizoram, for example, roughly 90 percent of people over the age of seven can read and write. Kolkata (Calcutta) and its immediate environs also stand out as a relatively well-educated area, despite the general distress of the lower Ganges Basin. The most pronounced discrepancies, however, are found in the southern reaches of South Asia. In health, longevity, and education, the extreme south far outpaces the rest of the region.

India has instituted several programs to reduce poverty and enhance health and education, especially among the most deprived segments of the population. These initiatives have been plagued, however, by high levels of corruption. To reduce fraud in both welfare programs and elections, India has created an ambitious personal identification system in which every citizen is to be given a unique ID number linked in a centralized database to such biometric sources of information as photographs, fingerprints, and iris scans. As of July 2013, almost 400 million Indians had been issued their identification numbers. Although many Indians have great hopes for the project, others see it is an unwarranted intrusion on personal privacy.

The Educated South Southern South Asia's relatively high levels of social welfare are clearly visible in Sri Lanka. Considering its meager economic resources and long-lasting civil war, Sri Lanka must be considered one of the world's great success

▲ **Figure 12.43 Education in Kerala** India's southwestern state of Kerala, which has virtually eliminated illiteracy, is South Asia's most highly educated region. It also has the lowest fertility rate in South Asia. Because of this, many argue that women's education and empowerment are the best and most enduring form of contraception.

stories of social development. It demonstrates that a country can achieve significant health and educational gains even in the context of an "undeveloped" economy. Sri Lanka's average longevity of 76 years and its literacy rate of more than 90 percent stand in favorable comparison with comparable figures in some of the world's industrialized countries. The Sri Lankan government has achieved these results by funding universal primary education and inexpensive medical clinics.

On the mainland, Kerala in southwestern India has achieved even more impressive results. Kerala is not a particularly prosperous state. It is extremely crowded, has a high rate of unemployment, and has long had difficulty feeding its population. Kerala's indices of social development, however, are the best in India, comparable to those of Sri Lanka (Figure 12.43). About 90 percent of Kerala's people are literate, the state's average life expectancy is 75 years, and several diseases, such as malaria, have been essentially eliminated. Since Kerala is poorer than Sri Lanka, its social accomplishments are all the more impressive.

Some observers attribute Kerala's social successes to its state policies. For most of the period since Indian independence, Kerala has been led by a socialist party that has stressed mass education and community health care. Although no doubt an important factor, this does not seem to offer a complete explanation. West Bengal, for example, also has a socialist political heritage, but it has not been nearly as successful in its social programs. Kerala's neighboring state of Tamil Nadu, on the other hand, also has made very rapid social progress in recent years despite having a different political environment. Some researchers suggest that one of the key variables for explaining the success of the far south is the relatively high social position of its women.

Gender Relations in South Asia It is often said that South Asian women are accorded a very low social position in both the Hindu and the Muslim traditions. In higher-class families of both religions throughout the Indus–Ganges Basin, women traditionally were secluded to a large degree, and their social relations with men outside the family were severely restricted. Throughout northern India, women traditionally leave their own families shortly after puberty to join the families of their husbands. As outsiders, often in distant villages, young brides have few opportunities, and it is not uncommon for them to be bullied by their parents-in-law. Widows in the higher castes, moreover, are not supposed to remarry and instead are encouraged to go into permanent mourning.

Several social indices show that women in the Indus–Ganges Basin still suffer pronounced discrimination. In Pakistan, Bangladesh, and such Indian states as Rajasthan, Bihar, and Uttar Pradesh, female levels of literacy are much lower than those of males. An even more disturbing statistic is that of gender ratios, the relative proportion of males and females in the population. A 2011 study found that for India as a whole, only 914 girls are born for every 1000 boys and that in parts of northern India the ratio is as low as 824 to 1000. An imbalance of males over females often results from differences in care. In poor families, boys typically receive better nutrition and medical care than do girls, which results in higher rates of survival. An estimated 10 million girls, moreover, have supposedly been lost in northern India over the past 20 years due to sex-selective abortion. Economics plays a major role in this situation. In rural households, boys are usually viewed as an asset because they typically remain with and work for the well-being of their families. In the poorest groups, elderly people (especially widows) subsist largely on what their sons provide. Girls, on the other hand, marry out of their families at an early age and must be provided with a dowry. They are thus seen as an economic liability.

Evidence suggests that the social position of women is improving, especially in the more prosperous parts of western India, where employment opportunities outside the family context are emerging. But even in many of the region's middle-class households, women still suffer major disadvantages. Dowry demands seem to be increasing in many areas, and there have been several well-publicized murders of young brides whose families failed to provide enough goods. In some parts of South Asia, gender ratios are growing even more male-biased now that technology allows gender-selective abortion. Laws against this practice have generally been ignored.

Although the social bias against women across northern South Asia is striking, it is much less evident in southern India and Sri Lanka. In Kerala especially, women have relatively high status, regardless of whether they are Hindus, Muslims, or Christians. Here the overall gender ratio shows the normal pattern, with a slight predominance of females over males. Female literacy is very high in Kerala, which is one reason the state's overall illiteracy rate is so low. Not surprisingly, the high social position of women in southwestern India has deep historical roots. Among the Nairs—Kerala's traditional military and land-holding caste—all inheritance up to the 1920s had to pass through the female line. By 2000, however, census data showed that even in Kerala the number of girls being born relative to the number of boys was declining, showing some evidence of sex-selective abortion.

REVIEW

12.13 How has the economy of India been transformed since the reforms of 1991?

12.14 Why do levels of social and economic development vary so much across South Asia?

12.15 How does the social position of women vary in different regions of South Asia?

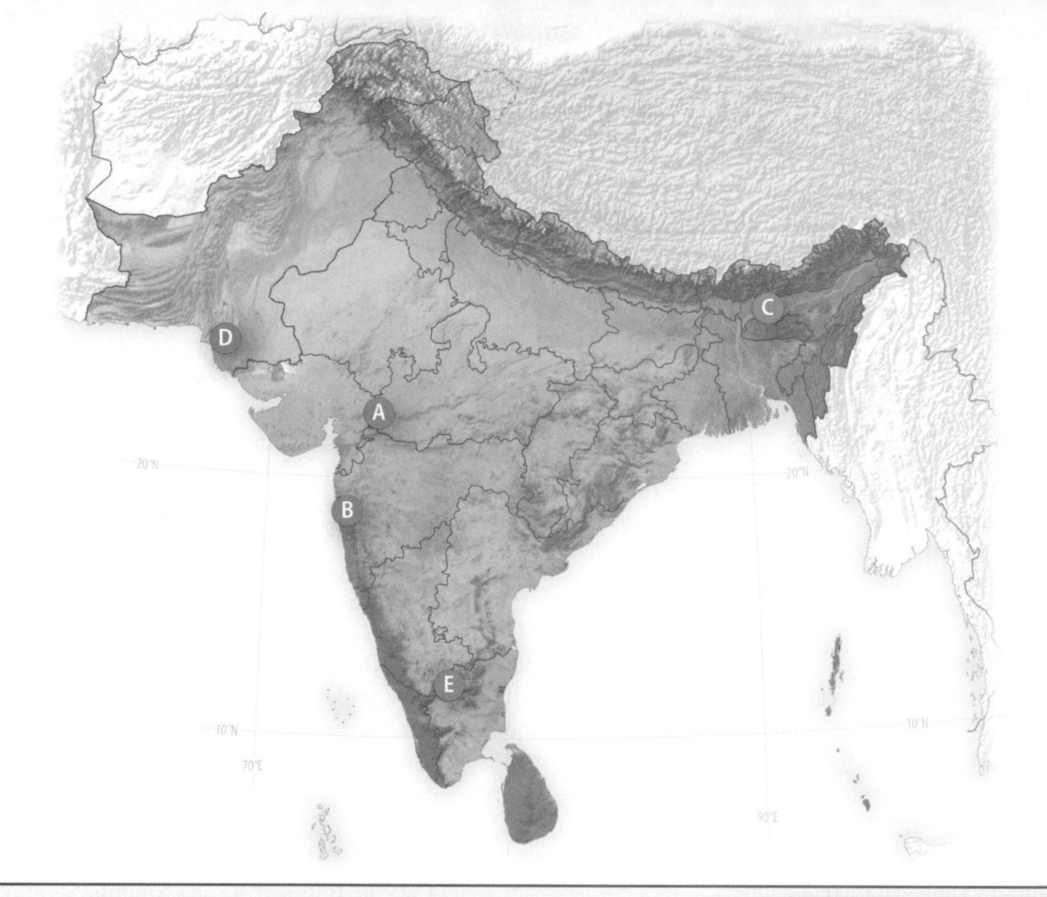

Physical Geography and Environmental Issues

Environmental degradation and instability pose particular problems for South Asia. The region's monsoon climate causes both floods and droughts to be more problematic here than in most other world regions. Rising sea level associated with global climate change directly threatens the low-lying Maldives, and changes in rainfall may play havoc with the monsoon-dependent agricultural systems of India, Pakistan, and Bangladesh.

12.1 Why is India so eager to build massive dams, such as the one visible here?

12.2 Why are some Indians deeply opposed to such dam- and canal-building projects, and what potential problems are encountered when irrigation water is brought into desert areas?

Population and Settlement

Continuing population growth in this already densely populated region demands attention. Although fertility rates have declined rapidly in recent years, Pakistan and northern India cannot easily meet the demands imposed by their expanding populations.

12.3 Why have shantytowns such as the one visible in this photograph grown so rapidly in and around the large cities of India in recent years?

12.4 How might the government of India reduce the problem of shantytown growth and create better living conditions for people residing in such slums?

Key Terms

British East India Company *(p. 581)*
caste system *(p. 569)*
cyclone *(p. 554)*
dalit *(p. 572)*
Dravidian *(p. 574)*
federal state *(p. 582)*
forward capital *(p. 568)*
Green Revolution *(p. 568)*
Hindu nationalism *(p. 569)*
Indian diaspora *(p. 591)*
Indian subcontinent *(p. 550)*
Jainism *(p. 572)*
linguistic nationalism *(p. 577)*
maharaja *(p. 583)*
micro-credit *(p. 590)*
monsoon *(p. 557)*
Mughal Empire (also spelled *Mogul*) *(p. 570)*
orographic rainfall *(p. 557)*
outsourcing *(p. 594)*
salinization *(p. 566)*
Sikhism *(p. 573)*
Tamil Tigers *(p. 585)*

In Review • Chapter 12
South Asia

Cultural Coherence and Diversity

South Asia's diverse cultural heritage, shaped by peoples speaking several dozen languages and following several major religions, makes for a particularly rich social environment. Unfortunately, cultural differences have often translated into political conflicts. Ethnically or religiously based separatist movements have severely challenged the governments of Pakistan, India, and Sri Lanka. In India, moreover, religious strife between Hindus and Muslims persists, whereas in Pakistan and Bangladesh Islamic radicals clash with the state.

12.5 What historical and geographical features help account for the fact that the northeastern part of India, visible in this map detail, has such linguistic and cultural diversity?

12.6 What kinds of problems are associated with the cultural diversity found in this part of India?

Geopolitical Framework

Geopolitical tensions within South Asia are particularly severe, again demanding global attention. The long-standing feud between Pakistan and India escalated dangerously in the late 1990s, leading many observers to conclude that this was the most likely part of the world to experience a nuclear war. Although tensions between the two countries have lessened, the underlying sources of conflict—particularly the struggle in Kashmir—remain unresolved. Pakistan, for its part, is increasingly troubled by religious and ethnic conflict. Its Shiite Muslim and Christian communities have suffered a number of attacks, leading to protests such as the one shown in this photograph.

12.7 What particular features of Karachi, Pakistan's largest city, have led to such intensive ethnic and religious tensions in recent years?

12.8 What policies might the city government of Karachi, or the national government of Pakistan, enact in order to reduce such tensions?

Economic and Social Development

Although South Asia remains one of the poorest parts of the world, much of the region has seen rapid economic expansion in recent years. Many argue that India in particular is well positioned to take advantage of economic globalization. Large segments of its huge labor force are well educated and speak excellent English, the major language of global commerce. But will these global connections help the vast numbers of India's poor or merely the small number of its economic elite? Advocates of free markets and globalization tend to see a bright future, while skeptics more often see growing problems.

12.9 Why is India investing such large amounts of money into a system that will provide unique identification numbers for all of its citizens, linked with such biological information as photographs, fingerprints, and iris scants?

12.10 What are some reasons why this program has been controversial, both in India and abroad?

MasteringGeography™

Looking for additional review and test prep materials? Visit the Study Area in **MasteringGeography™** to enhance your geographic literacy, spatial reasoning skills, and understanding of this chapter's content by accessing a variety of resources, including **MapMaster™** interactive maps, videos, RSS feeds, flashcards, web links, self-study quizzes, and an eText version of *Diversity Amid Globalization*.

Authors' Blogs

Scan to visit the author's blog for chapter updates

www.gad4blog.wordpress.com

Scan to visit the GeoCurrents blog

www.geocurrents.info

13 Southeast Asia

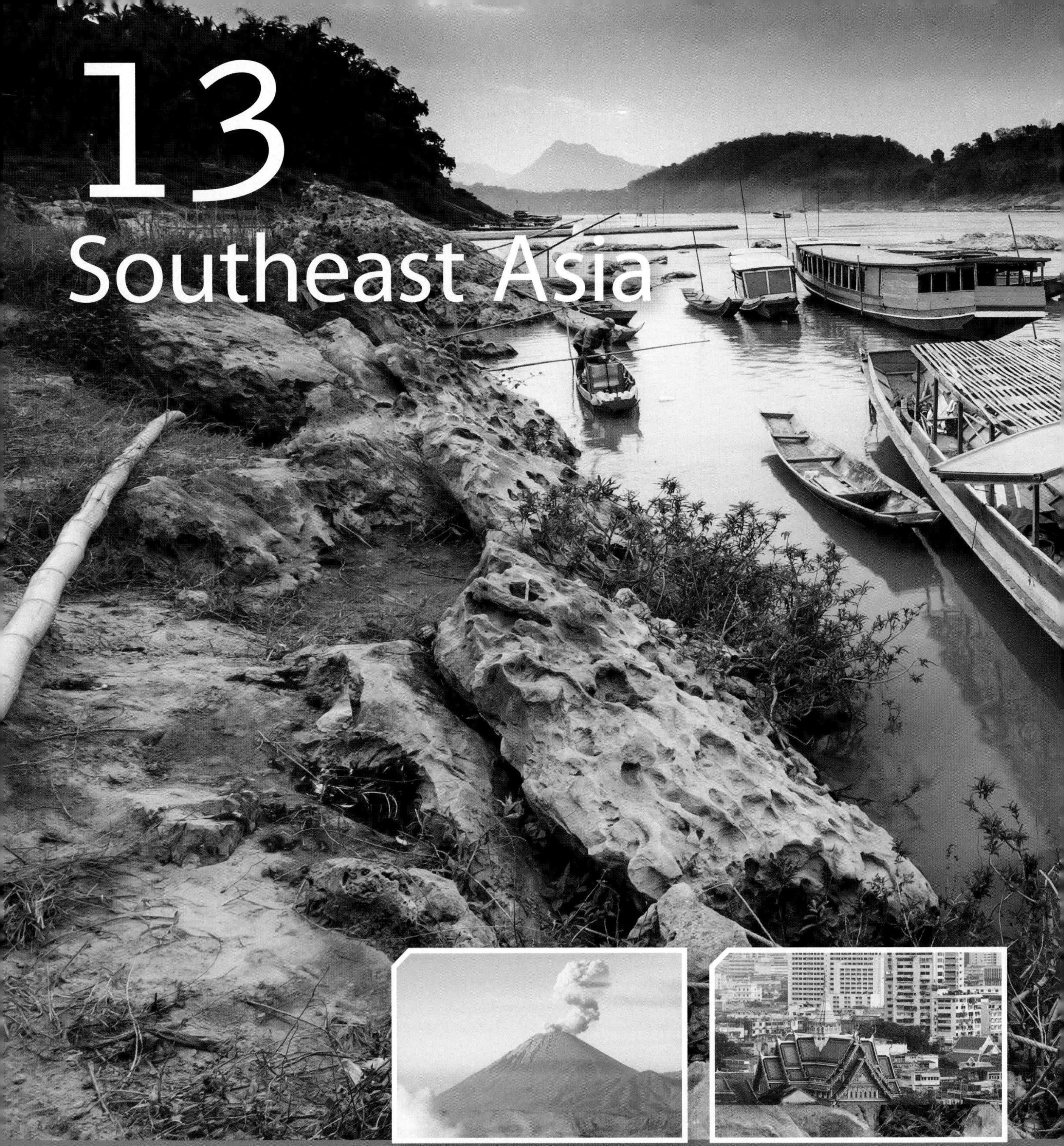

Physical Geography and Environmental Issues

Southeast Asia is divided into a mainland area, noted for its rugged mountains and broad river valleys, and an island region, famed for its rainforests, earthquakes, and volcanoes. Southeast Asian rainforests are vital centers of biological diversity, but they are rapidly diminishing due to commercial logging and agricultural expansion.

Population and Settlement

Southeast Asia has a particularly uneven pattern of population distribution, with some areas experiencing serious crowding and others noted for their sparse settlement.

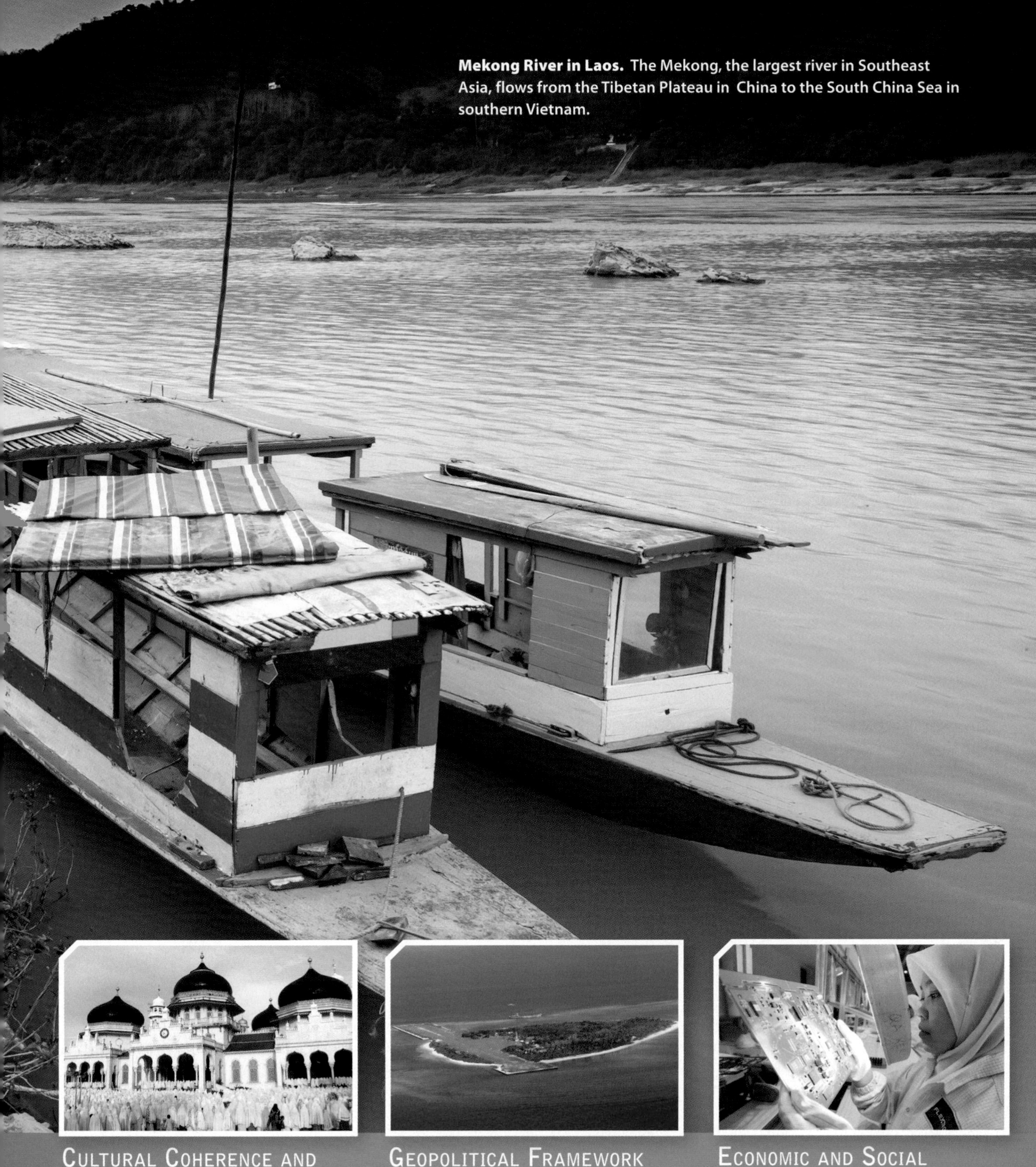

Mekong River in Laos. The Mekong, the largest river in Southeast Asia, flows from the Tibetan Plateau in China to the South China Sea in southern Vietnam.

Cultural Coherence and Diversity

Culturally, Southeast Asia is characterized by much more diversity than coherence, hosting significant areas of Muslim, Buddhist, and Christian religions, as well as a vast number of languages.

Geopolitical Framework

Southeast Asia is one of the most geopolitically unified regions of the world, with all but one of its countries belonging to the Association of Southeast Asian Nations (ASEAN). Ethnic conflicts and maritime tensions with China, however, generate geopolitical problems in several Southeast Asian countries.

Economic and Social Development

Southeast Asia as a whole contains some of the world's most globalized and advanced economies and some of its most isolated and impoverished; it has also experienced marked periods of boom and bust in recent decades.

Rugged, landlocked Laos has long been one of the poorest counties in Asia, with an economy relatively disconnected from the global commercial system. However, Laos is changing rapidly, attracting investment funds from other countries and experiencing an economic boom. In 2012, the Laotian economy expanded at a sizzling rate of 8.3 percent. Much of Laos's recent growth is attributable to its dam-building efforts, which have relied heavily on investments from China. Laos is a mountainous country with heavy precipitation, giving it a large potential for hydroelectricity generation. As demand for electricity in sparsely populated Laos is relatively small, much of the resulting power will be exported to neighboring countries, particularly Thailand. Laos currently plans to build at least 11 major dams by 2020.

Dam-building in Laos, as elsewhere, is very controversial, due both to local environmental impacts and to the fact that large numbers of people are often displaced by the resulting reservoirs. The most contentious dams are those planned for the Mekong, Southeast Asia's largest river. The Xayaburi Dam, currently under construction and slated for opening in 2019, has generated opposition not only from environmental and human rights groups, but also from neighboring countries. Vietnam is concerned that the dam could harm rice production in the Mekong Delta, while Cambodia is worried that it could devastate freshwater fisheries in the lower portions of the river. Some environmentalists have warned that as many as 20 fish species could be wiped out by the project, including the Mekong giant catfish. The world's largest freshwater fish, the Mekong catfish can weigh up to 660 pounds (300 kg).

The Laotian government has responded to such criticism by redesigning the dam to allow some fish migration and to reduce sedimentation, although it has not yet released all of its new plans to the public. But it is determined to proceed with its ambitious dam-building plans, claiming that this is the best way to lift the country out of poverty and to generate the electricity needed by Southeast Asia without producing massive quantities of greenhouse gases. Southeast Asian economies are growing strongly, increasing the region's energy demands at a rapid pace. If you have energy available to sell, you can make a good profit.

Southeast Asia as a whole contains some of the world's most globalized and advanced economies and some of its most isolated and impoverished; it has also experienced marked periods of boom and bust in recent decades.

The Southeast Asian World Region

Southeast Asia (Figure 13.1) consists of 11 countries that vary widely in spatial extent, population, cultural attributes, and levels of economic and social development. Geographically, the region is commonly divided into an Asian mainland zone and the islands, or the insular realm. The mainland includes Burma (Myanmar), Thailand, Cambodia, Laos, and Vietnam. Insular Southeast Asia includes the sizable countries of Indonesia, the Philippines, and Malaysia, as well as the small countries of Singapore, Brunei, and East Timor. Although classified as part of the insular realm because of its cultural and historical background, Malaysia actually splits the difference between mainland and islands. Part of its national territory is on the mainland's Malay Peninsula and part is on the large island of Borneo, some 300 miles (480 km) distant. Borneo also includes the Muslim sultanate of Brunei, a small, but oil-rich island country of 400,000 people covering an area slightly larger than Rhode Island. Singapore is essentially a city-state, occupying a small island just to the south of the Malay Peninsula.

Controversies surround the names of two Southeast Asian countries. Starting in 1989, the government of Burma has insisted that the country's English name is *Myanmar* (officially, the Republic of the Union of Myanmar). Both terms are indigenously used, although in Burmese they would be rendered

LEARNING OBJECTIVES

After reading this chapter you should be able to:

- Identify the key environmental differences between the equatorial belt of insular Southeast Asia and the higher-latitude zone of mainland Southeast Asia.
- Explain how environmental differences influenced human settlement and economic development.
- Describe the driving forces behind deforestation and habitat loss in the different regions of Southeast Asia.
- Explain how the interaction of tectonic plates and the resulting volcanism and seismic activity have influenced Southeast Asian history and development.
- Show how the differences among plantation agriculture, rice growing, and swidden cultivation in Southeast Asia have molded settlement patterns.
- Describe the role of primate cities and other massive urban centers in the development of Southeast Asia.
- Outline the ways in which religions from other parts of the world have spread through Southeast Asia, including how religious diversity has influenced the history of the region.
- Identify the controversies surrounding cultural globalization in Southeast Asia, explaining why some people in the region welcome the process, whereas others resist it.
- Trace the origin and spread of ASEAN and explain how this organization has influenced geopolitical relations in the region.
- Describe the major ethnic conflicts in Southeast Asia, showing why certain countries in the region have such deep problems in this regard.
- Explain why levels of economic and social development vary so widely across the Southeast Asian region.

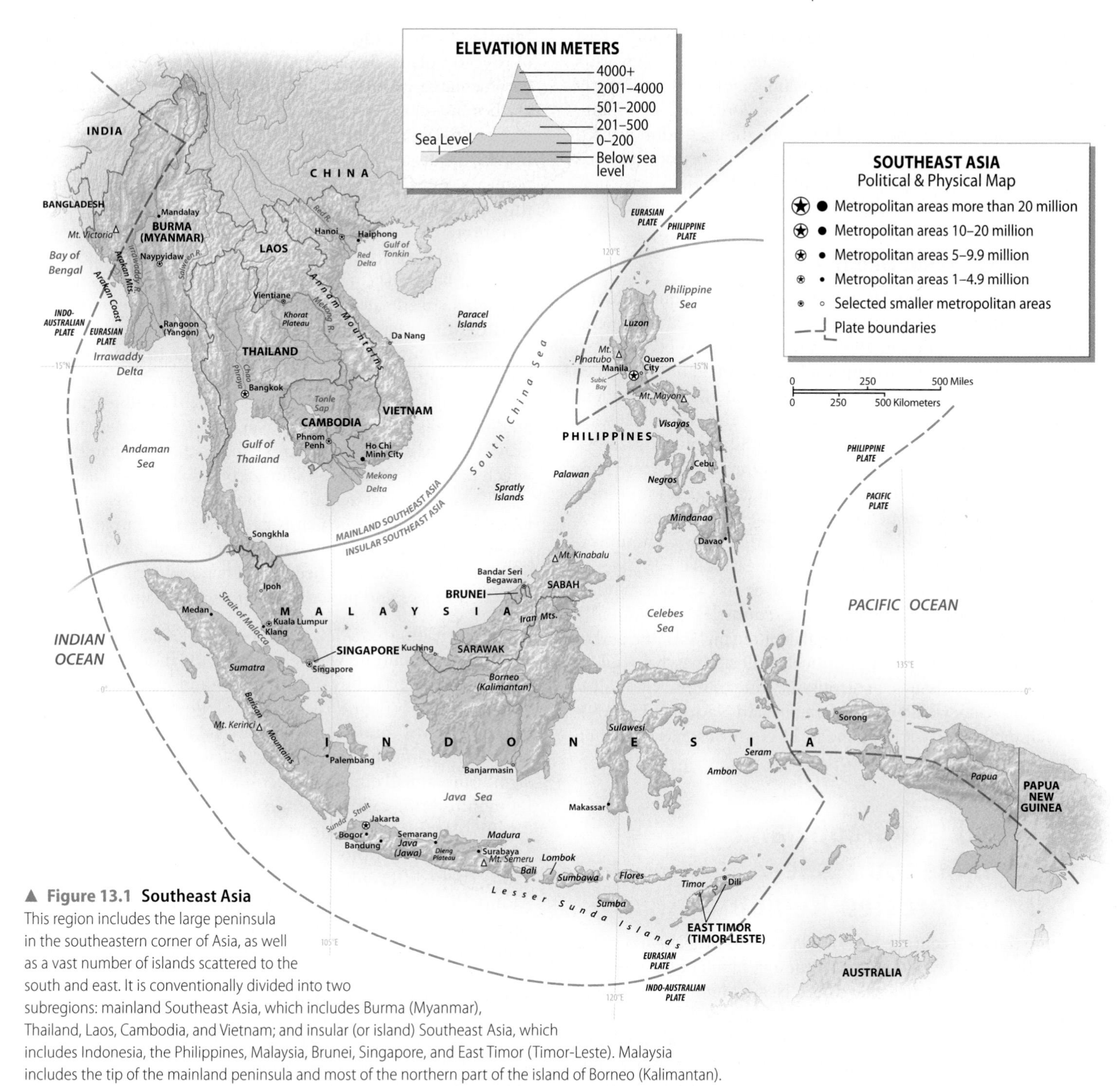

▲ **Figure 13.1 Southeast Asia** This region includes the large peninsula in the southeastern corner of Asia, as well as a vast number of islands scattered to the south and east. It is conventionally divided into two subregions: mainland Southeast Asia, which includes Burma (Myanmar), Thailand, Laos, Cambodia, and Vietnam; and insular (or island) Southeast Asia, which includes Indonesia, the Philippines, Malaysia, Brunei, Singapore, and East Timor (Timor-Leste). Malaysia includes the tip of the mainland peninsula and most of the northern part of the island of Borneo (Kalimantan).

as "Bama" and "Myanma." Owing to the repressive nature of the Burmese government, opposition groups within the country, as well as the governments of the Unites States, the United Kingdom, and Canada, have continued to use the term *Burma*, as does this book. If the government of Burma continues along its current path of reform, we may switch to *Myanmar* in future editions. Less controversial is the official name of East Timor: Timor-Leste. As *Leste* derives from the Portuguese word for "east," most English language sources, including this book, continue to use the more familiar term *East Timor.*

Southeast Asia occupies an important place in discussions of globalization. It includes some of the world's most globally networked countries, such as Singapore, as well as some of the countries most resistant to worldwide economic and cultural forces—most notably Burma. Debates about the benefits and drawbacks of economic globalization have often focused on Southeast Asia. Although human well-being has increased across most of the region, the sweatshops associated with low-cost global production are criticized for their low wages and harsh working conditions. Owing to its high levels of economic integration, Southeast Asia is also highly vulnerable to fluctuations in the global economy.

Southeast Asia's involvement with the larger world is not new. Chinese and especially Indian influences date back many centuries. Later, commercial ties with the Middle East opened the doors to Islam, and today Indonesia is the world's most populous country with a Muslim majority. More recently came the heavy imprint of Western colonialism, as Britain, France, the Netherlands, and the United States administered large Southeast Asian colonies. During this period, national territories were rearranged, populations relocated, and new cities built to serve trade and military needs.

Southeast Asian Geopolitics

Southeast Asia's resources and its strategic location made it a major battlefield during World War II. Yet long after the end of this global conflict in 1945, war continued to be waged in the region. Resistance to imperialism mounted, but even after the colonial regimes were replaced by newly independent countries, Southeast Asia was a zone of contention between the world powers and their different philosophical systems. In Vietnam, Laos, and Cambodia, communist forces, supported by China and the Soviet Union, struggled for control of local territory and people. Opposing this movement were the United States and its allies, concerned that communism would rapidly spread throughout Southeast Asia.

Ironically, while communist forces did prevail in Vietnam, Laos, and Cambodia, these countries subsequently opened their economies to global capitalism. With the end of the Cold War, competing political philosophies have taken a back seat to issues of economic integration. Geopolitically, the **Association of Southeast Asian Nations (ASEAN)**, which includes 10 of the region's 11 countries, has brought a new level of regional cooperation.

Physical Geography and Environmental Issues: A Once-Forested Region

Before the 20th century, Southeast Asia was probably the most heavily forested world region. Although much of the region is still wooded, most areas have been cleared for agriculture and human settlement. The commercial logging of tropical forests, however, remains a controversial issue in several Southeast Asian countries. To understand regional patterns of both forest preservation and destruction, it is necessary to examine differences in climates and landforms across the region.

Patterns of Physical Geography

The physical environments of insular Southeast Asia differ significantly from those of the mainland part of the region. The island belt is mostly situated in the equatorial zone, constituting one of the world's three main zones of tropical rainforest. Mainland Southeast Asia, on the other hand, is mostly located in the tropical wet-and-dry zone, noted for its large seasonal differences in rainfall. Distinguishing between the mainland and the islands is thus one of the keys for understanding the geography of Southeast Asia.

Mainland Environments Mainland Southeast Asia is an area of rugged uplands interspersed with broad lowlands and deltas associated with large rivers. The region's northern boundary lies in a cluster of mountains connected to the highlands of eastern Tibet and south-central China. In the far north of Burma, peaks reach 18,000 feet (5500 meters). From this point, a series of mountain ranges radiates out, extending through western Burma, along the Burma–Thailand border, and through Laos into southern Vietnam.

Several large rivers flow southward out of Tibet into mainland Southeast Asia. The longest is the Mekong, which flows through Laos, Thailand, and Cambodia before entering the South China Sea through an extensive delta in southern Vietnam (Figure 13.2). Second longest is the Irrawaddy, which flows through Burma's central plain before reaching the Andaman Sea. This river also has a large delta. Equally significant are two smaller rivers: the Red River, which forms a sizable and heavily settled delta in northern Vietnam, and the Chao Phraya, which has created the fertile alluvial plain of central Thailand.

The centermost area of mainland Southeast Asia is Thailand's Khorat Plateau, which is neither a rugged upland nor a fertile river valley. This low sandstone plateau averages about 500 feet (175 meters) in height and is noted for its thin soils. Water shortages and periodic droughts also plague the area.

Monsoon Climates Mainland Southeast Asia is affected by the seasonally shifting winds known as the monsoon. The climate is characterized by a distinct hot and rainy season from May to October (Figure 13.3). This is followed by dry, but still generally warm conditions from November to April. Only the central highlands of Vietnam and a few coastal areas receive significant rainfall during this period. In the far north, the winter months bring mild and sometimes rather cool weather.

Two tropical climate regions are found in mainland Southeast Asia. Although both are affected by the monsoon, they differ in the total amount of precipitation. Along the coasts and in the highlands, the tropical monsoon (Am) climate dominates. Rainfall totals for this climate usually register more than 100 inches (250 cm) each year.

▼ **Figure 13.2 Delta Landscape** The Mekong Delta in southern Vietnam encompasses a complex maze of waterways. Extensive tracts of fertile farmland lie between the canals and river channels. The waterways are used for transportation and provide large quantities of fish and other aquatic resources.

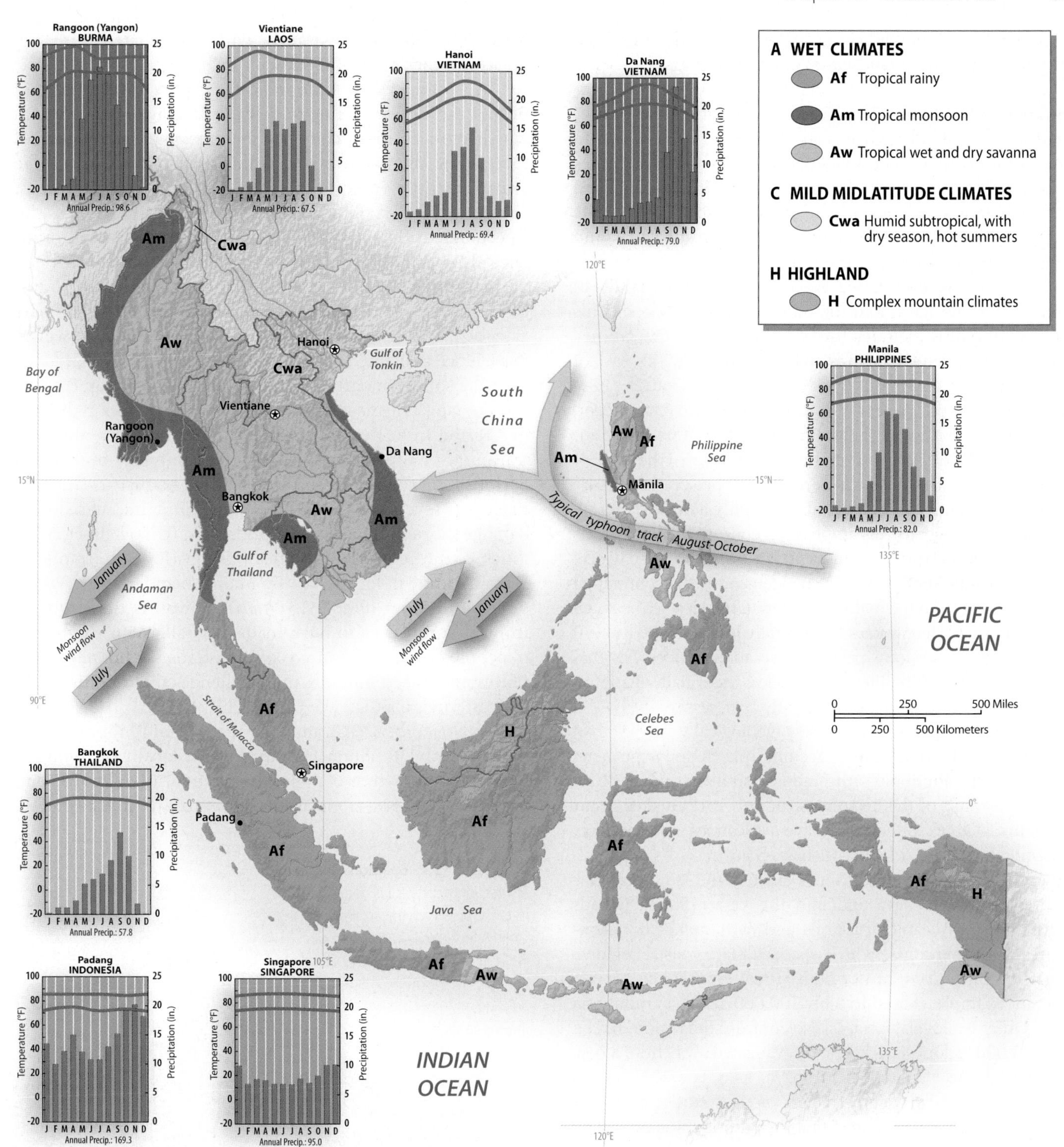

▲ **Figure 13.3 Climate of Southeast Asia** Most of insular Southeast Asia is characterized by the constantly hot and humid climates of the equatorial zone. Mainland Southeast Asia, on the other hand, has the seasonally wet and dry climates of the tropical monsoon and tropical savanna types. Only in the far north are subtropical climates, with relatively cool winters, encountered. The northern half of the region is strongly influenced by the seasonally shifting monsoon winds. Northeastern Southeast Asia—and especially the Philippines—often experiences typhoons from August to October.

The greater portion of mainland Southeast Asia falls into the tropical savanna (Aw) climate type. Here annual rainfall totals are about half as much. In most cases, this can be explained by interior locations sheltered from the oceanic source of moisture by mountain ranges. A good portion of Thailand receives less than 50 inches (125 cm) of annual rainfall. Much of Burma's central Irrawaddy Valley is almost semiarid, with rainfall totals below 30 inches (75 cm). Forests in these areas are especially vulnerable, being easily converted by fire and agriculture into rough landscapes of brush and grass.

Insular Environments The signal feature of insular Southeast Asia is its island environment. Indonesia alone is said to contain more than 13,000 islands, whereas the Philippines supposedly contains 7000. Borneo and Sumatra are the third and sixth largest islands in the world, respectively, while many thousands of others are little more than specks of land rising at low tide from a shallow sea.

Indonesia is dominated by the four great islands of Sumatra, Borneo, Java, and the oddly shaped Sulawesi. This island nation also includes the western half of New Guinea and the Lesser Sunda Islands, which extend to the east of Java. A prominent mountain spine runs through these islands as a result of tectonic forces. The two largest and most important islands of the Philippines are Luzon (about the size of Ohio) in the north and Mindanao (the size of South Carolina) in the south. Sandwiched between them are the Visayan Islands, which number roughly a dozen.

Closely related to this impressive collection of islands is one of the world's largest expanses of shallow seas. These waters cover the **Sunda Shelf**, which is an extension of the continental shelf stretching from the mainland through the Java Sea between Java and Borneo. Here waters are generally less than 200 feet (70 meters) deep. Some local peoples have adopted lifestyles that rely on the rich marine life of this region, essentially living on their boats and setting foot on land only as necessary.

Insular Southeast Asia is less geologically stable than the mainland. Four of Earth's tectonic plates converge here: the Pacific, the Philippine, the Indo-Australian, and the Eurasian. As a result of this tectonic structure, earthquakes occur frequently. Large, often explosive volcanoes are another consequent feature of the insular Southeast Asian landscape. A string of active volcanoes extends the length of eastern Sumatra across Java and into the Lesser Sunda Islands (Figure 13.4). Volcanic eruptions and earthquakes occasionally result in **tsunamis**, which can devastate coastal regions. A massive earthquake in northern Sumatra on December 26, 2004, for example, caused roughly 100,000 deaths in Indonesia alone. Other geological hazards also confront the region. In late 2006, 13,000 Indonesians had to be evacuated away from a new mud volcano near the city of Surabaya. As of late 2011, it was still oozing out 10,000 cubic meters of toxic sludge each day. An inquiry in 2012 determined that the eruption was caused by a poorly regulated natural-gas drilling operation.

Island Climates The climates of insular Southeast Asia are more varied than those of the mainland. Most of insular Southeast Asia, unlike the mainland, receives rain during the Northern Hemisphere's winter because the monsoon winds of this season cross large areas of warm equatorial ocean, absorbing moisture. On Sumatra and Java, where north winds blow between November and March, heavy rains occur on the northern side of these east–west-running islands. But during the May–September period, the heaviest rains are found on the southern flanks of these same islands because of the southwesterly winds associated with the Asian summer monsoon.

The climates of Indonesia, Singapore, Malaysia, and Brunei are heavily influenced by their equatorial location, which reduces seasonality, as does the maritime nature of the region. Temperatures remain elevated throughout the year, with little variation. The average high temperature in Singapore, for example, varies between 85°F (29.5°C) in December to 89°F (31.7°C) in April. The equatorial influence also brings high and evenly distributed rainfall to much of the insular realm. As a result, large areas of island Southeast Asia are placed into the tropical rainforest (Af) climate category. The southeastern islands of Indonesia, however, experience a distinct dry season from May to October.

The southern part of the Philippines also has an equatorial climate, but most of the country experiences a distinct dry period from December to April. The northern and central parts of the Philippines are frequently hit by tropical cyclones, or **typhoons**, especially from August to October. Each year several typhoons strike the Philippines with heavy damage through flooding and landslides. In November 2013, much of the central Philippines was devastated by Typhoon Haiyan, known locally as Typhoon Yolanda. With winds of 195 miles per hour (315 km/h), Haiyan is the strongest tropical storm ever to have made landfall. With thousands of people dead and hundreds of thousands left homeless, the Philippines faced monumental challenges that required substantial foreign assistance (Figure 13.5).

In November 2013, much of the central Philippines was devastated by Typhoon Haiyan, known locally as Typhoon Yolanda. With winds of 195 miles per hour (315 km/h), Haiyan is the strongest tropical storm ever to have made landfall.

▼ **Figure 13.4 Bromo Volcano** Insular Southeast Asia is noted for its widespread volcanism, which has created fertile soils in many areas, while also generating many natural hazards. The eruption of Bromo Volcano on the island of Java in Indonesia took place on January 22, 2011.

▲ **Figure 13.5 Typhoon Haiyan** The Philippines has suffered several devastating tropical storms in recent years, including Typhoon Bobha in 2012 and Typhoon Haiyan in 2013. **Q: What features of Philippines' physical and human geography make it particularly vulnerable to cyclone damage?**

With thousands of people dead and hundreds of thousands left homeless, the Philippines faced monumental challenges that required substantial foreign assistance. Nargis caused over 150,000 deaths, left as many as 1 million people homeless, and resulted in roughly $10 billion in damages. Casualty rates were particularly high because Burma's military government reacted slowly to the crisis, restricting the activities of international aid agencies.

The Deforestation of Southeast Asia

Deforestation and related environmental problems have long been major issues throughout most of Southeast Asia (Figure 13.6). Although colonial powers cut Southeast Asian

▼ **Figure 13.6 Environmental Issues in Southeast Asia** This was once one of the most heavily forested regions of the world. Most of the tropical forests of Thailand, the Philippines, peninsular Malaysia, Sumatra, and Java, however, have been destroyed by a combination of commercial logging and agricultural settlement. The forests of Borneo (Kalimantan), Burma (Myanmar), Laos, and Vietnam, moreover, are now being rapidly cleared. Water and urban air pollution, as well as soil erosion, is also widespread.

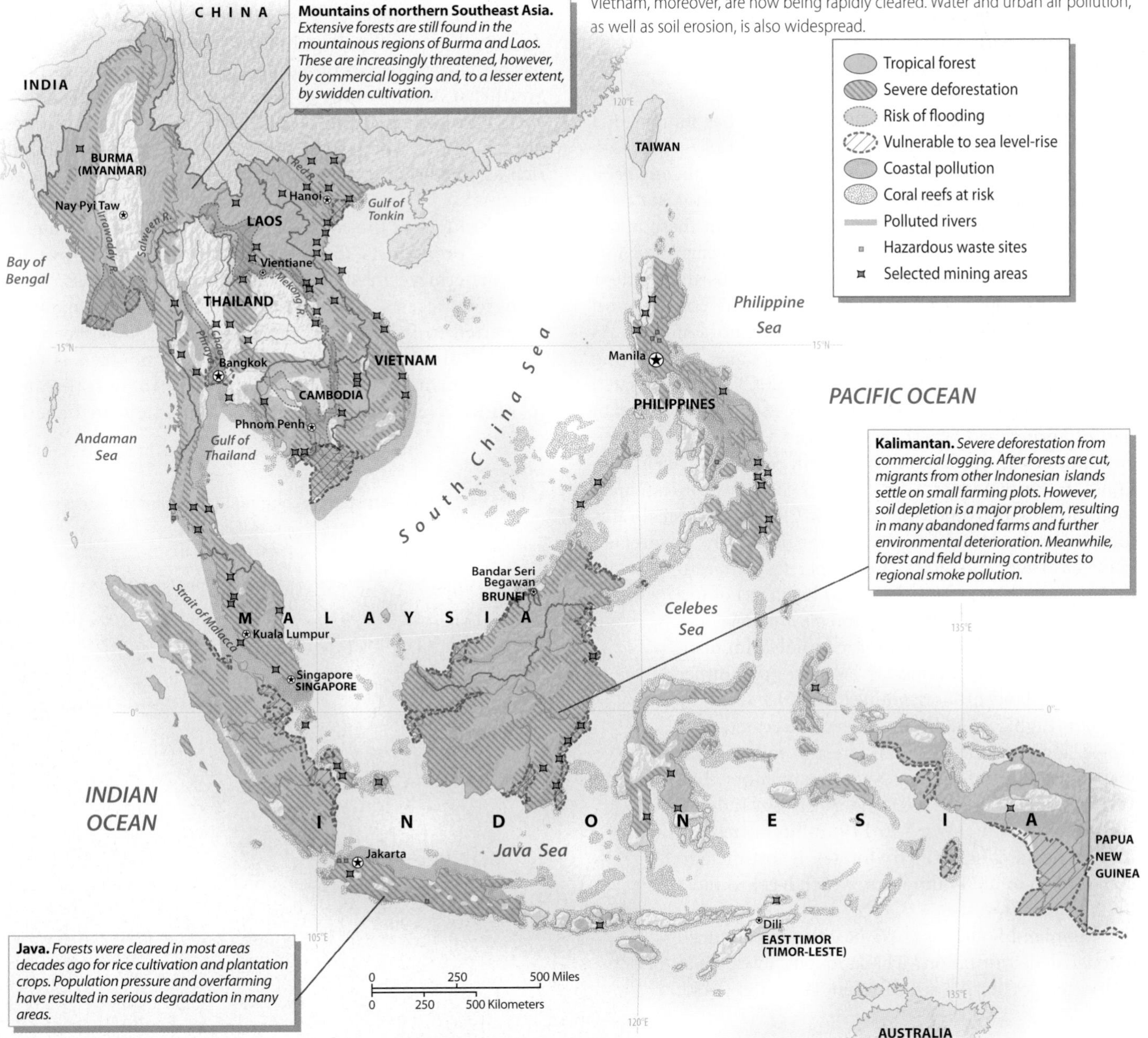

forests for tropical hardwoods and naval supplies and indigenous peoples have long cleared small areas of forest for agricultural use, rampant deforestation came only in the late second half of the twentieth century, with large-scale international commercial logging. This activity has largely been driven by the developed world's appetite for wood products such as tropical hardwood, plywood, and paper pulp. Increasingly, China is playing a major role as well. By 2011, it was estimated that more than half of all timber shipped worldwide was destined for China.

Although several Southeast Asian countries have been transforming forests into farmlands in order to increase food production and relieve population pressure in their more densely settled areas, agriculture and population growth are not the only cause of deforestation. Most forests are cut so that the wood products can be exported to other parts of the world. Subsequently, some of the logged-off lands are replanted (often with fast-growing, "weedy" tree species), while others are opened for agricultural settlement. In Malaysia and Indonesia, vast areas of former rainforest have been planted with African oil palms, which yield large quantities of inexpensive edible oil.

Local Patterns of Deforestation Malaysia has long been a leading exporter of tropical hardwoods. Most of the primary forests in western (or peninsular) Malaysia have already been cut, which has resulted in increased logging in the states of Sarawak and Sabah on the island of Borneo. In these areas, the granting of forestry concessions to Malaysian and foreign firms has harmed local tribal people by disrupting their traditional resource base. Although Malaysia's conservation policies look good on paper, they are often not enforced. As a result, the World Bank estimates that logging is occurring at several times the sustainable rate. A 2013 study indicated that more than 80 percent of the forests on the Malaysian part of Borneo had been heavily logged.

Indonesia, the largest country in Southeast Asia, contains an estimated two-thirds of the region's forest area, including about 10 percent of the world's true tropical rainforests. Its forest coverage, however, is deeply threatened; between 1990 and 2005, Indonesia lost an estimated 69 million acres (28 million hectares) of forest. Most of Sumatra's primary forests are gone, and those of Borneo (Kalimantan) are rapidly diminishing (Figure 13.7). Indonesia's last forestry frontier is on the island of New Guinea. A 2013 study by Human Rights Watch, moreover, indicated that illegal logging and general forestry mismanagement cost the Indonesian government more than US$7 billion between 2007 and 2011. But Indonesia, which is still covered in 119 million acres (48 million hectares) of primary forests, is fortunate in comparison to the Philippines, which has lost most of its original forests.

Mainland Southeast Asia has also experienced extensive deforestation. Thailand cut more than 50 percent of its forests between 1960 and 1980. Damage to the landscape was severe; flooding increased in lowland areas, and erosion on hillslopes led to the accumulation of silt in irrigation works and hydroelectric facilities. As a result, a series of logging bans in the 1990s severely restricted commercial forestry. Many of Thailand's cutover lands are being reforested with fast-growing Australian eucalyptus trees, a nonnative species that cannot support local wildlife. As the forests of Thailand disappeared, mainland Southeast Asia's logging frontier moved into Burma, Vietnam, Laos, and Cambodia. A 2013 study by the Worldwide Fund for Nature indicated that between 1973 and 2009, the countries of mainland Southeast Asia cleared almost a third of their forests for timber and agriculture, with losses in Vietnam amounting to 43 percent. The same report also claimed that the pace of clearing was accelerating so rapidly that the region risks losing a third of its remaining forests by 2030.

Such forest clearance figures can be misleading, however, since in many parts of Southeast Asia, logged-over lands and other degraded areas are gradually returning to tree coverage. As a result, some experts think that Vietnam's total forest area has actually increased in recent years. Indonesia and other Southeast Asian countries also support reforestation projects, which often try to recruit local people to nurture tree growth. The resulting secondary forests are not as biologically rich as primary forests, but over time their biodiversity does increase.

Throughout the coastal areas of Southeast Asia, a more specific problem is the destruction of the mangrove forests that thrive in shallow and silty marine areas. Often mangrove forests are burned for charcoal, but many are converted to fish and shrimp ponds, as well as rice fields and oil palm plantations. Mangrove forests serve as nurseries for many fish species, so their destruction threatens to undercut several important Southeast Asian fisheries.

▼ **Figure 13.7 Deforestation/Logging in Kalimantan** Extensive deforestation has occurred on the island of Borneo in recent decades. The line between forests and cleared lands is often quite sharp.

Mangrove forests also protect coastal areas from storm surges associated with the tropical cyclones that often strike the region. Such wetland forest zones are particularly widespread in coastal Borneo and on the east coast of Sumatra.

Protected Areas Despite rampant deforestation, Indonesia, like most other Southeast Asian countries, has created several large national parks and other protected areas. Kutai National Park in the province of East Kalimantan, for example, covers more than 741,000 acres (300,000 hectares). Southeast Asian rainforests are among the most biologically diverse areas on the planet, containing a large number of species that are found nowhere else. Conservation officials hope that protected areas will allow animals such as the orangutan, which now lives only in a small portion of northern Sumatra and in a somewhat larger area of Borneo (Kalimantan), a chance to survive in the wild. Others are less optimistic, noting that Southeast Asian national parks often exist only on paper, receiving little real protection from loggers or immigrant farmers. Much of Kutai National Park, for example, has been logged and burned. According to some sources, over 80 percent of logging activities in Indonesia are illegal in some manner. In Malaysia, however, conservation tends to be much more effective than in most other countries of the region (see *Working Toward Sustainability: Malaysia's National Physical Plan*).

Southeast Asia is also one of the world's main centers in the trade in endangered species and animal products. Many animals of the region's tropical forests and seas are in high demand in neighboring China, both for eating and for their alleged pharmaceutical properties. The pangolin, or scaly anteater, is particularly endangered by this trade.

▲ **Figure 13.8 Burning Peatlands** The soil of most wetland areas in Southeast Asia is composed largely of peat, an organic substance that can burn when dry. Draining for agricultural expansion, as well as drought, often results in extensive peat fires. Here a C130 airplane is dropping water on burning peat in Lampung, Sumatra.

Fires, Smoke, and Air Pollution

Logging operations typically leave large quantities of wood (small trees, broken logs, roots, branches, and so on) on the ground. Exposed to the sun, this remaining "slash" becomes highly flammable. Indeed, it is often burned on purpose in order to clear the ground for agriculture or tree replanting. Commercial forest cutting is responsible for most burning, even though the large logging firms commonly blame small-scale farmers. Wildfires also are common in other Southeast Asian habitats. The most thoroughly deforested areas are often covered by rough grasses. Such grasslands frequently are purposely burned every year so that cattle can graze on the tender shoots that emerge after a fire passes through. Grassland fires, in turn, prevent forest regeneration, further undercutting the region's biodiversity. In drained wetland areas, organic peat soils often burn, causing extensive air pollution (Figure 13.8). Such drained areas, moreover, are rapidly spreading due mainly to the planting of oil palm plantations.

In the late 1990s, wildfires associated with both logging practices and a severe drought raged so intensely across much of insular Southeast Asia that the region suffered from disastrous air pollution. During that period, a commercial airliner crashed because of poor visibility, countless road accidents resulted, two ferries collided in smoke-laden conditions, and hundreds of thousands of people were admitted to hospitals with life-threatening respiratory problems. The smoke crisis of the late 1990s led several Southeast Asian countries to devote more attention to air quality. Deforestation continues, however, and the fire threat remains. In June 2013, an estimated 800 fires resulted in such heavy levels of air pollution that the prime minister of Malaysia had to declare a state of emergency in several districts.

Efforts to protect Southeast Asia's air quality are also hampered by continuing industrial development, along with a major increase in vehicular traffic. Southeast Asia's large metropolitan areas have extremely unhealthy levels of air and water pollution. Several cities, including Bangkok and Manila, have built rail-based public transportation systems in order to reduce traffic and vehicular emissions. Bangkok, in particular, has substantially reduced its levels of ozone and sulfur dioxide, but particulate matter remains at unhealthy levels.

Climate Change in Southeast Asia

Most of Southeast Asia's people live in coastal and delta environments, making the region particularly vulnerable to the rise in sea level associated with global warming. Periodic flooding is already a major problem in many of the region's low-lying cities. Southeast Asian farmland is also concentrated in delta environments and thus could suffer from saltwater intrusion and heightened storm surges. A 2012 report by the World Bank estimated, through the use of advanced computer simulations, that the highly productive

Working Toward Sustainability

Malaysia's National Physical Plan

Malaysia has often been criticized by outsiders for its environmental record. It was one of the first countries in Southeast Asia to engage in massive commercial logging, and it pioneered the development of palm oil plantations. In 2010, however, Malaysia surprised some of its critics when it released its ambitious National Physical Plan, designed to foster both economic growth and social development, while attempting to safeguard the natural environment over the course of the next decade. Much of the plan focuses specifically on geographical issues, seeking to encourage residential, commercial, and industrial development within existing urban areas. The idea is to make more room available in rural areas for both natural habitat and agricultural and forestry production (Figure 13.1.1).

Environmental and Cultural Preservation "Sustainability" is one of the key words of the new plan. To enhance sustainability, the plan emphasizes the importance of "ecosystem services," such as the ability to provide clean drinking water and to decompose waste products. High levels of biological diversity, in turn, are considered necessary to maintain ecosystem services. The report also argues that biodiversity should be maintained at the genetic, species, and ecosystem levels. Such thinking reflects a high level of environmental knowledge on the part of the authors.

The Malaysian National Physical Plan seeks to maintain not only "pristine forests, hills and wetlands, habitats for the Malaysian wildlife," but also cultural and historical landscapes. The preservation ethos is supposed to extend down to the level of "individual buildings of architectural merits and historical interests." The goal here is both to protect Malaysia's heritage and to promote sites that might be attractive to tourists, both foreign and domestic.

Climate Issues Malaysia's new physical plan is also concerned about climate change. It seeks to lower greenhouse gas emissions by encouraging public transportation and dense urban settlement, characterized by mixed-use zoning. One specific goal is the development of "non-pollutive live-work-play activities in a [single] building." The plan further seeks to maintain "the open countryside and forested areas as carbon sinks in combating climate change." It also advocates sustainable forestry practices and integrated river basin management procedures.

Malaysia's National Physical Plan thus puts forward an ambitious and forward-thinking agenda for sustainable national development. Whether the plan will actually be followed remains to be seen. Malaysia does have a long way to go in creating an environmentally sustainable economic system. According to a World Resources Institution report, it has the world's fourth highest level of per capita greenhouse gas emissions, owing largely to deforestation and the draining of coastal wetlands. In June 2013, moreover, the country suffered from an extremely severe episode of air pollution; Malaysian officials, however, claimed that this was largely due to forest fires in neighboring Indonesia.

◄ **Figure 13.1.1 Taman Negara National Park, Malaysia** Malaysia's National Plan calls for urban growth, sustainable development, and the preservation of national parks such as Taman Negara.

Mekong Delta in Vietnam could experience a sea-level rise of almost 12 inches (30 cm) by 2040, which would, in turn, cause a crop production decline of about 12 percent.

Changes in precipitation across Southeast Asia brought about by global warming remain highly uncertain. Many experts foresee an intensification of the monsoon pattern, which could bring increased rainfall to much of the mainland. Enhanced precipitation would likely result in more destructive floods, but it could bring some agricultural benefits to dry areas, such as Burma's central Irrawaddy Valley. Complicating this scenario, however, is the prediction that climate change could intensify the El Niño effect, which would result in more extreme droughts. Some observers think that the prolonged drought that Indonesia experienced in the late 1990s was a harbinger of future conditions.

Greenhouse Gas Emissions All Southeast Asian countries have ratified the 1997 Kyoto Accord. But since they are officially classified as developing countries—even wealthy Singapore—none are obligated to reduce their emissions of greenhouse gases. Still, Southeast Asia's overall emissions from conventional sources remain low by global standards. Indonesia, Malaysia, and Vietnam, however, are planning to increase their reliance on coal-fired electricity generation, which would greatly increase the region's emissions.

When greenhouse gas emissions associated with deforestation are factored in, Southeast Asia's role in global climate change becomes much larger than it appears at first glance. By some estimates, Indonesia is the world's third largest contributor to the problem, following only China and the United States. Vast quantities of carbon dioxide are released into the atmosphere when slash is burned, a routine practice in many areas. A greater problem, however, is the release of carbon from peat, the partially decayed organic matter that accumulates in perennially saturated soils. The wetlands of coastal Indonesia hold 60 percent of the world's tropical peat, containing roughly 50 billion tons of carbon. During drought periods, peat soils sometimes burn, releasing the stored carbon. More worrisome is the fact that both Indonesia and Malaysia are actively draining coastal wetlands to make room for agricultural expansion, a process that results in the gradual oxidation of the peat.

▲ **Figure 13.9 Geothermal Plant in the Philippines** Owing to its geologically active location with numerous volcanos, the Philippines is one of the leading countries of the world in geothermal power production.

Energy in Southeast Asia Southeast Asia was one of the world's first major oil-exporting regions, with production concentrated in Indonesia, Malaysia, and Burma. Production failed to keep up with demand in most areas, however, and now only tiny Brunei remains a significant oil exporter relative to the size of its economy. Indonesia, Malaysia, and Vietnam, however, still have significant oil reserves. Natural gas deposits in the region are larger than those of oil, with Indonesia and Malaysia having the 13th and 15th largest proven reserves, respectively. New natural gas fields have recently been discovered offshore from East Timor, although tussles with Australia over ownership have slowed down their exploitation. Coal is an important source of energy in Vietnam, Thailand, and especially Indonesia. Indonesia is the world's seventh largest coal producer, and it exports much of its production, especially to India.

Most renewable energy in Southeast Asia comes from hydropower and geothermal plants. Laos gets 92 percent of its energy from hydro, one of the highest figures in the world. Hydropower is also important in Vietnam and Burma, and ambitious new dam-building projects promise to supply increasing amounts of electricity, much of which will be exported to neighboring countries. Environmental and cultural activists, however, generally oppose dam-building, which displaces indigenous peoples and destroys habitat. Geothermal power is concentrated in the Philippines and Indonesia, which are noted for their volcanoes and hot springs (Figure 13.9). These two countries rank second and third, respectively, in electricity from geothermal, trailing only the United States.

Southeast Asia does not produce nuclear power, although several countries in the region are planning to do so in the future. Here Vietnam is the clear leader, with 15 nuclear power plants currently in the planning stage. Indonesia, Malaysia, and Thailand also have plans for nuclear power plants. The Philippines, on the other hand, built a nuclear generator, but never brought it online, due to environmental and safety concerns. In 2011, its government decided to turn the mothballed Bataan Nuclear Power Plant into a tourist attraction.

REVIEW

13.1 Why do the mainland and insular regions of Southeast Asia have such distinctive climates and landforms, and how have these differences impacted the human communities of these two regions?

13.2 How have people changed the physical landscapes of Southeast Asia over the past 50 years?

13.3 What factors make climate change a particularly worrying matter in Southeast Asia?

Population and Settlement: Subsistence, Migration, and Cities

With just over 600 million inhabitants, Southeast Asia is not heavily populated by Asian standards. One of the reasons for this relatively low density is the extensive tracts of rugged mountains, which generally remain thinly inhabited. In contrast, relatively dense populations are found in the region's deltas, coastal areas, and zones of fertile volcanic soil (Figure 13.10). Many of the favored lowlands of Southeast Asia

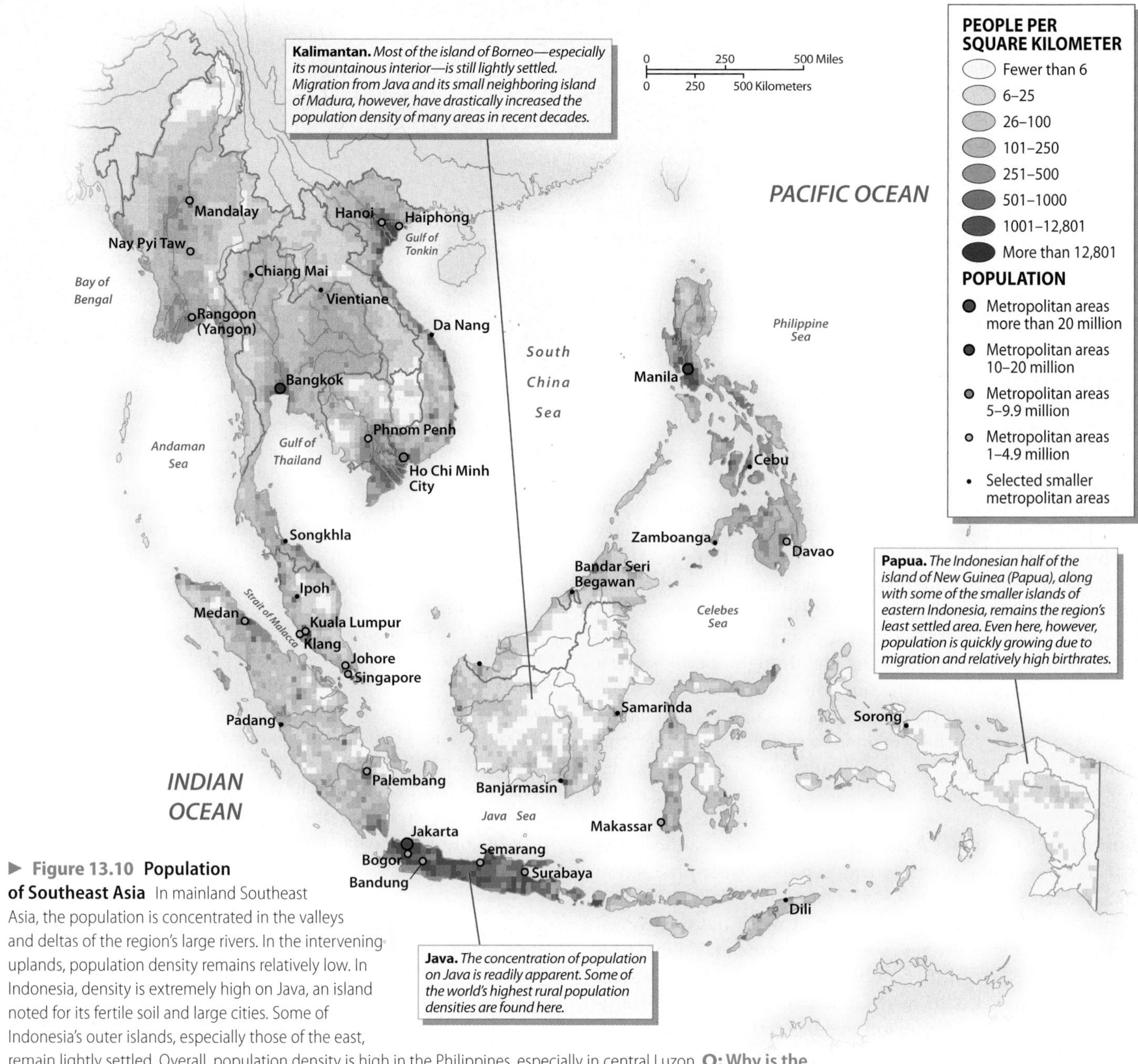

▶ **Figure 13.10 Population of Southeast Asia** In mainland Southeast Asia, the population is concentrated in the valleys and deltas of the region's large rivers. In the intervening uplands, population density remains relatively low. In Indonesia, density is extremely high on Java, an island noted for its fertile soil and large cities. Some of Indonesia's outer islands, especially those of the east, remain lightly settled. Overall, population density is high in the Philippines, especially in central Luzon. **Q: Why is the population of Vietnam so unevenly distributed? What are some of the political and economic consequences of such uneven distribution?**

have experienced striking population growth over the past half century. Demographic growth and family planning have thus become increasingly important concerns throughout much of the region. Several countries, especially Singapore and Thailand, have seen rapid reductions in birth rates, but large families are still common in Cambodia, Laos, and especially East Timor.

Settlement and Agriculture

Much of insular Southeast Asia has infertile soil, which is unable to support intensive agriculture and high rural population densities. The island rainforests, though lush and biologically rich, typically grow on a poor base. Plant nutrients are locked up in the vegetation itself, rather than being stored in the soil where they would easily benefit agriculture. The incessant rain of the equatorial zone also tends to wash nutrients out of the soil. Agriculture must be carefully adapted to this limited fertility by constant field rotation or the application of heavy amounts of fertilizer.

Some notable exceptions can be found to this generalization about soil fertility and settlement density in equatorial Southeast Asia. Unusually rich soils connected to volcanic activity are scattered through much of the region, particularly on the island of Java. Java, with more than 50 volcanoes, is blessed with highly productive agriculture that supports a large array of tropical crops and a very high population density. Approximately 138 million people live on the island—more than half the total population of Indonesia—in an area smaller than the state of Iowa. Dense populations also are found in pockets of fertile alluvial soils along the coasts of insular Southeast Asia, where people have traditionally supplemented land-based farming with fishing and other commercial activities.

The demographic patterns in mainland Southeast Asia are less complicated than those of the island realm. In all the mainland countries, population is concentrated in the agriculturally intensive valleys and deltas of the large rivers. The population core of Thailand, for example, clusters around the Chao Phraya River, just as Burma's is focused on the Irrawaddy. Vietnam has two distinct foci: the Red River Delta in the far north and the Mekong Delta in the far south. In contrast to these densely settled areas, the middle reaches of the Mekong River provide only limited lowland areas in Laos, which reduces its population potential. In Cambodia, the population historically is centered around Tonle Sap, a large lake with an unusual seasonal flow reversal. During the rainy summer months, the lake receives water from the Mekong drainage, but during the drier winter months it contributes to the river's flow.

Agricultural practices and settlement forms vary widely across the complex environments of Southeast Asia. Generally speaking, however, three farming and settlement patterns are apparent.

Swidden in the Uplands Swidden, also known as shifting cultivation or "slash-and-burn" agriculture, is practiced throughout the uplands of both mainland and island Southeast Asia (Figure 13.11). In the **swidden** system, small plots of tropical forest or brush are periodically cut or "slashed" by hand. The fallen vegetation is then burned to transfer nutrients to the soil before subsistence crops are planted. Yields remain high for several years and then drop off dramatically as the soil nutrients are exhausted and insect pests and plant diseases multiply. These plots are abandoned after a few years and allowed to revert to woody vegetation. The cycle of cutting, burning, and planting is then moved to another small plot not far away—thus the term *shifting cultivation*. Villages generally control a large amount of territory so they can rotate their fields on a regular basis. After a period of 10 to 75 years, farmers must return to the original plot, which once again has accumulated nutrients in the dense vegetation.

Swidden is sustainable when population densities remain low and when upland people control enough territory. Today, however, the swidden system is increasingly threatened. It cannot easily support the increasing population that has resulted from relatively high birth rates and, in some cases, migration. With greater population density, the rotation period must be shortened, undercutting soil resources. The upland swidden system is also often undermined by commercial logging. Road building presents another threat. Vietnam, for example, has recently built a highway system through its mountainous spine, designed to aid economic development and more fully integrate its national economy. Partly as a result of this activity, lowlanders are streaming into the mountains, disrupting local ecosystems and indigenous societies.

When swidden can no longer support the local population, upland people sometimes adapt by switching to cash crops that allow them to participate in the commercial economy. In the mountains of northern Southeast Asia, often called the **Golden Triangle,** one of the main cash crops is opium, grown by local farmers for the global drug trade. Drug eradication programs between 1998 and 2006 proved relatively successful, reducing the area of land devoted to opium production by 85 percent. Since then, however, opium growing in Burma has experienced resurgence and now accounts for roughly one-quarter of the global total. Methamphetamine manufacturing has also emerged as a major problem in much of the Golden Triangle.

Plantation Agriculture With European colonization, Southeast Asia became a center of plantation agriculture,

▼ **Figure 13.11 Swidden Agriculture** In the uplands of Southeast Asia, swidden (or slash-and-burn) agriculture is widely practiced. When done by tribal peoples with low population densities, swidden is not environmentally harmful. When practiced by large numbers of immigrants from the lowlands, however, swidden can cause deforestation and extensive soil erosion.

producing high-value specialty crops ranging from rice to rubber. Even in the 19th century, Southeast Asia was closely linked to a globalized economy through the plantation system. Forests were cleared and swamps drained to make room for these commercial farms. Labor was supplied, often unwillingly, by indigenous people or by contract laborers brought in from India or China.

Plantations are still an important part of Southeast Asia's geography and economy. Most of the world's natural rubber, for example, is grown in Malaysia, Indonesia, and Thailand, while sugarcane has long been a major plantation crop in the Philippines and Indonesia (Figure 13.12). More recently, pineapple plantations have spread in both the Philippines and Thailand, which are now the world's leading exporters of this fruit. Indonesia is the region's leading producer of tea, while Indonesia and Malaysia together produce more than 90 percent of the world's palm oil. Coconut oil and **copra** (dried coconut meat) are widely produced in the Philippines, Indonesia, and elsewhere. In the early 2000s, Vietnam rapidly emerged as the world's second largest coffee producer, its large harvests reducing global coffee prices.

Rice in the Lowlands The lowland basins and deltas of mainland Southeast Asia are largely devoted to intensive rice cultivation. Throughout most of Southeast Asia, rice is the preferred staple food. Traditionally, rice was mainly cultivated on a subsistence basis by rural farmers. But as the number of wage laborers in Southeast Asia has grown as a result of economic development, so has the demand for commercial rice cultivation. Until 2013, Thailand was the world's largest rice exporter, but it was surpassed in that year by both India and Vietnam. Thailand's government is currently subsidizing the rice market, which results in higher prices for Thai farmers, but makes exports more expensive. Across most of Southeast Asia, the use of agricultural chemicals and high-yield crop varieties, along with improved water control, has allowed rice production to keep pace with population growth.

In those areas without irrigation, yields remain relatively low. Rice growing on Thailand's Khorat Plateau, for example, depends largely on the uncertain rainfall, without the benefits of the more sophisticated water-control methods. In some lowland districts lacking irrigation, dry-field crops, especially sweet potatoes and manioc, form the staple foods of people too poor to buy market rice on a regular basis. In some of the poorest parts of Southeast Asia, population growth has combined with economic stagnation to force larger numbers of people into such meager diets. Elsewhere, economic growth and declining birth rates have significantly reduced the burden of poverty.

▼ **Figure 13.12 Rubber Plantation** Most of the world's natural rubber in produced on plantations in Southeast Asia. This photograph shows a rubber tree being tapped on the island of Sumatra in Indonesia

Recent Demographic Change

Most Southeast Asian countries have seen a sharp decrease in birth rates over the past several decades. Because the region is not facing the same kind of population pressure as East or South Asia, governments have promoted a wide range of population policies. In countries with a highly uneven population distribution, internal relocation away from densely populated areas to outlying districts is a common response.

Population Contrasts The Philippines, the second most populous country in Southeast Asia, has a relatively high fertility rate, at over three children per woman (Table 13.1). When a popular democratic government replaced a dictatorship in the 1980s, the Philippine Roman Catholic Church pressured the new government to cut funding for family-planning programs. As a result, many clinics that had dispensed family-planning information were closed. Although high birth rates are not associated with Catholicism in other parts of the world, the Church's outspoken stand on birth control in the Philippines has inhibited the dispersal of family-planning information. Still, fertility rates have been slowly declining, and the government has been responding. In 2013, for example, a new law required the establishment of family-planning clinics designed especially to serve the needs of the poorest communities in the country. In Southeast Asia's other predominately Roman Catholic country, East Timor, the fertility rate remains extremely high, at around 5.4 children per woman.

Laos and Cambodia, countries of Buddhist religious tradition, also have fertility rates above the replacement rate, although these rates have dropped rapidly in recent years. Thailand, which shares cultural traditions with Laos yet is considerably more developed, saw a much earlier drop in its birth rate. Here the total fertility rate (TFR) dropped

Table 13.1 Population Indicators

Country	Population (millions) 2013	Population Density (per square kilometer)	Rate of Natural Increase (RNI)	Total Fertility Rate	Percent Urban	Percent < 15	Percent > 65	Net Migration (Rate per 1000)
Burma (Myanmar)	53.3	79	1.0	2.0	31	28	5	−2
Brunei	0.4	71	1.4	1.6	72	25	4	3
Cambodia	14.4	80	1.8	2.8	20	34	4	−4
East Timor	1.1	74	2.7	5.7	30	41	5	−14
Indonesia	248.5	130	1.5	2.6	50	29	5	−1
Laos	6.7	29	2.0	3.2	27	36	4	−2
Malaysia	29.8	90	1.3	2.1	64	26	5	4
Philippines	96.2	321	1.5	3.0	63	33	4	−2
Singapore	5.4	7,971	0.6	1.3	100	16	10	19
Thailand	66.2	129	0.4	1.6	46	19	10	−0
Vietnam	89.7	270	1.0	2.1	32	24	7	−0

Source: Population Reference Bureau, *World Population Data Sheet, 2013.*

from 5.7 in 1970 to 1.56 currently, a figure that will soon bring population decline if it persists. Thailand has promoted family planning for both population and health reasons, including the relatively high incidence of AIDS in the country. But the Thai government is now concerned about future labor shortages and as a result is contemplating removing the mandatory retirement age.

The city-state of Singapore stands out on the demographic charts with its extremely low fertility rate of 1.3, which by some accounts is the lowest in the world. As a result, the population pyramid of Singapore has a completely different shape from that of the Philippines (Figure 13.13). If Singapore did not experience relatively high rates of immigration, it's population would be steadily dropping. As early as 1987, Singapore abandoned its earlier fertility limitation policy to institute its "Have Three or More (if you can afford it)" campaign, directed particularly at the most highly educated segment of its population. As a result, Singaporeans can receive more than $10,000 in direct government subsidies for having a third or fourth child (Figure 13.14). Singapore also encourages the immigration of highly skilled workers.

Indonesia, with the region's largest population at 248.5 million, also has seen a dramatic decline in fertility in recent decades, with its fertility rate now roughly at the replacement

▼ **Figure 13.13 Population Pyramids of Singapore and the Philippines** (a) Singapore has a very low birth rate and an aging population; as a result, its government encourages the immigration of skilled workers. (b) The Philippines, by contrast, has a high birth rate and a young population.

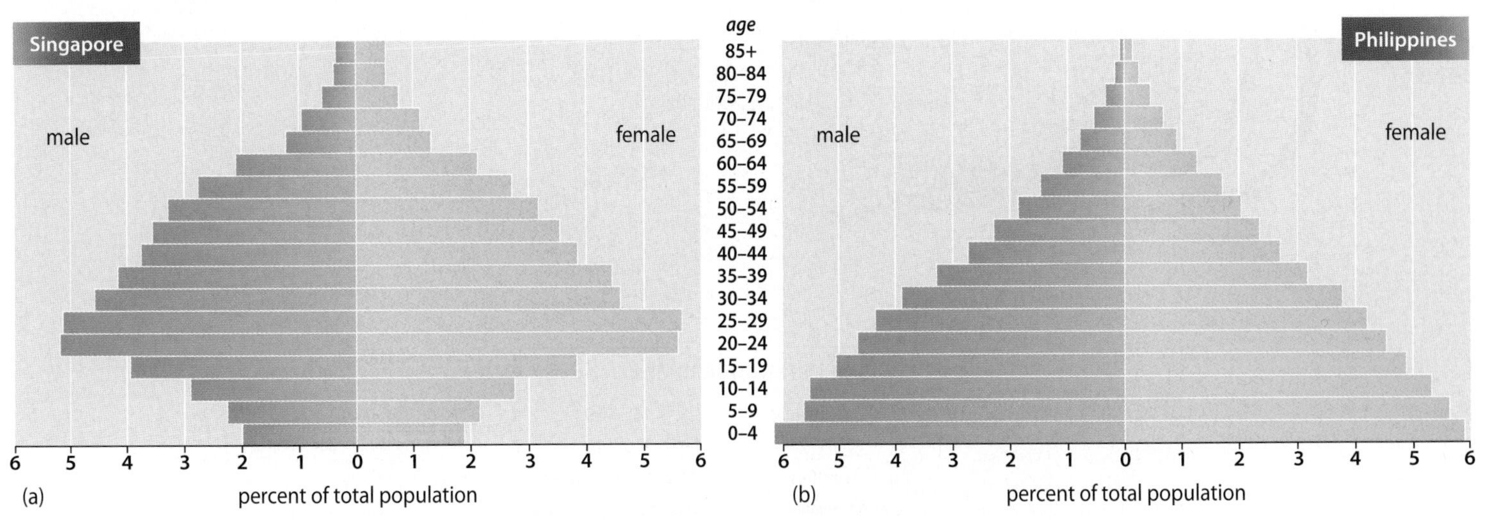

level. As with Thailand, this drop in fertility seems to have resulted from a strong government family-planning effort, coupled with improvements in education.

Growth and Migration Until recently, Indonesia had an official policy of **transmigration,** with the government helping people move from densely populated to lightly populated parts of the country (Figure 13.15). Primarily because of migration from Java and Madura, the population of the outer islands of Indonesia has grown rapidly since the 1970s. The province of East Kalimantan, for example, experienced an astronomical growth rate of 30 percent per year during the last two decades of the twentieth century. As a result of this shift in population, many parts of Indonesia outside Java now have moderately high population densities, although many of the more remote districts remain lightly settled.

High social and environmental costs often accompany these relocation schemes. Javanese peasants, accustomed to

▲ **Figure 13.14 Pro-Reproduction Advertisement in Singapore** Due to its extremely low birth rate, the population of Singapore will probably begin to decline soon. Shown here are perfumes created by local students and launched in a government-backed "romance" campaign.

▼ **Figure 13.15 Indonesian Transmigration** The distribution of population in Indonesia shows a marked imbalance: Java, along with the neighboring island of Madura, is one of the world's most densely settled places, whereas most of the country's other islands remain rather lightly populated. As a result, the Indonesian government has encouraged the resettlement of Javanese and Madurese people to the outer islands, often paying the costs of relocation. This transmigration scheme has resulted in a somewhat more balanced population distribution pattern, but has also caused substantial environmental degradation and intensified several ethnic conflicts.

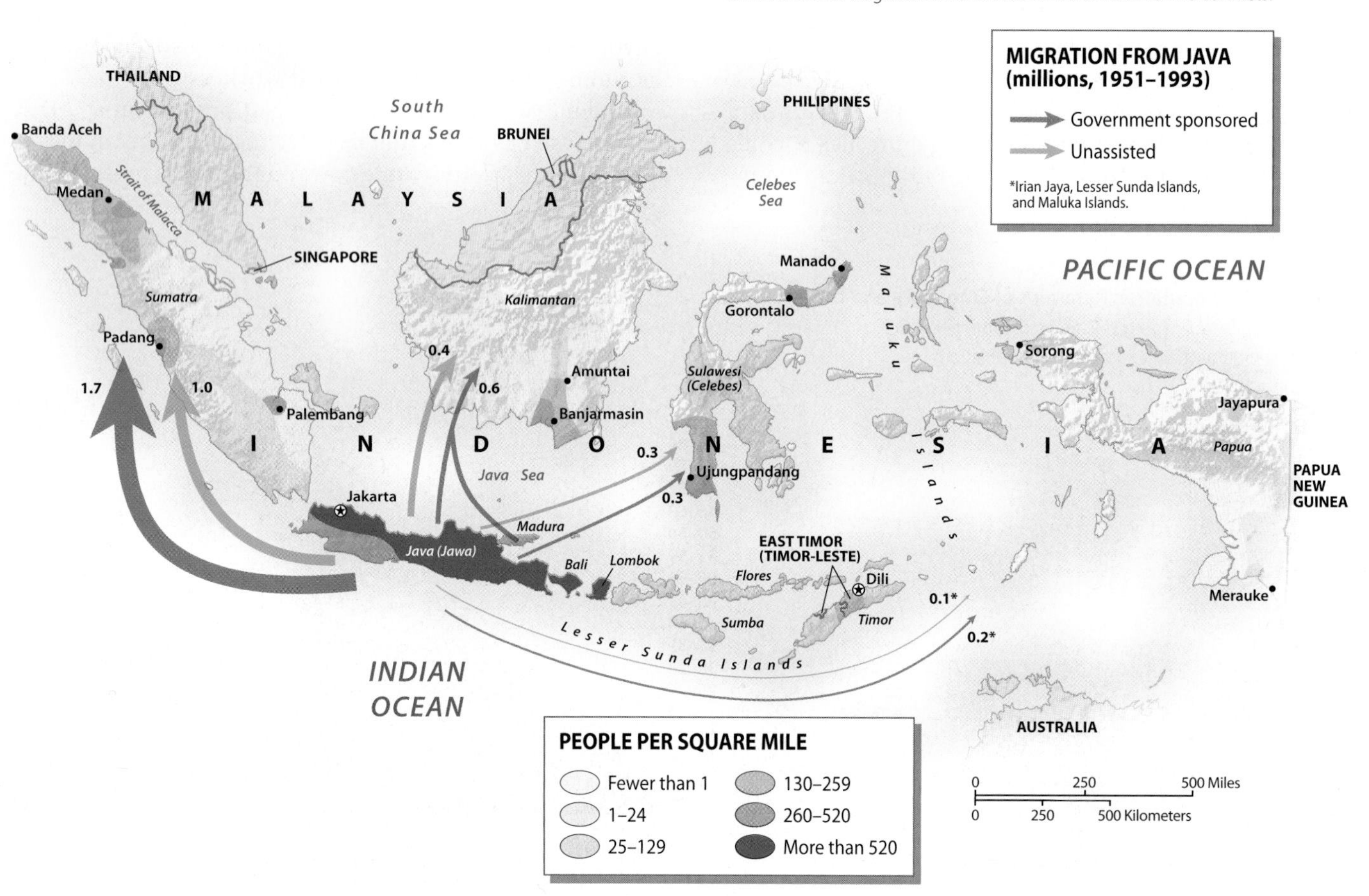

working the highly fertile soils of their home island, often fail in their attempts to grow rice in the former rainforest of Borneo (Kalimantan). Abandoned fields are common after repeated crop failures. In some areas, farmers have little choice but to adopt a semi-swidden form of cultivation, moving to new sites once the old ones have been exhausted. The term **shifted cultivators** is sometimes used for such displaced rural migrants. Partly because of these problems, the Indonesian government restructured and substantially reduced its transmigration program in 2000. It is currently creating, however, a huge new agricultural project in the Merauke region of south-central New Guinea, which could entail the transfer of over half a million people to the area (Figure 13.16). Huge sugarcane plantations are planned for the region, which could cover up to 62,000 acres (25,000 hectares).

The government of the Philippines has also used internal migration to reduce population pressure in the core areas of the country. Beginning in the late 19th century, people began streaming out of central and northwest-central Luzon for the frontier zones. In the early 20th century, this settlement frontier still lay in Luzon; by the postwar years, it had moved south to the island of Mindanao. Today, however, internal migration is less appealing, both because the population disparities between the islands have been reduced and because the south is politically unstable. The emphasis is now on international migration, resulting in a Filipino overseas population of roughly 8 million.

Population issues and policies thus vary greatly across Southeast Asia. Virtually every part of the region, however, has seen a rapid expansion of its urban population in recent decades.

Urban Settlement

Southeast Asia is not a heavily urbanized region. Even Thailand retains a rural flavor, which is somewhat unusual for a country that has experienced so much industrialization. Currently, however, the region's urban population is growing at 1.75 times the world average. As a result, Southeast Asia's urbanization rate is expected to increase from 45 percent in 2008 to 57 percent in 2030.

Several Southeast Asian countries have **primate cities**—single, large urban settlements that overshadow all others. Thailand's urban system, for example, is dominated by Bangkok, just as Manila far surpasses all other cities in the Philippines. Both have recently grown into megacities with metropolitan areas of more than 12 million residents. More than half of all city dwellers in Thailand live in the Bangkok metropolitan area. In both Manila and Bangkok, as in other large Southeast Asian cities, explosive growth has led to housing shortages, congestion, and pollution. In Manila, more than half of the city's population lives in squatter settlements, usually without basic water and electricity service. Both cities also suffer from a lack of parks and other public spaces, enhancing the appeal

▼ **Figure 13.16 Migrant Settlement in Indonesia** Over the past several decades, people have been moving in large numbers from the crowded island of Java to Indonesia's outer islands. The migrant settlement visible in this image is on the Indonesian portion of the island of New Guinea.

of massive shopping malls. Bangkok's Paragon Mall has recently emerged as a major urban focus, complete with a conference center and a concert hall. Although Bangkok is a modern metropolis, older and more traditional buildings, especially Buddhist temples, abound (Figure 13.17).

Thailand, the Philippines, and Indonesia are all making efforts to encourage growth of secondary cities by decentralizing economic functions. The goal is to stabilize the population of the primate cities. In the case of the Philippines, the city of Cebu has emerged in recent years as a relatively dynamic economic center, leading to hopes that a more balanced urban system may be emerging. Several former colonial "hill stations" (resort cities) are also emerging as major cities in their own rights (see *Cityscapes: Baguio, the "Summer Capital of the Philippines"*).

Urban primacy is less pronounced in other Southeast Asian countries. Vietnam, for example, has two main cities: Ho Chi Minh City (formerly Saigon) in the south, with a metropolitan population of 7.5 million people, and the capital city of Hanoi in the north, with 6.5 million residents in its metro area. Indonesia's Jakarta is a gargantuan city with a metropolitan population of more than 20 million, but the country has a host of other large and growing metropolitan areas, including Bandung and Surabaya. Rangoon (Yangon), until recently the capital city of Burma, has doubled its population in the last two decades to more than 5 million residents. In Cambodia, the capital of Phnom Penh is still a relatively small city of around 2.5 million. Vientiane, the capital and largest city of Laos, has only about 750,000 inhabitants.

Kuala Lumpur, the largest city in Malaysia with around 5 million inhabitants in its metro area, has received heavy investments from both the national government and the global business community. This has produced a modern city of grand ambitions that is free of most infrastructure problems plaguing other Southeast Asian cities. The Petronas Towers, owned by the country's national oil company, were the world's tallest buildings at almost 1500 feet (450 meters) when completed in 1996.

The independent republic of Singapore is essentially a city-state of 5.3 million people on an island of 240 square miles (600 square km) (Figure 13.18). Although space is at a premium, Singapore has successfully developed high-tech industries that have brought it great prosperity. Unlike other Southeast Asian cities, Singapore has no squatter settlements or slums. Only in the much-diminished Chinatown and historic colonial district does one find older buildings. Otherwise, Singapore is an extremely clean and orderly city of modern high-rise skyscrapers, apartment complexes, and space-intensive industry (Figure 13.19). But despite its crowded conditions, Singapore makes space available for parks and other green areas. Even though its population grew by 68 percent between 1986 and 2007, Singapore's green cover increased in the same period from 36 percent to 46 percent.

Singapore is unique in Southeast Asia not only because it is a city-state, but also because most of its people have a Chinese cultural background. But as we shall see in the following section, strong Chinese influences are also found in most of the other large cities of the region.

REVIEW

13.4 Why do different parts of Southeast Asia vary so much in regard to both population density and population growth?

13.5 Why is export-oriented plantation agriculture so widespread in Southeast Asia, and what are some of the problems associated with its spread?

13.6 Why are some economists and planners concerned about the phenomenon of urban primacy in such Southeast Asian countries as Thailand and the Philippines?

▼ **Figure 13.17 Bangkok** Bangkok saw the development of an impressive skyline during its boom years from the late 1970s through the late 1990s. Older and more traditional buildings, especially Buddhist temples, are still found in many areas.

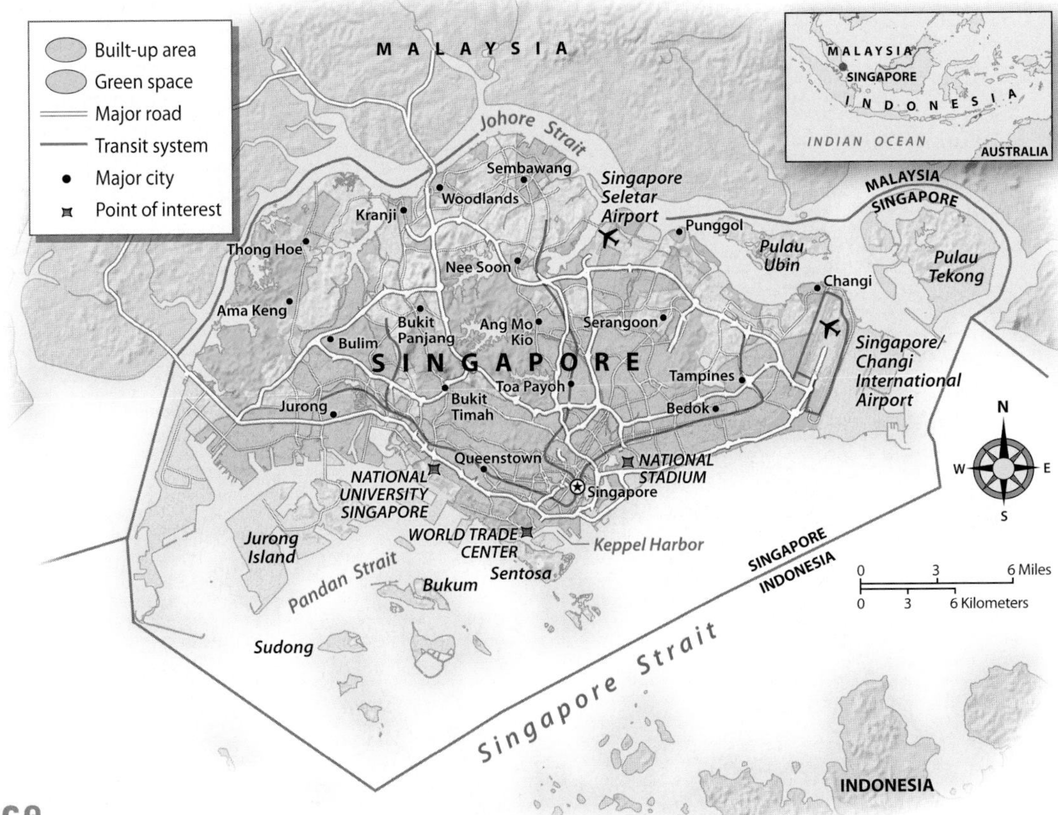

▶ **Figure 13.18 Singapore** The economic and technological hub of Southeast Asia, Singapore is a very small and extremely crowded country, forming one of the world's few city-states. Despite its high population density, Singapore has devoted almost half of its area to open spaces. The country consists of one main island and some 60 smaller islands.

Cultural Coherence and Diversity: A Meeting Ground of World Cultures

Unlike many other world regions, Southeast Asia lacks the historical dominance of a single civilization. Instead, the region has been a meeting ground for cultural traditions from South Asia, China, the Middle East, Europe, and even North America. Abundant natural resources, along with the region's strategic location on oceanic trade routes connecting major continents, have long made Southeast Asia attractive to outsiders. As a result, the modern cultural geography of this diverse region reflects a long history of borrowing and combining external influences.

The Introduction and Spread of Major Cultural Traditions

In Southeast Asia, contemporary cultural diversity is embedded in the historical influences connected to the major religions of Hinduism, Buddhism, Islam, and Christianity (Figure 13.20).

South Asian Influences The first major external influence arrived from South Asia roughly 2000 years ago when small numbers of migrants from South Asia helped local

◀ **Figure 13.19 Panoramic View of Singapore** Essentially a city state, Singapore is a densely populated country noted for its careful planning and modern architecture.

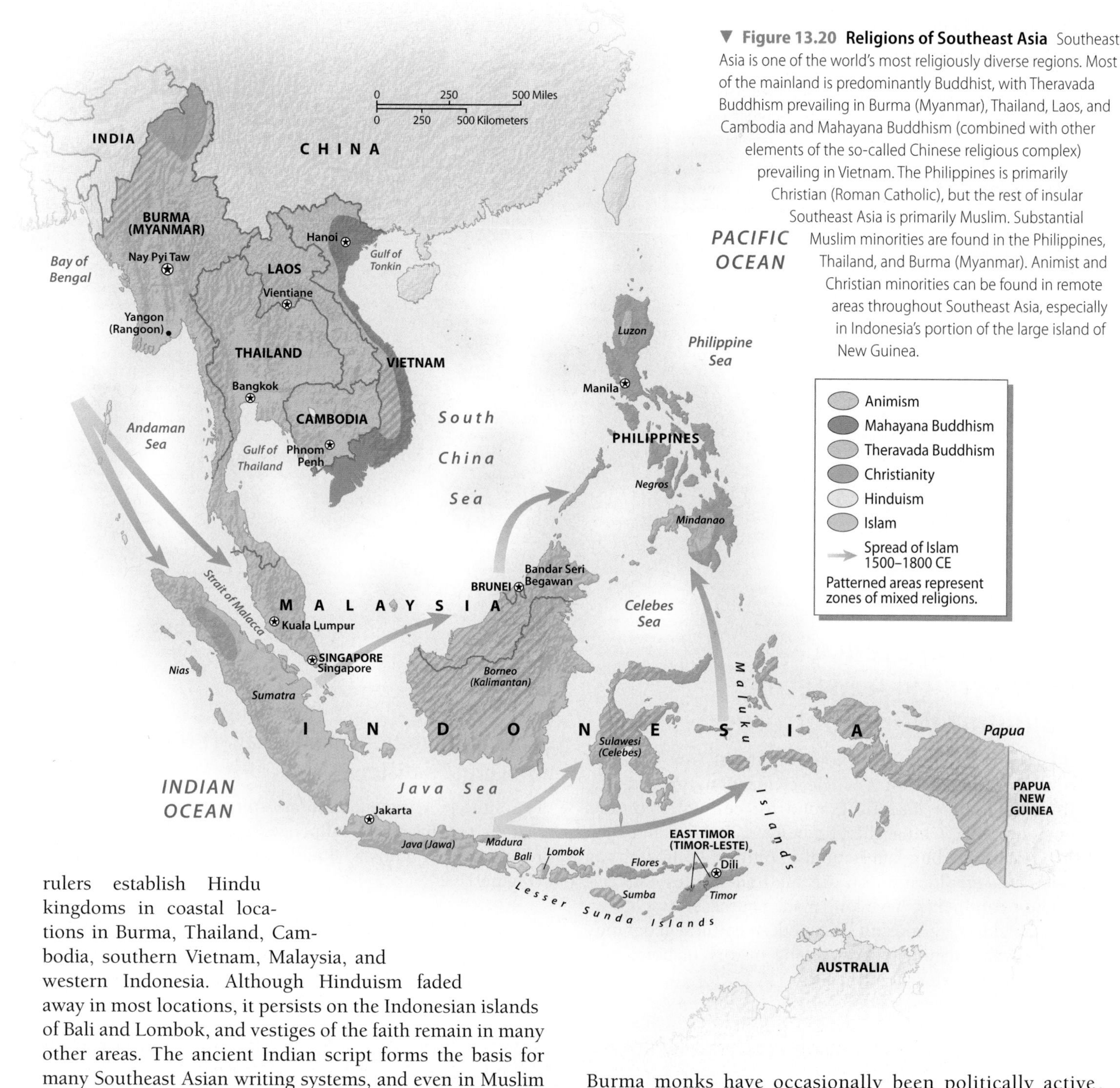

▼ **Figure 13.20 Religions of Southeast Asia** Southeast Asia is one of the world's most religiously diverse regions. Most of the mainland is predominantly Buddhist, with Theravada Buddhism prevailing in Burma (Myanmar), Thailand, Laos, and Cambodia and Mahayana Buddhism (combined with other elements of the so-called Chinese religious complex) prevailing in Vietnam. The Philippines is primarily Christian (Roman Catholic), but the rest of insular Southeast Asia is primarily Muslim. Substantial Muslim minorities are found in the Philippines, Thailand, and Burma (Myanmar). Animist and Christian minorities can be found in remote areas throughout Southeast Asia, especially in Indonesia's portion of the large island of New Guinea.

rulers establish Hindu kingdoms in coastal locations in Burma, Thailand, Cambodia, southern Vietnam, Malaysia, and western Indonesia. Although Hinduism faded away in most locations, it persists on the Indonesian islands of Bali and Lombok, and vestiges of the faith remain in many other areas. The ancient Indian script forms the basis for many Southeast Asian writing systems, and even in Muslim Java, the Hindu epic called the **Ramayana** remains a central cultural feature to this day.

A second wave of South Asian religious influence reached mainland Southeast Asia in the 13th century in the form of Theravada Buddhism, which is closely associated with Sri Lanka. Virtually all of the people in lowland Burma, Thailand, Laos, and Cambodia converted to Buddhism at that time. Today this faith forms the foundation for many of the social institutions of mainland Southeast Asia. Saffron-robed monks, for example, are a common sight in Thailand and Burma, where Buddhist temples abound. In Thailand, the country's revered constitutional monarchy remains closely connected with Buddhism, and in Burma monks have occasionally been politically active (Figure 13.21). Although Southeast Asian Theravada Buddhism shares many traits with the Mahayana Buddhism of East Asia, the two are different enough that they are mapped separately in Figure 13.20.

Chinese Influences Unlike most other mainland peoples, the Vietnamese were not heavily influenced by South Asian civilization. Instead, their early connections were with East Asia. Vietnam was actually a province of China until about a thousand years ago, when the Vietnamese established a kingdom of their own. But although the Vietnamese rejected China's rule, they retained many attributes of Chinese culture. The traditional religious and philosophical

Cityscapes

Baguio, the "Summer Capital of the Philippines"

▲ **Figure 13.2.1 Baguio City** The former colonial hill station of Baguio was designed by the noted American architect and city planner Daniel Burnham.

An important urban category in both South and Southeast Asia is that of the former colonial hill station. During the colonial era, imperial agents from Europe and North America often had a difficult time adjusting to the hot climates of the southern Asian lowlands and thus spent as much time as possible in areas of higher elevation. Some of their favored mountain resorts eventually grew into substantial cities. When the United States gained control of the Philippines at the beginning of the 20th century, it selected a small village at the southern end of the mountains of northern Luzon to become the official summer capital of the country. The famous American architect and city planner Daniel Burnham was selected to design a new city on the site. For several years in the early 1900s, the entire American government of the Philippines moved from Manila to Baguio from March to June, the hottest time of the year. The United States also established a military base, Camp John Hay, in the pine-clad hills around Baguio, which later became a major "rest and rehabilitation" center, noted for its golf course and scenic views.

▼ **Figure 13.2.2 Panorama of Baguio City** Known as the "city of pines," Baguio is noted for its scenery, resorts, and universities.

Continuing Growth Baguio City continued to grow after the Philippines gained independence in 1946 (**Figure 13.2.1**). Like the American administrators before them, the Filipino elite found much to admire in the cool mountain environment, and many built second homes in the area. The Philippine Military Academy was built near the city, and Camp John Hay was eventually transformed into a tourist resort. Several colleges and universities were also established, attracting students from all over the Philippines. By the beginning of the 2000s, international students—especially South Koreans eager to improve their English language skills—began to come to Baguio as well. South Korean companies followed along, making Baguio an increasingly cosmopolitan city. Although its population of about 350,000 is a small fraction of that of Manila, the mile-high (1600-meter-high) city of Baguio occupies a much more important place in the Philippine urban hierarchy and public imagination than such numbers would indicate (**Figure 13.2.2**).

Cultural Diversity Although Baguio is a cosmopolitan city favored by the Filipino elite, it also serves as the cultural capital of the indigenous peoples of the mountains of northern Luzon. These rugged highlands were never fully subdued by the Spaniards, and as a result, animism and other indigenous cultural practices persist to this day, both in rural areas of the highlands and in Baguio itself.

Baguio's explosive growth in recent years has generated serious overcrowding in many parts of the city. Squatter settlements are expanding as well, many of which are built on steep and unstable slopes. The city's 180 inches (450 cm) of annual rainfall lead to frequent landslides and mudflows, which often disrupt roads. Earthquakes are another hazard. A 1990 quake of 7.8 magnitude destroyed much of the historical downtown, necessitating extensive rebuilding. But despite such natural hazzards, people continue to flock to the Philippines' "City of Pines."

1. Why are most of the former colonial hill stations located in South and Southeast Asia?
2. What features of the Baguio region would have made it attractive to American colonial agents in the early 1900s?

Google Earth Virtual Tour Video

http://goo.gl/yjjbqq

beliefs of Vietnam centered around Mahayana Buddhism and Confucianism.

East Asian cultural influences elsewhere in Southeast Asia are linked more directly to immigration from southern China. Although people have been moving to the region from China for hundreds of years, migration reached a peak in the 19th and early 20th centuries. At first, most migrants were single men. Many returned to China after accumulating money, but others married local women and established mixed communities. In the Philippines, the elite population today is often described as "Chinese Mestizo," being of mixed Chinese and Filipino descent. In the 19th century, Chinese women begin to migrate in large numbers, allowing the creation of ethnically distinct Chinese settlements. Urban areas throughout much of Southeast Asia still have large and cohesive Chinese communities. In Malaysia, the Chinese minority constitutes a little less than one-third of the population, whereas in the city-state of Singapore three-quarters of the people are of Chinese ancestry.

In much of the region, relationships between the Chinese minority and the indigenous majority are strained. One source of tension is the fact that overseas Chinese communities tend to be relatively prosperous. Chinese emigrants frequently prospered by becoming merchants, a job that was often ignored by the local people. As a result, they exert a major economic influence on local affairs, which is often resented. In 1998 and again in 2001, anti-Chinese rioting broke out in several Indonesian cities. Since then, the situation of the Chinese minority in Indonesia has improved considerably.

▲ **Figure 13.21 Monks in Thailand** Protests by Buddhist monks and nuns have become common in Thailand and Burma. The monks in this photograph are protesting attacks on Buddhist temples in Bangladesh, carried out by Muslim extremists after a Buddhist man posted anti-Muslim photos on Facebook.

The Arrival of Islam Muslim merchants from India and Southwest Asia arrived in Southeast Asia more than a thousand years ago, and by the 1200s their religion began to spread. From an initial focus in northern Sumatra, Islam diffused into the Malay Peninsula, through the main population centers in the Indonesian islands, and east to the southern Philippines. By 1650, it had largely replaced Hinduism and Buddhism in Malaysia and Indonesia. The only significant holdout was the small, but fertile island of Bali. As Islam spread across Java, thousands of Hindu musicians and artists fled to Bali, giving the island an especially strong tradition of arts and crafts. Partly because of this artistic legacy, Bali is now one of the premier destinations of international tourism.

Today the world's most populous Muslim country is Indonesia, where some 87 percent of the people follow Islam (Figure 13.22). This figure, however, hides a significant amount of internal diversity. In some parts of Indonesia, especially northern Sumatra (Aceh), orthodox forms of Islam took root. In others, such as central and eastern Java, a more lax form of worship emerged that retained certain Hindu and even animistic beliefs. Islamic reformers, however, have long been striving to instill more orthodox forms of worship among the Javanese. Today the young people of the island are increasingly turning to mainstream Islam.

In Malaysia and parts of Indonesia, Islamic fundamentalism has recently gained ground. Whereas the current Malaysian government has generally supported the revitalization of Islam, it is wary of the growing power of the fundamentalist movement. Some Malaysian Muslims, moreover, accuse the fundamentalists of advocating social practices derived from the Arabian Peninsula that are not necessarily religious in origin. Adding to Malaysia's religious tension is the fact that almost all ethnic Malays, who narrowly form the country's majority population, follow Islam, whereas most members of the minority communities follow different religions. Religious tensions mounted in 2009, when the government banned Christians from using the term "Allah" to refer to God. In 2013, the Catholic Church's envoy to Malaysia was forced to apologize to the government for supporting the use of "Allah" in a Christian context, which hard-line Muslims viewed as an attempt to spread the Christian faith.

Islam was still spreading eastward through insular Southeast Asia when the Europeans arrived in the 16th century. When Spain claimed the Philippine Islands in the 1570s, it found the southwestern portion of the archipelago to be thoroughly Islamic. The Muslims resisted the Roman Catholicism introduced by the Spaniards, and much of the southwestern Philippines remains an Islamic stronghold.

The Philippines as a whole is currently about 85 percent Roman Catholic, making it, along with East Timor, the only predominantly Christian country in Asia. Several Protestant sects have been spreading rapidly in the Philippines over the past several decades, however, creating a more complex religious environment.

Christianity and Indigenous Cultures Christian missions spread through much of Southeast Asia in the late 19th and early 20th centuries, when European colonial powers controlled the region. French priests converted many people in Vietnam to Catholicism, but they had little influence elsewhere. Beyond Vietnam, missions failed to make headway in areas of Hindu, Buddhist, or Islamic heritage.

Missionaries were far more successful in Southeast Asia's highland areas, where they found a wide array of tribal societies that had never accepted the major lowland religions. These peoples retained their indigenous belief systems, which generally focus on the worship of nature spirits and ancestors. Although some modern tribal groups retain such animist beliefs, others converted to Christianity. As a result, notable Christian concentrations are found in the Lake Batak area of north-central Sumatra, the mountainous borderlands between southern Burma and Thailand, the northern peninsula of Sulawesi, and the highlands of southern Vietnam. **Animism** is still widespread in the mountains of northern Southeast Asia, central Borneo, far eastern Indonesia, and the highlands of northern Luzon in the Philippines. In Indonesia, however, animist practices are technically illegal, as monotheism is one of the country's founding principles. As a result, many Indonesian tribal groups have been gradually converting to either Islam or Christianity.

Indonesia has experienced pronounced religious strife over the past 10 years, especially between its Muslim majority and its Christian minority. Interfaith relations had been relatively good until the late 1990s. Indonesia, despite its Muslim majority, is a secular state that has always emphasized tolerance among its officially recognized religions (Islam, Christianity, Buddhism, Hinduism, and Confucianism). With the Indonesian economic disaster of the late 1990s, however, relations deteriorated. Religious conflicts broke out in many areas, especially in the Maluku Islands of eastern Indonesia and in central Sulawesi. Between 1999 and 2003, an estimated 2000 people were killed in Christian–Muslim clashes on the island of Sulawesi alone. Transmigration is also implicated in Indonesia's religious conflicts, as in many instances indigenous Christian and animist groups are now competing against Muslim immigrants from Java and Madura. In recent years, however, religious tensions have significantly decreased in most parts of the country.

Religion and Communism By 1975, communism had triumphed in Vietnam, Cambodia, and Laos. In all three countries, religious practices were strongly discouraged. At present, Vietnam's officially communist government is struggling against a revival of faith among the country's Buddhist majority and its 8 million Christians. Buddhist monks are frequently detained, and the government reserves for itself the right to appoint all religious leaders. Also expanding is Vietnam's indigenous religion of Cao Dai, a syncretic (or mixed) faith that venerates not only the Buddha, Confucius, and Jesus, but also the French novelist Victor Hugo. Between 3 and 8 million Vietnamese currently belong to this religious community.

Geography of Language and Ethnicity

As with religion, language in Southeast Asia expresses the long history of human movement. The linguistic geography of the region is complicated—too complicated, in fact, to be adequately conveyed in a single map. The several hundred distinct languages of the region can, however, be placed into five major linguistic families. These are *Austronesian,* which covers most of the insular realm; *Tibeto-Burman,* which includes most of the languages of Burma; *Tai-Kadai,* centered on Thailand and Laos; *Mon-Khmer* encompassing most of the languages of Vietnam and Cambodia; and *Papuan,* a Pacific language family found in eastern Indonesia (Figure 13.23).

The Austronesian Languages Austronesian is one of the world's most widespread language families, extending from

◄ **Figure 13.22 Indonesian Mosque** Indonesia is often said to be the world's largest Muslim country, as more Muslims reside here than in any other country in the world. Islam was first established in northern Sumatra, which is still the most devoutly Islamic part of Indonesia. Acehnese people are shown praying in front of the Baiturrahman Grand Mosque in the city of Banda Aceh. **Q: What are some of the political consequences of the high level of religious devotion found among most of the people of Aceh in northern Sumatra?**

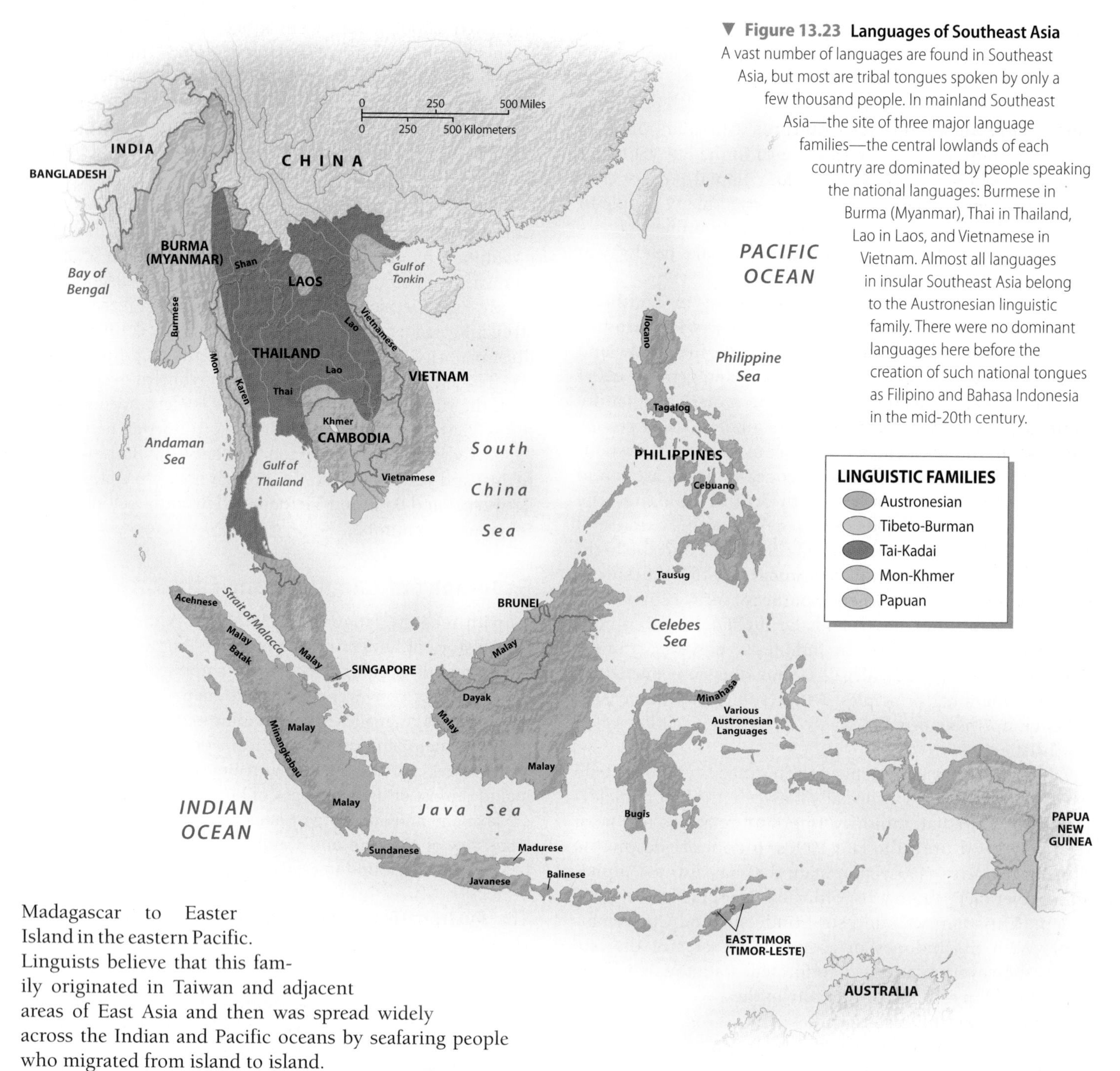

▼ **Figure 13.23 Languages of Southeast Asia** A vast number of languages are found in Southeast Asia, but most are tribal tongues spoken by only a few thousand people. In mainland Southeast Asia—the site of three major language families—the central lowlands of each country are dominated by people speaking the national languages: Burmese in Burma (Myanmar), Thai in Thailand, Lao in Laos, and Vietnamese in Vietnam. Almost all languages in insular Southeast Asia belong to the Austronesian linguistic family. There were no dominant languages here before the creation of such national tongues as Filipino and Bahasa Indonesia in the mid-20th century.

Madagascar to Easter Island in the eastern Pacific. Linguists believe that this family originated in Taiwan and adjacent areas of East Asia and then was spread widely across the Indian and Pacific oceans by seafaring people who migrated from island to island.

Today almost all insular Southeast Asian languages fall within the Austronesian family. This means that elements of grammar and vocabulary are widely shared across the islands. As a result, it is relatively easy for a person who speaks one of these languages to learn any other. But despite this linguistic commonality, more than 50 distinct Austronesian languages are spoken in Indonesia alone. And in far eastern Indonesia, a variety of languages falls into the completely separate family of Papuan, closely associated with New Guinea.

One language, however, overshadows all others in insular Southeast Asia: Malay. Malay is indigenous to the Malay Peninsula, eastern Sumatra, and coastal Borneo and was spread historically by merchants and seafarers. As a result, it became a common trade language, or **lingua franca,** understood and used by peoples speaking different languages throughout much of the insular realm. Dutch colonists in Indonesia eventually employed Malay as an administrative language, although they wrote it with the Roman alphabet, rather than in the Arabic-derived script used by native speakers. When Indonesia became an independent country in 1949, its leaders decided to use the lingua franca version of Malay as the basis for a new national language called *Bahasa Indonesia* (or more simply, *Indonesian*). Although Indonesian is slightly different from Malaysian, they essentially form a single, mutually intelligible language.

The goal of the new Indonesian government was to find a common language that would overcome ethnic differences throughout the far-flung state. In general, this policy has been successful. Indonesian is now used in schools, the government, and the media, and most of the country's people understand it. But regional languages, such as Javanese, Balinese, and Sundanese, continue to be primary languages in most homes. More than 100 million people speak Javanese, which makes it one of the world's major tongues.

The eight major languages of the Philippines also belong in the Austronesian family. Despite more than 300 years of colonization by Spain, Spanish never became a unifying force for the islands. During the American period (1898–1946), English served as the language of administration and education. After independence following World War II, Philippine nationalists decided to create a national language that could replace English and help unify the new country. They selected Tagalog, the language spoken in the Manila region. The first task was to standardize and modernize Tagalog, which had many dialects. After this was accomplished, it was renamed *Filipino* (alternatively, *Pilipino*). Today, mainly because of its use in education, television, and movies, Filipino has emerged as a unifying national language.

Tibeto-Burman Languages Each country of mainland Southeast Asia is closely identified with the national language spoken in its core territory. This does not mean, however, that all the inhabitants of these countries speak these official languages on a daily basis. In the mountains and other remote districts, other languages remain primary. This linguistic diversity reinforces ethnic divisions despite national educational programs designed to foster unity.

Burma provides a good example of ethnic and linguistic diversity. Its national language is Burmese, which is closely related to Tibetan and written in its own script (Figure 13.24). People who speak Burmese as their first language are considered to be members of the nationally dominant Burman ethnic group. Today some 32 million people fall into this category. Although the nationalistic government of Burma has attempted to force its version of unity on the entire population, a major schism developed with several non-Burman tribal groups inhabiting the rough uplands that flank the Burmese-speaking Irrawaddy Valley. Although most of these peoples speak languages in the Tibeto-Burman family, these languages are quite distinct from Burmese.

Tai-Kadai Languages The Tai-Kadai linguistic family probably originated in southern China and then spread into Southeast Asia starting around 1200 CE. Today closely related languages within the Tai subfamily are found through most of Thailand and Laos, in the uplands of northern Vietnam, in Burma's Shan Plateau, and in parts of southern China. Most Tai languages are quite localized, spoken by small ethnic groups. However, two of them, Thai and Lao, are national languages, spoken in Thailand and Laos, respectively.

Linguistic terminology here is complicated. Historically, the main language of the Kingdom of Thailand, called *Siamese* (just as the kingdom was called *Siam*), was restricted to the lower Chao Phraya Valley, which formed the national core. In the 1930s, however, the country changed its name to Thailand to emphasize the unity of all the peoples speaking closely related Tai languages within its territory. Siamese was similarly renamed *Thai*, and it has gradually become the unifying language for the country. There are still substantial variations in dialects, however, with those of the north often considered separate languages.

Lao, another Tai language, is the national tongue of Laos. More Lao speakers reside in Thailand than in Laos, however, where they form the majority population of the relatively poor Khorat Plateau of the northeast. The dialect of Lao spoken in northeastern Thailand, called Isan locally, seems to be declining in favor of standard Thai. Roughly a third of Thailand's population, however, still speaks Lao (Isan) rather than Thai as their first language.

Mon-Khmer Languages The Mon-Khmer language family probably covered almost all of mainland Southeast Asia 1500 years ago. It contains two major languages—Vietnamese (the national tongue of Vietnam) and Khmer (the national language of Cambodia)—as well as a host of minor languages spoken by hill peoples and a few lowland groups. Because of the historical Chinese influence in Vietnam, Vietnamese was usually written with Chinese characters until the French imposed the Roman alphabet, which

▼ **Figure 13.24 Burmese Script** Seen here on a Buddhist temple, the Burmese script has characters with round shapes, supposedly because the palm leaves that were originally used as writing surfaces would have been split by straight lines. Burma's script, like those of Thailand, Laos, and Cambodia, can be traced back to a writing system from ancient India.

remains in use. Khmer, like the other national languages of mainland Southeast Asia, is written in its own script, ultimately derived from India.

The most important aspect of linguistic geography in mainland Southeast Asia is probably the fact that in each country the national language is limited to the core area of densely populated lowlands, whereas the peripheral uplands are inhabited primarily by tribal peoples speaking separate languages. In Laos, up to 40 percent of the population is non-Lao. This linguistic contrast between the lowlands and the uplands poses an obvious problem for national integration. Such problems are most extreme in Burma, where the upland peoples are numerous and well organized—and have strenuously resisted the domination of the lowland Burmans. The Burmese government has, in turn, devoted much of its energies to resisting the influences of global culture.

▲ **Figure 13.25 Filipina Entertainer** Performers from the Philippines, who have long adopted global musical styles, are often in demand abroad. Charice (Charmaine Clarice Relucio Pempengco) is a young singer who rose to fame on the basis of her YouTube videos. She has performed extensively in other Asian countries and has appeared on the U.S. television show *Glee*.

Southeast Asian Culture in Global Context

The imposition of European colonial rule ushered in a new era of globalization in Southeast Asia, bringing with it European languages, Christianity, and new governmental, economic, and educational systems. This period also deprived most Southeast Asians of their cultural autonomy. As a result, with decolonialization after World War II, several countries attempted to isolate themselves from the cultural and economic influences of the emerging global system. Burma retreated into its own form of Buddhist socialism, placing strict limits on foreign tourism, which the government viewed as a source of cultural contamination. Although the door has opened substantially since the 1980s, the government of Burma remains wary of foreign practices and influences, an attitude that has become a major source of tension within the country. Burma's government is relatively tolerant of music, however, which has resulted in a surge of spontaneous concerts and underground music festivals. At these events, hip-hop and techo-beat styles are often mixed with more traditional Burmese musical forms.

Other Southeast Asian countries have been more receptive to foreign cultural influences. This is particularly true in the case of the Philippines, where American colonialism may have predisposed the country to many of the more popular forms of Western culture. As a result, Filipino musicians and other entertainers are in demand throughout East and Southeast Asia (Figure 13.25). Thailand, which was never subjected to colonial rule, is the most receptive country of the mainland to global culture, with its open policy toward tourism, mass media, and economic interdependence.

Cultural globalization has been recently challenged in some Southeast Asian countries. The Malaysian government, for example, is highly critical of many American films and television programs. Islamic revivalism, in both Malaysia and Indonesia, also presents a challenge to global culture. A recent poll, for example, found that 62 percent of Indonesians believe that U.S. culture is "disruptive" to Indonesian society. Singapore's leaders have also criticized cultural influences from the West, but the city-state remains committed to globalization. Singapore is also seeking to enhance tourism and bolster its economy by building Las Vegas–style casinos. A huge gambling complex called Resorts World Sentosa opened its doors in February 2010, welcoming 130,000 visitors in its first week of operation.

Language and Globalization As the global language, English causes ambivalence in much of Southeast Asia. On one hand, it is the language of questionable popular culture; yet on the other hand, citizens need proficiency in English if they are to participate in international business and politics. In Malaysia, widespread proficiency in English was challenged in the 1980s as nationalists stressed the importance of their native tongue. This movement distressed the business community, which considers English vital to Malaysia's competitive position, as well as the influential Chinese communities, for which Malaysian is not a native language.

In Singapore, the situation is more complex. Mandarin Chinese, English, Malay, and Tamil (from southeastern India) are all official languages. The local languages of southern China are also common in home environments, as 75 percent of Singapore's population is of southern Chinese ancestry. The Singapore government now encourages Mandarin Chinese and discourages southern Chinese dialects. It also supports English, while discouraging *Singlish,* an English-based dialect containing many southern Chinese words and grammatical forms. Singaporean officials have occasionally slapped restrictive ratings on local films for "bad grammar" if they contain too much Singlish dialogue.

In the Philippines, nationalists have long criticized the common use of English, even though widespread fluency has proved beneficial to the millions of Filipinos who have emigrated for better economic conditions or who work as crew members in the globalized shipping industry (see *Exploring Global Connections: Baltimore Turns to the Philippines to Recruit Teachers for Inner-City Schools*). The Philippine government has been gradually deemphasizing

Exploring **Global Connections**

Baltimore Turns to the Philippines to Recruit Teachers for Inner-City Schools

Many U.S. public schools, especially those in inner-city districts and rural areas, are having trouble recruiting teachers—particularly in math, science, and special education. Baltimore has pioneered a new strategy of hiring foreign educators. Most of those teachers come from the Philippines, a country whose public school system was established by American educators when the United States took possession of the Philippines. Today the Philippines has a large number of trained, motivated, English-speaking teachers, most of them women. Although salaries in U.S. inner-city school districts may be low by American standards, they can be as much as 25 times greater than those of the Philippines.

Baltimore's Gambit In 2005, Baltimore City Public Schools recruited 108 teachers from the Philippines to help meet staffing shortages in schools labeled "persistently dangerous." By 2009, more than 600 Filipino teachers, amounting to some 10 percent of the city's entire teaching force, were working in Baltimore (**Figure 13.3.1**). Several other school districts followed suit, including schools in Prince George County in Virginia and in other states, such as Texas, Georgia, New York, and California. The results have been quite impressive; for example, fourth-grade students in Baltimore improved 30 points in math and 11 points in reading between 2005 and 2010.

Cultural differences, however, along with students' poor, often-troubled backgrounds, have created numerous problems. Filipina teachers tend to approach their classrooms in what some American journalists have characterized as "an extraordinarily warm, disciplined and familial" manner. Some teachers have run into problems because they hugged their students too freely by American standards. Furthermore, pundits have criticized the recruitment process for making it too easy to avoid recruiting U.S. citizens for these jobs; instead of attending job fairs and trying to persuade American teachers and recent graduates in education to accept positions in troubled inner-city schools, Baltimore officials have simply turned to the Philippines.

Growing Controversies Due to such concerns, the recruitment of Filipina teachers to the United States has recently been curbed. Foreign educators typically enter the United States with a temporary, nonimmigrant H1-B visa, which allows American employers to temporarily employ foreign workers in specialty occupations. Usually, such visas are good for three years and can be renewed for another three years, during which time the visa-holder qualifies to apply for U.S. permanent residency. School authorities in Prince George County recently decided to renew the three-year visas for teachers in such "critical" areas as special education, English as a second language, and high school math and science, but not for teachers who specialize in "noncritical areas," such as pre-K education, home economics, music, and social studies. This decision prompted calls for the board of education to stop the dismissals, nonrenewal of visas, and impending terminations without valid reasons.

The program has also generated controversy in the Philippines, where educational authorities have expressed fear that the emigration of some of the best and brightest teachers will undermine domestic education. But considering the meager wages of teachers in the Philippines, such widespread desires to move abroad are hardly surprising. Many other former Filipina teachers, after all, now work as maids and nannies in Hong Kong, Singapore, and the Persian Gulf states.

▼ **Figure 13.3.1 Filipino Teachers in Baltimore** Several urban school districts in the United States have hired teachers from the Philippines, a country that sends many high-skilled emigrants to work abroad.

English in favor of Filipino in public schools, and, as a result, some fear that competence in English is slowly declining. Others counter that the spread of cable television, with its many U.S. shows, has enhanced the country's English language skills.

Sports and Globalization Southeast Asia is not particularly known for its athletic culture, although interest and participation have been growing, in part because of the Southeast Asian Games, an Olympic-style event that encompass all 11 countries of the region. As is true over most of the world, soccer is popular throughout Southeast Asia. Since 1996, ASEAN has held a biennial soccer championship for member countries; most of these events have been won by Singapore or Thailand. Basketball is also popular, and a Southeast Asian championship event is held every other year. In this sport, the Philippines is the acknowledged Southeast

▲ **Figure 13.26 Sepak Takraw** A fast-paced game that might be considered something of a cross between volleyball and soccer, sepak takraw originated in Thailand and Malaysia. The sport is now played throughout Southeast Asia and increasingly in other parts of the world as well.

Asian leader. Indonesia and Malaysia dominate the region in badminton, which is also popular in the region.

An indigenous Southeast Asian game that is gaining popularity both at home and abroad is *sepak takraw,* which is perhaps best described as a cross between volleyball and soccer (you cannot use your hands), played with a rattan ball. Top players must have gymnastic skill, as they often do complete flips in order to kick the ball with a downward motion over the net (Figure 13.26).

Martial arts also have widespread appeal across Southeast Asia. Most countries of the region have their own particular forms of fighting. In the Philippines, *arnis* (or *eskrima*) involves a well-developed set of techniques that generally involve the use of rattan sticks, although edged weapons are also used. *Pencak silat* is the umbrella term for martial arts in Indonesia and Malaysia. A wide array of weapons, both edged and blunt, are used in *Pencak silat* competitions. Although similar martial arts are found in Thailand, the country's best-known combat sport is *muay thai,* or Thai kickboxing. Competitors use a wide variety of punches, kicks, elbow jabs, and knee thrusts. Over the past several decades, *muay thai* has gained popularity across much of the world.

Although the diffusion of global cultural forms is beginning to merge cultures across Southeast Asia, religious revivalism operates in the opposite direction. This same combination of opposing forces is also found in the geopolitical realm.

REVIEW

13.7 What major world religions have spread into Southeast Asia over the past 2000 years, and in what parts of the region did they become established?

13.8 How have different Southeast Asian countries reacted to the challenges of cultural globalization?

13.9 How have such strongly multilingual countries as Indonesia, the Philippines, and Singapore dealt with the challenges posed by the lack of dominant language?

Geopolitical Framework: War, Ethnic Strife, and Regional Cooperation

Southeast Asia is sometimes defined as a geopolitical entity composed of the 10 different countries that form the Association of Southeast Asian Nations, or ASEAN (Figure 13.27). Today ASEAN, more than anything else, gives Southeast Asia regional coherence. East Timor—or Timor-Leste, as the country is officially called—gained independence only in 2002 and is not yet a member of ASEAN. In 2013, however, the ASEAN Coordinating Council Working Group announced that it would soon be reviewing the readiness of East Timor to join the regional associaiton. Despite the general success of ASEAN in reducing international tensions, several Southeast Asian counties are still struggling with serious ethnic and regional conflicts.

Before European Colonialism

The modern countries of mainland Southeast Asia were all indigenous kingdoms at one time or another before the onset of European colonialism. Cambodia emerged earliest, reaching its height in the 12th century, when it controlled much of what is now Thailand and southern Vietnam. By the 1300s, independent kingdoms had been established by the Burman, Siamese (Thai), Lao, and Vietnamese people in the major river valleys and deltas. These kingdoms were in a nearly constant state of war with one another, often fighting more for labor than for territory. Victors would typically take home thousands of prisoners to settle their lands, leading to considerable ethnic mixing.

The situation in insular Southeast Asia was different, with the premodern map bearing little resemblance to that of the modern nation-states. Many kingdoms existed on the Malay Peninsula and on the islands of Sumatra and Java, but few were territorially stable. In the Philippines, eastern Indonesia, and central Borneo, most societies were organized at the local level. Indonesia, the Philippines, and Malaysia thus owe their modern territorial configurations almost wholly to the European colonial influence (Figure 13.28).

The Colonial Era

The Portuguese were the first Europeans to arrive (around 1500), lured by the cloves and nutmeg of the Maluku region of eastern Indonesia. In the late 1500s, the Spanish conquered most of the Philippines, which they used as a base for their silver and silk trade between China and the Americas. By the 1600s, the Dutch had started staking out Southeast Asian territory, followed by the British. With superior naval weaponry, the Europeans were quickly able to conquer key

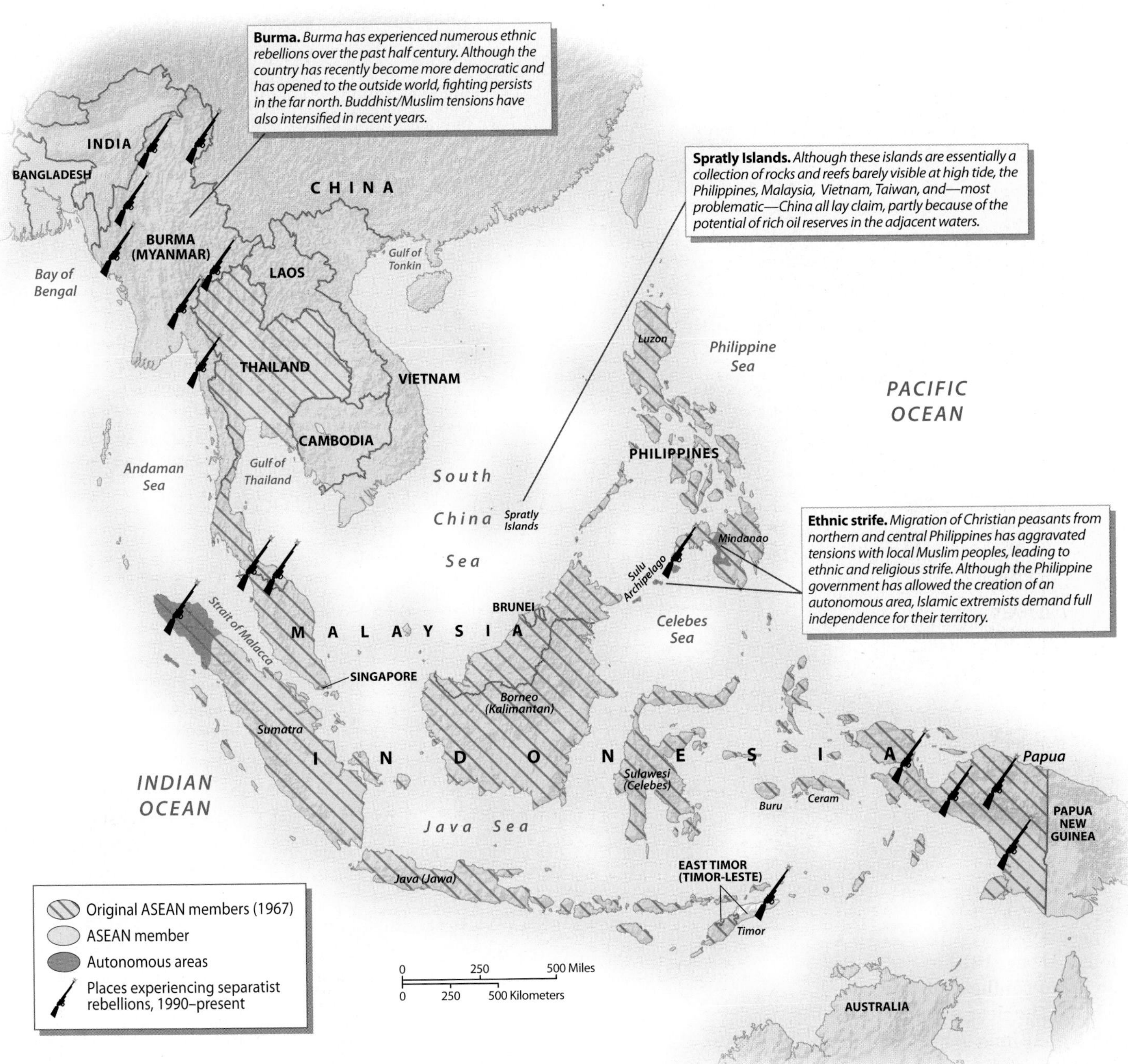

▲ **Figure 13.27 Geopolitical Issues in Southeast Asia** The countries of Southeast Asia have managed to solve most of their border disputes and other sources of potential conflicts through ASEAN (the Association of Southeast Asian Nations). Internal disputes, however, mostly focused on issues of religious and ethnic diversity, continue to plague several of the region's states, particularly Indonesia and Burma (Myanmar). ASEAN also experiences tension with China over the Spratly Islands of the South China Sea.

ports and strategic trading locales. Yet for the first 200 years of colonialism, except in the Philippines, the Europeans did not control large expanses of land in the region.

Dutch Power By the 1700s, the Netherlands had become the most powerful force in the region. As a result, a Dutch Empire in the "East Indies" (or Indonesia) began appearing on world maps. From its original base in northwestern Java, this empire continued to grow into the early 20th century, when the Dutch defeated their last main adversary, the powerful Islamic sultanate of Aceh in northern Sumatra. Later, the Dutch invaded the western portion of New Guinea in response to German and British advances in the eastern half. In a subsequent treaty, these imperial powers sliced New Guinea down the middle, with the Netherlands taking the west.

British, French, and U.S. Expansion The British, preoccupied with their empire in India, concentrated their attention on the sea-lanes linking South Asia to China. As a result, they established several fortified trading outposts along the vital Strait of Malacca, the most notable being on the island of

◀ **Figure 13.28 Colonial Southeast Asia** With the exception of Thailand, all of Southeast Asia was under Western colonial rule by the early 1900s. The Netherlands had the largest empire in the region, covering the territory that was later to become Indonesia. France maintained a substantial imperial realm in Vietnam, Laos, and Cambodia, as did Britain in Burma (Myanmar) and Malaysia (including Singapore and Brunei). The Philippines was colonized by Spain, but passed to the control of the United States in 1898. **Q: Note that Thailand was the only country in Southeast Asia that was not colonized by Western powers. How did Thailand avoid this fate, and what if any consequences did it have for the country's future development?**

Singapore, founded in 1819. To avoid conflict, the British and Dutch agreed that the British would limit their attention to the Malay Peninsula and northern Borneo. Britain allowed Muslim sultans to retain limited powers, and their descendants still enjoy token authority in several Malaysian states. When Britain left this area in the early 1960s, Malaysia emerged as an independent state. Two small portions of the former British sphere did not join the new country. In northern Borneo, the Sultanate of Brunei gained independence, backed by its large oil reserves. Singapore briefly joined Malaysia, but then withdrew and became fully independent in 1965. This divorce was carried out partly for ethnic reasons. Malaysia was to be a primarily Malay state, but with Singapore included its population it would have been almost half Chinese.

In the 1800s, European colonial power spread through most of mainland Southeast Asia. British forces in India fought several wars against the kingdom of Burma before annexing the entire area in 1885, including considerable upland territories that had never been under Burmese rule. During the same period, the French moved into Vietnam's Mekong Delta, gradually expanding their territorial control to the west into Cambodia and north to China's border. Thailand was the only country to avoid colonial rule, although it did lose substantial territories to the British in the Malay Peninsula and to the French in Laos. Thai independence was maintained partly because it served British and French interests to have a buffer state between their colonial realms.

The final colonial power to enter the area was the United States, which took the Philippines first from Spain and then, after a bitter war, from Filipino nationalists between 1898 and 1900. The U.S. army subsequently conquered the

Muslim areas of the southwest, most of which had never been fully under Spanish authority.

Growing Nationalism Organized resistance to European rule began in the 1920s in mainland countries, but it took the Japanese occupation during World War II to shatter the myth of European invincibility. After Japan's surrender in 1945, agitation for independence was renewed throughout Southeast Asia. As Britain realized that it could no longer control its South Asian empire, it also withdrew from adjacent Burma, which achieved independence in 1948. The Netherlands failed to reconquer Indonesia after World War II and was forced to acknowledge Indonesian independence in 1949. The United States granted long-promised independence to the Philippines on July 4, 1946, although it retained key military bases, as well as considerable economic influence, for several decades.

The Vietnam War and Its Aftermath

Unlike the United States, France was determined to maintain its Southeast Asian empire. Resistance to French rule was organized primarily by communist groups that were deeply rooted in northern Vietnam. As French forces returned to Southeast Asia in 1946, the leader of this resistance movement, Ho Chi Minh, organized a separatist government in the north. Open warfare between French soldiers and the communist forces went on for almost a decade. After a decisive defeat in 1954, France agreed to withdraw. An international peace council in Geneva decided that Vietnam would be divided into two countries. As a result, communist leaders came to power in North Vietnam, allied with the Soviet Union and China. South Vietnam simultaneously emerged as an independent, capitalist-oriented state with close political ties to the United States.

The Geneva peace accord did not end the war. Communist guerrillas in South Vietnam fought to overthrow the new government and unite it with the north. North Vietnam sent troops and war materials across the border to aid the rebels. Most of these supplies reached the south over the Ho Chi Minh Trail, an ill-defined network of forest passages through Laos and Cambodia that steadily drew these two countries into the conflict. In Laos, the communist Pathet Lao forces challenged the government, while in Cambodia the **Khmer Rouge** guerrillas gained considerable power. In South Vietnam, the government gradually lost control of key areas, including much of the Mekong Delta, a region perilously close to the capital city of Saigon.

U.S. Intervention By 1962, the United States was sending large numbers of military advisors to South Vietnam. In Washington, D.C., the **domino theory** became accepted foreign policy. According to this notion, if Vietnam fell to the communists, then so would Laos and Cambodia; once those countries were lost, Burma and Thailand, and perhaps even Malaysia and Indonesia, would join the Soviet-dominated communist bloc. In 1961, President John Kennedy sent 400 members of the U.S. Army Special Forces to train South Vietnamese troops, and the next year he promised additional assistance. The United States' involvement in the conflict seriously escalated in 1964, under the leadership of U.S. President Lyndon Johnson. One precipitating event was the so-called Gulf of Tonkin Incident, a naval clash between American and North Vietnamese forces. In response, the U.S. Congress gave the president wide powers to conduct military operations in Vietnam without the formal declaration of war.

After the Gulf of Tonkin Incident, U.S. efforts ramped up quickly. By 1968, over half a million U.S. troops were fighting a ferocious land war against the communist guerrillas (Figure 13.29). But despite superiority in arms and troops—and domination of the air—U.S. forces failed to gain control over much of the countryside. As casualties mounted and the antiwar movement back home strengthened, the United States began secret talks with North Vietnam in search of a negotiated settlement. By the early 1970s, U.S. troop withdrawals began in earnest.

Communist Victory With the withdrawal of U.S. forces and financial support, the noncommunist governments of the former French zone began to collapse. Saigon fell in 1975, and in the following year Vietnam was officially reunited under the government of the north. Reunification was a traumatic event in southern Vietnam. Hundreds of thousands of people fled to other countries, especially the United States. The first wave of refugees consisted primarily of wealthy professionals and businesspeople, but later migrants included many relatively poor ethnic Chinese. Most of these refugees fled on small, rickety boats; large numbers suffered shipwreck or pirate attack.

Vietnam proved fortunate compared with Cambodia. There the Khmer Rouge installed one of the most brutal regimes the world has ever seen. City dwellers were forced into the countryside to become peasants, and most wealthy and educated people were summarily executed. The Khmer Rouge's goal was to create a wholly new agrarian society by returning to what they called "year zero." After several years of bloodshed that took an estimated 1.5 million lives, neighboring Vietnam invaded Cambodia and installed a far less brutal, but still repressive regime.

▼ **Figure 13.29 U.S. Soldiers and Vietcong Prisoners** The United States maintained a substantial military presence in Vietnam in the 1960s and early 1970s. Although U.S. forces claimed many victories, they were ultimately forced to withdraw, leading to the victory of North Vietnam and the reunification of the country.

Fighting between different factions continued for more than a decade, but by the late 1980s the United Nations (UN) was able to broker a peace settlement. A parliamentary system of government subsequently brought peace to the shattered country. Although Cambodia is officially a constitutional monarchy with an elected government, corruption is widespread, and democratic institutions remain weak and unstable. In 2007, the UN set up a tribunal to try surviving Khmer Rouge officials who had allegedly been guilty of genocidal actions in the early 1970s. But as of July 2013, only five of the former top officials of the regime had been formally charged with crimes, and only one had been convicted.

Vietnam stationed significant numbers of troops in Laos after 1975. Large numbers of Hmong and other tribal peoples—many of whom had fought on behalf of the United States—fled to Thailand and the United States. The Communist Party still maintains a monopoly over political power in Laos, although much of the economy has been opened to private firms. Many Laotian Hmong refugees who fled to the United States after the war, however, continue to seek political change at home.

Geopolitical Tensions in Contemporary Southeast Asia

The most serious geopolitical problems in Southeast Asia today occur within countries rather than between them. In several areas, locally based ethnic groups struggle against centralized national governments that inherited territory from the former colonial powers. Tension also results when tribal groups attempt to preserve their homeland from logging, mining, or interregional migration. Such conflicts are especially pronounced in the large, multiethnic country of Indonesia.

Conflicts in Indonesia When Indonesia gained independence in 1949, it included all the former Dutch possessions in the region except western New Guinea. The Netherlands retained this territory, arguing that its cultural background distinguished it from Indonesia. Although Dutch authorities were preparing western New Guinea for independence, Indonesia demanded the entire territory. Bowing to pressure from the United States, the Netherlands relented and allowed Indonesia to take control in 1963. The Indonesian government formally annexed western New Guinea in 1969.

Tensions in western New Guinea increased in the following decades as Javanese immigrants, along with mining and lumber firms, arrived in the area. Faced with the loss of their land and the degradation of their environment, indigenous residents formed the separatist organization OPM (*Organisesi Papua Merdeka*) and launched a rebellion. Rebel leaders demanded independence, or at least autonomy, but they faced a far stronger force in the Indonesian army. The struggle between the Indonesian government and the OPM is currently a sporadic, but occasionally bloody guerrilla affair. Indonesia is determined to maintain control of the region in part because it is home to one of the country's largest taxpayers, the highly polluting Grasberg copper and gold mine run by the Phoenix-based Freeport-McMoRan Corporation (Figure 13.30).

A more intensive war erupted in 1975 on the island of Timor. The eastern half of this poor island had been a Portuguese colony (the only survivor of Portugal's 16th-century foray into the region) and had evolved into a mostly Christian society. The East Timorese expected independence when the Portuguese finally withdrew. Indonesia, however, viewed the area as its own territory and immediately invaded. A ferocious struggle ensued, which the Indonesian army won in part by starving the people of East Timor into submission.

After the economic crisis of 1997, Indonesia's power in the region slipped. A new Indonesian government promised an election in 1999 to see if the East Timorese wanted independence. At the same time, however, Indonesia's army began to organize loyalist militias to intimidate the people of East Timor into voting to remain within the country. When it was clear that the vote would be for independence, the militias began rioting, looting, and slaughtering civilians. Under international pressure, Indonesia finally withdrew, UN forces arrived, and the East Timorese began to build a new country.

▼ **Figure 13.30 Grasberg Mine** Grasberg, located in the Indonesian province of Papua, is the world's largest gold mine and third largest copper mine. Employing more than 19,000 people, Grasberg is of great economic importance to Indonesia. It is also highly polluting, generating 253,000 tons (230,000 metric tons) of tailings per day and is thus opposed by many local inhabitants.

Considering the devastation caused by the militias, the reconstruction of East Timor was not easy. Even selecting an official language proved difficult; after some deliberation, both Tetum (one of 16 indigenous languages) and Portuguese were selected. In 2006, ethnic rioting further damaged the country, requiring Australian intervention to reestablish peace. A relatively successful election in 2007 brought hope that East Timor could overcome its divisions, but in 2008 a well-coordinated attack by former soldiers and police officers came close to killing the country's president and prime minister. As a result of such instability, a UN peacekeeping force remained in the country until the end of 2012.

Secession struggles have occurred elsewhere in Indonesia. In the late 1950s, central Sumatra and northern Sulawesi rebelled, but they were quickly defeated and eventually reconciled to Indonesian rule. Another small-scale war flared up in the late 1990s in western and southern Borneo (Kalimantan). Here the indigenous Dayaks, a tribal people partly converted to Christianity, began to clash with Muslim migrants from Madura, a densely inhabited island north of Java. The Indonesian military has restored order in Kalimantan's cities, but the countryside remains troubled. The southern Maluku Islands, especially Ambon and Seram, have also seen fighting between Muslims and Christians. Even the area's cities are now divided into Muslim and Christian sectors, and crossing to the wrong side can prove dangerous.

Indonesia's most serious regional conflict has long been that of Aceh in northern Sumatra. Many Acehnese—in general, Indonesia's most devout Muslim group—have demanded the creation of an independent Islamic state. Although the government has allowed Aceh a high degree of autonomy as a "special territory of Indonesia," it has been determined to prevent actual independence. Ironically, the devastation caused by the December 2004 tsunami, which left over 500,000 Acehnese homeless, allowed a peace settlement to be finally reached, as the needs of the area were so great that the separatist fighters agreed to lay down their weapons. A 2006 election brought former rebel leaders into the heart of Aceh's government. Their use of strict Islamic law has proved very controversial. In 2013, a series of bills was put forth in the legislature of Aceh that would prohibit women from wearing tight clothing and allow the stoning of adulterers and the flogging of homosexuals.

Indonesia has obviously had difficulties creating a unified nation over its vast, sprawling extent and across its numerous cultural and religious communities. As a creation of the colonial period, Indonesia has a weak historical foundation. Some critics have viewed it as something of a Javanese empire, even though many non-Javanese individuals have risen to high governmental positions. Indonesia now has the highest degree of political freedom in Southeast Asia, but its politically influential military remains suspicious of movements for regional autonomy.

Regional Tensions in the Philippines The Philippines has long been struggling against a regional secession movement in its Islamic southwest. In the 1990s, it sought to defuse tensions by creating the Autonomous Region in Muslim Mindanao (ARMM), which would provide a degree of political and cultural autonomy to the main Muslim areas of the country. However, the more extreme Islamist factions continued to fight, demanding greater autonomy, if not outright independence. In 2012, a new peace treaty was signed based on the promise that autonomy would be enhanced, with the autonomous area renamed Bangsamoro. But despite these accommodations, the hard-line group Abu Sayyaf, reportedly linked to Al Qaeda, continues to demand full independence and continues to fight against the Philippine government. Several hundred U.S. military advisors are currently working with the Philippine military in the Mindanao region. As a result, some Filipino nationalists argue that their country is again falling under U.S. domination.

The Philippines' political problems are not limited to the south. A long-simmering communist rebellion has affected almost the entire country. In the mid-1980s, the New People's Army (NPA) controlled one-quarter of the country's villages scattered across all the major islands. Although the NPA's strength has declined since then, it remains a potent force in many areas. In May 2013, after the government suspended peace talks with the group, NPA rebels attacked an elite police commando unit, killiing seven troops and wounding eight more. The Philippine national government, although democratic, is itself not particularly stable, suffering from periodic coup threats, corruption scandals, and mass protests.

The Opening of Burma? Burma has also long been a war-ravaged and relatively isolated country. Its simultaneous wars of the 1970s, 1980s, and 1990s pitted the central government, dominated by the Burmans, against the country's varied non-Burman societies. Fighting intensified gradually after independence in 1948, until almost half of the country's territory had become a combat zone (Figure 13.31). Burma's troubles, moreover, are by no means limited to the country's ethnic minorities.

From 1988 until 2011, Burma was ruled by a repressive military regime bitterly resented by much of the population. The Burmese state at the time was also extremely suspicious of the outside world. In 2005, it mandated the creation of a new capital city at Nay Pyi Taw, located in a remote, forested area 200 miles north of the old capital of Rangoon (Yangon). The new capital, which was supposedly picked on the basis of astrological calculations and defense considerations, was to be largely closed to the outside world. Such actions, however, did not generate stability. In 2007, Burma was convulsed by massive pro-democracy protests against the government, led by Buddhist monks. Despite the religious nature of the protests, they were repressed harshly.

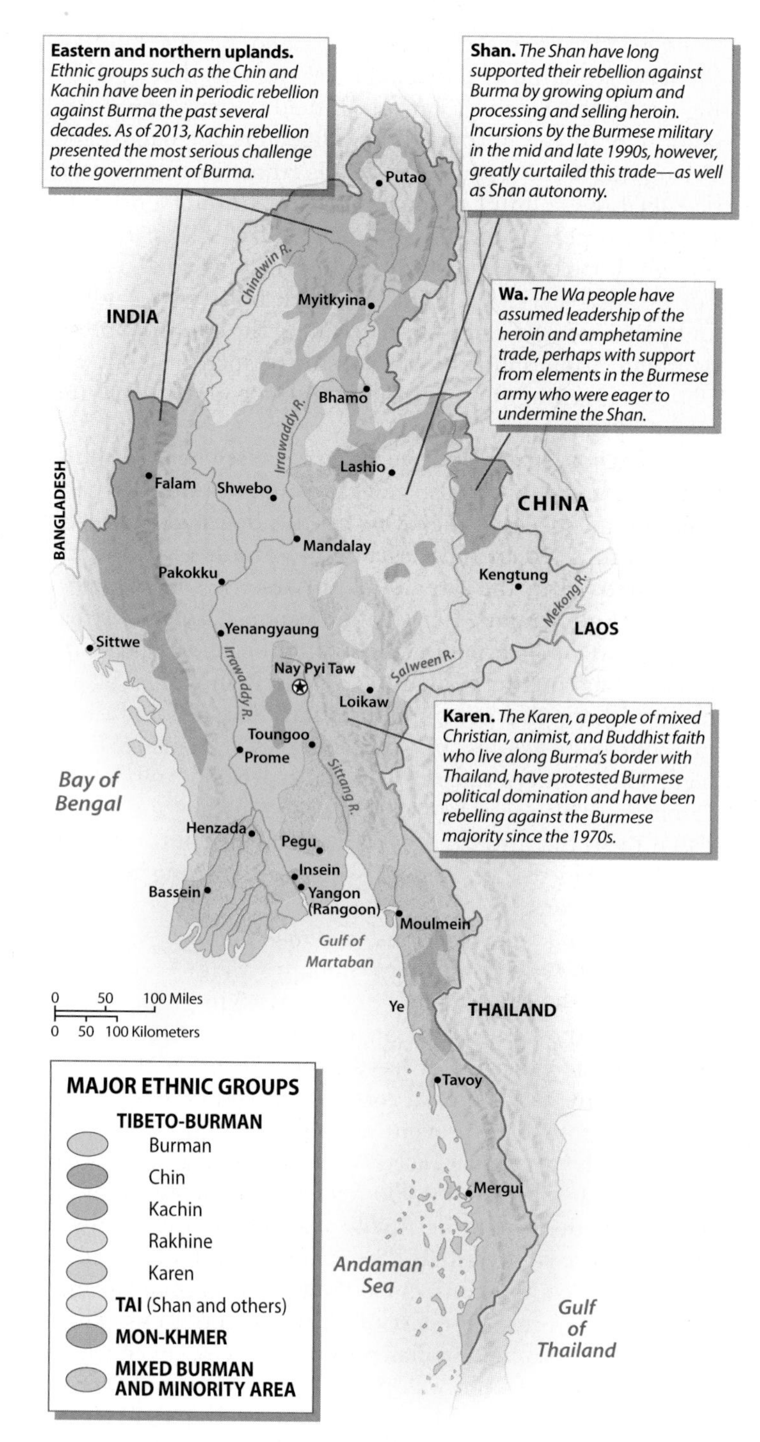

▶ Figure 13.31 Ethnic Groups and Conflicts in Burma (Myanmar) Although the central lowlands of Burma (Myanmar) are primarily populated by people of the dominant Burman ethnic group, the peripheral highlands and the southeastern lowlands are home to numerous non-Burman peoples. Most of these peoples have been in periodic rebellion against Burma (Myanmar) since the 1970s, owing to their perception that the national government is attempting to impose Burman cultural norms. Economic stagnation and political repression by the central government have intensified these conflicts.

More recently, however, Burma has undergone major changes, making progress toward democratization. Significantly, in 2011, its government released the democratic opposition leader and Nobel Peace Prize recipient Aung San Suu Kyi from house arrest, allowing her to participate in the political process; she has recently declared her intention of running for the presidency of Burma in 2015. Burma has also opened itself to the rest of the world to a significant extent, seeking better relations with the European Union, the United States, and other democratic states. But despite these changes, Burma continues to be a troubled country, burdened by ethnic conflicts and the lack of effective governmental institutions.

The discontented ethnic groups of Burma have sought to maintain authority over their own cultural traditions, lands, and resources. They generally see the national government as a Burman institution seeking to impose its language and, in some cases, its religion upon them. Most of the insurgent ethnic groups live in rugged and inaccessible terrain, but even some lowland groups have rebelled. Partly as a result of these problems, many Burmese people have migrated to Thailand, even though they generally face many difficulties there (see *People on the Move: Burmese Migrants in Thailand*).

The Muslim peoples of the Arakan coast, known as the Rohingyas in western Burma, have faced especially severe discrimination. The government regards them as recent immigrants from Bangladesh and therefore denies them citizenship. Many Rohingyas have been forced out of the country, and thousands more live in squalid refugee camps in Burma (Figure 13.32). In early 2013, radical Buddhist mobs attacked several communities of Rohingyas and other Burmese Muslims, creating a human rights catastrophe. As a result, UN Secretary-General Ban Ki-moon demanded that the Burmese government award citizenship to the Rohingyas and afford basic protection and human rights to all of the country's minority groups. This demand was repeated in August 2013, but so far without result.

Several of Burma's ethnic insurgencies have been financed largely by opium growing and heroin manufacture. The Shan, a Tai-speaking people inhabiting a plateau area within the Golden Triangle of drug production, formed a breakaway state in the 1980s and 1990s based largely on the narcotics trade. Their efforts faltered in the late 1990s after Burmese agreements with Thailand reduced the Shan heroin trade, just as military operations cut off the supply of raw opium reaching the Shan factories. The government's triumph over the Shan was also made possible by alliances with rival ethnic groups that subsequently moved into the drug trade. The most powerful of these groups is the United Wa State Army (UWSA), which controls several heroin and methamphetamine factories and maintains over 20,000 heavily armed soldiers. Over the past several years, Wa leaders have been resisting efforts by Burma to incorporate their militia into the Burmese army. In 2013, independent military analysts announced that China was supplying the UWSA with advanced weaponry, including missile-equipped helicopters. Such allegations, however,

▲ **Figure 13.32 Rohingya Refugee Camp** Tens of thousands of Muslim Rohingyas have been driven out of Burma. Most of these refugees live in grim camps in neighboring Bangladesh.

were denied by both China and Wa state officials. Peace talks between the government and the Wa began in July of the same year.

Other ethnic conflicts persist elsewhere in Burma. For many years, the struggle was most intensive along the border with Thailand, home to the several million members of the Karen ethnic group. Although Burmese military victories resulted in reduced conflict, the human rights situation in the region remains grim, and thousands of people continue to lose their legs, if not their lives, every year from land mines, which are still being scattered by the Burmese army. Since 2011, the most persistent fighting has been in northern Burma, where the mostly Christian Kachin ethnic group has been seeking both autonomy and the end of dam-building and resource-extraction projects. Reports from early 2013 claimed that the Burmese army was gaining ground on the poorly funded Kachin rebels.

Trouble in Thailand Compared to Burma's recent history, that of Thailand appears peaceful and stable. Thailand enjoys basic human freedoms and a thriving free press, although it does have a legacy of military takeovers followed by periods of authoritarian rule. In 2006, mass protests in Bangkok led to the resignation of the corruption-plagued prime minister Thaksin Shinawatra, who had remained very popular among poor Thai voters in rural areas. Several months later, the Thai army seized power in a bloodless coup, apparently with the blessing of King Bhumibol, Thailand's revered constitutional monarch. Although the new military leaders allowed a quick restoration of democracy, they also banned several political movements, substantially reducing the freedom of the Thai people. Huge protests and counterprotests followed for several years, causing major problems. In 2011, Thaksin's sister, Yingluck Shinawatra, became prime minister after winning a landslide victory. Although many observers predicted continuing troubles, more stable conditions have apparently returned to Thailand.

Conflict persists, however, in Thailand's far south, a primarily Malay-speaking, Muslim region. Minor rebellions have flared up in southern Thailand for decades, but the violence sharply escalated in 2004. The Thai government initially responded with harsh military measures, a strategy that many Thai military leaders viewed as counterproductive. But the more conciliatory stance taken by the Thai government after 2006 has been no more effective, as a shadowy Islamist group continues to attack Thai-speaking, Buddhist government officials, as well as ordinary civilians. From 2004 to July 2013, more than 4000 Thai citizens lost their lives in this conflict. The war has also raised tensions between Thailand and Malaysia, as Thailand accuses its southern neighbor of not doing enough to prevent militants from slipping across the border.

International Dimensions of Southeast Asian Geopolitics

As events in southern Thailand show, geopolitical conflicts in Southeast Asia can be complex affairs, involving several different countries, as well as nonstate organizations. Historically, some of the most serious tensions emerged when two countries claimed the same territory. More recently, radical Islamist groups have posed the biggest challenge.

Territorial Conflicts In previous decades, several Southeast Asian countries quarreled over their boundaries. The Philippines, for example, still maintains a "dormant" claim to the Malaysian state of Sabah in northeastern Borneo, based on the fact that in the 1800s the Islamic Sultanate of the Sulu Archipelago in the southern Philippines controlled much of its territory. With the rise of ASEAN, the Philippines government concluded that maintaining peaceful relations with neighbors were more important than pursuing claims to additional territory and let the issue go dormant. It was therefore deeply embarrassed in early 2013 when a group of 200 armed Filipinos, led by a descendant of the former Sultan of Sulu, landed in a Sabah town to claim it as their own. When Malaysian troops retook the town, more than 50 Philippine citizens were killed.

More difficult has been the dispute over the Spratly Islands in the South China Sea, a group of rocks, reefs, and tiny islands that may well lie over substantial undersea oil reserves (Figure 13.33). The Philippines, Malaysia, Vietnam, and Brunei have all advanced territorial claims over the Spratly Islands, as have China and Taiwan. In 2002, all parties pledged to seek a peaceful solution, but by 2010, military analysts highlighted a worrisome naval buildup in the area, particularly on the part of China, Indonesia, Malaysia, and Singapore. Over the next

People on the Move

Burmese Migrants in Thailand

For decades, Thailand has been a major destination for migrants from neighboring Burma. In the past, members of ethnic groups residing along the Thai-Burma border, such as the Karen, the Mon, and the Shan, often crossed the borders to visit friends, buy goods, or seek health-care services. In the 1980s, under the military regime in Burma, this temporary migration continued unofficially, even though border crossings were not officially allowed. A large number of asylum seekers fighting against the Burma government started to enter Thailand to take refuge in the same period. Since the 1990s, migrants from Burma have come to Thailand mostly for economic reasons.

The Lures and Challenges of Thailand In Burma, slow economic growth, high unemployment, and forced labor for government development projects have pushed workers to cross into Thailand for job opportunities and higher wages (Figure 13.4.1). Although the minimum daily wage is Thailand only US$10, such pay can be as much as 10 times higher than what workers can get in Burma, due in part to differences in currency strength. Most of the Burmese migrant workers are employed in unskilled occupations that are not desired by local Thai workers—particularly in agriculture, construction, fisheries, and domestic service.

Regardless of their ultimate occupation, most migrants from Burma come from farming backgrounds. As a result, they have to gain new skills in Thailand. Learning the Thai language is particularly important. Yet there are no official training programs for migrant workers, and the children of Burmese immigrants often receive no education at all. A new school recently opened near Bangkok, however, that will offer the children of Burmese workers a chance to study Burma's national curriculum in the Burmese language, opening up educational possibilities.

Legal and Political Issues Currently, over a million migrant workers from Burma are registered through Thailand's National Verification Program, aimed at giving such people some protection under Thai labor laws. Migrants from Burma constitute as much as 10 percent of Thailand's total workforce and over 80 percent of its legal foreign workforce (the remaining legal migrant workers in Thailand come chiefly from Cambodia and Laos). But many migrants cross the Burma–Thailand border illegally and avoid registering with the authorities. They do so primarily because Thai employers often try to recoup transport, visa, and other costs by making deductions from the workers' salaries or by holding their passports until they get their money back. Some sources estimate the number of illegal workers to be around 1 million, whereas others think that a figure of 3 million is more reasonable.

Many of the illegal migrants come with the help of human smugglers, often risking their lives in doing so. In a much-publicized incident in June 2013, at least 12 Burmese migrant workers drowned off Thailand's southwestern coast when the boats they were in sank. Many illegal migrants who do make it to Thailand are forced to live in slavelike conditions, working for up to 20 hours a day, often for months on end, and with little or no pay. Illegal migrants also face arrest and deportation. The fact that so many Burmese still cross the border into Thailand indicates how desperate the economic situation is in their homeland.

1. How are the different levels of economic development in Thailand and Burma evident in the Mae Sot/Myawaddy region?
2. Why does the boundary between the two countries deviate from the course of the Moei River in several places?

Google Earth Virtual Tour Video

http://goo.gl/NIMal0

▼ **Figure 13.4.1 Burmese Migrants in Thailand** Most Burmese migrants in Thailand work in dirty, dangerous, and difficult jobs that do not pay well. These Burmese workers are sorting squid in front of their apartment building in the Thai port of Mahachai.

several years, tensions escalated, especially between China and the Philippines. As a result, the Philippines has sought closer military ties with the United States. In 2013, the Philippine government requested UN arbitration with China under the 1982 UN Convention on the Law of the Sea, or UNCLOS. This maneuver, however, is unlikely to be successful, as UNCLOS has not been a very effective agreement and China has rejected arbitration.

Owing primarily to the South China Sea conflict, relations between China and ASEAN as a whole have deteriorated, although China retains close ties with Cambodia and Laos. Relations between Burma and China were quite close

◀ **Figure 13.33 The Spratly Islands** These tiny Islands are small and barely above water at high tide, but they are geopolitically important. Oil may exist in large quantities in the vicinity, heightening the competition over the islands. Southeast Asian countries have been especially concerned about China's military activities in the Spratlys.

until recently, but they cooled off considerably after 2011, when Burma cancelled a major Chinese-financed dam project and sought to open its economy to the rest of the world. Vietnam, on the other hand, has long been in a tense relationship with China, despite the fact that they have similar market-oriented communist governments. Vietnam is particularly angered by the fact that China controls the Paracel Islands, another archipelago in the South China Sea, which Vietnam claims as its own territory.

ASEAN and Global Geopolitics Within Southeast Asia itself, the development and enlargement of ASEAN reduced geopolitical tensions. Initially, ASEAN was an alliance of nonsocialist countries fearful of communist insurgency. Through the 1980s, the United States maintained naval and air bases in the Philippines, and U.S. military force bolstered the anticommunist coalition. In the early 1990s, however, the United States, under mounting pressure from Filipino nationalists, withdrew from the Philippines. By that time, the struggle between communism and capitalism was no longer an international issue. In 1995, communist Vietnam itself gained membership in ASEAN, followed by Laos and Cambodia.

Although ASEAN is on friendly terms with the United States, one purpose of the organization is to prevent the United States—or any other country—from gaining undue influence in the region. ASEAN leaders are thus keen to include all Southeast Asian countries within the association. ASEAN's ultimate international policy is to encourage conversation and negotiation over confrontation and to enhance trade. As a result, the ASEAN Regional Forum (ARF) was established in 1994 as an annual conference in which Southeast Asian leaders could meet with the leaders of both East Asian and Western powers to discuss issues and ease tensions. Seeking to further improve relations with East Asia, ASEAN leaders also established an annual ASEAN+3 meeting, where its foreign ministers confer with those of China, Japan, and South Korea. Economic integration within Southeast Asia is also a priority, as national leaders hope to create an "ASEAN Economic Community" by 2015 that would function as a single market. Progress toward these goals, however, has thus far been halting.

Global Terrorism and International Relations ASEAN has not defused all of the international political tensions in Southeast Asia. Most worrisome has been the rise of radical Islamic fundamentalism in the Muslim parts of the region. One group, Jemaah Islamiya (JI), which calls for the creation of a single Islamic state across Indonesia, Malaysia, southern Thailand, and the southern Philippines, has cooperated extensively with Al Qaeda. In 2002, JI agents blew up a tourist-oriented nightclub in Bali, killing 202 people. The group followed this by bombing an American-run hotel in Jakarta in 2003, setting off explosions in Australia's Indonesian embassy in 2004, and attacking three tourist restaurants in Bali in 2005.

By 2006, however, Indonesian authorities began to make headway against JI. Some of its leaders were arrested, and top-level operatives were killed in security raids; in 2012, the Philippines claimed to have taken out two of the group's highest-level leaders. As a result, some analysts believe that JI is no longer an effective fighting force. Most evidence shows that few Southeast Asian Muslims support radical groups such as JI. Since 2004, even mainstream Muslim political parties have not done particularly well in national elections in Indonesia and Malaysia. Although radical Islam has relatively few supporters in Southeast Asia, the people of the region have generally been wary of U.S. foreign policy initiatives designed to combat global terrorism. Resentment against the United States is especially pronounced in Indonesia and Malaysia, where most people see the conflicts in Iraq and Afghanistan as an assault on Islam.

REVIEW

13.10 How did European colonization influence the development of the modern countries of Southeast Asia, and how did that process differ in the insular and the mainland regions?

13.11 How did the emergence and spread of ASEAN reduce geopolitical tensions in Southeast Asia, and why was the ASEAN process unable to resolve some conflicts in the region?

13.12 Why has Burma suffered from so many ethnic conflicts over the past 50 years?

Economic and Social Development: The Roller-Coaster Ride of Developing Economies

In the 1980s and early 1990s, Southeast Asia was often held up as a model for a new globalized capitalism. With investment capital flowing from Japan, the United States, and other developed countries, Thailand, Malaysia, and Indonesia experienced impressive economic booms. In 1997, however, regional economies suffered a profound crisis, with the currencies of both Thailand and Indonesia being devalued almost 50 percent. Subsequent years have seen economic ups and downs, with expansion between 2002 and 2007, turning again to recession with the global economic crisis of 2008–2009, followed by renewed recovery in 2010.

Uneven Economic Development

Southeast Asia today is a region of strikingly uneven economic and social development. Some countries, such as Indonesia, have experienced both booms and busts; others, such as Burma, Laos, and East Timor, missed the expansion of the 1980s and 1990s and remain impoverished. Oil-rich Brunei and technologically sophisticated Singapore, on the other hand, remain two of the world's more prosperous countries (Table 13.2). In the Philippines, the situation is somewhat more complicated.

The Philippine Decline Sixty years ago, the Philippines was the most highly developed Southeast Asian country and was considered by many to have the brightest prospects in all of Asia. It boasted the best-educated populace in the region, and it seemed to be on the verge of sustained economic development. Per capita gross national income (GNI) in 1960 was higher in the Philippines than in South Korea. By the late 1960s, however, Philippine development had been derailed. Through the 1980s and early 1990s, the country's economy failed to outpace its population growth, resulting in declining living standards. The Philippine people are still well educated and reasonably healthy by world standards, but even the country's educational and health systems declined during the dismal decades of the 1980s and 1990s.

Why did the Philippines fail so spectacularly despite its earlier promise? Although there are no simple answers, it is clear that dictator Ferdinand Marcos (who ruled from 1968 to 1986) squandered billions of dollars, while failing to enact programs conducive to genuine development. The Marcos regime instituted a kind of **crony capitalism** in which the president's many friends were granted huge sectors of the economy, while those perceived to be enemies had their properties confiscated. After Marcos declared martial law in 1972 and suspended Philippine democracy, revolutionary activity intensified, and the country began to fall into a downward spiral. Although it is tempting to blame the failure of the Philippines on the Marcos regime, such an explanation is only partially adequate. Indonesia and Thailand also saw the development of crony capitalism, yet their economies have proved more competitive. Also, when the Marcos dictatorship was finally replaced by an elected democratic government in 1986, the Philippine economy continued to languish.

Table 13.2 Development Indicators

Country	GNI per capita, PPP 2011	GDP Average Annual % Growth 2000-11	Human Development Index (2011)[1]	Percent Population Living Below $2 a Day	Life Expectancy (2013)[2]	Under Age 5 Mortality Rate (1990)	Under Age 5 Mortality Rate (2011)	Adult Literacy (% ages 15 and older) Male/Female	Gender Inequality Index (2011)[3]
Burma (Myanmar)	–	–	.498	–	65	107	62	95/90	0.437
Brunei	49,910	1.2	.855	–	78	12	7	97/94	–
Cambodia	2,230	8.4	.543	49.5	62	117	43	83/66	0.473
East Timor	5,200	5.6	.576	–	66	180	54	64/53	–
Indonesia	4,500	5.4	.629	46.1	70	82	32	96/90	0.494
Laos	2,580	7.3	.543	66.0	67	148	42	88/75	0.483
Malaysia	15,650	4.9	.769	2.3	75	17	7	95/91	0.256
Philippines	4,140	4.9	.654	41.5	69	57	25	95/96	0.418
Singapore	59,380	6.0	.895	–	82	8	3	98/94	0.101
Thailand	8,360	4.2	.690	4.1	75	35	12	96/92	0.360
Vietnam	3,250	7.3	.617	43.4	73	50	22	95/91	0.299

[1]United Nations, *Human Development Report, 2013.*

[2]Population Reference Bureau, *World Population Data Sheet, 2013.*

[3]Gender Inequality Index—A composite measure reflecting inequality in achievements between women and men in three dimensions: reproductive health, empowerment and the labor market that ranges between 0 and 1. The higher the number, the greater the inequality.

Source: World Bank, *World Development Indicators, 2013.*

In the early years of the new millennium, however, the Philippine economy showed some signs of revival, and by 2012 it was growing at a relatively rapid rate of 6.6 percent. The government turned its attention to infrastructure problems, such as electricity generation, and foreign investments began to flow into the country. A vibrant local economy emerged in the former U.S. naval base of Subic Bay, which boasts both world-class shipping facilities and a highly competent local government (Figure 13.34). Cebu City, on the central Visayan island of Cebu, also has expanded quickly, giving rise to its local nickname of "Ceboom." The Philippines has also capitalized on language skills to develop a major "call center" industry, offering technical assistance and other services by telephone to customers in the United States and other English-speaking countries. Its banking system is now sound, and it has accumulated substantial foreign reserves, leading to a new sense of optimism.

▲ **Figure 13.34 Subic Bay, the Philippines** The former U.S. naval base in Subic Bay, the Philippines, is now a thriving industrial and export-processing center.

Despite its recent stability, the Philippine economy continues to be undermined by political and social problems. The Philippine political system, modeled on that of the United States, entails elaborate checks and balances between the different branches of government. Critics contend that such "checks" are so effective that it is difficult to accomplish anything. Another obstacle is the fact that the Philippines has the least equitable distribution of wealth in the region. Many members of the elite are fantastically wealthy by any measure, but roughly half of the country's people subsist on less than $2 a day. As a result, many Filipinos must seek employment abroad; women often work as nurses, maids, or nannies, whereas men often work in construction or on ships. Remittances sent by Filipinos working abroad have greatly helped the national economy, but this exodus of labor represents in many respects a tragedy for the country as a whole.

The Regional Hub: Singapore Singapore presents a profound contrast to the Philippines—and indeed to all other Southeast Asian countries. Over the past half century, this small city-state has been Southeast Asia's greatest developmental success. It has transformed itself from an **entrepôt** port city—a place where goods are imported, stored, and then transshipped—to one of the world's most prosperous and modern states. Singapore, a thriving high-tech manufacturing center, also stands as the communications and financial hub of Southeast Asia. The Singaporean government has played an active role in the development process, but has also allowed market forces freedom to operate. Singapore has encouraged investment by multinational companies (especially those involved in technology) and has invested heavily in housing, education, and some social services (Figure 13.35). Roughly 85 percent of Singapore's residents live in public housing provided by the country's Housing and Development Board.

Although Singapore has experienced strong, sustained economic growth for the past several decades, in recent years its economy has cooled down considerably. In 2012, for example, it grew at the anemic rate of 1.3 percent. Some observers worry that Singapore has become too expensive for manufacturing, which has been increasingly relocating to other countries. The Singaporean government has responded by encouraging high tech, particularly biotechnology. Banking is also promoted, and by some measures Singapore is the world's fourth largest financial center, trailing only London, New York, and Hong Kong. Critics contend, however, that Singapore tolerates too much banking secrecy, allowing tainted money to be safely stored in its accounts.

Singapore's political system has nurtured economic development, but it is somewhat repressive and by no means fully democratic. Although elections are held, the government manipulates the process, ensuring that the opposition never gains control of parliament. Freedom of speech, moreover, is limited by libel and slander laws that make it easy for the Singaporean government to successfully sue its critics. Many Singaporeans object to such policies, but others counter that they have brought fast growth, as well as a clean, safe, and remarkably corruption-free society with a negligible unemployment rate.

Although the Singaporean government has thus far been able to repress dissent, its authoritarian form of capitalism confronts a new challenge in the Internet. National leaders want the communication services that the Net provides, but they are worried about the free expression that it allows, fearing that it will lead to excessive individualism. In 2013, the U.S. government announced that it is "deeply concerned" about Singapore's establishment of new restrictive laws for the licensing of online news websites. It will be interesting to see how Singapore responds to this challenge, in part because some experts claim that the governments of China and Vietnam are following the technocratic, authoritarian capitalism pioneered by Singapore.

▲ **Figure 13.35 Housing in Singapore** Despite its free-market approach to economics, the government of Singapore has invested heavily in public housing. Many Singaporeans live in buildings similar to the ones depicted here. **Q: Singapore is in general a conservative country devoted to free-market economics. Considering this fact, why then has its government invested so heavily in public housing?**

Malaysia's Solid Economy Although not nearly as prosperous as Singapore, Malaysia has experienced rapid economic growth over the past several decades and is now generally classified as an upper-middle-income country. Development was initially concentrated in agriculture and natural resource extraction, focused on tropical hardwoods, plantation products (mainly palm oil and rubber), and tin. More recently, manufacturing, especially in labor-intensive high-tech sectors, has become the main engine of growth (Figure 13.36). As Singapore prospered, many of its enterprises moved their manufacturing operations into neighboring Malaysia. Increasingly, Malaysia's economy is multinational; many Western high-tech firms operate in the country, and several Malaysian companies have themselves established subsidiaries in foreign lands. As a result of its prosperity, Malaysia has attracted hundreds of thousands of illegal immigrants, mostly from Indonesia and the Philippines. Highly dependent on the export of electronic goods, the Malaysian economy remains vulnerable to fluctuations in the North American, European, and Japanese markets.

The economic geography of modern Malaysia shows large regional variations. Most industrial development has occurred on the west side of peninsular Malaysia, with most of the rest of the country remaining dependent on agriculture and resource extraction. More important, however, are disparities based on ethnicity. The industrial wealth generated in Malaysia has been concentrated in the Chinese community. Ethnic Malays remain less prosperous than Chinese Malaysians, and those of South Asian origin tend to be poorer still. Many of the country's tribal peoples have suffered as development proceeds and their lands are taken away.

The disproportionate prosperity of the local Chinese community is a feature of most Southeast Asian countries. The problem is particularly acute in Malaysia, however, simply because its Chinese minority is so large (some 30 percent of the country's total population). The government's response has been one of aggressive "affirmative action," by which economic clout is transferred through the so-called New Economic Policy (NEP) to the numerically dominant Malay, or **Bumiputra** ("sons of the soil"), community. This policy has been reasonably successful, although it has not yet reached its main goal of placing 30 percent of the nation's wealth in the hands of the Bumiputras. Because the Malaysian economy as a whole has grown rapidly since the 1970s, the Chinese community has thrived even as its relative share of the country's wealth declined. Considerable resentment, however, is still felt by the Chinese, many of whom believe that Malaysia's economic and educational systems are biased against them. Even many Malays think that the system has become too inefficient to persist. According to a 2008 survey, 71 percent of Malaysians as a whole agreed that the country's affirmative action program was obsolete.

As a result of such changing attitudes, affirmative action emerged as a central issue in Malaysia's 2013 general election, when it was challenged by the opposition party. The election itself proved highly controversial; although the opposition

▼ **Figure 13.36 High-Tech Manufacturing in Malaysia** Many foreign companies have established high-tech manufacturing facilities in Malaysia. Most of their factories are located in the western part of Peninsular Malaysia.

party won a majority of the popular vote, the ruling party took more parliamentary seats and thus remained in power. The defeated candidate, Anwar Ibrahim, denounced the results as fraudulent, and anger was particularly pronounced in the Chinese community.

Thailand's Ups and Downs Thailand also climbed rapidly during the 1980s and 1990s into the ranks of the world's newly industrialized countries, although it remained less prosperous than Malaysia. Thailand also experienced a major downturn in the late 1990s that undermined much of this development. Recovery began in earnest after 2000, but the military coup of 2006, coupled with the intensifying insurgency in the far south, resulted in another slowdown, as did the global recession of 2008. Export growth of both agricultural goods and machinery, however, remains strong, leading to solid economic growth in 2012.

Japanese companies were leading players in the Thai boom of the 1980s and early 1990s. As labor costs in Japan itself became too expensive for assembly and other manufacturing processes, Japanese firms began to relocate factories to such places as Thailand. They were particularly attracted by the country's low-wage, yet reasonably well-educated workforce. Thailand was also seen as politically stable, lacking the severe ethnic tensions found in many other parts of Asia. Although Thailand's Chinese population is large and economically powerful, relations between the Thai and the local Chinese have generally been good.

Thailand's economic expansion has by no means benefited the entire country to an equal extent. Most industrial development has occurred in the historical core, especially in the greater Bangkok metropolitan area. Yet even in Bangkok, the blessings of progress have been mixed. As the city begins to choke on its own growth, industry has begun to spread outward. The entire Chao Phraya lowland area shares to some extent in the general prosperity due both to its proximity to Bangkok and to its rich agricultural resources. In northern Thailand, the Chiang Mai area has profited from its ability to attract large numbers of international tourists.

Other parts of the country have not been so fortunate. Thailand's Lao-speaking northeast (the Khorat Plateau) is the country's poorest region. Soils here are too thin to support highly productive agriculture, yet the population is sizable. Because of the poverty of their homeland, northeasterners often are forced to seek employment in Bangkok. As Lao speakers, they often experience ethnic discrimination. Men typically find work in the construction industry; northeastern women more often make their living as prostitutes (Figure 13.37). Not coincidentally, northeastern Thailand has strongly supported ousted populist Prime Minister Thaksin Shinawatra. The Muslim area of southern Thailand also remains very poor, contributing to the region's insurgency.

Indonesian Economic Development At the time of independence (1949), Indonesia was one of the poorest countries in the world. The Dutch had used their colony largely as an extraction zone for tropical crops and mineral resources, investing little in infrastructure or education. The population of Java mushroomed in the 19th and early 20th centuries, leading to serious land shortages.

The Indonesian economy finally began to expand in the 1970s. Oil exports fueled the early growth, as did the logging of tropical forests. However, unlike most other petroleum exporters, Indonesia continued to grow even after prices plummeted in the 1980s. Production subsequently declined, and in 2004 Indonesia had to begin importing oil. But like Thailand and Malaysia, Indonesia proved attractive to multinational companies eager to export from a low-wage economy. Large Indonesian firms, some three-quarters of them owned by local Chinese families, also have capitalized on the country's low wages and abundant resources.

Despite rapid growth in the 1980s and 1990s, Indonesia remains a poor country. Until recently, its pace of economic expansion seldom matched that of Thailand, Singapore, and Malaysia, and it has remained more dependent on the unsustainable exploitation of natural resources. The financial crisis of the late 1990s, moreover, hurt Indonesia more severely than any other country, and political instability continues to hamper economic recovery. Yet Indonesia was one of the few large countries to avoid the recent global economic crisis, with its economy growing by more than 4 percent in 2009, and it has continued to expand at a healthy pace through 2013. But despite such growth, Indonesia's economy has been burdened by large fuel subsidies, a legacy of the time when it was an oil-exporting country. When the government cut such subsidies in June 2013, gasoline prices immediately rose by 44 percent, resulting in major protests in several Indonesian cities.

As in Thailand, development in Indonesia exhibits pronounced geographical disparities. Northwest Java, close to the

▼ **Figure 13.37 Sex Tourism in Thailand** Thailand has one of the highest rates of prostitution in the world. Most prostitutes cater to a local clientele, but those working in Bangkok's infamous Patpong district are usually hired by foreign men. Prostitution in Thailand is associated with high rates of HIV infection and with the brutal exploitation of women and girls.

capital city of Jakarta, has boomed, and much of the resource-rich and moderately populated island of Sumatra has long been relatively prosperous. Eastern Borneo (Kalimantan), another resource area, also is relatively well-off. In the overcrowded rural districts of central and eastern Java, however, many peasants have inadequate land and remain near the margins of subsistence. Far eastern Indonesia has experienced little economic or social development, and throughout the remote areas of the "outer islands," tribal peoples have suffered as their resources have been taken by and their lands lost to outsiders.

Vietnam's Uneven Progress The three countries of former French Indochina—Vietnam, Cambodia, and Laos—experienced only modest economic expansion during Southeast Asia's boom years of the 1980s and 1990s. This area endured almost continual warfare between 1941 and 1975, which persisted until the mid-1990s in Cambodia. Critics contend that the socialist economic system adopted by these countries prevented sustained economic growth. This debate is now moot, however, since a globalized capitalist model of development has largely been adopted in all three countries.

Vietnam's economy is much stronger than those of Cambodia and Laos. The country's per capita GNI, however, is still low by global standards. Postwar reunification in 1975 did not initially bring the anticipated growth, and economic stagnation ensued. Conditions declined in the early 1990s after the fall of the Soviet Union, Vietnam's main supporter

▼ Figure 13.38 Development Issues in Southeast Asia: Human Development Index Levels of health, education, and other aspects of human development vary widely across Southeast Asia. As the human development figures shown on the map indicate, Singapore, Brunei, and Malaysia have reached high levels of development, whereas Burma, Cambodia, and East Timor lag well behind.

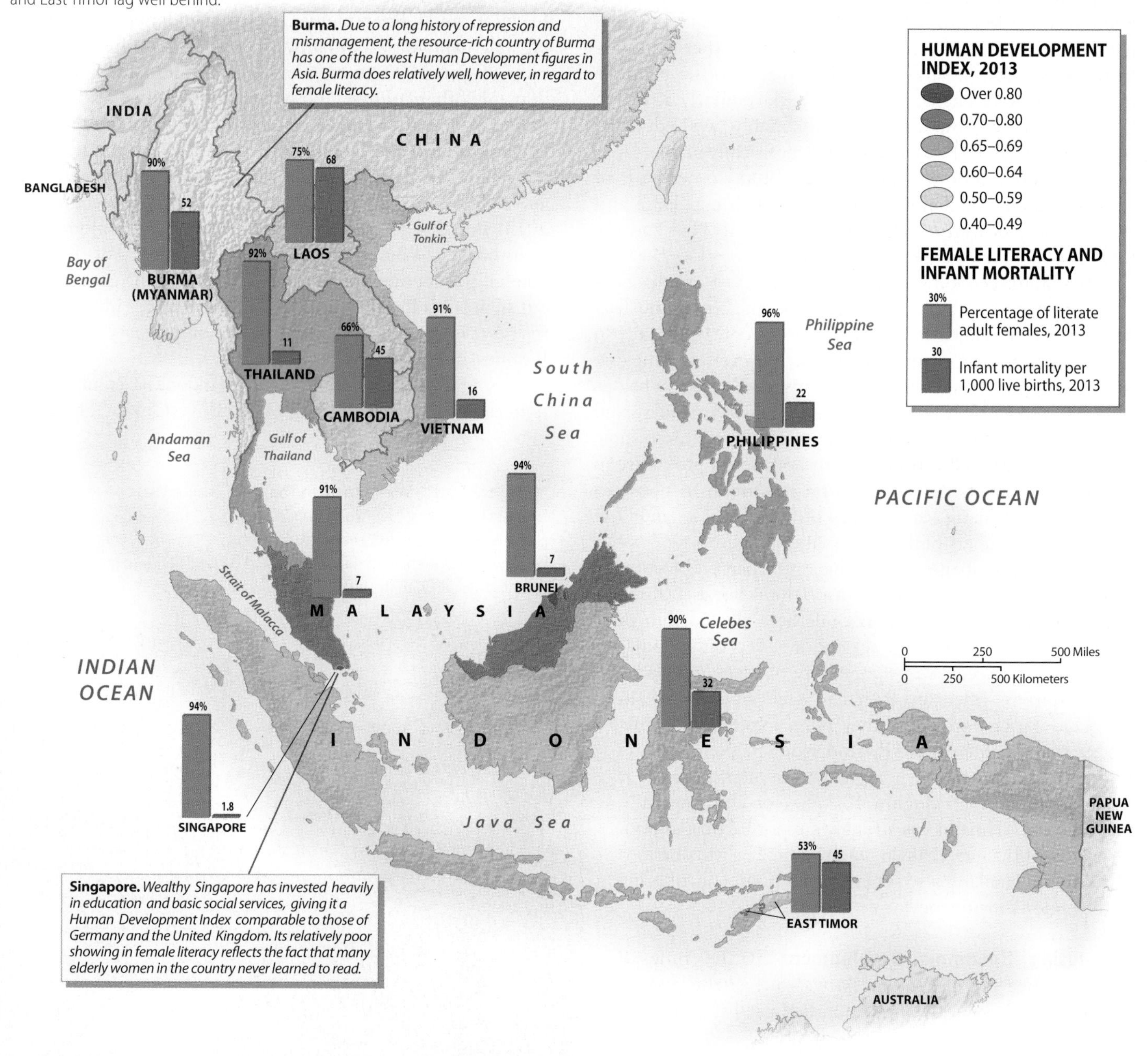

and trading partner. Until the mid-1990s, Vietnam remained under embargo by the United States. Frustrated with their country's economic performance, Vietnam's leaders began to embrace market economics, while retaining the political forms of a communist state. They have, in other words, followed the Chinese model. Vietnam now welcomes multinational corporations, which are attracted by the low wages received by its relatively well-educated workforce (Figure 13.38).

Such efforts seemingly began to pay off after 2000. By 2007, the Vietnamese economy was expanding at the rapid rate of about 8 percent a year. Foreign investment began to pour into the country, and exports—especially of textiles—began to surge. At roughly the same time, Vietnam joined the World Trade Organization, which encouraged exports, while ensuring the continuation of market-oriented reforms. More recently, however, growth has slowed, due in part to the country's poorly developed banking sector. In 2013, the Vietnamese government devalued its currency in order to boost exports and hence ramp up economic growth.

Vietnam's recent economic development has generated a significant degree of social tension. Many lowland peasants and upland tribal peoples have been excluded from the boom and are thus growing increasingly discontented. Tensions between the north (the center of political authority) and the more capitalistic south (the center of economic power) seem to be increasing. Trade disputes have arisen with the United States, which has accused Vietnamese exporters of dumping such products as farm-raised catfish and shrimp on American markets.

Rapid Growth in Impoverished Laos and Cambodia Laos and Cambodia face some of the most serious economic problems in the region. In Cambodia, the ravages of war, followed by continuing political instability, long undermined economic development. Laos faces special difficulties owing to its rough terrain and relative isolation. Both countries also are hampered by a lack of infrastructure; outside the few cities, paved roads and reliable electricity are rare. As a result, the economies of both Cambodia and Laos remain largely agricultural in orientation, also relying heavily on environmentally destructive logging and mining operations. In Laos, subsistence farming still accounts for more than three-quarters of total employment.

The Laotian government is pinning its economic hopes on dam- and road-building projects. The country is mountainous and has many large rivers, and it could therefore generate large quantities of electricity, in high demand in neighboring Thailand and Vietnam. Up to 100 new dam projects have been proposed for the Mekong River system, most of them to be located in Laos. The Asian Development Bank also is supporting an ambitious road-building program that would pass through Laos in order to link China to Thailand. As a result of both infrastructure development and environmentally problematic investments in mining and forestry, the Laotian economy has been growing at a very rapid rate in recent years, exceeding 8 percent in 2012.

Like Laos, Cambodia is a deeply impoverished country that has experienced a striking economic boom in recent years. Foreign investment has generated an expanding textile sector, tourism is thriving (Angkor Wat, the largest religious monument in the world, receives well over half a million visitors annually), mining is taking off, and in 2005 important oil and natural gas fields were discovered in Cambodian territorial waters. From 2004 to 2007, the Cambodian economy grew at an annual rate of over 10 percent, owing in large part to investment from and trade with China. But Cambodia still faces major economic challenges. A lack of both skills and basic infrastructure, as well as problems of political instability and corruption, could easily undermine its economic recovery. The recent economic boom has also led to an epidemic of "land-grabbing," in which politically connected elites take over the properties of impoverished peasants in order to develop them (Figure 13.39).

Despite their basic lack of development, Cambodia and Laos are not as poor as one might expect from the official economic statistics. Both countries have relatively low population densities and abundant resources. Their low per capita GNI figures, more importantly, partially reflect the fact that many of their people remain in a subsistence economy. Although the Laotian highlanders make few contributions

▼ **Figure 13.39 Land Clearance in Phnom Penh** Severe conflicts have emerged in Cambodia's booming capital city, Phnom Penh, as slum areas are cleared to make way for development projects. In this photo, a woman wearing a helmet tries to stop a bulldozer from demolishing her home, while her daughter tries to pull her away from the dangerous machine.

to GNI, most of them at least have adequate shelter and food.

Burma's Troubled Economy Burma stands near the bottom of the scale of Southeast Asian economic development. For all of its many problems, however, Burma is a land of great potential. It has abundant resources—including natural gas, oil, minerals, water, and timber—as well as a large expanse of fertile farmland. Its population density is moderate, and its people are reasonably well educated. The country, however, has seen little economic or social development.

Burma's economic woes can be traced in part to the continual warfare that it has experienced. Most observers also blame economic policy. Beginning in earnest in 1962, Burma attempted to isolate its economic system from global forces in order to achieve self-sufficiency under a system of Buddhist socialism. Its intentions were admirable, but the experiment was not successful. Instead of creating a self-contained economy, Burma found itself burdened by smuggling and black-market activities. As a result of Burma's authoritarian rule and human rights abuses, the United States, the European Union, and other countries imposed economic sanctions on the country.

In the early 1990s, Burma began to open its economy to market forces and increase its involvement in foreign trade. By 2000, however, the reform measures had stalled out, and in 2003 a banking crisis resulted in another round of economic havoc. High rates of inflation and worrisome fiscal deficits undermine business confidence, but probably the most damaging economic impediment is Burma's exchange-rate policy. In 2009, when the official exchange rate was 6.4 Burmese kyat to the U.S. dollar, you could get up to 1090 kyat to the dollar in street markets.

The dismal state of the Burmese economy was one of the main reasons why the country's government decided to open up to the rest of the world and begin to democratize in 2011. It was hoping to attract foreign investments, secure export markets, and end international sanctions. In 2012, the United States responded by easing investment and banking import restrictions. The Burmese economy has revived to some degree following these changes, but it remains to be seen whether solid growth will continue.

Hope for East Timor By most measures, East Timor is one of the poorest and least developed countries in Asia. Its violent separation from Indonesia, along with the ethnic unrest that hit the country in 2006, undermined its economy. But substantial offshore oil and especially natural gas deposits have recently been discovered and are now being brought into production, although disputes with Australia over their maritime boundaries have slowed down the process. While few jobs have been created in East Timor's energy sector, fossil fuel exports have helped the government establish a balanced budged and achieve several years of solid economic growth.

Globalization and the Southeast Asian Economy

As the previous discussion shows, Southeast Asia has undergone rapid, but uneven integration into the global economy (Figure 13.40). Singapore has thoroughly staked its future to the success of multinational capitalism, and several neighboring countries are following suit. According to one measurement, Malaysia and Singapore have, respectively, the fourth and fifth most trade-dependent economies in the world. Even communist Vietnam and once-isolationist Burma have opened their doors to international commerce, hoping to find advantages in economic globalization. The booming palm oil business in Indonesia and Malaysia has also attracted a lot of foreign investment (see *Everyday Globalization: The Southeast Asian Palm Oil Connection*).

Much debate has arisen among scholars over the roots of Southeast Asia's economic gains, as well as its more recent economic problems. Those who credit primarily the diligence, discipline, and entrepreneurial skills of the Southeast Asian peoples are optimistic about future economic expansion. Skeptics argue, however, that most of the region's growth has come from the application of large quantities of labor and capital unsupported by real advances in productivity. Regardless of how this debate turns out, it is clear that Southeast Asia's globalized economies depend heavily upon exports to the international market. In the 1990s, many observers thought that the booming economies of Southeast Asia had become overly dependent on exports to the United States, but in recent years the rise of China has resulted in a more balanced trading regime. In 2012, for example, 9.9 percent of Thailand's exports went to the United States, while 17.7 percent went to China (including Hong Kong) and another 10.2 percent went to Japan. Currently, the main concern in Southeast Asia is competition from China in exporting inexpensive consumer goods to the rest of the world. As is true in the United States, many people in the region believe that the Chinese currency is undervalued, giving China an unfair advantage in export markets.

Globalized industrial production in Southeast Asia remains controversial. Consumers in the world's wealthy countries worry that many of their basic purchases, such as sneakers and clothing, are produced under exploitative conditions. Movements have thus emerged to pressure both multinational corporations and Southeast Asian governments to improve the working conditions of laborers in the export industries. The multinational firms in question counter that such exploitative conditions occur not in their own factories, but rather in those of local companies that subcontract for them; activists, however, point out that multinationals have tremendous influence over their local subcontractors. Some Southeast Asian leaders object to the entire debate, accusing Westerners of wanting to prevent Southeast Asian development and protect their own home markets, under the pretext of concern over worker rights.

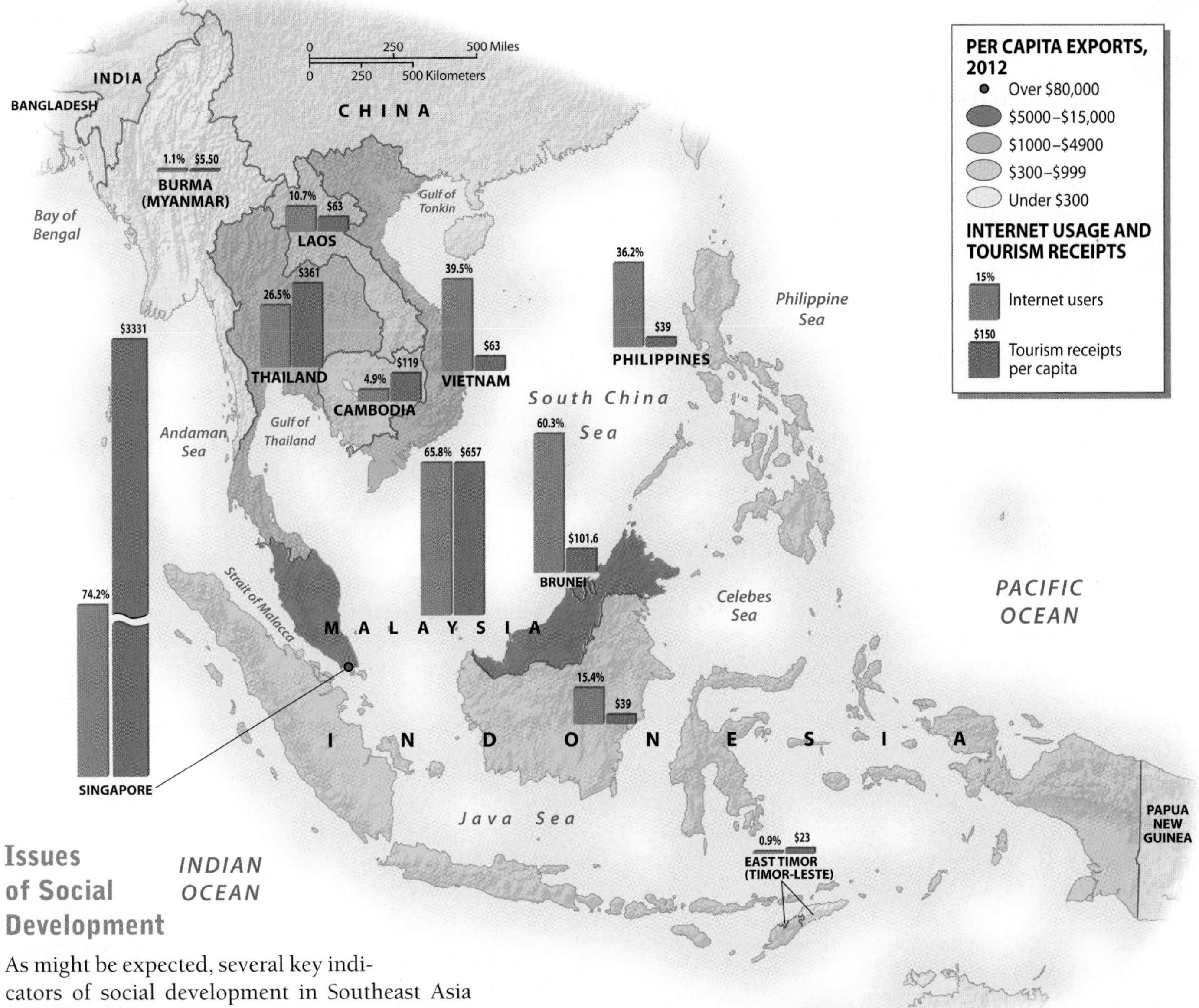

▲ **Figure 13.40 Global Linkages in Southeast Asia: Exports, Internet Usage, and Tourism** Levels of globalization vary tremendously across Southeast Asia. Whereas Singapore and Malaysia are closely linked to the global economy, Burma is still relatively isolated. With Burma's recent political transformation, however, it is beginning to experience more extensive globalization.

Issues of Social Development

As might be expected, several key indicators of social development in Southeast Asia correlate well with levels of economic development. Singapore thus ranks among the world leaders in regard to health and education, as does the small, yet oil-rich country of Brunei. East Timor, Laos, and Cambodia, not surprisingly, come out near the bottom for these measures. The people of Vietnam, however, are healthier and better educated than might be expected on the basis of their country's overall economic performance. Burma, on the other hand, has seen a decline in some social indicators. Its per capita health budget is one of the lowest in the world, whereas its military budget, as a percentage of its overall economic output, is one of the highest.

Education and Health With the exception of Laos, Cambodia, Burma, and East Timor, Southeast Asia has achieved relatively high levels of social welfare. But even the poorest countries of the region have made substantial gains over the past several decades. In East Timor, for example, average life expectancy at birth rose from 33 years in 1978 to 62 years in 2011. Improvements in mortality under the age of five have also been striking, as is evident from the figures in Table 13.2. Overall, progress has been more pronounced in prosperous countries such as Malaysia and Singapore, whereas war-torn Cambodia has made smaller gains. Thailand has recently begun to provide a social safety net for its citizens, hoping to raise its level of social development to that found in the world's wealthier countries. Social conditions are improving in East Timor in part due to funds from the country's rapidly growing oil and natural gas industries.

Most of the governments of Southeast Asia have placed a high priority on basic education. Literacy rates are relatively high in most countries of the region. Much less success, however, has been realized in university and technical education.

Everyday Globalization

The Southeast Asian Palm Oil Connection

The next time you eat a packaged snack food, look to see if the label lists "palm oil." The chances are good that it will.

Palm oil, derived from the fruit of the oil palm, is one of the world's most rapidly growing agricultural exports. It now accounts for roughly 30 percent of the total global production of edible oils and fats. Although the oil palm itself is native to West Africa, almost all commercial production now takes places in Southeast Asia. In 2013, Indonesia produced an estimated 31 million metric tons (MT), Malaysia 19 million MT, and Thailand 2.1 million MT, whereas the next largest producer, Colombia, produced only 1 million MT. International agricultural firms have invested heavily in Southeast Asian production. Minnesota-based Cargill Incorporated, for example, controls palm oil plantations in Sumatra, Borneo, and the Indonesian portion of New Guinea, and it operates refiners for the resulting harvests in Indonesia, Malaysia, and the Netherlands (Figure 13.5.1).

Advantages and Disadvantages Palm oil is inexpensive to produce, especially in the massive plantations pioneered in Southeast Asia, encouraging its consumption worldwide. It has a neutral flavor and works well in both baking and frying. Palm oil has also been widely used as a substitute for trans fats, which have been found to increase the risk of coronary diseases.

But despite its increasing popularity, palm oil is also one of the world's most criticized food products. Health advocates note that, unlike most vegetable oils, it is very high in saturated fats, which have also been linked to heart disease and other health problems. The environmental problems associated with palm oil are equally serious. Vast tracts of former rainforests and wetland swamp forests in Indonesia and Malaysia have been transformed into palm oil plantations over the past several decades. In the process, biological diversity is threatened, while vast amounts of greenhouse gases are emitted. Palm oil plantations are regarded as one of the main reasons for the endangerment of the Borneo orangutan and the critical endangerment of the closely related Sumatran orangutan. Indigenous peoples often lose their lands as well when new palm oil plantations are established.

Seeking Sustainability As environmental criticisms of palm oil have mounted, the industry has responded by seeking to establish less harmful production techniques. Created in 2004, the Roundtable on Sustainable Palm Oil seeks to establish global standards. This nonprofit organization brings together palm oil producers, processors, and traders, as well as food and soap manufacturing companies and environmental and social organizations. Reflecting the global nature of the palm oil business, the roundtable itself is based in Zurich, Switzerland, but it also has major offices in Malaysia and Indonesia.

Thus far, the record of the Roundtable on Sustainable Palm Oil has been mixed. Some environmental groups give it credit for trying to address the most critical problems, but others, such as Friends of the Earth, argue that it merely seeks to create a smokescreen, hiding inherently unsustainable practices. Some palm oil producers are also dissatisfied, arguing that they have gone to great lengths to have their harvests certified as sustainable, only to find that the market does not offer them any rewards for such efforts. As a result, critics think that a much more concerted effort must be made to reduce the destructive impacts of palm oil production.

1. Why are the geometrical patterns so regular in the oil palm plantation areas?
2. Why are there so few houses in the planation areas?

Google Earth Virtual Tour Video

http://goo.gl/rQsjdJ

Figure 13.5.1 Palm Oil Plantation Most of the world's palm oil is produced from plantations in Southeast Asia. The clearing of rain forests and the draining of swamps to make room for such plantations is highly controversial.

As Southeast Asian economies continue to grow, this educational gap is beginning to have negative consequences, forcing many students to study abroad. High levels of basic education, along with general economic and social development, also have led to reduced birth rates throughout much of Southeast Asia. With population growing much more slowly now than before, economic gains are more easily translated into improved living standards.

▲ **Figure 13.41 Minangkabau House** The Minangkabau of Western Sumatra are known for their their large traditional houses and for the high social position of women in their society.

Gender Relations Historically, Southeast Asia has been characterized by relatively high levels of gender equity. Some scholars contend that women had a higher social position, on average, in Southeast Asia than in any other region of the world. In traditional kingdoms of the region, women often played important economic roles as marketers, merchants, and financiers, and many reached high political positions as diplomats, translators, and even royal bodyguards. When the Spaniards first reached the Philippines, they noted that almost all of the women, but not the men, of the Manila area were literate in the indigenous Tagalog script. Even in today's world, some anthropologists have gone so far as to describe the Muslim Minangkabau people of western Sumatra as living in a "modern matriarchy," since Minangkabau women still have a great deal of authority over their large households, which are traditionally based on descent from female ancestors (Figure 13.41).

Historians have also argued, however, that the position of women in Southeast Asia began to decline as religious and philosophical belief systems from other parts of the world spread through the region, as they tended to put men in the dominant position. In some countries, women have been able to take advantage of modern education and changing ideas to reclaim positions of authority. According to the global Gender Gap Index of 2012, the Philippines has the eighth lowest gender gap in the world, trailing only Iceland, Finland, Norway, Sweden, Ireland, New Zealand, and Denmark. Other Southeast Asia countries, however, do not rank nearly so high. Thailand and Vietnam, for example, received only average rankings (65th and 66th, respectively, out of 135 countries), while Indonesia was placed in the 97th position and Cambodia in the 103rd.

Despite these generally positive tendencies, Southeast Asia is also the site of some of the world's most intensive sexual exploitation. Commercial sex is a huge business in Thailand. Despite its massive scope, prostitution is technically illegal in Thailand, which means that it is a major source of corruption. Other Southeast Asian countries, particularly the Philippines, Vietnam, and Cambodia, are also centers of a globally oriented commercial sex trade. Many workers in Southeast Asian brothels are underage, many have been coerced into the activity, and some are reportedly held as virtual slaves. Young women, girls, and boys are frequently trafficked from the poorer parts of the region, often in connection with the drug trade. Sexually transmitted diseases, including HIV-AIDS, are associated with this activity, although the Thai government has engaged in a relatively successful public health campaign focused on condom use. Poor women from Indonesia, moreover, are frequently trafficked into Malaysia, where they are promised good jobs, but often end up instead as underpaid maids or sex workers.

Two of the main centers of commercial sex in Southeast Asia developed around U.S. military bases during the Cold War: Pattaya in Thailand and Angeles City in the Philippines. Both cities have lost their military bases, but have expanded their economies by focusing on tourism, much of it sex-related. Pattaya now supports an estimated 20,000 sex workers. Here the high end of the business includes thousands of Russian and Ukrainian women. As a result, Pattaya now has a major Russian presence, attracting hundreds of thousands of Russian tourists every year, in addition to wealthy Russian investors. Russian organized crime now plays a major role in the city, illustrating one of the seamier aspects of globalization in modern Southeast Asia.

REVIEW

13.13 Why have some Southeast Asian countries experienced sustained economic growth and social development, whereas others have more generally experienced stagnation in the same period?

13.14 Why have the major Southeast Asian economies experienced such shared booms and busts over the past several decades?

13.15 Why does the position of women in Southeast Asia look favorable from some angles, but not from others?

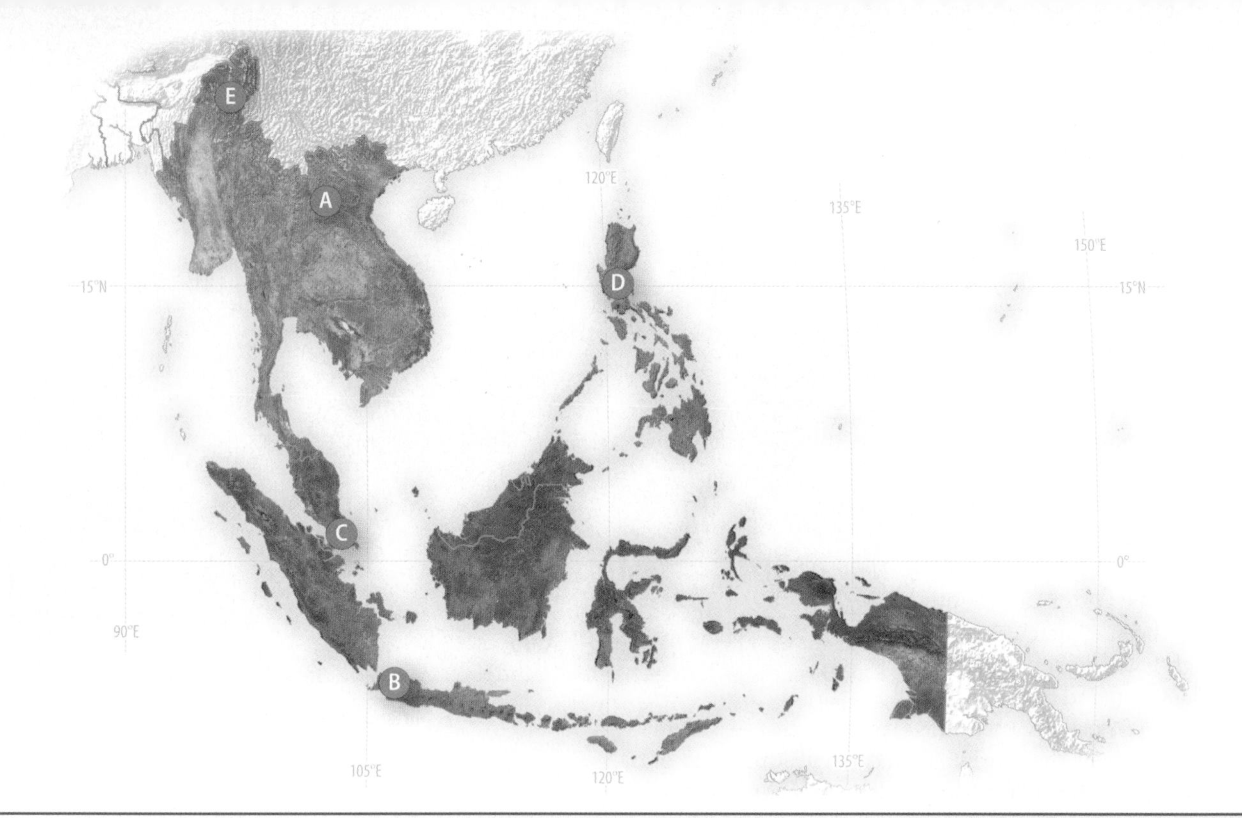

Physical Geography and Environmental Issues

Some of the most serious problems created by globalization in Southeast Asia are environmental. Commercial logging and the development of plantations have resulted in extensive deforestation. The draining of swamplands has led to massive forest and peat fires, creating severe air pollution. Dam-building has generated clean electricity, but at the cost of habitat destruction. Conservation efforts, however, are under way in most countries of the region.

13.1 Why is the Mekong River in Laos such a promising site for the generation of hydroelectricity?

13.2 Why is the burning and oxidation of peatlands a more serious problem in Southeast Asia than in any other region of the world?

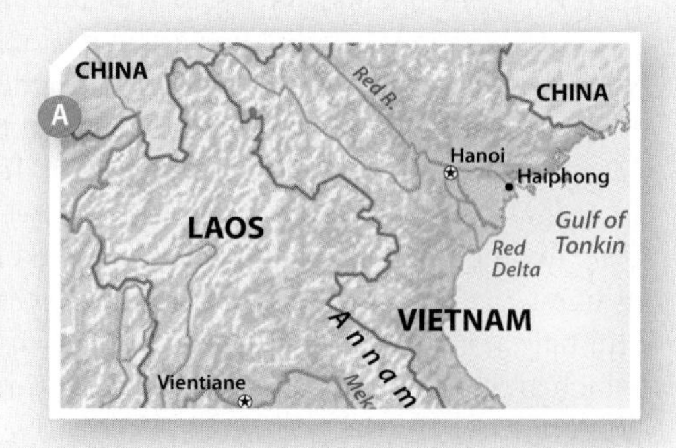

Population and Settlement

As people move from densely populated, fertile lowland areas into remote uplands, both environmental damage and cultural conflicts often follow. Population movements in Southeast Asia also have a global dimension. Although most countries of the region have seen major reductions in their birth rates, the Philippines and East Timor continue to experience rapid population gains. Cities are growing rapidly throughout the region, and the Jakarta, Bangkok, and Manila metropolitan areas are now some of the world's largest urban aggregations.

13.3 What do the patterns in this image tell us about the development of the Jakarta metropolitan area?

13.4 Why has Jakarta grown so much more rapidly over the past several decades than other Indonesian cities?

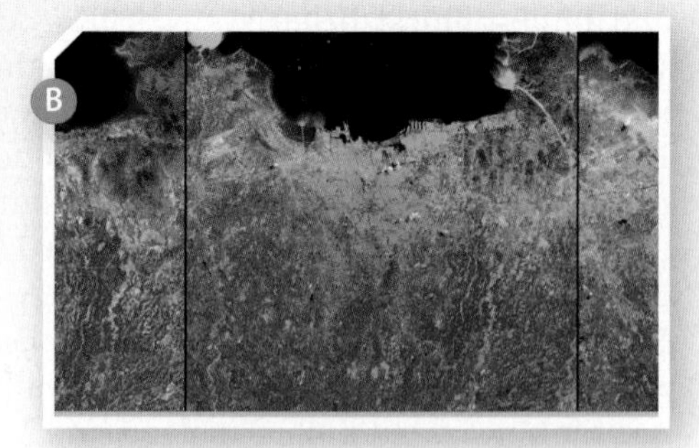

Key Terms

- animism *(p. 621)*
- Association of Southeast Asian Nations (ASEAN) *(p. 602)*
- Bumiputra *(p. 638)*
- copra *(p. 612)*
- crony capitalism *(p. 636)*
- domino theory *(p. 629)*
- entrepôt *(p. 637)*
- Golden Triangle *(p. 611)*
- Khmer Rouge *(p. 629)*
- lingua franca *(p. 622)*
- primate city *(p. 615)*
- Ramayana *(p. 618)*
- shifted cultivators *(p. 615)*
- Sunda Shelf *(p. 604)*
- swidden *(p. 611)*
- transmigration *(p. 614)*
- tsunami *(p. 604)*
- typhoon *(p. 604)*

In Review • Chapter 13
Southeast Asia

Cultural Coherence and Diversity

Southeast Asia is characterized today by tremendous cultural diversity. Most of the world's major religions, for example, are represented in the region. In recent decades, conflicts over language and religion have caused serious problems in several Southeast Asian countries. However, the region has found a new sense of regional identity as expressed through the Association of Southeast Asian Nations (ASEAN). Cultural globalization has also long been pronounced in many parts of Southeast Asia—particularly Singapore, Thailand, and the Philippines.

13.5 As Singapore is considered to be an English-speaking country, why is its government working hard to teach its citizens proper spoken English?

13.6 What geographical and historical factors have led Singapore to develop such a particular, and highly valued, local form of the English language?

Geopolitical Framework

The relative success of ASEAN has not solved all of Southeast Asia's political tensions. Many of its countries still argue about geographical, political, and economic issues, while insurgencies remain active in the Philippines and Thailand. Both the Philippines and Indonesia have established autonomous areas in order to reduce the desire for secession. Cambodia, Laos, and Burma have also been held back by repressive governments, although reforms have recently been enacted, especially in Burma.

13.7 Some of the signs carried by protestors in this rally demand that China stop "poaching" in Philippine waters. Why is concern about illegal fishing by other countries so pronounced in the Philippines?

13.8 Why are the tiny Spratly Islands such a controversial issue in Southeast Asia?

Economic and Social Development

Although ASEAN has played an economic as well as a political role, its economic successes have been more limited. Most of the region's trade is still directed outward toward the traditional centers of the global economy: North America, Europe, and East Asia. A significant question for Southeast Asia's future is whether the region will develop an integrated regional economy. A more important issue is whether social and economic development will be able to lift the entire region out of poverty instead of benefiting just the more fortunate areas.

13.9 What are some of the challenges involved in building natural gas and oil pipelines across Burma and into China?

13.10 Why is China so keen to build natural gas and oil pipelines in Burma?

Authors' Blogs

Scan to visit the author's blog for chapter updates

www.gad4blog.wordpress.com

Scan to visit the GeoCurrents blog

www.geocurrents.info

14 Australia and Oceania

Physical Geography

Very diverse environments characterize this huge region that includes a continent-sized landmass as well as thousands of small oceanic islands. While sea-level rise from global warming threatens many of these low-lying islands, Australia and New Zealand debate how best to reduce their CO_2 emissions.

Population and Settlement

Growing, dense cities punctuate the sparse rural settlement pattern of Oceania, with urban places as the magnets attracting migrants from both within and outside of the region.

The Moai. Huge monolithic human figures, called *moai* in Polynesian, and believed to represent clan leaders from the period 1250 to 1500 CE, stare inland across Rapa Nui (Easter) island in the far eastern corner of Oceania. Tourists at the base of the statues provide a sense of scale for these very large statues.

Cultural Coherence and Diversity

Both Australia and New Zealand, originally products of European culture, are seeing new cultural geographies take shape because of immigration from other parts of the world, as well as from their own native peoples, the Aborigines and Maori.

Geopolitical Framework

A heritage of colonial geographies overlaying native cultures is being replaced by contemporary power struggles between global powers, dominated by the tensions between China and the United States.

Economic and Social Development

While Australia and New Zealand are relatively wealthy because of world trade, most of island Oceania struggles economically. Even Hawaii has troubles during global downturns with its boom-or-bust tourist economy.

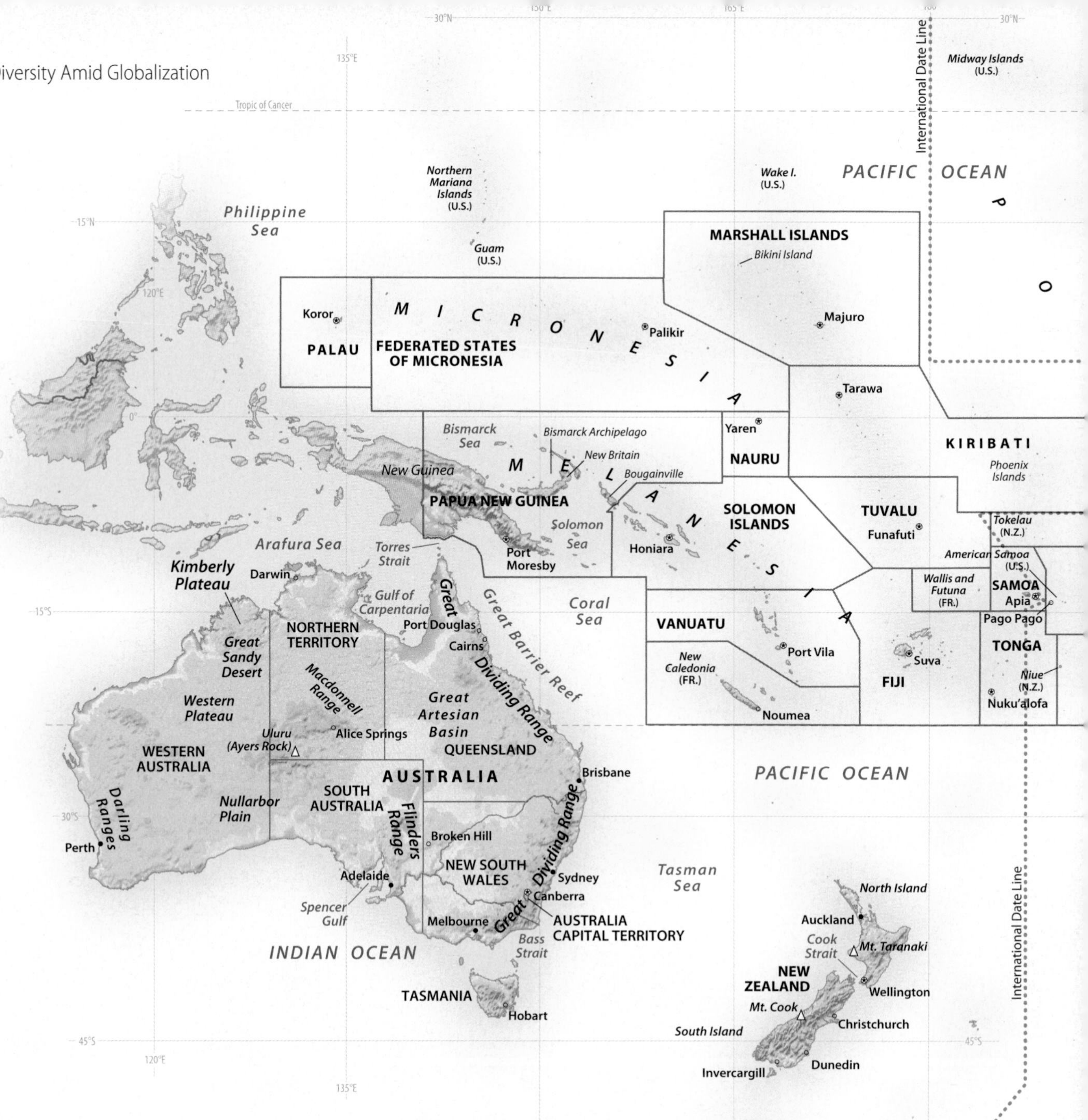

Oceania is a vast water world stretching from New Guinea to Hawaii, punctuated with some 25,000 islands. Images of tropical Pacific islands usually portray idyllic serenity under calm, sunny skies, but island people actually have lots to worry about: Will they be flooded by rising seas; washed away by storms; starved by crop failure; ruined by pestilence; attacked by strangers? Even on the continent-sized island of Australia, some worry about having too many people, while others say the population is too small. For people on low-lying Kiribati, the concern is about losing their island to rising sea level. On tourist-riddled Oahu, native Hawaiians worry about being driven from their ancestral lands by foreign real estate corporations.

Aiding and abetting these concerns about the sustainability of island life is the mystery of Easter Island (called *Rapa Nui* in Polynesian), a small island tucked away in the far southeastern corner of Oceania, closer to Chile in South America than to Australia. Like many Pacific Islands, it was discovered and settled long ago by Polynesian people who arrived in large sailing canoes. Those original settlers, however,

Although there's no shortage of opinions on the demise of the Easter Islanders, this mystery energizes worries about the sustainability of human populations and the vagaries of island life. Their fate represents a reality lying behind the postcard stereotype of Oceania.

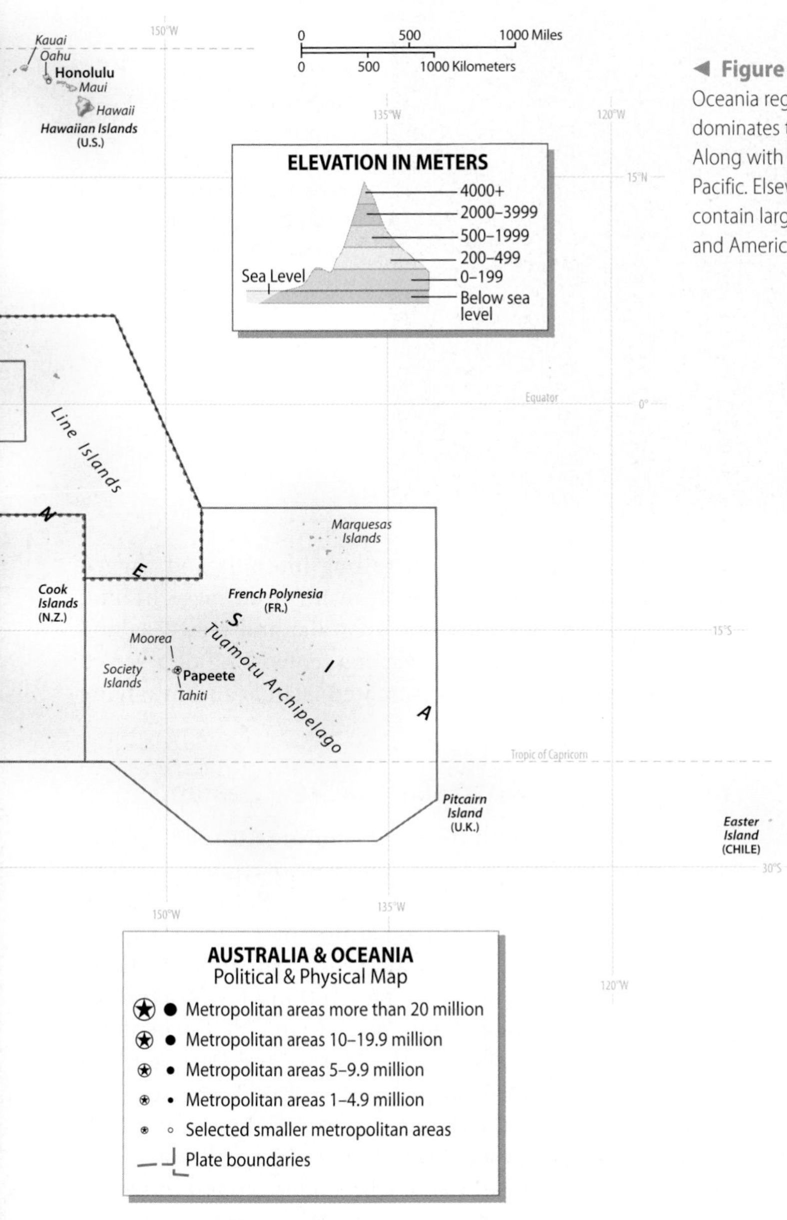

◀ **Figure 14.1 Australia and Oceania** More water than land, the Australia and Oceania region sprawls across the vast reaches of the western Pacific Ocean. Australia dominates the region, both in its physical size and in its economic and political clout. Along with New Zealand, Australia represents largely Europeanized settlement in the South Pacific. Elsewhere, however, the island subregions of Melanesia, Micronesia, and Polynesia contain large native populations that have mixed in varied ways with later European, Asian, and American arrivals.

are gone now, having fallen victim to some unknown disaster of island life. Perhaps the Easter Islanders were driven away by hostile strangers—or perhaps they ate themselves out of house and home. Maybe an epidemic of some sort wiped them out—a plague, perhaps, carried by rats and mice—or maybe rodents broke into the storehouses and ate all the seeds for next year's crops. No one knows for sure except the silent monoliths, the huge stone-faced statues (called *moai* in Polynesian) that, some say, may have been part of the problem.

Although there's no shortage of opinions on the demise of the Easter Islanders, this mystery energizes worries about the sustainability of human populations and the vagaries of island life. Their fate represents a reality lying behind the postcard stereotype of Oceania. Within this immense oceanic realm lies Australia, which because of its large size (approximately the same size as the continental United States) is often considered a continent, rather than an island. To its east, a three-hour flight from Australia, is New Zealand. This island country has a much smaller population—4.5 million people compared to Australia's 23.1 million—but the two nations are closely linked by shared historical ties to Britain. New Zealand, however, is usually considered part of **Polynesia** ("many islands") because of its native Maori people, who are distinctly different from the Aboriginal native people of Australia.

Hawaii, 4400 miles (7100 km) northeast of New Zealand, shares the same Polynesian heritage as New Zealand and can be thought of as the northeastern boundary of Oceania. Far to the southeast, more than 3000 miles (4400 km) from Hawaii, lie the Polynesian islands of Tahiti and Easter Island, marking the corner of Oceania closest to South America (Figure 14.1).

Four thousand miles (6400 km) west of Tahiti and the other islands of French Polynesia, well across the International Date Line, lies New Guinea, containing the confusing boundary between Oceania and Asia. Today an arbitrary political boundary bisects the island, dividing Papua New Guinea, the eastern half,

LEARNING OBJECTIVES

After reading this chapter you should be able to:

- Describe the geographic characteristics of the region known as Oceania.
- Identify the major environmental issues problematic to Australia and Oceania, as well as the pathways toward solving those problems.
- Explain how the Pacific Rim of Fire is linked to the landforms of Oceania.
- Describe the different sources of energy used in Australia and New Zealand and how this influences the amount and kind of greenhouse gas emissions produced in each country.
- Summarize the prehistoric peopling of the Pacific, as well as the colonial exploration and settlement of Australia and Oceania.
- Explain the changing migration patterns to and within postwar Australia and Oceania.
- Describe the historical and modern interactions between native peoples and Anglo-European migrants in Australia and Oceania.
- Describe the different pathways to independence taken by countries in Oceania.
- Summarize why and how Oceania has become a contested region between global superpowers.
- Describe the diverse economic geographies of Oceania.
- Explain the positive and negative interactions of Australia and Oceania with the global economy.

◀ **Figure 14.2 The Australian Outback** Arid and generally treeless, the vast lands of the Australian Outback resemble some of the dry landscapes of the U.S. West. This photo is near Pilbara, Western Australia.

which is usually considered part of Oceania, from Papua and West Papua, which, as part of Indonesia, are usually thought of as Southeast Asia. This western part of Oceania is often called **Melanesia,** meaning "dark islands," because early explorers considered local peoples to be darker-skinned than those in Polynesia.

Finally, the more culturally diverse region of **Micronesia** ("small islands") lies north of Melanesia and west of Polynesia. It includes microstates such as Nauru and the Marshall Islands, as well as the U.S. territory of Guam.

Physical Geography and Environmental Issues: Varied Landscapes and Habitats

Australia is made up of a vast semiarid interior—the **Outback**, a dry, sparsely settled land of scrubby vegetation—fringed by tropical environments in its far north and hilly, mountainous topography with summer-dry Mediterranean climates in the east, west, and south (Figure 14.2). In contrast, the two small islands that make up New Zealand are known for their landscapes of green rolling foothills and rugged snow-capped mountains, which result from more humid, cooler climates. Surrounding Australia and New Zealand is the true island realm of Oceania, consisting of a varied array of both high, volcanic-created islands and low-lying coral atolls.

▼ **Figure 14.3 The Great Dividing Range** This mountainous area is a dominant feature of eastern Australia, extending for over 2300 miles (3700 km) from north to south. This photo is from the Thompson Valley in the state of Victoria.

Topography of Australia and New Zealand

Three major topographic regions dominate Australia's physical geography (Figure 14.3). The vast, irregular Western Plateau, which averages only 1000 to 1800 feet (300 to 550 meters) in elevation, occupies more than half of the continent. To the plateau's east, the Interior Lowland Basins stretch for more than 1000 miles (1600 km), from the swampy coastlands of the Gulf of Carpentaria in the north to the valleys of the Murray and Darling rivers in the south, Australia's largest river system. Farthest east is the forested and mountainous country of the Great Dividing Range, which extends over 2300 miles (3700 km) from the Cape York Peninsula in northern Queensland to southern Victoria (Figure 14.3). Nearby, off the eastern coast of Queensland, the Great Barrier Reef offers a final dramatic subsurface feature: Over the past 10,000 years, one of the world's most spectacular examples of coral reef-building has produced a living legacy, now protected by the Great Barrier Reef Marine Park (Figure 14.4).

As part of the Pacific Rim of Fire, New Zealand owes its geologic origins to volcanic mountain-building, which produced its two rugged and spectacular islands. The North Island's active volcanic peaks, reaching heights of more than 9100 feet (2800 meters), and its geothermal features reveal the country's fiery origins. Even higher and more rugged mountains comprise the western spine of the South Island. Mantled by high mountain glaciers and surrounded by steeply

▲ **Figure 14.4 The Great Barrier Reef** Stretching along the eastern Queensland coast, the famed Great Barrier Reef is one of the world's most spectacular examples of coral reef-building. Threatened by varied forms of coastal pollution, much of the reef is now protected in a national marine park. **Q. How is global warming affecting tropical coral reefs like the Great Barrier Reef?**

sloping valleys, the Southern Alps are some of the world's most visually spectacular mountains, complete with narrow, fjord-like valleys that indent much of the South Island's isolated western coast (Figure 14.5).

Australia and New Zealand's Climates

As noted, zones of higher precipitation encircle Australia's arid center (Figure 14.6). In the tropical low-latitude north, seasonal changes are dramatic and unpredictable. For example, Darwin can experience drenching monsoonal rains in the Southern Hemisphere summer, December to March, followed by bone-dry winters, lasting from June to September (see the climograph for Darwin in Figure 14.6). Indeed, life across the region is shaped by this annual rhythm of what is called locally "the wet" and "the dry." By the end of the dry season, natural wildfires often dot the landscape of northern Australia, sometimes even threatening the suburbs of Sydney and Melbourne in the southeast (Figure 14.7, page 655).

Along the east coast of Queensland, precipitation is high (60–100 in., or 150–250 cm), but diminishes rapidly moving westward into the interior. Rainfall at interior locations, such as the Northern Territory's Alice Springs, averages less than 10 inches (25 cm) annually. South of Brisbane, more midlatitude influences dominate eastern Australia's climate. Coastal New South Wales, southeastern Victoria, and Tasmania experience the country's most dependable year-round rainfall, with averages of 40–60 inches (100–150 cm) of precipitation per year; winter snow frequently covers the nearby mountains. Farther west, summers are hot and dry in much of South Australia and in the southwest corner of Western Australia. These zones of Mediterranean climate produce the **mallee** vegetation, scrubby eucalyptus woodland.

Climates in New Zealand are influenced by three factors: latitude, the moderating effects of the Pacific Ocean, and proximity to local mountain ranges. Most of the North Island is distinctly subtropical; the coastal lowlands near Auckland, for example, are mild and wet year-round. On the South Island, conditions become distinctly cooler as you move closer to the South Pole. Indeed, the island's southern edge feels the seasonal breath of Antarctic chill, as it lies more than 46° south of the equator, at a latitude similar to Portland, Oregon, on the West Coast of the United States. Mountain ranges on New Zealand's South Island also display incredible local variations in precipitation: West-facing slopes are drenched with more than 100 inches (250 cm) of precipitation annually, whereas lowlands to the east average only 25 inches (65 cm) per year. The Otago region, inland from Dunedin, sits partially in the rain shadow of the Southern Alps, and its rolling, open landscapes resemble the semiarid expanses of North America's West (Figure 14.8).

▼ **Figure 14.5 The New Zealand Alps** The dominant topographic feature of the South Island, these picturesque mountains, referred to locally as the Southern Alps, rise to heights over 12,000 feet (3600 meters).

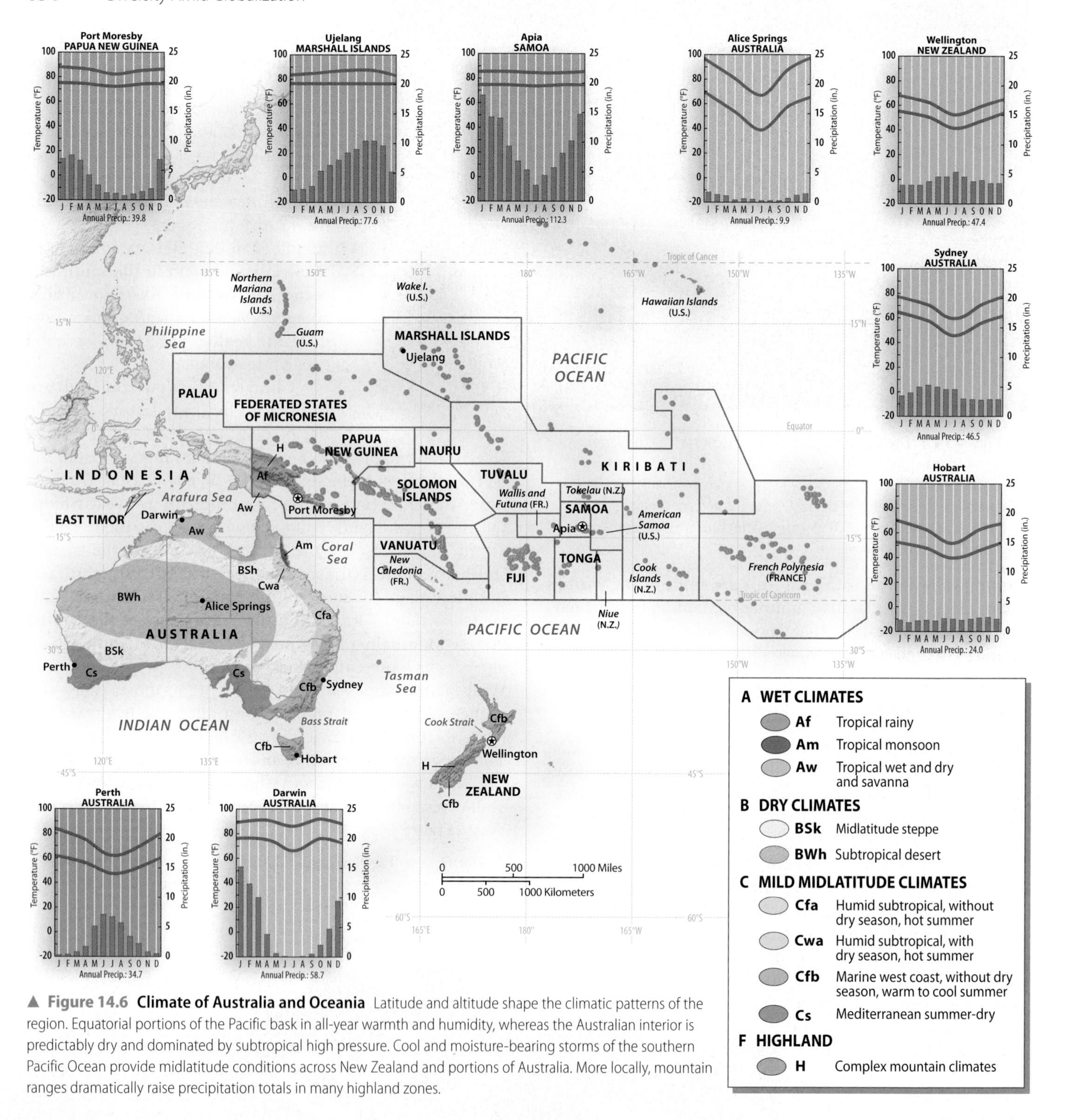

▲ **Figure 14.6 Climate of Australia and Oceania** Latitude and altitude shape the climatic patterns of the region. Equatorial portions of the Pacific bask in all-year warmth and humidity, whereas the Australian interior is predictably dry and dominated by subtropical high pressure. Cool and moisture-bearing storms of the southern Pacific Ocean provide midlatitude conditions across New Zealand and portions of Australia. More locally, mountain ranges dramatically raise precipitation totals in many highland zones.

Oceania's Diverse Environments

Given the vast area of the globe covered by Oceania, from Hawaii west to Asia and from the tropics south almost to Antarctica, its physical geography also varies greatly. Ocean currents and wind patterns, having served as natural highways of migration between these islands, also produce different kinds of weather and climate, while geologic forces have given rise to an array of island landscapes.

Island Landforms Much of Melanesia and Polynesia is part of the seismically active Pacific Rim of Fire. As a result, volcanic eruptions, major earthquakes, and tsunamis are common across the region. For example, in 1994, volcanic eruptions and earthquakes on the island of New Britain (Papua New Guinea) forced more than 100,000 people from their homes. Only four years later, a massive tsunami triggered by an offshore earthquake swept across the north coast of New Guinea, killing 3000 residents and destroying numerous villages. Such events are unfortunately a part of life in this geologically active part of the world.

▲ **Figure 14.7 Australian Wildfires** Huge and savage dry-season wildfires (known as bushfires in Australia) threaten both rural settlements and sprawling city suburbs in the southeast. This fire in February 2009, just 70 miles from the heart of Melbourne, was the worst fire disaster in 25 years and may be a harbinger of even more damaging fires accompanying global warming.

High and Low Islands Most of Oceania's islands were created by two distinct processes: either volcanic eruptions or, alternatively, coral reef-building. Those with a volcanic heritage are referred to as **high islands** because most of them rise hundreds and even thousands of feet in elevation above sea level. The Hawaiian Islands are good illustrations, with a volcanic mountain of more than 13,000 feet (4000 meters) on the Big Island of Hawaii. Tonga, Samoa, Bora Bora, and Vanuatu provide other examples of high islands (Figure 14.9, page 656). Even larger and more geographically complex are the *continental high islands* as found in New Guinea, New Zealand, and the Solomon Islands.

In contrast, **low islands**, as the name suggests, are formed from coral reefs, making the islands not just lower, but also flatter and usually smaller than high islands. Further, because the soil on these islands originated as coral, it is generally less fertile than the soil of high islands and supports less varied plant life. Low islands often begin as barrier reefs around or over sunken volcanic high islands, resulting in an **atoll** (Figure 14.10, page 656). The world's largest atoll, Kwajalein in Micronesia's Marshall Islands, is 75 miles (120 km) long and 15 miles (25 km) wide. Low islands dominate the countries of Tuvalu, Kiribati, and the Marshall Islands. Clearly, these low islands are the most vulnerable to rising sea levels associated with climate change.

Island Climates The Pacific high islands usually receive abundant precipitation because of the orographic effect, resulting in dense tropical forests and vegetation. On the island of Kauai in the state of Hawaii, Mt. Walaleale may be one of the wettest spots on Earth, receiving an average annual rainfall of 470 inches (1200 cm). In contrast to the high islands, low islands receive less precipitation, typically less than 100 inches (250 cm) annually. As a result, water shortages are common.

Australia's Unique Plants and Animals

Because of the Australian continent's long geologic history of separation and isolation from other landmasses, its bioregions contain a unique array of plants and animals

◄ **Figure 14.8 Central Otago, South Island** On New Zealand's South Island, the Southern Alps capture rainfall on the west coast, but leave areas to the east in a drier rain shadow. As a result, the Central Otago region has a semiarid landscape resembling portions of the U.S. West.

▲ **Figure 14.9 Bora Bora** The jewel of French Polynesia, Bora Bora displays many of the classic features of Pacific high islands. As the island's central volcanic core retreats, surrounding coral reefs produce a mix of wave-washed sandy shores and shallow lagoons.

found nowhere else in the world. More specifically, 83 percent of its mammals and 89 percent of its reptiles are unique to that country. Best known are the country's **marsupials** (mammals that raise their young in a pouch)—the kangaroo, koala, possum, wombat, and Tasmanian Devil (Figure 14.11). Fully 70 percent of the world's known marsupials are found in Australia. Also unique is the platypus, an egg-laying mammal.

Australia is also known for a high number of venomous species, including a large range of snakes, spiders, scorpions, jellyfish, stingrays, and octopi. Even the male platypus carries venom in its foot spurs.

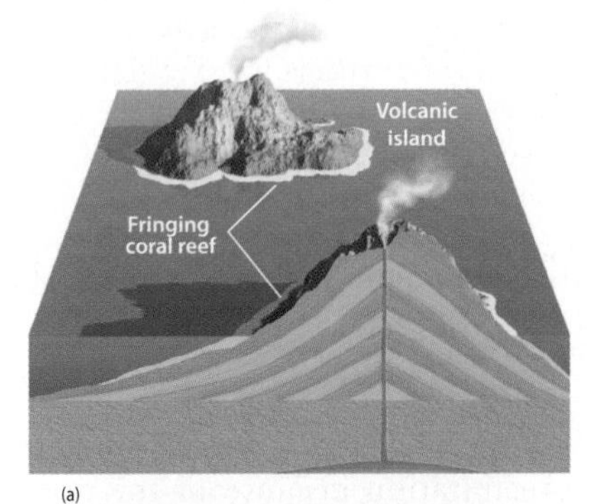

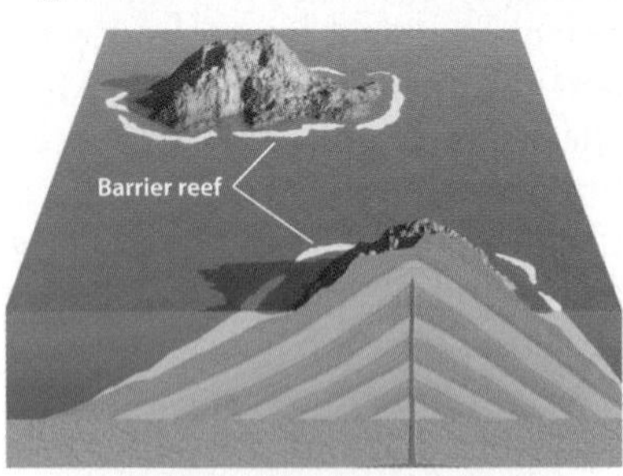

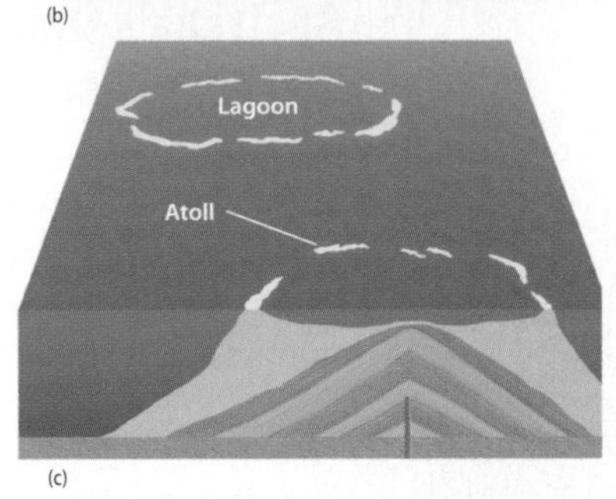

▲ **Figure 14.10 Evolution of an Atoll** Many Pacific low islands begin as rugged volcanoes (a) with fringing coral reefs. However, as the extinct volcano subsides and erodes away, the coral reef expands, becoming a larger barrier reef (b). (The term *barrier reef* comes from the hazards these features pose to navigation when approaching the island from the sea.) Finally, all that remains (c) is a coral atoll surrounding a shallow lagoon.

Exotic Species The introduction of exotic (nonnative) plants and animals has caused problems for endemic (native) species throughout the Pacific region. In Australia, for example, rabbits brought from Europe successfully multiplied in an environment that lacked the diseases and predators that normally checked their numbers. Before long, rabbit populations had reached plague-like proportions, stripping large sections of land of its vegetation. The animals were brought under control only through the purposeful introduction of the rabbit disease myxomatosis.

▼ **Figure 14.11 Kangaroo** Among Australia's unique flora and fauna is the iconic kangaroo. This photo is of a grey kangaroo, the largest member of the kangaroo family, of which there are four other species.

▲ **Figure 14.12 Island Pest** The brown tree snake, which arrived in Guam accidentally in the 1950s, has now taken over large parts of the island's forestlands and killed off most native bird species. Because these snakes, which reach 10 feet in length, climb along electrical wires, they frequently cause power outages throughout Guam.

The introduction of exotic plants and animals to Oceania's island environments has had similar effects. For example, many small islands possessed no native land mammals, and their native bird and plant species proved vulnerable to the ravages of introduced rats, pigs, and other animals. The larger islands of the region, such as those of New Zealand, originally supported several species of large, flightless birds. The largest of these, the moas, were substantially larger than ostriches. During the first wave of human settlement in New Zealand some 1500 years ago, moa numbers fell rapidly as they were hunted by humans and their eggs were consumed by invading rats. By 1800, the moas had become completely extinct.

The spread of nonnative species continues today, perhaps at an even greater pace. In Guam, the brown tree snake, which arrived accidentally by cargo ship from the Solomon Islands in the 1950s, has taken over the landscape (Figure 14.12). Forest areas contain more than 10,000 snakes per square mile, which have wiped out nearly all the native bird species. In addition, the snakes cause frequent power outages as they crawl along electrical wires. The brown tree snake has already done its damage to Guam, but it threatens other islands as well because it readily hides in cargo containers shipped to other islands.

Complex Environmental Issues

Globalization has exacted an environmental toll on Australia and Oceania (Figure 14.13). Specifically, the region's considerable

▼ **Figure 14.13 Environmental Issues in Australia and Oceania** Modern environmental problems belie the myth that the region is an earthly paradise. Tropical deforestation, extensive mining, and a long record of nuclear testing by colonial powers have brought severe challenges to the region. Human settlements have also extensively modified the pattern of natural vegetation. Future environmental threats loom for low-lying Pacific islands as sea levels rise from climate change.

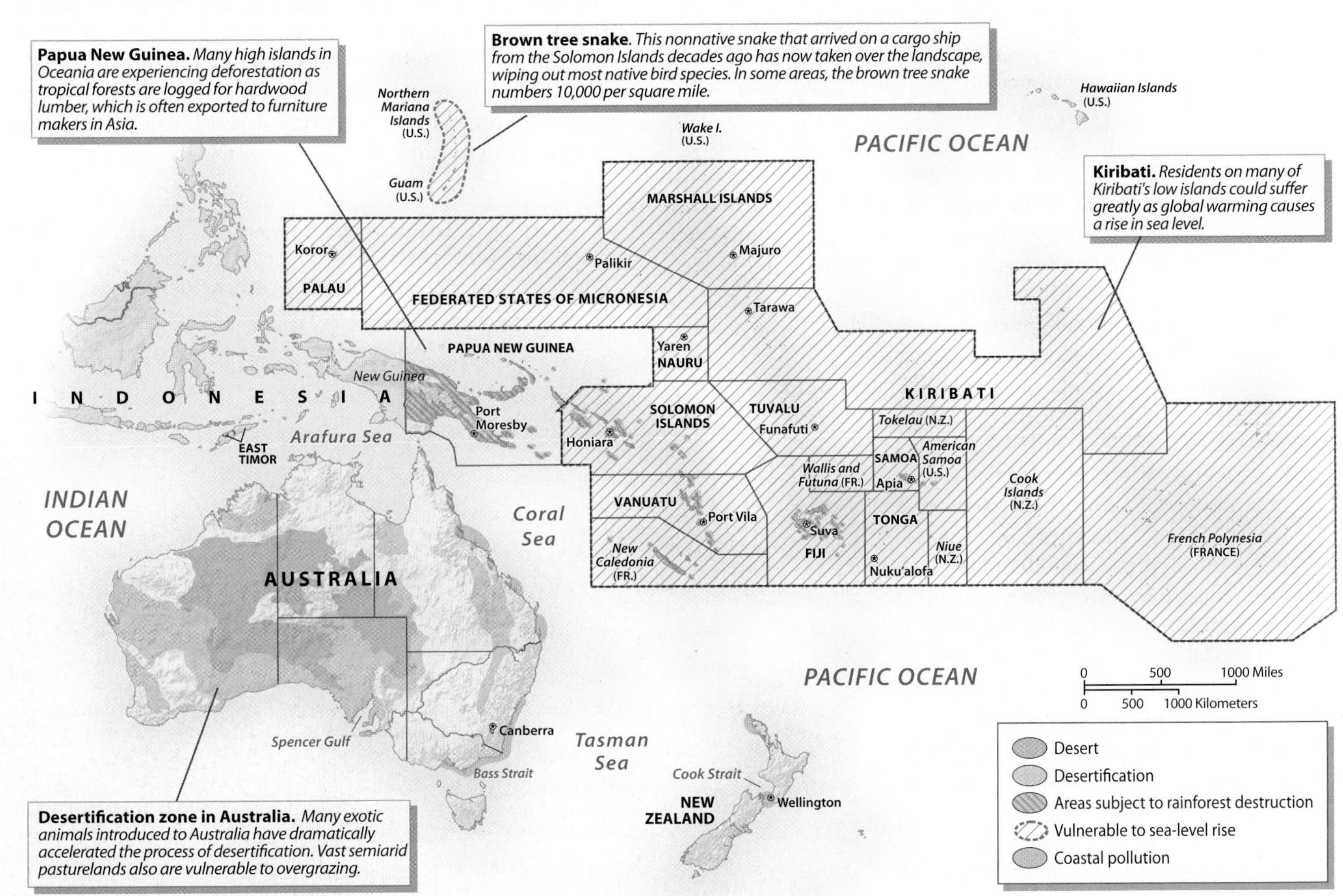

▲ **Figure 14.14 The Globalization of Nauru** A Nauruan local points to the scarred landscape that remains after decades of phosphate mining by Australian companies. With no mineral riches left, the islanders have tried several different schemes for finding their place in the global economy.

base of natural resources has been opened to development, much of it by outside interests. Although gaining from the benefits of global investment, the region has also paid a considerable price for encouraging development, and the result is an increasingly threatened environment.

Major mining operations have greatly affected Australia, Papua New Guinea, New Caledonia, and Nauru. Some of Australia's largest gold, silver, copper, and lead mines are located in sparsely settled portions of Queensland and New South Wales, polluting watersheds in these semiarid regions. In Western Australia, huge open-pit iron mines dot the landscape, unearthing ore that is bound for global markets, particularly China and Japan. To the northeast, gold mining is transforming the Solomon Islands, while an even larger gold-mining venture has raised environmental concerns on the island of New Guinea. Elsewhere, Micronesia's tiny Nauru has been virtually turned inside out as much of the island's jungle cover has been removed to get at some of the world's richest phosphate deposits (Figure 14.14).

Deforestation is another major environmental threat across the region. Vast stretches of Australia's eucalyptus woodlands have been destroyed to create pastures. In addition, coastal rainforests in Queensland cover only a fraction of their original area, although a growing environmental movement in the region is fighting to save the remaining forest tracts. Tasmania has also been an environmental battleground, particularly given the biodiversity of its midlatitude forest landscapes. The island's earlier European and Australian development featured many logging and pulp mill operations, but more than 20 percent of the land is now protected by national parks.

Many high islands in Oceania are also threatened by deforestation. With limited land areas, islands are subject to rapid tree loss, which, in turn, often leads to soil erosion. Although rainforests still cover 70 percent of Papua New Guinea, more than 37 million acres (15 million hectares) have been identified as suitable for logging (Figure 14.15). Some of the world's most biologically diverse environments are being threatened in these operations, but landowners see the quick cash sales to loggers as attractive, even though this nonsustainable practice is contrary to their traditional lifestyles.

Climate Change in Oceania

Even though Oceania contributes relatively little atmospheric pollution to the global atmosphere, the harbingers of climate change are widespread and problematic. New Zealand mountain glaciers are melting, and Australia suffers from frequent droughts and devastating wildfires. Warmer ocean waters have caused widespread bleaching of the Great Barrier Reef off Australia's coast as microorganisms die, and rising sea levels threaten low-lying island nations. United Nations (UN) projections for

▼ **Figure 14.15 Logging in Oceania** Much of Oceania's tropical forest is being destroyed by logging. In this photo hardwood tree trunks await loading for transfer to Asian mills where they will be made into furniture for the growing markets of China.

the future are also highly disturbing: Sea levels may rise 4 feet (1.4 meters) by century's end; stronger tropical cyclones could devastate Pacific islands, with widespread damage to land and life; and island inhabitants will suffer from changed coastal resources as the ocean continues to warm (see *Working Toward Sustainability: Sea-Level Rise and the Future of Low Islands*). In response to these threats, the actions taken and policies implemented by Oceania's countries vary considerably, depending on their susceptibility to climate change, the source and magnitude of their atmospheric emissions, and the state of local economies.

Power Generation and Emissions in Australia and New Zealand Australia, with its 22 million people, generates the most global-warming emissions by far in this large region, primarily because almost 80 percent of its electricity is generated from fossil fuels. Coal is the favored fuel, since Australia has immense deposits of it. Also linked to coal are the considerable emissions generated from the mining and transport of that fuel. Although use of wind and solar power has increased lately, currently less than 5 percent of Australia's energy is produced by renewables. Added to that mix is the 13 percent generated by hydroelectric power. Australia's emissions total 405 million metric tons of CO_2, making it the world's 17th largest atmospheric polluter.

In contrast, 57 percent of New Zealand's energy comes from hydroelectric power, primarily generated in the high and wet mountains of the South Island. Wind and solar—particularly wind—supply fully 13 percent of the country's power. Given this array of clean energy, it is no surprise that most of New Zealand's global-warming emissions result from methane produced by the country's large livestock population. As a result, New Zealand produces 40 million metric tons of CO_2 emissions, enough to rank it the world's 71st largest emitter.

Power and Emissions in Oceania Papua New Guinea has almost twice as many people as New Zealand, but because it has far fewer livestock, their emissions are only 5.3 million metric tons per year, ranking them121st in the world. This is despite the fact that most of their power comes from fossil fuels.

Among the island countries, a significant difference occurs in power generation between low and high islands. High islands use topography and higher rainfall to their advantage by generating hydroelectric power. Fiji, for example, generates 39 percent of its power from hydro and plans to rely even more on that source of energy as it builds an expanded system of dams and generators funded by China (Figure 14.16).

Without any topography to speak of, low islands such as Kiribati, Samoa, and the Solomon Islands must generate all of their power from imported oil and coal. Despite this dependence on fossil fuels, these islands produce miniscule amounts of global-warming emissions compared to Australia and New Zealand, due primarily to their small populations. The Solomon Islands, for example, with more than half a million people, produces only 359,000 metric tons of CO_2. Other small island nations produce similarly small amounts of emissions. Even so, many of the low islands are now experimenting with wind and solar renewables to reduce their dependence on imported fuel.

Emission Reduction Plans and Action Until a 2007 change of government, Australia was the only industrial country besides the United States to not ratify the Kyoto Protocol. According to many analysts, the fact that Australia is the world's largest exporter of coal influenced that decision. Although the former government rejected the Kyoto Protocol and downplayed the threat of global warming, the current government is committed to a 5 percent reduction in greenhouse gas (GHG) emissions by 2020. To achieve that goal, the government recently approved a carbon tax program that could be fully operational by 2015. Strong opposition to this plan, however, continues from three powerful components of the Australian economy: the coal industry, aluminum manufacturers (which use coal-based power plants to generate the energy needed to refine aluminum), and agriculture.

When New Zealand ratified the Kyoto Protocol, it committed itself to a 5 percent reduction in GHG emissions over its 1990 baseline; however, a booming economy

▼ **Figure 14.16 Hydroelectric Power in Fiji** To reduce its dependence on coal and oil, Fiji is increasing its hydroelectric capacity. This new dam was built with funding from China.

Working Toward **Sustainability**

Sea-Level Rise and the Future of Low Islands

Low islands have very little topographic relief, since they are basically coral atolls that have grown from the sea as barrier reefs. True, some low islands began life as volcanoes, but eons of erosion have worn them down to near nothing. The highest point in the Marshall Islands, for example, home to some 68,000 people, is about 30 feet (9 meters). Because the sustainability of Oceania's low islands appears questionable, given the array of problems challenging their future, Pacific islanders are working on what many consider inevitable—abandoning their homeland.

Rising Sea Level At the time of writing, the best estimates regarding global-warming-related sea-level rise are for a 4-foot (1.4-meter) increase by the year 2100 (**Figure 14.1.1**). These estimates, however, are at the conservative end of the scale. Some scientists think it could be considerably higher, perhaps closer to 10 feet (33 meters) by century's end. These are average figures for global sea-level rise; it could be higher or lower, depending on regional factors such as the ocean basin geology and local ocean temperatures. (Recall that warm water expands.) Additionally, this average sea-level rise does not take into account the tidal range for specific locations. In the South Pacific, seasonal extreme high tides can add another 10 feet (3.3 meters) to the average sea level.

More to the point, Pacific islanders say they are already experiencing the effects of climate change and it's causing numerous problems. Flooding is more common during high tides, and flooding brings related problems such as salt-water contamination of precious freshwater supplies and damage to low-lying taro fields. Also, warmer ocean temperatures are causing detrimental changes to local reef ecologies that are crucial for islanders' sustenance. Even the deep-water tuna are leaving as they seek cooler waters elsewhere.

▼ **Figure 14.1.1 Sea-level Rise and Pacific Low Islands** This is Funafuti Atoll, home to half the population of Tuvalu.

▼ **Figure 14.1.2 Flood Damage on Kiribati** This woman and grandchild mourn the loss of their home from storm surf on the island of Kiribati.

Aggravating these problems with sea-level rise and warming is the existing high population density that comes from decades of relatively high population growth and little vacant land. Basically, with a growing island population, there's nowhere go except to leave the island. Tragically, that's what many people are preparing for.

Migration with Dignity Kiribati, for example, has adopted a "migration with dignity" program that emphasizes education and vocational training so its people can find jobs elsewhere. A new maritime training college has been created to help locals gain jobs with global shipping firms. Australian aid has created a nurse-training program for islanders who might seek jobs off-island in that understaffed field. Kiribati has also developed a medical-training relationship with Cuba. (**Figure 14.1.2**)

These programs, however, are mainly designed for younger people, to give them the training and skills needed to compete in foreign lands. But what about the older people, those too old for retraining and less able to adapt to life in a foreign land? How do they maintain their dignity under those conditions? No one seems to have the answer to that question.

1. Which island (or parts of an island) do you think are most vulnerable to flooding from rising sea levels? Why?
2. What areas of the United States face similar problems from sea-level rise as Oceania?

Google Earth Virtual Tour Video

http://goo.gl/E48xOc

▲ **Figure 14.17 New Zealand's Greenhouse Gas Problem** Worldwide, cattle and sheep emit the greenhouse gas methane, but only in New Zealand are these livestock emissions greater than those produced by human activities. New Zealand has discussed a "flatulence tax" on sheep farms, but the country's scientists are also working on an antiflatulence inoculation that would reduce sheep emissions.

resulted in an unexpected 50 percent increase of emissions over the last decade, primarily because of an expanding export livestock industry. In response to this increase, the government is now proposing a 10–20 percent reduction in GHG emissions to be achieved by 2020. To address its livestock methane emissions issue, New Zealand is discussing a much publicized "flatulence tax," which would be levied on livestock operations (Figure 14.17). Livestock specialists are also experimenting with grass and grain fodder mixtures that could reduce livestock emissions.

In 2005, New Zealand committed itself to either a carbon tax or a cap-and-trade emissions plan, and it spent considerable effort in piecing together a master plan to reduce all categories of GHG emissions. However, as of this writing (late 2013), that plan is on hold because of fears by the current government that such a plan could stifle New Zealand's economy.

Many Pacific nations—most notably, Tuvalu, Kiribati, and the Marshall Islands—maintain they are already experiencing problems from sea-level rise, coral bleaching, and degraded fishery resources because of warming ocean waters related to climate change. As a result, these nations have banded together into a strident political union lobbying for a global solution to climate change. In the global-warming debate, these small island nations have been outspoken in their demands that developed nations such as the United States, Japan, and the countries of Western Europe provide the island nations with financial aid to mitigate damage from global warming.

Similarly, Hawaii was one of the first U.S. states to formally address global warming, starting with an analysis of the sources of its GHG emissions and then setting a target for reduction of those emissions. Since most of Hawaii's emissions result from oil- and coal-fired power plants, the state, like Australia and other countries, is actively promoting sustainable energy generation through wind, tidal, and solar power (Figure 14.18).

REVIEW

14.1 Why does Australia have such unusual fauna? Give examples.
14.2 What is the major source of GHG emissions in New Zealand, and what steps are being taken to reduce these emissions?
14.3 Describe the different climate regions found in Australia and New Zealand. What factors produce those regions?
14.4 How are high and low islands formed?

Population and Settlement: Migration, Cities, and Empty Spaces

Present-day population patterns across the region reflect the combined and interactive influences of indigenous and European settlement. In countries such as New Zealand and Australia, as well as in the Hawaiian Islands, Anglo-European migration has structured the distribution and concentration of contemporary populations. In contrast, elsewhere in Oceania population geographies are determined by the local needs and experiences of

▼ **Figure 14.18 Wind Power in Hawaii** Like many islands in Oceania, Hawaii relies heavily on imported oil and coal to meet its energy needs. To reduce these imports the state passed the Hawaii Clean Energy Initiative that has a goal of providing 70 percent of its energy needs from renewable energy by 2030. This wind farm is on South Point, Hawaii and is in the process of being rebuilt with modern turbines.

native peoples (Figure 14.19). Currently, migration is changing the human geography of Australia and Oceania with an increase of intraregional migration as people move about for a variety of reasons.

Contemporary Population Patterns

Despite the stereotypes of life in the Outback, modern Australia is one of the most urbanized populations in the world (Table 14.1). About 90 percent of the country's residents live within either the Sydney or the Melbourne metropolitan area. Perth, in Western Australia, however, has been the country's 21st-century boomtown because of its mineral resource trade with Asia (see *Cityscapes: Perth, Australia's Boomtown*).

Inland, in Australia's legendary Outback, population densities decline as rapidly as the rainfall. New South Wales, with its long Pacific coastline, is the country's most heavily populated state. Its sprawling capital city of Sydney (over 4 million people), focused around one of the world's most magnificent natural harbors, is the largest metropolitan area in the entire South Pacific. In the nearby state of Victoria, Melbourne (with 3.8 million residents) has long competed with Sydney for status as Australia's premiere city, claiming

▼ **Figure 14.19 Population of Australia and Oceania** About 37 million people occupy this world region. Although Papua New Guinea and many Pacific islands feature mainly rural settlements, most residents of the region live in the large urban areas of Australia and New Zealand. Sydney and Melbourne account for almost half of Australia's population, and most New Zealand residents live on the North Island, home to the cities of Auckland and Wellington.

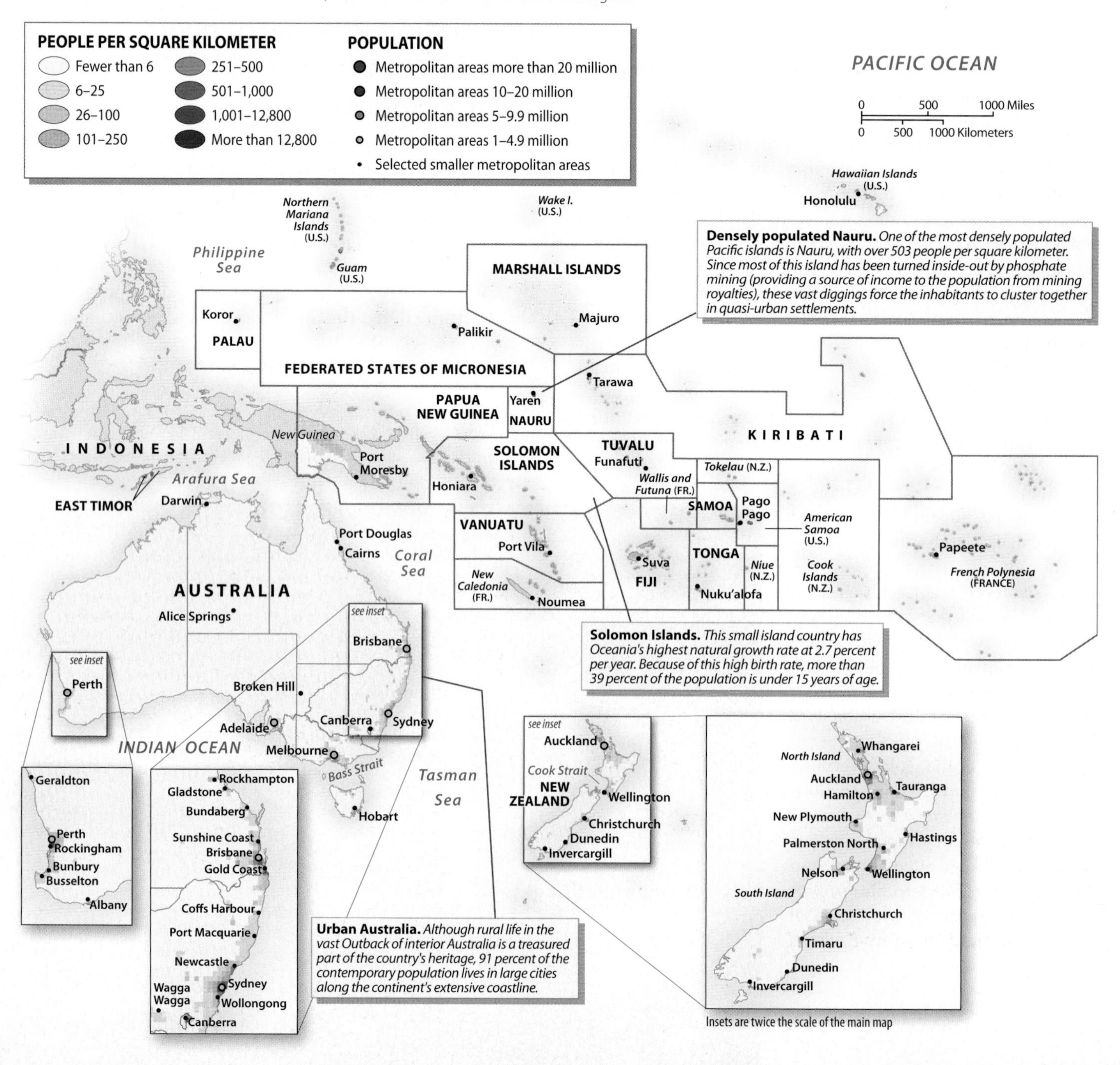

Table 14.1 Population Indicators

Country	Population (millions) 2013	Population Density (per square kilometer)	Rate of Natural Increase (RNI)	Total Fertility Rate	Percent Urban	Percent <15	Percent > 65	Net Migration (Rate per 1000)
Australia	23.1	3	0.7	1.9	82	19	14	10
Fed. States of Micronesia	0.1	152	1.9	3.5	22	35	4	−15
Fiji	0.9	47	1.2	2.6	51	29	5	−8
French Polynesia	0.3	70	1.1	2.1	51	27	5	−0
Guam	0.2	295	1.5	2.9	93	27	7	−10
Kiribati	0.1	146	2.0	3.6	54	34	4	−1
Marshall Islands	0.1	310	2.5	3.9	65	41	2	−18
Nauru	0.01	503	2.0	3.0	100	38	2	8
New Caledonia	0.3	14	1.2	2.2	58	26	7	4
New Zealand	4.5	16	0.7	2.0	86	20	14	−0
Palau	0.02	46	0.6	2.2	77	20	6	0
Papua New Guinea	7.2	16	2.1	4.0	13	38	3	0
Samoa	0.2	67	2.2	4.5	21	38	5	−17
Solomon Islands	0.6	20	2.7	4.6	20	39	3	0
Tonga	0.1	138	2.0	3.9	23	37	6	−17
Tuvalu	0.01	436	1.4	3.1	47	32	5	−9
Vanuatu	0.3	22	2.6	4.0	24	37	4	0

Source: Population Reference Bureau, *World Population Data Sheet, 2013.*

cultural and architectural supremacy over its slightly larger neighbor (Figure 14.20). In between these two metropolitan giants, the location of the much smaller federal capital of Canberra (population 325,000) represents a classic geopolitical compromise, similar to the founding of Washington, DC: locating a country's capital city near—but not in—two competing urban areas.

In New Zealand, more than 70 percent of the country's 4.4 million residents live on the North Island, with the Auckland region (over 1.1 million) dominating the metropolitan scene in the north and the capital city of Wellington (165,000) anchoring settlement along the Cook Strait in the south. Settlement on the South Island is mostly located in the somewhat drier lowlands and coastal districts east of the mountains, with Christchurch (340,000) serving as the largest urban center. Elsewhere, rugged and mountainous terrain on both the North and the South islands produces much lower population densities.

In Papua New Guinea, only 13 percent of the country's population is urban, with most people living in the isolated interior highlands. The nation's largest city is the capital, Port Moresby (around 200,000), located along the narrow coastal lowland in the far southeastern

▼ **Figure 14.20 Downtown Melbourne** Metropolitan Melbourne lies along the Yarra River. Capital of the Australian state of Victoria, Melbourne resembles many modern North American cities, with its high-rise office buildings, entertainment districts, and downtown urban redevelopment.

Cityscapes

Perth, Australia's Boomtown

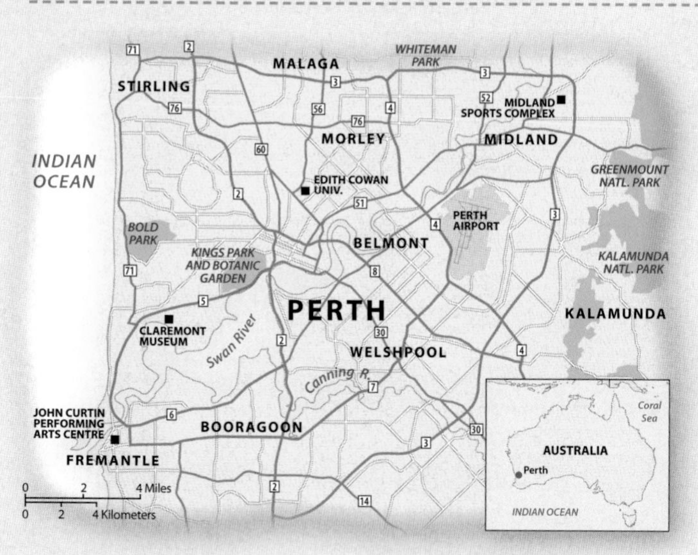

▲ **Figure 14.2.1 Perth** Located on the far west coast of Australia, Perth is currently booming because of nearby iron ore and coal mines that supply China.

Perth is an urban poster child for globalization, with its fortune closely linked to China's voracious appetite for oil, iron ore, and coal. Right now, business is booming in Western Australia, and Perth is exploding with action. But even though the supplies are bountiful, Perth's future could be shaky.

Located on the Swan River in the state of Western Australia, Perth is Australia's fastest-growing city, with suburbs now sprawling north and south of downtown along a narrow coastal strip extending some 70 miles (150 km) north to south (**Figure 14.2.1**). Current population estimates place 1.8 million people in this exploding metropolitan area, with projections that it will reach 2.2 million in the next decade. Besides fast growth, another statistic associated with Perth is that it has the country's lowest unemployment rate, as mining and oil companies attract workers from all over the world with outrageously high wages.

Downtown, the symbol of this boom is the new 45-story headquarters for BHP Billiton, one of the world's largest multinational mining and petroleum companies. Other new buildings, such as luxury boutiques and high-end restaurants, reveal the city's new wealth (Figure 14.2.2).

Two-Speed Economy However, not everyone is striking it rich; instead, people complain about Perth's two-speed economy of haves and have-nots. Those tied to oil and mining are making good money, but others, such as Perth's public workers—firefighters, police, teachers, and nurses—are being left in the dust as the cost of living skyrockets. Fierce competition for housing has driven up rents 80 percent in the last five years—so much so that Perth now has mortgage delinquency rates three times the national average. Bankruptcy rates, too, have risen dramatically, with many small firms being driven out of business as larger chain stores and multinational service industries move in for a piece of the action.

The cost of living in Perth is now higher than in the traditional eastern cities of Sydney and Melbourne. This has created a large class of FIFO (fly in, fly out) workers who work in Perth and Western Australia, but live elsewhere. Not uncommon in the mining and oil business are two-week work stints of 12-hour days, followed by two or three weeks off. This makes the FIFO lifestyle possible.

Perth's explosive growth has led to physical problems as well because it has outgrown an infrastructure of highways, transit systems, schools, and hospitals built earlier for a much smaller urban area. Traffic jams are notorious, leading to long home-to-work commute times. Pundits say that W.A., the common term for Western Australia, now means "wait awhile" because congestion is commonplace—not just on streets and highways, but also in supermarket and store checkout lines.

Thinking About the Future What about Perth's future? Even as urban planners draw up designs for rapid transit systems, higher density in-fill, and smart communities for home and work, they complain about the boomtown mentality that thinks more about getting rich today than addressing tomorrow's problems. Entrepreneurs hedge their bets about Perth's future not because they think China's appetite for oil, coal, and iron ore will lessen, but because they fear Perth and Western Australia may price themselves right out of the world market by paying high wages. The current Australian minimum wage is $16 an hour, but global investors know that African miners work for $2 a day. As is the case with many other facets of a globalized economy, work often flows to where labor is the cheapest. That fact could cloud Perth's future.

Google Earth Virtual Tour Video

http://goo.gl/E1I4lt

1. What are some of Perth's boomtown characteristics manifested in the landscape seen during your Google Earth tour?
2. Are there FIFO workers in any part of North America? Where, and why?

▼ **Figure 14.2.2 Downtown Perth along the Swan River** This tranquil photo masks the frenetic boom going on in Perth and its suburbs.

corner of the country. In stark contrast to Papua New Guinea, the largest urban area on the northern margin of Oceania is Honolulu (over 1 million), on the island of Oahu. Here rapid metropolitan growth has occurred since World War II because of U.S. statehood and the scenic attractions of Honolulu's mid-Pacific setting.

Historical Geography

The region's remoteness from the world's hearth areas in Africa and Europe meant that Oceania lay beyond the dominant migratory paths of earlier peoples. Even so, prehistoric settlers eventually found their way to the isolated Australian interior and even the far reaches of the Pacific. Much later, the pace of new in-migrations increased once Europeans explored the region and identified its resource potential.

Peopling the Pacific The large islands of New Guinea and Australia, with their nearness to the Asian landmass, were settled much earlier than the more distant islands of the Pacific. By 60,000 years ago, the ancestors of today's native Australian, or **Aborigine**, population were making their way out of Southeast Asia and into Australia (Figure 14.21). The first Australians most likely arrived using some kind of watercraft, but because such boats were probably not capable of more lengthy voyages, the more distant islands remained inaccessible to settlement for tens of thousands of years. During the last glacial period, however, sea levels were much lower than they are now, which would have allowed easier movement to Australia across relatively narrow spans of water from what is now Southeast Asia. It is not known whether the original Australians arrived in one wave of people or in many, but the available evidence suggests that they soon occupied large portions of the continent, including Tasmania, which was at that time connected to the mainland by a land bridge because of the lower sea level.

Eastern Melanesia was settled much later than Australia and New Guinea. By 3500 years ago, some Pacific peoples had mastered long-distance sailing and navigation, which eventually opened the entire oceanic realm to human habitation. In that era, people gradually moved east to occupy New Caledonia, the Fiji Islands, and Samoa. From there, later movements

▼ **Figure 14.21 Peopling the Pacific** Recent settlement of Pacific islands by Austronesian peoples from Southeast Asia shaped cultural patterns across the oceanic portions of the realm. Eastward migrations through the Solomon Islands, Fiji, and the Cook Islands were followed by later movements to the north and south.

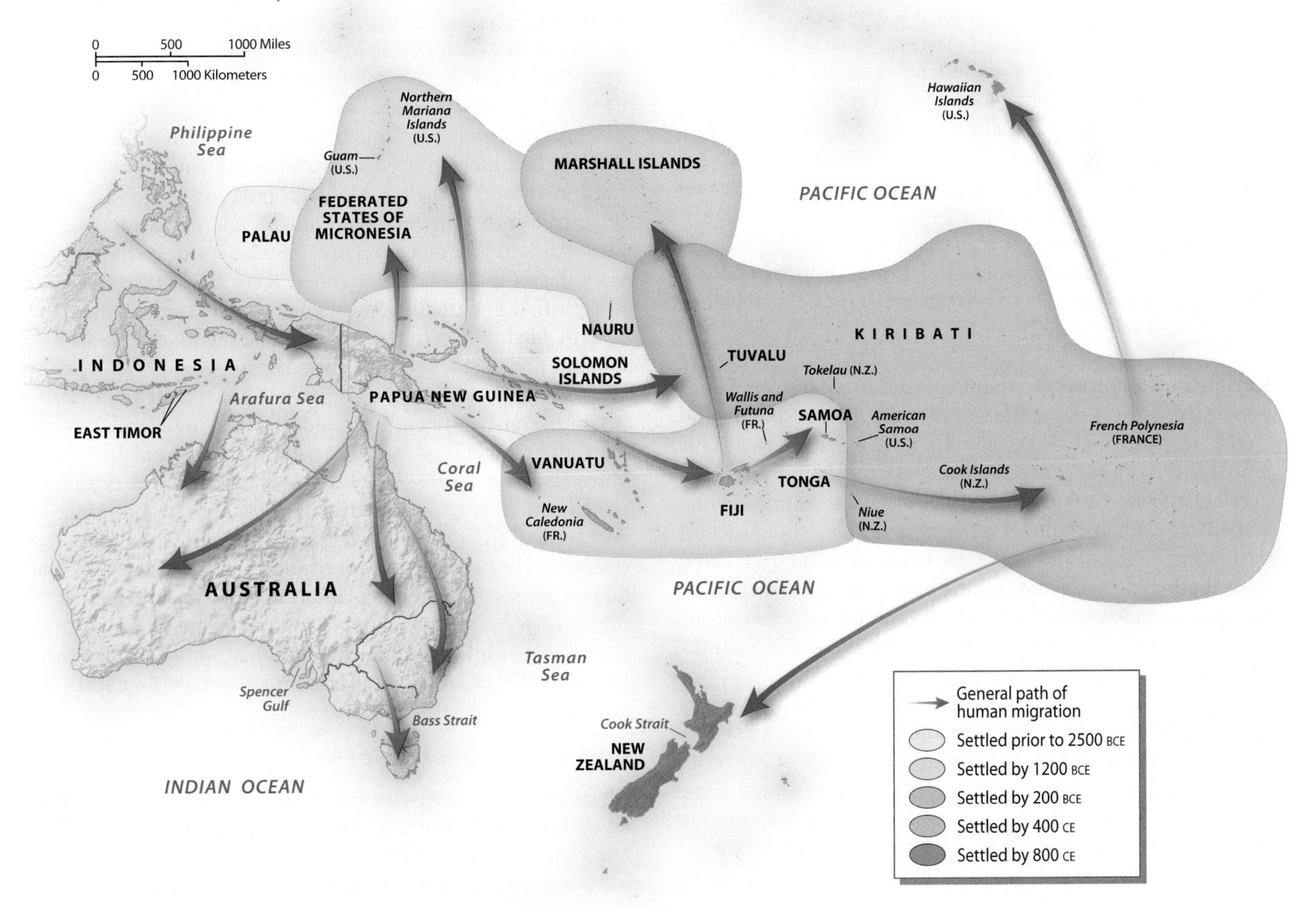

▲ **Figure 14.22 Polynesian Ocean-Voyaging Canoes**
In 2011, seven ocean-voyaging canoes were built using a combination of traditional and modern materials (the hulls were fiberglass instead of wood). They were then sailed between far-flung Polynesian islands to revive cultural traditions. **Q: How did the ancient Polynesians navigate as they crossed the South Pacific to outlying islands?**

took seafaring folk north into Micronesia, with the Marshall Islands occupied around 2000 years ago.

Continuing movements from Asia further complicated the story of these migrating Melanesians. Some of the migrants mixed culturally and eventually reached western Polynesia, where they formed the core population of the Polynesian people. By 800 CE, they had reached such distant places as Tahiti, Hawaii, and Easter Island. Prehistorians hypothesize that population pressures may have quickly reached crisis stage on the relatively small islands, leading people to attempt dangerous voyages to colonize other Pacific islands. Equipped with sturdy outrigger sailing vessels and ample supplies of food, the Polynesians were able to colonize most of the islands they discovered (Figure 14.22).

European Colonization About six centuries after the Maori people brought New Zealand into the Polynesian realm, Dutch navigator Abel Tasman spotted the islands on his global exploration of 1642, marking the beginning of a new chapter in the human occupation of the South Pacific. British sea captain James Cook surveyed the shorelines of both New Zealand and Australia between 1768 and 1780, with the belief that these distant lands might be worthy of European development. By 1800, other European expeditions were also exploring the Pacific, placing most of Oceania's island groups on European maps.

European colonization of the region began in Australia when the British needed a remote penal colony to which convicts could be exiled. The southeastern coast of Australia was selected as an appropriate site, and in 1788 the First Fleet arrived with 750 prisoners in Botany Bay, near what is now Sydney. Other fleets and more convicts soon followed, as did boatloads of free settlers. Before long, free settlers outnumbered the convicts, who were gradually being released after serving their sentences. The growing population of mainly English-speaking people soon moved inland and also settled other favorable coastal areas. British and Irish settlers were attracted by the agricultural and stock-raising potential of the distant colony, as well as by the lure of gold and other minerals (a major gold rush occurred in Australia during the 1850s). The British government also encouraged the emigration of its own citizens, often paying the transportation fare of those too poor to afford it themselves.

These new settlers came into conflict with the Aborigines immediately upon arrival. No treaties were signed, however, and in most cases Aborigines were simply displaced from their lands. In some places—most notably, Tasmania—they were hunted down and killed. In mainland Australia, the Aborigines were greatly reduced in numbers by diseases introduced by the foreign settlers. By the mid-19th century, Australia was primarily an English-speaking British colony with the native peoples driven into submission.

Somewhat later, the lush and fertile lands of New Zealand also attracted British settlers. European whalers and sealers arrived shortly before 1800, with more permanent agricultural settlement taking shape after 1840, when the British formally declared sovereignty over the region. As new arrivals grew in number and the scope of planned settlement colonies on the North and South islands expanded, tensions with the native Maori population increased. Well organized into kingdoms and chiefdoms, the Maori put up formidable resistance (Figure 14.23). Widespread Maori wars began in 1845 and engulfed New Zealand until 1870. The British eventually prevailed, however, and as was the case in Australia with the Aboriginals, the native Maori lost most of their land.

Similarly, native Hawaiians lost control of their lands to immigrants. Hawaii emerged as a united and powerful Polynesian kingdom in the early 1800s, and for many years its native rulers limited U.S. and European claims to their islands. However, increasing numbers of missionaries and settlers from the United States were allowed in, and by the late 19th century control of the Hawaiian economy had largely passed to foreign plantation owners. In 1893, U.S. interests were strong enough to overthrow the Hawaiian monarchy, resulting in formal political annexation to the United States in 1898, with Hawaii becoming a state in 1959.

Settlement Landscapes

The settlement geography of Australia and Oceania offers an interesting mixture of local and global influences. The contemporary cultural landscape still reflects the imprint of indigenous peoples in those areas where native popula-

◀ **Figure 14.23 Maori Warrior** Body ornamentation, including tattoos, was common in traditional Maori culture, particularly among the high-status warrior class. Before battle, the Maori warriors would perform a ceremonial dance, called the ***haka,*** that included fierce facial contortions with the tongue and eyes, as shown by this Maori dancer. *haka* dances (along with the facial contortions) are now a common part of New Zealand culture, particularly among sports teams.

tions remain numerically dominant. Elsewhere, patterns of recent colonization have produced a modern global scene mainly shaped by Europeans. The result includes everything from German-owned vineyards in South Australia to houses on New Zealand's South Island that appear to have been plucked directly from the British Isles. In addition, processes of economic and cultural globalization have resulted in urban forms that make cities such as Sydney or Auckland look strikingly similar to places such as San Diego or Seattle (Figure 14.24).

The Urban Transformation As mentioned, both Australia and New Zealand are highly urbanized, Westernized societies, and thus the vast majority of their populations live in city and suburban environments. As in Europe and North America, much of this urban transformation came during the 20th century as the rural economy contracted and as opportunities for urban manufacturing and service employment grew. As urban landscapes evolved, they initially took on many of the characteristics of their largely European models and then later blended in a strong dose of North American urban and suburban influences. The result is a landscape in which many North Americans are

▼ **Figure 14.24 Downtown Sydney** In the foreground is the distinctive opera house that opened in 1973 and is now a World Heritage Site because of its unique architecture. The park on the left of the photo is the Royal Botanic Gardens.

▲ **Figure 14.25 Port Moresby, Papua New Guinea** Urban poverty and high crime afflict the city of Port Moresby, the capital of Papua New Guinea. The city's slums, many built out on the water, reflect stresses of recent urban growth as rural residents emigrate from the even poorer nearby highlands.

quite comfortable, even though the varied local accents heard on the street and many features of the metropolitan scene are reminders of the strong and lasting attachments to British traditions.

The affluent Western-style urban environments in Australia and New Zealand offer a dramatic contrast with the urban landscapes found in less developed cities in the region. If you walk the streets of Port Moresby in Papua New Guinea, you see a very different urban landscape, which reveals evidence of the large gap between rich and poor within Oceania. Port Moresby is the country's political capital and largest commercial center. Rapid growth here has produced many of the classic problems of urban underdevelopment: Adequate housing is scarce, and the building of roads and schools lags far behind the need, while street crime and alcoholism rise (Figure 14.25). Elsewhere, urban centers such as Suva (Fiji), Noumea (New Caledonia), and Apia (Samoa) also reflect the economic and cultural tensions generated as local populations are exposed to Western influences. Rapid growth is a common problem in the smaller cities of Oceania because native people from rural areas and nearby islands gravitate toward the job opportunities available. In the past 50 years, the huge global growth of tourism in places such as Fiji and Samoa has also transformed the urban scene, replacing traditional village life with a landscape of souvenir shops, honking taxicabs, and crowded seaside resorts.

Rural Australia and New Zealand Rural landscapes across Australia and the Pacific region express a complex mosaic of cultural and economic influences. In some settings, Australian Aborigines or native Papua New Guinea Highlanders can still be found in their familiar homelands, their traditional way of life and settlements barely changed from pre-European times. Yet such settlement landscapes are becoming increasingly rare. Global influences penetrate the scene as the cash economy, foreign tourism and investment, and the currents of popular culture make their way from city to countryside.

Most of rural Australia is too dry for farming or serves as only marginally valuable agricultural land. Much of the remainder of the interior, however, features areas supporting range-fed livestock, areas without any agricultural potential, and isolated areas where Aboriginal peoples still pursue their traditional forms of hunting and gathering. Sheep and cattle dominate rural Australia's livestock economy. Many rural landscapes in the interior of New South Wales, Western Australia, and Victoria, for example, are oriented around isolated sheep stations—ranch operations that move the flocks from one large pasture to the next. Cattle can sometimes be found in these same areas, although many of the more extensive range-fed cattle operations are concentrated farther north, in Queensland.

Croplands also vary across the region. A band of commercial wheat farming mingles with the sheep country across southern Queensland; the moister interiors of New South Wales, Victoria, and South Australia; and a swath of favorable land east and north of Perth. Elsewhere, specialized sugarcane operations thrive along the narrow, warm, and humid coastal strip of Queensland. To the south and west, productive irrigated agriculture has developed in places such as the Murray River Basin, allowing for the production of orchard crops and vegetables. **Viticulture**, or grape cultivation, increasingly shapes the rural scene in places such as South Australia's Barossa Valley, the Riverina district in New South Wales, and Western Australia's Swan Valley. Indeed, the area under grape cultivation grew by 50 percent between 1991 and 1998 as the popular Chardonnay, Cabernet Sauvignon, and Shiraz varieties boosted revenues from wine production to more than $540 million per year (Figure 14.26).

Although much smaller in area than that of Australia, New Zealand's rural settlement landscape includes a variety of agricultural activities. Ranching clearly dominates the New Zealand countryside, with the vast majority of agricultural land devoted to livestock production, particularly sheep grazing and dairying. Commercial livestock outnumber people in New Zealand by a ratio of more than 20 to 1, and this is apparent everywhere in the countryside. Dairy operations are present mostly in the lowlands of the north, where they sometimes mingle with suburban landscapes in the vicinity of Auckland.

Rural Oceania Elsewhere in Oceania, varied influences shape the rural landscape. On high islands with more water, denser populations take advantage of diverse agricultural opportunities; on the more barren low islands, fishing is often more important. Several types of rural settlement can be identified across the island realm. In rural New Guinea, village-centered shifting cultivation dominates: Farmers clear a patch of forest and then, after a few years, shift to another patch, thus practicing a form of land rotation. Subsistence foods such as sweet potatoes, taro (another starchy root crop), coconut palms, bananas, and other garden crops are often found in the same field. Increasing numbers of planters also include commercial crops such as coffee (Figure 14.27).

▲ **Figure 14.26 Vineyards in Australia** Australia is the fourth largest wine exporter in the world. Much of these exports go to Asia, particularly to East Asia, where Australian wines are second only to those from France. This photo is from the Hunter Valley wine region, close to Sydney in New South Wales, where the oldest vineyards were planted in the early 19th century.

Commercial plantation agriculture has also made its mark in many more accessible rural settings. In these places, settlements consist of worker housing near crops that are typically controlled by absentee landowners. For example, copra (coconut), cocoa, and coffee operations have transformed many agricultural settings in places such as the Solomon Islands and Vanuatu. Sugarcane plantations have reshaped other island settings, particularly in Fiji and Hawaii.

Diverse Demographic Paths

A variety of population-related issues face residents of the region today. In Australia and New Zealand, although populations grew rapidly from natural increases in the 20th century, today's low birth rates produce a pattern similar to that of North America, where any population growth takes place only from immigration.

Very different demographic challenges grip many less-developed island nations of Oceania (Figure 14.28). High population growth rates of over 2 percent per year are not uncommon, causing crowding and land hunger. Tuvalu (north of Fiji), for example, has just over 12,000 inhabitants, but they are crowded onto a land area of about 10 square miles (26 square km), making it one of the world's most densely populated countries.

Out-migration from several island nations is very high. In Tonga and Samoa, for example, the lack of employment is a considerable push force. In contrast, Australia and New Zealand remain attractive to migrants, as does New Caledonia, mainly because of its recent mining boom.

▼ **Figure 14.27 Commercial Taro Cultivation** While taro is a traditional subsistence crop throughout Oceania, taro is also grown on plantations in Hawaii for commercial sales throughout the region. This taro plantation is on the island of Kauai.

Too Many People or Not Enough? According to politicians, talk radio pundits, and activists of different stripes, Australia is facing a severe population crisis. However, the kind of crisis is not clear. Australia either (a) already has too many people or (b) is not growing fast enough and has too small a population.

Groups like Sustainable Population Australia (SPA) call for an environmentally responsible population policy that restricts population size. They say that because so much of the country is either arid or semiarid, it has a limited rural carrying capacity. Global warming will make matters

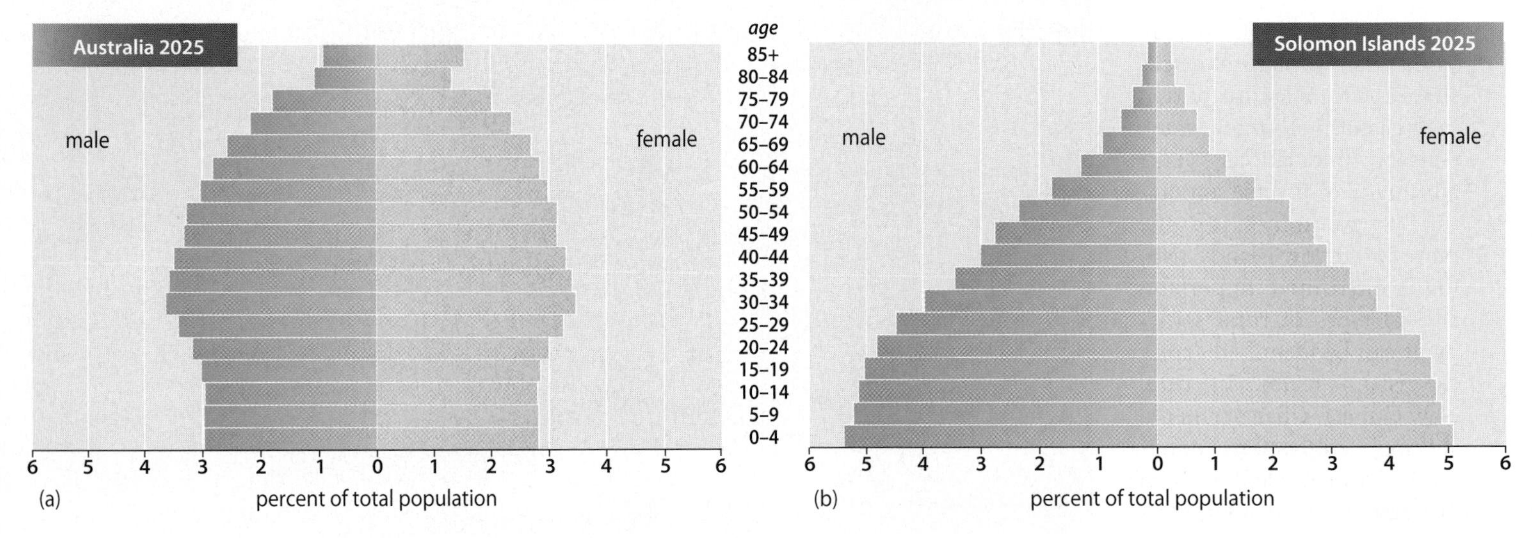

▲ **Figure 14.28 Population Pyramids of Australia and the Solomon Islands (2025)** (a) Like many developed countries, Australia has very low natural growth, as shown by the forecast for 2025. (b) In contrast, like many developing countries, the Solomon Islands forecast shows the classic pyramidal shape of a young and growing population.

even worse, they argue, expanding Australia's arid interior and forcing even more people into the crowded coastal urban strips, where seasonal brushfires are already a serious problem. Tim Flannery—scientist, author, activist, and Australian of the Year in 2007—argues that the country is already overpopulated by some 8 million people. Environmental groups advocating population limits often find their bandwagon crowded with an uneasy array of anti-immigration groups with their baggage of ethnic discrimination and racism (Figure 14.29).

On the other side of the controversy are those who say Australia's low birth rate will lead to a declining population unless immigration increases. For example, the Australian Labor Party's minister for family and community services has said that the very low birth rates currently observed could halve the population by the end of the century. To combat this so-called baby bust, the Labor Party and business coalitions such as the Australian Population Institute are calling for maternity leave, tax breaks, and, most of all, increased immigration to shore up Australia's population. Enhanced population growth, they argue, is absolutely necessary for the national economy, with a larger market, larger labor force, and broader base of taxpayers.

In a country basically the size of the mainland United States, yet with a total population only slightly larger than that of the greater Los Angeles metropolitan area, many find it difficult to believe Australia is already overpopulated.

▼ **Figure 14.29 Is Australia over- or underpopulated?** Australians debate whether their country has already reached its population limits and, thus, should close the door to immigrants or, alternatively, whether its population is too small and should therefore encourage more immigration. This photo is of a crowded area at a horse race outside of Melbourne.

This is particularly true for those living on densely populated Pacific islands, casting a wary eye at rising sea levels and a longing eye at Australia's empty spaces, which might offer refuge if the islanders should have to leave their flooded homes.

REVIEW

14.5 Compare the populations of Australia and New Zealand as to size, density, and level of urbanization.
14.6 Describe the prehistoric peopling of the Pacific.
14.7 Is Australia over- or underpopulated? What are the arguments for each point of view?

Cultural Coherence and Diversity: A Global Crossroads

The Pacific world offers excellent examples of how cultural geographies are shaped and transformed as different groups migrate to a region, interact with one another, and then change over time. As Europeans and other outsiders arrived in Oceania, colonization forced native peoples to resist or adjust. More recently, worldwide processes of globalization have also redefined the region's cultural geography, provoking fears of homogenization that have led to cultural preservation efforts as native groups attempt to protect their heritage.

Multicultural Australia

Australia's cultural patterns illustrate many of the fundamental processes of globalization at work. Today, although still dominated by its colonial European roots, the country's multicultural character is becoming increasingly visible as native people assert their cultural identity and, at the same time, varied immigrant populations play larger roles in Australian society.

Aboriginal Imprints For thousands of years, Australia's Aborigines—the indigenous people of Australia—dominated the continent. Prehistorically, they did not farm in any way, opting instead for a hunting-and-gathering way of life that persisted up to the time of the European conquest. Because of their foraging way of life in the semiarid Outback, settlement densities remained low. Tribal groups were often isolated from one another, and prehistorians estimate that their overall population never numbered more than 300,000 inhabitants. Because these people were clustered in many different areas, their language became fragmented. Although precise counts vary, there were probably 250 languages spoken at the time of European contact, and almost 50 indigenous languages can still be found today.

Radical cultural and geographic changes accompanied the arrival of Europeans, and Aboriginal populations were decimated in the process. The geographic results of colonization were striking: Aborigines were relocated to the sparsely settled interior, particularly in northern and central Australia, where fewer Europeans competed for land. From 1909 until 1969, large numbers of Aboriginal children were forcibly taken from their families and raised in government-run boarding schools with the ostensible goal of acculturating them to white Australian society. This controversial program produced what today is referred to as the "Stolen" or "Lost Generation" of Aborigines, who fit neither into their native cultural group nor into white society. More recently, the inhumane treatment of Aborigines was considered so despicable that in 2008 the Australian government issued a formal apology to their native peoples.

Nonetheless, despite the Aborigines' treatment by white Australia, their culture perseveres in Australia, and a native people's movement is growing (Figure 14.30). Indigenous people today account for approximately 2 percent (or 430,000) of Australia's population, but their geographic distribution has changed dramatically over the past century. Aborigines account for almost 30 percent of the Northern Territory's population (many of these near Darwin), and other large native reserves are located in northern Queensland and Western Australia. Most native people, however, live in the same urban areas that dominate the country's overall population geography. Indeed, more than 70 percent of Aborigines live in cities, and very few of them still practice traditional hunting-and-gathering lifestyles. Processes of cultural assimilation are clearly at work: Urban Aborigines are frequently employed in service occupations, Christianity has often replaced traditional animist religions, and only 13 percent of the native population still speaks a native language.

Still, forces of diversity are also operative, as evidenced by a growing Aboriginal interest in preserving traditional cultural values. Particularly in the Outback, several Aboriginal languages remain strong and have growing numbers of speakers. In addition, cultural leaders are preserving

▼ **Figure 14.30 Australian Aborigines.** The native people of Australia, who inhabited the continent before European colonization, the Aborigines continue to struggle for equal rights and decent living conditions. This family lives in a traditional community near Alice Springs in the Northern Territory where Aborigines make up 30 percent of the population.

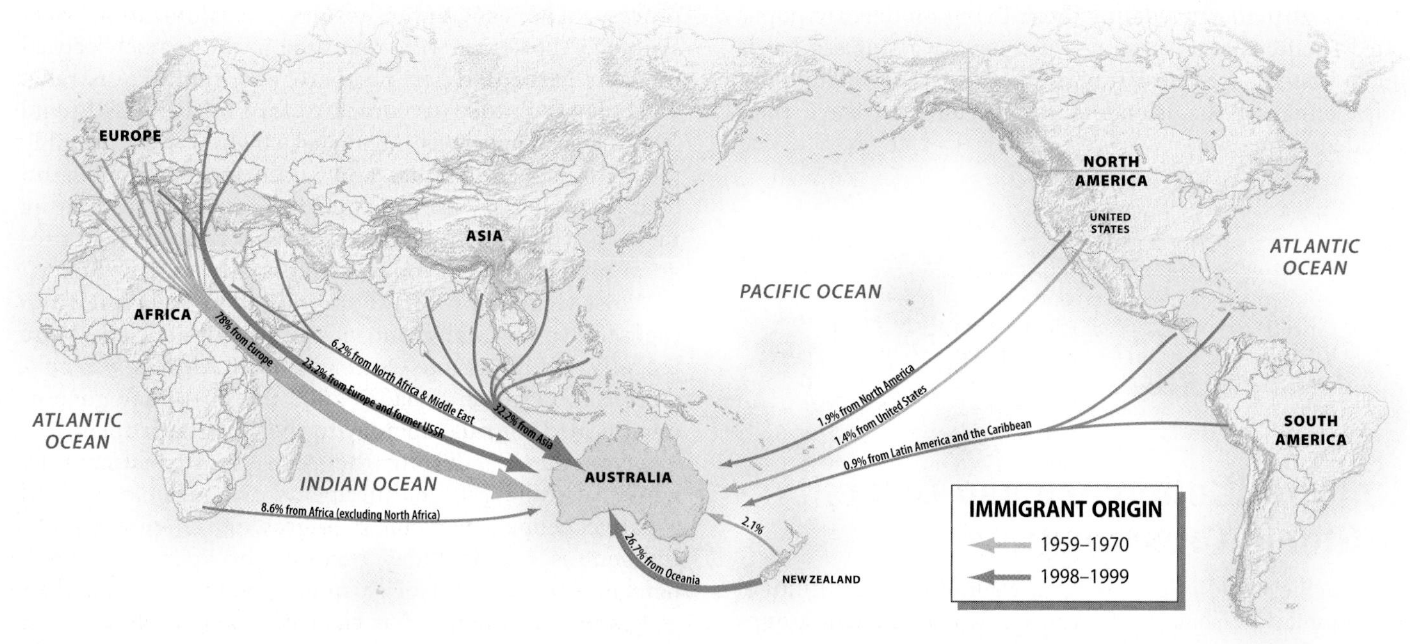

▲ **Figure 14.31 Changing Immigration to Australia** This map shows the dramatic changes in the source areas for immigrants to Australia after the White Australia Policy was abolished in 1973. **Q: What are the different factors that explain the changing patterns of immigration to Australia?**

some aspects of Aboriginal spiritualism, and these religious practices often link local populations to surrounding places and natural features that are considered sacred. Most notably, the sandstone rock formation known to white Australians as Ayers Rock, a tourist attraction located roughly in the middle of the country, is called Uluru by the Aborigines and is sacred to them. In fact, a growing number of these sacred locations are at the center of land-use controversies between Aboriginal populations and Australia's European majority.

The future of Aboriginal cultures remains unclear. Pressures for cultural assimilation are intense as many native people move to more Western-oriented urban settlements and lifestyles. At the same time, rapid rates of natural increase (almost twice the national average) and a growing cultural awareness of Aboriginal traditions act to preserve elements of the country's indigenous cultures.

A Land of Immigrants Most Australians reflect the continent's more recent European-dominated migration history, but even these patterns have become more complex as a rising tide of Asian cultures becomes increasingly important (Figure 14.31). Overall, more than 70 percent of Australia's population continues to reflect a British or Irish cultural heritage. These groups dominated many of the 19th- and early 20th-century migrations into the country, and as a result, close cultural ties to the British Isles remain strong.

A need for labor along the fertile Queensland coast caused European plantation owners to import inexpensive workers from the Solomons and New Hebrides in the late 19th century. These Pacific island laborers, known as **kanakas**, were spatially and socially segregated from their Anglo employers, but further diversified the cultural mix of Queensland's "sugar coast." Historically, however, nonwhite migrations to the country were strictly limited by what is usually called the **White Australia Policy**. These governmental guidelines promoted European and North American immigration at the expense of other groups. This remained national policy until 1973.

Recent migration trends have reversed this historical bias, and more diverse inflows of new workers and residents are adding to the country's multicultural character. Since the 1970s, the government's Migration Program has been dominated by a variety of people chosen on the basis of their educational background and potential for succeeding economically in Australian society. Thus, a growing number of families have come from places such as China, India, Malaysia, and the Philippines. Smaller numbers have arrived as refugees from troubled parts of the world, such as Central and South Asia (see *People on the Move: Australia's Asylum Problem*). The result of this legal and irregular immigration is a more diverse population. Indeed, today 25 percent of Australia's people are foreign-born, reflecting the country's global popularity as a migration destination. In the period 2000–2010, almost 40 percent of the settlers arriving in the country were from Asia. Major cities offer particularly attractive possibilities: Sydney's Asian population already exceeds 10 percent and is growing rapidly, while Perth's culture and economy are increasingly linked to Asian countries.

Cultural Patterns in New Zealand

New Zealand's cultural geography broadly reflects the patterns seen in Australia, although the precise cultural mix differs slightly. The native **Maori** people are more numerically important and culturally visible in New Zealand than their Aboriginal counterparts are in Australia. British colonization clearly mandated the dominance of Anglo cultural traditions by the late 19th century, but the Maori survived, even though they lost most of their land in the process. After the initial decline, the native population began rebounding in the 20th century, and today the Maori account for more than 8 percent of the country's 4 million residents. Geographically, the Maori remain most numerous on the North Island, including a sizable concentration in metropolitan Auckland. Although urban living is on the rise, many Maori, similar to the Aborigines, are committed to preserving their religion, traditional arts, and Polynesian culture (Figure 14.32). In addition, Maori joins English as the official languages of New Zealand.

Many New Zealanders still identify with their largely British heritage, but the country's cultural identity has increasingly separated from its British roots. Several processes have forged New Zealand's special cultural character. As Britain tightened its own links with the European continent after World War II, New Zealanders increasingly formed a more independent and diverse identity. In many ways, popular culture ties the country ever more closely to Australia, the United States, and continental Europe, a function of increasingly global mass media. Several major movies, for example, have been filmed in New Zealand, including *The Adventures of Tintin, The Hobbit, Avatar, The Lord of the Rings* trilogy, and *Whale Rider*.

▲ **Figure 14.32 Maori Artwork** Woodcarving is an active part of preserving Maori culture in New Zealand, as demonstrated by this carver at the Puia Cultural Center in Rotorua, North Island, New Zealand.

The Mosaic of Pacific Cultures

Native and colonial influences produced a variety of cultures across the islands of the South Pacific. In more isolated places, traditional cultures are largely insulated from outside influences. In most cases, however, modern life in the islands revolves around an intricate cultural and economic interplay of local and Western customs. The result illustrates the fact that the relative cultural insularity of the past is gone forever, and in its place is a Pacific realm rapidly adjusting to powerful forces of globalization.

Language Geography A modern language map reveals some significant cultural patterns that both unite and divide the region (Figure 14.33). Most of the native languages of Oceania belong to the **Austronesian** language family, which encompasses wide expanses of the Pacific, much of insular Southeast Asia, and, somewhat surprisingly, Madagascar. Linguists hypothesize that the first prehistoric wave of oceanic mariners spoke Austronesian languages and thus spread them throughout this vast realm of islands and oceans. Within this broad language family, the Malayo-Polynesian subfamily includes most of the related languages of Micronesia and Polynesia, suggesting a common cultural and migratory history for these widespread peoples.

Melanesia's language geography is more complex and still incompletely understood by outside experts. Although coastal peoples often speak languages brought to the region by the seafaring Austronesians, more isolated highland cultures, particularly on the island of New Guinea, speak varied Papuan languages. In fact, more than 1000 languages have been identified, creating such linguistic complexity that many experts question whether they even constitute a unified "Papuan family" of related languages. The New Guinea highlands may hold some of the world's few remaining **uncontacted peoples**—cultural groups that have yet to be "discovered" by the Western world.

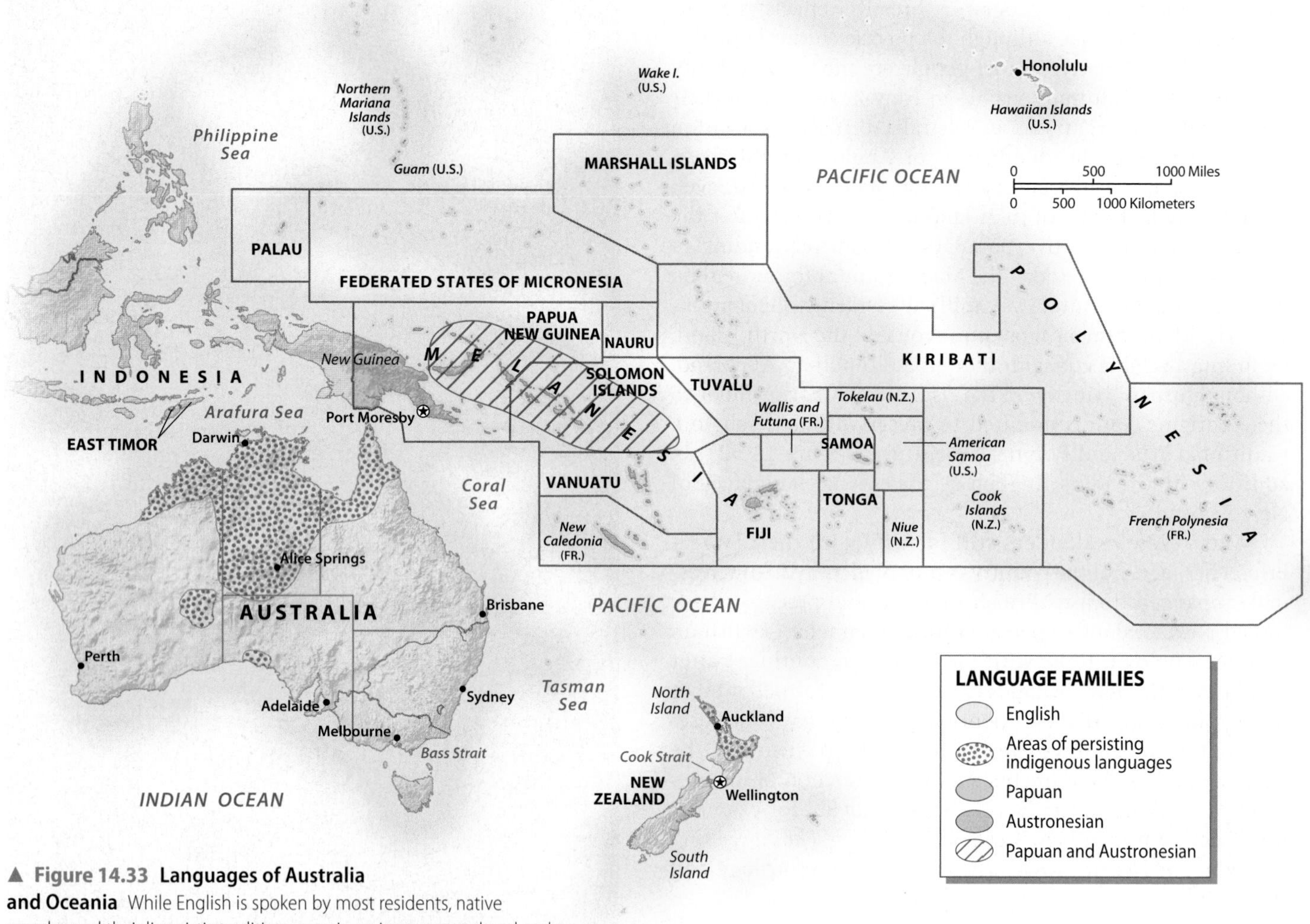

▲ **Figure 14.33 Languages of Australia and Oceania** While English is spoken by most residents, native peoples and their linguistic traditions remain an important cultural and political force in both Australia and New Zealand. Elsewhere, traditional Papuan and Austronesian languages dominate Oceania. The French colonial legacy also persists in some Pacific locations. Tremendous linguistic diversity has shaped the cultural geography of Melanesia, and more than 1000 languages have been identified in Papua New Guinea alone.

Given the frequency of contact between different island cultures, it is no surprise that people have generated new forms of intercultural communication. For example, several forms of **Pidgin English** (also known simply as *Pijin*) are found in the Solomons, Vanuatu, and New Guinea, where it is the major language used between different ethnic groups. In Pijin, a largely English vocabulary is reworked and blended with local grammar. Pijin's origin is commonly traced to 19th-century Chinese sandalwood traders ("pijin" is the Chinese pronunciation of the word for "business"). Today Pijin has become a globalized language of sorts in Oceania as trade and political ties develop between different native island groups. About 300,000 people in Oceania speak Pidgin on a regular basis (see *Everyday Globalization: Hawaiian Pidgin*).

Village Life Traditional patterns of social life are as complex and varied as the language map. In many cases, life revolves around predictable settings. For example, across much of Melanesia, including Papua New Guinea, most people live in small villages, often occupied by a single clan or family group. Many of these traditional villages contain fewer than 500 residents, although some larger communities may house more than 1000 people. Life often revolves around the growing and gathering of food, annual rituals and festivals, and complex networks of kin-based social interactions.

Traditional Polynesian culture also focuses on village life (Figure 14.34), although strong class-based relationships often exist between local elites (who are often religious leaders) and ordinary residents. Polynesian villages are also more likely linked to other islands by wider cultural and political ties. Despite the Western stereotype of Polynesian communities as idyllic and peaceful, violent warfare was actually quite common across much of the region prior to European contact.

People on the Move

Australia's Asylum Problem

In 2012, some 16 million people sought asylum from political and social persecution, according to the UN. Many of them left their homelands and crossed into a neighboring country seeking refuge. Others traveled longer distances, seeking out a specific country for their refuge. Of these homeless masses, each year about 10,000 reach Australia—or come close enough to its shores that they're rescued by the Australian Coast Guard. Once in the hands of Australian authorities, these refugees—most of whom ask for asylum—are housed in detention camps while their claims of persecution are validated (or not) and their requests for asylum processed.

Australia has a long tradition of taking care of its destitute and downcast. Perhaps the country's compassion stems from its own humble beginnings as a debtor's prison. Nevertheless, although many Australians are proud of its 21st-century asylum policies, others are less so, arguing that the world's riffraff are identifying Australia as an easy target for a life on welfare. As a result, today the asylum topic has become a full-blown controversy in Australia.

The Pacific Solution One of the most controversial aspects of the refugee-asylum topic is Australia's so-called Pacific Solution, a government policy designed to dissuade refugees from targeting Australia as their goal. Originally instigated in 2001, the Pacific Solution had three components: First, the Australian navy was charged with intercepting refugee boats before they could make landfall on Australian territory; second, detention centers were created on Nauru and Christmas Island (**Figure 14.3.1**) to keep refugees off Australian territory (and to deprive them of Australian law and civil rights) while their claims are processed; and, third,

▼ **Figure 14.3.1 Refugees and the Australian Navy** Charged with intercepting refugee boats on the high seas, the Navy then delivers the refugees to detention centers on Nauru and Christmas Island.

▲ **Figure 14.3.2 Refugees from Southeast Asia arrive in Australia** This boatload of refugees traveled first to Indonesia, where they then departed illegally for Australia.

and complementing the second point, Australia relinquished its ownership of small offshore islands where refugees often landed and where they could plead amnesty under Australian law. Also, a harder stance was taken in giving amnesty, granting it to only 40 percent of refugees.

The Pacific Solution was so controversial among Australians that it was shelved with a change of government in 2007. However, as the numbers of asylum seekers increased, the current Labor government began implementing the offshore detention component once again in August 2012, with the reopening of the processing center (opponents call it a prison) on Nauru Island.

Refugee Source Areas How do these asylum seekers get to Australia, and where do they come from? Most arrive in overloaded boats of questionable seaworthiness that probably departed most recently from some part of Indonesia, Australia's closest neighbor (**Figure 14.3.2**). These craft, however, may have started their voyages much farther away, in Sri Lanka, Bangladesh, Pakistan, or even China. Evidence for that comes from the nationalities of the refugees, which is where the story gets murky. Many reaching the shores of Australia are originally from Afghanistan, but have spent time in refugee camps in Pakistan, Iran, or Iraq. Because of their time in these other countries, the question of whether or not they've been persecuted becomes complex. Another complication is that, if the decision is made to return these refugees to their country of origin, which is done when there is no evidence of persecution, which country would that be—Pakistan or Afghanistan?

1. Because a key element in Australia's amnesty policy is whether or not refugees were "persecuted" in their homeland, make a list of the reasons Australian refugees might have been persecuted in these different countries—Afghanistan, Sri Lanka, Pakistan, and Iran.
2. What other countries and regions of the world that you've read about in this book are magnets for refugees and what sort of amnesty policies do they have?

Google Earth Virtual Tour Video

http://goo.gl/1kl02h

▲ **Figure 14.34 Native Village in Fiji** This reconstructed native village in the highlands of Fiji actually serves as an eco-lodge, providing tourists with basic accommodations while providing information on traditional Fijian culture.

Gender Geographies

It is usually unwise to generalize about gender in a large multicultural region; nevertheless, we can note several key points that complement our understanding of the cultural geography of Australia and Oceania.

Gender in Australia and New Zealand Both countries were early supporters of female suffrage, empowering women to vote in national elections in 1893 (New Zealand) and 1902 (Australia). Earlier, in some cases women were allowed to vote in local and state elections, although not until decades later could females actually hold office. Since that time, both countries have elected women as national leaders, with New Zealand's Helen Clark becoming prime minister in 1999 and Australia electing Julia Gillard to a similar office in 2010 (Figure 14.35).

Despite having women as their nation's political leaders, both countries still have considerable gender gaps in terms of female representation in government, employment salaries, and social support services. As a result, in the global gender index New Zealand ranks 18th and Australia is 25th, with both countries scoring lower than most western European countries but higher than the United States.

Colonial Influences on Maori Gender Roles Before European colonization, Maori women and men were equal in social status and power. This was a product of the overarching Maori principle of equity and balance in their nonhierarchical society. For example, the Maori language had no gender distinctions in personal pronouns (no "his" or "hers"); instead, all pronouns were gender-neutral. Further evidence for this gender equity comes from the prominent role women played in Maori proverbs and legends.

However, English colonial society was troubled by this gender-neutral native society and enacted legislation to deprive Maori women of their status. In 1909, New Zealand law required Maori women to undergo legal marriage ceremonies that emphasized male ownership of all property—not just of land and livestock, but also of the wife. Missionary schools for Maori women had also long reinforced English notions of female domesticity where women were secondary to male authority.

As a result, gender roles in contemporary Maori society now reflect these earlier colonial notions of male dominance. To further their European values, missionaries rewrote Maori proverbs and legends to emphasize male characters with heroic warrior attributes. Today Maori women never participate in the haka war dances; instead, female participation is limited to subservient songs and dances.

Gender in Aboriginal Society Traditional Aboriginal society has distinct gender roles, with clear-cut distinctions between what are referred to as "women's business" and that of men. Important is that these distinctions are played out not just in daily affairs, but also in the Dreamtime, an abstract parallel universe for Aboriginal people of central importance. In physical life as well as in Dreamtime, women are responsible for the vitality and resilience of family lives, while men's business centers on the larger group or tribe. These distinct gender roles also involve the landscape, with certain areas and locales linked closely to either men or women, but rarely to both. As a result, the Aboriginal territory also becomes highly gendered.

Sports in Oceania

Like all other aspects of culture, colonial influences have left their mark on Oceania's playing fields, particularly in those former English colonies where cricket, soccer, and

▼ **Figure 14.35 Julia Gillard, former Australian Prime Minister** Ms. Gillard was the first female elected prime minister of Australia and served from 2010 to 2013. In this photo she autographs a student's shirt at a cricket match between Sri Lanka and Australia.

Everyday **Globalization**

Hawaiian Pidgin

"Eh, howzit, brah. So what you like grind?"

"We no moa da kine. No worries, brah. I grind any kine."

Got it? If not, then you're not proficient in Hawaiian pidgin. That's probably OK unless you're a local, since Hawaiians hate having mainlanders talking their talk.

Hawaiian pidgin is also called Hawaiian Pidgin English or even Hawaiian Creole English (abbreviated by linguists as HCE). However, in the islands it is most commonly known simply as pidgin. Pidgin is the everyday language used by many nonwhite Hawaiian residents, but should not be confused with native Hawaiian, which is the traditional Polynesian language of the islands and is, along with English, an official language of the state.

Beginnings of Hawaiian Pidgin Hawaiian pidgin reportedly began as a local variant of Pacific Island pidgin and evolved during the 19th century as different ethnic groups communicated while working on Hawaii's sugar plantations. Like Pacific or Melanesian pidgin, it contains many Cantonese words and grammatical structures, but Hawaii pidgin is unique because of its heavy use of words and grammar from Portuguese, Japanese, Spanish, Korean, and Tagalog (the language spoken by Filipino sugar workers).

By the early 20th century, pidgin moved out of the sugarcane fields and into the towns and cities where different ethnic groups clustered. Children learned it from their school classmates and then fed new versions of the language back to their parents and grandparents as the language evolved. In the demographic fluidity of society after World War II, Hawaii pidgin and native Hawaiian started interacting and crossbreeding, just as did Hawaii's multicultural population (Figure 14.4.1). Consequently, today native Hawaiians commonly use Hawaii pidgin instead of Polynesian Hawaiian. In the last several decades, Hawaiian writers and artists have attempted written works in pidgin. Perhaps best known is a New Testament bible titled *Da Jesus Book.*

Unique Characteristics Although often sounding like English, Hawaii pidgin has key differences, such as replacing the *th* sound in words like *that* and *think* with a *d,* resulting in dat" and "dink." Also, words ending with an *l* sound, like *people,* are pronounced with an *o,* so that *people* become "peepo." Many other grammatical and pronunciation characteristics make Hawaiian pidgin unique and are known only to fluent speakers and academic linguists.

Once more we issue a friendly warning to mainlanders that just because you might understand some pidgin words, do not—repeat, do not—attempt speaking pidgin to a native islander. Minimally, you'll be given a bad case of "stink eye"; worse, you might hear something like "Eh, haole boy, you like beef? Kay den."

(Translation of opening dialog: "Hey, bro, what do you want to eat? Oh, I'll eat anything, no worry.")

▼ **Figure 14.4.1 Native Hawaiians** A group of Native Hawaiians gather on an island beach to sing and chat, much of it in Hawaiian pidgin.

rugby dominate the sporting scene. Netball, an English version of basketball played with seven players, is one of the most popular women's sports in both Australia and New Zealand, followed by field hockey. However, in New Zealand, where rugby is arguably the nation's favorite sport, the national team, the All Blacks (named for their uniform color, not their ethnicity), has integrated Polynesian warrior rituals and dances into its pre- and postgame activities (Figure 14.36).

Besides the traditional colonial sports, both Australia and New Zealand have become world powers in water sports such as sailing, surfing, and wind surfing.

Pacific Islanders and American Football Although rugby is played widely throughout Oceania, in recent years young men from the Pacific islands increasingly look to American football as their ticket to a better future. This future includes a U.S. college scholarship and perhaps even a chance to play in the National Football League (NFL), which currently has close to 30 Samoan players.

At the other end of this Pacific island–U.S. mainland pipeline are American college football teams that increasingly recruit Pacific islanders on scholarship. In fact, rare is a top-ranked college team that doesn't feature at least one player from Oceania (Figure 14.37). American Samoans make up

▲ **Figure 14.36 New Zealand All-Black Rugby** Before and after each rugby game, the All Blacks perform the Maori haka, a traditional ritual dance associated with war, contests, or tribal challenges.

the majority because they are American citizens and need no visas, although football players from Tonga and Fiji are also recruited because of similar Polynesian cultural and physical traits. Although a genetic heritage of large, strong males is important, coaches say that adding to the attraction is a Polynesian culture that emphasizes group activity and bonding, coupled with a strong work ethic. Some pundits add that a long history of tribal warfare doesn't hurt. Others say that the Polynesian tradition of male group dancing assures that even the largest men still have a remarkable agility that helps them on the football field.

Most sports historians credit Brigham Young University (BYU) in Utah as the first college to recruit Pacific islanders, back in the late 1950s, drawing on its ties to the Mormon Church and its strong missionary presence in Oceania. But once BYU started the process, the floodgates opened, with other western teams—most notably, Washington State University, the University of Washington, UCLA, and the University of Arizona—filling their rosters with Pacific islanders. Not all players were 300-pound linemen, for Samoans have played at other positions. Jack Thompson (his Anglicized name), the "Throwin' Samoan," was a star Washington State quarterback and the National Collegiate Athletic Association's most prolific passer in 1978, before beginning an NFL career. More recently, his nephew, Tavita Pritchard, played the same position at Stanford University. Other notable Samoan quarterbacks are Jeremiah Masoli (University of Oregon) and Marques Tavita Tuiasosopo (University of Washington).

Interactions with the Larger World

Although traditional culture persists in some areas, most Pacific islands have witnessed tremendous cultural transformations in the past 150 years (see *Exploring Global Connections: Hypercommunication Comes to Oceania*). Settlers from Europe, the United States, and Asia brought new values and technological innovations that have forever changed Oceania's cultural geography and its place in the larger world. The result is a modern setting where Pidgin English has replaced native languages, Hinduism is practiced on remote Pacific islands, and people from traditional fishing communities now work at resort hotels and golf course complexes.

Anglo-European colonialism transformed the cultural geography of the Pacific world by introducing new political and economic systems. In addition, the region's cultural makeup was changed by new people migrating into the Pacific islands. Hawaii illustrates the pattern. By the mid-19th century, Hawaii's King Kamehameha was already entertaining a varied assortment of whalers, Christian missionaries, traders, and navy officers from Europe and the United States. A small elite group of **haoles**, or light-skinned European and American foreigners, was profiting from commercial sugarcane plantations and Pacific shipping contracts. Labor shortages on the islands led to the importation of Chinese, Portuguese, and Japanese workers, who further complicated the region's cultural geography. By 1900, the Japanese had become a dominant part of the island workforce.

After the United States formally annexed the islands in 1898, the cultural mix revealed in the Hawaiian census of 1910 suggests the magnitude of culture change: More than 55 percent of the population was Asian (mostly Japanese and Chinese), native Hawaiians made up another 20 percent, and about 15 percent (mostly imported European workers) were white. By the end of the 20th century, however, the Asian population was less dominant, as about 40 percent of Hawaii's residents were white. In addition, the small number of remaining native Hawaiians had been joined by an increasingly diverse group of people

▼ **Figure 14.37 Samoan Football Players** Even though rugby is the most popular sport in Oceania, a large number of Samoans play football in the United States, at both college and professional levels. Here Joey Iosefa, running back for the University of Hawaii, carries the ball against San Jose State.

▲ **Figure 14.38 French Church Architecture in Tahiti** This countryside church in Moorea, Tahiti, in a style that is distinctly French, provides a strong clue to the strong cultural relationship between that island country and France.

from other Pacific islands. Cultural mixing between these groups has produced a rich mosaic of Hawaiian cultures, showing a unique blend of North American, Asian, Pacific, and European influences.

Hawaii's story has also played out in many other Pacific island locations. In the Mariana Islands, Guam was absorbed into the United States' Pacific empire at the conclusion of the Spanish-American War in 1898. Thereafter, not only did native people feel the effects of Americanization (the island remains a self-governing U.S. territory today), but also thousands of Filipinos were moved there to supplement its modest labor force. To the southeast, the British-controlled Fiji Islands offered similar opportunities for redefining Oceania's cultural mix. The same sugar-plantation economy that spurred changes in Hawaii prompted the British to import thousands of South Asian Indian laborers to Fiji. The descendants of these Indians (most practicing Hinduism) now constitute almost half the island country's population and often come into sharp conflict with the native Fijians. In French-controlled portions of the realm, small groups of traders and plantation owners filtered into the Society Islands (Tahiti), but a larger group of French colonial settlers (many originally part of a penal colony) had a major impact on the cultural makeup of New Caledonia. Still a French colony, New Caledonia's population is more than one-third French, Still today French architecture can be found in both New Caledonia and Tahiti (Figure 14.38).

REVIEW

14.8 Compare the situations of the Australian Aborigines and the New Zealand Maori.

14.9 Summarize the language geography of Oceania in terms of both native and colonial languages.

14.10 How and why have Maori gender roles changed since European colonization?

14.11 Describe Hawaiian cultural changes over the last century.

Geopolitical Framework: A Region of Dynamic Polities

Pacific geopolitics reflects a complex interplay of local, colonial-era, and global-scale forces (Figure 14.39). These complexities become apparent in the story of Micronesia's Marshall Islands. This sprinkling of islands and atolls (covering 70 square miles, or 180 square km, of land) historically consisted of many ethnic groups that made up small political units. In 1914, the Japanese moved into the islands, and the area remained under their control until 1944, when U.S. troops occupied the region. Following World War II, a UN trust territory (administered by the United States) was created across a wide swath of Micronesia, including the Marshall group. Demands for local self-government grew during the 1960s and 1970s, resulting in a new constitution and independence for the Marshall Islands in the early 1990s. Today, still benefiting from U.S. aid, government officials in the modest capital city on Majuro Atoll struggle to unite island populations, protect large maritime sea claims, and resolve a generation of legal and medical problems that grew from U.S. nuclear bomb testing in the region. Similar stories are typical across the region, suggesting a 21st-century political geography that is still very much in the making.

Roads to Independence

The newness and fluidity of the region's political boundaries are remarkable. The region's oldest independent states are Australia and New Zealand, and both were 20th-century creations that still discuss whether they want to complete their formal political separation from the British Crown. Elsewhere, political ties between colony and mother country remain close and, perhaps, more enduring. Even many of the newly independent Pacific **microstates**, with their tiny overall land areas, keep special political and economic ties to countries such as the United States.

Independent Australia (1901) and New Zealand (1907) gradually created their own political identities, yet both still struggle with the final shape of these identities. Although Australia became a commonwealth in 1901, it still acknowledges the British Crown as the symbolic head of its government. A national referendum in 1999 asked Australians to decide whether they would like to drop their remaining tie to Britain, with the country becoming a genuine republic with its own president replacing the British queen as head of state. Despite a strong movement toward complete independence from Britain, a slight majority (55 percent) voted to retain Australia's ties to the Crown; thus, they remain today still linked symbolically to Britain. In New Zealand, formal legislative links with Great Britain were not broken until 1947, and today New Zealand is pondering the same formal break with the British Crown being debated by the Australians.

Elsewhere in the Pacific, colonial ties were cut even more slowly; in fact, the process has not yet been completed.

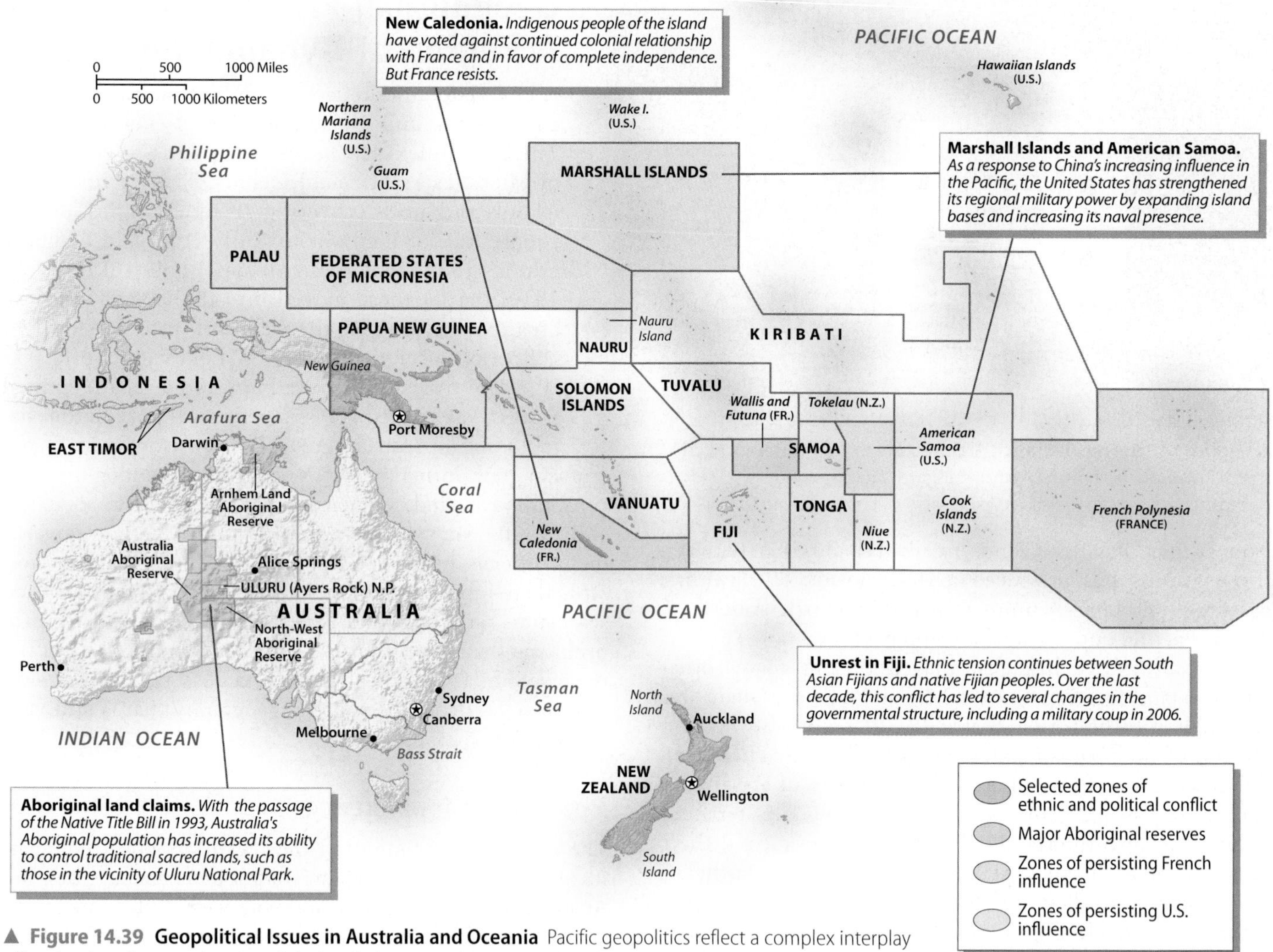

▲ **Figure 14.39 Geopolitical Issues in Australia and Oceania** Pacific geopolitics reflect a complex interplay of local, colonial-era, and current global-scale forces, resulting in a political geography that is still very much in the making today.

In the 1970s, Britain and Australia began giving up their colonial empires in the Pacific. Fiji (Great Britain) gained independence in 1970, followed by Papua New Guinea (Australia) in 1975 and the Solomon Islands (Great Britain) in 1978. The small island nations of Kiribati and Tuvalu (Great Britain) also became independent in the late 1970s.

The United States has recently turned over most of its Micronesian territories to local governments, while at the same time maintaining a large influence in the area (Figure 14.40). After gaining these islands from Japan in the 1940s, the U.S. government provided large monetary subsidies to islanders and also utilized a number of islands for military purposes. Bikini Atoll was destroyed by nuclear tests, and the large lagoon of Kwajalein Atoll was used until recently as a missile target. Further, a major naval base was established in Palau, the westernmost archipelago of Oceania.

By the early 1990s, both the Marshall Islands and the Federated States of Micronesia (including the Caroline Islands) had gained independence. Their ties to the United States, however, remain close. Several other Pacific islands remain under U.S. administration. Palau is a U.S. "trust territory," which gives Palauans some local autonomy. The people of the Northern Marianas chose to become a "self-governing commonwealth in association with the United States," a rather vague political position that allows them to become U.S. citizens. The residents of self-governing Guam and American Samoa are also U.S. citizens. As noted earlier, Hawaii became a full-fledged U.S. state in 1959, yet debate continues regarding native land claims and sovereignty.

Other colonial powers appear less inclined to give up their oceanic possessions. New Zealand still controls substantial territories in Polynesia, including the Cook Islands, Tokelau, and the island of Niue. France has even more extensive holdings in the region. Its largest maritime possession is French Polynesia, which includes a large expanse of mid-Pacific territory. To the west, France still controls the much smaller territory of Wallis and Futuna in Polynesia and the larger island of New Caledonia in Melanesia.

Exploring Global Connections

Hypercommunication Comes to Oceania

Oceania has 36 million people living in 17 political units located on 25,000 different islands spread out over 6000 miles (9600 km). Thus, it is no surprise that communication and isolation have long been issues. Until interisland air travel came along in the 1960s, visiting friends and family on other islands meant hours or even days on ferries, copra boats, or island traders. The rise of island airlines like Samoa Air and Fiji Airways helped some, but this service was—and still is—rather expensive.

Cell Phones and Internet Today, however, a communications revolution is in progress that bridges long distances and remote locations. Ryan Peseckas, an anthropologist doing fieldwork in three Fijian villages, wrote recently:

> Mobile phones now mediate every type of social interaction in Fiji. I witnessed a houseful of mourners participating in a funeral ceremony on a distant island by listening to the service via a mobile phone, set on "speakerphone." I heard sordid tales of marriages destroyed by text messages and of youths in the Viti Levu highlands dialing random numbers in the hope that a beautiful Lau island girl would pick up on the other end. Many romances in Fiji begin with a randomly dialed number, the lovers sometimes corresponding by phone for months or years before actually laying eyes upon one another.

Data collected on Internet usage tell a similar story. As of October 2012, it appeared that about 50 percent of Oceania's total inhabitants were Facebook users. On New Caledonia, for example, out of a population of 260,000, over 100,000 people were regularly checking their Facebook accounts. Similar figures come from French Polynesia and Guam. The pattern is spotty, however, because data show several other island countries with very little global connectivity. Why is that?

▲ **Figure 14.5.1 Cell Phones in Oceania** Earlier attempts to bring digital communications relied on satellite connections, which don't work as well as hardwire submarine cables.

Satellites and Submarine Cables Satellite communication first came to the remote islands of Oceania decades ago, bringing with it a smattering of global TV, irregular phone service, and slow, but welcome Internet service. Anyone in North America without access to broadband knows well the frustration of satellite connections.

However, the real communications revolution started in 2006 with the laying of the first modern submarine fiber-optic cables to Guam, Fiji, and French Polynesia (**Figure 14.5.1**). Along with the cables came broadband Internet and extensive cell phone networks, developed by a Jamaica-based firm experienced in island communications.

The main digital submarine cable routes are in the northern hemisphere, running directly east–west across the Pacific from North America to Asia (with Hawaii a major connecting hub). Direct lines to Australia and New Zealand were established a decade ago. Islands along this route, such as Samoa and Fiji, were the first to be connected; Guam was also connected early on as the submarine cable network was expanded between Hawaii and Hong Kong (**Figure 14.5.2**). However, islands not along these cable routes do not presently have the same connectivity.

Today international firms are finding it profitable to lay cable to most island nations in the South Pacific. With it comes a communications revolution that has forever changed the way Oceania's islands interact with one another, as well as with the larger world.

▼ **Figure 14.5.2 Map of South Pacific Submarine Cables** Cables from North America are connected via Hawaii, while those to Asia go through Guam.

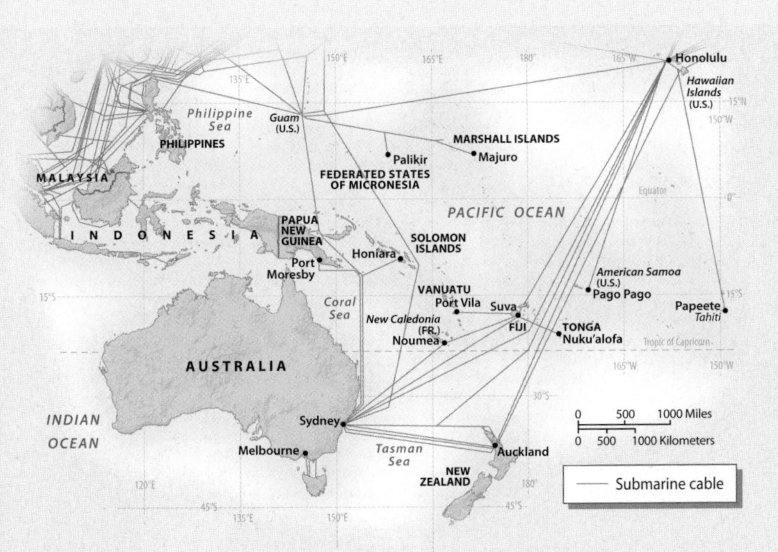

▲ **Figure 14.40 United States Military Bases in the South Pacific** This U.S. base is on Kwajalein Atoll in the Marshall Islands. During the European colonial period, Germany used Kwajalein as a copra trading center, but after Germany's defeat in World War I, the Japanese took over the island. It was subsequently taken over by the U.S. Army in 1944 and served as a command center for postwar American nuclear testing at other Marshall Island atolls. Today the Kwajalein base is part of the Ronald Reagan Missile Defense Test Site. **Q: Describe how the landscape may have been changed by different uses since 1941.**

Land and Native Rights Issues

British colonialists first interacted with native peoples in the North American colonies, and those experiences apparently formed the Empire's template as to how native peoples were to be treated in Australia and Oceania. The initial response, as in North America, was extermination. But when that proved unsuccessful, English Land Law (which can be traced back to the Romans in England) took over by expropriating native lands and converting them into private property held by Europeans. These past actions set the scene for legal, social, and ethnic struggles that continue still today.

Aboriginal and Maori Rights in Australia and New Zealand Indigenous peoples in both Australia and New Zealand have used the political process to gain more control over land and resources in their two countries. The strategies these native groups have used parallel similar efforts in North America and elsewhere. In Australia, Aboriginal groups are discovering newfound political power from both more effective lobbying efforts by native groups and a more sympathetic federal government, interested in rectifying historical discrimination that left native peoples with no legal land rights. In recent years, the Australian government established several Aboriginal reserves, particularly in the Northern Territory, and expanded Aboriginal control over sacred national parklands such as Uluru (called Ayers Rock by the British settlers) (Figure 14.41). Further concessions to indigenous groups were made in 1993, when the government passed the **Native Title Bill**, which compensated Aborigines for lands already given up and gave them the right to gain title

▼ **Figure 14.41 Aborigines at Uluru-Kata Tjuta National Park** Uluru, also known as Ayers Rock, is a large sandstone formation of great spiritual significance to the Anangu Aboriginal people who live in this part of Australia's Northern Territory. The rock and its surrounding area are not only a national park, but also a World Heritage Site.

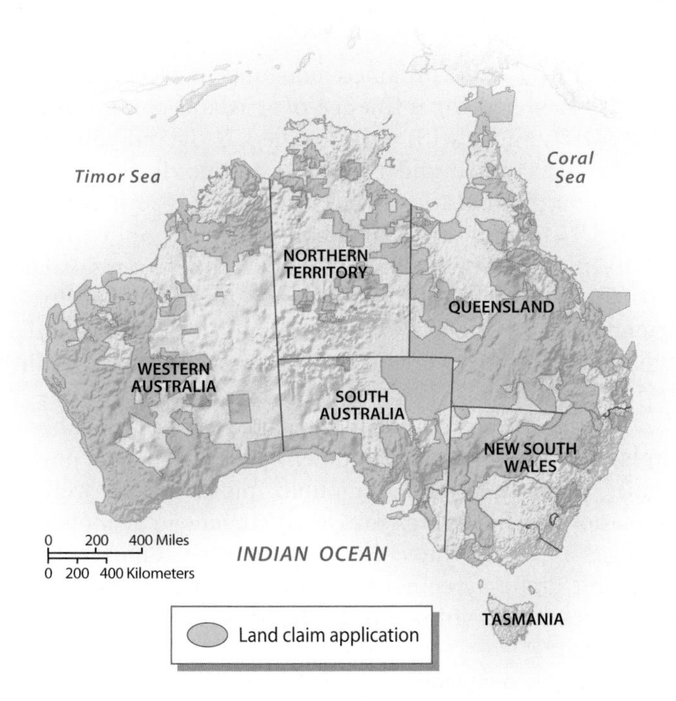

▲ **Figure 14.42 Applications for Native Land Claims in Australia** This map shows the applications for native land claims in Australia filed by different Aboriginal groups as of 2013. Note that these are applications only, not government-approved claims. Nevertheless, the widespread extent of the claims shows why the topic is so contentious and controversial.

to unclaimed lands they still occupied. The bill also provided them with legal standing to deal with mining companies in native-settled areas.

Efforts to expand Aboriginal land rights, however, have met strong opposition. In 1996, an Australian court ruled that pastoral leases (the form of land interest held by the cattle and sheep ranchers who control most of the Outback) do not necessarily negate or replace Aboriginal land rights. Grazing interests were infuriated, which led the government to respond that Aboriginal claims allow them to visit sacred sites and do some hunting and gathering, but do not give them complete economic control over the land (Figure 14.42).

In New Zealand, Maori land claims have generated similar controversies in recent years. Because the Maori constitute a far larger proportion of the overall population and because the lands they claim tend to be more valuable than rural Aborigine lands in Australia, the issues are even more complicated. Recent protests include civil disobedience, ever-increasing Maori land claims over much of the North and South islands, and a call to return the country's name to the indigenous *Aotearoa*, "Land of the Long White Cloud."

Native Rights in Hawaii As is the case with indigenous peoples in Australia, New Zealand, and North America, native Hawaiians have issues with their government about human rights, access to ancestral land, and the political standing of native people. Attempts are being made to resolve these contentious issues in Hawaii, but the path forward is uncertain.

Native Hawaiians, who call themselves *Kanaka maoli*, are descendants of Polynesian people who arrived in the Hawaiian Islands about a thousand years ago. They lived in an independent and sovereign state that was recognized by major foreign powers until the islands were annexed by the United States in 1898. The legality of this annexation, however, is still contested today and underlies native Hawaiian demands for a return of their historical sovereignty.

Giving credence to this demand is the U.S. admission—on two occasions—that the annexation process was illegal: initially in 1893, when the Hawaiian queen was displaced (President Cleveland opposed annexation of Hawaii because of his concern about its legality), and more recently in 1993, on the 100th anniversary of that earlier event, when President Clinton signed the Apology Bill. This bill, which was passed without debate by Congress, once again referred to the illegal overthrow of Hawaiian sovereignty in 1893.

Today many Hawaiian nationalists support a return to Polynesian sovereignty. One pathway might be similar to that of the Navajo nation in Arizona, Utah, and New Mexico. Extremists, however, reject the idea of native Hawaiians having the same standing as U.S. Indian tribes and instead demand that the UN invalidate the 1898 annexation, thereby granting native Hawaiians a return to complete international sovereignty (Figure 14.43). Were that to happen, native Hawaiian ancestral lands would be a separate country within the U.S. state of Hawaii. Understandably, this solution is resisted by private and corporate landowners in Hawaii because it could call into question the legality of their ownership of, for example, residential properties in the Honolulu suburbs or hotel parcels along Waikiki Beach.

Another land issue is the disposition of those lands historically administered by the Hawaiian monarch, lands that were ceded to the U.S. government in the 1898 annexation. These lands, which constitute almost half of the land area of the Hawaiian Islands, were transferred to the local government in 1959 upon statehood, with the stipulation that they be used for the benefit of native Hawaiians. Many argue that these lands should be turned over to the native Hawaiians to use as they please, but the state government resists this notion because these lands generate revenue through rental to public agencies (a portion of the Honolulu airport, for example, is built on ceded lands) and the U.S. military.

To resolve these issues, the U.S. Congress is pondering the Native Hawaiian Government Reorganization Act (commonly called the "Akaka Bill" after its author, former Senator Daniel Akaka). If passed, this law would give native Hawaiians legal standing similar to that of Native Americans. Currently, unlike Native Americans, native Hawaiians have no legal standing or recognition in U.S. courts, a legality that prevents

▲ **Figure 14.43 Native Hawaiian Nationalism** Thousands of red-shirted supporters of native Hawaiian sovereignty gather to march along Waikiki Beach in Honolulu in the annual Justice for Hawaiians rally.

them from seeking redress for acts of discrimination or settling land claims. Equally important, the law would, for the first time, provide a legal mechanism for interaction between native Hawaiians and the U.S. government, a legal necessity that currently does not exist.

The Strategic Pacific

As was the case during World War II, Oceania is once again a strategic global region where numerous countries are expanding their influence. The reasons vary widely, from engaging in dollar diplomacy that buys friends, to exploiting mineral and timber resources, to establishing strategically placed military bases. The major players include the reigning superpowers of the United States and China, with Russia also included because it needs to keep track of both superpowers; the Pacific Rim countries of Japan, South Korea, Taiwan, and Indonesia; France, as a former colonial master; and the two most powerful locals, Australia and New Zealand.

Tensions Between Taiwan and China For several decades now, disputes between Taiwan, also known as the Republic of China (ROC), and the giant People's Republic of China (PRC) have inflamed tensions in Oceania. Each country has vied for influence over island communities, particularly those with votes in the UN. Island countries like Samoa, Tonga, and Fiji, for example, have received vast amounts of aid from the PRC and have often voted in the UN to support China's controversial policy toward Tibet. In contrast, Taiwan, which currently counts Palau, Kiribati, and the Marshall Islands among its political friends (all of them recent recipients of Taiwan's aid), relies on their UN votes to neutralize China's ambitions to reclaim Taiwan.

Behind these political allegiances is a checkbook diplomacy that funds development projects such as power plants, schools, roads, and public buildings, as well as educational scholarships for island citizens to schools and colleges in either Taiwan or China. Also important are favorable trade agreements; in the last two years, China has increased its volume of trade 50 percent with Pacific island states.

A major recipient of China's aid has been the military government of Fiji, which seized power in 2006 and alienated both Australia and New Zealand. Not unimportant is that Fiji is home to some 20,000 overseas Chinese, a group that reportedly enjoys considerable influence with the Fijian military government. Although Fiji was once considered friendly to the United States, its movement closer to China and even North Korea, where it plans to open an embassy (with North Korea doing the same in Fiji), is clearly bothersome and problematic to the United States.

America's Asia Pivot Because of the PRC's growing influence in Oceania, coupled with North Korea's continued hostility toward the West, the United States is increasing its diplomatic and military presence in the South Pacific. The term *Asia Pivot* is commonly used to describe the shift in America's foreign and military policy away from Iraq and Afghanistan and toward the Asia–Pacific world. In late 2011, then Secretary of State Hillary Clinton wrote in a widely circulated article titled "America's Pacific Century" that "One of the most important tasks of American statecraft over the next decade will be to lock in a substantially increased investment—diplomatic, economic, strategic, and otherwise—in the Asia–Pacific region." When China protested that this U.S. shift was a thinly veiled policy of containment, Washington's reply was that, no, the new U.S. policy was simply one of "Asia management."

Semantics aside, the Asia Pivot includes the establishment of numerous small military bases scattered around the Pacific (and, for that matter, around the world), commonly called "lily-pad" outposts. The recent—and controversial—agreement with Australia to post 2500 U.S. Marines near Darwin in northern Australia is an example (Figure 14.44). Although the military understandably does not publicize its lily-pad bases, reportedly there are new or recently revitalized military bases in Australia's Cocos Islands, American

Samoa, Tinian Island, the Marshall Islands, and the Marianas. Andersen Air Force Base in Guam has undergone a recent expansion of its facilities, and similar expansion is also taking place on existing U.S. military bases in Japan, the Philippines, and South Korea. The geopolitical goals behind this Asia Pivot seem clear enough: to reestablish the United States as a major political and military player in Oceania.

Continuing Influence of New Zealand and Australia Not to be overlooked is that Australia and New Zealand still play key political roles in the South Pacific. Although these two countries sometimes disagree on strategic and military matters, their size, wealth, and collective political influence in the region make them important forces for political stability. Also, special colonial relationships still connect these two nations with many Pacific islands. Australia maintains close political ties to its former colony of Papua New Guinea, and New Zealand's continuing control over Niue, Tokelau, and the Cook Islands in Polynesia confirms that its political influence extends well beyond its borders. How these two countries will respond to China's growing influence, however, is unclear.

Australia's number one trade partner is China, yet it maintains unquestioned political ties to North America. Australia could be showing Oceania that a middle pathway is possible; that Pacific Island countries do not necessarily have to choose between the United States and China (or between China and Taiwan). Easy for Australia to do because of its economic clout, say the small island states, but hard for smaller, less fortunate countries to do when economic aid and low-interest loans are desperately needed, regardless of the source: China or the United States.

▼ Figure 14.44 U.S. Marines in Australia As part of the Asia Pivot, the United States reached an agreement with Australia to station up to 2500 Marines on the country's north coast near Darwin. Although the U.S. denies that the Marines are part of a strategy to contain China, and instead maintains northern Australia is simply a good place to train troops, world pundits think otherwise.

REVIEW

14.12 Describe the colonial history of the Pacific islands and contrast it to that of Australia and New Zealand.

14.13 What are the arguments for and against sovereignty for native Hawaiians?

14.14 Describe why and how China and Taiwan are causing geopolitical tensions in Oceania.

14.15 Explain the Asia Pivot.

Economic and Social Development: Increasing Ties to Asia

As with all other world regions, the Pacific realm contains economic situations of both wealth and poverty. Even within affluent Australia and New Zealand, pockets of pronounced poverty occur, and large economic disparities exist as well between those Pacific countries with global trade ties and those small island nations lacking resources and external trade. Tourism offers some relief from abject poverty, but the whims and economics of foreign tourists can be fickle, creating a Pacific version of boom-and-bust economies. Overall, the economic future of the Pacific realm remains uncertain and highly variable because of its small domestic markets, its peripheral position in the global economy, and its diminishing resource base (Table 14.2).

The Australian and New Zealand Economies

Much of Australia's past economic wealth has been built on the cheap extraction and export of abundant raw materials. Export-oriented agriculture, for example, has long been one of the key supports of Australia's economy. Australian agriculture is highly productive in terms of labor input, and it produces a wide variety of both temperate and tropical crops, as well as huge quantities of beef and wool for world markets. Although farm exports

Table 14.2 Development Indicators

Country	GNI per capita, PPP 2011	GDP Average Annual % Growth 2000–11	Human Development Index (2011)[1]	Percent Population Living Below $2 a Day	Life Expectancy (2013)[2]	Under Age 5 Mortality Rate (1990)	Under Age 5 Mortality Rate (2011)	Adult Literacy (% ages 15 and older) Male/Female	Gender Inequality Index (2011)[3]
Australia	38,610	3.1	.938	–	82	9	5	--/--	0.115
Fed. States of Micronesia	3,580	0.0	.645	44.7	68	56	42	--/--	–
Fiji	4,610	1.2	.702	22.9	69	30	16	--/--	–
French Polynesia	22,000*	–	–	–	76	–	–	98/98*	–
Guam	28,700*	–	–	–	78	–	–	98/98*	–
Kiribati	3,300	0.4	.629	–	65	88	47	--/--	–
Marshall Islands	8,800*	1.5	–	–	72	52	26	--/--	–
Nauru	5,000*	–	–	–	60	–	–	--/--	–
New Caledonia	37,700*	–	–	–	77	–	–	97/96	–
New Zealand	28,930	2.0	.919	–	81	11	6	--/--	0.164
Palau	11,080	6.3	.791	–	69	32	19	--/--	–
Papua New Guinea	2,570	4.3	.466	57.4	63	88	58	64/57	0.617
Samoa	4,270	2.6	.702	–	73	30	19	99/99	–
Solomon Islands	2,350	4.9	.530	–	67	42	22	--/--	–
Tonga	5,000	1.0	.710	–	72	25	15	99/99	0.462
Tuvalu	3,400*	1.1	–	–	65	58	30	--/--	–
Vanuatu	4,330	3.9	.626	–	71	39	13	84/81	–

[1]United Nations, *Human Development Report, 2013.*

[2]Population Reference Bureau, *World Population Data Sheet, 2013.*

[3]Gender Inequality Index—A composite measure reflecting inequality in achievements between women and men in three dimensions: reproductive health, empowerment and the labor market that ranges between 0 and 1. The higher the number, the greater the inequality.

Source: World Bank, *World Development Indicators, 2013.*

*Additional data from the *CIA World Factbook*, 2013.

◀ **Figure 14.45 Australian Trade with China** Containers from Asia testify to the recent explosion of two-way trade between Australia and China. Raw materials, mainly iron ore, are exported to China, and consumer goods flow from China into Australia, making that country a key beneficiary of China's recent economic growth.

▶ **Figure 14.46 Queensland's Gold Coast** Many of these luxury hotels in the Surfer's Paradise section of the Gold Coast are owned by Japanese firms specializing in accommodations for Asian tourists. **Q: What are the environmental implications of this Gold Coast development?**

are still important to the economy, the mining sector has grown more rapidly since 1970, making Australia one of the world's mining superpowers.

Mining's recent growth is primarily due to increased trade with China, which has made Australia the world's largest exporter of iron and coal. Besides these two resources, Australia produces an assortment of other materials—namely, bauxite (aluminum ore), copper, gold, nickel, lead, and zinc. As a result, the New South Wales–based Broken Hill Proprietary Company (BHP) is one of the world's largest mining corporations. Hardly affected by the global recession of 2008–2009, China's appetite for natural resources is huge—equaled, apparently, only by Australia's willingness to sell its resources (Figure 14.45). Even during the recession, China pumped $40 billion into the Australian economy. Besides natural resources, half a million Chinese tourists visited Australia and added to the local economy by keeping busy Australian lifeguards, blackjack dealers, and real estate brokers. In addition, 70,000 Chinese students are currently attending Australian universities.

With growing numbers of Asian immigrants and expanding economic links with Asian markets, promise is high for Australia's economic future. In addition, an expanding tourism industry is helping to diversify the economy. More than 7 percent of the nation's workforce is now devoted to serving the needs of more than 4 million tourists each year. Popular destinations include Melbourne and Sydney, as well as Queensland's resort-filled Gold Coast, the Great Barrier Reef, and the vast, arid Outback. Along the Gold Coast, most luxury hotels are owned by Japanese firms and provide a bilingual resort experience for their Asian clientele (Figure 14.46).

New Zealand is also a relatively wealthy country, although somewhat less well off than Australia. Before the 1970s, New Zealand relied heavily on exports to Great Britain, especially agricultural products such as wool and butter. Problems developed, however, with these colonial trade linkages in 1973 when Britain joined the European Union, with its strict agricultural protection policies. Unlike Australia, New Zealand lacked a rich base of mineral resources to export to global markets. As a result, the country had slipped into a serious recession by the 1980s. Eventually, the New Zealand government enacted drastic reforms, and the country that had previously been noted for its lofty taxes, high levels of social welfare, and state ownership of large companies changed dramatically. Privatization became the watchword, and most state industries were sold off to private parties. As a result, New Zealand has been transformed into one of the most market-oriented countries in the world and has largely recovered from its earlier recession.

Oceania's Economic Diversity

Varied economic activities shape the Pacific island nations. One way of life is oriented around subsistence-based economies, such as shifting cultivation or fishing. In other places, a commercial extractive economy dominates, with large-scale plantations, mines, and timber activities competing for land and labor with the traditional subsistence sector. Elsewhere, the huge growth in global tourism has transformed the economic geographies of many island settings, forever changing the way that people make a living. In addition, many island nations still benefit from direct subsidies and economic assistance from present and former colonial powers,

▲ **Figure 14.47 Nickel Mine in New Caledonia** New Caledonia has a quarter of the world's nickel resources, most of which are located in the area of the North Star Nickel Mine, the second largest nickel mine in the world. Mining is a mainstay of the New Caledonia economy, but fluctuations in the demand and price for nickel cause sporadic problems for the country. Even more worrisome is the fact that the fiscal future is not particularly bright because of dwindling reserves in the North Star mine.

designed to promote development and stimulate employment.

Melanesia is the least developed and poorest part of Oceania because these countries have benefited the least from tourism and from subsidies from wealthy colonial and ex-colonial powers. Today most Melanesians still live in remote villages that remain somewhat isolated from the modern economy. The Solomon Islands, for example, with few industries other than fish canning and coconut processing, has a per capita gross national income (GNI) of only $2350 per year. Similarly, Papua New Guinea's economy produces a per capita GNI of $2570. Although traditional exports such as coconut products and coffee have increasingly been supplemented with the rapid development of tropical hardwoods, the economic returns remain low. Further, gold and copper mining has dramatically transformed the landscape, although political instability has often interfered with mineral production in places such as Bougainville. In New Guinea's interior highlands, much village life is focused on subsistence activities. In contrast, Fiji is the most prosperous Melanesian country, with a per capita GNI of $4610, largely because of its tourist economy, reflecting its popularity with Chinese and Japanese tourists.

The Economic Impact of Mining Among the smaller islands of Melanesia and Micronesia, mining economies dominate New Caledonia and Nauru. New Caledonia's nickel reserves, the world's second largest, are both a blessing and a curse (Figure 14.47). Although they currently sustain much of the island's export economy, income from nickel mining will lessen in the near future as reserves dwindle. Dramatic price fluctuations for the industrial economy also hamper economic planning for the French colony. Other activities include coffee growing, cattle grazing, and tourism. To the north, the tiny, phosphate-rich island of Nauru also depended on mining; however, as noted earlier, that day is past, with mineral deposits exhausted and the future uncertain.

The South Pacific Tuna Fishery The South Pacific is home to the world's largest tuna fishery, and this resource contributes significantly to local island economies, both directly and indirectly. Tuna fishing and processing accounts for about 10 percent of all wage employment in southern Oceania, although 90 percent of the working tuna boats come from far-flung Pacific nations—namely, China, Japan, South Korea, and the United States. These foreign boats are charged access fees for fishing in each island's offshore territory (Figure 14.48).

▲ **Figure 14.48 Trouble in the Tuna Industry** These Samoan fishermen returned to traditional subsistence fishing in their outriggers after the economic collapse of the Samoan tuna fishing and canning industry. The collapse was due to competition from countries in East Asia that pay lower wages.

More specifically, international law allows the extension of sovereign rights 200 miles (320 km) beyond a country's coastline in an **Exclusive Economic Zone (EEZ)**. Within the EEZ, that country has economic control over resources such as fisheries, as well as retaining ocean-floor mineral rights. In Oceania, because island countries are often composed of a series of islands and because each country's EEZ is delimited from its outermost points, the South Pacific tuna fishery is composed of a patchwork of intersecting EEZs, each with different fishing regulations and fees. Complicating the matter is the fact that, as tuna become more scarce (and more valuable as a resource), many island nations have closed their EEZs to outside countries, reserving the tuna resource for local boats and processing. Whether this fragmentation will aid or injure the sustainability of the tuna fishery is not clear.

Micronesian and Polynesian Economies Throughout Micronesia and Polynesia, economic conditions depend on both local subsistence economies and economic linkages to the wider world beyond. Many archipelagos export a few food products, but native populations survive mainly on fish, coconuts, bananas, and yams. Some island groups enjoy large subsidies from either France or the United States, even though such support often comes with a political price. China is also increasingly involved with economic development plans that often imply political support.

Other Polynesian island groups have been completely transformed by tourism. In Hawaii, more than one-third of the state's economy flows directly from tourist dollars. With almost 7 million visitors annually (including more than 1.5 million from Japan), Hawaii represents all the classic benefits and risks of the tourist economy. Job creation and economic growth have reshaped the island realm, but congested highways, high prices, and

▼ **Figure 14.49 Asian Honeymooners in Guam** During the 1970s, Guam and other Pacific islands built a lucrative tourist business catering to Japanese honeymooners. However, today that industry is suffering because of the decreasing rate of marriage in Japan, China, and South Korea.

the unpredictable spending habits of tourists have put the region at risk for future problems. Elsewhere, French Polynesia has long been a favored destination of the international jet set. More than 20 percent of French Polynesia's GNI is derived from tourism, making it one of the wealthiest areas of the Pacific. More recently, Guam has emerged as a favorite destination of Japanese and Korean tourists, especially those on honeymoons (Figure 14.49).

Oceania in Global Context

Many international trade flows link the area to the far reaches of the Pacific and beyond. Australia and New Zealand dominate global trade patterns in the region. In the past 30 years, ties to Great Britain, the British Commonwealth, and Europe have weakened in comparison with growing trade links to Japan, East Asia, the Middle East, and the United States. Australia, for example, now imports more manufactured goods from China, Japan, and the United States than it does from Britain and Europe. Other global economic ties have come in the form of capital investment in the region. Today U.S. and Japanese banks and other financial institutions dot the South Pacific landscape from Sydney to Suva. Both Australia and New Zealand also participate in the Asia–Pacific Economic Cooperation Group (APEC), an organization designed to encourage economic development in Southeast Asia and the Pacific Basin.

To promote more economic integration within the region, Australia and New Zealand signed the **Closer Economic Relations (CER) Agreement** in 1982, which successfully slashed trade barriers between the two countries. As a result, New Zealand benefited from the opening of larger Australian markets to New Zealand exports, and Australian corporate and financial interests gained new access to New Zealand business opportunities. Since the CER Agreement's signing, trade between the two countries has expanded almost 10 percent per year. Today more than 20 percent of New Zealand's imports and exports come from Australia, and this pattern of regional free trade is likely to strengthen in the future.

Smaller nations of Oceania, while often closely tied to countries such as China, Taiwan, Japan, the United States, and France, also benefit from their proximity to Australia and New Zealand. More than half of Fiji's imports come from those two nearby nations, and countries such as Papua New Guinea, Vanuatu, and the Solomon Islands enjoy a similarly close trading relationship with their more developed Pacific neighbors.

Continuing Social Challenges

Australians and New Zealanders enjoy high levels of social welfare, but face some of the same challenges evident elsewhere in the developed world (see Table 14.2). Life spans average about 80 years in both countries, and rates of child mortality have fallen greatly since 1960. However, echoing patterns in North America and Europe, cancer and heart disease are leading causes of death, and alcoholism is a continuing social problem, particularly in Australia. Furthermore, Australia's rate of skin cancer is among the world's highest, the result of having a largely fair-skinned, outdoors-oriented population from northwest Europe in a sunny, low-latitude setting. Overall, Australia's Medicare program (initiated in 1984) and New Zealand's system of social services provide high-quality health care to their populations.

Not surprisingly, the social conditions of the Aborigines and Maoris are much less favorable than those of the population overall. Schooling is irregular for many native peoples, and levels of postsecondary education for Aborigines (12 percent) and Maoris (14 percent) remain far below the national averages (32–34 percent). Many other social measures reflect this same pattern. Less than one-third of Aboriginal households own their own homes, whereas more than 70 percent of white Australian households are homeowners. Furthermore, considerable discrimination against native peoples continues in both countries, a situation that has been aggravated and publicized with the recent assertion of indigenous political rights and land claims. As with African-American, Hispanic, and Native American populations in North America, simple social policies do not yet exist as solutions to these lasting problems.

Even in Hawaii, the social welfare of the native people is problematic. The proportion of native Hawaiians living below the poverty level is much higher than that of any other ethnic group (Figure 14.50). This group also has the shortest life expectancy and the highest infant mortality rate; rates of death from cancer and heart disease are almost 50 percent higher than for other groups in the United States; and native Hawaiian women have the highest rate of breast cancer in the world. Further, 55 percent of native Hawaiians do not complete high school, and only 7 percent have college degrees.

In other parts of Oceania, levels of social welfare are higher than you might expect based on the region's economic situation. Many of its countries and colonies have invested heavily in health and education services and have achieved considerable success. For example, the average life expectancy in the Solomon Islands, one of the world's poorer countries as measured by per capita GNI figures, is a respectable 67 years. By other social measures as well, the Solomon Islands and several other Oceania states have reached higher levels of human well-being than exist in most Asian and African countries with similar levels of economic output. This is partly a result of successful policies, but it also reflects the relatively healthy natural environment of Oceania because many of the tropical diseases that are so troublesome in Africa simply do not exist in the Pacific islands.

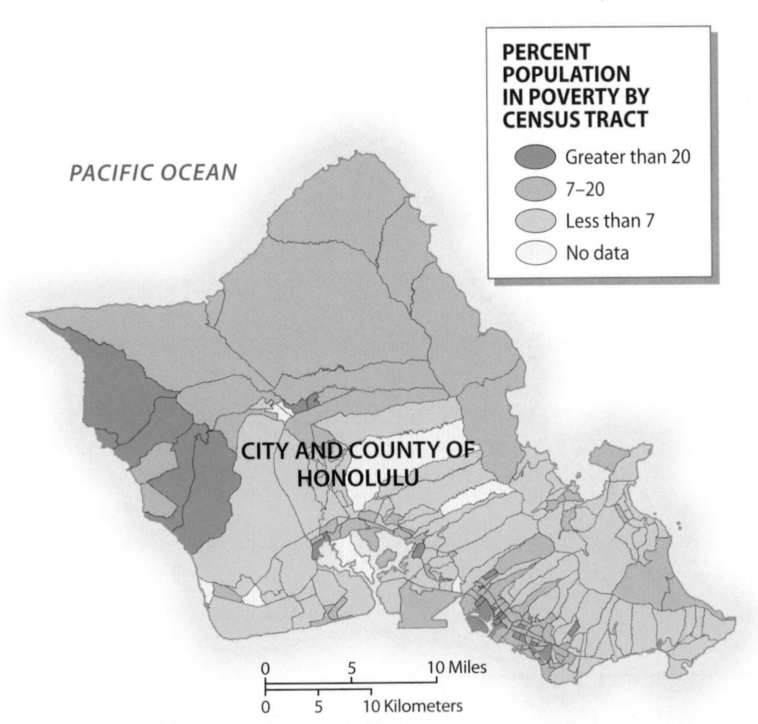

◀ **Figure 14.50 Poverty in Hawaii** The distribution of poverty on the Hawaiian island of Oahu shows that the highest levels of poverty are in the urban pockets around Honolulu in the southeast, as well as in the western census tracts, which have the highest number of native Hawaiians.

REVIEW

14.16 Explain what an Exclusive Economic Zone (EEZ) is and why it is important to Pacific island economies.

14.17 What is the Closer Economic Relations (CER) Agreement? Why is it important to understand the economics of Oceania?

14.18 What are the major social development challenges in Oceania?

14.19 How has Australia's foreign trade changed over the last 20 years in terms of its trading partners and the goods being exported and imported?

▼ **Figure 14.51 O'ahu** Sometimes called "The Gathering Place," O'ahu is the third largest of the Hawaiian Islands and home to Honolulu, the state's capital, as well as Pearl Harbor, site of a U.S. Naval Base. O'ahu has nearly a million residents—about 75 percent of Hawaii's population.

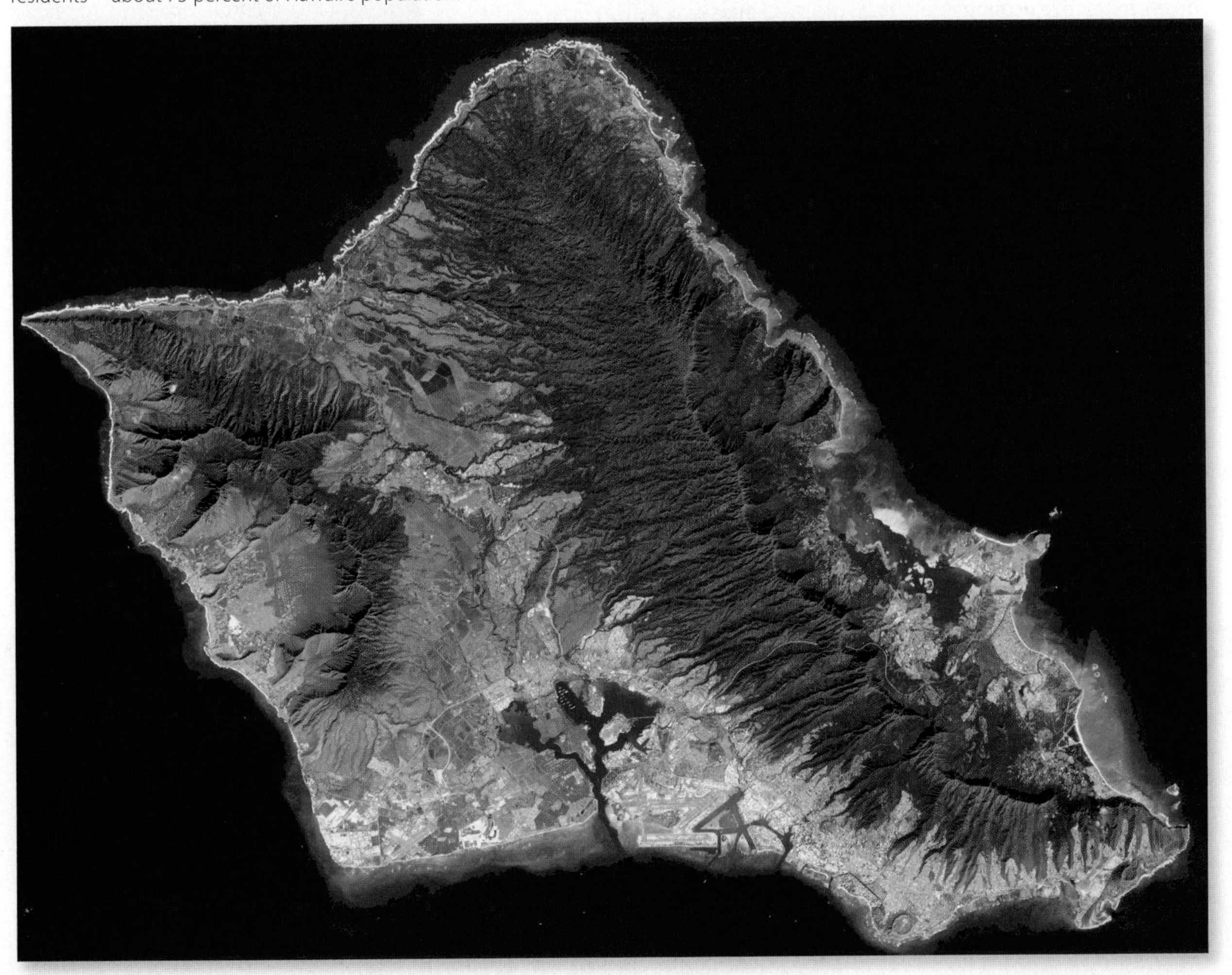

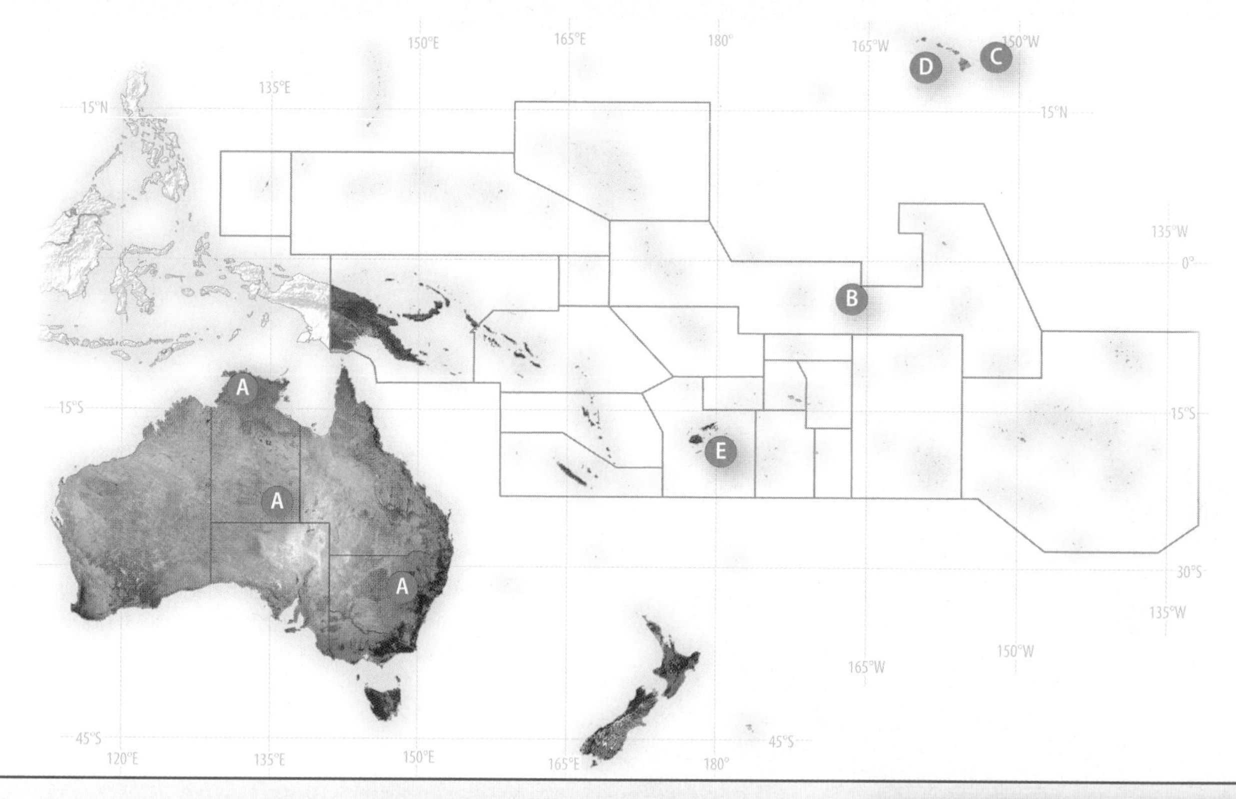

Physical Geography and Environmental Issues

The natural environment has witnessed accelerated change in the past 50 years. Urbanization, tourism, extractive economic activity, exotic species, and, most recently, global warming and climate change have altered the landscapes of Oceania.

14.1 Imagine a line from Darwin in the north through Alice Springs to Sydney in the southeast; then, drawing upon the climographs for these three places, describe in a narrative essay the different climates found along this transect in terms of rainfall, temperatures, and seasonality.

14.2 Given the different climates you've discussed, describe the different agricultural activities you might expect to find along this north-south transect.

Population and Settlement

Migration is a major theme in Oceania, beginning with the earliest Aboriginal peoples who came from the Asian mainland, to prehistoric island-hopping Polynesian people, to more recent European emigrants that populated Australia and New Zealand.

14.3 Make a list of what Polynesian sailors would have to take along on their voyages to set up new settlements on distant islands.

14.4 Today, because of sea-level rise from global warming, many low island people see an unsustainable future and are discussing migrating to Australia. But some Australian groups maintain their land is already overpopulated. Is it? Evaluate and discuss the arguments put forth that Australia is either over- or underpopulated.

Key Terms

Aborigine *(p. 665)*
atoll *(p. 655)*
Austronesian *(p. 673)*
Closer Economic Relations (CER) Agreement *(p. 690)*
Exclusive Economic Zone (EEZ) *(p. 689)*
haole *(p. 678)*
high island *(p. 655)*
kanaka *(p. 672)*
low island *(p. 655)*
mallee *(p. 653)*
Maori *(p. 673)*
marsupial *(p. 656)*
Melanesia *(p. 652)*
Micronesia *(p. 652)*
microstate *(p. 679)*
Native Title Bill *(p. 682)*
Oceania *(p. 650)*
Outback *(p. 652)*
Pidgin English *(p. 674)*
Polynesia *(p. 651)*
uncontacted people *(p. 673)*
viticulture *(p. 668)*
White Australia Policy *(p. 672)*

In Review • Chapter 14
Australia and Oceania

Cultural Coherence and Diversity

In Australia and New Zealand, the Aborigines and Maori work to preserve their minority cultures in the face of dominating Anglo influences. In different ways, the same is true of Oceania's peoples as tourists and foreign workers become increasingly common on once remote islands.

14.5 These tourists are watching Native Hawaiians roast a pig in the traditional manner. What other aspects of Polynesian culture have become tourist attractions in Hawaii? What about in New Zealand?

14.6 In a group of three, have one person each represent the issues facing Australian Aborigines, New Zealand Maoris, and native Hawaiians in their different countries. Then, as a group, compare the similarities and differences. Predict where each native group might be in 10 years.

Geopolitical Framework

Oceania has become entangled in the world's larger geopolitical tensions as China and the United States jockey for economic and political influence in the region. Some small island nations feel pressure to align themselves with one superpower or the other, but economically powerful Australia has forged a different solution, maintaining continued political allegiance to the United States, while continuing to develop economic ties to its major trading partner, China.

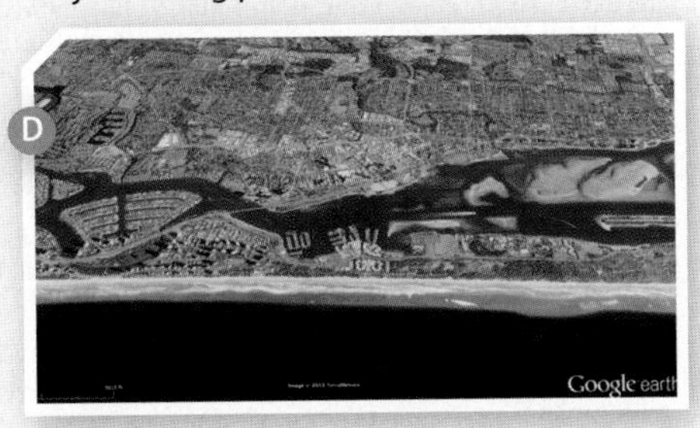

14.7 This is a Google Earth image of Pearl Harbor, Hawaii, which includes the huge U.S. naval base that is the headquarters for the Pacific fleet. Drawing upon what you've learned about tropical island environments in this chapter, make a list of how this coastal environment has been changed since 1900.

14.8 Working with several other students, each of whom will represent a specific country in Oceania (including Australia), discuss the geopolitical and economic ties and tensions between that country and China. How might these change in the next five years?

Economic and Social Development

Australia is the economic powerhouse of the region because of its rich mineral resources, with coal and iron dominating the list of trade commodities. In contrast, island nations, from Fiji to Hawaii, are largely at the mercy of global tourism, with boom-and-bust cycles linked to the vitality of the global economy.

14.9 Make a list of the economic and social advantages and liabilities for an island community of a large tourist resort such as the one pictured in Fiji (or Samoa).

14.10 You are part of a team responsible for creating a comprehensive economic and social development plan for a small South Pacific island. Choose a specific island, learn its needs and resources, define your goals, and describe a plan for achieving them within five years.

MasteringGeography™

Looking for additional review and test prep materials? Visit the Study Area in **MasteringGeography**™ to enhance your geographic literacy, spatial reasoning skills, and understanding of this chapter's content by accessing a variety of resources, including **MapMaster**™ interactive maps, videos, RSS feeds, flashcards, web links, self-study quizzes, and an eText version of *Diversity Amid Globalization*.

Authors' Blogs

Scan to visit the author's blog for chapter updates

www.gad4blog.wordpress.com

Scan to visit the GeoCurrents blog

www.geocurrents.info

Glossary

Aborigine An indigenous inhabitant of Australia.

acid rain Harmful form of precipitation high in sulfur and nitrogen oxides. Caused by industrial and auto emissions, acid rain damages aquatic and forest ecosystems in regions such as eastern North America and Europe.

adiabatic lapse rate The rate an air mass cools or warms with changes in elevation, which is usually around 5.5° F per 1000 feet (1°C per 100 meters). Contrast with environmental lapse rate.

African diaspora The forced removal of Africans from their native areas to localities around the globe, especially due to slavery.

African Union (AU) Founded in 1963, the organization grew to include all the states of the continent except South Africa, which finally was asked to join in 1994. In 2004 the body changed its name from the Organization of African Unity (OAU) to the African Union. It is mostly a political body that has tried to resolve regional conflicts.

agrarian reform A popular but controversial strategy to redistribute land to peasant farmers. Throughout the 20th century, various states redistributed land from large estates or granted land titles from vast public lands in order to reallocate resources to the poor and stimulate development. Agrarian reform occurred in various forms, from awarding individual plots or communally held land to creating state-run collective farms.

agribusiness The practice of large-scale, often corporate farming in which business enterprises control closely integrated segments of food production, from farm to grocery store.

agricultural density The number of farmers per unit of arable land. This figure indicates the number of people who directly depend upon agriculture, and it is an important indicator of population pressure in places where rural subsistence dominates.

alluvial fan A fan-shaped deposit of sediments dropped by a river or stream flowing out of a mountain range.

Altiplano The largest intermontane plateau in the Andes, which straddles Peru and Bolivia and ranges in elevation from 10,000 to 13,000 feet (3000 to 4000 meters).

altitudinal zonation The relationship between higher elevations, cooler temperatures, and changes in vegetation that result from the environmental lapse rate (averaging 3.5°F for every 1000 feet [6.5°C for every 1000 meters]). In Latin America, four general altitudinal zones exist: tierra caliente, tierra templada, tierra fria, and tierra helada.

animism A wide variety of tribal religions based on the worship of nature spirits and human ancestors.

anthropogenic Caused by human activities.

anthropogenic landscapes Landscapes that have been heavily transformed by humans.

apartheid The policy of racial separateness that directed separate residential and work spaces for white, blacks, coloreds, and Indians in South Africa for nearly 50 years. It was abolished when the African National Congress came to power in 1994.

Arab League A regional political and economic organization focused on Arab unity and development.

Arab Spring A series of public protests, strikes, and rebellions in the Arab countries, often facilitated by social media, that have called for fundamental government and economic reforms.

areal differentiation The geographic description and explanation of spatial differences on Earth's surface, including both physical as well as human patterns.

areal integration The geographic description and explanation of how places, landscapes, and regions are connected, interact, and are integrated with each other.

Association of Southeast Asian Nations (ASEAN) An international organization linking together the 10 most important countries of Southeast Asia.

atoll Low, sandy islands made from coral, often oriented around a central lagoon.

Austronesian A language family that encompasses wide expanses of the Pacific, insular Southeast Asia, and Madagascar.

autonomous area Minor political subunit created in the former Soviet Union and designed to recognize the special status of minority groups within existing republics.

autonomous region In China, province-level regions that have been given a degree of political and cultural autonomy due to the fact that they are inhabited in large part by minority groups.

Baikal-Amur Mainline (BAM) Railroad Key central Siberian railroad connection completed in the Soviet era (1984), which links the Yenisey and Amur rivers and parallels the Trans-Siberian Railroad.

balkanization Geopolitical process of fragmentation of larger states into smaller ones through independence of smaller regions and ethnic groups. The term takes its name from the geopolitical fabric of the Balkan region.

Berlin Conference The 1884 conference that divided Africa into European colonial territories. The boundaries created in Berlin satisfied European ambition but ignored indigenous cultural affiliations. Many of Africa's civil conflicts can be traced to ill-conceived territorial divisions crafted in 1884.

biodiversity The array of species, both flora and fauna, found in an ecosystem or bioregion.

biofuels Energy sources derived from plants or animals. Throughout the developing world, wood, charcoal, and dung are primary energy sources for cooking and heating.

bioregion A spatial unit or region of local plants and animals adapted to a specific environment such as a tropical savanna.

Bodhisattva In the religion of Mahayana Buddhism, a spiritual being that helps others attain enlightenment.

Bolsa Familia A successful conditional cash transfer program developed in Brazil to address extreme poverty. Poor families who qualify receive a monthly check from the government provided they keep their children in school and vaccinated.

Bolsheviks A faction within the Russian communist movement led by Lenin that successfully took control of the country in 1917.

boreal forest Coniferous forest found in high-latitude or mountainous environments of the Northern Hemisphere.

brain drain Migration of the best-educated people from developing countries to developed nations where economic opportunities are greater.

brain gain The potential of return migrants to contribute to the social and economic development of their country of origin with the experiences they have gained abroad.

British East India Company Private trade organization with its own army that acted as an arm of Britain in monopolizing trade in South Asia until 1857, when it was abolished and replaced by full governmental control.

bubble economy A highly inflated economy that cannot be sustained. Bubble economies usually result from the rapid influx of international capital into a developing country.

buffer zone An array of nonaligned or friendly states that "buffer" a larger country from invasion. In Eurasia, maintaining a buffer zone has been a long-term policy of Russia (and also of the former Soviet Union) to protect its western borders from European invasion.

Bumiputra The name given to native Malays (literally, "sons of the soil"), who are given preference for jobs and schooling by the Malaysian government.

Burakumin A group of people in Japan who suffer from discrimination due to the employment of their ancestors in "polluting" jobs.

capital leakage The gap between the gross receipts an industry (such as tourism) brings into a developing area and the amount of capital retained.

capitalism An economic system based on private property and the market exchange of goods and services.

carbon inequity The position taken by developing countries such as China and India, which argue that, because Western industrial countries in North America and Europe have been burning large amounts of fossil fuels since the mid-19th century and because CO_2 stays in the atmosphere for hundreds of years, these countries caused the global warming problem and therefore should fix it.

Caribbean Community and Common Market (CARICOM) A regional trade organization established in 1972 that includes mostly former English Caribbean colonies as its members.

Caribbean Diaspora The scattering of a particular group of people over a vast geographic area. Originally, the term referred to the migration of Jews out of their homeland, but now it has been generalized to refer to any ethnic dispersion.

caste system Complex division of South Asian society into different hierarchically ranked hereditary groups. Most explicit in Hindu society, the caste system is also found in other South Asian cultures to a lesser degree.

Central American Free Trade Association (CAFTA) A 2004 free trade agreement between the United States and five Central American countries—Guatemala, Nicaragua, El Salvador, Honduras, and Costa Rica—plus the Dominican Republic.

central place theory A theory used to explain the distribution of cities, and the relationships between different cities, based on retail marketing.

centralized economic planning An economic system in which the state sets production targets and controls the means of production.

chain migration A pattern of migration in which a sending area becomes linked to a particular destination, such as Dominicans with New York City.

chernozem soils A Russian term for dark, fertile soil, often associated with grassland settings in southern Russia and Ukraine.

China proper The eastern half of the country of China where the Han Chinese form the dominant ethnic group. The vast majority of China's population is located in China proper.

choke point A strategic setting where narrow waterways are vulnerable to military blockade or disruption.

choropleth map A thematic map in which areas use color shading or different patterns to depict differences in whatever is being mapped.

circular migration Temporary labor migration, in which an individual seeks short-term employment overseas, earns money, and then returns home.

clan A social unit that is typically smaller than a tribe or ethnic group but larger than a family, based on supposed descent from a common ancestor.

climate The average weather conditions for a place, usually based upon 30 years of weather measuresments.

climate change The measured change in climate from a previous state, contrasted with normal variability.

climographs Graph of average annual temperature and precipitation data by month and season.

Closer Economic Relations (CER) An agreement signed in 1982 between Australia and New Zealand designed to eliminate all economic and trade barriers between the two countries.

Cold War The ideological struggle between the United States and the Soviet Union that was conducted between 1946 and 1990.

colliding plate boundary A tectonic boundary where two different tectonic plates are moving against each other.

colonialism The formal and established rule over local peoples by a larger imperialist government.

coloured A racial category used throughout South Africa to define people of mixed European and African ancestry.

Columbian exchange An exchange of people, diseases, plants, and animals between the Americas (New World) and Europe/Africa (Old World) initiated by the arrival of Christopher Columbus in 1492.

Commonwealth of Independent States (CIS) A loose political union of former Soviet republics (without the Baltic states) established in 1992 after the dissolution of the Soviet Union.

communism A belief based on the writings of Karl Marx that promoted the overthrow of capitalism by the workers, the large-scale elimination of private property, state ownership and central planning of major sectors of the economy (both agricultural and industrial), and one-party authoritarian rule.

conflict diamonds Diamonds that are sold on the black market to fund armed conflict and civil war, especially in Sub-Saharan Africa.

Confucianism The philosophical system developed by Confucius in the 6th century BCE that stresses the creation of a proper social order.

connectivity The degree to which different locations are linked with one another through transportation and communication infrastructure.

continentality Inland climates, removed from the ocean, with hot summers and cold winters such as those found in interior North America and Russia.

copra Dried coconut meat.

core-periphery model According to this scheme, the United States, Canada, western Europe, and Japan constitute the global economic core, while other regions make up a less-developed economic periphery from which the core extracts resources.

Cossacks Highly mobile Slavic-speaking Christians of the southern Russian steppe who were pivotal in expanding Russian influence in 16th- and 17th-century Siberia.

counterinsurgency The suppression of a rebellion or insurgency by military and political means, which includes not just warfare but also winning the support of local peoples by improving community infrastructure (schools, roads, water supply, etc.).

creolization The blending of African, European, and even some Amerindian cultural elements into the unique sociocultural systems found in the Caribbean.

crisis mapping The use of open source and web-based technologies to create detailed maps of areas experiecning humanitarian crises, typically supported by volunteer mappers.

crony capitalism A system in which close friends of a political leader are either legally or illegally given business advantages in return for their political support.

cuentapropista The Spanish term for small scale entrepreneurs in Cuba. A socialist country, Cuba is experimenting with licensing small businesses to address the country's economic and unemployment problems.

cultural assimilation The process in which immigrants are culturally absorbed into the larger host society.

cultural homeland A culturally distinctive settlement in a well-defined geographic area, whose ethnicity has survived over time, stamping the landscape with an enduring personality.

cultural imperialism The active promotion of one cultural system over another, such as the implantation of a new language, school system, or bureaucracy. Historically, this has been primarily associated with European colonialism.

cultural landscape Primarily the visible and tangible expression of human settlement (house architecture, street patterns, field form, etc.), but also includes the intangible, value-laden aspects of a particular place and its association with a group of people.

cultural nationalism A process of protecting, either formally (with laws) or informally (with social values), the primacy of a specific cultural system against influences from another culture.

cultural syncretism or hybridization The blending of two or more cultures, which then produces a synergistic third culture or specific behavior that exhibits traits from all cultural parents.

culture Learned and shared behavior by a group of people empowering them with a distinct "way of life"; it includes both material (technology, tools, etc.) and non-material (speech, religion, values, etc.) components.

culture hearth An area of historical cultural innovation.

cyclones Large storms, marked by well-defined air circulation around a low-pressure center. Tropical cyclones are typically called hurricanes in the Atlantic Ocean and typhoons in the western Pacific.

Cyrillic alphabet Based on the Greek alphabet and used by Slavic languages heavily influenced by the Eastern Orthodox Church. Attributed to the missionary work of St. Cyril in the 9th century.

dacha A Russian country cottage used especially in the summer.

Dalai Lama The leader of the Tibetan Buddhist faith. The current Dalai Lama is an important advocate for Tibet rights.

dalits The so-called untouchable population of India; people often considered socially polluting because of their historical connections with occupations classified as unclean, such as leatherworking and latrine-cleaning.

de-territorialization The expansion of an activity (such as a sport) closely tied to a specific place or region to a non-place based global scale.

deciduous tree A tree that looses its leaves to inhibit or halt photosynthesis during the harsh season. While most commonly leaves fall before the cold winter, deciduousness can also take place as plants prepare for a hot and dry summer.

decolonialization The process of a former colony's gaining (or regaining) independence over its territory and establishing (or reestablishing) an independent government.

deforestation The loss of forest cover, usually due to logging followed by the spread of agriculture or grazing.

demographic transition model A four-stage scheme that explains different rates of population growth over time through differing birth and death rates.

denuclearization The process whereby nuclear weapons are removed from an area and dismantled or taken elsewhere.

dependency theory A popular theory to explain patterns of economic development in Latin America. Its central premise is that underdevelopment was created by the expansion of European capitalism into the region that served to develop "core" countries in Europe and to impoverish and make dependent peripheral areas such as Latin America.

desertification The spread of desert conditions into semiarid areas owing to improper management of the land.

desiccation The process of extreme drying; in desert areas, such processes can lead to the disappearance of lakes when the rivers that previously fed them are diverted for agriculture or other purposes.

devolution The breakng apart or separation within a political unit such as a nation-state.

diaspora The scattering of a particular group of people over a vast geographical area. Originally, the term referred to the migration of Jews out of their original homeland, but now it has been generalized to refer to any ethnic dispersion.

divergent plate boundary A place where two tectonic plates move away from each other. Here, magma often flows from Earth's interior, producing volcanoes on the surface such as in Iceland. In other divergent zones large trenches or rift valleys are created.

diversity The differences between cultures, ethnicities, economies, landscapes, and regions.

dollarization An economic strategy in which a country adopts the U.S. dollar as its official currency. A country can be partially dollarized, using U.S. dollars alongside its national currency, or fully dollarized, when the U.S. dollar becomes the only medium of exchange and a country gives up its own national currency. Panama fully dollarized in 1904; more recently, Ecuador became fully dollarized in 2000, followed by El Salvador in 2001.

domestication The purposeful selection and breeding of wild plants and animals for cultural purposes.

domino theory A U.S. geopolitical policy of the 1970s that stemmed from the assumption that if Vietnam fell to the communists, the rest of Southeast Asia would soon follow.

Dravidian A major family of related languages found mostly in southern India: Tamil, Telugu, Malayalam, and Kannada are the major Dravidian languages.

Eastern Orthodox Christianity A loose confederation of self-governing churches in eastern Europe and Russia that are historically linked to Byzantine traditions and to the primacy of the patriarch of Constantinople (Istanbul).

economic convergence The notion that globalization will result in the world's poorer countries gradually catching up with more developed economies.

Edge city Suburban node of activity that features a mix of peripheral retailing, industrial parks, office complexes, and entertainment facilities.

El Niño An abnormally warm current that appears periodically in the eastern Pacific Ocean and usually influences storminess along the western coasts of the Americas. During an El Niño event, which can last several years, torrential rains often bring devastating floods to the Pacific coasts of North, Central, and South America.

entrepôt A city and port that specializes in transshipment of goods.

environmental lapse rate The decline in temperature as one ascends higher in the atmosphere. On average, the temperature declines 3.5°F for every 1000 feet ascended, or 6.5°C for every 1000 meters.

ethnic religions A religion closely identified with a specific ethnic or tribal group, often to the point of assuming the role of the major defining characteristic of that group. Normally, ethnic religions do not actively seek new converts.

ethnicity A shared cultural identity held by a group of people with a common background or history, often as a minority group within a larger society.

European Union (EU) The current association of 27 European countries that are joined together in an agenda of economic, political, and cultural integration.

Eurozone The common monetary policy and currency of the European Union; those countries of Europe using the euro as its currency and who are members of the EU's common monetary system, contrasted to those countries having a national currency and monetary system. France is an example of the former, and the United Kingdom of the latter.

exclave A portion of a country's territory that lies outside its contiguous land area.

Exclusive Economic Zone (EEZ) An area of the ocean decreed by international law where one local country has more rights to fishing and mineral rights than do other countries.

exotic river A river that issues from a humid area and flows into a dry area otherwise lacking streams.

fair trade An international certification movement to identify primary commodities exported from the developing world in which farmers earn a better price for their product. Commodities such as coffee, tea, and forest products are certified "fair trade" when small-scale producers earn more for their product and production methods are viewed as environmentally and socially sustainable.

federal state A country in which the major territorial subdivisions have a significant degree of political autonomy, such as the United States, India, or Mexico; federal states are contrasted with unitary states, in which the central governments sets most policies.

Fertile Crescent An ecologically diverse zone of lands in Southwest Asia that extends from Lebanon eastward to Iraq and that is often associated with early forms of agricultural domestication.

fjord Flooded, glacially carved valley. In Europe, fjords are found primarily along Norway's western coast.

food insecurity When people lack physical or economic access to sufficient, safe, and nutrious food to meet their dietary needs for extended periods of time.

formal region A geographic concept used to describe an area where a static and specific trait (such as a language or a climate) has been mapped and described. A formal region contrasts with a functional region. A geographic concept of areal or spatial similarity, large or small.

forward capital A capital city deliberately positioned near a contested territory, signifying the state's interest and presence in this zone of conflict.

fossil water Water supplies that were stored underground during wetter climatic periods.

fracking (also called hydraulic fracturing) A set of drilling technologies that injects a mix of water, sand, and chemicals underground in order to release and enhance the removal of natural gas and oil.

free trade zone (FTZ) A duty-free and tax-exempt industrial park created to attract foreign corporations and create industrial jobs.

functional region A geographic concept used to describe the spatial extent dominated by a specific activity. The circulation area of a newspaper is an example, as is the trade area of a large city.

gender The social and cultural expressions of male- and femaleness, which contrasts with sex, which is the biological distinction between male and female.

gender equity A social condition, state, or goal of complete parity and equality between males and females.

gender gap A term often used to describe gender differences in salary, working conditions, or political power.

gender roles How female and male behavior differs in a specific cultural context.

gentrification A process of urban revitalization in which higher income residents displace lower-income residents in central-city neighborhoods.

geographic Information systems (GIS) A computerized mapping and information system that is able to analyze vast amounts of data through both in its totality but also through layers of specific kinds of information, such as microclimates, hydrology, vegetation, or landuse zoning regulations.

geography the spatial science that describes and explains physical and cultural phenomena on Earth's surface.

geomancy The traditional Chinese and Korean practice of designing buildings in accordance with the principles of cosmic harmony and discord that supposedly course through the local topography.

geopolitics The relationship between politics and space and territory.

glasnost A policy of greater political openness initiated during the 1980s by Soviet president Mikhail Gorbachev.

global positioning system (GPS) Originally used to describe a very accurate satellite-based location system, but now also used in a general sense to describe smartphone location systems that may use cell phone towers as a subsitute for satellites.

global warming The increase in global temperatures as a result of the magnification of the natural greenhouse effect due to human-produced pollutants. The resulting climate change could cause profound—and probably damaging—changes to Earth's environments.

globalization The increasing interconnectedness of people and places throughout the world through converging processes of economic, political, and cultural change.

Golden Triangle An area of northern Thailand, Burma, and Laos that is known as a major source region for heroin and is plugged into the global drug trade.

Gondwana The name given for the southerly supercontinent that existed over 200 million years ago that included Africa, South America, Antarctica, Australia, the Arabian Peninsula, and the Indian subcontinent.

graphic or linear scale A ruler-like symbol on a map that translates the map's cartographic scale into visual terms.

grassification The conversion of tropical forest into pasture for cattle ranching. Typically, this process involves introducing species of grasses and cattle, mostly from Africa.

Great Escarpment A landform that rims southern Africa from Angola to South Africa. It forms where the narrow coastal plains meet the elevated plateaus in an abrupt break in elevation.

Greater Antilles The four large Caribbean islands of Cuba, Jamaica, Hispaniola, and Puerto Rico.

Green Revolution A term applied to the development of new techniques, starting in the 1960s, that have transformed agriculture in India and many other developing countries; Green Revolution techniques usually involve the use of hybrid seeds that provide higher yields than native seeds when combined with high inputs of chemical fertilizer, irrigation, and pesticides.

greenhouse effect The natural process of lower atmosphere heating that results from the trapping of incoming and reradiated solar energy by water moisture, clouds, and other atmospheric gases.

gross domestic product (GDP)/gross national product (GNP) GDP is the total value of goods and services produced within a given country (or other geographical unit) in a single year. GNP is a somewhat broader measure that includes the inflow of money from other countries in the form of the repatriation of profits and other returns on investments, as well as the outflow to other countries for the same purposes.

gross national income (GNI) The value of all final goods and services produced within a country's borders (gross domestic product, or GDP), plus the net income from abroad (formerly referred to as gross national product, or GNP).

gross national income (GNI) per capita The figure that results from dividing a country's GNI by the total population.

Group of Eight (G-8) A collection of powerful countries that confer regularly on key global economic and political issues. It includes the United States, Canada, Japan, Great Britain, Germany, France, Italy, and Russia.

Gulag Archipelago A collection of Soviet-era labor camps for political prisoners, made famous by writer Aleksandr Solzhenitsyn.

Hajj An Islamic religious pilgrimage to Makkah. One of the five essential pillars of the Muslim creed to be undertaken once in life, if an individual is physically and financially able to do it.

haoles Light-skinned Europeans or U.S. citizens in the Hawaiian Islands.

high islands Larger, more elevated islands, often focused around recent volcanic activity.

Hindu nationalism A contemporary "fundamentalist" religious and political movement that promotes Hindu values as the essential—and exclusive—fabric of Indian society. As a political movement, it generally has less tolerance of India's large Muslim minority than other political movements.

homelands Nominally independent ethnic territories created for blacks under the grand apartheid scheme. Homelands were on marginal land, overcrowded, and poorly serviced. In the post-apartheid era, they were eliminated.

Horn of Africa The northeastern corner of Sub-Saharan Africa that includes the states of Somalia, Ethiopia, Eritrea, and Djibouti. Since the 1980s, drought, famine, ethnic conflict, and political turmoil have undermined development efforts in this area.

Human Development Index (HDI) For the past three decades, the United Nations has tracked social development in the world's countries through the Human Development Index (HDI), which combines data on life expectancy, literacy, educational attainment, gender equity, and income.

human trafficking A practice in which women are lured or abducted into prostitution.

hurricanes Storm systems with an abnormally low-pressure center sustaining winds of 74 mph or higher. Each year during hurricane season (July–October), a half dozen to a dozen hurricanes form in the warm waters of the Atlantic and Caribbean, bringing destructive winds and heavy rain.

hydraulic fracturing (fracking) A process for extracting oil from shale rock that involves injecting a fluid mixture into the ground that breaks apart (or fractures) the rock layer, thus making it easier to pump out the oil.

hydropolitics The interplay of water resource issues and politics.

ideographic writing A writing system in which each symbol represents not a sound but rather a concept.

indentured labor Foreign workers (usually South Asians) contracted to labor on Caribbean agricultural estates for a set period of time, often several years. Usually the contract stipulated paying off the travel debt incurred by the laborers. Similar indentured labor arrangements have existed in most world regions.

Indian diaspora The historical and contemporary movement of people from India to other countries in search of better opportunities. This process has led to large Indian populations in South Africa, the Caribbean, and the Pacific islands, along with Western Europe and North America.

Indian subcontinent The large Eurasian peninsula that extends south of the Himalayan Mountains and that encompasses most of South Asia.

Industrial Revolution That period of time in the 18th century when European factories first changed from using animate power (human and animals) to inanimate power (water and coal) to power machines.

informal sector A much-debated concept that presupposes a dual economic system consisting of formal and informal sectors. The informal sector includes self-employed, low-wage jobs that are usually unregulated and untaxed. Street vending, shoe shining, artisan manufacturing, and even self-built housing are considered part of the informal sector. Some scholars include illegal activities such as drug smuggling and prostitution in the informal economy.

insolation Incoming solar energy that enters the atmosphere adjacent to Earth.

insurgency A political rebellion or uprising.

internally displaced persons Groups and individuals who flee an area due to conflict or famine but still remain in their country of origin. These populations often live in refugee-like conditions but are harder to assist because they technically do not qualify as refugees.

Iron Curtain A term coined by British leader Winston Churchill during the Cold War that defined the western border of Soviet power in Europe. The notorious Berlin Wall was a concrete manifestation of the Iron Curtain.

irredentism A state or national policy of reclaiming lost lands or those inhabited by people of the same ethnicity in another nation-state.

Islamic fundamentalism A movement within both the Shiite and Sunni Muslim traditions to return to a more conservative, religious-based society and state. Often associated with a rejection of Western culture and with a political aim to merge civic and religious authority.

Islamism A political movement within the religion of Islam that challenges the encroachment of global popular culture and blames colonial, imperial, and Western elements for many of the region's problems. Adherents of Islamism advocate merging civil and religious authority.

isolated proximity A concept that explores the contradictory position of the Caribbean states, which are physically close to North America and economically dependent upon that region. At the same time, Caribbean isolation fosters strong loyalties to locality and limited economic opportunity.

Jainism A religious group in South Asia that emerged as a protest against orthodox Hinduism in the 6th century BCE. Jains are noted for their practice of nonviolence, which prohibits them from taking the life of any animal.

kanakas Melanesian workers imported to Australia, historically concentrated along Queensland's "sugar coast."

Khmer Rouge Literally, "Red (or communist) Cambodians." The left-wing insurgent group led by French-educated Marxists that ruled Cambodia from 1975 to 1979, during which time it engaged in genocidal acts against the Cambodian people.

kibbutzim Collective farms in Israel.

kleptocracy A state where corruption is so institutionalized that politicians and bureaucrats siphon off a large percentage of a country's wealth for personal gain.

laissez-faire An economic system in which the state has minimal involvement and in which market forces largely guide economic activity.

latifundia A large estate or landholding in Latin America.

less developed country (LDC) A country with an economy that is less well developed than industrial countries such as Japan, North America, and Europe.

Lesser Antilles The arc of small Caribbean islands from St. Maarten to Trinidad.

Levant The eastern Mediterranean region.

liberation theology A political movement within the Catholic church that began in the 1970s by a Peruvian priest, Fr. Gustavo Gutiérrez. It emphasizes the role of the church to address the needs of the poor brought on by unjust economic, political, and social conditions.

life expectancy The statistical average for how long a person within a specific data category is expected to live.

lingua franca An agreed-upon common language to facilitate communication on specific topics such as international business, politics, sports, or entertainment.

linguistic nationalism The promotion of one language over others that is, in turn, linked to shared notions of political identity. In India, some nationalists promote Hindi as the unifying language of the country, yet this is resisted by many non-Hindi-speaking peoples.

location factors The various influences that explain why an economic activity takes place where it does.

loess A fine, wind-deposited sediment that makes fertile soil but is very vulnerable to water erosion.

low island Flat, low-lying islands formed by coral reefs, and contrasting with high islands that were formed from volcanic eruptions.

Maghreb A region in northwestern Africa, including portions of Morocco, Algeria, and Tunisia.

maharaja Historical term for Hindu royalty, usually a king or prince, who ruled specific areas of South Asia before independence, but who was usually subject to overrule by British colonial advisers.

mallee A tough and scrubby eucalyptus woodland of limited economic value that is common across portions of interior Australia.

Mandarins A member of the high-level bureaucracy of Imperial China (before 1911). Mandarin Chinese is the official spoken language of the country and is the native tongue of the vast majority of people living in north, central, and southwestern China.

Maori Indigenous Polynesian people of New Zealand.

map projection The cartographic and mathematical solution to translating the surface of a founded globe (usually Earth) to a flat surface (usually a piece of paper) with a minimum of distortion.

map scale The relationship between distances on a mapped object such as Earth and depiction of that space on a map. Large scale maps cover small areas in great detail, where small scale maps depict less detail but over large areas.

maquiladoras Assembly plants on the Mexican border built and owned by foreign companies. Most of their products are exported to the United States.

maritime climate Climate moderated by proximity to oceans or large seas. It is usually cool, cloudy, and wet, and lacks the temperature extremes of continental climates.

Maroons Runaway slaves who established communities rich in African traditions throughout the Caribbean and Brazil.

marsupial A class of mammals found primarily in the Southern Hemisphere with the distinctive characteristic of carrying their young in a pouch. Kangaroos are perhaps the best-known marsupial, with wallabies, koalas, wombats, and the Tasmanian Devil also found in Oceania.

Marxism A term referring to the political philosophy developed by Karl Marx in the 1800s and based on the ideas of communism.

medieval landscape Urban landscapes from 900 to 1500 CE characterized by narrow, winding streets, three- or four-story structures (usually in stone, but sometimes wooden), with little open space except for the market square. These landscapes are still found in the centers of many European cities.

medina The original urban core of a traditional Islamic city.

megacities Urban conglomerations of more than 10 million people.

Megalopolis A large urban region formed as multiple cities grow and merge with one another. The term is often applied to the string of cities in eastern North America that includes Washington, DC; Baltimore; Philadelphia; New York City; and Boston.

Melanesia Pacific Ocean region that includes the culturally complex, generally darker-skinned peoples of New Guinea, the Solomon Islands, Vanuatu, New Caledonia, and Fiji.

Mercosur The Southern Common Market established in 1991 that calls for free trade among member states and common external tariffs for nonmember states. Argentina, Paraguay, Brazil, and Uruguay are full members and Venezuela's full membership is pending; the Andean countries of Colombia, Ecuador, Peru, Bolivia, and Chile are associate members.

meridians (lines of longitude) Meridians run north-south, from pole to pole, and measure distance east or west from the Prime Meridian (0 degrees) located near London, England.

mestizo A person of mixed European and Indian ancestry.

micro-credit An economic development strategy that entails the provision of small loans to poor people who want to start small-scale businesses.

Micronesia Pacific Ocean region that includes the culturally diverse, generally small islands north of Melanesia. Includes the Mariana Islands, Marshall Islands, and Federated States of Micronesia.

microstates Usually independent states that are small in both area and population.

mikrorayon Large, state-constructed urban housing project built during the Soviet period in the 1970s and 1980s.

Millennium Development Goals Part of a group of programs implemented since 2000 to foster development in the world's poorest countries. The Millennium Development Goals are part of a global United Nations effort to reduce extreme poverty by 2015.

minifundia A small landholding farmed by peasants or tenants who produce food for subsistence and the market.

monocrop production Agriculture based upon a single crop, often for export.

monotheism A religious belief in a single God.

Monroe Doctrine A proclamation issued by U.S. President James Monroe in 1823 that the United States would not tolerate European military action in the Western Hemisphere. Focused on the Caribbean as a strategic area, the doctrine was repeatedly invoked to justify U.S. political and military intervention in the region.

monsoon The seasonal pattern of changes in winds, heat, and moisture in South Asia and other regions of the world that is a product of larger meteorological forces of land and water heating, the resultant pressure gradients, and jet-stream dynamics. The monsoon produces distinct wet and dry seasons.

more developed country (MDC) A country with a well-developed industrial economy.

Mughal Empire (also spelled Mogul) The Muslim-dominated state that covered most of South Asia from the early 16th to late 17th centuries. The last vestiges of the Mughal dynasty were dissolved by the British following the rebellion of 1857.

nation-state A relatively homogeneous cultural group (a nation) with its own political territory (the state). While useful conceptually, the reality of today's globalized world is that there are very few countries that fit this simplistic definition because of the influx of migrants and/or the presence of minority ethnic groups.

Native Title Bill Australian legislation signed in 1993 that provides Aborigines with enhanced legal rights over land and resources within the country.

neocolonialism Economic and political strategies by which powerful states indirectly (and sometimes directly) extend their influence over other, weaker states.

neoliberism Economic policies widely adopted in the 1990s that stress privatization, export production, and few restrictions on imports.

neotropics Tropical ecosystems of the Americas that evolved in relative isolation and support diverse and unique flora and fauna.

net migration rate A statistic that depicts whether more people are entering or leaving a country through migration.

new urbanism An urban design movement stressing higher density, mixed-use, pedestrian-scaled neighborhoods where residents might be able to walk to work, school, and local entertainment.

non-renewable energy Those energy sources such as oil and coal with finite reserves.

nonmetropolitan growth A pattern of migration in which people leave large cities and suburbs and move to smaller towns and rural areas.

North American Free Trade Agreement (NAFTA) An agreement made in 1994 among Canada, the United States, and Mexico that established a 15-year plan for reducing all barriers to trade among the three countries.

North Atlantic Treaty Organization (NATO) Initially NATO was a group of North Atlantic and European allies who came together in 1949 to counter the Soviet threat to western Europe.

northern sea route An ice-free channel along Siberia's northern coast that will grow in importance given sustained global warming.

Oceania A major world subregion that is usually considered to include New Zealand and the major island regions of Melanesia, Micronesia, and Polynesia.

offshore banking Financial services offered by islands or microstates that are typically confidential and tax exempt. As part of a global financial system, offshore banks have developed a unique niche, offering their services to individual and corporate clients for set fees. The Bahamas and the Cayman Islands are leaders in this sector.

oligarchs A small group of wealthy, very private businessmen who control (along with organized crime) important aspects of the Russian economy.

Organization of American States (OAS) Founded in 1948 and headquartered in Washington, DC, the organization advocates hemispheric cooperation and dialogue. Most states in the Americas belong except Cuba.

Organization of the Petroleum Exporting Countries (OPEC) An international organization (formed in 1960) of 12 oil-producing nations that attempts to influence global prices and supplies of oil. Algeria, Gabon, Indonesia, Iran, Iraq, Kuwait, Libya, Nigeria, Qatar, Saudi Arabia, the United Arab Emirates, and Venezuela are members.

orographic effect The influence of mountains on weather and climate, usually referring to the increase of precipitation on the windward side of mountains, and a drier zone (or rain shadow) on the leeward or downwind side of the mountain.

orographic rainfall Enhanced precipitation over uplands that results from lifting (and cooling) of air masses as they are forced over mountains.

Ottoman Empire A large, Turkish-based empire (named for Osman, one of its founders) that dominated large portions of southeastern Europe, North Africa, and Southwest Asia between the 16th and 19th centuries.

Outback Australia's large, generally dry, and thinly settled interior.

outsourcing A business practice that transfers portions of a company's production and service activities to lower-cost settings, often located overseas

Palestinian Authority (PA) A quasi-governmental body that represents Palestinian interests in the West Bank and Gaza.

Pan-African Movement Founded in 1900 by U.S. intellectuals W. E. B. Du Bois and Marcus Garvey, this movement's slogan was "Africa for Africans," and its influence extended across the Atlantic.

parallels (lines of longitude) Parallels circle the mathematical globe in an east-west direction and measure a location north or south of the Equator, which is 0 degrees. When coupled with longitude, latitude provides a precise and unique mathematical address for any point on Earth.

pastoral nomadism A traditional subsistence agricultural system in which practitioners depend on the seasonal movements of livestock within marginal natural environments.

pastoralists Nomadic and sedentary peoples who rely upon livestock (especially cattle, camels, sheep, and goats) for their sustenance and livelihood.

perestroika A program of partially implemented, planned economic reforms (or restructuring) undertaken during the Gorbachev years in the Soviet Union designed to make the Soviet economy more efficient and responsive to consumer needs.

permafrost A cold-climate condition in which the ground remains permanently frozen.

physiological density A population statistic that relates the number of people in a country to the amount of arable land.

Pidgin English (Pijin) A version of English that also incorporates elements of other local languages, often utilized to foster trade and basic communication between different culture groups.

plantation America A cultural region that extends from midway up the coast of Brazil, through the Guianas and the Caribbean, and into the southeastern United States. In this coastal zone, European-owned plantations, worked by African laborers, produced agricultural products for export.

plate tectonics The theory that explains the gradual movement of large geological platforms (or plates) along Earth's surface.

podzol soils A Russian term for an acidic soil of limited fertility, typically found in northern forest environments.

pollution exporting The process of exporting industrial pollution and other waste material to other countries. Pollution exporting can be direct, as when waste is simply shipped abroad for disposal, or indirect, as when highly polluting factories are constructed abroad.

Polynesia Pacific Ocean region, broadly unified by language and cultural traditions, that includes the Hawaiian Islands, Marquesas Islands, Society Islands, Tuamotu Archipelago, Cook Islands, American Samoa, Samoa, Tonga, and Kiribati.

population density The number of people per areal unit, usually measured in people per square kilometer or per square mile.

population pyramid A graph representing the structure of a population, including the percentage of young and old. The percentages of all different age groups are plotted along a vertical axis that divides the population into male and female. In a fast growing population with a large percentage of young people and a small percentage of elderly, the graph will have a wide base and a narrow tip, giving it a pyramidal shape.

postindustrial economy An economy in which the tertiary and quaternary sectors dominate employment and expansion.

prairie An extensive area of grassland in North America. In the more humid eastern portions, grasses are usually longer than in the drier western areas, which are in the rain shadow of the Rocky Mountain range.

primate cities Massive urban settlements that dominate all other cities in a given country. A primate city is usually the capital of the country in which it is located.

prime meridian Zero degrees longitude, from which locations east and west are measured in a system of latitude and longitude. Currently, the most used Prime Merdian is that established in 1851 at the Naval Observatory in Greenwich, England (in southeastern London). Before that other countries and cultures established their own prime meridians upon which to base their maps and navigation systems.

pristine myth The idea put forward by geographer William Denevan that the Americas were more ecologically disturbed in 1492 than they were in the mid-1700s due to the demographic collapse of Amerindian peoples.

proctectorate During the period of Western global imperialism, a state or other political entity that remained autonomous but sacrificed its foreign affairs to an imperial power in exchange for "protection" from other imperial powers.

proven reserves the amount of a non-renewable energy source (oil, coal, and gas) still in the ground that is feasible to exploit under current market conditions.

purchasing power parity (PPP) An important qualification to GNI per capita data is the concept of adjustment through PPP, an adjustment that takes into account the strength or weakness of local currencies.

qanat system A traditional system of gravity-fed irrigation that uses gently sloping tunnels to capture groundwater and direct it to needed fields.

Quran (also spelled Koran) A book of divine revelations received by the prophet Muhammad that serves as a holy text in the religion of Islam.

rain shadow A drier area of precipitation, usually on the leeward or downwind side of a mountain range, that receives less rain and snowfall than the windward or upwind side. A rain shadow is caused by the warming of air as it descends down a mountain range; this warming increases the ability of an air mass to hold moisture.

Ramayana One of the two main epic poems of the Hindu religion, the Ramayana is also commonly performed in the shadow puppet theaters of the predominately Muslim island of Java.

rate of natural increase (RNI) The standard statistic used to express natural population growth per year for a country, region, or the world based upon the difference between birth and death rates. RNI does not consider population change from migration. Though most often a positive figure (such as 1.7 percent), RNI can also be expressed either as zero or even as a negative number for no-growth countries.

refugee A person who flees his or her country because of a well-founded fear of persecution based on race, ethnicity, religion, ideology, or political affiliation.

region A geographic concept of areal or spatial similarity, large or small.

regulatory lakes A term applied to a series of lakes in the middle Yangtze Valley of China. Regulatory lakes take excess water from the river during flood periods and supply water to the river during dry periods.

remittances Monies sent by immigrants working abroad to family members and communities in countries of origin. For many countries in the developing world, remittances often amount to billions of dollars each year. For small countries, remittances can equal 5 to 10 percent of a country's gross domestic product.

remote sensing A method of digitally photographing Earth's surface from satellites or high altitude aircraft so that the information captured can be manipulated by computers to translate information into certain electromagnetic bandwidths, which, in turn, emphasizes certain features and patterns on Earth's surface.

Renaissance-Baroque period A historical period, dated roughly from the 16th to 19th centuries centuries characterized by certain urban planning designs and architectural styles that are still found today in many European cities.

renewable energy Energy sources, such as solar, wind, and hydro, that are replenished by nature at a faster rate than they are used or consumed.

representative fraction The cartographic and mathematical expression of the relationship between distance on a map and that on Earth's surface, which is expressed as a fraction depicting the scale of the map. For example, a common scale is that of one inch on a map depicting one mile on the surface.

rimland The mainland coastal zone of the Caribbean, beginning with Belize and extending along the coast of Central America to northern South America.

rural-to-urban migration The flow of internal migrants from rural areas to cities that began in the 1950s and intensified in the 1960s and 1970s.

Russification A policy of the Soviet Union designed to spread Russian settlers and influences to non-Russian areas of the country.

rust belt Regions of heavy industry that experience marked economic decline after their factories ceased to be competitive.

Sahel The semidesert region at the southern fringe of the Sahara, and the countries that fall within this region, which extends from Senegal to Sudan. Droughts in the 1970s and early 1980s caused widespread famine and dislocation of population.

salinization The accumulation of salts in the upper layers of soil, often causing a reduction in crop yields, resulting from irrigation with water of high natural salt content and/or irrigation of soils that contain a high level of mineral salts.

samurai The warrior class of traditional Japan. After 1600, the military role of the samurai declined as they assumed administrative positions, but their military ethos remained alive until the class was abolished in 1868.

Schengen Agreement The 1985 agreement between some—but not all—European Union member countries to reduce border formalities in order to facilitate free movement of citizens between member countries of this new "Schengenland." For example, today there are no border controls between France and Germany, or between France and Italy.

sectarian violence Conflicts that divide people along ethnic, religious, and sectarian lines.

sectoral transformation The evolution of a labor force from being highly dependent on the primary sector to being oriented around more employment in the secondary, tertiary, and quaternary sectors.

secularism The term describes both the separation of politics and religion, as well as the non-religious segment of a population. An example of the first usage is the secularism of the United States constitution, which clearly separates State from Church; whereas in the second usage it is common to refer to the growing secularism of Europe, referring to the disinterest in religion of a large part of the population.

sediment load The amount of sand, silt, and clay carried by a river.

Shanghai Cooperation Organization (SCO) An international organization composed of China, Russia, Kazakhstan, Kyrgyzstan, Tajikistan, and Uzbekistan that aims to enhance security and economic cooperation in Central Asia.

Shi'a Islam The second largest denomination of Islam, whose adherents are called Shi'ites or Shias. While difficult to assess accurately, generally speaking 75–90 percent of the world's Muslims are Sunni, with Shias making up the remaining 10–25 percent.

shields Large upland areas of very old exposed rocks that range in elevation from 600 to 5000 feet (200 to 1500 meters). The three major shields in South America are the Guiana, Brazilian, and Patagonian.

shifted cultivators Migrant farmers who are either transplanted by government relocation schemes or forced to move on their own when their lands are expropriated.

Shiites Muslims who practice one of the two main branches of Islam; especially dominant in Iran and nearby southern Iraq.

shogunate The political order of Japan before 1868, in which power was held by the military leader known as the shogun, rather than by the emperor, whose authority was merely symbolic.

Sikhism A South Asian religion, concentrated in the Indian state of Punjab, that shows some similarities with both Islam and Hinduism.

Silk Road An historical trade route that extended across Central Asia, linking China with Europe and Southwest Asia.

siloviki Members of military and security forces.

Slavic peoples A group of peoples in eastern Europe and Russia who speak Slavic languages, a distinctive branch of the Indo-European language family.

social and regional differentiation "Social differentiation" refers to a process by which certain classes of people grow richer when others grow poorer; "regional differentiation" refers to a process by which certain places grow more prosperous while others become less prosperous.

socialist realism An artistic style once popular in the Soviet Union that was associated with realistic depictions of workers in their patriotic struggles against capitalism.

socialist state A state in which most key economic sectors are controlled by the government.

Spanglish A hybrid combination of English and Spanish spoken by Hispanic Americans.

Special Economic Zones (SEZs) Relatively small districts in China that were fully opened to global capitalism after China began to reform its economy in the 1980s.

spheres of influence In countries not formally colonized in the 19th and early 20th centuries (particularly China and Iran), limited areas called "spheres of influence" were gained by particular European countries for trade purposes and more generally for economic exploitation and political manipulation.

steppe Semiarid grasslands found in many parts of the world. Steppe grasses are usually shorter and less dense than those found in prairies.

structural adjustment programs Controversial yet widely implemented programs used to reduce government spending, encourage the private sector, and refinance foreign debt. Typically, these IMF and World Bank policies trigger drastic cutbacks in government-supported services and food subsidies, which disproportionately affect the poor.

subduction zone Areas where two tectonic plates are converging or colliding. In these areas, one plate usually sinks below another. They are characterized by earthquakes, volcanoes, and deep oceanic trenches.

subnational organizations Groups that form along ethnic, ideological, or regional lines that can induce serious internal divisions within a state.

Suez Canal Pivotal waterway connecting the Red Sea and the Mediterranean opened, in 1869.

Sunda Shelf An extension of the continental shelf from the Southeast Asia mainland to the large islands of Indonesia. Because of the shelf, the overlying sea is generally shallow (less than 200 feet [61 meters] deep).

Sunni Islam The major denomination of Islam, with its adherents comprising from 75 to 90 percent of the world's Muslims.

Sunnis Muslims who practice the dominant branch of Islam.

superconurbation A massive urban agglomeration that results from the coalescing of two or more formerly separate metropolitan areas.

supranational organization Governing bodies that include several states, such as trade organizations, and often involve a loss of some state powers to achieve organizational goals.

sustainable agriculture A system of agriculture where organic farming principles, a limited use of chemicals, and an integrated plan of crop and livestock management combine to offer both producers and consumers environmentally friendly alternatives.

swidden A form of tropical cultivation in which forested or brushy plots are cleared of vegetation, burned, and then planted in crops, only to be abandoned a few years later as soil fertility declines. Also called slash-and-burn agriculture or shifting cultivation.

syncretic religions The blending of different belief systems. In Latin America, many animist practices were folded into Christian worship.

taiga The vast coniferous forest of Russia that stretches from the Urals to the Pacific Ocean. The main forest species are fir, spruce, and larch.

Taliban A harsh, Islamic fundamentalist political group that ruled most of Afghanistan in the late 1990s. In 2001, the Taliban lost power, but it later regrouped in Pakistan and continues to fight against the Afghan government in southern and eastern Afghanistan.

Tamil Tigers The common name of the rebel forces in Sri Lanka (officially known as the Liberation Tigers of Tamil Eelam, or LTTE) that fought the Sri Lankan army from 1983 until their defeat in 2009.

terrorism The systematic use of terror to achieve political or cultural goals.

theocracy A government run by religious leaders.

theocratic state A political state led by religious authorities.

tonal language A language in which the meaning of a syllable varies in accordance with the tone in which it is uttered.

total fertility rate (TFR) The average number of children who will be borne by women of a hypothetical, yet statistically valid, population, such as that of a specific cultural group or within a particular country. Demographers consider TFR a more reliable indicator of population change than the crude birthrate.

townships Racially segregated neighborhoods created for nonwhite groups under apartheid in South Africa. They are usually found on the outskirts of cities and classified as black, coloured, or South Asian.

Trans-Siberian Railroad Key southern Siberian railroad connection completed during the Russian empire (1904) that links European Russia with the Russian Far East terminus of Vladivostok.

transform fault an earthquake fault where the ground on each side of the fault moves in opposite directions because of tectonic forces.

transhumance A form of pastoralism in which animals are taken to high-altitude pastures during the summer months and returned to low-altitude pastures during the winter.

transmigration The planned, government-sponsored relocation of people from one area to another within a state territory; the term is usually associated with Indonesia.

transnational migration Complex social and economic linkages that form between home and host countries through international migration. Unlike earlier generations of migrants, 21st century immigrants can maintain more enduring and complex ties to their home countries as a result of technological advances.

Treaty of Tordesillas A treaty signed in 1494 between Spain and Portugal that drew a north–south line some 300 leagues west of the Azores and Cape Verde islands. Spain received the land to the west of the line and Portugal the land to the east.

tribalism Allegiance to a particular tribe or ethnic group rather than to the nation-state. Tribalism is often blamed for internal conflict within Sub-Saharan states.

tribes A group of families or clans with a common kinship, language, and definable territory but not an organized state.

tropical savanna A grassland with few trees found in tropical climates with a distinct dry season.

tsar A Russian term (also spelled czar) for "Caesar," or ruler; the authoritarian rulers of the Russian empire before its collapse in the 1917 revolution.

tsetse fly A biting fly common in Africa that transmits sleeping sickness, which can be deadly to humans and *nagana*, which can kill livestock, especially horses, cattle, and camels.

tsunamis Very large sea waves induced by earthquakes.

tundra Arctic region with a short growing season in which vegetation is limited to low shrubs, grasses, and flowering herbs.

typhoons Large tropical storms, synonymous to hurricanes, that form in the western Pacific Ocean in tropical latitudes and cause widespread damage to the Philippines and coastal Southeast and East Asia.

uncontacted peoples Cultures that have yet to be contacted and influenced by the Western world.

Union of South American Nations (UNASUR) An intergovernmental organization proposed in 2008 to improve the economic integration of South American countries.

unitary state A political system in which power is centralized at the national level.

universalizing religions A religion, usually with an active missionary program, that appeals to a large group of people regardless of local culture and conditions. Christianity and Islam both have strong universalizing components.

urban decentralization The process in which cities spread out over a larger geographical area.

urban heat island An effect in built-up areas in which development associated with cities often produces nighttime temperatures some 9 to 14°F (5 to 8°C) warmer than nearby rural areas.

urban primacy The situation found in a country in which a disproportionately large city, such as London, Seoul, or Bangkok, dominates the urban system and is the center of economic, political, and cultural life.

urbanized population That percentage of a country's population living in settlements characterized as cities. Usually, high rates of urbanization are associated with higher levels of industrialization and economic development because these activities are usually found in and around cities. Conversely, lower urbanized populations (less than 50 percent) are characteristic of developing countries.

vernacular region An areal unit of similarity that is neither formal nor functional, but, instead, is more informal and cognitive, thus a region of the mind having imprecise boundaries. Vernacular regions can be thought of as spatial stereotypes used by humans to characterize an area generally. Examples abound: The Village in Manhattan; The Hill in Washington, DC; Watts in Los Angeles; and on and on.

viticulture Grape cultivation, usually for the purpose of making wine.

water stress An environmental planning tool used to predict areas that have—or will have—serious water problems based upon the per capita demand and supply of freshwater.

weather The day-to-day meterological phenomena, such as wind, rain, sun, humidity, on Earth's surface.

White Australia Policy Before 1975, a set of stringent Australian limitations on nonwhite immigration to the country. Largely replaced by a more flexible policy today.

World Trade Organization (WTO) Formed as an outgrowth of the General Agreement on Tariffs and Trade (GATT) in 1995, the WTO is a large collection of member states dedicated to reducing global barriers to trade.

Photo Credits

Chapter 1 Opening Photo David R. Frazier/Danita Delimont Photography/Newscom Chapter 1 Inset 1 Frans Lemmens/Getty Images Chapter 1 Inset 2 Chapter 1 Inset 3 FRILET Patrick/Hemis/Alamy Chapter 1 Inset 4 Danita Delimont/Gallo Images/Getty Images Chapter 1 Inset 5 Gaizka Iroz/AFP/Getty Images Chapter 1 Inset 6 Images & Stories/Alamy Figure 1.1 Purepix/Alamy Figure 1.2 Jon Arnold/Jon Arnold Images Ltd/Alamy Figure 1.1.1 Ton Koene/ZUMAPRESS/Newscom Figure 1.3 Bhandol/Alamy Figure 1.4 Frans Lemmens/Getty Images Figure 1.6 Interfoto/Alamy Figure 1.7Figure 1.8 Jerome Favre/Bloomberg/Getty Images Figure 1.10 Jim West/Alamy Figure 1.11 Andrew Fox/Alamy Figure 1.12 Figure 1.16 Westend61/Getty Images Figure 1.25 Tim Graham/Alamy Figure 1.2.1 Northfork/Alamy Figure 1.26 FRILET Patrick/Hemis/Alamy Figure 1.3.1 Stuart Price/AFP/Getty Images Figure 1.29 BILAWAL ARBAB/EPA/Newscom Figure 1.30a Craig Aurness/Corbis Figure 1.30b Tony Waltham/Robert Harding World Imagery Figure 1.4.1 BILAWAL ARBAB/EPA/Newscom Figure 1.31 YAY Media AS/Alamy Figure 1.32 Jens Benninghofen/Alamy Figure 1.5.1 Peter Stroh/Alamy Figure 1.33 travelbild.com/Alamy Figure 1.35 Jeff Greenberg/Alamy Figure 1.37 ThavornC/Shutterstock Figure 1.38 Danita Delimont/Gallo Images/Getty Images Figure 1.39 PHILIPPE LOPEZ/AFP/Getty Images/Newscom Figure 1.40 Wang Yue/ZUMA Press/Newscom Figure 1.41 Wesley Bocxe/Photo Researchers/Getty Images Figure 1.42 Gaizka Iroz/AFP/Getty Images Figure 1.45 Hubert Boesl/dpa/Corbis Figure 1.47 ROLEX DELA PENA/EPA/Newscom Figure 1.49 MIKE CLARKE/AFP/Getty Images/Newscom Figure 1.52 Terry Whittaker/Alamy Figure 1.53 Images & Stories/Alamy Chapter 1 In Review A Joerg Boethling/Alamy Chapter 1 In Review B Michele Falzone/JAI/Corbis Chapter 1 In Review D Les Rowntree Chapter 1 In Review E EPA/Alamy Chapter 1 In Review F Liba Taylor/Robert Harding Picture Library Ltd/Alamy

Chapter 2 Opening Photo Wicaksono Saputra/Alamy Chapter 2 Inset 1 KYODO Kyodo/ReutersFigure Chapter 2 Inset 2 Ashley Cooper pics/Alamy Chapter 2 Inset 3 David R. Frazier Photolibrary, Inc./Alamy Chapter 2 Inset 4 Ajay Verma/Reuters Chapter 2 Inset 5 Calvin Larsen/Science Source Figure 2.1 NASA Earth Observing System Figure 2.4 NHPA/SuperStock Figure 2.5 KYODO Kyodo/Reuters Figure 2.6 Arctic-Images/SuperStock Figure 2.7 Vacclav/Shutterstock Figure 2.1.2 Blickwinkel/Hummel/Alamy Figure 2.10 Vacclav/Shutterstock Figure 2.17 Tony Cunningham/Alamy Figure 2.19 Ashley Cooper pics/Alamy Figure 2.21 David R. Frazier Photolibrary, Inc./Alamy Figure 2.23 BL Images Ltd/Alamy Figure 2.2.1 Joerg Boethling/Alamy Figure 2.2.2 Roberto Schmidt/AFP/Getty Images/Newscom Figure 2.25 Ajay Verma/Reuters Figure 2.27 Michael Doolittle/Alamy Figure 2.28 louise murray/Alamy Figure 2.29 Fabio Pili/Alamy Figure 2.30 James P. Blair/Getty Images Figure 2.3.1 Mireille Vautier/Alamy Figure 2.31 MICHAEL REYNOLDS/EPA/Newscom Figure 2.32 Les Rowntree Figure 2.33 cappi thompson/Shutterstock Figure 2.34 H. Mark Weidman Photography/Alamy Figure 2.35 Figure 2.36 Julien Menghini/Science Source Chapter 2 In Review A KYODO Kyodo/Reuters Chapter 1 In Review B Todd Shoemake/Shutterstock Chapter 1 In Review C Jim West/Alamy Chapter 1 In Review D Ocean/Corbis Chapter 1 In Review E United States Geological Survey (USGS), National Aeronautics and Space Administration (NASA), Esri Inc.

Chapter 3 Opening Photo William Scott/AGE Fotostock Chapter 3 Inset 1 Caleb Foster/Fotolia Chapter 3 Inset 2 William Wyckoff Chapter 3 Inset 3 Alison Wright/National Geographic Image Collection/Alamy Chapter 3 Inset 4 Ian Shive/Aurora Photos/Alamy Chapter 3 Inset 5 Don Mason/Corbis/Glow Images Figure 3.2 Adrien Veczan/ZUMA Press/Newscom Figure 3.3 MICHAEL REYNOLDS/EPA/Newscom Figure 3.4 US Geological Survey/NASA Figure 3.5 Caleb Foster/Fotolia Figure 3.1.2 Northfork/Alamy Figure 3.8 Kevin Schafer/Documentary Value/Corbis Figure 3.10 Veronique de Viguerie/Getty Images News/Getty Images Figure 3.11a Karen Holzer/USGS Information Services Figure 3.11b Glacier Bay National Park and Preserve Figure 3.13 Songquan Deng/Shutterstock Figure 3.16 Mastering_Microstock/Shutterstock Figure 3.18 iStockphoto/Thinkstock Figure 3.19 William Wyckoff Figure 3.20 Craig Aurness/Corbis/Glow Images Figure 3.22 William Wyckoff Figure 3.2.2 John Mitchell/Alamy Figure 3.2.3 uaridh Stewart/ZUMApress/Newscom Figure 3.2.4 Photoshot/Alamy Figure 3.26a Alison Wright/National Geographic Image Collection/Alamy Figure 3.26b Figure 3.27b J. Emilio Flores/La Opinion/Newscom Figure 3.3.1 Amy Sussman/Corbis Figure 3.3.2 AP Images/Alex Katz Figure 3.30 ZUMA Press, Inc./Alamy Figure 3.4.2 Action Plus Sports Images/Alamy Figure 3.33 Pat and Rosemarie Keough/Corbis Figure 3.34 Ian Shive/Aurora Photos/Alamy Figure 3.36 Richard Hamilton Smith/Corbis Figure 3.37 Inga Spence/Photolibrary/Getty Images Figure 3.38 Jim Noelker/Age Fotostock Figure 3.39 Katja Kreder/imagebroker/Alamy Figure 3.40 Camilla Zenz/ZUMAPRESS/Newscom Figure 3.42 David South/Alamy Figure 3.43 Don Mason/Corbis/Glow Images Figure 3.5.2 ANGELA PETERSON/MCT/Landov Chapter 3 In Review A Chapter 3 In Review B Radius Images/Corbis Chapter 3 In Review C iStockphoto/Thinkstock Chapter 3 In Review E Frontpage/Shutterstock

Chapter 4 Opening Photo Dado Galdieri/Bloomberg via Getty Images Chapter 4 Inset 1 Rob Crandall Chapter 4 Inset 2 Brazil Photo Press/LatinContent/Getty Images Chapter 4 Inset 3 Rob Crandall Chapter 4 Inset 4 AFP/Getty Images Chapter 4 Inset 5 Bob Daemmrich/Corbis Figure 4.2 David Santiago Garcia/Aurora Photos/Alamy Figure 4.3 Ian Trower/Robert Harding/Newscom Figure 4.4 Danny Lehman/Corbis Figure 4.5 Rob Crandall Figure 4.7 EVARISTO SA/AFP/Getty Images/Newscom Figure 4.9 Rob Crandall Figure 4.11a Bernard Francou, IRD Figure 4.11b Bernard Francou, IRD Figure 4.13a NASA Figure 4.13b NASA Figure 4.14 Rob Crandall Figure Rob Crandall Figure 4.1.1 Fernando Vergara/AP Images Figure 4.1.2 Javier Galeano/AP Images Figure 4.16 Martin Bernetti/AFP/Getty Images/Newscom Figure 4.18a Rob Crandall Figure 4.18b Rob Crandall Figure 4.18c Rob Crandall Figure 4.18d Rob Crandall Figure 4.19 Brazil Photo Press/LatinContent/Getty Images Figure 4.2.1 Santiago Urquijo/Flickr/Getty Images Figure 4.2.2 Martin Alipaz/EFE/Newscom Figure 4.20 ENRIQUE CASTRO-MENDIVIL/Reuters/Landov Figure 4.23 Orlando Kissner/AFP/Getty Images Figure 4.24 Rob Crandall Figure 4.25 Rob Crandall Figure 4.3.2 AFP/Getty Images Figure 4.27 Alexandro Auler/STR/LatinContent WO/Getty Images Figure 4.28 CARLOS EDUARDO CARDOSO/CONTE/Newscom Figure 4.29 Christian Heeb/Aurora Photos, Inc. Figure 4.31 Rob Crandall Figure 4.4.1 Rob Crandall Figure 4.33 Felipe Trueba/EPA/Newscom Figure 4.35 Bob Daemmrich/Corbis Figure 4.36 Rob Crandall Figure 4.37 Paulo Fridman/Corbis Figure 4.5.2 Kaveh Kazemi/Getty Images Figure 4.38 Rob Crandall Figure 4.41 Carlos Jasso/Reuters Figure 4.42 Rob Crandall Chapter 4 In Review A iStockphoto/Thinkstock Chapter 4 In Review B luoman/E/Getty Images Chapter 4 In Review D US Geological Survey and NASA Chapter 4 In Review E Frontpage/Shutterstock

Chapter 5 Opening Photo Bildagentur-online/Schickert/Alamy Chapter 5 Inset 1 Grand Tour/Corbis Chapter 5 Inset 2 Rob Crandall Chapter 5 Inset 3 Rob Crandall Chapter 5 Inset 4 Laif/Frank Heuer/Redux Pictures Chapter 5 Inset 5 Photoshot/Newscom Figure 5.2 Rob Crandall Figure 5.3 Grand Tour/Corbis Figure 5.4 NASA Figure 5.5 Ho News/Stephane Corvaja/ESA/Reuters Figure 5.7 Alejandro Ernesto/EPA/Newscom Figure 5.11 Rob Crandall Figure 5.12 Rob Crandall Figure 5.16 Bettmann/Corbis Figure 5.17 ORLANDO BARRIA/EPA/Newscom Figure 5.2.1 Javier Galeano/AP Images Figure 5.2.2 Figure 5.18 Rob Crandall Figure 5.20 RANU ABHELAKH/Reuters/Corbis Figure 5.22Figure 5.24 Rob Crandall Figure 5.25 Rob Crandall Figure 5.3.1 Rob Crandall Figure 5.3.2 Rob Crandall Figure 5.4.1 Robin Utrecht/EPA/Alamy Figure 5.29 Rob Crandall Figure 5.31 Laif/Frank Heuer/Redux Pictures Figure 5.33 Photoshot/Newscom Figure 5.5.1 Huntstock/Disability Images/Alamy Chapter 5 In Review B

Chapter 6 Opening Photo Jake Lyell/Alamy Chapter 6 Inset 1 Chapter 6 Inset 2 Chapter 6 Inset 3 Chapter 6 Inset 4 Chapter 6 Inset 5 Figure 6.2 Jake Lyell/Alamy Figure 6.3 Rob Crandall Figure 6.4 AfriPics.com/Alamy Figure 6.5 Robert Caputo/Getty Images Figure 6.6 Marion Kaplan/Alamy Figure 6.8 Rob Crandall Figure 6.9 Chris Howes/Wild Places Photography/Alamy Figure 6.11 Daniel Berehulak/Getty Images Figure 6.1.1 Bruno D'Amicis/Nature Picture Library Figure 6.1.2 Mike Goldwater/Alamy Figure 6.1.3 Akintunde Akinleye/Reuters Figure 6.1.4 Rob Crandall Figure 6.17 Jake Lyell/Alamy Figure 6.19 RODGER BOSCH/AFP/Getty Images/Newscom Figure 6.21 Jon Hrusa/epa/Corbis Figure 6.22 Horizons WWP/Alamy Figure 6.2.1 John Warburton-Lee Photography/Alamy Figure 6.2.2 Jeremy Graham/dbimages/Alamy Figure 6.23 James Marshall/Corbis Figure 6.24 Max Milligan/AWL Images/Getty Images Figure 6.26 THEGIFT777/E+/Getty Images Figure 6.29 Rob Crandall Figure 6.30 Neil Cooper/Alamy Figure 6.32 Gavin Hellier/Alamy Figure 6.34 Amar Grover/AWL Images/Getty Images Figure 6.3.1 SHASHANK BENGALI/MCT/Newscom Figure 6.3.2 face to face/ZUMA Press/Newscom Figure 6.35 Streeter Lecka/Getty Images Figure 6.38 Ulrich Doering/Alamy Figure 6.4.1 SATourism/Greatstock Photographic Library/Alamy Figure 6.4.2 ALEXANDER JOE/AFP/Getty Images Figure 6.41 Mohamed Abiwahab/AFP/Getty Images/Newscom Figure 6.42 Simon Maina/AFP/Getty Images Figure 6.43 G P Bowater/Alamy Figure 6.44 Ahmed Jallanzo/Africa Media Online/Alamy Figure 6.5.1 Str/Epa/Newscom Figure 6.5.2 Stuart Price/AFP/Getty Images Figure 6.46 Siphiwe Sibeko/Reuters Figure 6.47 Heiner Heine/Alamy Chapter 6 In Review E Ines Gesell/Getty Images

Chapter 7 Opening Photo Duby Tal/Albatross/AGE Fotostock Chapter 7 Inset 1 De Agostini/Getty Images Chapter 7 Inset 2 Gavin Hellier/Robert Harding World Imagery Chapter 7 Inset 3 imago stock&people/Newscom Chapter 7 Inset 4 Shaam News Network/Handout/Reuters Chapter 7 Inset 5 Murad Sezer/Reuters Figure 7.2 Rex Features/AP images Figure 7.3 Jon Arnold Images Ltd/Alamy Figure 7.4 NASA Figure 7.5 Protasov AN/Shutterstock Figure 7.7 De Agostini/Getty Images Figure 7.8 Galyna Andrushko/Shutterstock Figure 7.9 Egmont Strigl/IMAGEBROKER/AGE Fotostock Figure 7.12 DAVID BUIMOVITCH/AFP/Getty Images/Newscom Figure 7.13 Worldspec/NASA/Alamy Figure 7.14 Patrick Syder/AGE Fotostock Figure 7.16 NASA Figure 7.17 Izzet Keribar/Lonely Planet Images/Getty Images Figure 7.19 Wigbert R√∂th Image Broker/Newscom Figure 7.20 Ben Curtis/AP Images Figure 7.1.1 Dereje Belachew/Alamy Figure 7.1.2 wael hamdan/Alamy Figure 7.1.3 Megapress/Alamy Figure 7.21 Jalil Bounhar/AP Images Figure 7.22 Marcia Chambers/dbimages/Alamy Figure 7.24 DarkGrey/Flickr Open/Getty Images Figure 7.25a Gavin Hellier/Robert Harding World Imagery Figure 7.25b Rob Crandall Figure 7.2.1 Abaca/Newscom Figure 7.2.2 Masdar-HO/AP Images Figure 7.29 imago stock&people/Newscom Figure 7.33 Chris Hondros/Staff/Getty Images Figure 7.34 Fadi Al-Assaad/Reuters Figure 7.36 Thierry GRUN/Alamy Figure 7.39 DEBBIE HILL/UPI/Newscom Figure 7.40 Xinhua/Newscom Figure 7.41 Shaam News Network/Handout/Reuters Figure 7.3.2 PixelPro/Alamy Figure 7.4.2 dpa/Mehmet Biber/Landov Figure 7.5.1 Dominique VERNIER/Fotolia Figure 7.5.2 Adam Reynolds/Bloomberg/Getty Images Figure 7.45 Murad Sezer/Reuters Figure 7.46 Louafi Larb/Reuters Figure 7.47 PATSTOCK/AGE Fotostock Chapter 7 In Review A Lukasz Janyst/Shutterstock Chapter 7 In Review C johnnydao/Shutterstock Chapter 7 In Review D Barry Gregg/Getty Images

Chapter 8 Opening Photo Katja Kreder/JAI/Corbis **Chapter 8 Inset 1** Les Rowntree **Chapter 8 Inset 2** Kacper Pempel/Reuters **Chapter 8 Inset 3** Ponizak/Caro/Alamy **Chapter 8 Inset 4** Peter Jordan/Alamy **Chapter 8 Inset 5** Nikolas Georgiou/Alamy **Figure 8.2** Sean Gallup/Getty Images **Figure 8.4** FRUMM John/Hemis/Alamy **Figure 8.5** Michele Boiero/Shutterstock **Figure 8.7** smart.art/Shutterstock **Figure 8.9** Magnus Qodarion/Alamy **Figure 8.10** SARAH LEEN/National Geographic Stock **Figure 8.1.1** Les Rowntree **Figure 8.13** Kacper Pempel/Reuters **Figure 8.2.1** Rob Few/Alamy **Figure 8.15** Les Rowntree **Figure 8.16** Les Rowntree **Figure 8.3.1** Les Rowntree **Figure 8.3.2** oern Haufe/Getty Images **Figure 8.17** Russell Gordon/Danita Delimont, Agent/Danita Delimont/Alamy **Figure 8.19** Peter Forsberg/EU/Alamy **Figure 8.21** Les Rowntree **Figure 8.4.1** Erika Rowntree **Figure 8.4.2** Magdalena Cooper **Figure 8.22** Ponizak/Caro/Alamy **Figure 8.23** Jonathan Larsen/Diadem Images/Alamy **Figure 8.24** kerkla/E+/Getty Images **Figure 8.27** Peter Jordan/Alamy **Figure 8.28a** Bettmann/Corbis **Figure 8.28b** picture-alliance/dpa/Newscom **Figure 8.30** FEHIM DEMIR/EPA/Newscom **Figure 8.31** Ann Pickford/Alamy **Figure 8.34** David Henderson/Photolibrary/Getty Images **Figure 8.35** Velko Angelov/Bloomberg/Getty Images **Figure 8.37** Nikolas Georgiou/Alamy **Chapter 8 In Review A** US Geoloagical Survey/NASA **Chapter 8 In Review C** Attila JANDI/Shutterstock **Chapter 8 In Review D** bernd.neeser/Shutterstock **Chapter 8 In Review E** Ingram Publishing/Thinkstock

Chapter 9 Opening Photo Konstantin Mikhailov/ZUMApress/Newscom **Chapter 9 Inset 1** PhotoXpress/ZUMAPRESS/Newscom Photo **Chapter 9 Inset 2** Cindy Miller Hopkins/Danita Delimont/Alamy Photo **Chapter 9 Inset 3** Jochem D Wijnands/Taxi/Getty Images Photo **Chapter 9 Inset 4** Sergei Chirikov/epa/Corbis Photo **Chapter 9 Inset 5** ITAR-TASS Photo Agency/Alamy **Figure 9.1** NATALIA KOLESNIKOVA/AFP/Getty Images/Newscom **Figure 9.3** Preusser Volker/WoodyStock/Alamy **Figure 9.6** Serguei Fomine/ZUMAPRESS/Newscom **Figure 9.7** Sergbob/Fololia **Figure 9.8** Masami Goto/Glow Images **Figure 9.1.1** Yekaterina Pustynnikova/AP Images **Figure 9.2.1** Krasilnikov Stanislav/ZUMA Press/Newscom **Figure 9.9** NASA **Figure 9.10** Irakli Gedenidze/AP Images **Figure 9.12** PhotoXpress/ZUMAPRESS/Newscom **Figure 9.2.1** Yul/Fotolia Figure 9.2.2 ITAR-TASS Photo Agency/Alamy **Figure 9.13** ITAR-TASS/Lev Fedoceyev/Newscom **Figure 9.16** Gavin Hellier/Jon Arnold Images Ltd/Alamy **Figure 9.17** Sergei Karpukhin/Reuters **Figure 9.3.1** WYSOCKI Pawel/hemis.fr/Alamy **Figure 9.3.3** Gavin Hellier/Jon Arnold Images Ltd/Alamy **Figure 9.19 Iain Masterton/Alamy Figure 9.4.1** Mara Zemgaliete/Fotolia **Figure 9.4.2** Misha Japaridze/AP Images **Figure 9.20 Cindy** Miller Hopkins/Danita Delimont/Alamy **Figure 9.21** ITAR-TASS Photo Agency/Alamy **Figure 9.24** Delaroche, Hippolyte (Paul) (1797-1856)/The Art Gallery Collection/Alamy **Figure 9.27** Gerner Thomsen/Alamy **Figure 9.29** Jochem D Wijnands/Taxi/Getty Images **Figure 9.30** Sergey Dubrovskiy/Alamy **Figure 9.31** Sergie Ilnitsky/Corbis **Figure 9.32** Artur Lebedev/Newscom **Figure 9.33** Sergei Chirikov/epa/Corbis **Figure 9.36** Uzakov Sergei/Corbis **Figure 9.37** AP Images/Sergey Ponomarev **Figure 9.38** Tom Schulze/Sovfoto **Figure 9.40** ITAR-TASS Photo Agency/Alamy **Figure 9.42** SERGEY DOLZHENKO/EPA/Newscom **Figure 9.5.1** Gene Blevins/LA DailyNews/Corbis **Figure 9.5.2** Astapkovich Vladimir Itar-Tass Photos/Newscom **Figure 9.44** ITAR-TASS Photo Agency/Alamy **Chapter 9 In Review A** Svetlana Bobrova/Shutterstock **Chapter 9 In Review C** ITAR-TASS Photo Agency/Alamy **Chapter 9 In Review D** ITAR-TASS Photo Agency/Alamy

Chapter 10 Opening Photo Pavel Parmenov/Fotolia **Chapter 10 Inset 1 Chapter 10 Inset 2 Chapter 10 Inset 3 Chapter 10 Inset 4 Chapter 10 Inset 5 Chapter 10** Bbbar/Fotolia **Figure 10.4** Deidre Sorensen/Photoshot/Newscom **Figure 10.6a** NASA/AFP/Newscom **Figure 10.6b** NASA/AFP/Newscom **Figure 10.6c** Japan Aerospace Exploration Agency **Figure 10.7** Ted Wood/Aurora Photos/Alamy **Figure 10.1.1** Nurlan Kalchinov/Alamy **Figure 10.9** TAO Images Limited/Getty Images **Figure 10.12** Ted Wood/Aurora Photos/Alamy **Figure 10.2.1** Zoonar GmbH/Alamy **Figure 10.14** GM Photo Images/Alamy **Figure 10.3.1** SYED JAN SABAWOON/EPA/Newscom **Figure 10.18** Tony Gervis/iStock/Getty Images **Figure 10.19** J Marshall/Tribaleye Images/Alamy **Figure 10.20** Ryumin Alexander Itar-Tass Photos/Newscom **Figure 10.21** Thomas L. Kelly/Alamy **Figure 10.24** Zarip Toroyev/AP Images **Figure 10.25** AP Images/Ng Han Guan **Figure 10.26** AP Images/Scott Nelson **Figure 10.27** NASA/ZUMAPRESS/Newscom **Figure 10.4.1** Michael Coyne/Glow Images **Figure 10.29** David Turnley/Corbis **Figure 10.30** Theodore Kaye/Alamy **Figure 10.31** Wang wei bj/AP Images **Figure 10.5.1** Ton Koene/ZUMAPRESS/Newscom **Figure 10.32 Figure 10.33** John Costello/MCT/Newscom **Chapter 10 In Review A** Cai Chuqing/Shutterstock **Chapter 10 In Review B Chapter 10 In Review C** The larger of Bamyan's 2 most famous Buddha alcoves following their destruction by the Taliban **Chapter 10 In Review D** Maximilian_Clarke/iStock/Getty Images **Chapter 10 In Review E** Michal Cerny/Alamy

Chapter 11 Opening Chapter Im ChanKyung/Topic Photo Agency/Corbis **Chapter 11 Inset 1** Photo Marco Brivio/Getty Images **Chapter 11 Inset 2** BL Images Ltd/Alamy **Chapter 11 Inset 3** Rob Crandall **Chapter 11 Inset 4** Yonhap/AP Images **Chapter 11 Inset 5** imago stock&people/Newscom **Figure 11.4** George F. Mobley/National Geographic Stock **Figure 11.5** AFP/Getty Images **Figure 11.6** Corbis **Figure 11.9** Bob Sacha/Documentary/Corbis **Figure 11.10** Michael Reynolds/Corbis **Figure 11.11** Marco Brivio/Getty Images **Figure 11.1.1** KAZUHIRO NOGI/AFP/GettyImages **Figure 11.13** Zhang Ling/ZUMAPRESS/Newscom **Figure 11.14a** Panorama Images/The Image Works **Figure 11.14b** Chinafotopress/ZUMApress/Newscom **Figure 11.15** China Daily China Daily Information Corp - CDIC/Reuters **Figure 11.16** Color China Photo/AP Images **Figure 11.17** Jim Richardson/Getty Images **Figure 11.21** Skye Hohmann Japan **Figure 11.22** Liu Xiaoyang/China Images/Alamy **Figure 11.23** Everett Kennedy Brown/epa/Corbis **Figure 11.24** BL Images Ltd/Alamy **Figure 11.25 BL** Images Ltd/Alamy **Figure 11.27** B.S.P.I./Documentary/Corbis **Figure 11.28** Iain Masterton/Alamy **Figure 11.29** Rob Crandall **Figure 11.30** RedChopsticks/Getty Images **Figure 11.31** Kyodo News/AP Images **Figure 11.33** Masa Uemura/Alamy **Figure 11.34** Olaf Schubert/imagebroker/Alamy **Figure 11.36** ZUMA Press, Inc./Alamy **Figure 11.38** Universal Images Group/SuperStock **Figure 11.40** Yonhap/AP Images Figure 11.41 Rob Crandall **Figure 11.3.1** SIPHIWE SIBEKO/Reuters/Landov **Figure 11.42** Kyodo/Newscom **Figure 11.43** imago stock&people/Newscom **Figure 11.4.1** Imaginechina/Corbis **Figure 11.44** Digital Globe/ZUMAPRESS/Newscom **Figure 11.5.1** Nayan Sthankiya/Corbis **Figure 11.45** Yuan shuiling/Imaginechina/AP Images **Figure 11.47** Lintao Zhang/Getty Images **Chapter 11 In Review B** ADRIAN BRADSHAW/EPA/Newscom **Chapter 11 In Review C** SeanPavonePhoto/Fotolia **Chapter 11 In Review D** Patrick Lin/AFP/Getty Images/Newscom **Chapter 11 In Review E** Chinafotopress/Ma Jian/ZUMApress/Newscom

Chapter 12 Opening Chapter Lynsey Addario/VII/Corbis **Chapter 12 Inset 1** Mufty Munir/EPA/Newscom **Chapter 12 Inset 2** Rob Crandall **Chapter 12 Inset 3** India Today Group/Getty Imag **Chapter 12 Inset 4** Danish Ismail FK/TW/Reuters **Chapter 12 Inset 5** Philippe Lissac/GODONG/picture-alliance/Godong/Newscom **Figure 12.3** SAM PANTHAKY/AFP/Getty Images **Figure 12.1.1** STEVENS FREDERIC/SIPA/Newscom **Figure 12.5** Mufty Munir/EPA/Newscom **Figure 12.6** Vivek Menon/Natural Picture Library **Figure 12.7** Aditya "Dicky" Singh/Alamy **Figure 12.10** AFP/Getty Images **Figure 12.15** Juergen Poehlitz/Glow Images **Figure 12.16** Earl & Nazima Kowall/Documentary Value/Corbis **Figure 12.17** Rob Crandall **Figure 12.18** dbtravel/Alamy **Figure 12.19** AAMIR QURESHI/AFP/GETTY IMAGES **Figure 12.20** NOAH SEELAM/AFP/Getty Images/Newscom **Figure 12.22** India Today Group/Getty Image **Figure 12.23** Money Sharma/EPA/Newscom **Figure 12.24** Margaret Bourke-White/Time & Life Pictures/Getty Images **Figure 12.26** Indranil Mukher Jee/AFP/Getty Images/Newscom **Figure 12.28** Olaf Kruger Image Broker/Newscom **Figure 12.3.1** AFP Photo/Indranil Mukherjee/Newscom **Figure 12.32** Rahul Talukder/Ap Images **Figure 12.4.1** STR/EPA/Newscom **Figure 12.34** Anupam Nath/Ap Images **Figure 12.36** Danish Ismail FK/TW/Reuters **Figure 12.37** Deshakalyan Chowdhur/AFP/Newscom **Figure 12.39** Hemis/Alamy **Figure 12.40** Philippe Lissac/GODONG/picture-alliance/Godong/Newscom **Figure 12.5.1** SOBERKA Richard/hemis.fr/Hemis/Alamy **Figure 12.41** John Henry Claude Wilson/Glow Images **Figure 12.43** Rob Crandall **Chapter 12 In Review A** Ocean/Corbis **Chapter 12 In Review B** Paul Biris/Flickr/Getty Images **Chapter 12 In Review D** REHAN KHAN/EPA/Alamy **Chapter 12 In Review E** HARISH TYAGI/EPA/Newscom

Chapter 13 Opening Chapter Francois Marclay/Getty Images **Chapter 13 Inset 1** Tuul/Robert Harding/Newscom **Chapter 13 Inset 2** Image Source/SuperStock **Chapter 13 Inset 3** Corbis Wire/epa/Corbis **Chapter 13 Inset 4** FRANCIS R. MALASIG/EPA/Newscom **Chapter 13 Inset 5** Jonathan Drake/Bloomberg/Getty Images **Figure 13.2** Julia Rogers/Alamy **Figure 13.4** Tuul/Robert Harding/Newscom **Figure 13.5** Kevin Frayer/Getty Images **Figure 13.7** Luca Tettoni/Corbis **Figure 13.8** Michael S. Yamashita/Terra/Corbis **Figure 13.1.1** Jochen Schlenker/Robert Harding World Imagery/Alamy **Figure 13.9** Darren Whiteside/Reuters **Figure 13.11 Figure 13.12** G P Bowater/Alamy **Figure 13.14** ED WRAY/AP Images **Figure 13.17** Image Source/SuperStock **Figure 13.19** Konstantin Kalishko/Zoonar GmbH/Alamy **Figure 13.2.1** joyfull/Shutterstock **Figure 13.21** CHRISTOPHE ARCHAMBAULT/AFP/Getty Images/Newscom **Figure 13.22** Corbis Wire/epa/Corbis **Figure 13.24** SuperStock/Getty Images **Figure 13.25** JEFF HAYNES/Reuters/Landov **Figure 13.3.1** Sarah L. Voisin/Washington Post/Getty Images **Figure 13.26** Vivek Prakash/ZUMA Press/Newscom **Figure 13.29** MPI/Getty Images **Figure 13.30** Tetra Images/Alamy Figure 13.32 Thierry Falise/LightRocket/Getty Images **Figure 13.4.1** ZUMA Press, Inc/Alamy **Figure 13.33** FRANCIS R. MALASIG/EPA/Newscom **Figure 13.34 PHC** Larry Foster/National Archives **Figure 13.35** ROSLAN RAHMAN/AFP/Getty Images/Newscom **Figure 13.36** Jonathan Drake/Bloomberg/Getty Images **Figure 13.37** STEPHEN SHAVER/AFP/Getty Images **Figure 13.39** Reuters/Corbis **Figure 13.5.1 Figure 13.41** Robin Laurance/LOOK Die Bildagentur der Fotografen GmbH/Alamy **Chapter 13 In Review C** AP Images/Wong Maye-E **Chapter 13 In Review D** Ezra Acayan/NurPhoto/Sipa USA/Newscom **Chapter 13 In Review E** Fan shaoguang/ICHPL Imaginechina/AP Images

Chapter 14 Opening Chapter Chris Schmid/Corbis **Chapter 14 Inset 1** Chad Ehlers/Photographer's Choice/Getty Images **Chapter 14 Inset 2** Lan Waldie/redbrickstock.com/Alamy**Chapter 14 Inset 3** Emily Riddell/AGE Fotostock **Chapter 14 Inset 4** Lucy Pemoni GAC/Reuters **Chapter 14 Inset 5** Torsten Blackwood/AFP/Getty Images **Figure 14.2** Paul Mayall Australia/Alamy **Figure 14.3** Radius Images/Corbis TIPS Images/Water Rights/Newscom **Figure 14.4** TIPS Images/Water Rights/Newscom **Figure 14.5** Maxine House/Alamy **Figure 14.7** Glenn Nicholls/AP images **Figure 14.8** Bon Appetit/Alamy **Figure 14.9** Chad Ehlers/Photographer's Choice/Getty Images **Figure 14.11** TOMO/Fotolia **Figure 14.12** Michael & Patricia Fogden/Getty Images **Figure 14.14** Torsten black wood/afp/Newscom **Figure 14.15** Friedrich Stark/Alamy **Figure 14.16** Jiang Xianming/ZUMA Press/Newscom **Figure 14.1.1** Torsten Blackwood/AFP/Getty Images/Newscom **Figure 14.1.2** Kyodo via AP Images **Figure 14.17** Michele and Tom Grimm/Alamy **Figure 14.18** Laszlo Podor/Alamy **Figure 14.20** Pictures Colour Library/Travel Pictures/Alamy **Figure 14.2.2** Andrew Watson/Getty Images **Figure 14.22** Douglas Peebles Photography/Alamy **Figure 14.23** Bill Bachmann/Science Source **Figure 14.24** Lan Waldie/redbrickstock.com/Alamy**Figure 14.25** Friedrich Stark/Alamy **Figure 14.26** Paul Kingsley/Paul Kingsley/Alamy **Figure 14.27** Mark A. Johnson/Alamy **Figure 14.29** Allan Symon Gerry/Alamy **Figure 14.30** Marianna Day Massey/ZUMA Press/Newscom **Figure 14.32** Emily Riddell/AGE Fotostock **Figure 14.3.1** Rick Rycroft/AP Images **Figure 14.3.2** YONGYOT PRUKSARAK/EPA/Newscom **Figure 14.34** Greg Vaughn/Alamy **Figure 14.34** Greg Vaughn/Alamy **Figure 14.35** AFP/Getty Images **Figure 14.4.1** Lucy Pemoni/AP Images **Figure 14.36** Phil Walter/Getty Images **Figure 14.37** Damon Tarver/Cal Sport Media/Newscom **Figure 14.38** Eric Nathan/Alamy **Figure 14.5.1 Figure 14.5.2 Figure 14.40** Andre Seale/Andre Seale/Alamy **Figure 14.41** Glenn Campbell/The Sydney Morning Herald/Getty images **Figure 14.43** Lucy Pemoni GAC/Reuters **Figure 14.44** Glenn Campbell/The Sydney Morning Herald/Fairfax Media/Getty Images **Figure 14.45** Torsten Blackwood/AFP/Getty Images **Figure 14.46** Brisbane Architectual and Landscape Photographer/Getty Images **Figure 14.47** Photononstop/SuperStock **Figure 14.48** Fred Krger/picture-alliance/dumont Bildar/Newscom **Figure 14.49** SuperStock **Chapter 14 In Review A Chapter 14 In Review B Chapter 14 In Review C Chapter 14 In Review D** Stocktrek Images/Getyy Images **Chapter 14 In Review E**

Text and Illustration Credits

Chapter 1 Text 1.1 page 10 PBS NewsHour, http://www.pbs.org/newshour/bb/business/july-dec05/borders_12-15.html; and http://www.bricklin.com/albums/fpawlf2000/friedman.htm **Table 1.1 page 22** Population Reference Bureau, World Population Data Sheet, 2012 **Table 1.2 page 44** World Bank, World Development Indicators, 2013 **Figure 1.51 page 46** Used by permission of the Center for International Earth Science Information Network (CIESIN), Columbia University, 2006. Where the Poor Are: An Atlas of Poverty (Palisades, NY: Columbia University). Available at: http://sedac.ciesin.columbia.edu/data/collection/povmap/povatlas **Text 1.2 page 23** UN World Commission on Environment and Development

Chapter 2 Figure 2.3 page 53 Data from McKnight and Hess, 2005, Physical Geography: A Landscape Appreciation, 8th ed., Upper Saddle River, NJ: Prentice Hall **Figure 2.26 page 72** Based on Clawson and Fisher, 2004, World Regional Geography, 8th ed., Upper Saddle River, NJ: Prentice Hall **Table 2.1 page 67** BP Statistical Review of World Energy 2013, BP p.l.c. Used by permission.

Chapter 3 Table 3.1 page 95 Population Reference Bureau, World Population Data Sheet, 2012; net migration rate data from United Nations, Population Division, World Population Prospects: The 2008 Revision Population Database. **Figure 3.25 page 107** Modified from Jordan, Domosh, and Rowntree, 1998, The HumanMosaic, Upper Saddle River, NJ:Prentice Hall **Figure 3.27a page 111** RUBENSTEIN, JAMES M., THE CULTURAL LANDSCAPE: AN INTRODUCATION TO HUMAN GEOGRAPHY, 9th Ed., ©2008. Reprinted and Electronically reproduced by permission of Pearson Educationi, Inc. Upper Saddle River, New Jersey. **Figure 3.28 page 113** Adapted from Jerome Fellman, Arthur Getis, and Judith Getis, Human Geography (Dubuque, Iowa: Brown and Benchmark, 1997), p.164. Used by permission of The McGraw-Hill Companies, Inc. **Text 3.1 page 116** President John F. Kennedy summarized the links in a speech to the Canadian parliament in 1962. **Table 3.2 page 123** World Bank, World Development Indicators, 2012; human development index data from United Nations, Human Development Report, 2009; life expectancy data from Population Reference Bureau, World Population Data Sheet, 2012 **Figure 3.35 page 123** Modified from Clawson and Fisher, 2004,*World Regional Geography, 8th ed.*, Upper Saddle River, NJ: Prentice Hall, and Howard Veregin, ed., 2010, *Goode's World Atlas, 22nd ed.*, Upper Saddle River, NJ: Prentice Hall

Chapter 4 Text 4.1 page 187 Calderón **Figure 4.21 page 161** Based on Clawson, 2000, Latin America and the Caribbean: Lands and People, 2nd ed., Boston: McGraw-Hill **Figure 4.43 page 189** World Bank Development Indicators, 2013. Used by permission.

Chapter 5 Box 5.1 page 198 Adapted from URL www.newswatch .nationalgeographic.com-How Crisis Mapping Saved Lives in Haiti, July 2, 2012 **Figure 5.6 page 200** Copyright © by Pearson Education, Upper Saddle River, NJ **Figure 5.8 page 202** Based on Aguado and Burt, 2004, Understanding Weather and Climate, 3rd ed., Upper Saddle River, NJ: Prentice Hall **Figure 5.9** Based on DK World Atlas, London: DK Publishing, 1997, pp. 7, 55 **Table 5.1 page 207** Population Reference Bureau, World Population Data Sheet, 2010;net migration rate data from United Nations, Population Division, World Population Prospects: The 2008 Revision Population Database; additional data from CIA World Factbook. 2010. **Figure 5.15 page 209** Copyright © by Pearson Education, Upper Saddle River, NJ **Figure 5.19 page 214** Data based on Philip Curtin, The Atlantic Slave Trade, A Census, Madison: University of Wisconsin Press, 1969, p. 268 **Figure 5.21 page 216** Robert Voeks, "African Medicine and Magic in the Americas," Geographical Review 83, no. 1, (1993): 66–78. Used by permission of the American Geographical Society. **Figure 5.26 page 221** Data from Bonham C. Richardson, The Caribbean in the Wider World, 1492–1992, Cambridge, Cambridge University Press, 1992, p. 56, reprinted with the permission of Cambridge University Press; and David P. Henige, Colonial Governors from the Fifteenth Century to the Present, Madison: University of Wisconsin Press, 1970 **Figure 5.27 page 223** Data from Barbara Tenenbaum, ed., Encyclopedia of Latin American History and Culture, 1996, vol. 5, p. 296, with permission of Charles Scribner's Sons; and John Allcock, Border and Territorial Disputes, 3rd ed., Harlow, Essex, UK: Longman Group, 1992 **Figure 5.28 page 225** N. Nerurkar and M.P. Sullivan. 2011. "Cuba's Offshore Oil Development: Background and U.S.Policy Considerations" Congressional Research Service, Nov. 28, 2011 **Box 5.4 page 226** Adapted from Mies van Niekerk, "Second Generation Caribbeans in the Netherlands: Different Migration Histories, Diverging Trajectories," Journal of Ethnic and Migration Studies 33, no. 7 (2007): 1063–1081; and Simon Romero, "With Aid and Migrants, China Expands Its Presence in a South American Nation," New York Times, April 10, 2011. **Table 5.2 page 228** World Bank, World Development Indicators, 2010; human development index data from United Nations, Human Development Index, 2009; life expectancy data from Population Reference Bureau, World Population Data Sheet, 2010; additional data from CIA World Factbook, 2010. **Figure 5.34 page 233** United Nations, Human Development Report, 2009 **Figure 5.5.2 page 235** Migration Policy Institute analysis of American Community Survey, 2010. Reproduced with permission of Migration Policy Institute in the format Republish in a book via Copyright Clearance Center

Chapter 6 Box 6.1 page 252 Based on Tina Rosenberg's "In Africa's Vanishing Forests, the Benefits of Bamboo" New York Times, March 13, 2012. **Text 6.2 page 253** Michael Watts **Box 6.5 page 289** Based on Africa in their Words: A Study of Chinese Traders in South Africa, Lesotho, Botswana, Zambia and Angola (2012) by Terence McNamee. Discussion Paper 2012/2013 for the Brenthurst Foundation, South Africa **Figure 6.15page 254** Famine Early Warning System (FEWS) Network, April to June 2013 Report, U.S. Agency for International Development **Figure 6.18 page 258** Data from the Population Reference Bureau, World Population Data Sheet, 2012 **Figure 6.28 page 267** From Newman, 1995, The Peopling of Africa, New Haven: Yale University Press, 141. Used by permission of Yale University Press. **Figure 6.31 page 269** From Phillips, 1984, The African Political Dictionary, Santa Barbara, CA: Clio Press Ltd., 196. **Figure 6.40 page 280** Data from UNHCR 2012 **Table 6.1 page 255** Population Reference Bureau, World Population Data Sheet, 2010 **Table 6.2 page 284** World Bank, World Development Indicators, 2010

Chapter 7 Figure 7.11 page 306 Modified from Soffer, Arnon, Rivers of Fire: The Conflict over Water in the Middle East, 1999, p. 180. Reprinted by permission of Rowman and Littlefield Publishers, Inc. **Figure 7.18 page 312** Modified from Clawson and Fisher, 2004, World Regional Geography,8th ed., Upper Saddle River, NJ: Prentice Hall, and Bergman and Renwick, 1999, Introduction to Geography, Upper Saddle River, NJ: Prentice Hall **Table 7.1 page 310** Population Reference Bureau, World Population Data Sheet, 2010. **Figure 7.28 page 322** Modified from Rubenstein, 2005, An Introduction to Human Geography, 8th ed., Upper Saddle River, NJ: Prentice Hall **Figure 7.30 page 323** Modified from Rubenstein, 2011, An Introduction to Human Geography, 10th ed., Upper Saddle River, NJ: Prentice Hall, and National Geographic Society, 2003, Atlas of the Middle East, Washington, DC **Figure 7.31 page 324** Reprinted from Rubenstein, 2011, An Introduction to Human Geography, 10th ed., Upper Saddle River, NJ: Prentice Hall **Figure 7.32 page 325** Modified from Rubenstein, 2011, An Introduction to Human Geography, 10th ed., Upper Saddle River, NJ: Prentice Hall and National Geographic Society, 2003, Atlas of the Middle East, Washington, DC **Figure 7.37 page 331** Modified from Rubenstein, 2011, An Introduction to Human Geography, 10th ed., Upper Saddle River, NJ: Prentice Hall **Figure 7.38 page 332** Modified from Rubenstein, 2011, An Introduction to Human Geography, 10th ed., Upper Saddle River, NJ: Prentice Hall **Table 7.2 page 338** World Bank, World Development Indicators, 2010. **Figure 7.44 page 340** Modified from Rubenstein, 2011, An Introduction to Human Geography, 10th ed., Upper Saddle River, NJ: Prentice Hall **Text 7.1 page 315** In Palace of Desire, by Naguib Mahfouz

Chapter 8 Table 8.1 page 361 Data from CIA World Factbook 2013 **Table 8.2 page 365** Population Reference Bureau, World Population Data Sheet, 2012; net migration rate data from United Nations, Population Division, World Population Prospects: The 2010 Revision Population Database. **Figure 8.12 page 366** http://populationpyramid.net/Germany/1950/ **Figure 8.29 page 385** Data from CIA World Factbook, 2010 **Table 8.3 page 387** World Bank, World Development Indicators, 2012; human development index data from United Nations, Human Development Report, 2011; life expectancy data from Population Reference Bureau, World Population Data Sheet, 2012.

Chapter 9 Table 9.1 page 415 Population Reference Bureau, World Population Data Sheet, 2010. Table 9.2 page 440 World Bank, World Development Indicators, 2010

Chapter 10 Text 10.1 page 455 Ban Ki-moon **Table 10.1 page 465** Population Reference Bureau, World Population Data Sheet, 2012; net migration rate data from United Nations, Population Division, World Population Prospects: The 2008 **Figure 10.15 page 470** ©2012 Pearson Education, Inc. **Table 10.2 page 485** World Bank, World Development Indicators, 2010; human development index data from United Nations, Human Development Index, 2009; life expectancy data from Population Reference Bureau, World Population Data Sheet, 2010

Chapter 11 Figure 11.1 page 511 Population Reference Bureau, World Population Data Sheet, 2010. **Figure 11.2 page 537** World Bank, World Development Indicators, 2010

Chapter 12 Table 12.1 page 561 Population Reference Bureau, World Population Data Sheet, 2010. **Text 12.1 page 562** Zillur Rahman **Table 12.2 page 589** World Bank, World Development Indicators, 2010

Chapter 13 Table 13.1 page 613 Population Reference Bureau, World Population Data Sheet, 2010; net migration rate data from United Nations, Population Division, World Population Prospects: The 2008 Revision Population Database **Table 13.2 page 636** World Bank, World Development Indicators, 2010; human development index data from United Nations, Human Development Report, 2009; life expectancy data from Population Reference Bureau, World Population Data Sheet, 2010

Chapter 14 Table 14.1 page 663 Population Reference Bureau, World Population Data Sheet, 2013; net migration rate data from United Nations, Population Division, World Population Prospects: The 2012 Revision Population Database; additional data from CIA World Factbook, 2013 **Text 14.1 page 684** Hillary Clinton, America's Pacific Century, 2011 **Table 14.2 page 686** World Bank, World Development Indicators, 2013; human development index data from United Nations, Human Development Report, 2011; life expectancy data from Population Reference Bureau, World Population Data Sheet, 2013; additional date from CIA World Factbook, 2013

Index

Note: Information found in tables is indicated by "t" following the page number and similarly, illustrations are indicated by "f" after the page number.

A

Aborigine, 665, 671–672, 671*f*
  gender issues among, 676
  native rights issues, 682–683
  social challenges for, 690–691
Abu Dhabi Future Energy Corporation, 319
Abulu, Tony, 273
Abu Sayyal (Philippines), 631
Acadiana (U.S. cultural region), 107*f*, 110
Accra, Ghana, 262, 264, 264*f*
Acid rain
  in Europe, 360
  in North America, 90*f*, 91
Aconcagua Peak, Chile, 142
Acquired immune deficiency syndrome (AIDS)
  in Africa, 24, 25
  in Caribbean population, 208–209
  in Sub-Saharan Africa, 256, 257–259, 259*f*
Adiabatic lapse rate, 60
Advance Brazil program, 148
Aerial photographs, 17–18
Afghanistan
  Al Qaeda in, 480–481
  birth rate, 465–466
  development indicators of, 485*t*
  economy of, 488, 488*f*, 489
  ethnolinguistic map of, 472*f*
  languages of, 472–473
  NATO in, 481
  population pyramid, 466*f*
  refugees in Iran, 483, 483*f*
  social conditions in, 489, 490
  Taliban in, 474, 482
  U.S. occupation of, 480–481
  women in, 488, 490, 491*f*
Africa. *See also individual countries*; North Africa
  climate categories, 18*f*
  common languages, 31
  cultural influences in colonial Latin America, 169
  human development index, 46
  migrant works, 6
  as plateau continent, 243
  religions of, Caribbean adoption of, 215–216, 216*f*
  rift valley, 54
  small scale solar energy, 70, 70*f*
  Southwest Asia/North Africa, 296–348
  Sub-Saharan Africa, 238–293
African Americans, 99–100
African Union (AU), 242, 277, 279*f*
Agribusiness, 124
  in Latin America, 181
  soy production in Latin America, 181, 181*f*
  toxic chemical use by, 150
Agriculture
  Columbian exchange, 165–166
  density of, 256
  expansion in Sub-Saharan Africa, 250
  "green revolution," 179
  monocrop, 210
  shifting cultivation, 611
  sustainable, 93
  swidden, 260, 611, 611*f*
AIDS. *See* Acquired immune deficiency syndrome
Ainu people, 523, 524*f*
Air pollution
  in China, 504, 505*f*
  in Latin America, 151–154
  in Mexico City, 153–154
  in North American cities, 91
  in Russian Domain, 410–411
  in Santiago, Chile, 153, 153*f*
  in Southeast Asia, 607
Alaska, United States, 94
al-Assad, Bashar Hafez, 333
Albania
  development indicators of, 387*t*
  population indicators of, 365*t*
Alcohol consumption, compared, 443
Algeria
  development indicators of, 338*t*
  economy of, 341
  population indicators of, 310*t*
  urban growth in, 316
  violence in, 330
  women in, 345, 346*f*
Algiers, Algeria, 329*f*
All Assam Students Union, 584
Alluvial fans, 464
Alps, 355
  skiers and climate change, 360
Al Qaeda, 481
  goals of, 41
  organizational structure of, 41
  Al-Shouf Biosphere Reserve, 305
Alsou, 432
Alternative energy
  in Japan, 506
  in North America, 93–94
Altiplano, 141, 141*f*
  lithium reserves in, 182
Amazon Basin, 143–144, 143*f*, 148
  as Brazilian frontier, 161–162
  deforestation of, 75, 150, 151*f*
  Puerto Maldonado in, 138, 160*f*
  violence against environmentalists, 162
Ambedkar, B. R., 572*f*
American Prairie Reserve (Montana), 92
Amerindians
  language of, 166
  Latin America, 164–166
    demographic collapse, 164–155
  Latin American presence of, 138
  political recognition of, 166
  poverty levels of, 188
  pristine myth of, 165
Anatolian Plateau, 301
Andes Mountains, 141–142
Angeles City, Philippines, 645
Angkor Wat monument, 641
Anglo America, 84, 86
Anglo-Zulu Wars, 277
Angola
  Chinese investment in, 535, 535*f*
  development indicators of, 284*t*
  economy of, 291
  independence, 278
  population indicators of, 255*t*
Anguilla
  development indicators of, 228*t*
  population indicators of, 207*t*
Animism, in Southeast Asia, 619, 621
Antarctica, territorial claims in, 57
Anthropogenic landscape, 513
Antigua and Barbuda
  development indicators of, 228*t*
  population indicators of, 207*t*
Antilles, 197–198
Apartheid, 264, 277, 278
Apennine Mountains, 355
Apple, 539
Arabian Desert, 302
Arabian Peninsula, 301, 334–335
Arab-Israeli conflict, 331–333
Arab League, 329*f*, 334
Arab Spring, 300, 300*f*
  causes of, 328
  cell phone use in, 327, 327*f*
  impact on tourism, 347
  women's participation in, 345
Aral Sea, 455–457, 458*f*
Arctic, ecological sensitivity of, 414
Argentina
  culture of, tango, 170, 171, 171*f*
  development indicators of, 178*t*
  energy resources of, 184
  Falkland/Malvinas conflict, 168, 174
  megacities in, 140, 140*f*
  migration from Bolivia and Paraguay, 163
  Pope Francis, 168, 168*f*
  population growth of, 154
  population indicators of, 156*t*
Armenia, 437
  development indicators of, 440*t*
  foreign capital in, 446
  population indicators of, 415*t*
Arctic Sea ice, 61*f*
  polar bears in Russian Domain, 414
ASEAN+3 meeting, 635
ASEAN Regional Forum (ARF), 635
Asia. *See also individual countries*
  Central Asia, 450–491
  East Asia, 494–545
  South Asia, 548–595
  Southeast Asia, 598–645
  Southwest Asia/North Africa, 296–347
Asian migration
  to Latin America, 163
  to North America, 105, 105*f*, 106
Asia-Pacific Economic Cooperation Group (APEC), 690
Assam Accords, 584
Association of Southeast Asian Nations (ASEAN), 602
  members, 626
  purpose of, 635
Astana, Kazakhstan, 450*f*–451*f*, 452
Ataturk, Kemal, 330
AU. *See* African Union
Aung San Suu Kyi, 632
Australia
  cities of
    Melbourne, 663*f*
    Perth, 664, 664*f*
    Sydney, 567*f*
  culture of, 671–672
  development indicators of, 686*t*
  economy of, 685, 686, 687
  environmental geography of
    climate of, 653, 654*f*
    Great Barrier Reef, 652, 653*f*
    Great Dividing Range, 652, 652*f*
    Outback of, 652, 652*f*
  geopolitics of, 685
  human development index, 46
  map of east coast, 17*f*
  plants and animals in, 655–656
  population indicators of, 663*t*
  population policy (Pacific Solution), 675
  population pyramid for, 670*f*
Australia and Oceania, 650–652. *See also individual countries*; Oceania
  atoll evolution in, 656*f*
  climate patterns in, 65*f*, 653
    global warming and, 658–659
  culture of, 673
    aborigines in, 671–672, 671*f*, 676
    gender issues, 676
    immigrants in, 672, 675
    language and, 673–674, 674*f*
    Maori in, 651
    native rights issues, 682–683, 683*f*
    village life in, 674, 676*f*
  development indicators of, 686*t*, 690
  economy of, 685, 687–688
    mining in, 658*f*, 687
    tourism in, 689–690
  emissions, 659, 661
  environmental geography of, 652
    Australian Outback in, 652, 652*f*
    climates of Australia and New Zealand, 653
    climates (map), 654*f*
    Great Barrier Reef in, 653*f*
    high islands in, 649
    issues in, 657*f*
    Melanesia in, 652
    Micronesia in, 651
    Polynesia in, 651
    topography of Australia and New Zealand, 652–653

Australia and Oceania (*Continued*)
geopolitics of
independence movements, 679–680
map of, 650*f*–651*f*, 680*f*
Pacific microstates in, 679
history of
European colonization in, 666
initial habitation in, 665–666, 665*f*
island, 654–655
maps (entire region)
climate, 654*f*
initial migration, 665*f*
language, 674*f*
political, 650*f*–651*f*
population, 662*f*
topographical, 650*f*–651*f*
population pyramids in, 670*f*
population/settlement patterns in, 661–663, 665, 666–667
debates, 669–671
density map of, 662*f*
historical, 665–666
indicators of, 663*t*
population indicators of, 663*t*
rural, 668–669
urban, 667–668
social challenges in, 690–691
Austria
development indicators of, 387*t*
population indicators of, 365*t*
Austronesian languages, 621–622, 673
Autonomous Region in Muslim Mindanao (ARMM), 631
Axum, Kingdom of, 274
Ayodhya Mosque controversy, 569*f*
Azerbaijan, 452, 453
cities in, 467
development indicators of, 485*t*
economy of, 486
energy resources of, 460–461
independence in, 479–480
population pyramid, 466*f*

B

Baghdad, Iraq, 316
Baguio, Philippines, 619, 619*f*
Bahamas
development indicators of, 228*t*
offshore banking in, 230
population indicators of, 207*t*
Bahrain
city of Manama, 317*f*
development indicators of, 338*t*
population indicators of, 310t
violence in, 334
Baikal-Amur Mainline Railroad, 417
Baikonur Cosmodrome, 482, 482*f*
Baku, Azerbaijan, 475*f*
Baku-Tbilsi-Ceyhan Pipeline, 489, 489*f*
Balkan countries, 384
economic transition of, 391, 393
ethnicities in, 385*f*
geopolitical change in, 382*f*
history of, 384–385
independence wars of, 384–385
map of, 385*f*
Balkanization, 384
Baltimore City Public Schools, 625
Bamboo
advantages of, 252
to reduce deforestation in Sub-Saharan Africa, 251, 252
Bananas, Caribbean production of, 228
Banda, Joyce, 292
Bangalore, India
call center industry in, 579
high tech industry in, 550
as Silicon Plateau, 592–593
Bangkok, Thailand, 615–616, 616*f*
economy of, 639
Bangladesh, 550
anti-blasphemy protests in, 582, 582*f*
cities of, 568, 568*f*
development indicators of, 44*t*, 589*t*
economy of, 590
independence of, 582
migration to Assam, India, 584
natural hazards in, 554–555
population of
indicators of, 22*t*, 561*t*
stabilization of, 562
poverty in, 588
remittances to, 594
Banking
in Caribbean, 196, 230
Japan's crisis in, 537
offshore, in Caribbean, 230
in Singapore, 637
Bantu, migrations of, 267, 267*f*
Barbados
development indicators of, 228*t*
population indicators of, 207*t*
Barefoot College, 70, 70*f*
Baseball, 219, 219*f*, 222
in Europe, 379
in Japan, 528
in South Korea, 528
Basketball. *See also* NBA (National Basketball Association)
in Europe, 379
in Southeast Asia, 625–626
Bay Area, California, map of, 14*f*
Beer, U.S. importation of, 114, 114*f*
Beijing, China, 516, 517*f*
Belarus
development indicators of, 440*t*
foreign investment in, 446
language patterns in, 428
links with Russia, 436
population indicators of, 415*t*
urban clusters in, 416
Belgium
development indicators of, 387*t*
population indicators of, 365*t*
Belize
development indicators of, 228*t*
ecotourism, 232
forest cover of, 204–205
housing in, 213*f*
nature preserves, 205, 205*f*
population indicators of, 207*t*
Benin
development indicators of, 284*t*
population indicators of, 255*t*
Berber language, 326
Berdymukkammedov, Gurbanguly, 480
Berlin Conference, 276–277, 276*f*
Berlin, Germany, Jewish synagogue, 376*f*
Berlin Wall, 383, 384*f*
Bermuda
development indicators of, 228*t*
population indicators of, 207*t*
reinsurance business in, 230, 230*f*
Bharatiya Janata Party (BJP), 569
Bhumibol (king of Thailand), 633
Bhutan
development indicators of, 589*t*
economy of, 589
hydroelectric exports of, 589
Bhutto, Benazir, 587
Bikila, Abeke, 272
Bill and Melinda Gates Foundation, 243
bin Laden, Osama, 335
Biodiversity
in Caribbean environments, 201, 202–203
in Costa Rica, 150
in Latin America, 141
Biodiversity, definition of, 72
Biofuel
alternatives in Africa, 252
bamboo as, 252, 252*f*
of Brazil, 184
deforestation's threat to, 250
as majority of energy production, 251, 251*f*, 253
Biomass fuel, 362
Bioregion
definition of, 72
desert, 75
grassland, 75
temperate deciduous forest, 76–77
temperate forest, 78*f*
tropical rainforest, 73–75, 73*f*
world bioregions, 72*f*–73*f*
Bison, reintroduction of, 92
Black Belt (U.S. cultural region), 107*f*, 110
Bogotá, Colombia
as bike-friendly, 152, 152*f*
urban transportation in, 152
Bolívar, Simón, 173, 173*f*
Bolivia
agrarian reform in, 160
agricultural frontier of, 160–161
Amerindian political pressure in, 166
development indicators of, 178*t*
Día del Mar, 174
education in, in native languages, 167
emigration, 163
global warming impact on glaciers, 147–148, 147*f*
La Paz and El Alto, 159, 159*f*
lithium reserves of, 182
mining in, 182
population indicators of, 156*t*
privatizing water, 72
racial designations, 188
Bollywood, 577
Bolshevism, 433
Bono, 243
Bora Bora, 655, 656*f*
Boreal forests
in North America, 89
in northern Russia, 406*f*
Bosnia and Herzegovina
development indicators of, 387*t*
population indicators of, 365*t*
Botswana
development indicators of, 284*t*
population indicators of, 255*t*
Boza, Mario, 150
Bracero program, 164
Brahmaputra River, 551, 553
Brahmin caste, 571
Brain drain
from Caribbean, 234, 235
from Egypt, 345
from Russian Domain, 421
Brazil
African cultural survivals in, 169–170
agricultural frontiers in, 160–162
soy, 181, 181*f*
Amerindians in, 166
carbon dioxide emissions from, 64
carnival traditions in, 169–170, 170*f*
China as trading partner, 172
culture of, soap operas, 170
deforestation in, 148, 150*f*
development indicators of, 44*t*, 178*t*
as economic engine, 178
energy, 184
environmental geography of, dam projects, 144, 144*f*
ethnicity in, color lines, 171
exports of, 181
"green city" in, 152
greenhouse gases from cattle, 76, 76*f*
history of, discovery and settlement, 172–173
impact of slavery in, 271
Interoceanic Highway in, 138
investment in North America, 132
Japanese descendants in, 163*f*
megacities in, 140
music in, 171
population indicators of, 22*t*, 156*t*
population of, 140
growth of, 154
poverty levels of, 187
racial issues, 187–188
rainforest destruction, 76, 76*f*
rural settlement in, 158
shield of, 142
soccer in, 170, 171
social development of, Bolsa Familia, 187
tropical savannas, 74–75, 74*f*
Brazos Wind Farm, 83*f*–84*f*, 85
Bretton Woods, 130
BRIC countries, 172
as strategic partnerships, 186
British East India Company, 578, 581
Bromo Volcano, Java, Indonesia, 604*f*
Brunei, 600
development indicators of, 636*t*
population indicators of, 613*t*
Bryant, Kobe, 528
Bubble, economic, 10
Buddhism, 33–34, 34*f*
landscapes, 34*f*
Mahayana, 521, 521*f*
in South Asia, 570, 573–574
in Southeast Asia, 618
Tibetan, 474–475
monastery, 474*f*
Buenos Aires, Argentina, 140*f*
Bombonera Stadium, 170
Bukhara, Uzbekistan, mosque in, 473*f*
Bulgaria
development indicators of, 387*t*
energy sources of, 361*t*
female CEOs in, 379
population indicators of, 365t
Buraku Liberation League, 523
Burakumin, 523
Burj Khalifa building, Dubai, 317, 318*f*
Burkina Faso
development indicators of, 284t
population indicators of, 255t
Burma (Myanmar)
Cyclone Nargis in, 604–605

democratization of, 632
development indicators of, 636t
drug trade, 6–7
economy of, 642
emigration to Thailand, 634, 634f
ethnic conflict in, 631–633
Karen people of, 633
language in, 623, 623f
name of, 600
repressive government of, 600–601
Rohingya persecution in, 632, 633f
war in, 632f
Burnham, Daniel, 619
Burundi
development indicators of, 284t
population indicators of, 255t
Bush, George W., 223

C

Cabral, Alvares, 172
Cairo, Egypt, 314–315
emergence as religious center, 316
Calderón, Enrique, 187
California
agriculture, 56
Mediterranean bioregion, 76, 77f
Call centers
in India, 579, 579f
in the Philippines, 579, 637
Cambodia
development indicators of, 636t
economy of, 641
impact of dams, 600
Khmer Rouge era, 629–630
prostitution in, 645
religion in, 621
Cameroon
development indicators of, 284t
population indicators of, 255t
Camp John Hay, Philippines, 619
Canada
Baffin Island fjord, 52f
Barbados as tax haven, 230
border issues, 119
border of, 118
conflicts with native populations in, 120–121
cultural identity of, 104
U.S. influences, 115
development indicators of, 123t
economic issues in, 118
as federal state, 120
foreign-owned corporations in, 131
immigration to, 106
NAFTA and, 118
Nunavut territory created, 120
oil reserves in, 94
political history of, 117, 118, 120
population indicators of, 95t
provinces of, assembly of, 118
tar sands in, 90f, 94, 94f
urban population of, 96
U.S.'s relationship with, 116–120
Cannabis, growing locations of, 7f
Cao Dai faith, 621
Cape Verde
development indicators of, 284t
music from, 272
population indicators of, 255t
Capitalism
in China, 497–498
crony, 636
described, 38
globalization as consequence of, 8
Capital leakage, from Caribbean economy, 232
Caracas, Venezuela, urban form, 157f
Carbon dioxide
emissions of
North America, 87
reduction efforts for, 64–65
sources, 64
as greenhouse gas, 63, 63f
increase in, 62–64
from tropical forest destruction, 75
Carbon inequity issue, 64
Cargill, Incorporated, 644
Caribbean, 194, 196. *See also* Latin America
brain gain of, 234
cities of, 211–212
climate patterns of, 199–201
climate change, 201–203
global warming and, 202–203
hurricanes in, 199–201
culture of, 212–217
Asian immigration and, 216–217
baseball in, 219, 219f, 222
Carnival in, 196f, 218
colonialism's impact on, 214–216
creolization of, 214, 217
maroon societies in, 215
music in, 217–219
neo-Africa in, 213, 214–215
plantation America and, 210–211
development indicators of, 227, 228t
economy of, 226–227
agriculture in, 227–228
assembly plants in, 229
banana production in, 228
capital leakage in, 232
coffee production in, 227–228
development indicators of, 228f
free trade zones in, 229–230
gambling in, 230–231
industrialization of, 224, 227, 229–230
offshore banking in, 230
remittances to, 233, 233f
sugar production in, 227
tourism in, 231–232, 231f
education in, 234
energy issues of, 229–230
environment of, 196–203
biodiversity in, 201, 202–203
deforestation in, 196, 203f, 204–205
forest in, 204–205
issues in, 196, 203–206
rimland in, 197
vegetation in, 199
geopolitics of, 222–223
colonialism in, 221f, 222
colonies in, present-day, 225
disputes in, 223f
independence movements in, 224
integration in, regional, 225–226
neocolonialism in, 222
U.S. occupations, 222
U.S. policies in, 222–223
indentured labor in, 216–217
isolated proximity of, 196
languages of, 217
maps of (entire region)
climate, 200f
emigration, 209f
environmental issues, 203f
HDI, 233f
language, 218f
political, 195f, 223f
population, 206f
topographical, 195f
tourism, 231f
population/settlement patterns of, 206–207, 206–212
demographic trends in, 207–208
diaspora of, 209, 209f
emigration from, 209–210, 234–235
fertility decline in, 208
HIV/AIDS prevalence in, 208–209
housing, 212, 213f
indicators of, 207t
rural, 210–211
subsistence farming in, 210, 211
urban, 205–206, 211–212
religion in, African, 215–216, 216f
slave trade in, 214–215, 214f
social development of, 228t, 232–225, 232–235
education and, 234
HDI in, 233f
women's status and, 234
soil quality, 196–197
Caribbean Community and Common Market (CARICOM)
on climate change, 203
creation of, 227
Caribbean Sea, 196, 197f
marine reserves, 205
Carnival
in Brazil, 169–170, 170f
in Caribbean, 196f, 218
Carpathian Mountains, 355
Caspian Sea, 457–458
Caste system, 569–570, 571
Castro, Fidel, 224
Castro, Raúl, 194, 224
Catherine the Great, 427
Catholicism
in Latin America, 138
liberation theology of, 168
in North America, 111, 113f
in Sub-Saharan Africa, 168
worldwide distribution of, 169f
Cattle
deforestation in Latin America, 148, 150
displaced by bison herds, 92
hamburgers and tropical rainforest destruction, 76, 76f
in Sub-Saharan Africa, 261–262
Masai, 262
overgrazing by, 250
Caucasus, languages of, 429f
Caucasus Mountains, 409, 409f
Cayman Islands
development indicators of, 228t
offshore banking, 230
population indicators of, 207t
Cell phones
Arab Spring, 327, 327f
in Oceania, 681
smartphones in East Asia, 539, 539f
in Southwest Asia/North Africa, 339
in Sub-Saharan Africa, 238f–239f, 240, 287, 287f
Census
of Canada, 96–97
of U.S., 96–97
Central African Republic
development indicators of, 284t
population indicators of, 255t
Central America
cities of, 154
history of, United Provinces era, 173
level of violence in, 178
maquiladoras, 179–180
remittances to, 186
rural settlement and migration, 158
uplands of, 142
U.S. intervention in, 122
Volcanic Axis in, 142
Central American Free Trade Association (CAFTA), 140, 172, 174f
origins of, 176
Central Asia, 452–453
climate patterns of
global warming in, 462–463
map of, 456f
culture of, 468, 475
economy of, 484
energy issues, 460–462, 489
foreign investment in, 489
global context of, 489
post-communist, 484–487
environmental geography of, 453
Aral Sea in, 455–457, 458f
desertification in, 459
exotic rivers in, 465
hydropolitics in, 455–457
lake fluctuation in, 457–458
geopolitics of, 476, 479
Afghanistan wars in, 480–481
China's role in, 478–479, 482–484
foreign military bases in, 482
independence in, 479–480
map of, 454f–455f, 477f
Pakistan's role in, 484
Russia's role in, 482–484
SCO in, 482–483, 482–484
Turkey's role in, 484
U.S. role in, 482
history of
communist rule in, 477–479
Mongol Empire in, 469, 470f
Silk Road in, 466, 468
Soviet control in, 477–478
language in, 466, 468, 470–473
maps of (entire region)
climate, 456f
environmental issues, 458f–459f
HDI, 485f
language, 470f–471f
oil and gas pipelines, 460f
political, 454f–455f, 477f
population, 462f–463f
topographical, 454f–455f
population/settlement in, 462f–463f
density of, 462f–463f
issues of, 465–466
lowland, 464–465
map of, 462f–463f
migration patterns in, 466
religion in, 473–475
Islamic, 473–474
Tibetan Buddhist, 474–475
social development of, 489–491
"bride kidnapping" in, 491
HDI map in, 485f
human trafficking in, 491
Central Asia-China Gas Pipeline, 460

Central Business District (CBD)
- in Latin America, 156, 157, 159
- in North America, 101
- in South Africa, 264

Central place theory, 516
CGIAR, 450
Chaco War, 174
Chad
- development indicators of, 284*t*
- physiological density in, 256
- population indicators of, 255t

Chain migration, 210
Charice (Charmaine Clarice Pempengco), 624*f*
Charlemange, Manno, 219
Chávez, Hugo, 224
Chechnya, 437–438, 438*f*
Chernobyl disaster, 411, 413
Cherulyot, Vivian Jepkemoi, 274*f*
Chesapeake Bay, 87*f*
Child labor
- in China, 539
- in India, 587*f*
- in South Asia, 588

Child mortality, 46. *See also* Infant mortality
- as social indicator, 187
- in Southeast Asia, 643
- in Southwest Asia/North Africa (map of), 339*f*
- in Sub-Saharan Africa, 256, 291

Chile
- air pollution in Santiago, 153*f*
- Andes Mountains in, 141–142
- climate of, 146
- deforestation in, 148
- development indicators of, 178t
- exports of, 181
- mining production in, 181–182
- Pinochet diaspora from, 163
- politics of student protests, 176, 176*f*
- population indicators of, 156*t*

China
- agriculture
  - global aspects of, 515
  - regions of, 513–514
- Australian trade with, 686*f*, 687
- basketball in, 115, 116–117, 117*f*
- carbon dioxide emissions from, 64, 65, 65*f*
- Central Asia and, 453
- cities in, 513
  - central place theory, 516
  - migration to, 545, 545*f*
- communism in, 497
- culture of
  - gender issues, 545
- desertification in, 77*f*
- economy of, 540
  - alternative energy in, 506–507
  - development indicators of, 44*t*, 537*t*
  - industrial reform in, 542
  - post-communist, 541
  - rust belt in, 543
  - SEZ in, 541
  - social/regional differentiation in, 541–542, 542*f*
  - western China's, 487–488
- emigration
  - to Southeast Asia, 620
  - to Suriname, 226
- energy futures, 69
  - pipelines in Central Asia, 489
- energy issues, 506–507
- energy use, 507*f*
- environmental geography of, 501–502
  - dune stabilization in, 504, 505*f*
  - flooding in, 508–509
  - loess plateau erosion in, 509, 510*f*, 513
  - pollution in, 504
  - reforestation in, 504
  - Three Gorges Dam in, 506, 507–508
- geopolitics of
  - African connection, 535
  - Central Asian relations in, 478–479, 482, 482–484
  - Hong Kong and, 534
  - military buildup in, 536
  - Taiwan and, 533–534
  - tensions in Oceania, 684
  - territorial issues in, 534
- globalization gains, 5
- history of
  - boundaries in, map of, 529*f*
  - division in, 533–534
  - early, 529–530
  - Manchu Qing Dynasty in, 530
  - modern, 530
  - Opium War, 530, 530*f*
- immigration, North Korean refugees in, 541, 541*f*
- investment
  - in North America, 132
  - in Sub-Saharan Africa by, 285, 288, 290
- Islamic issues in, 474
- language/ethnicity in, 525–526
- maps of (entire region), economic differentiation, 542*f*
- maquiladora relocation to, 179
- merchants in Africa, 289
- migration to
  - heartland, 528
  - Southeast Asian culture influenced by, 619
  - urban system of, 516
- population indicators of, 22t, 511t
- population planning, 21
- population pyramid, 512*f*
- religion in, Mahayana Buddhism in, 521, 521*f*
- renewable energy in, 68
- Russian boundary issues, 438
- social conditions in, 543–544

Chipko movement, 556
Chisinau, Moldova, 404*f*
Choropleth maps, 17, 18*f*
Christianity, 32, 111, 113, 113*f*, 321, 325
- in East Asia, 522
- Eastern Orthodox, 425
- in Eritrea, 268, 269*f*
- Great Schism of, 374
- in Russian Domain, 429
- Russian Orthodox, 430
- in South Asia, 574
- in Southeast Asia, 620–621
- in Southwest Asia/North Africa, 321
- in Sub-Saharan Africa, 268–269

Chui Sai On, 535
Churchill, Winston, 435
Circular migration, 210
Cities
- concentric zone model of, 101, 101*f*
- decentralization of, 101–102
- densities of, 27–28
- edge, 101, 102, 102*f*
- gentrification in, 102
- "green," 152
- mega-, 140
- primate, 615–616
  - Moscow, 416
- sprawl of, consequences of, 101–102
- squatter settlements in, 28
- urban functions, 28
- urbanism, 28
- urbanization and, 28–29
- urban morphology, 28

Clans, in Somalia, 283
Climate, 54–56
- continentality of, 58
- controls of
  - global pressure systems as, 59
  - global wind patterns as, 59
  - greenhouse effect as, 58
  - land-water interaction as, 58–59, 59*f*
  - latitude as, 58
  - solar energy as, 54, 56, 58, 58*f*
  - topography, 60
- definition of, 61
- El Niño in, 146–147, 148
- of Latin America, 144, 145*f*, 146–147
- maritime, 58–59
- of North America, 89*f*
- regions of, 60–61, 62*f*–63*f*
- topography and, 60

Climate change, 61–63
- in the Caribbean, 199–203
- in Central Asia, 462–463
  - Mongolia, 462
- in East Asia, 510
- in Europe, 360
- frequency of hurricanes and, 201
- in Latin America, 144, 146–148
  - in Bolivia, 147–148, 147*f*
- in North America, 94–95
- in Oceania, 658–659
- in Russian Domain, 413–414
- in South Asia, 550, 558–560
  - Pakistan, 560
- in Southeast Asia, 607–609
- in Southwest Asia/North Africa, 308–309
- in Sub-Saharan Africa, 254
- three areas of concern with, 71

Climograph, 61
Closer Economic Relations (CER) Agreement, 690
Clothing industry, in Bangladesh, 590
Coal, 66
- Black Thunder Mine (Wyoming), 125*f*
- coal train, 67*f*
- in Europe, 360–361

Coca cultivation, locations of, 7*f*
Cocaine, production centers of, 7*f*
Coffee
- Caribbean production of, 227–228
- Fair Trade certified, 156, 183
- global trends, 182*f*
- in Latin America, 182–183
  - labor-intensive, 183
- as Latin American export, 165, 181
  - by Brazil, 181
- Rwandan production of, 261
- shade grown and organic, 156
- Sub-Saharan African production of, 261

Cold War, 122
- Angola and Mozambique independence, 278
- buffer zone in, 383
- in East Asia, 527, 536
- end of, 7, 8*f*, 383–386
- in European geopolitics, 352, 380, 383
- India during, 586
- Pakistan as US ally, 586
- Soviet role in, 435
- U.S. Cuba policy, 224

Colombia
- Andes Mountains in, 141
- development indicators of, 178t
- emigration
  - to United States, 164
  - to Venezuela, 163
- floriculture in, 261
- guerrilla insurgents in, 177
- history of, Gran Colombia era, 173
- music in, 171
- population growth of, 154
- population indicators of, 156t
- urban transportation in, 152
- Venezuela boundary dispute with, 174
- volcano (1985), 55

Colonialism, 39
- in 1914, 40*f*
- in Australia and Oceania, 666, 679–680, 682
- in the Caribbean, 194
- in Caribbean, 214–216, 221*f*, 222
- decolonization and, 39–40, 277–278, 330
- as globalization, 4
- hill stations, 619
- influence on culture, 29, 30
- in Latin America, 36, 138, 155, 171
  - complex patterns, 164
- map of world (1914), 40*f*
- neo-, 40, 222–223
- North African urban landscapes and, 316
- in South Asia, 579, 580–581
- in Southeast Asia, 626–628, 628*f*
- in Southwest Asia/North Africa, 328–330
- in Sub-Saharan Africa, 242, 262, 274–276, 276*f*, 277–278

Columbian exchange, 165–166
Communication systems, global, 4, 5*f*
Communism, 433
- in Central Asia, 477–479
- in China, 497, 540
  - cultural features of, 528
- in Southeast Asia, 621
- in Vietnam, 629–630

Communist governments, 38
Comoros
- development indicators of, 284t
- population indicators of, 255t

Concentric zone model, 101, 101*f*
Conflict diamonds, 281, 282
Confucianism
- gender issues and, 545
- legacy of, 520
- modern role of, 520–521

Congo
- conflict in, 278–279
- development indicators of, 284*t*
- malaria deaths in, 259
- population indicators of, 255*t*

Congo River, 244, 245*f*
Congress Party, 569
Conservation
- in Niger, 250
- in Sub-Saharan Africa, 247, 250–251
- of wildlife, 253–254

Consumer culture, 4–5
North American exports of, 115–116
Continentality, of weather, 58
Contraceptives, in Bangladesh, 562
Convection cells, in Earth's interior, 50, 53*f*
Convention on International Trade in Endangered Species (CITES), 253
Core-periphery model, 42–43
Corporations, transnational
geography of, 131
home countries of, 131–132
power of, 4
Cossacks, 402*f*
privileges of, 426
Costa Rica
development indicators of, 178t
exports of, 181
high tech manufacturing in, 180
national parks of, 150, 151*f*
population indicators of, 156t
Counterinsurgency, 41
Counterterrorism, 41
Country, 38
LDC and MDC, 43
Creolization, 214, 217
Crimean Peninsula, 436
Criminal activity, global, 5–6
Crisis mapping, 198
Croatia
development indicators of, 387*t*
population indicators of, 365*t*
Crony capitalism, 636
Crosby, Alfred, 165
Crude death rate, 22
Cuba
agricultural system of, 212, 212*f*
development indicators of, 228t
farmlands, 197–198
fertility rates in, 208
health care in, 208
Hurricane Dennis, 200, 201*f*
offshore drilling near, 224, 225*f*
political history of, 224
population indicators of, 207*t*
population pyramid, 208*f*
small business (*cuentapropistas*) in, 194
tourism in, 231, 232
urban, 211–212, 212*f*
Cuban missile crisis, 122
Cultural assimilation, North America, 104
Cultural identity, Hispanic, 105*f*
Culture
change in, globalization and, 4–6, 29
collision of, 29–30
dimensions of, 29
global, 5, 6*f*
homogenization of, 11
hybridities of, 5, 31
imperialism in, 30
landscapes of, in geography, 14
language and, 30–32
learned, 29
nationalism in, 30
syncretism in, 31
Culture clash, 29–30, 29*f*
Culture hearth, 300
Curitiba, Brazil, 152
Cyclone, 554–555
Cyprus
development indicators of, 387*t*
fiscal crisis in, 6, 394
gas wars, 394–395
map of, 394*f*
population indicators of, 365*t*
women's political participation in, 379
Cyrillic alphabet, 374, 374*f*
Christian Orthodox use of, 376
Czech Republic
development indicators of, 387*t*
population indicators of, 365*t*

D

Dacha, 424
Dalai Lama, 474, 479
Dalits, 572, 572*f*
Dam
Belo Monte, Brazil, 144, 144*f*
in Bhutan, 555
in Laos, 641
on Mekong River, 641
Merowe Dam (Sudan), 307
Nurek Dam, 461, 487*f*
regulatory lakes, 508
removal of
Pacific Northwest, 93
politics of, 93
rationale for, 93
Sardar Sarovar (India), 553, 553*f*
South American, map of (entire region), 143*f*
in Southeast Asia, 600
in Tajikistan, 486
Tekeze Dam, 245
Three Gorges, 506, 507–508, 508*f*
Danube River, 358, 358*f*
Darfur region, Sudan, 331
Dari language, 473
Dead Sea, remote sensing of, 19*f*
Death rate, crude, 22
Debt relief, Sub-Saharan African nations, 290
Decolonization, 39–40, 277–278, 330
Deforestation
in Australia and Oceania, 658, 658*f*
biofuels threatened by, 250
in Brazil, 150*f*
in Caribbean, 196, 203*f*, 204–205
in Central Asia, 459–460
in Chile, 148
in China, 503–504, 504*f*
in East Asia, 504
from grassification, 148, 150
greenhouse gases associated with, 609
in Haiti, 204
in India, 556, 556*f*
in Indonesia, 606, 606*f*
in Latin America, 148–151
in Madagascar, 251
in Malaysia, 75*f*, 606
in Philippines, 606
in South Asia, 555–556
in Southeast Asia, 602, 605–607
in Southwest Asia/North Africa, 304–305
in Sub-Saharan Africa, 247, 250–251, 253
bamboo to reduce, 251, 252, 252*f*
in Thailand, 606
Delhi, India, 566
Demilitarized zone, 533, 533*f*
Democracies, 37–38
Democratic Republic of Congo
development indicators of, 284t
ethnic conflict in, 281
as kleptocracy, 287
population indicators of, 255*f*
Demographic transition model, 25–26, 25*f*
Dengue fever, global warming and, 148
Denmark, 361*f*
border controls of, 367
development indicators of, 387*t*
population indicators of, 365*t*
Dependency theory, 185
Desert, 75
in Central Asia, 455, 457*f*, 464–465
Kalahari, 248
in Latin America, 146
in Rajasthan, India, 591
Southwest Asia/North Africa, 302
in Sub-Saharan Africa, 248, 248*f*
Desertification, 75, 77*f*
in Central Asia, 455, 459
in China, 504
in East Asia, 503–504
in the Sahel, 248–250
in Sub-Saharan Africa, 248–250
Desiccation, in Central Asia, 455
Development
core-periphery model of, 42–43
economic
growth rates of, 45
indicators of, 43–45, 44t
inequalities in, 42
growth vs., 43–44
social indicators of, 45–47
world map of human development, 45*f*
Dhaka, Bangladesh, 568, 568*f*
Diamonds, conflict, 281, 282
Díaz-Canel, Miguel, 224
Dibaba, Tirunesh, 272, 274*f*
Diepsloot, South Africa, 264, 265*f*
Diet, globalization and, 5
Divergent plate boundary, 54
Diversity
economic unevenness, 11*f*
globalization and, 4, 11
Djibouti
development indicators of, 284t
population indicators of, 255t
Dollarization, 186–187
Dominica
development indicators of, 228t
population indicators of, 207t
Dominican Republic
baseball in, 219, 222
development indicators of, 228t
forest cover of, 204, 204*f*
free trade zones in, 229, 229*f*
independence of, 224
population indicators of, 207*t*
tourism in, 232
Domino theory, 629
Dravidian languages, 574–575, 576–577
Dresden, Germany, 371, 371*f*
Drug trade
Afghan poppy farming, 488, 488*f*
border security and, 121
countries active in, 7*f*
global routes of, 5–6, 7*f*
in Latin America, 177–178, 177*f*
in Mexico, 177–178
map of cartel control, 177*f*
Moroccan crops, 314–315, 316*f*
in Southeast Asia, 611, 632–633
Druze, 324
Dubai, United Arab Emirates, 317, 318*f*
foreign workforce of, 320
urbanization of, 317
Durban Agreement on emission reduction, 64–65
Dust Bowl, of 1930s, 75

E

Earthquakes, 55
Baguio City, Philippines, 619
in Concepción, Chile, 141
in East Asia, 500*f*, 501, 501*f*
Haiti (2010), 196, 197*f*
in South Asia, 551
Earth Summit, of 1992, 64
biodiversity project on Socotra, 304
East African Community (EAC), 278
East African economy, 291
East Asia, 496–497
climate patterns in
global warming in, 510
map of, 498*f*–499*f*
culture of, 519
Chinese heartland, 528
Confucianism in, 520–521
cosmopolitan fringe in, 527–528
writing systems in, 519–520
economy of, 498
agriculture in, 513
development indicators of, 537t
"Jakota triangle" in, 537
energy in, 506–507
environmental geography of, 498–499
geopolitics of, 529
China's growing power in, 536
global aspects of, 535–536
imperialism in, 530
map of, 496*f*–497*f*
North Korean crisis in, 533, 535–536
Senkaku/Diaoyu crisis, 534, 536*f*
language in, 522
maps of (entire region)
climate, 498*f*–499*f*
environmental issues, 503*f*
imperial, 531*f*
language, 524*f*–525*f*
political, 496*f*–497*f*, 529*f*
population, 512*f*
topographical, 496*f*–497*f*
physical geography of, 499–500, 499–503
population/settlement of, 496–497, 511
map of, 512*f*
urbanization in, 513
religion in, 519
Easter Island, 648*f*–649*f*, 650–651
Eastern Europe
Soviet economic planning in, 391
transition period, 388, 391–392
Western hypermarkets in, 391*f*
Eastern Orthodox Christianity, 425
East Timor
conflict in, 630–631
development indicators of, 636t
economy of, 642
language of, 631
name of, 501
population indicators of, 613t
Economic Community of Central African States (ECCAS), 278
Economic Community of West African States (ECOWAS), 278, 291
Economic development, uneven benefits of, 42
Economy
agglomeration, 126–127, 127*f*
bubble, 10

Economy (*Continued*)
capital investment's role in, 126
connectivity's role in, 125
development of
indicators of, 43–45, 44t
inequalities in, 42
global
elements of, 4–5
instability of, 10
government spending's role in, 126
growth rates of, 45
innovation and research's role in, 126
lifestyle amenities' role in, 127
market demand's role in, 126
natural resources' role in, 126
postindustrial, 84
productive labor's role in, 126
recession of 2008–2010, 5
sectoral transformation of, 125–126
trickle-down, 10
Ecotourism
in Bhutan, 555
in Kyrgyzstan, 489
Lake Baikal, 412
in Mongolia, 489
in Socotra, 304
Ecuador
Amerindian political pressure in, 166
development indicators of, 178t
dollarization in, 186, 187
native languages in, 168
population indicators of, 156t
Edge city, 101, 102, 102*f*
Education
adult literacy, 46–47
in Caribbean, 234
gender equity in, 47*f*
in North America, 129
in South Asia, 594–595
in Southeast Asia, 643, 645
in Sub-Saharan Africa, 292
Egypt
Alexandria beachside view, 308–309, 308*f*
brain drain from, 345
Cairo, 314–315
Islamic Cairo, 315
modern and old, 314*f*
New Cairo City, 315*f*
Coptic Christianity, 325, 331
development indicators of, 338*t*
Naama Bay, 346*f*, 347
population indicators of, 310*t*
population of
growth in, 321
pyramid of, 320*f*
poverty in, 345
recent era, 330–331
sea-level changes, impact of, 308–309
Tahrir Square rally, 300*f*, 327*f*
Elephant, 253–254
El Niño, 146–147, 148
in Southeast Asia, 608
El Salvador
agricultural frontier of, 161
development indicators of, 178t
dollarization in, 186
emigration from, 164
population indicators of, 156t
Emigration
from Caribbean, 209–210, 234–235
from Latin America, 163–164
Encomienda system, 160
Endangered species
CITES and, 253
in East Asia, 504–505
Energy
in Australia and Oceania, 659
in Central Asia, 460–462
definition of, 66, 66*f*
"dirty" and "clean," 66
in economy of Russian Domain, 440
European components, by country, 361*t*
future demand, 69
Latin American energy sector, 184, 184*f*
North American consumption of, 93*f*
North American production of, 124
solar, 56, 56*f*, 58
Japan, 506
in South Asia, 560
in Southeast Asia, 609
Sub-Saharan Africa shortages, 251–253, 251*f*
wind, Denmark's, 361*f*
English
as global lingua franca, 31
internationalization of, 5
pidgin, 674
in South Asia, 577
India, 577
in Southeast Asia, 624–625
teaching around the world, 115
English Channel, 358
Environment
globalization and, 8
transnational firms' impact on, 8
Environmentalists, Brazil, 162
Environmental lapse rate, 146
Equatorial Guinea
development indicators of, 284t
oil production in, 290
population indicators of, 255*f*
Eritrea, 268, 269*f*
development indicators of, 284t
independence, 283
population indicators of, 255*t*
Erosion, in Latin America, 150–151
Estonia
development indicators of, 387*t*
population indicators of, 365*t*
Ethiopia
Blue Nile, 245
development indicators of, 284*t*
distance running in, 272, 274*f*
Eritrean independence, 283
Italian incursions into, 277
population indicators of, 255t
population pyramid of, 256, 256*f*
Ethnic conflict
in Burma (Myanmar), 632–633, 632*f*
in Democratic Republic of Congo, 281
in Kashmir, 583–584
in Mali, 281
in South Asia's northeastern fringe, 584
in Sri Lanka, 585, 585*f*
in Sub-Saharan Africa, 281
Ethnicity
South African categories, 264
United States projected, 106*f*
Ethnic neighborhoods, in Europe, 378, 378*f*
Ethnic religions, 32
Ethnic separatism, 38*f*
Europe, 352
cities of, 371
history of, 369–370
medieval landscape in, 369–370
preservation of, 370*f*
Renaissance-Baroque period and, 370
climate patterns of, 356, 357*f*
continental, 356, 357*f*
global warming in, 360
map of, 357*f*
marine west-coast, 356, 357*f*
maritime, 357*f*
Mediterranean, 356
culture of, 372
global context of, 377
immigrant clustering and, 378
diversity of, 352
Romani peoples, 367, 367*f*
economy of, 386
current crisis of, 393–395
development indicators of, 387t
eastern European changes in, 388, 391
eastern European disparities in, 394*f*
economic disparities among, 392*f*
industrial centers in, 388, 389*f*
Industrial Revolution in, 386
integration of, 388, 391–392
postwar, 388
textile industry in, 386
environmental geography of, 352
acid rain in, 360
Alpine mountain system in, 354–355
central uplands in, 355
issues in, 358, 360
landforms in, 352, 354
lowlands in, 352, 354, 355*f*
rivers and ports in, 358
seas surrounding, 358
shield landscape in, 355–356
size and northerly location (map), 354*f*
western highlands in, 355–356
gender issues in, 379–380
geopolitics of, 380
Balkans in, 384–385
boundaries in, changing, 381*f*
Cold War in, 352, 380, 383
devolution in, 386
east-west division in, 380, 383
Eurozone, 393–395
Iron Curtain in, 383, 383*f*, 435
irredentism in, 380
issues in, 380
Scotland's devolution in, 386
world wars and, 380
immigration to, 368*f*, 369
Fortress Europe and, 369
leaky borders and, 369
Schengen Agreement and, 366–367
undocumented, 367
language in, 372
geographies of, 372–374
Germanic, 372
map of, 373*f*
Romance, 372, 374
Slavic, 374
maps of (entire region)
boundary change, 381*f*
climate, 357*f*
economic disparities, 392*f*
environmental issues, 359*f*
geopolitical, 381*f*
immigration, 368*f*
industrial region, 389*f*
language, 373*f*
political, 353
population, 364*f*
religion, 375*f*
topographical, 353
migration in, 366, 368*f*
from Southwest Asia/North Africa, 320
population/settlement patterns of, 363
demographic trends in, 363, 366*f*
growth rate of, 363
indicators of, 365t
map of, 364*f*
public transportation in, 370
religion in, 374
Christianity's schism in, 374
contemporary patterns of, 376, 378
immigrant influence on, 378
Islam in, 376, 378
Judaism in, 376
map of, 375*f*
Protestant revolt in, 376
soccer teams in, 36
social development of, 386
European Coal and Steel Community (ECSC), 388
European Economic Community (EEC), 388
European Higher Education Area (EHEA), 377
European Monetary Union (EMU), 393
current crisis and, 383–385
European Union (EU), 352
border security of, 369
controversy, 354*f*
emission trading scheme of, 363
expansion of, 388, 390*f*
current economic crisis and, 383–385
formation of, 388
gender discrimination in, 379, 379*f*
greenhouse gas reduction by, 363
Kyoto Protocol and, 360
links to Southwest Asia/North Africa, 347
map of, 390*f*
nation-states eclipsed, 39
preserving Socotra, 304
states weakened by, 8
Turkey and, 335, 347
Eurosports, 37
Eurotunnel, 358
Eurovision music awards, 432
Eurozone, 393
Evergreen forests, 77–78, 78*f*
Evora, Cesária, 272
Exclave, 439
in Kyrgyzstan, 478
Exclusive Economic Zone (EEZ), 689
Exotic rivers
in Central Asia, 465
in Southwest Asia/North Africa, 313

F

Fair Trade, coffee certified as, 156, 183
Falkland/Malvinas conflict, 174
Family planning, in Bangladesh, 562
Famine Early Warning System (FEWS), 254
Fast-food, traditional diets affected by, 5

Fatah political party, 333
Favela, 156–157, 158*f*
Federalism
in India, 582
in U.S. and Canada, 120
Federal state, 120
Federated States of Micronesia
development indicators of, 686t
population indicators of, 663t
Female circumcision, 292
FEMEN (Ukrainian feminist organization), 444, 444*f*
Feng shui, 521, 522*f*
Fenno-Scandian Shield, 355–356
Fergana Valley, 478, 478*f*
Fertile Crescent, 311
Fes, Morocco, 316, 317*f*
Fiji
development indicators of, 686t
GHG emissions in, 659, 659*f*, 661
population indicators of, 663*t*
Finland
development indicators of, 387*t*
migration to, 421
population indicators of, 365*t*
Fisheries, South Pacific, 688–689, 689*f*
Fisheries, of North America, 91
Fishing, in Japan, 514, 515*f*
Fjords, 355
in Norway, 356*f*
Flooding
in Bangladesh, 555–556, 556*f*
in East Asia, 508–509
in South Asia, 555–556
Floriculture
in Colombia, 261
in Kenya, 261, 261*f*
Florida, Richard, 12
Food insecurity, 254*f*
in Sub-Saharan Africa, 286
Food resources
in South Asia, 565–566
in Sub-Saharan Africa, 254
Foreign direct investment (FDI)
by emerging market nations, 132
in India, 594
in Latin America, 185*f*, 186
in North America, 131
in Saudi Arabia, 342
in Sub-Saharan Africa, 243, 288
Foreign workers, in Southwest Asia/North Africa, 318*f*, 320
Forest
boreal, in Russian Domain, 406*f*
in Caribbean, 204–205
clear cutting, 79*f*
in India, 555–556
in Iran, 305
Ituri, 244, 245*f*, 251
in mainland South Asia, 602
in South Asia, 555–556, 556*f*
in Sub-Saharan Africa, 244, 245*f*, 247
temperate, 76–77, 78*f*
tropical, 73–75, 73*f*
deforestation in Latin America, 148
large-scale infrastructure project impact on, 138
in Turkey, 305
Valdivian, 148
Forestry
in Japan, 514
in Latin America, 182–184
plantation, 183–184
Forward capital, 568
Fossil fuels
in Azerbaijan, 486
as "dirty" energy, 66
future demand for, 69
geography of, 339–341, 340*f*
in Kazakhstan, 484
of North America, 124
North American reserves, 93–94
percentage of power from, 66
production maps, 340*f*
proven reserves, 66–67, 67*t*
reserves of, location of, 340*f*
reserves, production, and consumption, 66–68, 67*t*
in Southeast Asia, 609
in Southwest Asia/North Africa, 340*f*
Fossil water, in Southwest Asia/North Africa, 307
Foster, Norman, 319
Foxconn, 539, 539*f*
Fracking, 68, 68*f*
environmental hazards of, 94
forecasts, 69
France
Basque separatism, 38*f*
colonies in Africa, 275
colonies in Caribbean, 225
colonies in India, 580–581
colonies in Southeast Asia, 629
colonies in Southwest Asia/North Africa, 328–329
cultural nationalism, 30
development indicators of, 387t
energy sources of, 361t
immigrant issues, 378
industrial region, 388
population indicators of, 365t
secularism, 34
in Southeast Asian history, 627–628
as unitary state, 120
Francis (Pope), 138, 168*f*
shifting Catholic demographics, 168
Franco, Francisco, dictatorship, 37
Frauenkirche Cathedral, 371, 371*f*
Free trade zones
in Caribbean, 229, 229*f*
in Dominican Republic, 229, 229*f*
French Guiana
development indicators of, 228*t*
Kourou space center, 199*f*
maroon societies of, 215, 215*f*
population indicators of, 207*t*
French Polynesia
development indicators of, 686t
population indicators of, 663*t*
Friedman, Thomas
"electronic herd" of traders, 9, 9*f*, 10
"flat world" concept, 9, 12
Fujimori, Alberto, 163
Fujisawa Sustainable Smart Town (SST), 506, 506*f*

G

G8 countries, 42–43
role of, 131
Gabon
development indicators of, 284t
oil wealth in, 290
population indicators of, 255t
Gambia
development indicators of, 284t
population indicators of, 255t
Gambling
in Caribbean, 230–231
as global business, 7
Gandhi, Mohandas, 571, 582
Ganges-Brahmaputra Delta, 551, 553
Ganges River, 551
Gaza and West Bank, 332, 332*f*
control of, 332
development indicators of, 338t
Israeli settlements in, 332, 332*f*
security barrier in, 332–333, 333*f*
Gazprom, 437
in Central Asia, 489
new ventures, 447
state control of, 442
GDP. *See* Gross domestic product
Gebrselassie, Haile, 272
Gelana, Tiki, 272, 274*f*
Gender
definition, 34
double day, 35
in Europe, 379–380
globalization and, 34–36
sex preferences and birth rates, 24
Gender equity, 47
in education, 47*f*
in Southeast Asia, 645
Gender Gap Index, 645
Gender issues. *See individual regions*
Gender roles, 34–35
traditional dress and, 35*f*, 345
Generic prescription drugs, from Israel, 344, 344*f*
Genghis Khan, 469
Gentrification, 102
Geographic information systems (GIS), 18–19
layers of information, 19*f*
Geographic specialization, 9
Geography, 12
areal differentiation in, 13, 13*f*
areal integration in, 13
of corporations, 131–132
cultural landscapes in, 14
environmental issues, 20
of fossil fuels, 339–341, 340*f*
of migration, 26
population trends in, 21–22
regions in, 13
formal, 13
functional, 13–14
vernacular, 14, 14*f*
of religion, 32–34, 34*f*–35*f*
scales in, global to local, 13
settlement, 21, 26–29
tools used in, 15
types of
human, 12
physical, 12
regional, 12
systematic, 12
thematic, 12
Geology, 52. *See also* Plate tectonics
convection cells and, 50, 53*f*
hazards from, 55
Geomancy, 521
Geopolitics, 7–8. *See also specific region or country*
definition, 37
Georgia
development indicators of, 440t
invasion from Russia, 437
politics of, 404, 436–437
population indicators of, 415t
subtropical agriculture in, 409*f*
withdrawal from CIS, 435
Geothermal energy, 55
Germanic languages, 372
Germany
carbon dioxide emissions from, 65*f*
colonies in Africa, 276, 276*f*, 277
development indicators of, 387t
Dresden symbolic landscape, 371, 371*f*
energy sources of, 361t
fossil fuel emissions, 360
low-carbon development for Guyana, 203
Nazi era, 380
population indicators of, 365t
population of, 24*f*
growth, 24*f*
renewable energy, 68
reunification of, 383
Ruhr industrial region, 388
universities, 377, 377*f*
Ghana
development indicators of, 284t
economy of, 291
population indicators of, 255t
Ginseng, production of, 124*f*
Glaciers
global warming in Bolivia, 147–148, 147*f*
retreat of, in Central Asia, 462–463, 462*f*
Glasnost, 435
Global communication systems, 4, 5*f*
MTV Russia, 431–432
sports TV and, 36–37
Global culture, soccer, 378–379
Globalization
in Australia and Oceania, 678–679
in Central Asia, 475, 476
colonial-era forms of, 4
complexity of, 12
cultural change and, 4–6, 29–30
cultural imperialism and, 30
customer call centers, 579
demographics of, 6
diffusion of U.S. culture and, 114–115
directionality of, 5
diversity and, 11, 11*f*
economic, 4–5
economic convergence, 9
environmental concerns with, 8
flat and spiky worlds, 12, 12*f*
gender and, 34–36
geopolitics and, 7–8
homogeneous landscapes, 14
illegal activity in, 6–7
of Indian economy, 593–594
inequality issue, 10
in Japan, views of, 526–527
of Korean culture, 527
language in Southeast Asia and, 624–625
of Latino culture, 170–171
nation-state concept and, 38
"naturalness" of, 10
of Nauru, 658*f*
in North America, 84, 88–89
of North American culture, 113–116
opinions on, 8–11
advocates of, 8–10
critics' of, 10
protests against, 8, 8*f*
recession and, 41–42
in Russian Domain, 401, 442, 444, 446
energy projects, 447
local impacts, 447
virtual world, 445, 445*f*
smartphones in East Asia, 539, 539*f*
social dimensions of, 8
in South Asia

Globalization (*Continued*)
cultural tensions and, 578
sports, 578
in Southeast Asia, 601–602, 624–626
economy affected by, 642
Southwest Asia/North Africa
cities affected by, 316–317, 326–328
culture affected by, 326–328
sports and, 36–37
in East Asia, 528
state power affected by, 7–8
Sub-Saharan African culture affected by, 271–273
terrorism as reaction against, 40
tourism in the Maldives, 592, 592*f*
ubiquitous nature of, 30
US sports, 115–116
Global positioning systems (GPS), 16
Global sports TV, 36–37
global media sports complex, 115
in Southwest Asia/North Africa, 328
Global warming, 61–63
anthropogenic, 61
causes of, greenhouse gases as, 62–63
climate effects from, 95
effects of, 63, 64*f*
El Niño and, 148
Kyoto Protocol on, 64, 360, 659
in India, 560
regional effects of
Caribbean, 202–203
Central Asia, 462–463
East Asia, 510
Europe, 360
Latin America, 147–148
North America, 94–95
Russian Domain, 413–414
South Asia, 550, 558
Southeast Asia, 607–609
Southwest Asia/North Africa, 308–309
Sub-Saharan Africa, 254
sea level rise from, 64*f*
temperature change map, 414
GNI. *See* Gross national income
Goa, 574, 583
religion in, 574
Gobi Desert, 455, 459, 459*f*, 464–465
Golan Heights, water availability in, 307
Gondwanaland, breakup of, 243
Google Android, 539
Google Earth, 539–540
Gorbachev, Mikhail, 435
on civil rights, 438
Government. *See* State; *specific region or country*
Grameen Bank, 590, 590*f*
Graphic scale, 17
Grasberg mine, 630*f*
Grassification, 148, 150
Gray, John, 10
Great Barrier Reef, 652, 653*f*
Great Britain. *See also* United Kingdom (UK)
colonies in Africa, 276
colonies in Caribbean, 225
colony in India, 569–570
cricket, 578
missionary efforts of, 574
Hong Kong as colony, 40
immigration from South Asia, 578*f*
opium wars in China, 530, 530*f*
protectorates in Southwest Asia/North Africa, 329–330
in Southeast Asian history, 627–628
Greater Antilles, 197–198
population of, 206–207
Greater Arab Free Trade Area (GAFTA), 347
Great Escarpment, 244
Great Rift Valley, 243, 244
Greece
development indicators of, 387t
leaky border of, 369
migration from, 366
origins of democracy, 38
population indicators of, 365t
Green Belt Movement, 250–251
Green Fund, 65
Greenhouse effect, 56*f*
Greenhouse gases (GHG), 62. *See also* Carbon dioxide
Caribbean measures for, 203
in China, 510, 510*f*
emission trading scheme, 363
EU's reduction of, 363
in Latin America, 147
limiting of, international debate on, 63–65
in Southeast Asia, 608–609
Greenland, as region, 86
Greenpeace, 319
Green Revolution, 562, 564–565, 565*f*
regional benefits, 591
Green Schools for Green Bhutan Movement, 555, 555*f*
Green technology, in Masdar City, UAE, 319, 319*f*
Grenada
development indicators of, 228*t*
population indicators of, 207t
Gross domestic product (GDP), 44
Gross national income (GNI), 43*f*, 44
Gross national income (GNI) per capita, 44
among Indian regions, 591
in the Philippines, 636
in Vietnam, 640
Group of Eight. *See* G8 countries
Growth, development vs., 43–44
Grozny, Chechneya, 438*f*
Guadeloupe
development indicators of, 228t
population indicators of, 207*t*
Guam, 657
development indicators of, 686*t*
population indicators of, 663*t*
Guatama, Siddhartha, 570
Guatemala
agricultural frontier of, 161
deportations from U.S., 175
development indicators of, 178*t*
emigration from, 164
grassification in, 148, 150, 150*f*
Mayan city of Tikal, 164, 165*f*
population growth of, 162, 162*f*
population indicators of, 156t
Guayana
forest cover of, 205
GHG amelioration aid, 203
Guinea
development indicators of, 284t
population indicators of, 255t
Guinea-Bissau
development indicators of, 284t
population indicators of, 255t
Gulag Archipelago, 420
Gulf Coast industrial region, 126
Gulf of Tonkin Incident, 529
Guna peoples, 166, 166*f*
Gutiérrez, Gustavo, 168
Guyana
development indicators of, 228t
foreign investment in, 205
population indicators of, 207*t*
South Asian population of, 217

H

H-1B visa, 132–133
Indian professionals with, 550
for teachers from the Philippines, 625
Haiti
development indicators of, 228t
earthquake (2010), 196, 197*f*
crisis mapping after, 198, 198*f*
Enriquillo Fault, 197, 197*f*
history of, 204
independence of, 224
population indicators of, 207*t*
population pyramid, 208*f*
rara music, 218–219, 219*f*
Voodoo dictators of, 216
Hajj, 322, 323*f*
Hakka, diaspora of, 526
Hamas political party, 333
Han people, 525–526
Havana, Cuba, 211–212, 212*f*
Hawaii, United States
annexation of, 678, 683
emission controls on, 661
ethnic/cultural mix in, 678–679
geography of, 86
haoles in, 678
as high islands, 655
language of, 677
nationalism in, 583, 584*f*
native rights in, 683–684
Polynesian heritage of, 651
rainfall in, 655
social welfare of native people in, 691
tourism in, 689–690
wind power in, 661*f*
HDI. *See* Human development index
Health care
air pollution and, 153
Caribbean migrants in United States and, 235
in North America, 129–130
in Russian Domain, 442–443, 443*f*
in Sub-Saharan Africa, 292
Hefajat-e-Islam group, 582
Heroin, production centers of, 7*f*
High tech industry
in Hong Kong, 540
in India, 548*f*–549*f*, 550, 592–593
in Japan, 537, 538*f*
in Malaysia, 638, 638*f*
in Nairobi, Kenya, 263
smartphones in East Asia, 539
in South Korea, 538
in Taiwan, 540
in Turkey, 343, 345*f*
Himalayan Range, 453, 551
Hindi language, 576
role of, 577
Hinduism, 32, 568–570
regional distribution of, 572
in Southeast Asia, 617–618
Hindu nationalism, 569
Hip-hop, in Southwest Asia/North Africa, 327
Hiragana, 520
Hispanic Borderlands (U.S. cultural region), 105–106, 107–108, 110
Hispanic population of United States, 105–106, 105*f*
Hitler, Adolf, 380
dictatorship, 37
HIV. *See* Human immunodeficiency virus
Ho Chi Minh Trail, 629
Homs, Syria, 333*f*
Honduras
Amerindians in, 166
deportations from U.S., 175
development indicators of, 178t
population indicators of, 156*t*
Hong Kong
autonomy of, 534
culture of, cosmopolitan, 527–528
development indicators of, 537t
economy of, 540, 542*f*
as Special Administrative Region (SAR), 516
verticality of, 518
Horn of Africa, 248, 261*f*
Huang He River, 508–509
Hugo, Victor, 521
Human development index (HDI), 46
of Caribbean countries, 233*f*
in Central Asia, 485*f*
Latin America, 178
rankings of, 46
in Southeast Asia, 640*f*
Human immunodeficiency virus (HIV)
in Caribbean population, 208–209
in Russian Domain, 443
in Southeast Asian sex trade, 645
in Sub-Saharan Africa, 256, 257–259, 259*f*, 291
Human smuggling, in Thailand, 634
Human trafficking
in Central Asia, 491
in Russian Domain, 444
Humboldt, Alexander von, 146
Hungary
development indicators of, 387t
population indicators of, 365t
Hurricane, 59
in Caribbean, 199–201, 201*f*
paths of, selected, 202*f*
tracking paths of hurricanes, 202*f*
Hurricane Sandy, 58*f*, 87, 87*f*
Hybridization, 5, 31
Hyderabad, India, 548*f*–549*f*, 550
as high-tech center, 593
Hydraulic fracturing ("fracking"), 68, 68*f*
environmental hazards of, 94
Hydropolitics
in Central Asia, 455–457
fossil-fuel poor nations in Central Asia, 461
of Israel, 307
of Southwest Asia/North Africa, 307

I

Iberian countries, Latin America's colonization by, 138
Ibrahim, Anwar, 639
Iceland
development indicators of, 387t
divergent plate boundary, 54, 55*f*
economic turmoil in, 10
energy sources of, 361t
population indicators of, 365t

renewable energy, 68
volcanoes in, 356
Ice sheets, in Antarctica, 57, 57*f*
Ideographic writing, 519
IMF. *See* International Monetary Fund
Immigration. *See also* Emigration; Illegal immigration; International migration
to Canada, 106
to Caribbean, 216–217
economies affected by, 119–120, 121
to Europe, 368*f*, 369
in Europe
cultural influence of, 378
gender issues in, 379–380
immigrant clustering and, 378
to Japan, 523
to Latin America, 138, 162–163
Mexico-U.S. border, 175
to North America, 5, 6, 104–106
North America, entrepreneurs, 133
to Russia, 420–421
United States
community gardens, 112
deportations, 175, 175*f*
undocumented, 175, 175*f*
to U.S., 104–106, 121–122, 132–133, 132*f*
Imperialism
cultural, 30
in East Asia, 530, 531*f*
of Japan, 532
Inca civilization, 164
Indentured labor, 216–217
India, 550. *See also* South Asia
anti-poverty programs, 594
Bodoland issue, 584, 584*f*
carbon dioxide emissions from, 64, 65, 65*f*
cities of, 566, 568
city growth, 28
cultural hybrids, 31
economic unevenness, 11*f*
economy of, 588
call centers, 579, 579*f*
development indicators of, 44t, 589t
development map of, 588*f*
foreign investment in, 594
globalization of, 593–594
less-developed areas', 591
software industry in, 593
textiles, 591, 592*f*
energy in, 560
environment of, 551, 553, 556
ethnic conflicts, 583–584, 583*f*
federalism in, 582
geopolitics of
international, 586–587
state structure of, 582–583
immigration in, from Bangladesh, 584
investment in North America, 132
maps of (entire region), economic, 593*f*
middle class in, 588
migration from Bangladesh, 584
border barrier, 585*f*
outsourcing to, 180
Pakistan's nuclear capacity and, 586
population of, 18*f*, 22*t*, 27, 27*f*, 561–562, 561*t*
poverty in, 550
reforestation in, 556
relations with China, 586–587
social development in, 595
women's roles, 595
Indian Subcontinent, 550. *See also* South Asia
Indo-European languages, 575–576
Indonesia, 604
conflict in, 630
culture of, foreign influences on, 624
deforestation in, 606, 606*f*
development indicators of, 44*t*, 636*t*
economy of, 639–640
regional disparities, 639–640
family planning in, 612
Islam in, 620, 621, 621*f*
political disunity in, 630–631
population indicators of, 22t, 613t
population of, 610–611, 613–614
religious strife in, 621
transmigration in, 614–615, 614*f*, 621
urban settlement in, 616
women in, 645
Indus-Ganges Plain, 551, 553
Indus River, 551
Industrialization
European Russia, 416
European urban landscapes and, 370
in India, 591–592, 592*f*
Puerto Rico, 224
in Sub-Saharan Africa, 286*f*
urbanization related, 29
Industrial Revolution, 386
Infant mortality, 46. *See also* Child mortality
as social indicator, 187
Informal sector
in Caribbean, 192*f*–193*f*
in Latin America, 157–158, 180, 180*f*
in Russian Domain, 442
Infosys corporate campus (India), 548*f*–549*f*, 550
Ingushetia, 438
Insolation, 58
Insurgency, 40
Intergovernmental Panel on Climate Change (IPCC), 463
Internally displaced persons (IDPs), 280–281, 280*f*
from Syria, 336–337, 337*f*
International government, UN as, 5
International Meteorite Collectors Association (IMCA), 408
International migration. *See also* Immigration; Migration
Philippines, 6
International Monetary Fund (IMF), 9
austerity programs, 10
creation of, 130
privatizing water, 71–72
structural adjustment by, 285
International Network for Bamboo and Rattan (INBAR), 252
Internet
access in North America, 114
censorship in China, 543
censorship in Turkmenistan, 480
in Iran, 339
leading global language, 115
in Oceania, 681
restrictions in Singapore, 637
Russian surveillance of, 438
in Sub-Saharan Africa, 287
Interoceanic Highway, 138
Inuit people, 120–121
Iran
Afghan refugees in, 483, 483*f*
aridity of, 302, 303*f*
development indicators of, 338*t*
economy of, 341–342
forests in, 305
geopolitics of, 335
Central Asia in, 484
immigration in, Afghan refugees, 483, 483*f*
language of, 326
mosque in Qom, 316*f*
population of
fertility decline in, 320–321
indicators of, 310t
pyramid of, 320
qanat water system, 306
women in, 345
Iranian Plateau, 301
Iraq
as contrived territory, 330
counter terror strategy in, 41
cultures in, 334
development indicators of, 338t
economy of, 341
energy futures, 69
historical urbanization of, 315
multicultural map of, 334*f*
political history of, 334
population indicators of, 310t
salinization in, 306
U.S. presence in, 334
Iraq war, 122
Ireland, 356
development indicators of, 387t
economic turmoil in, 10
fertility rate of, 363
IRA attacks, 40–41
Lithuanian migration to, 366
population indicators of, 365t
Iron Curtain, 383, 383*f*, 384*f*
as buffer zone, 383
creation of, 435
Irredentism, 380
Irrigation, in Central Asia, 456
Islam, 32–33
in Central Asia, 453, 473–474
diffusion of, 322*f*, 323
diversity in practices in, 321
in East Asia, 522
emergence of, 321–324
in Europe, 376, 378
head scarf in France, 378
internationalism of, 326
pillars of, 322
in Russian Domain, 430–431, 430*f*
Shiite-Sunni split in, 322, 324
in South Asia, 570–571, 572–573
in Southeast Asia, 620
in Southwest Asia/North Africa, 321–324
in Sub-Saharan Africa, 268, 269–270, 270*f*
in Thailand, 633
urbanization affected by, 316
Islamabad, Pakistan, 568, 568*f*
Islamic fundamentalism, 300, 327
in Aceh, Indonesia, 631
in Central Asia, 473, 482
in Pakistan, 569, 587
in the Philippines, 631
in Russian Domain, 431
in Southeast Asia, 620
Jemaah Islamiya (JI), 635
Islamic Movement in Uzbekistan (IMU), 482
Islamism, 300
Isolated proximity, 196
Israel
Arab-Israeli conflict and, 331–333
creation of, 331–332
development indicators of, 338t
economy of, 344, 344*f*
evolution of, 331*f*
immigration from Russia, 320, 421
kibbutzim in, 313
population indicators of, 310t
religion in, 324
security barrier in, 332–333, 333*f*
water management
hydropolitics of, 307
projects, 307, 307*f*
West Bank settlements of, 332*f*
women in, 347
Istanbul, Turkey, 311, 311*f*
urban growth in, 316
Italy
colonies in Africa, 277
development indicators of, 387t
energy sources of, 361t
leaky borders of, 369
population indicators of, 365t
urban landscapes of, 369*f*
Ituri rain forest, 244, 245*f*, 251
Ivory Coast
conflict in, 281
development indicators of, 284*t*
economy of, 291
population indicators of, 255*t*
Ivory, trade in, 253–254

J

Jainism, 574
Jalade-Ekeinde, Omotola, 273, 273*f*
Jamaica
brain drain from, 234
development indicators of, 228*t*
population indicators of, 207*t*
reggae music in, 219
James, LeBron, 528
Japan
Ainu people of, 523, 524*f*
culture of
cosmopolitanism of, 527–528
language in, 522–523
minority groups in, 523, 524*f*
national identity in, 522–523
Shinto in, 521
social conditions in, 537–538
sports, 528, 528*f*
earthquake and tsunami (2011), 54*f*, 55
economy of, 537
agriculture in, 511, 514
banking crisis in, 537
development indicators of, 44t, 537t
high tech aspects, 537, 538*f*
imports to, 513, 514
outsourcing to Thailand, 639
energy resources of, 507
environment geography of
physical, 500–501, 500*f*
pollution in, 504
environment sustainability in, 504
history of
closing and opening in, 532
early, 532
imperial period of, 532

Japan (*Continued*)
impact of recession, 4
living standards in, 537–538
Mahayana Buddhism in, 521
minority groups in, 523
physical environment of, 500–501, 500*f*
pollution exporting, 504
population indicators of, 22t
population pyramid, 512*f*
population/settlement patterns in, 511
population indicators of, 511*t*
superconurbation in, 517
urban, 516–518, 519*f*
urban-agricultural dilemma in, 511–512
settlement pattern, 27
smart city movement in, 506
territorial extent of, 500
Tohoku earthquake and tsunami, 54*f*, 55, 501, 501*f*, 510
women in, 545
writing system in, 519–520
Java, population of, 611
Jemaah Islamiya (JI), 635
Jerusalem, 296*f*–297*f*, 298, 324*f*
Jewish faith/Judaism, 113
Jewish Pale, 376
Johannesburg, South Africa, 264, 265*f*
Johnson-Sirleaf, Eileen, 292
Jonathan, Goodluck, 273
Jordan
development indicators of, 338t
population indicators of, 310t
Jordan River Basin water politics, 306*f*
Jordan River Valley, 301, 302*f*
irrigation in, 313
Judaism, 33, 321, 324, 325
in Europe, 376
in Russian Domain, 431
in Southwest Asia/North Africa, 321
synagogue in Berlin, 376*f*

K

Kabila, Joseph, 281
Kabila, Laurent, 281
Kabul, Afghanistan, 468, 469, 469*f*
Kaká, 170–171
Kalahari Desert, 248
Kaliningrad, Russia, 438–439, 439*f*
Kamchatka Peninsula, Russia, 407, 407*f*, 409
Kanaka, 672
Kanji, 519
Karachi, Pakistan, 567, 567*f*
map of, 567*f*
Urdu language in, 576
Karakalpak people, 457
Karakoram Range, 551
Karen people, 633
Karzai, Hamad, 481
Kashgar (Kashi), China, 466
Kashmir, India, 583–584, 583*f*, 586*f*
Katakana, 520
Kazakhstan, 452
Aral Sea recovery, 457
Astana, 450*f*–451*f*, 452
cities in, 467
desertification in, 459
development indicators of, 485t
economy of, 484
energy resources of, 461
evolution of apples, 461, 461*f*
fossil fuels in, 484
language issues, 472
population indicators of, 465t
Russian language issue, 475
women in, 491
Kennedy, John F., 116
Vietnam War role of, 629
Kenya
development indicators of, 284t
distance running in, 272, 274*f*
economy of, 291
floriculture in, 261, 261*f*
highway construction, 285*f*
population indicators of, 255*f*
Khmer Rouge, 629–630
Kibaki, Mwai, 271
Kibbutzim, 313
Kiev, Ukraine, 416, 418–419
map of, 419*f*
Podil district, 418, 419*f*
Kimberly Process, 281
Kim Il-Jong, 522
Kipyego, Sally Jepkosgei, 274*f*
Kiribati
development indicators of, 686*t*
population indicators of, 663*t*
Kleptocracy, 287
Kolkata, India, 556, 558
bookstore, 576*f*
education in, 594
as literary center, 576
slums in, 591
Korea
agriculture in, 514, 515
culture of, 524, 527
demilitarized zone in, 533, 533*f*
diaspora from, 524
division of, 533
environmental geography of, physical, 503
Koreans in Japan, 523
landscapes of, 503
language of, 522–523
map, 502*f*
population in, 514
urban primacy in, 516
wildlife in, 506
writing systems in, 519
Korean War, 122
Kosovo, 384–385, 385*f*
conflict in, 122
KRI (Russian Game Developers) Conference, 445, 445*f*
Kshatriya caste, 571
Kurdish people
in Iraq, 334
language of, 326
map of, 39*f*
as stateless nation, 39
Kuta, Feli, 272
Kuwait
development indicators of, 338t
migrant labor, 26
population indicators of, 310t
Kwan-il-so Labor Camp, 539–540, 539*f*
Kyoto, Japan, 517–518
Kyoto Protocol, 64, 659
EU targets, 360, 363
in India, 560
Kyrgyzstan
development indicators of, 485t
economy of, 486
ethnic violence in, 479
exclaves, 478
independence in, 479
language issues, 471–472
population indicators of, 465t
transhumance in, 464

L

Ladysmith Black Mambazo, 242
Lagos, Nigeria, 262, 264*f*
as film capital, 273
Lake Baikal, 412, 412*f*
Landsat satellite program, 17–18
Language. *See also* English; *individual languages*
of Afghanistan, 472–473
Amerindian, 167*f*
Arab dialects, 321
of Australia and Oceania, 673–674, 674*f*
Austronesian, 621–623
Berber, 326
branches and groups of, 31
of Caribbean, 217, 218*f*
of Caucasus, 429*f*
of Central Asia, 466, 469, 470, 470–473
Uyghur, 471
Uzbek, 471
of China, 525–526, 525*f*, 527
culture and, 30–32
Dravidian, 574–575, 576–577
of East Asia, 522, 524*f*–525*f*
of Europe, 372–374, 373*f*
families of, 32f–33*f*
families and subfamilies of, 31
Germanic, 372
homogenization of, 5
Indo-European, 575–576
international students in Europe, 377
of Japan, 522–523
of Korea, 522–523
of Kurdish people, 326
of Latin America, 138, 167–168, 167*f*
Mandarin Chinese, 525
of Moldova, 428
Mongolian, 468, 469
Mon-Khmer, 623–624
Parsi, 574, 574*f*
Romance, 372, 374
of Russian Domain, 427–430, 427*f*
Sinitic, 526
Slavic, 374
in South Asia, 574–575, 575–577
Bengali, 576
Hindi, 576
of Southwest Asia/North Africa, 325–326, 326*f*
of Sub-Saharan Africa, 265–268
Tai-Kadai, 623
of Taiwan, 527
Tajik, 471, 472
Tibetan, 471
Tibeto-Burman, 623
tonal, 526
Transcaucasian, 429–430
Turkic, 468, 471–472
Turkish languages, 326
of Ukraine, 428, 428*f*
Laos
communism in, 630
development indicators of, 636t
economy of, 600, 641
language in, 623
Pathet Lao era, 629
population in, 612, 613*t*
religion in, 621
La Paz, Bolivia, 159, 159*f*
Latifundia, 160
Latin America, 138. *See also* Caribbean
agricultural frontiers of, 181
quinoa, 181
Atacama Desert in, 146, 146*f*
cities of, 154–158
bike riding in, 152, 154
informal sector of, 157–158
megacities, 140, 140*f*
periférico in, 156
squatter settlements in, 158
urban form of, 156–158, 157*f*
urban frontier of, 156
urban primacy, 155
climate patterns of, 144, 146–147
altitudinal zonation in, 146, 147*f*
El Niño in, 148
global warming and, 147–148
culture of
blended religion in, 168–170
global reach of, 170–171
indigenous people in, 164–168, 167*f*
language in, 167–168, 167*f*
national identity in, 171
soccer and, 170–171
telenovelas and, 170
economy of, 140
agricultural production in, 181
development indicators of, 178t
development strategies for, 179–180
dollarization in, 186–187
energy sector of, 184, 184*f*
export dependency of, 180–184
foreign investment in, 179–180, 185*f*
forestry in, 182–184
global, 186
industrialization of, 179
informal sector of, 180
maquiladoras in, 179, 179*f*
mining in, 181–182
monetary policy in, 187
neoliberalism in, 138, 140, 176, 179
outsourcing to, 180
primary export, 140
remittances to, 185*f*, 186
trade blocs, 172, 174*f*, 176–177
education in, 188, 188*f*
emigration from, 163–164
environmental geography of, 141
agricultural land problems in, 150–151
air pollution in, 151–154
Andes Mountains in, 141–142
biodiversity in, 141
deforestation in, 148–151
extractive industries emphasis of, 140
grassification in, 148, 150
landforms in, 141–142
lowlands in, 143
river basins in, 143–144, 143*f*
shields in, 142
uplands in, 142
urban problems in, 151–154
water issues in, 154
ethnicity of
colonial categories, 166–167
mestizo peoples, 163, 166–167
geopolitics of, 171–172
border conflicts in, 173–175
boundaries in, 138
democracy in, 176
drug trade in, 177–178, 177*f*

history of, 171, 173
indigenous organization in, 176
insurgencies in, 176, 177
regional organizations in, 172
trade blocs in, 172, 174f
women's participation in, 189f
history of
Amerindian survival in, 165–166
colonial, 138, 155, 171, 173f
Columbian exchange in, 165–166
concept of, 138
demographic collapse and, 164–165
Iberian conquest in, 138, 171
independence in, 173
political, 173
revolution in, 173
immigration to, 162
Asian, 163
European, 162–163
informal sector of, 157–158, 180, 180f
Interoceanic Highway, 138
literature of, 170, 171
maps of (entire region)
climate, 89f, 145f
environmental issues, 149f
female politicians, 189f
foreign investment, 185f
language, 167f
migration, 161f
political, 139f, 172f, 174f
population, 155f
topographical, 139f
megacities of, 140
urban transport, 152
migration in, 161f, 163–164
rural-to-urban, 154, 158
mountain ranges in, 141–142
national parks in, 150
neotropics in, 141
physical geography of, 141–144
pop culture, 170
soap operas, 170
population/settlement patterns of, 138, 154–156, 155f
agrarian reform in, 160
agricultural frontiers in, 160–162
encomienda system in, 160
growth in, 162
indicators of, 156t
latifundia in, 160
minifundia in, 160
rural, 158, 160
rural landholdings in, 160
race and class in, 188
as region of emigration, 164
religion
Catholicism in, 168, 169f
syncretism in, 168–170
social development of, 187
poverty and, 187
race and, 187–188
women's status and, 188–189, 189f
squatter settlements in, 156–157, 157f
wars, 173–174
water issues, urban, 154
Latin American Free Trade Association (LAFTA), 174f, 176
Latitude, 15–16, 15f
climate affected by, 58, 58f
Latvia
development indicators of, 387t
population indicators of, 365t
LDC. *See* Less-developed country
Lebanon, 324, 330
development indicators of, 338t
population indicators of, 310t
Lenin, Vladimir Ilyich Ulyanov, 433–434
on culturally defined republics, 435
Lesotho
development indicators of, 284t
population indicators of, 255t
Less-developed country (LDC), 43
Lesser Antilles, 198–199
population of, 207
Levant, 301
Liberation theology, 168
Liberation Tigers of Tamil Ealam, 585
Liberia
cell phone use, 287f
conflict diamonds in, 281
development indicators of, 284t
ex-slaves in, 276
population indicators of, 255f
Libya
development indicators of, 338t
Europe colonialism and, 330
fragmentation of, 330
population indicators of, 310t
women in, 347
Liechtenstein
development indicators of, 387t
population indicators of, 365t
Life expectancy, 25
in Southeast Asia, 643
Sub-Saharan Africa, 291–292
Lifestyle migrants, 100
Lima, Peru, squatter settlements in, 158
Linear scale, 17
Lingua franca
English as, 31
Malay as, 622
Linguistic nationalism, 577
Literacy, adult, 46–47
Lithium metal, 182
Lithuania
development indicators of, 387t
migration to Ireland, 366
population indicators of, 365t
Loboda, Svetlana, 432, 432f
Loess
in Central Asia, 464
dwellings carved in, 514, 514f
in East Asia, 509, 510f, 513
Longitude, 15, 15f
Lord's Resistance Army, 240
Lowlands
in Latin America, 143, 146
city location, 154
oxisol degradation of, 151
Luanda, Angola, 262
Chinese real estate development in, 290, 290f
Lukashenka, Alyaksandr, 436
Lula da Silva, 148
Luxembourg
development indicators of, 387t
population indicators of, 365t

M

Maathai, Wangari, 250–251
Macau, China, 534
Macau Forum, 535
Macedonia
development indicators of, 387t
population indicators of, 365t
Madagascar
deforestation in, 251
development indicators of, 284t
population indicators of, 255f
poverty within, 46f
Maghreb, as French protectorate, 329
Maharaja, 583
Mahfouz, Naguib, 315
Mahmoud, Yuonis, 328f
Mahuad, Jamil, 187
Makkah, 322, 323f
Malaria
in Sub-Saharan Africa, 259
conquest and, 275
Malawi
development indicators of, 284t
population indicators of, 255t
Malay, 622
Malaysia
Bumiputra affirmative action policy, 638–639
culture in, foreign influences on, 624
deforestation in, 75f, 606
development indicators of, 636t
economy of, 638
language in, 624
local Chinese community in, 638
National Physical Plan, 608
population indicators of, 613t
religion in, 620
urban settlement in, 616
Maldives, 553–554
climate change in, 558
development indicators of, 589t
economy of, 591
tourism, 591, 592, 592f
population indicators of, 561t
Mali
development indicators of, 284t
ethnic conflict in, 281
music in, 272, 272f
population indicators of, 255t
Malta
development indicators of, 387t
population indicators of, 365t
Manama, Bahrain, 317, 317f
Manchu Qing Dynasty, 530
Manchuria, culture in, 526
Mandarin Chinese, 525
Mandarins, 520
Mandela, Nelson, 278
Mangroves, in Southeast Asia, 606–607
Manila, Philippines, 615–616
Maoism, in South Asia, 586
Maori people, 651, 673, 673f
culture of, 673, 673f
gender issues, 676
Mao Zedong, 541
influence in India, 586
Map projections, 16, 16f
Maps
choropleth, 17, 18f
patterns and legends, 17
scale, 16–17
Map scale, 16–17, 17f
Maquiladora, 179, 179f
Marcos, Ferdinand, 636
Marine west coast climate, 89, 90f
Maritime climate, 58–59, 356
Marley, Bob, 219
Maroon society, 215
Marshall Islands
development indicators of, 686t
population indicators of, 663t
Martial arts, in Southeast Asia, 626
Martinique
development indicators of, 228t
population indicators of, 207t
Marxism, 401
Chinese ideology and, 522
dynastic succession principle, 522
Masdar City, United Arab Emirates, 317, 319
Mauritania
development indicators of, 284t
population indicators of, 255t
Mauritius
development indicators of, 284t
population indicators of, 255t
Mauryan Empire, 570
Mbeki, Thabo, 278
MDC. *See* More-developed country
Medina, 316, 316f
Mediterranean climate
Atlas Mountains and Maghreb, 302, 304f
in Australia, mallee vegetation, 653
Levant, 302
Mediterranean shrub and woodland bioregion, 75–76
in California, 89, 90f
Megacities, 140, 140f
Megalopolis, 96
in Japan, 517
in Latin America, 155–156
in Mexico, 142
Mekong Delta, 600, 602f, 608
Mekong River, 641
Melanesia, 652
Mendes, Chico, 162
Mercator projection, 16, 16f
Mercosur, 140, 172, 174f
success of, 176–177
Meridians, 15
Mesopotamia, urbanization of, 315–316
Messi, Lionel, 170–171
Meteorite collecting, 408
Methane, as greenhouse gas, 63, 64
Mexican-American War, 173–174
Mexican Plateau, 142
Mexico
agrarian reform in, 160
agricultural frontier of, 161
air pollution, 151, 153–154
Amerindian movements in, 166
Aztecs celebrated in, 171
border with U.S., 175
cities of, 154
culture of, telenovelas, 170
development indicators of, 178t
drug cartels, 177–178, 177f
as economic engine, 178
emigration to U.S., 164
energy of, 184
femicide in Ciudad Juárez, 179
Indian cultures, 188
maquiladoras, 179, 179f
Mexican Plateau of, 141, 142
music in, 171
population indicators of, 156t
population of, 140
growth of, 154
poverty levels of, 187, 188
pre-Columbian civilizations in, 164–165
remittances to, 186
rural landscape, 158, 160
U.S. border of, 87, 121f
Mexico City, Mexico, 142
air pollution in, 151, 153
water issues, 154

Micronesia, 652
Midaltitude climates, Latin America, 146
Middle East. *See* Southwest Asia/North Africa
Middle East, use of term, 299
Middle East Peace Treaty (1979), 122
Migrant workers, 6
Persian Gulf states (map of region), 318*f*
Migration. *See also* Emigration; Illegal immigration; Immigration; International migration; *specific region or country*
in Australia and Oceania, initial, 665*f*
of Bantu, 267, 267*f*
Caribbean, 209–210, 209*f*
intraregional, 210
in Central Asia, 466
chain, 210
Chinese, into Sub-Saharan Africa, 289
circular, 210
in Europe, 360, 369
factors in, 26
in Indonesia, 614–615, 614*f*, 615*f*
shifted cultivators, 615
international, 6, 7*f*
in Latin America, 162, 163–164
rural-to-urban, 154, 158
lifestyle, 100
net rates of, 27
in North America, 97
nonmetropolitan, 100
northward, African American, 99–100
rural-to-urban, 100
southward, 100
westward, 97–98
from North Korea, 541
patterns of, 26–29
push vs. pull forces of, 26
rural to urban, 21
Latin America, 154
Southwest Asia/North Africa, 317–318
in Russian Domain, 418, 420*f*
Chinese, 421, 421*f*
eastward, 418–420
international, 420
political motives for, 419–420
Russification in, 420
undocumented immigration in, 421
urban, 421, 423
in South Asia, 562, 564*f*
in Southeast Asia, 614–615
in Southwest Asia/North Africa, 317–318, 320
in Sub-Saharan Africa, 258, 267, 267*f*
transnational, 210
Military, U.S., 122
Asia Pivot, 684–685, 685*f*
in Central Asia, 482
in Japan, 532
in South Pacific, 680, 682*f*
Millennium Development Goals, 287
Minh, Ho Chi, 629
Minifundia, 160
Mining
in Mongolia, 486–487
rare earth elements, 539
Mobutu, President, 287
Moldova
development indicators of, 440t
foreign capital in, 446
language patterns in, 428
political tensions in, 437
population indicators of, 415t
Monaco
development indicators of, 387*t*
population indicators of, 365*t*
Monarchies, 37
Mongol Empire, 469, 470*f*
map of, 470*f*
Mongolia, 455
cities in, 467–468
copper-gold mining in, 486–487
development indicators of, 485*t*
economy of, 486–487
languages of, 468, 471
nomadic pastoralism in, 465, 465*f*, 467, 467*f*
population indicators of, 465t
social conditions in, 490
U.S. ties, 482
women in, 490–491
Mon-Khmer languages, 623–624
Monocrop production, 210
Monroe Doctrine, 122, 222
Monsoon, 59, 59*f*, 557–558, 558*f*, 559*f*
in Southeast Asia, 602–604
tropical seasonal forest, 74
Montenegro
development indicators of, 387t
population indicators of, 365t
Montserrat
development indicators of, 228t
population indicators of, 207t
Moon, Ban-Ki, 455
Morales, Evo, 159, 160, 166, 181
More-developed country (MDC), 43
Morocco
Atlas Mountains, 300, 301*f*
development indicators of, 338t
drug crops, 314–315, 316*f*
population indicators of, 310t
Morsi, Mohamed, 331
Moscow International Business Center, 423
Moscow, Russia, 416, 416*f*
benefits of capitalism, 447
Central Asian communities, 422, 422*f*
core area, 423
downtown, 423*f*
St. Basil's Cathedral, 430*f*
Mount Kilimanjaro, 244, 244*f*
Mozambique
Chinese investment in, 535
development indicators of, 284t
independence, 278
population indicators of, 255t
Mt. Pinatubo, 55
Mubarak, Hosni, 345
Mugabe, Robert, 277
Mughal/Mogul Empire, 570–571
Red Fort of Delhi, 571*f*
rise and fall of, 580
Muhajir people, 567
Muhammad, 322
Multinational corporations, U.S. influence and, 115
Mumbai, India, 566
as economic center, 591
hutments in, 566, 566*f*
Muñoz Marín, Luis, 224
Myanmar. *See* Burma (Myanmar)

N

Naboa, Gustavo, 187
Nairobi, Kenya, 263
Namib Desert, 248, 248*f*
Namibia
development indicators of, 284t
independence, 278
population indicators of, 255t
Nasheed, Mohammed, 592
National Football League (NFL), 37, 115
in Europe, 379
Pacific Island players in, 677–678
National Hurricane Center (Miami), 200–201
Nationalism
cultural, 30
definition, 38
linguistic, 577
in Southeast Asia, 629
National parks
in Costa Rica, 150, 151*f*
in Russian Domain, 409
in Southeast Asia, 607
in Sub-Saharan Africa, 247
Nation-state
decentralization and devolution, 39
definition of, 38
weakening concept of, 38–39
Native Americans
bison's return, 92
challenges to federal power, 120–121
displacement of, 97–98
gaming issue among, 120
homelands of, 110, 110*f*
Navajo Reservation, 110
political power of, 120–121
territory controlled by, historical, 117
Native Hawaiians, 691
Natural gas
in Latin America, 184
politics of, 394–395
production/reserves map of, 340*f*
in Russian Domain, 446
use by Europe, 361
Natural resources
commodification of, 8
economic role of, 126
international financial agency strategies, 10
of North America, 122–124
of Russian Domain, 440
Nature, globalization of, 72–73
Nauru
development indicators of, 686t
globalization of, 658*f*
population indicators of, 663t
NBA (National Basketball Association), 36–37
followings abroad, 115
international players, 116*f*–117*f*
N'Dour, Youssou, 242
Nekrasov, Nikolai, 445
Neocolonialism, 40, 222
Neoliberalism, in Latin America, 138, 140, 176, 179
Neo-Nazis, 371
Nepal
constitutional struggles in, 586
development indicators of, 589*t*
economy of, 589–590
Maoist rebellion in, 586
population indicators of, 561*t*
poverty in, 588
tourism in, 589–590, 589*f*
Netherlands
development indicators of, 387*t*
emigration from Suriname, 226, 226*f*
population indicators of, 365*t*
sea-level rise in, 360, 360*f*
South Africa colonization, 275
Southeast Asian colonies of, 627
Netherlands Antilles, 225
development indicators of, 228t
population indicators of, 207t
Net migration rates, 27
New Caledonia
development indicators of, 686t
nickel mine in, 688*f*
population indicators of, 663*t*
New Delhi, India
Akshardham Temple, 570*f*
as British capital, 582
New Guinea, 615
conflict in, 630
New People's Army (Philippines), 631
New urbanism, 102
New York City, cultural diversity of, 112
New Zealand, 651
All-Black rugby team, 677, 678*f*
Central Otago region, 655*f*
culture of, 673
development indicators of, 686t
economy of, 687
environmental geography of
climate of, 652–653, 653*f*, 655*f*
Southern Alps, 653, 653*f*
topography, 652–653
geopolitics of, 685
greenhouse gases in, 659, 661, 661*f*
population indicators of, 663*t*
Nicaragua
agrarian reform in, 160
development indicators of, 178*t*
emigration from, 163–164
exports of, 180–181
population indicators of, 156*t*
rift valley of, 142
Niger
conservation in, 250
development indicators of, 284*t*
food insecurity in, 254
population indicators of, 255*f*
reforestation of, 250
Niger Delta, 253, 253*f*, 291
Nigeria
city growth, 28
development indicators of, 44*t*, 284*t*
Ibo yam cultivation in, 260
languages spoken, 266–267
malaria deaths in, 259
music in, 272
oil exploitation in, 253, 253*f*
oil wealth of, 291
population growth, 24*f*
population indicators of, 22*t*, 255*f*
urban traditions of, 262
Niger River, 245
Nile Delta, impact of climate change on, 308
Nile River, 244–245
Aswan Dam, 306–307
as exotic river, 313
Nile Valley, 310–311, 311*f*
farmers in, 313, 313*f*
Niyazov, Sparmurat, 480

Nkrumah, Kwame, 277
Non-renewable energy, 66. *See also* Fossil fuels; Petroleum
North Africa, 242
  geopolitics of, 330–331
  Maghreb region, 301
  population clusters in, 310
  religion in, 325
  use of term, 300
North America
  affluence and abundance of, 86
  birth rates in, 97
  cities of
    decentralization of, 100
    global, 130
    growth of, 101
    historical evolution of, 101
    impact of immigration, 132–133
    sprawl of, 101–102
  climate patterns of, 89, 90*f*, 94–95, 95
    global warming in, 94–95
    impact on glaciers, 95, 95*f*
  cultural characteristics, 86
  cultural identity of, 104
  culture of
    assimilation in, 104
    globalization of, 113–116
    identities in, 104, 107*f*
    persistent homelands in, 106–107
    regions of, 84
  economy of, 122, 130
    agriculture in, 123–124
    alternative energy in, 93–94
    fossil fuels in, 124
    global investment patterns of, 131–132
    history of, 124–125
    indicators of, 123t
    industrial raw materials in, 124
    location factors in, 126
    natural resource base of, 122–124
    postindustrial, 85
    regional patterns of, 126–127
    sectors of, 124
    trade patterns in, 131
    transportation and communication networks in, 125
    wealth distribution in, 127–129
  energy in
    consumption of, 93*f*
    renewable, 93
  environmental geography of, 86–87
    acid rain in, 90*f*, 91
    air pollution in, 91
    atmospheric alterations in, 91
    binational agreements, 118
    continuing conflicts, 118–120
    fisheries in, 91
    human modification of, 88, 90
    soil transformation in, 90
    vegetation changes in, 88
    water management and, 91
  ethnicity in, 104
  ethnic neighborhoods in, 108–109, 111
    Los Angeles, 111, 111*f*
    Toronto, 106
    Vancouver, Canada, 108–109
  Europeanization of, 98, 99*f*
  foreign direct investment (FDI) in, 131–132
  geographic region, 84
  geopolitics of
    political conflicts, 118–120, 119*f*
    U.S.-Canada relationship in, 116–120
  global diffusion of culture, 114–116
  globalization in, 84, 88–89
  immigration to, 104–106
    undocumented, 121–122
  maps of (entire region)
    climate, 89*f*
    cultural region, 107*f*
    environmental issues, 90*f*
    European settlement expansion, 99*f*
    geopolitical issues, 119*f*
    political, 85*f*
    population, 96*t*
    poverty, 128*f*
    topographical, 85*f*
  migration in, 97–98
    nonmetropolitan, 100
    northward, African American, 99–100
    rural-to-urban, 100
    southward, 100
    westward, 99
  physical geography of, 88–89
    Great Lakes, 118, 118*f*
  population/settlement patterns of, 95–97
    Asian immigration to, 105, 105*f*, 106
    density of, 95t
    historical, 97–98
    immigrant, 86*f*, 90
    indicators of, 95t
    rural, 102–103, 103
    Southern states, 100
    Sun Belt growth in, 100
    urban, 95–96, 100
  poverty levels in, 128–129, 128*f*
    children's, 128*f*
    Hawaiian, 691, 691*f*
  religion in, 111, 113, 113*f*
  retirement focus in, 129, 130*f*
  social issues in, 127–130
    aging as, 129
    education as, 129
    gender equity, 129–130
    health care as, 129–130
    race as, 127, 129
  vegetation, 89
  wealthier communities, 127, 127*f*
North American Free Trade Agreement (NAFTA), 118, 172, 174*f*
  history of, 176
  maquiladoras, 179
  Mexican deforestation and, 148
North Atlantic Treaty Organization (NATO), 122
  in Afghanistan, 481
  as Cold War creation, 383
  creation of, 122
  Russia on expansion of, 438
  Russia in, 436
North China Plain, 513
North Korea. *See also* Korea
  crisis in, 533, 535–536
  development indicators of, 537t
  economics, 10
  famine impact in, 541
  Google Earth monitoring, 539–540
  migration from, 541
  population indicators of, 511t
  self-sufficiency goal of, 515
  South Korean economic initiatives, 538
North Korea Economy Watch, 539
North Sea, 358
Norway
  aid to Guyana, 203
  development indicators of, 387t
  energy sources of, 361t
  fjords in, 356*f*
  human development index, 46
  population indicators of, 365t
  tundra, 79*f*
  women in workforce of, 379
Novosibirsk, Russia, 447, 447*f*
Nuclear power
  in Europe, 361
  German phase-out of, 362
  in Japan, 501, 507
  in South Korea, 507
Nuclear weapons/power, 435–436
  denuclearization of, 435–436
  pollution from, 411, 413
Nunavut, Canada
  creation of, 120–121
  cultural identity of, 110
Nuraliev, Boris, 445
Nyerere, Julius, 277, 277*f*

O

Oasis settlements, 313, 313*f*
Obama, Barack, 271
Obeah, 215
Oceania, 650
  climates of, 655
  economy of, 687–688, 687–690, 689*f*
  environmental geography of
    atoll formation, 655, 656*f*
    high islands, 655
    island climates, 655
    island landforms, 654–655
    low islands, 655
  ethnic/cultural mix of, 679
  hypercommunication in, 681
  invasive species in, 659
  island landforms, 654–655
  mining in, 688
  sports in, 676–678
Ogallala Aquifer, depletion of, 90*f*, 91
Oil spills, 91
Olympics, 432
  in China, 528
  in Russian Domain, 432
Oman
  development indicators of, 338t
  population indicators of, 310t
1C Company, 445
Open Street Map mapping platform, 198
Opium, growing locations of, 6–7
Organization of American States (OAS), 122, 171
Organization of the Petroleum Exporting Countries (OPEC)
  influence of, 347
  in Southwest Asia/North Africa, 300
  in Sub-Saharan Africa, 251
Orinoco Basin, 143*f*, 144
Orographic effect, 60, 60*f*
Orographic rainfall, 557
Orthodox Christians, 113
  Great Schism and, 374
Ottoman Empire, 323
  domination by, 328, 330
Otunbayeva, Roza, 491
Outsourcing
  in India, 594
  to Latin America, 180
  in North America, 133
  Uruguay, 180
Overgrazing, in Southwest Asia/North Africa, 304–305
Oxisols, 151
Ozone, as greenhouse gas, 63

P

Pahlavi, Mohammad Reza, 335
Pajitnov, Alexie, 445
Pakistan, 550
  China's military connection to, 586
  cities of, 567–568
  climate change in, 560
  economy of, 590
    development indicators of, 44t, 589*t*
  education in, 594
  energy in, 560
  family planning in, 562
  Federally Administered Tribal Areas (FATA), 594
  geopolitics of
    Central Asia in, 484
    international, 587
  India's nuclear capacity and, 586
  Islamic fundamentalism in, 569
  Kasmir conflict, 583–584
  partition era, 582
  population of, 22*t*, 561*t*, 561*f*, 562
  post-9/11 security crisis in, 587
  radical Islamists in, 587
  religion in, 573, 574
  Urdu language in, 576
  women in, 595
Palau
  development indicators of, 686t
  population indicators of, 663t
Palestine, 331–333
Palestinian Authority (PA), 332
Palm oil plantations, 75
  in Southeast Asia, 612, 642
    sustainability issue, 644, 644*f*
  in Sub-Saharan Africa, 261
Pan-African Green Belt Movement, 250–251
Pan-African Movement, 277
Panama
  Amerindians in, 166, 166*f*
  development indicators of, 178t
  dollarization in, 187
  educational access in, 188*f*
  Guna ban on cattle ranching in, 150
  population indicators of, 156t
Panama Canal, 138
  expansion of, 138, 186, 186*f*
Papua New Guinea
  boundary of, 651–652
  development indicators of, 686t
  New Britain volcanoes and earthquakes, 654
  population indicators of, 663t
  Port Moresby, 658*f*
  tsunami in, 654–655
  uncontacted peoples in, 673
Paraguay
  development indicators of, 178t
  emigration, 163

Paraguay (*Continued*)
immigration from Korea, 163
native languages in, 168
population indicators of, 156t
Parallels, 15
Paraná basalt plateau, 142
fertility of, 151
Paris, France, 370
Parsi, 574, 574*f*
Partition, 582
Pashto language, 473
Pastoralism
in Central Asia, 463
nomadic, 465
Pasture-User Groups (PUGs), 467
in Southwest Asia/North Africa, 312
in Sub-Saharan Africa, 261–262
Patagonian shield, 142, 142*f*
Pattaya, Thailand, 645
Peace activists, Germany, 371, 371*f*
Pelé, 170
Peña, Enrique, 178
Peñalosa, Enrique, 152
Perestroika, 435
Periférico (perimeter highway), 156
Permafrost, 407, 414
Persad-Bissessar, Kamla, 217, 217*f*
Persian Gulf, 317
fossil fuels, 341
numbers of migrants, 318, 318*f*, 320
Persian Gulf cities, 317, 317*f*
Perth, Australia, 664, 664*f*
Peru
development indicators of, 178*t*
informal sector of, 180*f*
Interoceanic Highway in, 138
population indicators of, 156t
Petroleum. *See also* Fossil fuels
Caribbean supplies of, 228–229
in Central Asia, 453, 460–462
map of pipelines, 460*f*
pipelines, 489
Cuba's offshore drilling, 224, 225*f*
in Equatorial Guinea, 290
Gulf Coast petroleum refining and, 126*f*
India's imports of, 560
in Indonesia, 639
in Latin America, 184
in Niger Delta, 253, 253*f*
in Nigeria, 291
OPEC and, 300, 347
pipelines, in Russian Domain, 446*f*
production/reserves map of, 340*f*
in Russian Domain, 446–447
in Southeast Asia, 609
in Southwest Asia/North Africa, 340*f*
in Sub-Saharan Africa, 290–291
in Xinjiang, 460
Philanthropy, 243
Philippines
Baguio City, 619, 619*f*
birth rate in, 613*f*
call centers in, 579
culture of, 624, 624*f*
deforestation in, 606
economy of
crony capitalism in, 636
decline in, 636
development indicators of, 636t
former bases in, 637, 637*f*
revival of, 637
emigration, 5
emigration to United States, 625, 625*f*
gender gap in, 645
languages in, 623, 624–625
migration, internal, 615
New People's Army in, 631
political system of, 637
political tensions in, 631
population indicators of, 613t
primate city of, 615–616
prostitution in, 645
territorial issues of, 633
terrorist links in, 635
Typhoon Bobha in, 604, 605*f*
Phnom Penh, Cambodia, 641*f*
Physiological density, 256
of Southwest Asia/North Africa, 310
Pidgin English, 674
Plantation
in the Caribbean, 210–211, 210*f*
forestry, 183–184
monocrop production in, 210
in Southeast Asia, 611–612
in Sub-Saharan Africa, 261
sugar, 210, 210*f*
Plantation America, 210–211, 210*f*
Plata Basin, 143*f*, 144
Plate tectonics, 52–55, 53*f*
Caribbean Plate, 199
colliding plate boundary and, 54
convection cells and, 53*f*
divergent plate boundary and, 54
global pattern of, 53*f*
insular Southeast Asia, 604
mountain formation from, 55
subduction zone and, 54
Poaching, in Sub-Saharan Africa, 253–254
Poland
border issues in, 366*f*
development indicators of, 387t
environmental issues in, 360
Jewish Pale, 376
population indicators of, 365t
spiky world of, 12*f*
Politics
of dam removal, 93
geo-, 7–8
and globalization, 7–8
of natural gas, 394–395
Pollution
in China, 504, 505*f*
in East Asia, 504, 505*f*
exporting of, 504
in Japan, 504
nuclear, 411, 413
in Russian Domain, 410–411
water, 410–411
Polynesia, 651
Population. *See also specific region or country*
age of, 22–24
of the Caribbean, 196
demographic transition model of, 25–26, 25*f*
density
international variation in, 27–28
urban vs. rural variation in, 27
of world, map of, 20*f*–21*f*
in geography, 21
growth
in Caribbean, 208
of cities, 28
crude death rate and, 22
of Latin America, 154
RNI and, 22, 25–26
variations in rates of, 21
of Latin America, 140
life expectancy and, 25
of North America, 95–97
census data, 96–97
farm, 103
planning, 21
pyramids, 24, 24*f*
TFR and, 22, 23*f*
trends in, 21
urbanized, 28–29
Population density, 27–28
in Caribbean region, 206–207
in Central Asia, map of, 462*f*–463*f*
in China, 545
of Latin America, 141
in Sub-Saharan Africa, 256
Population pyramids
China and Japan, 512*f*
Cuba and Haiti, 208*f*
of Ethiopia and South Africa, 256*f*
Guatemala and Uruguay, 162*f*
North America, 98*f*
Singapore and the Philippines, 613*f*
Sri Lanka and Pakistan, 561*f*
Pornography, as global business, 7
Portugal
colonies in Africa, 274, 278
development indicators of, 387*t*
Latin America's colonization by, 138
population indicators of, 365t
trade links to China, 535
women in workforce of, 379
Postindustrial economy, 84
Potato, 165
Poverty, 46
deep, 45–46
definition of, 42, 46
in Hawaii, 691, 691*f*
in Latin America, 187
in Madagascar, 46*f*
in North America, 128–129, 128*f*
slavery and, 285
in South Africa, 290
in South Asia, 551, 587–588, 587*f*
in Southwest Asia/North Africa, 343, 345
in Sub-Saharan Africa, 283, 285–287
World Bank definition of, 42, 42*f*
in Yemen, 345
PPP. *See* Purchasing power parity
Prairie, 75
in North America, 89
Primate cities
in Korea, 516
in Philippines, 615–616
in Southeast Asia, 615–616
in Taiwan, 516
in Thailand, 615–616
Prime meridian, 15
Project Tiger, 557, 557*f*
Prostitution
as global business, 7
in Southeast Asia, 639, 645
Protected areas, in Southeast Asia, 607
Protectorates, in Southwest Asia/North Africa, 329
Protestantism
in Europe, 376, 378
in North America, 111, 113, 113*f*
revolt, in Europe, 376
Protests
against globalization, 8, 8*f*
against Russian central government, 438
Chilean student, 176, 176*f*
Pussy Riot, 438, 439*f*
Public housing, in Singapore, 638*f*
Puerto Rico
commonwealth status of, 223–224
development indicators of, 228t
economy of, 223–224
forestry conservation, 204, 205*f*
population indicators of, 207*t*
Purchasing power parity (PPP), 44–45
in Latin America, 178
Pussy Riot, 438, 439*f*
Putin, Vladimir
anti-Americanism of, 439
background, 437
challenges to, 438
oligarchs and, 402
on Olympics in Sochi, 432, 433*f*
population programs, 424
religious affiliation of, 430
siloviki ties, 438
strong centralized rule of, 433, 433*f*, 437
Pyrenees, 354–355

Q

Qanat system, 306
Qatar
development indicators of, 338t
population indicators of, 310t
women in, 327
Qinghai-Tibet Railway, 487, 487*f*
Quebec, Canada
cultural identity of, 106–107
political status of, 120
Quran, 322. *See also* Islam

R

Race
in Latin America, 187–188
in North America, 127, 129
Rain shadow, 60
Ramadan, 322
Ramayan, 618
Rate of natural increase (RNI), 22, 25–26
in Caribbean, 208
in Europe, 363
Recession, global (of 2008—2010)
EU emission reductions, 363
European economic crisis, 393–395, 393*f*
globalization and, 5, 41–42
impact on economic activity, 42*f*
impact on immigration, 121
impact on North America, 127, 128
impact on oil and gas revenues, 341
impact on remittances, 164
impact on tourism, 591
instability of global markets, 10
Reducing Emissions from Deforestation and Degradation (REDD) program, 65
Norway aid to Guyana, 203
Reforestation
advantages of bamboo, 252
in Caribbean, 204–205
in China, 504
in India, 556
in Niger, 250
in Sub-Saharan Africa, 250
Refugees, 27*f*
Afghan refugees in Iran, 483, 483*f*
in Australia, 672, 675, 675*f*
in Central Asia, 483
Latin American, 164
migration data on, 26, 27

Palestinian, 337
from Southwest Asia/North Africa, 320, 334
in Sub-Saharan Africa, 280–281, 280*f*
from Syria, 320, 336–337, 336*f*, 337*f*
Region. *See also specific region*
formal, 13, 14*f*
functional, 13–14, 14*f*
in geography, 12–13
vernacular, 14, 14*f*
world, major, map of, 20*f*
Regional geography, 12, 19
Religion. *See also* Buddhism; Christianity; Islam; Judaism; Protestantism; *specific region or country*
ethnic, 32
geography of, 32–34, 34*f*–35*f*
of Latin America, 168–170, 169*f*
secularism and, 34
in Sub-Saharan Africa, 268–270
interaction among religions, 270
syncretic, 169
universalizing, 32
Remittances
in Bangladesh, 594
to Caribbean, 233, 235
from Filipino diaspora, 637
from Indian diaspora, 591
to Latin America, 164, 185*f*, 186
in Russian Domain, 421
Remote sensing, 17–18
Dead Sea, 19*f*
Renewable energy, 66
Caribbean, 229
China's support for, 504
German transformation, 362
issues with, 68
uses of, 68
Representative fraction, 17
Reunion
development indicators of, 284t
population indicators of, 255t
Rhine River Delta, 354
Rhinoceros, 253*f*
Rhône River, 358
Rice
in India, 562, 563, 565*f*
in Southeast Asia, 612
Rift valleys, 54, 142, 243, 244
Rim of Fire
Melanesia and Polynesia in, 654
New Zealand in, 652–653
Rio de Janeiro, Brazil
gondola lines over favelas, 156–157, 158*f*
Maracaña Stadium in, 170, 170*f*, 171
RNI. *See* Rate of natural increase
Robinson map projection, 16, 16*f*
Rocky Mountains, 88, 88*f*
formation, 55
Rohingyas, 632, 633*f*
Romance languages, 372, 374
Romania
development indicators of, 387*t*
population indicators of, 365*t*
Romero, Oscar (Archbishop), 168
Rouhani, Hassan, 335
Roundtable on Sustainable Palm Oil, 644
Rousseff, Dilma, 189
Royal Botanic Garden, Edinburgh, 304
Rubber plantations, 612, 612*f*
Rushdie, Salman, 577
Russia
carbon dioxide emissions from, 65*f*
in Central Asia's politics, 482–484
changes since 1991, 391
civil liberties in, 438
development indicators of, 44t, 440t
as diamond producer, 282
economy of, petroleum, 446–447
emigration from, 420–421
energy futures, 69
geopolitics in, 437–438
corruption in, 442
gas pipeline politics in, 394–395
territorial disputes of, 438–439
immigration to, 420–421
imperialism in China, 530, 531*f*
Kaliningrad exclave of, 438–439, 439*f*
language patterns in, 428–429
population indicators of, 22t, 415*t*
power of, reassertion of, 439
Siberian *taiga* forest, 78
uneven economic growth of, 43
women in, Southeast Asian sex trade, 645
Russian Domain, 400. *See also* Putin, Vladimir
agricultural regions in, 404, 406*f*, 407
air pollution in, 410–411
Armenia, 404
Belarus, 403
cities in, 423–424
suburbs, 424*f*
climate patterns in
global warming in, 413–414
map of, 405*f*
culture of, 425
global context of, 431
music in, 431–432
nationalism in, 432
Soviet-era, 431
western-oriented, 431
economy of, 440
corruption in, 442
development indicators of, 440t
foreign investment in, 444, 446
globalization in, 442, 444, 446
industrial zones in, 441*f*
natural resources in, 440
oil pipelines in, 446*f*
petroleum, 446–447
post-Soviet, 441–443
privatization in, 442
regional ties in, 442
Soviet legacy in, 440–441
state control of, 442
environmental geography of, 404–409
agricultural regions, 442
arctic wildlife in, 414
Caucasus Mountains in, 409, 409*f*
devastation in, 410
European west in, 404, 407
issues in, 410*f*
Norilsk pollution, 411, 411*f*
nuclear pollution in, 411, 413
permafrost in, 407, 414
post-Soviet challenges in, 413
Russian Far East in, 409
Siberia, 407, 409
taiga in, 407
Transcaucasia in, 409, 409*f*
Ural Mountains in, 407, 409
wildfires in, 414
wildlife challenges in, 413
geopolitics of, 433, 435–438
Baikonur Cosmodrome, 482, 482*f*
in Central Asia, 482–483
changing boundaries, 400
Commonwealth of Independent States in, 435–436
denuclearization in, 435–436
issues in, 436*f*
map of, 436*f*
southern, 436–437
Soviet, 433–435, 435
western, 436–437
Georgia, 400
issues with, 404
history of, 401–402, 425–427, 433–435
human trafficking in, 444
immigration, Central Asian, 421, 422*f*, 466
independent states in, 401
language in, 428*f*
former Central Asian republics, 475–476
map of, 427*f*
patterns in, 427–430
Transcaucasian, 429–430
Lena Pillars (Siberia), 399*f*, 400
maps of (entire region)
agricultural, 406*f*
climate, 405*f*
environmental issues, 410*f*
geopolitical, 436*f*
health care and alcoholism, 443*f*
industrial zones, 441*f*
language, 427*f*
natural resources of, 441*f*
political, 402*f*–403*f*
population, 415*f*
recent migration flows, 420*f*
Russian Empire, 426*f*
Soviet geopolitical system, 434*f*
topographical, 402*f*–403*f*
meteorite fragments at Chelyabinsk, 408, 408*f*
migration in, 420*f*
Chinese, 421, 421*f*
eastward, 418–420
international, 420
political motives for, 419–420
Russification in, 420
undocumented immigration in, 421
urban, 421, 423
Moldova, 403–404, 404*f*
population/settlement patterns of, 415
changing structure of, 424*f*
demographic crisis in, 424–425
distribution of, 415*f*
European Core, 416
indicators of, 415*t*
map of, 415*f*
Siberian hinterland, 416–417
urban, 421, 423–424
religion in, 429–430
Christian, 429
non-Christian, 429–430
Siberia, 407
Buryat peoples, 412, 412*f*
northern sea route (Murmansk), 413, 414*f*
Slavic Russia, 401
social development of, 440
health care and alcoholism, 442–443
problems in, 442
women's position in, 443–444
tensions with, 422
Ukraine, 403
U.S.'s parallels with, 401
water pollution in, 410–411
Russian Empire
growth of, 425–427, 426*f*
heritage of, 425
legacy of, 427
origins of, 425
Russian Federation Treaty (1992), 437
Russian Orthodox Church, 429
Russian Republic, 400
Russification, 420
Russo-Japanese War (1905), 426
Rwanda
coffee in, 261
development indicators of, 284*t*
population indicators of, 255*t*

S

Sábado Gigante TV program, 170
Sahara Desert, 302
The Sahel, 248–250
climate change and, 293
history of, 274
Saint Kitts and Nevis
development indicators of, 228t
population indicators of, 207t
Saint Lucia
development indicators of, 228t
population indicators of, 207t
Saint Vincent and the Grenadines
development indicators of, 228t
population indicators of, 207t
Sakhalin Island, 446, 447
Salafists, 324, 330, 331
in Russian Domain, 431
Salinization
in Central Asia, 455
in India, 566
Samarkand, Uzbekistan, 468*f*
Samoa
development indicators of, 686t
population indicators of, 663t
Samsung, 539
Samurai, 528
San Andreas fault zone, 54, 54*f*
San Juan, Puerto Rico, 211
history of, 220
Old San Juan, 220*f*
San Marino
development indicators of, 387*t*
population indicators of, 365*t*
Santería, 215
Santiago, Chile, air pollution in, 153*f*
Santo Domingo, Dominican Republic, 211, 211*f*
free trade zones near, 229, 229*f*
São Tomé and Principe
development indicators of, 284*t*
population indicators of, 255*t*
Sardar Sarovar Dam (India), 553, 553*f*
Satellite communications, in Oceania, 681
Saudi Arabia
absolute monarch in, 37
conservatism of, 335

Saudia Arabia (*Continued*)
development indicators of, 338t
energy futures, 69
gender inequality, 47
Hofuf oasis, 313
independence, 330
migrant labor, 26
population growth of, 321
population indicators of, 310t
Sunni Arab majority in, 335
women's rights, 345
Yanbu, 341, 342, 342*f*–343*f*
Savanna, in Sub-Saharan Africa, 247, 247*f*
Schengen Agreement, 366–367
Scotland, devolution from UK, 386
Sea level, rise of, 64*f*
in Australia and Oceania, 659
Caribbean, 201–202
in Mekong Delta, 608
in Netherlands, 360, 360*f*
in Russian Domain, 414
Southwest Asia/North Africa, 308–309
Secularism
in East Asia, 522
religion and, 34
Sediment load, 508
Senegal
development indicators of, 284t
population indicators of, 255t
Senkaku/Diaoyu crisis, 534, 536*f*
Seoul, South Korea, 516
Gangnam District, 494*f*–495*f*, 496
Separatism, ethnic, 38*f*
Sepoy Mutiny, 581
September 11, 2001 attacks, 41, 41*f*
Serbia
Balkan wars and, 384–385, 385*f*
development indicators of, 387*t*
energy sources of, 361t
population indicators of, 365*t*
Seth, Vikram, 577
Settlement patterns, 21, 26–28. *See also* Cities; Suburbs; *specific region or country*
oasis, 313, 313*f*
squatter, 28, 156–157, 157*f*, 159*f*, 264, 265*f*
township-and-range, 103
Seychelles
development indicators of, 284*t*
population indicators of, 255*t*
SEZ. *See* Special Economic Zone
Shanghai, China, 515–516, 516*f*
economic boom in, 544
Shanghai Cooperation Organization (SCO), 482–483, 482–484
Shenzhen, China, 542*f*
Shi'a Islam, 33
in Iraq, 334
Shields
Guiana Shield, 199
in Latin America, 141
South American, 142
Shifting cultivation, 611*f*
Shiite, 322, 324
Shinawatra, Thaksin, 633
Shinawatra, Yingluck, 633
Shinto, 521
Shogunate, 532
Sialkot, Pakistan, 590
Siberia
economic development of, 447, 447*f*
environment of, 407, 409
Evenki minority, 429, 429*f*
hinterland of, 416–417
lumber extraction in, 413
Mirny, 417*f*
Novosibirsk, 447, 447*f*
permafrost of, thawing of, 414
population of, 416
Siberian Gas Pipeline, 440, 446
Siberian Pacific Pipeline, 447
Siberian *taiga* forest, 78
Sierra Leone
conflict diamonds in, 281
development indicators of, 284t
population indicators of, 255t
Sikhism, 573
Silk Road, 466, 468
Sindhi people, 567
Singapore, 600, 616, 617*f*
culture of, Western influences, 624
development indicators of, 636*t*
economy of, 637
language in, 624
map of, 617*f*
population in, 613, 613*f*, 613*t*, 614*f*
indicators of
as port entrepôt, 637
Sinitic languages, 526
Six-Day War (1967), 331*f*, 332
Slash and burn agriculture, 611, 611*f*
Slavery
African poverty and, 285
Caribbean, 208, 210–211, 210*f*, 214–215
cultural impact of, 271–272
in Latin America, 138
cultural patterns, 164
syncretic religious practices, 159
legacy of, 242
Liberia and, 276
routes of, map of, 214*f*, 271*f*
in Sub-Saharan Africa, 240
Slavic languages, 374
Cyrillic alphabet, 374, 374*f*
Slavic peoples, 425
Slim, Carlos, 140
Slovakia
development indicators of, 387t
population indicators of, 365t
Slovenia
creation of, 384
development indicators of, 387t
least income disparity by gender, 379
population indicators of, 365t
Soccer
in Central Asia, 476
in Europe, 378–379, 378*f*
European teams, 36
globalization of, 115–116
in Latin America, 170–171
Latin Americans on U.S. teams, 171
in South Asia, 578–579
in Southeast Asia, 625
in Southwest Asia/North Africa, 327–338
World Cup, 171
Sochi, Russia, 432, 433*f*
games as symbol, 432
price tag for Olympics, 432
Socialism described, 38
Socialist state, 434
Socotra, unique environment of, 302, 304, 304*f*
Soil
Caribbean, quality of, 196–197
in Sub-Saharan Africa, 245, 247
transformation of, in North America, 90
types in Russian Domain
chernozem soils, 407
podzoil soils, 404
Solar energy, 68
in Germany, 362, 362*f*
in Japan, 506
small scale, 70
in Sub-Saharan Africa, 240
Solar intensity, and latitude, 58, 58*f*
Solomon Islands
development indicators of, 686*t*
population indicators of, 663*t*
population pyramid for, 670*f*
Somalia
clans in, 283, 283*f*
development indicators of, 284*t*
disintegration of, 281
political boundaries in, 283
population indicators of, 255*f*
Somaliland, 283
South Africa
apartheid in, 264, 278
black homelands in, 278
DeBeers diamond mines in, 282
development indicators of, 284t
Drakensberg Range, 244
economy of, 242
establishment of, 277, 278
ethnicity of, coloured, 264
languages spoken, 268, 268*f*
population indicators of, 255*f*
population pyramid of, 256–257, 256*f*
poverty in, 290
Sandton, Johannesburg, 264, 265*f*
squatter settlements of, 264, 265*f*
townships of, 264
urban industrialism in, 264
South Asia, 550–551
agriculture in, 562, 562–566, 564–565, 565*f*
Green Revolution in, 564–565, 565*f*
cities of, 566, 566–568
climate patterns of
global warming in, 558, 560
map of, 559*f*
monsoon, 557–558, 558*f*, 559*f*
orographic rainfall in, 557
culture of, 568–569
caste system in, 569, 570–571
global context of, 577–579
Southeast Asian culture influenced by, 617–618
economy of, 551, 587–588, 587*f*, 588–589, 589*t*, 589–593, 593*f*, 593–594
energy in, 560
environmental geography of, 551
cyclones in, 554–555
deforestation in, 555–556
flooding in, 555–556
Indus-Ganges-Brahmaputra lowlands in, 551, 553
regions in, 551
southern islands in, 553–554
wildlife in, 556–557
ethnic conflicts in, 583
Kashmir's, 583–584, 583*f*, 586*f*
northeastern fringe's, 584
Sri Lanka's, 585
food resource issues in, 565–566
geopolitics of, 551, 579
historical empires in, 580
independence in, 580–583
international, 586–587
Maoism in, 586
map of, 552*f*
history of, 569
British conquest in, 579, 580–581
Buddhism in, 570
colonial period in, 579, 580–581
early civilizations in, 569
Hindu civilization in, 569–570
independence in, 580–583
Islam's arrival in, 570–571
Partition in, 582
language in, 574–577
English, 577
Indo-European, 575–576
linguistic nationalism in, 577
map of, 575*f*
maps of (entire region)
climate, 559*f*
early civilization, 570*f*
environmental issues, 554*f*
geopolitical, 580*f*
language, 575*f*
migration, 564*f*
monsoon, 558*f*
political, 552*f*
population, 563*f*
religion, 573*f*
satellite photo, 553*f*
topographical, 552*f*
population/settlement patterns in, 561
diaspora of, 578*f*
indicators of, 561*t*
map of, 563*f*
migration in, 562, 564*f*
poverty in, 550
religion in, 569–570, 571, 570, 570–571, 571, 572, 573*f*, 573–574, 574, 574*f*
satellite photograph of, 553*f*
social development in, 594–595
education in, 594–595
gender ratios, 595
women's status and, 595
South Asia Terrorism Portal, 584
South China Sea conflict, 633–635, 635*f*
Southeast Asia, 600, 600–602
agriculture in, 610–612, 611, 612, 632–633
Asian mainland zone, 600
climate patterns in
global warming in, 607–609
map of, 603*f*
monsoon, 602–604
culture of, 617
Chinese influences in, 617, 619
global context of, 624–626
indigenous, 621, 626
South Asian influences in, 617–618
economy of, 636
competition with China, 642
development indicators of, 636t
differences in, 636
drug trade in, 611
palm oil production, 644, 644*f*
education and health, 642–643
energy in, 609
environmental geography of, 602
air pollution in, 607
deforestation in, 602*f*, 605–607
fire in, 607

protected areas in, 607
slash and burn agriculture in, 611, 611*f*
Sunda Shelf in, 604
ethnicity in, 621
geopolitics of, 626
ASEAN in, 602, 635
Cold War impact on, 602
communism in, 621
map of, 601*f*
nationalism in, 629
tensions in, 630
territorial conflicts in, 633–635
terrorism in, 635
history of
British expansion in, 627–628
colonial, 626–627, 628*f*
Dutch power in, 627
French expansion in, 627–628
nationalism in, 629
pre-colonial, 626
U.S. expansion in, 628–629
Vietnam War in, 629–630
insular realm (islands) in, 600, 604–605
language in, 621, 621–624
Austronesian, 621–623
map of, 622*f*
Mon-Khmer, 623–624
Tai-Kadai, 623
Tibeto-Burman, 623
maps of (entire region)
climate, 603*f*
colonial, 628*f*
economic, 643*f*
environmental issues, 605*f*
geopolitical, 627*f*
HDI, 640*f*
language, 622*f*
political, 601*f*
population, 610*f*
topographical, 601*f*
population/settlement patterns in, 609–610
contrasts in, 612–613
density map of, 610*f*
indicators of, 613*t*
migration patterns in, 614–615
primate cities in, 615
urban, 615–616
religion in, 619, 620–621
map of, 618*f*
social development of
HDI in, 640*f*
issues in, 643, 645
Southeast Asian Games, 625
Southern African Development Community (SADC), 278
Southern Cone, economies of, 178
South Korea. *See also* Korea
chaebol conglomerates, 538
culture of, 527–528
economy of, 537*t*, 538–539
immigration to Latin America, 163
North Korean refugees in, 541
population indicators of, 511t
Seoul's Gangnam District, 494*f*–495*f*, 496
South Sudan, 242
independence, 283
religious aspects, 270
Sudd wetlands in, 245
Southwest Asia/North Africa, 242, 298–301
agriculture in
domestication, 311
dryland, 314–315
regions of, 311–312
cities of, 310
Cairo, 314, 314*f*, 315*f*
globalization's effect on, 316–317
legacy of, 315–316
Persian Gulf, 317, 317*f*
climate change in, 308–309
climate patterns in, 301–302, 303*f*
global warming and, 308–309
map of, 303*f*
as culture hearth, 300
culture of, 321
conservative influences on, 327
globalization and, 326–328
music in, 327
television, 327
economy of, 341–343
development indicators of, 338t
generic prescription drugs from Israel, 344
global links to, 347
oil export in, 341–342
regional patterns in, 341–344
tourism in, 347
environmental geography of, 304–308
deforestation in, 304–305
fossil fuels in, 340*f*
issues in, 305
landforms in, 301
Levant in, 301
Maghreb in, 301
overgrazing in, 304–305
salinization of, 305–306
Islamic fundamentalism in, 300
Islamism in, 300
language in, 325–326
Berber, 326
Kurdish, 326
map of, 325*f*
Persian, 326
Semitic, 325–326
Turkish, 325
maps of (entire region)
agricultural, 312*f*
child mortality, 339*f*
climate, 303*f*
geopolitical issues, 329*f*
Islamic diffusion, 322*f*
language, 325*f*
political, 298*f*–299*f*
population, 309*f*
religion, 323*f*
topographical, 298*f*–299*f*
migration patterns of, 317–318, 320
politics of, 330
Arab-Israeli conflict in, 331–333
colonialism in, 328–330
decolonization in, 330
European protectorates in, 329
independence in, 330
issues in, 329*f*
North African instability in, 330–331
U.S. involvement in, 335–337
population growth of, 320
population/settlement patterns of, 309
demographic changes in, 320–321
exotic river, 313
geography of, 310–311
growth in, 321
indicators of, 310t
oasis, 313, 313*f*
pastoral nomadic, 312
physiological density of, 310
rural, 311–315
poverty in, 343, 345
religion in
diversity of, 324–325
Islam in, 321–324
Judeo-Christian tradition in, 321
map of, 323*f*
sectarian violence in, 300
technology and popular culture in, 326–327
water management in, 306–308
fossil water extraction in, 307
Great Man-made River and, 307
hydropolitics in, 307–308
qanat system of, 306
water as transport link in, 308
women's role in, 345–347
Soviet Union, 400
autonomous areas granted by, 434
Central Asia controlled by, 477–478
centralization of, 435
centralized economic planning, 440
dissolution of, 400, 401–402, 435
impact on Central Asia, 466
during World War II, 380
economic legacy of, 440–441
economic planning in Eastern Europe, 391
expansion of, 435
geographic scope of, 400
geopolitical structure of, 433
map of, 434*f*
republics controlled by, 434
socialist realist art in, 431
Warsaw Pact and, 383
Soybeans, rainforest destruction and, 76
Spain
Basque terrorism, 41
colonialism, 39
development indicators of, 387t
energy sources of, 361t
Latin America's colonization by, 138
mining resources of, 173
racial mixing, 166–167
leaky borders of, 369
migration from, 366
population indicators of, 365t
Spanglish, 114
Spanish-American War (1898), 222
Special Economic Zone (SEZ), 541
Specialization, geographic, 9
Sports
in Australia and Oceania, 676–678, 678*f*
globalization and, 36–37
in Central Asia, 476
in Russian Domain, 432
in Southeast Asia, 625–626
sepak takraw, 626, 626*f*
Southwest Asia/North Africa, 327–328
Spratly Islands, 633–634, 635*f*
Squatter settlements, 28
in Kabul, Afghanistan, 469, 469*f*
in Latin American cities, 156–157, 157*f*, 158*f*
in Mumbai (Bombay), 566
in Nairobi, Kenya, 263, 263*f*
in Peru, 158
of South Africa, 264, 265*f*
Sri Lanka, 553
conflict map of, 585*f*
development indicators of, 589*t*
economy of, 590
education in, 594–595
ethnic conflict in, 585
population of
indicators of, 561*t*
pyramid of, 561*f*
religion of, 570
Tamil language in, 576
Stalin, Joseph, 420
as chess player, 432
expansion of Soviet space, 435
political centralization of, 435
in textbooks, 432
State. *See also* Nation-state
definition of, 38
federal, 120
unitary, 120
weakening of, globalization's, 7–8
Steppe, 75
partitioning of, 476–477
pastoralism, 465, 465*f*
St. Petersburg, Russia, 416, 417*f*
core area, 423
Straits of Hormuz, 308*f*
Streetcar suburbs, 101
Structural adjustment programs, 285
Subduction zone, 54
Subic Bay, the Philippines, 637, 637*f*
Submarine cables, in South Pacific, 681, 681*f*
Subnational organizations, 176
Sub-Saharan Africa, 240
cell phone use in, 238*f*–239*f*, 240
Chinese merchants in, 289
civil society in, 293
climate of, 247
global warming and, 254
map of, 246*f*
colonization of, 262, 274–277
Berlin Conference and, 276–277, 276*f*
decolonization after, 277–278
map of, 276*f*
South Africa's establishment and, 277, 278
culture of, 265–266
complexity of, 242
globalization and, 271–273
language-ethnicity boundaries in, 268
music in, 271–272
running in, 272–273, 274*f*
sport in, 272–273, 274*f*
world music in, 242
economy of, 242–243, 281, 282
Chinese investment in, 285, 288, 290
conflict diamonds in, 281
debt forgiveness and, 243, 290
development indicators of, 284*t*
differentiation within, 290–291
East Africa in, 291
foreign aid in, 287–288, 288*f*
foreign investment in, 287, 288, 290
global links to, 287–290
growth, 285, 287
growth rates, 242
infrastructure and, 285
Millennium Development Goals and, 287

Sub-Saharan Africa (*Continued*)
- oil and mineral production in, 290–291
- paved roads, 285, 285*f*
- trade levels in, 287
- West Africa in, 291

emigration from, 272
energy shortages, 251–253, 251*f*
environment of
- basins in, 243–244
- conservation efforts in, 247, 253–254
- deforestation in, 250–251, 251–252
- desertification in, 248–250
- deserts in, 248, 248*f*
- Great Escarpment in, 244
- issues in, 248, 249*f*, 250–251
- plateaus in, 243
- poaching and, 253–254
- reforestation efforts in, 250
- savannas in, 247, 247*f*
- soils in, 245, 247
- tropical forests in, 244, 245*f*, 247, 250
- vegetation of, 247
- watersheds of, 244–245

export agriculture in, 260–261
food insecurity in, map of, 254*f*
foreign assistance impact on, 242–243
geopolitics of, 274
- boundary issues in, 279–280
- conflict in, 278–280
- corruption in, 286–287
- ethnic conflicts in, 281
- independence in, 277–278
- refugees and, 280–281, 280*f*
- secessionist movements in, 281–282
- supranational organizations in, 278, 279*f*
- tribalism in, 278–279
- women's participation in, 292, 293*f*

history of
- AU in, 279*f*
- disease in, 274–276
- European colonization, 274–277
- kingdoms in, 274
- prehistorical, 274

HIV/AIDS in, 256, 257–259, 259*f*, 291
Human Development Index (HDI) scores, 283
languages of, 265–267
- African, 266*f*, 267–268, 267*f*
- European, 268
- identity and, 268

malaria in, 259
maps of (entire region)
- climate, 246*f*
- environmental issues, 249*f*
- European colonization, 276*f*
- foreign aid, 288*f*
- historical states and kingdoms, 275*f*
- HIV prevalence, 258*f*
- Islamic extent, 269*f*
- language, 266*f*
- political, 241*f*
- population distribution, 260*f*
- refugee and displaced persons, 280*f*
- supranational organization, 279*f*
- topographical, 241*f*
- women in work force and politics, 293*f*

mountains in, 244, 244*f*
philanthropy in, 243
population/settlement patterns of, 240, 254, 256–259, 259–262
- agricultural density of, 256
- Bantu migration, 267, 267*f*
- demographic trends in, 256–257
- density of, 259, 260*f*
- distribution of, 260*f*
- family size and, 257, 257*f*
- indicators of, 255*f*
- pastoralism, 261–262
- physiological density of, 256
- plantation agriculture, 261
- subsistence farming, 259–260
- urban, 259, 262–264

poverty in, 283, 285–286, 290
- development policy failures and, 285–286
- slave trade's effect on, 285

religion in, 268
- Christian, 268–269
- conflict and coexistence among, 270
- Islamic, 268, 269–270, 270*f*

social development in, 291–292
- education and, 292
- health issues and, 291–292
- life expectancy and, 291–292
- women's status and, 292, 293*f*

Sudd wetlands in, 245–246, 246*f*
Victoria Falls in, 243, 243*f*

Subsistence farming
- in Caribbean, 210, 211
- in Laos and Cambodia, 641–642
- in Sub-Saharan Africa, 242, 242*f*, 259–260

Suburbs
- commuters from, 102
- downtowns of, 102
- streetcar, 101

Sudan, 242
- civil war in, 331
- development indicators of, 284t, 338t
- ethnic conflict in, 283
- petroleum, 341
- population indicators of, 255*f*, 310t
- poverty in, 343, 345
- religious conflict in, 270

Sudra caste, 571–572
Sufism, 324
Sugar production
- in the Caribbean, 227
- deforestation in Caribbean, 204
- in Latin America, 165

Sultan of Sulu, 633
Sunda Shelf, 604
Sunni Islam, 33, 322, 324
- in Iraq, 334
- in Russian Domain, 431

Superconurbation, 517
Supranational organizations, 176
Suriname
- Chinese immigration in, 226
- development indicators of, 228t
- emigration to Netherlands, 226, 226*f*
- forest cover of, 205
- maroon societies of, 215, 215*f*
- population indicators of, 207*t*
- South Asian population of, 216

Sustainability
- bamboo to reduce deforestation in Sub-Saharan Africa, 252
- bison return to North America, 92
- cities in United Arab Emirates, 319
- concept of, 23
- forest products in Latin America, 183
- Germany's move to renewable energy, 362
- green schools and eco-tourism in Bhutan, 555
- Japan's smart city movement, 506
- Lake Baikal recovery, 412, 412*f*
- Malaysia's National Physical Plan, 608
- Masdar City, United Arab Emirates, 319
- pastoralism in Central Asia, 467
- sea-level rise and the future of Oceania's low islands, 660
- small scale solar power, 70, 70*f*
- Southeast Asian palm oil plantations, 644
- urban gardens in Cuba, 212
- urban transport in Latin America, 152

Sustainable agriculture, 93
Sustainable development, 23
Swahili language, 267–268
Swaziland
- development indicators of, 284t
- population indicators of, 255t

Sweatshop conditions in Southeast Asia, 642
Sweden
- development indicators of, 387t
- energy sources of, 361t
- population indicators of, 365t
- women's political participation in, 379
- women in workforce of, 379

Swidden agriculture, 260, 611, 611*f*
Swiss Agency for Cooperation and Development (SDC), 467
Switzerland
- development indicators of, 387*t*
- population indicators of, 365*t*

Syncretism, cultural, 31
Syria
- civil war refugees, 320, 336–337, 336*f*, 337*f*
- development indicators of, 338*t*
- instability in, 333–334, 333*f*
- Maronite Christians in, 330
- population indicators of, 310*t*
- religion in, 324–325
- sectarian violence, 300

T

Tagalog/Filipino language, 623
Tagore, Rabindranath, 576
- Nobel prize winner, 577

Taiga, in Russian Domain, 407
Taiga forest, 78, 78*f*
Tai-Kadai languages, 623
Taiwan
- culture of, cosmopolitanism of, 527–528
- development indicators of, 537t
- diplomatic recognition of, 533
- economy of, 540
- environmental geography of, physical, 501
- geopolitics of, 684
  - China and, 533–534
- language in, 527
- physical environment of, 501
- population of, 511*t*, 514

Tajikistan
- birth rate, 466
- cities in, 468
- development indicators of, 485t
  - population indicators of, 465t
- economy of, 486
- independence in, 479
- Nurek Dam, 487*f*
- Russian language issue, 475–476
- social conditions in, 490

Tajik languages, 471, 472
Taliban, 474, 481
- opium proceeds, 488
- in Pakistan, 567
- women in Afghanistan, 490

Tamil language, 577
Tamil Tigers, 585
Tanzania
- development indicators of, 284t
- economy of, 291
- population indicators of, 255t
- subsistence farming in, 242*f*

Tarim Basin, population patterns in, 464, 464*f*
Tar sands, 94
- in Canada, 90*f*, 94*f*

Tatar ethnic identity and language, 428
Tatarstan Republic, Kazan Mosque, 430*f*
Tectonic plates. *See* Plate tectonics
Tehran, Iran, 311
Telenovela, 170
Televisa, 170
Temperate deciduous forests, 76–77, 78*f*
Temperature
- increase in, 63, 63*f*
- Southwest Asia/North Africa, 308
- in insular Southeast Asia, 604

Tenochtitlán, 164
Terrorism
- global, 40–41
  - globalization and, 40–41
  - military responses to, 41
- local, 40
- in Philippines, 631
- in Southeast Asia, 635

Teva Pharmaceutical Industries, 344, 344*f*
TFR. *See* Total fertility rate
Thailand
- border issues in, 633
- Burmese migrants to, 634
- deforestation in, 606
- development indicators of, 636t
- economy of, 639
- family planning in, 613
- language in, 623
- political turmoil in, 633
- population indicators of, 613t
- primate city of, 615–616
- prostitution in, 639, 639*f*, 645
- settlement patterns, 611

Theocratic state, 322, 474
Theravda Buddism, 618, 620*f*
Thermal inversion, 153, 153*f*
Third world countries, 43. *See also* Development
Three Gorges Dam, 506, 507–508, 508*f*
Tiananmen Square, 516, 517*f*
- protest and crackdown at, 544

Tibet
agriculture in, 463
Buddhism in, 474–475
China's relations with, 534
Chinese occupation of, 479
economy of, 487
environment in, 455, 463
government-in-exile, 574
history of, 469–470
languages of, 471, 623
shrinking glaciers in, 462–463, 462*f*
social conditions in, 490
Tibetan Government in Exile, 479
Tibetan Plateau, 453*f*, 454
Tibeto-Burman languages, 623
Tiger Protection Force, 557
Timber, sustained-yield forestry and, 23
Timor-Leste. *See* East Timor
Togo
development indicators of, 284t
population indicators of, 255t
Tokyo, Japan, 500, 517, 519*f*
Tonal languages, 526
Tonga
development indicators of, 686*t*
population indicators of, 663*t*
Total fertility rate (TFR), 22, 23*f*
in India, 561
Latin America, 162
in Southeast Asia, 612–613
in Sub-Saharan Africa, 257
Touré, Amadou Toumani, 281
Tourism
in Australia and Oceania, 687*f*, 689–690
in Bhutan, 589
in Caribbean, 196, 206, 231–232, 231*f*
cruise ships, 231, 232*f*
in Cuba, 231, 232
in Dominican Republic, 232
in Hawaii, 689–690
in Maldives, 591, 592, 592*f*
in Nepal, 589, 589–590, 589*f*
to Southeast Asia, 639
in Southwest Asia/North Africa, 347
in Thailand, 639
Township-and-range survey system, 103
Trade. *See also* Drug trade; *specific region or country*; *specific trade associations*
Fair Trade, 156
free zones of, 229, 229*f*
in ivory, 253–254
North America's patterns of, 131
Trade blocs
in Caribbean, 226–227
in Latin America, 172, 174*f*, 176–177
Transcaucasia, 409, 409*f*, 429–430, 429*f*
instability of, 437
Transcaucasian language, 429–430, 429*f*
Transhumance
in Central Asia, 464
in Southwest Asia/North Africa, 312
in Sub-Saharan Africa, 249
TransMilenio transport system, 152, 152*f*
Transnational migration, 210
Trans-Siberian Railroad, 417
modernized, 440
Treaty of Tordesillas, 172
Tribalism, in Sub-Saharan Africa, 278–279
Trickle-down economics, 10
Trinidad and Tobago
development indicators of, 228t
population indicators of, 207*t*
South Asian population of, 217, 217*f*
steel drums, 196*f*, 218
Tropical forest destruction, 75, 75*f*
Brazil, 76, 76*f*
Tropical savannas, 74–75, 74*f*
Tropical seasonal forests, 74, 74*f*
Tsarina Catherine the Great, 427
Tsar Peter the Great, 416
imperial expansion of, 426–427, 426*f*
Tsars, 418
Tsetse flies, 262
Tsunamis
in East Asia, 499, 501
in Oceania, 654
in Papua New Guinea, 654–655
in Southeast Asia, 604
Tohoku earthquake and tsunami, 501, 501*f*, 510
Tulu, Derartu, 272
Tundra, 79, 79*f*
in North America, 89, 95
in Russian Domain, 407
Tunisia, 327
Arab Spring in, 330
development indicators of, 338*t*
population indicators of, 310*t*
Turkey, 301
after World War I, 330
development indicators of, 338*t*
economy of, 343, 345*f*
EU and, 335
forests in, 305
geopolitics of, Central Asia in, 484
hydropolitics of, 307
Islamist pressures in, 327
languages of, 326
offshore gasfields held by Cyprus, 395, 395*f*
population indicators of, 310t
satellite view of, 302*f*
Turkic languages, 468, 471–472
Turkmenistan
development indicators of, 485*t*
economy of, 486
energy resources of, 461
independence in, 480
Karakum Desert of, 457*f*
population indicators of, 465*t*
Turks and Caicos
development indicators of, 228*t*
population indicators of, 207*t*
Tuvalu
development indicators of, 686*t*
population indicators of, 663*t*
sea level rise and, 655
Typhoon, 59
in Southeast Asia, 604–605
Philippines, 604

U

Uganda
development indicators of, 284t
economy of, 291
population indicators of, 255t
solar energy in, 238*f*–239*f*, 240
UK. *See* United Kingdom
Ukraine
development indicators of, 440t
EU and NATO, 437
foreign capital in, 446
language patterns in, 428, 428*f*
links with Russia, 436–437
population indicators of, 415*t*
religious split in, 430
Svetlana Loboda, 432, 432*f*
urban clusters in, 416
Ulaanbaatar Naadam festival, 476, 476*f*
UN. *See* United Nations
UN Commission on Environment and Development, sustainable development, 23
Undocumented immigration
in Europe, 367, 369
in Malaysia, 638
in North America, 119–120, 121–122
in Russian Domain, 421
in Thailand, 634
Unemployment, globalization and, 10*f*
Union of Arab Football Associations, 328
Union Carbide disaster in Bhopal, India, 594
Union of South American Nations (UNASUR), 172, 174*f*
as Brazil's initiative, 177
Unitary state, 120
United Arab Emirates
development indicators of, 338t
Dubai, 317, 318*f*
foreign workforce of, 320
population indicators of, 310*t*
population pyramid, 320*f*
sustainable cities in, 319
women's dress in, 327
United Kingdom (UK). *See also* Great Britain
development indicators of, 387*t*
industrialization, 386, 387*f*, 388, 388*f*
investment in Afghanistan, 489
population indicators of, 365*t*
Scotland's devolution from, 386
South Asia conquered by, 579, 580–581
in Southeast Asian history, 627–628
traditional currency, 393
United Nations (UN), creation of, 5
United Nations (UN) World Heritage Sites, Socotra, 304
United States (U.S.). *See also* Military, U.S.
African Americans in, 271
Asia Pivot, 684–685, 685*f*
Atlantic coastline, 87, 87*f*
carbon dioxide emissions from, 64, 65*f*
Census of, 96–97
colonies in Oceania, 680
economy of, development indicators of, 44*t*, 123*t*
energy futures, 69
ethnic composition of, projected, 106*f*
expansion of, 117, 122
expressions of globalization, outsourcing, 133
as federal state, 120
gender roles, 36
geopolitics of
Afghanistan occupation in, 480–481
Canada's relationship in, 116–120
Canadian border in, 118
Caribbean policies in, 222–223, 222*f*
Central Asia in, 482
domino theory, 629
Mexican border in, 87
Southwest Asia/North African in, 335–337
global culture in, 5, 6*f*
global reach of, 122
Gulf Coast of, 90*f*, 91
Hispanic population of, 164
human development index, 46
immigration to, 104–106
distribution of, 105*f*
from Mexico, 164
politics of, 121–122
skilled worker, 132–133, 132*f*
by year and origin, 104*f*
impact of recession, 4
isolationism of, 122
Mexico-U.S. border, 175
NAFTA and, 118
occupation of Caribbean nations, 222
occupation of Haiti, 204
occupation of Philippines, 122, 619, 623, 635
political roots of, 117–118
population density in, 27, 27*f*
population growth, 24*f*
population indicators of, 22t, 95*t*
regions of, 107*f*
renewable energy, 68
Russian Domain's parallels with, 401
in Southeast Asian history, 628–629
sports abroad, 36–37
summer monsoon in Southwest, 59*f*
trade deficit, container shipping to, 131*f*
trade deficit of, 131
United Wa State Army (UWSA), 632–633
Universalizing religions, 32
Univision TV network, 170
Untouchables, 572, 572*f*
Uplands, of Latin America, 142
Ural Mountains, 407, 409
formation, 55
Urban decentralization, 101–102
Urban gardens
in Cuba, 212–213
in United States, 112
Urban heat islands, 91
Urbanization, 28–29. *See also individual countries, cities of*
in Central Asia, 466–468
of Dubai, 317
in East Asia, 513, 515
Islam's effect on, 316
Latin America, 139
elite spine of, 156, 157, 159
North America, 95–96
New York skyline, 97*f*
in Soviet Union, 421, 423
mikrorayons, 423
ringlike morphology, 423
Urbanized population, 28–29
Urban primacy
in Latin America, 155
in Southeast Asia, 615–616
Uruguay
development indicators of, 178t
outsourcing in, 180
population indicators of, 156t
population pyramid, 162
U.S. *See* United States
U.S. Department of Energy, 319

Ushahidi mapping platform, 198, 198*f*
Uzbekistan
cities in, 466–467
development indicators of, 485*t*
economy of, 484, 486, 486*f*
independence in, 480
industrial pollution in, 455
Islamic Movement in Uzbekistan (IMU), 482
population indicators of, 465t
salinization in, 459
Uzbek language, 471

V

Vaishya caste, 571
Valdivian forest, 148
Vallejo, Camila, 176
Vancouver, Canada, cultural identity of, 108–109
Vanuatu
development indicators of, 686t
population indicators of, 663t
Venezuela
agrarian reform in, 160
boundary disputes with Colombia, 174
Colombian immigration to, 163
culture of, telenovelas, 170
development indicators of, 178t
petroleum, 184
population indicators of, 156t
Video-gaming industry
in Russian Domain, 445
South Korea, 528
Vieques Island, 223
Vietnam
deforestation in, 606
development indicators of, 636t
economy of, 640–641
impact of dams in, 600
as part of China, 618, 620
population indicators of, 613t
religion in, 621
reunification of, 629
settlement patterns, 611
urban settlement in, 616
Vietnam War, 122, 629–630
Vietcong prisoners, 629*f*
Vijayanagara kingdom, 580
Virgin of Guadalupe, 159
Virgin Islands, 222
Visa, H-1B, 132–133
Viticulture, 668, 669*f*
Vladivostok, Russia, 447
Volcanic Axis of Central America, 142
Volcanoes, 55, 56*f*
in East Asia, 500*f*
Indonesia, 50*f*–51*f*
in Southeast Asia, 604, 604*f*
Volga-Don Canal, 440, 441*f*
Voodoo, 215

W

Wahhabis, 324, 335
War
in Burma (Myanmar), 631–633
in Latin America, 173–174
Russo-Japanese War (1905), 426
in Sudan, 331
in Vietnam, 629–630
War of the Pacific (1879–1882), 173
War of the Triple Alliance, 174
within-country vs. between countries, 11
World (I and II), 122, 380
Water. *See also* Hydropolitics; *specific region or country*
access, 71–72
access to freshwater, 69, 71*f*
budget, global, 69
hydropolitics of
Central Asia's, 455–457
Israel's, 307
Mexico City, 154
pollution, 410–411
resource management, in North America, 91
sanitation, 71
scarcity, 71
in Southwest Asia/North Africa, 301, 306–308
stress, 71
women and, 71, 72*f*
West Africa
music in, 271
urban traditions of, 262–264
West-East Gas Pipeline (China), 460
Western culture, global communications and, 29
Western Sahara, 347
development indicators of, 338t
population indicators of, 310t
White Australia policy, 672
Wildfires, 90*f*
in Australia, 653, 655*f*
in Russian Domain, 414
Wildlife
conservation of, 253–254
endangered
CITES and, 253
in East Asia, 504–505
invasive, 657
Patagonian guanaco, 142*f*
reserves, 253
in South Asia, 556–557
Wind energy, 68, 69*f*
in China, 507, 507*f*
in Denmark, 361*f*
in Germany, 362
in North America, 83*f*–84*f*, 85
in Puerto Rico, 229, 229*f*
Wind, global patterns of, 59
Women. *See also gender issues by individual region*; gender roles
in Afghanistan, 488, 490, 491, 491*f*
in Caribbean, 234
political presence, 234
challenges of, in Qatar and UAE, 327
in China, 545
in Latin America, 188–189, 189*f*
political access in, 189, 189*f*
literacy, 47, 47*f*
in Mongolia, 490–491
in Russian Domain, 443–444
in South Asia, 595
in Southwest Asia/North Africa, 345–347
political participation of, 346*f*
in Sub-Saharan Africa, 292, 293*f*
and water, 71, 72*f*
World Bank, 9
creation of, 130
privatizing water, 71–72
resource export strategy, 10
structural adjustment by, 285
World Health Organization (WHO), 259
World Trade Organization (WTO), 131
functions of, 9, 10
member countries of, 9*f*
online gambling decision, 230–231
Russian Domain, in, 444, 447
states weakened by, 8
World War I, 380
World War II, 122, 380
World Wildlife Fund, 319
WTO. *See* World Trade Organization

X

Xinjiang
anti-Chinese conflict in, 474
economy of, 488–489
ethnic tension in, 480, 480*f*
petroleum in, 460
population patterns in, 464*f*
religion in, 473
social conditions in, 490
Xi Zinping, 289, 289*f*

Y

Yakuza, 523
Yanbu, Saudi Arabia, 341, 342, 342*f*–343*f*
Yangtze River, 507–508
Yanjmaa, Sükhbaataryn, 491
Yanukovich, Viktor, 437
Yao Ming, 116–117, 117*f*, 528
Yeltsin, Boris, 435
Yemen
Al Qaeda influence in, 335
development indicators of, 338*t*
population indicators of, 310*t*
poverty in, 345
Yokohama Smart City Project, 506
Yugoslavia, 384, 385*f*
breakup of, 383
Yunnan Province, China, 527*f*
Yushchenko, Viktor, 437

Z

Zaatari Refugee Camp (Jordan), 336, 336*f*
Zambezi River, 243, 245, 245*f*
Zambia
development indicators of, 284*t*
population indicators of, 255*t*
Zimbabwe
development indicators of, 284*t*
independence, 277
population indicators of, 255*t*
urban population, 29
Zoroastrianism, 574, 574*f*
Zuma, Jacob, 278

World – Physical

Great Basin	Land features
Caribbean Sea	Water bodies
Aleutian Trench	Underwater features